WATER–ROCK INTERACTION

PROCEEDINGS OF THE ELEVENTH INTERNATIONAL SYMPOSIUM ON WATER – ROCK INTERACTION
WRI-11, 27 JUNE – 2 JULY 2004, SARATOGA SPRINGS, NEW YORK, USA

Water–Rock Interaction

Edited by

Richard B. Wanty
U.S. Geological Survey, Denver, Colorado, USA

Robert R. Seal II
U.S. Geological Survey, Reston, Virginia, USA

Volume 1

A.A. BALKEMA PUBLISHERS LEIDEN / LONDON / NEW YORK / PHILADELPHIA / SINGAPORE

WATER–ROCK INTERACTION

INTERNATIONAL

Cover: The Boquet River which flows through the Adirondack Mountains in the high peaks of New York (Photo courtesy of Brady McTigue).

Published by: A.A. Balkema Publishers, a member of Taylor & Francis Group plc
www.balkema.nl and www.tandf.co.uk

For the complete set of volumes: ISBN 90 5809 641 6
For volume 1: ISBN 90 5809 643 2
For volume 2: ISBN 90 5809 644 0

Printed in The Netherlands

Table of Contents

Crustal fluid–rock interactions, mass transfer and cycling of volatiles

Water–rock interactions in groundwater systems and sedimentary basins

Carbon dioxide and hydrogen sulfide sequestration

Advances in spectroscopic and microscopic techniques for the study of water–rock interactions

Radionuclide interactions with minerals and microbes

Complexity of mineral surfaces: experimental and theoretical studies

Volume 2

Geochemical modeling from molecular to global scales

Reactivity of organic compounds during water–rock interactions

Geomicrobiology: a symposium in honor of Henry Ehrlich

Iron biogeochemistry

Aqueous geochemistry and biogeochemistry

Environmental geochemistry

Preface

The 11th International Symposium on Water–Rock Interaction was held in Saratoga Springs, New York, USA from June 27 to July 2, 2004. The Symposium addressed all aspects of Water–Rock Interaction including recent advances in biogeochemistry, weathering system science, crustal fluids, geochemical modelling, radioactive nuclide chemistry, magmatic processes and degassing, environmental geochemistry, geochemical spectroscopies and microscopies, and isotope geochemistry.

The symposium was sponsored by the Energy and Environment Directorate, Lawrence Livermore National Laboratory; Worldwide Universities Network; the U.S. Department of Energy; the Geology and Paleontology Program and the Office of International Science and Engineering, National Science Foundation; the National Research Program, Water Resources Division, U.S. Geological Survey; the Water–Rock Interaction Working Group of the IAGC; Lawrence Berkeley National Laboratory-Earth Sciences Division; Rensselaer Polytechnic Institute; Shell International Exploration and Production; the International Association of Geochemistry and Cosmochemistry; Thermo Electron Corporation; The Geochemical Society; GV Instruments Inc.; Blackwell Publishing; LI-COR Biosciences, Inc.; Teledyne Leeman Labs; Kluwer Academic/Plenum Publishers; and the Penn State University Center for Environmental Chemistry and Geochemistry.

The volumes include over 330 manuscripts submitted by scientists from over 35 countries. The papers include theoretical, analytical, laboratory, and field-based studies of processes spanning from high- to low-temperature and from abiotic to biotic processes. The volume is organized according to session at the symposium, generally from high-temperature and pressure to low-temperature and pressure environments as follows:

1. Volcanic-geothermal water–rock processes and degassing: a tribute to Donald White
2. Crustal fluid–rock interactions, mass transfer, and cycling of volatiles
3. Water–rock interactions in ground-water systems and sedimentary basins
4. Carbon dioxide and hydrogen sulfide sequestration
5. Advances in spectroscopic and microscopic techniques
6. Radionuclide interactions with minerals and microbes
7. Complexity of mineral surfaces: experimental and theoretical studies
8. Weathering studies at all space and time scales
9. Geochemical modeling from molecular to global scales
10. Reactivity of organic compounds
11. Geomicrobiology: a tribute to Henry Ehrlich
12. Iron biogeochemistry
13. Aqueous geochemistry and biogeochemistry
14. Environmental geochemistry

The interdisciplinary nature of water–rock interactions is evident in the considerable overlap in many of the sessions. We were fortunate to be addressed by both plenary and keynote speakers; abstracts for the plenary presentations are presented in the first section of the book. For those sessions with keynote presentations, the keynote papers lead each section of the book, followed by all other contributed papers in alphabetical order by first author's family name.

All of the manuscripts have been reviewed for scientific content by peers. This volume would not have been possible without the insightful and timely efforts of our colleagues listed below. Our sincere thanks go to:

Teofilo Abrajano
Greg Arehart
Robert Ayuso
Laurie Balistrieri
Gray Bebout
Deborah Bergfeld
JK Böhlke
Mike Borda
David Borrok
George Breit
Bob Burruss
Chunfang Cai
Vyllinniskii Cameron
Carlo Cardellini
Ian Cartwright
Lisa Cave
Dornadula Chandrasekharam
Thomas Chapin
Jon Chorover
I-Ming Chou
Oleg Chudaev
Dave Cole
George Darling
Lou Derry
Dr. Hailang Dong
Greg Druschel
Tatsuji Ebihara

Carrick Eggleston
Annette Summers Engel
Bill Evans
Jaisong Fang
Jeremy Fein
Nora Foley
David Fowle
Gustavo Garzon
Susan Glasauer
Larry Gough
Johnson Haas
Jane Hammarstrom
Dan Hayba
Janet Herman
Joann Holloway
Evangelos Kakouros
Andreas Kappler
Ken Kemner
Yousif Kharaka
Allan Kolker
Kurt Konhauser
David Kreller
Amy Kroenke
Jim Kubicki
Lee Kump
Jen Lewicki
Laura Liermann
Gail Lipfert
Mario Lorrai
Tim Lowenstein
Don Macalady
Kevin Mandernack
Robert Mariner
Ross McCartney
Jeff McLean
Jude McMurry
Peter Moeller
Joel Moore
Charles Myers
Heino Nitsche
Jim Palandri
Gene Parkin
Nadine Piatak
Geoff Plumlee
Mike Reddy
Rich Reeder
Don Rimstidt
Jennifer Roberts
Eric Roden
Marjorie Schulz
Jill Scott
Pat Shanks
Mark Stanton
Carl Steefel
Lisa Stillings
Yuri Taran
Kevin Taylor
Hsin-Yi Tseng
Benjamin Turner
David Vance
Thijs VanSoest
Joop Varekamp
Philip Verplanck
Greg Wandless
Bronwen Wang
Lois Wardell
Tamie Weaver
Alan Welch
Art White
Laurie Wirt
Scott Wood
Nathan Yee
Bob Zielinski
Min Zhang
Wenbin Zhou

The advice of Yousif Kharaka and Rosa Cidu from their past experiences as WRI editors is greatly appreciated. The co-editors also thank Debra Bandy and Dorothy Trujillo for their technical assistance. In addition, we thank Nadine Piatak for her extraordinary efforts at reviewing and formatting numerous manuscripts, particularly some of the more challenging ones. Andy Zimmerman helped us greatly at the beginning to "debug" some of our instructions to the authors. The co-editors also thank Kathleen Johnson, Mineral Resources Program Coordinator for the U.S. Geological Survey, for her support of our editing activities.

Susan L. Brantley
Secretary General
WRI-11

Richard B. Wanty
Co-editor
WRI-11 Proceedings

Robert R. Seal, II
Co-editor
WRI-11 Proceedings

Organization

IAGC Executive Committee

President: Eric M. Galimov (Russia)
Vice President: John Ludden (France)
Secretary: Mel Gascoyne (Canada)
Treasurer: David T. Long (USA)
Past President: Gunter Faure (USA)
Council Members: A. Demeny (Hungary)
J. Gurney (South Africa)
R. Harmon (USA)
J. Hoefs (Germany)
M. Javoy (France)
J. Kramers (Switzerland)
G. Kurat (Austria)
N.V. Sobolev (Russia)

Water–Rock Working Group:

Yousif Kharaka (USA; Chairman)
Tomas Paces (Czech Republic)
Yves Tardy (France)
Brian Hitchon (Canada)
Hitoshi Sakai (Japan)
Halldor Armansson (Iceland)
Mike Edmunds (UK)
Oleg Chudaev (Russia)
Brian Robinson (New Zealand)
Luca Fanfani (Italy)
Susan Brantley (USA)

WRI-11 Organizing Committee

Secretary General: Sue Brantley (Penn State)
Assistant to Secretary General: Denise Kowalski (Penn State)
Conference Coordinator: Barb Impelliteri (Penn State)
Editors, Abstract Volume: Richard B. Wanty (U.S. Geological Survey), Robert R. Seal II (U.S. Geological Survey)

Scientific Program: Martin Schoonen, Chair (SUNY Stony Brook), Jun Abrajano (Rensselaer Polytechnic Institute), Jay Ague (Yale University), Ariel Anbar (University of Rochester), Gray Bebout (Lehigh University), Michael J. Borda (University of Delaware), Patrick Brady (U.S. Department of Energy), Jon Chorover (University of Arizona), Dave Cole (Oak Ridge National Laboratory) Louis Derry (Cornell University), Bill Evans (U.S. Geological Survey), Jeremy Fein (University of Notre Dame), A.J. Francis (Brookhaven National Laboratory), Janet Herman (University of Virginia), James Kubicki (Penn State), Lee Kump (Penn State), Jennifer Lewicki (Lawrence Berkeley Laboratory), Ann Maest (Buka Environmental), Patricia Maurice (University of Notre Dame), Kathryn Nagy (University of Illinois at Chicago), Jennifer Rogers (University of Kansas), Wayne Nesbitt (University of Western Ontario), Brian Phillips (SUNY Stony Brook), Rich Reeder (SUNY Stony Brook), Everett Shock (Arizona State University), Carl Steefel (U.S. Department of Energy)

Poster Program: Tom Bullen (U.S. Geological Survey)

Fund Raising: Sue Brantley (Penn State), Ann Maest, Funding Chair (Buka Environmental), Patricia Maurice (University of Notre Dame), Jim Boles (University of California, Santa Barbara)

Student and Scientist Funding: Louise Criscenti, Chair (Sandia National Laboratories), Robyn Hannigan (Arkansas State University), Carla Koretsky (Western Michigan University)

Accompanying Guest Program: Becky Seagle, Chair (Penn State), Denise Kowalski (Penn State), Barbara Impellitteri (Penn State)

Social Program: Sue Brantley (Penn State), Ann Maest (Buka Environmental)

European Coordinator: Vala Ragnarsdottir (University of Bristol, UK)

Local Organizing Committee: Jun Abrajano, Chair (Rensselaer Polytechnic Institute),
John Arnason (SUNY Albany), Doug Burns (USGS, Albany District), Kyle Nichols (Skidmore College)

Field Trips: Gene Ulmer, Chair (Temple University), Richard Bopp (Rensselaer Polytechnic Institute), Doug Burns (U.S. Geological Survey), Dave Grandstaff (Temple University), Janet Herman (University of Virginia), William Kelly (NY State Museum), Lee Kump (Penn State), James McLelland (Colgate University), Don Siegel (Syracuse University), Lara Storm (Rensselaer Polytechnic Institute), David Wark (Rensselaer Polytechnic Institute), Andy Zimmerman (University of Florida, Gainesville).

Plenary presentations

Water–Rock Interaction, Wanty & Seal II (eds)
© 2004 Taylor & Francis Group, London, ISBN 90 5809 641 6

The silicification of microorganisms: a comparison between *in situ* experiments in the field and laboratory

Liane G. Benning
School of Earth Sciences, University of Leeds, Leeds, UK

Bruce W. Mountain
Institute of Geological and Nuclear Sciences, Wairakei Research Centre, Taupo, New Zealand

ABSTRACT: The mechanisms that drive and control silicification in active geothermal systems at the Earth surface are still poorly understood. To explore the reactions that control inorganic and microbial silicification rates in such systems we have carried out *in situ* experiments (a) in natural geothermal pools in New Zealand (multi-species system) and (b) in the laboratory (single-species system). The selected field sites exhibit contrasting chemistry and they generate markedly different sinter growth rates, but despite the ubiquitous presence of microorganisms, these silicification rates are primarily controlled by the amorphous silica supersaturation and the pH of the effluent fluids. In the *in situ* laboratory experiments the textures growing in the presence of bacteria are similar to those observed in the field experiments, but they are not markedly different from experimental abiotic textures. The differences and similarities between these two settings and with literature data of natural and laboratory silicification experiments will be discussed.

1 INTRODUCTION

The growth rate of silica sinters in active geothermal hot springs as seen in New Zealand or Yellowstone National Park is controlled both by inorganic and biogenic factors. However, the mechanisms that lead to the precipitation of amorphous silica from geothermal waters, or the silicification of the ubiquitously present microorganisms, are still controversial. Geothermal fluids contain high concentrations of silicic acid (H_4SiO_4) derived from water-rock interaction at depth. Upon reaching the surface, geothermal fluids cool and silicic acid begins to polymerize. Experimental evidence shows that the ratio of monomeric to polymeric silica plays an important role in silica precipitation rates (Carroll et al. 1998). It has also been recognized that mesophilic and thermophilic organisms can act as substrates for silica precipitation (e.g. Cady & Farmer 1996; Mountain et al. 2003 and references therein). Nevertheless, the question as to whether microorganisms play an active or passive role in silica precipitation is still unresolved. Walter et al. (1972, 1976) concluded that silicification of microbial material was a passive process. Later studies have postulated that microbial material, such as cell walls and extracellular polysaccharides, can provide preferred sites for silica nucleation and that some bacteria can be preserved (Ferris et al. 1986; Konhauser & Ferris 1996; Phoenix et al. 2000, 2001; Konhauser et al. 2001). Whether silicification is an entirely passive or microbially-promoted process remains to be answered and is currently a subject of continued study (Benning et al. 2004a).

Textural studies of New Zealand sinters have demonstrated the importance of microorganisms, both prokaryotes and eukaryotes (Campbell et al. 2002; Jones et al. 1997, 1999, 2000, 2001; Mountain et al. 2003). Several laboratory-based studies of silicification of microorganisms (*Bacillus subtilis* and *Calothrix sp*) have been reported recently (Fein et al. 2002; Yee et al. 2003). However, these studies were focused on silica-microbe interactions in batch reactors and they have shown that these microorganisms have a low affinity for silica. Few experimental studies on the contribution of thermophilic microorganisms to the microscopic and macroscopic textural development of geothermal deposits are available (Inagaki et al. 1997). Field-based studies of sinter growth rates and textural development are also rare (Schultze-Lam et al. 1995; McKenzie et al. 2001; Phoenix 2001), largely due to slow growth rates and the need to make regular visits to geographically inconvenient sample localities.

In this paper we discuss *in situ* growth experiments performed both in the field (New Zealand, up to 2 years) and in the laboratory. These experiments had the goal to evaluate the processes and mechanisms that govern silicification in inorganic and mixed inorganic/microbial systems.

2 METHODS

2.1 In situ *field experiments and sample characterization*

The *in situ* precipitation rates of silica from geothermal solutions were determined from glass slides that were placed in various hot springs in the Taupo Volcanic Zone in New Zealand (Fig. 1).

At each locality, the pH, T, dissolved O_2 and conductivity were measured prior to slide deployment. Water samples from each site were collected and analyzed for anions (ion chromatography), cations (atomic absorption and inductively coupled plasma optical emission spectroscopy, AAS/ICP-OES), and $H_2S_{(aq)}$ and HCO_3^- (manual titration). These results are shown in Table 1. At specific time intervals, slides from each locality were retrieved. Temperature and pH were re-measured and sample material was stored in 2.5% glutaraldehyde solution. Mineralized/silicified structures were characterized using scanning and transmission electron microscopy

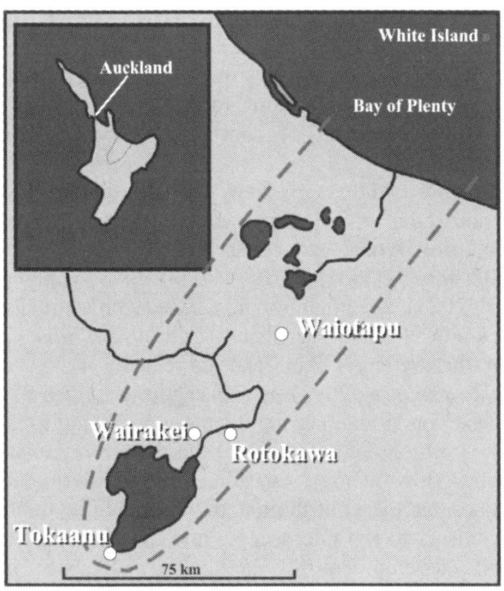

Figure 1. Location of study areas in the Taupo Volcanic Zone (TVZ). Dashed line is the approximate boundary of the TVZ (after Mountain et al. 2003).

(SEM/TEM). In addition, small quantities of the surface layers were scraped off and prepared for synchrotron-based Fourier transform infrared micro-spectroscopy (SR-FTIR). Samples from Champagne Pool were used to ascertain the infrared characteristics of naturally mineralized material (WT1) and fresh *in situ* grown material on the deployed slides (WT2, WT3). Infrared spectra were measured on station 1.4.3 at the Advanced Light Source at the Lawrence Berkeley National Laboratory using a Nicolet NicPlan infrared microscope fitted with a 730 mid-IR spectrometer with the light derived from an external synchrotron source. Further details on SR-FTIR can be found in Benning et al. (2004a).

Results from four localities chosen for their extremes in pH and temperature are discussed. These are: Rotokawa (Pool RK2), Waiotapu (Champagne Pool, N-side), Wairakei (Main Drain) and Tokaanu (Healy Bore 2). A more detailed description of these sites of the sampling methods and analytical techniques can be found in Mountain et al. 2003.

2.2 In situ *laboratory experiments*

The placement of substrates into natural geothermal fluid emanations is one method of measuring sinter growth rates, however, this approach is not conducive to determining species-dependent textural development. This is because it is not possible to grow isolated species within an experimental set-up placed in a natural environment. A laboratory approach is a better option for the simulation of sinter growth in the presence of a single species of microorganism. The growth of monospecies sinters allows us to observe the effect of different species of microorganism on silica deposition and sinter development. The presence of silica in the experimental setup creates

Table 1. Water composition (mg kg^{-1}) and main effluent temperatures from the geothermal localities studied.

	Wairakei	Rotokawa	Waiotapu	Tokaanu
T/°C	62	85	75	93
pH	8.23	3.01	5.46	7.54
Na	1127	824	1134	1274
K	155	88	161	100
Li	10.7	6.8	8.7	16.6
Ca	19.6	31	33	55
Al	0.57	4.9	0.19	<0.1
Fe	0.05	0.33	0.04	0.08
Mg	<0.01	3.6	0.04	0.70
Mn	0.005	0.58	0.08	0.1
SiO_2	559	333	430	290
As	4	0.64	4.7	6
Cl	1858	1198	1926	2104
HCO_3	9.3	<1	153	25
H_2S	<0.2	16	7	0.3

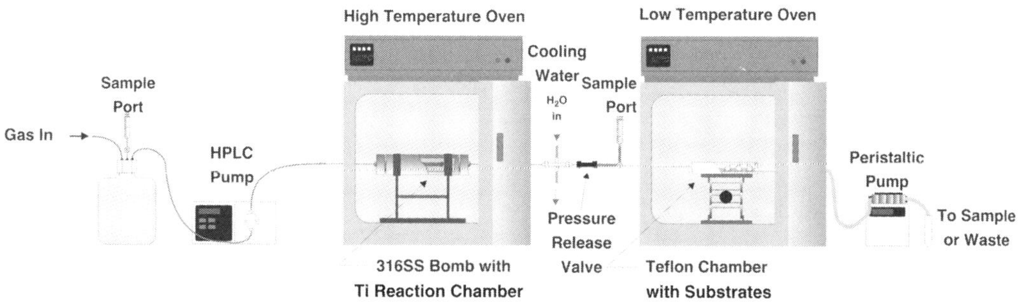

Figure 2. Experimental apparatus for simulation of sinter formation/microbe interaction.

additional difficulty because at near-neutral to alkaline pH, silica polymerizes at temperatures below about 100°C.

Experimental solutions created at ambient temperatures from sodium metasilicate are unsuitable for simulations as they do not represent the true monomeric/polymeric silica ratios found in natural settings. In order to obviate this problem, the laboratory apparatus used consisted of a dual-oven flow through setup similar to that shown in Figure 2.

A synthetic SiO_2-free solution of Wairakei main drain composition and pH was passed through a Ti reaction chamber enclosed within a 316 SS autoclave maintained at 250°C. The autoclave was partially filled with liquid water to maintain saturated vapor pressure thus preventing rupture of the Ti chamber. The chamber was filled with silica gel (SA = $650\,m^2/g$) and the synthetic solution was passed through the chamber via a peristaltic pump. Equilibration of the fluid with the silica gel is rapid and at temperature, the solution is wholly monomeric. This approach mimics the situation occurring in a natural hot spring environment upon issuance from depth. The resulting solution then passes into a second oven maintained at 60°C containing a Teflon tray holding glass slides arranged in a similar configuration to those deployed at the field sites. During cooling from 250°C to 60°C, polymerization of the monomeric silica begins. Abiotic silica deposition was simulated by sterilizing the Teflon tray and slides. Silicification in the presence of bacteria was simulated by using slides placed in a culture of *Bacillus flavothermus (BF)* overnight. BF is a thermophilic facultative anaerobe isolated from the main drain at Wairakei Power Station (Mountain et al. 2003). Silicification was observed with no added media or with additional media added using a second peristaltic pump that metered out small quantities of concentrated tryptic soya broth. This allowed simultaneous silicification and growth of the microorganisms. The change in total silica, and the monomeric vs. polymeric fraction in the effluent solution were monitored using standard methods (ICP-OES and spectrophotometric molybdate yellow method). Furthermore, the growth and structure of the silica mineralization and the preservation of microbes were characterized using the microscopic methods described above.

3 RESULTS AND DISCUSSION

3.1 *Field observations*

The growth structures observed on the deployed slides were composed in most cases of poorly ordered opaline silica nanospheres or nanosphere aggregates (between 25–500 nm in Ø). These are intermixed with non-mineralized microbial filaments or biofilms or they enclose mineralized microorganisms. The rates of silica deposition varied dramatically between the four sites. The highest rate was observed at Wairakei (T = 60°C, pH = 8.2) where the slides were fully immersed in the effluent stream. A longitudinal growth rate of 10 mm/day was measured and an overall growth rate of $350\,kg\,yr^{-1}\,m^{-2}$ of drain wall was estimated. The resulting structures formed thick porous amorphous silica deposited as large fan structures (<0.5 meters).

At Waiotapu (T = 75°C, pH = 5.5) the slides were protruding ~0.5 cm above the water-air interface and after 30 days, silica deposition at the tip of the slides lead to the formation of small spicules (~1 mm high and at 1 mm spacing). These spicules continued to grow and after 1.5 years spicule arrays as high as 7 mm have formed (Fig. 3a). A rate of $0.023\,mm\,day^{-1}$ was calculated from the measurements. The spicules consisted of porous, microbe-rich layers of amorphous silica that contained many microorganisms that were fully encased in nanoparticulate silica (Fig. 3b).

The infrared spectra for the Champagne pool samples are shown in Figure 4. The frequencies of the main features in the samples were used to derive the spectral band assignments by comparison and

Figure 3. *In situ* silicification experiments at Champagne Pool, Waiotapu. T = 75°C, pH = 5.7. The measured vertical growth rate is 0.023 mm/day. (a) Total experimental time is 1.5 years. (b) SEM microphotograph of silicified bacterial colonies on a slide.

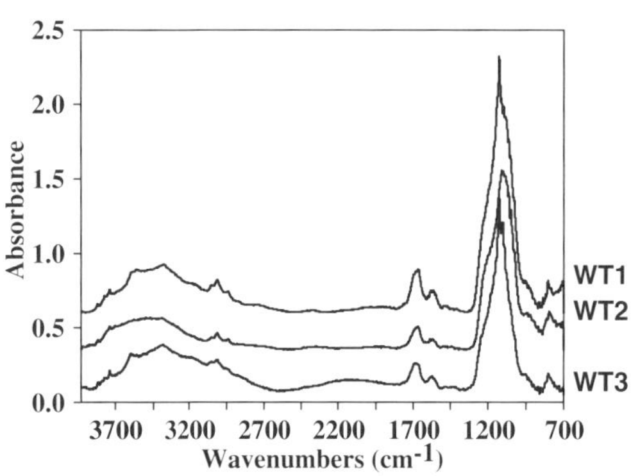

Figure 4. Infrared absorbance spectra of Champagne Pool samples. The numbers correspond to band # in Table 2.

cross-referencing against literature data on other bacterial cells, amorphous silica, or other macromolecules (Fig. 4; Table 2). At high wavenumbers, specific features for silica (band 1), OH (band 2) and lipids (band 3) are found. At lower frequencies, the characteristic features for bacterial cells are manifested in the strong amide I and amide II bands (bands 4 and 5). In addition, less well-defined features from lipids and carboxylic acid (band 6) and polysaccharides (band 7) are found. For silica, the main siloxane and silane frequencies, which partially overlap with the carbohydrate frequencies are found between 1200 and 800 cm^{-1} (band 7).

At lower frequencies two silica vibrations (bands 8 and 9) are characteristic for silane and siloxane frequencies. These vibrations do not interfere with the organic vibrations in bacterial cells and they are representative of inorganic hydrated silica vibrations. Extensive details and discussions on all these vibrations and their characteristics can be found in Benning et al. (2004a). All samples exhibited a mixture of bacterial cell components (proteins, lipids and carbohydrates), and amorphous silica peaks (Table 2). These data corroborate the observations from the SEM and TEM work that show strongly silicified cells with silica partly reacting with the polysaccharides while the proteins and lipids remain relatively unaffected. The silicification degree varies only slightly between samples and the intensity of the

Table 2. Frequencies and band assignments for absorptions confirmed in the IR spectra of the Waiotapu samples.

Band #	Wavenumbers (cm^{-1})	Band assignment[a,b]
1	~3700	Si–OH, surface hydroxyls(1)
2	3250–3500	νO–H and νN–H of water, amide A(2)
3	2980–2845	$\nu_{as}CH_3$, $\nu_{as}CH_2$, $\nu_s > CH_2/-CH_3$; lipid methyl and methylene (2,3)
4	1650	νC=O; Amide I(2–5)
5	~1530	δN–H and ν C–N; Amide II(2–5)
6	1450–1370	$\delta_{as}CH_2/\delta_{as}CH_3$; Lipids, ν_sC–O; Carboxylic acid(2,4,6)
7	1200–950	νC–O–C/νSi–O–Si; Carbohydrates/ Siloxanes(6–8)
8	~ 950	$\nu Si-H_n/\nu Si-OH$; Silanes/ Silanols(8)
9	~800	$\nu Si-O-Si/SiO_4$ rings/Siloxanes(9)

[a]ν = stretching; δ = bending; as = asymetric, s = symetric.
[b]Sources: 1: Tripp & Hair (1991); 2: Nauman et al. (1996); 3: Stuart & Ando (1997); 4: Zeroual et al. (1995); 5: Williams & Fleming (1996); 6: Nelson (1991); 7: Wong et al. (1991); 8: Liquier & Taillandier (1996); 9: Nakamoto (1986); 10: Lippencot et al. (1958).

Si–O–Si band near 1000–1100 cm^{-1} indicates a strong dominance of Si–O–Si bonds due to their more ionic character. The presence of the bands at 950 and 800 cm^{-1} proves that the silicification process is occurring independently of the organic functional groups that represent the cells. This is in accordance with the observations that silica polymerizes inorganically from the supersaturated solutions and is not induced or controlled by the presence of the microorganisms.

At Rotokawa and Tokaanu the deposition of silica occurred much more slowly. At Rotokawa (T = 60°C, pH = 3) the slides were deployed about 1 m from the spring RK2 and protruded about 0.5 cm above the air water-interface. After nine months only thin silica layers and small (<1 mm) spicules were observed on the sides of the slides and a growth rate of ∼0.004 mm day^{-1} was estimated. The slow growth rate is attributed to the very low pH (∼3.0) which inhibits silicic acid deprotonation and polymerization.

The lowest growth rate (0.002 mm/day) was observed at Tokaanu (T = 45–55°C, pH = 7.5), where the slides were placed subaqueously. The growth structures contained alternate cyanobacteria and/or phototrophic bacterial layers that were silicified. Variations in the density and morphology of microorganisms is attributed to large variations in temperature and the non-constant water flow over the experimental period (2 years).

Modeling the dissolved silica saturation indices in the fluids from these sites shows that the rate of subaqueous precipitation is governed by the amorphous silica supersaturation and pH of the thermal fluid (Fig. 5). Despite the ubiquitous presence of microorganisms in these hot springs, there is no evidence that they induce or prevent silica precipitation, at least on a macroscopic scale. The relationship between silica spherule size and the degree of preservation of microbial material can be explained by the effect of Ostwald ripening of the silica aggregates.

Microscopic observations of slides from all sites (Fig. 3b; more photographs in Mountain et al. 2003) showed that despite the large variation in deposition rates, resulting textures were invariably made up of a mixture of inorganic silica deposition (nanospheres, aggregates, thin films or layers, etc.) and microorganisms that were partly or fully silicified.

3.2 *Laboratory observations*

In abiotic *in situ* laboratory experiments, silicification was solely a function of supersaturation; within hours the formation of silica nanospheres and aggregates on the slide surface was observed at the air–water interface. In the experiments carried out in the presence of

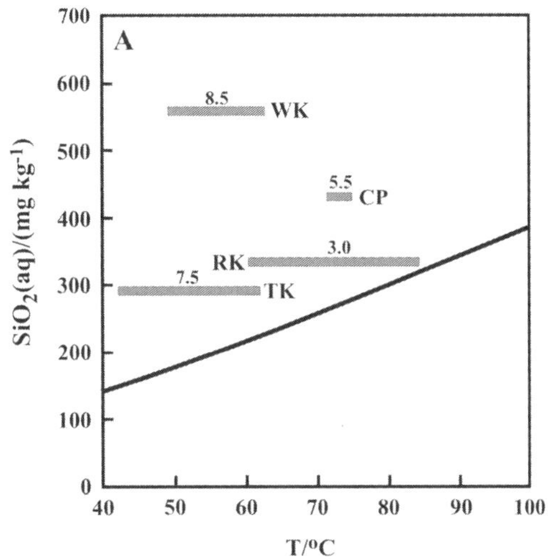

Figure 5. Saturation state of thermal waters versus temperature for with the solid black line representing the equilibrium amorphous silica solubility calculated from Gunnarsson and Arnórsson (2000). Gray bars show where amorphous silica is undersaturated (below) and supersaturated (above) with respect to amorphous silica solubility for each locality. Numbers above bars show the measured pH. WK, Wairakei; RK, Rotokawa; CP, Champagne Pool; TK, Tokaanu (Mountain et al. 2003).

B. flavothermus silica–microbe interactions were similar to those observed in field experiments. Microscopic observations showed that although *BF* becomes encased in silica, it continues to metabolize in the highly supersaturated silica solution. The bacteria became silicified but inorganic polymerization and aggregation of nanospheres dominated the process. When the experiments were carried out with added media, thin films of silica formed at the air-water interface and nanospheres or aggregates of silica deposited among the un-mineralized bacteria (Fig. 6a). These observations are similar to those observed from field experiments showing EPS or an organic-rich film that appeared to encourage silica aggregation.

In experiments where the microbes were previously grown on the slides most microbes were fully mineralized. Nevertheless, non-mineralized microbes also co-existed throughout the experiments (Fig. 6b). The silicification rates of *B. flavothermus* in all the experiments were fast (<8 h), however, the rates at which these processes occurred could not be quantified. These experiments show that inorganic processes dominate the polymerization and deposition of silica from solution. The presence of bacterial colonies may enhance the precipitation rate only as far as they may provide a more active surface where silica

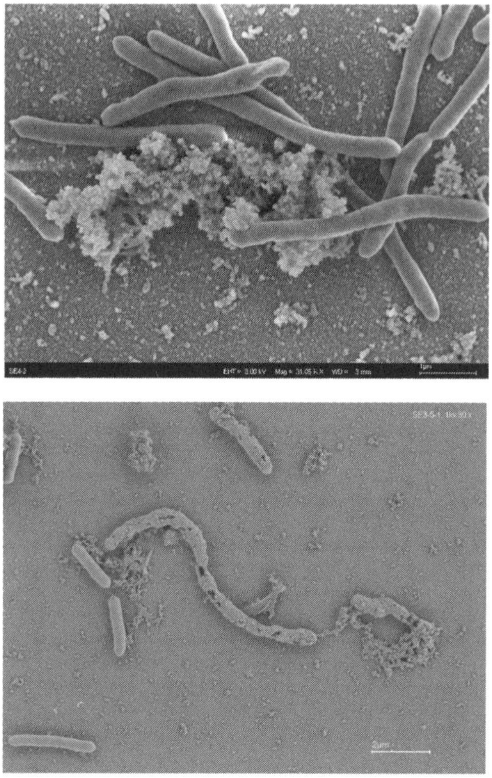

Figure 6. Microphotographs from the *in situ* laboratory silicification of *B. flavothermus* (a) live bacteria together with aggregates of silica nano-spheres (silicified for 2 hours, with added media; bar = 1 μm) (b) live and templated microorganisms (silicified for 8 hours with the bacteria grown on the slides prior to silicification; bar = 2 μm).

nanosphere aggregation can occur. Further experiments, to elucidate the rates of these processes and the effect of nutrient and various bacterial components, are underway.

The growth of silica sinters in modern or ancient hot springs follows a multistep process governed by inorganically driven silica poly-condensation that may be affected by an increase in bacterial EPS. From the *in situ* field experiments presented here, the role of the organic components of the microbial biomass could not be resolved. Laboratory silicification batch experiments carried out previously with cyanobacteria (*Calothrix sp.* Yee et al. 2003) showed that in single step silicification experiments the affinity of silica to bind to the bacterial surface is low. In sequential experiments (Phoenix et al. 2000; Benning et al. 2004a, b) it has been shown that the silicification is governed by an initial increase in EPS in the bacterial sheath with concomitant silica polycondensation and nanosphere formation. This followed by EPS aided aggregation of silica nanospheres (stickiness of the EPS sugars). Benning et al. (2004b) showed that silica polymerization and subsequent silica nanosphere nucleation in the presence of cyanobacteria is not enhanced but that in a first stage cyanobacteria react by increasing the amount of EPS in their sheath. This process encourages the accumulation of inorganically nucleated silica nanoparticles into larger silica aggregates, which upon further addition of silica will lead to cyanobacterial fossilization.

Experimental and theoretical evidence indicates that the polymerization of silica monomers leads to the formation of branching clusters that eventually collapse to form a spherical particle. Although, the parameters controlling this collapse are unclear, some evidence indicates that it may be the dehydroxylation of silanol or silane clusters, which leads to the formation of the first nanospheres.

4 CONCLUSIONS

Both laboratory and field experiments have shown that silica precipitation can occur subaqueously and subaerially without the influence of bacterial surfaces. Measured growth rates and fluid chemistry from field studies has shown that the rate of silica precipitation depends partly on the degree of supersaturation with respect to amorphous silica and partly on solution pH. The presence of microbes is more likely to have an effect on the macroscopic rate of silica precipitation. The rate per unit area does not change but the overall amount of silica precipitated is greatly increased. The laboratory experiments also suggest that other sites of silica polymerization and growth include the air–water interface. This may partly explain the finely-laminated textures observed in microspicules from Champagne Pool.

ACKNOWLEDGEMENTS

LGB acknowledges funding from The Leverhulme Trust (Ref. #: F/00122/F), the United Kingdom Natural Environment Research Council Grant No. GR9/04623, as well as funding from the Envirosync Program sponsored by the UK Central Laboratories of the Research Councils, Direct Access grants no: 37059 and 38463. The Advances Light Source is supported by the Director, Office of Sciences, Office of Basic Energy Sciences, Materials Sciences Division of the U.S. Department of Energy under contract No. DE-AC03-76SF00098 at the Lawrence Berkeley National Laboratory. BWM acknowledges funding from the Foundation for Research in Science and Technology (Contract C05X0201: Geothermal for the New Millenium) and from the NSOF Extremophiles Programme. We are indebted to Eric Condliffe (School of Earth Sciences, Leeds) and Adrian Hick (School of Biology, Leeds) for help with the SEM and TEM work and to Mark Tobin for support during the synchrotron IR work. Sampling permission in New Zealand was granted by the Department of Conservation, Tongariro and Bay of Plenty Conservancies, New Zealand.

REFERENCES

Benning, L.G., Phoenix, V., Yee, N. & Tobin, M.J. 2004a. Molecular characterization of cyanobacterial cells: A synchrotron-based infrared study. *Geochim. Cosmochim. Acta* 68(4): 729–741.

Benning, L.G., Phoenix, V.R., Yee, N. & Konhauser, K.O. 2004b. The dynamics of cyanobacterial silicification: An infrared micro-spectroscopic investigation. *Geochim. Cosmochim. Acta* 68(4): 729–741.

Cady, S.L. & Farmer, J.D. 1996. Fossilization processes in siliceous thermal springs: trends in preservation along thermal gradients. In G.R. Brock & J.A. Goode (eds.), *Evolution of Hydrothermal Ecosystems on Earth (and Mars?)*: 150–173. Chichester: John Wiley.

Carroll, S., Mroczek, E.K., Alai, M. & Ebert, M. 1998. Amorphous silica precipitation (60 to 120°C); comparison of laboratory and field rates. *Geochim. Cosmochim. Acta* 62: 1379–1396.

Campbell, K.A., Rodgers, K.A., Brotheridge, J.M.A. & Browne, P.R.L. 2002. An unusual modern silica-carbonate sinter from Pavlova spring, Ngatamariki, New Zealand. *Sedimentol.* 49, 835–854.

Fein, J.B., Scott, S. & Rivera, N. 2002. The effect of Fe on Si adsorption by *Bacillus subtilis* cell walls: Insights into non-metabolic bacterial precipitation of silicate minerals. *Chem. Geol.* 182: 265–273.

Ferris, F.G., Beveridge, T.J. & Fyfe, W.S. 1986. Iron-silica crystallite nucleation by bacteria in a geothermal sediment. *Nature* 320: 609–611.

Gunnarsson, I. & Arnórsson, S. 2000. Amorphous silica solubility and the thermodynamic properties of $H_4SiO_4^\circ$ in the range of 0° to 350°C at P_{sat}. *Geochim. Cosmochim. Acta* 64: 2295–2307.

Inagaki, F., Hayashi, S., Doi, K., Motomura, Y., Eiji, I. & Ogata, S. 1997. Microbial participation in the formation of siliceous deposits from geothermal water and analysis of the extremely thermophilic bacterial community. *FEMS Microbiol. Ecol.* 24: 41–48.

Jones, B., Renaut, R.W. & Rosen, M.R. 1997. Biogenecity of silica precipitation around geysers and hot spring vents, North Island, New Zealand. *J. Sediment. Res.* 67: 88–104.

Jones, B., Renaut, R.W. & Rosen, M.R. 1999. Actively growing siliceous oncoids in the Waiotapu geothermal area, North Island, New Zealand. *J. Geol. Soc. London* 156: 89–103.

Jones, B., Renaut, R.W. & Rosen, M.R. 2000. Stromatolites forming in acidic hot-spring waters, North Island, New Zealand. *Palaios* 15: 450–475.

Jones, B., Renaut, R.W. & Rosen, M.R. 2001. Microbial construction of siliceous stalactites at geysers and hot springs: Examples from Whakarewarewa geothermal area, North Island, New Zealand. *Palaios* 16: 73–94.

Konhauser, K.O. & Ferris, F.G. 1996. Diversity of iron and silica precipitation by microbial mats in hydrothermal waters, Iceland: Implications for Precambrian iron formations. *Geol.* 24: 323–326.

Konhauser, K.O., Phoenix, V.R., Bottrell, S.H., Adams, D.G. & Head, I.M. 2001. Microbial-silica interactions in Icelandic hot spring sinter: possible analogues for some Precambrian siliceous stromatolites. *Sedimentol.* 48: 415–433.

Lippencot, E. R., Van Valkenburg, A., Weir, C.E. & Bunting, E.N. 1958. Infrared studies of polymorphs of silica and germania. *J. Res. Nat. Bur. Stand.* 61: 61–70.

Liquier, J. & Taillandier, E. 1996. Infrared spectroscopy of nucleic acids. In H.H. Mantsch & D. Chapman (eds.), *Infrared Spectroscopy of Biomolecules:*131–158. Chichester: Wiley.

McKenzie, E.J., Brown, K.L., Cady, S.L. & Campbell, K.A. 2001. Trace metal chemistry and silicification of microorganisms in geothermal sinter, Taupo Volcanic Zone, New Zealand. *Geothermics* 30: 483–502.

Mountain, B.W., Benning, L.G. & Boerema, J. 2003. Experimental Studies on New Zealand Hot Spring Sinters: Rates of Growth and Textural Development. *Can. J. Earth Sci.* 40: 1643–1667.

Nakamoto, K. 1986. Infrared and raman spectra of inorganic and coordination compounds. New York: Wiley.

Naumann, D., Schultz, C. & Helm, D. 1996. What can infrared spectroscopy tell us about the structure and composition of intact bacterial cells? In H.H. Mantsch & D. Chapman (eds.), *Infrared Spectroscopy of Biomolecules*: 279–310. New York: Wiley-Liss, Inc.

Nelson, W.H. 1991. *Modern techniques for rapid microbiological analysis*. New York: VCH Publishers.

Phoenix, V.R., Adams, D.G. & Konhauser, K.O. 2000. Cyanobacterial viability during hydrothermal biomineralization. *Chem. Geol.* 169: 329–338.

Phoenix, V.R. 2001. *Microbial-biomineral interactions and their significance for the formation of chemical sediments*. PhD thesis. University of Leeds.

Schultze-Lam, S., Ferris, F.G., Konhauser, K.O. & Wiese, R.G. 1995. *In situ* silicification of an Icelandic hot spring microbial mat: implications for microfossil formation. *Can. J. Earth Sci.* 32: 2021–2026.

Stuart, B. & Ando, D.J. 1997. *Biological applications of infrared spectroscopy*, ACOL publication (University of Greenwich) Chichester: Wiley.

Tripp, C.P. & Hair, M. 1991. Reaction of chloromethylsilanes with silica: A low-frequency infrared study. *Langmuir* 7: 923–927.

Walter, M.R. 1976. Geyserites of Yellowstone National Park: An example of abiogenic stromatolites. In M.R. Walter (ed.), *Stromatolites. Developments in Sedimentology* 20: 87–112.

Walter, M.R., Bauld, J. & Brock, T.D. 1972. Siliceous algal and bacterial stromatolites in hot spring and geyser effluents of Yellowstone National Park. *Science* 179: 402–405.

Williams, D.H. & Fleming, I. 1996. *Spectroscopic methods in organic chemistry (5th edn.)*. London: McGraw-Hill International Ltd.

Wong, P.T.T., Wong, R.H., Caputo, T.A., Godwin, T.A. & Rigas, B. 1991. Infrared-spectroscopy of exfoliated human cervical cells – Evidence of extensive structural changes during carcinogenesis. *Proc. Natl. Acad. Sci. U.S.A.* 88: 10988–10992.

Yee, N., Phoenix, V.R., Konhauser, K.O., Benning, L.G. & Ferris, F.G. 2003. The effect of cyanobacteria on Si precipitation kinetics at neutral pH: Implications for bacterial silicification in geothermal hot springs. *Chem. Geol.* 199: 83–90.

Zeroual, W., Manfait, M. & Choisy, C. 1995. FT-IR spectroscopy study of perturbations induced by antibiotic on bacteria *Pathologie et Biologie* 43(4): 300–305.

Water–Rock Interaction, Wanty & Seal II (eds)
© 2004 Taylor & Francis Group, London, ISBN 90 5809 641 6

Control of dilation and collapse during weathering and soil formation on Hawaiian basalts

O.A. Chadwick & R.D. Goldstein
Department of Geography, University of California, Santa Barbara, CA, USA

ABSTRACT: Weathering is a fundamental process that involves swapping rock-derived elements for atmosphere-derived components. Strain is a fundamental property of the resulting soil that can be interpreted in the context of additions and losses from weathering profiles. Here we present strain data from three soil chronosequences in Hawaii that span from arid to hyperhumid environments and interpret them with respect to additions of C and water and losses of Si and Al during the weathering process. It takes tens to hundreds of thousands of years for soils to shift from dilation to collapse as elemental leaching slowly mitigates the effect of C added as organic matter and water incorporated into clay minerals. Soils forming in humid environments that have unimpeded drainage show the most rapid shift to collapse whereas soils in hyperhumid environments take longer because pores become plugged and water moves through them slowly, and soils in arid environments are exposed to far less leaching power.

1 INTRODUCTION

Weathering is a process of mineral conversion with soluble byproducts that are easily lost through leaching. Our concepts often involve an implicit assumption that weathering and soil formation involve losses and hence collapse of the weathering material (Brimhall & Dietrich 1987). The reality is that different parent materials weather differently, different climates have different efficiencies in byproduct removal, and additions from atmospheric sources tend to mitigate losses and collapse (Brimhall et al. 1992; Vitousek et al. 1997; Capo & Chadwick 1999). Calculation of strain, or the amount of collapse/dilation of a weathering profile, involves use of a relatively immobile element as an indicator element. As other elements are added or lost from the weathering volume, the index element will be diluted or concentrated accordingly.

Changes in strain imply specific additions and losses that provide greater understanding of interactions occurring at the interface of the lithosphere, hydrosphere, biosphere and atmosphere. For instance, on arkosic sands collapse is related to Si leaching losses in humid environments; in the same profiles, horizons showing expansions exhibit enhanced levels of C added as organic matter (Chadwick et al. 1990; Brimhall et al. 1992). C has also been implicated in expansion of subsurface horizons where calcium carbonate accumulates in desert soils. The Ca in calcretes is usually delivered to soil profiles dissolved in rainwater and it combines with atmospheric C reduced via photosynthesis to form soil horizons that show clear physical as well as chemical evidence of expansion (Capo & Chadwick 1999). Aeolian addition and subsequent incorporation of chemically mature elements such as Al and Fe can mitigate collapse caused by leaching of more soluble components or even drive expansion in their accumulation zones (Brimhall et al. 1992).

Strain calculations provide a basis for interpretation of the long-term effects of mass transfers from rock to water and from atmosphere to soil. Their interpretation usually benefits from multiple ancillary lines of evidence. Knowledge of soil–landscape relationships is very important as well as use of microscopic and isotopic evidence. Here we analyze profile strain and the gains of C and water and contrast them with losses related to leaching as basalt weathers and soil develops in different climate zones in Hawaii. We interpret these results in the context of a considerable body of knowledge about the properties of Hawaiian soils.

2 METHODS

We sampled three chronosequences on lava flows that have been influenced by 500 mm, 2500 mm and 4000 mm of annual rainfall. Hawaii provides an ideal setting for weathering research on basalt because the parent material composition varies little while age varies across the islands (MacDonald et al. 1983; Moore & Clague 1992; Wolfe & Morris 1996), and rainfall varies on individual volcanoes relative to the asymmetric impact of the northeast trade winds (Giambelluca et al. 1986). Soil profiles were excavated by hand or backhoe, described using standard USDA procedures and sampled by horizon. Bulk density was measured using the clod and volume displacement method and rock fragments were estimated on a volume basis (Soil Survey Staff 1992). Horizon samples were characterized in a number of different ways (Chadwick et al. 2003), but here we present data derived from loss on ignition (weight loss on heating to 500°C) and x-ray fluorescence (XRF; major element analysis).

Mass fraction lost or gained for each element was calculated as follows (Brimhall & Dietrich 1987; Chadwick et al. 1990; Brimhall et al. 1992):

$$\tau_{j,w} = ((P_w C_{j,w})/(P_p C_{j,p})(\varepsilon_{i,w} + 1)) - 1 \qquad (1)$$

where P is soil bulk density, C_j is the concentration of the element of interest, w represents weathered horizons and p represents parent material and:

$$\varepsilon_{i,w} = ((P_p C_{j,p})/(P_w C_{i,w})) - 1 \qquad (2)$$

where C_i represents the concentration of an immobile element. Here we use Nb measured by XRF as the "immobile element" (Kurtz et al. 2000).

Expansion and collapse result from a balance of losses and additions to soil profiles. Losses occur as elements are leached and as for most soils, Hawaii soils lose elements in the following order: base cations, Si, Al, Fe (although this can be modified under reducing conditions). The fact that Si composes nearly 50% of the non-oxygen portion of basalt means that its loss strongly influences overall loss vectors. Here we focus on both Si and Al as indicators of loss from profiles. Gains in soil profiles that are not strongly influenced by aeolian sources, which is the case in Hawaii, generally come from additions of C and water. C comes from the biological cycle and rainwater is incorporated into clay structures during weathering. Here we use loss on ignition as a combined measure of C and water added to soil profiles. The indicators of addition and loss are combined into a single parameter as follows:

$$\text{Net Gain/Loss} = \text{LOI} + (\tau_{Si,w} + \tau_{Al,w}) + 1 \qquad (3)$$

Data for 47 soil profiles are presented in this paper; they were measured on a horizon basis with as many as 11 horizons per profile. For presentation the data have been summarized in several ways: (1) the profile data has been averaged on a depth-weighted basis to provide one value per profile, (2) in plots of strain as a function of soil age, profiles sampled on the same age flows have been averaged, (3) in plots of strain vs. net gain and loss of element we present all the depth-weighted profile data (i.e. 47 data points in all).

3 RESULTS

The chronosequences characterize weathering and soil development in three soil environments: arid with no appreciable deep leaching; humid with appreciable deep leaching, and relatively little Fe reduction; and hyperhumid with impeded drainage developing over time leading to Fe reduction and loss as an important soil-forming factor.

3.1 *Arid chronosequence*

The arid chronosequence was sampled on lava flows ranging from 2 ky to 350 ky. Within the timeframe of this chronosequence, we are quite certain that the climate has always been dry (Chadwick et al. 2003; Ziegler et al. 2003). Present vegetation is bunch grass and land use is low-intensity cattle pasture. Arid soils in Hawaii accumulate carbonate (Whipkey et al. 2001), but at a rate that is much slower than that occurring on continents because there is far less mineral aerosol and dissolved Ca being delivered to these soils. The soils also exhibit a mineral transformation trend that indicates rapid transition from primary minerals to poorly crystalline allophane with a slower transition to halloysite (Ziegler et al. 2003).

Depth-weighted average strain for the arid soils indicates >100% expansion of the profiles for the first 30 ky followed by a decline that leaves the 170 ky profiles only slightly expanded (Fig. 1, top). The 350 ky profiles show a small amount of collapse. Soils that are 30 ky and younger gain organic matter and water to a far greater extent than they lose Si and Al (Fig. 1, bottom). The older soils have lost more Si but little Al and importantly more of their mineral assemblages have transformed from allophane, a strongly hydrated mineral, to halloysite that has far less water. As a consequence, losses nearly balance gains and the profiles show minimal strain.

3.2 *Humid chronosequence*

The humid chronosequence was sampled on lava flows ranging from 0.3 ky to 4,100 ky old (Crews et al. 1995). While the sites are all at 16°C and 2500 mm of rain, they have been influenced by differences in climate primarily due to differences in the altitude of the tropical inversion layer controlled by glacial-interglacial cycles (Hotchkiss et al. 2000). The youngest soil is still composed of primary minerals mixed with organic matter, the middle aged soils,

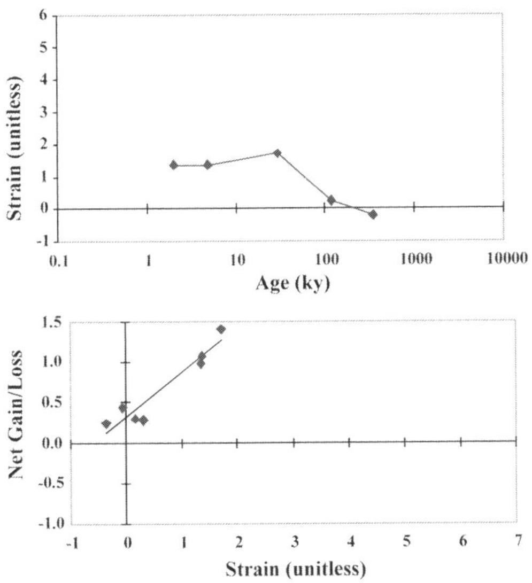

Figure 1. Depth-weighted average strain for the arid chronosequence as a function of soil age based on lava flow age (top) and strain as a function of the depth-weighted average net gain/loss (bottom).

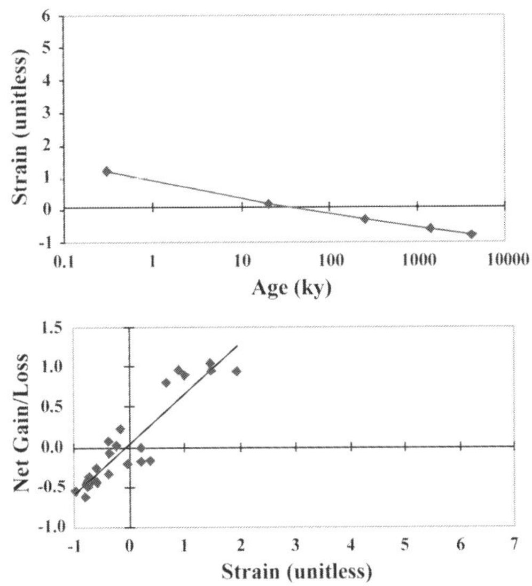

Figure 2. Depth-weighted average strain for the humid chronosequence based on lava flow age (top) and strain as a function of the depth-weighted average net gain/loss (bottom).

20 to 150 ky are dominated by non-crystalline allophane and ferrihydrite, and the two oldest soils are dominated by crystalline halloysite, gibbsite and hematite (Vitousek et al. 1997; Chorover et al. 1999). The intermediate age soils have gained considerable amounts of organic matter that is chemically linked with allophane and ferrihydrite, leading to slow turnover of the C compounds (Torn et al. 1997).

Depth-weighted average strain for the humid soil indicates >100% expansion in the youngest soil but after 20 ky nearly most of the expansion has been lost and by 150 ky the soils show considerable collapse (Fig. 2, top). The 4,100 ky soils show an average of 80% collapse. The youngest soils have gained much organic matter and water while not losing virtually any Si and Al (Fig. 2, bottom). The intermediate age soils have gained even more organic matter and water but have lost nearly 90% of the original Si and 30–40% of the original Al. Thus the net gain and loss values range around 0 and so do the strain values. The oldest soils exhibit the greatest amount of collapse and the balance between addition and loss is strongly shifted to the negative side (Fig. 2, bottom). The decline in both parameters is due to a transition from allophane and ferrihydrite to halloysite, gibbsite, and hematite. This transition lowers the structural water content as well as the chemical stability of organic matter leading to a decline in organic C levels. Even as organic matter and structural water declines in these old soils, Si and Al continue to be leached.

3.3 *Hyperhumid chronosequence*

The hyperhumid soils were sampled on lava flows ranging from 0.2 ky to 4,100 ky old. They are on windward exposures of the volcanoes where the trade winds bring moisture during most days of the year. The young soils (0.2 and 6 ky) are dominated by primary minerals but about a quarter of the mineral assemblage in the 6 ky soils is composed of allophane and ferrihydrite. The 40 ky soils shows a dramatic decrease in particle size compared to the younger soils and all the primary minerals have weathered to allophane and ferrihydrite. The finer-grained soil has smaller pores and water tends to pond during heavy rainfall events. The 40 ky soils have horizons where Fe has been redistributed leading to even greater plugging of soil porosity. The older soils are poorly drained and they are anaerobic much of the time leading to considerable loss of Fe through reduction (Miller et al. 2001). The mineralogy of the 150 and 400 ky soils is dominated by allophane but the 1,400 and 4,100 ky sites have halloysite as its major mineral component with gibbsite as a secondary component. These soils have thick O-horizons because of decreased C cycling efficiency (Schuur et al. 2001). We excluded O-horizons and started the analysis at the top of the mineral soil.

Depth-weighted average strain rises rapidly to nearly 600% dilation by 6 ky and then declines, but it remains over 100% dilation as long as 400 ky before dropping to nearly 90% collapse in the 1,400 ky and

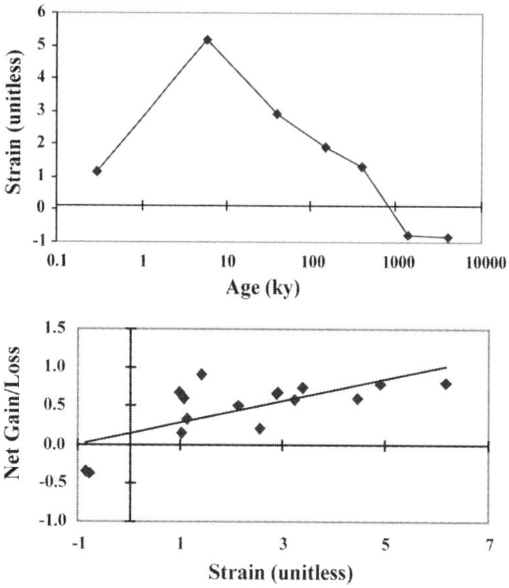

Figure 3. Depth-weighted average strain for the hyperhumid chronosequence based on lava flow age (top) and strain as a function of the depth-weighted average net gain/loss (bottom).

4,100 ky soils (Fig. 3, top). The youngest soils gained large amounts of organic matter and by 6 ky the soils are composed primarily of organic matter and allophane. The bulk densities of these soils are only about 0.2–0.3 g/cm^3 due to the massive increase in structural water and organic matter. These young soils have lost little Si and Al and therefore the dominance of C and structural water leads to high levels of dilation (Fig. 3, bottom). The relationship among strain and gains and losses of components is less clear in Figure 3, bottom, probably because Fe represents a major loss vector that is not accounted for in our calculations. The 150 and 400 ky soils have lost most of their Fe because the soils have been subjected to long-term reducing conditions. Addition of Fe to the calculations would pull those points to lower net gain/loss values and likely reduce the noise in the analysis. Fe loss is not significant in the other two chronosequences.

4 CONCLUSIONS

Soil formation is driven by reactions that transfer atmospherically and biologically derived acids into secondary soil minerals during weathering of primary minerals inherited in parent material or dust. Initially this process leads to dilation of soil profiles as C accumulates as humus and water accumulates in secondary clays. Eventually, organic matter and clay accumulation reach steady state, but leaching of elements continues to occur. Leaching of Si, in particular, and Al to a lesser extent, counteracts the accumulation of C and water leading to long-term collapse of soil profiles. One of the interesting results from Hawaii is the length of time over which soils remain dilated. The transition from dilation to collapse occurs between 20 and 150 ky for the humid chronosequence, 170 and 350 ky for the arid chronosequence and 400 and 1,400 ky for the hyperhumid chronosequence.

The reasons for differences in the transition point are complex, having to do with amount of effective moisture and efficiency of leaching, but also related to mineral dehydration and transformation during soil development, and the interaction of C and water with minerals at different stages of soil development. It is evident that the quantifiable property of strain is an emergent property of soil that can be explained by considering the interactions among atmospheric additions to and hydrologic losses from soil profiles as well as internal transformations during pedogenesis. Quantified strain and its explanatory variables allow us to consider weathering in a far broader context, one that explicitly focuses on sources and sinks in Earth surface chemical reactions.

REFERENCES

Brimhall, G.H. & Dietrich, W.E. 1987. Constitutive mass balance relations between chemical composition, volume, density, porosity, and strain in metasomatic systems: results on weathering and pedogenesis. *Geochim. Cosmochim. Acta* 51: 567–587.

Brimhall, G.H., Chadwick, O.A., Lewis, C.J., Compston, W., Williams, I.S., Danti, K.H., Dietrich, W.E., Power, M.E., Hendricks, D.M. & Bratt, J. 1992. Deformational mass transport and invasive processes in soil evolution. *Science* 255: 695–702.

Capo, R.C. & Chadwick, O.A. 1999. Sources of strontium and calcium in desert soil and calcrete. *Earth Planet. Sci. Lett.* 170: 61–72.

Chadwick, O.A., Brimhall, G.H. & Hendricks, D.M. 1990. From a black box to a grey box – a mass balance interpretation of pedogenesis. *Geomorph.* 3: 369–390.

Chadwick, O.A., Derry, L.A., Vitousek, P.M., Heubert, B.J. & Hedin, L.O. 1999. Changing sources of nutrients during four million years of ecosystem development. *Nature* 397: 491–497.

Chadwick, O.A., Gavenda, R.T., Kelly, E.F., Ziegler, K., Olson, C.G., Elliott, W.C. & Hendricks, D.M. 2003. The impact of climate on the biogeochemical functioning of volcanic soils. *Chem. Geol.* 202: 195–223.

Chorover, J., DiChiaro, M.J. & Chadwick, O.A. 1999. Structural charge and cesium retention in a chronosequence of tephritic soils. *Soil Sci. Soc. Am. J.* 63: 169–177.

Crews, T.E., Kitayama, K., Fownes, J.H., Ripley, R.H., Herbert, D.A., Mueller-Dombois, D. & Vitousek, P.M.

1995. Changes in soil phosphorus and ecosystem dynamics along a long soil chronosequence in Hawaii. *Ecology* 76: 1407–1424.

Giambelluca, T.W., Nullet, M.A. & Schroeder, T.A. 1986. *Rainfall Atlas of Hawaii.* Hawaii Department of Land and Natural Resources, No. R76, Honolulu, HI.

Hotchkiss, S.C., Vitousek, P.M., Chadwick, O.A. & Price, J. 2000. Climate cycles, geomorphological change, and the interpretation of soil and ecosystem development. *Ecosystems* 3: 522–533.

Kurtz, A.C., Derry, L.A., Chadwick, O.A. & Alfano, M.J. 2000. Refractory element mobility in volcanic soils. *Geol.* 28: 683–686.

Macdonald, G.A., Abbott, A.T. & Peterson, F.L. 1983. *Volcanoes in the Sea. 2nd edition.* Honolulu: University of Hawaii Press.

Miller, A.J., Schuur, E.A.G. & Chadwick, O.A. 2001. Redox control of phosphorus pools in Hawaiian montane rainforest soils. *Geoderma* 102: 219–237.

Moore, J.G. & Clague, D.A. 1992. Volcano growth and evolution of the island of Hawaii. *Geol. Soc. Am. Bull.* 104: 1471–1484.

Schuur, E.A.G., Chadwick, O.A. & Matson, P.A. 2001. Carbon cycling and soil carbon storage in mesic to wet Hawaiian montane forests. *Ecology* 82: 3182–3196.

Soil Survey Staff. 1992. *Soil Survey Laboratory Methods Manual.* Soil Survey Investigations, Report No. 42. v. 2.0, USDA, Lincoln, NE.

Torn, M.S., Trumbore, S.E., Chadwick, O.A., Vitousek, P.M. & Hendricks, D.M. 1997. Mineral control of soil carbon storage and turnover. *Nature* 389: 170–173.

Vitousek, P.M., Chadwick, O.A., Crews, T.E., Fownes, J.H., Hendricks, D.M. & Herbert, D. 1997. Soil and ecosystem development across the Hawaiian Islands. *GSA Today* 7: 1–8.

Whipkey, C.E., Capo, R.C., Hsieh, J.C.C. & Chadwick, O.A. 2002. Development of magnesian calcite and dolomite in Quaternary soils on the island of Hawaii. *J. Sediment. Res.* 72: 158–165.

Wolfe, E.W. & Morris, J. 1996. *Geologic map of the Island of Hawaii.* US Geological Survey Map I-2524-A.

Ziegler, K., Hsieh, J.C.C., Chadwick, O.A., Kelly, E.F., Hendricks, D.M. & Savin, S.M. 2003. Halloysite as a kinetically controlled end product of arid-zone basalt weathering. *Chem. Geol.* 202: 461–478.

Water–Rock Interaction, Wanty & Seal II (eds)
 ISBN 90 5809 641 6

Chemical weathering in steady-state orogens – examples from the Himalaya and Southern Alps

C.P. Chamberlain & J.R. Waldbauer
Department of Geological and Environmental Sciences, Stanford University, Stanford, California, USA

ABSTRACT: We present an analytical model for steady-state weathering that captures two elements of steady-state orogens, continual supply of fresh rock to the weathering regime and the formation of easily weathered calcite in active hydrothermal systems. This one-dimensional model accounts for chemical weathering of constituent minerals and the release of cations from the weathering profile as material moves from the bottom of the weathering zone to the surface at constant velocity. This model is parameterized in terms of "effective surface age" which is defined as the amount of time it takes for rock to move from the lower boundary of the weathering profile to the surface. Comparison of model results with data sets throughout the world show that our model reproduces observed trends in chemical weathering. Chemical weathering rates are closely linked with tectonism and decline sharply with increasing effective surface age.

1 INTRODUCTION

The weathering of silicate minerals has long been recognized as one of the major long-term controls on the levels of atmospheric CO_2. Because of the link between chemical weathering and the carbon budget of Earth there have been numerous studies that quantify chemical weathering rates. These studies have shown that there are a large number of variables that exert influence on rates of rock weathering. Mineralogy of the bedrock is of primary importance, as the types of minerals present and their cation content will determine how much CO_2 will be consumed by their weathering. The effect of climate on weathering rates is also significant. White & Blum (1995) found that an increase in temperature and precipitation results in an increase in weathering rates, a trend that has been incorporated into global CO_2 models (Berner 1995). The other major factor in determining long-term, large-scale weathering rates is tectonic uplift. Tectonic uplift exposes fresh minerals to weathering and, therefore, determines the total amount of weatherable material on the Earth's surface. Recently, Riebe et al. (2001) found that tectonics rather than climate is the dominant factor controlling long-term weathering rates. The role of uplift in setting weathering rates, though incompletely quantified, is nevertheless essential to the global carbon cycle, as shown by the strong correlation of chemical and physical weathering rates in many world rivers (Gaillardet et al. 1999).

In this paper, we present an analytical model for steady-state chemical weathering with an emphasis on the effect of uplift on the total rate of weathering in a depth profile. This model is generally applicable to steady-state active orogens where denudation is balanced by rock uplift or in ancient orogens where denudation is compensated by isostatic uplift. The model is parameterized in terms of the "effective surface age", which allows comparisons of widely disparate tectonic regimes. From this model, we determine chemical weathering rates as a function of effective surface age, thereby showing how tectonism is related to chemical weathering. We also model the effect of the presence of disseminated hydrothermal calcite on the Sr isotope composition of weathered material, as disseminated calcite is a ubiquitous feature of active orogens as a result of active hydrothermal systems.

2 MODELLING CHEMICAL WEATHERING AS A FUNCTION OF UPLIFT

We developed a model for steady-state chemical weathering for the case where uplift provides a constant supply of fresh rock to the bottom of the weathering zone and physical erosion removes material from the surface at an equivalent rate. These conditions produce a constant weathering profile that is balanced by the uplift of fresh rock which together results in a stable flux of cations from the chemical weathering of parent rock. This balance of erosion and uplift occurs in active collisional tectonic environments. In these mountain belts, known as steady-state orogens, the mean surface elevation does not change; it is balanced by rock uplift and erosion (Koons 1990; Willet et al. 2001). Typically in active steady-state orogens rock moves through into the weathering zone at relatively high velocities (on the order of millimeters per year), partially chemically altered in the weathering zone, and is physically removed at the surface quickly enough to maintain a constant weathering profile. Our model should also be applicable to inactive mountain belts where erosion is compensated by isostatic uplift (Matmon et al. 2003). Here the downward propagation of a weathering front would equal the upward movement of fresh rock.

We modeled the chemical weathering rate of constituent minerals of unweathered rock and the total release of cations from the weathering profile as material moves from the bottom of the weathering zone to the surface at constant velocity. We assume semi-neutral conditions (pH $\sim$ 5–8), chemical weathering rate is independent of pH and the kinetics of dissolution are zeroth order with respect to water. We assume that fresh rock is supplied to the lower boundary at the same rate it is removed from the surface. The solution to this differential equation is as follows, and its derivation is given in Waldbauer & Chamberlain (in press):

$$R_i = \frac{q_i^0}{\tau}\left[1 - \exp\left(-\frac{6k_iSR_i}{\rho_i d_i}\tau\right)\right]$$

In this equation R_i is the weathering rate of mineral i in mol/(m$^3 \cdot$ s), q_i^0 is the concentration of the mineral in the unweathered rock, k_i is the dissolution rate (mol/m^2/s), ρ_i is the molar density of the mineral, d_i is the grain size (diameter of an assumed spherical grain), and SR_i is the surface roughness. We define an "effective surface age," τ, which is the amount of time it takes for rock to move from the lower boundary of the weathering profile to the surface, as:

$$\tau = \frac{Z}{u}$$

where Z is the depth of the weathering zone (m) and u is the uplift rate (m/s). The notion of effective surface age is a useful one, as it allows the weathering rate to be expressed as a function of a single variable that encompass measurable quantities related to tectonic and weathering processes in an orogenic belt. Effective surface ages vary greatly in differing tectonic environments. In an active collisional orogen like the Himalayas or Andes, uplift rates can be as high as 10 mm/yr and weathering zones are about 2 m deep, resulting in an effective surface age of 200 years. By contrast, a tectonically inactive craton in a tropical climate can be uplifting at only 0.01 mm/yr or less, while the weathering zone might be 30 m or deeper, so effective surface ages in these environments are in the 3–10 Myr range. This five-orders-of-magnitude range in effective surface ages is attributable more to the range of uplift rates across the globe ($\sim$4 orders of magnitude) than to variations in the depth-to-bedrock ($\sim$2 orders of magnitude); hence effective surface age is more a measure of tectonic uplift than of weathering profile depth.

This model is admittedly simplified in several points. First, we assume that the weathering profile is in a steady state, and does not change significantly in depth. This assumption certainly holds in the steady-state orogen described above, and may be essentially valid for a broader range of circumstances (Heimsath et al. 1997). Second, we ignore the production of secondary minerals, such as clays, during the weathering of silicates. A third simplifying assumption is that all minerals begin weathering at the same depth and mineral surface reactivity remains essentially constant, which may not hold true in some environments (White et al. 2001).

3 RESULTS

3.1 *Chemical weathering rates and CO_2 drawdown in tectonic regimes*

Figure 1a shows the results of this steady-state weathering model for the weathering of a typical granitic rock. The chemical weathering rate of the granitic substrate declines with increasing surface age, dropping by more than an order of magnitude between typical active orogen effective surface ages and temperate craton conditions. For areas with $\tau > 10^5$ yr, the chemical weathering rate is very low, and for the slowest-uplifting, thickest-mantled tropical cratons, the rate is $\sim 10^{-11}$ mol/(m$^3 \cdot$ s). These results are in general agreement with published measurements of weathering rates, as shown in Figure 1b, which shows the observed annual Si discharge from a range of world rivers against effective surface age.

Higher weathering rates of low effective surface age environments means that weathering in tectonically active areas should draw down significantly more CO_2 than weathering in tectonically inactive areas. Hence, it is in the building of large mountain belts that the greatest amount of atmospheric CO_2 can be sequestered in marine carbonate. This result of the steady-state chemical weathering model is in good accord with the recent detailed field study of

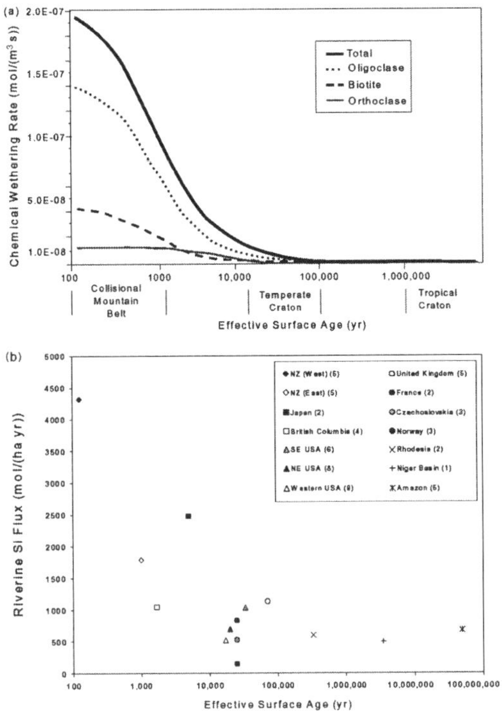

Figure 1. (a) Results of steady-state chemical weathering of granitic material. Ranges of effective surface ages for different tectonic regimes are indicated. (b) Riverine Si fluxes as a function of effective surface age for a number of world rivers.

Mortatti & Probst (2003), who found that the Andes region accounts for 78% of CO_2 drawdown by silicate weathering in the Amazon basin. These calculations are also consistent with the idea that large uplifted plateaus, such as the Himalayas, have had a significant impact on the CO_2 budget of the Earth (Raymo & Ruddiman 1992).

3.2 *Interpretation of the marine Sr isotope record; the role of calcite weathering*

In order to reconstruct the history of atmospheric CO_2, it is necessary to have some information about the past continental rock weathering rates, as the weathering of silicate minerals serves as a major sink for atmospheric CO_2. One proxy that has been used to estimate past continental rock weathering is the strontium isotope composition of marine carbonates. The past 40 million years have seen a marked increase in the seawater $^{87}Sr/^{86}Sr$ ratio; a trend attributed to the India–Asia collision and consequent uplift of the Himalayas. The uplift of the Himalaya exposed large amounts of fresh granitic rock with ^{87}Sr-enriched feldspar, which when weathered possibly sequestered large amounts of atmospheric CO_2 in marine sediments (Edmond 1992; Richter et al. 1992). This interpretation, however, is complicated by the fact that hydrothermal systems are a ubiquitous feature in active collisional mountain belts, as a result of topographic and buoyancy driven fluid flow (Koons & Craw 1991; Chamberlain et al. 1995; Templeton et al. 1998). These hydrothermal systems deposit ^{87}Sr-enriched calcite, which has no net effect in the long-term drawdown of CO_2 (Blum et al. 1998). This easily weathered ^{87}Sr-enriched calcite can significantly affect the interpretation of the strontium seawater curve, and, thus, it is important to assess the importance of carbonate weathering.

The strontium isotope composition of weatherable material will be strongly influenced by the abundance of calcite and the uplift rate. Using the following equation it is possible to model the strontium isotopic composition of minerals in the weathering zone as a function of uplift:

$$\left(\frac{^{87}\mathrm{Sr}}{^{86}\mathrm{Sr}}\right)_{\mathrm{surface}} = \left(\frac{^{87}\mathrm{Sr}}{^{86}\mathrm{Sr}}\right)_0 + \left(\frac{^{87}\mathrm{Rb}}{^{86}\mathrm{Sr}}\right)_0 \left(e^{\lambda(15\mathrm{km}/u)} - 1\right)$$

where the subscript 0 indicates the isotopic ratios at the closure temperature and u is the uplift velocity in km/yr. With knowledge of the initial ^{87}Sr and ^{87}Rb content of the rock, this equation provides a value for the strontium isotope composition of material weathering at the surface.

As shown in Figure 2a,b even a small amount of calcite has a strong influence on the strontium flux and strontium isotopes derived from weathering of granitic composition rocks. Secondary calcite dominates the strontium flux from low effective surface ages and holds the strontium isotope composition relatively constant over a wide range of effective surface ages. Since calcite has such a profound influence in tectonically active areas, and these areas have the highest weathering rates, this complicates the interpretation of the strontium seawater curve. Moreover, if our interpretation is correct that uplift rates as expressed by the effective surface age exerts strong control over the strontium isotopic composition of the ocean than the Sr flux should be inversely correlated and well coupled. This observation is precisely what has been found in a recent paper that measured Sr concentrations and Sr isotopes of benthic foraminifera for the early Cenozoic (Lear et al. 2003).

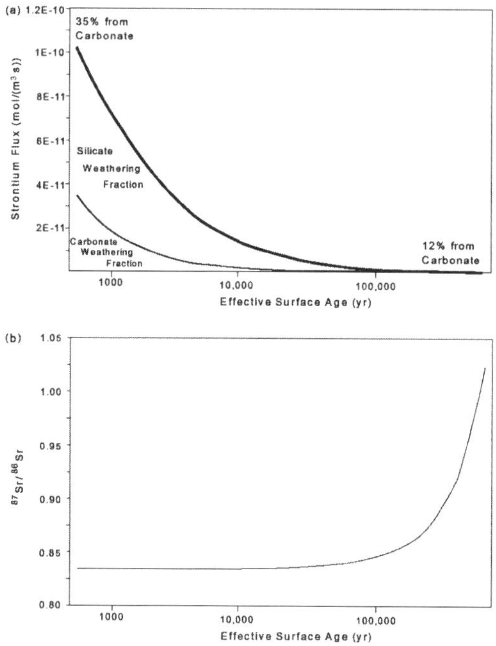

Figure 2. (a) Calculated Sr flux from weathering of granite with 3% disseminated calcite. (b) Calculated strontium isotopic ratios of weathered material. The weathering of 3% calcite holds the isotopic composition essentially constant over a large range of effective surface ages, yet the total chemical weathering rate drops significantly over the same range.

4 CONCLUSIONS

The analytical model presented in this paper quantifies the role of uplift in setting bulk weathering rates, and demonstrates that areas with low effective surface ages such as tectonically active areas are likely to play the largest role as land-based CO_2 sinks. This is important for reconstructing the history of atmospheric CO_2. An interesting consideration in the discussion of the relationship between tectonics and weathering is the question of what it means for a landscape to be in a steady state. We recognize that there must be a scale dependence in steady state chemical weathering, but as yet this has not been quantified.

REFERENCES

Berner, R.A. 1995. Chemical weathering and its effect on atmospheric CO_2 and climate. *Rev. Mineral.* 31: 565–584.

Blum, J.D., Gazis, C.A., Jacobson, A.D. & Chamberlain, C.P. 1998. Carbonate versus silicate weathering in the Raikhot watershed within the High Himalayan Crystalline Series. *Geol.* 26: 411–414.

Chamberlain, C.P., Zeitler, P.K., Barnet, D.E., Winslow, D., Poulson, S.R., Leahy, T. & Hammer, J.E. 1995. Active hydrothermal systems during the recent uplift of Nanga Parbat. Pakistan Himalaya. *J. Geophys. Res.* 100: 439–453.

Edmond, J.M. 1992. Himalayan Tectonics, Weathering processes, and the strontium isotope record in marine limestones. *Science* 258: 1594–1597.

Gaillardet, J., Dupre, B., Louvat, P. & Allegre, C.J. 1999. Global silicate weathering and CO_2 consumption deduced from the chemistry of large rivers. *Chem. Geol.* 159: 3–30.

Heimsath, A.M., Dietrich, W.E., Nishiizumi, K. & Finkel, R.C. 1997. The soil production function and landscape equilibrium. *Nature* 388: 358–361.

Koons, P.O. 1990. The two-sided orogen: Collision and erosion from the sand box to the Southern Alps, New Zealand. *Geol.* 18: 679–682.

Koons, P.O. & Craw, D. 1991. Evolution of fluid driving forces and composition within collisional orogens. *Geophys. Res. Let.* 280: 935–938.

Lear, C.H., Elderfield, H. & Wilson, P.A. 2003. A Cenozoic seawater Sr/Ca record from benthic foraminiferal calcite and its application in determining global weathering fluxes. *Earth Plan. Sci. Let.* 208: 69–84.

Matmon, A., Bierman, P.R., Larsen, J., Southworth, S., Pavish, M., Finkel, R.C. & Caffe, M. 2003. Erosion of an ancient mountain range, the Great Smokey Mountains, North Carolina and Tennessee. *Am. J. Sci.* 303: 817–855.

Mortatti, J. & Probst, J.L. 2003. Silicate rock weathering and atmospheric/soil CO_2 uptake in the Amazon basin estimated from river water geochemistry: seasonal and spatial variations. *Chem. Geol.* 197: 177–196.

Raymo, M.E. & Ruddiman, W.F. 1992. Tectonic forcing of late Cenozoic climate. *Nature* 359: 117–122.

Richter, F.M., Rowley, D.B. & DePaolo, D.J. 1992. Sr isotope evolution of seawater: the role of tectonics. *Earth Plan. Sci. Let.* 109: 11–23.

Riebe, C.S., Kirchner, J.W., Granger, D.E. & Finkel, R.C. 2001. Strong tectonic and weak climatic control of long-term chemical weathering rates. *Geol.* 26: 511–514.

Templeton, A.S., Chamberlain, C.P., Koons, P.O. & Craw, D. 1998. Stable isotopic evidence for mixing between metamorphic and surface-derived waters during uplift of the Southern Alps, New Zealand. *Earth Plan. Sci. Let.* 154: 73–92.

Waldbauer, J.R. & Chamberlain, C.P. in press. Influence of uplift, weathering and base cation supply on past and future CO_2 levels. A history of atmospheric CO_2 and its effects on plants, animals and ecosystems. Springer Verlag.

White, A.F. & Blum, A.E. 1995. Effects of climate on chemical weathering rates in watersheds. *Geochim. Cosmochim. Acta* 59: 1729–1747.

White, A.F., Bullen, T.D., Schulz, M.S., Blum, A.E., Huntington, T.G. & Peters, N.E. 2001. Differential rates of feldspar weathering in granitic regoliths. *Geochim. Cosmochim. Acta* 65: 847–869.

Willett, S.D., Slingerland, R. & Hovius, N. 2001. Uplift, shortening, and steady state topography in active mountain belts. *Am. J. Sci.* 301: 455–485.

Water–Rock Interaction, Wanty & Seal II (eds)

Origin of the carbon dissolved in the groundwater and derivation of Earth diffuse emission of CO_2: the case of the Italian Peninsula

G. Chiodini & S. Caliro
Osservatorio Vesuviano INGV, Napoli, Italy

C. Cardellini
Dipartimento Scienze della Terra Università di Perugia, Perugia, Italy

ABSTRACT: Diffuse Earth degassing is the main source for the carbon dissolved in the groundwaters of Apennines, which host the main aquifers of the Italian Peninsula. We used data from about 200 springs to derive a regional map of CO_2 Earth degassing from central and south Italy. A globally significant amount of deeply derived CO_2 (~10% of global CO_2 emitted from subaerial volcanoes) is released by two large areas located in western Italy. The map shows that the anomalous flux of CO_2 suddenly disappears in correspondence of a narrow band where most of the Italian seismicity concentrates. Here, at depth, the gas accumulates in crustal traps generating CO_2 overpressured reservoirs which favor seismogenesis.

1 INTRODUCTION

Recently the first quantitative estimations of present-day CO_2 emission from volcanoes were published (Gerlach 1991; Allard et al. 1991; Marty & Tolstikhin 1998) but very little data exists on non-volcanic CO_2 degassing (not directly from craters or flanks of volcanoes). As reviewed by Kerrick et al. (2001), non-volcanic degassing can provide a globally-significant amount of CO_2 to the atmosphere. In areas where high CO_2 fluxes directly affect regional aquifers, most or part of the deeply-generated gas can be dissolved by groundwaters because of the relatively high solubility of CO_2 in water. However, little data exists on the amount of deeply derived CO_2 dissolved by groundwaters. We will show here both the methods used to investigate the origin of the carbon dissolved by Apennine groundwaters and the results obtained in our investigations.

The Apennine chain divides central-southern Italy into two geographical regions characterized by distinctive geological, geophysical and geothermal features: the Tyrrhenian hinterland to the west and the Adriatic foreland to the east. The Tyrrhenian hinterland is characterised by a crust thickness of 20–25 km, high heat flow (locally higher than 200 mWm^{-2}, Della Vedova et al. 1984), positive magnetic anomalies (Arisi Rota & Fichera 1985), shallow earthquakes and a positive regional gravity anomaly. From the upper Miocene to the recent Quaternary, crustal extension occurred in the Tyrrhenian hinterland, causing widespread magmatic activity, crustal thinning and graben development (Barberi et al. 1971; Serri et al. 1993). Extensive Quaternary volcanics and Neogene units with low permeability overlie the compressional structures made up mainly of Mesozoic carbonate formations. The Tyrrhenian hinterland is also the seat of many drilled geothermal fields (e.g. Larderello, Monte Amiata, Latera, Torre Alfina, Cesano, Mofete) and of evident degassing of CO_2 from deep sources as manifested by numerous CO_2-rich gas emissions (Fig. 1).

The Adriatic foreland, in contrast, is characterized by a crustal thickness of approximately 30–40 km, normal heat flow (30 mWm^{-2}; Barchi et al. 1998), negative gravity anomalies and by the absence of CO_2-rich gas emissions.

These two areas, characterised by a very different geological framework, are separated by the Apennine chain mainly made up of permeable Mesozoic limestones and dolostones which host the major aquifers of the Italian peninsula. The groundwaters circulating in the Apennines have been studied in detail with the aim both to highlight a possible input of a gas phase of deep provenance and to quantify the amount of earth degassing in the region.

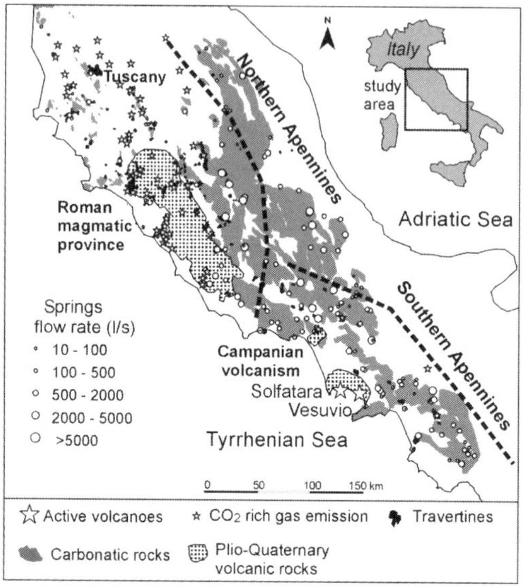

Figure 1. Location map of the studied springs and main evidence of earth degassing processes in central south Italy.

2 CHEMICAL COMPOSITION OF APENNINE GROUNDWATERS

2.1 *Sampling and analysis*

The data base is made up of the chemical and isotopic compositions of 139 springs of central and southern Apennine sampled during the last five years. The springs were selected based on their high flow rates in order to have a data set highly representative of the groundwater circulation of the region. A total flow rate of 2.21×10^6 l/s (58% of the total water discharge of the Apennine) has been sampled and analyzed for major chemical components, for the isotopic composition ($\delta^{13}C_{DIC}$) of dissolved inorganic carbon (DIC) and, in the springs of the southern Apennines (77 samples), for the dissolved gas composition (CO_2, N_2, O_2, Ar, CH_4, He).

2.2 *Chemical composition of groundwaters*

Most of the samples have a Ca(Mg)-HCO_3 composition while a minor amount of samples is characterised by either Ca-Mg-SO_4 or Na-Cl compositions. The saturation indexes of the samples with respect to the main minerals of the hosting rocks suggest that a major process controlling groundwater composition is the dissolution of calcite at near-equilibrium conditions ($SI_{cal} = -0.08 \pm 0.2$). In some cases (~30%) the waters also are close to equilibrium with dolomite while all the groundwaters are undersaturated with respect to gypsum (or anhydrite). However, dissolution of gypsum (or anhydrite) is evident in some springs showing SO_4 concentrations more than one order of magnitude higher than the springs of other aquifers. With dissolution of $CaSO_4$, dolomite dissolves and calcite precipitates to maintain water-mineral equilibrium (dedolomitization), shifting ground water toward Ca-Mg-SO_4 composition. Additional processes locally affecting the groundwater composition are dissolution of halite from evaporite formations and mixing of meteoric water with seawater in the aquifers located near the Tyrrhenian coast.

DIC was calculated for each sample with the aqueous speciation model *PHREEQC* (Parkhurst 1995). DIC molality ranges from 2.4 to 73 mmol/kg with an average value of 12 mmol/kg. The $\delta^{13}C$ values of the DIC range from −15‰ to +5‰ (PDB) where the more negative values characterise samples with the lowest DIC contents while the highest values are found in water with the highest DIC. This wide range of $\delta^{13}C$ values and the large variations observed in DIC values, suggest that inorganic carbon in the studied aquifers is derived from more than one source.

3 ORIGIN OF THE DIC

In order to investigate the effects of the input of a CO_2-rich gas phase into groundwater, we developed a specific computation procedure, named IGP (input of a gas phase), which computes the variations on the chemical composition of dissolved species, including gases, and on the $\delta^{13}C$ of the DIC.

3.1 *Simulation of groundwater composition during the IGP rich in CO_2*

Variable amounts of a gas phase of known composition and $\delta^{13}C$ value of CO_2, are added in a multistep process to 1 kg of solution at a fixed temperature T. The solution dissolves the gases until the total pressure of dissolved gases, P_{tot}, reaches a selectable threshold value (P_{deg}). When during a step P_{tot} becomes higher than P_{deg}, the solution is degassed to $P_{tot} = P_{deg}$, assuming ideal gas behavior according to Henry's law. At each step the solution is maintained at equilibrium with the calcite, dissolving or precipitating the necessary amount of the mineral. The computations were made by means of the code *PHREEQC* (Parkhurst 1995) integrated with two routines which compute at each step the carbon isotopic composition of DIC (Carbon Isotopic Composition, CIC) and the dissolved and degassed gas phase composition (Dissolved Degassed Gas, DDG).

The routine CIC uses the approach of Wigley et al. (1978); it is assumed that isotopic equilibrium exists between all dissolved carbon species in solution and between all outputs and the parent solution. It is also assumed that no fractionation occurs during the addition of any carbon to the solution. Isotopic fractionation therefore only takes place during the output of carbon from the parent solution. In detail, CIC solves the two following mass–balance equations:

$$\Delta(DIC) = \sum_{i=1}^{N} \Delta I_i - \sum_{i=1}^{M} \Delta O_{i1} \tag{1}$$

and

$$\delta^{13}C + \Delta(\delta^{13}C) = \frac{(\delta^{13}C + 1000)\cdot(DIC - \sum_{i=1}^{M} \alpha_{i-s}\Delta O_i)}{DIC + \Delta(DIC)} + \frac{\sum_{i=1}^{N}(\delta^{13}C_i^* + 1000)\cdot \Delta I_i}{DIC + \Delta(DIC)} - 1000 \tag{2}$$

where I_i and O_i are respectively the *i*th input and the *i*th output of carbon in the solution, respectively (expressed in molality); $\delta^{13}C$ is the bulk isotopic composition of the solution; $\delta^{13}C_i^*$ is the isotopic composition for the *i*th input species; α_{i-s}, is the fractionation factor between the *i*th output and the solution; N and M are the total numbers of incoming and outgoing carbon phases. In the cases applied the terms ΔO_i (carbon outputs in moles) include the CO_2 degassed and the calcite precipitated. In order to evaluate the effect of these outputs on the isotopic composition, the fractionation factors α_{i-s} of CO_2 and calcite with respect to carbon in solution have been computed assuming isotopic equilibrium at the temperature of the solution between CO_2, calcite, H_2CO_3, HCO_3 and CO_3 from:

$$a_{CO_2-S} = \frac{a_{CO_2} - H_2CO_3[H_2CO_3] + a_{CO_2-HCO_3}[HCO_3]}{TDIC} + \frac{a_{CO_2-CO_3}[CO_3]}{TDIC} \tag{3}$$

and

$$\frac{a_{CaCO_3-S} = a_{CO_2-S}}{a_{CO_2-CaCO_3}} \tag{4}$$

where the values of α are from Deines et al. (1974).

The routine DDG computes the composition of dissolved and degassed gas phases. In many problems the initial gas composition of the solution is assumed to be in equilibrium with the atmosphere while the composition of the incoming gas phase is suitably chosen. If P_{tot} becomes higher than P_{deg} then the solution is degassed to $P_{tot} = P_{deg}$. Single-step degassing was modeled starting from the compositions of the undegassed aqueous phase, referring to the following mass–balance of the generic *i*th gaseous species, which is expressed by the following relationship (Chiodini et al. 1996):

$$C_{O,i} = x_{CO_2} \times C_{v,i} + (1 - x_{CO_2}) \times C_{l,i} \tag{5}$$

where $C_{o,i}$, $C_{V,i}$ and $C_{l,i}$ are ratios of the concentration of the ith species and concentration of CO_2 in the undegassed liquid, in the degassed gas and in the residual liquid, respectively, while x_{CO_2} is the fraction of CO_2 degassed during the process. Accounting for the distribution coefficient with respect to CO_2, between the liquid and the gaseous phases,

$$BC_i = C_{V,i}/C_{V,i}C_{l,i} = K_{H,i}/K_{H,i}/K_{H,CO_2} \tag{6}$$

(computed at any temperature from Henry's constants, $K_{H,j}$), equation 5 may be rewritten as follows:

$$C_{l,i} = C_{0,i}/(1 + x_{CO_2} \times (BC_i - 1)) \tag{7}$$

The computation of the dissolved gas phase, and of the degassed one, is done iteratively changing the value of x_{CO_2} in equation 7 and computing the concentration of any gaseous species in the degassed liquid and its partial pressure P_i until $P_{tot} = P_{deg}$.

With this tool we investigated the processes occurring in a groundwater which infiltrates in an Apennine aquifer and which is subsequently affected by the input of a deep gas phase.

3.2 *Origin of the carbon dissolved in Apennine groundwaters*

During groundwater circulation, a common source of carbon is organic matter of soil. This source is active during infiltration of waters in recharge areas. The DIC and $\delta^{13}C$ values of a water which infiltrates in a carbonatic aquifer was simulated with the above IGP model choosing typical rain water, i.e. water in equilibrium with atmospheric gases and a $\delta^{13}C$ value of −5.5‰ as starting solution and setting $P_{deg} = 1$ bar. The incoming gas was chosen as pure CO_2 with carbon isotopic composition typical of the organic matter of the soil ($\delta^{13}C$ from −18‰ to −28‰), and the isotopic composition of the carbon from calcite dissolution was set at +2.21‰. The last is the average value of 567 samples of Apennine carbonate rocks (practically all the available data), which are characterised by $\delta^{13}C$ values from −2.44‰ to 4.40‰, mean of 2.21‰ and σ of 0.66‰ (Ferreri et al. 1997; Chiodini et al. 2000; Ghisetti et al. 2001; Morettini et al. 2002). An amount of 0.01 mol of incoming gas was added to the initial solution in 200 steps. The

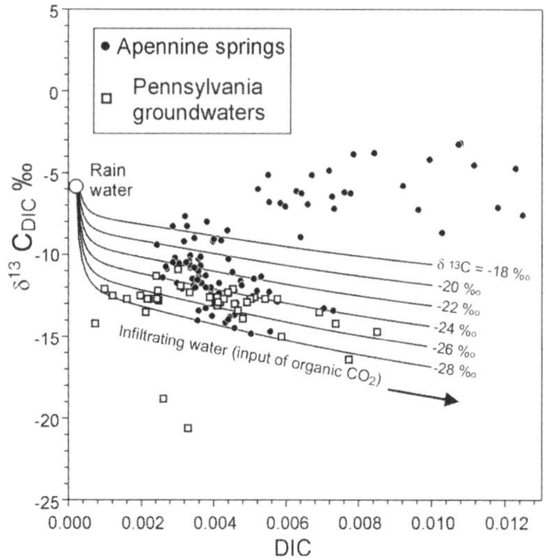

Figure 2. DIC vs $\delta^{13}C_{DIC}$ diagram. Analytical compositions of Apennine springs and of groundwaters circulating in Pennsylvanian carbonate aquifers are compared with the simulated compositions of infiltrating waters.

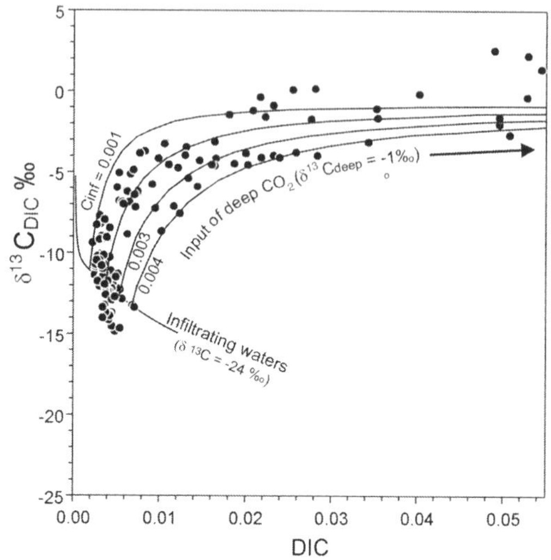

Figure 3. DIC vs $^{13}C_{DIC}$ diagram. Compositions of Apennine springs are compared with the simulated compositions resulting from addition of "inorganic" CO_2 to infiltrating waters.

results obtained for DIC and $\delta^{13}C$ values are shown in Figure 2 as "infiltrating waters" lines. Each of these curves shows that $\delta^{13}C$ values of groundwaters become more negative with increasing DIC contents.

The model explains the compositions of some Apennine springs and other groundwaters from carbonate aquifers. In particular the DIC-$\delta^{13}C$ compositions of groundwaters circulating in Pennsylvanian carbonate aquifers (U.S.A.; data from Deines et al. 1974), plot in the field of simulated infiltrating waters (Fig. 2). Here the carbon is derived solely from two sources: carbonate minerals and organic CO_2 in soil. Instead, most Apennine springs plot outside the field of simulated infiltrating waters because they show increasing $\delta^{13}C$ values with DIC contents. Apennine groundwaters fit the DIC and $\delta^{13}C$ simulated values of a second model shown in Figure 3. In this case the initial solutions were assumed as infiltrating waters which underwent the addition of different amounts of organic carbon (0.001, 0.002, 0.003 and 0.004 mol/kg) from the soil which, for reference, was chosen with a ^{13}C value of −24‰. The degassing pressure was set at 1 bar and the incoming gas was chosen as pure CO_2 with carbon isotopic composition ($\delta^{13}C = -1$‰) typical of deep inorganic carbon sources active in the Italian peninsula, such as volcanoes, geothermal systems and cold gas emissions. The good agreement of measured and simulated DIC and $\delta^{13}C$ compositions suggests that the Apennine groundwaters, besides the addition of carbon from a biologic source, active during the infiltration of the groundwaters, are affected by additional input of carbon from an inorganic source characterised by higher ^{13}C values.

3.3 *Quantification of carbon sources active in Apennine aquifers*

The total amount of carbon transported in Apennine groundwaters (TC) is computed as 1.17×10^{11} mol/y. This value is the sum of the carbon transported by each spring (computed by multiplying DIC by flow rate) divided by 0.58, which is the sampled fraction of Apennine groundwaters.

A fraction of this amount of carbon derives directly from dissolution of carbonate rocks, mainly limestone with some dolostone, which comprise the Apennine aquifers. Considering that calcite (or dolomite) dissolution leads to an equimolar increment of DIC and Ca (or Ca + Mg) in solution, and that minor amounts of Ca derive from gypsum dissolution, locally present in the aquifers, the fraction of DIC derived from carbonate dissolution (C_{carb}) is:

$$C_{carb} = Ca + Mg - SO_4 \quad (8)$$

where Ca, Mg and SO_4 are the concentrations expressed in molality. Figure 4 compares the measured DIC and ($Ca + Mg - SO_4$) of the samples, with theoretical DIC and C_{carb} values simulated for a solution in equilibrium with calcite, to which were added, in multiple steps, 0.1 mol/l of CO_2. The good fit between simulated and measured values confirms

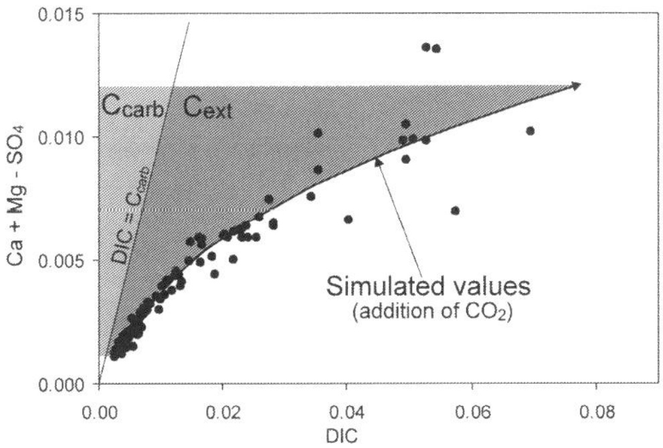

Figure 4. DIC vs ($Ca + Mg - SO_4$) of Apennine groundwaters compared with the values simulated for a solution in equilibrium with calcite, at which were added, in a multi step process, 0.1 moles/liter of CO_2 (see text).

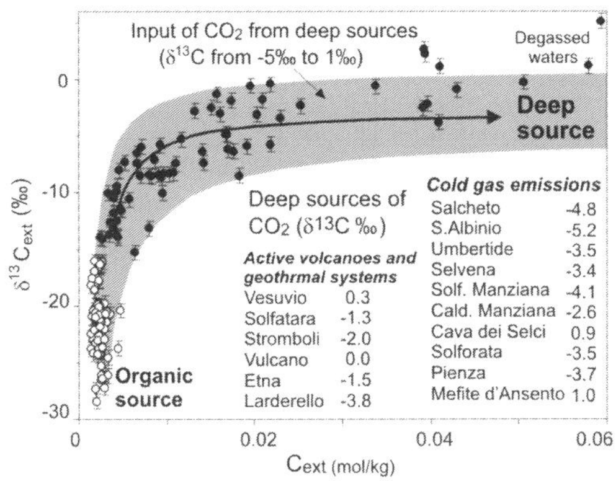

Figure 5. Diagram of $\delta^{13}C_{ext}$ ‰ vs C_{ext} of Apennine groundwater. Error bars indicate the uncertainty in the computed $\delta^{13}C_{ext}$ value which derives from the variability of the isotopic carbon composition of carbonate mineral. Open circles refer to groundwater with a C_{ext} and $\delta^{13}C_{ext}$ compatible with dissolution of biological CO_2 only; solid dots refer to groundwater with a high C_{ext} and $\delta^{13}C_{ext}$ compatible with the further input of deeply derived CO_2. The carbon isotopic compositions of CO_2 released by Italian active volcanoes, geothermal systems and cold gas emissions are shown for comparison.

that carbonate mineral dissolution in Apennine aquifers is controlled by external sources of CO_2.

The two gray areas of Figure 5 are proportional to the carbon derived from carbonate minerals (light gray) and from external sources. The total amount of carbon derived from carbonate mineral dissolution (TC_{carb}) results in 3.86×10^{10} mol/y. This value has been computed with the same method used for the computation of TC but using C_{carb} instead of DIC. The amount of carbon deriving from external sources (TC_{ext}) is 7.84×10^{10} mol/y, i.e. the difference $TC - TC_{carb}$.

For each spring the concentration of the externally derived carbon, C_{ext}, and its isotopic composition ($\delta^{13}C_{ext}$) were computed from measured values by the equations:

$$C_{ext} = DIC - C_{carb} \tag{9}$$

$$\delta^{13}C_{DIC} \times DIC = \delta^{13}C_{ext} \times C_{ext} + \delta^{13}C_{carb} \times C_{carb} \tag{10}$$

setting the isotopic composition of the carbon from carbonate mineral dissolution ($\delta^{13}C_{carb}$) at 2.21‰. In the case that the water was affected before sampling by CO_2 degassing and/or carbonate mineral precipitation, equation 9 underestimates C_{ext} and equation 10 overestimates $\delta^{13}C_{ext}$ because of isotopic fractionation due to the carbon sinks from the solution. However these uncertainties are considered negligible over the range of DIC and PCO_2 values of most of the samples in this study (Appendix I in Chiodini et al. 2000).

In Figure 5 the new variables C_{ext} and $\delta^{13}C_{ext}$ clearly indicates the presence of two carbon sources. The samples of low carbon content (open circles in Fig. 5) have an isotopic composition ($\delta^{13}C_{ext} = -21.4 \pm 2.9$‰) very similar to that of soil CO_2 in vegetated areas (Deines et al. 1974; Cerling et al. 1991; Robinson & Scrimgeour 1995). The first source is therefore the biogenic CO_2 of the soil which is dissolved during infiltration of meteoric waters in recharge areas. The samples of relatively high DIC (closed circles) fit a hyperbolic trend. It is worth noting that these samples fall in the compositional field of infiltrating waters to which are added variable amounts of CO_2 with $\delta^{13}C$ from −5‰ to +1‰, i.e. the isotopic signature of the CO_2 of deep provenance emitted by active Italian volcanoes (Vesuvio, Solfatara, Etna, Vulcano, and Stromboli), geothermal fields (Larderello), and many cold gas emissions in the western part of the region (Fig. 5). The second source is therefore, a deep source of inorganic CO_2 and affects most groundwaters circulating in the Apennines.

Further evidence for the double origin of C_{ext} in Apennine groundwaters is given by the probability plots of $\delta^{13}C_{ext}$ (Fig. 6a) and log C_{ext} (Fig. 6b). Both diagrams show the existence of two populations: the first, population A, is represented by the samples where the "external" carbon derives only from biogenic CO_2 (background), while the second, population B, is represented by samples in which variable amounts of deeply derived CO_2 are added during groundwater circulation. In addition, Fig. 6b highlights a very low variance of C_{ext} of population A suggesting that the biogenic source, which is active in

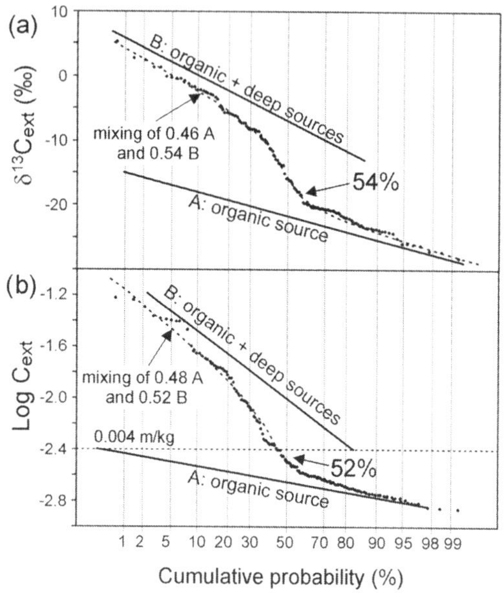

Figure 6. Probability plots of (a) $\delta^{13}C_{ext}$ and (b) log C_{ext} of Apennine groundwater. In both plots the measured values fit curves with an inflection point, which are typical of bimodal distributions originated by the partial overlapping of two normal (log–normal) populations.

the soil during infiltration of recharge waters, is quite constant in the Apennines. Consequently its mean C_{ext} value of 0.0022 mol/kg can be assumed as a representative value for all the recharge waters of the Apennines and can be used to compute the total amount of carbon derived from the organic source, which results to 2.70×10^{10} mol/y. Subtracting this amount from TC_{ext} (see above) we obtain a value of 5.14×10^{10} mol/y which represents the amount of carbon derived from inorganic deep sources.

Summing up, the total amount of carbon transported by Apennine groundwaters is 1.17×10^{11} mol/y. Minor amounts of this carbon derive from dissolution of carbonate minerals (33.2%) and from the organic matter of the soils (23.5%) while the greatest part derives from a deep inorganic source (43.3%), pointing out the relevance of non-volcanic earth degassing processes in the study area.

4 MAP OF CO_2 EARTH DEGASSING OF CENTRAL-SOUTHERN ITALY

For the elaboration of a general map of CO_2 earth degassing in central-southern Italy we used each spring as a measuring point and we applied the method of sequential Gaussian simulations (Deutsch & Journel 1998), that recently has been applied to

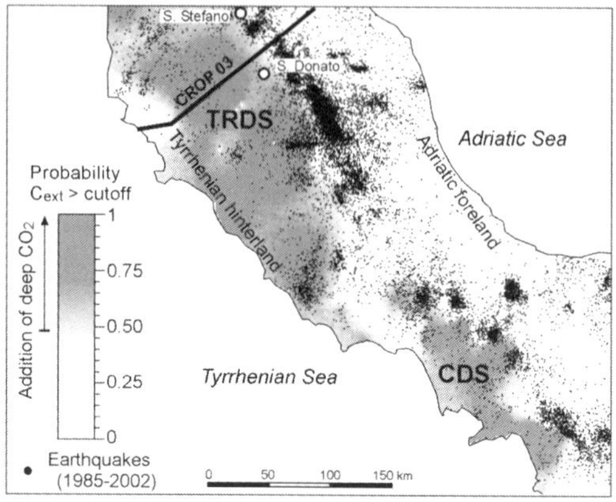

Figure 7. Map of the probability that C_{ext} > cutoff. The map, which represents the probability that a deep source of carbon is active at any location (see text), is compared with the recent seismic activity recorded by the National Seismic Network (INGV, 2003).

investigate soil CO_2 diffuse degassing in volcanic areas (Cardellini et al., 2003). The aim was to produce a map that represents the probability that a deep source of CO_2 is active. The data set on which the geostatistical approach is based has been integrated with data from literature relative to 60 other springs, most of which are from smaller carbonate aquifers of western Italy. In each case the selection has been based on the high flow rates of the springs (flow rates >10 l/s). In the probability plot of log C_{ext} (Fig. 6b) we highlighted the value of 0.004 m/kg which separates the background values from those affected by the input of a deeply derived carbon. Above this value the probability that C_{ext} derives only from the biogenic source is very low (<1%), i.e., above this value a deep source of CO_2 is active in the hydrogeological basin of the springs. We selected 0.004 m/kg as the cutoff value for the elaboration of the C_{ext} probability map of Figure 7 which derives from the post processing of the result of the stochastic simulation procedure. The C_{ext} values were simulated according to a grid of 16000 squared cells of 4 km^2. The C_{ext} distribution was transformed in to a standard normal distribution (normal scores of C_{ext} data) that is used for the simulation. The normal scores were simulated at each unsampled location by random sampling of a Gaussian conditional cumulative distribution function defined on the basis of original data and of previously simulated data within its neighborhood. The Gaussian cumulative distribution function is defined from the simple kriging estimate and variance computed for each location on the basis of the variogram model of the normal scores. Afterwards, the simulated normal scores are back transformed in to C_{ext} simulated values applying the inverse of the normal score transform. Three-hundred alternative simulations were performed and 300 equiprobable realisations were drawn. The set of realisations was post processed drawing a map of the probability that at any location the C_{ext} is above the cutoff value assumed as an upper limit for the C_{ext} derived only from dissolution of biogenic CO_2. Because the cutoff is the maximum value expected for dissolved carbon, if only a biological carbon source is present, Figure 7 represents the probability that a deep source of CO_2 is active at any location.

The highest probability (>0.5) defines two large anomalies located on the Tyrrhenian side of the Italian peninsula. These are two regional structures degassing deeply derived CO_2 as evidenced also by the widespread occurrence of strong gas emissions (Rogie et al. 2000) and Quaternary travertines deposited by CO_2-rich groundwater (Fig. 1). The northern structure partially overlaps the Tuscany, Roman magmatic province (TRDS, Tuscan Roman degassing structure) while the southern structure relates to Campanian volcanism (CDS, Campanian degassing structure). The two volcanic provinces are characterized by quaternary potassic and ultrapotassic magmas rich in fluids with high CO_2/H_2O ratios (Foley 1992). Geochemical features of the magmas are consistent with the melting of a mantle source metasomatized by crustal material (Peccerillo 1999). We suggest that the TRDS and CDS reflect the degassing of this metasomatized, uprising mantle. The map highlights an intriguing coincidence among seismicity and the eastern boundary of the CO_2 degassing areas. The location of earthquake epicenters forms a belt that separates two distinct crustal domains: the Tyrrhenian hinterland, where both the TRDS and the CDS are located, and the Adriatic foreland where deeply derived CO_2 flux is virtually absent.

The total amounts of deeply derived CO_2 released from TRDS and CDS have been estimated to be 1.4×10^{11} mol/y and 0.7×10^{11} mol/y respectively. The computation of the flux of CO_2 of deep provenance has been done by (i) calculating the specific flux of C_{ext} affecting the hydrogeological basin of each spring, (ii) using these data to simulate the C_{ext} flux over the entire area, and (iii) subtracting the contribution from soil CO_2.

5 N_2, Ar AND He DISSOLVED IN APENNINE GROUNDWATERS

We investigated in detail the groundwaters of the Campania region testing the possibility of a common source for the gas dissolved in the groundwaters and for the gas emitted by the main manifestations of the region, i.e. the active volcanoes Vesuvio and Solfatara (Campi Flegrei) and the cold gas emission Mefite

Table 1. Chemical and isotopic composition of the gases emitted by Campanian volcanoes and by Mefite.

	Solfatara	Vesuvio	Mefite
CO_2	987600	983800	980000
N_2	2798	1855	13850
Ar	3.2	5.7	11.5
He	8.2	5.3	14.9
$\delta^{13}C_{CO2}$	−1.25	0.35	1.01
R/Ra	2.85	2.60	2.44

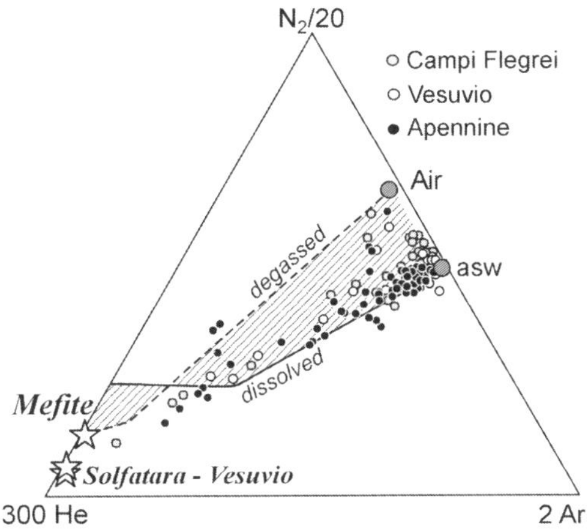

Figure 8. He–Ar–N_2 diagram for Campanian groundwaters and main gas emissions (Mefite, Solfatara and Vesuvio).

(Fig. 1). Mefite cold gas emission is an impressive manifestation, located in a non-volcanic area of the Apennines, which is characterized by a very high CO_2 flow rate (>300 t/d, Rogie et al. 2000) similar to those which characterize Vesuvio and Solfatara (150 t/d and 1500 t/d of deeply derived CO_2 respectively; Frondini et al. 2004; Chiodini et al. 2000). The less reactive species (N_2–Ar–He, Fig. 5) and the isotopic composition of the carbon of CO_2 and of He are here considered in order to investigate the relations between volcanic and non-volcanic degassing. In spite of the different geological framework the 3 gas emissions show similar He and carbon isotopic signatures and similar N_2–Ar–He relative contents (Table 1, Fig. 8).

The data of Mefite, Solfatara and Vesuvio have been used as incoming gas in the IGP procedure in order to investigate the origin of the gases dissolved in the groundwaters circulating in the main carbonate aquifers of Apennine and in the volcanic complex Vesuvio and Campi Flegrei. The He–Ar–N_2 compositions of the dissolved and degassed gas phases, which have been simulated by adding a gas with the same composition of Mefite to air saturated water (asw) and setting $P_{deg} = 1$ bar, are shown in Figure 8. Most of the groundwater of Apennine, as well as the groundwaters of Vesuvio and Campi Flegrei, fit the line of the "dissolved gas" simulated compositions. Similar results are obtained using as incoming gas the fluids discharged by the hydrothermal systems of Vesuvio and Solfatara. If we use as incoming gas other fluids richer in N_2, typical of the TRDS, the field of simulated compositions is different and doesn't fit the data analyzed in the Campania groundwaters. Concluding, the good fit between simulated and analytical data observed in Figure 8 is not trivial and reinforces the hypothesis that a common, deep source of gas generates the CO_2 anomaly of CDS.

6 CONCLUSIONS

The carbon isotopic signature of DIC as well as the composition of the gases dissolved in groundwaters suggests that a large process of diffuse Earth degassing affects the Apennine. This process is the main source for the carbon dissolved in the groundwaters of Apennines which host the main aquifers of the Italian Peninsula. We used data from about 200 springs to draw a regional map of Earth degassing. The map highlights the presence in the Tyrrhenian sector of Italy of two degassing structures, named TRDS and CDS. The total CO_2 released by TRDS and CDS (estimated as 2.1×10^{11} mol/y) is globally significant, being $\sim$10% of the present-day total CO_2 discharge from subaerial volcanoes of the Earth (Williams et al. 1992; Brantley & Koepenick 1995; Marty & Tolstikhin 1998; Kerrick 2001). This result suggests an underestimation of CO_2 globally released by the Earth, because unquantified processes of CO_2 Earth degassing from non-volcanic environment affects almost all the tectonically active areas of the world.

A conceptual model to explain Earth degassing in central Italy is proposed on the basis of the 40 km deep seismic reflection profile CROP03 (Fig. 9, Barchi et al. 1998).

In our opinion the source of the gas is the upwelled mantle located in the Tyrrhenian hinterland. In this zone, mantle fluids may enter the ductile lower crust at near lithostatic pressure, infiltrate upwards through the interconnected network of extensional fractures and normal faults and generate the high CO_2 flux domains observed at the surface.

Following the information from the CROP03 profile, we hypothesize that CO_2 from the mantle intrudes the deeper, now inactive thrusts affecting the Moho and the upper crustal levels at the boundary between the Tyrrhenian hinterland and the Adriatic foreland. In this zone, the arrangements of the deeper thrusts and low-angle normal faults describe outward verging structures which act as traps (structural seal)

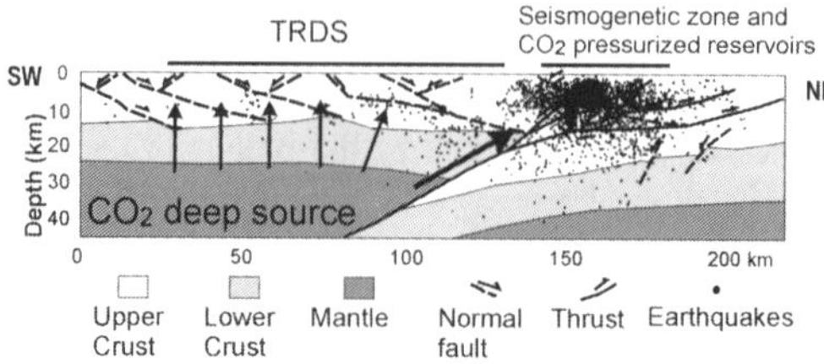

Figure 9. Conceptual model of CO_2 Earth degassing in central Italy.

in which CO_2 may accumulate and generate over-pressured reservoirs. These over-pressured structures fed by mantle derived CO_2 are, in our opinion, the primary cause of Apennine earthquake activation processes. We suggest that mantle degassing processes have an active primary role in the geodynamics of the region.

REFERENCES

Allard, P. et al. 1991. Eruptive and diffuse emissions of CO_2 from Mount Etna. *Nature* 351(6325): 387–391.

Arisi Rota, F. & Fichera, R. 1985. Magnetic interpretation connected to geomagnetic provinces: the Italian case history. *Tectonophysics* 138: 176–196.

Barberi, F., Innocenti, F. & Ricci, C.A. 1971. Il magmatismo nell Appennino centro-settentrionale: La Toscana Meridionale. *Rend. Soc. It. Miner. Petrol.*: 169–210.

Barchi, M., Minelli, G. & Pialli, G. 1998. The CROP03 profile: A synthesis of results on deep structures of the Northern Apennines. *Mem. Soc. Geol. It.*: 383–400.

Brantley, S.L. & Koepenick, K.W. 1995. Measured Carbon-Dioxide Emissions from Oldoinyo-Lengai and the Skewed Distribution of Passive Volcanic Fluxes. *Geology* 23(10): 933–936.

Cardellini, C., Chiodini, G. & Frondini, F. 2003. Application of stochastic simulation to CO_2 flux from soil: mapping and quantification of gas release. *J. Geophys. Res.* 108(B9): 2425, doi:10.1029/2002JB002165.

Cerling, T.E., Solomon, D.K., Quade, J. & Bowman, J.R. 1991. On the Isotopic Composition of Carbon in Soil Carbon-Dioxide. *Geochi. Cosmochim. Acta* 55(11): 3403–3405.

Chiodini, G., D'Alessandro, W. & Parello, F. 1996. Geochemistry of gases and waters discharged by the mud volcanoes at Paternò, Mt. Etna (Italy). *Bull. Volcanol.* 58: 51–58.

Chiodini, G., Frondini, F., Cardellini, C., Parello, F. & Peruzzi, L. 2000. Rate of diffuse carbon dioxide Earth degassing estimated from carbon balance of regional aquifers: The case of central Apennine, Italy. *J. Geophys. Res.* 105(B4): 8423–8434.

Deines, P., Langmuir, D. & Harmon, R.S. 1974. Stable carbon isotope ratios and the existence of a gas phase in the evolution of carbonate ground waters. *Geochim. Cosmochim. Acta* 38: 1147–1164.

Della Vedova, B., Pellis, G., Foucher, J.P. & Rehault, J.P. 1984. Geothermal structure of Tyrrhenian. *Sea. Mar. Geol.* 55: 271–289.
Deutsch, C.V. & Journel, A.G. 1998. GSLIB: Geostatistical software library and users guide. Applied geostatistical series. New York: Oxford University Press.
Ferreri, V., Weissert, H., Dargenio, B. & Buonocunto, F.P. 1997. Carbon isotope stratigraphy: A tool for basin to carbonate platform correlation. *Terra Nova* 9(2): 57–61.
Foley, S.F. 1992. Petrological characterization of the source components of potassic magmas: geochemical and experimental constraints. *Lithos*. 28: 187–204.
Frondini, F., Chiodini, G., Caliro, S., Cardellini, C., Granieri, D. & Ventura, G. 2004. Diffuse CO_2 degassing at Vesuvio, Italy. *Bull.Volcanol.* (in press).
Gerlach, T. 1991. Volcanology – Etnas Greenhouse Pump. *Nature* 351(6325): 352–353.
Ghisetti, F., Kirschner, D.L., Vezzani, L. & Agosta, F. 2001. Stable isotope evidence for contrasting paleofluid circulation in thrust faults and normal faults of the central Apennines, Italy. *J. Geophys. Res.* 106(B5): 8811–8825.
INGV. 2003. Catalogo della sismicità strumentale, INGV.
Kerrick, D.M. 2001. Present and past nonanthropogenic CO_2 degassing from the solid Earth. *Reviews in Geophysics* 39(4): 565–585.
Marty, B. & Tolstikhin, I.N. 1998. CO_2 fluxes from mid-ocean ridges, arcs and plumes. Chemical *Geology* 145 (3–4): 233–248.
Morettini, E. et al. 2002. Carbon isotope stratigraphy and carbonate production during the Early-Middle Jurassic: examples from the Umbria-Marche-Sabina Apennines (central Italy). *Palaeogeography Palaeoclimatology Palaeoecology* 184(3–4): 251–273.
Parkhurst, D.L. 1995. User guide to PHREEQC: A computer program for speciation, reaction-path, advective-transport, and inverse geochemical calculations. *U.S. Geol. Surv. Water Resour. Invest. Rep.* 6: 95–4227.
Peccerillo, A. 1999. Multiple mantle metasomatism in central-southern Italy: Geochemical effects, timing and geodynamic implications. *Geology* 27(4): 315–318.
Robinson, D. & Scrimgeour, C.M. 1995. The contribution of plant C to soil CO_2 measured using $\delta^{13}C$. *Soil. Biol. Biogeochem.* 27(12): 1653–1656.
Rogie, J.D., Kerrick, D.M., Chiodini, G. & Frondini, F. 2000. Flux measurements of nonvolcanic CO_2 emission from some vents in central Italy. *J. of Geophys. Res.* 105(B4): 8435–8445.
Serri, G., Innocenti, F. & Manetti, P. 1993. Geochemical and petrological evidence of the subduction of delaminated Adriatic continental litosphere in the genesis of the Neogene-Quaternary magmatism of Central Italy. *Tectonophysics* 223: 117–147.
Wigley, T.M.L., Plummer, L.N. & Pearson, F.J.J. 1978. Mass transfer and carbon isotope evolution in natural water systems. *Geochim. Cosmochim. Acta* 42: 1117–1139.
Williams, S.N., Schaefer, S.J., Calvache, M.L. & Lopez, D. 1992. Global carbon-dioxide emission to the atmosphere by volcanoes. *Geochim.Cosmochim. Acta* 56(4): 1765–1770.

Water–Rock Interaction, Wanty & Seal II (eds)
© 2004 Taylor & Francis Group, London, ISBN 90 5809 641 6

Chlorine stable isotopes in sedimentary systems: does size matter?

Max Coleman
Jet Propulsion Laboratory, Caltech, Pasadena, California, USA

ABSTRACT: Stable isotope abundances vary because of size (atomic mass) differences. The Cl stable isotope system was one of the first described theoretically, but had a slow, disappointment-strewn development, relative to others. Analytical improvements reduced uncertainty from >10‰ to <0.3‰ but gave only small differences (~1‰) in compositions of geological materials. Eventually, brines and groundwater chlorides gave larger differences (~5‰). Physical processes (e.g. diffusion, adsorption, probably are the main controls of groundwater compositions. In contrast, the processes producing brine compositions are still enigmatic and need further work for full understanding. Anthropogenic groundwater contaminants (chlorinated aliphatic hydrocarbons and perchlorate) show variations resulting from manufacturing processes; implying possibilities of tracing sources. However, they are subject to microbial degradation, producing much larger fractionations (≤15‰), but therefore offering better possibility of monitoring natural attenuation. So atomic size is important to give isotopic fractionation and the size of fractionation matters, allowing observation of otherwise undetectable processes.

1 INTRODUCTION

This paper aims to give a short and selective history of the frustrating development of Cl stable isotope geochemistry. Readers may be depressed by the apparently small return of science from the immense investment of effort expended by the pioneers in this area, but please do read on because there really is a happy ending.

I include a brief description of the technology of analysis, which has been and still is one of the key factors both allowing and restraining progress. Some examples of its application will indicate the scope of the method to produce large and characteristic isotopic fractionation values. Because of its relatively low chemical reactivity in sedimentary systems, Cl demonstrates fewer effects of normal isotopic fractionation but therefore has the power to record physical processes.

2 HISTORY

2.1 *Early history*

In the late 1850's Cannizzaro resurrected the idea, put forward by Avogadro 50 years before, that equal volumes of gas contained equal numbers of molecules. This allowed him to refine estimates of molecular and atomic weights by weighing the gases and relating them to the weight of hydrogen. All atomic weights were integer multiples the weight of hydrogen, with the exception of Cl, 35.5. It was not until Aston developed what was effectively the mass spectrograph, a modification of Thompson's Positive Ray apparatus, that the non-integer atomic weight was confirmed to be the result of the abundances of two isotopes, ^{35}Cl ~75% and ^{37}Cl ~25% (Aston 1919).

2.2 *Later history*

In a classic theory paper, Urey (1947) calculated the result of the equilibrium isotopic fractionation between perchlorate and hydrogen chloride molecules to be a massive isotopic difference in $^{37}Cl/^{35}Cl$ of 9.2% (more usually expressed now in parts per thousand as 92‰). This should have been measurable by the instruments of the day which could analyze less than 1% differences, but was not found. In fact the major advance in analytical capability came from the various developments from two laboratories, those of Nier & McKinney in the period 1946–1955. That work produced the double collector mass spectrometer, allowing stable isotope measurements to be made routinely to a precision of better than 1‰ (Nier 1947; McKinney et al. 1950). However, no one was able to measure any significant natural variations in Cl stable isotope abundance ratios. Hoering & Parker

(1961) analyzed 81 samples, including natural perchlorate and chloride from Chile, again with no detectable variation.

Further development of the method by the research group in the University of Arizona, finally improved the analytical precision to about 0.25‰ (Kaufmann 1984; Kaufmann et al. 1984). This gave hope that Cl stable isotope geochemistry might start to follow the examples of the other light stable isotope systems, like C or S, and show significant variations in ocean waters of different geological ages as well as characterizing geochemical processes. In the next ten years, although the Arizona group measured the first natural variations to be observed, the range of Cl isotope ratios in more than 200 samples of all geological ages was disappointingly small. Most samples were near to the value for modern seawater, adopted as the reference standard. The total range extended from 1.5‰ less than seawater chloride to just over 2‰ more.

Subtle further refinements of the technique by Eggenkamp (1994) now allow an analytical precision of about 0.05‰. Despite the increased data quality, analysis of more than 200 other geological samples by 1994 had produced a range from 2.5‰ less than seawater chloride to just over 1.5‰ more.

3 MEASUREMENT TECHNIQUES

The methods referred to above all measure the sample in the form of a gas. However Xiao & Zhang (1992) pioneered use of a technique, thermal ionization mass spectrometry (TIMS), previously used for measurement of radiogenic isotope ratios. Both of these commonly used methods for measurement of Cl isotope ratios require that the sample to be converted into an appropriate chemical form from which it is ionized so that the different masses may be separated in a magnetic sector mass spectrometer. Results are reported as the difference of the $^{37}Cl/^{35}Cl$ ratio in the sample from that ratio in a standard reference material, $\delta^{37}Cl$, in parts per thousand, ‰.

$$\delta^{37}Cl = \frac{R_{sample} - R_{standard}}{R_{standard}} \times 1000 \qquad (1)$$

where $R = {}^{37}Cl/{}^{35}Cl$. Positive, negative and zero δ values indicate respectively, greater, less and the same relative abundance of the minor, heavier isotope, ^{37}Cl, compared with the standard.

This formulation evolved from one of the methods by which isotope measurements are made but is used for both. In fact, there is no single standard curated and distributed by a single agency, as is the case for H and O. However, Kaufmann et al. (1984) showed that all ocean water chloride had the same isotopic composition within the limits of measurement precision and defined the standard as SMOC, standard mean ocean chloride, analogous with SMOW, standard mean ocean water, the standard for H and O.

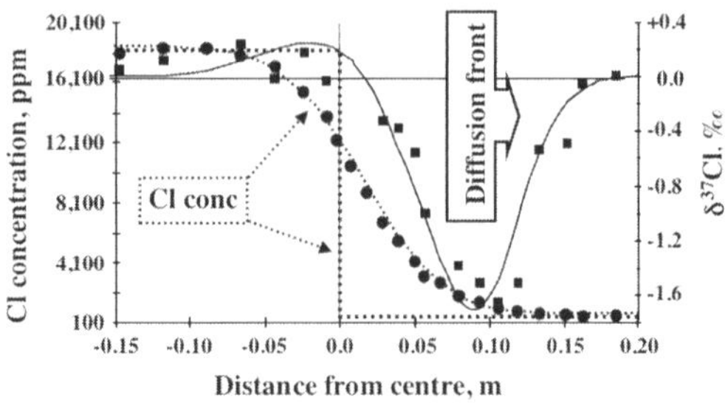

Figure 1. Laboratory experiments to measure the differential diffusion of ^{35}Cl and ^{37}Cl to produce a large change in $\delta^{37}Cl$.

3.1 *Sample preparation agonies*

For both types of measurement Cl has to be in a specific, chemically pure form, methyl chloride, for dual inlet stable isotope ratio mass spectrometry (SIRMS) and cesium chloride for TIMS. Without going into details, both approaches require analyst time, skill and patience.

There are frustrations associated with both techniques. In the case of analysis of chlorinated aliphatic hydrocarbons, commonly analyzed materials, organic chloride has to be separated as inorganic chloride ion and then converted into CsCl for TIMS analysis. The situation is even worse for SIRMS, where the organic Cl is separated by conversion to an inorganic chloride, precipitated as insoluble AgCl which is then converted back to organic methyl chloride. The main features of each of the isotopic analytical techniques are described very briefly below.

3.2 *Comparison of SIRMS and TIMS*

For SIRMS the sample is in gaseous form, methyl chloride. After introduction to the mass spectrometer, it is ionized by electron bombardment. Introduction of the sample gas is alternated repeatedly with reference gas of known isotopic composition. This allows multiple direct estimates to be made of their difference in isotope ratio in about 15 minutes. Minimum sample size is $\leq$4 micromole. Reproducibility of analysis for the whole analytical procedure is $\leq$0.10‰ (2σ).

In TIMS the CsCl solution, prepared from the sample Cl, is evaporated on a filament to form a solid that is ionized by resistive heating of the filament in the mass spectrometer. Multiple readings of the sample isotope ratio are taken over a period of hours. Standards can be run as samples to confirm the calibration of the mass spectrometer. Precision for a single run is $\leq$0.2‰ but to determine the value relative to a standard, which also has the same uncertainty, the overall uncertainty is about 0.3‰ (2σ). To compensate for this, sample size can be as little as 0.3 micromole or less.

Godon et al. (2004) showed that a large variety of different seawater samples all gave the same result within error, thus confirming the use of SMOC as a standard. They also showed that both methods give comparable mean values for samples near to that of the standard but that more extreme values may not. However, part of the discrepancy may be due to the fact that in TIMS analysis the result may be changed by the amount of sample loaded onto the filament.

To summarize the points made above, whichever method is used, a lot of effort and time is needed to make any measurements and those made on sedimentary materials showed very little variation in comparison to the larger ones of other stable isotope systems.

4 ISOTOPIC FRACTIONATION PROCESSES

4.1 *Crystallization*

Eggenkamp (1994) performed careful experiments crystallizing Na-, K-, and Mg-chlorides from aqueous solutions, but only NaCl gave a fractionation factor significantly larger than the analytical uncertainty. The solid had a value +0.24‰ (0.07 1σ). The implication of this result is that during continued evaporation of seawater, the first-formed halite will have a positive value relative to SMOC. The residual seawater will have lost more ^{37}Cl relative to ^{35}Cl to the precipitate and consequently will be relatively depleted in the former, giving it a slightly negative δ^{37}Cl value. Continued evaporation will produce more negative values for the residual brine and a succession of precipitates which will track those values, but 0.24‰ more positive. Thus, evaporite salts may even have negative SMOC values. The systematic shift of isotope values depends on the extent of completion of a reaction, and can be modeled as a Rayleigh fractionation process. Eggenkamp (1994) used this result to interpret data on halite and other evaporite minerals from cores of Late Permian Zechstein salts. Successive analyses traversing vertically upwards in a core showed repeated sequences of more negative δ^{37}Cl, which then reverted to a more positive value in a relatively short vertical distance. Each sequence represents evaporation of a batch of marine brine terminated by recharge from a volume of new seawater. The sequences correspond to Zechstein evaporite cycles defined by changes in mineral composition.

4.2 *Diffusion*

Variations of chloride concentration in pore-waters correlated with δ^{37}Cl could be interpreted as preferential diffusion of ^{35}Cl relative to ^{37}Cl (Desaulniers et al. 1986; Eggenkamp et al. 1994). Although the effects were quite small and the uncertainties relatively large, the trends were persuasive, if not quantitative. The theory of diffusion is well established, but predicting isotopic diffusion coefficients also requires knowledge of the chloride ion's physical properties and especially its interaction with its local water molecules. Consequently, Coleman et al. (2001) had to resort to laboratory experiments to determine the ratios of Cl isotope diffusion coefficients under various conditions of temperature and chloride concentration. There are two points of significance in the results (e.g. Fig. 1). They show large and consistent relative depletion of ^{37}Cl of up to approximately 2‰ in the diffusion front; and a corresponding, small relative enrichment in ^{37}Cl in the residual brine, from which ^{35}Cl has diffused faster. These data were used to estimate times of large-scale diffusion in natural systems (Section 7).

4.3 *Microbial processes*

There has been some effort to measure isotopic fractionation associated with microbial degradation of chlorinated hydrocarbon solvents and of perchlorate. This is because of their significance as environmental pollutants. In both cases, for either natural or engineered bioremediation, the δ^{37}Cl values of the residual contaminant and of the reaction product can indicate the extent of completion of the process. If the fractionation factor is constant, the reaction progress may be modeled as a Rayleigh fractionation process. This is best exemplified by the data for microbial perchlorate reduction shown in Figure 2 (Coleman et al. 2003). In this case reduction of perchlorate produces chloride approximately 15‰ more negative than the starting material. The residual perchlorate shows an increase in δ^{37}Cl as a precise function of the extent of completion of the reduction. Equally, the accumulated chloride independently can give the same information. Together, the chloride and perchlorate data at any stage can be checked for Cl mass and isotopic balance to validate the interpretation. The dramatically large isotopic fractionation factor allows the data for such processes to be interpreted even if there is presence of background values of natural chloride.

Microbial dechlorination of chlorinated aliphatic hydrocarbon (CAH) solvents like tetrachloroethene (perchloroethene or PCE) is more complex since it may proceed in stages not necessarily mediated by the same type of microbe. Each stage may remove

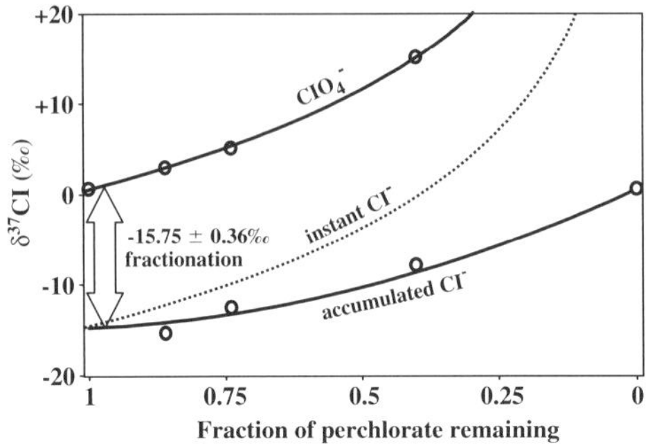

Figure 2. Microbial reduction of perchlorate to chloride, showing Rayleigh fractionation behavior.

just one or maybe two Cl atoms from the molecule as dissolved chloride producing successively trichloroethene (TCE), dichloroethene (DCE) and vinyl chloride. The results of Numata et al. (2002) showed a large range of isotopic fractionation values for dechlorination of TCE to cis-DCE and PCE to cis-DCE. However, their method did not permit separation of different Cl-bearing molecules and rigorous checking by mass balance calculations. Their results gave consistent fractionation factors with chloride from TCE 5‰ to 6‰ more negative than the original material. Their results for dechlorination of PCE were very variable.

5 OIL-FIELD BRINES

5.1 *Waters in clastic oil-field reservoirs*

The first really sizeable Cl isotope variations observed consistently were values ranging from near zero to −4.2‰ in samples of oil-field brines from the Forties Field, reported at the WRI-7 meeting (Coleman 1992). Despite the time since then, there is still no clear understanding of how these values are generated. Similarly, negative values down to −5‰ occur in waters from Oseberg Field, too (Ziegler et al. 2001). These data give information on routes and processes of mixing of distinct fluids.

Despite the technical difficulty of making Cl isotope measurements, Eggenkamp & Coleman (2000) resurrected a chemical procedure published in 1895 that allowed separation of Br from Cl and made δ^{81}Br measurements on the Oseberg samples. The results are shown in Figure 3. So we are now in the situation where we can explain neither the Cl nor Br isotope compositions of these waters and their negative correlation implies at least a mixture of two components where different processes control the isotopic fractionations of the two elements.

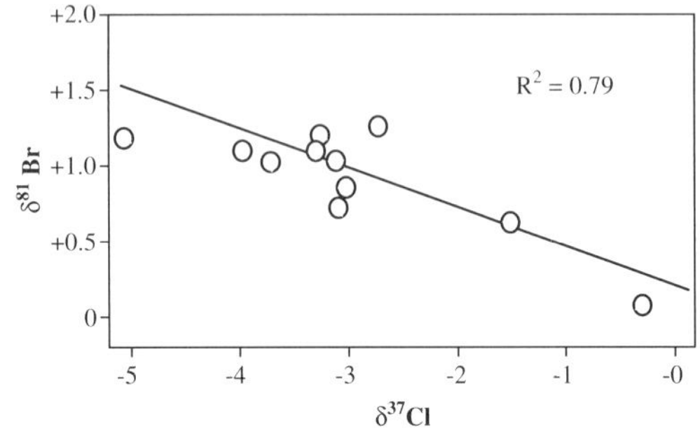

Figure 3. Negative correlation of Cl and Br stable isotope compositions from Oseberg oil-field waters.

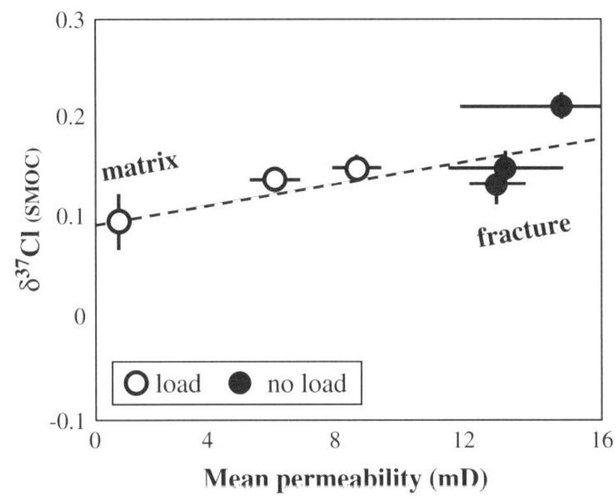

Figure 4. Relationship of δ^{37}Cl and permeability of chalk cores subjected to controlled imposed stress.

5.2 *Water in carbonate oil-field reservoirs*

A recent study of a large carbonate reservoir aimed to address not only the problem of showing admixture of different fluids, but also aimed to use the intrinsic information in the data to identify whether the water came from fracture or matrix permeability. The work involved not only measurements of field samples but also laboratory simulations of compaction and stress-induced fracture permeability. The other self-imposed difficulty was to use only the trace water extracted from produced oil. Despite the complexity of the data set, Poszwa et al. (2003) produced the results shown in Figure 4.

The effects are small in the experiments but relate qualitatively to the expected direction of fractionation, where processes like diffusion and adsorption will slightly enhance the relative abundance of ^{35}Cl in the micro-permeability of the matrix. The field data fortunately showed a much larger range of δ^{37}Cl values (−0.9 to −1.4‰), that were independent of brine chemical compositions but related to the

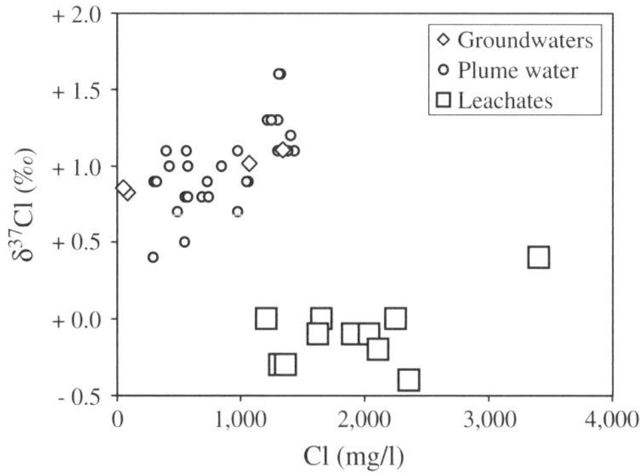

Figure 5. Chloride concentrations and isotopic compositions from a land-fill site and nearby locations.

permeability in the same way expected from the laboratory simulation experiments.

6 CONTAMINATED LAND

6.1 *Contaminated land – chloride pollution*

There are number of cases in the UK where there is a plume of chloride in groundwater down-gradient from a landfill site, despite carefully engineering a very low permeability liner to retain the waste and its aqueous leachate. Analysis of leachate, nearby and control groundwaters produced a wide range of values for Cl concentrations and isotopic compositions (Loomis et al. 1998). It is clear from Figure 5 that the high values of chloride in the plume are not the result of admixture leachate with groundwaters.

In this case there is a rational explanation for the results. The landfill site is a former quarry where clay was extracted for brick making. The clay is organic rich and also contains minor amounts of sulfide. The process of clay extraction exposed the sulfide to aerobic oxidation, which resulted in a local low-pH environment, allowing leaching of chloride from the clay.

6.2 *Contaminated land – CAH*

The microbial dechlorination fractionation process described in Section 4.3 indicated that successive Cl atoms were stripped from the CAH. An attempted application of this approach to a field site resulted in a small but useful data set (Fig. 6).

There were two surprises from these data: they appear to show a relationship between C and Cl isotope values and the isotope data relate to the concentrations. Although unexpected in terms of the microbial process, it seems that there is complete rather than step-wise dechlorination occurring

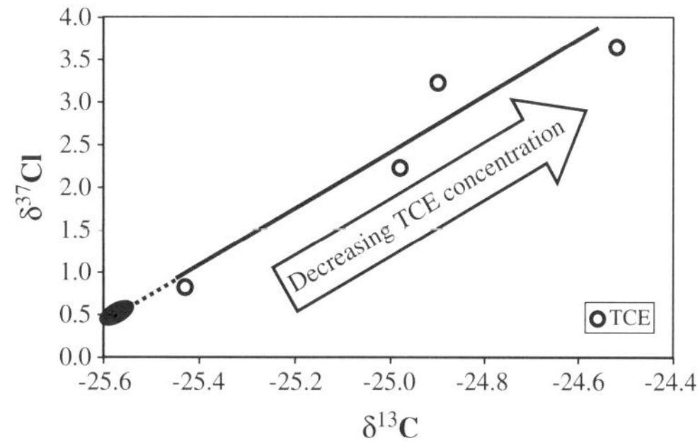

Figure 6. Carbon and chlorine isotopic values of TCE from a contaminant plume. The solid ellipse at the bottom left of the line indicates a possible, original value for the TCE.

indicated by the increase in C isotope values as the process progresses. Although, the fractionation factor has not been calibrated in the laboratory, the results are consistent with a Rayleigh fractionation process and extrapolation back may constrain the original composition of the TCE.

7 RADIOACTIVE WASTE SITES

Two case studies are considered that show the value of the Cl isotopes in obtaining both positive and negative information about the potential suitability of a site for construction of an underground test laboratory to investigate radioactive waste storage problems. The deliberately indirect phrasing in the previous sentence indicates the degree of public sensitivity to such issues and that not only geo-scientific criteria are important in such decisions. The two case studies were part of a larger program of research of the French organization ANDRA. The aim in both cases was to define the hydrodynamic stability of the sub-surface waters, since a site where there had been little flow in the past offered the promise for future lack of disturbance too.

7.1 *Vienne Site, France*

The rocks are fractured granite with pore-waters that show a concentration profile implying a possible diffusive gradient (Fig. 7). However, δ^{37}Cl data indicate that there has been no detectable diffusion.

The reason for this definitive negative conclusion is obvious by reference to Figure 1. As well as the negative δ^{37}Cl at the diffusion front, caused by faster migration of the ^{35}Cl, there is a complementary increase in δ^{37}Cl in the residual source material. It is clear that this is not the case here and there is no evidence of diffusion in the Vienne Site and more energetic advective mixing is operating.

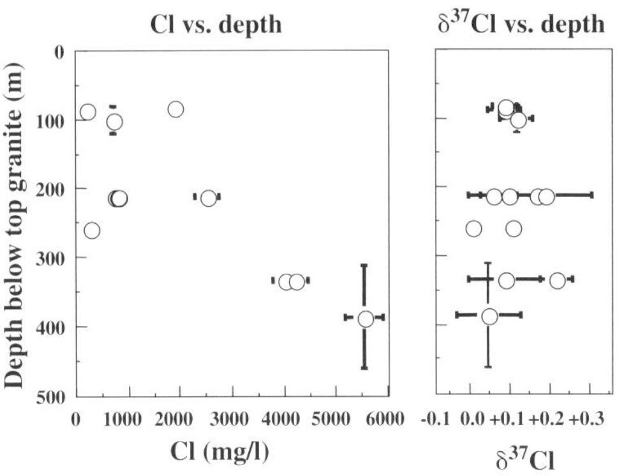

Figure 7. Variation of chloride with depth at the Vienne Site, but without corresponding variation of δ^{37}Cl values.

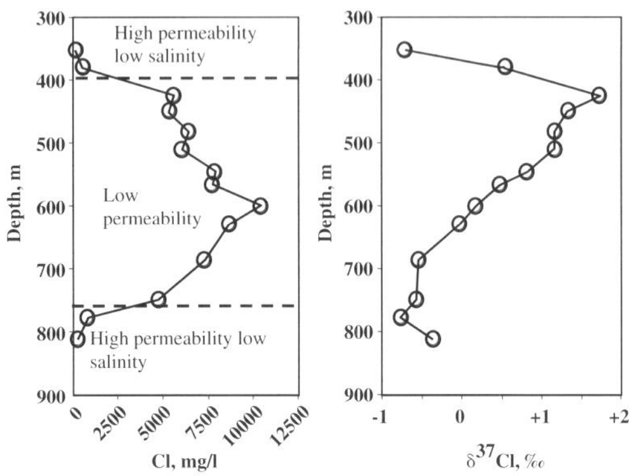

Figure 8. Cl and δ^{37}Cl variation with depth at the Gard Site, showing evidence of large-scale natural diffusion.

7.2 *Gard Site, France*

In contrast to the Vienne Site, the Gard Site is a marine sedimentary sequence of very low permeability fine-grained calcareous mud-rocks. Above and below the 400-meter thick mudrock succession there are highly permeable marine sands. During the Oligocene the system suffered tectonic activity and a lateral extension of the system was folded allowing erosion of a fold limb and ingress of meteoric water to the permeable sandstones. This activity replaced the marine pore-fluids above and below the mudrocks with fresh water, thus establishing two salinity gradients. Unlike the Vienne Site, in this case it appears that diffusion has really occurred (Coleman et al. 2001). The salinity profile and the associated δ^{37}Cl data are shown in Figure 8.

The salinity profile shows clearly a double diffusion system, with simultaneous migration of the assumed original seawater pore-fluid upward and downward from the mid-depth of the mudrock. If the pore-waters' original salinity was similar to the modern value of just over 20,000 mg/l there has been considerable loss even from the center of the formation. This is supported by the isotopic evidence, which again on the assumption that the marine pore fluid had an isotopic composition like modern seawater, was originally 0‰. The more positive values in the center and negative ones at the two diffusion fronts are strong evidence of diffusion. However, the lack of vertical coincidence of the salinity and δ^{37}Cl maxima implies occurrence of another process, probably related to mudrock compaction prior to diffusion. Using a normal Fick's law diffusion model and the diffusion coefficients mentioned in Section 4.2, it is possible to compute a diffusion time of 15.5 million years, indicating a remarkably stable hydrodynamic system. It is also gratifying to see the extrapolation of a two-week diffusion experiment over a length scale of 40 cm extrapolated to 15 million years and 400 meters.

8 CONCLUSIONS

From the above case studies it is apparent that size does matter, not only in causing isotopic fractionation but also in producing effects sufficiently large to be significantly above measurement uncertainties. However, in most of the examples above it is also clear that size is not enough and what one does with the data is equally important. In particular there are now a number of very large and reproducible effects for which there is either no or insufficient explanation in terms of basic understanding of relevant processes. However, as in the earlier days of this branch of science, it will be easier to make advances in understanding, if the means of making high precision measurements becomes faster and easier.

REFERENCES

Aston, F.W. 1919. The constitution of the elements. *Nature* 104: 393.

Coleman, M.L. 1992. Water composition variation within one formation. In Y. Kharaka & A.S. Maest (eds.) *Water–Rock Interaction – Proceedings of the 7th international symposium on water–rock interaction*: 1109–1112. Rotterdam: Balkema.

Coleman, M.L., Eggenkamp, H.G.M. & Aranyossy, J.F. 2001. Chlorine stable isotope characterisation of solute transport in mudrocks. *ANDRA: Actes des journées scientifiques, 1999 Chapter 10*: 155–175, EDP Sciences, France.

Coleman, M., Ader, M., Chaudhuri, S.K. & Coates, J.D. 2003. Microbial isotopic fractionation of perchlorate chlorine. *Appl. Env. Microbiology* 69: 4997–5000.

Desaulniers, D.E., Kaufmann, R.S., Cherry, J.A. & Bently, H.W. 1986. ^{37}Cl-^{35}Cl variations in a diffusion-controlled groundwater system. *Geochim. Cosmochim. Acta* 50: 1757–1764.

Eggenkamp, H.G.M. 1994. *δ^{37}Cl, the geochemistry of chlorine isotopes.* Ph.D. thesis, Utrecht University, 150 pp.

Eggenkamp, H.G.M., Middelburg, J.J. & Kreulen, R. 1994. Preferential diffusion of ^{35}Cl relative to ^{37}Cl in sediments of Kau Bay, Halmahera, Indonesia. *Chem. Geol. (Isotope Geosci. Sect.)* 116: 317–325.

Eggenkamp, H.G.M. & Coleman, M.L. 2000. Rediscovery of classical methods and their application to the measurement of stable bromine isotopes in natural samples. *Chem. Geol.* 167: 393–402.

Godon, A., Jendrzejewski, N., Eggenkamp, H.G.M., Banks, D.A., Ader, M., Coleman, M.L. & Pineau, F. 2004. A cross calibration of chlorine isotopic measurements and suitability of seawater as the international reference material. *Chem Geol.* (in press).

Hoering, T.C. & Parker, P.L. 1961. The geochemistry of the stable isotopes of chlorine. *Geochim. Cosmochim. Acta* 23: 186–199.

Kaufmann, R.S. 1984. *Chlorine in ground water: stable isotope distribution*. Ph.D. thesis, University of Arizona: Tuscon, 137 pp.

Kaufmann, R.S., Long, A., Davis, S. & Bentley, H. 1984. Natural variations of chlorine stable isotopes. *Nature* 309: 338–340.

Loomis, J.L., Coleman, M. & Joseph, J. 1998. Use of stable chlorine isotopes to evaluate the origin of a Cl-rich plume in the Oxford Clay, England. Goldschmidt Conference, Toulouse, *Mineral. Mag.* 62A: 901–902.

McKinney, C.R., McCrea, J.M., Epstein, S., Allen, H.A. & Urey, H. 1950. Improvements in mass spectrometers for the measurement of small differences in isotope abundance ratios. *Rev. Sci. Instrum.* 21:724–730.

Nier, A.O. 1947. A mass spectrometer for isotope and gas measurement. *Rev. Sci. Instrum.* 18:398–411.

Numata, M., Nakamura, N., Koshikawa, H. & Terashima, Y. 2002. Chlorine isotope fractionation during reductive dechlorination of chlorinated ethenes by anaerobic bacteria. *Environ. Sci. Technol.* 36: 4389–4394.

Poszwa, A.C., Coleman, M.L., Pouya, A., Ader, M. & Bounenni, A. 2003. Variation of brine compositions resulting from flow from matrix or fracture permeability, investigated by high-pressure laboratory experiments. *Geophys. Res. Abs.* 5:4597.

Urey, H. 1947. The thermodynamic properties of isotopic substances. *J. Chem. Soc.* 562–581.

Xiao, Y.-K., Zhang, C.-G. 1992. High precision isotopic measurement of chlorine by thermal ionization mass spectrometry of the Cs_2Cl^+ ion. *Int. J. Mass Spectr. Ion Proc.* 116: 183–192.

Zeigler, K., Howarth, R.J. & Coleman, M.L. 2001. Palaeohydrodynamics of fluids in the Brent Group (Oseberg Field, Norwegian North Sea) from chemical and isotopic compositions of formation waters. *Appl. Geochem.* 16: 609–632.

Water–Rock Interaction, Wanty & Seal II (eds)
© 2004 Taylor & Francis Group, London, ISBN 90 5809 641 6

Arsenic mobilization from contaminated sediments: a full-scale experiment in progress

J.G. Hering, S. Dixit & K. Campbell
Environmental Science & Engineering, California Institute of Technology, Pasadena, CA, USA

P.A. O'Day
School of Natural Sciences, University of California, Merced, CA, USA

ABSTRACT: The mobilization of arsenic (As) was examined in a system where the deposition of Fe and As have been substantially modified by large-scale manipulations. This engineering practice was designed to decrease As concentrations in water supplied to the City of Los Angeles. Accomplishing this objective, however, has resulted in significant accumulation of As and Fe in the sediments of a reservoir on the Los Angeles Aqueduct. As and Fe are released into the porewater at depth in the sediment, consistent with reductive dissolution of Fe^{III} oxyhydroxides. Factors influencing the possible re-sorption of As onto residual Fe^{III} oxyhydroxides solids have been examined. Reduction of As^{V} to As^{III} alone cannot account for As mobilization since As occurs in the solid phase as As^{III} well above the depth at which it is released into the porewater. Competition from other porewater constituents could suppress re-sorption of As released by reductive dissolution.

1 INTRODUCTION

The mobilization of As from aquifer sediments poses a serious threat to human health in many parts of the world, particularly in South Asia (Nordstrom 2002). Such mobilization can result in the occurrence of As in groundwater at concentrations of hundreds to thousands of µg/L.

As in alluvial sediments is often associated with Fe oxyhydroxides and is mobilized by conditions that destabilize the solid carrier phase or that promote desorption of As from the solid (Smedley & Kinniburgh 2002; Welch et al. 2000). Manipulation experiments (e.g., a push–pull experiment in which an organic substrate is injected into an aquifer) can demonstrate the effects of environmental conditions on As mobility (Harvey et al. 2002). However, it is often infeasible to perform large-scale biogeochemical experiments.

In some cases, large-scale manipulations of environmental systems are engineered for specific goals such as ecosystem restoration. A useful opportunity to study the biogeochemical controls on As mobility is provided by a full-scale engineering project that is currently being implemented by the Los Angeles Department of Water and Power (LADWP).

2 THE INTERIM ARSENIC MANAGEMENT PLAN FOR LOS ANGELES

The Los Angeles Aqueduct (LAA) transports water from the eastern Sierra Nevada to Los Angeles for municipal water supply. Because of geothermal inputs, As concentrations in the LAA water supply are naturally elevated with a historical annual average of 20 µg/L. In 1996, an interim As management plan was implemented to lower the concentration of As in the water delivered by the LAA (Stolarik & Christie 1999). Since 1996, ferric chloride and polymer have been added to the aqueduct at the Cottonwood Treatment Plant, an existing facility originally used to control turbidity. Ferric chloride addition results in the formation of an Fe^{III} oxyhydroxide floc that is transported as suspended load in the aqueduct and deposited in North Haiwee Reservoir as the flow velocity decreases. Ferric chloride doses are adjusted to remove sufficient As from the dissolved phase that the As concentrations below Haiwee Reservoir are <10 µg/L. This practice has resulted in significant accumulation of As and Fe in the Haiwee sediments since 1996.

3 SIGNAL AND IMPACT OF ARSENIC AND IRON DEPOSITION

Sediments and porewaters have been sampled in the inlet channel where the LAA feeds into North Haiwee Reservoir (Kneebone et al. 2002). Sampling has thus far been limited to locations very near the bank because of restrictions on human contact with the LAA water supply. Thus it cannot be claimed that the samples obtained are entirely representative of conditions throughout the reservoir.

Sediment cores collected from the inlet channel were found to contain elevated concentrations of Fe, As, and Mn relative to sediments at a control site. Sediment concentrations of these elements remain elevated throughout the core length sampled (ca. 4% Fe, 600 μg/g Mn, and 200 μg/g As). Porewater profiles reveal that Mn, Fe, and As are mobilized at depth in the sediment with Mn released into the porewaters at a more shallow depth than either Fe or As. As concentrations in porewaters at depth in the sediment reached values >1000 μg/L (Kneebone et al. 2002). The depth at which increased porewater concentrations are observed varied in sampling conducted in different seasons but, in general, Mn, As, and Fe porewater concentrations were all low in the surficial sediments and overlying water. At depth in the sediments, As and Fe concentrations in the porewaters were always strongly correlated (Dixit, unpubl.; Campbell, unpubl.).

From the water supply perspective, the interim As management plan is fulfilling its objectives. As concentrations in water delivered to the LAA Filtration Plant in Sylmar, CA are consistently below 10 and often below 5 μg/L. Thus, the As deposited to the sediments in association with the Fe^{III} oxyhydroxide floc must be almost entirely retained within the sediment. The low porewater concentrations of As in the surficial sediment, despite elevated concentrations at depth in the sediment, indicate that As diffusing upward in the sediment is sequestered into the solid phase before reaching the sediment–water interface. Nonetheless, numerous questions remain regarding the controls on the rate and extent of As mobilization in this system.

4 BIOGEOCHEMICAL CONTROLS ON ARSENIC MOBILITY

Even though elevated concentrations of As are observed at depth in the porewaters of the Haiwee sediments, the As concentration in the sediments is fairly constant with depth. The amount of As that would need to be released from the sediments in order to support the observed porewater concentrations can be estimated by assuming reasonable values for the porosity of the sediment and the density of the Fe^{III} oxyhydroxide floc. This estimate suggests that <2.5% of the As and <1% of the Fe in the sediments is released into the porewaters. Notably the molar ratios of Fe:As are similar in the sediments and porewaters and in the calculated values for the original floc (Hering & Kneebone 2001).

These observations raise the question of what factors control the rates of release of As and Fe. For Fe, mobilization into the porewaters can be attributed to reductive dissolution but, for As, other processes (e.g., competitive desorption) could also play some role. A related question is whether the Fe and As released by reductive dissolution are partly repartitioned (i.e., by sorption) to the solid phase, which contains the bulk of the deposited oxyhydroxide floc. And, if such re-sorption does not occur, whether that is due to alteration of the solid phase or because the porewater composition does not favor As sorption.

Several hypotheses can be stated regarding the biogeochemical processes controlling As mobility in Haiwee sediments:

1. *As oxidation state hypothesis*. It may be hypothesized that reduction of As^{V} to As^{III} is responsible (in whole or in part) for As mobilization from the sediments into the porewater.
2. *Solid diagenesis hypothesis*. The Fe^{III} oxyhydroxide floc could have undergone some alteration such that the residual solid in the sediment is not as good a sorbent for As as the original floc.
3. *Porewater composition hypothesis*. Re-sorption of As onto the residual floc may not be favored under the chemical conditions prevailing in the porewater. In particular, re-sorption of As could be inhibited by elevated concentrations of Fe^{II}, phosphate, or dissolved organic matter (DOM) in the porewater.

4.1 *Arsenic oxidation state hypothesis*

A substantial body of work on As removal from drinking water by coagulation with hydrolyzing Fe^{III} salts demonstrates that, under conditions relevant to water treatment, better removal efficiencies are achieved with As^{V} than with As^{III} (Shen 1973; Sorg & Logsdon 1978; Jekel 1994; Hering et al. 1996). Similar results have been observed for adsorption of As^{III} and As^{V} onto pre-formed Fe^{III} oxyhydroxides at comparable sorbate and sorbent concentrations (Wilkie & Hering 1996). Such studies, however, focus on low sorbent concentrations (typically $[Fe]_T < 120$ μM corresponding to a solids concentration of ca. 12 mg/L) and sorbate–sorbent ratios at which As^{V} removal from solution is nearly complete.

In contrast, As sorption studies conducted with Fe^{III} oxyhydroxides at higher sorbate and sorbent concentrations and higher sorbate-to-sorbent ratios have demonstrated comparable sorption of As^{III} and As^{V} at circumneutral pH and preferential sorption of As^{III} at higher pH values (Manning et al. 1998; Raven et al. 1998; Dixit & Hering 2003). However, the relative affinity of the Fe^{III} oxyhydroxide surface for As^{III} and As^{V} is affected not only by pH but also by the presence of competing sorbates such as phosphate. In the presence of phosphate, the extent of sorption of As^{III} and As^{V} onto hydrous ferric oxide (HFO) are still comparable at pH ~8 but sorption of As^{III} is significantly depressed at lower pH values (Fig. 1).

In Haiwee sediments, examination of cores using X-ray Near Edge Spectroscopy (XANES) indicated that As in the solid phase was present as As^{III} throughout the sediment core; As^{V} was detected only in the surficial sediment (Kneebone et al. 2002). Since As occurs as As^{III} in the sediment well above the depth at which As is released into the porewater, it is apparent that reduction of As^{V} to As^{III} is not, in itself, sufficient to account for the mobilization of As at depth.

4.2 *Solid diagenesis hypothesis*

At the Cottonwood Treatment Plant, the Fe^{III} oxyhydroxide floc is formed under conditions that would produce an amorphous solid. This solid material might become increasingly crystalline as it aged in the sediment. Amorphous Fe^{III} oxyhydroxides can also undergo transformation to the mixed Fe^{II}, Fe^{III} solid magnetite during reductive dissolution (Benner et al. 2002).

The bulk Fe signature in a sediment core taken at Haiwee reservoir in December 2001 was examined by Extended X-ray Absorption Fine Structure (EXAFS) Spectroscopy. Iron EXAFS spectra were collected at the Stanford Synchrotron Radiation Laboratory using a Lytle fluorescence detector at room temperature. Spectra were fit assuming that Fe-bearing phases in the sediment are primarily a mixture of detrital phyllosilicate minerals (indicated by X-ray diffraction analyses of a control core) and ferrihydrite floc. Sediment spectra are similar but not identical to a reference spectrum of synthetic ferrihydrite (Fig. 2), consistent with the presence of additional detrital phyllosilicate minerals in the core.

Spectra for surficial and deeper sediments were remarkably similar to each other and least-squares fits indicated no major structural change with depth. These spectra show no evidence of Fe reduction or conversion to magnetite. The possibility of ferrihydrite conversion to goethite is not excluded on the basis of EXAFS alone because the local atomic structure of both phases is nearly identical (O'Day et al. in press). However, bulk X-ray diffraction analysis of sediment cores collected in 2003 has indicated an absence of crystalline Fe^{III} oxide or oxyhydroxide phases.

Laboratory studies of As^{III} sorption onto HFO and goethite indicate comparable sorption for both solids when sorbed As^{III} is normalized to surface site densities (Dixit & Hering 2003). In this study, the maximum sorption density for As^{III} on HFO was ca. 20-fold higher than on goethite. This would imply that the capacity of the surface for As sorption would decrease with alteration of the solid to goethite.

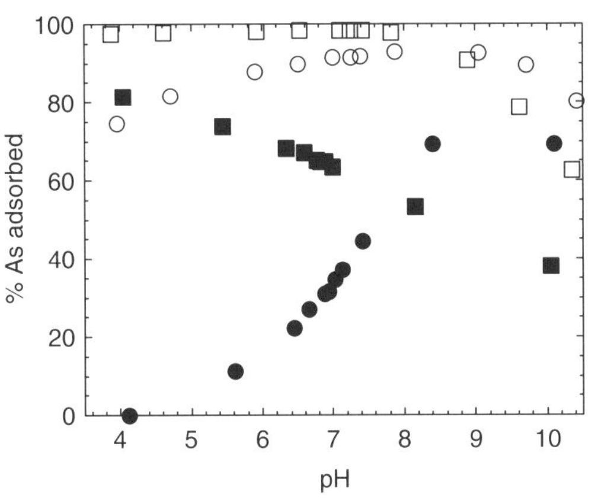

Figure 1. Sorption of As^{III} (circles) and As^{V} (squares) on HFO in the presence (closed symbols) or absence (open symbols) of phosphate. Experimental conditions: 0.03 g/L HFO, $[As]_T = 10\,\mu M$, $[PO_4]_T = 100\,\mu M$, 0.01 M $NaClO_4$. Reprinted with permisision from Dixit & Hering (2003). Copyright American Chemical Society.

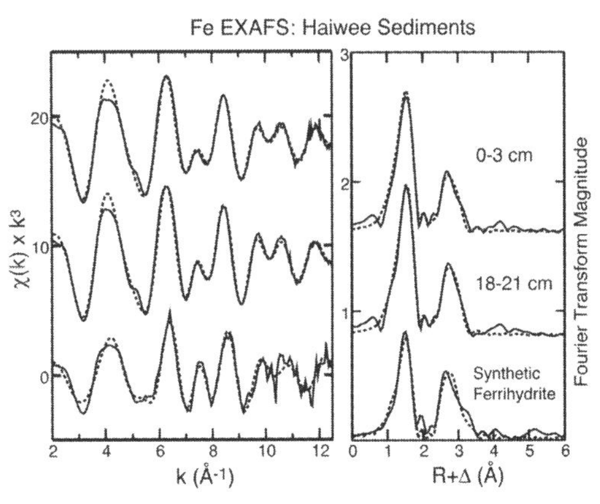

Figure 2. Iron EXAFS and Fourier transforms of bulk sediments from a Haiwee core at two intervals, 0–3 cm and 18–21 cm, compared to a reference sample of synthetic ferrihydrite. Solid lines are experimental data; dashed lines are non-linear least-squares fits.

However, this difference largely reflects the lower surface area of synthetic goethite (relative to HFO). If alteration to goethite in the sediments occurred without substantial decrease in specific surface area, then the sorption capacity of the solid would not be affected. A recent laboratory study showed that aging of HFO did not result in the desorption of As^{V} (Ford 2002).

4.3 *Porewater composition hypothesis*

The composition of the porewater changes with depth in the sediment as a result of the respiration and mineralization of organic matter. In the Haiwee sediments, pH measured with a micro-electrode in September 2003 decreased only moderately (from 7.8 to 6.7) with depth (Dixit, unpubl.). Total dissolved P concentrations in the porewater (like those of As, Fe, and Mn) were elevated only deeper in the sediment column. It is possible that the release of phosphate associated with the remineralization of organic matter could lead to competitive desorption of As from the Fe^{III} oxyhydroxide floc. Consistent with this hypothesis, it was observed that a significant fraction (ca. 80%) of the As associated with the sediment was released by leaching with 1 M phosphate at pH 4. The presence of elevated phosphate in the porewaters may also inhibit the re-sorption of As (released by reductive dissolution) onto the residual Fe^{III} oxyhydroxide solids in the sediment. Note that, in this study, only total dissolved P (not phosphate, *per se*) was measured in the porewater. In addition, it has not yet been possible (given the small volumes of porewater recovered) to distinguish between As^{III} and As^{V} in the porewaters though the preponderance of As^{III} in the solid phase suggests that As^{III} may also dominate the dissolved pool.

DOM has not yet been measured in the porewaters of Haiwee sediments but DOM concentrations are often elevated in sediment porewaters. If present at sufficient concentrations, DOM could promote desorption of As and/or inhibit As re-sorption onto the Fe^{III} oxyhydroxide floc (Redman et al. 2002).

Elevated concentrations of dissolved Fe have been measured in Haiwee sediments. As with As and P, the speciation of Fe in the porewaters has not yet been determined but it may be assumed that inorganic Fe^{II} is predominant (van der Zee et al. 2002). A laboratory study of competitive sorption of Fe^{II} and As^{III} was conducted using goethite as the sorbent. Sorption of Fe^{II} was shown to be reversible. In single sorbate experiments, sorption of both Fe^{II} and As^{III} was observed. In the presence of both sorbates, however, competition between Fe^{II} and As^{III} was not observed (Fig. 3; Dixit, unpubl.). Although the potential for

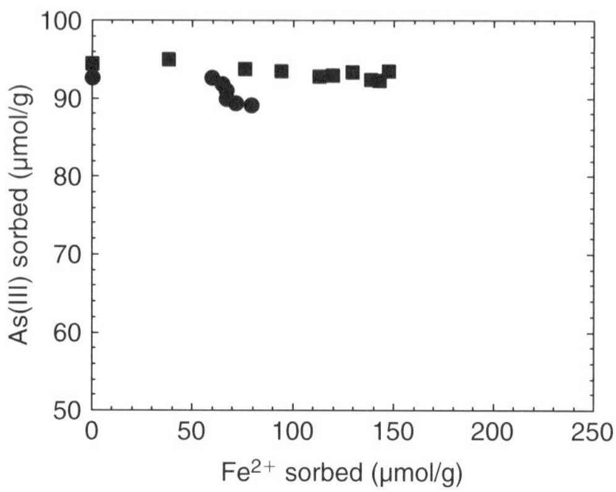

Figure 3. Sorption of As^{III} and Fe^{II} onto goethite at pH 6.5 (circles) and 7.5 (squares). Experimental conditions: I = 0.01, 2.5 g/L goethite, $[As^{III}]_T = 250\,\mu M$, $[Fe^{II}]_T = 0$–1.5 mM.

competition between Fe^{II} and As^{III} needs to be assessed for conditions relevant to Haiwee reservoir, these results suggest that Fe^{II} is not likely to suppress As^{III} sorption dramatically.

5 *IN SITU* SORPTION EXPERIMENTS

Because of the difficulty of obtaining sufficient volumes of porewaters for sorption experiments (or even for complete analytical characterization), an alternative approach was adopted in which freshly-prepared HFO could be exposed to *in situ* conditions with minimal disturbance of the sediments. The gel probe sampler (Kneebone et al. 2002; Krom et al. 1994), which was used to obtain porewater profiles, was modified for this purpose. In the original gel probe sampler, polyacrylamide (clear) gel slabs held in a Plexiglas ladder were introduced into the sediment and allowed to equilibrate with the porewater. For the modified sampler, the gels were doped with HFO before being loaded into the sampler. The sampler was loaded with both HFO-doped and clear gels. It was hoped that the clear gels could be used (as previously) to obtain porewater profiles of various elements and that the HFO-doped gels would provide information on the extent of sorption of those elements in response to their porewater concentrations.

In obtaining porewater profiles using the (clear) gel probe sampler, it can be assumed that the equilibration of the clear gels does not perturb the ambient porewater composition significantly. This assumption is reasonable because the volume of water in the gel slabs is quite small and the porewater constituents are not concentrated in the (clear) gel. This is not the case, however, with the HFO-doped gels.

Since significant concentration of porewater constituents onto the HFO-doped gels (i.e., by sorption) may be expected, the concentration of these species in the porewater in contact with the sampler could be substantially depleted (Zhang et al. 1995; Ernstberger et al. 2002; Zhang et al. 2002). Thus, the accumulation of sorbates in the HFO-doped gels and the concentrations of solutes in the clear gels will reflect the ambient porewater composition only if the sorbing constituents are resupplied to the porewater from the sediments on the timescale of equilibration of the (mixed clear and HFO-doped) gel probe sampler. Preliminary deployment of the sampler (for several hours) suggested some artifacts associated with kinetic limitations of either As resupply from the sediment or uptake onto the HFO-doped gels or both. Further work is in progress to assess performance of the HFO-doped gel probe sampler.

6 CONCLUDING COMMENTS

The deposition of As-enriched Fe^{III} oxyhydroxide floc to the sediments of Haiwee Reservoir provides a unique setting for the study of As mobilization. Sediment deposition patterns in Haiwee reservoir have been significantly perturbed by the addition of ferric chloride to the LAA. Because this manipulation is designed to substantially decrease dissolved As concentrations in the LAA water supply, it has necessarily been carried out at a large scale. Geochemical studies in this system, in which As and Fe inputs are relatively well defined, provide insight into the biogeochemical controls on As mobility.

REFERENCES

Benner, S.G., Hansel, C.M., Wielinga, B.W., Barber, T.M. & Fendorf, S. 2002. Reductive dissolution and biomineralization of iron hydroxide under dynamic flow conditions. *Environ. Sci. Technol.* 36(8): 1705–1711.

Dixit, S. & Hering, J.G. 2003. Comparison of arsenic(V) and arsenic(III) sorption onto iron oxide minerals: Implications for arsenic mobility. *Environ. Sci. Technol.* 37: 4182–4189.

Ernstberger, H., Davison, W., Zhang, H., Tye, A. & Young, S. 2002. Measurement and dynamic modeling of trace metal mobilization in soils using DGT and DIFS. *Environ. Sci. Technol.* 36(3): 349–354.

Ford, R.G. 2002. Rates of hydrous ferric oxide crystallization and the influence on coprecipitated arsenate. *Environ. Sci. Technol.* 36(11): 2459–2463.

Harvey, C.F., Swartz, C.H., Badruzzaman, A.B.M., Keon-Blute, N., Yu, W., Ali, M.A., Jay, J., Beckie, R., Niedan, V., Brabander, D., Oates, P.M., Asfaque, K.N., Islam, S., Hemond, H.F. & Ahmed, M.F. 2002. Arsenic mobility and groundwater extraction in Bangladesh. *Science* 298(5598): 1602–1606.

Hering, J.G., Chen, P.Y., Wilkie, J.A., Elimelech, M. & Liang, S. 1996. Arsenic removal by ferric chloride. *J. Am. Water Works Assoc.* 88(4): 155–167.

Hering, J.G. & Kneebone, P.E. 2001. Biogeochemical controls on arsenic occurrence and mobility in water supplies. In: W.T. Frankenberger (ed.), *Environmental Chemistry of Arsenic*: 155–181. New York: Marcel Dekker.

Jekel, M.R. 1994. Removal of arsenic in drinking water treatment. In: J.O. Nriagu (ed.), *Arsenic in the Environment. Part I: Cycling and Characterization*: 119–132. New York: Wiley-Interscience.

Kneebone, P.E., O'Day, P.A., Jones, N. & Hering, J.G. 2002. Deposition and fate of arsenic in iron- and arsenic-enriched reservoir sediments. *Environ. Sci. Technol.* 36: 381–386.

Krom, M.D., Davison, P., Zhang, H. & Davison, W. 1994. High-resolution Pore-water sampling with a gel sampler. *Limnol. Oceanogr.* 39: 1967–1972.

Manning, B.A., Fendorf, S.E. & Goldberg, S. 1998. Surface structures and stability of arsenic(III) on goethite: spectroscopic evidence for inner-sphere complexes. *Environ. Sci. Technol.* 32(16): 2383–2388.

Nordstrom, D.K. 2002. Public health – Worldwide occurrences of arsenic in ground water. *Science* 296(5576): 2143–2145.

O'Day, P.A., Rivera, N., Root, R. & Carroll, S.A., in press. X-ray absorption spectroscopic study of iron reference compounds for the analysis of natural sediments. *Am. Mineral.*

Raven, K.P., Jain, A. & Loeppert, R.H. 1998. Arsenite and arsenate adsorption on ferrihydrite: Kinetics, equilibrium, and adsorption envelopes. *Environ. Sci. Technol.* 32: 344–349.

Redman, A.D., Macalady, D.L. & Ahmann, D. 2002. Natural organic matter affects arsenic speciation and sorption onto hematite. *Environ. Sci. Technol.* 36(13): 2889–2896.

Shen, Y.S. 1973. Study of arsenic removal from drinking water. *J. Am. Water Works Assoc.* 65(8): 543–548.

Smedley, P.L. & Kinniburgh, D.G. 2002. A review of the source, behaviour and distribution of arsenic in natural waters. *Appl. Geochem.* 17(5): 517–568.

Sorg, T.J. & Logsdon, G.S. 1978. Treatment technology to meet the interim primary drinking water regulations for inorganics: Part 2. *J. Am. Water Works Assoc.* 70(7): 379–393.

Stolarik, G.F. & Christie, J.D. 1999. Interim arsenic management plan for Los Angeles, *Proc. Amer. Water Works Assoc. Annual Conference*, Chicago, IL.

van der Zee, C., van Raaphorst, W. & Helder, W. 2002. Fe redox cycling in Iberian continental margin sediments (NE Atlantic). *J. Marine Res.* 60(6): 855–886.

Welch, A.H., Westjohn, D.B., Helsel, D.R. & Wanty, R.B. 2000. Arsenic in ground water of the United States: occurrence and geochemistry. *Ground Water* 38: 589–604.

Wilkie, J.A. & Hering, J.G. 1996. Adsorption of arsenic onto hydrous ferric oxide: Effects of adsorbate/adsorbent ratios and co-occurring solutes. *Colloids and Surfaces*

A-Physicochemical and Engineering Aspects 107: 97–110.
Zhang, H., Davison, W., Miller, S. & Tych, W. 1995. *In-Situ* high-resolution measurements of fluxes of Ni, Cu, Fe, and Mn and concentrations of Zn and Cd in porewaters by dgt. *Geochim. Cosmochim. Acta* 59(20): 4181–4192.
Zhang, H., Davison, W., Mortimer, R.J.G., Krom, M.D., Hayes, P.J. & Davies, I.M. 2002. Localised remobilization of metals in a marine sediment. *Sci. Total Environ.* 296(1–3): 175–187.

Water–Rock Interaction, Wanty & Seal II (eds)
© 2004 Taylor & Francis Group, London, ISBN 90 5809 641 6

Polymeric silicate complexing in aqueous fluids at high pressure and temperature, and its implications for water–rock interaction

C.E. Manning
Dept. of Earth & Space Sciences, University of California, Los Angeles, CA, USA

ABSTRACT: The solubility of silicate minerals in H_2O typically increases strongly as pressure (P) and temperature (T) rise along crustal geotherms. This simple observation has profound implications for material transfer in a wide range of geologic environments, but the controls on the P-T dependence of solubility remain poorly understood. New experimental studies suggest that solubility enhancement arises in large part from the increasing stability of silica-rich polymers at high P and T, e.g. $Si_2O(OH)_6$ dimers and $AlSi_2O_2(OH)_8^-$ and $NaAlSi_2O_2(OH)_8$ trimers. Although such complexes are well known in surface waters, their role in the aqueous solutions associated with magmatic and metamorphic processes is only now being appreciated. For example, SiO_2 dimers are abundant in quartz-saturated solutions in the system SiO_2-H_2O at >5 kbar and >500°C. In more complex systems, silica multimers incorporate Al, alkalis, and, potentially, other rock-forming constituents. In the deep crust and mantle, these complexes may predominate relative to monomeric species and control the chemical interaction of fluids with crustal and mantle rocks. Silicate polymers increase chemical potential gradients, and therefore enhance mass transfer per unit fluid mass relative to polymer-free solutions. The abundance of silicate polymers also modifies the transport properties of dense aqueous fluids.

1 AQUEOUS SILICA POLYMERIZATION IN PURE H_2O

Silica polymers have long been recognized in aqueous solutions at low P and T. Early studies demonstrated their stability at high concentration in basic solutions or at strong quartz oversaturation (e.g. Lagerstrom 1959; Ingri 1959; Weill & Bottinga 1970; Busey & Mesmer 1977; Crerar et al. 1981; Cary et al. 1982; Alvarez & Sparks 1985). Nevertheless, thermodynamic analyses typically assumed that these complexes were not significant in equilibrium quartz-saturated H_2O at near-neutral pH because of lower aqueous silica concentration (e.g. Walther & Helgeson 1977). However, the concentration of aqueous silica in quartz-saturated H_2O increases dramatically with increasing P and T (Anderson & Burnham 1965; Manning 1994). Recent experimental studies show that this solubility enhancement is accompanied by increased extent of polymerization. Silica concentrations in equilibrium with quartz-undersaturated mineral assemblages at near-neutral pH are lower than predicted if monomers are the only silica species (Zhang & Frantz 2000; Newton & Manning 2002) and Raman spectroscopy of high-P-T quartz-saturated H_2O yields direct evidence for silica dimers (Zotov & Keppler 2000, 2002).

The identification of aqueous silica dimers is consistent with our knowledge of the structure and phase relations of hydrous silicate melts. In many silicate-H_2O systems at elevated P, the H_2O-saturated melt and the melt-saturated aqueous phase approach each other in composition with isobarically increasing T. The compositional difference between the two disappears along a critical curve, above which there is a single fluid whose composition varies without phase change from pure H_2O to relatively H_2O-poor, hydrous melt on the liquidus. In the system SiO_2-H_2O, the structure of supercritical hydrous SiO_2 melt is strongly polymerized at <10 wt% H_2O, such that bridging-oxygen rich species – Q_3 ($Si_2O_5^{2-}$) and Q_4 (SiO_2) – predominate (Farnan et al. 1987). Thus, in the supercritical region, an increase in SiO_2 concentration relative to H_2O is accommodated by the progressive increase in the abundance of Si–O–Si bonding; i.e. polymerization. The system SiO_2-H_2O has a second critical end point at ~10 kbar and 1060°C (Kennedy et al. 1962). The comparatively low P and T of this feature indicate that there should be significant capacity for silica polymerization in fluids at geologically accessible conditions.

The significant equilibrium concentration of silica dimers in H_2O-rich solutions well below the hydrous melting temperature of quartz was not predicted by any theory of aqueous silica. This lack of prediction is in part attributable to a conceptual framework within which the thermodynamics of aqueous solutions are developed with reference to a state of infinite dilution. The analogy between hydrous melts and polymerized aqueous solutions implies that it is perhaps more useful to view the mixture of aqueous silica monomers and multimers at crustal metamorphic conditions as precursors to the polymerized hydrous melt produced at the water-saturated quartz solidus. Within such a framework, the structural chemistry of silicate melts (e.g. Hess 1971, 1995) will provide the best guide to speciation of dense aqueous fluids.

Initial steps toward developing new thermodynamic models of aqueous silica are under way. Newton & Manning (2003) investigated the thermodynamic properties of polymerized silicate solutions by constraining the activity coefficient of aqueous silica at 800°C and 12 kbar through measurement of the concentration of total dissolved aqueous silica ($SiO_{2(t)}$) in H_2O in equilibrium with silica-buffering mineral assemblages. For the reaction 2 SiO_2 monomers (m) = 1 Si_2O_4 dimer (d), the equilibrium constant (K_{md}) is $K_{md} = X_d/X_m^2$, where X represents the mole fraction of $SiO_{2(t)}$ occurring as the subscripted species, and mixing is assumed to be ideal. This formulation is valid irrespective of the hydration states of the silica species because of the small silica concentration at these conditions (≤2.5 mol%). For a standard state of unit activity of the hypothetical pure-monomer solution at the P and T of interest, the monomer–dimer model leads to $\gamma = X_m/X_s$ and

$$K_{md} = \frac{(1 - \gamma_s)}{2\gamma_s^2 X_s} \quad (1)$$

where γ_s and X_s are the activity coefficient and mole fraction of total silica. The data yield $K_{md} = 155$ at 800°C and 12 kbar, in excellent agreement with results from *in situ* Raman spectroscopy of a quartz-saturated solution at the same P-T (Zotov & Keppler 2002). Thermodynamic modeling indicates that higher polymers are unlikely to be significant at this P-T. In the system SiO_2-H_2O, a solution in equilibrium with quartz at 800°C, 12 kbar, contains 2.5 mol% silica, of which 70% occurs in dimers, and γ_s is small (0.30). Even at the low concentration (0.1 mol%), the activity coefficient is only 0.75 and the dissolved silica is substantially polymerized (25%).

2 Al AND Al-Si COMPLEXES

Like quartz, the solubility of corundum (Al_2O_3) and diaspore (AlO(OH)) in H_2O increase with P and T (Becker et al. 1983; Lin 2001; Tropper & Manning, in press). At 10 kbar, corundum solubility increases from 1 millimolal at 700°C to 20 millimolal at 1100°C. At all T, Al solubility in H_2O in equilibrium with diaspore or corundum at 10 kbar is at least ten times than at 1 kbar. Multimeric Al clusters have been observed in concentrated solutions at low P and T (e.g. Öhman & Sjoberg 1981; Bottero et al. 1987; Furrer et al. 1992, 2002; Boudot et al. 1996; Gérard et al. 2001). Though no direct observation has yet been made at elevated P and T, it is possible than, as with silica, the enhanced solubility of Al arises in part from increased stability of multinuclear Al clusters.

Aqueous fluids in the crust and mantle coexist with silicate minerals, so it is of fundamental importance to characterize solution and transport mechanisms of Al in Si-bearing solutions. Our experiments on the solubility of corundum (Al_2O_3) + kyanite (Al_2SiO_5) in H_2O at 700°C, 10 kbar, yield higher Al solubility (6.17 ± 0.55 mmol/kg H_2O) relative to corundum-only in H_2O at the same conditions. Because pH is near neutral, the elevation of Al concentration above that in the system corundum-H_2O implies formation of Al-Si multimers.

Al-Si complexing in aqueous solutions is a subject of some dispute. The formation of Al-Si complexes involving a bridging O may form in a wide variety of stoichiometries (e.g. Mueller et al. 1981). Browne & Driscoll (1992) reported that at 25°C, 1 bar, 95% of aqueous Al and Si were bound in Al-Si complexes, primarily $AlOSi(OH)_3^{2+}$ formed via

$$Al^{3+} + H_4SiO_4 = AlOSi(OH)_3^{2+} + H^+ \quad (2)$$

This result was challenged by Farmer & Lumsdon (1994), who found that minimal pH differences between Al solutions with and without Si were inconsistent with extensive Al-Si polymerization. Pokrovski et al. (1996) obtained a similar result at 25°C. Significantly, however, Pokrovski et al. (1996) and Salvi et al. (1998) showed that $AlOSi(OH)_3^{2+}$ and other Al-Si complexes become progressively more stable with increasing T along the steam saturation curve. Thus, although Al-Si complexes are likely unimportant at equilibrium at 25°C in natural waters, they may predominate in high-T environments such as hydrothermal systems. Combination of these observations with the new results on corundum-kyanite solubility suggests that enhanced stability of Al-Si complexes persists to the highest P and T yet studied, which represent deep crustal and mantle metasomatic environments.

3 Na-Al-Si COMPLEXES

A simple compositional model for the continental crust is feldspar + quartz. Thus, the extension of results in simple systems described above logically involves addition of alkalis. Antignano & Manning (2003) measured the solubility of albite + quartz in H_2O at 580°C, 5–12 kbar. Albite exhibited incongruent dissolution, yielding a husk of paragonite ($NaAl_3Si_3O_{10}(OH)_2$) that mantles the albite grain. This is important because it produces a compositionally invariant solution at equilibrium and fixed *P-T*. The concentration of total dissolved solids (TDS) reaches 1.0 molal at 12 kbar. Si, Na, and Al solubilities all increase strongly with *P*, such that at 12 kbar molalities are: Si, 0.80; Na, 0.15; Al, 0.08. The aqueous Si concentrations in the albite-only experiments are higher than those in fluid equilibrated with quartz (Manning 1994) at the same conditions. This is consistent with the data of Anderson & Burnham (1983) on albite-H_2O at low *P*. Enhanced solubility of silica in the presence of albite + quartz, relative to quartz alone, suggests complexing of Si with Na and Al. Strong support for the formation of polymers lies in the fact that pH is near-neutral, so deprotonated silicic acid monomers are not abundant in these solutions. Formation of multimeric complexes involving Si, Al, and Na also explains the enhancement of Al solubility relative to that in equilibrium with corundum in H_2O. These complexes probably include a mixture of Na-Al, Na-Si, and Na-Al-Si molecules. Alkali-Al and Alkali-Si complexes are stable over a wide range of *P* and *T*, and may predominate under appropriate conditions (Anderson et al. 1987; Pascal & Anderson 1989; Anderson 1995; Pokrovskii & Helgeson 1995, 1997; Diakanov et al. 1996; Tanaka & Takahashi 1999). Na-Al-Si complexing has not been characterized in sufficient detail, but the stability of Si-Al trimers with associated Na^+ in aluminosilicate glasses and melts suggests that such structures are likely stable in aqueous solution as well (e.g. de Jong et al. 1981; Kubicki & Sykes 1995). In addition, the relatively low *P* and *T* of the albite-H_2O critical curve (Shen & Keppler 1997) are consistent with extensive Na-Al-Si complexing in the supercritical fluid phase.

4 IMPLICATIONS

The formation of polymeric silicate complexes in crustal and mantle fluids has important consequences for mass transfer and fluid properties. Viscosity and diffusivity will be significantly different in a polymer-bearing fluid than in a simple electrolyte solution. The large compositional dependence of silica polymerization (Newton & Manning 2003) yields a greater contrast in chemical potentials between different silica-buffering assemblages. For example, transport from silica-saturated to silica-undersaturated rocks (e.g. crustal to mantle lithologies in subduction zones) will require less fluid to attain the same compositional shift in the presence of polymers than in their absence (Newton & Manning 2002). The substantial enhancement of Al solubility through formation of Al-Si and Na-Al-Si complexes provides a simple explanation for the long-standing problem of precipitation of abundant aluminosilicate minerals from aqueous fluids in a wide range of metamorphic environments (e.g. Kerrick 1988; Widmer & Thompson 2001; Sepahi et al. 2004). A more general problem in crustal metamorphism is that many metamorphic rocks show evidence for significant compositional modification by flowing fluids (e.g. Ague 1997), but the fluid volumes required to produce the effect are often unreasonably high. Polymer formation could ameliorate this dilemma because it provides a mechanism for more efficient mass transfer. Thus, the recognition of polymeric silicate complexes represents an important step toward understanding water–rock interaction in metamorphic and igneous environments of the crust and mantle.

REFERENCES

Ague, J.J. 1997. Crustal mass transfer and index mineral growth in Barrow's garnet zone, northeast Scotland. *Geol.* 25: 73–76.

Alvarez, R. & Sparks, D.L. 1985. Polymerization of silicate anions in solutions at low concentrations. *Nature* 318: 649–651.

Anderson, G.M. 1995. Is there alkali-aluminum complexing at high temperatures and pressures? *Geochim. Cosmochim. Acta* 59: 2155–2161.

Anderson, G.M. & Burnham, C.W. 1965. The solubility of quartz in supercritical water. *Am. J. Sci.* 263: 494–511.

Anderson, G.M. & Burnham, C.W. 1983. Feldspar solubility and the transport of aluminum under metamorphic conditions. *Am. J. Sci.* 283-A: 283–297.

Anderson, G.M., Pascal, M.L. & Rao, J. 1987. Aluminum speciation in metamorphic fluids. In H.C. Helgeson (ed.), *Chemical Transport in Metasomatic Processes*: 297–321. Dordrecht: Reidel.

Antignano, A. & Manning, C. 2003. Solubility of albite + paragonite ± quartz in H_2O at 1 GPa, 580°C: Implications for metamorphic fluids. *Eos* 84(46): Abstract V22D-0613.

Becker, K.H., Cemic, L. & Langer, K.E.O. 1983. Solubility of corundum in supercritical water. *Geochim. Cosmochim. Acta* 47: 1573–1578.

Bottero, J.-Y., Axelos, M., Tschoubar, D., Cases, J.-M., Fripiat, J.-J. & Fiessinger, F. 1987. Mechanism of formation of aluminum trihydroxide from Keggin Al_{13} polymers. *J. Colloid Interf. Sci.* 117: 47–57.

Boudot, J.-P., Maitat, O., Merlet, D. & Rouiller, J. 1996. Occurrence of non-monomeric species of aluminium in undersaturated soil and surface waters: consequences for the determination of mineral saturation indices. *J. Hydrol.* 177: 47–63.

Browne, B.A. & Driscoll, C.T. 1992. Soluble aluminum silicates: stoichiometry, stability, and implications for environmental geochemistry. *Science* 256: 1667–1670.
Busey, R.H. & Mesmer, R.E. 1977. Ionization equilibria of silicic acid and polysilicate formation in aqueous sodium chloride solution to 300°C. *Inorg. Chem.* 16: 2444–2450.
Cary, L.W., de Jong, B.H.W.S. & Dibble, W.E., Jr. 1982. A ^{29}Si NMR study of silica species in dilute aqueous solution. *Geochim. Cosmochim. Acta* 46: 1317–1320.
Crerar, D.A., Axtmann, E.V. & Axtmann, R.C. 1981. Growth and ripening of silica polymers in aqueous solutions. *Geochim. Cosmochim. Acta* 45: 1259–1266.
de Jong, B.H.W.S., Keefer, K.D., Brown, G.E. & Taylor, C.M. 1981. Polymerization of silicate and aluminate tetrahedral in glasses, melts, and aqueous solutions III. Local silicon environments and internal nucleation in silicate glasses. *Geochim. Cosmochim. Acta* 45: 1291–1308.
Diakonov, O. Pokrovski, G., Schott, J., Castet, S. & Gout, R. 1996. An experimental and computational study of sodium-aluminum complexing in crustal fluids. *Geochim. Cosmochim. Acta* 60: 197–211.
Farmer, V.C. & Lumsdon, D.G. 1994. An assessment of complex formation between aluminum and silicic acid in acidic solutions. *Geochim. Cosmochim. Acta* 58: 3331–3334.
Farnan, I., Kohn, S.C. & Dupree, R. 1987. A study of the structural role of water in hydrous silica glass using cross-polarisation magic angle spinning NMR. *Geochim. Cosmochim. Acta* 51: 2869–2873.
Furrer, G., Trusch, B. & Müller, C. 1992. The formation of polynuclear Al_{13} under simulated natural conditions. *Geochim. Cosmochim. Acta* 56: 3831–3838.
Furrer, G., Phillips, B.L., Ulrich, K.-U., Pöthig, R. & Casey, W.H. 2002. The origin of aluminum flocs in polluted streams. *Science* 297: 2245–2247.
Gérard, F., Boudot, J.P. & Ranger, J. 2001. Consideration on the occurrence of the Al_{13} polycation in natural soil solutions and surface waters. *Appl. Geochem.* 16: 513–529.
Hess, P.C. 1971. Polymer model of silicate melts. *Geochim. Cosmochim. Acta* 35: 289–306.
Hess, P.C. 1995. Structure, dynamics and properties of silicate melts. *Revs. Mineral.* 32: 145–189.
Ingri, N. 1959. Equilibrium studies of polyanions IV. Silicate ions in NaCl medium. *Acta Chem. Scand.* 13: 758–775.
Kennedy, G.C., Wasserburg, G.J., Heard, H.C. & Newton, R.C. 1962. The upper three-phase region in the system SiO_2-H_2O. *Am. J. Sci.* 260: 501–521.
Kerrick, D.M. 1988. Al_2SiO_5-bearing segregations in the Lepontine Alps, Switzerland: Aluminum mobility in metapelites. *Geol.* 16: 636–640.
Kubicki, J.D. & Sykes, S. 1995. Molecular orbital calculations on the vibrational spectra of Q^3-T(OH) species and the hydrolysis of a three-membered aluminosilicate ring. *Geochim. Cosmochim. Acta* 59: 4791–4797.
Lagerstrom, G. 1959. Equilibrium studies of polyanions III. Silicate ions in $NaClO_4$ medium. *Acta Chem. Scand.* 13: 722–736.
Lin, H.A. 2001. *Aqueous geochemistry at subduction zone conditions: experimental constraints.* Unpublished Ph.D. Thesis, University of California, Los Angeles, 247 p.
Manning, C.E. 1994. The solubility of quartz in H_2O in the lower crust and upper mantle. *Geochim. Cosmochim. Acta* 58: 4831–4839.
Mueller, D., Hoebbel, D. & Gessner, W. 1981. ^{27}Al NMR studies of aluminosilicate solutions. Influences of the second coordination sphere on the shielding of aluminum. *Chem. Phys. Lett.* 84: 25–29.
Newton, R.C. & Manning, C.E. 2002. Solubility of silica in equilibrium with enstatite, forsterite, and H_2O at deep crust/upper mantle pressures and temperatures and an activity–concentration model for polymerization of aqueous silica. *Geochim. Cosmochim. Acta* 66: 4165–4176.
Newton, R.C. & Manning, C.E. 2003. Activity coefficient and polymerization of aqueous silica at 800°C, 12 kbar, from solubility measurements on SiO_2-buffering mineral assemblages. *Contrib. Mineral. Petrol.* 146: 135–143.
Öhman, L.-O. & Sjoberg, S. 1981. Equilibrium and structural studies of silicon(IV) and aluminum(III) in aqueous-solution. 1. The formation of ternary mononuclear and polynuclear complexes in the system Al^{3+}-gallic acid-OH^- – a potentiometric study in 0.6 m NaCl. *Acta Chem. Scand. Series A*: 201–212.
Pascal, M.L. & Anderson, G.M. 1989. Speciation of Al, Si and K in supercritical solutions: Experimental study and interpretation. *Geochim. Cosmochim. Acta* 53: 1843–1855.
Pokrovski, G.S., Schott, J., Harrichoury, J.-C. & Sergeyev, A.S. 1996. The stability of aluminum silicate complexes in acidic solutions from 25 to 150°C. *Geochim. Cosmochim. Acta* 60: 2495–2501.
Pokrovskii, V.A. & Helgeson, H.C. 1995. Thermodynamic properties of aqueous species and the solubilities of minerals at high pressures and temperatures: The system Al_2O_3-H_2O-NaCl. *Am. J. Sci.* 295: 1255–1342.
Pokrovskii, V.A. & Helgeson, H.C. 1997. Thermodynamic properties of aqueous species and the solubilities of minerals at high pressures and temperatures: the system Al_2O_3-H_2O-KOH. *Chem. Geol.* 137: 221–242.
Salvi, S., Pokrovski, G.S. & Schott, J. 1998. Experimental investigation of aluminum-silica aqueous complexing at 300°C. *Chem. Geol.* 151: 51–67.
Sepahi, A.A., Whitney, D.L. & Baharifar, A.A. 2004. Petrogenesis of andalusite-kyanite-sillimanite veins and host rocks, Sanandaj-Sirjan metamorphic belt, Hamadan, Iran. *J. Metamorph. Geol.* 22: 119–134.
Shen, A. & Keppler, H. 1997. Direct observation of complete miscibility in the albite-H_2O system. *Nature* 385: 710–712.
Tanaka, M. & Takahashi, K. 1999. The identification of chemical species of silica in sodium hydroxide, potassium hydroxide and sodium chloride solutions by FAB-MS. *Anal. Sci.* 15: 1241–1250.
Tropper, P. & Manning, C.E. 2004. The solubility of rutile and corundum in H_2O at high P and T: constraints on Ti and Al mobility during high-P metamorphism. EMPG Conference Abstracts, in press.
Walther, J.V. & Helgeson, H.C. 1977. Calculation of the thermodynamic properties of aqueous silica and the solubility of quartz and its polymorphs at high pressures and temperatures. *Am. J. Sci.* 277: 1315–1351.
Weill, D. & Bottinga, Y. 1970. Thermodynamic analysis of quartz and cristobalite solubilities in water at saturation vapor pressure. *Contrib. Mineral. Petrol.* 25: 125–132.

Widmer, T. & Thompson, A.B. 2001. Local origin of high pressure vein material in eclogite facies rocks of the Zermatt-Saas zone, Switzerland. *Am. J. Sci.* 301: 627–656.
Zhang, Y.-G. & Frantz, J.D. 2000. Enstatite-forsterite-water equilibria at elevated temperatures and pressures. *Am. Mineral.* 85: 918–925.
Zotov, N. & Keppler, H. 2000. *In-situ* Raman spectra of dissolved silica species in aqueous fluids to 900°C and 14 kbar. *Am. Mineral.* 85: 600–603.
Zotov, N. & Keppler, H. 2002. Silica speciation in aqueous fluids at high pressures and high temperatures. *Chem. Geol.* 184: 71–82.

Volcanic-geothermal water–rock interactions and degassing: a symposium in memory of Donald White

Water–Rock Interaction, Wanty & Seal II (eds)
© 2004 Taylor & Francis Group, London, ISBN 90 5809 641 6

Yellowstone caldera inflation–deflation and hydrothermal Cl flux revisited

Robert O. Fournier
U.S. Geological Survey (retired), Portola Valley, California, USA

ABSTRACT: Geodetic leveling documented a change from uplift to subsidence within the Yellowstone caldera in 1984. This change was coincident with seismicity that appears to have been the result of movement of hydrothermal fluid radially outward from the caldera. Renewed uplift of a portion of the caldera, beginning in 1996 or 1997, was first detected by satellite interferometric synthetic aperture radar techniques. Modeling of that data suggests that hydrothermal fluid movement may have been important at the start, but that injection of magma may be responsible for a large new region of uplift centering on the northern rim of the caldera. Monitoring of chloride flux in rivers draining Yellowstone Park has shown that there are large seasonal variations and year to year variations that correlate with precipitation. However, there have been no detected changes in chloride flux from Yellowstone's hydrothermal system that can be attributed to variations in the amount of magmatic fluid that may have leaked upward from a deep, high-pressure system.

1 INTRODUCTION

Donald E. White provided leadership and inspiration for widely ranging research on the hydrothermal system at Yellowstone National Park (YNP) that was initiated in the 1960s, and continues to the present. An important component of that effort was an assessment of a minimum rate of heat discharge by the hydrothermal system (Fournier et al. 1976). The flux of Cl^- into the rivers originating in YNP and draining to the east of the continental divide was measured, and then heat flux was calculated using an enthalpy- Cl^- model developed for that purpose. Subsequently Norton & Friedman (1985, 1991) and Friedman & Norton (2000) expanded the coverage to include the rivers draining YNP west of the continental divide, and monitored the Cl^- flux out of YNP for the water years 1983–1999. The minimum hydrothermal heat flux value derived from these Cl^- flux measurements has proved to be an important parameter in the development of alternative models of dynamic processes that may be operating within the Yellowstone magmatic-hydrothermal system (Fournier 1989).

One intent of the ongoing effort to monitor Cl^- flux is to provide information about possible leakages of magmatic fluids into the overlying hydrostatically-pressured system that might correlate with observed changes in direction of vertical movement of the caldera floor described by Dzurisin et al. (1999, and references therein). For various reasons, discussed below, observed changes in Cl^- flux from the hydrothermal system have not correlated with cycles of inflation and deflation of the Yellowstone caldera, although information obtained from seismic and vertical ground motion monitoring suggests that movement of magmatic fluid is likely to be involved in some of the deformation.

2 UPLIFT AND SUBSIDENCE

In 1976 it was discovered that the Yellowstone caldera had undergone significant uplift since the first leveling survey in 1923 (Pelton & Smith 1979, 1982). Subsequent leveling surveys confirmed that the caldera continued to undergo uplift until 1984 (Dzurisin et al. 1994). The center of that uplift was along an axis running from the Sour Creek resurgent dome (SC) in the northeast region of the Yellowstone caldera, to the Mallard Lake resurgent dome (ML) in the southwest region of the caldera (Fig. 1).

In 1985 the largest historic swarm of earthquakes recorded in Yellowstone occurred outside the caldera to the west of Norris Geyser Basin (Waite & Smith 2002). Waite & Smith (2002) concluded that the temporal shift of that seismicity could be explained by migration of hydrothermal fluids radially outward from the caldera, following rupture of a sealed hydrothermal system within the caldera. Coincident with the start of the 1985 seismicity, the caldera floor began to subside and continued to do so until 1995, when it

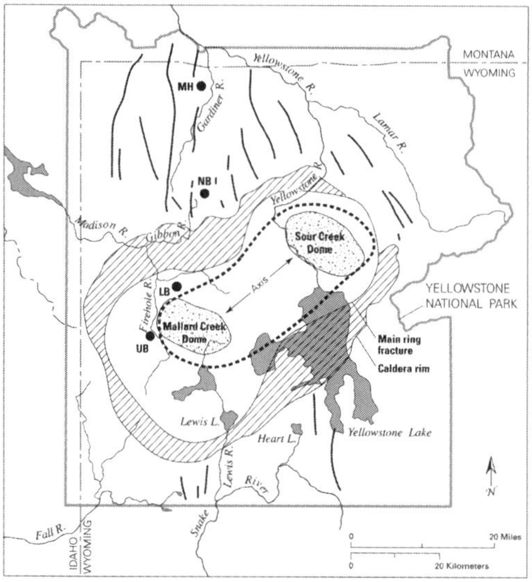

Figure 1. Map of Yellowstone National Park, Wyoming, showing 0.6 Ma Yellowstone caldera, two resurgent domes, active and inactive hydrothermal features (solid black) major faults (solid lines outside caldera region), and zone of major uplift and subsidence prior to 1997 (area within dashed line) MH is Mammoth Hot Springs, NB is Norris Geyser Basin, LB is lower Geyser Basin, and UB is Upper Geyser Basin (see text for discussion).

again began to uplift (Wicks et al. 1998; Dzurisin et al. 1999).

The renewal of uplift in 1995 is particularly noteworthy because it was detected by the satellite interferometric synthetic aperture radar (InSAR) method (Wicks et al. 1998). This method allowed deformation of the entire caldera and the surrounding region to be determined for precise time intervals between satellite images, rather than at selected points over the time period required to make a leveling traverse on the ground. In addition, subsidence within the caldera from 1992 to 1995 was found to migrate from the SC dome to the ML dome (Wicks et al. 1998), and the ensuing inflation after 1995 also started beneath the SC dome and migrated to the ML dome. That deformation is consistent with migration either of hydrothermal fluid or magma into two sill-like bodies that are about 8 kilometers directly beneath these domes (Wicks et al. 1998). Additional InSAR images for the years 1997–2001 showed a major change in the pattern of deformation in 1997 or 1998, in which a broad region, centered on the north caldera rim (Fig. 2), began to rise (Wicks et al. 2002). Because of its proximity to the Norris Geyser Basin it has been called the Norris Uplift Anomaly (NUA) by Wicks et al. (2002).

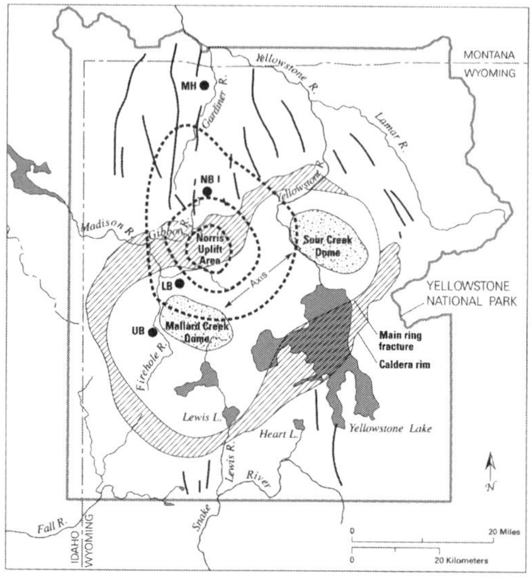

Figure 2. Map of Yellowstone National Park, Wyoming, showing the location of the Norris Uplift Anomaly (denoted schematically by concentric dashed lines). Other features and notation are the same as for Figure 1.

The NUA was preceded by a period of active seismicity that started in 1994 and continued through 2002. It is interesting that the center of the NUA roughly corresponds with the epicenter of a M 6.1 earthquake that occurred on 30 June 1975 (Pitt et al. 1979). Diagrams in a poster session presented by Wicks et al. (2002) showed a broad region of uplift roughly 30 km wide and 40 km long extending northward, as shown approximately in Figure 2, with a likely depth to the source of about 10 to 19 km. However, they also determined that the observed uplift could result from emplacement of an inclined dike extending upward from about 19 km to within 14 km of the surface. Diagrams in the poster session presented by Wicks et al. (2002) also showed that as the NUA began to rise, the rest of the Yellowstone caldera began to subside.

In addition to the geodetic measurements, tilted terraces and other geomorphic indicators show that there have been several cycles of significant uplift and subsidence within the caldera during postglacial time (Pierce et al. 1997). Two end-member models have been proposed that can account for these cycles of caldera inflation–deflation (Fournier & Pitt 1985; Fournier 1989; Dzurisin et al. 1990). In the first model uplift occurs when brine that is evolved from crystallizing rhyolitic magma is trapped at lithostatic pressure in hot (>400°C) plastic rock, and subsidence occurs when there is leakage of that magmatic fluid through a self-sealed zone separating the lithostatically-pressured region from the surrounding hydrostatically-pressured region; no basalt need be directly involved. In the second model uplift occurs during times when pulses (slug flow) of upward moving basalt from the mantle are injected into (or at the base of) the overlying silicic system, and subsidence occurs as a result of continuing crystallization and cooling of basaltic and rhyolitic magma between pulses of basalt injections. It seems likely that both mechanisms are occurring at Yellowstone.

At this time it appears that at least some of the inflation that was measured from 1923 to 1984 was the result of trapping of magmatic fluid at lithostatic pressure in plastic rock. Episodic leakage of some of this fluid through a self-sealed zone into cooler brittle rock during the period of deflation (1984–1995) is consistent with the model derived from seismicity. The return to uplift within the caldera in late 1995 may have resulted in part from resealing by mineral deposition of a relatively narrow zone separating the hydrostatic and lithostatic hydrothermal systems. However, it seems likely that a renewal of injection of basalt from the mantle into the shallow rhyolitic system also has been important in returning the system to inflation.

It is commonly assumed that the buoyant upflow of basaltic magma through solid crust stops, and ponding occurs, at the base of a large silicic magmatic chamber because rhyolitic magma is less dense than basaltic magma. At Yellowstone there has been no basalt erupted within the Yellowstone caldera, whereas basalt has erupted repeatedly to the surface in the region surrounding the caldera (Christiansen 1984). The implication is that molten or partly molten silicic material underlies most of the Yellowstone caldera. However, a three-dimensional analysis of the P and S velocity structure indicates the presence of molten or partly molten silicic material within the caldera only under the resurgent domes, and at a depth of about 8 km (Miller & Smith 1999). Elsewhere within the caldera the seismic and gravity data are consistent with the presence of quasi-plastic subsolidus granite in the depth range 6 to 12 km (Miller & Smith 1999). The most plausible explanations for the lack of basaltic surface eruptions inside of the caldera are: (1) rhyolitic magma underlies the entire Yellowstone caldera at depths too great to be detected by the method employed by Miller & Smith (1999); (2) pockets of semi-molten material too small to detect are present within the regions where only quasi-plastic subsolidus granite is indicated by the P and S velocity structure; and (3) basaltic magma that moves buoyantly upward within the caldera tends to follow a path through a pipe-like region beneath the resurgent domes where the pipe is filled with hotter and less viscous material than is present in the surrounding quasi-plastic rock.

Another possible mechanism for interrupting the upward movement of basaltic magma through the crust is diversion of that magma into sills where horizontal brine-filled lenses are present at lithostatic pressure in quasi-plastic rock. Bailey (1990) suggested that aqueous fluids in the deep crust, in general, reside in horizontal lenses, and Fournier (1999) suggested that magmatic fluids trapped at lithostatic pressure in relatively shallow hot rock that behaved plastically also would generally reside in relatively thin horizontal lenses. Upward flowing basalt that encountered such a lens would likely spread out along the base of the lens (a sill injection) while the thermal expansion of the brine (and possible separation of a gas phase) would provide additional energy to uplift overlying rock and cause lateral hydraulic opening and extension of the fracture. Thus, the presence of a relatively small horizontal pocket of brine could aid in the rapid lateral movement of the basaltic magma into a much larger sill. After emplacement of the sill, the subsequent cooling and thermal contraction of the brine (and possible condensation of water-rich gas) would contribute to a relatively rapid reversal from inflation to deflation.

3 CHLORIDE FLUX VARIATIONS

Fournier (1989) estimated that mixing about 0.2–0.4% magmatic fluid with 99.6–99.8% meteoric water could account for the concentration of Cl^- in the deepest circulating Yellowstone thermal water. According to that estimate, a very small change in the amount of magmatic brine that mixes into the hydrostatically-pressured hydrothermal system would likely cause a significant change in the Cl^- discharged from the system. Stream flow and Cl^- flux data for the four rivers draining Yellowstone, presented in Friedman & Norton (2000) for the years 1983–1999, clearly show that there are, indeed, large variations from year to year in the total amount of Cl^- carried away by the rivers that drain Yellowstone Park. The data also show that within any given water year there also are very large seasonal variations in this Cl^- flux, with much higher flux during the large run-off of snowmelt in the spring. Figure 3a shows variations in the instantaneous flow rate of the Firehole River throughout the calendar year 1986, and calculated instantaneous rates of flow of the deep component of thermal water into that river at indicated times. The water year 1986 (October 1, 1985–September 30, 1986) was relatively wet, and the run-off in the spring of the year was correspondingly high. The calculated rate of flow of hot spring water began to increase at about the same time that the run-off of snowmelt began, but peaked significantly before the peak in flow rate of the river. Figure 3b shows similar data for the calendar year 1988. The water year 1988 was relatively dry compared to 1986, and

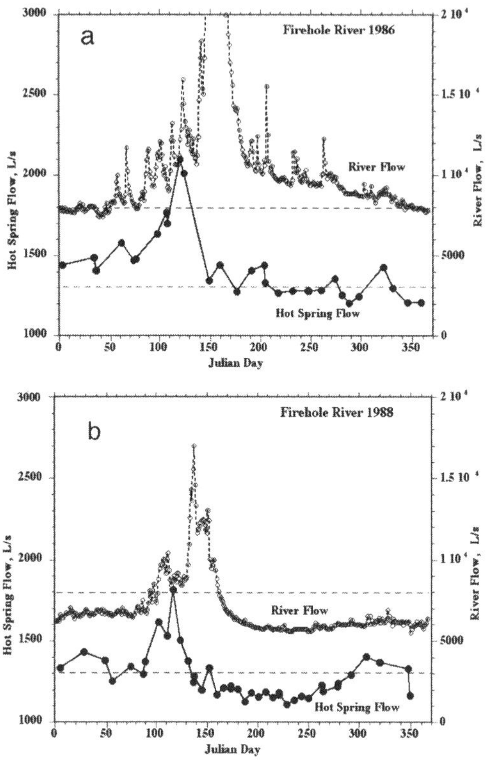

Figure 3. Instantaneous rates of discharge of water in the Firehole River 4.2 km upstream from Madison Junction, and calculated rates of instantaneous discharge of thermal water into the Firehole River determined by the Cl^- flux method during (a) 1986 and (b) 1988, assuming 400 mg/kg Cl^- in hot spring water, and 2.0 mg/kg background Cl^- in non-thermal water (see text for discussion).

the river flow and calculated hot spring flows in 1988 were correspondingly low. Again, the peak in calculated hot spring flow preceded the peak in river flow. Note that in spite of the large difference in the total flow of river water in those years, calculated rates of hot spring discharge into the Firehole River during the late summer and fall were not significantly different in 1986 and 1988. The data used by Fournier et al. (1976) to estimate the amount of hot spring water and heat discharged from the Yellowstone hydrothermal system were collected mostly in the late summer and early fall. The method used by Fournier (1989) to compare variations in the amount of hot spring water and heat discharged from the Yellowstone hydrothermal system in different years relied heavily on late summer and fall data, and discounted data collected during the run-off of snowmelt in the spring.

There are several factors that contribute to uncertainties in the determination of the total Cl^- discharged each year into each of the four rivers draining the park. These include errors in the measurement of total stream discharge and Cl^- concentration in the river water, and the integration, over a year's time, of a series of instantaneous Cl^- flux measurements. Friedman & Norton (1990 & in press) addressed each of these issues, claiming $\pm 5\%$ accuracy for the river flow measurements, and ± 1 to 2% for the Cl^- analyses. They concluded that there has been a decline of 10% in the annual thermal Cl^- flux from the Yellowstone hydrothermal system over the past 19 years, after correcting for climatic factors. However, there are several reasons why this conclusion might be viewed with skepticism.

Ingebritsen et al. (2001) used the Friedman & Norton (2000) Yellowstone data, and assumed a constant value of 0.7 mg/L for the concentration of non-thermal Cl^- in the river waters to calculate decadal-scale trends in values for "excess Cl^-" (Cl^- contributed by hot springs) in Yellowstone's river waters. They concluded that there might have been a statistically significant (at the 5% level) decline in excess Cl^- in the Firehole River, but found no long-term trends for the Yellowstone, Snake, and Fall Rivers. The major correlation that Ingebritsen et al. (2001) found was between excess Cl^- and stream discharge. They found no obvious correlation with earthquakes or inflation/deflation cycles.

A major problem in the estimation of the total flux of hydrothermal Cl^- integrated over an entire water year is dealing with rapidly changing conditions during the spring runoff. During that period some river discharge determinations may have uncertainties significantly greater than $\pm 5\%$ because sand and gravel are quickly redistributed at river high flow rates, causing changes in cross-sections of river channels. Thus, rating curves used to estimate flow between actual field measurements are less accurate compared to periods of low flow rates. An even greater uncertainty is in the assignment of a value for background (non-thermal) Cl^- for waters in different rivers, and for water in the same river during different seasons of the year. Friedman & Norton (1990, 2000, & in press) and Ingebritsen et al. (2001) assumed a constant value of 0.7 mg/L for background Cl^-, whereas Fournier et al. (1976) used a value of 2.0 mg/L. However, it seems unlikely that the background Cl^- remains constant throughout the year in any particular river because, as the season progresses, different proportions of water enter the rivers from snowmelt, rain, and discharge of groundwater that picks up non-hydrothermal Cl^- as a result of rock weathering. The amount of non-hydrothermal Cl^- from rock weathering is likely to be different for each river system because of differences in the types of rock that are weathered. Christiansen (2001) reports Cl^- concentrations ranging from 270 to 7400 parts per million in Yellowstone rhyolitic pumice and obsidian.

Uncertainties in the concentration of background Cl^- are most important for those river waters containing less than about 20 mg/L Cl^-. When the concentration of Cl^- in a river water is more than about 20 mg/L, it is unimportant in the determination of the volume of the thermal water component (assumed to contain 400 mg/L Cl^-) whether one uses 0.7 or 2.0 mg/L for the background Cl^-. This situation generally prevails for the Gibbon, Firehole, and Madison Rivers. It seldom is attained in the Yellowstone, Snake, and Fall Rivers. In those rivers small differences between the assumed and the actual background Cl^- during the spring runoff can result in very large errors in the calculated flux of hydrothermal Cl^- and hot spring water. In addition, transient storage of Cl^- in the Yellowstone and Shoshone Lakes results in very strong correlations between lake level and Cl^- flux into the Yellowstone and Snake Rivers draining those lakes (Fournier et al. 1976; Ingebritsen et al. 2001).

4 CONCLUSIONS

Inflation within the caldera prior to 1985 appears to have been due, in part, to the accumulation of magmatic fluid at lithostatic pressure below a self-sealed zone. Seismic evidence suggests that at least some of the deflation starting in 1985 and extending to 1995 is the result of leakage of hydrothermal fluids through a self-sealed zone into brittle rock to the north of the caldera. Vertical movements since 1995 appear to be the result of a combination of hydrothermal processes and upward injection of magma (possibly basaltic) to relatively shallow levels beneath the Yellowstone caldera. However, to date there have been no unambiguous correlations of changes in Cl^- flux from the hydrothermal system with observed changes in direction of vertical movements within the caldera.

There are various factors that make such a correlation difficult: (1) seasonal variation in Cl^- flux with changes in the water table mask small changes in Cl^- flux that might be associated with leakages of magmatic fluids; (2) it may take more than a year (and different times after different deformation events) for a change in composition of the deeply circulating hydrothermal fluid to be reflected in the composition of the surface discharge; (3) the leaked magmatic fluid might be a very high-density brine that ponds beneath the overlying dilute hydrothermal system, with relatively slow upward diffusion of Cl^- across the boundary; (4) most of a leaked, highly saline brine may move outward to the side of the caldera and then downward into cooler rock; and (5) most of the leaked fluid may be a gas, separated during decompressional "boiling" of a deep, high-temperature (>400°C) brine, and containing only a few hundred mg/kg Cl^-. The possible leakage of a gas-rich fluid is supported by tomographic imaging of the Vp and Vp/Vs structure beneath Yellowstone (Husen & Smith, in press) that can be best modeled by the presence of gas within pore spaces to the northwest of the caldera at 2.0 km depth (b.s.l.).

Seasonal variation in Cl^- flux with changing water table can occur for various reasons. In some places Cl^- in transient storage at shallow levels may be flushed less efficiently from the system when the water table is lowered (Friedman & Norton, in press, Ingebritsen et al. 2001). The drop-off in Cl^- flux before the peak in River discharge shown for the Firehole River suggests that flushing of Cl^- in transient storage at shallow levels may occur relatively efficiently as the water table rises. In addition, in some places the Cl^- flux may decrease with decreasing water table because a lowering of hydrostatic pressure within a reservoir where boiling conditions are present results in a slightly superheated condition. Excess boiling occurs (evaporative or non-adiabatic) utilizing heat stored in the rock. Consequently, there is an increase in the proportion of steam to water in upflow channels. The result is a decrease in the rate of discharge of Cl^--rich water and a great increase in the rate of discharge of steam. However, because of the increased boiling, the Cl^- concentration in the discharged liquid fraction increases. Fournier et al. (2002) showed that this process appears to occur every year at Norris Geyser Basin, preceding a basin-wide hydrothermal disturbance that usually occurs in the late summer or early fall. They collected and analyzed water samples from eight widely distributed hot springs and geysers at Norris during the summer season of 1995. The springs at Norris which discharged from the deepest and hottest (~300°C) subsurface reservoir all showed a systematic increase in Cl^- concentration as the summer progressed, while the ratios of dissolved constituents in those waters remained constant, indicating an increase in the ratio of steam to liquid water discharged from those springs.

Conditions within the Yellowstone hydrothermal system are constantly changing as a result of (1) variations in input of meteoric recharge water, (2) episodic regional and local tectonic movements, with associated earthquake activity, and (3) likely changing conditions at the magmatic-hydrothermal interface. At this time the influence of seasonal changes in precipitation appears to be the dominant factor. The unequivocal determination of effects of local and regional tectonic movements has proved to be very difficult. Conditions at the hydrothermal-magmatic interface are still mainly conjectural.

REFERENCES

Bailey, R.C. 1990. Trapping of aqueous fluids in the deep crust. *Geophys. Res. Letters* 17: 1129–1132.

Christiansen, R.L. 1984. Yellowstone magmatic evolution: Its bearing on understanding large-volume explosive volcanism. In *Explosive volcanism: inception, evolution, and hazards:* 84–95.Washington, DC: National Academy Press.

Christiansen, R.L. 2001. The Quaternary and Pliocene Yellowstone Plateau volcanic field of Wyoming, Idaho, and Montana. *U.S. Geol. Survey Prof. Paper*729-G.

Dzurisin, D., Savage, J.C. & Fournier, R.O. 1990. Recent crustal subsidence at Yellowstone caldera, Wyoming. *Bull. Volcanol.* 52: 247–270.

Dzurisin, C.D., Wicks, C. Jr. & Thatcher, W. 1999. Renewed uplift at the Yellowstone caldera measured by leveling surveys and satellite radar interferometry. *Bull. Volcanol.* 61: 349–355.

Dzurisin, C.D., Yamashita, K.M. & Kleinman, J.W. 1994. Mechanisms of crustal uplift and subsidence at the Yellow stone caldera, Wyoming. *Bull. Volcanol.* 56: 261–270.

Fournier, R.O. 1989. Geochemistry and dynamics of the Yellowstone National Park hydrothermal system, Wyoming. *Ann. Rev. Earth Planet. Sci.*17: 13–53.

Fournier, R.O. 1999. Hydrothermal processes related to movement of fluid from plastic into brittle rock in the magmatic-epithermal environment. *Econ. Geol.* 94: 1193–1212.

Fournier, R.O. & Pitt, A.M. 1985. The Yellowstone magmatic-hydrothermal system, U.S.A. *Geothermal Resources Council 1985 Int. Sym. on Geothermal Energy*. International Volume: 319–327.

Fournier, R.O., Weltman, U., Counce, D., White, L.D. & Janik, C.J. 2002. Results of weekly chemical and isotopic monitoring of selected springs in Norris Geyser Basin, Yellowstone National Park during June-September, 1995. *U.S. Geol. Survey Open File Report 02–344*.

Fournier, R.O., White, D.E. & Truesdell, A.H. 1976. Convective heat flow in Yellowstone National Park. In *Proceedings, UN Symp. on Development and Use of Geothermal Resources*, San Francisco. 1975 (I): 731–739. Washington, DC: U.S. Government Printing Office.

Friedman, I. & Norton, D.R. 1990. Anomalous chloride discharged from Yellowstone National Park. *J. Volc. Geotherm. Res.* 42: 225–234.

Friedman, I. & Norton, D.R. 2000. Data used for calculating chloride flux out of Yellowstone National Park for the water years 1983–1999. *U.S. Geol. Survey Open File Report 00–194.*

Friedman, I. & Norton, D.R. in press. Is Yellowstone losing its steam? Chloride flux out of Yellowstone National Park. In L.A. Morgan (ed.), *Geoscience studies in the greater Yellowstone area: volcanic, hydrothermal, and tectonic processes in the Yellowstone Geoecosystem.* U.S. Geological Survey Prof. Paper.

Husen, S. & Smith, R.B. in press. Evidence for gas and magmatic sources beneath the Yellowstone volcanic system from tomographic imaging. *J. Volc. Geotherm. Res.*

Ingebritsen, S.E., Galloway, E.M., Colvard, E.M., Sorey, M.L. & Mariner, R.H. 2001. Time-variation of hydrothermal discharge at selected sites in the western United States: implications for monitoring. *J. Volc. Geotherm. Res.* 111: 1–23.

Miller, D.S. & Smith, R.B. 1999. P and S velocity structure of the Yellowstone volcanic field from local earthquake and controlled-source tomography. *J. Geophys. Res.* 104: 15105–15121.

Norton, D.R. & Friedman, I. 1985. Chloride flux out of Yellowstone National Park. *J. Volc. Geotherm. Res.* 26: 231–250.

Norton, D.R. & Friedman, I. 1991. Chloride flux and surface discharge out of Yellowstone National Park, 1982–1989. *U.S. Geol. Survey Bull. 1959.*

Pelton, J.R. & Smith, R.B. 1979. Recent crustal uplift in Yellowstone National Park. *Science* 206: 1179–1182.

Pelton, J.R. & Smith, R.B. 1982. Contemporary vertical surface displacements in Yellowstone National Park. *J. Geophys. Res.* 87: 2745–2761.

Pierce, K.L., Cannon, K.P. & Meyer, G. 1997. Yellowstone Caldera "heavy breathing" based on Yellowstone Lake and River changes in post-glacial time (asbtract). *EOS Trans. AGU Fall Meet. Suppl.* 78: F802

Pitt, A.M., Weaver, C.S. & Spencer, W. 1979. The Yellowstone Park earthquake of June 30, 1975. *Bull. Seismol. Soc. Am.* 69(1): 187–205.

Waite, G.P. & Smith, R.B. 2002. Seismic evidence for fluid migration accompanying subsidence of the Yellowstone caldera. *J. Geophys. Res.*107(B9). 2177, doi: 10.1029/2001JB000586.

Wicks, C. Jr., Thatcher, W. & Dzurisin, D. 1998. Migration of fluids beneath the Yellowstone caldera inferred from satellite radar interferometry. *Science* 282: 458–462.

Wicks, C. Jr., Thatcher, W. & Dzurisin, D. 2002. Satellite InSAR reveals a new style of deformation at Yellowstone caldera (asbtract). *EOS Trans. AGU Fall Meet. Suppl.* 83: T12A-1291.

Water–Rock Interaction, Wanty & Seal II (eds)
© 2004 Taylor & Francis Group, London, ISBN 90 5809 641 6

Oxidation reactions for reduced Fe, As, and S in thermal outflows of Yellowstone National Park: biotic or abiotic?

D.K. Nordstrom, J.W. Ball & R.B. McCleskey
U.S. Geological Survey, Boulder, Colorado

ABSTRACT: Oxidation reactions for S, As, and Fe have been measured downstream in several hot spring discharges at Yellowstone National Park. Oxidation of dissolved H_2S to S_2O_3 is the fastest reaction and occurs abiotically and nearly instantaneously on exposure to air. In high pH overflows (pH > 7), S_2O_3 shows little tendency to oxidize further. In low pH overflows, S_2O_3 disproportionates, producing elemental sulfur and sulfite. Solid sulfur is then oxidized more slowly to sulfuric acid by microbial catalysis. Oxidation of dissolved As^{III} to As^{V} and Fe^{II} to Fe^{III} in acid waters (pH = 2.7) requires microbial catalysis and proceeds at rates comparable to those in acid mine drainage. The oxidation rate for Fe^{II} is about 2 mmol/L/h and for As^{III} is about 0.07 mmol/L/h, both of which are many orders of magnitude faster than the abiotic rate.

1 INTRODUCTION

Geothermal waters rising from depths contain reduced forms of iron [Fe^{II}], arsenic [As^{III}], and sulfur [S^{II-}]. Dissolved sulfur can occur as both H_2S and SO_4^{2-} but dissolved iron and arsenic occur only in their reduced forms until they are exposed to oxygen by mixing with air or with oxygenated ground waters. Oxidation sets up strong redox gradients and affects the chemical behavior of the redox elements. In thermal overflows the oxidation gradient is often the zone of abundant microbial growth. Rates of oxidation-reduction reactions are commonly controlled by microbial processes rather than by abiotic processes but distinguishing between the two types can be difficult. This study describes examples of oxidation of Fe^{II}, As^{III}, and S^{II-} in Nymph Creek and Ojo Caliente thermal overflows in Yellowstone National Park, discusses their rates, and abiotic vs. biotic oxidation mechanisms, and discusses similar mechanisms deduced from data from other springs.

2 FIELD SITES AND METHODS

2.1 *Nymph Creek*

Nymph Creek is located along the roadside just north of Norris Geyser Basin, where there are several small hot springs (Fig. 1). The main source of Nymph Creek thermal water comes from a small discharge in a crack in silica sinter at a pH of 2.7–3.0, a temperature of 60–62°C; the main constituents are SiO_2 (240 mg/L), SO_4 (300 mg/L), Na (70 mg/L), K (45 mg/L) and Cl (35 mg/L). Dissolved sulfide is rarely detectable, Fe^{T} is about 3 mg/L and As^{T} is about 0.1 mg/L. A transect was chosen in the overflow to define the rate of iron and arsenic

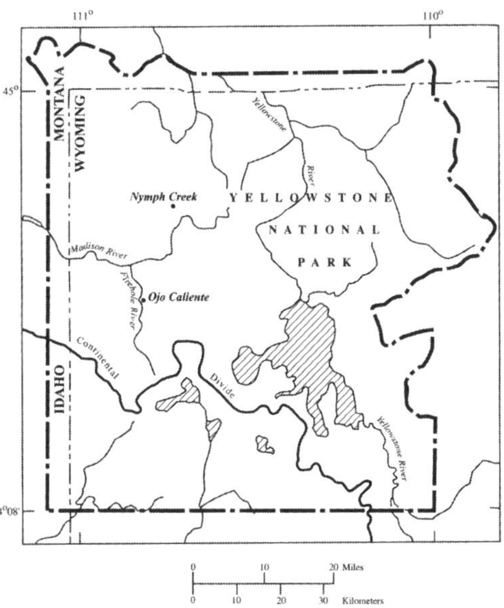

Figure 1. Location of sampling areas in Yellowstone Park.

oxidation from the gradient in concentration and travel time.

2.2 *Ojo Caliente*

Ojo Caliente is located next to the Firehole River in Lower Geyser Basin (Fig. 1). It discharges into a large pool at the boiling temperature (~93°C at that altitude) and drains into 4 distinct channels. Two of these channels were chosen to represent the range of flow from fastest to slowest. The slowest flow channel is also the longest. The pH at Ojo begins at about 7.5 and increases to about 8.5 due to CO_2 degassing. The water composition is dominated by Na (330 mg/L), Cl (300 mg/L), HCO_3 (235 mg/L) and SiO_2 (240 mg/L). Concentrations of Fe are usually at or below detection (<0.002 mg/L) and H_2S concentrations vary from 0.3 to 1 mg/L. Concentrations of dissolved As are about 1 mg/L.

2.3 *Analytical methods*

A mobile laboratory truck containing an ion chromatograph, UV-visible spectrophotometer, autotitrator, and reagent-grade water system was parked close to each sampling site so that unstable species such as redox species of Fe and intermediate sulfur oxyanions could be determined as soon as possible after collection. Samples for the determination of As^{III} and As^T, Fe^{II} and Fe^T, and intermediate sulfur species were filtered by pumping from the source with a portable peristaltic pump fitted with medical-grade silicone tubing through a 142-mm diameter all-plastic filter holder (Kennedy et al. 1976) containing a 0.1-μm mixed-cellulose-ester filter membrane.

Stabilizing reagents for intermediate sulfur species were put into the sample bottle before the sample was collected. Samples for H_2S determination were collected directly into a plastic syringe to minimize the inclusion of atmospheric air, then forced through a syringe-mounted 0.45-μm membrane filter. Samples were filtered directly into a measuring cuvette in which the appropriate reagents were quickly mixed. Total lag time between sampling and analysis was less than 20 seconds.

Concentrations of H_2S were determined by measuring absorbance using a Hach DR/2000 field spectrophotometer using Hach Method 8131, which is based on the APHA (1985) methylene blue method. Thiosulfate and S_xO_6 were determined (after removal of H_2S with addition of $ZnCl_2$ and NaOH) by syringe-filtering the sample directly into the ion chromatograph on-site in the mobile laboratory. Iron(total) and Fe^{II} concentrations were determined using a colorimetric method based on the FerroZine method of Stookey (1970), as modified by To et al. (1999). Arsenic(total) and As^{III} concentrations were determined by hydride generation atomic-absorption spectrometry using a Perkin-Elmer AAnalyst 300 atomic absorption spectrometer in absorption mode with a FIAS-100 flow injection analysis system hydride generator, quartz cell, and furnace.

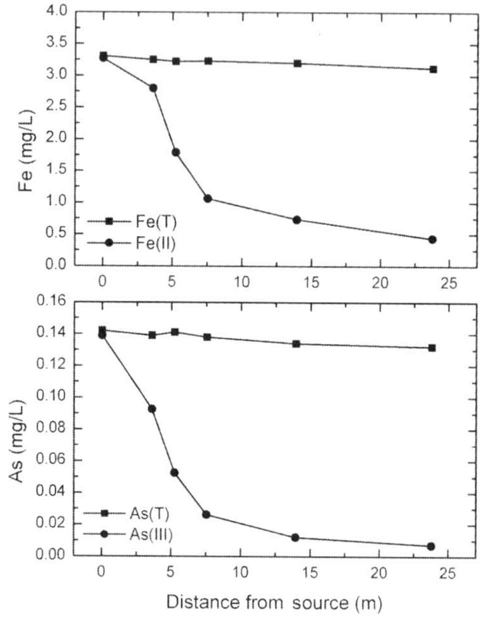

Figure 2. Iron and arsenic oxidation with distance along Nymph Creek.

3 RESULTS

3.1 *Nymph Creek*

Both Fe and As were found to oxidize rapidly upon exposure to the atmosphere (Fig. 2). The transect was resampled after dark. Similar results were obtained, indicating that photocatalysis was not significant in the first 15 m.

3.2 *Ojo Caliente*

At Ojo Caliente, H_2S was found to oxidize and volatilize rapidly upon exposure to the atmosphere in the overflow (Fig. 3). Dissolved oxygen increases from below detection (<0.1 mg/L) to more than 3.5 mg/L in the longer channel (that also reaches lower temperatures of about 50°C before entering the Firehole River). Thiosulfate concentrations increase quantitatively with the decrease in H_2S. Most other major and trace constituents undergo little change other than small evaporative concentration effects. Arsenic undergoes some oxidation in the longer channel but does not oxidize in the faster one.

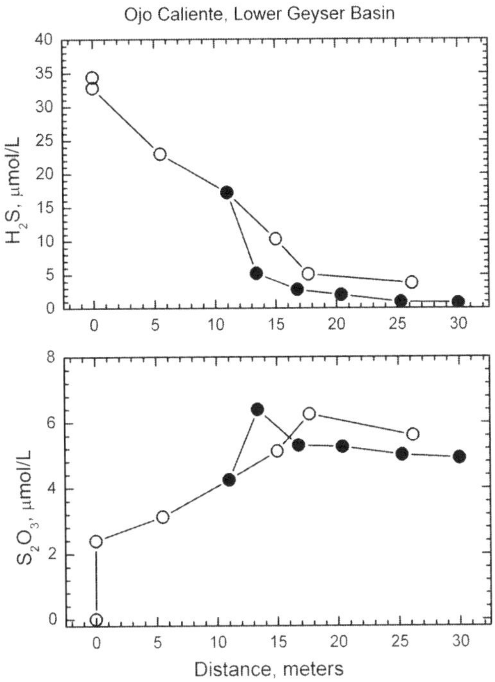

Figure 3. Oxidation of H_2S and formation of S_2O_3 with distance in 2 channels of Ojo Caliente overflow. Closed symbols show the separation of values for channel with slower and longer flow.

Measurements of H_2S oxidation in acid hot springs, such as Frying Pan Spring, have shown that thiosulfate does form but is quickly degraded to elemental sulfur making the pool turbid (Ball et al. 1998; Xu et al. 1998). Oxidation of the resultant colloidal sulfur forms sulfuric acid, keeping Frying Pan Spring at a pH of about 2.3.

4 DISCUSSION

The steady-state redox profiles for the oxidation of iron, arsenic, and sulfur species in thermal overflows can be combined with travel times to determine oxidation rates. The theoretical framework is the reactive-advective equation for steady flow:

$$\left(\frac{\partial c}{\partial t}\right)_x = 0 \quad \text{and} \quad \left(\frac{v\partial c}{\partial x}\right)_t = r$$

where c is concentration of a constituent, t is time, x is distance, v is velocity, and r is reaction rate. Although the concentration does not change at one location with time, it does change from one location to the next such that by finite difference:

$$\frac{v\Delta c}{\Delta x} = \frac{\Delta x \Delta c}{\Delta t \Delta x} = \frac{\Delta c}{\Delta t} = r$$

Hence, by measuring the travel time from station to station (=velocity) and the concentration from station to station in an overflow transect, the oxidation rate can be calculated for any given redox species. This method also assumes zero-order reaction rate, which is not exactly valid. The difference between linear and non-linear rates, however, is not large when comparing microbial and abiotic rates. Applying this method to the arsenic and iron oxidation in Nymph Creek, the oxidation rate of Fe^{II} is about 2 mmol/L/h. For As^{III}, the oxidation rate is about 0.07 mmol/L/h. These rates compare to the inorganic rates (at 25°C) of 10^{-5} mmol/L/h for Fe^{II} and $<10^{-8}$ mmol/L/h for As^{III} (see Nordstrom 2003). Hence, microbial catalysis increases iron oxidation rates in the field by more than 5 orders of magnitude and increases arsenic oxidation by more than 9 orders of magnitude.

Other measured field rates for microbial iron oxidation are similar. The Fe^{II} oxidation rates are 1–12 mmol/L/h for *in situ* and they are even higher for engineered bioreactors (Nordstrom 2003). Microbial arsenite oxidation has more commonly been reported as doubling time and less commonly as a half-life for As^{III}, the appropriate form for exponential (or first order) change. Nevertheless, Gihring & Banfield (2001) observed arsenite oxidation in the lab by a Yellowstone isolate, *Thermus* HR 13, at about 0.12 mmol/L/h, very similar to the rates reported here.

These rates of iron and arsenic oxidation are clearly microbially mediated. Oxidation of H_2S, however, appears to be dominated by abiotic mechanisms until elemental S or thiosulfate is formed. The rate of thiosulfate formation for the data from Ojo Caliente is 0.16 mmol/L/h and for lab studies the value is about 0.003 mmol/L/h at 25°C (Chen & Morris 1972; O'Brien & Birkner 1977). Although the *in situ* rate is about 50 times faster in the field, the temperature is also nearly 70° higher. At temperatures where few microbes can survive, the oxidation of H_2S to S_2O_3 is instantaneous. At Angel Terrace our data shows instantaneous formation of 4 µmol/L S_2O_3 from H_2S in less than a second. Data on H_2S oxidation in abiotic lab studies show a substantial increase in rate constants with increasing temperature that can account for the increase observed for our Yellowstone field rates (Zhang & Millero 1994).

For Ojo Caliente an estimate of the quantity of H_2S oxidized *vs.* the amount volatilized can be made. Of the original 30 µmol/L H_2S in the Ojo pool, 10 µmol/L was converted to S_2O_3 only. The remaining 20 µmol/L must have been volatilized. Hence, for a hot spring at boiling temperature and with a pH of about 7.7, about 2/3 of the original H_2S is volatilized and about 1/3 stays in the water as S_2O_3.

5 CONCLUSIONS

Both biotic and abiotic mechanisms are operative during the oxidation of redox-sensitive elements in thermal overflows. For iron and arsenic, oxidation in acid springs is catalyzed by iron and arsenic oxidizing microbes that can increase the oxidation rate by many orders of magnitude. For H_2S oxidation, rapid abiotic oxidation to thiosulfate and elemental sulfur occurs followed by the slower oxidation of sulfur and thiosulfate to sulfuric acid. For one circumneutral pH hot spring, Ojo Caliente, the first estimate has been made of the quantity of H_2S volatilized (67%) and the remaining dissolved H_2S (33%) quantitatively converts to S_2O_3.

REFERENCES

APHA, American Public Health Association, 1985. Method 428C. Methylene blue method for sulfide. In *Standard Methods for the Examination of Water and Wastewater*, 14th ed.: 403–405.

Ball, J.W., Nordstrom, D.K., Cunningham, K.M., Schoonen, M.A.A., Xu, Y. & DeMonge, J.M. 1998. Water-chemistry and on-site sulfur-speciation data for selected springs in Yellowstone National Park, Wyoming, 1994–95. *U.S. Geol. Survey Open-File Report* 98–574.

Chen, K.Y. & Morris, J.C. 1972. Kinetics of oxidation of aqueous sulfide by O_2. *Environ. Sci. Tech.* 6: 529–537.

Gihring, T.M. & Banfield, J.F. 2001. Arsenite oxidation and arsenate respiration by a new *Thermus* isolate. *FEMS Microbiology Letters* 204: 335–340.

Kennedy, V.C., Jenne, E.A. & Burchard, J.M. 1976. Backflushing filters for field processing of water samples prior to trace-element analyses. *U.S. Geol. Survey Open-File Report* 76–126.

Nordstrom, D.K. 2003. Effects of microbiological and geochemical interactions in mine drainage. In J.L. Jambor, D.W. Blowes, and A.I.M. Ritchie (eds.), *Environmental aspects of mine wastes, Mineralogical Association of Canada Short Course* 31: 227–238.

O'Brien, D.J. & Birkner, F.B. 1977. Kinetics of oxidation of reduced sulfur species in aqueous solution. *Environ. Sci. Tech.* 11: 1114–1120.

Stookey, L.L. 1970, FerroZine – a new spectrophotometric reagent for iron. *Anal. Chem.* 42: 779–781.

To, T.B., Nordstrom, D.K., Cunningham, K.M., Ball, J.W. & McCleskey, R.B. 1999. New method for determination of dissolved Fe(II) concentrations in acid mine waters. *Environ. Sci. Tech.* 33: 807–813.

Xu, Y., Schoonen, M.A.A., Nordstrom, D.K., Cunningham, K.M. & Ball, J.W. 1998. Sulfur geochemistry of hydrothermal waters in Yellowstone National Park, Wyoming, USA: 1. Origin of thiosulfate in selected hot spring waters. *Geochim. Cosmochim. Acta* 62: 3729–3743.

Zhang, J.-Z. & Millero, F.J. 1994. Kinetics of oxidation of hydrogen sulfide in natural waters. In C.N. Alpers & D.W. Blowes (eds.), *Environmental Geochemistry of Sulfide Oxidation; American Chemical Society Symposium Series* 550: 393–409. Washington, D.C.: American Chemical Society.

Water–Rock Interaction, Wanty & Seal II (eds)
© 2004 Taylor & Francis Group, London, ISBN 90 5809 641 6

Chemical aspects of exploration of the Theistareykir high-temperature geothermal area, N-E Iceland

H. Ármannsson
ÍSOR Iceland GeoSurvey, Orkugarður, Reykjavík, Iceland

ABSTRACT: Surface exploration was carried out in the Theistareykir high-temperature geothermal area, NE-Iceland in 1972–1974 and 1981–1984, and the area monitored intermittently from 1991–2000. The results of surface exploration suggested that the temperature of the fluid was at least 280°C and that it was drawn from a single reservoir with good permeability. The recharge is from the southeast, and the origin of the fluid probably far to the south. The fluid may be 100 years old or more. The geothermal area has been divided into five sub-areas, three of which are expected to be productive. Wells have been drilled in two of these in 2002 and 2003, and drilling results seem to confirm the predictions made during surface exploration.

1 INTRODUCTION

Theistareykir is a high-temperature geothermal area in NE-Iceland (Fig. 1). For centuries it hosted the main sulphur mine in Iceland, providing the Danish king with raw material for gun powder. Early records tell of prospecting for sulphur, Bemmelen & Rutten (1955) mapped interglacial lavas and palagonite, and Kjartansson (1972) reported on exploration for clay. The first geothermal exploration was carried out in 1972–1974 (Grönvold & Karlsdóttir 1975), and a major geothermal assessment was made in 1981–1984 (Layugan 1981, Gíslason et al. 1984, Ármannsson et al. 1986, Darling and Ármannsson 1989). The area was monitored intermittently from 1991–2000 (Ármannsson et al. 2000), and Ármannsson (2001) reviewed research in the area up to the year 2000.

Gautason et al. (2000) suggested drill sites based on available knowledge. The first well was drilled in 2002 and the second in 2003. The purpose of this paper is to reflect on the results of drilling with reference to information gained from surface exploration with special emphasis on chemistry.

2 RESULTS OF SURFACE EXPLORATION

Ármannsson et al. (1986) divided the area into five sub-areas (Fig. 2), three of which (A, C and D) appeared promising for drilling. Gas geothermometers gave the temperature ranges shown in Table 1. Darling and Ármannsson (1989) concluded from isotope values for fumarole steam that, in area D (Tjarnarás), the steam had been condensed to a fraction of 0.15 to 0.25 of the original steam at temperatures in the range 130–200°C, and that gas geothermometer temperatures were probably too high. Their interpretation of isotope values for area C (Theistareykjagrundir) was that the steam was mostly secondary steam and that the geothermometer

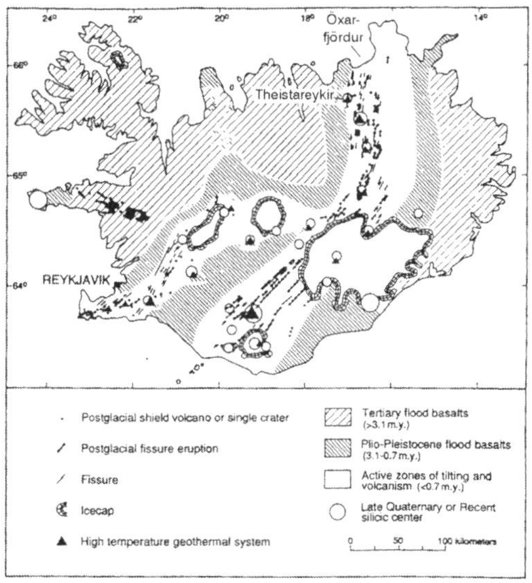

Figure 1. Iceland. Tectonic map (Sæmundsson 1986). Location of Theistareykir.

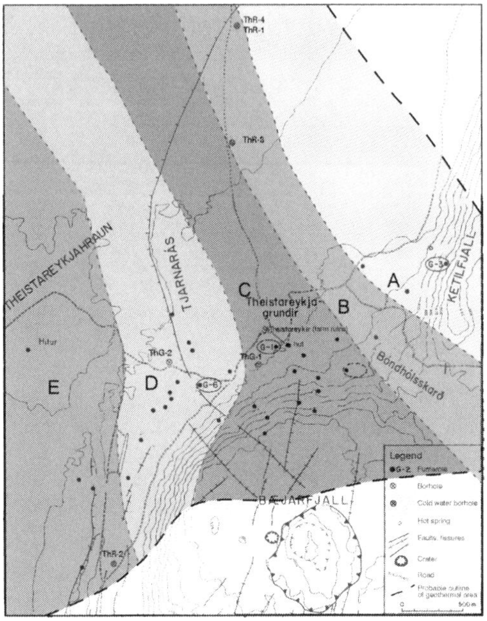

Figure 2. Theistareykir. Division into five subareas (Ármannsson et al. 1986). Geothermal and cold water boreholes in and north of the geothermal area are shown as well as three fumaroles sampled during monitoring 1991–2000.

Table 1. Gas geothermometer temperatures in A, C and D (°C).

Subarea	1980s °C	Fumarole	1991 °C
A	272–315	G-3	289
C	232–271	G-1	284
D	274–309	G-6	263

temperatures were probably too low. They concluded, however, that the steam rising from area A (Ketilfjall) was undisturbed and that the geothermometer temperatures were close to true values. Ármannsson et al. (1986) proposed a deep inflow to the area from the southeast with area C closest to the source. Thus area C was considered promising even though the gas geothermometer values seemed rather low. A relatively large, cool, shallow flow was predicted through Bóndólsskarð (Fig. 2), preventing primary steam from rising to the surface in area B. Areas C and D are more accessible than area A, and therefore the suggestion was that the first drillholes be situated in these two subareas. The dissolved solids content of the steam was very low suggesting that the reservoir fluid was dilute. Low radon

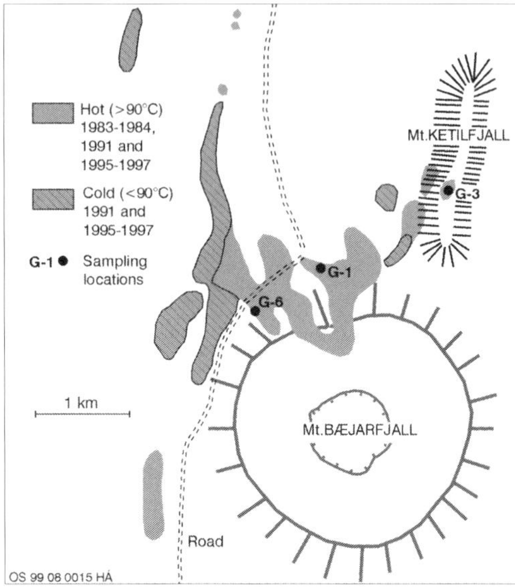

Figure 3. Theistareykir. Changes in surface manifestations 1983–1984 to 1991 (Ármannsson et al. 2000). Fumarole sampling locations used during monitoring shown.

concentrations were interpreted to suggest good permeability especially in area D.

3 MONITORING AND HEAT LOSS

The area was visited again in 1991 at the beginning of a monitoring program in which some unexploited high-temperature areas were to be monitored to establish the extent of natural changes in geothermal areas as opposed to changes due to production (Ármannsson et al. 2000). Changes in surface manifestations were mapped and steam samples collected from 3 fumaroles, one from each of areas A, C and D. At that time considerable changes in surface manifestations were observed, mainly cooling in area D (Fig. 3). There was information from local people that following earthquakes in 1958 the surface activity in the area had increased drastically but had been declining since. The gas geothermometer temperature for the fumarole from area D (G-6, Figs. 2 and 3) had decreased drastically, but little change or a slight increase were recorded for gas geothermometer temperatures for fumarole steam from areas A (G-3, Figs. 2 and 3) and C (G-1, Figs. 2 & 3; Table 1). It is possible that the secondary effects suggested by Darling & Ármannsson (1989), condensation and formation of secondary steam, were less pronounced this time. The area has been visited a few more times, but it has remained relatively unchanged after 1991.

Table 2. Heat loss from the Theistareykir area.

Inflow to Lón, Öxarfjörður	$20\,m^3/s$
Local groundwater ambient temperature	3.7°C
Inflow to Lón, mean temperature	7.2°C
Excess temperature	3.5°C
Heat power of area	$3.5 \times 20 \times 42 \approx 300$ MW
Areal extent of Theistareykir	$15\,km^2$
Heat loss from area	$300/15 = 20\,MW/km^2 = W/m^2$

Using information from Hafstad (1989 and personal communication) about the Lón estuary in Öxarfjörður, 20 km to the north of the Theistareykir area, Ármannsson (2001) calculated the heat loss from the Theistareykir area (Table 2). This estuary is believed to receive solely subsurface water from the Theistareykir area. The values are minimum but they suggest a total output of 300 MW. Therefore a powerful geothermal system with a temperature of about 280°C recharged with dilute, probably relatively old, water from far south and with an isotope signature of $\delta D = -100‰$ and $\delta^{18}O = -12‰$ was predicted prior to drilling.

4 RESULTS OF DRILLING

Prior to deep drilling, four shallow "cold water" wells, ThR-1-4 (Fig. 2) were drilled to obtain drilling fluid. Well ThR-2, which is 102 m deep, just reached the groundwater table, and the temperature of the water proved close to boiling. In ThR-3 the groundwater table is also close to 100 m depth. The temperature was 66°C there, but 90°C at 140 m depth. The groundwater table was at about 100 m depth in wells ThR-1 and ThR-4, the former is 128 m deep and the temperature of the water was 26–28°C, but the latter is 150 m deep, with water temperature 26–35°C (Hafstad 2000). Water from these two wells was used as drilling fluid for the two deep geothermal wells.

Well ThG-1 was drilled in area C (Theistareykjagrundir; Fig. 2) in Autumn 2002 to a depth of 1953 m, with a casing to 614 m depth. The major inflows are at 620–640 m depth and 1620–1640 m depth. Other inflows are observed at 710, 860–880, 1050, 1230–1240, and 1350 m depths and possibly at 1780–1800 and 1900–1910 m depths. An overpressured aquifer was observed at 212 m depth (Gudmundsson et al. 2002). The well started discharging in late October 2002 and two samples for chemical analysis have been collected from it, one in November 2002 and a second one in July 2003. The well discharged until August 2003 when it was closed and logged.

Table 3. Chemical composition of deep water and deep steam in fluid from well ThG-1, Theistareykir at 280°C.

Constituent	Water phase	Steam phase
pH	7.96	
CO_2 mg/l	37.53	1463
H_2S mg/l	30.65	265
B mg/l	1.29	
SiO_2 mg/l	574.3	
Na mg/l	89.44	
K mg/l	18.36	
Mg mg/l	0.003	
Ca mg/l	0.38	
F mg/l	0.79	
Cl mg/l	80.81	
SO_4 mg/l	7.06	
Al mg/l	1.57	
Fe mg/l	0.0055	
Mo mg/l	0.0144	
Mn mg/l	0.0014	
Zn mg/l	0.0050	
As mg/l	0.0057	
H_2 mg/l		25.3
N_2 mg/l	0.10	33.8

Table 4. Chemical geothermometer results for fluid from well ThG-1, Theistareykir.

Type	Quartz[1]	Na/K[2]	CO_2[3]	H_2S[3]	H_2[3]	H_2/Ar[4]
t°C	276	279	244	237	295	302

[1] Arnórsson et al. 1983, [2] Fournier 1977, [3] Arnórsson et al. 1998, [4] Giggenbach 1991.

The results of chemical analysis suggest that in November 2002 the fluid was still contaminated by drilling fluid and thus the results from July 2003 (Table 3) are used for the purpose of interpretation. The measured enthalpy at the surface was 2180 kJ/kg and the total flow from the well 16–17 kg/s. The calculated steam fraction at depth at 280°C was 0.611.

The stable isotope composition for the total fluid was $\delta D = -108‰$ and $\delta^{18}O = -12.7‰$, or a little lower than predicted. Thus the fluid probably originated as precipitation far to the south of the area, probably as far south as Vatnajökull, and it may also be 100 years old or more.

Geothermometer temperatures were calculated in six ways with the results reported in Table 4. As is apparent, the results come in pairs, and there is a significant difference between the pairs. It had been suggested that, in this part of the area, some secondary steam was formed in fumaroles. Using Arnórsson et al's (1998) calibrations for the CO_2, H_2S and H_2 geothermometers, the same kind of pattern emerges from the results for the fumaroles, i.e. CO_2

and H_2S temperatures around 240°C, but H_2 temperatures of about 300°C or higher. As H_2 and Ar are relatively insoluble at these conditions they may not be affected by the secondary steam formation whereas CO_2 and H_2S will dissolve in the water phase and be released again at a lower temperature.

One model that explains these geothermometer temperatures assumes that the main inflows are at 620–640 m and 1620–1640 m, that the upper inflow is essentially liquid phase at 280°C but the deeper one essentially vapor phase at 300°C or higher. The logged temperatures of well ThG-1 are 270–280°C at 6–700 m and 300–305°C at 1600–1700 m. The temperature of the liquid phase at 280°C is reflected in the results for the solute geothermometers, whereas the temperature of the steam phase is approximated by the H_2 and H_2/Ar temperatures. CO_2 and H_2S travel some way after being dissolved in the water phase and before the fluid flashes at a lower temperature.

Well ThG-2 was drilled October to December 2003. After some circulation losses suggesting aquifers in the top layers, the well was cased to a depth of about 260 m. Shortly after drilling started again, a total loss of circulation was encountered and persisted. Logging revealed a well diameter in excess of 30- over a 10 m interval. An attempt to seal off this interval by gradually pouring 120 m^3 of gravel and sand into the well and cementing with 140–150 m^3 of concrete failed to stop the circulation loss. Drilling was resumed, a total loss of circulation was encountered again at 327 m depth, but drilling was still continued with total loss to 617 m depth where it was cased again. The final depth was 1720 m, drilled again with a total loss of circulation (>50–55 l/s) from 657 m depth. It was finished with a slotted liner. Pumping tests suggested very high permeability, but the well is still warming up now (December 2003), so it is not clear what the final temperatures are.

Extremely strong flows are inferred from 260 m depth and considerably below this. There is a possibility that this constitutes a large cave or some such feature whose temperature has been estimated at a little over 200°C, and that this large flow might cause cooling of the rock over some distance from the cave, possibly sufficient to cause partial condensation of steam rising to the surface in the vicinity. It is suggested that this may be the mechanism responsible for the condensation of steam in area D suggested by Darling & Ármannsson (1989).

5 CONCLUSIONS

The results of surface exploration in the Theistareykir area suggested that it encompassed 3 distinct subareas (A, C and D, Fig. 2) suitable for drilling, drawing fluid from a single base reservoir with a temperature of at least 280°C. The fluid originates far to the south of the area, and is probably more than 100 years old. Results of stable isotopes in fumarole steam suggested that in area D, there was condensation of fumarole steam during its passage from the reservoir to the surface, but that in area C the fumarole steam was in some cases secondary steam.

During drilling of wells ThG-1 and 2 large inflows of relatively shallow cool water, which can explain condensation and secondary steam formation in area C and D, were encountered. Good permeability was predicted for area D from results of surface exploration. Thus the preliminary results of drilling do not contradict those of surface exploration.

REFERENCES

Ármannsson, H. 2001. Theistareykir. A review of research, exploration and its cost (In Icelandic). Orkustofnun OS-2001/035, 24 pp.

Ármannsson, H., Gíslason, G. & Torfason, H. 1986. Surface exploration of the Theistareykir high-temperature geothermal area, Iceland, with special reference to the application of geochemical methods. *Appl. Geochem.* 1: 47–64.

Ármannsson, H., Kristmannsdóttir, H., Torfason, H. & Ólafsson, M. 2000. Natural changes in unexploited high-temperature geothermal areas in Iceland. *Proc. World Geothermal Congress 2000*, 521–526.

Arnórsson, S., Fridriksson, T. & Gunnarsson, I. 1998. Gas chemistry of the Krafla geothermal field, Iceland. In G.B. Arehart & J.R. Hulston (eds.) *Water-rock Interaction*, 613–616. Rotterdam: Balkema.

Arnórsson, S., Gunnlaugsson, E. & Svavarsson, H. 1983. The chemistry of geothermal waters in Iceland III. Chemical geothermometry in geothermal investigations. *Geochim. Cosmochim. Acta* 47: 567–577.

Bemmelen, R.W. & Rutten, M.G. 1955. Table mountains of Northern Iceland. Leiden. E.J. Brill.

Darling, W.G. & Ármannsson, H. 1989. Stable isotope aspects of fluid flow in the Krafla, Námafjall and Theistareykir geothermal systems of northeast Iceland. *Chem. Geol.* 76: 197–213.

Fournier, R.O. 1977. Chemical geothermometers and mixing models for geothermal systems. *Geothermics* 5: 41–50.

Gautason, B., Ármannsson, H., Árnason, K., Sæmundsson, K., Flóvenz, Ó.G. & Thórhallsson, S. 2000. Thoughts on the next steps in the exploration of the Theistareykir geothermal area (In Icelandic). Orkustofnun BG-HÁ-KÁ-KS-ÓGF-STh-2000/04.

Giggenbach, W.F. 1991. Chemical techniques in geothermal exploration. In F. D'Amore (ed.) *Applications of geochemistry in geothermal reservoir development*: 119–142. Rome: UNITAR/UNDP.

Gíslason, G., Johnsen, G.V., Ármannsson, H., Torfason, H. & Árnason, K. 1984. Theistareykir. Surface exploration of the high-temperature geothermal area (In Icelandic). Orkustofnun OS-84089/JHD-16, 134 pp + 3 maps.

Grönvold, K. & Karlsdóttir, R. 1975. Theistareykir. An interim report on the surface exploration of the geothermal area (In Icelandic). Orkustofnun JHD-7501, 37pp.

Gudmundsson, Á., Gautason, B., Thordarson, S., Egilson, T. & Thorisson, S. 2002. Exploration drilling at Theistareykir. Well Th.G-1. 3rd stage: Drilling of production part to 1953 m depth (In Icelandic). Orkustofnun OS-2002/079, 59 pp.

Hafstad, T.H. 1989. Öxarfjördur. Groundwater investigations 1987–1988. A contribution to a special project on fish farming (In Icelandic). Orkustofnun OS-89039/VOD-08 B, 25 pp.

Hafstad, T.H. 2000. Theistareykir: The drilling of a fresh water well (In Icelandic). Orkustofnun ThH-00/14, 6 pp.

Kjartansson, H. 1972. Clay formations in Dalasýsla and Thingeyjarsýsla districts (In Icelandic). Orkustofnun JKD, 83 pp.

Layugan, D.B. 1981. Geoelectrical soundings and its application in the Theistareykir high-temperature area. United Nations University. Geothermal Training Programme Report 1981–1985, 101 pp.

Saemundsson, K. 1986: Subaerial volcanism in the western North Atlantic. In Vogt, PR. and Tucholke, B.E.(eds.) *The Geology of North America, M. The Western North Atlantic Region*. 69–86. Boulder, Colorado: Geological Society of America.

Water–Rock Interaction, Wanty & Seal II (eds)

The source and relative age of non-thermal and up to 90°C ground waters in the Skagafjördur tholeiite flood basalt province, N-Iceland

Stefán Arnórsson & Árný E. Sveinbjörnsdóttir
Science Institute, University of Iceland, Reykjavík, Iceland

ABSTRACT: Combined data on hydrogeology and the natural tracers δ^2H, $\delta^{18}O$, ^3H, Cl and B have been used to identify the origin and relative age of ground waters and <90°C geothermal waters in the basalt province of the Skagafjördur valley system in N-Iceland and to delineate flow directions. Ground waters and thermal waters in the relatively permeable Quaternary formations in the interior highlands are of local origin and a few years to a few decades old. Ground waters in the less permeable Mid-Tertiary flood basalts north of the highlands are older. Under the steep mountains on the east side of the Skagafjördur valley they represent precipitation that has fallen on high ground in these mountains and are 100's to 1000's of years old. On the west side of the Skagafjördur valley the geothermal waters are a mixture of recent local precipitation and a Pre-Holocene precipitation.

1 INTRODUCTION

Natural tracers have been extensively used to identify the origin of ground waters, in particular geothermal waters, and together with artificial tracers, they have been used to map both local and regional directions of flow underground (e.g. Craig et al. 1956; Árnason 1976; 1977; Giggenbach 1992; Adams 1995; Arnórsson 1995). In this contribution it is demonstrated how combined data on hydrogeology and many natural isotopic (^3H, ^2H, ^{18}O) and chemical (Cl, B) tracers can be used with more confidence than the use of one or a few tracers to delineate flow directions and the relative age of thermal waters and non-thermal ground waters in a Tertiary-Quaternary basalt province in north Iceland.

2 GEOLOGICAL FEATURES

The study area covers the watershed of the river of Héradsvötn, the main river of the Skagafjördur valley in central north Iceland. The Skagafjördur valley system dissects a monotonous Mid-Tertiary flood basalt sequence that generally dips a few degrees towards south and southwest. The basalts are tholeiites, slightly evolved relative to MORB (Arnórsson et al. 2002). Quaternary hyaloclastites and basalt lavas overlie the Tertiary formations in the interior highlands to the south of the Skagafjördur valley system. Sporadic outcrops of Quaternary formations, which rest unconformably upon the Tertiary formations, occur in several places.

Thermal springs are widespread on low ground in the study area, particularly in the central part of the Skagafjördur valley. They are concentrated where a recent NNE trending fracture zone crosses the valley. Temperatures reach 90°C.

The upper part of the flood basalt succession in the mountains on either side of the Skagafjördur valley is unaltered and free from vesicle minerals. In outcrops down in the valley the basalts are also quite fresh but partly vesicle filled. The vesicle minerals include chalcedony, calcite and the low temperature (<100°C) zeolites chabazite, thomsonite and stilbite. The sparse vesicle fillings and the small size of their crystals indicate that the exposed basalt lavas have never been deeply buried (see Walker 1960; Neuhoff et al. 1999). The present lava tops of the 1100–1300 m high mountains on the east side of the Skagafjördur valley are considered to be close to the original lava plateau surface (Jóhannesson 1991).

In the oldest rocks exposed in the Skagafjördur area, which occur on the headland on the east side of the fjord of Skagafjördur, the degree of alteration and the extent of vesicle fillings are more extensive than in younger rocks farther south. Basalts with similar degree of alteration are considered to exist at 1–2 km depth below the valley floor in central Skagafjördur where the geothermal activity is most concentrated.

3 GEOHYDROLOGY

The primary permeability of the basalt lava successions in Iceland is considered to be negatively related to the degree of alteration and the extent of vesicle fillings and, therefore, depth of burial (see Neuhoff et al. 1999). No deep drillholes have been sunk in the Skagafjördur area to provide data on permeability. It is, however, considered that data on global permeability from monitoring of 1–3 km deep hot water production wells in the Valley of Eyjafjördur, about 50 km to the east of Skagafjördur, give a good idea of what the global permeability may be in Skagafjördur. The two valleys occur in the same Mid-Tertiary tholeiite flood basalt succession and they have had similar history subsequent to their formation. According to Björnsson et al. (1990), the global permeability in these formations lies in the range 0.4 10^{-15} to $10^{-14}\,m^2$. These numbers are considered to be applicable to the lava succession underlying the valley floor of Skagafjördur but on higher ground permeability will be higher, and particularly in the interior highlands, most likely by two to three orders of magnitude as deduced from well testing data from different parts of Iceland, presented by Saemundsson & Fridleifsson (1980).

For the lower global permeability value, it would take 25°C ground water 2400 years to flow a distance of 5 km if the hydraulic gradient was 1 in 10. The corresponding value for the higher permeability value is 100 years. An average hydraulic gradient of 1/10 is considered to be a reasonable value for the slopes and mountains on the east side of the Skagafjördur valley. Down in the valley and on its west side the gradient will be less. It is, therefore, concluded that waters infiltrating the mountains and emerging in springs in the lowlands of the Skagafjördur valley are hundreds to thousands of years old. In the interior highland springs waters are expected to be much younger, even if the lateral flow distance is greater and the hydraulic gradient not as steep. Flow over a distance of 10 km along a gradient of 1/100 would take 1–10 years.

4 TRITIUM

Ground waters on low ground in Skagafjördur generally do not contain detectable tritium (Fig. 1). When this is not the case, relatively low in situ pH and departure from calcite saturation (Fig. 2) suggest that the cause is mixing of the deep ground water with ^3H bearing shallow water. Contrary to the ground water on low ground, non-thermal ground waters on high ground contain rather similar ^3H as surface waters, or 7–12 TU as compared to 10–14 TU, and thermal ground waters have 4–5 TU. Waters containing 4–5 TU could either be a few years old or around 40 years (Fig. 3). Alternatively, the ^3H content could be the consequence of mixing of water with different ^3H content. The ground waters on low

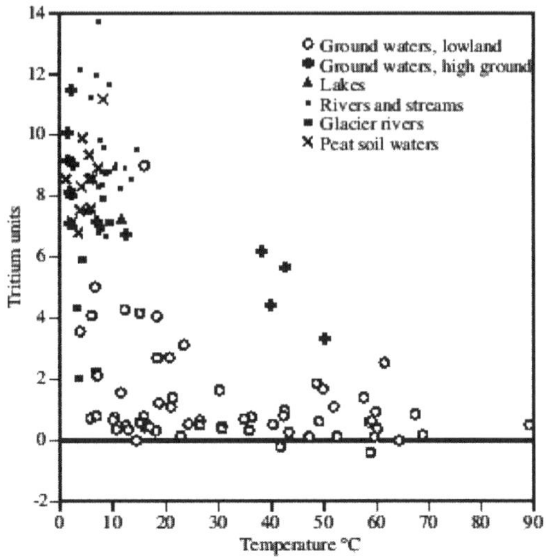

Figure 1. The tritium content of natural waters in Skagafjördur, northern Iceland.

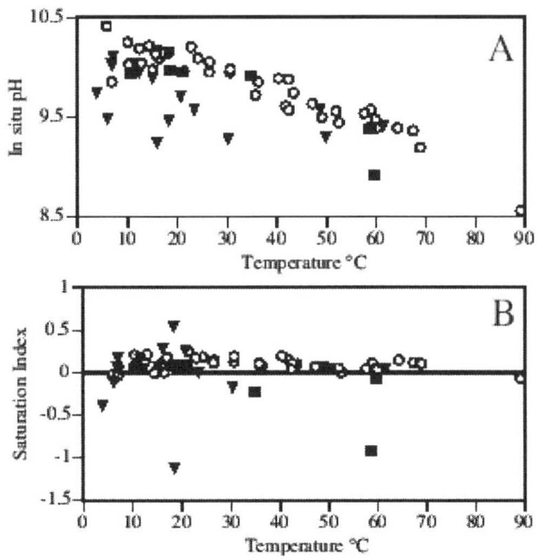

Figure 2. *In situ* pH (A) and calcite saturation (B) of ground waters from lowland areas in Skagafjördur area. Samples with >1.5 TU are specifically labeled (triangles) as wells as samples where the chalcedony equilibrium temperatures are significantly higher than discharge temperatures (squares).

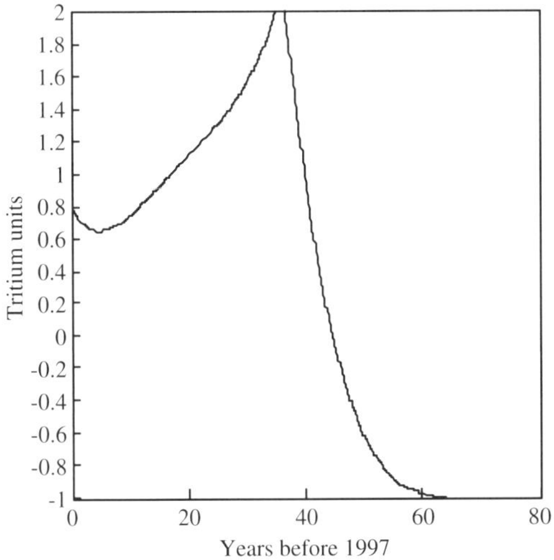

Figure 3. Tritium content versus age of ground waters.

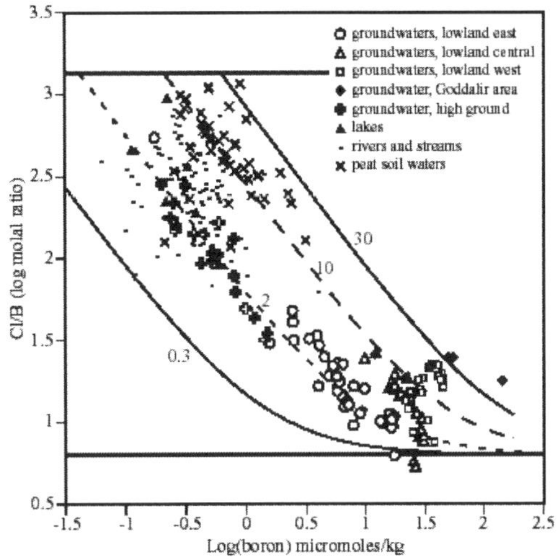

Figure 4. Relationship between Cl/B ratios and B concentrations in natural waters from Skagafjördur. The lines represent Cl/B ratios of waters that have dissolved variable amounts of Cl and B from the rock (in the molal ratio of 7) and with selected initial sea salt Cl concentrations (ppm) as indicated by the numbers by each line.

ground are pre-bomb; they do not contain a significant component that is <50–60 years (Fig. 3).

5 CHLORIDE AND BORON

It is well established that Cl acts as a conservative component. In the basalt-water environment in Iceland this is also the case for B (Arnórsson & Andrésdóttir 1995). In the study area surface waters on low ground contain 10–20 ppm Cl but on high ground, farther inland, it is 1–2 ppm. The source of the Cl in these waters is dominantly sea salts. The Cl/B ratios of these waters are similar to that of seawater or lower, due to rock leaching. Cl and B are leached from the basalt in the ratio of 7 (molal). As a consequence, the Cl/B ratios of surface- and ground waters decreases towards this value with progressive rock dissolution (Fig. 4). The Cl/B ratios of the water and their B (or Cl) content can be used to retrieve the B (or Cl) sea salt concentration in the parent precipitation (see Arnórsson & Andrésdóttir 1995).

Ground waters on high ground contain similar sea salt Cl as surface waters and are therefore of local origin. The sea salt Cl in ground waters on the east side of the Skagafjördur valley is similar to that of surface waters high ground. Together with geohydrological considerations, this indicates that these waters originated as precipitation in the mountains on the east side of the valley. By contrast, ground waters, all of which are thermal, emerging on the west side of the Skagafjördur valley, contain sea salt Cl in concentrations similar to that of precipitation falling on low ground, or somewhat higher. These waters, therefore largely represent local precipitation.

6 δ^2H AND $\delta^{18}O$

The δ^2H and $\delta^{18}O$ values of the waters in the study area are −75 to −120‰ and −8.5 to −14.5‰, respectively. Most surface waters from rivers and streams and <50°C ground waters plot close to the global meteoric line. By contrast, many peat soil- and lake waters are enriched in ^{18}O, which is considered be due to evaporation/evapotranspiration.

Surface waters, ground waters and thermal waters in the interior highlands possess similar isotopic composition indicating that the source of both the ground waters and the thermal waters is local. Ground waters merging in the eastern part of the Skagafjördur valley have isotopic signatures similar to surface waters on high ground. As based on hydrogeological considerations, it is considered that the source of these waters is in the mountains to the east of the valley. Thermal springs emerging in the valley bottom tend to have a little more negative δ^2H-values and those on the western side even more negative. The last mentioned group of waters display a significant oxygen shift. Their δ^2H-values are more negative than that of any surface waters (today's precipitation) in the watershed of Héradsvötn today, in fact more negative than that of any precipitation in Iceland today (see Árnason 1977). These waters must contain a component that has fallen as precipitation when the climate was colder than today (>10,000 years). The oxygen-shift observed for these waters,

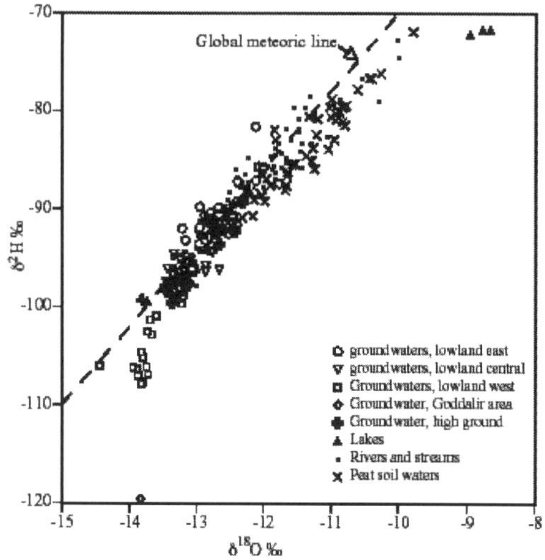

Figure 5. δ^2H versus $\delta^{18}O$ in natural waters from Skagafjördur.

which reflects extensive water–rock interaction, is consistent with an "old" water component in these waters.

7 CONCEPTUAL HYDRO-GEOCHEMICAL MODEL OF GROUND WATER FLOW

Thermal and non-thermal ground waters in the interior highlands south of the Skagafjördur valley are considered to be of local origin because their sea salt Cl content and δ^2H-values are similar to those of local precipitation. Ground water emerging on the east side of the Skagafjördur valley are 100's to 1,000's of years old as inferred from available global bedrock permeability data. They originate as precipitation on the high mountains to the east of the valley as deduced from hydrogeology and indicated by their content of sea salt derived Cl and δ^2H-values. In view of the low temperatures of these waters their flow underground must be driven by hydraulic head. Thermal waters emerging on the west side of the Skagafjördur valley contain a Pre-Holocene water component as indicated by their δ^2H-values. The calculated sea salt derived Cl is similar to that of local precipitation today or slightly higher. The sea salt Cl content of these waters is thought to be the consequence of the sea salt Cl in the parent precipitation and a tiny sea water component which owes its existence to seawater infiltration at the beginning of the Holocene when the lowland areas were transgressed by the ocean. Whichever the relative contribution from the two sources, precipitation or infiltrated seawater, the parent water must be water infiltrating the bedrock on low ground, i.e. local. Temperatures of the thermal waters on the west side of the Skagafjördur valley are sufficiently high for the water circulation to be density driven in the Recent fractures. The isotopically light Pre-Holocene water component in these waters may have been residing in the bedrock prior to the formation of the fractures but mobilized to flow into these fractures subsequent to their formation. Thermal waters emerging in the valley floor have δ^2H-values, which are intermediate between those on either side of the valley. Their sea salt Cl content is also intermediate between that of local precipitation ($\sim$10 ppm) and highland precipitation (1–2 ppm). These waters most likely represent a mixture of local water and distantly derived water but they may also contain a small Pre-Holocene component. The distantly derived water is mostly likely derived from the river of Héradsvötn. It feeds the alluvial sediments covering the valley floor and some of this water may have infiltrated fractures in the underlying bedrock.

REFERENCES

Adams, M.C. 1995. Vapor, liquid and two phase tracers for geothermal systems. *World Geothermal Congress, Florence Italy:* 1875–1880.

Árnason, B. 1976. Groundwater systems in Iceland traced by deuterium. *Soc. Sci. Islandica* 42: 236 p.

Árnason, B. 1977. Hydrothermal systems in Iceland traced by deuterium. *Geothermics* 5: 125–151.

Arnórsson, S. 1995. Geothermal systems in Iceland: Structure and conceptual models. II. Low-temperature areas. *Geothermics* 18: 603–629.

Arnórsson, S. & Andrésdóttir, A. 1995. Processes controlling the distribution of boron and chlorine in natural waters in Iceland. *Geochim. Cosmochim. Acta* 59: 4125–4146.

Arnórsson, S., Gunnarsson, I., Stefánsson, A., Andrésdóttir, A. & Sveinbjörnsdóttir, Á.E. 2002. Major element chemistry of surface- and ground waters in basaltic terrain, N-Iceland. I. Primary mineral saturation. *Geochim. Cosmochim. Acta* 64: 4015–4041.

Björnsson, A., Axelsson, G. & Flóvenz, Ó.G. 1990. The nature of hot spring systems in Iceland. *Náttúrufraedingurinn* 60: 15–38 (In Icelandic with an English summary).

Giggenbach, W.F. 1992. Isotopic shifts in waters from geothermal and volcanic systems along convergent plate boundaries. *Earth Planet. Sci. Lett.* 113: 495–510.

Craig, H., Boato, G. & White, D.E. 1956. Isotope geochemistry of thermal waters. In: *Proc. Second Conference on Nuclear Processes in Geologic Settings. National Academy of Sciences – National Research Council, Paper* 400: 29–38.

Jóhannesson, H. 1991. The mountains west of Eyjafjördur II. Reykjavík, *The Icelandic Tourist Association Yearbook 1991*, 246 p (In Icelandic).

Neuhoff, P.S., Fridriksson, Th., Arnórsson, S. & Bird, D.K. 1999. Porosity evolution and mineral paragenesis during low-grade metamorphism of basaltic lavas at Teigarhorn, eastern Iceland. *Am. J. Sci.* 299: 467–501.

Saemundsson, K. & Fridleifsson, I.B. 1980. Application of geology in geothermal studies in Iceland. *Náttúrufraedingurinn* 60: 157–188 (In Icelandic with an English summary).

Walker, G.P.L. 1960. Zeolite zones and dike distribution in relation to the structure of the basalts of eastern Iceland. *J. Geol.* 68: 515–528.

Water–Rock Interaction, Wanty & Seal II (eds)

Modeling hydrothermal alteration in the Pinatubo geothermal field, Philippines

M.I.R.D. Balangue, M.G.B. Collantes & M.T. Mamitag
National Institute of Geological Sciences, University of the Philippines, Diliman, Quezon City, Philippines

ABSTRACT: Geothermal waters discharged by deep wells drilled at the Pinatubo geothermal field prior to the 1991 eruption was titrated with a pre-eruption andesite to simulate water–rock interaction in the system, Results of computer modeling of water–rock interaction at 300°C in the Pinatubo geothermal field predicted a series of hydrothermal mineralization consisting of early albite + Fe-chlorite + clinozoisite + hematite + muscovite + quartz. With decreasing water-rock ratio, occurrence of antigorite + diopside + epidote + pyrite + wairakite was calculated. With decreasing water–rock ratio, the andesite undergoes alteration ranging from sericitic to propylitic types. There is close similarity with the results of the modeling exercise and observed hydrothermal alteration at 300°C for the neutral-pH alteration suite. The "dry" environment resulting from the low water/rock ratio may explain the predicted occurrence of pyroxenes (diopside-hedenbergite) over the observed actinolite. The formation of some minerals (pyrrhotite, anhydrite) is probably controlled by the composition of the fluid.

1 INTRODUCTION

Water–rock interaction in the Pinatubo Geothermal Field was simulated to determine whether the observed hydrothermal mineral phases are in equilibrium with the present-day geothermal fluids.

2 PINATUBO AS A GEOTHERMAL SYSTEM

Prior to the 1991 eruption of Mt. Pinatubo, an active hydrothermal system manifested at the surface by thermal springs and solfataras was observed. Most of the springs were concentrated along a narrow, ~25-km-long belt north-northwest of the summit at altitudes of 900–1180 m (Delfin 1984). This pre-eruption hydrothermal system was a source of elemental sulfur and had attracted geothermal exploration, including the drilling of three deep wells (Fig. 1). Drilling encountered acidified reservoir fluids (pH to 2.3, temperatures to 336°C; PNOC 1990). Chemical and isotopic (helium, oxygen and carbon) analyses of well fluids indicate the presence of magmatic components such as carbon and sulfur gases, as well as heat in the Pinatubo reservoir (Ruaya et al. 1992).

Subsequent drilling of three deep wells revealed a pile of andesitic to dacitic volcanic rocks, 1,300 to more than 2,000 m thick underlying the area. The upper few hundred meters of this unit are largely dacite and andesite breccias and tuffs believed to be erupted from the modern Pinatubo. This sequence is underlain by dacitic lava flows and by a thicker sequence of water-lain (?) pyroclastics presumed to be deposits of the ancestral Pinatubo. Dioritic dikes of uncertain affinity crosscut the volcanics generally below 200 m in altitude (Delfin et al. 1996)

Hydrothermal alteration in the Pinatubo geothermal field results mainly from the reaction between the reservoir rock and neutral pH chloride fluids. The

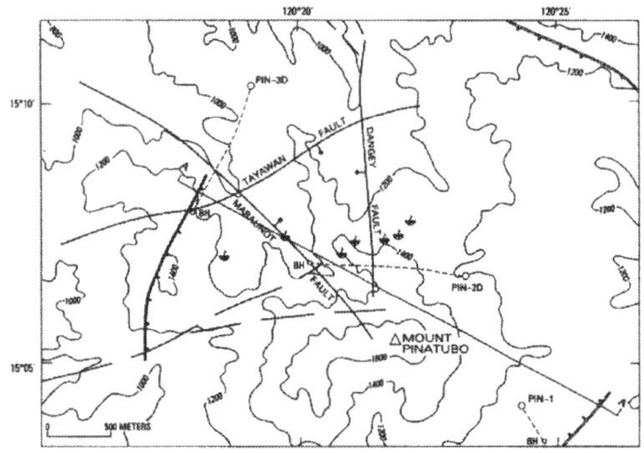

Figure 1. Well location map of Mount Pinatubo exploratory well Pin-1 had a total vertical depth of 2,733 m; well Pin-2D, 2,216 m; well Pin-3D, 2,190 m. (Delfin et al. 1996).

Table 1. Selected chemical analyses of discharge and downhole waters from Pinatubo wells. [WHP = wellhead pressure; H = enthalpy. Concentrations in milligrams per kilogram.] Modified from Delfin et al. 1996.

Well no.	pH (at 25°C)	Na	K	Ca	Mg	Fe	Cl	SO_4	SiO_2
Pin-1	5.04	5,417	531	92.2	206	114	8,778	1,532	418
Pin-2D	2.32	2,717	492	39.2	233	407	4,768	2,627	1,129
Pin-3D	4.19	17,400	3,540	2,207	133	525	37,760	500	1,660

observed clay zonation in other Philippine geothermal areas was not found at Mount Pinatubo. (Reyes 1990). Persistence of the smectite and illite-smectite zones to depths where temperatures are over 300°C has been interpreted to suggest existence of several regimes of hydrothermal activity. Both smectite and illite-smectite zones are stable only up to temperatures of 180°C and 230°C, respectively (Reyes 1990).

Despite the acidic fluids discharged by the wells (pH = 2.18–3.18 at the weir box and well pin-2D; Pagado & Villarosa 1989), acid alteration assemblages are restricted to major permeable zones associated with fault intersections. Acid minerals in the drillholes include alunite, diaspore, pyrophyllite, dickite, anhydrite, and pyrite. These occur at depths as shallow as 165 m to as deep as –514 m, where temperatures range from 164°C to 302°C. The limited distribution of acid minerals, despite the widespread occurrence of acid fluids in the boreholes, implies restricted permeability within the hydrothermal system.

3 GEOCHEMICAL MODELLING

The calculations were done at 300°C, which is close to the maximum reservoir temperature (maximum measured temperature in Pin-2D is 338°C). The calculation was executed by titrating small quantities of andesite into the aqueous phase one after the other to simulate the reaction of host rock as geothermal waters permeate the country rocks from fractures. The titration was done using the speciation program SOLVEQ and the reaction path program CHILLER (1998 version; Reed & Spycher 1984). The database for these programs is SOLTHERM (Spycher & Reed 1989).

The water chemistry used in the modeling consisted of the resultant waters after isoenthalpically boiling the well discharges (Table 1) from 300 to 100°C using CHILLER. The aqueous components at 300°C were reacted with a surface pyroxene andesitic rock (Bernard et al. 1996) assumed to be equivalent to the reservoir rock. Composition of the andesite used for the simulation is given in Table 2. The reaction was stopped after titrating about 15 kg of andesite into one kilogram of fluid.

Table 2. Composition of andesite used in the simulation (Bernard et al. 1996). Components are in weight %.

SiO_2	60.36
FeO	15.96
Al_2O_3	6.10
CaO	6.98
MgO	4.38
Na_2O	3.84
K_2O	1.59

4 RESULTS AND DISCUSSIONS

Figure 2 shows the minerals calculated as oversaturated with the geothermal fluids from Pinatubo. Simulation of hydrothermal alteration predicted formation of predominantly silicate minerals. At 300°C the andesite is predicted to alter to early albite + Fe-chlorite + clinozoisite + hematite + muscovite + quartz. With decreasing water-rock ratio, occurrence of antigorite + diopside + epidote + hedenbergite + pyrite + wairakite was calculated. The "dry" environment resulting from the low water/rock ratio may explain the predicted occurrence of pyroxenes (diopside-hedenbergite) over the observed actinolite.

There is marked similarity in most of the minerals (albite + anhydrite + antigorite + Fe-chlorite + clinozoisite + diopside + epidote + hedenbergite + hematite + pyrite + quartz + wairakite. Certain minerals are, however, predicted for one well but not in the other wells. For example, paragonite is calculated for Pin-2D but not for the other 2 wells. Diaspore and pyrrhotite are calculated for Pin-1 and Pin-2D only. The amount of dissolved Na in the waters may not be the primary factor controlling the formation of paragonite. Pin-2D has the lowest Na content of the three wells but it is the only one where paragonite is predicted to form. Other factors will have to be considered for the formation of these minerals. The dissolved H_2S probably controls the formation of pyrrhotite as suggested by the relationship of relatively higher H_2S with prediction of pyrrhotite formation (Table 1 and Fig. 2).

The occurrence of anhydrite appears to be determined by the SO_4 contents of the fluid with the Pin-1

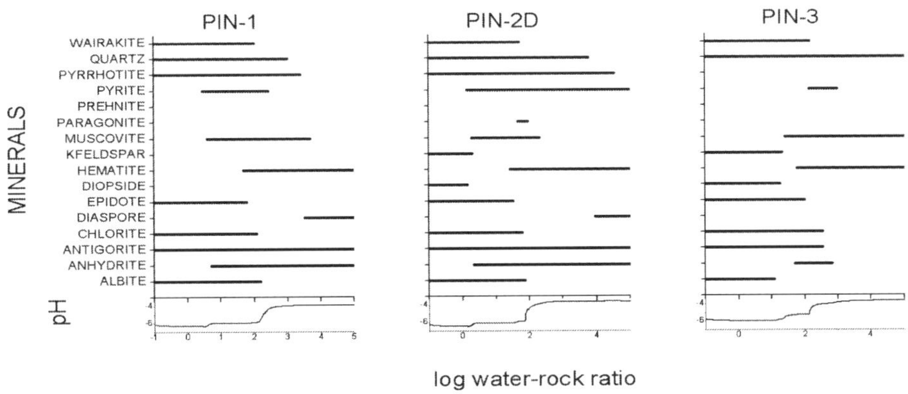

Figure 2. Paragenetic-style diagrams showing principal alteration minerals predicted to form as a result of the reaction of "boiled" Pinatubo geothermal waters with pre-eruption andesite at 300°C.

Table 3. pH Mineral buffer pairs.

Well	Mineral Pair	pH
Pin-1	Muscovite-albite	5.9
	Pyrite-pyrrhotite	5.9
Pin-2D	Muscovite-albite	5.9
	Pyrite-pyrrhotite	6.2
	Albite-Kspar	6.2
Pin-3D	Pyrite-pyrrhotite	5.5
	Microcline + Albite	5.8

and Pin-2D discharges having higher SO_4 concentrations (Table 1) and anhydrite predicted over longer water–rock ratio (Fig. 2). Whereas, Pin-3D waters have relatively lower SO_4 and anhydrite calculated only at the start of water–rock interaction.

Water–rock interaction appears to be effective in the progressive neutralization of the acidic waters. The overall increase in pH is interrupted by mineral buffers as shown by pH curve plateaus (Fig. 2, lower plot). Table 3 lists down the mineral buffer pairs for the three wells and corresponding pH.

The good correlation between predicted and observed replacement minerals indicates that local chemical equilibrium exists between neutral-pH alteration minerals and the present hydrothermal fluids at 300°C. Additional work is needed to simulate the occurrence of the acid alteration assemblage.

ACKNOWLEDGEMENTS

The results presented in this paper are made possible with financial assistance from the University of the Philippines-Diliman Office of the Vice-Chancellor for Research and Development. The authors would also like to thank the National Institute of Geological Sciences-University of the Philippines and the PNOC-EDC Geothermal Division.

REFERENCES

Bernard, A., Knittel, U., Weber, B., Weis, D., Albrecht, A., Hattori, K. & Oles, D. 1996. Petrology and geochemistry of the 1991 eruption products of Mount Pinatubo (Luzon, Philippines). In C.G. Newhall & R.S. Punongbayan (eds.), *Fire and Mud: Eruptions and Lahars of Mount Pinatubo, Philippines*. Seattle: University Of Washington Press.

Delfin, F.G., Jr. 1984. Geology and geothermal potential of Mt. Pinatubo: Philippine National Oil Company, unpublished report, 36 p.

Delfin, F.G, Villarosa, H.G., Layugan, D.B., Clemente, V.C., Candelaria, M.R. & Ruaya, J.R. 1996. Geothermal Exploration of the pre-1991 Mt. Pinatubo Hydrothermal System. In C.G. Newhall & R.S. Punongbayan (eds.), *Fire and Mud: Eruptions and Lahars of Mount Pinatubo, Philippines*. Seattle: University Of Washington Press.

PNOC-EDC 1990. Mt. Pinatubo resource assessment report: Philippine National Oil Company-Energy Development Corporation (PNOC-EDC) internal report, 46 p.

Reed, M.H. & Spycher, N.F. 1984. Calculation of pH and mineral equilibria in hydrothermal waters with application to geothermometry and studies of boiling and dilution: *Geochim. Cosmochim. Acta* 48: 1479–1492.

Reed, M.H. & Spycher, N.F. 1993. SOLTHERM: Data base of equilibrium constants for aqueous-mineral-gas equilibria.

Reyes, A.G. 1990. Petrology of Philippine geothermal systems and the application of alteration mineralogy to their assessment: *J. Volcanol. Geotherm. Res.* 43: 279–309.

Ruaya, J.R., Ramos, M.N. & Gonfiantini, R. 1992. Assessment of Magmatic Components of the Fluids at the Mt. Pinatubo Volcanic Geothermal System, Philippines, from Chemical and Isotopic Data: *Rept. Geol. Surv. Japan No.* 279: 142–152.

Water–Rock Interaction, Wanty & Seal II (eds)
© 2004 Taylor & Francis Group, London, ISBN 90 5809 641 6

Sulfur stable isotope ratios of the Mt. Apo geothermal fluid, Philippines

F.E.B. Bayon & B.M.G. Sambrano
PNOC-EDC, Fort Bonifacio, Makati City, Philippines

ABSTRACT: Sulfur isotope ratios in the Mt. Apo geothermal system effectively define neutral and acid zones in the field. Near the upflow region, S isotopes are in equilibrium, as predicted temperatures from sulfur pairs agree with measured downhole temperatures. Here, $\delta^{34}S_{SO_4}$ ratios are typically +17 to +20‰, and $\delta^{34}S_{H_2S}$ is around −3‰. In the neutral Marbel corridor, $\delta^{34}S_{SO_4}$ ratios are characteristically between +10 and +15‰ while $\delta^{34}S_{H_2S}$ remains at about −3‰. Sulfur isotopes from this sector onward along the outflow path, S-isotope temperatures do not coincide with measured well temperatures. Anhydrite derives its sulfur from the deep, $\delta^{34}S$-enriched SO_4 dissolved in fluid. Pyrite may have a sulfur source apart from H_2S, as isotope ratios in pyrite are higher (>0‰) than those in H_2S (−3‰). Magmatic SO_2 may be this other sulfur source. Sulfur isotope mineral geothermometry gives temperatures much higher than present day fluid temperatures.

1 INTRODUCTION

The study of sulfur and oxygen isotopes in sulfate in fluid and minerals of the Mt. Apo geothermal field forms part of the IAEA-assisted research program RC 9719/RO. The project is aimed primarily at evaluating the applicability of established sulfur isotope geothermometers, and characterizing sulfur-bearing species in acid and neutral fluids in Philippine geothermal systems. This paper presents data gathered for Mt. Apo, and its evaluation and interpretation.

Subsurface geology is composed of young extrusive rocks on top, underlain by older volcanics, clastic sediments and a monzodiorite intrusive unit in Sandawa Collapse, the center or upflow zone of the hydrothermal system. This sector also corresponds to the occurrence of acid-SO_4-Cl fluids, with SO_4 concentrations above 100 mg/kg and Cl ranging between 5000 and 6000 mg/kg. Fluid temperatures here are at least 270°C. Away from the upflow, within the Marbel Corridor, the fluid evolves into neutral-pH type, with reservoir Cl levels of around 2000–5000 mg/kg and temperature between 240 and 260°C.

2 RESULTS OF ISOTOPE ANALYSES

Fluid samples were gathered from wells TM-1D, SK-3D, SK-4B, SP-4D, APO-2D, APO-3D, KN-3D, SK-1D and TM-4D, and surface thermal features Imba, Sisiman, Marbel, Mainit, Agco, Lakes Venado, Jordan and Macadac. Anhydrite and pyrite grains were picked from cuttings in wells KN-1D, -2D, -3D, APO-2D, SK-2D and TM-1D. Not all acidic wells were sampled since most of these were already shut during the course of sample collection.

2.1 $\delta^{34}S$ ratios in sulfates

Dissolved SO_4 isotope ratios are grouped in 2 clusters: −5 to +5‰ CDT and +10 to +20‰ (Fig. 1). Surface thermal features show an almost uniform ratio of around −3‰, and these make up the low ratio group. Neutral well $\delta^{34}S$ ratios fall between +10 and +15‰, while high temperature acid wells (TM-2D, TO-1D, KN-3) have more enriched ratios (+17 to +20‰). Sulfur isotope ratios in Mt. Apo neutral wells are lighter compared with those in other geothermal fields, where $\delta^{34}S$ ratios are in the vicinity of +17‰ (i.e., Mahanagdong, Palinpinon, Bacon-Manito; Salonga et al. 1999; Bayon 1996). Nonetheless, these ratios are far more enriched compared with that of magmatic dissolved sulfur (0±3‰; Ohmoto 1986).

Anhydrite samples from neutral well cores and cuttings are in the range of +10 and +20‰. In the acid wells, ratios are again more enriched, approaching +20‰. Values are consistent with those of dissolved sulfate, and closer to those obtained from Mt. Pinatubo hydrothermal anhydrite (+20‰; McKibben et al. 1997) than primary magmatic anhydrite (+5 to +11‰; ibid.).

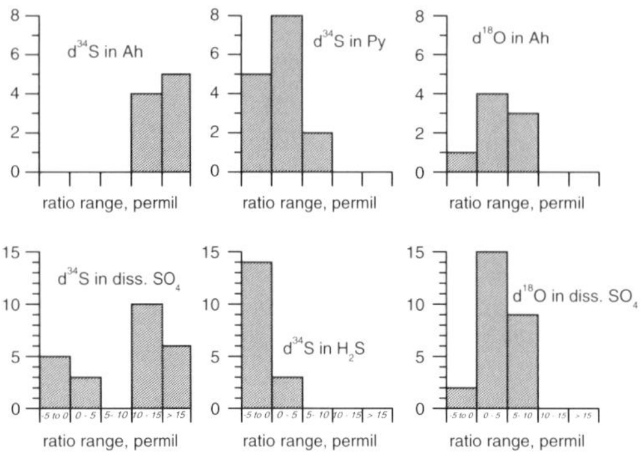

Figure 1. Frequency distribution of S isotope data.

2.2 $\delta^{34}S$ *in sulfides*

Sulfur isotope ratios in H_2S gas of wells are dominantly between −5‰ and 0‰ (about −3 to −4‰), except for APO-3D and Apo solfatara where $\delta^{34}S$ are >0‰. These ratios are akin to the depleted $\delta^{34}S$ ratios in dissolved SO_4 of thermal waters, and slightly depleted compared with the average magmatic H_2S ratio of about 0‰.

Pyrite samples, on the other hand, have slightly more enriched values between 0 and +5‰. The Apo solfatara pyrite sample is the lightest (−3‰), and this is unusual considering it owns the heaviest $\delta^{34}S_{H_2S}$. The pyrite ratios in Mt. Apo are comparatively higher than the close to 0‰ of pyrite crystals in other fields (Bayon 1996; Salonga et al. 1999), which have close affinity with pyrites in porphyry copper environments (Ohmoto & Rye 1979).

3 ISOTOPE GEOTHERMOMETRY

Estimated fluid temperature based on the difference in sulfur isotope ratios between pairs of dissolved SO_4 and H_2S, and Ah and Py, were computed and compared with T_{qtz} (representative of well measured temperatures). Geothermometry equations are given elsewhere (see references).

The fluid sulfur geothermometer gives a temperature range of 275°C (TO-1D) to 415°C (SK-4D), while the mineral isotope geothermometer has a range of about 300 to 650°C. Only wells TO-1D and TM-2D, both acidic and with high fluid temperatures, give relatively good agreement between isotope and quartz geothermometer (Fig. 2). The other samples reflected much higher sulfur isotope temperatures. It appears that it is only in the high temperature acid well reservoir fluid in Mt. Apo where sulfur isotopes are in equilibrium, and thus reflect current-day well temperatures. As for the mineral geothermometer, the even higher temperature range may be taken as an

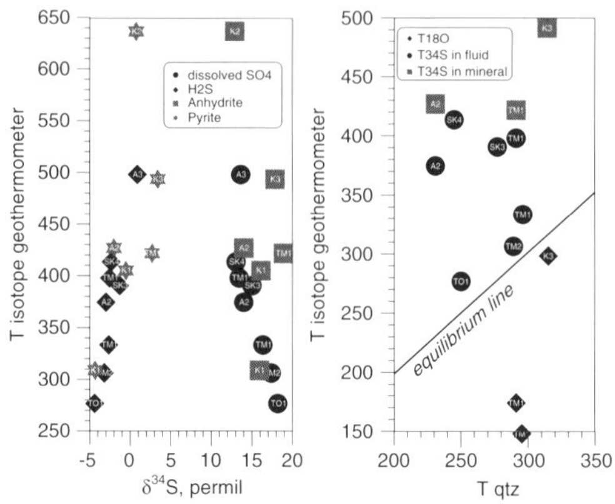

Figure 2. S isotope vs. temperature cross plots.

indication of deeper equilibration. However, though the estimated mineral temperatures approach magmatic levels, S isotopes in primary magmatic environments are much more depleted (i.e., +7‰ in Ah; McKibben et al. 1997) compared with those measured in well Ah and Py crystals.

4 CORRELATION OF ISOTOPE RATIOS

4.1 $\delta^{34}S$ *in fluid*

Three major factors control sulfur isotope equilibration in hydrothermal fluids – pH, S concentration and temperature. In theory, equilibration is faster at lower pH, greater total S concentration and higher temperature.

For dissolved SO_4, both fluid pH and temperature vaguely correlate with $\delta^{34}S$ (Fig. 3, filled circles). The acid wells (KN-3, TO-1D) are at the high ratio, low pH/high T end of the plots, the thermal springs at the opposite end, and the neutral wells in the middle. Dilution lines may be drawn to connect these data points (solid lines) in the plots, but these connections are highly speculative. Furthermore, there are quite a few data points that plot off these inferred dilution lines.

For H_2S (Fig. 3, filled squares), gas S concentration appears to have a degree of correlation with isotope ratio, but again this is speculative.

Although vague trends can be inferred from the cross plots, there is no clear, consistent and coherent relationship between chemistry and sulfur isotope ratios, suggesting a general state non-equilibrium of sulfur isotopes in the present geothermal environment. The influence of pH, sulfur concentration and temperature are evidently not enough to force exchange reactions and bring the sulfur isotopes to equilibrium, except perhaps in the upflow region.

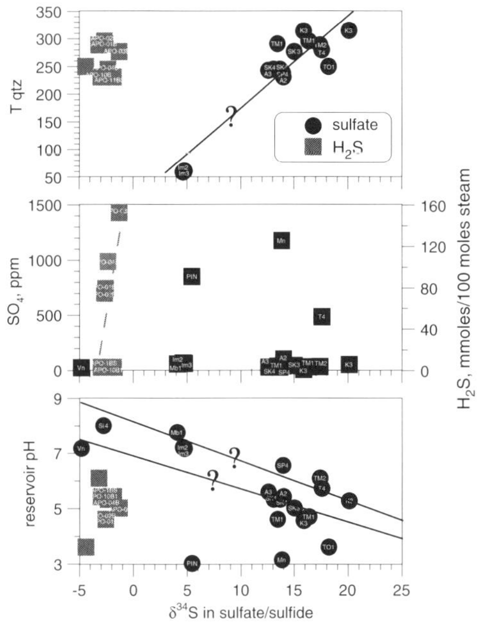

Figure 3. $\delta^{34}S$ in sulfate vs. chemistry cross plots.

Deeper and hotter equilibration can explain the lighter $\delta^{34}S_{SO_4}$ and higher S isotope temperatures, but this is unlikely if correlated with the hydrological flow model of the system. Sulfur isotopes in the acid bores (the upwelling zone of the system) appear to be in equilibrium at present fluid temperatures. The fluid that gets to the Marbel corridor, therefore, should have "starting" $\delta^{34}S$ similar to those of acid wells (i.e., >15‰ for SO_4 and −5‰ for H_2S). With the decrease in temperature and increase in pH going into the Marbel corridor and further northwest, there should be very little change in $\delta^{34}S$ ratios as isotope exchange reactions are not favored here. Thus, neutral well $\delta^{34}S$ should be similar to those of acidic wells. But this is not the case; there is a marked difference in ratios between neutral and acid wells. It is more likely that the depleted ratios in neutral wells are due to mixing between the deep, acidic reservoir fluid (>+15‰ $\delta^{34}S_{SO_4}$ and −5‰ $\delta^{34}S_{H_2S}$) and a shallow, isotopically-depleted fluid (≈−5 to 0‰ $\delta^{34}S$ in SO_4, perhaps −5‰ in H_2S). This shallow fluid may be condensed steam (where H_2S is converted to SO_4 but retains the $\delta^{34}S_{H_2S}$). The existence of a natural state steam cap in Mt. Apo makes this theory more feasible, and is perhaps the simplest and most logical explanation for the isotopic characteristic of the Mt. Apo neutral well waters. The mixing theory is also supported by the decrease in $\delta^{34}S_{SO_4}$ with increasing distance from the volcanic edifice, taken to represent the position of the upflow zone.

The degree of mixing can perhaps be roughly estimated based on the deviation of $\delta^{34}S_{SO_4}$ from the acid fluid ratio – the more depleted the ratio, the larger the proportion of steam condensate in the well fluid. Assuming +20.1‰ and −5‰ as heavy and light end-members, respectively, the amount of "steam dilution" in the samples are between 15 and 30% for neutral wells, 61 to almost 100% for thermal areas, and nil to 11% in acid wells.

Ratios in dissolved sulfate of thermal waters are commonly <0‰, and this suggests that dissolved sulfate here are derived from the conversion of H_2S to SO_4. The depleted ratios are retained by the resultant SO_4.

Assuming that KN-3 fluid is most representative of the upflow water (+20.1‰ $\delta^{34}S_{SO_4}$, ≈ −3‰ $\delta^{34}S_{H_2S}$), the $\delta^{34}S$ ratio of total sulfur and magmatic SO_2 (which produces dissolved SO_4 upon disproportionation) can be roughly estimated. Using the equation:

$$\delta^{34}S_{\Sigma S} = \delta^{34}S_{SO_4} * X_{SO_4} + \delta^{34}S_{H_2S} * X_{H_2S}$$

(Martinez Serrano et al. 1996), where $X_A = (m\Sigma A)/(m\Sigma A + m\Sigma B)$ and m is molality, $\delta^{34}S_{\Sigma S}$ is about 0‰. If SO_2 disproportionation results in as much as 70% enrichment in $\delta^{34}S_{SO_4}$ (Thode 1963), $\delta^{34}S_{SO_2}$ is roughly +3.9‰.

4.2 *$\delta^{34}S$ in Ah and Py*

Anhydrite minerals in TO-1D and APO-2D have isotope ratios very close to those of dissolved sulfate, and it is logical to invoke that the anhydrite precipitated from the present-day fluid and retained the sulfur isotope signature of dissolved sulfate from which it was derived. There is a slight difference in Ah and dissolved SO_4 ratios in KN-3, but these may be attributed to kinetic effects. In TM-1D, however, the mineral ratio is significantly heavier compared to the fluid ratio. In this case, the present TM-1D fluid is not the source of Ah. There may have been an "early" TM-1D fluid – the unmixed deep reservoir fluid with an enriched $\delta^{34}S_{SO_4}$ – which produced the Ah alteration mineral. Thus, the current well fluid is not the original fluid of the well; it now discharges the mixed fluid with characteristically lighter S isotope ratio. In general, however, it can be assumed that Ah ratios reflect the ratio in the source dissolved sulfate; where Ah and dissolved SO_4 ratios are similar, the source fluid for SO_4 in Ah is the original well fluid. On the other hand, heavier $\delta^{34}S$ ratios in Ah (compared with $\delta^{34}S$ in dissolved SO_4) signify fluid evolution, from the original $\delta^{34}S$-enriched brine to a mixed fluid type.

Pyrite s isotope ratios are considerably more enriched compared to those of H_2S, so that a similar inference cannot be invoked. The sulfur derived by the pyrite may be from a different source. This would be consistent with the absence of any relationship between the Mt. Apo gases and the pyrite-magnetite and pyrite-pyrrhotite mineral systems (Siega et al. 1999). From rough calculations on $\delta^{34}S_{\Sigma S}$ and $\delta^{34}S_{SO_2}$ (previous section), the pyrite ratios are closer to that of SO_2. Thus, pyrite might be directly derived from SO_2.

5 CONCLUSIONS

Sulfur isotopes in dissolved sulfate of the Mt. Apo geothermal fluid lie within the range of +17 and +20‰ CDT for acid bores in the upflow zone, +10 to +15‰ in neutral wells in the Marbel Corridor and outflow sectors, and −5 to +5‰ for thermal springs. The heavy ratios represent the deep reservoir fluid of the system. The ratios in neutral wells at Mt. Apo, lower compared with similar fluids from other fields, result from the mixing of the deep reservoir fluid and steam condensate.

Sulfur isotope ratios in anhydrite reflect the composition of the source dissolved sulfate. Where ratios differ between the 2 sulfate species, evolution of well fluid is inferred.

For H_2S in wells, the ratios are close to −3‰; this is more depleted relative to other geothermal areas in the Philippines, where ratios approach 0‰. Pyrite samples yield more enriched ratios of from 0‰ to +5‰. The difference in ratios between H_2S and pyrite may indicate that H_2S is not the source of sulfur for secondary pyrite minerals. Pyrite sulfur might be directly derived from SO_2.

Sulfur isotope geothermometers (for fluid and minerals) are not applicable to the Mt. Apo geothermal system, except in acidic, high temperature wells. The isotopes are not in equilibrium in the neutral sector of the present-day system. Since fluids here are mixed, S isotope geothermometer temperatures do not reflect deeper and hotter equilibration. In contrast, the isotopes attained a high degree of equilibration under acidic condition as shown by the relatively close agreement between isotope and quartz estimated temperatures.

ACKNOWLEDGEMENTS

This project was made possible through the support of the International Atomic Energy Agency in Vienna, Austria. The authors also wish to thank PNOC-EDC for permission to publish this paper.

REFERENCES

Bayon, F.E.B. 1996. Sulphur isotope applications in two Philippine geothermal systems. *Geothermal Resources Council Transactions* 20: 651–660.

Delfin, F.G. & Castro, C.C. 1984 (unpublished). Preliminary Evaluation of the Mt. Apo Geothermal Prospect.

Giggenbach, W.F. 1992. Isotopic and chemical composition of water and gas discharges from the Zunil geothermal system, Guatemala. In: *Geothermal investigation with isotope and geochemical techniques in Latin America. IAEA TECDOC*. 245–278. Vienna: International Atomic Energy Agency.

McKibben, M.A., Eldridge, C.S. & Reyes, A.G. 1997. Sulfur isotope systematics of the June 1991 Mount Pinatubo eruptions: A SHRIMP ion microprobe study. In C.G.Newhall & R.S. Punongbayan (eds.): *Fire and Mud Eruptions and Lahars of Mount Pinatubo, Philippines*: 825–844. University of Washington Press.

Martinez Serrano, R.G., Jacquier, B. & Arnold, M. 1996. The $\delta^{34}S$ composition of sulfates and sulfides at the Los Humeros geothermal system, Mexico and their application to physicochemical fluid evolution. *J. Volcanol. Geotherm. Res.* 73: 99–118.

Ohmoto, H. 1986. Stable isotope geochemistry of ore deposits. In J.W.Valley, H.P.Taylor & J.R.O'Neil (eds.), *Stable isotopes in high temperature geological processes. Reviews in Mineralogy* Vol. 16: 491–556. Washington DC: Mineralogical Society of America.

Ohmoto, H. & Rye, R.O. 1979. Isotopes of sulfur and carbon. In H.L. Barnes (ed.), *Geochemistry of Hydrothermal Ore Deposits, Second Edition*: 509–567. New York: John Wiley and Sons.

PNOC-EDC 1994 (unpublished). Mindanao I Geothermal Project Resource Assessment Update.

Salonga, N.D. 1995 (unpublished). Fluid and mineral equilibria in acid NaCl(+SO_4) reservoir: the case of Sandawa Collapse, Mt. Apo hydrothermal system.

Salonga, N.D., Aragon, G.M., Nogara, J.B. & Sambrano, B.G. 1996. Applications of stable isotopes in hydrological studies of Mt. Apo geothermal field. *Proceedings, 17th Annual PNOC-EDC geothermal conference, Manila*. 15–25.

Salonga, N.D., Bayon, F.E.B. & Martinez, M.M. 2000. The implication of ^{34}S and ^{18}O isotope systematics on the sulfur cycle in Mahanagdong hydrothermal system (Philippines). *Proceedings of the World Geothermal Congress 2000*, Japan. 2845–2850.

Siega, F.L., Salonga, N.D. & D'Amore, F. 1999. Gas equilibria controlling H_2S in different Philippine geothermal fields. *Proceedings, 20th Annual PNOC-EDC geothermal conference, Manila*. 29–36.

Thode, H.G. 1963. Sulphur isotope geochemistry. In D.M. Shaw (ed.), *Studies of Analytical Geochemistry*. The Royal Society of Canada Special Publication no. 6.

Water–Rock Interaction, Wanty & Seal II (eds)

Diffuse CO_2 degassing from summit areas on Mt. Hood, Oregon, USA

D. Bergfeld
U.S. Geological Survey, Menlo Park, Calif., USA

J.F. Howle
U.S. Geological Survey, Carnelian Bay, Calif., USA

M.E. Schmidt
OR State Univ., Corvallis, Oregon, USA

ABSTRACT: Results from a survey of diffuse CO_2 flux in the Devils Kitchen area near the summit of Mt. Hood in northern Oregon show that intense hydrothermal alteration impedes diffuse gas flux and CO_2 is emitted predominately from small gas vents and microfractures. Results indicate that at Devils Kitchen 0.78–0.94 of a tonne of CO_2 per day is emitted from a $\approx$5000 m^2 area. This suggests that a total of 6–7 tonnes of CO_2 per day could be emitted from the $\approx$37,000 m^2 of mostly inaccessible thermal ground around the crater.

1 INTRODUCTION

Studies of volcanic degassing have only recently begun to examine the location and amounts of diffuse CO_2 emissions on quiescent volcanoes (Chiodini et al. 1998; Giammanco et al. 1998; Gerlach et al. 2001). The results have been used with other data to estimate thermal emissions, evaluate geothermal reservoirs, track thermal fluids, identify faults and fractures, in addition to the general goal of estimating total diffuse CO_2 emissions. In areas where CO_2 is of magmatic origin, or emissions are directly related to magmatic processes, then research on diffuse degassing can also serve as a monitoring tool. Volcanoes with large areas of steaming ground provide excellent targets for diffuse degassing studies since CO_2 emissions are often strongly coupled with areas of high heat flow. This paper presents the first results from the investigation of diffuse CO_2 emissions from thermal areas near the summit of Mt. Hood, a volcano deserving diligent monitoring because of its proximity to major cities.

2 FIELD LOCATION

Mt. Hood is an andesitic to dacitic composite volcano in the High Cascades of northern Oregon. The most recent eruptive period was around 200–300 years ago (Crandell & Rubin 1977) producing a hornblende andesite dome below the summit (Wise 1968). The remnant of this dome, known as Crater Rock (CR), is partly surrounded by thermal ground containing numerous gas vents. Other thermal areas on Mt. Hood include the Steel Cliffs, a recent intracanyon flow which forms the east wall of the crater (Wise 1968; Friedman et al. 1982), and Devils Kitchen (DK) located below and east of CR (Fig. 1).

The DK is an area of steaming ground on the south side of the crater below the terminus of the Coalman Glacier. Snow covers the area above 3000 m in winter. In spring, the high heat flow exposes two sections of DK earlier than other parts of the mountain. During a reconnaissance survey on June 7, 2003, a patch of ground >1400 m^2 was already exposed. There are no high-velocity fumaroles on DK but numerous small gas vents with sub-boiling temperatures are scattered over the area. Yellow and white sublimate crusts surround the vents and also delineate microfractures that function as additional sites for gas emissions. Much of the host rock at DK has been hydrothermally altered to clay, creating a hard-packed, relatively impermeable, moist surface.

In early August 2003 the snow had melted from DK and we constructed a 10 m sampling grid across most of the accessible steaming ground (Fig. 1). The grid was sited over steeply-sloping ground that was bordered by large blocks of talus shed from the Steel Cliffs on the east, the glacier on the north, and was bisected by runoff from the glacier on the west. Due to the rough terrain, total grid coverage was irregular

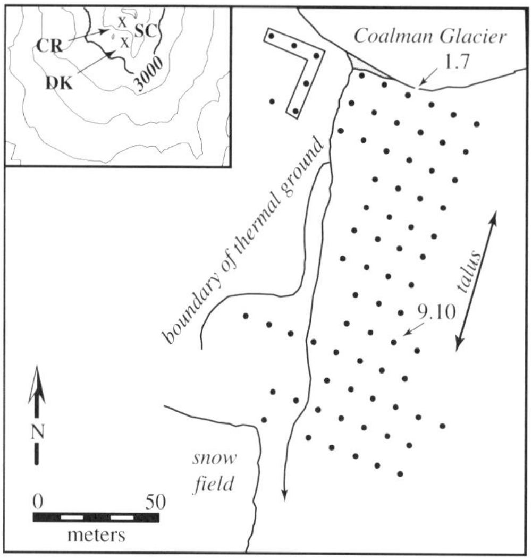

Figure 1. Location map showing DK grid. The t-shaped area encloses background sites. Inset shows summit of Mt. Hood. Locations of the DK grid and CR traverse are marked with an x. SC = Steel Cliff, contour interval = 200 m

in shape. We obtained flux measurements at 71 grid locations covering $\approx$5,000 m^2. We also measured fluxes at 14 sites along a single linear traverse above CR (Fig. 1). The traverse was located near several high-velocity gas vents and was bounded in part by the permanent snow field in the bowl of the crater. Crevasses and steep terrain prohibited extensive investigation of the CR area.

3 METHODS

3.1 *Flux measurement*

Diffuse CO_2 fluxes were measured using the accumulation chamber method with a West Systems flux instrument equipped with a 3 L chamber and a LI-COR 820 infrared gas analyzer (IRGA). Tests under ideal laboratory conditions show that measured fluxes may under-represent actual values by 12% (Evans et al. 2001), or may vary by $\pm$10% (Chiodini et al. 1998). Throughout this paper flux is reported in units of grams CO_2 m^{-2} d^{-1}. Soil temperatures were collected adjacent to each flux site, typically at a depth of 10 cm. The reconnaissance survey indicated that we would encounter large fluxes so we configured the IRGA to allow measurement of CO_2 concentrations up to 10,000 ppm. This configuration prohibited data collection at some low-flux grid sites west of the outflow creek. During our investigation we sometimes deviated from typical flux measurement protocol which minimizes soil disturbance so as not to enhance gas flux. The repercussions of this will be discussed later.

3.2 *Emissions estimates*

Estimates of the average CO_2 flux from sites in the DK grid were made using two methods commonly employed in flux studies. These include a minimum variance unbiased (MVU) estimator as outlined in Gilbert (1987) and a graphical statistical approach (GSA) following the methods of Chiodini et al. (1998). The MVU method analyzes the dataset to determine the type of distribution and then employs an appropriate set of equations to estimate the mean and associated confidence intervals for the mean. Total CO_2 emissions for the grid are determined using the estimated mean multiplied by the grid area. GSA methods also examine the type of distribution but use cumulative probability plots to divide the data into populations of similar flux values. For lognormal populations the means are determined using Sichel's T-estimator. Total CO_2 emissions are calculated using the population means multiplied by the percentage of grid area associated with those data.

4 RESULTS AND DISCUSSION

4.1 *Measured CO_2 emissions*

The values of CO_2 fluxes at the DK grid ranged from <8 to over 1100 g m^{-2} d^{-1}, and associated soil temperatures at 10 cm depth ranged from 4.7 to 87.9°C (Figs 2a, b, c). Background soil temperatures at 5 locations outside of the thermal area (Fig. 1) were between 10 and 15°C. Fluxes and temperatures along the CR traverse were 25–1800 g m^{-2} d^{-1}and 12.9–87.9°C, respectively, and are comparable to those at DK (Fig. 2d). Flux and temperature may be correlated at CR, but this is difficult to discern given the limited data set. At DK, flux and temperature data are scattered. The lack of correlation likely arises from the exceedingly compact ground that restricts rising thermal fluids to discrete pathways and the influence of the glacier and long periods of snow and ice cover that depress soil temperatures.

Results using the D'Agostino test on the DK data set indicate for a 95% confidence interval the data are lognormally distributed. Histogram and cumulative probability plots for fluxes at DK show the distribution is positively skewed with three flux populations (Figs 3a, b). The low and high flux populations are not well defined as they each are represented by only a few measurements that include about 4% of the data. For low fluxes this is due in part to our inability

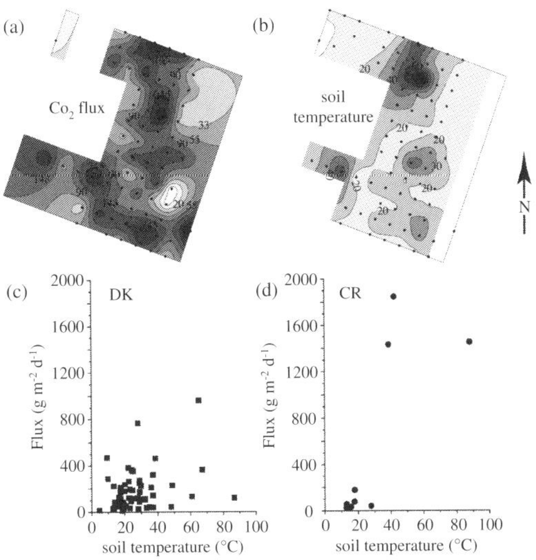

Figure 2. Flux and temperature data from the DK grid and CR traverse. Contour plots (a & b) are for DK only, white areas lack data. Data in b, c & d are only from sites collected at 10 cm.

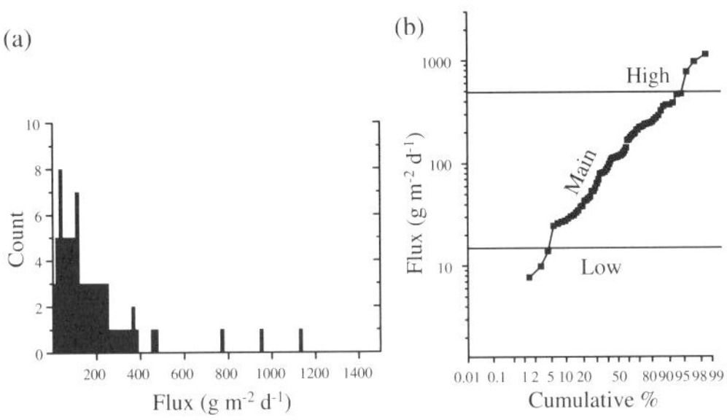

Figure 3. Histogram and cumulative probability plot for flux data at DK.

to measure flux at background sites. The kink in the probability plot around $470\,g\,m^{-2}\,d^{-1}$ defines the upper limit for the main population of data that includes the remaining 92% of the measurements.

Mean fluxes determined from estimator equations are shown in Table 1. Because of limited data for low and high flux populations it was not possible to calculate confidence intervals around those means. The data suggest that the high flux population contributes around 20% of total emissions from DK, illustrating the significance of contributions from gas vents. Total grid emissions calculated using GSA and MVU methods are similar and indicate that between 0.91 and 0.94 of a tonne of CO_2 per day ($t\,d^{-1}$ issues from the thermal ground. These values are small compared with diffuse emissions reported for areas on Nisyros volcano, Greece, Mammoth Mt., USA, and Solfatara of Pozzuoli crater, Italy, which range from 84 to $1513\,t\,d^{-1}$ (Cardellini et al. 2003).

4.2 *Error estimates*

Because of time constraints we were unable to investigate all aspects of measurement repeatability at sites on Mt. Hood. In other work we have examined measurement variation by making rapid successive measurements at a site. Triplicate measurements of degassing rates between 3 and 100 ppm s^{-1}, similar to rates at DK, show an increase in the standard error of the means as degassing rates rise (Table 2). Similarly, the coefficient of variation, the standard deviation of the sample population divided by the mean, increases to a value of around 12% at high CO_2 emission rates. We assume that similar variations in emission rates can be expected for measurements on Mt. Hood (Table 1).

Hard ground and the presence of buried rocks at DK made it difficult to obtain a tight seal between the accumulation chamber and the surface. It was often necessary to remove rocks in order to seat the chamber. This type of soil disturbance is undesirable as it can result in a temporary increase in surface flux.

Table 1. Results from statistical calculations for grid fluxes at Devils Kitchen. CI = confidence interval for the sample means. Total emissions reported in the last column are estimated for $37000\,m^2$ of thermal ground from the Hot Rocks, Steel Cliffs and DK areas.

Method	Area (m^2)	Mean flux ($g\,m^{-2}\,d^{-1}$)	Expected variance ($g\,m^{-2}\,d^{-1}$)	Std. error of mean	Lower CI (95%)	Upper CI (95%)	Grid emissions (tonne/day)	Grid emissions 95% CI (tonne/day)	Total emissions (tonne/day)
GSA high	200	953	105	–	–	–	0.19	–	–
GSA main	4600	155	17	–	129	198	0.71	0.6–1.0	–
GSA low	200	10	0.4	–	–	–	0.002	–	–
GSA total	5000	–	–	–			0.91	–	6.7
MVU	5000	188	21	29	149	249	0.94	0.7–1.2	6.9
MVU corrected	5000	155	17	24	123	206	0.78	0.6–1.0	5.7

Table 2. Statistical results for triplicate measurements of emissions from thermal areas at Long Valley caldera, CA. Rates are comparable to those at DK.

Location	Average rate ($ppm\,s^{-1}$)	Std. error of mean	Coeff. variation (%)
FVN 7	3.1[a]	0.04	1.7
BC 5	10.4	0.4	5.4
BC 27	14.0	0.3	3.2
DTH 33	23.6	0.2	1.7
BF 17	40.0	0.8	2.7
BC 9	41.7	2.8	11.8
BC 14	42.3	0.6	2.6
BC 1	56.8	0.6	1.9
BF 10	81.3	6.2	10.8
BF 9	92.3	5.5	10.4

[a] only 2 measurements taken.

During the DK investigation 52% of our measurements were made on the same day that the site was located. To evaluate the impact of soil disturbance we repeated flux measurements at two sites on successive days. The first measurement occurred soon after the site was leveled. Additional measurements were made after the site had stabilized for at least 24 hours. At site 1.7 we found that the flux decreased 22% on the second day and was essentially unchanged on the third day. The flux at site 9.10 decreased 32% on the second day and there was no opportunity for a third measurement. If we correct for a maximum 32% decrease in flux from newly prepared sites, then our emissions estimates for DK are lowered to $0.78\,t\,d^{-1}$ (Table 1).

4.3 *Fluxes near the summit*

A problem in determining a total diffuse emissions estimate for Mt. Hood is that much of the crater area is inaccessible. During our investigation a fortuitous nighttime snowfall highlighted the inaccessible thermal area known as Hot Rocks on the west side of CR. We mapped the snow-free area to obtain a size estimate of the thermal ground. By including this new area with the DK grid and mapped areas around the Steel Cliffs, we estimate that a minimum of $37{,}000\,m^2$ of thermal ground is present on Mt. Hood. If fluxes over this area are similar to those at DK, as was shown for sites on the CR transect, then total diffuse CO_2 emissions from the south summit Mt. Hood are around 6 to $7\,t\,d^{-1}$ (Table 1).

5 CONCLUSIONS

Compared to other recently active volcanoes, diffuse CO_2 emissions from the thermal areas around the summit of Mt. Hood are low. At DK, hydrothermal alteration reduces permeability and focuses rising thermal fluids along distinct paths. As such, diffuse soil emissions are probably a minor component of total CO_2 discharged from the crater, and gas vents provide the main source of CO_2. These results highlight a need to characterize advective vent emissions on Mt. Hood in order to provide a better estimate of total CO_2 discharged from the volcano. Future studies to obtain flux measurements on snow may provide a means to quantify fluxes from high volume point-sources.

REFERENCES

Cardellini, C., Chiodini, G. & Frondini, F. 2003. Application of stochastic simulation to CO_2 flux from soil: Mapping and quantification of gas release. *J. Geophys. Res.* 108: 3-1–3-13.

Chiodini, G., Cioni, R., Guidi, M., Raco, B. & Marini, L. 1998. Soil CO_2 flux measurements in volcanic and geothermal areas. *Appl. Geochem.* 13(5): 543–552.

Crandell, D.R. & Rubin M. 1977. Late-glacial and postglacial eruptions at Mt. Hood, Oregon. *Geol. Soc. Am. Abstr. Prog.*9(4): 406.

Evans, W.C., Sorey, M.L., Kennedy, B.M., Stonestrom, D.A., Rogic, J.D. & Shuster, D.L. 2001. High CO_2 emissions through porous media: transport mechanisms and implications for flux measurement and fractionation. *Chem. Geol.*177(1–2): 15–29.

Friedman, J.D., Williams, D.L. & Frank, D.L. 1982. Structural and heat-flow implications of infrared anomalies at Mt. Hood, Oregon, 1972–1977. *J. Geophys. Res.* 87: 2793–2803.

Gerlach, T.M., Doukas, M.P., McGee, K.A. & Kessler, R. 2001. Soil efflux and total emission rates of magmatic CO_2 at the Horseshoe Lake tree kill, Mammoth Mountain, California, 1995–1999. *Chem. Geol.* 177(1–2): 101–116.

Giammanco, S., Inguaggiato, S. & Valenza M. 1998. Soil and fumarole gases of Mount Etna; geochemistry and relations with volcanic activity. *J. Volcanol. Geotherm. Res.*81(3–4): 297–310.

Gilbert, R.O. 1987. In: *Statistical methods for environmental pollution monitoring*. New York: Van Nostrand Reinhold.

Wise, W.S. 1968. Geology of the Mt. Hood Volcano: In A.R. McBirney (ed.), *Andesite conference guidebook*. Bull. OR Dept. Geol. Miner. Ind. 62: 81–89.

Water–Rock Interaction, Wanty & Seal II (eds)
 ISBN 90 5809 641 6

Geochemical evolution of the young crater lake of Kelud volcano in Indonesia

A. Bernard & A. Mazot
BRUEGEL (Brussels unit for Environmental, Geochemical and Life Sciences Studies), Université Libre de Bruxelles, Brussels, Belgium

ABSTRACT: After the February 1990 plinian eruption of the Kelud volcano, a new lake rapidly filled its crater. This young lake offers the opportunity to study the evolution and re-equilibration of a volcanic-hydrothermal system after a magmatic eruption. Geochemical and thermal changes of the crater lake have been recorded during the period 1993–2003 in which a decrease in temperature and ionic concentrations of the lake was observed. The initial lake chemistry (1993–1997) was dominated by Na-K chloride waters. Today, Ca-Mg sulfate waters are the main component in the lake. The temporal evolution in the chemistry of the lake waters suggests the presence of two distinct hydrothermal systems feeding the lake: a deep system at high temperature (250°C) with neutral alkali-chloride fluids and a shallow aquifer at lower temperature dominated by Ca/Mg-sulfate waters.

1 INTRODUCTION

Kelud is one of the most active volcanoes of Java with a record of 29 eruptions since 1311 and with repose intervals between eruptions ranging from 1 to 75 years. After the last eruption that occurred in February 1990, a new lake rapidly filled the crater of this volcano. Persistent degassing from subaqueous fumaroles and hot springs discharging into the lake maintain water temperatures between 31–50°C, higher than the ambient temperature of 19°C. The lake is shallow with a maximum depth of 34 m and a volume estimated at 2.1 million $\cdot$ m^3.

A tunnel was drilled through the crater wall in 1926 to drain the lake and keep its volume constant. High overflows through the drainage tunnel (between 300 and 500 $kg \cdot s^{-1}$) are typically observed even during the dry season. This high overflow combined with the small volume of the lake make the residence times of the elements in the lake particularly short, around 80–100 days. The temperature and geochemistry of the lake has been regularly monitored since 1993 in order to understand the processes leading to the re-equilibration of the lake-hydrothermal system after the February 1990 magmatic eruption.

2 GEOCHEMISTRY OF THE LAKE WATERS

The composition of the lake is rather unusual for an active volcano and corresponds to a near-neutral pH (~6) relatively diluted water with a TDS of a few $g \cdot kg^{-1}$ (Table 1).

This neutral composition contrasts with the highly acidic waters (pH = 0–1) most frequently observed in active crater lakes where the discharge of magmatic volatiles (SO_2, HCl, HF) directly into the lake or into the sub-surface hydrothermal system produces acid

Table 1. Selected compositions of Kelud crater lake. Data are in $mg \cdot kg^{-1}$.

Date:	KCL3 18 Dec 93	KCL9 1 Aug 94	KCL37 24 Sept 02	KCL56 4 Sept 03
Temperature (°C)	42.8	42.1	33.2	30.7
Depth (m)	15	32	0	0
pH	5.9	6.3	6.5	6.5
TDS (g/kg)	3.2	3.7	2.2	2
Na	700	1020	342	271
K	92	102	39	30
Ca	105	130	135	147
Mg	55	67	80	78
Si	109	133	129	111
B	11	14	4	3
F	6	7	<1	<1
Cl	1300	1300	290	200
SO_4	631	692	670	679
HCO_3^-	238	207	435	472

sulfate-chloride waters with elevated Al and Fe contents.

The evolution of temperature and geochemistry during the period 1993–2003 is shown on Figure 1. The general trend is toward a decrease in temperature and TDS, the lake becoming more diluted with time. The main geochemical change observed is a decrease in Na + K chlorides, sulfates and Ca/Mg remaining almost constant during this period (Table 1).

With the exception of the 1993 data, there are positive correlations between chloride and the alkali metals or with B and Li but no correlation exists with Ca or Mg (Figs. 2 and 3).

During the 10-year monitoring period, two notable events were observed in 1996 and 2001 when a sudden increase in lake temperatures and bubbling occurred. In both events, the lake temperature peaked at 50°C. Heating episodes, sometimes cyclic, are relatively frequent in some crater lakes and reflect changes in the flow rate or in the enthalpy of hot fluids entering the lake. These heating episodes always represent an alarming situation because an increasing lake temperature can be a precursory signal for the renewal of magmatic activity as was observed 3 months before the 1990 eruption (Vandemeulebrouck et al. 2000).

The temperature increases observed in 1996 and 2001 were accompanied by rapid changes in lake-water composition with a net increase in Na + K chlorides, B and Li. These heating episodes could be

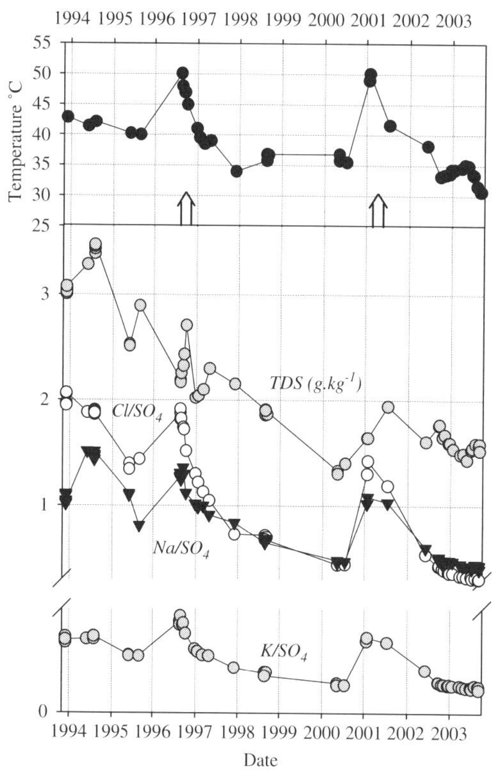

Figure 1. Temporal evolution of the temperature and chemistry of the lake during 1993–2003. Arrows represent the start of a heating episode. TDS is the Total Dissolved Solids.

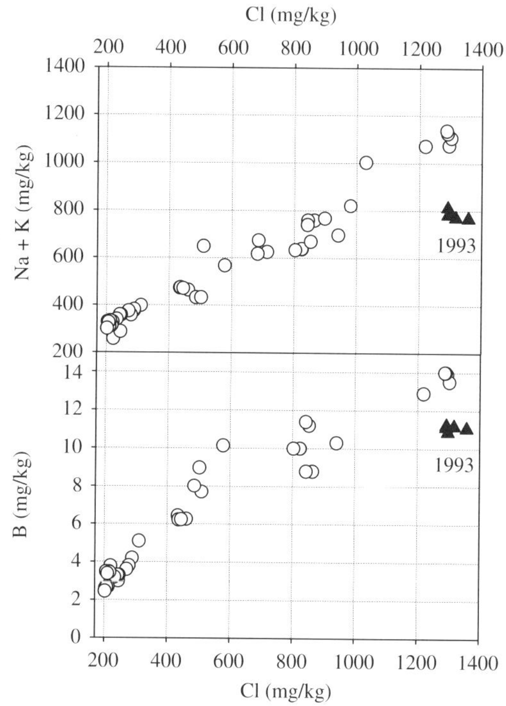

Figure 2. Evolution of chloride versus Na + K and boron during 1993–2003.

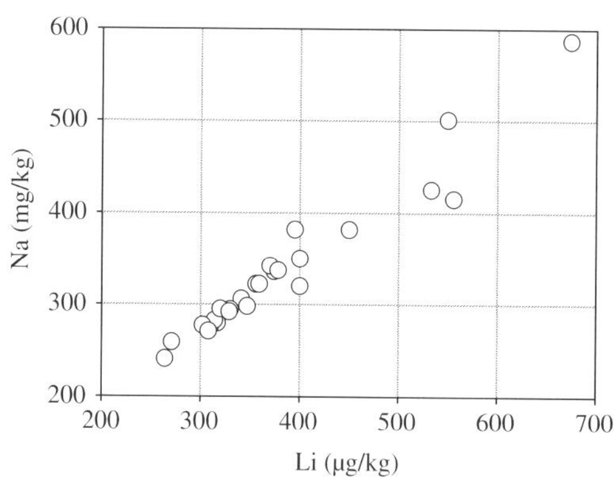

Figure 3. Evolution of Li versus Na during 1993–2003.

the consequence of the (re)opening of fractures at depth leading to an increase in the contribution from a deep hydrothermal aquifer. This evolution suggests that two distinct hydrothermal systems are feeding the crater lake. The first aquifer has neutral alkali chloride waters enriched in B and Li typical of the deepest part of hydrothermal systems and the second shallow system is dominated by Ca-Mg sulfate waters.

3 TEMPERATURE ESTIMATION OF THE DEEP HYDROTHERMAL SYSTEM

The presence of a deep alkali chloride system suggests the use of Na-K and K-Mg as geoindicators. Equilibrium temperatures for Na/K and K/Mg and based on Giggenbach (1988) are presented in Figure 4.

The lake water compositions are far from the equilibrium with a mineral assemblage as expected for waters resulting from the mixing between two types of fluids. However, the Na/K remained almost constant during the entire period despite a large decrease in their concentrations. This suggests that the composition of the deep neutral-chloride hydrothermal system remained stable during this period. The deep hydrothermal system most probably existed before the 1990 eruption and its composition was not significantly affected by the relatively brief (1 hour) plinian eruption. The average log Na/K value around 0.9 could reflect equilibrium temperatures close to 250°C in the deep hydrothermal system. On Figure 5, these Na/K equilibrium temperatures are compared to the Na-Li geothermometer of Verma & Santoyo (1997) assuming a saline brine.

4 STABLE ISOTOPE COMPOSITIONS

The hydrogen (D/H) and oxygen ($^{18}O/^{16}O$) isotope ratios also show changes clearly correlated with changes in temperature or chloride contents. Figure 6 shows the strong positive correlation between chloride and oxygen isotopic composition, except for the 1993 data. The same positive trend exists between chloride and (D/H). This trend can be interpreted as a mixing line between two end members. The first is represented by the deep neutral-chloride fluids with a heavier isotopic composition and the second by the shallow sulfate fluids with an isotopic composition identical to meteoric fluids (spring).

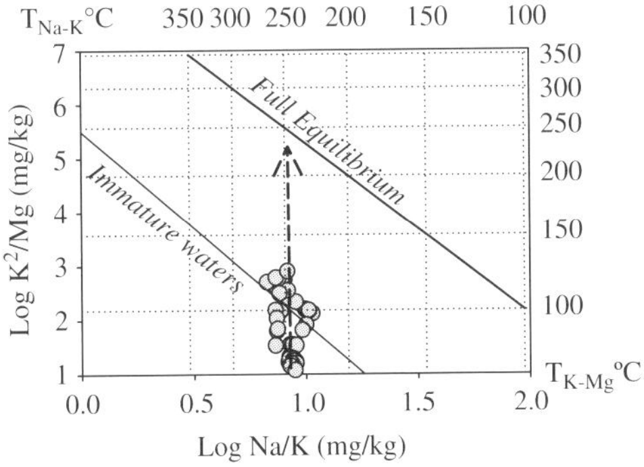

Figure 4. K-Mg and Na-K equilibrium temperatures.

5 THERMAL REGIME

Like many other active crater lakes, Kelud acts as a calorimeter trapping the heat supplied into the lake by subaqueous fumaroles and hot springs. A heat and mass balance model was used to evaluate the changes in the heat supplied to the lake by the hydrothermal fluids. The model is derived from Stevenson (1992) and assumes that the lake is in steady-state equilibrium, i.e. that the lake temperature is constrained

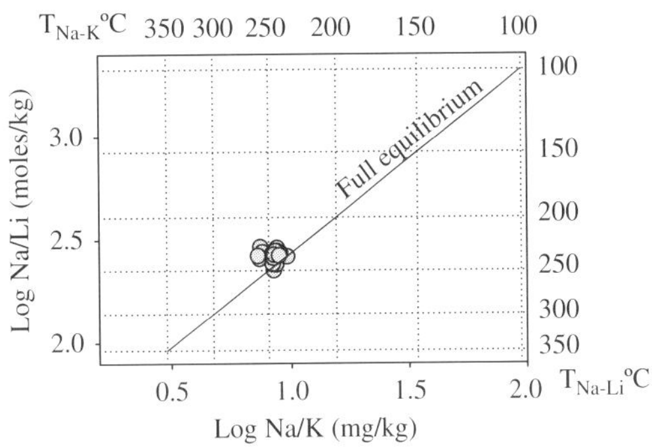

Figure 5. Na-Li and Na-K equilibrium temperatures.

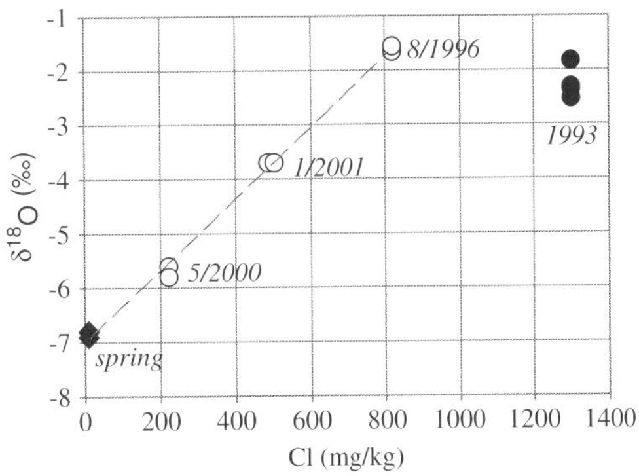

Figure 6. Evolution of the $\delta^{18}O$ (‰) isotopic composition versus chloride for the lake.

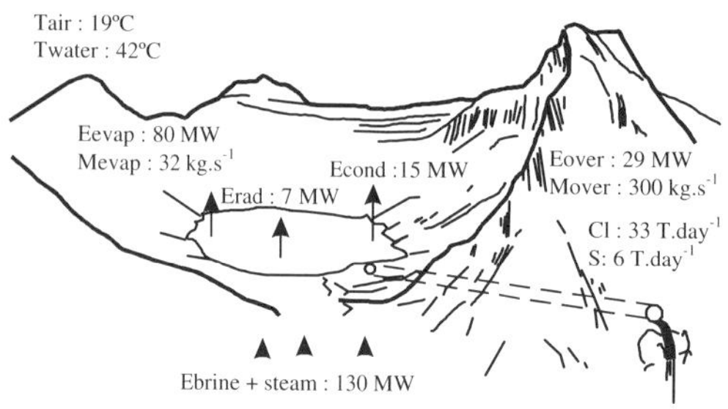

Figure 7. Heat and mass balance model calculated for a lake temperature of 42°C.

by the balance between heat input and output. Heat is derived from the enthalpy of hydrothermal fluids (*brine* + *steam*) entering and mixing within the lake and from solar and atmospheric radiation (*rad*). Heat is lost by evaporation (*evap*), conduction (*cond*), radiation (*rad*) and by the overflow (*over*) of the hot waters through the drainage tunnel. The lake surface radiated a thermal power estimated at 130 MW for the period 1993–1995 (Fig. 7), peaked to 250 MW in 1996 and 2001 when the lake surface reached 50°C and is today close to 80 MW.

6 CO_2 DEGASSING

The CO_2 flux emitted by the surface of the lake was measured during 2001–2003 by IR spectrophotometry using a Dräger Polytron. We modified the technique initially developed for soil gas flux monitoring by Chiodini et al. (1996) in order to work on a crater lake by using a floating accumulation chamber. During each field campaign, about 260 measurements were obtained to cover the entire lake surface (105,000 m^2). Results show a decrease in the CO_2 flux from 30,000 T/year in 2001 to 13,000 T/year in 2003 following the temperature decrease. These results compared to the CO_2 lost as HCO_3^- by the overflow of the lake waters through the drainage tunnel show that most of the CO_2 degasses from the lake surface and escapes to the atmosphere. Only a small fraction (<15%) is trapped and converted as HCO_3^- in the lake waters.

7 CONCLUSION

In the case of Kelud, the lack of acidity and low fluoride content suggest that the input of magmatic volatiles to the hydrothermal system is low or limited. Only the samples collected in 1993 are slightly different in Cl and stable isotopic compositions. This could suggest that a small input of magmatic volatiles was still present 3 years after the eruption.

The emission of neutral-chloride fluids at the top of a volcanic edifice is a relatively rare situation. These neutral-chloride fluids are generally discharged at low elevation on the flanks of the volcanic edifice or are only accessible by well-drilling. For Kelud, the 1990 eruption cleared the vent and permitted the ascent of fluids from the deep neutral-chloride hydrothermal system. Today, this channel opens towards the surface, is progressively clogged and the contribution from the deep Na/K chloride fluids to the lake is decreasing.

REFERENCES

Chiodini, G., Frondini, F. & Raco, B. 1996. Diffuse emission of CO_2 from the Fossa crater, Vulcano Island (Italy): *Bull. Volcanol.* 58: 41–50.

Giggenbach, W.F. 1988. Geothermal solute equilibria. Derivation of Na-K-Mg-Ca geoindicators. *Geochim. Cosmochim. Acta* 52: 2749–2765.

Stevenson, D.S. 1992. *Heat transfer in active volcanoes: models of crater lake systems.* PhD thesis, The Open University, 235p.

Vandemeulebrouck, J., Sabroux, J.C., Halbwachs, M., Surono, Poussielgue, N., Grangeon, J. & Tabbagh, J. 2000. Hydroacoustic noise precursors of the 1990 eruption of Kelut volcano, Indonesia. *J. Volcanol. Geotherm. Res.* 97: 443–456.

Verma, S.P. & Santoyo, E. 1997. New improved equations for Na/K, Na/Li and SiO_2 geothermometers by outlier detection and rejection. *J. Volcanol. Geotherm. Res.* 79: 9–23.

Water–Rock Interaction, Wanty & Seal II (eds)
© 2004 Taylor & Francis Group, London, ISBN 90 5809 641 6

Evidence of mantle derived fluid contributions to the thermal basins of Western Sicily: geotectonic and geodynamic implications

A. Caracausi, R. Favara, F. Italiano, A. Paonita & A. Rizzo
Istituto Nazionale Geofisica e Vulcanologia, sezione di Palermo

P.M. Nuccio
Dept. C.F.T.A., Università di Palermo

ABSTRACT: Geochemical investigations have been carried out over the seismic area of Western Sicily (Italy). The area, crossed by a major seismogenic structure, was severely damaged by the last destructive seismic sequence (Ml 6.4; January 1968). Three hydrothermal basins, located across this seismic belt, were investigated for the chemical and isotopic characteristics of natural waters and dissolved gases. The $^{3}He/^{4}He$ ratios ranging between 0.7 and 2.8 Ra highlight variable contributions of "mantle-derived" He into the different basins, which increase southwards. Although the investigated thermal basins are located over a continental area characterized by thrust tectonics, both the anomalous He isotopic ratios and the estimated fluxes of mantle-derived He reveal the large-scale motion of mantle-derived fluids in the upper crust. The results allow a better understanding of the deep hydrothermal systems, and have improved the knowledge of the local geotectonic and geodynamic framework, which may lead to a deeper insight of the development of local seismogenic processes.

1 INTRODUCTION

Geochemical investigations have been carried out during the last three years, on three hydrothermal systems of Western Sicily (Fig. 1). These systems have anomalously high heat flow. The seismicity of the area is caused by the deep structures crossing Western Sicily, which gave rise to the last destructive seismic crisis of January 1968, known as the Belice Valley earthquake (max Ml 6.4). As physico-chemical variations in some thermal springs of the area were recorded during that swarm of seismic activity (Favara et al. 2001), recent geochemical investigations focused their attention on the geochemistry of the thermal waters and the hydrologic characterization of the hydrothermal basins (Favara et al. 1998, 2001).

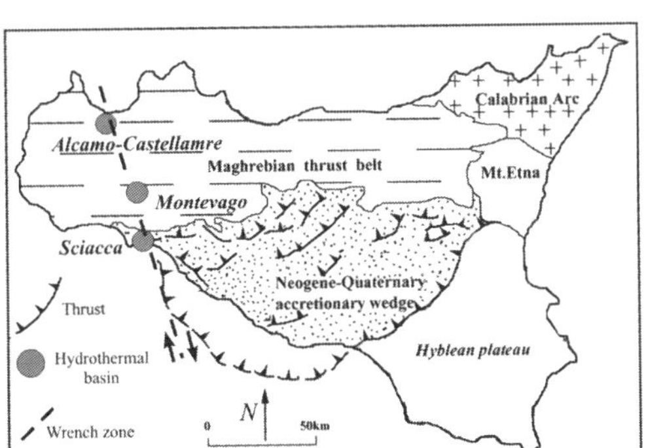

Figure 1. Sketch map of Sicily and sampling sites location.

We have studied both the chemical composition and the He isotopic signatures of the dissolved gases in the sampled thermal waters of three hydrothermal basins named Castellamare-Alcamo, Montevago and Sciacca basins.

Here we present the results of over three years of geochemical investigations, which demonstrate a considerable mantle degassing in all the basins and the relationship of that degassing with the geotectonic framework of the area.

2 GEOLOGICAL SETTING

Western Sicily is a complex geologic area, made by a segment of the "Apennine-Maghrebid Chain" that developed near the boundary between the European and African plates, with a southward vergence

(Catalano et al. 2000). In this area two different structural elements are well identified: Neogene-Pleistocene foredeep basins of Castelvetrano and Caltanissetta, and a foreland area localized from the south-eastern part of Sicily to the Sicily Channel. The Western Sicily chain is characterized by a complex deformational history, recorded by several tectonic events (Catalano et. al. 1995). The chain is characterized by a complex structural framework of thrust systems and the deformation gave rise to a pile of different structures with duplex geometry (Catalano et al. 1995). This area is seismically active, and the occurrence of recent earthquakes, such as Valle del Belice seismic crisis of 1968, is probably related to the transcurrent zone in southwest Sicily and their indirectly connected second-order tectonic element (Michetti et al. 1995).

Three hydrothermal basins are located close to this regional tectonic discontinuity: the Montevago basin crossed by the NWW trending faults system; the Castellammare-Alcamo basin located to the North and the Sciacca basin, to the South. Recent geological-structural findings reveal probable relationships between the structural discontinuities of the Sciacca area and those related to the recent volcanism of the Sicily Channel.

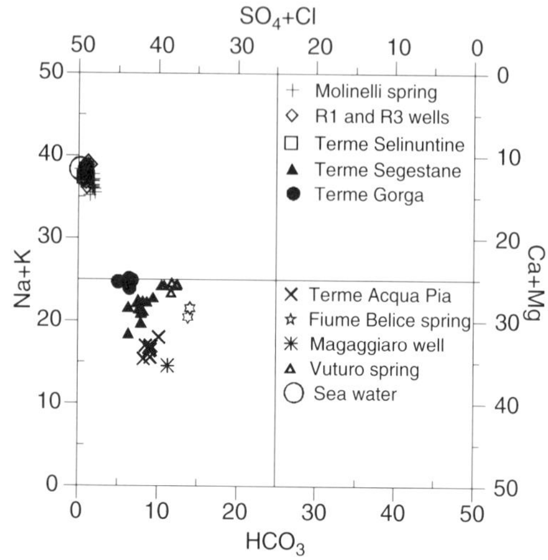

Figure 2. Langelier and Ludwing diagram.

3 FLUID GEOCHEMISTRY

During the last three years, waters of the three hydrothermal basins of Sciacca, Montevago and Castellamare-Alcamo, have been sampled monthly (Fig. 1). The chemical composition of both the thermal waters and dissolved gases, together with the isotopic signature of the dissolved He, allowed a detailed geochemical characterization of the hydrothermal systems.

For a preliminary classification of the waters, the concentrations of the major components are plotted on the Langelier–Ludwing diagram of Figure 2 (Langelier & Ludwing 1942). According to previous studies (Alaimo et al. 1984; Favara et al. 1998), two different families of groundwaters can be identified: chlorine-sulphate alkaline-earth waters (Alcamo and Montevago thermal basins), and chlorine-sulphate alkaline waters (Sciacca thermal basin).

The Na and Cl contents of the Sciacca thermal waters (Fig. 3) allow the different springs to be separated into groups based on proposed variable mixing percentages between sea water and meteoric waters.

Favara et al. (2001) identified two different hydrogeochemical circuits for the above-mentioned hydrothermal systems. The first may explain the Alcamo and Montevago thermal systems, suggesting a deep carbonate reservoir recharged by both carbonatic and selenitic waters in a mixing percentage of

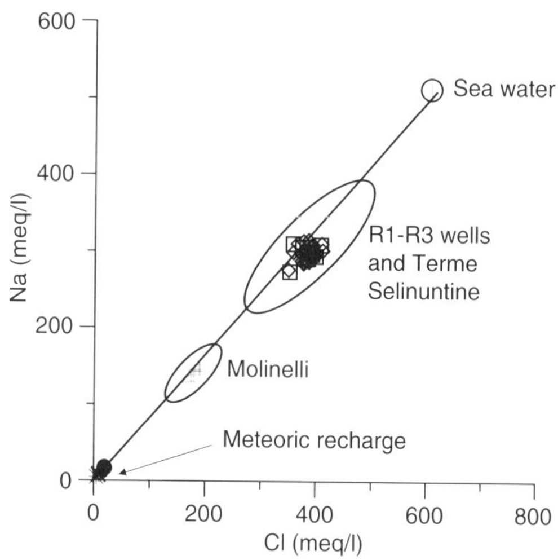

Figure 3. Based on NaCl contributions, the waters of the Sciacca hydrothermal basin show mixing between sea water and groundwaters.

70% and 30% respectively. The other, proposed for the Sciacca thermal system, is hypothesized as a deep carbonate reservoir fed by a mixture of sea water and waters equilibrated with carbonate rocks in the same percentage.

The CO_2-N_2-He diagram (Fig. 4) allows the dissolved gases in the three hydrothermal basins to be differentiated. All the sampled waters contain dissolved gases from two main sources: a shallow source directly linked to the atmosphere (as shown by the O_2

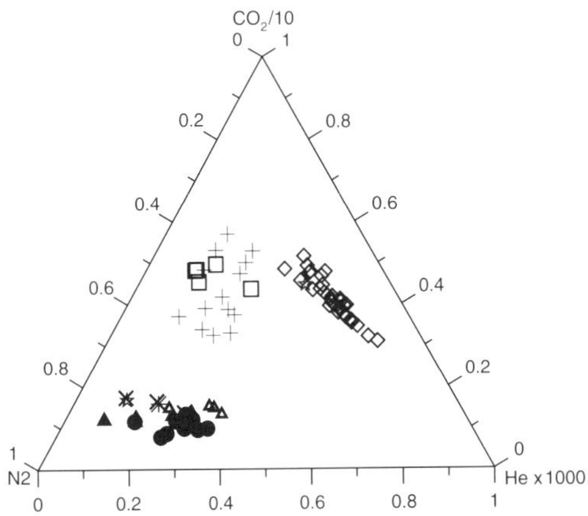

Figure 4. Characterization of the chemical composition of the dissolved gases in the thermal waters.

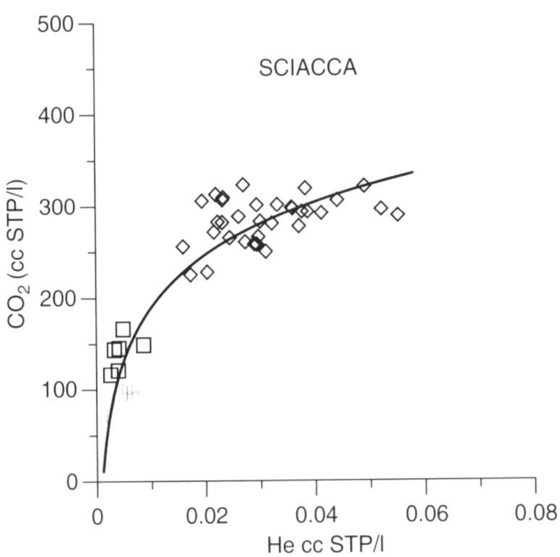

Figure 5. Several samples of Sciacca dissolved gases display He-CO_2 contents that fall along a degassing trend line.

and N_2 concentrations), and a deeper source, marked by the presence of CO_2 or N_2-dominanted gaseous components.

The analytical results highlight how waters from the thermal basins of Western Sicily have a relevant concentration of dissolved He, up to 2–3 orders of magnitude higher than the Air Saturated Water (ASW) standard.

Data collected from thermal springs sampled over the Sciacca basin show a decrease in the abundance of both dissolved He and CO_2 from the R1–R3 wells up to Molinelli hot spring, highlighting the occurrence of degassing at shallow depth (Fig. 5). Some of

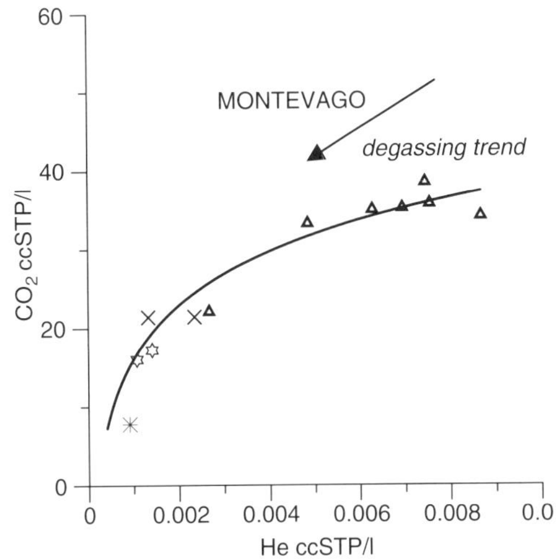

Figure 6. Analogous to Figure 5, samples of dissolved gases from the Montevago area show He-CO_2 contents that fall along a degassing trend line.

the thermal waters released over the Montevago basin also show evidence of degassing (Fig. 6).

4 HELIUM ISOTOPE CHARACTERIZATION

It is well known that the $^3He/^4He$ ratio is a powerful geochemical tool able to recognize possible contributions of mantle He in a continental area (O'Nions & Oxburgh 1983). The collected data allowed us to distinguish He coming from the atmosphere (1 Ra; Ra = $^3He/^4He$ ratio in air) from radiogenically produced He (0.02–0.03 Ra) and He coming from the mantle (about 8 Ra).

The isotopic composition of He dissolved in the studied thermal waters ranges from 0.7 to 2.8 Ra. The high $^4He/^{20}Ne$ ratios demonstrate that air contamination of the sampled hydrothermal waters was negligible. As a consequence it was possible to distinguish the three hydrothermal systems in terms of He mainly provided by only two sources: mantle and crust (Fig. 7). The $^3He/^4He$ isotopic ratio increases from the North (Alcamo area) to the South (Sciacca area). The relatively high $^3He/^4He$ isotope ratios measured in the basins suggest a surface connection with the mantle. Taking into account the general geodynamic framework of the area, marked by compressive tectonics, we suggest that the transcurrent zone (Fig. 1) may provide the link between the local subcontinental mantle and the host rocks of these thermal basins.

Under steady state conditions, the measured R/Ra values can be converted into continental flux of

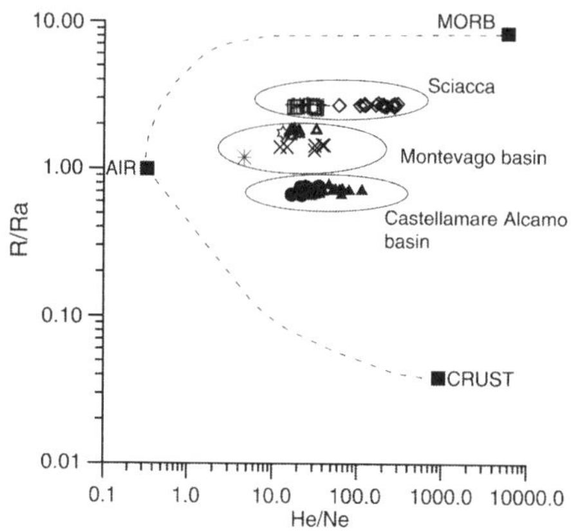

Figure 7. R/Ra versus He/Ne ratios show the variable contributions of atmospheric, radiogenic and mantle He sources.

mantle-derived He following the relation of O'Nions & Oxburgh (1988). The mantle He fluxes in the hydrothermal basins calculated in this way are up to two orders of magnitude higher than those of the typical continental crust, suggesting a probable presence of mantle-derived degassing melts possibly intruded in the crust. On the other hand, magmas intruded through lithospheric discontinuities could represent the major source providing other mantle volatile components such as CO_2, normally the main component among the gases dissolved in the investigated thermal basins.

5 CONCLUSIONS

Geochemical investigations of the thermal basins have revealed information concerning the genesis of thermal fluids discharged in Western Sicily. The hydrothermal systems are characterized by contributions of fluids of deep origin. Both the high He isotope ratio of the dissolved gases and the high values of the estimated mantle He flux allowed us to identify the occurrence of mantle-derived fluids. The high mantle He fluxes cannot be supported just by diffusion, implying an advective transport of He (i.e., gas transport through faults), strongly suggesting the presence of tectonic features.

These results imply the presence of a wrench zone, which probably extends from the north toward the Sicily Channel, possibly providing new tools for better understanding the development of local seismogenic processes.

REFERENCES

Alaimo, R., Carapezza, M., Dongarrà, G. & Hauser, S. 1984. Geochimica delle sorgenti termali siciliane. in *Risorse termali della sicilia e delle isole minori*: 211–225. Regione Siciliana Assessorato Territorio e Ambiente.

Catalano, R., DiStefano, P. & Vitale, F.P. 1995. Structural trends and paleogeography of the central Sicily belt: new insights. *Terranova* 7: 189–199.

Catalano, R., Franchino, A., Merlini, S. & Sulli, A. 2000. Central western Sicily structural setting interpreted from seismic reflection profiles. *Mem. Soc. Geol. It.* 55: 5–16.

Favara, R., Grassa, F., Inguaggiato, S. & D'Amore, F. 1998. Geochemical and hydrogeological characterization of thermal springs in Western Sicily, Italy. *J. Volcanol. Geotherm. Res.* 84: 125–141.

Favara, R., Grassa, F., Inguaggiato, S. & Valenza, M. 2001. Hydrogeochemistry and stable isotopes of thermal springs: earthquake-related chemical changes along the Belice Fault (Western Sicily). *Appl. Geochem.* 16: 1–17

Langelier, W.F. & Ludwing, H.F. 1942. Graphical methods for indicating the mineral character of natural waters. *J. Am. Water Works Assoc.* 34, 335

Michetti, A.M., Brunamonte, F. & Serva, L. 1995. Paleoseismological evidence in the Epicentral Area of the January 1968 Earthquakes, Belice, Southwestern Sicily. in *Perspectives in Paleoseismology, Association of Engineering Geologists*, Special Publication 6: 127–139.

O'Nions, R.K. & Oxburgh, E.R. 1983. Heat and helium in the Earth. *Nature* 306: 429–431.

O'Nions, R.K. & Oxburgh, E.R. 1988. Helium, volatile fluxes and the development of continental crust. *Earth Plan. Sci. Lett.* 90: 331–347.

Water–Rock Interaction, Wanty & Seal II (eds)
© 2004 Taylor & Francis Group, London, ISBN 90 5809 641 6

Experimentation of an automatic continuous monitoring station in a thermal well of Stromboli volcanic island (Italy)

M.L. Carapezza
Istituto Nazionale di Geofisica e Vulcanologia, sezione di Roma 1, Italy

R. Cioni, M. Guidi & A. Scozzari
Istituto di Geoscienze e Georisorse, CNR, Pisa, Italy

L. Pruiti
Istituto Nazionale di Geofisica e Vulcanologia, sezione di Catania, Italy

ABSTRACT: Chemical–physical modifications of groundwater in active volcanic areas provide important insights on magma ascent leading to volcanic eruptions. To fully exploit the great potential of fluid geochemistry, the traditional discontinuous sampling with subsequent laboratory analysis should be replaced by automatic continuous data recording stations. Such a station, that measures every second pH, Eh, conductivity, T and dissolved CO_2 and CH_4 contents, has been experimentally tested in a shallow thermal well of Stromboli for the first time. A strong explosive paroxysm occurred during the test. Before the event variations in T, pH and dissolved CO_2 indicative of an increasing gas influx were recorded by the station.

1 INTRODUCTION

In quiescent volcanoes the transition from a repose period to an eruption, when magma rises toward the surface, is accompanied by relevant geophysical and geochemical changes (i.e. anomalous seismicity, ground uplift, variations in the gravimetric, magnetic and electric fields, increase in the gas output, chemical modifications and temperature increase in the fumaroles and thermal waters; Scarpa & Tilling 1996). Fluid geochemistry is very promising in forecasting eruptions, as gases initially dissolved into the magmatic melt are released because of depressurization during magma ascent. Being much more mobile than the magma releasing them, gases reach the surface well before the eruption onset, producing early geochemical anomalies that precede the appearance of geophysical anomalies, as observed during volcanic unrest at Campi Flegrei (Barberi & Carapezza 1996) and Colima (Taran et al. 2002). The full exploitation of the geochemistry potential in volcanic monitoring is limited by the fact that, whereas geophysicists have long since developed continuous automatic recording systems, current geochemical methods are still largely based on discontinuous sampling of gases and waters or remote sensing measurements. A remarkable exception is the semi-continuous control of CO_2 soil flux that has been introduced in the monitoring system of several volcanoes (Badalamenti et al. 1994; Bergfeld et al. 2001). Here we present the results of a 2.5-month experiment, in a shallow thermal well of Stromboli volcano, of a continuous station that automatically records several physical-chemical parameters of the thermal water. An important explosive paroxysm occurred during the experiment that was heralded by some significant geochemical changes in the water recorded by the station.

2 GEOCHEMICAL BACKGROUND OF STROMBOLI

Stromboli volcano in the Aeolian archipelago (Southern Tyrrhenian Sea, Italy) is famous for its persistent mild explosive activity (e.g. Strombolian activity) characterized by very frequent low-energy explosions (on average one every 10–20 minutes). This normal activity is episodically interrupted by more violent and hazardous explosions whose ejecta and glowing avalanches also may affect the settled areas of the island. Lava effusion and tsunami generated by flank collapses also frequently occur during the most energetic eruptions (Barberi et al. 1993), the last of which began on 28 December 2002 and ended on 6 July 2003 (Bonaccorso et al. 2003).

A huge quantity of gas is continuously emitted from the summit craters. Total gas output has been estimated to vary from 6000 to 12,000 tons/day depending on the level of activity. The gas plume is mostly made of steam (over 50 wt %), followed by CO_2, SO_2 and minor HCl and HF (Allard et al. 1994). Zones of strong soil degassing have been identified by a CO_2 soil flux survey (Carapezza & Federico 2000). They occur both in the crater area rim and at the NE base of the cone, near the coast (Pizzillo zone). In both sites an automatic station recording the CO_2 soil flux (one measurement every hour) has been installed (Carapezza et al. 2002). A crater fumarole with a temperature of 95°C, the steam condensation point for the elevation, provided information on the magmatic gas composition. It has a CO_2 content (dry gas) of 98 vol %, appreciable contents of H_2 (6–700 ppm) and He (5–20 ppm), $^3He/^4He$ (R/Ra) of 2.7–3.5, $\delta^{13}C_{CO_2}$ of −2‰ (Carapezza & Federico 2000, Carapezza et al. 2002). In January 2003, gas at 240°C was sampled from a newly formed fracture in the crater area. The isotopic analysis of $^3He/^4He$ gave a value of 4.2 (R/Ra).

Thermal waters are found in shallow wells (depths of 5–16 m) in the NE coastal area (Stromboli village). Water temperature ranges from 25 to 42°C. Chemistry indicates that these waters are the result of mixing, in various proportions, of sea and meteoric waters (Carapezza & Inguaggiato 2001; Fig. 1). Evidence of a strong interaction of the thermal aquifer with a deep CO_2-rich gas of magmatic origin is provided by the high dissolved CO_2 content (Fig. 1), the isotopic composition of dissolved He (R/Ra up to 4.2) and of dissolved carbon that, as dissolved CO_2 content increases, tends to approach the isotopic composition of the CO_2 crater fumarole (Carapezza & Federico 2000; Carapezza et al. 2004; Capasso et al. 2004). These shallow thermal wells are ideal sites to test an automatic monitoring station. The privately owned Fulco well was selected for this study. It is located near the anomalous degassing site of Pizzillo; it has the greatest depth (16 m), is less contaminated by sea water and shows the highest dissolved CO_2 content (Fig. 1).

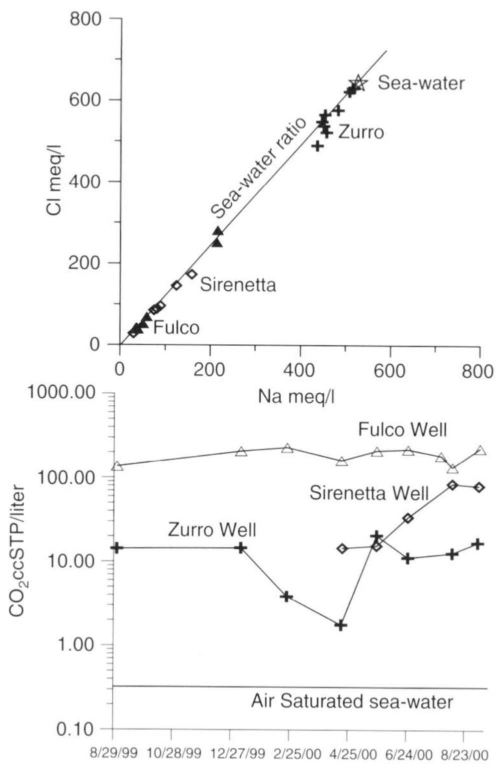

Figure 1. Na/Cl ratio and dissolved CO_2 in the thermal wells of Stromboli (after Carapezza & Inguaggiato 2001).

CO_2 and CH_4 dissolved in water (see Table 1) were monitored. During the experiment, the CH_4 always has been below the detection limit. Particular attention has been devoted to determination of dissolved CO_2 content. CO_2 is in fact, after water, the main volatile component of magma. It has a low solubility in magmatic melts (Stolper & Holloway 1988; Papale 1999) and is then released before other more soluble gas species when magma moves toward the surface. An increase of CO_2 output from depth, as indicated by anomalously high soil flux values, or by an increase of CO_2 dissolved in thermal water, could, therefore, provide early evidence of an ongoing magma ascent.

The monitoring station is made of two separated units dedicated respectively to the measurement of physical–chemical water parameters and of dissolved gas concentrations. It operates with flowing water (about 5 litres per minute) that, extracted from the well by a submersible pump, is sent to the two units by separate paths. In the first unit there are two cells where conductivity (first cell), T, pH and Eh (second cell) are measured (Fig. 2). In the cell of unit 2 (Fig. 3), gas is equilibrated by circulation in the headspace

3 THE MONITORING STATION

The station has been designed by the Institute of Geoscience and Earth Resources, CNR, Pisa. The Stromboli experiment, described here, is its first test in an active volcanic area. Temperature (T), pH, conductivity, redox potential (Eh) and the content of

Table 1. Measurement sensors.

Parameter	Sensor	Analytical range	Accuracy
pH	HA405-DXK-8/120 comb. Xerolit glass electrode (Ag/AgCl ref.)	2–14	0.03*
Eh	Pt4805-DXK-8/120 comb. platinum electrode	−1250 to +1250 mV	1.5* mV
Cond.	HT8936 four electrode configuration (Pt rings)	0–19.99 mS · cm^{-1}	2%
T	PT100 class A (BS1904 standard)	0–100°C	0.15°C at 0°C
CO_2	IR cell	0–10 vol %	2%
CH_4	IR cell	0–10 vol %	2%

* maximum deviation from the room temperature calibration.

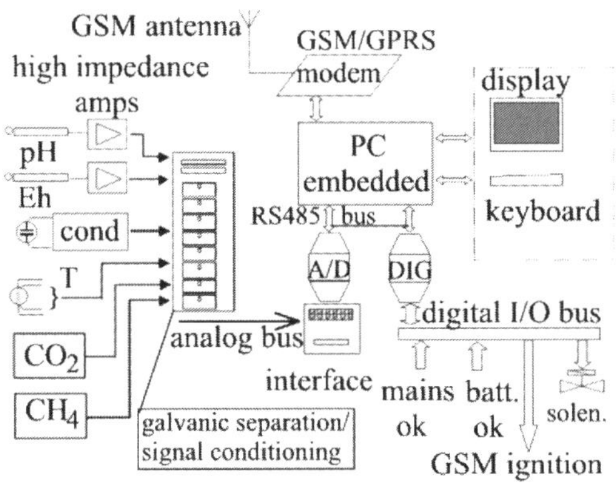

Figure 2. Simplified block diagram of the station.

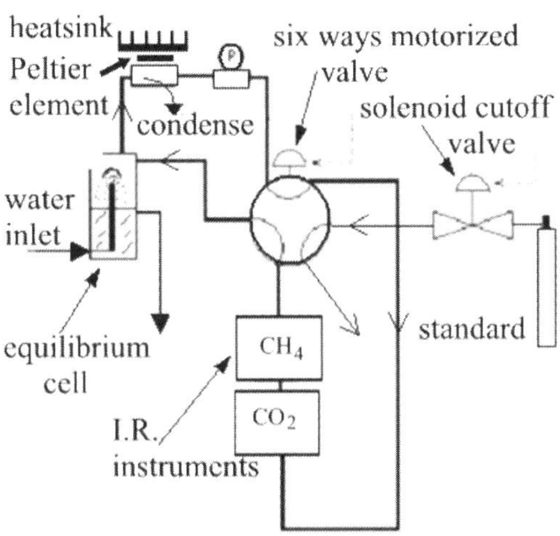

Figure 3. Measurement loop for dissolved gases.

above the water level inside the cell, within a closed loop which includes two I.R. spectrometers and a low-flow pump, which ensures the necessary gas circulation in the loop. The system is calibrated using a standard gas. A six-way valve is configured in such a way that the measurement loop is not interrupted, either during the calibration procedure, or during the measurements. A solenoid valve is added to cut the standard gas off, while the system is in measurement configuration.

The submersible pump has been chosen to withstand the temperature and the chemical aggressiveness of the water: wet parts are made of a plastic self-lubricating material able to work up to 90°C and suitable delivery/head curve has been selected to obtain the required water flow without creating overheating problems for the engine. Water adduction tubes and cells are thermally insulated. A simplified block diagram of the electronics appears in Figure 2.

All the signals are isolated and conditioned via a fifth order low pass filter with a high DC accuracy, and come from the sensors to the signal conditioning block either directly or via analog front-end. For the high-impedance electrodes (pH and Eh measurement), a specifically designed low-input loss-amplifier is used, which provides three port isolation and operates over a wide temperature range. Particular care has been taken to minimise the effect of ambient temperature drifts on pH measurement. The core of the data logging system is a Personal Computer in PC/104 format, which runs a program that controls all the activities of the station via the onboard serial ports. An RS485 port is used to communicate with the two data conversion devices, which perform all the input/output activities; one device (A/D) performs analog to digital conversion at the input signals, while the other one (DIG) manages the digital input/output signals to control loads such as solenoid and six-way valves, to switch the modem on and to check the main power line and batteries state. Data can be downloaded both onsite and remotely, via a GSM/GPRS modem connected directly to the PC; the remote link also allows us to see real-time values for all the parameters and to interact with the station in order to check and modify its internal parameters. Data are acquired every second; the average value and variance over 5 minutes are computed and recorded in a local removable non-volatile memory (Compact Flash card). The embedded PC performs a daily calibration opening the cut-off valve and switching the six-way valve, so that the standard gas enters into the measuring devices. Liquid water, in the form of small droplets or formed by condensation, can never enter the optical measurement cells, as it would induce large errors and eventually damage the cells. To prevent this problem, the gas temperature is maintained below the minimum expected temperature within the cells by a Peltier element (solid state heat pump) placed on the cold plate of a small separator (Fig. 3). This device allows separation of condensed water before the optical measurements cells.

4 RESULTS

The station was installed on 4 February 2003 and after two days of calibration it operated at full capacity until 10 April. Then the station had to be removed at the request of the well owner and was reinstalled a few months later into a dedicated thermal water well drilled on public land for this study. The February–April data are shown in Figure 4.

During this period a lava flow was continuously emitted from the volcano and on 5 April a violent explosive paroxysm occurred. The station worked nicely, regularly recording the five chemical–physical parameters. The main technical problem was related to the deposition of Fe-hydroxide from the circulating water. This modified the cell constant of the sensor utilized to measure the conductivity, affected the response of the Eh electrode and the measurements of dissolved CO_2 content, as the water flow into the cell was progressively reduced by the scaling. The pH electrode response was not influenced by the Fe-hydroxide deposition, nor was the temperature. Data reported in Figure 4 are the values corrected upon calibration and real water flow. All signals show a periodic component with a clear 12-hour period, likely due to tide phenomena affecting the near-shore aquifer.

It is interesting to observe that in the period preceding the 5 April paroxysm, T and disolved CO_2 increased and pH decreased without any conductivity variations. These trends reversed soon after the explosion, as confirmed also by discontinuous analyses (Carapezza et al. 2004). These variations indicate that the monitored thermal water was affected, before the explosion, by an increased input of deep acid gas, likely released by an ascending body of gas-rich magma that explosively erupted on 5 April.

In conclusion, the Stromboli test proved the reliability of the continuous station described here and

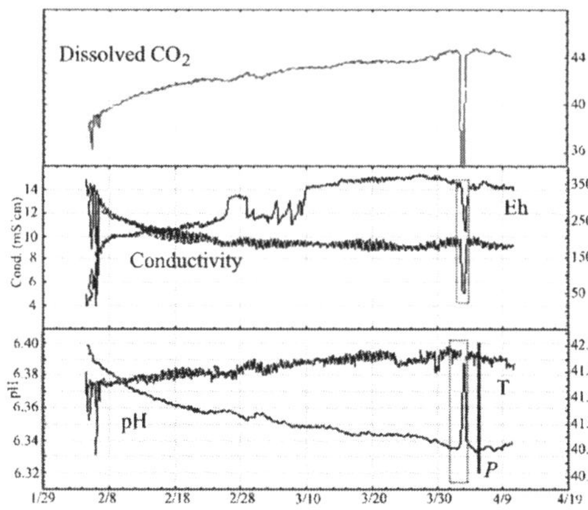

Figure 4. Variation with time (m/d) of the recorded parameters. *P* indicates the explosive paroxysm. Peaks within the dotted areas are due to calibration.

confirms the importance of developing automated geochemical tools for volcano (and seismic) monitoring.

REFERENCES

Allard, P., Carbonnelle, J., Metrich, M., Loyer, H. & Zettwoog, P. 1994. Sulphur output and magma degassing budget of Stromboli volcano. *Nature* 368: 326–330.

Badalamenti, B., Gurrieri, S. & Valenza, M. 1994. Continuous monitoring (temperature, CO_2 in soil gases and reducing capacity; Vulcano). In: Data related to eruptive activity, unrest phenomena and other observation on the Italian active volcanoes in 1992. *Acta Vulcanol.* 6: 46–48.

Barberi, F., Rosi, M. & Sodi, A. 1993. Volcanic hazard assessment at Stromboli based on review of historical data. *Acta Vulcanol.* 3: 173–187.

Barberi, F. & Carapezza, M.L. 1996. The problem of volcanic unrest: the Campi Flegrei case history. In: R. Scarpa & R. Tilling (eds.), *Monitoring and Mitigation of Volcano Hazards*: 771–786. Berlin Heidelberg: Springer Verlag.

Bergfeld, D., Goff, F. & Allard, P. (eds.) 2001. High CO_2 flux measurements in volcanic and geothermal areas, methodologies and results. *Chem. Geol.* 177.

Bonaccorso, A., Calvari, S., Garfì, G., Lodato, L. & Patanè, D. 2003. Dynamics of the December 2002 flank failure and tsunami at Stromboli volcano inferred by volcanological and geophysical observations. *Geophys. Res. Lett.* 30(18): 1941 doi: 10, 1029/2003GLO17702.

Capasso, G., Carapezza, M.L., Federico, C., Inguaggiato, S. & Rizzo, A. 2004. Carbon and helium isotopic variations in Stromboli fluids: early evidences of magma upraise in the 2002–2003 eruption. *Bull. Volcanol.*(submitted).

Carapezza, M.L. & Federico, C. 2000. The contribution of fluid geochemistry to the volcano monitoring of Stromboli. *J. Volcanol. Geotherm. Res.*95: 227–245.

Carapezza, M.L., Guidi, M. & Inguaggiato, S. 2002. Soil CO_2 degassing on Stromboli Island: influence of environmental parameters and insights on volcanic activity. *Geophys. Res. Abstract* EGS 27th General Assembly. Nice.

Carapezza, M.L. & Inguaggiato, S. 2001. Interaction between thermal waters and CO_2-rich fluids at Stromboli volcano (Italy). *Proc. WRI-10*, 2: 791–794. Rotterdam: Balkema.

Carapezza, M.L., Inguaggiato, S., Brusca, L. & Longo, M. 2004. Geochemical precursors of Stromboli 2002–2003 eruptive events. *Geophys. Res. Lett.* (submitted).

Papale, P. 1999. Modelling of the solubility of two-component H_2O-CO_2 fluid in silicate liquids. *Am. Mineral.* 84: 477–492.

Scarpa, R. & Tilling, A. (eds.) 1996. *Monitoring of volcanic hazards*. Berlin Heidelberg: Springer-Verlag.

Stolper, E. & Holloway, J.R. 1988. Experimental determination of the solubility of carbon dioxide in molten basalt at low pressure. *Earth Planet. Sci. Lett.* 87: 397–408.

Taran, Y., Gavilanes, J.C. & Cortés, A. 2002. Chemical and isotopic composition of fumarolic gases and the SO_2 flux from Volcàn de Colima, México, between the 1994 and 1998 eruptions. *J. Volcanol. Geotherm. Res.* 197: 105–119.

Water–Rock Interaction, Wanty & Seal II (eds)
© 2004 Taylor & Francis Group, London, ISBN 90 5809 641 6

Diffuse soil CO_2 degassing at Salcheto (Tuscany) and Lipary (Sicily): a probabilistic approach to mapping and quantification of gas release

C. Cardellini, F. Frondini & N. Morgantini
Dipartimento di Scienze della Terra, Università di Perugia, Perugia, Italy

G. Chiodini, R. Avino, S. Caliro, D. Granieri & M. Russo
Osservatorio Vesuviano – INGV, Napoli, Italy

ABSTRACT: Conditional sequential Gaussian simulations (sGs) have been applied to the study of soil diffuse degassing from a volcanic area of Lipari (Sicily) and a non-volcanic cold gas emission at Salcheto (Tuscany). The sGs algorithm was used to generate 100 realizations of CO_2 flux for each area. Simulation results were used to: (i) map the degassing structures, (ii) calculate the total CO_2 emission rate, and (iii) quantify the uncertainty of the estimation. The sGs is a suitable tool to model soil diffuse degassing, producing realistic images of the distribution of the CO_2 fluxes that honor the statistics of the original data. Previously unknown tectonic structures degassing moderately large quantities of CO_2 were discovered within both study areas.

1 INTRODUCTION

Recently, a great interest has been addressed to the study of earth CO_2 degassing. There are many objectives to such studies; e.g. the definition of the relations between the flux and the tectonic structures (Etiope et al. 1999; Lewicki & Brantley 2000), the quantification of deeply derived CO_2 release to the atmosphere in the framework of the carbon global budget (Allard et al. 1991; Brantley & Koepenick 1995; Kerrick et al. 1995; Seward & Kerrick 1996; Marty & Tolstikhin 1998), and the study of volcanic degassing. In particular, numerous studies have been focused on the diffuse soil CO_2 degassing from quiescent active volcanoes (e.g. Farrar et al. 1995; Brombach et al. 2001; Chiodini et al. 1998, 2001; Hernandez et al. 1998; Gerlach et al. 2001; Salazar et al. 2001). The mapping of diffuse degassing structures and the quantification of the amount of CO_2 released and the uncertainty associated with this estimate are common objectives of all of these studies. Mapping of soil CO_2 fluxes has mainly been performed using interpolation algorithms, usually kriging, that produce a set of estimated values at the unsampled locations. The variogram associated with these values does not match that of the original data and smoothes out the extreme extrapolated values. Also, the total CO_2 emission rate from an area has usually been calculated by (1) multiplying the arithmetic mean value of CO_2 fluxes by the surveyed area, (2) applying volume and area integration algorithms to the grids produced to contour the CO_2 flux, or (3) a graphical statistical approach independent from any interpolation technique (GSA, Chiodini et al. 1998). Together with the quantification of total CO_2 release, the definition of the related uncertainty is essential, especially in volcanic surveillance where the recognition of anomalous states of degassing is a goal. All of the commonly used methods provide an incomplete measure of this estimation uncertainty.

An alternative approach for mapping and estimating the gas release, based on a sequential Gaussian simulation algorithm (Deutsch & Journel 1998), was recently proposed by Cardellini et al. (2003). This approach (sGs method) is a suitable tool to model soil diffuse degassing, producing realistic images of the distribution of the CO_2 fluxes that honor the histogram and variogram of the original data. Furthermore, sGs allows one to estimate the uncertainty associated with the total CO_2 emission rate calculated.

The aim of this work is the application of the sGs method to soil gas flux data from the Faurdo area, located in the volcanic island of Lipari (Sicily), and from a non-volcanic cold gas emission at Salcheto (Tuscany). We map the degassing areas and evaluate the total CO_2 emitted from these areas and the uncertainty associated with these estimates.

2 METHODS

2.1 *Experimental data sets*

Four hundred and eleven soil CO_2 flux measurements were made at 10-m spacing along a grid at the Faurdo site. At the Salcheto site, 153 measurements were made along a grid at 15-m spacing.

The soil CO_2 flux was measured with the accumulation chamber "time 0" method (Chiodini et al. 1996; Chiodini et al. 1998) using a prototype chamber with a diameter of 20 cm equipped with a LI-COR LI-800 infrared CO_2 analyzer with a resolution of 0–20,000 ppm. The sensitivity of this equipment, in terms of CO_2 fluxes, is better than $0.1\,g\,m^{-2}\,d^{-1}$.

2.2 *Stochastic simulation*

SGs operates considering CO_2 fluxes from soil as the realization of a stationary multivariate Gaussian random function. Soil CO_2 fluxes are simulated at locations defined by a grid covering the area of interest. The simulation is conditional and sequential, i.e., the variable is simulated at each unsampled location by random sampling of a Gaussian conditional cumulative distribution function defined on the basis of original data and of previously simulated data within its neighborhood. As the sGs procedure needs a multigaussian distribution, CO_2 fluxes have to be transformed into a normal distribution (i.e., normal-score transform). The transformed data are then used in the simulation procedure. Simple kriging estimate and variance, computed according to the semivariogram model of normal scores, are used to define at each location a Gaussian conditional cumulative distribution function. A random value is drawn from the conditional cumulative distribution as one "reasonable" simulated value for that location. Once a value is simulated, it is added to the data set and can be used together with the original data to estimate the variable at the next locations of the grid. The simulation proceeds to the next grid location and loop until all nodes are simulated. Afterwards, the simulated normal scores are back transformed into values expressed in the original data unit, applying the inverse of the normal score transform. Changing the random path of grid nodes visited, by changing the starting point of simulation procedure (random-number seed), N alternative simulations can be performed and N equiprobable realizations can be drawn, each honoring the sampled data at their locations and reproducing the univariate statistics and bivariate properties of the data, within reasonable ergodic fluctuations (Deutsch & Journel 1998; Cardellini et al. 2003). The total CO_2 emission rate is computed for each realization by multiplying the simulated flux for each grid cell by the cell area and summing these products. The mean and the standard deviation of the total CO_2 emission rates computed for the N realizations are assumed to be the characteristic CO_2 emission rate for the study area and its uncertainty, respectively. The set of realizations is also used to draw the map of the CO_2 values "expected" at the grid cells (E-type estimates) obtained by a point-wise linear average of all of the realizations.

3 APPLICATIONS AND RESULTS

3.1 *Faurdo area*

Soil CO_2 fluxes measured in the Faurdo area ranged from 10 to $12{,}300\,g\,m^{-2}\,d^{-1}$. For this area, 100 simulations were performed using a simulation grid with 1-m spacing. The simulations were performed considering a spherical variogram model defined on the basis of the experimental variogram of normal scores (Fig. 1).

The map of the mean CO_2 flux values (Fig. 2) shows the presence of two well-defined anomalous degassing areas oriented ENE-WSW. Hot soils are present here and the chemical composition of soil gases is similar to the composition of Lipari fumaroles (Table 1), suggesting a hydrothermal origin of Faurdo soil gases. The total CO_2 released by the studied area is $6.6 \pm 0.38\,t\,d^{-1}$.

3.2 *Salcheto area*

Soil CO_2 fluxes measured in the Salcheto area ranged from 0.6 to $2500\,g\,m^{-2}\,d^{-1}$. One hundred simulations were performed using a simulation grid with 2-m spacing. Figure 3 shows the variogram model used for the simulations. In this area, the CO_2 fluxes show the presence of a degassing structure larger than the small area were the gas vents are located (Fig. 4). The chemical composition of the gas and the isotopic composition of CO_2 emitted by a vent located in the study area (Table 1) suggest a deep origin for the gas released by the soil.

The total CO_2 released by the studied area is $36 \pm 5.1\,t\,d^{-1}$.

4 CONCLUSIONS AND DISCUSSION

The diffuse degassing of CO_2 from two previously unknown degassing structures, one located in a

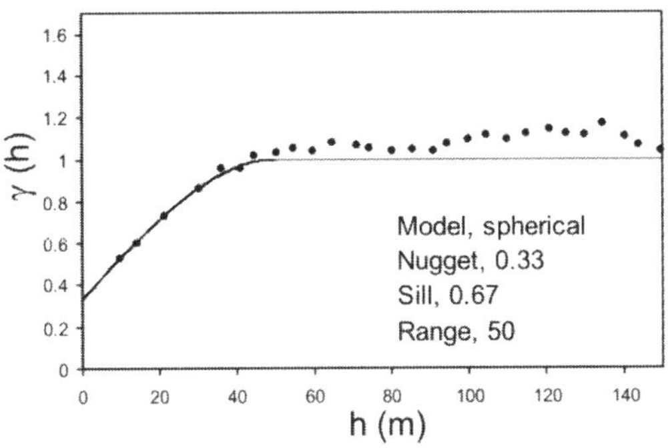

Figure 1. Experimental variogram and variogram model of the normal scores of the Faurdo CO_2 flux data.

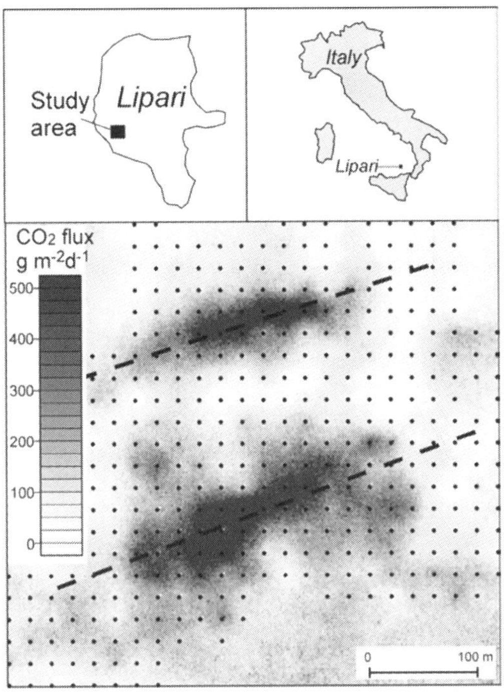

Figure 2. Map of CO_2 flux at Faurdo (Lipari). The gas of deep origin is released by two fractures (dashed lines) oriented ENE-WSW associated to a well known regional NNW-SSE strike-slip fault.

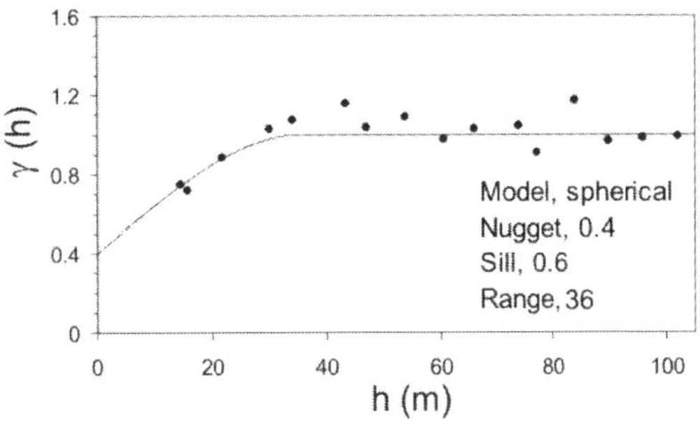

Figure 3. Experimental variogram and variogram model of the normal scores of the Salcheto CO_2 flux data.

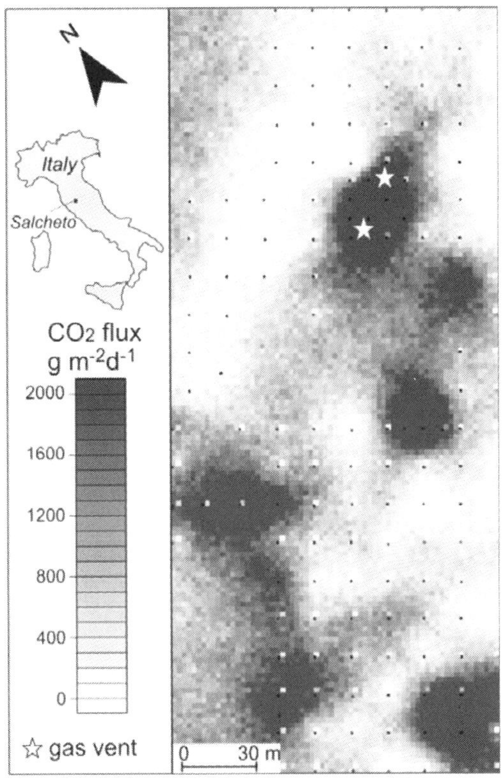

Figure 4. Map of CO_2 flux at Salcheto (Tuscany).

non-volcanic environment in Tuscany and the other

Table 1. Chemical and isotopic composition of fumaroles, vents and soil gasses (μmol/mol).

Sample	CO_2	H_2S	Ar	N_2	CH_4	H_2	He	$\delta^{13}C_{CO_2}$
Faurdo (soil)	935,000	0	150	18,500	45,000	5.9	2.76	n.a.
Lipari (fumarole)	94,600	884	105	13,500	38,700	609	2.37	n.a.
Salcheto (vent)	970,600	75.9	7.96	24,000	5330	6.72	8.75	−4.7

at Lipari volcano, was studied by means of detailed measurement campaigns performed with the accumulation chamber method. The results were treated with the stochastic simulation methodology (sequential Gaussian simulation, sGs) which was recently applied for the first time to the study of diffuse CO_2 flux from soils (Cardellini et al. 2003). This method allowed us to define and characterize the diffuse degassing structures affecting the Faurdo and Salcheto areas.

At Faurdo (Lipari), the data treated with sGs highlighted the presence of two linear features, likely faults or fractures, whose location was previously unknown. The total emission rate of CO_2 here is estimated to be $6.4\,t\,d^{-1}$, with a standard deviation of only $0.4\,t\,d^{-1}$ (6%). This low uncertainty of the total CO_2 emission rate, computed on the basis of sGs algorithms, depends on the detailed sampling design adopted at Faurdo. In the other study area (Salcheto, Tuscany) the total amount of gas released is estimated to be $36\,t\,d^{-1}$ with a standard deviation of $5.1\,t\,d^{-1}$ (14%). This estimated uncertainty is higher than at Faurdo due to the different sampling design employed at Salcheto, i.e., a lower number of samples and sampling density.

The presence of areas of anomalous CO_2 degassing, also if there are not macroscopic evidences (e.g., no fumaroles, no bare soils), may indicate the occurrence of a regional scale diffuse degassing. The magnitude of this degassing process is still largely unknown also because the numerous "small" degassing structures, like Faurdo and Salcheto, have not been studied. Although the amount of CO_2 released from the two areas is not itself "globally" significant, compared with other known degassing structures (e.g. Rogie et al. 2000; Chiodini et al. 2001; Cardellini et al. 2003), the study of such areas is at least "locally" important. Throughout central Italy, a relevant gas hazard is present; CO_2 degassing areas can constitute lethal traps for humans and nimals when high concentrations of CO_2 build in topographic depressions (Rogie et al. 2000). As a result, a better knowledge of the locations of these degassing areas and of the magnitude of CO_2 emissions from these areas are important tools to mitigate the gas hazard. Moreover, in the framework of the volcanic surveillance, areas like Faurdo can represent suitable areas for the monitoring of volcanic activity.

REFERENCES

Allard, P. et al. 1991. Eruptive & diffuse emissions of CO_2 from Mount Etna. *Nature* 351(6325): 387–391.

Brantley, S.L. & Koepenick, K.W. 1995. Measured carbon-dioxide emissions from Oldoinyo-Lengai and the skewed distribution of passive volcanic fluxes. *Geol.* 23(10): 933–936.

Brombach, T., Hunziker, J.C., Chiodini, G., Cardellini, C. & Marini, L. 2001. Soil diffuse degassing and thermal energy fluxes from the southern Lakki plain, Nisyros (Greece). *Geophys. Res. Lett.* 28(1): 69–72.

Cardellini, C., Chiodini, G. & Frondini, F. 2003 Application of Stochastic simulation to CO_2 flux from soil: Mapping and quantification of gas release *J. Geophys. Res.* 2425: doi: 10.1029/2002JB002165.

Chiodini, G., Cioni, R., Guidi, M., Raco, B. & Marini, L. 1998. Soil CO_2 flux measurements in volcanic and geothermal areas. *Appl. Geochem.* 13(5): 543–552.

Chiodini, G. et al. 2001. CO_2 degassing and energy release at Solfatara volcano, Campi Flegrei, Italy. J. *Geophys. Res.* 106(B8): 16213–16221.

Deutsch, C.V. & Journel, A.G. 1998. GSLIB: *Geostatistical software library and users guide. Applied geostatistical series.* New York: Oxford University Press.

Etiope, G. et al. 1999. Structural pattern and CO_2-CH4 degassing of Ustica Island, Southern Tyrrhenian basin. *J. Volcanol. Geotherm. Res.* 88(4): 291–304.

Farrar, C.D. et al. 1995. Forest-killing diffuse CO_2 emission at Mammoth Mountain as a sign of magmatic unrest. *Nature* 376(6542): 675–678.

Gerlach, T.M., Doukas, M.P., McGee, K.A. & Kessler, R. 2001. Soil efflux and total emission rates of magmatic CO_2 at the Horseshoe Lake tree kill, Mammoth Mountain, California, 1995–1999. *Chem. Geol.* 177(1–2): 101–116.

Hernandez, P.A. et al. 1998. Diffuse emission of carbon dioxide, methane, and helium-3 from Teide volcano, Tenerife, Canary Islands. *Geophys. Res. Lett.* 25(17): 3311–3314.

Kerrick, D.M., McKibben, M.A., Seward, T.M. & Caldeira, K. 1995. Convective hydrothermal CO_2 emission from high heat-flow regions. *Chem. Geol.* 121(1–4): 285–293.

Lewicki, J.L. & Brantley, S.L. 2000. CO_2 degassing along the San Andreas Fault, Parkfield, California. *Geophys. Res. Lett.* 27(1): 5–8.

Marty, B. & Tolstikhin, I.N. 1998. CO_2 fluxes from mid-ocean ridges, arcs and plumes. *Chem. Geol.* 145(3–4): 233–248.

Rogie, J.D., Kerrick, D.M., Chiodini, G. & Frondini, F. 2000. Flux measurements of nonvolcanic CO_2 emission from some vents in central Italy, *J. Geophys. Res* 105 (B4): 8435–8445.

Salazar, J.M.L. et al. 2001. Diffuse emission of carbon dioxide from Cerro Negro volcano, Nicaragua, Central America. *Geophys. Res. Lett.* 28(22): 4275–4278.

Seward, T.M. & Kerrick, D.M. 1996. Hydrothermal CO_2 emission from the Taupo Volcanic Zone, New Zealand. *Earth Planet. Sci. Lett.* 139(1–2): 105–113.

Water–Rock Interaction, Wanty & Seal II (eds)

Chemical composition and formation of thermal waters in Kuril's Islands (Far East Russia)

O.V. Chudaev & V.A. Chudaeva
Far East Geological Institute, Pacific Institute of Geography FEB, RAS, Vladivostok, Russia

K. Sugimori
School of Medicine, Toho University, Tokyo, Japan

A. Kuno & M. Matsuo
Department of Chemistry, Tokyo University, Tokyo, Japan

D.K. Nordstrom
United States Geological Survey, Boulder, USA

ABSTRACT: The purpose of this study is to provide geochemical data of thermal waters in Kuril's Islands (Mendeleev, Golovnina, and Ebeko volcanoes). Among the thermal water types three main groups can be distinguished: sodium-chloride, acid sulphate and chloride-sulphate-bicarbonate. Content and behavior of siderophile, chalcophile, lithophile, and rare-earth elements are discussed. These data, together with the results of isotopic studies, allow the authors to discuss the origin of these waters.

1 INTRODUCTION

The Kuril Islands are representative of modern volcanic arcs, which have experienced volcanic activity through the whole of Cenozoic time. Hot springs and mud pots are widespread in this area. Many researchers studied them in the 1970–1980s (Zelenov 1972; Baskov & Surikov 1975; Markhinin & Stratula 1977; Lebedev 1979; Kononov 1983; Menyailov et al. 1988), but information on the chemical elements is limited and the most reasonable results have been obtained only for the main ions. In this paper we present results of major, trace and rare earth elements as well as data on H, O, S and He isotope ratios for hot springs of Kunashir and Paramushir Islands.

2 GEOLOGICAL SETTING

We have studied thermal waters (Fig. 1) related with Mendeleev and Golovnina Volcanoes (Kunashir I.) and Ebeko Volcano (Paramushir I.). On Kunashir Island, Mendeleev and Golovnina Volcanoes are now at the stage of hydrothermal activity. Mendeleev is a volcano of the central type, and it is composed of interbanded andesite-basalts and pyroclastic formations. The extrusive dome, well defined in the relief, is composed of dacites. On the volcano slopes, the horizon of fresh infiltration waters is thin and strongly depends on the amount of precipitation. Golovnina Volcano is a truncated cone with a caldera inside. A large portion of the caldera is occupied by a

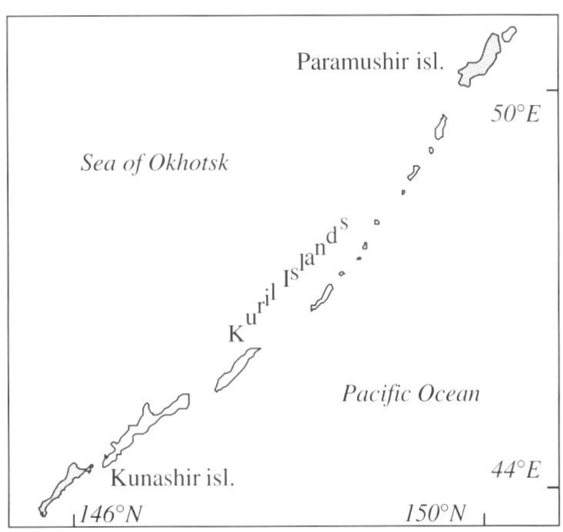

Figure 1. Location of the study areas.

lake. In the center of the Caldera there are two domes composed of andesite-dacites. Active solfatara fields are confined to the domes.

On Paramushir Island, Ebeko Volcano is incorporated into the volcano chain that forms the Vernadsky Ridge and is the most active volcano. It is composed of young lavas of two-pyroxene andesites. At present, the volcano demonstrates its high activity by abundant fumaroles and hot springs. A hydrothermal-magmatic system was formed in Pliocene-Holocene time (Rychagov et al. 2002). Several groups of fumaroles and thermal springs have been distinguished on Ebeko volcano, and the most intense are the fumaroles of the eastern slope.

3 RESULTS

3.1 *Main ions*

Among the thermal water types studied, three main groups can be distinguished. The first group includes waters in which SO_4^{2-} ion predominates. The water temperature on the surface reaches 97°C, and TDS varies from a few g/l to 27.4 g/l. These waters are acid to superacid with pH values ranging between 0.6 to 2.3, and their discharge, as a rule, is located within the fumarole fields of Mendeleev, Golovnina, and Ebeko Volcanoes. The main cation of these waters is Na. The second group is Na–Cl waters. The temperature of steam-gas mixtures in wells can reach 140°C with a maximum TDS of 14.2 g/l. These waters occur at depth, as a rule, and are tapped by wells. The waters are neutral or alkaline; pH varies from 6.9 to 8.2. Cl-ion predominates among the main anions. Bicarbonate and sulphate concentrations vary widely from 44 to 2300 mg/l and from 280 to 2460 mg/l, correspondingly. Na is the predominant cation with a maximum concentration of 4200 mg/l. The third group is dominated by sodium-chloride-sulphate-bicarbonate. Stolbovskye springs, located on the periphery of Mendeleev Volcano, are representative of this type. The contents of Cl-ion are slightly greater than the concentration of sulphate ion while the bicarbonate-ion is less than half the levels of the other two. Na is the main cation; its concentration reaches 500 mg/l. Ca is the second most abundant element, its concentration varies from 5 to 70 mg/l. The pH of these waters is close to neutral with temperatures of 76–78°C. TDS of the waters is 1.9 g/l.

Mendeleev and Ebeko display a classic vertical geochemical zoning of thermal waters. The fumarole fields are dominated by acid sulphate waters, and hypsometrically below, neutral or alkaline thermal waters of Na–Cl composition are common. The Stolbovskye springs are derivates of Na–Cl waters due to their dilution by the waters of the subsurface formation.

3.2 *Siderophile elements (Fe, Co, Ni, Mo)*

In the sulphate waters, Fe concentration is >10 mg/l. The highest content (144 mg/l) has been found in one of the boiling pots of Ebeko Volcano. Co and Ni contents are also increased. As a whole, Co and Ni concentrations in acid sulphate waters exceed 1 μg/l whereas Mo content is less than 1 μg/l. In thermal Na–Cl alkaline waters, the content of siderophile elements is less than that in sulphate waters with the exception of Mo, which has concentrations greater than 1 μg/l, because Mo accumulates predominantly in alkaline waters. A high concentration of Fe was defined in Well 101. In the Stolbovskye springs the contents of siderophile elements are extremely low. Co and Ni are below the limits of detection even with ICP-MS. Concentration of Fe is a few 10's of μg/l.

3.3 *Chalcophile elements (Cu, Zn, Pb, Sb, As)*

In the sulphate waters, the contents of these elements vary significantly. Stable high concentrations of these elements were defined in one of the boiling pots on Mendeleev Volcano. The content of Zn reaches a few mg/l and concentration of Pb is measured in 100's of μg/l. The maximum content of Cu is 39 μg/l. All these samples have a high concentration of As, up to 1.5 mg/l. Examination of the As oxidation state showed that it is present predominantly in the oxidized form (+5). In some manifestations in Ebeko's crater, As^{III} reaches 30% of the total of $As^{V} + As^{III}$. In alkaline Na–Cl waters, the chalcophile element concentrations are rather low with the exception of Cu. The contents of Sb do not exceed 1 μg/l, and Zn and Pb concentrations are below the limits of detection by ICP-MS. On Mendeleev Volcano, Sb concentration reaches 33 μg/l, Pb – 0.11 μg/l, and Zn – 4.8 μg/l. The content of As in these waters is rather stable – a few hundred μg/l. As is present in these waters in the reduced form (As^{III}). Cu concentration is 100's of μg/l and it is higher that in other types of waters. In the Stolbovskye springs (sodium-chloride-sulphate-bicarbonate), Cu content is about 37 μg/l. Pb and Zn are beyond the limits of detection. As contents in these waters are rather low – a few μg/l.

3.4 *Lithophile elements (Al, Li, Rb, Sr, Ba)*

In sulphate waters the contents of dissolved Al are high on Ebeko Volcano – 210 mg/l. In other manifestations they are a few 10's or 100's of mg/l. Li contents are, as a rule, 10's of μg/l, with the exception of the springs of the Mendeleev Volcano, where its concentrations reach a few 100's of μg/l. In the same springs, Rb and Ba concentrations are increased. The highest Sr contents in this water type characterize Ebeko Volcano. In Na–Cl waters, Al contents vary from 10's to 100's of μg/l. An abnormally high concentration of Al is found in Well 101. The waters of this well have increased concentrations of Rb and Li. The maximum content of Sr is 2.4 mg/l, and may be due to an influence from seawater. Among the Stolbovskye springs, Li contents vary insignificantly and are 260–280 μg/l. Al concentration varies from 62–150 μg/l. Contents of Rb, Sr, and Ba vary insignificantly within this water group.

3.5 *Rare-earth elements (REE)*

A general regularity of REE distribution is observed in the waters studied. In Figure 2 REE profiles normalized to North-American shale are shown. For comparison, the REE values for the surface-stream flows, atmospheric precipitation, and seawater in this region are given. REE concentrations in the alkaline thermal waters are the lowest and similar to those in the atmospheric precipitation and surrounding seawaters. The highest level (more than an order of magnitude) is found in the acid sulphate waters. The high TDS of waters shows a smoothed profile of REE. The surface-stream flows occupied an intermediate position. Generally, we observed similar profiles of REE in sulphate waters of Mendeleev and Golovnina. In Ebeko Volcano, however, the contents of LREE are higher and HREE are similar or lower. So, pH of waters controls the level of REE. In acid springs, more complete leaching of REE from the country rocks takes place. The Eu and Ce anomalies are smoothed out due to a valence change that results in a change of their mobility. In alkaline springs REE are not accumulated, and they remain at the level of atmospheric precipitation or seawater. In neutral surface-stream flows, concentrations of REE are controlled by the sorption-desorption process between water and the solids in suspension, so REE levels are higher than in atmospheric precipitations.

3.6 *Stable isotopes*

To reveal the source of waters, the O and H isotope ratio is often used (Craig 1961). Figure 3 portrays the O and H isotope ratios in the waters.

For comparison, data for atmospheric precipitations, surface-stream flows, and seawater also are given. Most of the studied springs are located along the meteoric water line, demonstrating the predominantly atmospheric origin of a water component. The shift towards the oxygen loading for acid sulphate waters is explained by an active interaction of high-temperature solutions with the country rocks.

For the hot springs of Mendeleev and Golovnina Volcanoes, we have determined the $\delta^{34}S$ values. On the fumarole fields of Mendeleev Volcano the $\delta^{34}S$ value is close to −3.2‰. In the Golovnina Caldera this value varies between +1.9‰ to +2.5‰. These values indicate abyssal (magmatic) origin of the S. At the same time, in the Na–Cl waters of the "Goryachie plyazh" (Mendeleev Volcano), $\delta^{34}S$ value is +20‰, and for the Stolbovskye springs this value is +22.1‰. Such an essential enrichment in ^{34}S is the result of influence of marine sediments, which are common among the country rocks.

In acid sulphate hot springs the $^3He/^4He$ ratio was obtained (Fig. 4). The highest values (10.6–10.8) are typical for springs of Ebeko Volcano. These high values mean very high percentage of mantle He. The lowest ratio (5–7) was determined for Golovnina Caldera. Mendeleev Volcano occupies an intermediate position ($^3He/^4He = 8–9.8$). These data correspond to geological and geophysical observation about volcanic activity of these volcanoes. At present, Ebeko Volcano is more active than Golovnina.

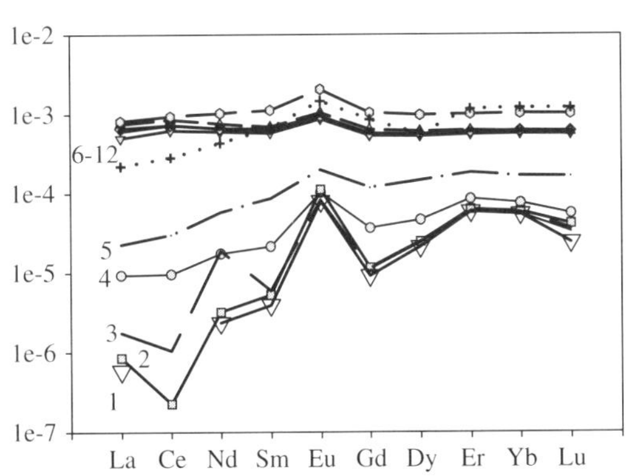

Figure 2. REE profile for waters of Ebeko volcano. 1 – sea water (Sea of Okhotsk); 2 – Na–Cl waters; 3 – rain; 4 – fresh water; 5 – Kuzminka River; 6–12 – acid sulphate waters.

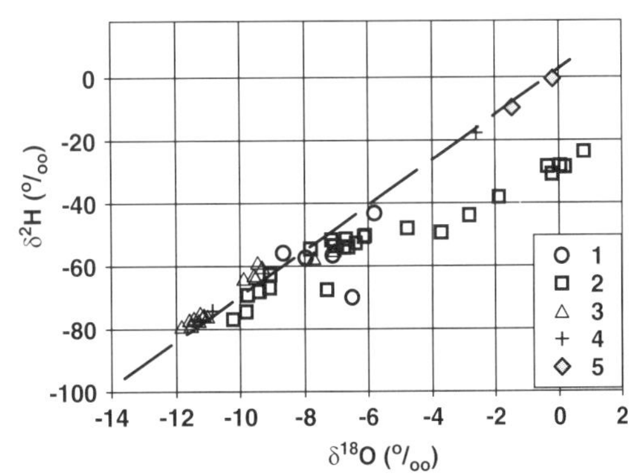

Figure 3. O vs. H isotopes in studied waters. Types of waters: 1 – chloride-sulphate; 2 – sulphate; 3 – fresh surface; 4 – atmospheric precipitations; 5 – sea water.

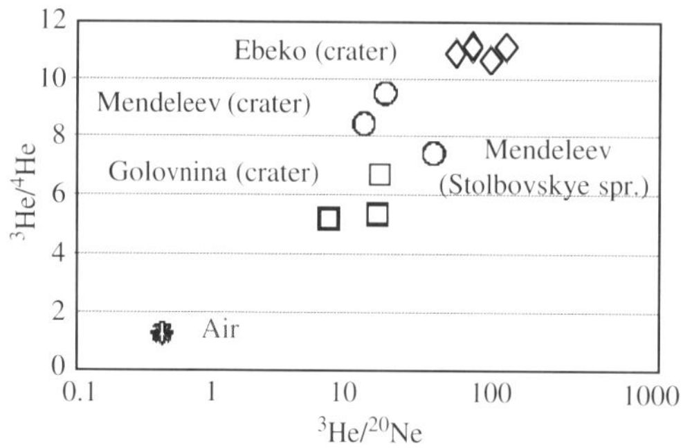

Figure 4. $^3He/^4He$ vs. $^3He/^{20}Ne$ in studied waters.

3.7 *Modeling of the water-rock interaction*

Using silica and sodium-potassium geothermometers it is possible to calculate the "base temperatures" of waters at depth. For Na–Cl waters of Mendeleev Volcano (Well 101) this temperature is 270°C, for Ebeko Volcano it is 127°C and for Stolbovskye springs it is 105°C. The index of mineral saturation calculated by SOLMINEQ indicates that Na–Cl waters are oversaturated relative to calcite, dolomite, adularia, albite, aragonite, saponite (Ca–Mg–Na) smectites, kaolinite, and a group of zeolites including laumontite. For Na–Cl waters along with the ion form of the elements, chloride complexes of Na, K, Ca are typical also. The waters of the Stolbovskye springs are oversaturated relative to the minerals of a kaolinite group, zeolites (clinoptilolite and mordenite), and saponite. In the superacid sulphate waters, oversaturation occurs only in the case of silica, and most of the elements are combined with oxidized S.

4 DISCUSSION & CONCLUSIONS

The results obtained indicate that the three groups of thermal waters, distinguished by the main ions, differ in many physical–chemical parameters: pH, TDS, and concentration of microelements including REE. Correspondingly they are characterized by different forms of elements in solution, by different associations of equilibrium minerals formed from solutions, as well as different extent of deep fluid participation. The chemical composition of the thermal waters is formed at the expense of both reworking of the country rocks and the supply material at depth. The flow of gaseous fluid is composed essentially of CO_2, H_2S, HF, HCl and other components (Menyailov et al. 1988), which, combining with infiltration waters, not only bring the additional components, but also change the physical–chemical parameters of the solution influencing significantly the reaction run in the water–rock system. The results obtained on the ratio of O and H isotopes suggest a predominantly meteoric origin of all the studied springs. They are in agreement with the data obtained by us (Chudaev et al. 2000) and other researchers (Rychagov et al. 2002) for the thermal waters of Kamchatka. Suggestion of significant seawater participation in the formation of the Na–Cl waters of Mendeleev Volcano (Baskov & Surikov 1975; Markhinin & Stratula 1977) have not been supported by recent data. We have observed the seawater influence only in the coastal springs discharging into seawater (Goryachie plyazh, Kunashir Island).

ACKNOWLEDGEMENTS

This work was supported by RFBR (grant # 01-05-64608). The authors wish to express their most sincere gratitude to Prof. K. Nagao (Tokyo University) and M. Kusakabe (Okayma University) for determination of helium and sulfur isotopes.

REFERENCES

Baskov, E.A. & Surikov, S.N. 1975. *Hydrotherms of the Pacific segment of the Earth.* Moscow: Nedra.

Craig, H. 1961. Isotopic variations in meteoric waters. *Science.* 133: 1702–1703.

Chudaev, O.V., Chudaeva, V.A., Karpov, G.A., Edmunds, M. & Shand, P. 2000. *Geochemistry of waters of the main geothermal areas of Kamchatka.* Vladivostok: Dalnauka.

Kononov, V.I. 1983. *Geochemistry of thermal waters in modern volcanism areas.* Moscow: Nauka.

Lebedev, L.M. 1979. *Minerals of modern hydrotherms.* Moscow: Nedra.

Markhinin, E.K. & Stratula, D.C. 1977. *Hydrotherms of the Kuril Islands.* Moscow: Nauka.

Menyailov, I.A., Nikitina, L.P. & Shapar, V.N. 1988. Features of chemical and isotopic composition of fumarole gases during inter-eruptive periods of Ebeko Volcano activity. *Volcanol. Seismol.* 4: 21–36.

Rychagov, S.N., Koroleva, G.P. & Stepanov, I.I. 2002. Ore elements in hypergenesis zone of steam-hydrotherms deposit: distribution, migration forms, sources. *Volcanol. Seismol.* 2: 37–58.

Zelenov, K.K. 1972. *Volcanoes as sources of ore-forming components of sedimentary formations.* Moscow: Nauka.

Water–Rock Interaction, Wanty & Seal II (eds)
© 2004 Taylor & Francis Group, London, ISBN 90 5809 641 6

Major, trace, and rare earth elements in the surface waters of two areas of the Kuril Islands

V.A. Chudaeva
Pacific Institute of Geography, Far East Branch Russian Academy of Science, Russia

O.V. Chudaev
Far East Geological Institute, Far East Branch Russian Academy of Science, Russia

K. Sugimory
Department of Biology, Toho University School of Medicine, Tokyo, Japan

M. Matsuo & A. Kuno
Department of Chemistry, Tokyo University, Japan

ABSTRACT: The purpose of this investigation is to evaluate the distribution of chemical elements in waters in the vicinity of hydrothermal activity. The primary area of research was surface waters, associated with the hydrothermal activity of the Kuril Islands in Russia. A smaller component of this study was connected with the Hakone hydrothermal area, Japan. The volcanic and hydrothermal activity significantly influences the chemical composition of surface water in the studied areas greatly elevating levels of many chemical elements (Zn, Cd, Pb, Li, Be, V, Cr, Mn, REE and some others). The distribution of elements in the surface waters mainly depends on the content of elements in the nearest hydrothermal systems.

1 INTRODUCTION

The Kuril Islands have an abundance of volcanoes and are associated with both acidic and alkaline hydrothermal features. There are numerous studies on the hydrothermal activity of the Kuril Islands conducted by Russian scientists. The surface waters of this region however, have been studied less (Nikitina 1978; Chertkova et al. 1986; Fazlulin 2000). Clearly it would be useful to investigate this specific natural environment. Our primary interest was to study the surface water distribution of chemical elements associated with hydrothermal activity on Kunashir and Paramushir Islands of Kuril ridge (Far East of Russia; Fig. 1) and in the Hakone area of Japan where data on trace and REE elements distribution have not been published.

2 METHODS

Together with a study of hydrothermal features, the researched objects were the Kislaya River (which originates from Mendeleev Volcano and flows into the Lesnaya River), streams and Goriachee Lake in Golovnina Caldera. Several other rivers, less associated with the hydrothermal activity and background waters, also were studied for comparison. In addition, several rivers on Paramushir Island were sampled: Kuzminka River (which flows from Ebeko volcano) and other rivers that originate on the north-east side of this volcano and flow into the Pacific Ocean (Fig. 1). The samples were also taken from a stream near Hakone volcano as well as one mud pot and one spring from this fumarole area.

Measurements of pH, Eh, conductivity, HCO_3 and dissolved oxygen were made in situ. All samples were filtered immediately through a 0.45 µm pore filter. The contents of trace elements in samples were analyzed using ICP-MS method. Major ion concentrations were evaluated with colorimetric, AAS and ICP-AES methods.

3 RESULTS AND DISCUSSION

The surface waters studied are mostly acidic (pH up to 2.5–1.8). Water which has not been influenced by

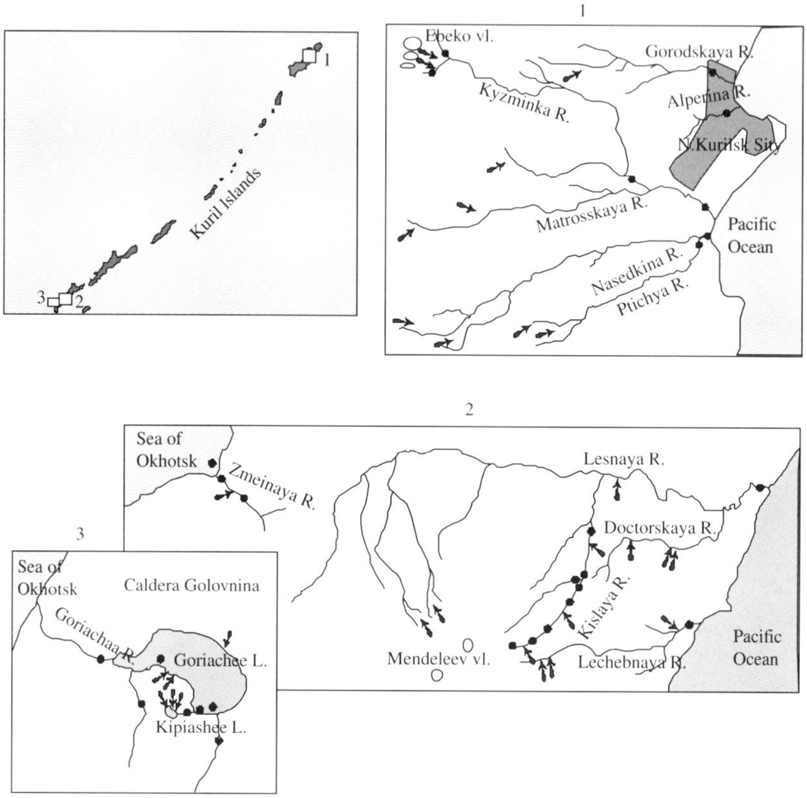

Figure 1. Scheme of sampling of the surface waters of the Paramushir (1) and Kunashir (2, 3) Islands. Circles – points of sampling; arrows show the main hot springs.

hydrothermal activity was closer to neutral (pH 5.5–7.5). The upper portion of the Kislaya River, originating from the Ebeko volcano (Paramushir Island) has a pH of 6.85, however the pH sharply decreases to pH 2.5–1.8 because of the influence of the fumarole and hydrothermal springs. The pH then increases to 3.2 when it is joined by the neutral water from left tributary. Even the Lesnaya River, which accepts water from the Kislaya River has a higher pH of 3.17 near its delta. The pH values of the Goriachee Lake of Golovnina Caldera are also no higher than 3. River water in the northeastern part of Paramushir Island is mostly acidic (pH 6.6–3.6) and in a stream of the NE fumarole's field of the Ebeko volcano the pH was as low as 0.82.

The low pH of surface waters was accompanied by a low oxygen concentration (up to 2–3 mg/l) and high Eh (up to 400 mv and higher). The aggressive acidic surface water resulted in the high mineralization. The water conductivity on Kunashir Island was up to 2–3 mS/cm, and the amount of major ions was some hundreds or thousands of mg/l higher in contrast to the background water samples, where the amount of major ions was less than 100 mg/l. In the stream flowing from the northeastern fumarole's field of Ebeko Volcano (Paramushir Island), the concentration of main ions was 16.6 g/l.

The acidic surface water was mostly sulfate or sulfate-chlorine (Golovnina Caldera), and either of mixed or predominantly Na cation composition. The exception was the Zmeinaya River (Kunashir Island) with another type of hydrothermal activity in the area. At a sampling point downstream from the hot springs there was water of Na-Cl composition with a pH close to neutral. There is some difference between Si and Al concentrations in the rivers of two islands (Fig. 2).

Acidic sulfate hydrothermal activity and underground water discharge in areas with fumaroles significantly change the surface water composition and introduce a wide variety of elements, particularly trace elements and REE. The distribution of trace elements varies greatly between tested waters.

The chemical composition of the Kislaya River (flows from the Mendeleev Volcano on Kunashir Island) was extensively studied and significant differences were found in the chemical composition in

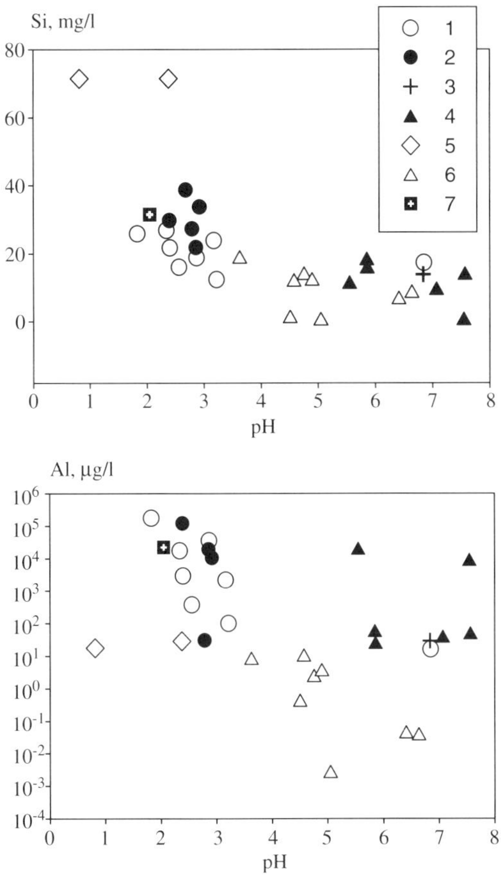

Figure 2. Relations of Si and Al in the surface waters of Paramushir and Kunashir Islands. 1 – Kislaya River; 2 – surface waters of Golovnin Caldera; 3 – Zmeinaya River; 4 – background rivers of Kunashir Island; 5 – streams of fumarole's field of Ebeko volcano; 6 – rivers of Paramushir; 7 – small river or creek flowing from fumarole s field on Hakone volcano.

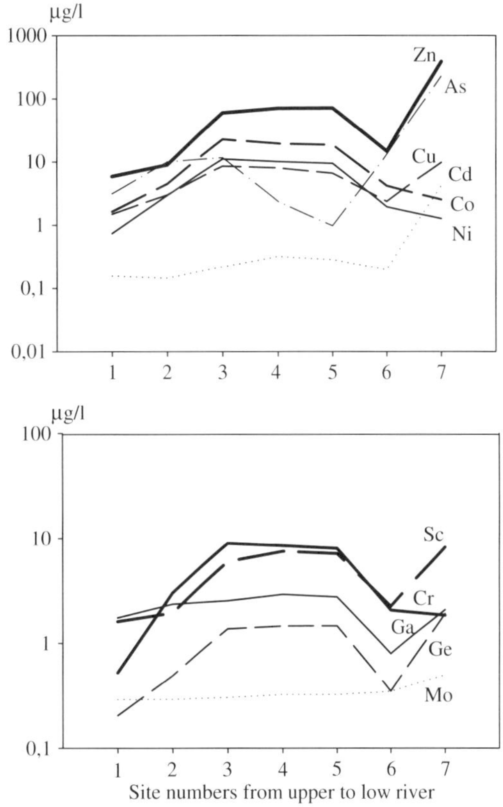

Figure 3. Variations of element concentrations in the Kislaya River.

different parts of the river (Fig. 3). The upper river (point 3), which is under the influence of the Upper Group of hot springs, has higher concentrations of Co (up to 20 µg/l) and Ni (up to 11 µg/l) than other points. The part of the river influenced by the Lower Group of hot springs have higher concentrations of Zn (up to 390 µg/l), Cd (>4 µg/l) and Pb (78 µg/l), as well as Li, Be, As, Rb, Cs, Tl, In, Ba and Hg. The latter three elements and Se are the most characteristic for Kislaya River. The highest concentration of Zn was found in Goriachee Lake of Golovnina Caldera (1400 µg/l). This lake also has a high concentration of Mn (up to 2600 µg/l).

The highest concentrations of dissolved Cu (>20 µg/l), V (>400 µg/l), Cr (19 µg/l), Fe (49 mg/l) and Mn (3.2 mg/l) were found in a stream of the north-eastern field of Ebeko volcano. High levels of Si, Al, Sr, Ge, Sc, Ga, Th and U also were common there.

The neutral water of the Stolbovskye hydrothermal springs influences the composition of the Zmeinaya River. There we found high concentrations of Li, Sr, Cu, Fe, Rb, Cs, Ba, Ga, Sc and Zn, but not Mn.

Golovnina Caldera has a unique hydrochemical system with fumaroles and various manifestations of hot water mostly with very low pH values and high levels of many elements. There is a two-lake system where a small lake, Kipiaschee (pH 2.7), flows into Goriachee Lake (pH 2.5–2.9). The stream that flows from the Goriachee lake (Fig. 1) into the Sea of Okhotsk, carries relatively high concentrations of Zn, V, Mn, Si, Ge, Li, Ga, As, Rb, Cs, Ba and Th.

The rivers in the northern part of Paramushir Island have extremely variable distributions of trace elements (sometimes two or three orders of magnitude),

Table 1. Some REE concentrations in the acid hot springs and pots and in the surface waters of two Kuril Islands, μg/l.

Sampling area*	La	Ce	Nd	Eu	Dy	Er	Lu
1	0.7–3.2	2.7–9.1	0.3–9.5	1.0–1.5	0.3–9.7	0.3–6.5	0.6–1.3
2	0.5–1.2	0.5–2.7	0.2–3.0	0.6–1.0	0.2–2.5	0.5–1.6	0.6–0.8
3	2.2–5.1	4.8–13.9	4.1–13.5	0.3–1.9	2.0–6.8	1.5–4.6	0.8–1.6
4	0.6–0.8	0.9–1.8	0.9–2.0	0.2–0.6	2.0–8.6	1.3–5.6	1.0–1.2
5	7.5–27.9	19.0–62.7	13.1–31.0	1.0–2.4	2.5–4.9	1.5–3.1	0.2–0.5
6	1.9–20.4	5.4–47.7	5.3–19.8	1.1–1.2	2.6–3.9	1.6–2.6	0.2–0.4
7	0.8–2.1	1.4–5.8	1.7–6.5	0.3–0.9	1.3–3.9	0.9–2.6	0.7–0.9
8	0.7	1.2	1.3	0.25	1.4	1.0	0.7

* 1 – springs and pots of the Mendeleev Volcano; 2 – Kislaya River; 3 – springs and pots of the Golovnin Caldera; 4 – surface waters of Golovnin Caldera; 5 – springs and pots of the Ebeko Volcano; 6 – streams on the fumarole's field of Ebeko Volcano; 7 – spring and pot of the Hakone Volcano; 8 – stream on the fumarole's field of Hakone Volcano.

yet the dissolved concentrations are lower than the streams originating from the fumarole's field.

The concentrations of REE are very different in different groups of springs and mud pots of the Kuril Islands and reach more than 100 μg/l total dissolved REE. Maximum contents are typical for acid springs with the lowest pH values (about 1). At the time of study the highest REE contents (>100 μg/l) for acid waters were found in a group of hot springs on Ebeko's north-eastern fumarole fields. Concurrently, the mud pots of Mendeleev volcano on Kunashir Island had total REE concentrations from 30 to 40 μg/l. In the Ebeko area, the concentrations of light REE are higher (up to La 28 μg/l; Ce 62.7; Nd 31; Sm 6.5; Gd 5.3 μg/l). In springs of the Mendeleev Volcano area, which, as noted above, have lower total REE concentrations, the heavier REE are higher (up to Gd 6.2, Dy 9.7, Er 6.5, Yb 6.1 μg/l). In the acid springs of the Golovnina Caldera, Kunashir Island, the light REE are more common than the heavy REE: Ce up to 13.9 and Nd up to 13.5 μg/l. For the Hakone area of Japan, REE concentrations were lower than those measured at Kuril (Table 1).

The streams located close to active areas have essentially different concentrations of REE. Moreover, the REE concentrations vary in light and heavy REE in different areas. Maximum concentrations are typical in the stream draining the northeastern fumarole field of Ebeko Volcano. The lowest concentrations are typical to the streams of Hakone area. In Golovnina Caldera the importance of light REE (La-Nd) decreases from the waters of springs and mud pots to the lake and stream waters. At the same time heavy REE concentrations in the lake and stream are close to it in the thermal waters. In the area of the Kislaya River the REE distribution exhibits some differences: decreasing of heavy REE contents in the river water compared to the springs. In the stream with the low pH level on the slope of Ebeko Volcano the REE concentrations were found in the same high levels as in the springs, but the REE concentrations in other rivers were essentially lower.

4 CONCLUSION

In the area studied hydrothermal activity significantly influences the chemical composition of surface water. The surface waters in areas of hydrothermal activity change from neutral to acidic (pH up to <3) and become mostly SO_4-Na with high concentration of elements. The distribution of trace elements and REE in the surface waters mostly depends on their distribution in the nearest hydrothermal systems. The REE concentrations vary widely in the groups of springs and pots of the Kuril Islands and concentrations of total dissolved REE exceed 100 μg/l. Depending on the proximity of the stream to active areas the waters have essentially different concentrations of REE which vary in light or heavy group contents.

ACKNOWLEDGEMENTS

The work has been done with financial support of the RFBR (project 01-05-64608) and Far East Branch Russian Academy of Sciences (project 03-1-09-004).

REFERENCES

Chertkova, L.V., Gavrilenko, G.M. & Eropheeva, E.A. 1986. Natural model of mixer of acidic river and sea water (Urieva River, Paramushir, Kuril Islands) *Geology of oceans and seas. Abstracts of conference.* V. 3: 148–149. (*In Russian*).

Nikitina, L.P. 1978. Metal migration from active volcano in sedimentation basin. 79p. Moscow: Nauka. (*In Russian*).

Fazlulin, S.M. 2000. Geochemical system of Uryev River *Volcanology and Seismology* 1: 36–45 (*In Russian*).

Water–Rock Interaction, Wanty & Seal II (eds)
© 2004 Taylor & Francis Group, London, ISBN 90 5809 641 6

Helium and carbon isotopic evidence for the relationship between outgassing and structure in the East African Rift System

W.G. Darling
British Geological Survey, Wallingford, UK

ABSTRACT: A significant amount of He and C isotopic data for fluids in the rifts of Kenya, Ethiopia and Djibouti now exists. Cross-axis measurements of both $^3He/^4He$ and $\delta^{13}C$ show that mantle outgassing is most readily identifiable at axial volcanic centres, but that some must also be occurring at rift boundaries, where structural style may be a control, particularly in the Kenya rift. Along-axis sections show that there is some variation in $^3He/^4He$ with crustal thickness although, perhaps contrary to expectations, maximum $^3He/^4He$ values are not associated with the thinnest crust. Values of $\delta^{13}C$ remain similar from northern Tanzania to Afar, and there is also evidence for mantle-influenced compositions more than 100 km from the main Kenya rift at Homa Mountain, Mt Kenya, and the Chyulu Hills.

1 INTRODUCTION

The East African Rift System (EARS) comprises eastern and western branches, apparently only linked in the southern Tanzania-Malawi area. The style of these branches is quite different. For example, the eastern branch exhibits major hydrothermalism, while the western branch has much less of the volcanicity or extension which promotes such activity. Since it is the existence of this hydrothermalism that provides the opportunity to characterise the deeper fluids of the EARS, this study is concerned mostly with the eastern branch together with Afar, including the active Asal rift in Djibouti. The intention is to examine possible links between outgassing and rift structure.

2 THE RIFT VALLEYS

2.1 *Background*

The eastern branch of the EARS forms a reasonably well-defined rift valley ("Eastern Rift") with fairly evenly-spaced late Quaternary volcanic centres along the axial line (Fig. 1). In the north, however, the Main Ethiopian Rift opens out into the Afar Triangle, a triple plate junction from which the active Asal rift heads off to the Gulf of Aden. In the south, the Eastern Rift splays out in central Tanzania to form "riftlets" which die out within a short distance.

The Eastern Rift consists of a series of separate, broadly aligned rift zones. For the purposes of this study, three main areas are identified: the Kenya Rift Valleys (KRV), the Main Ethiopian Rift (MER), and southern Afar. Each of these areas has differences in origin (e.g. Braile et al. 1995). The KRV is the result of an apparently relatively shallow mantle upwelling, whereas the MER is a consequence of surface impingement of a major mantle plume. Although this plume spreads beneath Afar, this is now an area of extension with a thin crust probably of oceanic origin.

As a result of this, there are profound differences in helium isotope characteristics between the MER plus Afar, and the KRV (Marty et al. 1993; Darling et al. 1995; Scarsi & Craig 1996). The areas can be regarded as having the helium isotope characteristics of mantle plume (hotspot) and mid-ocean ridge (MORB) basalts respectively. It is necessary to be aware of these different sources before comparing the helium isotope data between areas.

2.2 *Central volcanism vs. boundary faults*

Most late-Quaternary volcanism in the Eastern Rift is manifested as a chain of volcanic centres from Oldoinyo Lengai in the south to Ayelu in the north (Fig. 1), all more-or-less associated with the central axis of the rifts. Not every volcano with fumarolic activity has been investigated for He isotope ratios (the MER in particular has gaps), but there are

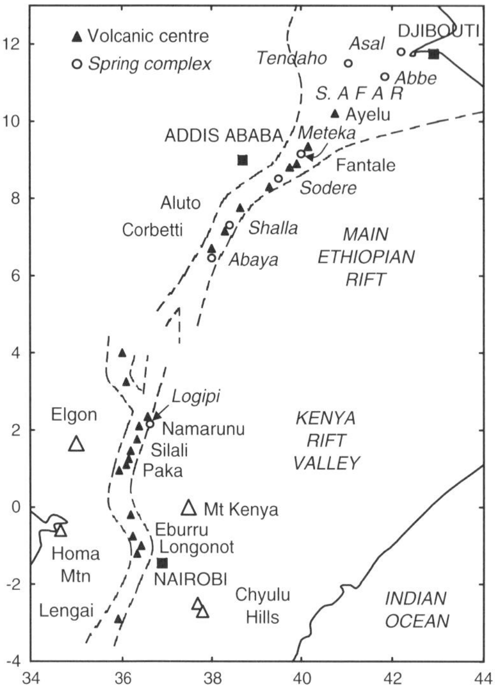

Figure 1. Map (lat-long) of the Eastern Branch of the East African Rift System. Boundaries of the rifts in Kenya, Ethiopia and southern Afar are shown, plus the locations of the main volcanic centres and spring complexes. Off-axis centres are also shown.

sufficient data to indicate that magmatism is a highly-important route for the outgassing of mantle He in the EARS (Craig et al. 1977; Darling et al. 1995).

Values of $\delta^{13}C\text{-}CO_2$ for the volcanic centres are very uniform, averaging out at around –4‰ (Darling 1998). There is no significant variation between the KRV and the MER suggesting that, whatever the difference in He source, the C source is the same. The value of –4‰ is similar to the general mantle signature of –5‰ (Deines 2002); there is no carbonate sedimentation in the KRV or MER to make a metamorphic origin possible. The CO_2 isotopic evidence therefore supports the He isotopic indications that magmatic activity promotes outgassing.

Outgassing presumably occurs over the whole width of rifted areas, but is often likely to be too diffuse to be easily detected. However, some concentration might be expected at the boundaries. Models based on KRISP geophysical studies (Prodehl et al. 1997) indicate that the main rift boundary faults (RBFs), at least in the KRV, are basically normal and may penetrate a considerable proportion of the crust. It seems probable that they form conduits for the transport of mantle fluids to the surface. However, by their sheer size and continuity they would have to be releasing a high volume of gas to reach the same concentrations as found on the discrete volcanic centres. Not only are the RBFs seldom associated with the strength of convective hydrothermal activity typically found above the axial magma chambers feeding the volcanic centres, but they are also more likely to be permeated by atmospheric gases. The effects of groundwater transport on dissolved He also cannot be ignored; most groundwater flow is directed into the rifts so will tend to dilute rather than enhance boundary fault concentrations.

Figure 2 shows that peak $^3He/^4He$ values (MORB or hotspot as appropriate) are only reached along the rift axes. However, even at the rift boundaries virtually all samples have R/R_A values >1, indicating a proportion of mantle He. Even where the rift boundary is ill-defined, notably adjacent to Addis Ababa, which is itself not regarded as being within the MER, there is evidence of a mantle He component ($5\,R_A$ in a thermal borehole in the city) at a considerable distance (~100 km) from the rift axis. There is also a tendency (not shown) for boundary He concentrations to be lower than axial concentrations.

Carbon isotopic evidence from the KRV (Fig. 2) partially supports the indications from He isotopic data, with sites having a range of dissolved inorganic carbon (DIC) and gas compositions from non-magmatic at around –15‰ (fresh waters) to magmatically affected at ~–2‰ (gases) and ~0‰ (high-alkalinity waters resulting from silicate hydrolysis). While the most consistently high-^{13}C values are associated with the axis of the KRV, the overall distribution is less clear-cut than for He isotopes. Lowest $\delta^{13}C$ values occur at the eastern rift boundary, though there are also some low values west of the central axis which mainly represent shallow groundwaters with low DIC. In the MER, for which there are fewer data, there appears to be no overall pattern.

2.3 *Nature of rift boundaries*

Some of the observed variations in isotope distributions across the rifts may be explicable in terms of rift boundary structure. Half-grabens seem to have been important in the development of the EARS, and it appears that major faulting in the KRV has occurred mainly on the western side of the rift (e.g. Macdonald et al. 1994). This might be expected to have an effect on the composition of fluids sampled at the surface, at or near the rift margins. In the KRV, the most depleted He and C isotopic results at or near a boundary (Fig. 2) seem to be associated with the eastern side. The fewer sites at or near the western boundary are comparatively enriched. This may be a legacy of the half-graben faulting, which would presumably tend to facilitate gas transport from

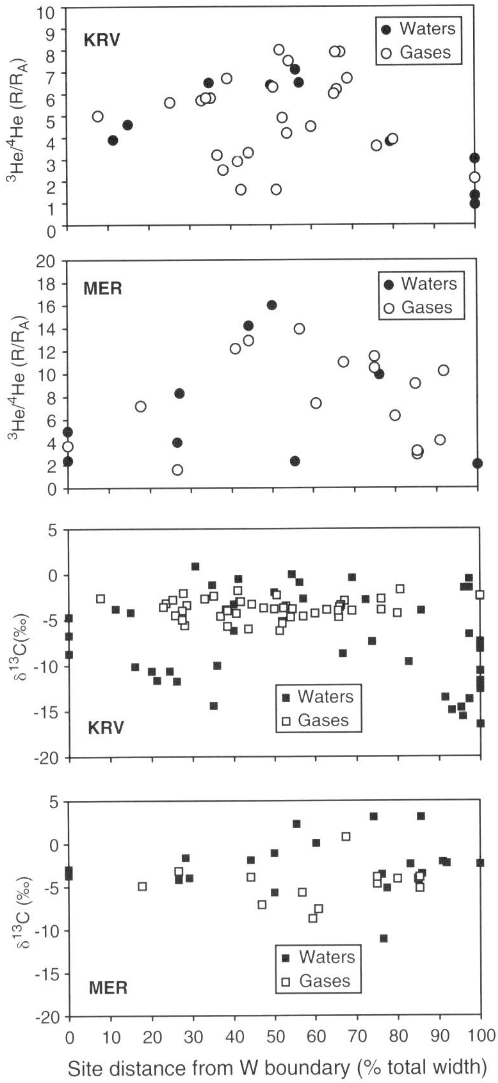

Figure 2. Helium and carbon isotope variations across the inner troughs of the KRV and MER. Because of variations in valley width, sites are located in proportional rather than absolute terms. Data from Darling et al. (1995) (KRV), Darling (1998) (KRV and MER) and Craig et al. (1977) (MER).

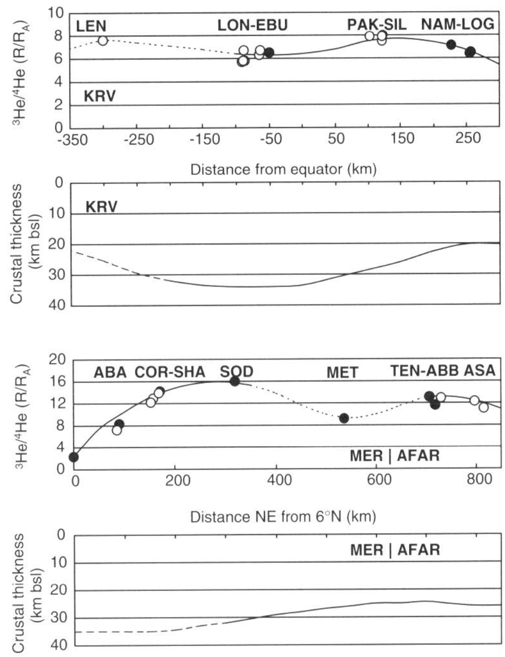

Figure 3. He isotope and crustal thickness variations along the axes of the KRV and MER. Distances are measured N (+ve) and S (−ve) of the equator for the KRV, and NE from Lat 6°N for the MER. He isotope data from Craig et al. (1977), Javoy et al. (1989), Darling et al. (1995), Darling (1998) and Scarsi & Craig (1996). The maximum $^3He/^4He$ curve is dotted where most speculative. Crustal thickness curve is based on data summarised in Braile et al. (1995), dashed where uncertain. Symbols as for Fig. 2. KRV: LEN-Lengai, LON-Longonot, EBU-Eburru, PAKPaka, SIL-Silali, NAM-Namarunu, LOG-Logipi. MER: ABA-Abaya, COR-Corbetti, SHA-Shalla, SOD-Sodere, MET-Meteka, TEN-Tendaho, ABB-Abbe, ASA-Asal.

beneath the rift. However, there are also ^{13}C-enriched sites on the eastern boundary: it is possible that these are associated with the NW-SE basement shear zones identified by Smith & Mosley (1993) as important structural elements. The MER is thought to have a more mature full graben structure and the data, though sparse, support this by showing fewer signs of lateral asymmetry (Fig. 2).

2.4 *Longitudinal changes with crustal thickness*

Darling et al. (1995) showed that in the KRV, lower maximum $^3He/^4He$ ratios are found near the centre of updoming in the Eburru-Longonot area, than to the N in the Paka-Silali area. Values decline again further N of Silali. This may be related to a combination of crustal thickness and distance from the centre of mantle upwelling. Figure 3 shows an N-S axial cross-section with a tentative curve of maximum $^3He/^4He$ values together with the depth of crust based on KRISP seismic reflection data (Braile et al. 1995). Although $CO_2/^3He$ ratios show little evidence for crustal production of He (Darling et al. 1995), the thicker crust under Eburru-Longonot may contribute enough ^4He to lower the $^3He/^4He$ ratio by ~1.5 R$_A$. The higher $^3He/^4He$ values of Paka-Silali would then represent a balance between crustal thickness and distance from the centre of the mantle upwelling, with values gradually declining again in the Namarunu-Logipi area with increasing distance from the centre.

Oldoinyo Lengai is the only data point available significantly to the south of the Eburru-Longonot area (Fig. 3). Little is known of the crustal thickness in northern Tanzania, and the curves for both this and maximum $^3He/^4He$ are speculative for this area. However, the Lengai value of 7.6 R_A (Javoy et al. 1989) is significantly higher than for the centre of the Kenya dome, and therefore could indicate a thinner crust. The beginnings of such a trend south of the thickest crust around Eburru-Longonot were noted by Braile et al. (1995). Alternatively, because Lengai is an active volcano it may simply be more efficient at bringing mantle He to the surface.

In the MER and southern Afar, both seismic reflection and He isotope data are more limited. Figure 3 shows an axial cross-section running approximately NE from Lat 6°N to western Djibouti. A progressive thinning of crust from the MER into Afar takes place. He isotope values reach their maximum at Sodere where the crust is thinning but not at its thinnest (cf. the KRV). The dip in $^3He/^4He$ between Sodere and the Tendaho-Abbe region is based on one data point and therefore speculative, but there does seem to be a real if small decline between the lakes area and Afar.

Values of $\delta^{13}C$-CO_2 along the the axes of both the KRV and MER change very little (Darling 1998).

3 OFF-AXIS CENTRES OF THE KRV

The major off-axis volcanic centres of the main KRV are Mount Elgon, Homa Mountain, Mount Kenya (all Tertiary) and the Chyulu Hills (Quaternary; Fig. 1). Some carbon isotope data are available for Homa Mt, Mt Kenya and the Chyulus, and He data for Homa Mt. A $\delta^{13}C$-CO_2 value of −2‰ and a $^3He/^4He$ value of 1.5 R_A were measured for vigorously-outgassing springs at Bala on Homa Mt, a carbonatite centre. He concentrations were very high, suggesting that MORB-like He is being diluted by inbred ^4He. On the northeastern flanks of Mt. Kenya at Meru, gas seeps have a $\delta^{13}C$-CO_2 value of −3.5‰, indicating that mantle outgassing is still taking place in the area. The high-output Mzima Springs, which emerge from a major lava flow at the southern end of the Chyulu Hills, give a $\delta^{13}C$-DIC value of −10.3‰ and a ^{14}C activity of 70 pmc. On hydrogeological evidence these waters are likely to be modern, so the elevation of ^{13}C and lowering of ^{14}C compared to unmodified modern KRV waters (Darling 1996) suggests a minor addition of mantle CO_2, which would indicate that there is some outgassing in this area also.

4 AFAR

Southern Afar has hotspot $^3He/^4He$ values only slightly lower than those of the MER (Fig. 3). Values of $\delta^{13}C$-CO_2 are also similar to those in the MER and KRV (Darling 1998). Even in the Asal area, where active rifting is taking place, there is little change in either He or C isotopic compositions of deep fluids. The consistency in $^3He/^4He$ values is attributed to the pervasive influence of the Afar mantle plume.

5 CONCLUSIONS

While outgassing occurs mainly via hydrothermal activity along the rift axes, there is evidence from fluid isotopic compositions at the rift margins that a proportion occurs via major faulting.

The elevated $^3He/^4He$ values in gases are undoubtedly derived from the mantle, but the highest values found so far do not coincide with the thinnest crust.

Although they lie more than 100 km away, major off-axis volcanic centres appear to have some relationship with the KRV proper, though this requires better characterisation.

ACKNOWLEDGEMENTS

I thank Halldór Ármannsson and Berhanu Gizaw for help with sampling, and Erika Griesshaber and Stephan Weise for the helium isotope analyses.

REFERENCES

Braile, L.W., Keller, G.R., Wendlandt, R.F., Morgan, P. & Khan, M.A. 1995. The East African Rift System. In K.H. Olsen (ed). *Continental Rifts: Evolution, Structure, Tectonics*: 213–231. Amsterdam: Elsevier.

Craig, H., Lupton, J.E. & Horowitz, R.M. 1977. Isotopic geochemistry and hydrology of geothermal waters in the Ethiopian Rift Valley. *Rep. Scripps Inst. Oceanogr.* 77–14.

Darling, W.G. 1996. *The geochemistry of fluid processes in the eastern branch of the East African Rift System*. Ph.D. thesis, Open University, 249p.

Darling, W.G. 1998. Hydrothermal hydrocarbon gases: 2, Application in the East African Rift System. *Appl. Geochem.* 13: 825–840.

Darling, W.G., Griesshaber, E., Andrews, J.N., Armannsson, H. & O'Nions, R.K. 1995. The origin of hydrothermal and other gases in the Kenya Rift Valley. *Geochim. Cosmochim. Acta* 59: 2501–2512.

Deines, P. 2002. The carbon isotope geochemistry of mantle xenoliths. *Earth-Sci. Revs.* 58: 247–278.

Javoy, M., Pineau, F., Staudacher, T., Cheminee, J.L. & Kraft, M. 1989. Mantle volatiles sampled from a

continental rift: the 1988 eruption of Oldoinyo Lengai (Tanzania). *Terra Abstrs.* 1: 324.
Macdonald, R., Williams, L.A.J. & Gass, I.G. 1994. Tectonomagmatic evolution of the Kenya rift valley: some geological perspectives. *J. Geol. Soc.* 151: 879–888.
Marty, B., Appora, I., Barrat, J.A., Deniel, C., Vellutini, P. & Vidal, P. 1993. He, Ar, Sr, Nd and Pb isotopes in volcanic rocks from Afar: evidence for a primitive mantle component and constraints on magmatic sources. *Geochem. J.* 27: 219–228.
Prodehl, C., Fuchs, K. & Mechie, J. 1997. Seismic refraction studies of the Afro-Arabian rift system – a brief review. *Tectonophys.* 278: 1–13.
Scarsi, P. & Craig, H. 1996. Helium isotope ratios in Ethiopian Rift basalts. *Earth Plan. Sci. Lett.* 144: 505–516.
Smith, M. & Mosley, P. 1993. Crustal heterogeneity and basement influence on the development of the Kenya Rift, East Africa. *Tectonics* 12: 591–606.

Water–Rock Interaction, Wanty & Seal II (eds)

A gas transfer study at Ojo Caliente, Yellowstone National Park, USA

B.R. Druschel, M.J. Borda & M.A.A. Schoonen
Department of Geosciences, Stony Brook University, Stony Brook, New York, USA

ABSTRACT: A detailed study was conducted at Ojo Caliente hot spring in Yellowstone National Park, WY, to characterize water flow regimes and to determine how it affects the kinetics of abiotic processes, in particular gas transfer. Hot spring drainages may contain many microbial communities which play a role in the dissipation of chemical and thermal processes. The dissolved gases CO_2, CH_4, H_2S, and O_2 were analyzed, to study gas exchange kinetics. All measured gases follow first-order rates of transfer that can be divided into two separate rates based on the change in hydrodynamics of the drainages. A study of H_2S at three varying light intensities shows that the rate of depletion is approximately twice as fast in bright daylight as compared to low light indicating the occurrence of photochemical and/or biological reactions.

1 INTRODUCTION

1.1 *Study goals*

Waters discharging in hot springs are thermally and chemically out of equilibrium with the surrounding atmosphere. The thermal and chemical disequilibrium in the hot spring environment is dissipated via abiotic and biotic processes. Examples of abiotic processes driven by the disequilibrium include cooling (evaporative and conductive cooling), mineral precipitation (i.e., travertine and geyserite formation), gas transfer (in/out gassing of volatile compounds), and electron transfer reactions (e.g., oxidation of hydrogen sulfide). The chemical disequilibrium in hot springs sustains microbial ecosystems that often include chemoautotrophic, heterotrophic, and photosynthetic microorganisms organized along chemical and thermal gradients. The microbial activity in hot spring systems assists in dissipating the chemical disequilibrium via metabolic processes, and the presence of biomats and other microbe-induced structures can affect the kinetics of abiotic processes. For example, *Phormidium tenue* var. *granuliferum* builds small mounds on top of microbial mats which increases the turbulence in the water as it flows over the mounds (Walter et al. 1976). This enhanced turbulence is thought to increase the rate of gas transfer reactions, which may be beneficial to the microbial community. While it has been widely recognized that the organization of microbial ecosystems is largely driven by chemical gradients and that these gradients will be affected by the presence of a microbial system, it remains unresolved how important this biological feedback is on the overall chemistry in hot spring environments. In order to address this question, it is imperative that the controls on the rates of the major abiotic processes are understood. There are two possible approaches to meet this objective, either a laboratory-based experimental study that simulates the hot spring environment or a detailed field study of hot springs where microbial activity is limited.

1.2 *Description of field site*

Here, we report the results of a detailed two-year study of the kinetics of gas transfer in the hot spring system Ojo Caliente located in the Lower Geyser Basin of Yellowstone National Park (YNP), WY. After an initial study in 2001, a second more thorough study was conducted in 2002. Ojo Caliente consists of a source pool and a well-defined run-off drainage that flows into the Firehole River. Ojo Caliente is part of the River Group, but it is isolated and there is no evidence for any other hydrothermal activity in its immediate surrounding. A large meadow with a smooth gradient and abundant flora surrounds Ojo Caliente and the ground is stable.

The drainage is approximately 24 m long. The channel begins wide (121.9 cm) and well defined, after about 4 m the channel becomes narrower. There are several forks in the drainage, however for simplicity only one fork was chosen to study (previous studies on the chemistry of these other forks show no significant deviation; Ball et al. 2001). The temperature of the source pool is ~92°C. At the overflow of the pool, station 1, the temperature is ~90°C and the end of the drainage is 67°C. The spring has essentially a constant discharge rate, chemical composition, and temperature at the time scale of our studies (up to 72 hours). Within the drainage there are changes in flow rate and turbulence as a result of roughness of the channel. Although there is microbial activity in the drainage, there are no extensive biomats nor microbe-induced mineral precipitates, such as travertine, which may significantly affect the flow regime. It should be noted however that after the water temperature drops below 80°C the presence of a red filamentous biomaterial is present in the drainage.

1.3 *Importance of research*

The variability in flow regime along the flow path coupled with the constant input composition of the solution entering the drainage make these springs natural laboratories to study the processes that control gas transfer. In particular, we wish to determine how the rate of gas transfer is affected by flow regime. Laboratory studies as well as studies on natural systems, such as oceans and lakes, indicate that the rate of gas transfer increases significantly with the transition from laminar to turbulent flow (Liss & Slater 1974; Brezonik 1994). Gas transfer kinetics have been studied for lakes and oceans, but there has been no work on the kinetics of gas exchange in hot springs. Gas transfer and heat transfer are the most important abiotic processes that place constraints on the development of microbial ecosystems in hot spring environments. Competition between the rate of gas transfer and the rate of microbial metabolism of those gases becomes more significant at higher temperatures due to accelerated gas transfer kinetics, and this competition may limit the temperature range in which microorganisms are able to subsist in any environment.

A better understanding of heat and gas transfer in hot springs is of importance, as the exploration of Mars for evidence for fossil life becomes a reality in the foreseeable future. High-resolution imagery of landforms on Mars suggests that there are ancient hydrothermal areas (Farmer 1996). These areas are prime targets for exploration. The distribution of microbial life in these systems (if life ever evolved) would have been constrained by the same processes that we can study in hot springs on Earth. The constraints on the distribution of life in the Martian systems would have been different from the constraints present on Earth due to differences in atmospheric pressure and temperature, which affect the boiling point and rate of evaporative cooling and gas transfer. Detailed studies such as the one reported here provide a basis for modeling the constraints placed on a Martian system for a given set of surface conditions (P, T).

2 THEORY OF GAS TRANSFER

Gas transfer between the hydrosphere and atmosphere plays a critical role in regulating the chemistry of natural waters (Brezonik 1994). The controlling factor in gas exchange is the partial pressure gradient between that of the component A in equilibrium with the water, P_A^{fluid} and that of the component in the atmosphere, P_A^{atm}. If the P_A^{fluid} is greater than the P_A^{atm}, the fluid will tend to degas.

The solubility of gases is defined by Henry's Law, which relates the partial pressure of the component A in the atmosphere to the activity, a, of A in the fluid:

$$K_H = a_{Aaq}/P_A^{atm} \qquad (1)$$

where K_H is the Henry's Law constant and a_A is the activity of A in the fluid. K_H is dependent on temperature. Many K_H values have been determined for most major constituents over a wide range of temperatures (Schoonen & Barnes 1997). Because the activity coefficients for dissolved molecular species are very close to unity in dilute solutions, it is acceptable to replace the activity of the dissolved molecular species with its molar concentration (i.e. $a_A = C_A$ in Equation 1).

From Fick's Law it can be derived that the flux of gas transfer can be determined from the following equation:

$$F_A = k_A(C_A - C_A^{eq}) \qquad (2)$$

where F_A is the mass transfer flux for component A (mol cm^{-2} s^{-1}), k_A is the mass transfer coefficient for A (cm/s), C_A is the concentration of A in the water (mol cm^{-3}), and C_A^{eq} is concentration of A in water in equilibrium with the atmosphere (mol cm^{-3}). Two conceptual models exist for the transfer of gas molecules. The first model is the two-film model, which assumes that gas transfer across the gas-water interface is rate determined by one or both films at the interface. Using this model, k_A (Equation 2) can be determined from the following:

$$k_A = D_A/z \qquad (3)$$

where D_A (cm^2/s) is the diffusion coefficient for component A in water and z (cm) is the boundary layer thickness in the water film. The boundary layer thickness is dependent on the hydrodynamics of the system. The two-film model is widely used to model gas transfer in large bodies of water (e.g., lakes, oceans, rivers; Liss & Slater 1974; Gislason 1989; Brezonik 1994). To implement this model it is necessary to determine a value for the boundary layer thickness. While z values have been reported for oceans and lakes at various degrees of wave action, there are no data for the types of systems we are interested in here.

The surface-renewal model is the second conceptual model. In this model the surface film is continually renewed as new parcels of water are brought to the surface by the effects of turbulent mixing (Danckwerts 1951; O'Connor & Dobbins 1958). These parcels, once at the surface, are assumed to equilibrate instantaneously. Therefore, the rate at which gas transfers from solution is solely dependent upon how fast the surface is renewed (i.e. the more turbulent the water, the faster the gas transfer). The k_A is dependent on the product of the square roots of D_A, the flow rate V (cm/s), and the water depth, d (cm):

$$k_A = D_A^{1/2} V^{1/2} d^{-3/2} \quad (4)$$

Both models yield different expressions for k_A. The two film model seems to work well in conditions found in bodies of water such as lakes and oceans, while the surface renewal model appears to work better for conditions where there is flowing water (e.g. rivers). It should be pointed out that none of the studies have measured k_A values for hot spring drainages, like the ones in this study (Holley 1976; Schnoor 1996). Our detailed study of the hot spring drainage, allows determination of the mass transfer flux and the transfer coefficient that will provide a better understanding of the gas transfer process in these environments.

If it is possible to accurately quantify k_A for a non-reactive gas in a drainage with varying degrees of turbulence, then one can evaluate the effects of the flow regime on gas transfer. Since the diffusion coefficients for CO_2, H_2S, O_2, and CH_4 are similar, D_A varies approximately $\pm 1 \times 10^{-5} cm^2/s$ at any given temperature, this would imply the k_A value for each would be nearly the same. Once the k_A value is known for a gas which is expected to undergo only degassing, then it can be used to approximate (for other constituents) the amount of gas that is involved in other chemical reactions. The main purpose of this gas study is to find a gas that can be used to determine a k_A value that is strictly due to degassing, determine if there is a link between the rate of degassing and the hydrodynamic conditions of the drainage and to use that value to further understand how gases are being consumed in other reactions.

3 RESULTS OF GAS ANALYSIS

The hydrodynamics of the drainage were studied both years by taking width and depth measurements, video recordings of the water flow in the drainage, and the turbulence was quantified using an *in situ* gypsum dissolution method to calculate a boundary layer thickness (Dreybrodt et al. 1992; Opdyke et al. 1987). The results of this study indicate a trend that the drainage is more turbulent in the latter part of the drainage. This would imply that gas and heat transfer rates are the fastest in the second part of the drainage.

CO_2 is chosen as the preferred reference gas in this study because it is the most abundant gas, it is not affected by travertine deposition at Ojo Caliente, and microbial consumption is thought to be insignificant compared to the total concentration. Figure 1 shows the first order rate of transfer, k_1, for CO_2 for the 2001 and 2002 study. For both years the rate of degassing is faster in the second part of the drainage. There is a significant difference in the rate in the first part of the drainage between the two years but this could be explained by the fact that in 2002 the initial measured concentration was approximately four times higher than in 2001. CH_4 follows the expected trend of faster degassing in the latter part of the drainage (Fig. 1), in fact there is little difference between the transfer rates of CO_2 and CH_4 (recall the notion that since the diffusion coefficients are similar we expect the rates of gas transfer to be similar if the only process occurring is gas transfer). The CH_4 rate of degassing in the first part of the drainage, however, is approximately twice as fast as the CO_2 rate of degassing. One possible explanation of this is biological activity.

Looking at O_2 ingassing rates (Figure 2), dissolution occurs more quickly in the second part of the drainage for both the 2001 and 2002 data sets. Focusing briefly on the 2002 data, the rate of O_2 ingassing in the first section is slower than the rate of degassing of CO_2 perhaps indicating H_2S oxidation.

In the 2002 study a more extensive study was conducted on H_2S concentrations in the drainage. Samples were taken at three varying light intensities throughout the day, early afternoon (70,600 lux),

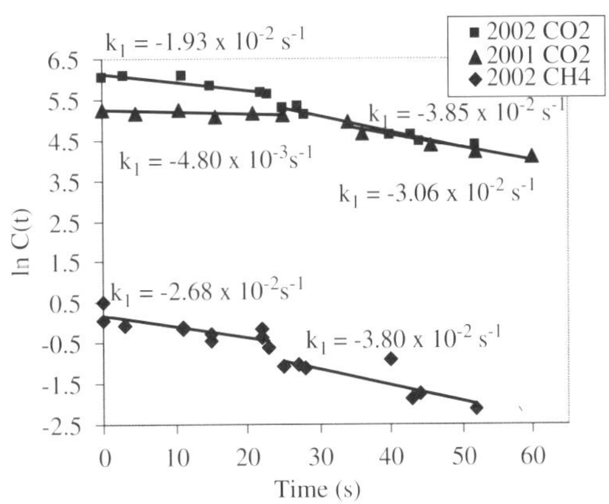

Figure 1. Plot showing first order rates of depletion of CO_2 and CH_4 for Ojo Caliente, YNP.

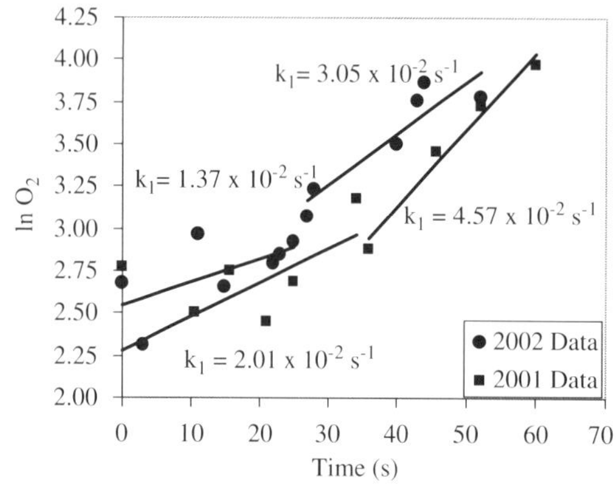

Figure 2. Ingassing trends and first order rates of reaction for O_2 from the 2001 and 2002 studies at Ojo Caliente, YNP.

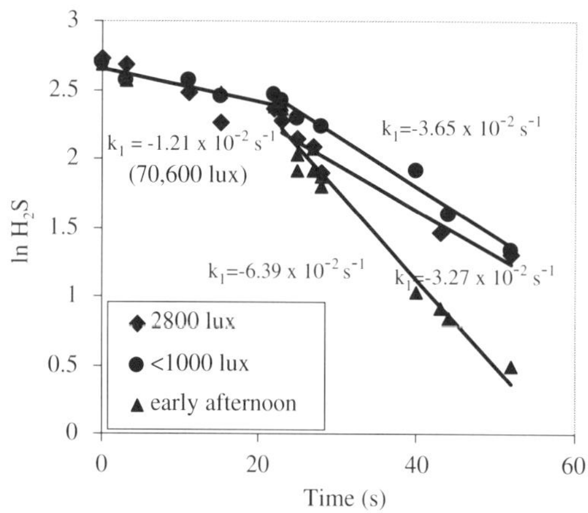

Figure 3. First order reaction rates for H_2S at three different levels of light at Ojo Caliente studied in 2002.

dusk (2800 lux), and dark (<1000 lux). Figure 3 shows that there is little difference in rates in the beginning of the drainage. Of particular interest is the second section of the drainage where the rates of depletion of H_2S become distinctly different especially for the early afternoon versus low light cases. Comparing the transfer rates of CH_4 to the rates obtained for CO_2 in 2002 indicates the low light cases are similar to the rates obtained for CO_2 in both sections of the drainage. The early afternoon case shows a rate almost twice as fast as the other light intensities in the second portion. This implies that there are some photochemical and/or biological reactions occurring. Note that in this part of the drainage, red filamentous biomaterial was observed.

4 CONCLUSIONS

Understanding and being able to quantify heat and gas transfer is a fundamental part of being able to model the abiotic processes in a hot spring. From this study at Ojo Caliente it is important to notice that all the abiotic processes follow the same trend (Figures 1–3). Having a conservative reference gas, CO_2, allows us insight on the abiotic processes controlling gas transfer between the hot spring and the atmosphere. It is also apparent that other processes such as oxidation of H_2S and biological consumption contribute to the dissipation of disequilibrium in these types of systems. This work can assist in understanding the constraints that might be placed on microbial ecosystems in similar drainages and in turn it can be determined how much of an effect biological communities can have on chemical processes.

ACKNOWLEDGEMENT

This project was supported by NSF-EAR.

REFERENCES

Ball, J.W., Nordstrom, D.K., Cunningham, K.M., Schoonen, M.A.A., Xu, Y. & DeMonge, J.M. 2001. Water-chemistry and on-site sulfur-speciation data for selected springs in Yellowstone National Park, Wyoming, 1994–1995. *U.S. Geol. Survey Open File Report* 01-49.

Brezonik, P.L. 1994. *Chemical kinetics and process dynamics in aquatic systems.* Boca Raton: Lewis Publishers.

Danckwerts, P.V. 1951. Significance of liquid-film coefficients in gas absorbtion. *Ind. Eng. Chem.* 43: 1460–1467.

Dreybrodt, W., Buhmann, D., Michaelis, J. & Usdowski, E. 1992. Geochemically controlled calcite precipitation by CO2 outgassing: Field measurements of precipitation rates in comparison to theoretical predictions. *Chem. Geol.* 97: 285–294.

Farmer, J.D. 1996. Hydrothermal systems on Mars: an assessment of present evidence. In M.R. Walter (ed.), *Evolution of hydrothermal ecosystems on Earth (and Mars?)*: 273–299. Chichester: Wiley.

Gislason, S.R. 1989. Kinetics of water-air interactions in rivers: A field study in Iceland. In: D.L. Miles (ed.), *Water-Rock Interaction-6:* 263–266. Rotterdam: Balkema.

Holley, E.R. 1976. Oxygen transfer at the air-water interface. In R.J. Gibbs & R.P. Shaw (eds.), *Transport processes in lakes and oceans.* New York: Plenum.

Liss, P.S. & Slater, P.G. 1974. Flux of gases across the air-sea interface. *Nature* 247: 181–184.

O'Connor, D.J. & Dobbins, W.E. 1958. Mechanisms of reaeration in natural waters. *Trans. Am. Soc. Civil Eng.* 153: 641–684.

Opdyke, B.N., Gust, G. & Ledwell, J.R. 1987. Mass transfer from smooth alabaster surfaces in turbulent flow. *Geophys. Res. Lett.* 14(11): 1131–1134.

Schnoor, J.L. 1996. *Environmental modeling.* New York: Wiley-Interscience.

Schoonen, M.A.A. & Barnes, H.L. 1997. Chemical and physical data for hydrothermal systems. In H.L. Barnes (ed.), *Geochemistry of Hydrothermal Ore Deposits*: 937–962. New York: Wiley.

Walter, M.R., Bauld, J. & Brock, T.D. 1976. Microbiology and morphogenesis of columnar stromatolites (Conophyton, Vacerrilla) from hot springs in Yellowstone National Park. In M.R. Walter (ed.), *Developments in Sedimentology, 20*: 275–310. Amsterdam: Elsevier.

Water–Rock Interaction, Wanty & Seal II (eds)
© 2004 Taylor & Francis Group, London, ISBN 90 5809 641 6

Transport of injected isobutane by thermal groundwater in Long Valley caldera, California, USA

W.C. Evans, T.D. Lorenson, M.L. Sorey & D. Bergfeld
US Geological Survey, Menlo Park, CA, USA

ABSTRACT: Isobutane was inadvertently injected into a thermal aquifer at the Casa Diablo power plant in the Long Valley caldera, California, beginning in January 1993. It was detectable in surface features ≤4 km away in about 2 years and reached Hot Creek gorge, 8 km away, within 4 years. The isobutane analyses show that tracer fronts travel east at ~2 km/yr and also reveal a westward flow of thermal water from Casa Diablo.

1 INTRODUCTION

The Long Valley caldera in eastern California formed 760,000 years ago, but recent and recurrent magmatism supports an active hydrothermal system (Sorey et al. 1991). The hot water feeds a geothermal power plant and many natural hot springs (Fig. 1). Some of the hot springs and thermal wells respond to seismicity (Roeloffs et al. 2003), and many thermal features are monitored for changes in chemistry, flow, or water level that might reflect intrusive processes beneath the caldera. Linking observed changes to the correct cause requires some understanding of the thermal-fluid flow regime, a topic discussed here using isobutane as a tracer.

2 HYDROLOGY

Water levels and temperature logs from thermal wells, geophysical measurements, and fluid geochemistry have all been incorporated into models of the hydrothermal system (Sorey 1985; Blackwell 1985; Sorey et al. 1991). The main features of these models can be summarized as follows. Precipitation on the west rim of the caldera recharges a geothermal reservoir beneath the West Moat near the young (<1–40 ka) Inyo Volcanic Chain. A plume of 215–240°C water rises from this reservoir and flows laterally southeastward through permeable layers toward the Casa Diablo power plant and around the south side of the resurgent dome (Fig. 1).

The elevation of the caldera floor drops steadily along this flow path, causing a decrease in depth to the thermal aquifer from west to east (Farrar et al. 2003). Vapors from the aquifer begin to find permeable conduits to the surface a few km west of Casa Diablo, feeding weak fumaroles (steam vents) at sites SRF, BF, and CHS. Vigorous venting occurs around the plant, where production wells tap the main thermal aquifer at ~150 m depth. Beneath the plant, there is a deeper and slightly cooler thermal aquifer at ~600 m depth (the injection zone) that is separated from the shallow aquifer by less permeable rocks. Whether these two flow systems merge away from Casa Diablo is unclear, but connections between them must exist as discussed later. Thermal waters appear at Hot Bubbling Pool (HBP), several km east of Casa Diablo. The main discharge of hot water occurs several km further east at springs in Hot Creek gorge (e.g., HC, HCW, HCFW), but a few smaller, sub-boiling springs (CC, SHSU, SHSL) that are probably part of the same flow system discharge even further east. Based on gaging and indirect geochemical methods, Sorey et al. (1991) estimated the total flow of hot water to be ~370 L/s.

Several questions about the hydrothermal system remain unresolved. The fluid velocity in the main thermal aquifer is not well known, although Blackwell (1985) modeled thermal profiles in wells to obtain natural-state velocity estimates of 104–228 m/yr prior to geothermal development. The flow regime beneath the resurgent dome is poorly resolved. Temperature reversals with depth in some wells suggest that the regional west-to-east flow also occurs beneath the southern half of the resurgent dome with eventual fluid discharge at Little Hot Creek (LHC). Other evidence from well logs, geophysical studies, and fluid geochemistry of the springs argues against such flow (Farrar et al. 2003, Pribnow et al. 2003).

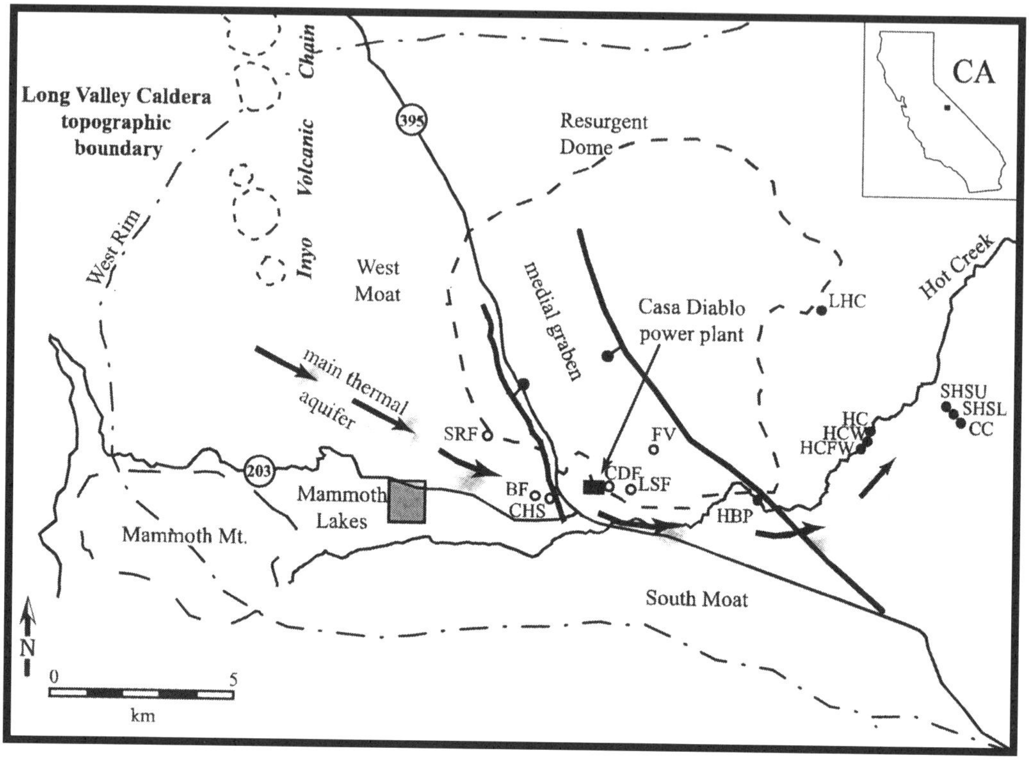

Figure 1. Map of the Long Valley caldera with locations of steam vents (open circles) and hot springs (filled circles) monitored for isobutane. Medial-graben boundary faults shown as heavy lines with ball and bar; arrows represent main thermal water flow path.

Finally, the effects of power plant operations on the hydrothermal system are not fully understood.

3 POWER PLANT OPERATIONS AND LEAKS

The power plant at Casa Diablo produces electricity using a binary generation process. Water at ~170°C is pumped up and used to heat isobutane, the working fluid that drives the turbines. The cooled water at ~85°C is injected into a deeper zone in order to prevent rapid cooling of the production zone. Occasional leaks in the heat exchangers allow high-pressure isobutane to enter the water stream and be pumped down into the injection zone. The spread of isobutane through the hydrothermal system can be traced by collecting gas samples at hot springs and steam vents.

A schematic cross section at Casa Diablo is shown in Figure 2. Production wells are located toward the west side of a fault-bounded graben and injection wells toward the east. Power production began at Casa Diablo in 1985 and was increased from 10 to 40 MWe in 1990. The total pumping rate is currently

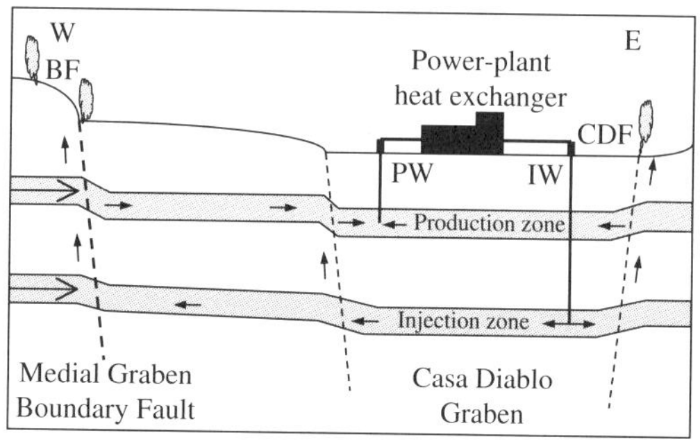

Figure 2. Schematic cross-section through the Casa Diablo power plant showing the two thermal aquifers and hypothesized fluid flow paths. Large arrows show natural west to east flow, and small arrows show flow induced by the production (PW) and injection (IW) wells.

about 850 L/s, considerably more than the estimated natural through-flow rate of 370 L/s in the aquifer, and plant operations have thus had definite impacts on the hydrothermal system even though the water is later injected. Steam venting and ground-temperature increases have occurred in several areas near the plant, and the ground surface within the graben has subsided by tens of cm relative to the surrounding caldera floor (Howle et al. 2003).

Company personnel confirm 10 isobutane leaks to date (Fig. 3a). Some leaks were quite large. For example, in the first leak, an estimated 45,000 kg of isobutane were injected into the deep aquifer during the last three weeks of January 1993, and ~18,000 kg were injected during the second leak in May 1994.

4 ISOBUTANE IN THE HYDROTHERMAL SYSTEM

Isobutane is a non-toxic gas (B.P. −11.7°C) with a very low solubility in water, comparable to that of nitrogen or helium (Wilhelm et al. 1977), so it will strongly fractionate into the vapor phase. Our study therefore focused mainly on sampling fumaroles and the gas phase in bubbling hot springs. This protocol can easily trace the spread of isobutane but does not directly provide information on the concentration of isobutane in the thermal water at a given site.

Samples were collected either in pre-evacuated tubes or in screw-cap septum bottles by water displacement. The results from sites that were repeatedly sampled are shown in Figure 3, expressed as parts-per-million by volume (ppmv) of the gas phase. Correction for air contamination in the samples is desirable but cannot be done using a consistent method for the entire dataset. However, the gas at all of the sites is strongly dominated by CO_2, and the variability induced by air contamination was eliminated by adjusting all analyses to 100% CO_2 and increasing the measured isobutane concentrations proportionately. Errors introduced by this adjustment are small for the samples in Figure 3. A few samples (~2%) with major air contamination or other problems were removed from the dataset.

Despite the fact that the isobutane is injected into the deeper zone beneath the main producing aquifer, it reaches the surface very quickly in the vicinity of the power plant. Two days after the end of the first leak in January 1993, a high concentration of isobutane was detected in the Casa Diablo fumarole (CDF) on the east side of the field (Fig. 3a). It seems likely that the near-vertical faults bounding the field provide a connection between the two thermal zones. The isobutane record at CDF shows that, in general, the concentration rises rapidly after a leak and then falls off exponentially with time.

For a few years following the first leak, sampling was limited to CDF and other sites near the power plant (not shown). By the time sampling was extended to sites away from the plant in 1995 (Fig. 3b), isobutane had already spread as far east as HBP and as far west as Basalt Fumarole (BF).

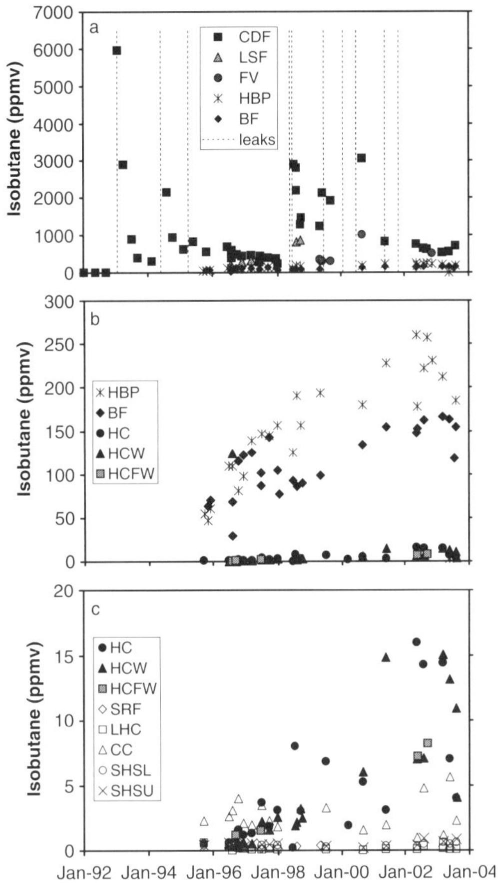

Figure 3. Adjusted isobutane concentrations at sites shown in Figure 1 (each value was multiplied by 100 and divided by the %-CO_2 measured in same sample). The timing of isobutane leaks is shown by dashed lines in part a. Open symbols in part c denote sites that have shown no apparent increase with time in response to the leaks.

The appearance of isobutane at BF was at first surprising, given that thermal groundwater is assumed to flow mainly from west to east. As noted earlier, however, operations at the power plant have affected the local hydrology, and it seems plausible that the high rate of injection has reversed the natural gradient in the deeper aquifer. Water in this aquifer may currently flow both west and east from the plant. Some of the water flowing westward must be able to rise back up into the shallow aquifer along faults beneath the BF area where limited boiling allows isobutane to reach the surface (Fig. 2).

The isobutane records at both HBP and BF show a noisy but general increase in concentration with time (Fig. 3b), not the clear, spike-like response to individual leaks that is seen at CDF (Fig. 3a). None of the leaks that occurred from May 1998 to October 2001 produced a recognizable jump at HBP or BF, and concentrations seem to have increased most rapidly between late 1995 and early 1998, a period of time when no leaks occurred. Extrapolating these early buildup curves down to zero (both sites had <5 ppmv isobutane in 1989) suggests that isobutane reached both places around January 1995.

The more gradual buildup of isobutane at HBP and BF, relative to CDF, can probably be attributed to dispersion of the tracer during flow. The seemingly random noise in the records is not understood but could reflect variations in the temperature at which the gases are collected into the sampling bottles. Temperature variations would affect the relative proportions of CO_2, a soluble gas, and the highly insoluble isobutane in the gas phase at the time of collection. Methane, another sparingly soluble gas and minor constituent in this system, shows temporal variations comparable to those of isobutane.

Isobutane is monitored at three sites in Hot Creek gorge (HC, HCW, HCFW) as shown in Figure 3c. The record is again noisy, due in part to detection limits, but all three sites show clear isobutane increases with time beginning about January 1997 or slightly earlier. To date, isobutane increases with time have not been detected at sites east of Hot Creek or at SRF, the westernmost site. One site (CC) does show isobutane concentrations of a few ppmv, but on average, the values have held steady over the period of record, indicating that this is the natural background level for this spring. The high concentrations of isobutane measured at FV (Fig. 3a) show that the main thermal aquifer does pass under the very southern part of the resurgent dome. However, isobutane from the power plant is not detectable (no increase with time) at LHC (Fig. 3c), which may be part of a different, smaller system.

The isobutane data provide new information on fluid velocities. Isobutane reached HBP, 4 km away from Casa Diablo, about 2 years after the first leak, and it reached Hot Creek gorge, another 4 km away, 2 years later. These arrivals yield tracer-front velocities of ~2 km/yr to both sites, an order of magnitude faster than the average fluid velocity modeled by Blackwell (1985). The disparity may result from pronounced channelization of flow through fractures in the eruptive units that constitute the thermal aquifer. Whatever the reason, this finding certainly has relevance to other geochemical monitoring that focuses on temporal responses. Calculating average fluid velocity based on isobutane is difficult with the available data and will always be imprecise due to the complex history of the leaks. If one assumes that the multiple leaks resemble continuous injection, and that the isobutane at HBP (Fig. 3b) has leveled off ~10 years after the first leak, the average fluid velocity would be ~400 m/yr, only a factor of 2 higher than obtained from thermal profile modeling. We also note that the modeling applied to conditions prior to development. However, it seems unlikely that power plant operations would have greatly altered velocities in the far field, because the water is reinjected and the two aquifers may merge east of Casa Diablo, where observed changes in the depths to thermal water are minimal.

5 CONCLUSIONS

Isobutane not only serves to generate electricity at the Casa Diablo geothermal plant, but also provides a useful way to track fluid flow in the hydrothermal system at Long Valley caldera. Its accidental release into the system has provided unique information on present-day flow directions and velocities. Isobutane has several distinct advantages as a tracer. It partitions strongly into the vapor phase and therefore reveals fluid flow beneath steam vents, something non-volatile tracers, such as fluorescent dyes, cannot do. It is a highly stable molecule with a symmetric structure that is not likely to adsorb strongly on substrates, and microbial production or degradation seems unlikely under geothermal conditions. Because many modern geothermal plants are binary, it may have a fairly broad application.

ACKNOWLEDGEMENTS

We thank Ted Courteposse and Bob Sullivan of Mammoth Pacific Co. for providing information on the isobutane leaks. Jennifer Dougherty helped with isobutane analyses.

REFERENCES

Blackwell, D.D. 1985. A transient model of the geothermal system in the Long Valley caldera. *J. Geophys. Res.* 90: 11229–11241.

Farrar, C.D., Sorey, M.L., Roeloffs, E., Galloway, D.L., Howle, J.F. & Jacobson, R. 2003. Inferences on the hydrothermal system beneath the resurgent dome in Long Valley Caldera, east-central California, USA, from recent pumping tests and geochemical sampling. *J. Volcanol. Geotherm. Res.* 127: 305–328.

Howle, J.F., Langbein, J.O., Farrar, C.D. & Wilkinson, S.K. 2003. Deformation near the Casa Diablo geothermal well field and related processes Long Valley caldera, Eastern California, 1993–2000. *J. Volcanol. Geotherm. Res.* 127: 365–390.

Pribnow, D.F.C., Schütze, C., Hurter, S.J., Fleshsig, C. & Sass, J.H. 2003. Fluid flow in the resurgent dome of Long Valley Caldera: implications from thermal data and deep electrical sounding. *J. Volcanol. Geotherm. Res.* 127: 329–345.

Roeloffs, E., Sneed, M., Galloway, D.L., Sorey, M.L., Farrar, C.D., Howle, J.F. & Hughes, J. 2003. Water level changes induced by local and distant earthquakes at Long Valley caldera, California. *J. Volcanol. Geotherm. Res.* 127: 269–303.

Sorey, M.L. 1985. Evolution and present state of the hydrothermal system in Long Valley caldera. *J. Geophys. Res.* 90: 11219–11228.

Sorey, M.L., Suemnicht, G.A., Sturchio, N.C. & Nordquist, G.A. 1991. New evidence on the hydrothermal system in Long Valley caldera, California, from wells, fluid sampling, electrical geophysics, and age determinations of hot spring deposits. *J. Volcanol. Geotherm. Res.* 48: 229–263.

Wilhelm, E., Battino, R. & Wilcock, R.J. 1977. Low-pressure solubility of gases in liquid water. *Chem. Rev.* 77: 219–262.

Water–Rock Interaction, Wanty & Seal II (eds)
© 2004 Taylor & Francis Group, London, ISBN 90 5809 641 6

Relationships between structural boundaries, fluid circulation and permeability changes during the 2002–2003 Stromboli eruptive crisis

A. Finizola
Osservatorio Vesuviano, Istituto Nazionale di Geofisica e Vulcanologia, Napoli, Italy

F. Sortino
Istituto Nazionale di Geofisica e Vulcanologia, Palermo, Italy

ABSTRACT: Reiterative Self-Potential (SP), temperature (T) and CO_2 soil gas measurements have been taken in the summit of Stromboli volcano before and during the 2002–2003 eruptions. These fluid measurements allowed an assessment of: (1) the role of previously identified structural boundaries (Large Fossa crater) in the opening of new fractures; (2) the increase in permeability in the summit area during the eruptions related to the SW crater, Large Fossa crater and Sciara del Fuoco sector collapse and boundaries; and (3) the best strategy to install a permanent fluid monitoring system in the area most sensitive to fluid-flow variations.

1 INTRODUCTION

In 1994 we began a fluid circulation survey near the summit of Stromboli volcano coupling Self-Potential (SP), temperature (T) and soil gas measurements. The aim of this multidisciplinary approach, involving geoelectrical and geochemical techniques, is to develop a detailed fluid circulation model for the summit crater area. The corresponding three papers (Finizola et al. 2002, 2003; Revil et al. 2004) discuss the structural and fluid circulation system before the Stromboli 2002–2003 eruptive crisis. In the first paper, the authors identified one hydrothermal system subdivided into three main convective cells intersected by two regional faults. In the convective cell located in the summit (Fossa-Pizzo) area, close to the active craters, SP, T and CO_2 measurements displayed anomalies related to the present-day (NE and SW) active craters and other anomalies associated with older crater boundaries (called Large Fossa crater and Pizzo crater) representing the preferential drains for hot fluid migration in the summit area (Finizola et al. 2003).

Since January 2003 we started a measurement campaign to assess the disruptions that occurred in the fluid circulation in the summit area, during the period of the eruptive crisis. This seismic activity, which began on 28 Dec. 2002, involved several important morphological changes:

- lava flow from the Sciara del Fuoco area from a eruptive fissure associated with partial collapse of the northern flank of the NE crater,
- 2 collapses in the submarine and aerial parts of the edifice ($\sim 20 \times 10^6\,m^3$), and complete refilling of the subaerial part by lava flows during the seven months of effusive activity,
- collapse of the summit craters, which reached the union of Central and NE, and also partial collapses inside the SW crater, and
- development of a set of fresh fractures cutting the entire Fossa area, some of them associated with vapor emission.

The main problem for the scientific community during this eruptive cycle was to assess the flank stability of the volcanic edifice, and in particular the possible reactivation of the old sliding plane of the Sciara del Fuoco sector collapse area ($\sim 1\,km^3$). On 5 April 2003, a large paroxysmal eruption damaged one house 2 km away from the active vents, but did not affect the stability of the flank edifice. At the end of July 2003, normal typical mild explosive activity began again in the summit craters, and the effusive rate in the Sciara del Fuoco area ceased.

The three main topics that will be discussed in this paper are: (1) What structural and related fluid circulation changes occurred during the eruptive crisis, (2) What is the importance of these changes for volcanic hazard assessment, and (3) What strategy can be developed to install a permanent fluid monitoring system at Stromboli.

2 DATA ACQUISITION

During the 2002–2003 Stromboli eruptions, three field campaigns were carried out in January, March and April 2003, to (1) map the new opened fractures in the summit area, and their temporal evolution and (2) reiterate SP, T and CO_2 soil gas measurements along a strategic North-South profile, the location of which was based on previous work.

SP data were acquired every meter using a high-impedance voltmeter, a pair of $Cu/CuSO_4$ non-polarizing electrodes, and an insulated electric cable. The reference potential was taken in the Fossetta area, 200 meters away from the huge fluid migration system of the Fossa area. T measurements were acquired by thermal probes and digital thermometer every 2.5 meters and at a precise depth of 30 cm. Measurements of soil CO_2 concentration were acquired with the same spacing as the T measurements. Gas was pumped through a copper tube 2 mm in diameter, inserted in the soil to a depth of 0.5 m, and analyzed in the field by infrared spectrometry.

3 RESULTS

3.1 *Structural changes in the Fossa area*

Detailed cartography of opened fractures that appeared at the beginning of the eruptive crisis superimposed on the thermal map constructed in August 1999 (Finizola et al. 2003), allow us to distinguish two kind of fractures (Fig. 1): (1) a NW-SE oriented set (fracture A-B-C-D-E), superimposed on the maximum thermal anomalies identified in 1999 (F1-F3 anomalies; Finizola et al. 2003), and (2) a set oriented approximately NE-SW, not related to previously defined thermal anomalies. The reiteration of the fracture cartography in April 2003 shows a lateral extension of about 10 meters on numerous fractures between January and April 2003 (Fig. 1). We also measured temperature and CO_2 concentration at 5 meter spacings along these fractures.

Between 1999 and 2003 there was a significant increase in temperature at several points inside the Large Fossa area, from 45–55° to more than 90°C respectively, while the cold zone defined in 1999 called Small Fossa crater (Finizola et al. 2003) did not display any variation during the 2002–2003 eruptive crisis (Fig. 2).

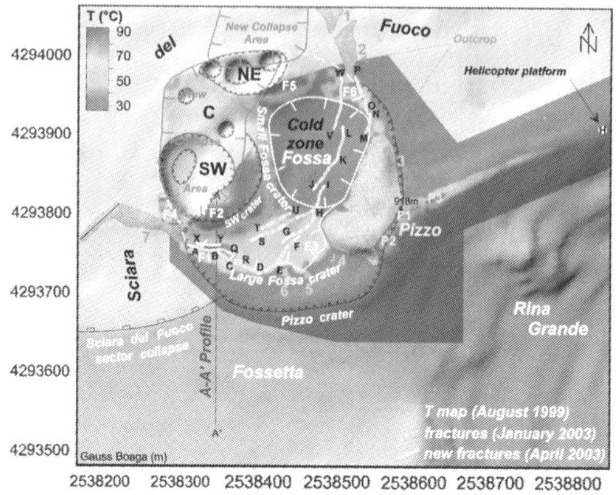

Figure 1. Fracture system map in January and April 2003 superimposed on the thermal map at 30 cm depth carried out in August 1999 with structural boundaries inferred by SP, T and CO_2 measurements (Finizola et al. 2003, modified). F1, F2, ... and P1, P2, ... represent the maximum of the thermal anomalies encountered in 1999 respectively in the Fossa and Pizzo area. Numbers 1–7 are outcrops (hatched pattern).

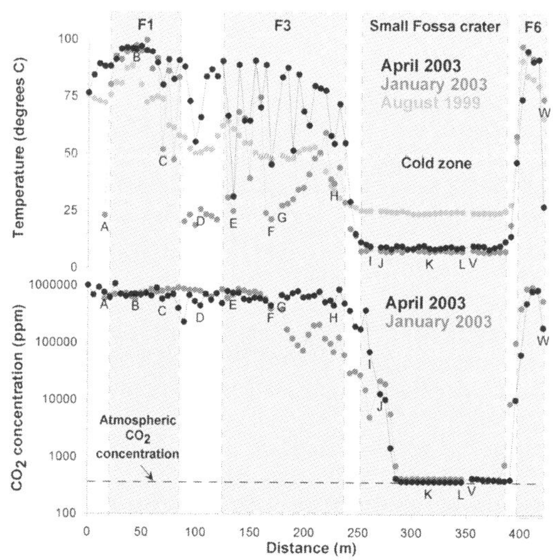

Figure 2. Reiteration in January and April 2003 of temperature and CO_2 soil concentration measurements along the new opened fractures crossing the entire Fossa area.

Along the fracture system crossing all the Fossa area from F1 to F6 anomaly, an increase in temperature was observed during the eruptive cycle between January and April 2003 (Fig. 2).

A permanent temperature monitoring system installed near the F1 anomaly, from March to May

2003, corroborated the general increase of temperature until the paroxysmal eruption of 5 April 2003. After the paroxysmal event, and until May, the temperature decreased slightly.

The CO_2 soil concentrations measured along the fracture system shows high CO_2 concentration in all the Fossa area since January 2003, except in the cold zone, which remained impermeable to gas ascent.

As heat is mainly transported by steam derived from energy accumulation through the latent heat of vaporization, the results presented above could be interpreted considering that the ascending steam took probably more time than the CO_2 propagation to migrate along the fracture system of the Fossa area.

3.2 *Reiterative A-A' profile*

Along the A-A' profile (Fig. 1), SP, T and CO_2 soil concentration measurements have been acquired since August 1994, August 1995, and October 2002 respectively, crossing the structural boundaries of SW crater, Large Fossa crater, Pizzo crater and the Sciara del Fuoco sector collapse (Figs. 1 & 3). This multidisciplinary approach allows a comparison of the suitability of these methods to assess permeability changes and related disruptions in fluid flow transport during this eruptive crisis.

3.2.1 *SP measurements*

Figure 3 shows SP data along the A-A' profile from August 1995 to January 2003. A clear increase was seen in the amplitude of the anomalies in the SW crater and the Large Fossa crater boundaries, respectively. The F2 and F1 anomalies, recorded in January and March 2003, reached greater values than ever recorded. This result can be interpreted as a major water flux generating a higher streaming potential signal, in good agreement with the general temperature increase of the Fossa area as seen before.

3.2.2 *T measurements*

Temperature measurements display (1) a reduction in the width of the F1 and F2 anomalies between 1994–1995 and 2002–2003, and (2) no significant change in the amplitude of the anomalies. This result is not in contradiction with the SP signal: (1) the reduction in the width of the F1 and F2 anomalies can be explained by the increase in permeability induced by recent fracturing, facilitating the migration of hot fluid along more opened structural boundaries and (2) the absence of increase in the amplitude of T anomalies is linked to the location of the Fossa hydrothermal system between the magma body and the surface. The latter allows the amplitude of T anomalies to stabilize at temperatures not exceeding the boiling point of water.

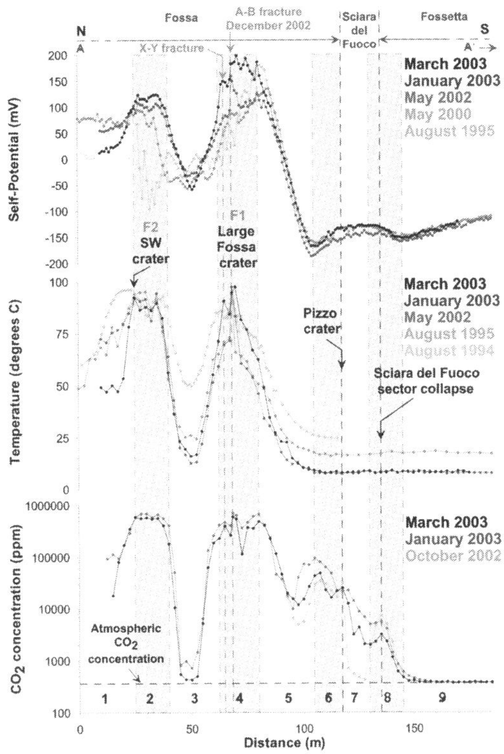

Figure 3. Reiteration of SP, T and CO_2 soil concentration measurements along the A-A' profile before and during the eruptive crisis.

3.2.3 *Soil CO_2 measurements*

Along the A-A' profile, an interesting change occurred in soil gas concentration between the Sciara del Fuoco sector collapse and Pizzo crater. In October 2002, before the eruptions, two distinct CO_2 peaks were identifiable in coincidence with the Sciara del Fuoco sector collapse and Pizzo crater. These two peaks were separated by very low CO_2 concentrations, showing a preferential gas migration along these two structural boundaries. In January 2003 a sharp increase in CO_2 concentrations characterized the spacing between the two structural boundaries, removing the previous recognition of two distinct anomalies. This transition in CO_2 concentrations between the Sciara del Fuoco sector collapse and Pizzo crater can be explained by an increase in permeability between the Sciara del Fuoco collapse and the Pizzo crater. This means that the southern Sciara del Fuoco sector collapse boundary probably has been affected slightly by the collapse events that occurred at the beginning of the eruptive cycle, increasing the permeability inside the Sciara del Fuoco sector collapse area. The data acquired in March 2003 display a

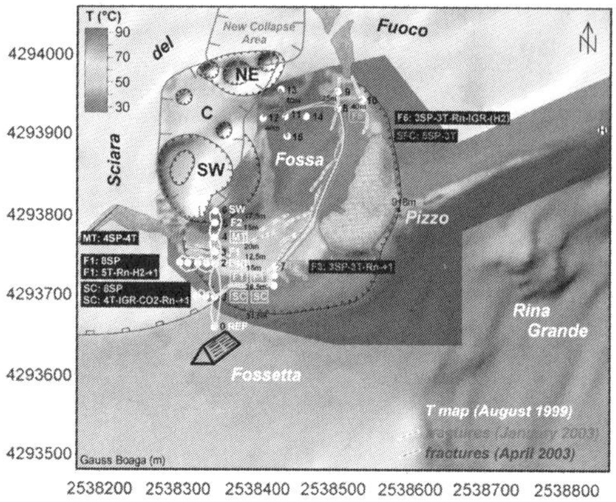

Figure 4. Fracture map of the Fossa area showing the location of the 16 sites (numbers 0 to 15) which have been chosen to install permanent SP, T and gas monitoring.

general decrease in CO_2 concentration along the entire profile, which can be interpreted as a lower input of CO_2 coming from depth.

We also observed a sharp decrease of all three investigated parameters inside the SW crater during the eruptive cycle (between January and March 2003). This result is probably related to cooling inside the SW crater, due to descent of the magma (the heat source) inside the conduits to sustain lower-elevation lava flows in the Sciara del Fuoco area.

4 STRATEGIES TO INSTALL A FLUID MONITORING SYSTEM

The use of different fluid measurement techniques during the Stromboli 2002–2003 eruptive crisis pointed out the different sensitivity of these techniques to assess fluid-flow variations. Moreover, the reiteration of SP, T and soil gas measurements along the A-A' profile before and during the eruptive cycle precisely defined areas where changes were observed in the signal prior to eruptions but where no significant variation occurred during the eruptive cycle. Based on these results, 16 sites have been chosen to install a permanent SP, T and soil-gas monitoring system in 2004. Figure 4 displays the location of these sites including:

- A reference site (no. 0 in Fig. 4) where no significant variation has been observed for all the parameters during the eruptive cycle.
- A control in several points of the new fracture system that appeared in 2002–2003 (no. 3, 7, 8, 9 & 10 in Fig. 4).
- A point of thermal and SP vs. depth gradient measurement where the temperature is hot (30–70°C) but not saturated by the boiling point of water (no. 2 in Fig. 4).
- A control of gas migration close to the Sciara del Fuoco sector collapse to assess permeability disruptions linked to possible reactivation of the gliding plane of Sciara del Fuoco sector collapse (no. 1 in Fig. 4).
- Measurements closer to the active craters to assess changes in heat transfer more influenced by the magma body of the SW and NE active craters (no. 5, 6, 12 and 13 in Fig. 4).

5 CONCLUSION

The combined SP, T and soil gas measurements carried out during the 2002–2003 Stromboli eruptive cycle show that this approach constitutes a powerful tool to assess (1) the locations of drains for fluid migration, which constitute points of weakness of a volcanic edifice that must be considered in volcanic hazard maps and (2) the timing of fluid flow and permeability changes along these structural boundaries. During a volcanic cycle, these measurements can provide important additional information to better assess the status of fracture propagation inside a volcanic edifice.

REFERENCES

Finizola, A., Sortino, F., Lénat, J-F. & Valenza, M. 2002. Fluid circulation at Stromboli volcano, (Aeolian Islands, Italy) from self-potential and soil gas surveys. *J. Volcanol. Geotherm. Res.* 116(1–2): 1–18.

Finizola, A., Sortino, F., Lénat, J-F., Aubert, M., Ripepe, M. & Valenza, M. 2003. The summit hydrothermal system of Stromboli: New insights from self-potential, temperature, CO_2 and fumarolic fluid measurements, with structural and monitoring implications. *Bull. Volcanol.* 65: 486–504, doi:10.1007/s00445-003-0276-2.

Revil, A., Finizola, A., Sortino, F. & Ripepe, M., in press. Geophysical investigations at Stromboli volcano, Italy. Implications for ground water flow and paroxysmal activity. *Geophys. J. Internat.*

Water–Rock Interaction, Wanty & Seal II (eds)
 ISBN 90 5809 641 6

Thermal springs at Colombian active volcanoes

G. Garzón, S.P. Salazar & D.Y. Serna
Instituto de Investigación e Información Geocientífica, Mineroambiental y Nuclear – INGEOMINAS, Manizales, Colombia

L. Bobadilla & L.E. Lesmes
INGEOMINAS, Bogotá D.C., Colombia

J.C. Diago
INGEOMINAS, Popayán, Cauca, Colombia

J. Mojica
INGEOMINAS, Cali, Valle, Colombia

ABSTRACT: In the past 18 years, 75 thermal springs were sampled and analyzed as part of a physicochemical monitoring project of Colombian active volcanoes. Chemical, mineralogical and isotopic studies were carried out with the purpose of classifying thermal spring discharges and understanding water-rock interactions. Acid sulfate waters were identified in the upper flanks of active volcanoes, whereas chloride and bicarbonate waters were identified in the lower regions in valleys, on active normal faults, and on intersections of faults. The chemistry of only one thermal spring indicates that full equilibrium has been established between water and host rocks. The chemistry of five thermal springs indicates partial equilibrium between waters and host rocks.

1 INTRODUCTION

The volcanic surveillance in Colombia was formally initiated after reactivation of Nevado del Ruiz volcano was recognized in 1985. The first thermal water samplings with the purpose of volcanic activity monitoring began in spring discharges located in the edifice of Nevado del Ruiz volcano (Giggenbach et al. 1990). In the course of reactivation of Galeras and Puracé volcanoes, these studies were extended to other five active Colombian volcanoes.

From 1996 to 2001 were carried out a detailed study of all thermal springs inventoried around 14 active Colombian volcanoes, in order to study its genesis, understand water-rock interaction and their possible use as a tool in the volcanic monitoring.

We present a physicochemical characterization of 75 thermal spring discharges (Garzón 1997; Garzón 2001) associated with active volcanoes in the northern, central and southern volcanic segments of Colombia (Fig. 1).

2 DISTRIBUTION AND TEMPERATURES OF SPRINGS

Thermal spring discharges at Colombian active volcanoes are located in altitudes between 1720 and 4700 m elevation. In the northern volcanic segment, two thermal springs with temperatures of 24 and 37°C are located close to Cerro Bravo volcano, 14 thermal springs (28–91°C) are located around Nevado del Ruiz, five thermal springs (22–84°C) are located on the Paramillo de Santa Rosa, four thermal springs (24–52°C) are located around Nevado del Tolima volcano, five thermal springs (28–92°C) are located on Cerro Machín. In the central volcanic segment, three thermal springs (29–32°C) are located near Nevado del Huila; 15 thermal springs (24–90°C) are located around Los Coconucos volcanic chain (Fig. 2), and two thermal springs with temperatures of 33 and 36°C are located on Sotará. In the southern volcanic segment, four thermal springs (26–62°C) are located near Doña Juana volcano, six thermal springs

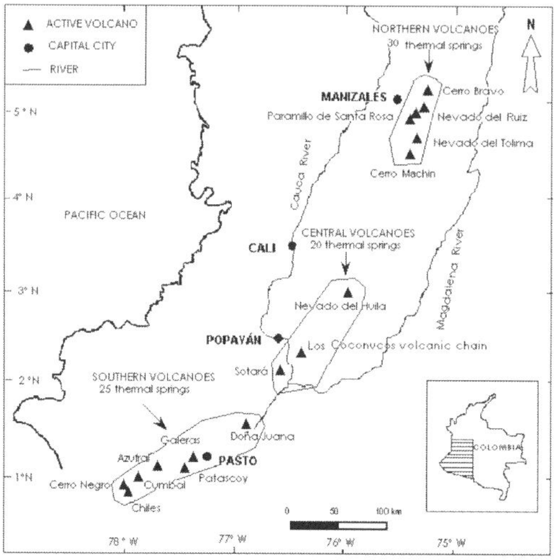

Figure 1. Colombian active volcanoes located in three main segments.

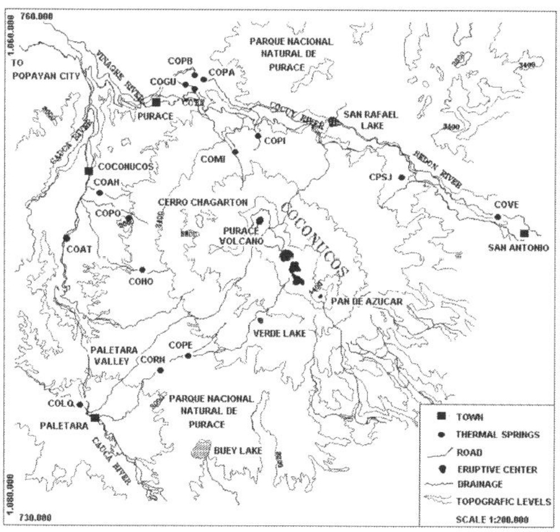

Figure 2. Location of thermal springs at Los Coconucos volcanic chain.

(22–45°C) are located near Galeras volcano, four thermal springs (25–76°C) are located near Patascoy volcano, five thermal springs (23–58°C) are located on Azufral volcano, four thermal springs (27–33°C) are located near Cumbal volcano, and two thermal springs (42 and 46°C) are located on Chiles volcano.

3 SAMPLING AND ANALYTICAL METHODS

Thermal spring waters were collected following the techniques described by Giggenbach & Goguel (1989). A *Barnant, K* type thermocouple thermometer was used to measure the temperature of spring discharges. The pH of springs was measured using a *Hanna* pH-meter. Cations in the waters were analyzed by a *Perkin Elmer 3110* atomic absorption spectrophotometer in the laboratory of gases and waters at INGEOMINAS, Bogotá. Anions were analyzed in the laboratory at Volcanological and Seismological Observatory at INGEOMINAS, Manizales by classical methods (e.g., volumetrically, titration, electrochemically). Unfiltered water samples were collected in glass bottles and oxygen and hydrogen isotopic compositions were determined with a *Balzers Prisma* quadrupole mass spectrometer in the isotope laboratory at Bundesanstalt für Geowissenschaften und Rohstoffe – BGR in Hannover, Germany. Mineralogical analyses were carried out for thermal spring host rocks by X-ray powder diffraction in a *Rigaku rint 2200* diffractometer with Bragg-Brentano geometry and optical microscopy of polish thin sections.

4 RESULTS AND DISCUSSIONS

4.1 *Major ions of thermal springs*

The major anion (SO_4^{2-}, Cl^-, HCO_3^-) and cation (Ca^{2+}, Mg^{2+}, Na^+, K^+, Si^{4+}) compositions were determined for 75 thermal spring waters. Thermal springs can be classified based on relative concentrations of the major anions SO_4^{2-}, Cl and HCO_3^- (Ellis & Mahon 1977; Lewicki et al. 2000).

In the northern volcanic segment, the two thermal springs at Cerro Bravo volcano are both bicarbonate in composition with pH of 6.0 and 6.3. Nevado del Ruiz volcano has two thermal springs classified as neutral chloride; eight thermal springs classified as acid sulfate with pH between 1.6 and 3.8, and four springs of bicarbonate composition (pH = 6.2–6.8). Paramillo de Santa Rosa has one neutral chloride thermal spring; one sulfate thermal spring and three bicarbonate thermal waters. Nevado del Tolima volcano has two bicarbonate thermal waters and two sulfate-bicarbonate waters. All five thermal springs at Cerro Machín volcano are bicarbonate-rich with pH between 6.2 and 8.3.

In the central volcanic segment, Nevado del Huila volcano has two neutral chloride thermal springs in the lower part of the edifice and one sulfate thermal spring in the upper flank. The Los Coconucos volcanic chain has five thermal springs with sulfate-chloride composition and ten with acid sulfate composition (Fig. 3). The two thermal springs at Sotará volcano have neutral chloride in composition.

In the southern volcanic segment, all four thermal springs at Doña Juana volcano are neutral chloride in

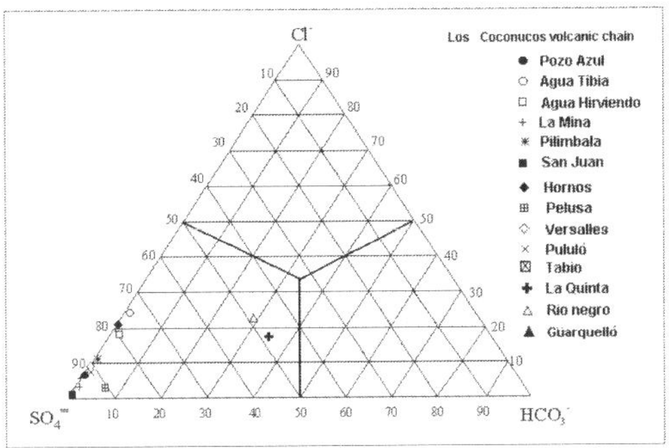

Figure 3. Cl-SO_4-HCO_3 diagram for thermal springs at Los Coconucos volcanic chain.

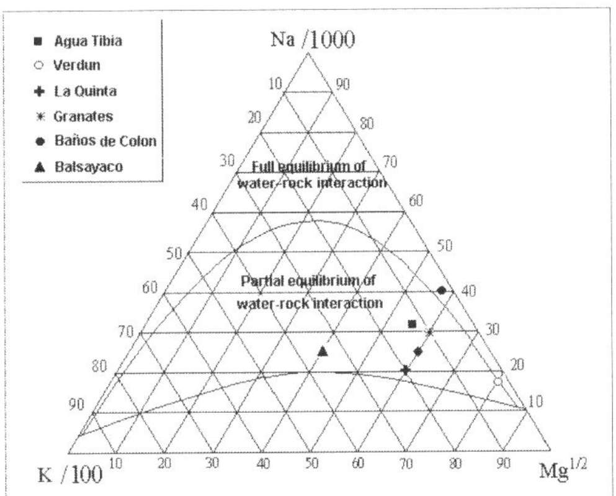

Figure 4. Na-K-Mg diagram for some thermal springs at Colombian active volcanoes.

composition. At Galeras volcano, four thermal springs are sulfate in composition and two bicarbonate in composition. Patascoy volcano has two thermal springs with neutral chloride waters and two with sulfate-chloride waters. Azufral volcano has two thermal springs with neutral chloride waters, two with sulfate waters and one with bicarbonate waters. All four thermal springs at Cumbal volcano have sulfate waters. And finally, Chiles volcano has two thermal springs that are sulfate in composition.

Waters from neutral chloride thermal springs originated from deep geothermal wells tend to establish a dynamic equilibrium with host rocks (Giggenbach 1988). Based on the relative concentrations of major cations, most thermal springs associated with Colombian active volcanoes are in disequilibrium with host rocks. Only Baños de Colon thermal spring located in the lower flanks of Patascoy volcano, established full equilibrium with a thermodynamically stable mineral assemblage of host rocks (Fig. 4). Five other neutral chloride thermal springs (La Quinta and Granates at Nevado del Ruiz, Agua Tibia at Los Coconucos volcanic chain, Verdún at Nevado del Huila, and Balsayaco at Patascoy) are in partial equilibrium with host rocks (Fig. 4).

4.2 *Minor ions of thermal springs*

The minor cation (Fe^{2+}, Al^{3+}, Li^{+}, Rb^{+}, Mn^{2+}, Zn^{2+}) compositions were determined for 75 thermal spring waters. In acid sulfate waters (pH = 1.6–2.8), Fe is in a range from 6 to 228 mg/l; Al concentration is from 11 to 935 mg/l; Li and Rb concentrations are less than 0.5 mg/l; Mn concentration is from 0.5 to 38 mg/l; and a chalcophile element like Zn is in concentrations from 0.2 to 8.5 mg/l. The highest concentrations of these elements were measured in RUAC thermal spring at Nevado del Ruiz volcano, which has pH of 1.6 and TDS of 16.7 g/l. In neutral chloride waters (pH = 6.4–7.4), Fe and Li concentrations were measured from 0.2 to 5.0 mg/l, whereas Mn concentration varies in a narrow range from 0.2 to 0.8 mg/l. In alkaline bicarbonate waters (pH = 6.2–8.3), Fe concentration is less than 22 mg/l and Li concentration is less than 2.9 mg/l.

4.3 *Mineralogical composition of thermal springs*

As a first attempt to understand water-rock interaction and its relation with water chemistry and minerals precipitated around spring discharges, selected rocks and mineral alterations were analyzed. Based on mineralogical analyses of thermal spring host rocks, two types of mineral alteration are indicated. These include (1) acid sulfate alteration due to advanced argillic alteration with alunite, amorphous silica and cristobalite formation, in the acid sulphate thermal spring COMI at Los Coconucos volcanic chain, which has pH = 2.0 and temperature of 46°C; and (2) intermediate to alkaline alteration with illite and carbonate (e.g., calcite and aragonite) precipitation, in the neutral chloride thermal spring DJTA at Doña Juana volcano, which has pH = 6.6 and temperature of 62°C.

4.4 *Isotopic composition of thermal springs*

$\delta^{18}O$ and δD values of some thermal springs, fumaroles and cold waters at Colombian active volcanoes are given in Table 1 and shown in Figure 5.

Also shown on Figure 5 is the global meteoric water line (Craig 1961). The sample of rain water collected in Manizales city lies on the global meteoric water line. Neutral and alkali thermal waters as well cold groundwater are close to global meteoric water line. Oxygen and hydrogen isotopic compositions of condensates from Deformes fumarole at Galeras volcano are a mixture of meteoric water and water

Table 1. Isotope composition of some thermal springs, fumaroles and cold waters.

Sample location	Sample type	Date (mm,yy)	Volcano	$\delta^{18}O$ ‰	δD ‰
RUHT	TS*	09-01	Ruiz	−13,5	−111,7
SREB	TS*	06-01	PSR**	−13,5	−101,8
SRSV	TS*	06-01	PSR**	−9,4	−86,4
SRRO	TS*	06-01	PSR**	−11,5	−88,0
SREC	TS*	06-01	PSR**	−11,4	−89,6
HUNE	TS*	06-01	Huila	−10,6	−91,6
HUVE	TS*	06-01	Huila	−11,5	−76,0
COMI	TS*	06-01	Ccs***	−10,2	−81,1
Deformes	Fumar.	06-01	Galeras	−6,0	−69,8
Deformes	Fumar.	07-01	Galeras	−6,1	−66,3
COPsam	Cold water	07-01	–	−9,8	−75,2
M/zales	Rainy water	09-01	–	−8,0	−55,1

* TS, Thermal Spring.
** PSR, Paramillo de Santa Rosa volcano.
*** Ccs, Coconucos volcanic chain.

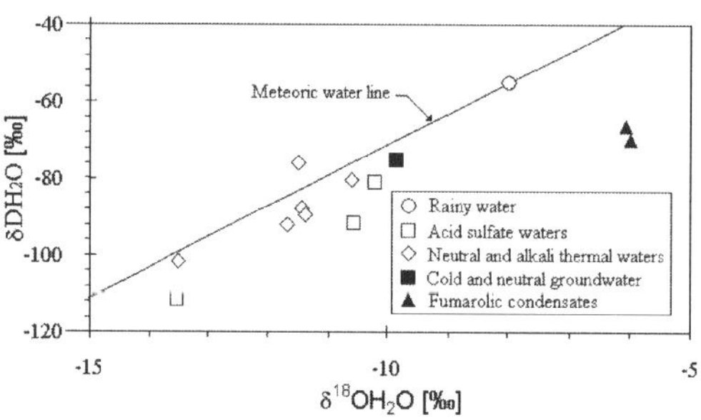

Figure 5. Oxygen and hydrogen isotopic composition of some thermal springs, fumarolic condensates and cold waters.

degassed from magmas. All acid sulfate thermal springs such as: RUHT (pH = 1.8), COMI (pH = 2.0) and HUNE (2.8) are enriched in ^{18}O relative to meteoric water line as a result of water-rock interactions.

4.5 *Spatial distribution of thermal springs*

Some thermal spring groups show particularities related to tectonic features. It is the case of a group of six thermal springs located around Los Coconucos volcanic chain. These thermal springs are acid sulfate in composition and are located on Los Coconucos geological fault, a NW-SE direction across Los Coconucos volcanic chain (Fig. 2). Based on data from Table 2, a dilution process is observed from COMI thermal spring in the upper flank of volcanic

Table 2. Physical and chemical data of some thermal springs at Los Coconucos volcanic chain.

Thermal spring	pH	T°C	Na	K	Ca	Mg	Al	TDS mg/l
COMI	2.0	46	290	68	213	304	113	13,6
COPI	2.3	28	130	27	94	121	48	7,1
COTA	2.2	28	118	24	89	108	62	8,6
COPA	2.3	27	117	23	87	108	64	7,8
COPB	2.4	25	71	14	53	66	40	5,7
COGU	2.5	24	76	14	57	67	42	5,9

chain to COGU thermal spring in the lower flank of volcanic edifice. COMI thermal spring has a higher temperature in its discharge due to its relative vicinity to volcanic conduit. For COPI thermal spring Koller & Aucott (1986) reported a surface temperature of 32°C. Lower temperatures are measured in COPI from 1994, perhaps due to structural changes in its water reservoir. It is possible a mixture of warm and cold reservoirs which was facilitated for some earthquakes registered in this zone by the Volcanological and Seismological Observatory at INGEOMINAS Popayán.

5 CONCLUSIONS

Acid sulfate thermal spring discharges are located in the upper flanks of volcanoes, whereas bicarbonate and neutral chloride thermal spring discharges in the lower flanks of volcanic edifices. Each type of thermal spring has distinguished physicochemical parameters, such as pH and defined range of ion concentrations. Relative concentrations of major cations in thermal springs indicate that most waters are far from equilibrium with host rocks. Three studied acid sulfate thermal springs are enriched in ^{18}O relative to meteoric water line as a result of water-rock interactions. Earthquakes can induce mixtures of reservoirs and modify temperatures and ion contents in the thermal springs.

ACKNOWLEDGEMENTS

The authors are grateful to Dr. Eckhard Faber and his collaborators at BGR, Hannover, Germany for performing isotopic analyses. We thank Dr. Jennifer Lewicki at Lawrence Berkeley National Laboratory, California, USA and an anonymous reviewer for their helpful suggestions and careful reading of the manuscript. This work was financially supported by INGEOMINAS, Colombia, Projects P97Q02 and C01R01.

REFERENCES

Craig, H. 1961. Isotopic variations in meteoric waters. *Science*, 133: 1702–1703.

Ellis, A.J. & Mahon, W.A.J. 1977. *Chemistry and geothermal systems*. Academic Press, New York, pp. 1–392.

Garzón, G. 2001. Catálogo de Fuentes Termales del Parque Nacional Natural Los Nevados. INGEOMINAS, Manizales, Colombia, 105pp. (in Spanish)

Garzón, G. 1997. Catálogo de Fuentes Termales del Suroccidente Colombiano. INGEOMINAS, San Juan de Pasto, Colombia, 212pp. (in Spanish)

Giggenbach, W.F. 1988. Geothermal solute equilibria. Derivation of Na-K-Mg-Ca geoindicators. *Geochim. Cosmochim. Acta*, 52: 2749–2765.

Giggenbach, W.F. & Goguel, R.L. 1989. Collection and analyses of geothermal and volcanic water and gas discharges. *Dep. Sci. Ind. Report CD-2401*, Petone, New Zealand, Chem. Div. DSIR Rep., pp. 1–53.

Giggenbach, W.F., García, P.N., Londoño, C.A., Rodríguez, V.L., Rojas, G.N. & Calvache, M.L. 1990. The chemistry of fumarolic vapor and thermal-spring discharges from the Nevado del Ruiz volcanic-magmatic-hydrothermal system, Colombia. *J. Volcanol. Geotherm. Res.*, 42: 13–39.

Koller, B. & Aucott, J. 1986. Algunas observaciones sobre los posibles cambios morfológicos del volcán Puracé y geoquímica de sus aguas termales después del sismo del 31 de marzo de 1983. INGEOMINAS, Popayán, 22pp. (in Spanish)

Lewicki, J.L., Fischer, T. & Williams, S.N. 2000. Chemical and isotopic compositions of fluids at Cumbal volcano, Colombia: evidence for magmatic contribution. *Bull. Volcanol.*, 62: 347–361.

Water–Rock Interaction, Wanty & Seal II (eds)

A hydrologic-hydrochemical model for the Troitsky crater lake, Maly Semyachik, Kamchatka

G.M. Gavrilenko
Institute of Volcanology, RAS, Petropavlovsk-Kamchatsky, Russia

P.G. Gavrilenko
Geology Department, Lomonosov Moscow State University, Moscow, Russia

ABSTRACT: A hydrologic-hydrochemical model is suggested for a drainless lake in the Troitsky crater of Maly Semyachik volcano based on hydrologic, chemical and thermal measurements. This model was used to calculate parameters that characterize the state of the lake during different phases of the volcano's activity.

1 INTRODUCTION

Maly Semyachik volcano (elevation = 1560 m) is an active volcano located in the Eastern Kamchatka Ridge (57°07′N and 159°53′E) 20 km from the Pacific Ocean coast. Semyachik's volcanic structure formed by three closely situated and sequentially formed strato volcanoes and constitutes a 3-km long ridge. The active SW part of this structure has an apical active crater (Troitsky Crater) which was formed by an explosive eruption about 400 years ago (Selyangin 1987).

At present the crater is filled with an acid thermal lake (Troitsky lake; Fig. 1). The lake is round, with a diameter of about 500–600 m and an area of more than 2×10^5 m^2. Maximum depths of the lake, which were measured from 1971 to 2000, vary from 140 to 117 m (Gavrilenko 2000; Gavrilenko et al. 1996; Gavrilenko et al. 1993). From 1971 to the present time its main crater bottom relief have been relatively constant due to the stagnant form of the crater bottom. Therefore, the observed shifts in lake depth are assumed to be due to changes in lake level, and therefore, water mass.

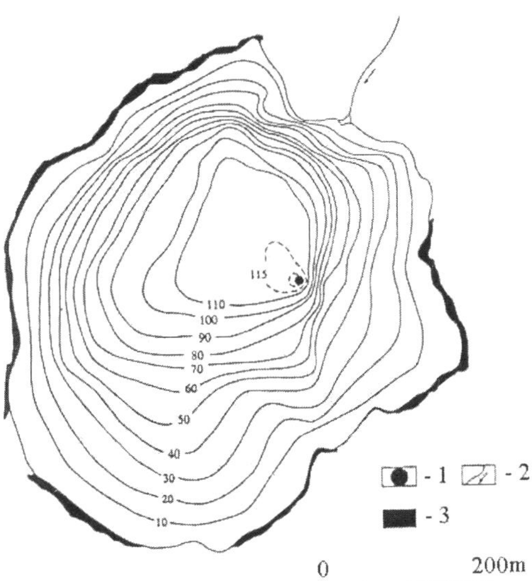

Figure 1. Bathymetry of the Troitsky crater lake (in meters). 1 – hydrothermal vent; 2 – streamlet; 3 – rock outcrops.

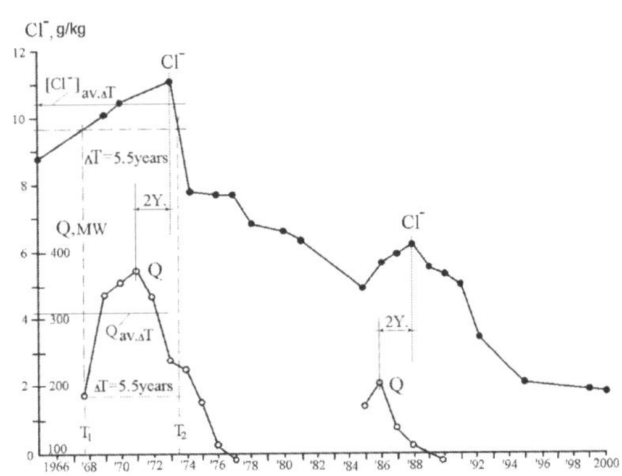

Figure 2. Variations of Cl$^-$ anion concentrations of the water Troitsky crater lake and its power output changes.

Hydrological, hydrochemical and thermal parameters of Troitsky Lake were monitored from 1965 to 2000 and observed to vary greatly (Gavrilenko et al. 1993; Gavrilenko et al. 1996; Gavrilenko 2000).

2 TROITSKY LAKE: A HYDROLOGICAL MODEL

Data gathered on the world's active volcano crater lakes point to considerable shifts in their levels that are unrelated to seasonal variations. Such shifts were also observed in the crater lake level of Maly Semyachik. These fluctuations can be explained based on the mathematical analysis of the water and thermal balances.

Two phases of Maly Semyachik peak activity in 1971 and in 1986 (heat powers up to 375 MW and 200 MW, respectively) and a decrease of activity after 1987 were analyzed by two independent methods. Based on these analyses, it was determined that there is hidden outflow from Troitsky lake by seepage through the lake bottom. The outflow was ~160 kg/s in 1971 and ~130 kg/s in 1986. The hidden outflow in the active 1986 phase was characteristic of that during the inactive period (1991–1992). Also, lake water volumes were equivalent during these time periods ($(10.4–10.7) \times 10^6$ m^3; Gavrilenko 2000). The equality in volumes of Troitsky crater lake waters in 1987 and 1991–1992 suggests stable lake level and bottom surface area. In 1971, when the volcano was more active, the lake was characterized by greater volume ($\sim 15 \times 10^6 m^3$), level, maximum depth (140 m), and seepage.

According to D'Arcy's law, the lake's hidden outflow is directly proportional to its bottom surface area. Therefore, knowing Troitsky's bottom area (for the period of 1992) and its hidden outflow volume, it is possible to calculate the filtration coefficient of lake-bottom rocks. This coefficient is 0.045 m/day, which corresponds to lake water filtration through water-resistant rocks.

Based on analysis of thermal and water balances for Troitsky's lake at different periods of activity, five stages of activity can be defined: (I) inactive phase, (II) primary phase of volcano activation, (III) continuation of active phase in stable regime, (IV) recession of volcano activity (but not cessation), (V) activity is practically fully stopped. Such real sequence of events in the life of closed lakes in volcano craters made it possible to determine that the more thermal power and the longer the active phase of the volcano, the more volume is reached by lacustrine waters, thus the higher the lake level. Moreover, during the active phase the lake's level increases due to different rates of water supply with depth fluid and evaporation which is less intensive being on average 1/2 (from 1/3 to 2/3; Fig. 3) (Gavrilenko, 2000).

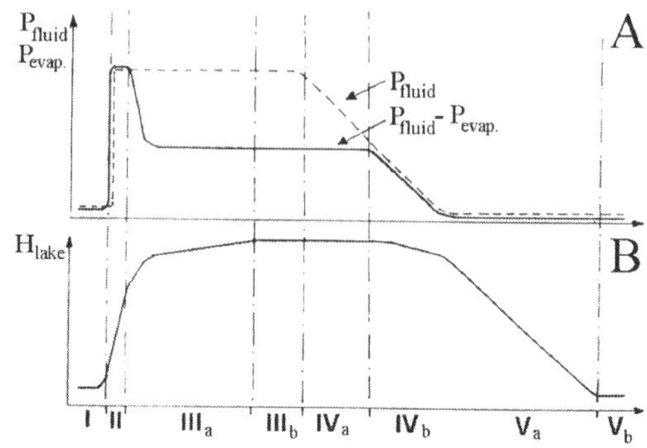

Figure 3. Idealized temporal plots a number of parameters of the Troitsky crater lake. Temporal phases: I, II, III, IV and V – see text; A – variations P_{fluid}(ascending volcanic fluid) and $P_{evap.}$ (water lost by evaporation). B – changes of H_{lake}(greatest depth of the lake).

Table 1. Relation between parameters characteristic for the present state of the Troitsky crater lake under different activity phases of Maly Semyachik, Kamchatka (Gavrilenko 2000).{a – measured parameters; b – parameters received through calculation; Q ($\approx Q_{fluid}$) – the total lake power (energy) output; – greatest recorded depth; V_{lake} – the lake volume; ΔV_{lake} – the rate of increase (or decrease) in the lake water volume; P_{seep} – the water lost by seepage through the lake bottom.}

	Phase			
	I	II	III	
Parameters	1946	1971	1986	1991–1992
Q	a	375 ± 15%	201 ± 15%	
MW	b~1100	374	207	<100
T,°C	a 66	42	31	17
H, m	a	140	132	117
	b ~154			
V_{lake},	a	15		10,4
10^6 m^3	b ~19		~10,7	
ΔV_{lake},	a	±0	+13	≤ −15
l/s	b			− 30
P_{seep}	a	162	130	
l/s	b	165	133	≤ 130

For example, in 1946 when the volcano power exceeded 1000 MW the lake's depth was ~154 m (Gavrilenko 2000). During Maly Semyachik's activation period in the beginning of 70's, its power was 375 MW and the lake's maximum depth was 140 m. And in the middle of 80's, when the volcano power reached 200 MW, the lake maximum depth was 132 m. When volcano's substantially dropped in 1992, the measured maximum depth became even less: – 117 m (Table1; Gavrilenko 2000; Gavrilenko et al. 1996; Gavrilenko et al. 1993). During inactive phase practically only hidden outflow compensates for the supply of precipitation which falls on the lake's water collecting area. During such periods hidden outflow is minimal and its volume approximately corresponds to intensity of atmospheric moisture supply. Accordingly, the lake depth is at a minimum.

If the lake water balance is disturbed, then the compensating processes of evaporation from the lake surface and hidden outflow stabilize the water masses. Hidden outflow is the most significant process as it occurs during all phases of volcano activity.

The data gathered on Troitsky Lake helps to clarify the dependence between several elements of water balance and morphometric parameters of the lake (depth, volume and bottom area). Examination of this lake geometrical analogue allowed the determination of possible quantitative connections between these parameters. When the lake's water mass increases, its bottom area and its level increase. However, these parameters change irregularly. The relation being 7:4:2.5 respectively. According to D'Arcy's law, when the bottom area increases, the lacustrine waters filtration speed increases, which reflects the hidden outflow intensity; its increase relates to the rise of the lake's level as 4 to 2.5 (Gavrilenko 2000). All the above calculations were made without taking into account existing seasonal variations of the lake level (volume). After this factor had been taken into account, it became clear that endogenic processes, that is intensity of inflow of water with fluid, are the result of the level increase in the Troitsky lake; activation periods of which have considerably greater time exposure (several years) than half-yearly seasonal processes. So, for example, during 5 years if the volcano power is 200 MW, the lake level will increase by 14.5 m and if it is 375 MW, the level will increase by almost 20 m when seasonal variations are ±2 m (Gavrilenko 2000).

3 VARIATIONS OF THERMAL AND HYDROCHEMICAL PARAMETERS OF TROITSKY LAKE

A correlation has been established between variations in the chemical composition and thermal power of lakes and the state of volcanic activity. (Gorshkov et al. 1975; Giggenbach & Glover 1975). Consequently, changes in lake parameters have been used to evaluate the state of volcanic activity. Importantly, however, hydrochemical changes have been observed to occur following thermal variations. Due to the large volume of the lake relative to the small outflow, attainment of chemical equilibrium is slow following change in the supply of mineralized fluids with variation in volcanic activity. Thermal inertia of the lake is based on the relationship between thermal reserves of the lake and its thermal power and is considerably less. This fact was confirmed by scientists who utilized mathematical calculations and thermo-physical and thermo-chemical laws. The delay for each lake varies and primarily depends on the lake water mass. For Troitsky Lake this time is approximately two years (see Fig. 2).

Correlation of chemical and thermal changes in crater lakes is due to the presence of a deep fluid that provides a common source of heat and mineralized fluids. In particular, for Troitsky Lake this fact is obvious at vapor phase with a heat content of about 2.8 MJ/kg and mineralization which exceeds dissolved substances in lake waters, i.e. during the activation periods their mineralization increases. Based on the relation:

$$m_{fluid} \times i = Q \qquad (1)$$

the probabilistic content of chloride-ion in this fluid ($[Cl^-]_{fluid}$ g/kg) can be estimated, where m_{fluid} is the fluid supply into the lake (in kg/s) with the enthalpy of $i = 2.8$ MJ/kg and Q equals the volcano thermal power measured at the lake. Because chemical and thermal changes in crater lakes are not synchronous in time (Fig. 2), it is more appropriate to compare their integral values for a period of time that encompasses the beginning and the end of active volcano phases and overlaps the extremes of thermal power and chemical content. For Troitsky Lake, the calculations were as follows: Two equations were equaled. The first equation having the mass flow of chloride-ion with a depth fluid in a time interval and the second equation having the quantity of chloride-ion outflow in a closed lake in the same time interval. Time interval had the same concentrations of chloride-ion, be it in the beginning or in the end. Having used data on Troitsky Lake during its activation phases from the end of 1960's until the beginning of the 1970's and in the middle of 1980's we calculated that in the first case, the average chloride-ion concentration in the fluid was 15 g/kg and in the second, it was 16 g/kg. The concentrations were calculated by using the following:

$$[Cl^-]_{fluid} = \left([Cl^-]_{\frac{av \cdot \Delta T}{Q_{av \cdot \Delta T}}}\right) \times i \times P_{seep} \qquad (2)$$

where $[Cl^-]_{av \cdot \Delta T}$ – average chloride-ion concentration of the lacustrine waters in active phase period, $Q_{av \cdot \Delta T}$ – average thermal power in active phase period, i – the enthalpy of the fluid and P_{seep} – the water lost by seepage through the lake bottom.

4 APPENDIX

Based on data gathered for Troitsky Lake from the late 1960's to the mid 1970's, it follows (see

Fig. 2): During $\Delta T = T_2 - T_1$ activation period, the following quantity of chloride ion entered the lake

$$\Sigma_{+}Cl^{-}(g) = m_{fluid} \times [Cl^{-}]_{fluid} \times \Delta T \quad (3)$$

If we input the equation (1), where $i = 2.8$ MJ/kg, and $Q_{av \cdot \Delta T}$ – average thermal power during time ΔT, equation (3) will look as follows:

$$\Sigma_{+}Cl^{-}(g) = (Q_{av \cdot \Delta T}/i) \times [Cl^{-}]_{fluid} \times \Delta T \quad (4)$$

During the same period, the following quantity of chloride ion left the lake due to the outflow:

$$\Sigma_{-}Cl^{-}(g) = P_{seep} \times [Cl^{-}]_{av \cdot \Delta T} \times \Delta T \quad (5)$$

where P_{seep} – the hidden outflow from the; $[Cl^{-}]_{av \cdot \Delta T}$ – average concentration of chloride in the lake during this period.

T_1 will correspond to the beginning of the time interval, and T_2 will correspond to the end of interval, when the quantity of chloride in the lake was equal; i.e. its concentrations were equal when the lake's volume was constant. Thus, it becomes obvious, that during a given time period due to the inflow of chloride with fluid, its quantity and concentration in the lake increased. Meanwhile, its outflow from the lake was constant as well. In the beginning it was less intensive than the inflow of Cl^{-} with fluid, but later when the volcanic activity diminished, Cl^{-} intensified. As a result Cl^{-} mass started to drop and its concentration started to decrease as well. At some point, T_2 the Cl^{-} concentration became exactly the same as T1. This means that the whole inflow mass of Cl^{-} with fluid during ΔT period was carried out during the same period through a hidden outflow. Therefore, we can equate equations (4) and (5) ΔT under the above-described conditions equaled 5.5 years for a given activation phase of Maly Semyachik volcano. During this period, average thermal power was 310 MW, and average Cl^{-} concentration was 10.4 g/kg. Since the level and volume of the lake were more or less the same, so was the outflow (~160 kg/s). If we input these values in equation (5), we will find that the average concentration of Cl^{-} ion in the deep fluid during the given period was approximately 15 g/kg. Analogous calculations for the volcano's active period in the mid 1980's produced $Cl^{-} = 16$ g/kg.

ACKNOWLEDGEMENTS

This work was supported by RFBR grant No. 02–05-64979 and FP Integration Grant Э0334/946.

REFERENCES

Gavrilenko, G.M. 2000. A Hydrological Model for a Lake in the Maly Semyachik Crater. *Volcanology and Seismology* 6: 21–31. (in Russian).

Gavrilenko et al. 1993. The Present-Day State of Maly Semyachik Volcano, Kamchatka. *Volcanology and Seismology* 15(2): 129–134.

Gavrilenko, G.M., Egorov, Yu, O. & Osipenko, A.B. 1996. Recent changes in the physico-chemical characteristics of the crater lake of Maly Semyachik volcano, Kamchatka. In S.J. Freeth (ed.), *Newsletter of the IAVCEI Commission on Volcanic Lakes. No. 9*: 11–13.

Giggenbach, W.F. & Glover, R.V. 1975. The use of chemical indicators in the surveillance of volcanic activity affecting the crater lake on Mount Ruapehu, New Zealand. *Bull. Volcanol.* 39: 70–81.

Gorshkov A.P. 1975. The deep structure of Maly Semyachik Volcano, Kamchatka and implications for the surface activity of the volcano. *Bull. Volcanol. Stations* 51: 50–59.

Water–Rock Interaction, Wanty & Seal II (eds)
© 2004 Taylor & Francis Group, London, ISBN 90 5809 641 6

Nitrogen transformations in hot spring runoff, Yellowstone National Park, USA

JoAnn M. Holloway
U.S. Geological Survey, Denver, CO, USA

Richard L. Smith & D. Kirk Nordstrom
U.S. Geological Survey, Boulder, CO, USA

ABSTRACT: Total nitrogen concentrations in hot springs at Yellowstone National Park vary widely in concentration, from <0.3 μM to 42.7 mM as N, with ammonium often occurring as the dominant nitrogen species. Two chemically distinct hot springs in Yellowstone National Park, acidic Nymph Creek and circumneutral Washburn Inkpot were investigated to determine variations in nitrogen concentration and speciation in hot-spring runoff. Ammonium concentration decreased by 41 percent (92 to 54 μM N-NH_4^+) over 148 m along Nymph Creek. There was no readily discernable decrease in ammonium in the drainage from the Washburn Inkpot (15.8 mM N-NH_4^+) over approximately 5 minutes of travel time along the 26 m reach. Nitrite concentration increased downstream in Nymph Creek from <0.2 to 1 μM N-NO_2^- and increased from 0.6 to 8.1 μM N-NO_2^- in the Washburn drainage. While there is some evidence from δ^{15}N-NH_4^+ measurements and denitrification assays for microbial nitrogen transformations in Nymph Creek and the Washburn drainage, nitrification of ammonium to nitrite and nitrate may be a rate-limiting step in nitrogen cycling for these hot spring drainages.

1 NITROGEN AND VOLCANISM

Nitrogen associated with volcanism has atmospheric, mantle, and sedimentary sources (Allen & Day 1935; Lowenstern et al. 1999; Goff & Janik 2002; Mariner et al. 2003), with nitrogen occurring as aqueous (e.g. NH_4^+, NO_3^-, NO_2^-), and gaseous (e.g. NH_3, N_2, N_2O) species in thermal waters and fumaroles.

Hot springs and geysers in Yellowstone National Park exhibit a range of nitrogen concentrations and speciation with some regional patterns suggesting an association between bedrock geology or tectonic setting and nitrogen chemistry. Historic data (Thompson & DeMonge 1996; Thompson et al. 1975; Ball et al. 1998; Ball et al. 2002) along the Firehole River, and Yellowstone Lake show generally low nitrogen concentrations (<0.3 to 50 μM N) with much higher and variable concentrations in the north-central region of the park (<0.3 μM to 42.7 mM), dominantly as ammonium. The highest naturally-occurring ammonium concentrations measured (14.9 to 42.7 mM NH_4-N L^{-1}; Gooch & Whitfield 1888; Ball et al. 1998) are associated with the Washburn Hot Springs complex near the Grand Canyon of the Yellowstone (Fig. 1).

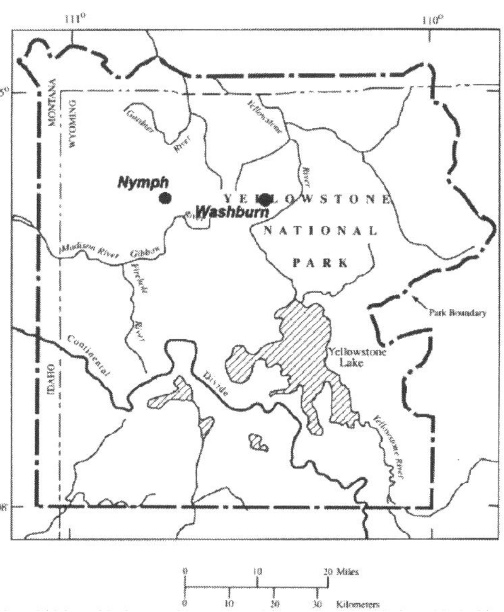

Figure 1. Yellowstone National Park, showing locations for Nymph Creek and Washburn Hot Springs.

In this study, we examine nitrogen concentration and speciation resulting when reduced ground waters with elevated ammonium concentrations come into contact with the atmosphere and microbial consortia in runoff channels. Microbial processes (e.g. denitrification and nitrification) active in hot spring runoff channels were inferred using a combination of $\delta^{15}N\text{-}NH_4^+$ values, concentrations of relatively unstable, intermediate forms of nitrogen (NO_2^-, N_2O), and denitrification potential assays.

2 NITROGEN SPECIATION IN HOT SPRING DRAINAGES

2.1 *Site selection and sample collection*

Two hot springs were selected for study based on location, degree of runoff channelization, and range of pH and nitrogen concentration. Nymph Creek is acidic (pH 2.5 to 3) with an extensive mat of green algae, thought to be *Cyanidium*, and is fed by a few small vents. We selected a single vent with sulfate-dominant water near the top of a well-channelized section of the creek, which is located north of Norris Geyser Basin. Nymph creek has moderate nitrogen concentrations (92 μM $N\text{-}NH_4^+$). A neutral (pH 6.4) sulfate spring in the lower Washburn Inkpot group was selected as a spring with high nitrogen concentration (15.8 mM $N\text{-}NH_4^+$). The Washburn group is located south of Mount Washburn near the Grand Canyon of the Yellowstone River.

Samples were filtered through a 0.1 μm membrane polycarbonate filter mounted in a filter plate or through a 0.45 μm syringe-tip mounted filter. Subsamples collected for ammonium and $\delta^{15}N\text{-}NH_4^+$ were preserved using sulfuric acid. The pH was calibrated and measured in the field using a flow-through cell to minimize temperature effects. Ammonium and nitrate were measured using ion chromatography (detection limits 2 μM $N\text{-}NH_4^+$ and 2 to 5 μM $N\text{-}NO_3^-$). Nitrite was analyzed using colorimetric and chemoluminescent methods (detection limit 0.2 μM $N\text{-}NO_2^-$). Dissolved nitrous oxide was determined by gas chromatography (detection limit 0.2 nM $N\text{-}N_2O$) on samples collected in serum bottles preserved with potassium hydroxide prior to sealing. Flow rate was estimated by timing the transport of a biodegradable dye through sample stations.

Laboratory incubations of sediment from Washburn Hot Springs and Nymph Creek were conducted at 40°C using an acetylene block technique (modified from Tiedje 1994). Runoff waters and stream-bed sediment were placed under anoxic conditions with 71 μM $N\text{-}NO_3^-$ added to stimulate denitrification.

2.2 *Nymph Creek*

Nymph Creek originates from small, hypoxic hydrothermal vents, with dissolved oxygen increasing downstream from 6 to 161 μM O_2. Travel time was approximately 5 minutes for the first 25 m of Nymph Creek, with a decrease in flow rate as the water reached the thick algal mat at 25 m downstream. Nymph Creek became somewhat more acidic (2.87 to 2.73 pH) downstream from the source vent (Fig. 2), with temperature decreasing from 62°C to 25°C. Ammonium concentrations in Nymph Creek decreased from 92 to 74 μM $N\text{-}NH_4^+$ from the source vent to 25 m downstream, with ammonium further decreasing to 54 μM $N\text{-}NH_4^+$ over the 148 m reach. Nitrite and nitrous oxide increased in concentration to 1.0 μM $N\text{-}NO_2^-$ and 0.4 nM $N\text{-}N_2O$, in the first 25 meters of the runoff transect before decreasing downstream. These trace nitrogen species are indicative of microbial activity, although they may also reflect abiotic oxidation of ammonium as water flows downstream from the source vent. Nitrate concentrations were below detection (5 μM $N\text{-}NO_3^-$) throughout Nymph Creek. The $\delta^{15}N\text{-}NH_4^+$ increased

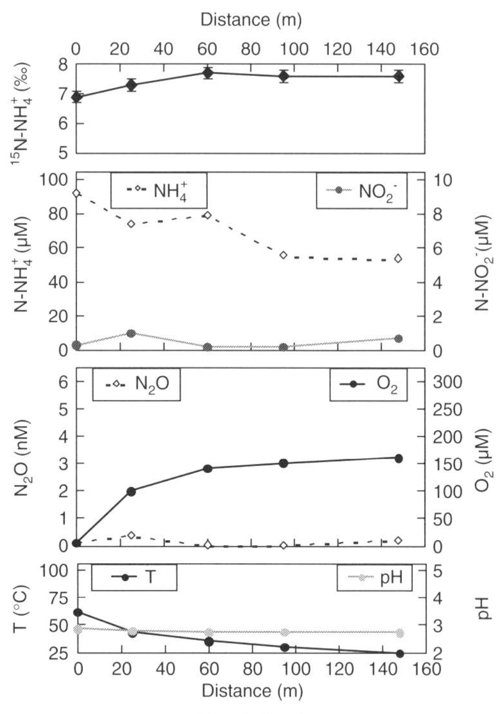

Figure 2. Transect for Nymph Creek, north of Norris Geyser Basin, showing nitrogen speciation, dissolved oxygen concentration, temperature and pH.

slightly from the vent (+6.9‰) to the sampling point 25 m downstream (+7.3‰). This slight enrichment was consistent with the uptake of ammonium.

2.3 *Washburn neutral inkpot*

The neutral Washburn inkpot is notable for naturally occurring millimolar concentrations of ammonium (Fig. 3). Travel time from the source along the 22.3 m transect was approximately three minutes. The pH in the outflow increased from 6.45 to 7.97 downstream from CO_2 degassing.

Ammonium concentrations vary between 14.5 and 16.1 mM NH_4^+ over the entire drainage, with no systematic changes in concentration. However, an increase in $\delta^{15}N\text{-}NH_4^+$ from 22.1 to 24.8‰ along the drainage may reflect microbially-mediated oxidation of ammonium in concentrations that are not measurable relative to the millimolar-scale ammonium concentration. Nitrite and nitrous oxide both increase downstream as the dissolved oxygen concentration increased from 41 to 234 μM O_2. Nitrate concentrations were below detection. Prior to entering Sulfur Creek, a third-order stream that drains into the Yellowstone River, nitrite concentrations increased to 8.1 μM $N\text{-}NO_2^-$. Nitrous oxide concentrations increase to 4.7 nM $N\text{-}N_2O$ prior to the runoff entering Sulfur Creek.

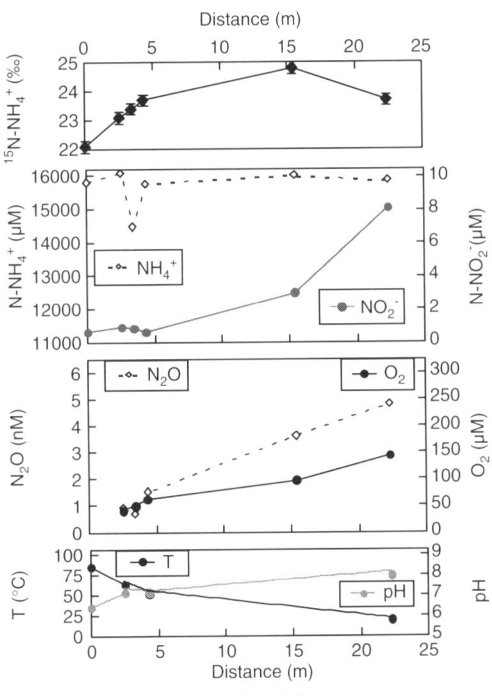

Figure 3. Transect of a neutral inkpot in the lower Washburn group, showing nitrogen speciation, dissolved oxygen concentration, temperature and pH.

3 MICROBIAL NITROGEN TRANSFORMATIONS

While there is only a slight decrease in ammonium concentration in Nymph Creek, there is marked increase in both nitrite and nitrous oxide 25 m downstream from the source vent. There was no measurable nitrate, even though there was complete oxidation of iron and arsenic within a few meters of the vents. The drainage from the neutral Washburn inkpot shows remarkable increases in trace nitrogen species, suggesting active microbial communities oxidizing ammonium. There were, however, no measurable nitrate concentrations.

Denitrification incubations helped our interpretation of the lack of nitrate in the overflow channels at Nymph Creek and Washburn. Formation of nitrous oxide was used to determine the potential for denitrification. Nymph Creek sediment had a delayed response to the onset of anoxia, with the formation of 8 nM N_2O per g sediment in 26 hr in sediment collected from below the algal mat. Denitrification-potential assays for Washburn sediments showed a strong and immediate response to the onset of anoxic conditions, with the concentrations of nitrous oxide produced at 26 hours ranging from 17 to 29 nM N_2O per g sediment. Faster rates were achieved with sediments collected further downstream from the spring. This increase in nitrous oxide production may be related to increased colonization of the channel sediments by denitrifying microbial communities.

The denitrification potential assays suggest microbial communities that reduce nitrate to nitrogen gas are present in both the Nymph and Washburn drainages. What little ammonium does oxidize to nitrate could thus be rapidly oxidized.

4 CONCLUSIONS

The distribution and concentration of nitrogen species in Nymph Creek and the Washburn drainage include ammonium as the dominant species of nitrogen, with the increase of trace nitrogen species nitrous oxide and nitrite within the first 25 m of each drainage. There were no measurable nitrate concentrations downstream from the hypoxic hot spring source waters. Our data support the interpretation that the accumulation of nitrite and nitrate in hot spring runoff channels is limited by low rates of nitrification relative to denitrification, and insufficient travel time to allow for more extensive oxidation of ammonium.

ACKNOWLEDGEMENT

J.K. Böhlke assisted with this study by providing ammonium isotope data.

REFERENCES

Allen, E.T. & Day, A.L. 1935. Hot springs of the Yellowstone National Park. *Carnegie Inst. Washington Publ.* 378.

Ball, J.W., McCleskey, R.B., Nordstrom, D.K., Holloway, J.M. & Verplanck, P.L. 2002. Water-chemistry data for selected springs, geysers, and streams in Yellowstone National Park, Wyoming, 1999–2000. *US Geol. Survey Open-File Report* 02-382.

Ball, J.W., Nordstrom, D.K., Cunningham, K.M., Schoonen, M.A., Xu, Y. & DeMonge, J.M. 1998. Water-chemistry and on-site sulfur speciation data for selected springs, geysers, and streams in Yellowstone National Park, Wyoming, 1994–1995. *US Geol. Survey Open-File Report* 98-574.

Goff, F. & Janik, C.J. 2002. Gas geochemistry of the Valles caldera region, New Mexico and comparisons with gases at Yellowstone, Long Valley and other geothermal systems. *J. Volcanol. Geotherm. Res.*116: 299–323.

Gooch, F.A. & Whitfield, J.E. 1888. Analyses of waters of the Yellowstone National Park with an account of the methods of analysis employed. *US Geol. Survey Bull.* 47.

Lowenstern, J.B., Janik, C., Fahlquist, L. & Johnson, L.S. 1999. A compilation of gas geochemistry and isotopic analyses from The Geysers geothermal field: 1978–1991. *Trans. Geotherm. Resources. Council* 23: 383–390.

Mariner, R.H., Evans, W.C., Presser, T.S. & White, L.D. 2003. Excess nitrogen in selected thermal and mineral springs of the Cascade Range in northern California, Oregon, and Washington: sedimentary or volcanic in origin? *J. Volcanol. Geotherm. Res.*121: 99–114.

Thompson, J.M. & DeMonge, J.M. 1996. Chemical analyses of hot springs, pools, and geysers from Yellowstone National Park, Wyoming, and vicinity, 1980–1993. *US Geol. Survey Open-File Report*, 96-68.

Thompson, J.M., Presser, T.S., Barnes, R.B. & Bird, D.B. 1975. Chemical analysis of the waters of Yellowstone National Park, Wyoming from 1965 to 1973. *US Geol. Survey Open-File Report* 75-0025.

Tiedje, J.M. 1994. Denitrifiers. In *Methods of Soil Analysis Part 2. Microbiological and biochemical properties*: 245–267. Madison, WI: Soil Science Society of America.

Water–Rock Interaction, Wanty & Seal II (eds)
 ISBN 90 5809 641 6

The geochemistry of geothermal waters in the Alvord Basin, southeastern Oregon

A.K. Koski & S.A. Wood
University of Idaho, Moscow, ID, USA

ABSTRACT: Waters from the Alvord, Mickey and Borax Lake thermal areas in the Alvord Basin, southeastern Oregon, have been analyzed for their major-element, trace-element, and stable isotope compositions. These waters are sodium-bicarbonate-chloride-type with a significant proportion of sulfate. The waters are also high in As (1–6 mg L^{-1}) and B (10–27 mg L^{-1}). Waters from different springs within each thermal area exhibit remarkably coherent chemical characteristics, in spite of variations in emergence temperatures of up to 30°C. Some components, such as bicarbonate, show little variation in composition over the course of a year, but others, such as arsenic and chloride, exhibit important temporal variations. Determination of As speciation indicates that As(V) dominates in most springs, even close to the vents. However, at Alvord there is evidence that the water emerges with As(III) dominant, but the arsenic is rapidly oxidized to As(V), suggesting mediation by microbial activity. Oxygen and hydrogen isotope compositions are similar for waters in all three thermal areas and are consistent with meteoric water that has undergone O-isotopic exchange during water–rock interaction. Values of sulfate $\delta^{34}S$ and bicarbonate $\delta^{13}C$ show a very narrow range both within and among the three thermal areas.

1 INTRODUCTION

The Alvord Basin is a north-northeast-trending graben in the northeastern Basin and Range Province, Harney County, southeastern Oregon. Three main groupings of thermal springs, Borax Lake, Mickey and Alvord hot springs, occur within the valley (Fig. 1; Brown & Peterson 1980; Cummings et al. 1993). The graben is bounded on the west by tilted fault-block mountains of the Steens and Pueblo Ranges, which attain elevations up to 2975 m. It is bounded on the east by the Sheepshead and Trout Creek Mountains. A large playa (the Alvord desert) covers approximately 250 km^2 in the north-central part of the valley. The basement in the area consists of Permian and Triassic metamorphic rocks (quartzite, greywacke, greenstone, sericite schist, quartz-muscovite schist, argillite and minor marble) and Cretaceous granitic rocks (gneissic granodiorite and quartz diorite) which intrude the older metamorphic rocks. These are overlain by Miocene and younger volcanic rocks successively comprising rhyolite and dacite flows and tuffs, andesite and basalt flows, and finally minor rhyolite and dacite flows and tuffs. The basin is filled with Plio-Pleistocene and Holocene alluvium. It is believed that the area underwent regional extension during the Cenozoic. Geophysical data suggest that each hot spring grouping is associated with faults. The system is probably recharged by meteoric water along the Steens Mountain-Pueblo Mountain fault block (Cummings et al. 1993).

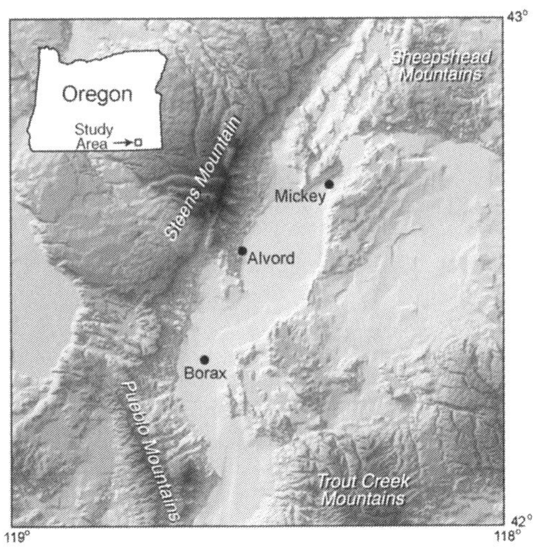

Figure 1. Map showing the location of thermal areas within the Alvord Basin, southeastern Oregon.

There have been a number of reports on the geochemistry of the thermal fluids feeding Borax Lake, Mickey and Alvord hot springs (Brown & Peterson 1980; Cummings et al. 1993; Schneider & McFarland 1995; Wood & Shannon 2003). However, questions remain as to the source of the waters and their dissolved constituents, and the geochemical processes occurring in these geothermal systems. Of particular interest is that these waters are comparatively high in As (1–6 mg L^{-1}) and B (10–27 mg L^{-1}). Our study is part of a larger interdisciplinary study of the microbial diversity in these hot springs. The goal of the study reported here is to characterize the major-element, trace-element, and stable isotopic compositions of these waters with a view towards determining the sources of the water and dissolved constituents, as well as the main processes of water-rock interaction.

Table 1. Selected average chemical characteristics of waters from the Alvord Basin – July 2002.

	Cold springs	Borax Lake	Alvord	Mickey
pH range	7.15–8.01	5.88–7.74	7.67–7.80	8.04–8.73
HCO_3^-	288(85)	427(27)	1258(60)	828(15)
Cl^-	5.1(5.4)	269(18)	858(28)	264(5)
SO_4^{2-}	6.3(4.2)	333(21)	227(10)	239(4)
As	0.01(0.01)	1.1(0.1)	4.5(0.2)	0.9(0.0)
B	ND	13.6(0.9)	27.5(1.1)	10.3(0.3)
Ca	36(11)	16.5(4.2)	11.4(0.5)	0.5(0.1)
K	ND	29.1(1.9)	71.9(4.8)	41.4(1.4)
Mg	26(3)	0.36(0.22)	1.78(0.05)	ND
Na	22(19)	535(22)	1240(52)	610(12)
Si	19(4)	65(3)	53(2)	95(3)

Average (standard deviation). All data except pH in mg L^{-1}. ND = not detected.

2 METHODS

Water samples were procured for major- and trace-element analysis in June, July and October 2002, and March, June and July 2003. At each site, temperature, conductivity and pH were measured in the field. Samples for determination of major- and trace-element cations were filtered through a 0.45-μm membrane and acidified with ultrapure HNO_3 in the field, and analyzed by inductively coupled plasma-atomic emission spectrometry (ICP-AES) upon return to the laboratory. Samples for anion analysis and alkalinity titrations were filtered through a 0.45-μm membrane but left unacidified. Alkalinity determinations were obtained via Gran titration and anions were determined via ion chromatography. The speciation of As in selected samples was determined using hydride-generation ICP-AES. A small number of samples were submitted to the University of Calgary for oxygen, hydrogen, carbon and sulfur isotope analysis.

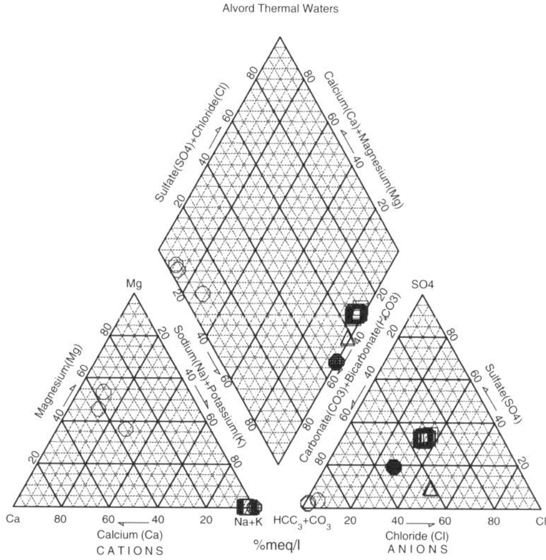

Figure 2. Piper diagram showing compositions of waters from the Alvord Basin (July 2002). Open circles = cold ground waters; closed circles = Mickey Hot Springs; open triangles = Alvord hot springs; open squares = Borax Lake hot springs.

3 RESULTS AND DISCUSSION

The thermal springs from the Alvord Basin discharge sodium-bicarbonate-chloride-type water with a significant proportion of sulfate (Table 1, Fig. 1). On the other hand, cold springs in the area discharge magnesium-calcium-bicarbonate-type waters. The total dissolved solids contents of the springs follow the order: cold springs < Borax Lake < Mickey < Alvord. Table 1 also shows that, whereas the cold springs contain negligible As and B, the thermal waters are enriched in these elements. As expected, the thermal waters also contain higher concentrations of Si than the cold waters, owing to the greater solubility of silica at elevated temperatures.

In each of these thermal areas, hot water issues from a number of individual vents, ranging from 3 vents at Alvord to more than 190 vents at Borax lake. Table 1 and Figure 2 demonstrate that the waters issuing from these vents are remarkably consistent in chemical composition within a thermal area, but the three thermal areas differ from one another. It is interesting that the chemical composition within a thermal area varies over only narrow limits, whereas the emergence temperatures of neighboring springs can vary by as much as 30°C (Fairley et al. 2003).

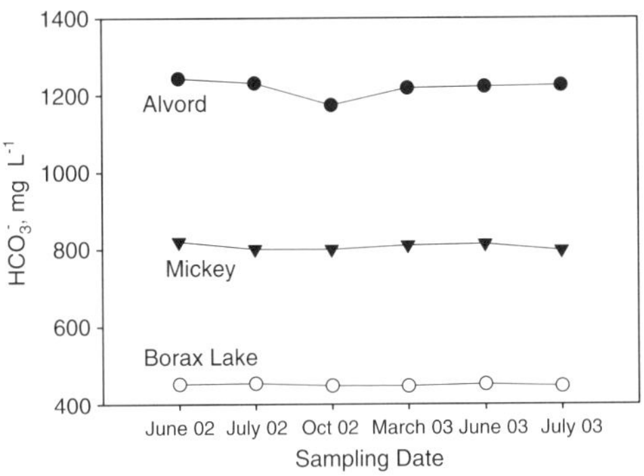

Figure 3. Bicarbonate concentration in thermal waters of the Alvord Basin as a function of time.

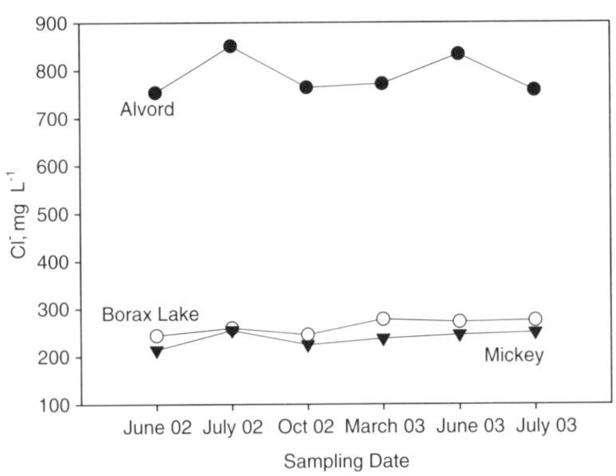

Figure 4. Chloride concentration in thermal waters of the Alvord Basin as a function of time.

The coherent chemistry within each thermal area, and the very different chemistry of cold ground waters in the area, suggest that the temperature variations cannot be a result of mixing with shallow groundwater.

Figure 3 shows that some constituents, such as bicarbonate, exhibit relatively minor temporal variations over more than a year.

However, some components, such as Cl^- and As (Figs. 4 & 5), show significant temporal variations, at least at Alvord. The reasons for these variations are unclear.

Chemical geothermometry (quartz, Na-K, and Na-Ca-K), based on the analytical data collected, yields temperatures of last equilibration of 150 to 225°C. These suggest minimum reservoir temperatures on the order of 200°C, in agreement with Cummings et al. (1993).

The results of speciation determinations for As are given in Table 2. At Borax Lake, 4 of 5 springs sampled were dominated by As(V) in both June and

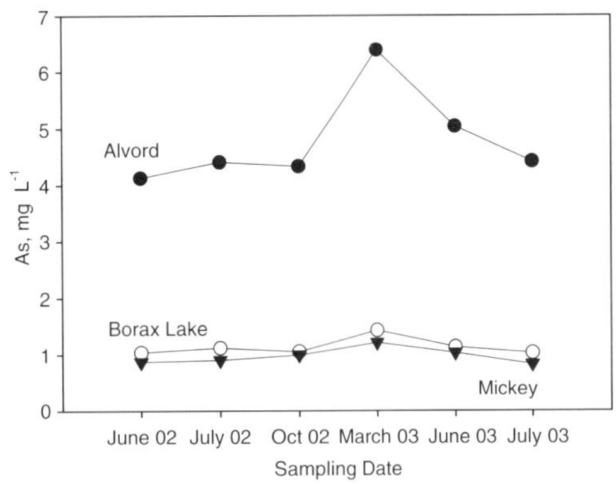

Figure 5. Arsenic concentration in thermal waters of the Alvord Basin as a function of time.

Table 2. Ratios of As(III)/As(V) in waters from the Alvord Basin.

Hot spring (number)	June 2003	July 2003
Borax (B 1060)[1]	0.05	0.01
Borax (B 1030)[1]	0.06	0.01
Borax (B 0440)[1]	0.07	0.01
Borax (B 0170)[1]	0.04	0.01
Borax (B 0070)[1]	7.5	2.0
Alvord[1]	4000	2.1
Alvord[2]	0.17	0.04
Alvord[3]	0.05	0.02
Mickey (M 0020)[1]	0.04	1.2×10^{-5}
Mickey (M 0020)[3]	0.04	NA
Mickey (M 0380)[1]	0.15	0.05
Mickey (M 0380)[3]	0.18	NA
Mickey (M 0430)[1]	55	2.0

[1] Sample taken from the vent of the spring; [2] Sample taken from head of outflow channel of spring; [3] Sample taken from several meters along the outflow channel.

July 2003. However, in one spring, As(III) was dominant. In all cases at Borax Lake, the samples were taken from within the vent. At Alvord, samples from within the vent were dominated by As(III), but As(V) became increasingly dominant as the vent fluid spilled out along the outflow channel. In two of the three springs sampled at Mickey, As(V) dominated in both the vent and in the outflow channels. However, in the third spring sampled at Mickey, As(III) was predominant. Taken together, these results suggest that As(III) is probably the dominant form of arsenic in the deep geothermal fluids, but that in many cases, oxidation to As(V) occurs very quickly as the fluid emanates from the vent. According to Gihring et al. (2001), the rate of inorganic oxidation of As(III) under these conditions is too slow to account for the observed rapid As(III)

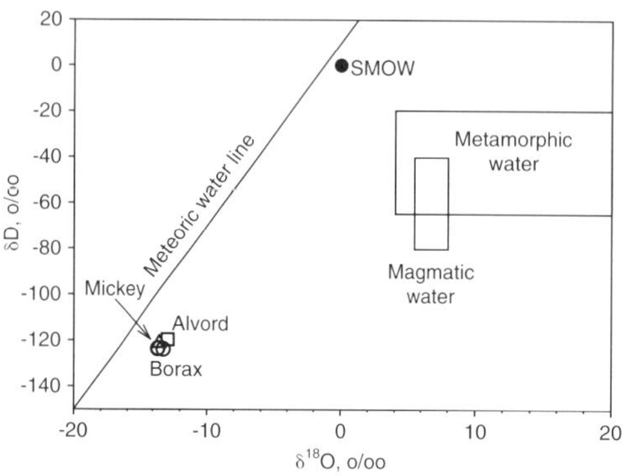

Figure 6. Hydrogen and oxygen isotopic composition of thermal waters from the Alvord Basin.

oxidation. These findings implicate microbial catalysis. However, additional investigation is required to fully understand the controls on As speciation in these waters.

Oxygen and hydrogen isotope data for the thermal waters are shown in Figure 6. These data show that the isotopic compositions of waters from all three thermal areas are very similar, and are consistent with meteoric water that has acquired slightly higher $\delta^{18}O$ values through water–rock interaction. This agrees with the findings of Cummings et al. (1993).

The isotopic compositions of sulfate in thermal waters from three different springs at Borax Lake are identical within analytical uncertainty. These values in turn differ by only approximately 1‰ from those at Alvord and Mickey. The narrow range of $\delta^{34}S$ observed indicates a lack of biogenic fractionation of sulfur isotopes in the thermal system at Alvord.

Table 3 indicates that the $\delta^{13}C$ values of bicarbonate in the thermal waters also exhibit very little variation within and between thermal areas. In one of the springs from Borax Lake, spring B 0070, the $\delta^{13}C$ value is slightly different from those of the other springs. Although the difference is small, it is outside the analytical uncertainty. It is interesting that B 0070 is the only one of three springs at Borax Lake in which a ratio As(III)/As(V) greater than unity was measured. The connection between these variables is under investigation.

Thus, one of the most important features of the geothermal system in the Alvord Basin is that, within the three main thermal areas, variations in major-element, trace-element, and stable isotopic compositions from spring to spring are quite small. Major- and trace-elemental compositions show some variation among the three thermal areas, but the stable isotope compositions of waters from all three areas are very similar. The same can be said about the Sr isotopic compositions of waters from the three thermal areas (Cummings et al. 1993). Thus, the sources of water and dissolved components must be similar for all three sites. Additional trace-element (including REE), stable (B, Li) and radiogenic (Nd, Pb) isotopic measurements are in progress, as are reaction path modeling calculations designed to elucidate further the water-rock interaction processes controlling the Alvord geothermal system.

Table 3. Sulfur isotopic composition of sulfate and carbon isotopic composition of bicarbonate for thermal waters from the Alvord Basin.

Hot spring (number)	$\delta^{34}S$ (‰)	$\delta^{13}C$ (‰)
Borax Lake (B 0070)	11.3	−4.9
Borax Lake (B 0170)	11.0	−4.2
Borax Lake (B 1030)	11.3	−4.2
Alvord	12.6	−4.2
Mickey (M 0380)	12.4	−4.3

ACKNOWLEDGMENTS

We thank Nick Raterman for his help with the As speciation determinations. The work described is supported by the NSF-Idaho EPSCoR Program and by the National Science Foundation under award number EPS-0132626.

REFERENCES

Brown, D.E. & Peterson, N.V. 1980. Preliminary geology and geothermal resource potential of the Alvord Desert Area, Oregon. Oregon Department of Geology and Mineral Industries Open-File Report 0-80-10, 57 p.

Cummings, M.L., St. John, A.M. & Sturchio, N.C. 1993. Hydrogeochemical Characterization of the Alvord Basin geothermal area, Harney County, Oregon, USA. Proc. 15th New Zealand Geothermal Workshop, 119–124.

Fairley, J., Heffner, J. & Hinds, J. 2003. Geostatistical evaluation of permeability in an active fault zone. *Geophys. Res. Lett.* 30, 1962.

Gihring, T.M., Druschel, G.K., McCleskey, R.B., Hamers, R.J., & Banfield, J.F. 2001. Rapid arsenite oxidation by *Thermus aquaticus* and *Thermus thermophilus*: Field and laboratory investigations: *Environ. Sci. Technol.* 35, 3857–3862.

Schneider, T.R. & McFarland, W.D. 1995. Hydrologic data and description of a hydrologic monitoring plan for the Borax Lake Area, Oregon. U.S. Geological Survey Open-File Report 95-367, 24 p.

Wood, S.A. & Shannon, W.M. 2003. Rare-earth elements in geothermal waters from Oregon, Nevada, and California. *Jour. Solid State Chem.* 171, 246–253.

Water–Rock Interaction, Wanty & Seal II (eds)
© 2004 Taylor & Francis Group, London, ISBN 90 5809 641 6

The scaling relationship between self-potential and fluid flow on Masaya volcano, Nicaragua

J.L. Lewicki
Earth Sciences Division, Lawrence Berkeley National Laboratory, Berkeley, CA, USA

G.E. Hilley
Department of Earth and Planetary Sciences, University of California, Berkeley, CA, USA

C. Connor
Department of Geology, University of South Florida, Tampa, FL, USA

ABSTRACT: The concurrent measurement of self-potential (SP) and soil CO_2 flux ($F_s^{CO_2}$) in volcanic systems may be an important tool to monitor intrusive activity and understand interaction between magmatic and groundwater systems. However, quantitative relationships between these parameters must be established to apply them toward understanding processes operating at depth. Power-law scaling exponents calculated for SP and $F_s^{CO_2}$ measured along a fault on the flanks of Masaya volcano, Nicaragua indicate a nonlinear relationship between these parameters. Scaling exponents suggest that there is a diminishing increase in SP with a given increase in $F_s^{CO_2}$, until a threshold (log $F_s^{CO_2} \approx 2.5\,g\,m^{-2}\,d^{-1}$) above which SP remains constant with increasing $F_s^{CO_2}$. Implications for subsurface processes that may influence SP at Masaya are discussed.

1 INTRODUCTION

SP anomalies have been observed within many volcanic and hydrothermal regions, and SP variations have been linked to changes in volcanic activity and hydrothermal fluid flow (Corwin and Hoover 1979; Fitterman and Corwin 1982; Massenet and Pham 1985; Hashimoto and Tanaka 1995; Zlotnicki et al. 2003). Monitoring these anomalies on active volcanoes may therefore be a tool to forecast eruptions and understand interaction between magmatic and groundwater systems. SP anomalies observed in volcanic systems have been primarily attributed to electrokinetic (EK) processes involving transport of charge contained in the electrical diffuse layer at the pore surface by electrolyte solution flow (Anderson and Johnson 1976; Massenet and Pham 1985; Revil et al. 1999; Ishido and Pritchett 1999) or rapid fluid disruption (RFD), a process whereby charge separation can be produced by rapid disruption or vaporization of liquid water by high heat and/or gas flow (Johnston et al. 2001). While EK potentials are produced by capillary flow of aqueous solutions, RFD may produce positive SP anomalies in thermal areas far above the groundwater table as charge is carried by water droplet and/or vapor flow (Johnston et al. 2001).

Lewicki et al. 2003 investigated the spatial relationship between SP, $F_s^{CO_2}$, and soil temperature (T_s) and the mechanism(s) that may produce SP anomalies on the flanks of Masaya volcano, Nicaragua (Figure 1). $F_s^{CO_2}$ and T_s up to $5.0 \times 10^4\,g\,m^{-2}d^{-1}$ and 80°C, respectively, and positive SP anomalies were measured within an area surrounding an inferred normal fault, adjacent to Comalito cinder cone (Figure 2). Lewicki et al. 2003 showed that SP, $F_s^{CO_2}$ and T_s were spatially correlated and that trends in $F_s^{CO_2}$ and T_s were consistent with advective transport of heat and CO_2 with steam along a fault zone. Also, a model calculation (Babu 2003) showed that polarization must be shallow (<10 m) to produce the high horizontal SP gradients and short wave lengths observed at the study site. Based on the presence of mechanisms to produce RFD (i.e., subsurface temperatures above liquid boiling points, high gas flow rates), a thick unsaturated zone (100 to 150 m), and relatively high resistivities, Lewicki et al. 2003 suggested that SP anomalies at Masaya are mainly produced by RFD.

While clear spatial correlation was observed between SP and $F_s^{CO_2}$, Lewicki et al. 2003 did not investigate the quantitative relationship between these parameters. Such a link is important to establish, to better understand the influence of fluid flow processes operating at depth on SP anomalies measured at the surface. Here, we establish power-law scaling relationships for measured values and discuss the subsurface processes that may influence SP at Masaya.

Figure 1. Areal photograph showing study site location adjacent to Comalito cinder cone on the flanks of Masaya volcano. Inset (A) shows locations of SP and $F_s^{CO_2}$ transects C1–C4, areal $F_s^{CO_2}$ measurements (dots), and inferred normal fault (dashed line).

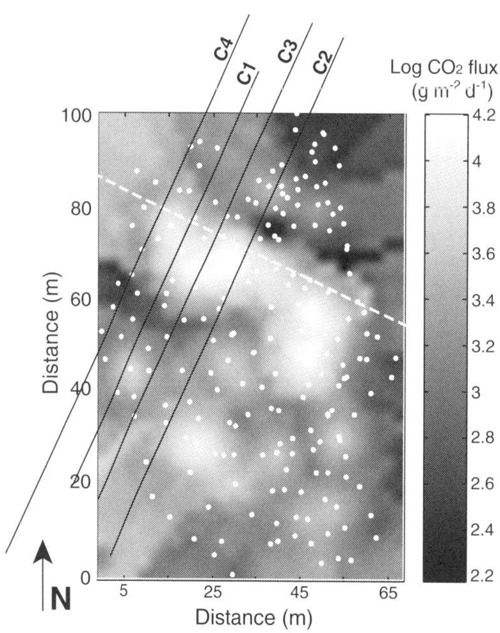

Figure 2. Interpolated image map of log $F_s^{CO_2}$ measured adjacent to Comalito cinder cone (Lewicki et al. 2003). Dashed line and dots show locations of inferred normal fault and $F_s^{CO_2}$ measurements, respectively. Also shown are locations of transects C1–C4.

2 DATA ANALYSIS

Scaling relationships between SP and $F_s^{CO_2}$ were calculated by linear regression of the log-transformed data. These relationships were determined separately for each traverse (C_1–C_4) because SP measurements along each traverse were referenced to the background base station at the southwest end of each traverse (Lewicki et al. 2003). To remove background effects not attributable to processes within the fault zone, we subtracted $F_s^{CO_2}$ measured at each background base station from all measurements along each traverse. These values are reported as "adjusted" $F_s^{CO_2}$. Since SP represents the electric potential difference between the base station and a given measurement along the traverse, SP adjustment was unnecessary.

3 RESULTS

Self-potential and log $F_s^{CO_2}$ measured along transects C_1–C_4 (Figure 3) are moderately positively correlated (correlation coefficient, $C = 0.67$, 0.71, 0.59, and 0.70 for transects C1, C2, C3, and C4, respectively). A plot of log SP versus log adjusted $F_s^{CO_2}$ for

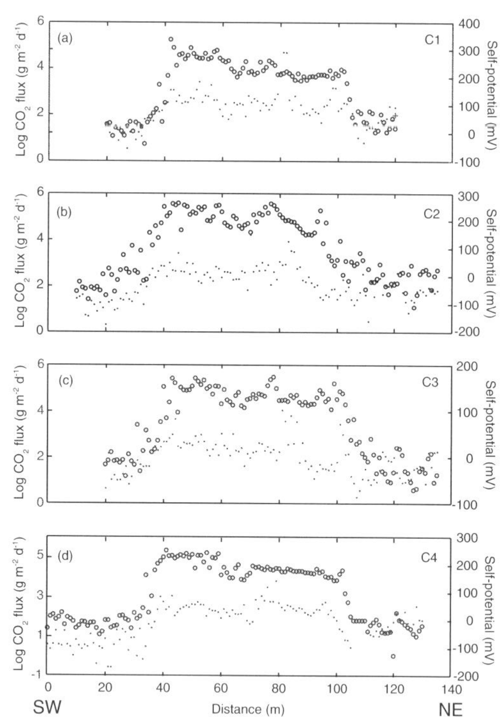

Figure 3. Plots of log $F_s^{CO_2}$ (dots) and SP (open circles) versus distance along transects (a) C1, (b) C2, (c) C3, and (d) C4 (modified from Lewicki et al. 2003).

transects C_1–C_4 (Figure 4) shows a sharp change in correlation of these parameters at $\log F_s^{CO_2} \approx 2.5\,g\,m^{-2}\,d^{-1}$. Log SP and log adjusted $F_s^{CO_2}$ are moderately positively correlated ($C = 0.44$ to 0.56 for C_1–C_4) at $\log F_s^{CO_2} \leq 2.5\,g\,m^{-2}\,d^{-1}$ and poorly correlated ($C = -0.25$ to 0.08 for C_1–C_4) at $\log F_s^{CO_2} > 2.5\,g\,m^{-2}\,d^{-1}$.

Linear regression of log-transformed values of SP and adjusted $F_s^{CO_2}$ was used to calculate the powerlaw scaling relation for SP versus adjusted $F_s^{CO_2}$:

$$SP = b(F_s^{CO_2})^a \quad (1)$$

(Figure 4). In the log-transformed space, the regressed slope yields the exponent value (a) and the intercept yields the scaling constant (b). Table 1 shows calculated power-law coefficients for transects C_1–C_4.

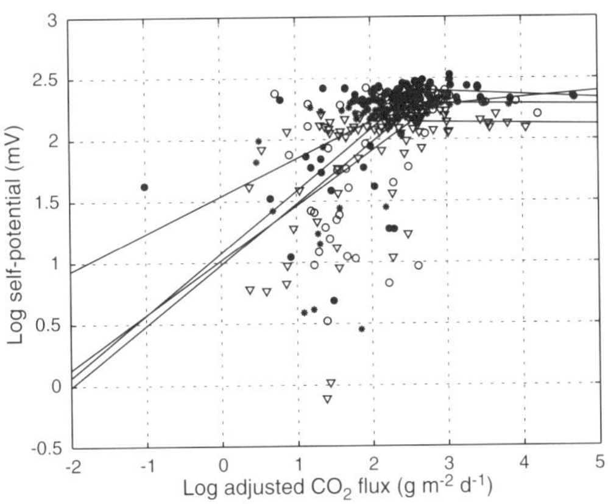

Figure 4. Plot of log SP versus log adjusted F_s^{CO2} for transects C1 (dots), C2 (circles), C3 (triangles) and C4 (stars). Best-fit lines determined by linear regression are shown for log adjusted $F_s^{CO2} \leq 2.5\,g\,m^{-2}\,d^{-1}$ and log adjusted $F_s^{CO2} > 2.5\,g\,m^{-2}\,d^{-1}$.

Table 1. Parameters a and b calculated for best-fit lines to log SP versus log adjusted $F_s^{CO_2}$. The mean (μ) and standard error (σ) of a and b are shown.

	$SP = b(F_s^{CO_2})^a$			
	$\log F_s^{CO_2} \leq 2.5$		$\log F_s^{CO_2} > 2.5$	
Traverse	a	b	a	b
C1	0.30	1.5	$-2.9e^{-2}$	2.5
C2	0.50	0.98	$5.2e^{-2}$	2.1
C3	0.44	1.0	$-1.0e^{-2}$	2.2
C4	0.50	1.1	$-6.8e^{-3}$	2.3
μ	0.44	1.2	$1.4e^{-3}$	2.3
σ	$9.2e^{-2}$	0.26	$3.5e^{-2}$	0.16

4 DISCUSSION AND CONCLUSIONS

We assume that rapid disruption (D) and/or vaporization (V) of groundwater by high gas and/or heat flow derived from a cooling intrusion is primarily responsible for the observed SP anomalies at Masaya (Lewicki et al. 2003). The magnitude of a given SP anomaly here should be proportional to the sum of the fluxes of positively charged water molecules produced by these processes ($F_t^{H_2O} = F_V^{H_2O} + F_D^{H_2O}$). Scaling exponents suggest that there is a diminishing increase in SP with a given increase in $F_s^{CO_2}$, until a threshold at $\log F_s^{CO_2} \approx 2.5\,g\,m^{-2}\,d^{-1}$ above which SP remains relatively constant with increasing $F_s^{CO_2}$.

Lavas and scoria at the Masaya study site are porous and highly fractured. We observe the highest rates of CO_2 (and steam) degassing along linear trends that correspond to the location/trend of the inferred normal fault (Figure 2). Elevated $\log F_s^{CO_2}$ values ($>2.5\,g\,m^{-2}\,d^{-1}$) along transects C_1–C_4 also correlate with fault/fracture locations (Figure 3). These observations indicate distinct gas transport mechanisms, where the highest fluxes are likely associated with advective transport along fractures and the moderate to low fluxes are associated with diffusive transport through the porous matrix. Relative to $F_s^{CO_2}$, positive SP anomalies show little variability over short spatial scales (Figure 3) and do not reflect the presence or absence of fractures. Hence, at elevated $F_s^{CO_2}$ values, log SP becomes poorly correlated with log adjusted $F_s^{CO_2}$.

One explanation for the observed scaling relationship between SP and adjusted $F_s^{CO_2}$ may be the presence of two sources of CO_2 degassing at depth. Based on carbon isotope compositions of soil CO_2 at Masaya, the CO_2 is ultimately derived from deep magmatic/marine carbonate sources (Lewicki et al. 2003). However, CO_2 exsolved from the cooling intrusion may either be dissolved in and subsequently degas from the groundwater system due to changes in subsurface temperatures, or the CO_2 may rise directly to the surface. In the former case, we would expect a positive (although not necessarily linear) correlation between CO_2 flux and SP, as areas of high heat flow will cause both CO_2 degassing from the groundwater and vaporization of groundwater yielding $F_V^{H_2O}$. The charged water vapor and CO_2 may then rise through the porous medium, creating relatively broad surface anomalies in SP and $F_s^{CO_2}$. In the latter case, CO_2 (and steam) degassed from the intrusion may rise directly to the surface along highly permeable fractures with minimal interaction with groundwater, producing the highest soil gas fluxes. If relatively low $F_V^{H_2O}$ is consequently produced during this process, then we would not expect SP to be correlated with $F_s^{CO_2}$ over the high $F_s^{CO_2}$ range. It is also unlikely that steam exsolved from the high-temperature intrusion would produce an SP anomaly, due to the absence of $F_V^{H_2O}$ production at temperatures $>\sim 400°C$ (Blanchard 1964). Therefore, the scaling relationships observed in Figure 4 may be related to two CO_2 flux populations, a low-flux population associated with groundwater degassing/vaporization and positive correlation of SP and $F_s^{CO_2}$, and a high-flux population associated with direct gas flow from depth along fractures and poor SP-$F_s^{CO_2}$ *correlation.*

The scaling relationship observed between SP and adjusted $F_s^{CO_2}$ may alternatively be explained by the influence of CO_2 flow on the SP anomaly by fluid disruption. With RFD, we would expect positive correlation between bulk gas flux from depth and SP; however, there may be a non-linear relationship between this gas flux and the number of water molecules ripped from the fluid ($F_D^{H_2O}$). This may explain the scaling relationship between adjusted $F_s^{CO_2}$ and SP at adjusted $F_s^{CO_2} \leq$ the threshold value. Above the $F_s^{CO_2}$ threshold, we observe a plateau in SP with adjusted $F_s^{CO_2}$ which may be due to local removal and depletion of the fluid source for RFD from the rock.

While intriguing trends are observed in SP and $F_s^{CO_2}$ data at Masaya, a range of processes may explain these trends. To use concurrent measurements of SP and $F_s^{CO_2}$ to better understand processes operating at depth in the volcanic system, additional field, laboratory, and numerical experiments are required to define the relationships between these parameters under a range of hydrogeologic and fluid/heat source conditions.

ACKNOWLEDGEMENTS

This work was supported by the U.S. Department of Energy under Contract No. DE-AC03-76SF00098. We thank D. Bergfeld and G.M. Hoversten for constructive review of the manuscript and L. Connor, E. Mayorga-Caldera, W. Sanford, W. Spinner, and W. Strauch for helpful assistance in the field.

REFERENCES

Anderson, L. A. and Johnson G. R. (1976). Application of the self-potential method to geothermal exploration in Long Valley, California. *J. Geophys. Res. 81*, 1527–1532.

Babu, R. (2003). Relationship of gravity, magnetic, and self-potential anomalies and their application to mineral exploration. *Geophysics 68*, 181–184.

Blanchard, D. C. (1964). Charge separation from saline drops on hot surfaces. *Nature 201*, 1164–1166.

Corwin, R. F. and D. B. Hoover (1979). The self-potential method in geothermal exploration. *Geophysics 44*, 226–245.

Fitterman, D. V. and R. F. Corwin (1982). Inversion of self-potential data from the Cerro-Prieto geothermal field, Mexico. *Geophysics 47*, 938–945.

Hashimoto, T. and Y. Tanaka (1995). A large self-potential anomaly on Unzen volcano, Shimabara peninsula, Kyushu island, Japan. *Geophys. Res. Lett. 22*, 191–194.

Ishido, T. and J. W. Pritchett (1999). Numerical simulation of electrokinetic potentials associated with subsurface fluid flow. *J. Geophys. Res. 104*, 15,247–15,259.

Johnston, M. J. S., J. D. Byerlee, and D. Lockner (2001). Rapid fluid disruption: A source for self-potential anomalies on volcanoes. *J. Geophys. Res. 106*, 4327–4335.

Lewicki, J. L., C. Connor, K. St-Amand, J. Stix, and W. Spinner (2003). Self-potential, soil CO_2 flux, and temperature on Masaya volcano, Nicaragua. *Geophys. Res. Lett. 30*, 1817.

Massenet, F. and V. N. Pham (1985). Mapping and surveillance of active fissure zones on a volcano by the self-potential method, Etna, Sicily. *J. Volcanol. Geotherm. Res. 24*, 315–338.

Revil, A., H. Schwaeger, L. M. C. III, and P. D. Manhardt (1999). Streaming potential in porous media 2. Theory and application to geothermal systems. *J. Geophys. Res. 104*, 20,033–20,048.

Zlotnicki, J., Y. Sasai, P. Yvetot, Y. Nishida, M. Uyeshima, F. Fauquet, H. Utada, Y. Takahashi and G. Donnadieu (2003). Resistivity and self-potential changes associated with volcanic activity: The July 8, 2000 Miyake-jima eruption (Japan). *Earth Planet. Sci. Lett. 205*, 139–154.

Water–Rock Interaction, Wanty & Seal II (eds)
© 2004 Taylor & Francis Group, London, ISBN 90 5809 641 6

The feasibility of using yttria-stabilized ZrO_2 sensors for pH measurements in geothermal field and industrial applications

Mark F. Manna, D.E. Grandstaff & G.C. Ulmer
Department of Geology, Temple University, Philadelphia, PA, USA

Edward P. Vicenzi
National Museum of Natural History, Smithsonian Institution, Washington D.C., USA

ABSTRACT: Zirconia sensors may be used for high-temperature (>90°C) pH sensing in such hostile environments as fumaroles, geothermal wells, and industrial applications, such as in boilers, exhaust stacks and high temperature and pressure aqueous systems. A commercially-available zirconia ceramic cell shows impressive durability during testing at 91°C over the pH range from 1.6–9.6. Current experiments have now surpassed more than 8000 hours of continuous operation in a laboratory flow-through test at 10^5 Pa, which demonstrates near-Nernstian behavior with rapid response to external pH change, on the order of minutes, which is largely limited by geometric mixing parameters. This zirconia sensor is constructed with Hg-HgO and has a resolution of 66 mV/pH unit at 91°C, allowing fairly precise measurements of pH in a very wide range of fluids.

1 INTRODUCTION

The pH is an important parameter affecting equilibrium and reaction rates in hydrothermal and industrial solutions. Yttria-stabilized zirconia (YSZ) sensors have been developed for measurement of pH in laboratory environments (Niedrach 1980; Bourcier et al. 1987; Macdonald et al. 1988; Lvov et al. 2003). Lvov et al. (2003) recently reviewed design parameters of YSZ pH sensors, providing the background to warrant our continued research to extend the current laboratory designs (Fig. 1) into geothermal field and industrial applications. In previous laboratory applications, despite observed near-Nernstian behavior, the durability of the YSZ membrane appeared to be a major limiting factor (Manna et al. 2001; Manna 2002). For example, other research teams have seen membrane life time limited to only a few hundred hours (Lvov et al. 2003) at temperatures and pressures of 340°C and 24.8 MPa and up to 1500 hrs (Ding et al. 2001) at 400°C and 40 MPa. While this is a significant achievement over the typical durability of glass electrodes (only a few hours) at 90°C or above, the field-practicality of YSZ pH-sensors is still in question. For example, the difficulty of emplacing or replacing sensors in harsh environments such as industrial geothermal power plants (Salton Sea, California; or geothermal wells in Hawaii or Iceland), volcanic vents (such as Volcano or Stromboli) or seafloor hydrothermal vents (e.g., Ding et al. 2001) requires that that sensor life be maximized.

Previous experiments have shown that YSZ sensor durability may be limited by thermal shock, chemical leaching or insufficiency of internal buffer (Manna et al. 2001; Manna 2002). In this paper we report recent experiments which have been conducted to test the long-term durability of the YSZ sensor relative to the durability issues created by internal buffer exhaustion and chemical leaching of the ceramic.

2 EXPERIMENTAL METHODS

YSZ pH sensors were constructed employing a commercially-available ZrO_2 ceramic membrane stabilized with 8 mole% Y_2O_3, generally following Bourcier et al. (1987), Macdonald et al. (1992) and Manna (2002) with minor modifications. However, we have increased the buffering capacity of the cell by filling it with two to three times as much Hg-HgO paste as in previous experiments. An intimate mixture of Hg/HgO was accomplished by careful employment of an automatic mortar and pestle using 99% pure (remaining 1% is listed as ignition residue by Arcos Chemicals) HgO powder and liquid mercury. Our custom-designed flow-through Pyrex glass

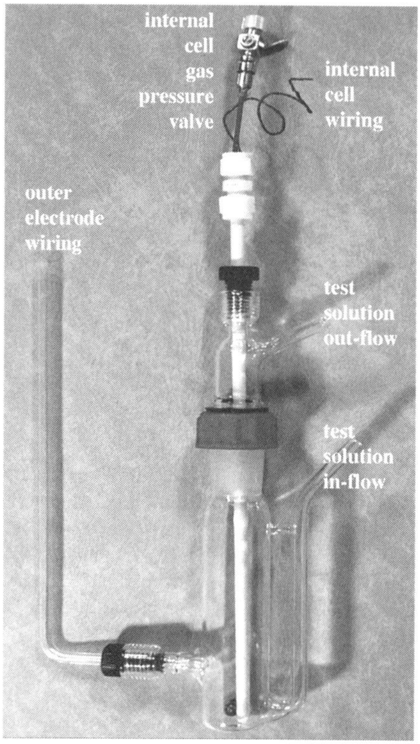

Figure 1. Custom-designed flow-through system (Pyrex glass) is mounted in a temperature-controlled water bath.

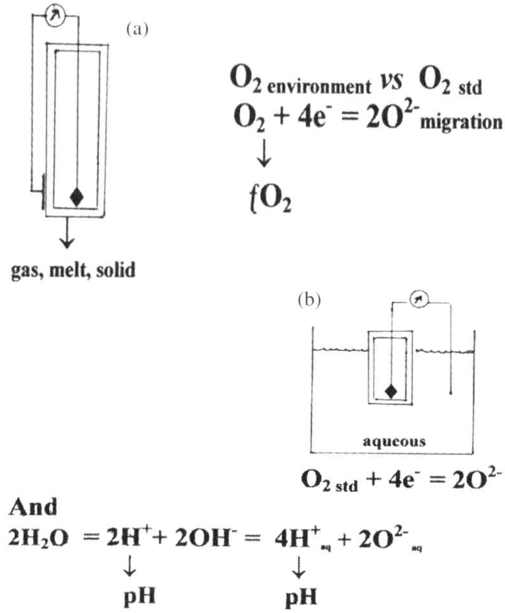

Figure 2. (a) and (b) show the two wiring methods that will allow the ZrO_2 membrane to sense either O_2 (a) or pH (b).

system (Fig. 1), which has a dead volume of 110 ml, is mounted in a temperature-controlled water bath. EMF data are automatically sampled every 5 minutes employing a Keithley© 6517A (high impedance electrometer with an internal resistance of 200 teraohms) and an Intel© Pentium III based computer running custom-written software. Tests have been conducted at 91°C, near the minimum cell operating temperature of 83°C. We are using pH buffer solutions having high-temperature pH values of 1.6 and 9.6 and varying the flow rate from 5 to 8 ml/min from experiment to experiment, but with a constant flow rate during any given experiment. Sensor response can be dependent on flow rate and sensor geometry. However, for this design, the response is constant over this range of flow rates.

The YSZ sensors are specific ion sensors and can be hard-wired to function as either an oxygen or hydrogen ion sensor. By hard wiring the sensor to preferentially detect an oxygen ion, the sensor can be made to detect the fugacity of oxygen in solid, melt, or vapor environments (e.g., Sato 1971; Fig. 2a here in).

By choice of different circuitry (e.g., Niedrach 1980), the YSZ sensor can also be used to measure hydrogen ion in a wide variety of geothermal and industrial aqueous solutions (Fig. 2b). However, in either configuration, the sensing mechanism is due to charge carrier migration by O^{2-} within the YSZ membrane (Bourcier et al. 1987; Macdonald et al. 1988; Lvov et al. 2003). In our research we have employed the latter configuration (Fig. 2b) by designing the sensor to preferentially detect the hydrogen ion and sense pH.

3 RESULTS

Data from a short-term experiment (~4 days) at 91°C for YSZ ceramic exposed to $pH_{T=91°C} = 1.60$ acidic buffer for ~23 hrs and then $pH_{T=91°C} = 9.61$ buffer for ~24 hrs. The buffer was then cycled back to the $pH_{T=91°C} = 1.60$ buffer to show repeatability and response time but the experiment ended due to a building power failure (Figure 3). The *emf* difference between the two buffers (~525 mV) is approximately what would be expected for Nernstian behavior (a slope of 66 mV/pH measured vs. 72 mV/pH ideal).

Although the slope of the data is near-Nernstian, all of the data in Figure 3 have a negative offset of ~380 mV from the theoretical *emf* for these pH values. This offset, which results from cell geometry and resistance, can be compensated by calibration and/or cell design. Electrode response time for the HgO/Hg paste is rapid, even at 91°C, and response is

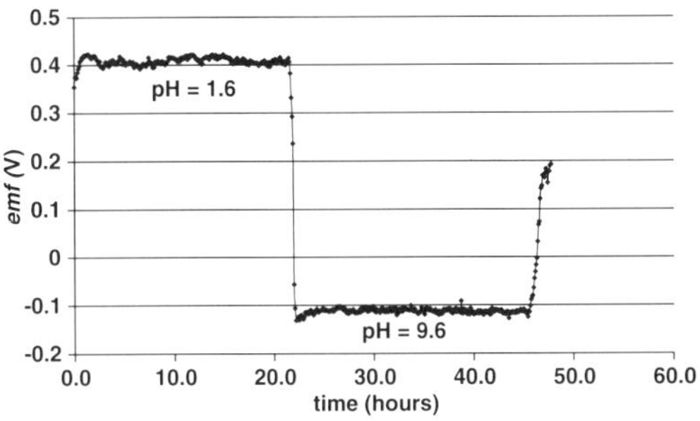

Figure 3. Graph showing data for a short-term experiment (~4 days) of a ZrO_2-Y_2O_3 ceramic constantly exposure to a pH = 1.6 acidic buffer ~23 hrs then a pH = 9.6 buffer for ~24 hrs. The buffer was changed in an attempt to cycle back to a pH = 1.6 buffer but the power was lost ending the experiment. The temperature of the solution was 91°C and a flow rate of 7.4 ml/min was used. The data shows ±20 mV of noise. There are no variations (excursions) in the data that could be associated to textural degradation or local induced factors.

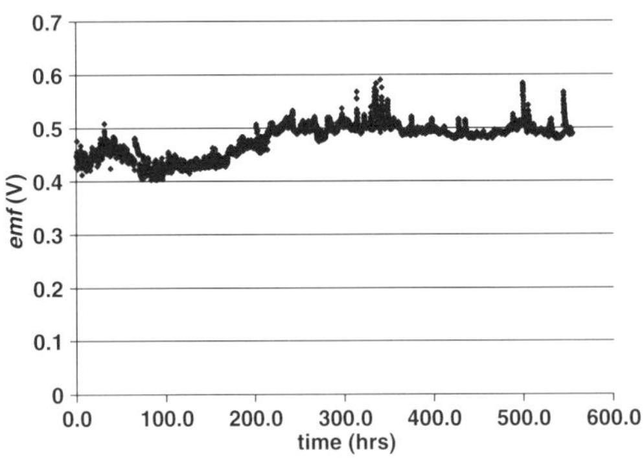

Figure 4. 550 hours of long term data from a YSZ ceramic membrane that was running in pH = 1.6 buffer with a flow rate of 5.8 ml/hr. The data between 50 and 200 hours may be affected by electrical issues that have been confirmed by other instruments in the building and/or stagnation of the pH buffer infiltration into the outer skin of the ceramic. These data contain, not only the ~±20 mV "noise" seen in other data (Fig. 3), but also ~100 mV transients which are related to the electrical variations present in the building, as discussed below. This illustrates the need for a stable *emf* environment to obtain high-quality data.

limited by mixing rate of the buffer solutions in the reaction chamber. The data show approximately ±20 mV of total noise. This noise may result mostly from local factors and electrode resistivity as discussed later.

Figure 4 shows data for another 91°C 10^5 Pa long-term experiment that is currently in progress. The graph shows approximately 550 hours of data which displays more variability than was seen in the short-term data set (Fig. 3).

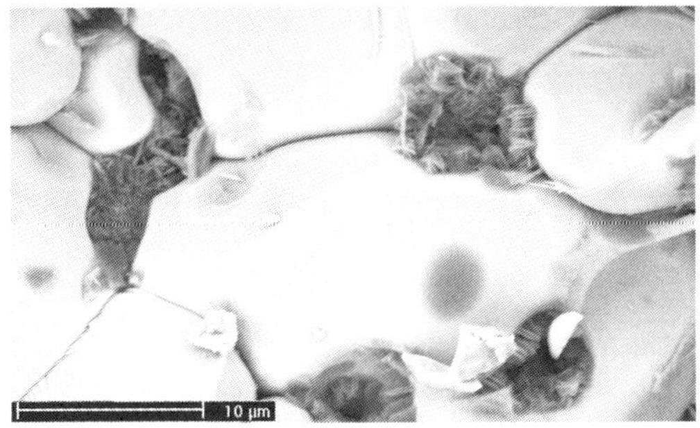

Figure 5. A scanning electron image (50% secondary/50% backscattered electron signal) of a ZrO_2 ceramic stabilized with 8 mol% Y_2O_3 and contains a continuous feldspathic glass that while increasing the overall bulk conductivity will over time zeolitize and mechanically and electronically degrade the ceramic. In this image the light gray grains are the YSZ ceramic and the finely crystalline areas in the cell triple junctions are the zeolitized glass.

The long-term data (Fig. 4) have more variability (excursions). The excursions take the form of dips or spikes. These excursions may be due either to variations in local factors (as discussed later) or changes in the ceramic. We have (Manna et al. 2001; Manna 2002) previously proposed that the texture of the ceramic plays a large role in the response time and the quality of the readings. This ceramic is characterized by a continuous inter-granular Ca-Al-silicate feldspathic glass phase that has been shown to dissolve and alter to zeolite over time and exposure to aqueous solutions (Manna et al. 2001; Manna 2002). Scanning electron microscope images disclose a porous outer skin in the used ceramic due to dissolution of the feldspathic glass. Fluids may become "stagnant" within these pores. These fluids also react with the interstitial glass to form zeolites (Fig. 5) within the porous layer. The crystals are too small to characterize precisely; however, the bladed or plate-like crystal form and calcic composition suggest they are of the heulandite group. When sufficient through-put has occurred this texturally trapped infiltration is replenished with new test pH buffer solution and the sensor starts to react to the new pH buffer. Long-term data will be needed to prove this phenomenon is the cause of the dips in the data, such as that at the beginning of Figure 4. The saturation of the outer porous areas of the ceramic does not appear to affect the overall reading quality of the ceramic in a solution that does not greatly vary in pH (for example does not change from greatly alkaline to greatly acidic). Figure 4 shows electrode response at pH 1.6 during a part of the long-term experiment.

As discussed above, the electrode signal is not entirely constant, even though the flow-through solution composition is constant. Long-term voltage "excursions" seen in the graph may be largely due to RF noise, as is further discussed below. Such excursions have been observed by other electrodes in the building and our own previous fO_2 measurements (Manna 2002). The issue of pore saturation in the outer areas of the ceramic, as it ages, may also be involved. Longer term experimentation should allow a more accurate model to be created. At press time this sensor has more than 8000 hrs of continuous operation at 91°C with similar results as shown in the Figure 4.

4 CONCLUSIONS

As suggested in our past fO_2 experiments (Manna et al. 2001) as the temperature increases, the electrode bulk resistivity will decrease and response time and signal to noise ratio will also decrease, thereby increasing response and data quality. This is important because, at the lowest temperatures, only one of the three commercially-available ceramics that we have investigated has a bulk resistivity low enough to be used. The other two ceramics do not contain the continuous feldspathic phase glass; therefore, their bulk resistivity is too great. In higher temperature applications the bulk resistivity problems of the non-glass phase containing ceramics should be acceptable, because the bulk resistivity should decrease by several orders of magnitude. At higher temperatures the response time of the buffer will also decrease and the overall offset of the readings from ideal Nernstian behavior will decrease, increasing reading quality.

The internal cell buffer can also play a role in the overall response time. Liquid based internal buffers (Hg/HgO for this project) will provide quicker response time because their contact resistance, compared to hard contact (solid Cu/Cu_2O) buffers, which will decrease the overall bulk resistance and increase response time. The buffer can also be varied for the pH of the sensing environment such that the output *emf* can be maximized by the choice of the proper buffer to avoid a null voltage (isoelectric point).

In designing these cells our several years of research have disclosed that there are factors that must be considered and these include: (1) susceptibility to impedance/capacitor behavior, (2) local factors relating to power supplies, and (3) textural degradation. Factors in the first category are exacerbated by the high resistivity at low temperatures which makes the cell circuit susceptible to external inductive fields, much like a metal-detector circuit. Since the cell resistance drops by about an order of magnitude with each 100°C increase, this type of problem becomes less severe at higher temperature. In regards to the second type of design factor, the circuit may be influenced by stability of the power supply to local devices (for example, DC rectifiers, NMR super-conductivity charging, or other typical laboratory RF noise), especially if a perfect electrical ground is not available. Certainly in field operations, the location of the cell circuit would sensibly avoid locations near generator turbines or transformers. We have already discussed the reality of the textural degradation and the need to be aware of this factor.

The current sensor design has been perfected to the limit of what is allowed in the laboratory environment, thus creating the necessity of field-based testing. At a temperature of about 90°C, pH values in natural geothermal fluids range from ca. 7 to 10 (Grandstaff et al. 1990). Under those conditions, the current sensor design would generate a range of about 200 mV, making the sensor suitable for differentiating well fluids within a manifolded set of wells or tracking daily variations in a single well or spring. Properly constructed for the operational environment, a sensor can be constructed to operate in a very wide range of pH's and temperature applications.

REFERENCES

Bourcier, W.L., Ulmer, G.C., & Barnes, H.L. 1987. Hydrothermal pH sensors of ZrO_2, Pd hydrides, and Ir oxides. In Ulmer, G.C. and Barnes, H.L. (eds.) *Hydrothermal Experimental Techniques.* 157–188. Wiley-Interscience: New York.

Ding, K., Seyfried, W.E., Tivey, M.K., & Bradley, A.M. 2001. In situ measurement of dissolved H_2 and H_2S in high-temperature hydrothermal vent fluids at the Main Endeavour Field, Juan de Fuca Ridge. *Earth Plan. Sci. Let.* 186: 417–425.

Grandstaff, D.E., Ulmer, G.C. & Kacandes, G.H. 1990. Prediction of near-field fluid compositions in high-level nuclear waste repositories: data from geothermal and experimental analogs. In Mellinger, G.B. (ed.) *Nuclear Waste Management III.* 347–362. Ceramic Transactions 9. Westerville, Ohio: Am. Ceram. Soc.

Lvov, S.N., Zhou, X.Y., Ulmer, G.C., Barnes, H.L., Macdonald, D.D., Ulyanov, S.M., Benning, L.G., Grandstaff, D.E., Manna, M., & Vicenzi, E. 2003. Progress on Yttria-stabilized zirconia sensors for hydrothermal pH measurements. *Chem. Geol.* 198: 141–162.

Manna, M.F. 2002. Zirconium oxide-yttrium oxide fO_2/pH electrode behavior: the role of temperature on acid durability, charge carrier type and Nernstian response. *M.S. Thesis*, Department of Geology, Temple University, Philadelphia, PA, 169 p.

Manna, M.F., Grandstaff, D.E., Ulmer, G.C., & Vicenzi, E.P. 2001. The chemical durability of yttria-stabilized zirconia pH and O_2 geothermal sensors. In R. Cidu (ed.) *Proceedings of the Tenth International Symposium on Water–Rock Interactions.* 295–299. Rotterdam: Balkema.

Macdonald, D.D., Hettiarachchi, S., & Lenhart, S. 1988. Thermodynamic viability of Yttria-stabilized pH sensors in high-temperature solutions. *J. Soln. Chem.* 17: 719–732.

Macdonald, D.D., Hettiarachchi, S., Song., H., Makela, K., Emerson, R., Haim, M., 1992. Measurements of pH in subcritical and supercritical aqueous systems. J. Solution Chem. 21, 849–881.

Niedrach, L.W. 1980. A new membrane-type pH sensor for use in high-temperature water. *J. Electrochem. Soc.* 127: 2122–2130.

Sato, M. 1971. Electrochemical Measurements and Control of Oxygen fugacities and other gaseous fugacities with solid electrolyte sensors. In Ulmer, G.C. (ed.) *Research Techniques for High Temperature and High Pressure.*43–99. New York: Springer Verlag.

Water–Rock Interaction, Wanty & Seal II (eds)

Insight into the hydrothermal system of Papandayan volcano, Indonesia

A. Mazot & A. Bernard
BRUEGEL (Brussels unit for Environmental, Geochemical and Life Sciences Studies), Universite Libre de Bruxelles, Brussels, Belgium

ABSTRACT: Since the last magmatic eruption that occurred in 1772, Papandayan volcano is in a state of passive degassing. A solfatara with medium-temperature gases (300°C) and acid sulfate-chloride hot springs have been observed for many years in the crater suggesting the presence of a large and permanent volcanic-hydrothermal system at shallow depth. Since 1994, geochemical monitoring of these acid springs has been carried out. In November 2002, phreatic activity of short duration erupted ash plumes to altitudes of about 5 km. New craters were formed and the composition of the acid springs changed reflecting the opening of new fractures at the top of the volcanic edifice. The volcanic material ejected during the eruption is primarily composed of deeply altered volcanic rocks from within the hydrothermal system. The alteration mineralogy is composed of pyrophyllite, alunite, quartz, pyrite, marcasite, kaolinite and anhydrite and is a typical assemblage of high-sulfidation environments.

1 INTRODUCTION

During quiescent period of a volcano, the geochemical survey of thermal springs can give useful information on the evolution of its activity (Taran et al. 1998; Federico et al. 2002).

Since 1994, sampling of hot springs in the crater of Papandayan volcano, Indonesia has been performed to give a description of the fluids associated with the hydrothermal system. The recent eruptive activity (November 2002) ejected ash and altered rocks from within the hydrothermal system, offering an opportunity to improve our knowledge of the alteration processes occurring at depth (Christenson 2000).

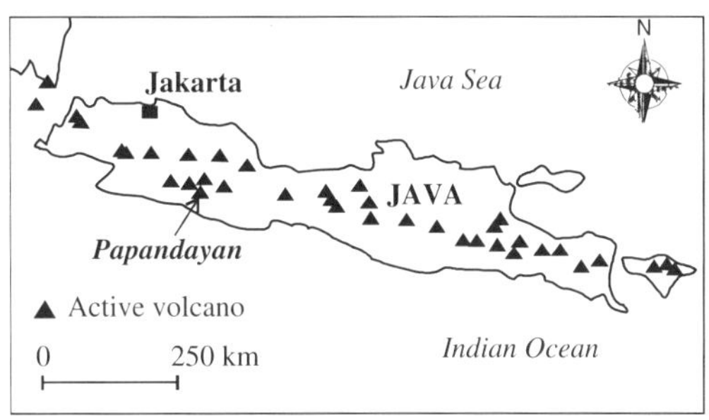

Figure 1. Location map of Papandayan volcano. Modified from Moore et al. (2002).

2 GEOLOGIC BACKGROUND AND RECENT ERUPTIVE ACTIVITY

Papandayan is a 2665 m high stratovolcano located 180 km southeast of the capital Jakarta (Fig. 1). During its historic activity, the eruptive centers moved to the Northeast with the formation of successive new craters. The last magmatic eruption of Papandayan volcano occurred in 1772 and created a horseshoe-shaped crater open to the Northeast, Papandayan volcano has an elliptical depression (1 × 1.5 km) and until November 2002, its activity was centered in two craters with multiple fumaroles (stable maximum temperatures of about 300°C), sulfur-mud pools and hot springs.

3 GEOCHEMISTRY OF HOT SPRINGS AND MINERALOGY OF ERUPTIVE PRODUCTS

3.1 *Composition of hot springs*

Based on chemical composition, two groups of hot springs can be distinguished. The first group of springs has outlet temperatures between 40 and 94°C and pH varying from 1.6 to 3.8 (Table 1). Their compositions are typical of acid sulfate-chloride waters. These fluids are likely to be formed by the direct absorption of magmatic gases such as SO_2 and HCl into the hydrothermal aquifer.

The second group is acid sulfate waters that contain high SO_4 but low amounts of chlorides. The pH of these springs ranges from 1.6 to 2.5 and temperatures range from 22 to 91°C (Table 1). The composition of these fluids is typical of steam condensates that are formed by a boiling acid aquifer at depth.

To understand the origin of these fluids, we studied the $\delta^{34}S$ of dissolved sulfates. Figure 2 shows $\delta^{34}S$ of sulfate versus the SO_4 weight fraction ($100 \times cSO_4/(cSO_4 + cCl)$). A positive correlation is observed between $\delta^{34}S$ values and the chloride contents. $\delta^{34}S$ tends to decrease with Cl concentration.

The high $\delta^{34}S$ values (9–15‰) observed in acid sulfate-chloride waters strongly suggest that dissolved sulfates are mainly formed by the disproportionation of magmatic SO_2 (Ohmoto et al. 1997). Kusakabe et al. (1992) showed that the disproportionation of SO_2 leads to a large isotopic fractionation between dissolved sulfates and H_2S or elemental sulfur. On the other hand, the low $\delta^{34}S$ values observed in chloride-depleted waters suggest that the origin of dissolved sulfates for these waters is the surficial oxidation of hydrogen sulfide (Kiyosu et al. 1983). These waters are likely to be formed by the condensation of steam rising from a boiling aquifer.

3.2 *Mineralogy of ashes and altered products*

Products of the 2002 eruption consist of grey ashes and altered blocks. These products are derived from highly altered lavas of andesitic to dacitic compositions. Some samples show complete silicification.

The mineral phases were studied by X-Ray diffraction (Fig. 3) and SEM/EDS. They are composed of amorphous silica, cristobalite, quartz, pyrite, marcasite, pyrophyllite ($AlSi_2O_5OH$) and alunite ($KAl_3(SO_4)_2(OH)_6$). Anhydrite or gypsum and kaolinite ($Al_2Si_2O_5(OH)_4$) are sporadic phases. Pyrite is very widespread and is well crystallized in altered rocks.

Table1. Chemical and isotopic composition of hot springs.

Site	Date	Temperature (°C)	pH	Na (mg/kg)	K (mg/kg)	Ca (mg/kg)	Mg (mg/kg)	Al (mg/kg)	Fe (mg/kg)	Cl (mg/kg)	SO_4 (mg/kg)	SO_4/Cl	$\delta^{34}S_{SO_4}$ (‰)
Papan1	July 1994	63	2.5	24	11	33	13	31	0.4	216	222	1	n.a.*
Papan2	08/23/1998	71	2.5	16	9	23	9	16	0.8	159	191	1.2	n.a.*
Papan3	02/05/2000	86	1.9	37	24	52	18	107	13	606	605	1	8.7
Papan4	02/05/2000	23	1.6	34	23	70	31	313	248	27	4905	182	−1
Papan5	07/31/2001	94	1.7	26	22	37	13	114	7	919	417	0.5	14.5
Papan6	09/14/2002	90	1.7	24	25	33	10	160	7	1422	790	0.6	n.a.*
Papan7	01/31/2003	40	3.8	30	7	99	20	7	1	46	359	7.8	n.a.*
Papan8	06/27/2003	92	2.5	41	8	177	77	81	556	14	2886	206	n.a.*
Papan9	06/27/2003	88	1.9	84	26	209	92	172	561	1192	2769	2.3	n.a.*

* not analyzed.

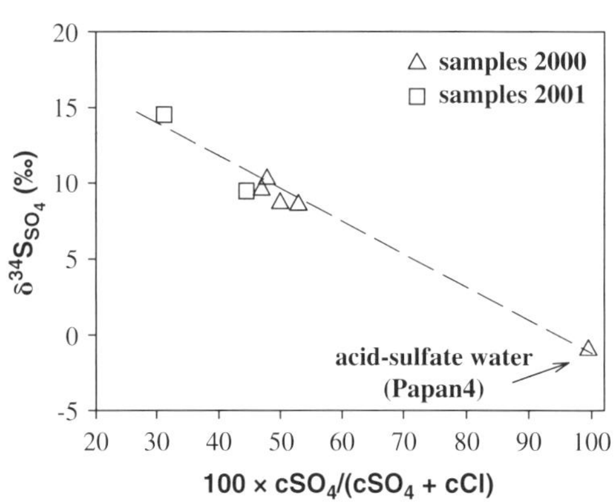

Figure 2. Plot of $\delta^{34}S_{SO_4}$ versus $100 \times cSO_4/(cSO_4 + cCl)$ of hot springs from Papandayan volcano. C_i are concentration in mg/kg.

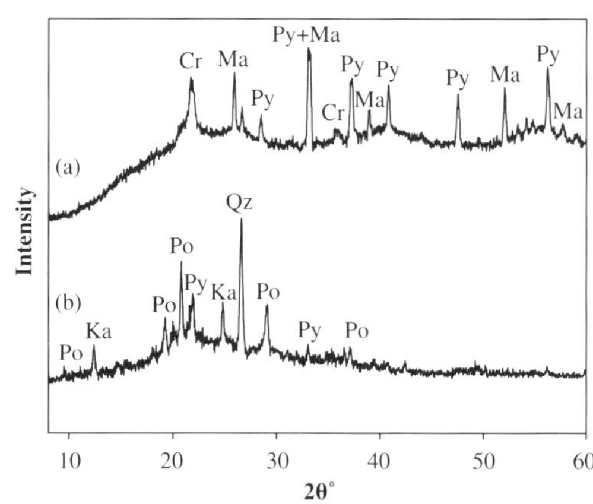

Figure 3. X-Ray diffraction patterns of (a) an altered block and (b) ash. Cr: cristobalite, Ma: marcasite, Py: Pyrite, Po: Pyrophyllite and Ka: kaolinite.

4 DISCUSSION

4.1 *Water–rock interaction*

The chemical composition of acid springs collected during the period 1994–2002 remained relatively constant. Even the samples collected two months before the November, 2002 eruption show no detectable changes, suggesting that no significant modification occurred in the pre-eruptive degassing. However, the samples collected after the eruption in January and July, 2003 reveal notable changes. The most significant changes are an increase in the SO_4/Cl ratios and Mg concentrations (Fig. 4 and Table 1). The increase in Mg could be explained by enhanced water–rock interactions.

To test this hypothesis, we evaluated the degree of fluid–rock interaction by calculating the percentage of residual acidity (PRA) recently introduced by Varekamp et al. (2000) for crater lake studies. The PRA is defined as the residual acidity left after the neutralization of a fluid by water-rock (WR) interaction. Since acidity is consumed by hydrolysis reactions during WR interaction, PRA reflects the rate of acid volatile input relative to the rate of WR interaction. PRA is independent of dilution or boiling.

The pure acid or initial acidity was computed with PHREEQC for each sample based on their Cl and SO_4 concentrations.

Relative to samples collected preceding the 2002 eruption, the samples collected in 2003 show a net decrease in the residual acidity and an increase in the relative Mg concentrations (Fig. 5). This trend reflects an increase in the neutralization of the acid springs that may be related to enhanced WR interaction following the eruption. This could result from the intrusion of fresh magma at shallow depth, but the same trend is observed for acid sulfate waters (i.e. steam condensates), which are likely to have been generated at a very shallow depth. A more likely explanation for the shift observed in the 2003 data is the opening of new fractures where unaltered (or less altered) volcanic rocks were brought in contact with the ascending acid waters.

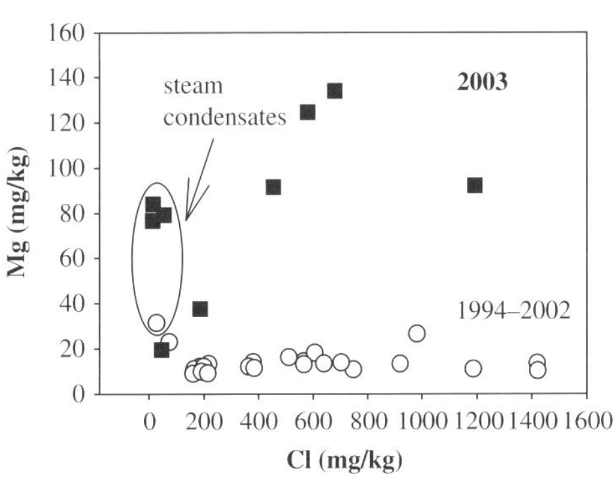

Figure 4. Mg versus Cl concentration of acid waters.

4.2 *Advanced argillic alteration*

The assemblage pyrophyllite, quartz, pyrite, alunite, kaolinite, and anhydrite found in the altered products can be explained by the intense reactions of acid fluids with rocks (Stoffregen 1987; Giggenbach 1992). A model based on the stability of these minerals as a function of KCl/HCl mole ratio versus temperature is shown in Figure 6.

Along reaction path (1), pyrophyllite can occur in association with kaolinite and alunite in a typical advanced argillic assemblage at 250–300°C (Hemley et al. 1980). Along this path, ascending magmatic vapors are directly absorbed by groundwater to

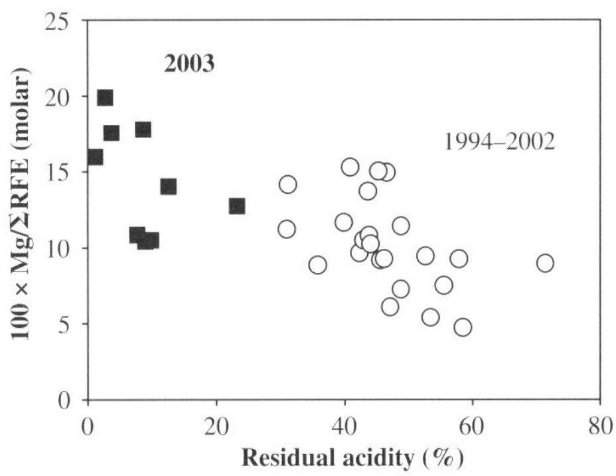

Figure 5. Plot of the percentage of residual acidity versus $100 \times Mg/\Sigma RFE$ (sum of major rock-forming elements $= Na + Ca + Mg + K + Al + Fe$).

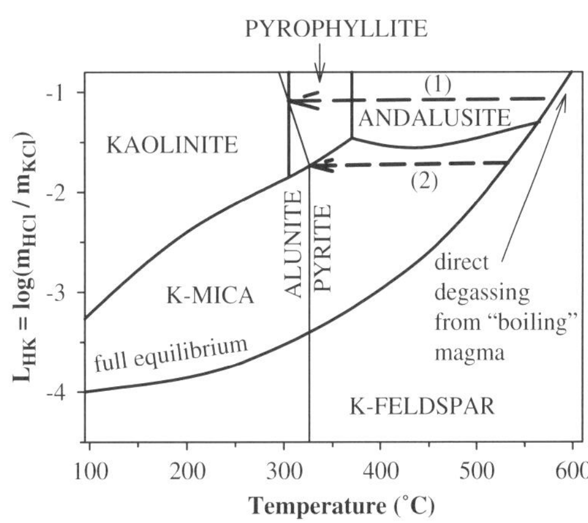

Figure 6. Stability of Al-silicates as a function of m_{HCl}/m_{KCl} and temperature. Dashed lines represent solution paths as discussed in text. Modified from Giggenbach (1992).

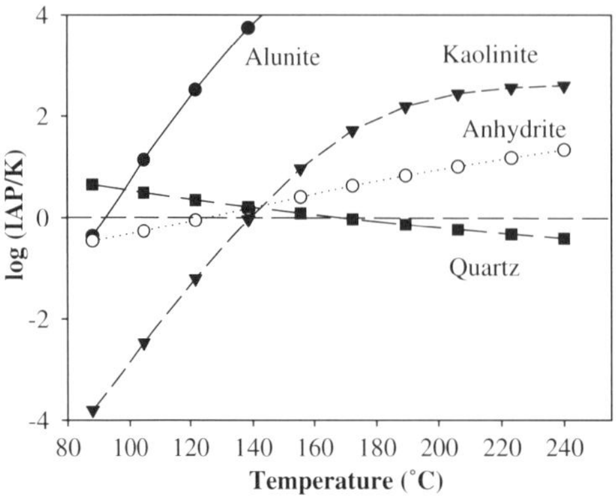

Figure 7. Saturation index of alteration minerals as a function of temperature for sample Papan9.

produce acid waters in the hydrothermal system. Ascending magmatic vapors following a less-direct path (2) will cross the stability field of K-mica and then cut the stability fields of pyrophyllite and kaolinite. Both reaction paths are possible for Papandayan waters.

4.3 *Saturation state of the acid springs*

An ion-association aqueous model PHREEQC (Parkhurst et al. 1999) with the thermodynamic database Wateq4f (Ball et al. 1991) was also used to model the saturation state of the acid-waters at higher temperatures in the subsurface hydrothermal system. The analytical data of acid springs collected after the 2002 eruption were used for the calculations and the temperature was increased by small increments up to 250°C. In Figure 7, the saturation index (log IAP/K) of minerals is plotted as a function of temperature. IAP is the ion activity product and K is the dissolution constant for the mineral phase considered. Waters are supersaturated with respect to alunite at 90°C. Kaolinite, quartz and anhydrite precipitate at higher temperatures (140°C).

5 CONCLUSIONS

The chemical and isotopic compositions of acid hot springs from the crater of Papandayan volcano suggest that contribution of magmatic volatiles (SO_2, HCl) to the hydrothermal system is likely. The 2002 eruption produced notable changes in the chemistry of these hot springs. Most of these chemical changes may be explained by modification of the dynamic circulation of hydrothermal fluids to the surface by the opening of new fractures. No evidence of the intrusion of fresh magma into the upper part of the hydrothermal system can be found. The altered mineral assemblage identified in the rocks ejected during the eruption and coming from within the hydrothermal system is typical of an advanced argillic alteration.

REFERENCES

Ball, J.W. & Nordstrom, D.K. 1991. WATEQ4F – User's manual with revised thermo-dynamic data base and test cases for calculating speciation of major, trace and redox elements in natural waters. *U.S. Geol. Survey Open-File Report* 90–129: 185 p.

Christenson, B.W. 2000. Geochemistry of fluids associated with the 1995–1996 eruption of Mt. Ruapehu, New Zealand: signatures and processes in the magmatic-hydrothermal system. *J. Volcanol. Geotherm. Res.* 97: 1–30.

Federico, C., Aiuppa, A., Allard, P., Bellomo, S., Jean-Baptiste, P., Parello, F. & Valenza, M. 2002. Magma-derived gas influx and water–rock interactions in the volcanic aquifer of Mt. Vesuvius, Italy. *Geochim. Cosmochim. Acta* 66: 963–981.

Giggenbach, W.F. 1992. Magma degassing and mineral deposition in hydrothermal systems along convergent plate boundaries. *Econ. Geol.* 87: 1927–1944.

Hemley, J.J., Montoya, J.W., Marinenko, J.W. & Luce, R.W. 1980. Equilibria in the system Al_2O_3-SiO_2-H_2O and some general implications for alteration/mineralization processes. *Econ. Geol.* 75: 210–228.

Kiyosu, Y. & Kurahashi, M. 1983. Origin of sulfur species in acid sulfate-chloride thermal waters, northeastern Japan. *Geochim. Cosmochim. Acta* 47: 1237–1245.

Kusakabe, M. & Komoda, Y. 1992. Sulfur isotopic effects in the disproportionation reaction of sulfur dioxide at hydrothermal temperatures. *Report, Geol. Survey of Japan* 279: 93–96.

Moore, J.N., Christenson, B.W., Browne, P.R.L. & Lutz, S.J. 2002. The mineralogic consequences and behavior of descending acid-sulfates waters: an example from the Karaha-Telaga Bodas geothermal system, Indonesia. *Proceedings, Twenty-Seventh Workshop on Geothermal Reservoir Engineering Stanford University, California, January* 28–30: 10 p.

Nicholson, K. 1993. *Geothermal fluids: Chemistry and exploration techniques*. Berlin: Springer-Verlag.

Ohmoto, H. & Goldhaber, M.B. 1997. Sulfur and carbon isotopes. In H.L. Barnes (ed.), *Geochemistry of hydrothermal ore deposits, 3rd ed.*: 517–611. New York: John Wiley & Sons.

Parkhurst, D.L. & Appelo, C.A.J. 1999. User's guide to PHREEQC – A computer program for speciation, batch-reaction, one-dimensional transport, and inverse geochemical calculations. *U.S. Geol. Survey Water-Resources Investigations Report* 99–4259: 312 p.

Stoffregen, R. 1987. Genesis of acid-sulfate alteration and Au-Cu-Ag mineralization at Summitville, Colorado. *Econ. Geol.* 82: 1575–1591.

Taran, Y., Fischer, T.P., Pokrovsky, B., Sano, Y., Aurora Armienta, M. & Macias, J.L. 1998. Geochemistry of the volcano-hydrothermal system of El Chichon Volcano, Chiapas, Mexico. *Bull. Vol.* 59: 436–449.

Water–Rock Interaction, Wanty & Seal II (eds)
© 2004 Taylor & Francis Group, London, ISBN 90 5809 641 6

Magmatic and fossil components of thermal and mineral waters in the Eger River continental rift (Bohemian massif, central Europe)

T. Paces & V. Smejkal
Czech Geological Survey, Praha, Czech Republic

ABSTRACT: The Eger River continental rift strikes SW-NE along the Czech Republic/German border. The rift produced Tertiary and Pleistocene volcanoes and deep reaching marginal and central faults. Its recent activity includes seismic swarms, increased heat flow and abundant discharge of CO_2 with traces of He. The isotopic composition of He indicates that over 50% is primordial, probably magmatic. The CO_2 flux is accompanied by convective ground-water flow. Isotopic composition of water indicates its meteoric origin. The dissolved components cannot be fully explained by water–rock interaction. Cl^-, SO_4^{2-} and Na^+ ions were introduced to the hydrologic system during Tertiary and Pleistocene volcanism. The salts were oxidized and concentrated by evaporation in a Tertiary playa lake. Brines percolated from cold playa lake towards hot volcanic spots. While water molecules evaporated due to magmatic heat and partly dissolved in the upwelling magma, dissolved salts were deposited in pores and fractures of wall rocks. Present meteoric water dissolves the fossil salts (Na_2SO_4 and NaCl) preserved at depth. Magmatic CO_2 acidifies the brines so that additional hydrolysis of rock can continue and the present composition of dissolved solids is formed. A uniform isotopic composition of sulfates and occurrence of a closed pocket of concentrated sodium-sulfate brine in mica schists indicate that the evolution of the geothermal and mineral waters has involved meteoric water–rock interaction, input of magmatic CO_2 and dissolution of fossil salts. S and Cl in the fossil salts are derived from volcanic volatiles. The volatiles dissolved and were oxidized in the playa lake and SO_4^{2-} and Cl^- ions were concentrated by evaporation. The direction of water percolation reversed during the evolution. The Tertiary trend in percolation from the playa lake to the volcanic centers changed to a recent direction from the elevated margins of the rift graben to lower elevations inside the graben and to places where CO_2 exsolves from the uprising thermal and mineral waters.

1 INTRODUCTION

Mineral and thermal waters in the western Bohemian massif in Central Europe (Czech Republic) have been investigated since medieval times. Payer (1522) wrote the oldest known scientific report about this topic. The modern continuous interest in thermal and mineral waters and in geological and geochemical origin and evolution of the waters is documented in numerous publications (Rosiwal 1894; Posepny 1895; Suess 1902; Kampe 1922; Michler 1932; Hynie 1949, 1958; Paces 1962, 1987; Vylita et al. 1991; Smejkal & Paces 1992; Dvorak 1998). A significant influence on the interpretation of the chemical and hydrogeological data have had publications by D.E. White, who showed the connection between meteoric, magmatic and connate waters and epithermal mineralizations (White 1955; White et al. 1956; White 1957a, 1957b, 1974). Data on stable isotopes of H, O, S (Smejkal 1978, 1979; Smejkal & Jetel, 1990; Smejkal & Paces 1992) and isotopic data on He (O'Nions & Oxburgh 1988; Weinlich et al. 1999), indicate that White's ideas are useful when we attempt to explain the origin and evolution of thermal and mineral waters in connection with the tectonic pattern and geological history of the Eger River continental rift.

2 GEOLOGICAL SETTING

The Eger River rift is a 160-km long graben striking SW-NE mostly in the Czech Republic along the border with Germany. The rift was active from Oligocene to Pleistocene (Kopecky 1978). The morphologically dominant Krusne Hory fault define the NW margin of the rift. Several transverse faults cross the rift in its SW part. The dominant transverse faults

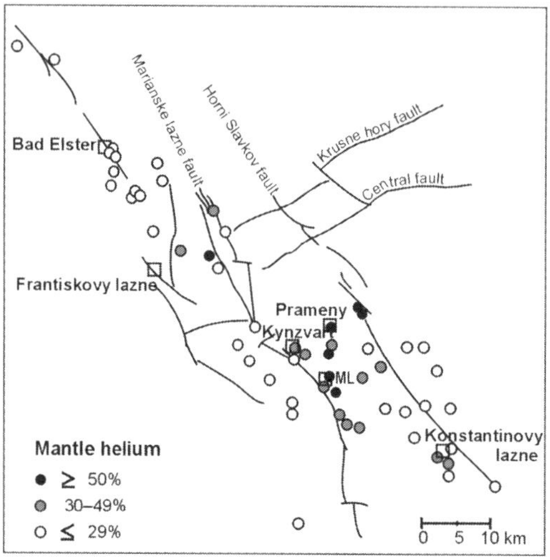

Figure 1. Tectonic sketch of the crossing of the Eger River rift (Krusne Hory fault and Central fault) with the transverse Marianske Lazne and Horni Slavkov faults. Points represent sampled springs and gas manifestations and their relative content of mantle He (data in Weinlich et al. 1999 and new unpublished data).

are the Marianske Lazne fault and Horni Slavkov fault. These crossings and less dominant associated transverse faults coincide with the youngest volcanism (0.2–0.9 Ma; Wagner et al. 1998) and recent occurrence of numerous thermal and mineral springs charged with CO_2 (Fig. 1). The crossings of the faults coincide with contemporary earthquake swarms (up to 3 M_L; Spicak & Horalek 2001). The rift valley is filled with Miocene and Oligocene lacustrine sediments. The following indicators suggest that the sediments were deposited in a large saline playa lake with high concentrations of Na_2SO_4: (1) occurrence of secondary sulfate minerals (gypsum and thenardite) in lacustrine sediments with isotopic composition indicating crystallization from saline solutions (Smejkal,1978); (2) occurrence of carbonates enriched in heavy isotopes of oxygen and carbon (Smejkal, 1984); (3) occurrence of fossil halophytes (Limnocarpus, similar to modern Ruppia) (Buzek & Holy, 1981); (4) sudden death loss of fish documented by peleontological records (Kvacek et al. in Paces et al. 1981); (5) geological reconstruction of sedimentation space.

3 GEOCHEMICAL DATA

The chemical and isotopic composition of selected springs of thermal and mineral water and gas are in Tables 1, 2 and 3.

Table 1. Chemical composition of thermal and mineral water from the Eger River rift (locality shown in Table 2 footnote).

Locality	1	2	3	4	5
Temp. C	39	73	7	13.5	20
pH	6.85	6.6	6.11	6.1	7.5
(mg/L)					
Na	230	1600	2600	6800	47000
K	12	99	46	110	290
Li	0.28	2.9	5	13	33
NH_4	0.029	0.36	0.26	1.8	35
Ca	43	127	148	417	455
Mg	9.5	44	93	167	278
Sr	0.19	2	0.85	3.2	4.7
Ba	0.09	0.05	0.04	0.000005	
Fe	0.02	1.03	28.4	7	83
Mn	0.023	0.16	0.58	0.19	1.5
Cl	50.1	613	858	2700	28700
Br	0.18	1.6			
I	0.0027	0.021			1.2
F	7.5	6.2	0.3		1.2
HCO_3	560	2110	2770	3310	1400
NO_3	7		0	82	85
SO_4	125	1400	2950	9810	54200
SiO_2	67	71	76		16
CO_2	$0^{\#}$	380	2900	1300	$0^{\#\#}$
TDS	1130	5970	9490	23400	132000

$^{\#}$Partial pressure of CO_2 0.0094 MPa, $^{\#\#}$Partial pressure of CO_2 0.046 MPa.

Table 2. Isotopic composition of H_2O, dissolved CO_2 and SO_4 in thermal and mineral waters from the Eger River rift.

Locality	1	2	3	4	5
(‰)					
$\delta^{18}O$ (SMOW)	−9.8	−9.6	10.51	−10.5	−10.4
δ D (SMOW)	−69.2	−69	66.4		−55
$\delta^{13}C$ (PDB)	−5.3	−6.3	−2.5	−3.4	
$\delta^{34}S$ (CDT)		6.3	6.4	6.2	6.4
$\delta^{18}O$–SO_4		6	5.1	6.1	6.7

1 – Teplice, source Pravridlo, 2 – Karlovy Vary, source Vridlo II, 3 – Marianske Lazne, source Krizuv spring, 4 – Frantiskovy Lazne, source Glauber IV, 5 – Odrava, drill hole HV 18.

The isotopic composition of H_2O in all thermal and mineral waters along the Eger River rift is on the meteoric line indicating that groundwater in the springs originates from local atmospheric precipitation (Fig. 2). The only significant shift is observed in the most concentrated brine from drill hole HV-18 (Odrava). The unusual shift is to the upper field from the meteoric line in a similar direction as was observed in brines from mines in the Canadian Shield by Frape & Fritz (1982, 1987).

Table 3. Chemical composition of gas from thermal and mineral (saline) water from the Eger River rift.

Locality	1	2	3	4	5
Total gas (mL/L)	14.5	[2.7*]	1190	1210	0
(% by volume)					
CO_2	99.7			97.88	99.5
H_2	0.0037	0.0028	0.01	0.001	
O_2	3.32	0.046	0.06	0.021	
N_2	94.72	0.168	0.38	0.78	
Ar	2.01	0	0.008	0.064	
He	0.0038	0	0.0003	0.00001	
CH_4	0.0003			0	0.01

1 – Teplice, source Pravridlo, 2 – Karlovy Vary, source Vridlo II, 3 – Marianske Lazne, source Ferdinand II, 4 – Frantiskovy Lazne, source Glauber III, 5 – Odrava, drill hole HV 18 [* Volume ratio CO_2 gas/thermal water].

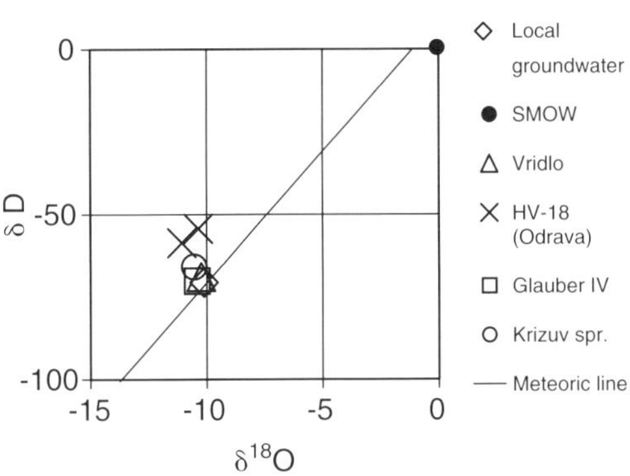

Figure 2. Isotopic composition of water in groundwater and thermal and mineral waters in the Eger River rift.

The major chemical characteristic of the waters in the Eger River rift is the high content of Na_2SO_4 followed by NaCl. The components cannot be derived from granite and mica schist in the basement of the tertiary sediments. This is demonstrated by a simple mass–balance calculation in Table 4.

The high proportion of Na indicates that there must be another source of Na besides rock-forming minerals. The disproportion in the mass balance is much higher for S and Cl. The salts cannot be derived from the lacustrine Miocene sediments because of the deep circulation of thermal water in the basement crystalline rocks.

The isotopic composition of SO_4^{2-} is almost identical in the waters from the area of the Tertiary playa lake while the waters outside the rift and in brines from Cenomanian and Permocarboniferous aquifers of the Bohemian massif are different and vary widely (Fig. 3).

Table 4. Calculated total consumption of granite in km^3 to supply major cations to the Karlovy Vary thermal springs during their 250,000 year history, recorded by calcareous sinters (Paces 1962).

Na	5.0	Ca	0.34
K	0.20	Mg	0.59

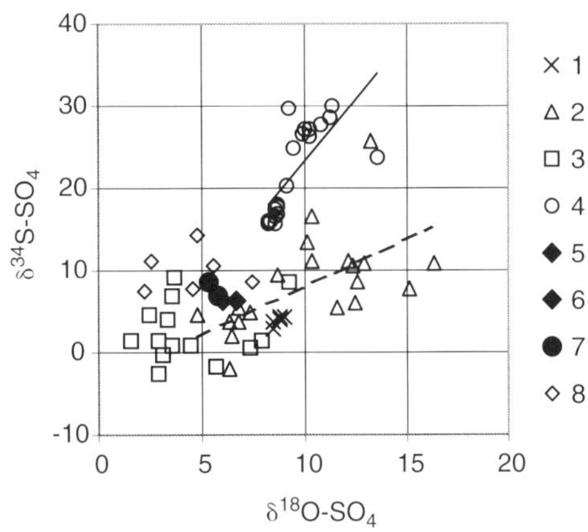

Figure 3. Isotopic composition of dissolved sulfates in groundwater and brines in the Bohemian massif (data by V. Smejkal). 1 – Precipitation, 2 – Brines from Cenomanian and Permocarboniferous aquifers, 3 – Shallow groundwater near the rift, 4 – Ca-Cl brines in granodiorite near but outside the rift, 5, 6, 7 – Thermal and mineral waters in the rift and under the Tertiary sediments (Karlovy Vary, Odrava, Marianske Lazne), 8 – Mineral water near Karlovy Vary without fossil salts (sulfates derived from local pyrite).

The origin of CO_2 cannot be clearly identified from the isotopic composition of C. Most data on $\delta^{13}C$ in dissolved CO_2 and gaseous CO_2 lay between −1.6 and −3‰. This C can be either from metamorphosed marine limestones (Barnes & O'Neil 1976) or from a magmatic melt. The geological setting of the Eger River rift makes highly improbable the possibility that the origin of CO_2 is in a current thermal decomposition of marine carbonates. The magmatic source of CO_2 was tested by isotopic analyses of traces of He in the gas (O'Nions & Oxburgh 1988, Weinlich et al. 1999).

It has been shown that He in the Eger River rift at the crossings with Marianske Lazne fault and Horni Slavkov fault have a very high content of mantle He, reaching 60% (Weinlich et al. 1999). The distribution of the mantle He in the traces of He in the CO_2 gas is shown in Figure 1. There is a statistically significant linear correlation between the content of mantle He and the logarithm of the content of CO_2 in water (Fig. 4).

The correlation, together with the high proportion of mantle He, suggests that the origin of CO_2 in the Eger River rift is magmatic rather than sedimentary or metamorphic.

There is no statistically significant correlation between CO_2 content and dissolved Na, SO_4^{2-} and Cl^-. In fact, the most concentrated Na-SO_4-Cl brine (TDS 130 g/L) from drill hole HV-18 in Odrava (Table 1, analysis 5) does not contain detectable CO_2 (calculated P_{CO_2} is 0.046 MPa). The brine was tapped at depth of 600 m in mica schist under the Tertiary sediments. It is not possible to explain such a high quantity of SO_4^{2-} by the oxidation of volcanic SO_2 and H_2S without an input of molecular O_2 at depth. The high concentration of SO_4^{2-} requires an oxidation of reduced sulfuric species in contact with atmospheric O_2 followed by an extensive concentration of the oxidized solution by evaporation. We suggest that the Na-SO_4-Cl brines in the Eger River rift have a complex but unified history in which volcanic gases produced by the Tertiary volcanism dissolved in the playa lake, the reduced volcanic species of S were oxidized by atmospheric O_2, the sulfuric and hydrochloric acids in lake water attacked detrital material from local granite and dissolved Na from feldspars. The decomposed feldspars were turned into large kaolinite deposits. The resulting solution was concentrated by evaporation. Surges of low pH due to the input of volcanic volatiles occasionally killed fish populations. Shores of the saline lake were inhabited by halophytes similar to the present gender of Ruppia. As suggested by the sketch in Figure 5, the concentrated brine infiltrated to fractures in the lake bottom and percolated towards the source of heat along the thermodynamic gradient. The source of heat was the basaltic magma feeding the local volcanoes. Water was partly assimilated by the basaltic melt and partly evaporated while salts were left behind in pores and fractures of wall rocks.

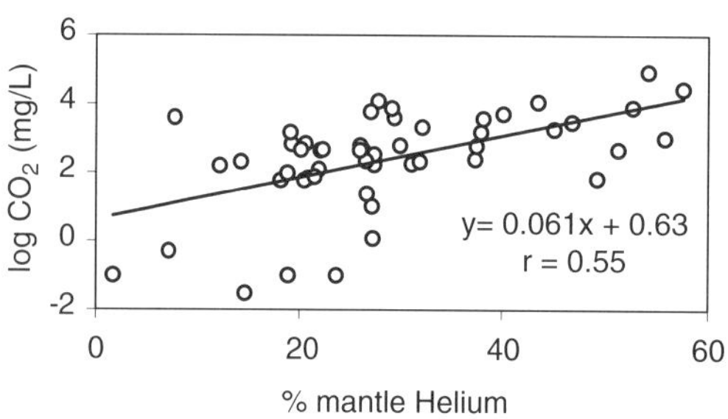

Figure 4. Correlation between content of free CO_2 gas in springs and fraction of mantle He in traces of He in CO_2 gas.

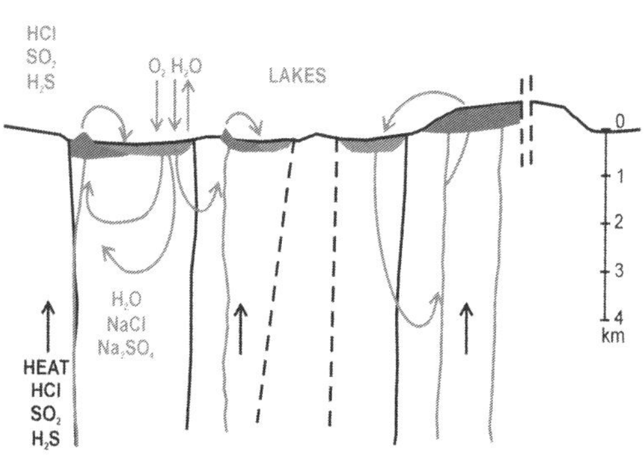

Figure 5. Sketch of the Eger River rift during Miocene volcanic activity. Light gray arrows represent percolation direction of water with dissolved solids and gases; darker gray represents basaltic lava chimneys and volcanoes with volcanic volatiles and heat. Dashed lines are faults and distance separators.

4 EVOLUTION OF THE HYDRODYNAMIC AND CHEMICAL SYSTEM

The concept of the evolution of present thermal and mineral waters in the Eger River rift is illustrated in the two sketches in Figure 5 and 6. The situation in Miocene time is depicted in Figure 5 when the volcanic activity generated heat and brought magmatic volatiles to the atmosphere and to the volcanic playa lake. Oxidation of reduced S species took place in the playa lake due to the input of atmospheric O_2. Very acid water dissolved detrital material derived from surrounding granitic rocks. The solution containing predominantly H^+, Na^+, SO_4^{2-} and Cl^- was concentrated by intense evaporation. The brine in the

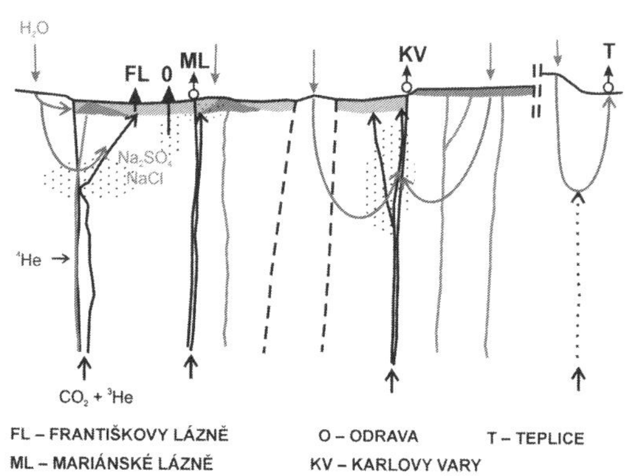

Figure 6. Sketch of Eger River rift at present. Light gray are tertiary lacustrine sediments, dark gray are basalts. Light gray dots represent assumed fossil salts, gray arrows indicate direction of present flow of groundwater, black arrows indicate flux of CO_2 with primordial ^3He from depth (asthenosphere?) and flux of radiogenic ^4He from granitic wall rocks.

lake percolated from the lake to the uprising lava. Water of the brine evaporated due to volcanic heat and was probably partly assimilated by dry lava. Sulfate and chloride salts precipitated and stayed in the fractured basement rocks.

The present situation is illustrated in Figure 6. The direction of the water percolation is opposite to the situation in Miocene time. Atmospheric precipitation infiltrates on elevated margins of the rift graben. Groundwater percolates under pressure to lower piezometric levels inside the graben. Groundwater meets upcoming CO_2 gas of magmatic origin at various depths. Acidified water dissolves selectively minerals of granite, basalt, amphibolite, and gneiss, depending on local geological conditions. The resulting mineral waters are lower in the content of sulfates and chlorides and high in the content of bicarbonates. The ratio between major cations is variable according to the type of dissolving rock.

The exsolution of CO_2 below the earth surface makes the column of water lighter and brings the thermal and mineral waters to the surface. Water with low TDS such as in Teplice (Table 1, locality 1) forms by the water–rock interaction only. More concentrated mineral waters are formed where the uprising groundwater, charged with CO_2, meets pockets of fossil salts, such as in the spring Vridlo in Karlovy Vary or water tapped by drill hole Glauber IV in Frantiskovy Lazne spa or the spring Krizovy in Marianske Lazne spa (Tables 1–3, localities 2–4). The most concentrated SO_4^{2-} brine in Odrava was tapped by the drill hole HV-18 in an isolated fractured nest in mica schist without access of CO_2 (Tables 1–3, locality 5). Therefore, the fossil salts were never brought to surface and represent probably the original fossil brine. It is probable that the original water of the brine has been exchanged with a more recent meteoric water during the 20 million years since the Miocene volcanic and playa lake event.

5 CONCLUSION

The complex history of the Na-SO_4-Cl brines in the Eger River rift fully agree with the conclusions of Donald E. White when he wrote in 1957: "Although some major types of deep water seem to be recognizable, there is much danger of oversimplifying the problems. Many waters are no doubt mixtures of different types, and some of the high salinity results from dissolution of salts by meteoric water."

ACKNOWLEDGEMENT

This paper has been influenced by ideas of Don White. I express my gratitude to Don, who predetermined my scientific and to a large extent my personal life by helping me to come to the USA from Czechoslovakia during difficult times of the Communist era. Not only to me, but to many of our mutual friends, his ideas on the complex nature of thermal and mineral waters and epithermal ore deposits remain alive. Tom Paces.

REFERENCES

Barnes, I. & O'Neil, J.R. 1976. Metamorphic reactions in flysch rocks. In: J. Cadek & T. Paces (eds.): *Proceedings International Symposium on Water–Rock Interaction, Czechoslovakia 1974*: 309–316. Praha: Ustredni ustav geologicky

Buzek, C. & Holy, F. 1981. Fruits of halophilous water plants (Limnocarpus C. Reid, Potamogetonaceae Dumotier) from the Cypris Formation of the Cheb and Sokolov Basins. *Jour. Geol. Sciences, Palaentology* 24: 163–177. Prague.

Dvorak, J. 1998. Hydrogeology and the genesis of mineral waters of the Carlsbad type in western Bohemia. *Münchner Geol. Hefte* B8: 63–69.

Frape, S.K. & Fritz, P. 1982. The chemistry and isotopic composition of saline groundwaters from the Sudbury Basin, Ontario. *Canad. J. Earth Sci.* 19: 645–661.

Frape, S.K. & Fritz, P. 1987. Geochemical trends for groundwaters from the Canadian Shield. In: P. Fritz & S.K. Frape (eds.) *Saline Water and Gases in Crystalline Rocks: 19–38.* GAC Special Paper 33, Geological Assoc. Canada.

Hynie, O. 1949. Geology of mineral springs in Bohemia and Moravia (in Czech). *Geotechnika* 7: Praha.

Hynie, O. 1958. Hydrogeology of mineral springs in spas of Wastern Behemia (in Czech). *Casopis pro mineralogii a geologii*: 3(2), Praha.

Kampe, R. 1922. Die Karlsbader Thermen. *Karlsbader ärzliche Vorträge*: Band 3: Jena 1927.

Kopecky, L. 1978. Neoidic taphrogenic evolution and young alkaline volcanism of the Bohemian Massif, Czechoslovakia. *J. Geol. Sci. Geology* 31: 91–107.

Michler, O. 1932. Wie alt ist es Karlsbader Sprudel? (without a reference) In: Paces. T., 1962.

O'Nions, R.K. & Oxburgh, E.R. 1988. Helium, volatile fluxes and the development of continental crust. *Earth Planet. Sci. Lett.* 90: 331–347.

Paces, T. 1962. Baitrag zum Geochemischen Studium der Karlsbader quellen. *Casopis pro mineralogii a geologii* 7: 418–427, Prague.

Paces, T. 1987. Hydrochemical evolution of saline waters from crystalline rocks of the Bohemian Massif. In: P. Fritz & S.K. Frape (eds.) *Saline Water and Gases in Crystalline Rocks:* 145–156. *GAC Special Paper 33.* Geol. Assoc. Canada.

Paces, T., Smejkal, V., Pazdera, A., Kobrova, M. & Barnes, I. 1981. A unique type of brine in the basement of the Cheb basin (in Czech). *Geologicky Pruzkum* 1981: 196–198.

Payer, V. 1522. Tractatus de termis Caroli Quarti. Printed in *Lipsiae, ex aedibus Valentini Schumanni.* In: Doktor Vaclav Payer z Lokte, Pojednani o Karlovych Varech z r.

1522, Translator: B. Ryba, *Publishing House Avicenum*: Praha, 1984.
Posepny, F. 1895. Über die Genesis der Erzlagerstätten. Berg. u. Hüttenmänn. *Jb. montan. Hochsch.* 431–524.
Rosiwal, A. 1894. Über neue Massnahmen zum Schutze der Karlsbader Thermen. *Jb. Geol. Reichsanst.*
Smejkal, V. 1978. Isotopic geochemistry of the Cypris Formation in the Cheb Basin, West Bohemia. *Vestnik Ustredniho ustavu geologickeho* 53: 3–18.
Smejkal, V. 1979. Oxygen isotopic composition of sulphates from some mineral waters and mine waters in Western Bohemia. *Isotope Hydrology 1978*: v. I, 83–98, Vienna: Int. Atom. Energ. Agency.
Smejkal, V. 1984. Isotopic composition of carbonated and differences in deposition environment during the Miocene lacustrine sedimentation in the Krusne Hory graben. In: 3. Arbeitslagung "Isotope in der Natur" *Berichte des Zentralinstituts fur Isotopen- und Strahlenforschung, ZFI – Mitteilungen* 84: 372–379. Leibzig.
Smejkal, V. & Jetel, J. 1990. Isotopic and geochemical indications of a drainless sulphate lake in the Permo-Carboniferous of the Krkonose-piedmont Basin. *Vestnik Ustredniho ustavu geologickeho* 65: 339–352.
Smejkal, V. & Paces, T. 1992. Origin of mineral waters of the Karlovy Vary type (in Czech). *Geologicky Pruzkum* 34: 33–37.
Spicak, A. & Horalek, J. 2001. Possible role of fluids in the process of earthquake swarm generation in the West Bohemia/Vogtland seismoactive region. *Tectonophys.* 36: 151–161.
Suess, E. 1902. Ueber heissen Quellen. 74. Vers. Deutsch Natur. u. Aerzte zu Karlsbad. In: *Library of the Czech Geological Survey*, Prague
Vylita, B., Bodlak P., Domecka K., Pecek J. & Zeman J. et al. 1991. New information on the Karlovy Vary spring structure (Nove poznatky o karlovarske zridelni struktuře) (in Czech). *Ustredni ustav geologicky*: ISBN 80-7075-021-9, 107 p.: Prague.
Wagner, G.A., Goegen K., Jonckheere R., Kappf H., Wagner I. & Woda C. et al. 1998. The age of Quaternary volcanoes Zelezna Hurka and Komorni Hurka (Western Eger Rift). Czech Republic: alpha-recoil track, TL, ESR and fission track chronometry. In: *Magma and Rift Basin Evolution – Excursion Guide. Abstracts.* Czech Republic, Liblice, Sept. 7–11, 1998, Czech Geological Survey: 95–96: Prague.
Weinlich, K.H., Bräuer, K., Kämpf, H., Strauch, G., Tesar, J. & Weise, S.M. 1999. An active subcontinental mantle volatile system in the western Eger rift, Central Europe: Gas flux, isotopic He, C and N and compositional fingerprints. *Geochim. Cosmochim. Acta* 63: 3653–3671.
White, D.E. 1955. Thermal springs and epithermal ore deposits. *Econ. Geol.*: 50th Anniversary Volume 99–154.
White, D.E. 1957a. Magmatic, connate, and metamorphic waters. *Geol. Soc. Am. Bull.* 68: 1659–1682.
White, D.E. 1957b. Outline of thermal and mineral waters as related to origin of Mississippi Valley ore deposits. In: Genesis of Stratiform Lead – Zink – Barite – Fluorite Deposits (J.S. Brown, ed.), *Economic Geology*, Monograph 3.
White, D.E. 1974. Diverse origin of hydrothermal fluids. *Econ. Geol.* 69: 954–973.
White, D.E., Craig, H. & Boato, G. 1956. Isotopic geochemistry of thermal waters. In: Proceedings of Second Conference on Nuclear Processes in Geological Settings. *National Academy of Sciences – National Research Council:* Publ. 400: 29–38, Washington D.C.

Water–Rock Interaction, Wanty & Seal II (eds)
© 2004 Taylor & Francis Group, London, ISBN 90 5809 641 6

Hydrothermal activity at Tacaná volcano, Mexico–Guatemala

D. Rouwet & Y. Taran
Instituto de Geofísica, UNAM, México

S. Inguaggiato
Istituto Nazionale di Geofisica e Vulcanologia, Sezione di Palermo, Italy

N. Varley
Universidad de Colima, Colima, México

ABSTRACT: The Tacaná volcano (4100 m a.s.l.) on the Mexico–Guatemala border is the northernmost extreme of the Central American Volcanic Arc. On 8 May 1986, after a minor phreatic explosion, a low-temperature fumarole field was formed at an altitude between 3200 and 3600 m a.s.l. on the northwest flank of the volcano. Numerous CO_2-rich thermal springs discharge on its slopes close to the contact between the volcanic edifice and granitic basement. Chemical composition of all the springs can be interpreted as a mixture of a concentrated deep component and dilute shallow water. The deep component is mainly meteoric water recharged at high elevations with dissolved magmatic volatiles whose C:S:Cl are close to that of magmatic gases. He isotopic ratios (5.7–6.2 R_a) and $C/^3He$ ($\sim 10^{10}$) in gases are typical for the Central American Volcanic Arc. The low-temperature fumarolic vapor is essentially fractionated meteoric water. A non-typical negative δD-Cl correlation for spring waters is caused by the special hydrological structure of the volcano.

1 INTRODUCTION

Tacaná volcano (4100 m a.s.l., 15°8′N, 92°9′W) is shared by two countries and is the northernmost extreme of the Central American Volcanic Arc (Fig. 1). Tacaná is a dome complex with a central andesitic dome in its main crater. The historical information about the Tacaná volcanic activity can be found in Macías et al. (2000). The last phreatic activity occurred in 1986 and was preceded by a seismic swarm with epicenters on the southwest flank of the volcano. According to Macías et al. (2000) the volcano is built within an old 9-km caldera rising more than 2000 m over the Paleozoic granitic and Late Tertiary ignimbrite basement.

Tacaná volcano is located in a very wet area with an annual precipitation rate more than 4 m. Low-temperature fumarolic activity is concentrated along a fracture between 3100 and 3600 m a.s.l., and most of thermal springs are located on the NW slopes close to the contact between basement and the edifice at elevations of about 2000 m (Fig. 1). One group of warm springs (La Calera, 27°C) is located on the southern slope at lower elevations. A very limited set of water analyses was reported by Medina (1986), and one chemical analysis of fumarolic vapor collected in 1986 was published by Martini et al. (1987).

We report here results of a complex geochemical study of thermal waters, free, dissolved and fumarolic gases of Tacaná volcano that includes δD, $\delta^{18}O$ in waters and fumarolic steam, $\delta^{13}C$ of free and dissolvd CO_2, $^3He/^4He$ and He/Ne in bubbling and dissolved gases.

2 CHEMICAL AND ISOTOPIC COMPOSITION OF WATER AND GAS DISCHARGES

2.1 *Water isotopes*

Isotope data are presented in Table 1. Local shallow and surface waters at Tacaná at ~2000 m a.s.l. are characterized by δD = −65% and $\delta^{18}O = -9\%$ vs. V-SMOW. All thermal waters are slightly isotopically lighter, with δD values in the range of −80 to −69‰ vs. V-SMOW and $\delta^{18}O$ between −10.5 and −9.5‰ vs. V-SMOW. Fumarolic condensates are strongly enriched in the light isotopes with values of δD from −116 to −109‰ vs. V-SMOW. The boiling temperature of water at 3500 m above sea level is near 87°C, and the difference between the water and steam isotopic compositions corresponds to the isotopic fractionation at this temperature.

2.2 *Chemistry of waters and solute geothermometry*

All thermal waters at Tacaná are near neutral and of a rare SO_4-HCO_3-Cl type with high HCO_3 and SO_4 contents, up to 1200 mg/l (Table 1). Temperature is positively correlated with TDS and varies from 27 to 63°C. On the triangle diagram (Fig. 3a) all points cluster together except for the La Calera spring with much higher bicarbonate concentrations. The average C-S-Cl composition of thermal waters plots between compositions of magmatic gases from Momotombo volcano, Nicaragua, and Colima volcano, Mexico. All waters are mixtures of two end-members: a shallow, dilute Na-HCO_3 water and a deep, SO_4-HCO_3-Cl water with an essentially "immature" cation composition (Fig. 3b). It can be suggested from the high Mg-content that the major part of the cations were leached from the andesitic edifice, not from the granitic basement.

There is almost a one-to-one correspondence between sampling temperature and a temperature estimated from the solubility of amorphous SiO_2.

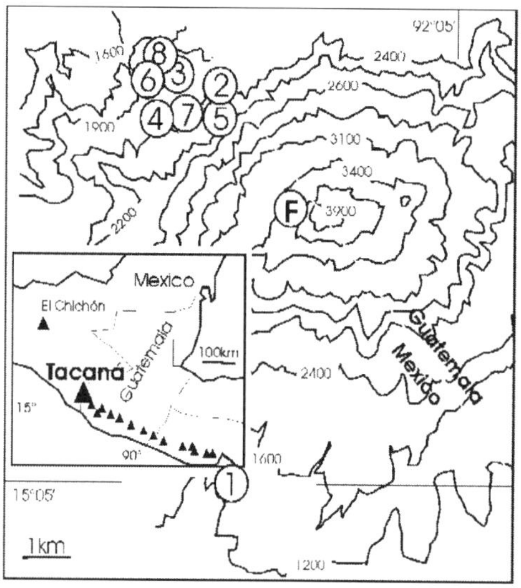

Figure 1. Location map of sample sites at Tacaná Volcano: 1 – La Calera spring, 2 – Toquián spring, 3 – Agua Caliente spring, 4 – Agua Tibia spring, 5 – Barillas spring, 6 – Agua Orlando spring, 7 – Zarco spring, 8 – Zacarías spring, F- fumarolic field. Inset: Location map of Tacaná Volcano in the Central American Volcanic Arc and El Chichón volcano.

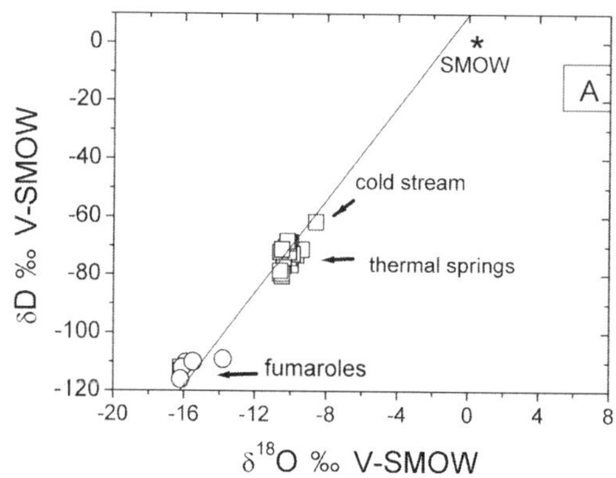

Figure 2. Craig diagram for spring waters and fumarole condensates. The isotope signature of "andesitic water" (A) is taken from Taran et al. (1989).

Table 1. Chemical and stable isotope composition of Tacaná thermal spring waters. Ion concentrations are expressed in mg/l, δD and $\delta^{18}O$ in ‰ vs V-SMOW, $\delta^{13}C$ of CO_2 in ‰ vs V-PDB. Numbers of samples correspond to numbers in Figure 1.

	1	2	3	4	5	6	7	8
t(°C)	27	63	51	32	39	44	45	39
pH	5.8	6.4	6.3	6.1	6.5	6.4	6.2	6.1
HCO_3	759	1110	1025	229	314	824	342	271
SO_4	282	1203	987	177	278	840	454	217
Cl	21	543	371	35	65	297	100	40
Na	124	631	452	84	130	374	188	107
K	17	61	59	13	19	54	28	18
Ca	137	342	309	40	64	250	82	46
Mg	74	142	151	29	39	121	64	32
SiO_2	106	200	216	108	121	164	143	128
δD	−70.4	−80.5	−79.6	−68.6	−74.2	−76.6	−76.9	−72.8
$\delta^{18}O$	−10.3	−10.5	−10.6	−10.0	−10.1	−10.0	−10.4	−9.9
$\delta^{13}C$	−2.8	−1.5	−2.1	−3.0	−1.4	−2.1	−1.7	−1.9

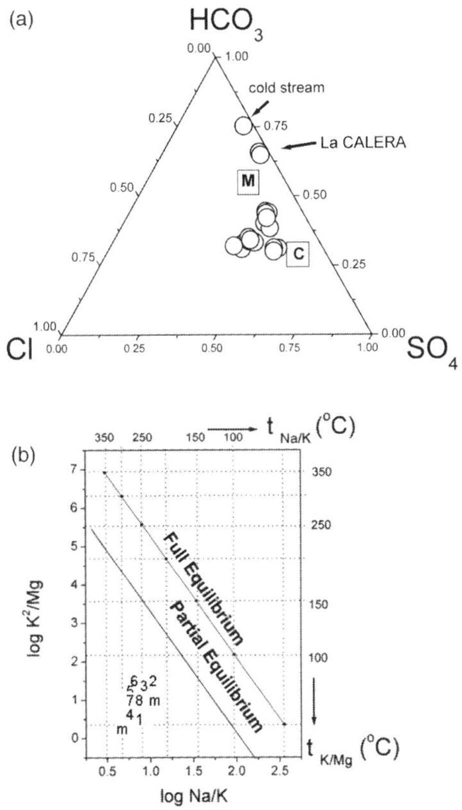

Figure 3. (a) Major anion composition of Tacaná thermal spring waters. M: composition of Momotombo volcanic gases, C: composition of Colima volcanic gases (Taran et al. 2002). (b) Major cation composition of Tacaná thermal spring waters, (modified from Giggenbach 1988). Numbers of samples correspond to numbers in Figure 1. m: meteoric water.

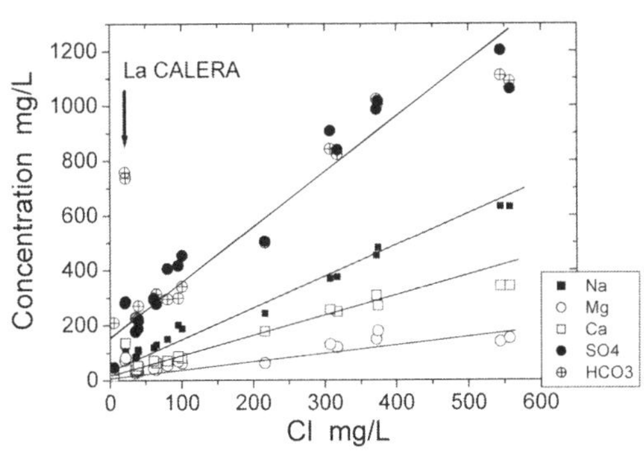

Figure 4. Mixing diagram for cations, sulphate and bicarbonate.

This means that the SiO_2 content is controlled by precipitation of amorphous silica and that the silica geothermometer cannot be used to estimate deep temperatures. The K/Mg temperatures are also close to the measured ones, and only the Na/K geothermometric temperatures (~250°C) can reflect in some degree the conditions of deep hydrothermal fluids (Fig. 3b).

2.3 *Mixing*

The mixing diagram (Fig. 4) shows an excellent linear mixing relationship between all species within the whole range of concentrations. The main and unusual feature of the Tacaná thermal waters is a negative correlation between major anions and δD (Fig. 5). This contrasts with the common positive correlation between Cl and δD in "volcanic" waters, which reflects mixing between Cl-free meteoric water and Cl-rich and D-enriched "andesitic" water (Taran et al. 1989, Giggenbach 1992). In the Tacaná case, the Cl-rich endmember is meteoric water that infiltrated down from much higher elevations than the discharge level. Because of the low proportion of the magmatic contribution (less than 6% comparing with ~10,000 ppm of Cl in a typical magmatic gas), the third, magmatic endmember, cannot be distinguished using the δD-chemistry correlation.

2.4 *Gas composition*

All gases from springs are CO_2-rich. On the air-free basis (after subtraction of O_2 and corresponding N_2 and Ar concentrations) the gases have high N_2/Ar ratios. He concentrations (tens of ppm) are typical for CO_2-rich gases related to magma degassing. The highest $^3He/^4He$ of $6.6R_a$, (where R_a is atmospheric value of 1.4×10^{-6}) was measured in fumarolic gases. Gases from springs are characterized by an R/R_a between 5.7 and 6.2. All He-isotope ratios are typical for the Central American Volcanic Arc (Hilton et al. 2002). Carbon isotopic composition of CO_2 from free gases varies from −1.5 to −3‰ V-PDB which is again typical for the region and is related to a large

Table 2. Composition of bubbling gas from thermal springs at Tacaná, $^3He/^4He$ (R/R_a) and some important ratios.

Spring No.	CO_2	N_2	Ar (ppm)	CH_4 (ppm)	He (ppm)	N_2/Ar	He/Ne	3/4
1	99.6	0.4	72	100	5.5	58	10	6.2
2	97.4	2.6	110	360	3.4	240	4.2	5.7
3	99.4	0.6	55	230	na	110	na	na
6	89.4	10	900	1500	na	110	na	na

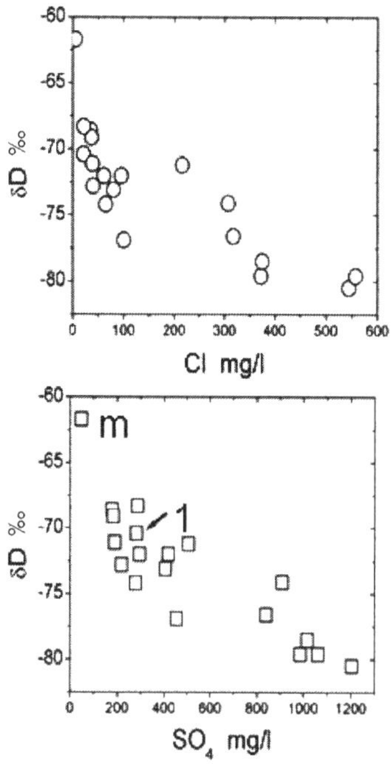

Figure 5. A negative correlation between δD and concentrations of Cl and SO_4 in thermal and cold waters discharging at Tacaná volcano. 1. La Calera spring, m: meteoric water, cold stream.

proportion of marine carbonates subducted beneath the Central American Volcanic Arc (Hilton et al. 2002).

2.5 *Volatile output from Tacaná*

The total steam discharge from fumarolic vents of Tacaná is low. However, several streams on the NW slope of the volcano drain thermal springs and have relatively high salinity. For example, Río Toquián, with the flow rate close to 1 m^3/s contains >400 mg/l of SO_4. It corresponds to ~20 t/d of the SO_2 flux assuming that all SO_4 is derived from magmatic input. Taking into account that there are at least 2 more streams with similar salinity and flow rates, the total S-output can be estimated at close to 50 t/d of SO_2, which is on the same level as many passive degassing volcanoes of the world.

3 CONCLUSIONS

The chemical and isotopic composition of thermal waters discharging at Tacaná volcano, mainly at the contact between basement and volcanic edifice at ~2000 m a.s.l., is very uniform and can be interpreted as a mixture of a shallow ground water with a deep SO_4-HCO_3-Cl water. The isotopic composition of the deep water corresponds to meteoric water infiltrated down inside the volcanic edifice from a much higher elevation than the level of discharge. This type of mixing is responsible for the unusual negative δD-Cl and δD-SO_4 correlation in thermal waters of the volcano.

Carbon and He-isotopic composition of gases from springs and fumaroles are in the range of values typical for the Central American Volcanic Arc. The total output of volatiles, despite weak fumarolic activity, can be estimated at ~50 t/d of SO_2, which is a typical value for a passively degassing volcano.

ACKNOWLEDGEMENTS

This work was supported by grant from DGAPA-UNAM # IN100802. The authors are especially grateful to W.C. Evans for enhancing this paper.

REFERENCES

Giggenbach, W.F. 1988. Geothermal solute equilibria: Derivation of Na-K-Mg-Ca geoindicators. *Geochim. Cosmochim. Acta* 52: 2749–2765.

Giggenbach, W.F. 1992. Isotopic shift in waters from geothermal and volcanic systems along convergent plate boundaries and their origin. *Earth Plan. Sci. Let.* 113: 495–510.

Hilton, D.R., Fischer, T.P. & Marty, B. 2002. Noble gases and volatile recycling at subduction zones. In: D. Porcelli, C.J. Ballentine, R. Wieler (eds), *Noble gases in geochemistry and cosmochemistry.Reviews in mineralogy and geochemistry*, 47: 319–371.

Martini, M., Capaccioni, B. & Giannini, L. 1987. Ripresa dell attivita sismica e fumarolica al Vulcano di Tacana (Chiapas, Messico) dopo un quarantennio di quiescenza. *Estrato da BOLLETTINO del Grupo Nazionale per la Vulcanologia*: 467–470.

Macías, J.L., Espíndola, J.M., García-Palomo, A., Scott, K.M., Hughes, S. & Mora, J-C. 2000. Late Holocene Peléan-style eruption at Tacaná volcano, Mexico and Guatemala: past, present and future hazards. *Geol. Soc. Am. Bull.* 112: 1234–1249.

Medina, A. 1986. Geoquímica de aguas y gases del volcán Tacaná. *Geotermia, Rev. Geotermia Geoenergía* 2(2): 95–110.

Taran. Y.A., Pokrovsky, B.G & Dubik, Y.M. 1989. Isotopic composition and origin of water from andesitic magma. *Dokl. Ac. Sci. USSR*, 304: 440–443.

Taran, Y.A., Gavilanes, J.C. & Cortés, A. 2002. Chemical and isotopic composition of fumarolic gases and the SO_2 flux from Volcán de Colima, Mexico, between the 1994 and 1998 eruptions. *J. Volcanol. Geotherm. Res.* 117: 105–119.

Water–Rock Interaction, Wanty & Seal II (eds)

Fluid geochemistry data processing for integrated assessment of Japanese geothermal resources using GIS: a case study for the Hohi area, Kyushu

H. Shigeno
Geological Survey of Japan, AIST, Tsukuba, Ibaraki, Japan

ABSTRACT: Discovery, evaluation and exploitation of geothermal energy resources in the world have been greatly enhanced by various applications of the geochemical techniques initiated in the 1950–60s especially by the late Donald E. White, US Geological Survey, and expanded and improved since then by geochemists working in many countries. Integrated analyses of fluid geochemistry data with various kinds of data from diverse disciplines on a GIS basis will make the processes more systematic, efficient and fruitful. Recent studies on fluid geochemistry database and data processing for the latest integrated assessment program of Japanese geothermal resources are reported, specifically as a case study for the Hohi area, Kyushu.

1 INTRODUCTION

Geochemical techniques have contributed to understanding the nature of geothermal systems (e.g. White 1970; D'Amore ed. 1992), and their combination with geological, geophysical, and geoengineering methods has promoted worldwide development of geothermal resources. This would be greatly enhanced by integrated approaches using GIS (geographic information systems) which will be able to manipulate various three-dimensional geo-sciences and engineering data more easily in the future.

In Japan, geothermal power generation had grown to ~530 MWe by 1996 (e.g. Shigeno 1995) through cooperative efforts of governments, companies and research institutes after the World Energy Crisis in 1973. But, it has been stagnant since then due to its slightly low economic feasibility under very high exchange rates of yen, regulations for its exploitation in and around the most promising National Park and hot-spring resorts, among other reasons. To contribute to better planning and long-term optimization of the resource development, environment protection, and hazard mitigation at volcanic and geothermal areas, a new research program for geothermal resources assessment using GIS and multi-disciplinary data was started at Geological Survey of Japan, AIST, in 2001 (Shigeno & Sakaguchi 2002).

In this article, the essential parts of the fluid geochemistry database and data processing methods in the assessment system (Shigeno 2003) will be outlined. Then, a case study result using hot-water chemistry data from the Hohi geothermal area, central Kyushu (Shigeno et al. 1985), will be discussed.

2 GEOTHERMAL RESOURCES ASSESSMENT PROGRAM USING GIS

In Japan, the Geographical Survey Institute has been systematically publishing digital elevation data on CD-ROM in three formats of hierarchical mesh sizes: so-called 1 km (80 × 80 points for 1° (E-W) × 40′ (N-S) quadrangle sheet-map area), 250 m (320 × 320 points for the 1° × 40′ area), and 50 m (200 × 200 points for 7′30″ × 5′ quadrangle area). Various governmental organizations have been conducting or planning publications of many kinds of electronic geographical data, including social, environmental and meteorological ones. Most of their data formats are relatively easily transformed to those of the above mesh system.

The basic concept of the geothermal resources assessment system, using GIS on PC, at the Geological Survey of Japan is shown in Figure 1. In this system, acquiring, managing, manipulating and analyzing data, and publishing the results are expected to be conducted and improved in a spiral manner by repeating the following processes.

1. Various kinds of data, preferably in electronic forms, are collected, transformed mostly to the above mesh formats, and stored as databases.

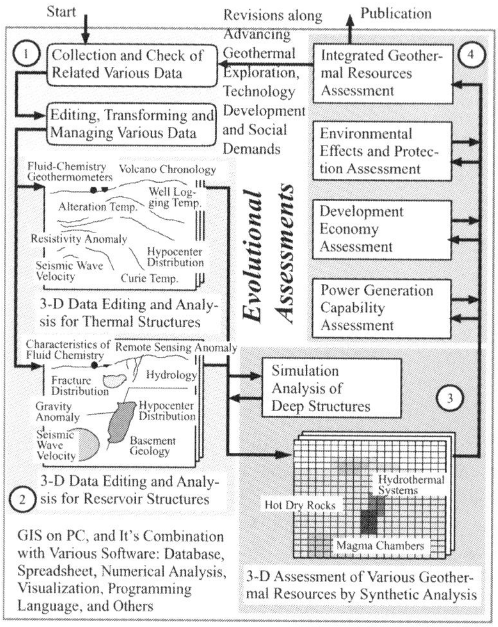

Figure 1. Conceptual model of geothermal resources assessment system flexibly using GIS at Geological Survey of Japan.

2. Geosciences and geoengineering data are transferred to layered mesh forms with appropriate depth intervals, and three-dimensional thermal and reservoir structure modeling are conducted.
3. Types, characteristics and amounts of geothermal resources with their reliabilities are estimated for each block in the layered mesh system by integrated analysis of the above two structures.
4. Capability of power generation using the resources, its economy, and problems for its exploitation are estimated for the blocks using various social, economical and environmental data.
5. Results of the assessments and their base data are appropriately summarized and published.

3 FLUID GEOCHEMISTRY DATABASE AND DATA PROCESSING

As a part of the assessment system, the steps and methods for processing various kinds of fluid geochemistry data for an area are as follows.

1. Data are transferred to fluid geochemistry databases with semi-standardized formats using spreadsheet software (e.g. Microsoft Excel).
2. Various kinds of figures for geothermal-fluid geochemistry (e.g. D'Amore ed. 1992), including those on maps, are drawn as standardized-format raster using a software, GPGG (Grouped programs for geothermal-fluid geochemistry data, tentative version), which has been developed with a BASIC language (e.g. Microsoft Visual Basic) and works under its interpreter environment.
3. Various kinds of mesh-format data are also visualized as standardized-format raster maps using basic functions of GIS software.
4. The above figures and maps are overlaid, if necessary, and manipulated using graphic software (e.g. Adobe Photoshop and Illustrator).
5. More advanced analysis and processing of the data (e.g. multivariate statistical analysis, and assigning parameter values estimated from geochemical data to the layered mesh system) are relatively easily conducted using GIS software, other kinds of commercial package software and newly developed programs for these purposes.

4 A CASE STUDY FOR THE HOHI AREA

4.1 *Data processing*

By using the above methods, a case study has been conducted for the chemical-analyses of hot-spring and geothermal waters reported by Shigeno et al. (1985) from the Hohi area, which is one of the most promising high-temperature hydrothermal areas to develop in Japan. In this area, five geothermal power plants have been operating with a total capacity of ca. 124 MWe.

Table 1 shows the data items for the hot-water geochemistry database. Figures 2, 3 and 4 are processed results of Stiff diagrams overlaid on a topographic map, Na-K-Mg geothermometer diagram (Giggenbach 1988), and logarithmic Cl vs. B concentration diagram (e.g. Shigeno 1995), respectively, for the hot waters from the area. Figure 5 shows a conceptual model of geothermal systems overlaid on Bouguer gravity anomaly map, as a result of integrated analysis using various kinds of data.

4.2 *Geological setting*

The Hohi area is located in a large graben system extending from the Beppu to Unzen areas in central Kyushu where E-W trending normal faults are widely distributed. High-temperature hot-spring and fumarole areas, and geothermal wells are, however, distributed on a SE-NW trending zone which corresponds to a steep-slope part of the Bouguer anomalies and a regional strike-slip fault zone (obliquely crossing the central part of Figures 2 and 5).

Two late-Quaternary volcanic complexes composed mainly of andesitic lava domes are distributed in this area. One is active Kuju complex, with the Iwoyama fumarole area (ca. 600°C) at the central part

Table 1. Data items of hot-water geochemistry database with semi-standardized format for geothermal resources assessment program using GIS at Geological Survey of Japan.

01. Sample number	16. Free CO_2 concent. (mg/l)
02. Location name	17. HCO_3^- concent. (mg/l)
03. Spring or well name	18. $CO_3^=$ concent. (mg/l)
04. Longitude (°, ×××.××××)	19. Na^+ concent. (mg/l)
05. Latitude (°, ××.××××)	20. K^+ concent. (mg/l)
06. Altitude (m, a.s.l.)	21. Ca^{++} concent. (mg/l)
07. Discharge mode (code)	22. Mg^{++} concent. (mg/l)
08. Well depth (m)	23. SiO_2 concent. (mg/l)
09. Discharge rate (l/min)	24. B concent. (mg/l)
10. Sampling date (yy:mm:dd)	25. $\delta D(H_2O)$ (‰)
11. Temperature (°C)	26. $\delta^{18}O(H_2O)$ (‰)
12. pH	27. –34. Additional items 1–8*
13. Total solved matter (mg/l)	35. + More additional items**
14. Cl^- concent. (mg/l)	
15. $SO_4^=$ concent. (mg/l)	

*e.g. concentrations of other dissolved components, isotopic compositions of dissolved components, scores of multivariate analysis, group codes, data reference codes, and others. ** More items could be added by expanding this format, but some modifications to GPGG and others would be necessary.

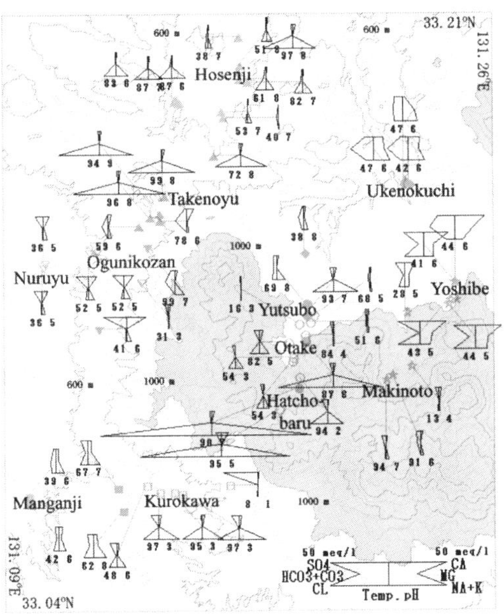

Figure 2. Stiff diagrams of hot waters for the Hohi area. Topographic map (ca. 15.9 km × 18.9 km) based on 50-m mesh elevation data by Geographical Survey Institute is overlaid.

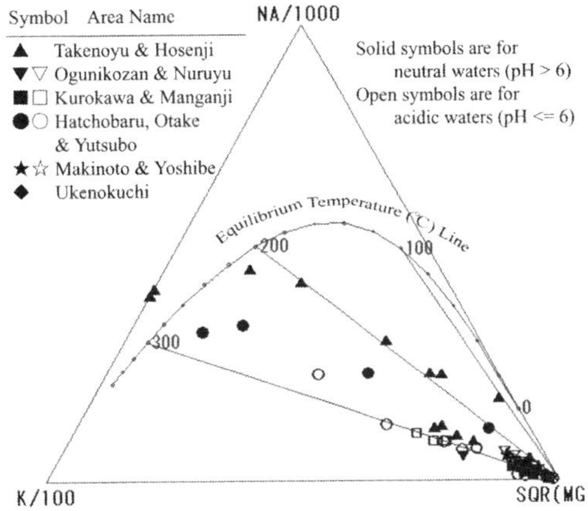

Figure 3. Na-K-Mg geothermometer diagram (Giggenbach, 1988) applied to hot waters for the Hohi area. Refer to Figure 2 for spatial distributions of the areas shown in the legend.

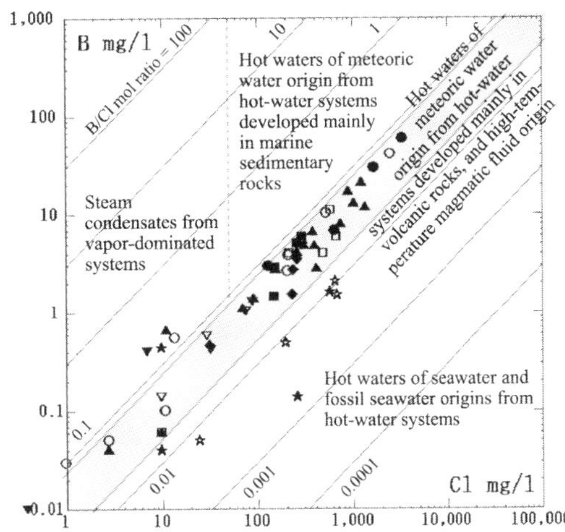

Figure 4. Cl vs. B concentration diagram of hot waters for the Hohi area. Refer to Figure 3 for the sample symbols. Domains of fluid origins and system conditions are after Shigeno (1995).

where a steam-blast eruption occurred in 1995 (Fig. 5). The other is older (ca. 0.4 Ma) and more eroded Waita–Hitome complex, which lacks fumarolic activity at the central part. The high-temperature hot-spring areas are distributed mainly around the Waita–Hitome complex (Figs. 2 & 5).

4.3 *Conceptual modeling of geothermal systems*

The older magma chamber (or hot igneous body) of the Waita–Hitome complex has probably been

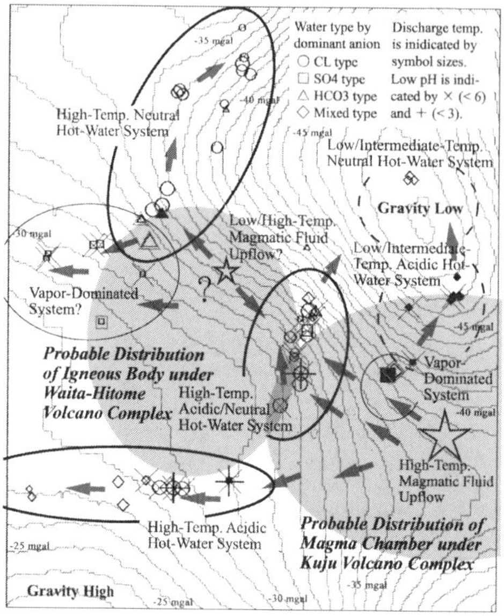

Figure 5. Integrated conceptual model of geothermal systems for the Hohi area. Classified B/Cl molar ratios of hot waters are shown with the water-type symbols in the legend (trimmed open ones, >0.08; solid ones, <0.02). Bouguer gravity anomaly map ($\rho = 2.67\,g/cm^3$) modified from CD-ROM data by Geological Survey of Japan is overlaid. Arrows indicate estimated geothermal-fluid flows at relatively shallow depths. See the text for more details referring to Figures 2, 3, and 4.

contributing mainly as a conductive heat source to the surrounding hydrothermal systems with dilute hot waters of meteoric water origin (Fig. 5). The neutral Na-Cl type hot-water system of the Takenoyu–Hosenji area is the typical case. At the Ogunikozan-Nuruyu area, however, development of a kind of vapor-dominated system is estimated on the basis of distributions of the weakly acidic SO_4-type waters with the relatively high B/Cl ratios.

The younger magma chamber of the Kuju complex, in contrast, is estimated to have been contributing mainly as a convective heat source supplying magmatic fluids in and around the Iwoyama area on the basis of the low pH values of the hot waters discharging near the complex (Fig. 5). Very limited distributions of the high-temperature hot-spring and fumarole areas around the Kuju complex, except toward W and NW directions, are probably caused by the two reasons: (1) insufficient heatsupply by thermal conduction from the young magma chamber distributed at the depth, and (2) dominant near-surface hydrology of groundwater of meteoric water origin flowing from the higher eastern part to the lower western part of the complex, through the Iwoyama fumarole area, along the active E-W fracture system and the SE-NW strike-slip fault zone.

The high-temperature geothermal areas, Iwoyama, Makinoto, Hatchobaru–Otake–Yutsubo, and Takenoyu, aligned on the SE-NW strike-slip fault zone, have probably been produced by the combination of the above two kinds of heating mechanisms, thermal conduction and magmatic fluid migration, from the depths of the two volcanic complexes. The geothermal areas away from the zone are estimated to be less promising. Development of a large-scale hydrothermal system is probably difficult for the southwestern part of the Hohi area, where pre-tertiary crystalline basement rocks are distributed at the shallow depth. The northeastern part of the Hohi area, where a large negative Bouguer anomaly possibly corresponding to a deep caldera structure is distributed, could be more promising. At the Yoshibe and Ukenokuchi areas, the mixed-type hot-spring waters with the immature Na-K-Mg temperatures, and the specific low B/Cl ratios suggest contributions of a kind of fossil seawater and/or steam-loss hot waters from a vapor-dominated system distributed at the Makinoto area.

5 SUMMARY

Fluid geochemistry data provide unique information on conditions and formation mechanisms of geothermal systems (e.g. White 1970; D'Amore ed. 1992). Integrating these analyses with multidisciplinary data produces more comprehensive understanding and evaluation of the systems. Various kinds of geographic (including geosciences and geoengineering) data have been published progressively in electronic forms, and they could be easily and economically used for the purpose with the help of GIS, especially by using standardized mesh formats.

The new research program for Japanese geothermal resources assessment using GIS at Geological Survey of Japan is still at the designing and trial stages. Case studies have been conducted for a couple of regional and promising geothermal areas using published nation-wide electronic data, especially of topography, geology, gravity, hypocenter distribution and thermal gradient, and some local fluid geochemistry, well logging and other data newly edited in the program. Concrete assessment work will be conducted at the second stage (2005–2008?) (Shigeno & Sakaguchi 2002; Shigeno 2003).

The present article outlines the part of fluid geochemistry data processing in the above program at the occasion of special symposium of WRI-11 in memory of the late Donald E. White.

ACKNOWLEDGEMENT

I thank Bill Evans for reviewing, and Rich Wanty for editing this paper with the traditional generosity and refreshing spirit of US Geological Survey.

REFERENCES

D'Amore, F. (ed.) 1992. *Applications of Geochemistry in Geothermal Reservoir Development*. New York: UN Institute for Training and Research.

Giggenbach, W.F. 1988. Geothermal solute equilibria. Derivation of Na-K-Mg-Ca-geoindicators. *Geochim. Cosmochim. Acta*, 52: 2749–2765.

Shigeno, H., Abe, K. & Noda, T. 1985. Conceptual hydrothermal system model for the Hohi area based on geochemistry of geothermal fluids. *Rept. Geol. Surv. Japan*, no. 264: 285–302. (In Japanese with English abstract)

Shigeno, H. 1995. Estimating deep environments of Japanese hydrothermal systems based on geochemical data from geothermal power plants. *Proc. World Geothermal Congress 95, Florence 1995*: 1019–1024.

Shigeno, H. & Sakaguchi, K. 2002. New program of Japanese geothermal resources assessment using GIS. *Chishitsu Nyusu (Geology News)*, no. 574: 24–45. (In Japanese with English title)

Shigeno, H. 2003. Simple database and synthetic analysis of geothermal-fluid chemistry data – A case study for the Hohi area, Japan, in geothermal resources assessment program using GIS. *Chishitsu Nyusu (Geology News)* (in press). (In Japanese with English title)

White, D.E. 1970. Geochemistry applied to the discovery, evaluation, and exploitation of geothermal energy resources, Rapporteur's report. *Proc. UN Symp. on the Development and Utilization of Geothermal Resources, Pisa 1970 (Geothermics, Spec. Iss., 2)*, 1: 58–80.

Water–Rock Interaction, Wanty & Seal II (eds)
© 2004 Taylor & Francis Group, London, ISBN 90 5809 641 6

Transition metals distribution in the rocks from Sierra-Leone Fracture Zone (Central Atlantic)

V.A. Simonov, Yu.P. Kolmogorov, A.S. Lapukhov & G.A. Tretyakov
Institute of Geology, Siberian Branch of Russian Academy of Sciences, Russia

ABSTRACT: From SRXRF data, it is established that geochemical characteristics of the majority of rare elements and noble metals in basalts of the Sierra-Leone Fracture Zone correspond to normal rocks of mid-ocean ridge basalts (MORB). In sites where hydrothermal ore-forming processes are active, enrichment in Cu, Sr, Ti and V were observed, as well as a sharp depletion in Cr and Ni. The geochemical features of basalts are appreciably influenced by the geodynamic processes of fracture zone development.

1 INTRODUCTION

Recent research on the World Ocean bottom has concentrated mainly in areas of modern hydrothermal ore-forming systems such as black smokers. During cruise 22 of R/V "Akademik Nikolaj Strakhov" (2000) in the Sierra-Leone Fracture Zone (Central Atlantic), traces of hydrothermal activity with displays of sulphidic mineralization have been discovered (Peyve et al. 2000; Mazarovich et al. 2001). Our earlier research on the rock material collected during this expedition has shown that both hydrothermal and magmatic systems could be responsible for ore formation (Mazarovich et al. 2001; Simonov et al. 2001). Nevertheless, questions of interrelation of magmatic and hydrothermal ore-forming systems in the given region remain open, and the essential distinction can be resolved by research on rare elements and noble metal geochemistry with the use of SRXRF methods.

The samples were collected with the help of dredge stations in the Sierra-Leone Fracture Zone along the Mid-Atlantic Ridge (MAR) rift valley in an interval 6°–7°10′ N (Fig. 1). Dredge station S2230 is located in the southern part of the study area, where features of hydrothermal activity have been observed (Mazarovich et al. 2001). Of special interest is material dredged at station S2234, on the northern slope of Sierra-Leone fracture. Here hydrothermal activity accompanied by a sulphidic mineralization was found. Basalts of stations S2238, S2239 and S2244, found only in the rift valley, are representatives of modern normal MAR magmatism. Dredge station S2246 is located near where the 7°10′ Fracture Zone crosses the MAR rift.

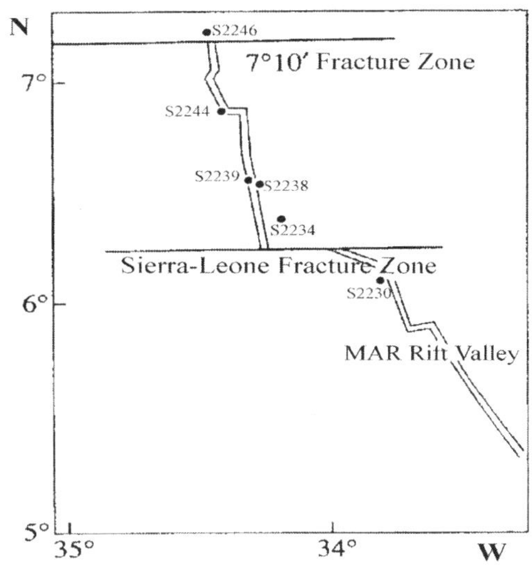

Figure 1. Location of dredge stations in the Sierra-Leone Fracture Zone.

2 METHODS

Chemical analyses of samples were carried out at the Laboratory of X-ray Fluorescence Element Analysis VEPP-3 in the Center of Synchrotron Radiation of the Institute of Nuclear Physics, Siberian Branch of the Russian Academy of Sciences. The principles of the instrumental analysis used in the present work are described in Kolmogorov et al. (1989). The laboratory spectrometer (from Oxford Instruments, Great Britain) consists of: the detector silicon–lithium

(SiLi) with the power range 5.9–133 keV, pulse processor XP-3, and peak analyzer software on a personal computer. Excitation of emission spectra of research samples was made by a beam of polarized monochromatized radiation at an energy of 25 keV. Samples were prepared by pressing powder into 30 mg tablets 6 mm in diameter. Fluorescence spectra were processed by an "external" standard method. The following standards of rocks were used as "external" standards: basalt BCR-1 (Incorporated Bureau of Standards of the European Economic Community, Brussels); diabases DNC-1, W-2; basalts BIR-1, BHVO-1 (US Geological Survey, USA). Noble metals in the samples were extracted by acid decomposition and concentration by organic extractants, then analyzed in Analytical Center UIGGM SB RAS by atomic absorption spectrometer (Perkin-Elmer model 3030Z with Zeeman background correction) with electrothermal atomization (HGA-600). All methods of the analysis correspond to requirements of System of accreditation of analytical laboratories (centers) of Russia Standards (Certificate of Accreditation #ROSS RU. 0001. 510590, valid until 07/07/2004).

3 RESULTS

The XRF-SR method allows us to investigate the geochemistry of a large number of elements (V, Cr, Mn, Fe, Ni, Cu, Zn, Ga, Rb, Sr, Y, Zr, Nb, Mo, As, Pb, Ag, Cd, Sn). In total, 950 element analyses were made. In most cases, levels of element contents for basalts, gabbroic, and ultrabasic rocks correspond to Clarke values (Vinogradov 1990). Among basalts and gabbroic groups of rocks, some elements were consistently below (Mn, Cu, Zn, Pb, Rb, Sr, Nb) or above (Zr, Y, Mo) their Clarke values. Essential differences are established for ultrabasic rocks, for which unprofitable relative Clarke contents are V, Fe, Mn, and superfluous – Zn, Pb, Mo, As. Zones where hydrothermal ore-forming processes are active display enrichment in Cu, Sr, Ti and V and sharp depletion in Cr and Ni.

3.1 *Rare earth elements*

Y and Zr are indicator elements, which are unaffected by secondary processes. Within the Sierra-Leone Fracture Zone, the Y/Zr ratio in basalts forms a linear trend close to the distribution found in chondrites (Fig. 2). Basalts of the most northern and southern sites of the study area (stations S2246 and S2230) are most enriched in Zr. Rocks of these stations are plotted on the Zr-Sr diagram (Fig. 3) where they form precise linear trends with a maximum in Zr. Because station S2246 is located where the 7°10′ fracture zone

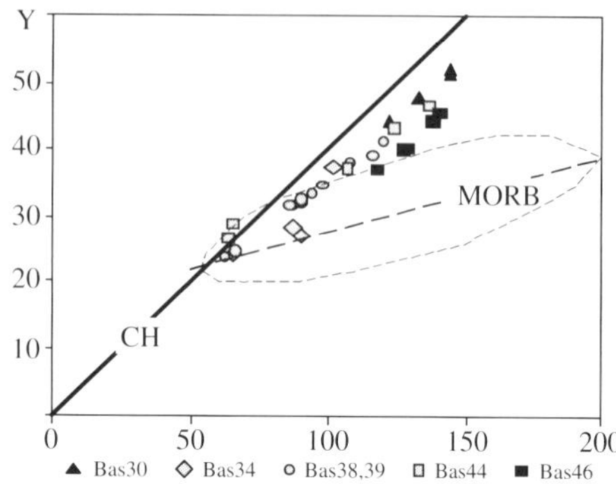

Figure 2. Diagram Y-Zr (ppm) in basalts of the Sierra Leone Fracture Zone. CH – chrondritic trend; Bas30 – data on basalts of corresponding stations (e.g. S2230).

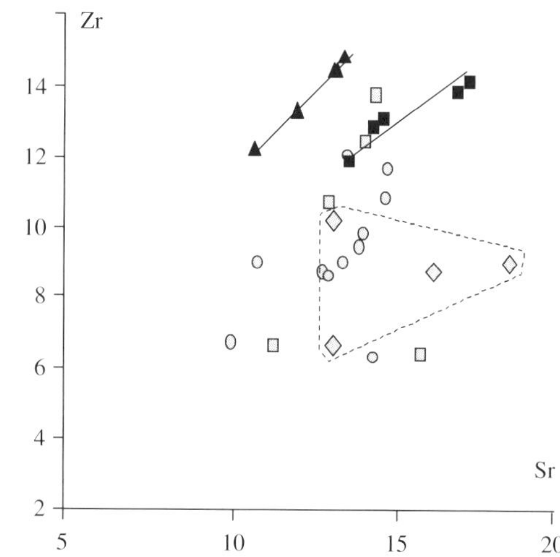

Figure 3. Diagram Zr-Sr (ppm) in basalts of the Sierra-Leone Fracture Zone. Symbols as in Figure 2.

crosses the MAR, and station S2230 is connected structurally to the Sierra-Leone Fracture Zone, it is possible to assume that enrichment of these two magmatic systems in Y and Zr is connected to the transform fracture development typical of Mid-Oceanic Ridges. Basalts of station S2234, where the sulphide mineralization was found, show a range of moderate Zr values with relatively increased contents of Sr. This feature is even more apparent in the context of element distributions along the MAR rift valley. There is also a generally decreasing trend of Sr values from north to south with a peak (up to 186 ppm) at station S2234. Taking into account that the basalts of this station are altered substantially, enrichment in Sr is most probably caused by development of hydrothermal systems.

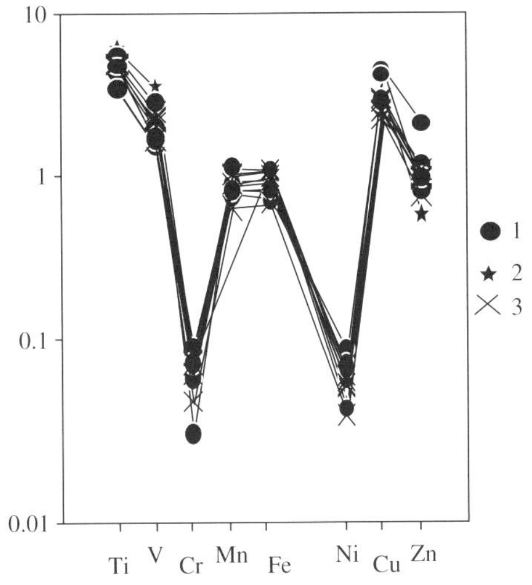

Figure 4. Distribution of transition metals in basalts. 1 – basalts from station S2234; 2 – a trend of normal mid-ocean ridge basalts (MORB); 3 – basalts from station S2239. Values of elements are normalized to a primitive mantle content according to Rollinson (1993). Figure is constructed according to authors' data.

3.2 *Transition metals*

A significant number of transition elements, Sc, Ti, V, Cr, Mn, Fe, Co, Ni, Cu, Zn (Rollinson 1993) are sufficiently mobile during various geological processes, and active hydrothermal activity in the Sierra-Leone Fracture Zone is accompanied by redistribution of these metals. In particular, Cu enrichment in basalts at station S2234 is likely connected with development of hydrothermal systems.

Plots of the transition metals normalized to chondrite and to primitive mantle show similar features. Therefore, only diagrams with the data normalized to primitive mantle were used. In comparison with primitive mantle, basalts of mid-ocean ridges are sharply depleted in Cr and Ni. Compared to these strong depletions, the distribution of other metals is not so distinct. Therefore the comparative analysis is more conveniently carried out without these elements. Figure 4 shows that Mn, Fe, and Zn values in basalts with sulphidic mineralization (station S2234) are close to mantle values. At the same time, enrichment in Ti, Cu, and to a lesser degree V, is observed. As a whole, patterns of transition metals in the investigated rocks are similar to normal mid-ocean ridge basalts (MORB).

3.3 *Noble metals*

Noble metal contents were analyzed on representative samples of basalts, gabbros and ultrabasic rocks from the Sierra-Leone Fracture Zone; 120 analyses of noble metals were performed. In most cases, concentrations of these elements are close to their Clarkes in corresponding types of rocks. Ultrabasic rocks are a little enriched in platinum and palladium. In all samples of basalts and gabbros, no significant differences in contents of platinum-group metals was observed. Basalts from sites near transform fractures (stations S2230 and S2246) are enriched in gold in comparison with rocks far from faults. Distribution of silver in basalts is extremely variable and changes from 0.04 up to 0.47 ppm. Contents of gold and silver decrease in the order: basalts > gabbro > ultrabasic rocks. In comparison with similar rocks from the 15°20′ fracture zone (Simonov et al. 1999), the samples of the Sierra-Leone Region have contents of platinum metals that are much lower, and gold contents the same or higher, which probably is a result of intense hydrothermal processes and the overprinted sulphide mineralization.

4 CONCLUSIONS

1. Basalts of the Sierra-Leone Fracture Zone, according to geochemical characteristics of the majority of rare elements and noble metals, are similar to normal mid-ocean ridge basalts (MORB).
2. In sites where hydrothermal processes and sulphide mineralization occur in basalts, enrichment in Cu, Sr and V is accompanied by depletion in Cr and Ni.
3. Compared to typical rocks from rift valleys, geochemical features of rare earth elements, transition metals and gold distribution in basalts from sites near faults are caused by the effects of transform fracture development on MAR magmatic systems.

ACKNOWLEDGEMENT

Work is carried out with the financial support of the Russian Foundation of Basic Research (project number01-05-65234, 02-05-64046).

REFERENCES

Kolmogorov Yu.P., Lapukhov A.S., & Melnikova R.D. 1989. *Nuclear Instruments and Methods in Physics Research,*A282, P. 692–694.

Mazarovich A.O., Simonov V.A., Peyve A.A., Kovyazin S.V., Tretyakov G.A., Raznitsin Yu.N., Saveljeva G.N., Skolotnev S.G., Sokolov S.Yu., & Turko N.N. 2001. Hydrothermal mineralization of the Sierra Leone Fracture Zone (Central Atlantic). *Lithology and minerals.* No. 5. P. 526–533.

Peyve A., Bonatti E., Brunelli D., Chilikov A., Cipriani A., Dobrolubova K., Efimov V., Erofeev S., Ferrante V.,

Gasperini L., Hekinian R., Ligi M., Maurizio G., Mazarovich A., Perfiliev A., Raznitsin Y., Savelieva G., Sichler B., Simonov V., Skolotnev S., Sokolov S., & Turko N. 2000. New data on some major MAR structures: preliminary results of R/V Akademik Nikolaj Strakhov 22 cruise *Inter Ridge News*. V. 9 (2).. P. 28.

Rollinson, H.R. 1993. *Using Geochemical Data: Evaluation, Presentation, Interpretation.* Longman Scientific and Technical. London. P. 352.

Simonov, V.A., Kolobov, V.Yu., Peyve, A.A. 1999. *Petrology and geochemistry of geodynamic processes in the Central Atlantic.* Novosibirsk: Publishing House of the Institute of Geology, Siberian Branch of Russian Academy of Sciences. P. 224.

Simonov V.A., Mazarovich A.O., Kovyazin S.V., Tretyakov G.A. 2001. Features of ore-forming processes in the Sierra Leone Fracture Zone (Central Atlantic) *Geology of the seas and oceans*. M.: GEOS, V. 1. P. 259–260.

Vinogradov, A.P. 1990. *Reference book on geochemistry.* Moscow: Nedra.

Water–Rock Interaction, Wanty & Seal II (eds)
© 2004 Taylor & Francis Group, London, ISBN 90 5809 641 6

Kr-81 and water–rock interactions on a million-year time scale

N.C. Sturchio & L.J. Patterson
University of Illinois at Chicago, Chicago, IL, USA

Z.T. Lu, X. Du & P. Mueller
Physics Division, Argonne National Laboratory, Argonne, IL, USA

R. Purtschert & B.E. Lehmann
Institute of Physics, University of Bern, Bern, Switzerland

M.I. Sultan
State University of New York at Buffalo, Amherst, NY, USA

B. El Kaliouby, Y. Dawood & A.M. Abdallah
Department of Geology, Ain Shams University, Cairo, Egypt

ABSTRACT: Measurements of ^{81}Kr and ^{36}Cl in deep groundwater from the Nubian aquifer were performed to determine the residence time of the groundwater. ^{81}Kr ages range from $\sim 2 \times 10^5$ to $\sim 1 \times 10^6$ yr and are correlated with ^{36}Cl/Cl ratios. Cosmogenic isotope ages are in fairly good agreement with predicted hydrodynamic residence times. Groundwater age distribution reflects distant lateral flow of deep groundwater, with local recharge and discharge of shallow groundwater responding to periodic climate fluctuations. Groundwater ages in the context of aquifer outcrop areas and drainage patterns indicate that recharge of deep water occurred in SW Egypt. Chemical evolution of deep groundwater scaled to ^{81}Kr age indicates water-rock interaction kinetics.

1 INTRODUCTION

The importance of studying groundwater in arid regions has increased in recent decades due to larger human populations in desert areas. At the same time the development of technologies for measuring both stable and rare radioactive isotopes has improved our understanding of groundwater systems. Cosmogenic nuclides have been useful in studies of several large aquifers containing very old water (>100,000 yr) such as the Great Artesian Basin, Australia (e.g. Bentley et al. 1986, Torgerson et al. 1991, Bethke et al. 1999) and the Milk River Aquifer, Canada (Phillips et al. 1986, Lehmann et al. 1991). In this paper, we discuss results obtained by using measurements of the cosmogenic isotopes ^{81}Kr and ^{36}Cl to determine residence time of groundwater in the Nubian aquifer of the Western Desert of Egypt. Knowledge of residence time allows consideration of long-term water-rock interaction kinetics by placing a time scale on the chemical evolution pathway.

1.1 *Geologic setting*

The Nubian aquifer, underlying the northeastern portion of the Sahara Desert, covers approximately 2 million km^2 and has been estimated to contain approximately 50,000 km^3 of groundwater (Thorweihe 1990) (Fig. 1). It rests unconformably upon a Proterozoic granite basement that outcrops to the west, south, and east, and slopes northward. The basement is elevated in the center creating two basins: the Dakhla Basin underlies the Western Desert of Egypt and eastern Libya, whereas the Kufra Basin underlies southeastern Libya, northeastern Chad and northwestern Sudan. The Dakhla Basin is bounded to the north by the saline waters of the Northwestern Basin and to the east by the Kharga Uplift that separates it from the Upper Nile Platform (UNP), a hydraulically unconnected, uplifted area with a different sedimentary history (Thorweihe 1990).

The water-bearing portion of the Dakhla Basin comprises four hydraulically-connected "horizons"

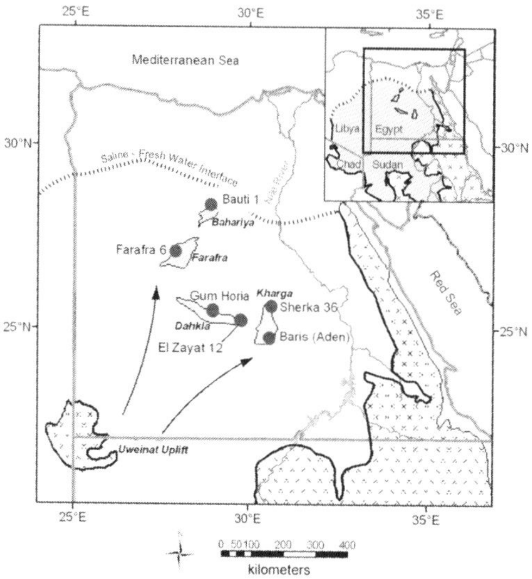

Figure 1. Map showing locations of wells sampled for groundwater analysis of ^{81}Kr and ^{36}Cl in the Nubian aquifer, and general groundwater flow direction.

of Paleozoic and Mesozoic epicontinental sandstone (the "Nubian sandstone") separated by thin layers of shale. The aquifer ranges from 400 to nearly 3000 m in thickness (Hesse et al. 1987) and has fair to good permeability overall, although the confining layers thicken northwards and restrict groundwater flow (Brinkmann et al. 1987). The piezometric surface indicates that groundwater flows roughly southwest to northeast (Ball 1927). The Nubian sandstone outcrops across the southern portion of the basin, but north of 25°N thick marine shales of Paleocene age confine the aquifer and create artesian conditions. Oases are topographic depressions where discharge occurs, and have developed where faults and folds in the basement and aquifer horizons have caused localized uplift and erosion of the confining beds. The climate of the Western Desert is hyperarid, with annual rainfall less than 20 mm/yr and less than 5 mm/yr over much of the area. No continuous surface recharge of the Dakhla Basin occurs today although it may receive baseflow from adjacent groundwater in the Kufra Basin. However, a wetter climate prevailed at various times throughout the Quaternary. Several intermittent periods of wetter conditions may have occurred during at least 450 kyr B.P. with the most recent recharge occurring between 9.5 to 4.5 kyr B.P. and 25 to 40 kyr B.P. (Pachur et al. 1987, Hesse et al. 1987). During pluvial periods, recharge occurred in the Nubian Sandstone outcrops – particularly at wadis (Hesse et al. 1987) and in mountainous areas – and as leakage through the confining layers north of the 25th parallel. Thus both local and large-scale regional flow occurs in the Nubian aquifer. Widespread local recharge is represented in the shallow zones of the aquifer where measurable ^{14}C exists with no indication of decay along the flowpath (Brinkmann et al. 1987). In contrast deeper waters represent baseflow from remote recharge areas to the south, have longer residence times, and dominate groundwater movement during arid periods (Brinkmann et al. 1987).

1.2 *Groundwater dating methods*

Groundwater age (or residence time) is defined as the amount of time some unit of water has been isolated from the atmosphere. Both hydrodynamic and isotopic methods have been devised for dating groundwater. The radioactive tracer ^{36}Cl ($t_{\frac{1}{2}}1/2 = 301 \pm 2$ kyr) has been used for dating very old groundwaters since the 1980s. It is produced in the atmosphere by spallation of Ar ($^{40}Ar(p,n\alpha)^{36}Cl$), falls out primarily as wet deposition along with stable Cl from marine aerosols, and after concentration by evapotranspiration, enters the groundwater system. Cl is considered a conservative groundwater tracer and groundwater age can be calculated from the decay of the atmospheric ^{36}Cl; however, in practice several factors complicate its use. The initial ratio of $^{36}Cl/Cl$ in recharge waters varies with latitude, meteorological conditions, distance between the recharge area and marine coastline, and time (Phillips 2000). Direct measurements of ^{36}Cl in modern local recharge are often used as initial values, but they do not address the problem of differential ^{36}Cl deposition in paleorecharge.

Another complication arises from the in situ production and behavior of ^{36}Cl. For water containing as little as 75–150 mg/L Cl^- there may be sufficient subsurface ^{36}Cl production from the neutron flux caused by the decay of U and Th series elements in the rock matrix to invalidate the method (Park et al. 2002). The mixing of two end members – young, ^{36}Cl-enriched water and old, ^{36}Cl-depleted water – can cause an apparent decrease in $^{36}Cl/Cl$ ratios that might otherwise be attributed to simple decay (Park et al. 2002). Furthermore, mass transport across formations with higher equilibrium $^{36}Cl/Cl$ ratios, higher salinities, or higher concentrations of U and Th can cause increases in aquifer ^{36}Cl concentrations. The lack of comparable modern recharge, the uncertainties in climate, extent, location and time of paleorecharge, the effects of aridity, the partial confinement of the aquifer, and the effects of cross-formational flow are all complicating factors in using ^{36}Cl to date Nubian groundwaters.

^{81}Kr ($t_{1/2} = 2.29 \times 10^5$ yr) is a cosmogenic isotope long recognized for its ideal physical and chemical properties for dating very old groundwaters and ice cores in the 5×10^4–10^6 yr age range (Loosli &

Oeschger 1969). It is produced in the upper atmosphere by cosmic-ray induced spallation and neutron activation of stable Kr isotopes. As a result of its long lifetime and chemical inertness in the atmosphere, ^{81}Kr is expected to have a constant and well-constrained atmospheric source term with negligible subsurface sources or sinks other than radioactive decay (Lehmann et al. 1991). The difficulty in analyzing ^{81}Kr, owing to low abundance ($^{81}Kr/Kr \sim 10^{-12}$), has been its main drawback. Recent developments in the Atom-Trap Trace Analysis (ATTA) method have enabled the analysis of ^{81}Kr in Kr gas extracted directly from ~2-ton groundwater samples using a tabletop apparatus, thus providing a more practical approach for dating very old groundwater (Du et al. 2003).

2 METHODS

2.1 *Field methods*

Water samples were taken from established wells at East Uweinat and the Dakhla, Kharga, Baris, Farafra and Bahariya oases in 1994 and 2002, and in the Upper Nile Platform in 2000 (Fig. 1). Stagnant wells were pumped or flowed (if artesian) until water temperature stabilized before being sampled. Two 250-mL polyethylene bottles of each sample were filled for ionic analysis, one of which was filtered with a 0.45 μm polypropylene syringe filter in the field. Dissolved gases were extracted from several tons of groundwater at the field sites by a method developed by colleagues at the University of Bern (Collon et al. 2000). This process involves spraying the groundwater into an evacuated plexiglass container and collecting the extracted gas in a gas cylinder. The gas cylinders were transported to the University of Bern for Kr separation.

2.2 *Laboratory methods*

Chloride concentrations were measured by ion chromatography. Samples were prepared for accelerator mass spectrometry (AMS) analysis of ^{36}Cl by precipitation and purification of AgCl. ^{36}Cl was measured at the PRIME Laboratory of Purdue University. Kr was separated by cryogenic and gas chromatographic techniques at the University of Bern (Collon et al. 2000) and $^{81}Kr/Kr$ was measured at Argonne National Laboratory (Du et al. 2003).

3 RESULTS AND DISCUSSION

Results of ^{81}Kr measurements are expressed in terms of the air-normalized ratio, $R/R_{air} = (^{81}Kr/Kr_{sample})/(^{81}Kr/Kr_{air})$, where R_{air} is the modern atmospheric $^{81}Kr/Kr$ ratio measured by ATTA, 1.10 $(\pm 0.05) \times 10^{-12}$ (Du et al. 2003). R/R_{air} values range from 4.8% to 52.6%. Using the simple expression for radioactive decay, and the ^{81}Kr decay constant $\lambda_{Kr} = 3.03\ (\pm 0.14) \times 10^{-6}\,yr^{-1}$, the age t_{Kr} of a groundwater sample is given by $t_{Kr} = -1/\lambda_{Kr} \ln (R/R_{air})$. The range of ages thus derived is 0.212–1.005×10^6 yr; these are apparent ages because the extent of water mixing within the sampled wells is unknown. Ages increase progressively from southwest to northeast. This age progression supports flow vectors predicted by numerical hydrodynamic models (Brinkmann et al. 1987), verifying distant lateral flow of deep groundwater toward the northeast, with from a recharge of deep groundwater occurring southwest of Dakhla. The general correspondence of ^{81}Kr age with predicted hydrodynamic age indicates negligible input of ^{81}Kr from subsurface sources, in agreement with conclusions reached from ^{81}Kr measurements of groundwater from the Milk River aquifer in Canada (Lehmann et al. 1991) and the Great Artesian Basin in Australia (Collon et al. 2000, Lehmann et al. 2003). The estimated diffusive loss of ^{81}Kr to aquitards was limited to $\leq 20\%$ for the Great Artesian Basin (Lehmann et al. 2003), and is expected to be even lower within the Nubian aquifer because of the smaller proportion of aquitard formations. ^{81}Kr ages increase progressively with distance to the north and east of Dakhla; their spatial distribution indicates relatively high flow velocities (~2 m/yr) from Dakhla toward Farafra, and lower velocities (~0.2 m/yr) from Dakhla toward Kharga and Baris and from Farafra to Bahariya. These observations are consistent with the areal distribution of hydraulically conductive sandstone within the aquifer and the increasing thickness of the aquifer in the downgradient direction and they provide support to some of the existing hydrodynamic models (Brinkmann et al. 1987, Hesse et al. 1987). Southwestward extrapolation of the ~2 m/yr flow rate inferred from the difference in ^{81}Kr ages for Dakhla and Farafra is consistent with recharge in the area of the Uweinat Uplift near the Egypt-Sudan border (where samples of groundwater were collected from the E. Uweinat #4 well). In this area, there is a topographic depression in which outcrops of the Nubian sandstone are exposed (or buried beneath sand dunes and sheets) at elevations between 200 and 600 m above sea level over a wide area, forming a broad catchment for recharge of the Nubian Aquifer.

There is a good correlation between $^{36}Cl/Cl$ ratios and ^{81}Kr ages (Fig. 2), in samples for which both measurements were made, providing a firm basis for calibrating apparent ^{36}Cl ages to an appropriate value of $[^{36}Cl/Cl]_{initial}$ in the Nubian aquifer. A best-fit curve through the data using the function

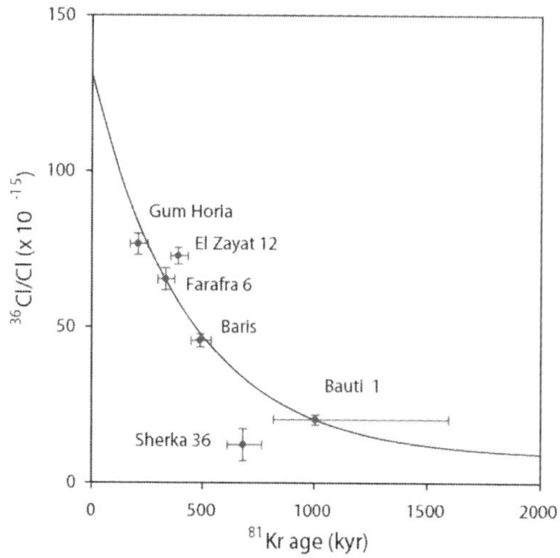

Figure 2. ^{36}Cl/Cl vs. ^{81}Kr age for Nubian aquifer groundwater samples ($\pm 1\sigma$error bars), showing best-fit exponential decay curve of ^{36}Cl. Intercept on y-axis represents $R_{initial}$, the initial ^{36}Cl/Cl ratio of groundwater, which is $131(\pm 11) \times 10^{-15}$ when $8(\pm 3) \times 10^{-15}$ is assumed for R_{seq}, the secular equilibrium value of ^{36}Cl/Cl in the sandstone.

$[^{36}Cl/Cl]_{measured} = [^{36}Cl/Cl]_{seq} + ([^{36}Cl/Cl]_{initial} - [^{36}Cl/Cl]_{seq}) \times \exp(-t_{Kr} \times \lambda_{Cl})$, where t_{Kr} is the ^{81}Kr-age in years, $\lambda_{Cl} = 2.30(\pm 0.02) \times 10^{-6}\,yr^{-1}$ is the ^{36}Cl decay constant, and $[^{36}Cl/Cl]_{seq} = 8(\pm 3) \times 10^{-15}$ is the secular equilibrium in-situ production value for the aquifer (assuming constant Cl concentration in water) gives $[^{36}Cl/Cl]_{initial} = 131(\pm 11) \times 10^{-15}$, which agrees well with the initial value of $125(\pm 10) \times 10^{-15}$ estimated for the Great Artesian Basin (Lehmann et al. 2003) and reasonably well with the measured value of $156(\pm 7) \times 10^{-15}$ for water sampled from a shallow well (E. Uweinat #4) in the unconfined portion of the Nubian aquifer within the central part of the inferred recharge area. ^{81}Kr and ^{36}Cl ages for all low-Cl (Cl < 35 mg/L) samples agree within error, but two samples exhibit discordance – sample El Zayat 12 (Cl = 59 mg/L) has a ^{36}Cl age of 2.79×10^5 yr that is significantly younger than its ^{81}Kr age of 3.91×10^5 yr, indicating that it may have had a higher value of $[^{36}Cl/Cl]_{initial}$; whereas for sample Sharka-36 (Cl = 95 mg/L) the ^{36}Cl age of 1.47×10^6 yr is significantly older than the ^{81}Kr age of 6.78×10^5 yr, although this discrepancy disappears if Sharka-36 water has experienced addition of subsurface Cl having low [^{36}Cl/Cl]. Such Cl could be obtained either by dissolution of Cl-bearing minerals or by diffusion from saline pore fluids in aquitards.

The intercomparison of the two methods in the Nubian aquifer system indicates that the ^{81}Kr method is more robust than the ^{36}Cl method for determining fluid residence time because the source and initial value of ^{81}Kr is better constrained and ^{81}Kr is insensitive to the Cl concentration of the water. This attribute is likely to impart great utility of ^{81}Kr as a tracer of fluid residence time in systems of moderate to high salinity, such as sedimentary basins and submarine hydrothermal systems. Further development and application of ^{81}Kr measurements in such water–rock systems is recommended for the insights that may be gained regarding time scales and mechanisms of water–rock interaction.

ACKNOWLEDGEMENTS

This work was supported mainly by the U.S. National Science Foundation (Grant EAR-0126297) and the U.S. Department of Energy, Nuclear Physics Division (Contract W-31-109-Eng-38). Partial support was provided by NASA's Land-Cover Land-Use Changes program (Grant NAG5-12409); the U.S. Department of Energy, Office of Basic Energy Sciences (Contracts W-31-109-Eng-38 and DE-AC03-76SF00098); the Swiss National Science Foundation; and the University of Bern.

REFERENCES

Ball, J. 1927. Problems of the Libyan Desert. *Geograph. J.* 70: 21–38, 105–128, 209–224

Bentley, H.W., Phillips, F.M., Davis, S.N., Habermehl, M.A., Airey, P.L., Calf, G.E., Elmore, D., Gove, H.E. & Torgersen, T. 1986. Chlorine 36 dating of very old groundwater 1. The Great Artesian Basin, Australia. *Water Resources Research* 22: 1991–2001.

Bethke, C.M., Zhao, X. & Torgersen, T. 1999. Groundwater flow and the ^{4}He distribution in the Great Artesian Basin of Australia. *J. Geophys. Res.* 104: 12999–13011.

Brinkmann, P.J., Heinl, M., Hollander, R. & Reich, G. 1987. Retrospective simulation of groundwater flow and transport in the Nubian Aquifer system. In E. Klitzsch & E. Schrank (eds.), *Berliner Geowissenschaftlich Abhandlungen*, ReiheA/Band 75.2: 465–516. Berlin: Verlag Von Dietrich Reimer.

Collon, P., Kuschera, W., Loosli, H.H., Lehmann, B.E., Purtschert, R., Love, A., Sampson, L., Anthony, D., Cole, D., Davids, B., Morrissey, D.J., Sherrill, B.M., Steiner, M., Pardo, R.C. & Paul, M. 2000. ^{81}Kr in the Great Artesian Basin, Australia: a new method for dating very old groundwater. *Earth Plan. Sci. Lett.* 182: 103–113.

Du, X., Purtschert, R., Bailey, K., Lehmann, B.E., Lorenzo, R., Lu, Z.-T., Mueller, P., O'Connor, T.P., Sturchio, N.C. & Young, L. 2003. A new method of measuring Kr-81 and Kr-85 abundances in environmental samples. *Geophys. Res. Lett.*30: 2068–2071.

Hesse, K.H., Hissene, A., Kheir, O., Schnacker, E., Schneider, M. & Thorweihe, U. 1987. Hydrogeological investigations in the Nubian aquifer system, Eastern Sahara. In E. Klitzsch & E. Schrank (eds.), *Berliner*

Geowissenschaftlich Abhandlungen, ReiheA/Band 75.2: 397–464. Berlin: Verlag Von Dietrich Reimer.
Lehmann, B.E., Loosli, H.H., Rauber, D., Thonnard, N. & Willis, R.D. 1991. ^{81}Kr and ^{85}Kr in groundwater, Milk River aquifer, Alberta, Canada. *Appl. Geochem.* 6: 419–424.
Lehmann, B.E., Love, A., Purtschert, R., Collon, P., Loosli, H.H., Kutschera, W., Beyerle, U., Aeschbach-Hertig, W., Kipfer, R., Frape, S.K., Herczeg, A., Moran, J., Tolstikhin, I.N. & Groning, M. 2003. A comparison of groundwater dating with Kr-81, Cl-36 and He-4 in four wells of the Great Artesian Basin, Australia. *Earth Plan. Sci. Letters* 211: 237–250.
Pachur, H.-J., Röper, H.-P., Kröpelin, S. & Goschin, M. 1987, Late Quaternary hydrography of the eastern Sahara. In E. Klitzsch & E. Schrank (eds.), *Berliner Geowissenschaftlich Abhandlungen*, ReiheA/Band 75.2: 331–384. Berlin: Verlag Von Dietrich Reimer.
Park, J., Bethke, C.M., Torgersen, T. & Johnson, T.M. 2002. Transport modeling applied to the interpretation of groundwater ^{36}Cl age. *Water Resources Research* 38: 10.1029/2001WR000399.
Phillips, F.M. 2000. Chlorine-36. In P.G. Cook & A.L. Herczeg (eds.), *Environmental tracers in subsurface hydrology*: 299–348. Norwell, Massachusetts: Kluwer Academic Publishers.
Phillips, F.M., Bentley, H.W., Davis, S.N., Elmore, D. & Swanick, G.B. 1986. Chlorine 36 dating of very old groundwater 2. Milk River Aquifer, Alberta, Canada. *Water Resources Research* 22: 2003–2016.
Thorweihe, U. 1990. Nubian Aquifer system. In R. Said (ed.), *The Geology of Egypt*: 601–611. Rotterdam: Balkema.

Water–Rock Interaction, Wanty & Seal II (eds)

Evidence for metastable equilibrium between hydrocarbons in volcanic gases

Y. Taran
Institute of Geophysics, UNAM, Coyoacan, Mexico D.F., Mexico

W.F. Giggenbach
Institute of Nuclear and Geological Sciences, Lower Hutt, New Zealand (Deceased 1997)

ABSTRACT: Despite a wide discussion about the relative importance of kinetics and thermodynamics in governing the chemical and isotopic composition of natural hydrocarbons, there was no direct evidence that thermodynamic equilibrium can be attained between hydrocarbons in any natural process. Here we report our data on the hydrocarbon composition of volcanic gases (C_1–C_4 and benzene) from White Island, New Zealand. We found that ratios of ethane/ethylene, propane/propylene and ethylene/benzene are close to their equilibrium values at temperatures of a fumarolic vent and at concentrations of hydrogen controlled by the H_2S–SO_2 "gas" buffer. This result suggests that the redox processes $C_nH_{2n} + H_2 \leftrightarrow C_nH_{2n+2}$ and $3C_2H_4 \leftrightarrow 3H_2 + C_6H_6$ at high temperature in volcanic vapors are reversible and fast in both directions, similar to reactions of thermal decomposition of water and the reduction–oxidation $CO_2 \leftrightarrow CO$.

1 INTRODUCTION

The concentration ratios H_2/H_2O and CO/CO_2 in volcanic gases are close to their equilibrium values at temperature of a fumarolic vent if it is assumed that the process of reduction–oxidation of sulfur species (SO_2 and H_2S) controls the redox-potential of the gaseous mixture (Giggenbach 1987). Light hydrocarbons including methane are always present in volcanic gases in very low but detectable concentrations (Capaccioni et al. 1995; Kiyosu & Asada 1995). Their molecular mass distribution is far from equilibrium and corresponds to the products of thermal degradation (pyrolysis) of sedimentary high-molecular-weight organic material, but not to a synthesis from CO_2 (CO) and H_2. Light olefins, ethene and propene, were reported in high-temperature (>400°C) volcanic gases by Capaccioni et al. (1995).

As was shown by Seewald (1994, 2001) in a series of laboratory experiments at 300–350°C and 350 bars pressure, low molecular weight aqueous hydrocarbons interact with water and Fe-bearing mineral assemblages under controlled redox-conditions within the time scale from hundreds to thousands of hours. An important finding from these experiments was that alkane–alkene pair and some other reduced-oxidized pairs with the same carbon number attained metastable thermodynamic equilibrium states. They react reversibly according to general simple redox reaction:

$$C_nH_{2n} + H_2 \leftrightarrow C_nH_{2n+2} \qquad (1)$$

For the dehydrogenation (oxidation) of alkanes and formation of alkenes, more oxidized conditions, higher temperature and lower pressure are preferable. It is possible that under appropriate conditions this type of temperature-controlled reaction can be found in nature.

2 ALKANE–ALKENE METASTABLE EQUILIBRIUM IN THE WHITE ISLAND VOLCANIC GASES

This study is based on a data set obtained during 1975–1989 by one of the authors (WFG) on White Island – a small volcanic island located in the Bay of Plenty, close to the northern coast of the North Island of New Zealand. Temperatures of fumaroles varied from 100 to 840°C covering almost the entire range of historically measured temperatures of volcanic gases. The ratios C_2H_6/C_2H_4 and C_3H_8/C_3H_6 (Fig. 1) plotted versus sampling temperature have clearly expressed temperature dependence and thus the relative amount of light saturated and unsaturated hydrocarbons with the same carbon number is

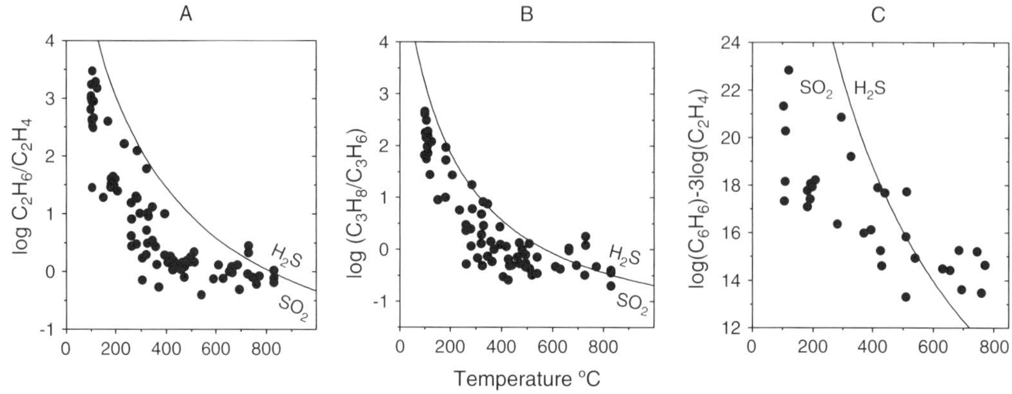

Figure 1. Variation in the concentration ratios of hydrocarbons with temperature of a fumarolic vent of White Island volcano, New Zealand: **A** = log(ethane/ethene); **B** = log(propane/propene); **C** = log(benzene)-3log(ethene). Equilibrium lines correspond to a situation where the redox state is controlled by reaction $SO_2 + 3H_2 = H_2S + 2H_2O$ at 1 bar total pressure.

apparently controlled by temperature of a fumarolic vent. In other words, reactions (1) in volcanic gases seem to be very fast like the thermal decomposition of water that controls the H_2/H_2O ratio. The mass action equation and redox-control for the equilibrium (1) can be written in form:

$$\log(X_{C_nH_{2n+2}}/X_{C_nH_{2n}}) = -\log K_n + R_H + \log f_{H_2O} \quad (2)$$

where X = mole fractions, $R_H = \log f_{H_2}/f_{H_2O}$, f = fugacities, which are very close to partial pressures, and K_n is the equilibrium constant. Volcanic gases consist mostly of water vapor, therefore the total pressure is close to the partial pressure of water vapor, and at the surface, close to 1 bar. Redox control, or H_2 concentration, in high-temperature volcanic gases is fixed at a given temperature and pressure by reduction–oxidation of the major sulfur species H_2S and SO_2 and thus, by the gas phase itself (Giggenbach 1987). The observed trend for the ethane/ethene ratio (Fig. 1A) passes below and subparallel to the gas buffer line and approaches equilibrium values at high temperatures. The temperature trend for the propane–propene ratio (Fig. 1B) is closer to the gas buffer line, and again, very close to the equilibrium values at high temperature.

This type of temperature-controlled behavior of hydrocarbons has never been observed before because of the lack of a large set of hydrocarbon data for volcanic gases. These trends indicate that chemical reactions among saturated and unsaturated hydrocarbons with the same carbon number are fast, with reaction rates close to those of the thermal decomposition of water, CO–CO_2 or H_2S–SO_2 redox reactions in the gas phase. They are most probably catalyzed by oxidized sulfur species, as shown by the hydrothermal experiments of Seewald (2001).

In spite of the near-equilibrium concentration ratios of alkanes and alkenes, their absolute concentrations in volcanic gases are too high and very far from equilibrium even at high temperature. This may indicate that the source of hydrocarbons in volcanic gases is buried organic matter of different degrees of maturity. Thermal degradation of these organics produces light hydrocarbons with a disequilibrium distribution and concentrations much higher than at equilibrium.

However, under high-temperature magmatic-hydrothermal conditions, light alkanes reversibly disproportionate to alkenes and hydrogen, but without breakage and rearrangement of the C–C bonds. Any reactions involving the breakage of C–C bonds may be fast enough to produce lower carbon number hydrocarbons and eventually CH_4, H_2 and CO_2, but there is no time to equilibrate the initial disequilibrium distribution inside a very dynamic and geologically short-lived system like an active volcano. The formation of new C–C bonds requires very specific conditions and catalysts and in the natural high-temperature magmatic-hydrothermal environment is highly unlikely.

3 BENZENE IN THE WHITE ISLAND VOLCANIC GASES

Benzene and more complicated volatile aromatics like toluene and ethylbenzene were analyzed in volcanic and hydrothermal gases on the ppm and ppb level in dry gas. Aromatics are an important part of oil and sedimentary organic material, and the aromatic ring itself (benzene) is stable at high temperature even under highly oxidizing conditions. It may be that there is geochemical importance in the fact that benzene can be synthesized in an empty (catalyst-free) reactor at 950°C from a CH_4-air mixture with a yield of 2.5% (York et al. 1996). This so-called oxidative aromatization of methane is the object of many studies in the catalytic chemistry of hydrocarbons. Methane can be converted to benzene in the presence of oxides like MoO_3 or WO_3 and the mechanism of the aromatization includes formation of ethane and ethene as intermediate reactants (Arutunov & Krylov 1998 and references therein). The presence of CH_4 and alkenes in volcanic gases may control the concentration of benzene by some mechanism. The thermodynamic constraints can be obtained from the reaction:

$$3C_2H_4 \rightarrow C_6H_6 + 3H_2 \quad (3)$$

The curve on Fig. 1C represents the equilibrium $X_{C_6H_6}/X^3_{C_2}H_4$ ratios expected for the gas buffer of hydrogen fugacity at 1 bar total pressure. The data for White Island fumaroles show a clear temperature-dependent trend in a "correct" direction along the gas buffer line. At high temperature, points are clustered closer to the gas buffer line. A decrease in the C_6H_6 absolute concentrations with temperature suggests that benzene is irreversibly consumed (reduced?) and decomposes, converting partially to alkenes, and that under high-temperature volcanic-hydrothermal conditions new C–C bonds do not form. In other words, the only oxidative or reductive degradation reactions are mechanisms of equilibration of hydrocarbons and CH_4 in the high-temperature crustal environment. Relationships between concentrations of alkanes and alkenes with the same carbon number, or between ethane and benzene, showing near-equilibrium ratios of concentrations most probably correspond to metastable equilibrium. Kinetic barriers for reactions, without forming or rearranging C–C bonds ("waves in the metastability pond", Giggenbach 1997), are much lower than for the full disproportionation of hydrocarbons to CO_2, H_2 and H_2O.

REFERENCES

Arutunov, V.S. & Krylov, O.V. 1998. *Oxidative Conversions of Methane*. Moscow: Nauka.

Capaccioni, B., Martín, M. & Mangani, F. 1995. Light hydrocarbons in hydrothermal and magmatic fumaroles. Hint of catalytical and thermal reactions. *Bull. Volcanol.* 56, 593–600.

Giggenbach, W.F. 1987. Redox-processes governing the chemistry of fumarolic gas discharges from White Island, New Zealand. *Appl. Geochem.* 2: 143–161.

Giggenbach, W.F. 1997. Relative importance of thermodynamic and kinetic processes in governing the chemical and isotopic composition of carbon gases. *Geochim. Cosmochim. Acta* 61: 3763–3785.

Kiyosu, Y. & Asada, N. 1995. Light hydrocarbons in volcanic gases from the Japanese Island arc. *Geochem. J.* 29: 231–142.

Seewald, J.S. 1994. Evidence for metastable equilibrium between hydrocarbons under hydrothermal conditions. *Nature* 370: 285–287.

Seewald, J.S. 2001. Aqueous geochemistry of low molecular weight hydrocarbons at elevated temperatures and pressures: Constraints from mineral buffered laboratory experiments. *Geochim. Cosmochim. Acta* 65: 1641–1664.

York, A.P.E., Claridge, J.B., Green, M.L.H. & Tsang, S.L. 1996. A study of the catalytical oxidative oligomerization of methane to aromatics. *Catalysis Lett.* 40: 207–214.

Water–Rock Interaction, Wanty & Seal II (eds)
© 2004 Taylor & Francis Group, London, ISBN 90 5809 641 6

Intrusion, degassing and hydrothermal systems in submarine felsic volcanic centers

B.E. Taylor & A. Timbal
Geological Survey of Canada, Ottawa, Canada

G.J. Holk
California State University, Long Beach, California

ABSTRACT: O isotope mapping of five Archean to Paleoproterozoic submarine felsic volcanic centers in Canada and Sweden documents a chronology of hydrothermal activity and massive sulfide formation of comparable complexity to the intrusive history of each center. Paired high- and low-temperature reaction zones represent the culmination of hydrothermal activity and can survive regional metamorphism, but reflect any host rock strain.

1 INTRODUCTION

Integration of whole-rock O isotope data, geologic mapping and U-Pb geochronology in selected Archean and Paleoproterozoic submarine volcanic centers has placed some constraints on the relative timing of magma emplacement, active and passive degassing, and hydrothermal activity. Four syn-volcanic intrusive complexes of variable size and complexity were selected in Canada (Noranda, Quebec: 625 km^2; >600 samples; 1198 including Beaty & Taylor 1982; Kennedy 1984; Cathles 1993; Paradis et al. 1993; Clifford-Ben Nevis, Ontario: 160 km^2; 251 samples; Snow Lake, Manitoba, 84 km^2; 575 samples; and Sturgeon Lake, Ontario: 525 km^2; 452 samples), plus one in Sweden (Kristineberg: 90 km^2; 304 samples). A four-year GSC-CAMIRO project focused on the relationship between regional hydrothermal alteration and volcanic-hosted massive sulfide (VMS) deposits. The Clifford-Ben Nevis area constituted an unmineralized control case.

2 O ISOTOPE MAPPING

Whole-rock O isotope analysis provides the basis for mapping regional variations in the distribution of heat and fluid in paleo-hydrothermal systems. Partially asynchronous, paired zones of ^{18}O depletion and enrichment were mapped using contour intervals of $\delta^{18}O_{WR} \leq 6.0, \geq 9.0$, and ≥ 12.0‰. These intervals afforded clear indication of high-temperature reaction zones (e.g. $\geq 300°C$; $\delta^{18}O_{WR} \leq 6.0$‰) and zones of lower-temperature alteration (ca. $\leq 250°C$; $\delta^{18}O \geq 9.0$‰) in marine hydrothermal systems. Sample densities varied from ca. 1.5 (Clifford-Ben Nevis) to 10.6 (Snow Lake) samples/km^2, with higher densities in areas of greater regional deformation. Close-spaced sampling across syn-volcanic faults, lithologic contacts, and fold hinges clarified the extent and relative timing of focused fluid flow in many cases.

VMS deposits occur stratigraphically between paired high-temperature and low-temperature zones of isotopic alteration in a single-event hydrothermal system, but may be found within alteration zones of multiple-event systems. A high-temperature reaction zone ($\delta^{18}O_{WR} \leq 6.0$‰) is commonly found about the locus of heat (e.g. sub-volcanic intrusion or intrusive complex) and is synchronous with the deposition of massive sulfide at or close to the paleo-sea floor (Fig. 1A,B). A zone of low-temperature alteration ($\delta^{18}O \geq 9.0$‰) may develop in the hanging wall rocks subsequent to mineralization, as sedimentation and/or volcanism continue during cooler, waning hydrothermal flow. Figure 1A–C shows schematically, features documented in several areas during isotopic mapping under the GSC-CAMIRO project. For example, detailed studies in the Noranda camp (Fig. 2) indicate that the high-temperature zones represent the effects of several hydrothermal systems, but low-temperature zones primarily reflect the last principal hydrothermal event.

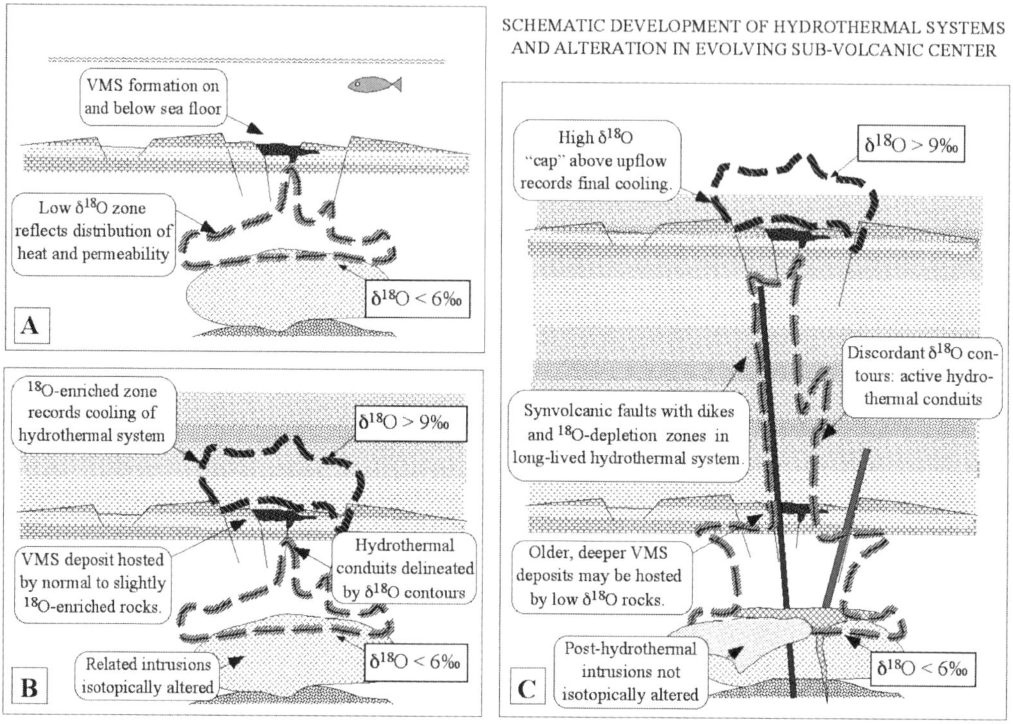

Figure 1. Schematic cross sections of subvolcanic intrusions showing development of O isotope zoning in (A-B) a single stage hydrothermal system; and (C) a multi-stage intrusive center with several episodes of magmatism and hydrothermal activity.

Many sub-volcanic intrusions (e.g. Flavrian, Noranda) are composite in nature, comprised of one or more sills. Contact relationships and textures (e.g. mega-jointing, vacuoles and pipes; Galley 2002) suggest magma-fluid interaction or vapor-saturation. In some areas (e.g. Noranda; Snow Lake; Kristineberg; Fig. 2), intrusions have invaded and assimilated hydrothermally altered rocks and VMS deposits. Even later intrusions in the same center may bear no evidence of isotopic alteration. Subaerial emergence may have played a role in some cases (e.g. Sturgeon Lake), but in other instances, isotopically unaltered intrusions (e.g. Lac Dufault gabbro, not shown in Fig. 2; Flavrian leucotonalites, Noranda) are spatially, but not temporally, related to earlier hydrothermal activity. Any direct magmatic contributions to mineralization appear limited to passive degassing.

Terranes metamorphosed regionally to the greenschist facies have mineral assemblages similar to those formed by hydrothermal alteration. O isotope techniques are particularly useful for mapping the limits of paleo-hydrothermal systems, and approaches incorporating isotopic mapping offer more to exploration of large areas.

2.1 *Magmatic and hydrothermal events*

Individual magmatic and hydrothermal events appear to be short-lived relative to the lifespan of a sub-volcanic center, allowing the distinction of individual magmatic/hydrothermal associations (or, "events") on the basis of O isotope systematics. Similar U-Pb ages (Davis & Trowell 1982; Davis et al. 1985; Galley 2002) and REE compositions of the Biedelman Bay intrusive complex (BBIC) and "Middle L" rhyolites indicate that they represent intrusive and extrusive magmatism subsequent to formation of the major Matabi (and smaller "F Group") VMS deposits. O isotope mapping of wall rocks and of fresh BBIC tonalite (average $\delta^{18}O_{WR}$ = ca. 7.5‰; 6–9‰ range) indicate emplacement postdated hydrothermal alteration. The tonalite is in abrupt intrusive contact with Darkwater basalt exhibiting $\delta^{18}O_{WR}$ of 1.0–2.7‰, and a variation of 7‰ elsewhere. The abrupt isotopic "discontinuity" indicates that emplacement of the BBIC at the present level postdated hydrothermal alteration of immediate wall rocks. The shallow depositional environment of the volcaniclastic and pyroclastic rocks (e.g. Morton

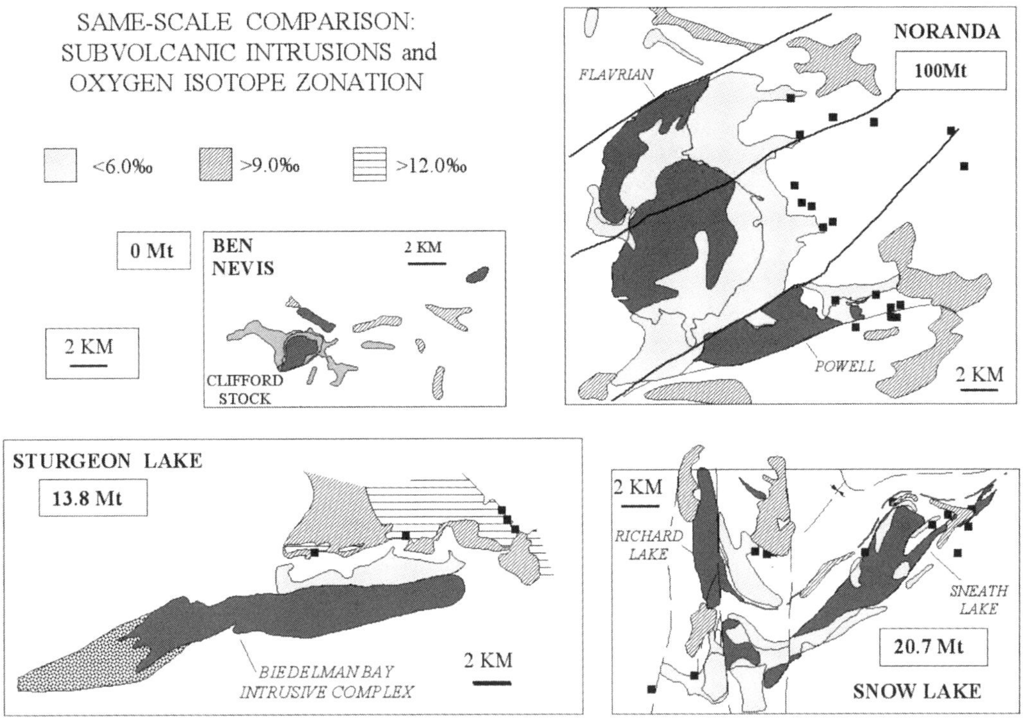

Figure 2. Maps compare O isotope zoning about subvolcanic intrusive centers. VMS deposits: solid black squares located in isotopically "normal" rocks, stratigraphically between zones of high- and low-$\delta^{18}O$ (total tonnage, Mt, is indicated for each area). In all maps, dark gray pattern (and stippled pattern on Sturgeon Lake map) represents sub-volcanic intrusion or intrusive complex.

et al. 1990; Hudak 1996) and isotopic data are consistent with an emerging complex.

Geologic and O isotope data for the Flavrian intrusive complex (FIC; Fig. 3A–C) illustrate a superposition of multiple magmatic and hydrothermal events. The FIC intruded the base of the Noranda cauldron, which eventually filled with volcanic rocks during 20,000 feet or more of subsidence, altered in deep-water (Spence & Rosen-Spence 1975). Hybrid leucotonalite and xenolith-bearing tonalite of the FIC intruded altered wall rocks and VMS deposits low in the intra-cauldron sequence, feeding eruption of overlying rhyolite well above most of the VMS deposits (Goldie 1978; Kennedy 1984; Paradis et al. 1988; Gibson 1989; Setterfield et al. 1995; Fig. 3A). Comparison of REE compositions of the FIC, leucotonalite feeder dikes and overlying rhyolite (Galley 2002), further supports syn- as well as post-VMS magmatism at Noranda. O isotope mapping (Fig. 3A–B) indicates a broad high-temperature zone that partially encloses and cross-cuts the FIC. The low-$\delta^{18}O$ zone includes the Early Leucotonalite (noted by Kennedy 1984, from $\delta^{18}O$ data) where it appears to have engulfed tonalite and xenolithic tonalite. Although the tell-tale (non-equilibrium) effects of hydrothermal alteration can be seen in Figure 3C, O isotope relationships are consistent with some ^{18}O-depletion of tonalite magma due to incorporation of previously-altered, low $\delta^{18}O$ wall rocks.

In contrast, the Late Leucotonalite of the FIC bears virtually no O isotope evidence of hydrothermal alteration. Sedimentological and volcanological evidence of succession in the cauldron east of the FIC, does not support a shallow or emergent complex (Spence & Rosen-Spence 1975; Caumartine & Calle 1990). Perhaps insufficient thermal energy was available, as local mega-columnar jointing of late-intruded marginal phases of the FIC (Taylor, unpubl.) suggests rapid cooling (aided by low-temperature fluids?). Apparently low VMS tonnage higher in the volcanic sequence underscores the importance of a sustained hydrothermal system and high-temperature reaction zone for the formation of VMS deposits.

2.2 *Effects of metamorphism and deformation*

Isotopic mapping in five study areas representing a marked variation in regional metamorphism (to

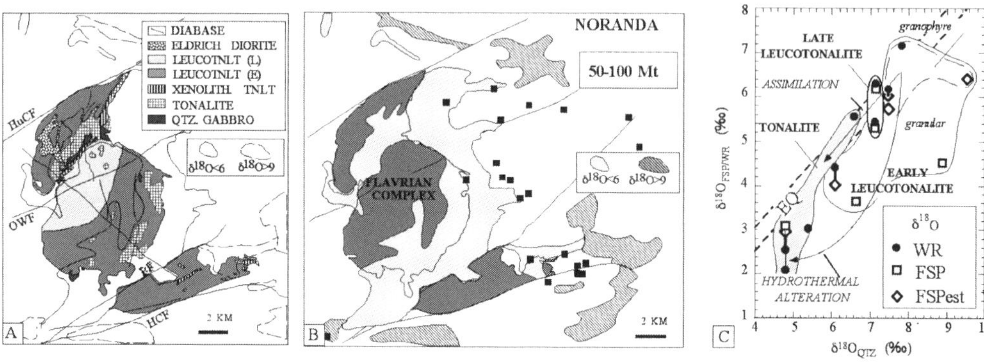

Figure 3. A–B. Geologic and O isotope zoning maps for the Flavrian intrusive complex, Noranda, Quebec; VMS deposits, solid black squares; OWF, Old Waite Paleofissure; BF, Beauchatel Fault; HuCF Hunter Creek Fault; HCF, Horne Creek Fault; C. Plot of $\delta^{18}O_{QTZ}$ (quartz) vs $\delta^{18}O_{WR/FSP}$ (whole-rock or feldspar) for various mapped facies of the FIC illustrates variations due to both hydrothermal alteration and assimilation of hydrothermally-altered (and ^{18}O-depleted) rocks.

sillimanite-zone) consistently indicate that $\delta^{18}O_{WR}$ values survive regional and contact metamorphism, in spite of intra-mineral re-equilibration. Shapes and sizes of the enrichment/depletion zones typically reflect the host rock strain, but the $\delta^{18}O_{WR}$ values have changed little, if at all.

2.3 *Terrane assessment and origin of deposits*

A positive correlation found between total aggregate tonnage of sulfide, mapped area of related intrusion (proxy for magmatic heat), and zones of O isotope depletion facilitates terrane assessment and exploration. The correlation supports the notion of ore genesis largely via leaching.

3 CONCLUSIONS

O isotope mapping provides a robust means to map submarine hydrothermal systems, to assess VMS potential, and to distinguish syn- vs. post-mineralization magmatic and hydrothermal events, despite regional metamorphism. Direct magmatic contribution to VMS formation is limited to passive degassing, and in some areas, magmas were emplaced without significant hydrothermal activity or linkage to VMS deposits.

REFERENCES

Beaty, D.W. & Taylor, H.P. Jr. 1982. Some petrologic and oxygen isotope relationships in the Amulet Mine, Quebec, and their bearing on the origin of Archean massive sulfide deposits. *Econ. Geol.* 77: 95–108.

Cathles, L.M. 1993. Oxygen isotope alteration in the Noranda mining district, Abitibi greenstone belt, Quebec. *Econ. Geol.* 88: 1483–1511.

Caumartine, C. & Calle, M-F. 1990. Volcanic stratigraphy and structure of the Mobrun Mine. *CIM Special. Volume* 43: 133–142.

Davis, D.W. & Trowell, N.F. 1982. U-Pb zircon ages from the eastern Savant Lake-Crow Lake area, Wabigoon subprovince, northwest Ontario. *Can. J. Earth Sci.* 19: 868–877.

Davis, D.W., Krogh, T.E., Hinzer, J. & Nakamura, E. 1985. Zircon dating of polycyclic volcanism at Sturgeon Lake and implications for base metal mineralization. *Econ. Geol.* 80: 1942–1952.

Galley, A.G. 2002. Characteristics of composite subvolcanic intrusive complexes associated with Precambrian VMS districts, in CAMIRO Project 94E-07: Interrelationships between subvolcanic intrusions, large-scale alteration zones and VMS deposits: *Open File 4431.* Ottawa: Geological Survey of Canada.

Gibson, H.L. 1989. *The mine sequence of the central Noranda volcanic complex: Geology, alteration, massive sulphide deposits and volcanological reconstruction.* PhD thesis (unpubl.). Ottawa: Carleton University.

Goldie, R.J. 1978. Magma mixing in the Flavrian pluton, Noranda Area, Quebec. *Can. J. Earth Sci.* 15: 132–144.

Hudak, G.J. III. 1996. *The physical volcanology and hydrothermal alteration associated with late caldera volcanic and volcaniclastic rocks and volcanogenic massive sulfide deposits in the Sturgeon Lake region of northwestern Ontario, Canada.* PhD thesis (unpubl.). Duluth: University of Minnesota.

Kennedy, L.P. 1984. *The geology and geochemistry of the Archean Flavrian pluton, Noranda, Quebec.* PhD thesis (unpubl.). London: University of Western Ontario.

Morton, M.L, Hudak, G.J., Walker, J.S. & Franklin, J.M. 1990. Physical volcanology and hydrothermal alteration of the Sturgeon Lake caldera complex. In J.M. Franklin, B.R. Schnieders &. E. Koopman (eds.): 74–94. *Geological Survey of Canada Open-File 2164*, Mineral Deposits in Western Superior Province, Ontario.

Paradis, S., Ludden, J. & Gelinas, L. 1988. Evidence for contrasting compositional spectra in comagmatic intrusive and extrusive rocks of the late Archean Blake River Group, Abitibi, Quebec. *Can. J. Earth Sci.* 25: 134–144.
Paradis, S., Taylor, B.E., Watkinson, D.H. & Jonasson, I.R. 1993. Oxygen isotope zonation and alteration in the northern Noranda district, Quebec: Evidence for hydrothermal fluid flow. *Econ. Geol.* 88: 1512–1525.
Setterfield, T.N., Hodder, R.W., Gibson, H.L. & Watkins, J.J. 1995. The McDonald-Despina fault set, Noranda, Quebec: Evidence for fault-controlled volcanism and hydrothermal fluid flow. *Explor. Mining Geol.* 4: 381–394.
Spence, C.D. & de Rosen-Spence, A.F. 1975. The place of sulfide mineralization in the volcanic sequence at Noranda, Quebec. *Econ. Geol.* 70: 90–101.

Water–Rock Interaction, Wanty & Seal II (eds)
© 2004 Taylor & Francis Group, London, ISBN 90 5809 641 6

Hydrochemical evolution of the Na-HCO_3-type groundwater from northern Sikhote-Alin (Far East Russia)

N.A. Tchepkaia, G.A. Chelnokov & O.V. Chudaev
Far East Geological Institute Rus. Acad. of Sci., Russia

V.A. Chudaeva
Pacific Institute of Geography Rus. Acad. of Sci., Russia

ABSTRACT: The Muhen River catchment, located in the northern end of Sikhote-Alin mountain system, is host to Ca-Mg-HCO_3- and Na-HCO_3-type cold CO_2-rich groundwaters issuing from closely spaced wells (~60 m intervals). Ca-Mg-HCO_3-type fluids are characterized by low TDS (300–1700 mg/L) and by enriched Fe, Mn, Ba, SiO_2. In contrast, the Na-HCO_3-type fluids have very high TDS (up to 14000 mg/L) and elevated contents of Li, B, Sr, Br and I. This paper demonstrates the origin and evolution of both types of water using detailed descriptions of geological and hydrogeological setting of the area, composition data for major, trace and rare earth elements in the waters, and thermodynamic modeling.

1 INTRODUCTION

In the Russian Far East (Khabarovsky krai), two different types of cold (7°C) CO_2-rich groundwater (Ca-Mg-HCO_3 and Na-HCO_3) issue from drilled wells. Both types of water are used by local spas. Ca-Mg-HCO_3-type fluid is from shallow depths (100 m), but the Na-HCO_3-type fluid source is deeper (150 m). The purpose of this study is to identify the processes that control the chemical evolution Na-HCO_3 groundwater and Ca-Mg-HCO_3 groundwater in the Punchi River catchment area of the northern Sikhote-Alin Mountains in Russia's Far East.

2 GEOLOGY AND HYDROGEOLOGY

The Muhen River catchment lies in the northern part of the Sikhote-Alin mountains system on the boundary zone of the Sikhote-Alin faulted belt and the Middle-Amur Cenozoic depression, at the north-eastern end of the Pereslavian graben.

The rocks exposed in the study region range in age from Mesozoic to Quaternary (Fig. 1). The Mesozoic sedimentary rocks outcropping to the east of the study area include sandstone, siltstone and shale. The sandstone is generally quite porous and permeable but commonly pore spaces have been filled with clay and carbonate minerals (Batukov 1974). The deep aquifer

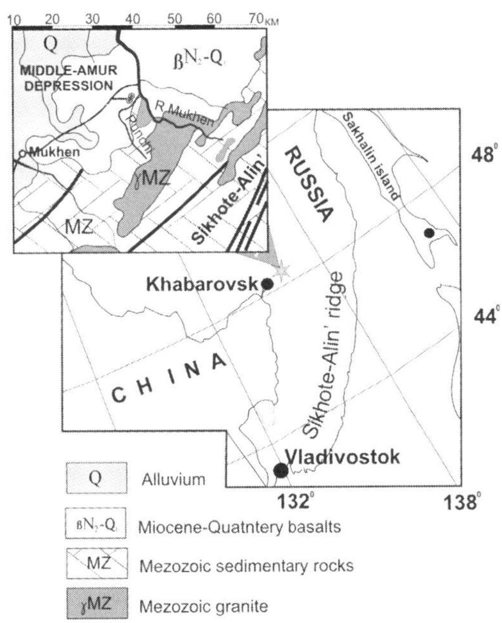

Figure 1. Geological map of the Muhen River catchment modified from Batukov 1974. Details are described in the text.

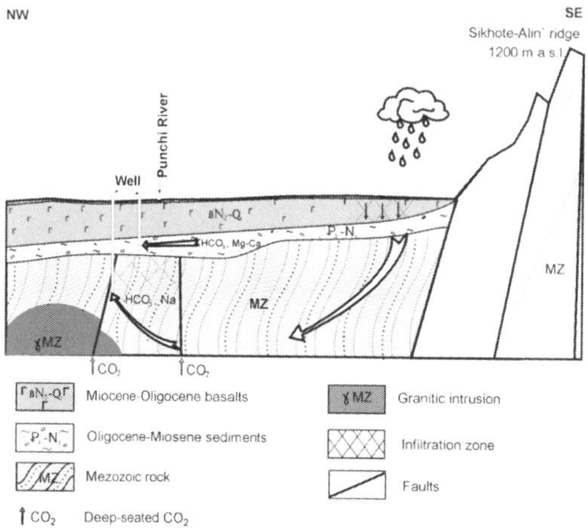

Figure 2. Hydrogeological cross-section through study area.

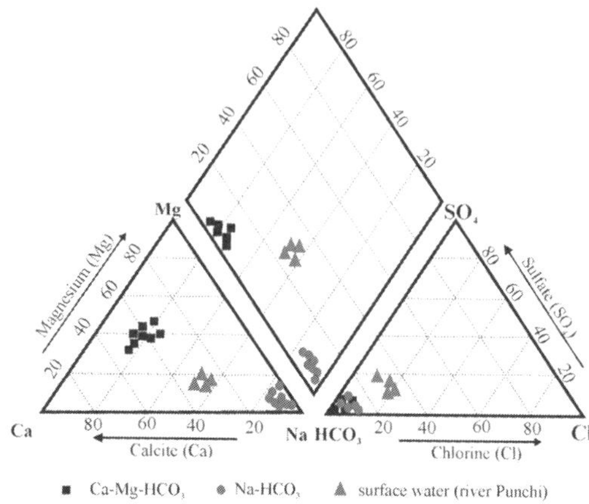

Figure 3. Trilinear diagram for analyses of surface and groundwater, illustrating the principal water types.

is hosted in the upper fractured zone of this sediment (Fig. 2).

The Oligocene–Miocene rocks consist of ~40 m of moderately sorted sand, gravel and cobbles, with coal beds and bodies of dolerite and basalt, and host shallow Ca-Mg-HCO_3-type fluids (Fig. 2). The main recharge zone for this water occurs in the north of basin where the sediment complex outcrops.

The sedimentary rocks are covered by 50–400 m of Miocene–Quaternary basalts. The basalts exhibit slight weathering and are characterized by very low permeability, with limited water circulation. The dominant minerals are plagioclase, pyroxene and olivine.

Alluvium from the Muhen River system is widespread in this area. It consists of poorly sorted sand and gravel with stone, siltstone and clay and reaches a thickness up to 15 m.

Numerous tectonic faults are spread across the study area, of which the Punchi fault is the largest. It is located under the cap of the Miocene–Quaternary basalts and trends southeast. The width of the fault zone is not less then 150–200 m. Published data (Batukov 1974) indicate that CO_2 gas migrates from the depths to the surface via the fault zone.

3 GROUNDWATER CHEMISTRY

Groundwater samples were collected over a three year period from the Punchi River and boreholes. Additionally, hydrochemical data from earlier investigations have been taken into consideration. Two chemically distinct groundwater types occur in the study area, as shown in Figures 2 and 3: Ca-Mg-HCO_3 and Na-HCO_3 waters. Sampling procedures are described in detail by Chudaeva et al. (1999). In all of the samples analyzed, the charge balance was better than ±5%. Representative analyses are shown in Table 1.

Table 1. Representative groundwater and surface water chemical analyses. Units for dissolved components are in mg/L.

Parameter	River Punchi	Na-HCO_3 groundwater	Ca-Mg-HCO_3 groundwater
pH	6.1	7	5.3
TDS	89.1	10100	330
Na	12.9	2400	17
Mg	1.8	57.3	20
K	2.9	46.4	0.93
Ca	6.5	113	29
SO_4	6	3	3.5
Cl	7.1	109	1.2
HCO_3	51.9	7300	229
F	NA	0.4	0.1
Al	0.001	0.002	0.078
B	NA	66	0.16
SiO_2	23	9.77	18.5
Fe	1.5	0.08	11.5
Mn	0.01	0.007	0.26
Ba	0.019	0.006	0.068
Sr	0.09	0.03	0.11
Li	0.005	2.1	0.006
Cu	0.008	0.142	0.002
Rb	0.002	0.07	0.001
Br	NA	0.146	0.07
I	NA	0.042	0.003

NA – not analysed.

Shallow samples are Ca-Mg-HCO_3-type water and the deeper samples are Na-HCO_3-type water. The shallow water typically has a total dissolved solid

(TDS) content of less than 400 mg/L, pH 4.9–5.8 and molar Ca/Mg is about 0.94. This type of water is characterized by high concentration of Fe^{2+} (up to 13 mg/L), Mn^{2+} (up to 0.6 mg/L), Ba^{2+} (up to 0.2 mg/L) and SiO_2 (at about 80 mg/L). Na-HCO_3 groundwater occurs beneath Ca-Mg-HCO_3 water and has very high TDS, up to 14,000 mg/L, and pH of 7.2–7.5. This groundwater is enriched in Li (to 3.5 mg/L), B (50–60 mg/L), Sr (to 0.3 mg/L), Br (0.15 mg/L) and I (~0.04 mg/L).

TDS increases with depth, from surface water to Ca-Mg-HCO_3 water and then Na-HCO_3 water, and depends primarily on the amount of dissolved sodium and bicarbonate. Positive correlations between K–Li ($r = 0.993$), Ca–Mg ($r = 0.97$), SiO_2–Fe ($r = 0.97$), K–Ca ($r = 0.92$) and negative correlation between SiO_2 and major cations (Ca, Na, K) are observed for both types of groundwater. Cl/Br, Na/Cl, and Br/I indicate that both types of groundwaters are meteoric in origin.

Molar Sr^{2+}/Ca^{2+} in the studied groundwaters range from 0.006 to 0.024 for Ca-Mg-HCO_3 and Na-HCO_3 water respectively. These ratios are close to molar Sr^{2+}/Ca^{2+}(0.007) of average shale (Krauskopf 1979) and sodium-rich feldspars (0.01–0.02; Smith 1974). The increase in Sr^{2+}/Ca^{2+} with evolution of Ca-Mg-HCO_3 to Na-HCO_3 water can be explained by the dissolution of plagioclase and the precipitation of carbonate minerals.

4 RARE EARTH ELEMENTS

Concentrations of rare earth elements (REE) for both types of groundwater and surface water were determined first from filtered samples and then from unfiltered samples. The REE abundances in the studied waters are listed in Table 2. Shale-normalized plots of the REEs in these waters are shown in Figure 4. For comparison, REE concentration data for the Miocene basalts (Martynov 1999) were included on this plot. We normalized the data to North American Shale Composite (NASC) data taken from Hannigton & Shelkovitch 2001. Shale-normalized REE plots for all water samples are flat to slightly enriched in the HREEs.

Unfiltered samples have higher REE concentrations than filtered ones, with larger variations for the HREE. In all samples the REE concentration becomes greater with increasing atomic number. The observed enrichment in HREEs compared to LREEs may result from formation of stronger carbonate complexation with increasing atomic number (Lee & Berne 1993). Cerium anomalies were clearly found in the REE patterns of the filtered samples (Fig. 4), whereas slight negative Ce anomalies are present in surface and unfiltered samples. In the analyses of basalt samples (Martynov 1999) Ce anomalies are also absent. This reflects the reducing conditions of the subsurface environment where both types of groundwater evolved. Positive Eu anomalies are evident for all water and basalt samples.

The REE concentration and shale-normalized signatures for both types of groundwater are similar and closely resemble the REE signature of basaltic rocks from northern Sikhote-Alin. This is likely related to a common path of groundwater evolution via basalts.

Table 2. REE abundance (ppb) in the water samples from the Muhen river catchment.

	Na-HCO_3		Ca-Mg-HCO_3		Muhen river
	1	2	1	2	2
La	0.49	0.025	0.54	0.345	0.53
Ce	0.469	0.001	0.58	0.435	0.517
Pr	0.502	0.001	0.52	0.087	0.52
Nd	0.157	0.082	0.487	0.3	0.222
Sm	0.047	0.025	0.133	0.099	0.062
Eu	0.6	0.093	0.61	0.127	0.597
Gd	0.132	0.049	0.198	0.179	0.15
Tb	0.601	n.d.	0.61	n.d.	0.601
Dy	0.105	0.103	0.213	0.17	0.118
Ho	0.53	0.008	0.54	0.032	0.527
Er	0.162	0.157	0.235	0.21	0.164
Tm	0.588	n.d.	0.598	0.0079	0.586
Yb	0.148	0.207	0.276	0.221	0.215
Lu	0.598	0.013	0.61	0.027	0.597
Y	0.454	0.009	0.92	0.73	0.532

Note: 1 – unfiltered samples; 2 – filtered samples.

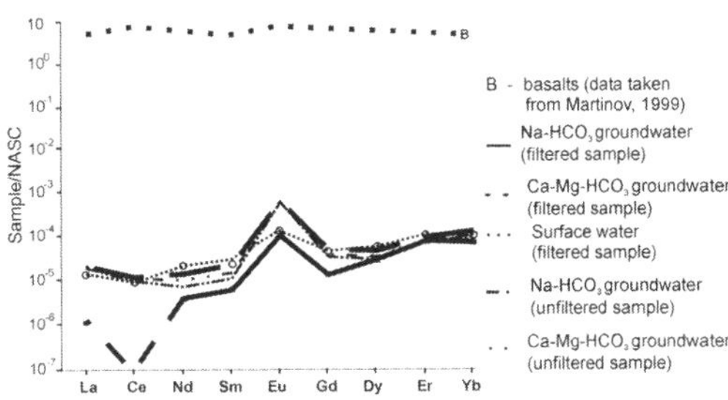

Figure 4. REE profiles of investigated waters.

5 THERMODYNAMIC CALCULATIONS

Chemical speciation modeling using WATEQ4F (Ball & Nordstrom 1991) indicate that Ca-Mg-HCO_3 waters are undersaturated with respect to the following minerals: all carbonates excluding siderite, gypsum, anhydrite, plagioclase and K-feldspar. The

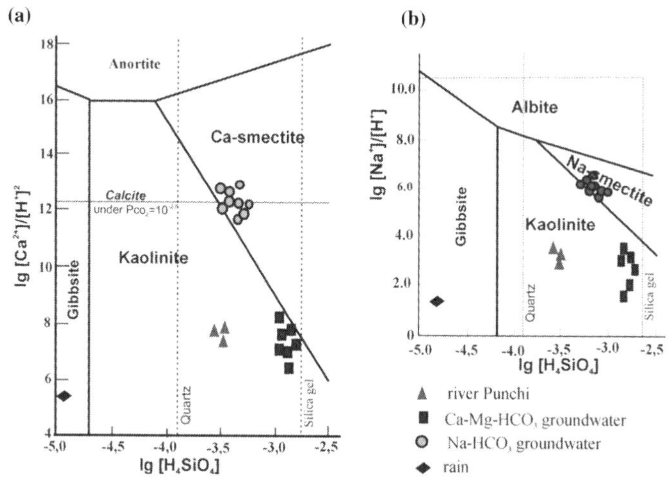

Figure 5. Stability diagrams for anorthite (a) and albite (b) and its possible weathering products gibbsite, kaolinite, and Ca- and Na-smectite.

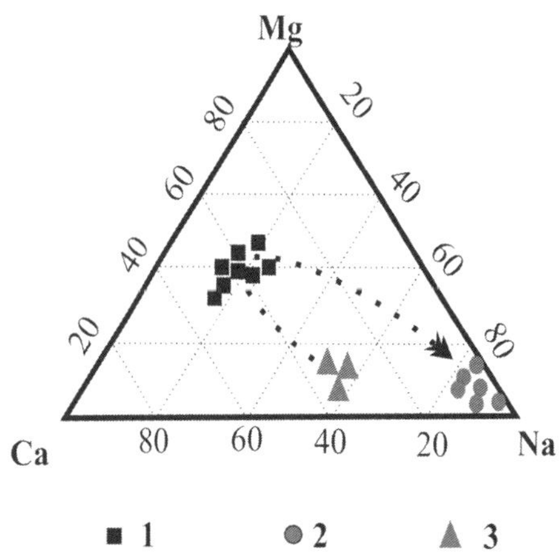

Figure 6. Evolution path of meteoric water (3) to Ca-Mg-HCO_3 water (1) and then to Na-HCO_3 water (2).

waters are supersaturated with quartz and clay minerals (kaolinite and smectite). Minor supersaturation with respect to siderite is caused by high concentration of Fe^{2+} in the groundwaters. Na-HCO_3 waters are saturated with respect to quartz and undersaturated with all carbonate minerals, albite, and clay minerals. This water is supersaturated with respect to sulfate minerals and anorthite.

Thermodynamic data indicate that both types of groundwater are undersaturated with respect to primary aluminosilicates and are in equilibrium with clay minerals. However, the Na-HCO_3 waters have already come to equilibrium with carbonate minerals. Therefore, Ca^{2+} does not accumulate in these waters and is removed from solution, forming secondary carbonates. Na^+, however, is leached from primary aluminosilicates.

To verify the possibility of the silicate mineral dissolution and identify the weathering products, ion activities of samples were plotted on the activity diagrams (Fig. 5). Na-HCO_3 type groundwater samples are plotted in the field of smectite, suggesting that primary silicate minerals can be weathered to smectite. However, surface water and Ca-Mg-HCO_3 groundwater samples are in the kaolinite field. In this case, weathering of primary silicates leads to precipitation of kaolinite. This is also in agreement with the mineralogical composition of rocks in the study area (Martynov 1999).

6 CONCLUSIONS

Detailed description of the hydrogeological setting of the area, together with chemical analyses of surface and groundwater provide information about the origin and evolution of two types of bicarbonate-rich groundwater found in the Muhen River catchment. Based on these results, we demonstrate that both types of groundwater are of similar origin. The source for both types of groundwater is meteoric water, while the residence times are different.

The water chemistry of Ca-Mg-HCO_3 results from the interaction of meteoric water with the basaltic cap rock. Enrichment of this groundwater by Sr, Ba, Mn, SiO_2, Al, Fe is caused by the leaching of these elements from aluminosilicate minerals. More than one source of calcium exists; the primary source is aluminosilicate and an additional source is carbonate cement in sedimentary rocks. This type of groundwater has a short residence time.

Na-HCO_3 water evolves from Ca-Mg-HCO_3 (Fig. 6) with increasing residence time. Our data indicate that Ca-Mg-HCO_3 waters are supersaturated with calcite and smectite. Therefore, Ca^{2+} is removed from water, precipitating in alteration minerals, whereas Na^+ is accumulated in solution. Cation exchange is also possible, however, it plays a minor role in evolution of the chemical composition of the groundwaters. The decrease in SiO_2 concentration during groundwater evolution is a result of clay mineral precipitation.

Drastically increasing TDS with depth is induced by diffusion of deep CO_2 gas via faults into groundwater. Accordingly, the groundwater becomes more acidic with depth, thereby promoting water–rock interaction.

ACKNOWLEDGEMENTS

We acknowledge financial support from the Russian Foundation for Basic Research (Project 03-05-06480) and from the Scientific School (1566.2003.05).

REFERENCES

Ball, J.W. & Nordstrom, D.K. 1991. User's manual for WATEQ4F with revised thermodynamic data base and test cases for calculating speciation of minor, trace and redox elements in natural waters. *U.S. Geol. Surv. Open File Rep.*, 91–183.

Batukov, S.I. 1874. Final report about exploring of the Punchi high pCO_2 mineral water deposit. Khabarovsk. 75 p.

Chudaeva, V.A., Chudaev, O.V., Chelnokov, A.N., Edmunds, W.M. & Shand, P. 1999. Mineral waters of Primorye (chemical aspect), Vladivostok, Dalnauka, 156 p.

Hannigan, R.E. & Sholkovitch, E.R. 2001. The development of middle rare earth elements enrichments in freshwaters weathering of phosphate minerals. *Chem. Geol.* 175: 495–508.

Krauskopf, K.B. 1979. *Introduction to geochemistry*. New York: McGrawHill.

Lee, J.H. & Byrne, R.H. 1993. Complexation of trivalent rare earth elements (Ce, Eu, Ga, Tb, Yb) by carbonate ions. *Geochim. Cosmochim. Acta* 57: 295–302.

Martynov, Y.A. 1999. Geochemistry of basalts of active continental margin and mature island arcs (North-West Pacific). Vladivostok, Dalnauka, 215 p.

Smith, J.V. 1974. Feldspar mineralogy, vol.2. Chemical and textural properties. Springer, Berlin Heidelberg New York.

Water–Rock Interaction, Wanty & Seal II (eds)
© 2004 Taylor & Francis Group, London, ISBN 90 5809 641 6

Chloride in hot springs of the Cascade volcanic arc – the source puzzle

M.C. van Soest
Center for Isotope Geochemistry, Lawrence Berkeley National Laboratory, Berkeley, CA, USA

W.C. Evans & R.H. Mariner
United States Geological Survey, Menlo Park, CA, USA

M.E. Schmidt
Department of Geosciences, Oregon State University, Corvallis, OR, USA

ABSTRACT: Relationships between Cl and ^3He concentrations, $^3He/^4He$ ratios and Br in Oregon Cascade volcanic arc hot and cold springs provide new evidence for the origin of the Cl in these springs. The relationship of Cl with ^3He indicates that a significant part of the Cl in the hot and cold springs can be derived from a magmatic source located beneath the crest of the volcanic arc. Evidence for a second Cl component in the hot springs is clearly present, but hard to distinguish given the resolution that the data provide. The most likely source for this component is Tertiary marine sediments that the thermal fluids encounter as they travel along long flow paths in deep aquifers to their discharge points away from the volcanic crest.

1 INTRODUCTION

Many geothermal systems in volcanic areas are characterized by high-chloride fluids, but the source of the chloride is often uncertain. Possible sources include (i) the magma (source) itself, (ii) Cl-rich crustal rocks (marine sediments) interacting with the magma, and (iii) Cl-rich aquifer rocks interacting with geothermal fluids. A further complication in the case of arc volcanoes is that the magmas can contain components derived from subducted (sedimentary) materials. Chloride isotopes do not provide a unique solution to this source question due to the very limited variation in chloride isotope ratios for the different sources. Therefore the problem of identifying chloride sources in volcanic areas has been addressed by (systematic) relationships with other (volatile) geochemical parameters (e.g. Giggenbach 1995).

2 OREGON CASCADE HOT SPRINGS

The Cl-rich hot springs in the Oregon Cascades clearly illustrate this problem. These springs discharge 10–40 km west of the arc crest, but are linked to recharge areas near the crest by stable isotopes and heat flow models (Ingebritsen et al. 1994). The hot springs all have temperatures $>40°C$ and contain magmatic helium, which indicates the presence of a magma-derived volatile component, but the discharge areas are likely underlain by Tertiary marine sediments that could supply part if not all of the chloride observed in these springs.

The hot springs are rich in N_2 and have total dissolved N_2/Ar ratios in excess of air saturated water (ASW – Mariner et al. 2003). Based on (i) $\delta^{15}N$ values up to +5.3‰, (ii) the presence of CH_4 with $\delta^{13}C$ values near –37‰, and (iii) the fact that the highest excess N_2 concentrations correlate with calculated aquifer temperatures near 120°C (the optimal temperature for N_2 release from sedimentary formations), Mariner et al. (2003) concluded that the most likely origin for the nitrogen was the Tertiary marine sediments. A correlation between Cl and N_2/Ar ratios in these springs (Fig. 1; Mariner et al. 2003) suggests that the source of the Cl may also be the Tertiary marine sediments that form or more likely underlie the aquifer.

However, our recent work on low-temperature springs near the crest of the arc in the Three Sisters area in the central Oregon Cascades provides some additional information on potential Cl sources.

3 THREE SISTERS AREA COLD SPRINGS

An episode of uplift, attributed to magmatic intrusion, began near the Three Sisters volcanoes in 1998

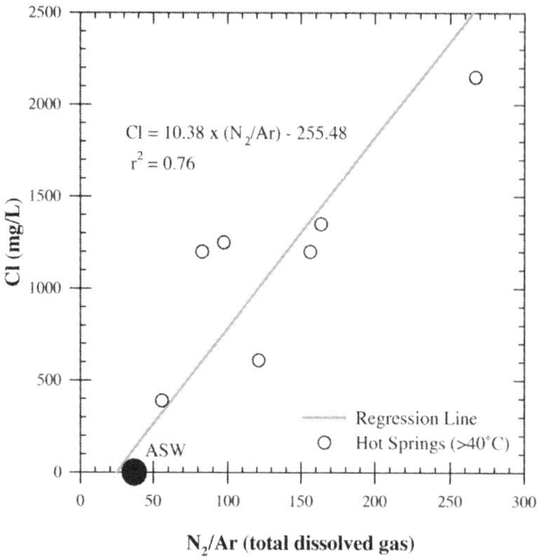

Figure 1. Cl vs. N_2/Ar for Oregon Cascade hot springs after Mariner et al. (2003).

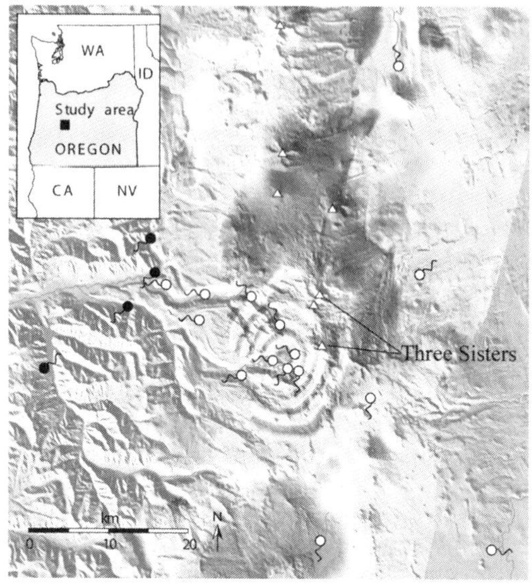

Figure 2. Map of the Three Sisters area with sample locations of cold springs (white symbols) and hot springs (black symbols). The area of uplift is indicated by the bulls-eye pattern. Several hot springs are located outside the area of this map. Their locations are given in Mariner et al. (2003).

(Wicks et al. 2002). To determine if this ongoing uplift has had any effects on the chemistry of local groundwater, water and gas samples have been collected from hot and cold springs in the Three Sisters area. The detailed results of this monitoring study can be found elsewhere (Evans et al. in press). The locations of these springs and their relation to the center of uplift and the Three Sisters are shown in Figure 2. As is characteristic for the Oregon Cascades the closest hot spring is over 25 km away from the volcanic edifice at the Cascade crest that forms the surface expression of the apparent heat source.

The cold springs that discharge in the area of the uplift all have temperatures <15°C but show slight and correlated anomalies in Cl concentration and temperature. Several have the highest $^3He/^4He$ ratios found anywhere in the Cascades (van Soest et al. 2002). These two observations suggest that there is a geothermal component in these springs and indicate that the source of the chloride could be magmatic. This notion is further reinforced by the fact that, in contrast to the hot springs to the west, these springs are located in a wide, arc-centered rift, where the presence of underlying marine sediments as an alternative source for the chloride is uncertain. In addition, Ingebritsen et al. (2002) have pointed out that the total discharge of Cl by these springs is compatible with the amount of Cl that could be supplied by magma at the long-term Cascade intrusion rate assuming a magmatic Cl concentration of 0.1 wt-%, implying that a sedimentary source is not necessary to explain the anomalous Cl discharge.

4 DIFFERENT CHLORIDE SOURCES?

The hot- and cold-spring characteristics each suggest a different chloride source, while the water, recharge derived from local precipitation (marked ASW in all figures), derives from the same source area in the high Cascades. To study this in more detail we combine our measurements of helium concentrations and isotopes in the cold springs with the hot-spring data from Mariner et al. (2003). Figure 3 shows that Cl concentrations in the cold springs correlate with ^3He concentration. Since the source of the ^3He is definitely magmatic, this supports the possibility of magma-derived chloride in the cold springs. The regression line was calculated for the cold-spring data and passes through the cluster of hot-spring data. If a magmatic Cl-source is invoked for the cold springs near the crest, which we think is most likely, the correlation line in Figure 3 indicates a magmatic Cl-source for at least part of the Cl budget at the hot springs. It is worth noting that both axes of Figure 3 are log scale, and therefore significant variations can be easily obscured. However, based on the regression line and the fact that all ^3He is magmatic, we can calculate that 54–114% of the Cl in the hot springs is magma-derived, with all but two samples falling in the range $100 \pm 20\%$. Given the large scatter in the data and the significant uncertainties associated with

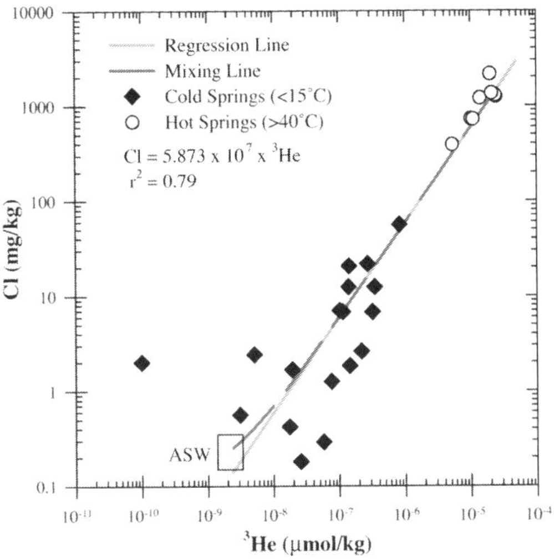

Figure 3. Cl vs. ^3He for Oregon Cascade hot springs and Three Sisters area cold springs.

the ^3He calculations these values should not be taken as absolute, but indicative that most hot springs carry a significant magma derived Cl component.

5 A SINGLE CHLORIDE SOURCE?

To better assess the implications derived from the observations made in Figure 3, we looked at Cl-$^3He/^4He$ and Br–Cl relations. The Br–Cl correlation for the cold springs is very tight ($R^2 > 0.99$ – Fig. 4 inset) and, when extrapolated to higher Cl and Br, passes through the hot-spring data (Fig. 4). This suggests that the halides in both fluid groups are derived from the same source and is consistent with the observations made from Figure 3. Although the slope of the regression line is not much different from the seawater ratio, the consistent offset of the data from the seawater trend would require some form of systematic elemental fractionation to occur between the time of their incorporation into the sediments and their discharge in the thermal fluids. Mariner et al. (1989) pointed out that Br is not as conservative as Cl in the thermal waters, so some form of fractionation process is possible. Similarly magmatic heating of marine sediments could result in differential halogen release, but this would require that the sediments occur in close proximity to the magma chamber. The presence of such sediments under the volcanic crest is uncertain, but the trend observed in our data could be indicative that they are there.

The Cl-$^3He/^4He$ plot shows an altogether different picture (Fig. 5). The cold-spring data trend is

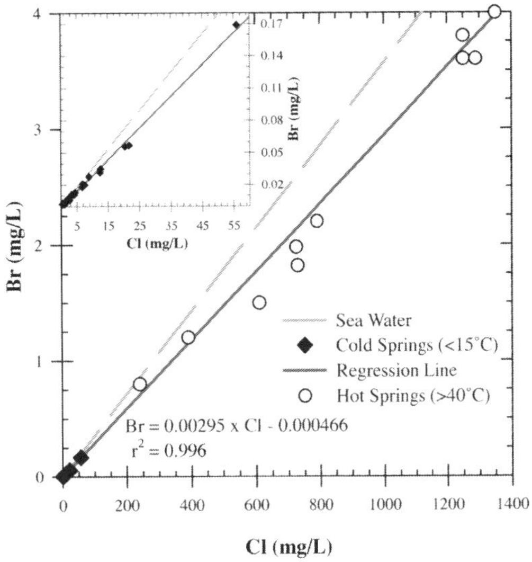

Figure 4. Br vs. Cl. The regression line is calculated through the cold-spring data. Details for the cold-spring data are shown in the inset. Both in this figure and Figure 5 there was more data available for the hot-spring group compared to Figure 3.

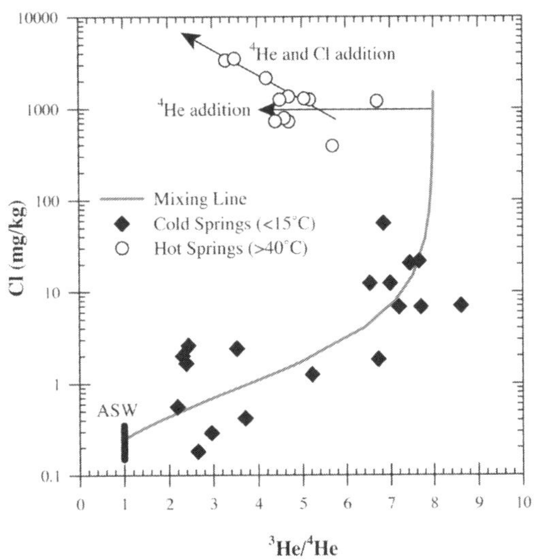

Figure 5. Cl vs. $^3He/^4He$, showing the mixing relationship for the cold springs and scenarios for the hot springs to evolve by interaction with aquifer rocks.

consistent with mixing a magmatic geothermal source with ASW and, therefore, also consistent with the Cl-^3He systematics, shown in Figure 3. The main difference in Figure 5 is that the hot-spring data no longer plot along a common mixing trend with the cold spring data. In Figure 3 the magmatic end-member has been chosen arbitrarily at the high end of the hot spring cluster for the Cl and ^3He concentrations. In Figure 5, however, the end-member is well defined by the upper mantle $^3He/^4He$ ratio of $8R_A$, consistent with the Cascades, in general. Differences in the hydrologic setting for the cold and hot springs are probably the main reason why the hot springs do not fall on the mixing line. The cold springs mainly reflect fairly shallow, local circulation of young groundwater that retains a clear but dilute magmatic signal in both chemistry and temperature. The hot springs are fed by ground water from the same recharge area that has infiltrated deeper and attained a stronger magmatic signal in both chemistry and temperature. The deep flow paths from recharge to discharge area probably shielded the fluids from interacting with large amounts of shallow ground waters of ASW composition and temperature.

The long and deep flow paths at high temperatures enhanced fluid interaction with the aquifer rocks, adding significant amounts of radiogenic ^4He to the fluids and lowering the $^3He/^4He$ ratio. Addition of ^4He along the flow path is consistent with a systematic decline of $^3He/^4He$ with distance from the volcanic crest (Mariner et al. 2002; Evans et al. in press) and is observed at many other volcanic systems in the world (e.g. Marty et al. 1989; van Soest et al. 1998). This is indicated in Figure 5 by the horizontal arrow extending from the mixing line towards lower $^3He/^4He$ ratios. The diagonal arrow shown in Figure 5 suggests that the water–rock interaction process that provides the radiogenic ^4He may also provide a significant amount of Cl. This may reflect leaching of the Tertiary marine sedimentary source (Mariner et al. 2003) and would be in concordance with the Cl-N_2/Ar correlation (Fig. 1).

The source for the magmatic Cl could be a combination of upper mantle and subduction derived components or crustal materials that interact with the magma as it moves through the crust. The high mantle like $^3He/^4He$ ratios suggests little or no addition of radiogenic ^4He as would be expected with the addition of crustal material. In combination with the uncertainty about the presence of sediments in the crust near the volcanic crest, a mantle/subduction origin for the Cl in the magma and the cold springs seems most likely.

6 CONCLUSIONS

Our work raises the strong possibility of a common magmatic Cl component as a significant source of elevated Cl in the hot and cold springs along the west side of the Cascade Range. However, we also show that the geothermal fluids that feed the hot springs interact with aquifer rocks along their deep flow paths from recharge to discharge and acquire radiogenic ^4He released from the rocks. There is good evidence that Cl is added to the thermal fluids in a similar manner. Given the resolution of our data and uncertainties about the Cl and He composition of each end-member it is difficult to establish the relative importance of each Cl source.

The most likely source for the Cl and radiogenic ^4He in the aquifers is Tertiary marine sediments that are estimated to both be a significant source for halides and radiogenic ^4He. Any variability in Cl source among the hot springs could be due to the distribution of Cl-bearing sediments along the flow paths of the fluids. On the other hand the source of Cl associated with the magma and cold springs is unclear, but based on the occurrence of mantle like helium ratios and uncertainty regarding the presence of marine sediments in the rift zone underlying the volcanic crest an upper mantle/subduction origin seems likely.

Our data also provide strong additional evidence for the hypothesis that the hot springs are sourced near the Cascade crest.

ACKNOWLEDGEMENTS

We would like to acknowledge B. Mack Kennedy, Steve E. Ingebritsen, and W.G. Darling for helpful reviews of this manuscript. Mark Huebner is acknowledged for performing the anion analyses. We would like to acknowledge UNOCAL for granting us permission to use hot spring helium isotope data contained in the report of their exploration effort in the Oregon Cascades. Pat Dobson is thanked for his assistance in obtaining access to the UNOCAL data. During this study MCvS was supported by the Assistant Secretary for Energy Efficiency and Renewable Energy, U.S. Department of Energy under contract number: DE-AC03-76SF00098.

REFERENCES

Evans, W.C., van Soest, M.C., Mariner, R.H., Hurwitz, S., Ingebritsen, S.E., Wicks, C.W. Jr. & Schmidt, M.E. in press. Magmatic intrusion west of Three Sisters, central Oregon, USA: The perspective from spring geochemistry. *Geology*.

Giggenbach, W.F. 1995. Composition of fluids in geothermal systems of the Taupo Volcanic Zone, New Zealand, as a function of source magma. In: Y.K. Kharaka & O.V. Chudaev (eds), *Water–Rock Interaction-8*: 9–12. Rotterdam: Balkema.

Ingebritsen, S.E., Mariner, R.H. & Sherod, D.R. 1994. Hydrothermal systems of the Cascade Range. *USGS Prof. Pap.* 1044-L: 86 p.

Ingebritsen, S.E., Evans, W.C., Mariner, R.H., Schmidt, M. & Hurwitz, S. 2002. Long- and short-term intrusion rates in the Three Sisters region, central Oregon Cascade Range. *GSA Abstr. with Prog.* 34, 6: 526–527.

Mariner, R.H., Presser, T.S., Evans, W.C., and Pringle, M.K.W. 1989. Discharge rates of thermal fluids in the Cascade Range of Oregon and Washington and their relationship to the geologic environment, in Muffler, L.J.P, Weaver, C.S., & Blackwell, D.D. (eds), *Proceedings of Workshop XLIV, Geological, Geophysical, and Tectonic Setting of the Cascade Range. USGS Open-File Report* 89–178: 663–694.

Mariner, R.H., Evans, W.C., Presser, T.S., & White, L.D. 2003. Excess nitrogen in selected thermal and mineral springs of the Cascade Range in northern California, Oregon, and Washington: Sedimentary or volcanic in origin? *J. Volcanol. Geotherm. Res.* 121: 99–114.

Marty, B., Jambon, A. & Sano, Y. 1989. Helium isotopes and CO_2 in volcanic gases of Japan. *Chem. Geology* 76: 25–40.

Wicks, C.W. Jr., Dzurisin, D., Ingebritsen, S., Thatcher, W., Lu, Z. & Iverson, J. 2002. Magmatic activity beneath the quiescent Three Sisters volcanic center, central Oregon Cascade Range, USA: *Geophys. Res. Lett.* 29(7): DOI: 10.1029/2001GL014205.

van Soest, M.C., Hilton, D.R. & Kreulen, R. 1998. Tracing crustal and slab contributions to arc magmatism in the Lesser Antilles island arc using helium and carbon relationships in geothermal fluids. *Geochim. Cosmochim. Acta* 62(19/20): 3323–3335.

van Soest, M.C., Kennedy, B.M., Evans, W.C. & Mariner, R.H. 2002. Mantle helium and carbon isotopes in Separation Creek geothermal springs, Three Sisters area, central Oregon: Evidence for renewed volcanic activity or a long term steady state system? *GRC Transactions* 26: 361–366.

Water–Rock Interaction, Wanty & Seal II (eds)

The magmato-hydrothermal system at Copahue volcano, Argentina

J.C. Varekamp
Earth & Environmental Sciences, Wesleyan University, Middletown CT

A.P. Ouimette
Volcano Hazards Division, U.S. Geological Survey, Menlo Park, CA

R. Kreulen
ISOLAB, Neerijnen, The Netherlands

ABSTRACT: Copahue Volcano (Argentina) has a hot acid crater lake and a series of acid hot springs. Chemical and isotopic data were collected from these waters from 1997–2003. The data indicate that both the lake and the springs are fed by a brine from a common, deeper reservoir. Acid fluids initially form at about 300°C in a deep zone above degassing magma and are later diluted with glacial meltwater to about 175°C. The acid fluids dissolve the surrounding rocks congruently. Mineral saturation calculations suggest that anhydrite and silica phases were the main secondary minerals that precipitated prior to the 2000 eruptions. In 2000, the major element concentrations in the hot spring fluids increased strongly, and alunite and clays became stable phases. We surmise that the intrusion and subsequent dissolution of chilled magma in the hydrothermal system saturated the fluids with new minerals.

1 INTRODUCTION

Copahue Volcano (37.85°S, 71.15°W, Province of Neuquen, Argentina) is about 3 km high and has an active volcano-hydrothermal system with a crater lake, hot springs and steam vents. The volcano had phreato-(magmatic?) eruptions in 1992 and 1995, during which large quantities of liquid sulfur were expulsed (Bermudez & Delpino 1995; Delpino & Bermudez 1993). The crater lake re-established itself after each explosive period and we started sampling the waters in the spring of 1997. The crater lake (~200 m diameter, 30–40 m deep, CPL samples) sits in an old explosion crater with walls of phreatic debris and glacier ice. The lake is sometimes deep green, sometimes milky white and turbulent, and temperatures ranged from 21 to 90°C. Local ambient temperatures are around freezing, indicating the input of a high-temperature fluid into the lake (Pasternack & Varekamp 1997). Some fine sulfur may float on the lake and glacial meltwater pours in from the sides. The water is acid (pH < 0.5) and the lake level varied by more than 10 m between sampling periods. Acid hot springs (CP samples), with temperatures of 63 to 83°C that differ strongly from that of the lake waters, emerge about 100 meters below the lake on the eastern mountain flank. The springs feed an acid river (Upper Rio Agrio, URA samples), which discharges into a large (~0.5 km^3) glacial lake (Lake Caviahue, pH ~ 2, CVL samples). The lake has a summer thermocline but is fully mixed during the winter season. The Lower Rio Agrio is the overflow of Lake Caviahue and this large stream leaves the Caviahue caldera through a V shaped opening and flows east to merge with the Rio Neuquen. The acid character of the stream is evident from red precipitates on river boulders more than 100 km from the source.

The north flank of the mountain has a geothermal system with a large thermal health spa (Figure 1). Many pools of green and dark coloured hot water are used for bathing, and the geothermal system has been drilled for heat extraction. The Las Machinas and Las Machinitas steam vents have acid condensation pools with multi-coloured algal colonies.

In July 2000, the volcano became active again with initial blasts ejecting lake sediment and hydrothermal debris, followed by magmatic eruptions, with ash dispersal up to 250 km from source. The 2000 magma was a basaltic andesite that showed some evidence of interaction with hydrothermally altered rocks (Goss 2001), and some of the ash was finely divided hydrothermal silica. Some altered boulders with

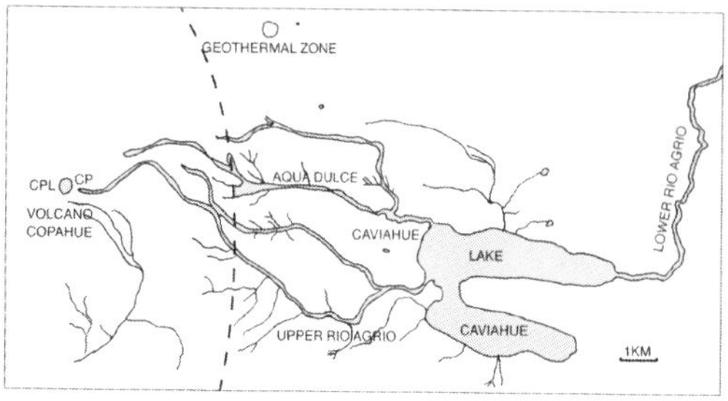

Figure 1. Hydrothermal surface features of Copahue volcano.

silica phases and large round blobs of native sulfur were ejected as well, which we interpret as fragments of the hydrothermal reservoir rock (Varekamp et al. 2001). During 2001 the crater lake restored itself while the springs had kept flowing during the eruptive period, and the system stabilized during 2002 and 2003.

We sampled and analyzed the waters to determine the character of this hyper-acid hydrothermal system, the ultimate origin of some of the gaseous components (e.g., Fehn et al. 2002), the processes of water–rock interaction inside the volcano (Varekamp et al. 2001), to assess its potential for ore metal transport (Delmelle & Bernard 2000), and to study relations between volcanic activity and water composition (e.g., Rowe et al. 1992a; Christenson 2000).

2 SAMPLING AND ANALYSIS

We analyzed waters for the volcanic elements (Cl, S) by ion chromatography and for major elements by ICP-AES. Minor and trace elements were analyzed on a subset of samples by ICP-MS. The pH values above 1 were measured with a digital pH meter; lower pH values were calculated based on charge balance, using speciation calculations at assumed pe values (Reed & Spycher 1984; Ouimette 2000). Stable isotope analyses of sulfur species were carried out at the University of Waterloo, Canada and for waters at ISOLAB. We also determined the Upper Rio Agrio water fluxes from measured cross sections and water flow rates, and with water analyses calculated element and heat fluxes. Lake Caviahue ($\sim$100 m deep) was sampled at 10 m depth intervals.

3 RESULTS

The crater lake and hot springs carry extremely concentrated acid Cl-SO_4 brines and are comparable in chemistry to the most concentrated crater lakes described worldwide (e.g., Rowe et al. 1992b; Varekamp et al. 2000; Delmelle & Bernard 2000). The CPL and CP are rich in both volcanogenic elements as well as in the rock-forming elements (Al, Fe, Mg, Ca, Na, K, Si – RFE) with total dissolved solids reaching over 90,000 ppm in some samples. Chemical data for CPL and CP waters in 1997–1999 show slightly variable concentrations (Table 1), with higher concentrations of the RFE in the hot springs relative to the lake. The Cl/S values were relatively stable in 1997–2000, but very different from S/Cl in glass inclusions of the Copahue magmas (Goss 2001). The S/Cl ratios declined strongly after the eruption. Lake and hot spring waters have calculated pH's approaching 0.3, and the Upper Rio Agrio and Lake Caviahue have pH values of approximately 1 and 2, respectively. The differences in temperature, RFE concentrations and $\delta^{18}O$ (see below) between the springs and crater lake suggest that the springs are not simple lake seepage. The springs tap the top of the deep and hot hydrothermal system, while the crater lake fluids are modified by meltwater input and evaporation. Element ratios in the lake and spring fluids are very similar to those in Copahue rocks, suggesting close to congruent rock dissolution. Detailed water profiles of Lake Caviahue show that the lake is chemically homogeneous, but thermally stratified in summer (Varekamp et al. 2001; Varekamp 2003). River water concentrations vary with the seasons, and the URA samples have chemical influences from multiple tributary inputs and dilution (Ouimette 2000).

Element ratios are not sensitive to dilution with melt water and are good indicators of changing water rock interaction during the eruptive period. The Mg/Cl values in the Upper Rio Agrio decreased between 1997 and 1999, and then rapidly increased during the eruptive period. The Mg/Cl values stayed well above pre-eruptive levels after the eruptions.

Table 1. Chemical data from the Copahue hot springs (CP samples) from 1997–2003 (in ppm, Mg/Cl × 100).

Sample	3/97	1/99	6/00	2/03
pH	0.4	0.3	0.4	0.6
T°C	68	83	70	58
Cl	10600	10900	8700	5700
SO_4	64300	65500	49300	11800
Na	940	890	1370	1220
Mg	640	520	1880	1590
Al	3870	3230	5140	2510
K	820	820	690	280
Ca	1100	820	670	1130
Fe	2190	2120	3380	2230
Mg/Cl	6.1	4.8	21.6	28
S/Cl	2.0	2.0	1.9	0.7

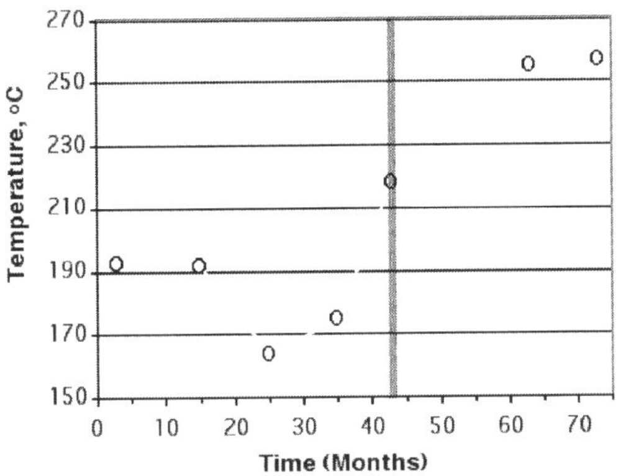

Figure 2. Quartz temperatures of the Copahue hot spring fluids showing the temperature increase with the onset of the 2000 eruption (Varekamp et al., 2003). Time axis: January 1997 = 0, eruption started at 43 months (dashed line).

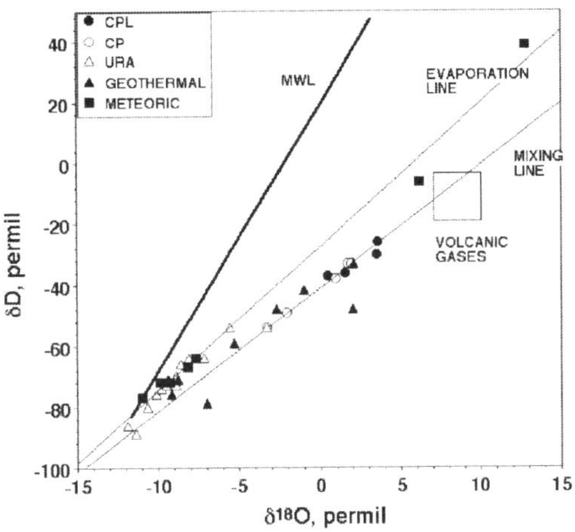

Figure 3. Stable isotope data of Copahue fluids. Evaporation line is calculated after Varekamp & Kreulen (2000).

3.1 *Mineral saturation calculations and thermometry*

The deep hydrothermal fluids are presumably saturated with elemental sulfur, given the pyroclastic sulfur eruptions of the past. We modeled the speciation and saturation conditions of the pre-eruptive fluids (1997–1999) with the program SOLVEQ (Reed & Spycher 1984), using the $HSO_4^- - S^0$ boundary to define the pe values in the system (Ouimette 2000). These calculations indicated that the fluids were saturated with anhydrite, several silica phases and a suite of Cu-sulfides. This is a mineral assemblage not unlike that found in high-sulfidation Cu deposits.

Silica thermometry of the CP fluids showed temperatures of fluid in equilibrium with quartz (Verma 2000) of about 175–185°C. During the eruptive period these values were higher and stayed around 250°C in the next few years (Fig. 2). The mineral saturation state of the fluids became much more complex during and after the eruption, with a multitude of minerals saturated in 2000 and 2002/2003.

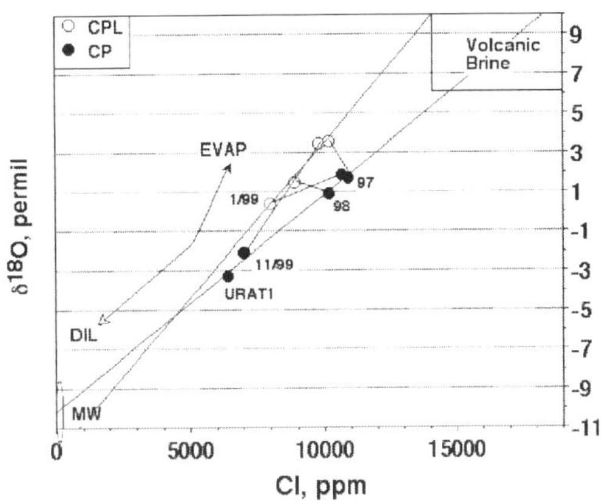

Figure 4. $\delta^{18}O$-Cl mixing diagram for the Copahue hot springs and crater lake. The crater lake data points plot above those of the springs as a result of evaporation processes in the lake. Approximate evaporation and dilution vectors are shown.

3.2 *Stable isotope studies*

The stable isotope measurements for a suite of water samples from Caviahue-Copahue, when plotted on a δD-$\delta^{18}O$ diagram, show the local meteoric water endmember, the effects of evaporation on surface waters (evaporation line in Fig. 3) and a suite of acidic and concentrated samples that are enriched in heavy isotopes.

The local meteoric waters evolve along an evaporation line that is very similar to a mixing line between arc volcanic gases (Giggenbach 1992) and local meteoric waters. As a result, the very dilute warm geothermal pools plot in a similar array as the crater lake fluids and hot springs. A $\delta^{18}O$ versus Cl diagram (Fig. 4) shows that the CP fluids plot on a mixing line between a magmatic endmember (1.5–2.0% Cl, $\delta^{18}O \sim +6$ to $+10$ *per mil*) and meteoric waters, whereas the lake waters plot above the line as a result of in-lake evaporation processes (as indicated by the approximate vectors in Fig. 4). The geothermal fluids plot in a cluster near the meteoric waters. The hot spring fluids, according to this diagram, represent up to 70% magmatic water during the most concentrated periods (no water isotope data of the later fluids were available at this time).

The sulfur isotope data on the HSO_4 and native sulfur ejecta from the 1995 eruptions, as well as sulfur froth from the lake, show heavy sulfate ($\sim +14$ *per mil*) and light elemental sulfur (–11 to –7 *per mil*). If we assume that these phases were in equilibrium in the hydrothermal reservoir, we obtain a temperature of isotopic equilibration of about 300°C.

4 CONCLUSIONS

The Copahue hydrothermal fluids are samples from a deep volcano-hydrothermal system that is fed by magmatic gases rich in SO_2. The hydrothermal system serves as an efficient scrubber for the volcanic gases (Symonds et al. 2001). We assume that the SO_2 disproportionated into elemental sulfur and HSO_4^- at about 300°C and that the top of the system is diluted with glacial melt water (about 30%) to a temperature of 180°C. This system was disturbed by an intrusion of magma in early-mid 2000, leading to higher temperatures in the top of the system, increased RFE element concentrations, and saturation with a multitude of minerals, which include alunite. The Mg/Cl ratios in the system increased several fold during the intrusive event, and the Mg/Cl ratio in the river may be a good precursor to monitor for future eruptions. Copahue represents a rare example of the occurrence of hydrothermal fluids on the surface of the earth from a deep source with a strong magmatic signal.

ACKNOWLEDGEMENTS

Funding for this study was provided by NSF (INT- 9813912 and INT-9704200) and the National Geographic Society (grant 7409-03). We thank the Wesleyan University students Scott Herman, Conor Gately, and Kathryn Flynn, and our Argentinan colleagues Daniel Delpino and Adriana Bermudez for their contributions to this project.

REFERENCES

Bermudez, A. & Delpino, D. 1995. Mapa de los peligros potenciales en el área delVolcán Copahue – sector Argentino. Volcanic Hazard map published by: the Geological Survey of the Province of Neuquen, Argentina.

Christenson, B.W. 2000. Geochemistry of fluids associated with the 1995–1996 eruption of Mt. Ruapehu, New Zealand: signatures and processes in the magmatic hydrothermal system. *J. Volc. Geotherm. Res.* 97: 1–30.

Delmelle, P. & Bernard, A. 2000. *Volcanic lakes. Encyclopedia of Volcanology*, Academic press, p. 877–896.

Delpino, D. & Bermudez, A. 1993. La actividad del Volcan Copahue durante 1992. Erupcion con emisiones de azufre piroclastico. Provencia del Neuquen-Argentina: XII Congreso Geologico Argentino y II Congreso de Exploracion de Hidrocarburos Actas T IV p. 292 – 301.

Fehn, U., Snijder, G. & Varekamp, J.C. 2002. Detection of recycled marine sedimentcomponents in crater lake fluids using ^{129}I: *J. Volc. Geotherm. Res.* 115: 451–460.

Giggenbach, W.F. 1992. Isotopic shifts in waters from geothermal and volcanic systems along convergent plate boundaries and their origin. *Earth Plan. Sci. Let.* 113: 495–510.

Goss, A.R. 2001. Magmatic evolution of Volcan Copahue, Neuquen, Argentina. Undergraduate senior thesis, Wesleyan University, Middletown CT USA, 209 p.

Ouimette, A.P. 2000. Hydrothermal processes at an active volcano, Copahue, Argentina. MA thesis, Wesleyan University, Middletown CT pp.220.

Pasternack, G.B. & Varekamp, J.C. 1997. Volcanic lake systematics I. Physical constraints: *Bull. Volcanol.* 58: 528–538.

Reed, M.H. & Spycher, N.F. 1984. Calculation of pH and mineral equilibria in hydrothermal waters with application to geothermometry and studies of boiling and dilution: *Geochim. Cosmochim. Acta* 48: p.1479–1492.

Rowe, G.L., Jr, Ohsawa, S., Takano, B., Brantley, S.L., Fernandez, J.F. & Barquero, J. 1992a. Using crater lake chemistry to predict volcanic activity at Poás volcano, Costa Rica: *Bull. Volcanol.* 54: 494–503.

Rowe, G.L., Jr, Brantley, S.L., Fernandez, M., Fernandez, J.F., Borgia, A. & Barquero, J. 1992b. Fluid-volcano interaction in an active stratovolcano: the crater lake system of Poas volcano, Costa Rica: *J. Volcanol. Geotherm. Res.* 49: 23–51.

Symonds, R.B., Gerlach, T.M. & Reed, M.H. 2001. Magmatic gas scrubbing: implications for volcano monitoring. *J. Volcanol. Geotherm. Res.* 108: 303–341.

Varekamp, J.C. 2003. Lake contamination models for evolution towards steady state. *J. Limnol.* (in press).

Varekamp, J.C. & Kreulen, R. 2000. The stable isotope geochemistry of volcanic lakes: examples from Indonesia. *J.Volcanol. Geotherm. Res.* 97: 309–327.

Varekamp, J.C., Pasternack, G.B. & Rowe, G.L. 2000. Volcanic lake systematics. II. Chemical constraints: *J. Volcanol. Geotherm. Res.* 97: 161–179

Varekamp, J.C., Ouimette, A.P., Herman, S.W., Bermudez, A. & Delpino, D. 2001. Hydrothermal element fluxes from Copahue, Argentina: a "beehive" volcano in turmoil. *Geol.* 29: 1059–1062.

Varekamp, J.C., Flynn, K., Ouimette, A.P., Herman, S., Delpino, D. & Bermudez, A. 2003. Compositional changes in hydrothermal fluids from Copahue volcano, Argentina, during the 2000 eruptions. *Pageophs* (in review)

Verma, M.P. 2000. Revised quartz solubility temperature dependence equation along the water vapor saturation curve. *Proceedings World Geothermal Congress 2000*, Kyushu - Tohoku, Japan, 1927–1932.

Water–Rock Interaction, Wanty & Seal II (eds)
© 2004 Taylor & Francis Group, London, ISBN 90 5809 641 6

Measurement of trace gas species at White Island volcano, New Zealand, using chemical traps

L.J. Wardell
Department of Earth & Planetary Sciences, McGill University, Montreal, Canada

P.R. Kyle
New Mexico Institute of Mining and Technology, Socorro, NM, USA

D. Counce
Los Alamos National Laboratories, Los Alamos, NM, USA

ABSTRACT: Measurements of trace element emissions from volcanoes increases the understanding of magmatic systems, effects of volcanism on atmospheric chemistry, as well as the role of magma degassing in ore genesis. Volcanic emission rates of Cl, F, Pb, As, Hg Sb and Se were determined at White Island, New Zealand, using chemical traps containing 4 N NaOH solutions. Traps were located in both the fumarole area and at the crater rim where they would be exposed to the main volcanic gas plume. The chemical ratios relative to sulfur of the fumarole samples were distinctly different from the ratios of samples exposed to the gas plume. The trace metal fluxes were determined using the species to S ratios in the solutions combined with independently determined SO_2 emission rates measured by correlation spectrometer. At White Island, chemical traps yielded flux values of $Cl = 84$, $F = 0.8\,Mg \cdot d^{-1}$(tonnes per day) and $Pb = 28$, $As = 17$, $Hg = 0.84$, $Sb = 1.8$ and $Se = 1.3\,kg \cdot d^{-1}$.

1 INTRODUCTION

New data on trace element emissions increases an understanding of magmatic systems, effects on atmospheric chemistry, as well as the role of magma degassing in ore genesis (Symonds et al. 1992). A large variety of trace elements (approximately 45 species) have been detected in volcanic emissions; however, for most species, correlation between changes in volcanic activity and trace element emissions is relatively rare (Mambo & Yoshida 1993). Since uncertainty still exists concerning the sources of trace metals in the atmosphere, there is a need to understand volcanic trace metal species and the magnitude of their contribution to the global atmosphere (Hinkley et al. 1997).

The intent of this study is to report on the use of chemical traps to sample volcanic emissions at White Island, New Zealand with the prime objectives of determining the flux rates of Cl, F, As, Hg, Pb, Sb, and Se and to compare trace element results for the main gas plume to emissions from a fumarolic area. A sample approximately 500 m from the main vent was also collected for comparison to samples collected close to the vent.

White Island volcano, New Zealand, is an andesitic composite volcano located off the coast of the North Island at the northeastern end of the Taupo Volcanic Zone. It has undergone many eruptive cycles since 1976 (Houghton & Narin 1989) and is New Zealand's most continuously active volcano. The activity of White Island was steady during the sampling period and was at the lower designated Volcanic Hazard Alert Level (IGNS 2002). The main gas plume emanated from the inner crater where several vents appeared above the crater lake. Emissions of F, Cl, S and other gaseous species have been previously measured (Giggenbach 1976; Giggenbach 1982; Giggenbach 1983; Giggenbach & Matsuo 1991; Giggenbach & Sheppard 1989; Le Cloarec et al. 1992; Love et al. 1999; Marty & Giggenbach 1990; Rose et al. 1986; Symonds et al. 1988; Tedesco & Toutain 1991; Wardell et al. 2001). However, there are few reports of metal emission rates. The majority of gas geochemistry studies at White Island are based on measurements from fumaroles, with the exception of SO_2 measurements by COSPEC (Rose et al. 1986; Wardell et al. 2001), airborne CO_2 measurements (Wardell et al. 2001), and SO_2, H_2O, and HCl flux measurements by FTIR (Love et al. 1999).

2 METHODS

At White Island, 300 ml of 4 molar NaOH solution was placed in 500 ml Nalgene jars with an opening and solution exposure surface of 52.8 cm^2. To hinder particulate matter from entering the caustic solution, the reaction vessels were placed in plastic buckets with lids (approximately 2 liters in size). The buckets were perforated with 0.5 cm holes to allow the volcanic gases to flow through the trap system and reach the reaction vessel sitting inside the bucket. The space between the bucket and the reaction vessel was filled with glass wool to further filter out particles. The traps were fastened to wood stakes and positioned approximately 0.5 to 1 m above the soil surface. Clean NaOH solution in closed jars traveled with the samples as blanks from the volcano to the laboratory to monitor for contamination.

In March 2001, traps were placed in triplicate at three locations around the same active crater as shown in Figure 1 (samples 1, 2, and 3). Other samples included two single traps located in areas of fumaroles (samples 5 and 6) and one trap situated on top of a mound, approximately 500 m from the active vent (sample 7). Fumarole areas existed along both sides of the crater wall. The area sampled consisted of low temperature fumaroles with a large amount of sulfur precipitation on the soil surface as well as some elemental sulfur fumarole structures. Gases were sampled by traps over a one-week period.

Analytical procedures were similar to those used by Goff et al. (1998) at Popocatèpetl. All alkaline solutions and blanks were analyzed at Los Alamos National Laboratory. Solutions were analyzed for S, Cl, F, Br, As, Sb, Se, Hg, and Pb. Ion chromatography (Dionex Model 4000i) was used for S, Cl, F, and Br analytes with all S species being converted to $SO_4^=$ prior to analyses. Bromine was not observed in any of the samples. Hydride atomic absorption (Perkin Elmer 5500 Atomic Absorption Spectrometer) was employed for the analyses of As, Sb and Se in the samples. Mercury was determined using the cold vapor atomic absorption technique (Perkin Elmer 5500 Atomic Absorption Spectrometer).

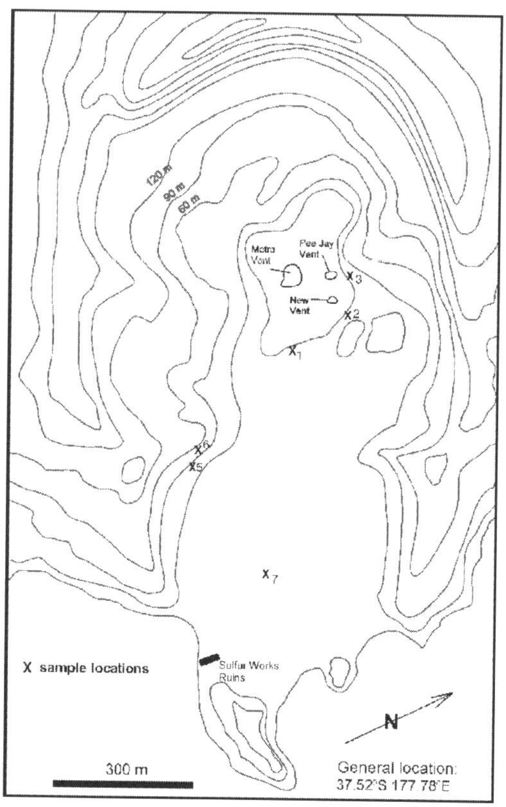

Figure 1. Sample locations at White Island volcano, New Zealand. The majority of samples are located along the edge of the inner crater that contains several active vents. The major portion of the volcanic plume originated from the Pee Jay vent during 2001.

3 RESULTS

Results for all samples in Table 1 are reported in ratios relative to S concentration on a weight basis. The ratios were scaled by the SO_2 flux average (1986 to 1999; Wardell, et al. 2001) to derive the trace species fluxes in Table 2. The chemistry of the two sets of triplicate samples exposed to the Pee Jay vent emissions (samples 1 and 3) agreed closely with each other and were averaged to produce the plume ratio values (Table 1). These plume ratios were used to

Table 1. Chemical Ratios from Trap Solutions, White Island, New Zealand.

Weight basis	Plume	Inner crater fumarole	Fumarole area	Down-wind sample
Cl/S	0.39	0.17	.16	–
F/S	0.0039	0.0013	1.4×10^{-4}	.0036
As/S	8.1×10^{-5}	2.3×10^{-5}	6.4×10^{-6}	9.6×10^{-5}
Hg/S	3.9×10^{-6}	9.9×10^{-7}	–	6.0×10^{-6}
Pb/S	1.3×10^{-4}	4.6×10^{-5}	1.0×10^{-5}	2.1×10^{-4}
Sb/S	8.3×10^{-6}	2.9×10^{-6}	2.1×10^{-6}	1.2×10^{-5}
Se/S	6.2×10^{-6}	2.9×10^{-6}	1.4×10^{-6}	6.0×10^{-6}

calculate the flux in Table 2 for White Island. Similarly, the two single samples placed by the fumarolic area (samples 5 and 6) agreed closely and were averaged together to represent the fumarole emissions. In addition, the middle sample located along the inner crater (sample 2) showed lower ratio values relative to the two other locations at the inner crater (samples 1 and 3). Results for sample 2 are more closely related to ratios for the two samples in the fumarole area near the crater wall. This sample therefore was not averaged with samples 1 and 3 and is presented as a separate fumarole area in Table 1. The chemistry of the single downwind sample (sample 7) agrees well with the crater rim plume samples for most species (Table 1).

4 DISCUSSION

Since the majority of the plume originated at the Pee Jay vent, the triplicate sample closest to this vent (sample 3) should best reflect that of the volcanic gas plume. It was expected that the chemistry of all three inner crater samples would correlate. However, the middle sample (sample 2) showed low ratio values relative to the other two crater samples. This sample location was selected as it seemed to be exposed to a large amount of gas. Results from this sample are more closely related to ratios derived from the two samples in the fumarole area near the crater wall (samples 5 and 6). It is therefore likely that the emissions at this location mainly originated from fumarolic activity from the inner crater wall and not the larger vents below. These observations indicate that sampling locations around the inner crater must be chosen carefully when trying to obtain samples from the plume.

Ratios from samples at the fumarole area along the main crater wall (Figure 1) are lower than the average value of the crater rim samples. A higher concentration of S in the caustic solution of the fumarole samples resulted in the lower ratios. It is possible that the elemental sulfur deposits in the fumarole area could have contaminated the trap solutions causing the higher sulfur concentrations and therefore, lower ratio values. However, since the fumarole area samples correlate with the set of samples taken at the inner crater, these results may reflect the actual emissions.

Table 2. Flux Values for White Island, New Zealand.

		This study	Previous work
Cl	Mg/d	84	70–340(HCl)[2,4,5]
F	Mg/d	0.80	<1–5(HF)[2,4,5]
As	Kg/d	17	14[4]
Hg	Kg/d	0.84	
Pb	Kg/d	28	9.3, 125[3,4]
Sb	Kg/d	1.8	4[4]
Se	Kg/d	1.3	

1. Wardell et al. 2001; 2. Love et al. 1999; 3. Le Cloarec et al. 1992; 4. Tedesco & Toutain 1991; and 5. Rose et al. 1986.

The single downwind sample agrees well with the crater rim plume samples for most species. This is particularly encouraging as it was only a single sample and being located downwind, was exposed to a more dilute plume. The caustic solution had a lower concentration of absorbed species but still maintained a close ratio with the crater rim plume samples, showing that plume sampling downwind is a possibility when going close to the inner crater is not desirable. An exposure time of one week was sufficient to collect detectible concentrations in the reported trace elements, except for F, which was below detection in the downwind sample.

Previous measurements of trace element fluxes (Table 2) vary, likely due to variability in sampling time and method. Results from this study tend to agree with the lower flux vales that were previously reported. Although it is a crude method, chemical traps are an alternative means of volcanic monitoring when faced with hindrances concerning costs, equipment availability or safe access to degassing areas. Chemical traps are a way to assess multi-component variations in volcanic plume fluxes when direct sampling is not possible (Goff et al. 1998).

5 CONCLUSIONS

The chemical trap method is a simple technique for monitoring trace species at active volcanoes. By comparing results to data acquired by other methods, it was shown that this method, although crude, yields comparable results. Short exposure times (1 week) can be sufficient and likely less time is needed for the more abundant chemical species. At White Island, chemical traps were able to distinguish between fumarolic emissions and vent emissions. Also, care is needed in locating sample sites around the inner crater to distinguish between plume and wall fumarole emissions. The sample placed downwind of the active crater gave similar results to the crater rim samples demonstrating that it is not necessary to place sample containers near the vent when it is not practical or safe.

ACKNOWLEDGEMENTS

This work was supported by grants from the NSF Office of Polar Programs and fellowship support from the DOE Global Change Education Program. A special thanks goes to IGNS-Wairakei Research Centre, Volcan Helicopters, and White Island Tours for their kind assistance.

REFERENCES

Giggenbach, W.F. 1976. Variation in the carbon, sulfur and chlorine contents of volcanic gas discharges from White Island, New Zealand. *Bull. Volcanol.* 39: 15–27.

Giggenbach, W.F. 1982. The chemical and isotopic composition of gas discharges from New Zealand andesitic volcanoes. *Bull. Volcanol.* 45(3): 253–255.

Giggenbach, W.F. 1983. Chemical surveillance of active volcanoes in New Zealand. In: H. Tazieff and J.C. Sabroux (Editors), *Forecasting Volcanic Events*. Elsevier, Amsterdam, pp. 311–322.

Giggenbach, W.F. & Matsuo, S. 1991. Evaluation of results from Second and Third IAVCEI Field Workshops on Volcanic Gases, Mt Usu, Japan and White Island, New Zealand. *App. Geochem.* 6: 125–141.

Giggenbach, W.F. & Sheppard, D.S. 1989. Variations in the temperature and chemistry of White Island fumarole discharges 1972–1985. In: B.F. Houghton and I.A. Nairn (Editors), *The 1976–82 eruption sequence at White Island volcano (Whakaari), Bay of Plenty, New Zealand. New Zealand Geological Survey*, Rotorua, 119–126.

Goff, F., Janik, C.J., Delgado, H., Werner, C., Counce, D., Stimac, J.A., Siebe, C., Love, S.P., Williams, S.N., Fischer, T. & Johnson, L. 1998. Geochemical surveillance of magmatic volatiles at Popocatepetl volcano; Mexico. *Geol. Soc. Am. Bull.* 110(6): 695–710.

Hinkley, T., Wilson, S.J., Lamothe, P.J., Landis, G.P., Finnegan, D.L., Gerlach, T.M. & Thornber, C.R. 1997. Metal emissions from Kilauea–proportions, source strength, and contribution to current estimates of volcanic injection to the atmosphere. *EOS, Transactions*, 79(45, Suppl.): 803–804.

Houghton, B.F. & Narin, I.A. 1989. A model for the 1976–82 phreatomagmatic and strombolian eruption sequence at White Island volcano, New Zealand. In: B.F. Houghton and I.A. Narin (Editors), *The 1976–82 eruption sequence at White Island volcano (Whakaari), Bay of Plenty, New Zealand. New Zealand Geological Survey*, Rotorua, pp. 127–137.

IGNS 2002. Current alert status of NZ's volcanoes. Institute of Geological and Nuclear Sciences. http://www.gns.cri.nz/earthact/volcanoes/status.html.

Le Cloarec, M.F., Allard, P., Ardouin, B., Giggenbach, W.F. & Sheppard, D.S. 1992. Radioactive isotopes and trace elements in gaseous emissions form White Island, New Zealand. *Earth Plan. Sci. Lett.* 108: 19–28.

Love, S., Schmidt, S.C., Goff, F., Counce, D., Pettit, D., Christenson, B.W. & Siebe, C. 1999. Passive infrared spectroscopic remote sensing of volcanic gases: Ground-based studies at White Island and Ruapehu, New Zealand, and Popocateptl, Mexico. In: *Remote Sensing of Active Volcanism, Geophysical Monograph* 116: 117–138.

Mambo, V.S. & Yoshida, M. 1993. Behavior of Arsenic in Volcanic Gases. *Geochem. J.* 27(4–5): 351–359.

Marty, B. & Giggenbach, W.F. 1990. Major and rare gases at White Island volcano, New Zealand: origin and flux of volatiles. *Geophys. Res. Lett.* 17(3): 247–250.

Rose, W., Chuan, R., Giggenbach, W., Kyle, P. & Symonds, R. 1986. Rates of sulfur dioxide and particle emissions from White Island volcano, New Zealand, and an estimate of the total flux of major gas species. *Bull. Volcanol.* 48: 181–188.

Symonds, R., Reed, M. & Rose, W. 1992. Origin, speciation, and fluxes of trace-element gases at Augustine volcano, Alaska: Insights into magma degassing and fumarolic processes. *Geochim. Cosmochim. Acta.* 56: 633–657.

Symonds, R.B., Rose, W.I. & Reed, M.H. 1988. Contribution of Cl- and F-bearing gases to the atmosphere by volcanoes. *Nature* 334: 415–418.

Tedesco, D. & Toutain, J. 1991. Chemistry and emission rate of volatiles from White Island volcano (New Zealand). *Geophys. Res. Lett.* 18(1): 113–116.

Wardell, L., Kyle, P., Dunbar, N. & Christenson, B. 2001. White Island, New Zealand: Carbon dioxide and sulfur dioxide emissions rates and melt inclusion studies. *Chem. Geol.* 177: 187–200.

Water–Rock Interaction, Wanty & Seal II (eds)
© 2004 Taylor & Francis Group, London, ISBN 90 5809 641 6

Monitoring CO_2 emissions at White Island volcano, New Zealand: evidence for total decreases in magmatic mass and heat output

C. Werner, B.W. Christenson, B.J. Scott, K. Britten & G. Kilgour
Institute of Geological and Nuclear Sciences, New Zealand

ABSTRACT: CO_2 and SO_2 emissions at White Island volcano in March 2003 (940 and 225 td^{-1}, respectively) were similar to previous estimates of long-term (20-yr) emission rates, but decreased significantly following abrupt decreases in 2–5 Hz volcanic tremor. Decreasing plume emission rates were coincident with the unprecedented formation of a crater lake. Such changes point to an overall decrease in magmatic mass and heat output from the volcano over a period of five months. Chemical modeling of crater lake waters suggests that decreases in permeability of the conduit was likely due to precipitation of sulfur in the vent region. Crater floor emissions of CO_2 (99 td^{-1}) were estimated to be >11% of the total CO_2 output from the volcano in March–April, 2003 and were found to be spatially consistent with IR imaging from previous studies.

1 INTRODUCTION

White Island is an andesitic composite volcano located ~50 km off the north coast of New Zealand. White Island occupies ~3.3 km^2 (~2×1.5 km) and the maximum elevation is 321 m above sea level. The main crater (1.2×0.4 km) is a complex of three volcanic subcraters oriented NW-SE at the southwestern corner of the island (Houghton & Nairn 1989). Over the last 28 years the volcano has experienced a number of eruptions from the westernmost subcrater including $10^6 m^3$ of andesite magma (1976–82), several smaller phreatic and phreato-magmatic eruptions between 1986 and 1991 (Nishi et al. 1996), and again in 2000. Magmas feeding these eruptions are thought to originate from shallow magma chambers located ~ 2 to 7 km beneath the central portion of the volcano (Houghton & Nairn 1989).

White Island hosts an active hydrothermal system within the main crater that is isolated from the sea through a chemically sealed zone around the volcano (Giggenbach 1987). Surface expressions of the hydrothermal system include steaming ground areas, small hot springs, and fumaroles around the main crater edges (100–150°C since 2001). The majority of volcanic-tectonic earthquakes at White Island occur at depths <1 km in these regions and are thought to be due to rapid changes in pore fluid pressure (Nishi et al. 1996).

White Island has a fairly long history of emissions measurements. Based on airborne correlation spectrometry (COSPEC) and fumarole chemistry from direct sampling, long-term emission rates for SO_2 and CO_2 were estimated at ~ 350 and 900–1000 td^{-1}, respectively (Rose et al. 1986). From 1986–99 the SO_2 emission rate was measured 17 times and varied between 171 and 900 td^{-1}, averaging 470 ± 70 td^{-1} (Wardell et al. 2001). The CO_2 emission rate was measured directly in January 1998 at ~ 2600 td^{-1}, and the CO_2 emissions from the crater floor were estimated to be minimal (~ 9 td^{-1}, n = 8; Wardell et al. 2001). Since the 2000 eruption the SO_2 emissions have declined steadily back to long-term levels. However, in 2003, dramatic changes occurred in the state of White Island including a reduction in 2–5 Hz volcanic tremor recorded at the volcano and the formation of an unprecedented crater lake. This paper presents three airborne measurements of the SO_2 and CO_2 emissions made over 5 months in 2003 and a CO_2 flux survey of the crater floor in context with these historically significant changes.

2 METHODS

2.1 *Aerial plume measurements*

SO_2 and CO_2 emission rates were on 4 March, 13 May, and 3 July, 2003 (Table 1). SO_2 emission rates were measured using a COSPEC mounted inside an aircraft with the range of view pointing toward the sky. On each flight 5–10 passes were completed beneath the plume at a height of 60 m above the sea surface normal to the wind direction.

Table 1. CO_2 and SO_2 emissions at White Island.

Date	CO_2 t/d cm	SO_2 t/d	Wind m/s
January 1998*	2600	225	
4 March 2003	842 ± 60	225 ± 67	9 ± 1
13 May 2003	338 ± 40	<8	10 ± 1
17 July 2003	275 ± 44	nd	7 ± 0.7

Wardell et al. 2001.

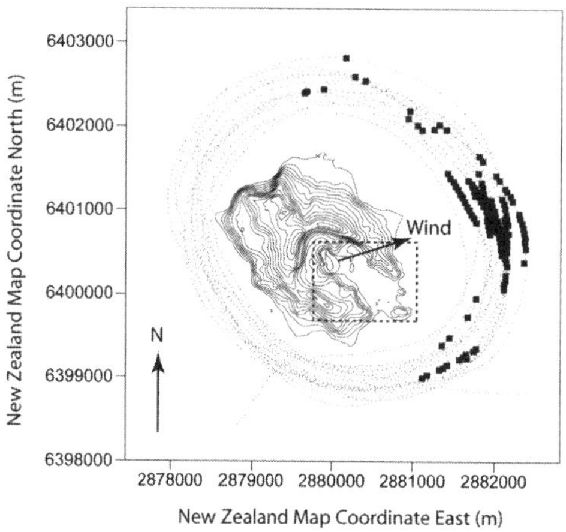

Figure 1. Flight path of CO_2 measurement around White Island on March 4, 2003. Similar flight paths were used for the March and July, 2003 measurements. The arrow indicates the wind direction on March 4, and the dark points on the flight path indicate positions where CO_2 concentrations in the air exceeded 352 ppm. The square shows the region of the main crater at White Island and the area is shown in Figure 2.

CO_2 emission rates were measured using a LICOR LI–7000 infrared gas analyzer and GPS logging once per second. The LICOR was calibrated prior to each flight using a zero and span gas (496 ± 5 ppm CO_2). The instruments were mounted in a twin-engine aircraft with an open-flow sampling of the external air upstream of the exhaust of the aircraft as previously described (Gerlach et al. 1997). Flights were performed during similar wind conditions (6–10 ms^{-1}), under clear skies in March and May, and hazy skies in July. Each flight consisted of flying circular orbits counterclockwise at 2 km from the central vent in the main crater at speeds between 50 and 60 ms^{-1} (100–110 knots; Fig. 1).

The orbits were flown between 60 and 800 m elevation with 30–60m elevation change between orbits, and measurements were corrected for the circular path. Wind speeds were measured on each orbit by continuously recording GPS, and by performing wind measurements at the height of the main plume

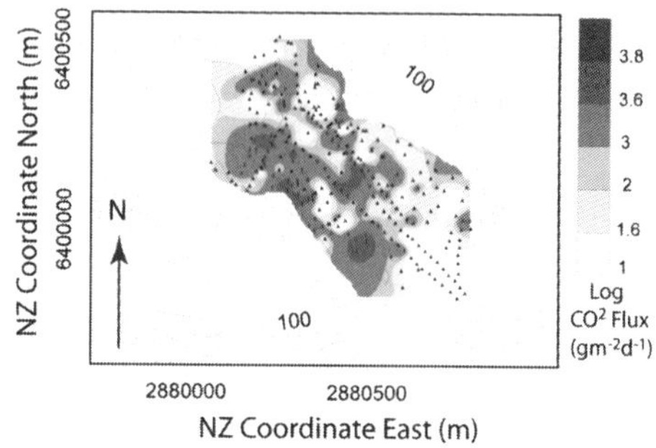

Figure 2. Map of CO_2 flux in the crater of White Island volcano (spatially interpolated by kriging). Dark grey indicates areas of high flux, and the points (n = 268) indicate the measurement locations.

(e.g. Doukas 2002). CO_2 emission rates in metric td^{-1} were calculated as

$$E = .457329 \frac{ASPXv}{T} \quad (1)$$

where A is the plume cross-sectional area normal to the wind direction in m^2, S is the mean wind speed in ms^{-1}, P is the atmospheric pressure in pascals, Xv is the average plume concentration, and T is the temperature in Kelvin. The constant (units are s t K $Pa^{-1} m^{-3} d^{-1}$) is derived from the formula weight of CO_2, the conversion factors $8.64 \times 10^4 \, sd^{-1}$ and $10^3 \, kgt^{-1}$, and the universal gas constant (8.314510 Pa $m^3 \, mol^{-1} K^{-1}$; Gerlach et al. 1997).

Plume concentrations were determined by subtracting the concentration of air sampled upwind of the volcano (background) from the remaining data for each elevation. The average plume concentration in (1) was calculated by dividing the volume of the contoured plume concentrations (kriged Surfer®) by the area using 1.5σ of average background concentration as a cutoff. (0.3, 0.9, and 0.3 ppm in the March, May, and July, respectively). The cutoff ensures that anomalies detected in the natural variability of the atmosphere are not included in the total area of the volcanic plume in (1).

2.2 *Crater floor measurements*

CO_2 fluxes were measured at 268 locations across the crater floor (~0.345 km^2) over three days (27 November and 17th December, 2002, and 8th April, 2003) using a West Systems accumulation chamber and a LICOR LI–820 infrared gas analyzer. Soil temperatures were measured at 10 cm depth at 210 sites. Measurement spacing varied between ~25 and 50 m. Measurement techniques have been described elsewhere (Werner et al. 2000).

3 RESULTS

3.1 *Aerial plume measurements*

Anomalous CO_2 was detected downwind of the volcano on every flight; however, SO_2 was only prevalent in March. On the second flight SO_2 was at the detection limit, and by the July flight no SO_2 was detected. CO_2 background measurements at given elevation varied by <1 ppm and typically <0.3 ppm. However background CO_2 concentrations changed with altitude from 356 ppm at 60 m to 348 ppm at ~800 m in March, from 359 ppm at 60 m to 353 ppm at 650 m in May, and from 376.5 ppm at 60 m to 379.7 ppm at 650 m in July. Maximum CO_2 anomalies in the plume ranged from >6 ppm (March) to <4 ppm (July). CO_2 anomalies were detected at 60 m above sea surface in each measurement.

The vertical extent of the plume was 750 m in March and 635 m in May and July. The width of the plume was ~ 2 km in March and May, and 1.5 km in July. The CO_2 emission rates calculated from (1) and SO_2 emission rates decreased significantly between the March and July measurements (Table 1).

3.2 *Crater floor CO_2 fluxes*

CO_2 fluxes and soil temperatures ranged from 1 to 7670 $gm^{-2}d^{-1}$ and 17 to 61.8°C, respectively. CO_2 fluxes had a coefficient of variation of 2.4 and were distributed between a high-flux population (HFP), which was determined to be lognormal at the 95% confidence interval using the D'Agostino's goodness of fit test, and the low-flux population (LFP) that was skewed, but not strictly lognormal. The break between the HFP ($n = 81$, backtransformed mean = 950 $gm^{-2}d^{-1}$) and LFP ($n = 187$, backtransformed mean = 14.5 $gm^{-2}d^{-1}$) occurred at $\log(CO_2\ flux) = 2.5$ or ~310 $gm^{-2}d^{-1}$. The arithmetic mean of all measurements was 492 $gm^{-2}d^{-1}$ and the declustered mean was 428 $gm^{-2}d^{-1}$, suggesting the spatial distribution of measurements did not largely affect the statistics of the population. Weighting the backtransformed means of HFP and LFP (the LFP was considered more lognormal than normal) by their proportion of the sample resulted in a weighted mean of 297 $gm^{-2}d^{-1}$. Multiplying the area (0.345 km^2; Fig. 2) by the weighted mean resulted in an emission rate of ~99 td^{-1}. Measurements of soil temperature at individual sites were not well correlated with log fluxes; however the mean temperature of HFP sites was higher (34 ± 1.5°C) than LFP sites (29 ± 0.7°C).

CO_2 fluxes were spatially variable across the crater floor but generally higher on the southwestern side of the main crater. A few high measurements were detected on the northeast side of the crater near the crater walls as well as in the central portions of the main crater near steaming ground.

4 DISCUSSION

4.1 *Aerial plume measurements*

The CO_2 and SO_2 emission rates measured in March 2003 (842 and 225 td^{-1}, respectively) were considered typical of normal background emissions and are similar to the long-term emission rates estimated by (Rose et al. 1986). The ratio of CO_2 to SO_2 emission (3.7 by wt) was very similar to the CO_2/SO_2 ratio expected for fumarole samples at White Island (3.43, (Rose et al. 1986). This suggests that very little of the SO_2 was being scrubbed from the plume during this measurement. It is also worth noting, however, that CO_2/SO_2 measured in fumaroles has varied over time between 1.4 and 8.2 (by wt, (Rose et al. 1986), and there is no confirmation of the CO_2/SO_2 ratio during these measurements. By May the CO_2 emissions had reduced to 338 td^{-1} and SO_2 had decreased to near detection limits ($CO_2/SO_2 = 42$). Coincident with these reductions was an increase in level of the lake in the 1978/90 crater such that a large solfatara region on the crater floor was now under water. By July all of the large fumarolic regions in the crater were drowned, no SO_2 emissions were detectable, and CO_2 remained low (275 td^{-1}).

The observations reveal a decrease in the total magmatic mass and heat output below levels typical of the 20 years potentially facilitated by a reduction in the near-surface permeability in the crater region due to mineral precipitation. The reduction in CO_2 emission suggests the thermal power output reduced from ~225 to 74 MW between the March and July measurements based on the magmatic H_2O/CO_2 ratio and 3.9 MJ/kg steam at 700°C. Furthermore, beginning in February, 2003 there was a dramatic reduction in 2–5 Hz volcanic tremor that had been prevalent at the volcano since March 2002. Such tremor is thought to be related to subsurface magma degassing or movement. Thus, the reduction of CO_2 emissions and the establishment of the lake (ie, a loss of steam/heat flux) in March 2003 is consistent with seismic evidence beginning a month prior for decreases in magmatic degassing or convection.

Chemical modeling of the crater lake waters sampled in December 2002 and July 2003 showed that elemental sulfur (as well as cristobalite) was supersaturated in both samples. Precipitation of such minerals could cause a reduction in permeability in the vent area. The lake chemistry suggests that sulfur became supersaturated at a temperature of 175°C thereby reflecting higher temperatures in the vent system. We observed CO_2 emissions in every measurement, but only significant SO_2 emissions in March. Thus, if we assume a constant CO_2/SO_2 ratio of 3.43 for magmatic gases and that the temperature of the crater was >116°C (the temperature needed to maintain liquid sulfur), we expect that ~50 td^{-1} of S could have collected at the bottom of the lake given a CO_2 emission rate of 338 td^{-1}. This equates to about 23.7 m^3d^{-1}, or for example, 0.24 cm d^{-1} over an arbitrarily chosen 100×100 m area at the bottom of the crater. If the temperature was <116°C then elemental S could have been precipitating in similar amounts (27 m^3d^{-1}); however, fumarole temperatures in the main crater did not suggest average temperatures below 116°C. Using the July CO_2 emission as a guide, 40 td^{-1} of S could have been accumulating at that time. In a study by Hurst et al. (1991) it was thought that accumulations liquid S and the changes in viscosity of S with temperature controlled permeability at the top of the volcanic conduit at Ruapehu volcano (thereby controlling the thermal and mass output by over periods of months). It is possible that the changes of viscosity with temperature of liquid S could also be important in controlling permeability and volcanic output at White Island beginning with the drowning of major fumaroles in March 2003. By October 2003 the lake volume was estimated to be increasing by 2700 m^3d^{-1} due to mainly input from precipitation, the lake temperature had stabilized at ~50°C, and seismicity remained low.

4.2 *Crater floor measurements*

The highest CO_2 fluxes measured in the crater of White Island were consistent with the magnitude of fluxes measured in volcanic regions worldwide (e.g., Werner et al. 2000). The crater floor emissions (99 td^{-1}) were an order of magnitude higher than estimated previously (9 td^{-1}, Wardell et al. 2001), but were lower, for example, than emissions measured at thermal areas of Yellowstone (e.g. 0.4 km^2, 290 td^{-1}; Werner et al. 2000).

The total CO_2emission from the volcano based on the March plume measurement and the crater floor fluxes was ~940 td^{-1}. Using the March plume measurement as a guide, the crater floor contributed approximately 11% to the total emissions at the volcano. The crater lake had not risen by the time of the flux measurements in April and no significant changes in volcanic/seismic activity occurred during that period; thus, we expect the crater floor emissions remained at steady state between Dec. 2002 and April 2003.

Although measurements of temperature and CO_2 flux were not well correlated, the spatial distribution of the crater floor fluxes was generally consistent with the distribution of thermal regions as indicated from thermal IR imaging (Mongillo 1993). IR images also suggest that the crater floor contribution to the total emissions reported here is underestimated because thermal areas exist on the flanks of the volcano that have not yet been measured for CO_2 flux.

5 CONCLUSIONS

CO_2 and SO_2 emission measurements at White Island during 2003 suggest that the volcano is emitting at or below long-term predicted levels, and that the plume component is highly sensitive to magmatic changes at depth. SO_2 and CO_2 emissions decreased concurrent with increasing lake levels suggesting a decrease in the mass and heat flux from the volcano. Furthermore, accumulations of elemental S at the bottom of the crater could potentially decrease conduit permeability and obstruct flow of magmatic volatiles and heat. However, more continuous monitoring is needed to fully understand the temporal variability of plume emissions as observed in 2003. Crater fluxes contributed ~11% of the total CO_2 emissions ($\leq$ 940 td^{-1}), and were spatially consistent with thermal IR imagery.

REFERENCES

Doukas, M.P. 2002. A new method for GPS-based wind speed determinations during airborne volcanic plume measurements, U.S. Geological Survey, David A. Johnston Cascades Volcano Observatory.

Gerlach, T.M., Delgado, H., McGee, K.A., Doukas, M.P., Venegas, J.J. & Cardenas, L. 1997. Application of the LI-COR CO_2 analyzer to volcanic plumes: A case study, volcan Popocatepetl, Mexico, June 7 and 10, 1995. *J. Geophys. Res.* 102(B4): 8005–8019.

Giggenbach, W. 1987. Redox processes governing the chemistry of fumarolic gas discharges from White Island, New Zealand. *Appl. Geochem.* 2: 143–161.

Houghton, B.F. & Nairn, I.A. 1989. A model for the 1976–82 phreatomagmatic and strombolian eruption sequence at White Island volcano, New Zealand, *N.Z. Geological Survey Bulletin* 103: 127–136.

Mongillo, M.A. 1993. A thermal infrared survey of White Island volcano. *Institute of Geological & Nuclear Sciences science report, 93/10*, 32 p.

Nishi, Y., Sherburn, S., Scott, B.J. & Sugihara, M. 1996. High-frequency earthquakes at White Island volcano, New Zealand: Insights into the shallow structure of a volcano- hydrothermal system. *J. Volcanol. Geotherm. Res.* 72(3–4): 183–197.

Rose, W.I., Chuan, R.L., Giggenbach, W., Kyle, P.R. & Symonds, R.B. 1986. Rates of sulfur dioxide and particle emissions from White Island volcano, New Zealand, and an estimate for the total flux of major gaseous species, *Bull. Volcanol.* 48: 181–188.

Wardell, L.J., Kyle, P.R., Dunbar, N. & Christenson, B. 2001. White Island volcano, New Zealand: carbon dioxide and sulfur dioxide emission rates and melt inclusion studies, *Chem. Geol.* 177(1–2): 187–200.

Werner, C., Brantley, S.L. & Boomer, K. 2000. CO_2 emissions related to the Yellowstone volcanic system 2. Statistical sampling, total degassing, and transport mechanisms, *J. Geophys. Res.* 105(B5): 10831–10846.

Water–Rock Interaction, Wanty & Seal II (eds)
© 2004 Taylor & Francis Group, London, ISBN 90 5809 641 6

Modeling study of WRI in the Bakki low-temperature geothermal field, south-west Iceland

Z.S. Zhang[1,2]
[1] *East China Institute of Technology, Fuzhou, Jiangxi Province, P.R China*
[2] *Nanjing University, Nanjing, Jiangsu Province, P.R China*

H. Ármannsson & H. Kristmannsdóttir
Orkustofnun, Reykjavik, Iceland

ABSTRACT: The Bakki geothermal field, with a maximum measured temperature of 137°C, is a low-temperature geothermal system. A relict high-temperature alteration is evidenced by the occurrence of epidote. Reservoir rocks are basalt lavas and basaltic hyaloclastites. A modeling study of water-rock interaction (WRI) at high temperature has been performed by the CHILLER program. Most of the important alteration minerals such as calcite, laumontite, stilbite, clinochlore, epidote-ord, clinozoisite, wairakite and muscovite were both identified in the field and by the geochemical modeling.

1 INTRODUCTION

The Bakki geothermal field, which is utilized for a district heating system, is located on the flank of the western active volcanic zone in southern Iceland and is a low-temperature geothermal system with maximum measured temperatures varying from 120°C to 137°C in the wells. But an earlier high temperature alteration is revealed by the occurrence of epidote (Kristmannsdóttir 1992; 1996). It can therefore be considered as a border-case between low- and high-temperature geothermal areas. Two water-rock interaction modeling runs have been made. One at 140°C is close to the present temperature as measured; the other is at a higher temperature of 250°C, which is suggested by the occurrence of epidote. The computer program CHILLER that operates for geothermal systems performed all the calculations. However it must be realized that many reactions that take place in the model may or may not occur in the natural systems. An essential component of effective use of the computer models is to compare model results with their natural counterparts to identify similarities and differences. This type of analysis and comparison greatly enhances understanding of the natural processes as indicated by Reed (1982).

2 MODELING STUDY

The main aim of the modeling study is to simulate the interaction between the reservoir rocks- basalt glass- and the groundwater that can be inferred by the isotopic data (Zhang 2001). The compositions of the basalt glass (Trønnes 1990) and the groundwater are listed in Tables 1 and 2, respectively. During the modeling processes, 0.01 or 0.1 grams of rock were titrated into 1 kg groundwater at each time step until completion, making it convenient to control the water/rock ratio. In the present study, the main emphasis is on comparison of the secondary minerals and the evolution of the fluid, and also to try to determine the main mineral-controlling reactions.

2.1 *Modeling at present temperature*

Figure 1A shows the evolution of the fluid during WRI at 140°C.The elements behave differently

Table 1. Chemical composition (%) of basalt glass used for the water-rock interaction (from Trønnes 1990).

SiO_2	Al_2O_3	FeO	Fe_2O_3	MnO	MgO	CaO	Na_2O	K_2O
48.1	13.9	8.1	4.8	0.24	6.58	12.0	2.52	0.29

Table 2. Groundwater compositions (mg/l) in the Bakki geothermal field (Zhang 2001).

Data	No.	MT	pH/T	CO_2	H_2S	B	SiO_2	Na	K	Mg	Ca	F	Cl	SO_4	Al	Fe	δ^2H	$\delta^{18}O$
9207	929018	4.4	8.56/25.5	16.6	0.02	<0.01	16.37	11.55	0.61	1.86	5.11	0.05	15.88	3.06	0.01	0.0011	−56.1	−8.25

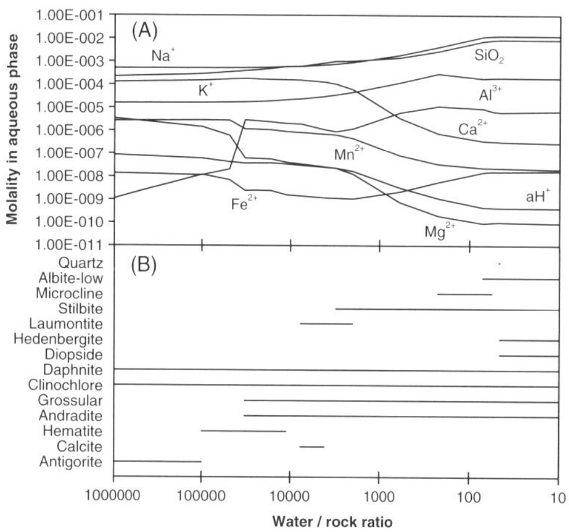

Figure 1. Water rock interaction at 140°C. A. The evolution of the cations in the fluid. B. The deposition of secondary minerals.

during WRI. The major cations, such as Na^+, Al^{3+}, K^+ and Fe^{2+} increase in concentration with a decrease in the water/rock ratio (W/R), which means they are dissolved from the rock during the reaction. However, other cations like Ca^{2+} and Mg^{2+} will be included into the crystal structure of secondary minerals such as calcite, laumontite and clinochlore during the reaction, so their concentration decreased when W/R decreased. SiO_2, the main component of the rocks, dissolves during reaction, so its concentration in the aqueous phase increases when W/R decreases.

Figure 1B shows the minerals present in the matrix during WRI at 140°C. Clinochlore and daphnite are precipitated throughout all values of W/R, from high to low; antigorite is formed at very high W/R from 1,000,000 to 100,000, and then dissolved. Hematite is formed with the W/R between 100,000 and 10,000; while calcite is precipitated mainly during the W/R range of 8000–6000. Laumontite is precipitated during W/R at the range of 8000–2000. Quartz only deposited at W/R around 45. Other minerals such as albite-low, diopside and hedenbergite, are deposited at low W/R. At W/R = 10, minerals present in the matrix are clinochlore, daphnite, andradite, grossular, stilbite, albite-low, diopside and hedenbergite.

Some minerals have the same behavior of deposition due to their similar chemical composition and therefore have the identical index in the SOLTHERM, database file of CHILLER (Reed & Spycher 1989). Clinochlore with daphnite, andradite with grossular, clinozoisite with epidote-ord, diopside with hedenbergite are examples of such pairs.

2.2 *Modeling at high temperature*

Due to the change from a high-temperature environment to a low-temperature one, the observed alteration is a mixture of relict alteration and that formed by the present temperature condition. Therefore another model run was executed imitating the high-temperature condition at 250°C to try to understand the evolution processes of this field.

The fluid evolution is more or less similar to that at low temperature. Na^+, Al^{3+}, K^+, Fe^{2+} and SiO_2 increase with decreasing W/R, as they are dissolved from the rocks. Ca^{2+}, Mg^{2+} and Mn^{2+} on the other hand, are fixed into the crystal structure of the secondary minerals such as calcite, epidote-ord, clinozoisite and clinochlore during the reaction, so their concentration is decreased by decreasing W/R (Fig. 2A). The anions behave more or less similar as at the lower temperature (Fig. 2B), as neither andradite nor hematite was deposited. Therefore the concentrations of HS^- and SO_4^{2-} are positive, whereas they were negative in the low-temperature run. The use of the negative concentrations of HS^- and SO_4^{2-} is a numerical method used by CHILLER to treat the oxidation-reduction processes, the increasingly negative value of M_{HS}^{-} or $M_{SO_4}^{2-}$ could be understood in terms of the dissolution of oxide rock components as used by Reed 1982 (Reed 1982; 1984). Figure 2C shows the alteration minerals that are deposited in the processes at 250°C. As would be expected, some of the typical relict minerals are formed at that temperature, such as epidote-ord and wairakite which are deposited in the matrix. Clinozoisite was also formed at this temperature. Muscovite, instead of only microcline, is deposited now, while some of the low temperature zeolites, like laumontite and stilbite are not deposited at this temperature.

The simulation studies show that at 250°C, the main secondary mineral assemblages are epidote-ord, clinozoisite, wairakite, calcite, clinochlore, muscovite and albite-low. Of those at least the secondary

Table 3. Mineral controlling reactions which were used in this modeling study.

Minerals	Mineral controlling reactions equations and its sequences	
Albite-low	$Alb\text{-}low + 4H^{+} = Na^{+} + Al^{3+} + SiO_2(aq) + 2H_2O$	(1)
Andradite	$Andradite + 0.25HS^{-} + 9.75H^{+} = 3Ca^{2+} + 3SiO_2(aq) + 5H_2O + 0.25SO_4^{2-} + 2Fe^{2+}$	(2)
Antigorite	$Antigorite + 6H^{+} = 3Mg^{2+} + 2.125SiO_2(aq) + 4.938H_2O$	(3)
Calcite	$Calcite + H^{+} = Ca^{2+} + HCO_3^{-}$	(4)
Clinochlore	$Clchl7A + 16H^{+} = 5Mg^{2+} + 2Al^{3+} + 3SiO_2(aq) + 12H_2O$	(5)
Clinozoisite	$Clinozoisite + 13H^{+} = 2Ca^{2+} + 3Al^{3+} + 3SiO_2(aq) + 7H_2O$	(6)
Daphnite	$Dap14/2 + 8H^{+} = 2.5Fe^{2+} + Al^{3+} + 1.5SiO_2(aq) + 6H_2O$	(7)
Diopside	$Diopside + 4H^{+} = Ca^{2+} + Mg^{2+} + 2SiO_2(aq) + 2H_2O$	(8)
Epidote-ord	$Epid\text{-}ord + 11.875H^{+} = 2Ca^{2+} + 2Al^{3+} + 3SiO_2(aq) + 0.125SO_4^{2-} + Fe^{2+} + 6.5H_2O$	(9)
Grossular	$Grossular + 12H^{+} = 3Ca^{2+} + 2Al^{3+} + 3SiO_2(aq) + 6H_2O$	(10)
Hedenbergite	$Hedenbergite + 4H^{+} = Ca^{2+} + Fe^{2+} + 2SiO_2(aq) + 2H_2O$	(11)
Hematite	$Hematite + 3.75H^{+} + 0.25HS^{-} = 2H_2O + 0.25SO_4^{2-} + 2Fe^{2+}$	(12)
Laumontite	$Laumontite + 8H^{+} = Ca^{2+} + 2Al^{3+} + 4SiO_2(aq) + 8H_2O$	(13)
Microcline	$Micr\text{-}max + 4H^{+} = K^{+} + Al^{3+} + 3SiO_2(aq) + 2H_2O$	(14)
Muscovite	$Muscovite + 10H^{+} = K^{+} + 3Al^{3+} + 3SiO_2(aq) + 6H_2O$	(15)
Quartz	$Quartz = SiO_2(aq)$	(16)
Stilbite	$Stilbite + 6H^{+} = Ca^{2+} + 2Al^{3+} + 7SiO_2(aq) + 10H_2O$	(17)
Wairakite	$Wairakite + 8H^{+} = Ca^{2+} + 2Al^{3+} + 4SiO_2(aq) + 6H_2O$	(18)

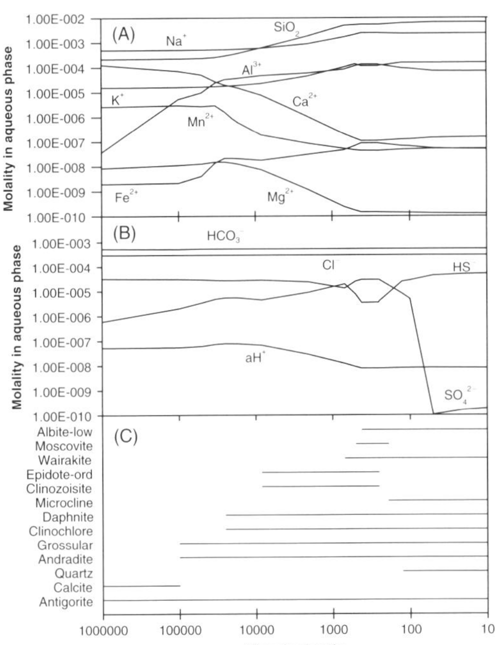

Figure 2. Water rock interaction at 250°C. A. The evolution of the cations in the fluid. B. The evolution of the anions in the fluid. C. The deposition of secondary minerals.

mineral assemblages of epidote and chlorite are found in the two studied wells. Other minerals like clinozoisite and wairakite were found in other wells in the Ölfus area.

If we combine the present temperature model with the high temperature model, most of the important minerals were identified both in the actual field and by the geochemical modeling, showing good agreement.

3 DISCUSSIONS AND CONCLUSIONS

The modeling study suggests that the formation of the alteration minerals depends on the temperature. Table 3 shows the main controlling reactions in the water-rock interaction. Table 4 shows minerals and equilibria controlling the composition of specific components of the fluid generated by water-rock interaction at different temperatures. For the major cation Na^{+}, albite-low formation reaction is the only reaction that controls the concentration in the fluid at both the modeling temperatures. For K^{+}, the controlling reaction is the formation of microcline at low temperature, while the formation of both microcline and muscovite controls at high temperature. The concentration of Ca^{2+} is controlled by the following mineral formation: andradite, calcite, diopside, grossular and laumontite at low temperature, andradite, calcite, clinozoisite, epidote-ord, grossular and wairakite at high temperature. Mg^{2+} is controlled by the formation of antigorite and clinochlore both at low and high temperature, and at low temperature also by diopside. The concentration of Fe^{2+} is controlled by the formation of andradite, daphnite and hematite at low temperature, but by andradite, daphnite and epidote at high temperature. Al^{3+} is controlled by the formation of albite, clinochlore,

Table 4. Elements and mineral phases controlling them at different temperatures.

	Temperature (°C)			
	140		250	
Component	Minerals	Reactions	Minerals	Reactions
Na^+	Albite-low	1	Albite-low	1
K^+	Microcline	14	Microcline	14
			Muscovite	15
Ca^{2+}	Andradite	2	Andradite	2
	Calcite	4	Calcite	4
	Diopside	8	Clinozoisite	6
	Grossular	10	Epidote-ord	9
	Hedenbergite		Grossular	10
	Stilbite	13	Wairakite	18
	Laumontite	17		
Mg^{2+}	Antigorite	3	Antigorite	3
	Clinochlore	5	Clinochlore	5
	Diopside	8		
Fe^{2+}	Andradite	2	Andradite	2
	Daphnite	7	Daphnite	7
	Hedenbergite	11	Epidote-ord	9
	Hematite	12		
Al^{3+}	Albite-low	1	Clinochlore	5
	Clinochlore	5	Clinozoisite	6
	Laumontite	13	Daphnite	7
	Daphnite	7	Epidote-ord	9
	Grossular	10	Grossular	10
	Stilbite	12	Muscovite	15
	Microcline	17	Wairakite	16
$SiO_2(aq)$	Albite-low	1	Albite-low	1
	Andradite	2	Andradite	2
	Antigorite	3	Antigorite	3
	Clinochlore	5	Clinochlore	5
	Daphnite	7	Clinozoisite	6
	Diopside	8	Daphnite	7
	Grossular	10	Epidote-ord	9
	Hedenbergite		Grossular	10
	Stilbite	11	Quartz	16
	Laumontite	17	Wairakite	18
	Quartz	13		
HS^-	Andradite	2	Andradite	2
	Hematite	12	Hematite	12
HCO_3^-	Calcite	4	Calcite	4
SO_4^{2-}	Andradite,	2	Andradite	2
	Hematite	12	Epidote-ord	9

laumontite, daphnite, grossular and microcline at low temperature, and clinochlore, clinozoisite, daphnite, epidote, grossular, muscovite and wairakite at high temperature. The concentration of SiO_2 is controlled by all the mineral formation reactions besides calcite and hematite. The formation of andradite and hematite is controlled by the concentration of HS^- and SO_4^{2-} in the fluid. The formation of calcite is controlled by the concentration of HCO_3^- in the geothermal fluid.

ACKNOWLEDGEMENTS

This paper was a part of research work as the UNU/GTP of the first author. The first author would like to express her gratitude to the government of Iceland, United Nations University and Dr. Ingvar Birgir Fridleifsson, Mr. Lúdvík S. Georgsson, Mrs. Gudrún Bjarnadóttir for their fellowship award and excellent guidance; the efficient help and kindness during the first author's stay in Iceland.

REFERENCES

Kristmannsdóttir, H. & Sveinbjörnsdóttir, A.E. 1992. Changes of stable isotopes and chemistry of fluids in the low-temperature geothermal field at Bakki-Thóroddstadir, Ölfus, SW-Iceland. 951–954 In Y.K. Kharaka & A.S. Maest (eds.) *Proceedings of the 7th International Symposium on Water-Rock Interaction*. Rotterdam: Balkema.

Kristmannsdóttir, H. & Ármannsson, H. 1996. Chemical monitoring of Icelandic geothermal fluids during production. *Geothermics*, 25(3): 349–364.

Reed, M.H. 1982. Calculation of multicomponent chemical equilibria and reaction processes in systems involving minerals, gases, and an aqueous phase. *Geochim. Cosmochim. Acta*, 46: 513–528.

Reed, M.H. & Spycher, N.F. 1984. Calculation of pH and mineral equilibria in hydrothermal water with application to geothermometry and studies of boiling and dilution. *Geochim. Cosmochim. Acta*, 48: 1479–1490.

Reed, M.H. & Spycher, N.F. 1989. SOLTHERM: Data base of equilibrium constants for aqueous-mineral-gas equilibria. Eugene Oregon: Department of Geological Sciences, University of Oregon.

Trønnes, R.G. 1990. Basaltic melt evolution of the Hengill volcanic system, SW Iceland, and evidence for clinopyroxene assimilation in primitive tholeiitic magmas. *J. Geophys. Res.*, 95(10): 15893–15910.

Zhang, Z.S. 2001. Water-Rock Interaction. In The Bakki Low-Temperature Geothermal Field, South-West Iceland. Report 17 in: *Geothermal training in Iceland 2001*. UNU G.T.P., Iceland: 405–434.

Water–Rock Interaction, Wanty & Seal II (eds)
© 2004 Taylor & Francis Group, London, ISBN 90 5809 641 6

Stable chlorine isotopic composition of geothermal waters from Yellowstone National Park

M. Zhang, M.Y. Hobbs & S.K. Frape
Department of Earth Sciences, University of Waterloo, Canada

D.K. Nordstrom, J.W. Ball & R.B. McCleskey
U.S. Geological Survey, Boulder, Colorado, USA

ABSTRACT: Sixteen spring-water samples from four different thermal areas in Yellowstone National Park (YNP) have been analyzed for their stable chlorine isotopic signatures with a standard deviation of 0.1‰. The $\delta^{37}Cl$ values of the waters ranged from −0.13 to +0.42‰. The most enriched $\delta^{37}Cl$ values were measured in Norris Geyser Basin, with an average value of +0.36‰. Within the Gibbon Geyser Basin, waters with pH values between 2.5 and 5.5 had the most positive $\delta^{37}Cl$ signatures (≈0.3‰), while an acid-sulfate water had the most depleted signature (−0.13‰). Examination of the $\delta^{37}Cl$ signatures of all the spring waters versus their $\delta^{18}O$ signatures shows two possible groups of samples, with those waters that are enriched in $\delta^{18}O$ also having more positive chlorine isotopic signatures. Chloride in deep reservoir water(s) at YNP could be derived from mixing of a small amount of magmatic brine, from the leaching of rhyolitic wall rocks, or both. Although the slightly positive signatures measured for most of the hot spring waters are consistent with a magmatic brine source for chloride, the dissolved salts in these waters are believed to be derived from leaching of the rhyolitic wall rocks. In this case, measurements of the wall rock $\delta^{37}Cl$ signature could be used to test this hypothesis. More extensive surveys of the $\delta^{37}Cl$ signatures of geothermal waters from YNP may expand the range of $\delta^{37}Cl$, and verify the possible relationship between chlorine isotopic composition and other geochemical parameters.

1 INTRODUCTION

Stable chlorine isotopes, when combined with other geochemical parameters, can be an effective tool for identifying potential sources of chloride and/or processes affecting fluid evolution. Physical processes such as diffusion and ion filtration can cause fractionation of Cl isotopes during transport (Kaufmann et al. 1988). Other processes that may result in fractionation of Cl isotopes include mineral precipitation and dissolution. The heavier isotope (^{37}Cl) is normally incorporated into the crystal lattice of minerals, whereas lighter ^{35}Cl is preferentially leached during water-rock interactions. Mixing of different waters also may result in a change in the isotopic signature of Cl, if there is a pronounced difference in the Cl signatures of the waters.

Chemical, stable and radiogenic isotopic compositions have been measured in geothermal waters from Yellowstone National Park (YNP). In general, the waters found in YNP are believed to be related to a single parent water located in a deep reservoir, with a temperature of ~350°C (Fournier 1989; Truesdell et al. 1977). An alternative model involving a separate magmatic body (Kharaka et al. 2000) may imply the existence of more than one reservoir. The hydrogen and oxygen isotopic signatures suggest that meteoric recharge is the main source of the parent water, and that contributions from brines of magmatic origin are of minor importance, accounting for less than a few percent (Truesdell et al. 1977). Solutes from the wall rock are leached into these recharge waters along the flow path.

The final chemical and isotopic composition of the thermal spring waters is likely determined by processes such as subsurface boiling and mixing that affect the parent water during ascent to the surface. Dramatic modifications to the chemical and isotopic composition of the waters can occur as a result of subsurface boiling. Condensates from steam are normally high in SO_4 and low in Cl^-, whereas dissolved salts are concentrated in residual liquid water. During ascent, the parent water is commonly mixed with shallow cold groundwater. As a consequence, the concentrations of dissolved salts decrease. Mixing between parent water and steam condensates can also occur, resulting in acidic Na–Cl water.

A reconnaissance study of spring waters from YNP was conducted to determine if there are observable differences in the stable chlorine isotopic signatures of different water types, or between waters from different locations. Observed differences may then be used to obtain further insight into the source(s) of chloride, and/or the dominant processes affecting chloride in the spring waters of YNP.

2 METHODOLOGY

Samples were collected from the Norris, Gibbon and Heart Lake Geyser basins and from the Nymph Creek area in Yellowstone National Park by the U.S. Geological Survey during field trips in 1998 and 2000. Detailed sampling locations are given by Ball et al. (2001, 2002). Samples from springs and geysers were collected as close to the source of each feature as possible. Changes in temperature, pH and water chemistry during sampling were minimized by collecting samples from the middle of pools using insulated sampling equipment, and by on-site analyses of unstable species in a mobile laboratory (Ball et al. 2001, 2002).

All water samples were analyzed for stable chlorine isotopes using the CH_3Cl procedure modified from Long et al. (1993). Chloride in solution is reacted with $AgNO_3$ to quantitatively precipitate Cl^- as AgCl. The AgCl is then reacted with CH_3I to form CH_3Cl, which is purified using gas chromatography and cryogenics. The ratio of ^{37}Cl to ^{35}Cl in the CH_3Cl gas is measured against Standard Mean Ocean Chloride (SMOC) using a mass spectrometer. All analyzes were conducted at the University of Waterloo Environmental Isotope Lab (EIL). The analytical precision was 0.1‰ (1σ), based on repeated measurements of SMOC.

3 RESULTS

The chlorine isotopic compositions of spring waters from YNP are presented in Table 1, along with their chemical compositions, δ^2H and $\delta^{18}O$. Consistent chlorine isotopic values were measured in three samples from the Norris Geyser Basin, with an average $\delta^{37}Cl$ of 0.36‰, which is the most enriched chlorine signature determined in this study. The three samples are all alkali-chloride-type waters with acidic pH values. Relative to chloride, the sulfate concentration was low (<100 mg/L). Although δ^2H values were identical within analytical precision in these samples, the $\delta^{18}O$ values of the waters ranged from −12.3 to −14.6‰.

A large variation in chlorine isotopic compositions (from −0.13 to 0.35‰) was measured for waters from the Gibbon Geyser Basin. This group of samples contained a number of different water types. Figure 1 shows the relationship between the chlorine signature and pH for waters from this Basin. Four samples

Table 1. Chemical and isotopic compositions of thermal waters from YNP. Concentrations are in mg/L; isotopes are in permil relative to SMOC for chlorine and VSMOW for hydrogen and oxygen.

Locality	Sample no.	T(°C)	pH	Ca	Mg	Na	K	Cl	SO_4	HCO_3	$\delta^{37}Cl$	δ^2H	$\delta^{18}O$
Norris Geyser Basin	98WA118	84	4.32	6.3	0.017	430	65	670	43	–	0.42	−133	−12.3
	98WA119	76	5.88	2.8	0.13	380	68	580	70	21.2	0.30	−133	−13.4
	98WA138	36.1	3.23	5.2	0.24	380	50	550	85	–	0.36	−137	−14.6
Gibbon Geyser Basin	98WA122	89	7.28	5.1	0.05	390	31	460	140	17.9	0.26	−133	−12.5
	98WA123	93	5.56	1.9	0.06	230	12	180	250	5.0	0.32	−127	−12.3
	98WA125	93.8	8.16	0.45	<0.001	510	25	420	130	337	0.12	−141	−15.6
	98WA126	75	2.52	2.3	0.13	160	15	100	470	–	0.32	−129	−12.7
	98WA127	92	8.77	3.8	0.003	380	16	380	140	109	0.11	−142	−15.6
	98WA128	78.7	4.41	3.7	<0.1	270	15	320	160	–	0.35	−122	−10.5
	98WA129	82	7.16	4.1	<0.1	460	51	520	160	63.4	0.11	−124	−10.1
	98WA130	68	1.84	10	0.43	330	36	390	1700	–	−0.13	−117	−6.6
Heart Lake Geyser Basin	98WA131	92	4.53	0.32	<0.001	190	16	38	330	–	0.01	−131	−14.3
Nymph Creek Area	00WA163	62	2.87	6.2	2.3	65	38	33	290	–	0.23	−143	−18.1
	00WA164	40.7	2.77	6.4	2.40	68	38	32	310	–	0.09	−142	−17.6
	00WA165	69.5	6.48	1.4	0.02	380	7.4	430	70	170	0.21	−131	−12.4
	00WA166	68.9	4.49	5.7	0.70	140	48	130	180	<1	0.10	−140	−16.2

(98WA122, 98WA125, 98WA127, and 98WA129) are alkali-chloride-sulfate waters, with sulfate concentrations of approximately 150 mg/L. These samples have neutral to slightly alkaline pH values. Similar chlorine signatures were measured in these samples, with an average of 0.15‰. An enrichment of the chlorine signature was measured in samples 98WA123, 98WA126 and 98WA128, where the pH values ranged between 2.5 and 5.5. Sample 98WA130 is quite unique in both chemical and isotopic compositions. This sample is an acid-sulfate water, with a sulfate concentration of 1700 mg/L. This water has a negative chlorine isotopic signature, and is enriched in both ^{2}H and ^{18}O.

The sample from the Heart Lake Geyser Basin had a chlorine signature of SMOC. This is also a sulfate-dominated water with a low concentration of dissolved salts.

Sample 00WA163 and 00WA164 from the Nymph Creek area are very similar in water chemistry. Both are acid-sulfate waters that are low in dissolved salts. Their chlorine isotopic compositions were 0.23 and 0.09‰. Sample 00WA166 is moderate in dissolved salts, with a $\delta^{37}Cl$ value of 0.1‰. The water chemistry of sample 00WA165 is similar to those of the Norris Geyser Basin, but the $\delta^{37}Cl$ value was lower (0.21 vs. 0.36‰).

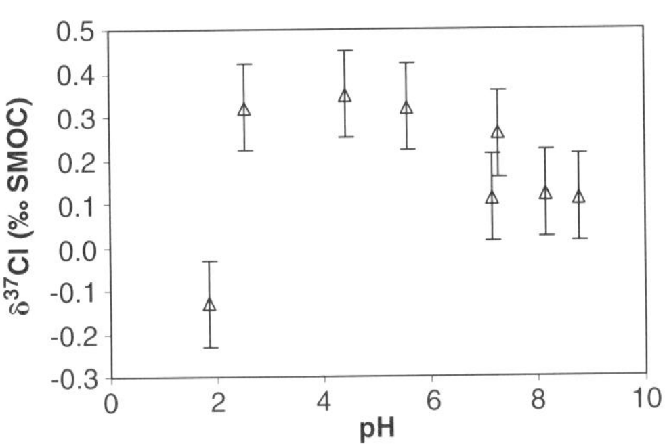

Figure 1. $\delta^{37}Cl$ vs. pH of the hot spring samples from the Gibbon Geyser Basin.

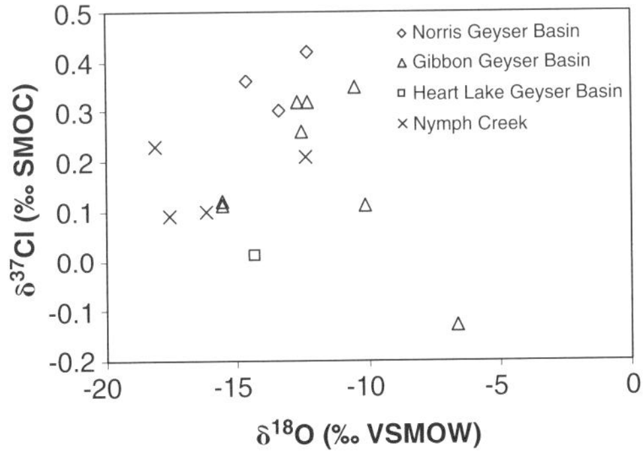

Figure 2. $\delta^{37}Cl$ vs. $\delta^{18}O$ of the thermal waters from the Yellowstone National Park.

Figure 2 shows $\delta^{37}Cl$ plotted against $\delta^{18}O$. If 98WA130 is excluded, there appear to be two groups of samples, in which spring waters with more positive $\delta^{37}Cl$ values are more enriched in $\delta^{18}O$.

4 DISCUSSION

The spring water samples from YNP show an overall enrichment in their chlorine isotopic signatures relative to SMOC. As the hot water ascends to the surface from the reservoir, it loses heat through processes such as conduction to surrounding wall rocks, decompressional boiling, or mixing with cold dilute water. Of these processes, subsurface boiling and mixing with cold dilute water have been identified as the dominant processes affecting geothermal waters at YNP (Truesdell et al. 1977).

During subsurface boiling, steam may separate from water. This process could produce chlorine isotope fractionation if the original parent water was acidic. Under such conditions, ^{35}Cl is preferentially incorporated into the vapor phase as HCl and the residual water becomes enriched in ^{37}Cl (Eggenkamp & Schuiling 1995). Although acid waters are often found associated with subsurface boiling, their acidity is a result of oxidation of H_2S to SO_4, which occurs on or near the surface.

A change in the chlorine isotopic signature could occur as a result of mixing between the parent water, which contains 300–500 mg/l Cl^- (Fournier 1989), and cold meteoric waters. However, the positive chlorine isotopic signatures measured probably reflect the parent water, because the contribution of chloride from the dilute water would be minimal. Chloride in the parent water could be derived from a brine of magmatic origin, from leaching of the rhyolitic wall rocks, or a combination of these sources.

Magmatic fluid accounts for less than a few percent by volume of the parent water (Truesdell et al. 1977). However, it could account for a substantial portion of the total chloride in the parent water, because the salt concentration of the magmatic brine is greater than 12–25% (Fournier 1989). Banks et al. (2000) reported a $\delta^{37}Cl$ value of +1.8‰ (±0.09) for a magmatic fluid. Therefore, the positive values for hot spring waters from YNP are consistent with a contribution of chloride from a magmatic fluid. However, Fournier (1989) suggests that dissolved salts in waters from YNP are mainly derived from leaching of wall rocks. In this case, the source of the slightly enriched chlorine signatures measured in the spring waters is the wall rocks through which the waters are transported. Measurements of the $\delta^{37}Cl$ signature of the rhyolitic wall rocks would allow further testing of this hypothesis.

The range in stable chlorine isotope values (−0.13 and 0.42‰) measured in the current study for hot spring waters from YNP is similar to the range of $\delta^{37}Cl$ signatures of between −0.16 and +0.23‰ measured for hot spring waters from the islands of Java, Bali, Flores and Pantar in Indonensia by Eggenkamp (1994). Kaufmann (1986) measured the stable chlorine isotope ratio of geothermal waters from springs, geysers, and wells in Iceland. The range of isotopic signatures was significantly larger, with $\delta^{37}Cl$ values between +1.60 and −0.98‰ (±0.24).

5 SUMMARY

In this initial study, the stable chlorine isotopic signatures of 16 hot spring waters from the Norris, Gibbon and Heart Lake Geyser Basins and from the Nymph Creek area in YNP were analyzed. The range of $\delta^{37}Cl$ values measured varied between −0.13 and +0.42‰.

The most enriched chlorine signatures were measured in samples from the Norris Geyser Basin (average +0.36‰). Of the spring waters from the Gibbon Geyser Basin, those with pH values between 2.5 and 5.5 had the most positive $\delta^{37}Cl$ values (≈0.3‰), whereas a sulfate-acid water from the same basin had a negative chlorine signature (−0.13‰). When the chlorine isotopic signatures of all 16 waters are examined versus their $\delta^{18}O$ signatures, the waters appear to fall into two groups, such that those enriched in $\delta^{18}O$ also have more positive $\delta^{37}Cl$ signatures.

Chloride in the parent water could be derived by mixing with a small amount of brine of magmatic origin, from leaching of the rhyolitic wall rocks, or both. The positive $\delta^{37}Cl$ values measured for most of the hot spring waters is consistent with a contribution of chloride from a magmatic fluid. However, if the dissolved salts in these waters are derived mainly from leaching of the rhyolitic wall rocks as suggested by Fournier (1989), then the wall rocks could be the source of the slightly enriched $\delta^{37}Cl$ signatures in these waters. Measurements of chlorine isotopic signature of the wall rocks could be used to further test this hypothesis.

In the future, more extensive surveys of the $\delta^{37}Cl$ signatures of geothermal waters from YNP may expand the range of $\delta^{37}Cl$, and verify the possible relationship between chlorine isotopic composition and other geochemical parameters.

REFERENCES

Ball, J.W., McCleskey, R.B., Nordstrom, D.K., Holloway, J.M. & Verplank, P.L. 2002. Water-chemistry data for selected springs, geysers, and streams in Yellowstone National Park, Wyoming, 1999–2000. *U.S. Geol. Surv. Open-File Rep.* 02–382.

Ball, J.W., Nordstrom, D.K., McCleskey, R.B., Schoonen, M. A.A. & Xu, Y. 2001. Water-chemistry and on-site sulfur-speciation data for selected springs in Yellowstone National Park, Wyoming, 1996–1998. *U.S. Geol. Surv. Open-File Rep.* 01–49.

Banks, D.A., Green, R., Cliff, R.A. & Yardley, B.W.D. 2000. Chlorine isotopes in fluid inclusions: Determination of the origins of salinity in magmatic fluids. *Geochim. Cosmochim. Acta* 64: 1785–1789.

Eggenkamp, H.G.M. & Schuiling, R.D. 1995. $\delta^{37}Cl$ variations in selected minerals: a possible tool for exploration. *J. Geochem. Explor.* 55: 249–255.

Eggenkamp, H.G.M. 1994. The geochemistry of chlorine isotopes. Ph.D. Thesis. Universiteit Utrecht., The Netherlands.

Fournier, R.O. 1989. Geochemistry and dynamics of the Yellowstone National Park hydrothermal system. *Ann. Rev. Earth Planet. Sci.* 17: 13–53.

Kaufmann, R.S., Long, A. & Campbell, D.J. 1988. Chlorine isotope distribution in formation waters, Texas and Louisiana. *Am. Assoc. Petrol. Geol. Bull.* 72: 839–844.

Kaufmann, R.S. 1986. $^{37}Cl/^{35}Cl$ ratios in Icelandic geothermal waters. *Int. Symp. Water-Rock Interaction, 5th, Reykjavik, Iceland: 325–328.*

Kharaka, Y.K., Sorey, M.L. & Thordsen, J.J. 2000. Large-scale hydrothermal fluid discharges in the Norris-Mammoth corridor, Yellowstone National Park, USA. *J. Geochem. Explor.* 69–70: 201–205.

Long, A., Eastoe, C.J., Kaufmann, R.S., Martin, J.G., Wirt, L & Finley, J.B. 1993. High-precision measurement of chlorine stable isotope ratios. *Geochim. Cosmochim. Acta*, 57: 2907–2912.

Truesdell, A.H., Nathenson, M. & Rye, R.O. 1977. The effects of subsurface boiling and dilution on the isotopic compositions of Yellowstone thermal waters. *J. Geophys. Res.*, 82: 3694–3704.

Water–Rock Interaction, Wanty & Seal II (eds)
© 2004 Taylor & Francis Group, London, ISBN 90 5809 641 6

Experimental study of water–rock interaction in active geothermal fields: Los Azufres, Mexico

W.B. Zhou*, Z.S. Zhang & M.G. Li
East China Institute of Technology, Fuzhou, Jiangxi Province, P.R. China
** present address: Nanchang University, Nanchang, Jiangxi Province, P.R. China*

ABSTRACT: The Los Azufres Geothermal Field in Mexico is a typical high-temperature hydrothermal system. High-temperature-pressure experiments were set up to study water–rock interaction in this system. Natural rock samples which had been cut and polished into small rock cylinders with a diameter around 4.0 mm were used as the solid phase in the experiments. Mineral alteration and the evolution of the equilibrium solution composition due to WRI processes were explored in this study. A conceptual model of the origin and evolution of the fluids in Los Azufres hydrothermal system is presented. These results might have great significance to the study of moderate temperature mineralization, exploitation of the geothermal field and the study of the migration of radionuclides in high-level waste repositories.

1 INTRODUCTION

Los Azufres geothermal field is located in central Mexico, approximately 200 km northwest of Mexico City. It is one of several Pleistocene silicic volcanic centers with active geothermal systems in the Mexico Volcanic Belt (MVB). It was the first geothermal field in Mexico to generate electricity from fluids in volcanic rocks, and represents the second most important geothermal field in Mexico after Cerro Prieto (Tello 1997). Geologically, the volcanic rock at Los Azufres can be divided into two principal units (Torres-Alvarado 2000): (1) A silicic sequence of rhyodacites, rhyolites and dacites with ages between 1.0 and 0.15 m.y. and a thickness up to 1000 m. (2) A 2700 m thick interstratification of lava flows and pyroclastic rocks of andesitic to basaltic composition with ages between 18 and 1 m.y., forming the local basement.

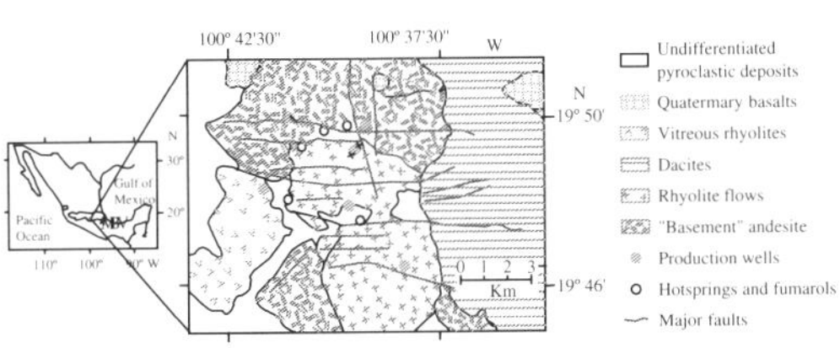

Figure 1. Simplified geological map of the Los Azufres geothermal field (after Dobson & Mahood 1985).

This unit constitutes the main aquifer with flow through fractures and faults. Three different fault systems, NE-SW, E-W and N-S can be distinguished in the field. The E-W system is the most important one for geothermal fluid circulation (Fig. 1).

2 EXPERIMENTAL METHODS

The experiments were designed to simulate water–rock interaction and chemical evolution of thermal fluids in geothermal systems. The distinguishing feature of this experimental study was that rock samples which had been cut and polished into small rock cylinders of about 4 mm diameter were used rather than the rock powders used in most experiments (Xiong 1998; Zhou & Zhang 2000). The advantage of this method is that water–rock interactions took place on the mineral surface, especially at the interface of different minerals, which was helpful in the study of surface interactions and alteration of minerals. At the same time it had the disadvantage that it was very difficult to interpret the chemical composition of the resulting fluid, as the rocks were usually heterogeneous materials (Zhang 2000). The experiments were carried out in the National Key Laboratory on Mineral Deposit Research, Nanjing University. Detailed experimental procedures were similar to the ones performed by Zhou & Zhang (2000).

The major-element compositions were measured by the analysis center in ECIT. An Elan 6000 ICP-MS in the Institute of Geochemistry, Chinese Academy of Sciences, Guangzhou, measured the compositions of solutions from the experiments. A JEOL JXA-8800M Electronic Probe X-Ray Microanalyzer (EPMA) in the National Key Laboratory on Mineral Deposit Research, Nanjing University, measured the mineral compositions before and after experiments. The measured conditions of the EPMA included accelerating voltage of $12 \sim 15$ KV, beam current of 2×10^{-8} A, and beam spot of $0 \sim 2\,\mu m$. ZAF correction was employed.

3 EXPERIMENTAL RESULTS AND DISCUSSIONS

Rock cylinders of M2 andesite from Los Azufres geothermal field were selected as the solid phase for the reaction. The liquid phase, MF, was prepared according to the composition of well waters in Los Azufres. The Water/Rock ratio (W/R) was set to 1, the temperature and pressure were chosen as 350°C and 50 MPa separately, and the duration of the reaction was set from 48 to 720 hours (See Table 1).

3.1 *Characteristics of the solid phase before and after experiments*

After slicing, the solid phase (rock cylinder) was studied before and after experimentation by means of microscope and electron microprobe analysis. Using a microscope, an obvious discolored alteration of feldspar and lots of little secondary crystals around the porphyritic crystals of pyroxene were observed in the rock slice after the experiment. More detailed electron microprobe analytical results are described below.

3.1.1 *Feldspar before and after experiment*

In the M2 rock cylinder from Los Azufres, two kinds of feldspar, porphyritic crystal and micro-crystal in matrix, can be distinguished. The component of An in feldspar in the matrix varies from 48.5 to 51.9%, and from 36 to 42% in the porphyritic crystal, corresponding to andesine-labradorite and andesine respectively. Both feldspars had normal girdle banding. In M3, a naturally altered rock sample, there are two kinds of feldspars, K-feldspar and albite, which are the products of water–rock interaction of andesine in the original andesite rock with fluid enriched in Na and K.

After the experiment, the secondary feldspar crystals around the porphyritic crystals were K-feldspar and albite, just like the natural altered rock sample. But in the leaked sample, the component of An in feldspar increased and the component of Ab decreased. The classification of feldspar before and after experiment is shown in Figure 2.

3.1.2 *Pyroxene before and after experiment*

In the M2 rock sample most pyroxenes were augite before the experiment, but during the experiment, most of them changed into diopside, actinolite and actinolite-Hb (Figs. 3, 4) and epidote. In the natural altered rock sample M3, all augite had changed into epidote and actinolite.

Electron microprobe analysis distinguished very little secondary minerals around the porphyritic crystal; these were hematite, chlorite, sphene, ilmentite, SiO_2 and calcite. Except for that, the typical mineral paragenesis was actinolite + epidote + chlorite + albite.

3.2 *Factors affecting the water–rock interaction*

The analytical results of the solution after the water–rock interaction experiment provide useful

Table 1. Parameters for the experiments.

No.	Rock	Fluid	W/R	T(°C)	P (MPa)	Duration (hours)
SY-02	M2	MF	1	350	50	144
SY-04	M2	MF	1	350	50	720
SY-08	M2	MF	1	350	50	48
SY-09	M2	MF	1	250	50	48
SY-10	M2	MF	1	150	50	48
SY-12	M2	Distilled water	1	350	50	48
SY-17	M2	MF	1	350	50	480

M2: rock sample; MF = 0.1 M NaCl + 0.02 M KCl.

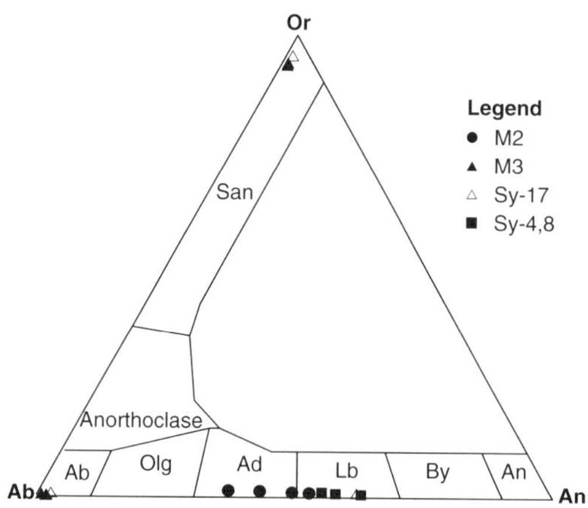

Figure 2. The classification of feldspar before and after experiment.

information on how the fluids were affected by the heterogeneous rock cylinder.

3.2.1 *Temperature*

Temperature had an important effect on water–rock interaction. The evolution of the elements when the interaction between MF and M2 andesite cylinder occurred at different temperatures (Fig. 5) show that high-temperature (350°C) was more favorable for water–rock interactions. At 350°C, most major elements were less than at 250°C and 150°C, except for Mg and Ca, but minor elements were higher. This result shows that the dominant process was rock dissolution at low-temperature, but at high-temperature, along with the rock dissolving (which was stronger than at low-temperature), some secondary minerals might precipitate.

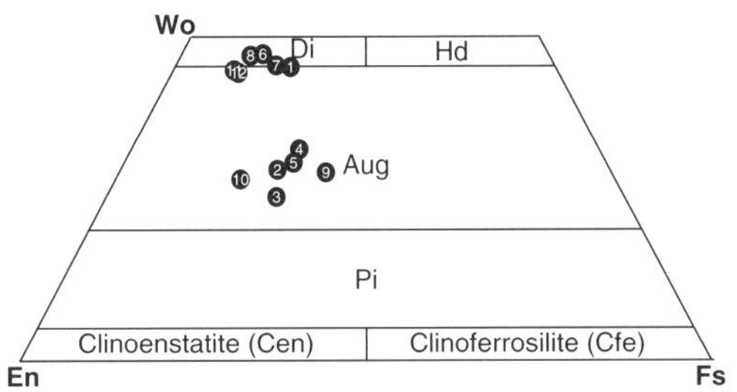

Figure 3. The classification of pyroxene before and after experiment: *The classification based on Morimoto (1989). No: 2,3,4,5 – M2 rock cylinder before experiment; 9- the remainder of SY-4; 10 – from M3; 1,6,7,8,11,12 – from SY-2,17.

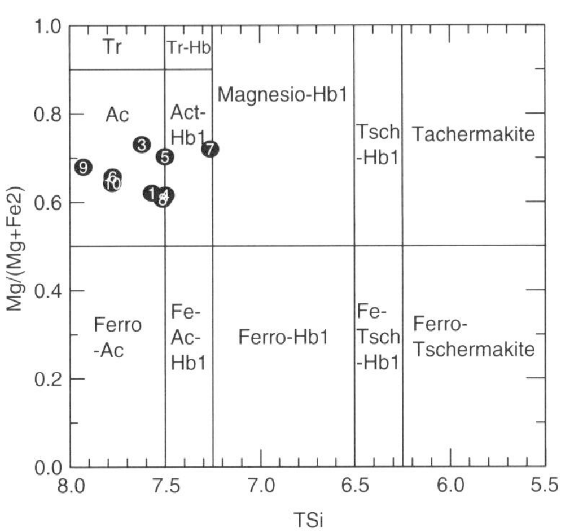

Figure 4. Classification of amphibole after experiment: * The classification according to Hawthorne (1981). No. 1, 2 from M3 rock; 3–10 from SY-4.

3.2.2 *The fluid composition*

After the interaction between distilled water with M2 andesite cylinder, Na, K, Si, Al, Ca, & Fe were enriched in solution (Fig. 6), much like the thermal fluid from the wells in the field. That indicated that the local meteoric water had added into the fluid through water–rock interaction when it infiltrated the andesitic rock zone.

4 ORIGIN AND EVOLUTION OF THE GEOTHERMAL FLUID

The origin and evolution of the thermal fluid in a geothermal system is always one of the most important and disputed questions. Through the study of rock geochemistry along with an experimental study, combined with the isotopic data, a conceptual model of the origin and evolution of the thermal fluid in Los Azufres geothermal system has been set up as follows (Fig. 7). First, heat, Na, Cl, H_2S, and CO_2, contained in andesitic water and magmatic water, ascended from a magma chamber. Local meteoric water infiltrated through the faults in pyroclastic rock and mixed with the andesitic water flowing up in the andesitic rock zone, and formed the original thermal

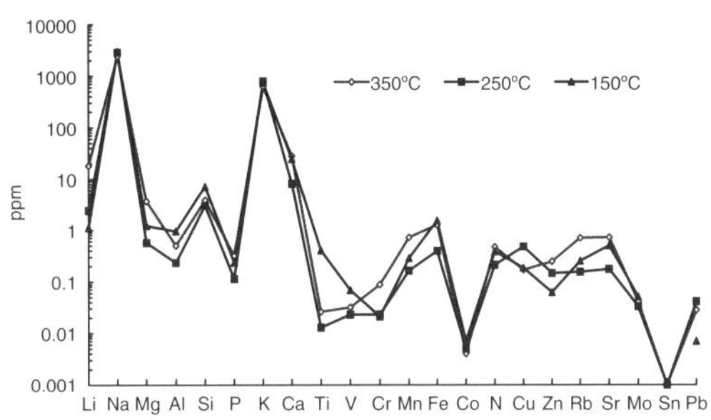

Figure 5. Experimental study of MF-M2 interaction: Effect of temperature on the evolution of elements.

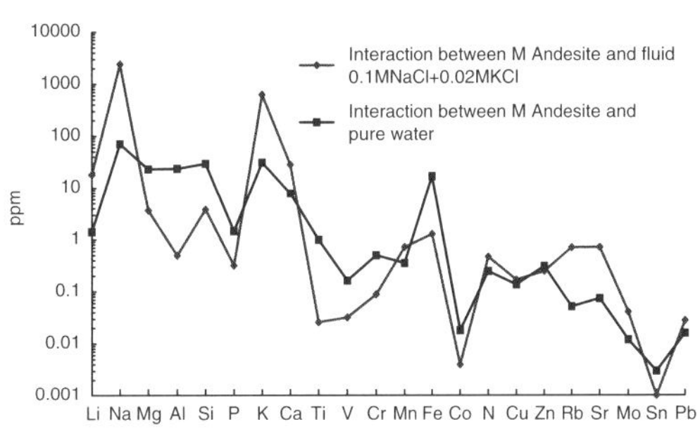

Figure 6. Experimental study of water-M2 interaction: Evolution of elements in the solution.

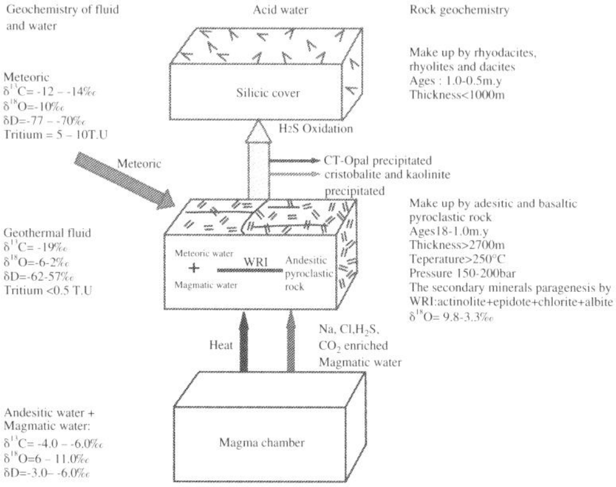

Figure 7. Model of origin and evolution of the geothermal fluid in Los Azufres geothermal field.

fluid. Then interaction between this original fluid and andesite took place, and oxygen isotopes exchanged between the water and rock, resulting in $\delta^{18}O$ decrease in rocks and $\delta^{18}O$ increase in water. At the same time the main paragenesis: actinolite + epidote + chlorite + albite formed in the altered rock, which made the contents of Si and Ca decrease in the altered rocks, but increase in the fluids. Silica precipitated in sequence as cristobalite and CT-opal as the temperature declined. H_2S in the fluid changed into sulfur or SO_4^{2-} by oxidation, forming an acid fluid.

5 CONCLUSIONS

Water-rock interactions at Los Azufres could be divided into two stages: a deeper high-temperature water–rock interaction stage and a near-surface water–rock interaction stage. These interactions had different characteristics and different products. During the deeper high-temperature water–rock interactions, porphyritic crystals of pyroxene and feldspar were changed into diopside, actinolite, epidote, actinolite-Hb, K-feldspar and albite, the contents of SiO_2 and CaO in the altered rock decreased, and Al_2O_3, Fe_2O_3, K_2O, Na_2O, & TiO_2 increased. In the near-surface water–rock interactions, the influential factors would be more complex than those of the deeper ones, but the products were simpler, include complex mixtures of CT-opal and cristobalite, and well-crystallized kaolinite.

Electron microprobe analysis indicated that the main dark porphyritic crystals in the original experimental rock sample from the Los Azufres geothermal field were augite and hematite; light porphyritic crystals were andesine or andesine to labradorite having normal girdle banding. But in the rock cylinder after experiment, the main slightly altered minerals around the porphyritic crystals were diopside, actinolite, epidote, albite, K-feldspar, hematite, chlorite, sphene, SiO_2 and actinolite-Hb.

The modeling study of MF-M2 andesite interaction indicates that the evolution of the elements in the solution were different. Contents of Na, Si, & Ca increased continuously, Cl, B, C, Li, & Ti were almost fixed, and Mg & Fe decreased. The contents of K and Al reached a peak during 1–10 days, and then declined 10 days later. The regular pattern of K, Na, & Ca was in accord with the experimental study.

ACKNOWLEDGMENTS

This paper was a part work of the project, which was financially supported by the International Atomic Energy Agency Research Contract No. CPR-9716. Special thanks to Ms. Jane Gerardo-Abaya and T. Ann Benson for their help in the application and implementation of the project. Special thanks also to our Mexican colleagues for their help in field investigation and sample collection. The authors greatly appreciate the help and supported from the National Key Laboratory on Mineral Deposit Research, Nanjing University.

REFERENCES

Dobson, P.F. & Mahood, G.A. 1985. Volcanic stratigraphy of the Los Azufres geothermal area, Mexico. *J. Volcanol. Geotherm. Res.* 25: 273–287.

Hawthorne, F.C. 1981. Amphibole spectroscopy. In D.R. Veblen (ed.), *Amphiboles and Other Hydrous Pyriboles – Mineralogy, Rev. Mineral 9A*: 103–139. Washington, DC: Mineralogical Society of America.

Tello, H.E. 1997. Geochemical model update of the Los Azufres, Mexico, geothermal reservoir. *GRC Transactions* 21: 441–448.

Torres-Alvarado, I.S. 2000. Mineral chemistry of hydrothermal silicates in Los Azufres geothermal field, Mexico. *Proceedings World Geothermal Congress*, 2000, Kyushu-Tohoku, Japan, May 28–June 10: 1861–1866.

Xiong, X.L. 1998. Experiment on the fluid/melt partition of fluorine in the system albite granite-H_2O-HF. *Geochem.* 27: 66–73 (in Chinese).

Zhang, Z.S. 2000. Study of water–rock interaction in active hydrothermal system–with case study of Los Azufres in Mexico and Tongonan in the Philippines. *Thesis for Master Degree of East China Geological Institute.*

Zhou, W.B. & Zhang, Z.S. 2000. Experimental study and modeling of water-rock interaction in Xiangshan fossil geothermal field, South-eastern China. *Proceedings World Geothermal Congress 2000*, Kyushu- Tohoku, Japan, May 28–June 10: 2319–2324.

Crustal fluid–rock interactions, mass transfer and cycling of volatiles

Water–Rock Interaction, Wanty & Seal II (eds)
 ISBN 90 5809 641 6

The isotopic heterogeneity of C in metamorphic complexes

O.V. Avchenko, S.N. Lavrik, I.A. Alexandrov & T.A. Velivetskaya
Far East Geological Institute, Far East Branch of Russian Academy of Sciences, Vladivostok, Russia

ABSTRACT: It has been discovered that metavolcanic rocks of the Kholodnikanski and Stanovoi complexes from the Aldan-Stanovoi Shield can be divided into two groups according to the inferred $\delta^{13}C_{PDB}$ values of metamorphic fluid. The first group is isotopically light (crustal), where $\delta^{13}C_{PDB}$ varies from −27.2 to −21.4‰, and the second group – isotopically heavy (mantle), where $\delta^{13}C_{PDB}$ varies from −7.2 to −13.6‰. The first group is found mainly by samples from the Stanovoi complex, whereas the second one is found in metavolcanic rocks of the Kholodnikanski complex. The difference in the C isotopic composition correlates with the primary composition of the respective complexes. Thus, mantle and crust signatures of the fluid correlate their protoliths rather than fluid-rock interaction processes during metamorphism.

1 INTRODUCTION

In general, the C of metamorphic rocks can have diverse sources. C can be derived from the mantle or lower levels of the Earth's crust with metamorphic fluid during the process of metamorphism. Alternatively, C may be inherited from the protoliths before their metamorphism. We can suggest that in cases where there is a sufficiently large resource of external C-bearing fluid, or if there is a high fluid/rock ratio during metamorphism, the C isotopic composition of various rocks will be similar, on the average, regardless of the origin of the rock. However, in the case of metamorphic rocks with distinctly different C isotopic compositions, which can be certainly related to their lithologic composition, where there are limited external C-bearing fluids or the fluid/rock ratio was small, then the primary heterogeneity of the C isotopic composition of the protoliths would be little changed. We have examined the question of C sources and amounts of C-bearing fluid associated with metamorphism using of two metamorphic complexes in the eastern Siberia.

2 SCIENTIFIC APPROACH

We have studied the C isotopic composition of metamorphic rocks that did not include C-bearing minerals, graphite or carbonates, visible using an ordinary petrographic microscope. We had reason to believe that the C we studied was contained in rocks as a fine-dispersed carbonaceous phase or that it occurred in fluid inclusions (Avchenko et al. 1998; Avchenko et al. 2002).

All analytic investigations of the C isotopic composition of metamorphic rocks were carried out at the Analytical Centre of the Far East Geological Institute of Far Eastern Branch of Russian Academy of Sciences using a Finnigan MAT 252 mass spectrometer. To prepare samples for the analysis, we converted C to CO_2 using black copper oxide as a source of oxygen. Samples for analysis were ground up and mixed with black copper oxide powder that had been previously calcined at 900°C for 30 minutes. The mixture was put into a quartz tube with a sealed end. Then water vapor and adsorbed gases were evacuated from the tube at a temperature of 200°C until pressure dropped to less than 0.001 mm of mercury. Conversion of C-containing compounds to CO_2 was accomplished at a temperature of 800°C for 20 minutes. Reaction products were frozen at −196°C, then cryogenic separated. The C isotope composition of the purified CO_2 was measured with an observed accuracy (95% confidence level) of ±0.1‰.

3 OBJECTS

We have studied metamorphosed magmatic rocks from two metamorphic complexes in the southern part of the Aldan Shield in the eastern Siberia – the Kholodnikanski and the Stanovoi complexes. These complexes differ from each other in their lithologic composition, history and metamorphic conditions.

The Kholodnikanski greenstone belt is situated in the southern margin of the Aldan Shield. At the base

Table 1. Chemical composition (weight %) and $\delta^{13}C_{PDB}$ value (‰) of the metamorphic rocks.

		SiO_2	TiO_2	Al_2O_3	Fe_2O_3	FeO	MnO	MgO	CaO	Na_2O	K_2O	P_2O_5	LOI	Rock name	Protolith composition	Series	$\delta^{13}C_{PDB}$
Kholodnikanski complex	z-1-v	66.07	0.43	17.20	2.47	0.80	0.05	2.66	1.23	4.52	2.74	0.09	1.74	muscovite crystal shale	Dacite	CAS*	–9.9
	z-2-e	62.66	0.86	15.26	3.28	2.06	0.10	3.07	3.66	5.82	2.11	0.34	0.78	chlorite-epidote crystal shale	andesite	CAS	–9.2
	z-3-a	47.62	0.98	13.10	4.97	6.54	0.18	11.92	8.65	2.89	1.02	0.07	2.06	amphibole crystal shale	komatiite basalt	TKS**	–7.5
	z-3-b	44.50	0.97	15.02	5.90	6.54	0.21	12.52	9.01	1.90	0.65	0.08	2.70	amphibole crystal shale	komatiite basalt	TKS	–9.2
	z-43-a	58.89	0.88	15.75	4.18	3.77	0.17	4.83	3.66	3.37	3.92	0.32	0.82	amphibole crystal shale	andesite	CAS	–21.4
	z-43-g	49.00	1.28	16.10	4.35	7.20	0.12	5.74	7.33	2.15	4.47	0.07	2.19	amphibole crystal shale	tholeiite basalt	TKS	–7.2
Stanovoi complex	OG-5	52.73	0.46	12.86	2.71	7.79	0.15	9.10	10.14	2.41	0.75	0.24	0.18	amphibole gneiss	tholeiite basalt	TKS	–23
	OG-5-1	52.45	0.64	11.24	3.32	7.50	0.36	10.49	9.17	2.15	1.24	0.26	0.60	amphibolite	tholeiite basalt	TKS	–8.7
	OG-6	51.81	0.67	15.00	2.10	9.08	0.00	8.48	8.99	1.72	0.61	0.19	0.35	garnet amphibolite	tholeiite basalt	TKS	–27.2
	OG-6-2	50.14	0.84	13.98	3.06	10.11	0.45	6.65	10.21	1.98	0.99	0.19	1.00	amphibolite	tholeiite basalt	TKS	–24.4
	OG-6-7	51.22	0.68	15.21	1.76	7.08	0.10	8.81	10.25	2.26	0.80	0.15	0.60	amphibolite	tholeiite basalt	TKS	–22.8
	OG-6-9	51.52	0.90	14.52	1.23	8.74	0.34	8.39	9.93	1.91	0.81	0.31	0.80	garnet amphibolite	tholeiite basalt	TKS	–23.8
	OG-9	50.50	0.42	14.77	3.66	7.60	0.00	7.33	10.41	2.76	0.98	0.18	0.64	amphibole gneiss	tholeiite basalt	TKS	–24.2
	OG-11-2	50.40	1.84	15.69	1.46	10.40	0.26	5.25	9.53	3.68	0.55	0.27	0.00	amphibole gneiss	tholeiite basalt	TKS	–22.5
	OG-15	59.42	0.93	17.80	1.49	3.20	0.08	2.54	8.06	3.97	1.01	0.60	0.25	garnet amphibolite	dacite	CAS	–13.6
	OG-15-2	47.33	0.92	15.13	2.66	9.96	0.18	6.88	11.95	1.84	1.05	0.24	0.80	amphibolite	tholeiite basalt	TKS	–9.8
	OG-18	56.20	0.42	12.46	4.90	4.15	0.21	7.86	10.13	2.12	1.07	0.17	0.00	amphibolite	tholeiite basalt	TKS	–23.4

* CAS – calc-alkaline series.
** TKS – tholeiite-komatiite series.

of the Kholodnikanski complex, there is an amphibolite series interbedded with amphibole-epidote shales. Amphibole-epidote and muscovite-epidote shales constitute the middle part of the sequence. The upper part consists of a series of biotite, muscovite and chlorite shales containing intercalations of quartz shales and quartzites in subordinate amounts. The total thickness of the sequence is about 2.5 km. The metamorphic peak conditions correspond to the epidote-amphibolite facies ($T = 550°C$; $P = 5$ kbar, THERMOCALC computer program). Determination of the premetamorphic nature of metamorphic rocks made it possible to distinguish two series in the observed sequence: a lower series that is volcanogenic, and an upper series that is volcanogenic-terrigenous dominated by volcanogenic rocks. Basic and ultra-basic volcanic rocks of the tholeiite-komatiite series were protoliths for the lower strata; basic, intermediate and acidic volcanic rocks of the calc-alkaline series and terrigenous rocks were protoliths for the upper strata (Table 1). The Sm-Nd isochron age of Kholodnikanski metamorphic complex protolytes is 2.41 Ga. (Lavrik et al. 2002).

The Stanovoi complex is situated in the south-eastern part of the Aldan-Stanovoi Shield. It consists mainly of metamorphic rocks of the amphibolite facies ($T = 550$–$690°C$; $P = 7$–8 kbar, THERMOCALC computer program), specifically biotite-hornblende, garnet-biotite, garnet-biotite-hornblende, epidote-biotite plagiogneisses and shales, amphibolites. Nutman et al. (1992) discovered two groups of zircons during an investigation of tonalite gneisses and crosscutting amphibolites in the area of the Stanovoi complex rocks. An age of 2.75 Ga was obtained for zircons taken from tonalite gneisses. This age, in our opinion, is that of the protolith of the Stanovoi gneisses. The second, younger group of zircons, has an age of 1916–1962 Ma and is only associated with green hornblende – plagioclase – quartz, the mineral paragenesis that is typical of the Stanovoi complex. Therefore, we propose that age of 1916–1962 Ma is the time of amphibolite facies metamorphism of the Stanovoi complex.

It is concluded that the metamorphic rocks of the Stanovoi complex were probably volcanogenic and volcanogenic-sedimentary rocks with the latter predominating, and are represented by biotite-amphibole gneisses.

4 DISCUSSION

The results of C isotopic analysis form two well defined groups on the basis of the $\delta^{13}C_{PDB}$ values: the first group is isotopically light and $\delta^{13}C_{PDB}$ varies from -27.2 to -21.4‰; the second group is isotopically heavy with $\delta^{13}C_{PDB}$ ranging from -13.6 to -7.2‰ (Table 1). Most of the samples from the Stanovoi complex belong to the first group, whereas the second includes mainly metavolcanic rocks from the Kholodnikanski complex. For the first group, C has an obvious biogenic or "crustal" nature, whereas the second group exhibits abiotic or "mantle" characteristics. It should be stressed that the C isotopic composition of the second group cannot be explained by fluid contamination with CO_2 from carbonate rocks due to their absence in the surrounding stratigraphy of the metamorphic series. Thus, we can consider the heterogeneous nature of metamorphic fluid C with "mantle" and "crustal" constituents. Heterogeneity of the C isotopic compositions is locally observed for samples taken within the limits of a single outcrop. For example, samples OG-5 and OG-5-1 from the Stanovoi metamorphic rocks (Table 1) are from the same outcrop and have different isotopic compositions and fall into two different genetic groups. Moreover, correlation of isotopic composition with protolith composition of the metamorphic complexes is strongly suggested. The greater abundance of Relatively high saturation of volcanogenic-sedimentary and sedimentary rocks in the Stanovoi complex with, compared to the Kholodnikanski complex, is consistent with the lighter C isotopic compositions from the former. It is probable the isotopic composition is mainly formed during the formation of metamorphic complex protoliths and not from exotic fluids infiltrating the rocks during metamorphism. In other words, the volume of metamorphic fluid is insufficient for buffering the isotopic composition. This conclusion is reinforced by the more intensive metamorphism of the Stanovoi gneisses containing dominantly "crustal" C. It may seem that if mantle component of C is received with metamorphic fluid, that the ratios should be reversed, that is, higher temperature and pressure metamorphism of the Stanovoi rocks should have resulted in a greater "mantle" character in the fluid C.

5 CONCLUSION

The present investigation demonstrated that "mantle" and "crustal" isotopic signature of the metamorphic fluid C is probably inherited from the protoliths of the metamorphic complexes. This result suggests that metamorphic fluid interacted with the rocks under the condition of the low fluid/rock ratio.

ACKNOWLEDGEMENTS

This study was supported by an integral grant from the Far Eastern and Siberian Branches of Russian Academy of Sciences, № 03-2-0-00-001.

REFERENCES

Avchenko, O.V., Khudolozhkin, V.O., Konovalova, N.P. & Barinov, N.N. 1998. Carbon-rich reduced fluids of the Sutam metamorphic complex. *Geochem. Int.* 36: 742–751.

Avchenko, O.V., Lee Dong Woo & Sapin, V.I. 2000. Interaction of metamorphic rocks with reduced fluid: mineralogical evidence. *Geochem. Int.* 38: 535–541.

Lavrik, S.N., Mishkin, M.A., Moiseenko, V.G. & Zhouravlev D.Z. 2002. First data on Sm-Nd isotope systematization of metavulcanites from the Kholodnikan Greenstone Belt, Southern Aldan shield. *Doklady Earth Sci.* 382: 31–34.

Nutman, A.P., Chernyshev, I.V., Baadsgaard, H. & Smelov, A.P. 1992. The Aldan Shield of Siberia, USSR: the age of its Archaen components and evidence for widespread reworking in the mid-Proterozoic. *Precambrian Res.* 54: 195–210.

Water–Rock Interaction, Wanty & Seal II (eds)
© 2004 Taylor & Francis Group, London, ISBN 90 5809 641 6

Dating fluid infiltration using monazite

J.C. Ayers, M. Loflin & C.F. Miller
Vanderbilt University, Nashville, Tennessee, USA

M.D. Barton
Univ. of Arizona, Tucson, Arizona, USA

C. Coath
Dept. of Earth Sciences, Univ. of Bristol, Bristol, UK

ABSTRACT: Thermodynamic calculations at temperatures from 25–200°C and experimental measurements at 1.0 GPa and 1000°C show that the solubility of monazite ($REEPO_4$) is low in near-neutral pH aqueous fluids but much higher in acidic and alkaline fluids. Acidic or alkaline fluids can recrystallize preexisting monazites or precipitate new monazites in contact metamorphic aureoles. Isotopic and geochronologic evidence from two aureoles suggests that monazite is more susceptible to hydrothermal alteration than zircon and may be used to map the extent and date the timing of fluid infiltration during contact metamorphism.

1 MONAZITE SOLUBILITY EXPERIMENTS

1.1 *Introduction and methods*

The solubility of monazite in pure H_2O is low. At 25°C the concentration of Ce in a $CePO_4$-saturated solution at pH = 7 is only $\sim 10^{-14}$ molal. Even at 1.0 GPa and 1000°C the solubility is only ~0.007 molal (Ayers 1991). At 25°C monazite solubility is known to increase as fluid acidity increases.

Because REE behave as hard acids, we would expect them to complex with hard acids, such as OH^-, Cl^- and F^- (Langmuir 1997). Fitting equations of state to experimental data to estimate stability constants of REE complexes led Haas et al. (1995) to conclude that these would be the dominant ligands of REE, and that most dissolved REE would be complexed at elevated temperatures. We therefore conducted a series of experiments to characterize the effect of OH^- and H^+ concentration on the solubility of monazite in order to identify compositions of fluids that could mobilize monazite.

Experimental measurement of the solubility of natural monazites in $H_2O \pm HCl \pm NaOH$ fluids used the single crystal mass loss method used previously by Ayers & Watson (1991, 1993). Cubes ~1–2 mm on edge were sawed from natural monazite crystals for use in solubility experiments. In terms of atoms per formula unit compositions ranges were Si 0.01–0.09, Th + U = 0.005–0.11, and Ca 0.002–0.025; Si and Th + U span the range of common monazite compositions, but Ca concentrations fall within the low end of that range. All experiments were conducted in a piston cylinder apparatus at 1.0 GPa and 1000°C. Cold-sealing capsules of the type described by Ayers et al. (1992) were used to encapsulate samples. The solubility experiments were not designed for the extraction of thermodynamic data: pure end-member monazite compositions were not used, but as shown below, measurements using four different monazite compositions revealed no measurable difference in solubility, suggesting that the solubility of natural monazite does not depend strongly on its composition.

1.2 *Results and discussion*

Because we could not measure pH in-situ, we express pH relative to neutral at 25°C. This is equivalent to assuming that the added acid HCl or base NaOH completely dissociates at experimental conditions (the values of the dissociation constants of these salts at experimental conditions are unknown).

The amount of monazite dissolved in the fluid at run conditions is represented by the mass loss = initial crystal mass – final crystal mass. The mass of fluid = final mass of fluid + mass of dissolved monazite. The mass % solubility s = 100 × (mass loss xt)/(mass loss xt + mass fluid).

The time required to reach equilibrium was evaluated by conducting experiments of varying duration. Experiments run for 24 h and 48 h in pure H_2O yield solubilities of 0.006 and 0.007 molal (assuming g.f.w of $CePO_4$) that are the same within error. Solubilities measured in 2 m NaOH were 0.085 m at 12.4 h and 0.083 m at 24 h. Solubility in 0.1 m HCl was 0.002 m at 24 h and 0.001 m at 53 h. We conclude that 24 h and possibly just 12 h are required to reach equilibrium.

Measured solubilities appear to be independent of monazite composition within the range investigated (Fig. 1). In 2 m NaOH, measured solubilities were 0.083 m for Brazil monazite and 0.085 for Arendal monazite. In H_2O, solubilities were 0.006 m and 0.007 m for Raade monazite and 0.005 m for Brazil monazite. In 0.01 m HCl, solubilities were 0.002 molal for EBay monazite and 0.001 m for Brazil monazite.

Measured solubilities demonstrate that the solubility of monazite at 1.0 GPa and 1000°C is very low, with $m_{REEPO_4} < 0.01$ over a wide range of pH values. However, between $pH = n - 6$ (neutral – 6 pH units) and $n - 7.3$, the solubility increases by roughly two orders of magnitude from $m_{REEPO_4} \sim 0.001$–0.1. Likewise, solubility is enhanced in strongly alkaline solutions, increasing from $m_{REEPO_4} \sim 0.001$–0.1 when pH increases from $pH = n + 6$ to $n + 7.3$. This results in a U-shaped solubility curve characteristic of amphoteric solids. The solubility minimum occurs at near-neutral pH values where neutral complexes such as $REE(OH)_3^\circ$ predominate. At pH values $< n - 6$ the activity of H^+ becomes high enough that it undergoes stepwise association with the hydroxyl ligands to form H_2O: $REE(OH)_3^\circ + H^+ = REE(OH)_2^+ + H_2O$, and the ligand number decreases with decreasing pH until it equals zero and REE^{3+} is the dominant species. At high pH values $> n + 6$ the activity of OH^- becomes high enough to form negatively charged complexes such as $REE(OH)_4^-$. The formation of charged complexes enhances the solubility of monazite at low and high pH values because their concentrations are added to the constant (pH-independent) concentration of the neutral complex $REE(OH)_3^\circ$. This type of pH-dependence of the solubility of REE compounds was demonstrated recently for $Nd(OH)_{3(xt)}$ (Wood et al. 2002).

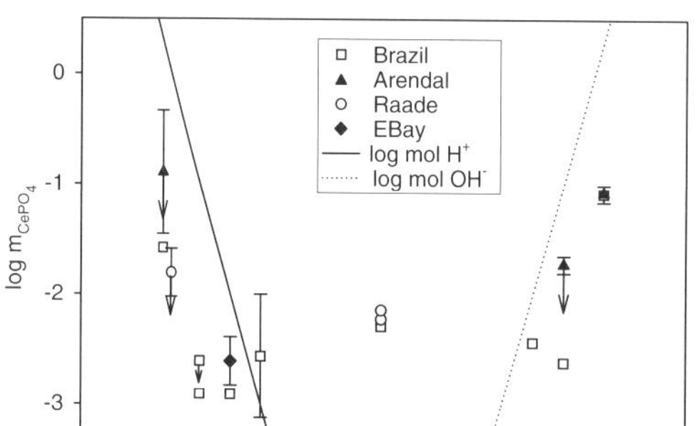

Figure 1. Measured aqueous solubility of monazite as a function of the concentration of H^+ in the initial fluid at 25°C. Concentrations of H^+ and OH^- at 25°C are shown for comparison.

Neutral complexes are favored when the dielectric constant of water ε_{H_2O} is low. A very rough extrapolation suggests that $\varepsilon_{H_2O} \sim 10$ at the conditions of this study compared with a value of ~80 at ambient conditions. This may explain the inference from the shape of the solubility curve that $REE(OH)_3^\circ$ predominate to 6 pH units below neutrality. However, an alternative explanation is that the range of pH values experimentally investigated does not extend over 12 orders of magnitude because the added salts did not completely dissociate at the experimental conditions. Just as it favors formation of neutral REE hydroxide complexes, a low value of ε_{H_2O} inhibits dissociation of HCl and NaOH, i.e. HCl may behave as a weak acid and NaOH as a weak base at experimental conditions. Experimentally measured values of electrical conductance of dilute HCl-bearing solutions allowed (Frantz & Marshall 1984) to estimate the value of the dissociation constant of HCl from 100–700°C and pressures up to 4000 bars; using the equation fit to their experimental data and a density of 0.792 g/cm^3 estimated from the EOS of (Kerrick & Jacobs 1981) yields $\log K_{HCl} = -3.87$. This results in calculated pH values ranging from 3.53 at $m_{HCl} = 0.001$ to 1.79 for $m_{HCl} = 2$. Plotting the data using these estimated pH values shows that the solubility remains constant, and by implication $REE(OH)_3^\circ$ remain the dominant species, until pH drops below 2. Unfortunately no data on the dissociation constant of NaOH at elevated pressures and temperatures are available. We conclude that significant amounts of monazite may dissolve and precipitate in the presence of acidic and alkaline fluids.

2 STUDIES OF MONAZITE IN CONTACT METAMORPHIC AUREOLES

2.1 *Introduction and methods*

We chose the contact metamorphic aureoles of two granitic plutons to investigate how monazite responds to infiltration of fluids in which it is soluble. Both aureoles contain monazite and display evidence of alteration by acidic magmatic fluids (sericitization). Contact metamorphism occurred at conditions under which monazite in granitic systems has been shown to be susceptible to hydrothermal alteration

(Poitrasson et al. 2000): mildly acidic (and for Birch Creek oxidizing and saline) fluids at temperatures of ~250–400°C and pressures of ~0.15–0.4 GPa.

Monazites were separated from rock samples using standard methods, then mounted in epoxy and polished. The Cameca IMS 1270 Multi-Collector High Resolution Ion Microprobe at UCLA was used for in-situ analysis of zircon U-Pb ages and monazite $\delta^{18}O$ and Th-Pb ages using backscattered electron images to guide our choice of analysis spots. For geochronology, spots on individual grains were analyzed using standard procedures outlined in Quidelleur et al. (1997) and Miller et al. (2000).

Most monazite and zircon analyses were corrected using ^{204}Pb. Reported zircon ages are $^{206}Pb/^{238}U$ for concordant and $^{207}Pb/^{206}Pb$ for discordant zircons. For oxygen isotope analysis a Cs^+ beam was used to sputter the sample and $^{18}O^-$ and $^{16}O^-$ beams collected simultaneously in two Faraday cups (Fayek et al. 2002). Analytical precision was typically 0.2‰ or better. Conventional mass spectrometric oxygen isotope analysis of monazite standards was performed by Bruce Taylor of the Geological Survey of Canada using a fluorination procedure followed by conventional gas source mass spectrometric analysis.

We used the measured conventional values for monazite standards $\delta^{18}O$ Brazil = 1.43‰ and 554 = 7.54‰ as the "true" values to calculate the average values of correction factor CF = (true − measured) for each of our two IMP sessions. In a given session standards 554 and Brazil have nearly identical correction factors even though their respective ThO_2 concentrations of ~3.7% and ~6.8% are very different, suggesting that composition does not affect the value of $\delta^{18}O$ measured on the IMP. Over two sessions the average of the standard deviation for multiple analyses of 554 was 0.46‰ (Fig. 2).

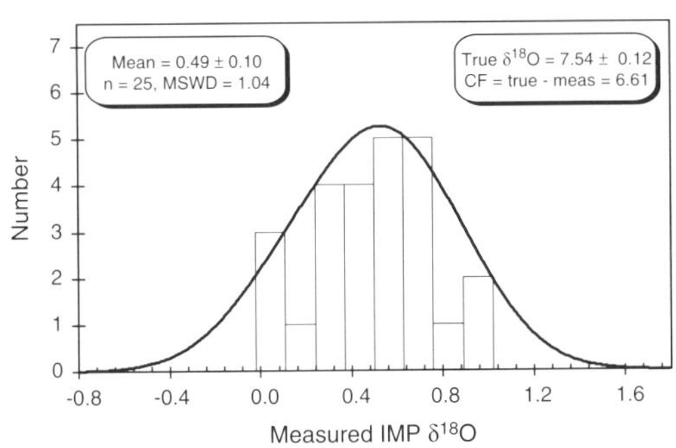

Figure 2. Histogram of $\delta^{18}O$ values measured on standard 554 during one ion microprobe session. Mean values given with 95% confidence limits.

2.2 *Results and discussion*

Monazites from Ireteba are most intensely altered near the contact with the Searchlight pluton, but nearly all monazites contain altered secondary zones and unaltered primary zones, resulting in a bimodal Th-Pb age distribution with age ranges of 14–18 and 60–65 Ma corresponding to intrusion of the subjacent Searchlight pluton and primary crystallization of the Ireteba pluton (Townsend et al. 2000). In contrast, primary and secondary zones show no significant difference in corrected $\delta^{18}O$ values (Fig. 3). If fluids caused the monazite to partially recrystallize they did not significantly change the oxygen isotope compositions of the secondary zones, suggesting that fluids were in equilibrium with respect to oxygen isotope exchange. Monazites from the Ireteba pluton display two generations of monazite with very different ages and compositions, but identical $\delta^{18}O$ distributions. This would occur if the fluid/rock ratio was low so that the oxygen isotope composition of the fluid was buffered by the rock, consistent with the low degree of hydrothermal alteration in the Ireteba pluton. In this situation altered zones date the timing of fluid infiltration, but oxygen isotopes do not fingerprint the fluid source because the $\delta^{18}O$ of the infiltrating fluid was buffered by rock.

The Birch Creek pluton is a two-mica granite in the White Mountains of eastern California (Barton 2000). Monazites from the pluton show concentric euhedral magmatic zoning, mean $\delta^{18}O = 8.2 \pm 0.2$‰ (Fig. 4) and mean Th-Pb age of 78.0 ± 0.7 Ma (n = 28). Monazites in the early Cambrian Deep Springs quartzite ~0.3 km from the contact display patchy chaotic zoning but have similar ages (78.3 ± 1.6 Ma, n = 17) and $\delta^{18}O$ values (9.0 ± 0.5‰), and ~1 km from the contact show concentric zoning,

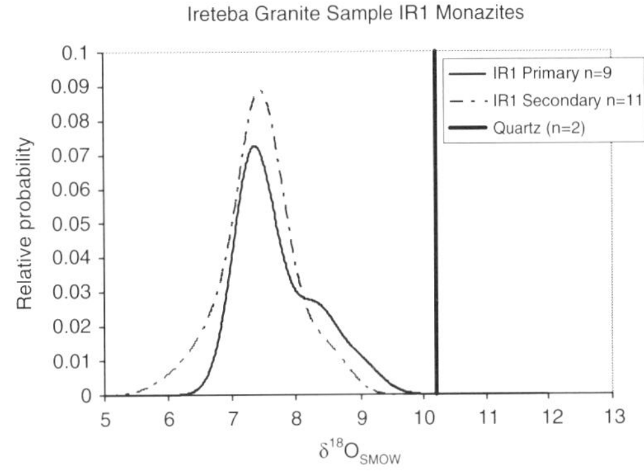

Figure 3. Cumulative histograms of corrected $\delta^{18}O$ values of primary and secondary zones in monazites from sample IR1 measured on the IMP. Values for quartz from the same sample measured by conventional methods shown for comparison.

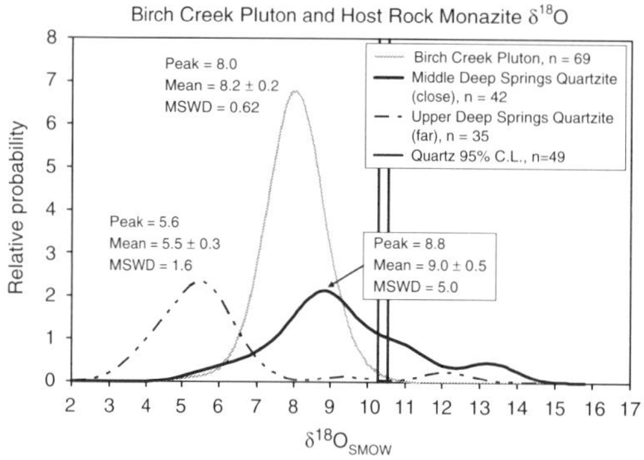

Figure 4. Cumulative histograms of corrected $\delta^{18}O$ values measured using the IMP. 95% confidence limits on the mean value for quartz from the same units measured using conventional methods (Barton, unpub. data) shown for comparison.

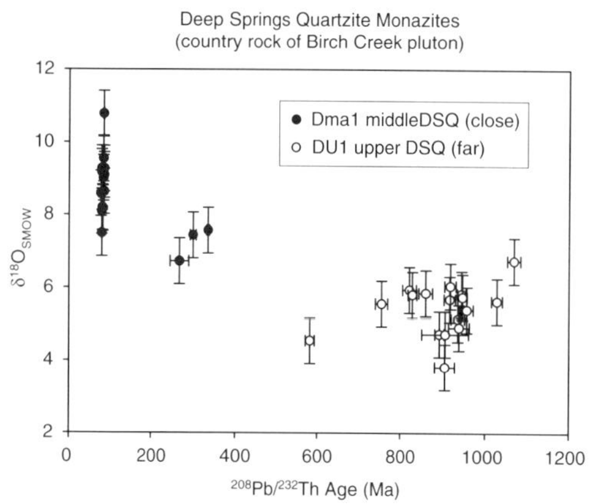

Figure 5. Distinct differences in Th-Pb ages and oxygen isotope compositions for monazites from the Deep Springs Quartzite ~0.3 km (close) and ~1 km (far) from the contact with the Birch Creek pluton.

$\delta^{18}O = 5.5 \pm 0.3$‰ and detrital ages (with some possibly low due to Pb loss) from 583–1069 Ma. Figure 5 shows the distinct difference in age and oxygen isotope compositions of altered monazites close to the contact, unaltered monazites far from the contact, and of three monazites close to the contact that plot as mixtures of altered and unaltered monazite. Monazites close to the contact completely recrystallized or were newly precipitated during fluid infiltration, while monazites ≥1.0 km away were unaffected. In contrast, zircons in the host rocks did not recrystallize, but preserved their magmatic zoning and Cambrian-Precambrian U-Pb ages. Zircon is most reliable for dating events prior to hydrothermal activity, whereas monazite, being more susceptible to alteration by acidic or alkaline fluids, is useful for mapping the extent and timing of fluid infiltration events.

REFERENCES

Ayers, J. C. & Watson, E. B. 1991. Solubility of apatite, monazite, zircon and rutile in supercritical aqueous fluids with implications for subduction zone geochemistry. *Phil. Trans. Roy. Soc. Lond. A* 335: 365–375.

Ayers, J. C., Brenan, J. B., Watson, E. B., Wark, D. A. & Minarik, W. G. 1992. A new capsule technique for hydrothermal experiments using the piston cylinder apparatus. *Amer. Min.* 77: 1080–1086.

Ayers, J. C. & Watson, E. B. 1993. Rutile solubility and mobility in supercritical aqueous fluids: Contrib. Mineral. Petrol. 114: 321–330.

Barton, M. D. 2000. Overview of the lithophile element-bearing magmatic-hydrothermal system at Birch Creek, White Mountains, CA; Field trip day one, Birch Creek, White Mountains, CA. In *SEG Guidebook Series* 32: 9–44.

Fayek, M., Harrison, T. M., Ewing, R. C., Grove, M. & Coath, C. D. 2002. O and Pb isotopic analyses of uranium minerals by ion microprobe and U-Pb ages from the Cigar Lake deposit. *Chemical Geology* 185(3–4): 205–225.

Frantz, J. D. & Marshall, W. L. 1984. Electrical conductances and ionization constants of salts, acids, and bases in supercritical aqueous fluids: I. Hydrochloric acid from 100° to 700°C and at pressures to 4000 bars. *American Journal of Science* 284: 651–667.

Haas, J. R., Shock, E. L. & Sassani, D. C. 1995. Rare earth elements in hydrothermal systems: Estimates of standard partial molal thermodynamic properties of aqueous complexes of the rare earth elements at high pressures and temperatures. *Geochimica et Cosmochimica Acta* 59: 4329–4350.

Kerrick, D. M. & Jacobs, G. K. 1981. A modified Redlich-Kwong equation for H_2O, CO_2, and H_2O-CO_2 mixtures at elevated pressures and temperatures. *American Journal of Science* 281: 735–767.

Langmuir D. 1997. *Aqueous Environmental Geochemistry*. Prentice Hall.

Miller, C. F., Hatcher, R. D., Jr., Ayers, J. C., Coath, C. D. & Harrison, T. M. 2000. Age and zircon inheritance of eastern Blue Ridge plutons, southwestern North Carolina and northeastern Georgia, with implications for magma history and evolution of the Southern Appalachian Orogen. *American Journal of Science* 300: 142–172.

Poitrasson, F., Chenery, S. & Shepherd, T. J. 2000. Electron microprobe and LA-ICP-MS study of monazite hydrothermal alteration: Implications for U-Th-Pb geochronology and nuclear ceramics. *Geochim. Cosmochim. Acta* 64: 3283–3297.

Quidelleur, X., Grove, M., Lovera, O. M., Harrison, T. M., Yin, A. & Ryerson, F. J. 1997. Thermal evolution and slip history of the Renbu Zedong thrust, southeastern Tibet. *Jour. Geophys. Res.* 102: 2659–2679.

Townsend, K. J., Miller, C. F., D'Andrea, J. L., Ayers, J. C., Harrison, T. M. & Coath, C. D. 2000. Low temperature

replacement of monazite in the Ireteba granite, Southern Nevada: geochronological implications. *Chem. Geol.* 172: 95–112.
Wood, S. A., Palmer, D. A., Wesolowski, D. J. & Benezeth, P. 2002. The aqueous geochemistry of the rare earth elements and yttrium. Part XI. The solubility of $Nd(OH)^3$ and hydrolysis of Nd^{3+} from 30 to 290°C at saturated water vapor pressure with in-situ pH measurement. In R. Hellmann & S. Wood (eds.), *Water–Rock Interactions, Ore Deposits, and Environmental Geochemsitry: A Tribute to David A. Crerar*, *Special Publication* 7: 229–256. The Geochemical Society.

Water–Rock Interaction, Wanty & Seal II (eds)
 ISBN 90 5809 641 6

High-precision isotopic analysis of nanomole quantities of silicate N

G. Bebout, B. Idleman & L. Li
Earth & Environmental Sciences, Lehigh University, Bethlehem, PA, USA

A. Hilkert
Thermo Electron (Bremen), Finnigan Advanced Mass Spectrometry, Barkhausenst, Bremen, Germany

ABSTRACT: We have developed a system for analyzing nanomole-sized quantities of silicate N by carrier-gas methods, through combination of a metal high-vacuum extraction line fabricated at Lehigh University and a commercially available continuous-flow, gas chromatography interface (Finnigan MAT Gas Bench II). Early testing has concentrated on analyses of aliquots of standard tank gases, and application of sealed-tube methods in the production of extremely small N_2 samples from silicate standards. Procedural blanks of this method are presently on the order of 0.5–1.0 nanomoles. Practical minimum sample size, taking into account blanks and other factors affecting N_2 transfer, is now ~70 nmoles, more than one order of magnitude smaller than that previously possible in our laboratory using dual-inlet microvolume methods. These methods will allow analyses of microgram-sized samples, thus affording higher spatial resolution and opening up many new avenues of investigation previously impeded by the absence of sufficiently N-rich materials.

1 INTRODUCTION

The nitrogen isotope system has shown exciting potential to trace interactions among Earth's major reservoirs, in particular providing a unique tracer of the transfer of organic components through a wide range of biogeochemical processes. Whereas the application of N isotopes has been extensive in studies of the biosphere (examining, for example, primary productivity, anthropogenic contributions, diagenesis, to name only a few areas), applications of N to trace the transfer of organic components into the deeper Earth have lagged behind, partly related to the trace quantities of N in many silicate materials (commonly <100 ppm, rarely >1000 ppm) and related analytical difficulties. Some of the most extensive use of N isotopes for study of the deeper Earth has been on mantle-derived materials, including diamonds and mantle-derived igneous rocks. Many of the laboratories pursuing such work have been those employing highly sensitive static mass spectrometry (analyzing into the picomole range with precision near ±0.5‰), and a number of these labs also devote attention to studies in cosmochemistry. Another category of research, again employing static mass spectrometry, has been that coupling study of N cycling with analyses of noble gases to assess deep Earth degassing and extents of crustal recycling at convergent margins, and in broader considerations of Earth and extraterrestrial volatiles evolution.

The great potential of the N isotope system as a tracer of flux of organic components into the solid Earth, well-synthesized recently by Boyd (2001), has been extremely under-explored, with only a few laboratories conducting research aimed at transfer of N into the crust and mantle via organic processes, diagenesis, and low- to high-grade metamorphism (and melting). Much of the pathway illustrated in Figure 1 requires additional work for thorough understanding of the important processes and variables.

2 EXISTING METHODS AND LIMITATIONS

Over the last decade or so, we have developed techniques for routine, quantitative extractions and isotopic analyses of trace amounts of N in micas and whole-rock silicate samples. Until recently, most of this work has involved sealed-tube extraction techniques and standard off-line cryogenic purification methods (i.e. using separate glass vacuum lines) and dynamic-vacuum, dual-inlet, viscous-flow mass spectrometry (Bebout & Fogel 1992, Sadofsky & Bebout 2000, Bebout & Sadofsky, in press). These methods are adaptations, with some significant

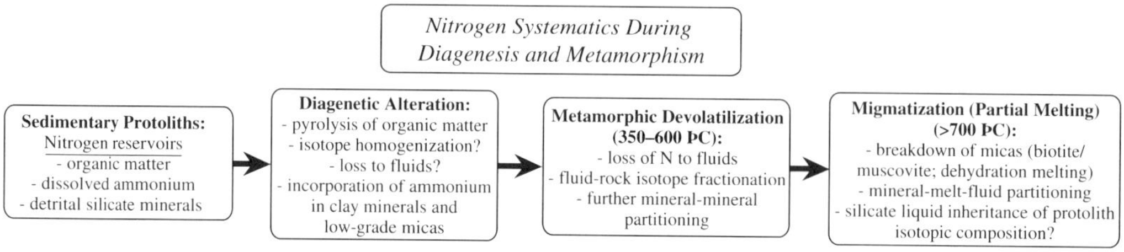

Figure 1. Illustration of pathway of organic N into lithosphere.

modification, of the sealed-tube combustion techniques initially used in biogeochemical studies (Macko 1981). They involve heating of mineral separate and whole-rock samples to 910–1250°C in sealed quartz tubes with CuO wire, Cu metal, with or without CaO (Bebout & Fogel 1992, Sadofsky & Bebout 2000). The N_2 gas samples analyzed using these methods are in the size range of 1–50 μmoles, a size range that can easily be handled using the variable volume (bellows) and microvolume (cold finger) inlets on a viscous-flow mass spectrometer (Finnigan MAT 252 at Lehigh Univ.). These methods yield highly accurate and precise $\delta^{15}N$ and N concentration data for samples with relatively high N concentrations; however, large amounts of sample are needed for materials with <100 ppm N (see Fig. 2, which illustrates the frustration of being limited to analyses of ≥1 μmole using a cold-finger microvolume or bellows). For $\delta^{15}N$ values of samples with >100 ppm N, 1σ for 2–10 duplicate analyses typically approaches 0.1‰, and N concentration data have <5% uncertainty (data for muscovites and biotites in Sadofsky & Bebout 2000). Blanks of these methods are <0.025 μmole N_2, affording analyses of N_2 samples to ~1.0 μmole. The precision of these analyses is typically higher than that obtainable by static mass spectrometry, with the latter methods commonly producing data with 1σ of 0.2–0.5‰.

The requirement of viscous-flow conditions (in turn requiring maintenance of sufficient inlet pressure) and the geometry of the sealed-tube experiments (placing limits on maximum sample sizes that can be fused without tube rupture) have imposed a practical limit of ~10 ppm as the minimum N concentration that can reliably be analyzed by these techniques (for samples containing <100 ppm N, 1σ ≤0.30‰). For samples with the lower N, one is forced to heat samples of >1 gram to satisfy requirements for the dual-inlet mass spectrometry (which of course wastes most of the sample because of the need to maintain sufficient inlet pressure).

As one example of how the minimum sample size requirement of ~1 mole can limit the ability to address a problem, a study by our group of Vermont metamorphic rocks (Sadofsky & Bebout 2000) detected differences in the $\delta^{15}N$ of coexisting biotite and white mica; however, because of minimum sample size limitations, it was necessary to separate coexisting white mica and biotite from "fist-sized" rock volumes, which sampled multiple foliation (and likely equilibrium) domains. Reduction in sample size, by over an order of magnitude (Fig. 2) would allow N-isotope analyses of far smaller domains on the scale of individual foliation selvages, perhaps sufficiently small to overcome the mixing problem caused by the coarser sampling. A number of other studies of fluid–rock interactions, in which N largely

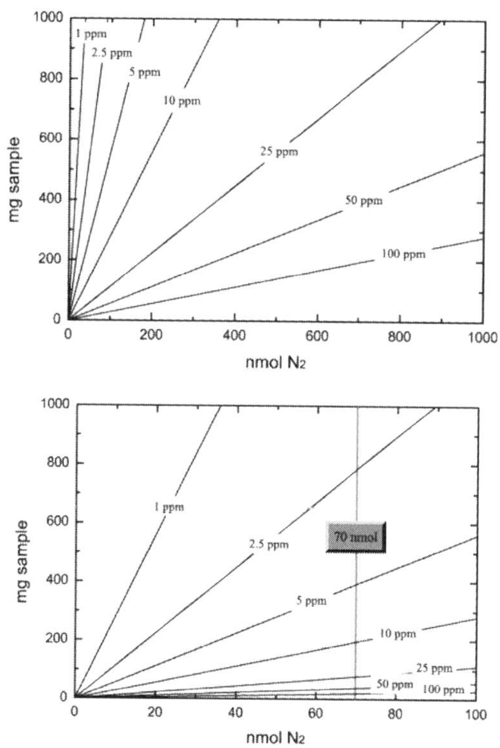

Figure 2. Relationships between sample size, N concentration, and moles (in nanomoles) of N_2 (and present 70 nmol limit).

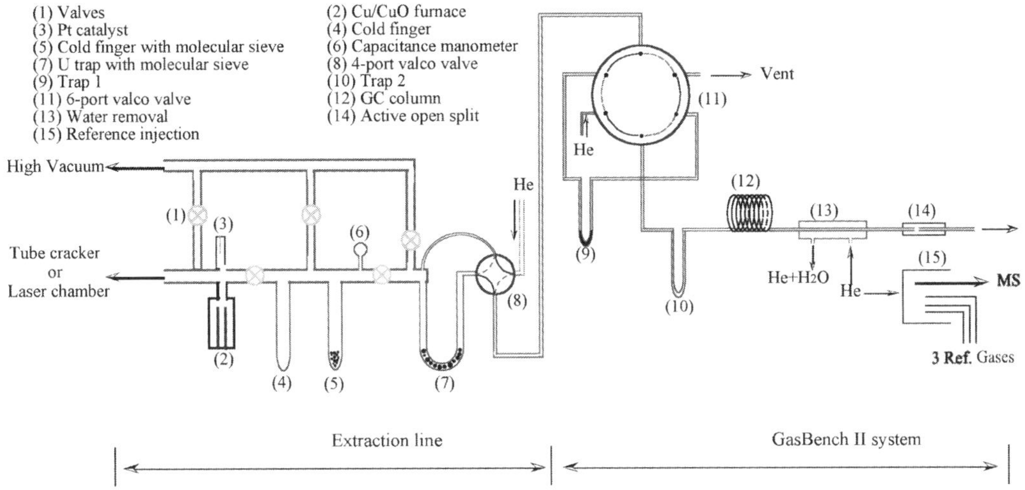

Figure 3. Schematic illustration of the Lehigh University high-vacuum line and interface with the Gas Bench II, and critical components of the GB II on the right.

resides in micas, would similarly benefit from the large reduction in sample size. For relatively low-N minerals (i.e. with <50 ppm N), separations of sufficient materials for a dual-inlet run are cumbersome and can require large volumes of rock.

Also at Lehigh, work underway on variably hydrothermally altered oceanic crust samples from ODP Leg 185 (Izu-Bonin-Mariana) has determined that the N concentrations are extremely low in these rocks (<10 ppm), requiring a more sensitive method. It has been obvious for some time that more routine applications of gas chromatography–isotope ratio monitoring techniques (Merritt & Hayes 1994, Brand 1995), perhaps ultimately involving laser ablations, constitute the next analytical frontier for N-isotope analyses of silicates (see recent studies employing element analyzer(EA)-GC-IRMS methods; Kyser et al. 2000, Jia et al. 2003).

3 THE NEW CARRIER GAS METHODS

Toward analyses of smaller N_2 samples, we have coupled a Finnigan MAT Gas Bench II (GBII) carrier gas system with a metal, low-blank vacuum extraction/purification line built at Lehigh. The initial design of the system, including the interface we built for transfer of small samples of N_2 into the GBII in a He stream, is illustrated schematically in Figure 3. This interface handles the awkward transition from the high-vacuum section of the system (the metal high-vacuum line) and the section of the system that "sees" both high-vacuum and considerable He flow depending on the gas transfer mode. The critical components of the interface are a 4-way Valco valve (Valve 8; rated for high-vacuum use) for manipulation of the He gas stream, and a U-trap (7) containing 5A molecular sieve. Several key components of this interface have seen significant modification since the initial version. The system is similar in a number of ways to those constructed by Young et al. (1998) for analyses of nanomole quantities of O_2, in their case liberating silicate O by UV-laser heating in the presence of F_2. We have modified the configuration of the GBII somewhat to allow the use of two U-traps (9 and 10 above), the first (9) for transferring over N_2 from our extraction line and vacuum-line-He-flow interface into the GC, and the second (10) initially included as a means of cryofocussing the N_2 before release into the GC.

4 PROGRESS ON SMALL-SAMPLE ANALYSES USING THE NEW SYSTEM

Peak m/z 28 beam intensities indicate that the total system blanks from sealed tubes (6 mm outside diameter) are on the order of 0.5–1.0 nanomole, at least 25 times smaller than the blank for our dual-inlet, sealed-tube analyses (Sadofsky & Bebout 2000, Bebout & Sadofsky, in press) and lower than the blanks reported in the recent applications of EA-GC-IRMS to silicate N isotope analyses (Kyser et al. 2000, Jia et al. 2003). In addition to testing on aliquots of standard gases, we have focused attention on analyses of two silicate internal laboratory standards, WS-1, a whole-rock metasedimentary sample, and SL-4, a fuchsite (Cr-rich white mica), containing

620 ppm and 1900 ppm N, respectively, and having $\delta^{15}N_{air}$ of 2.9‰ and 1.9‰, respectively (data for WS-1 in Bebout & Sadofsky, in press; data for SL-4 in Bebout 1997). For these two standards, 1–12 mg are analyzed (the larger amounts for the lower-N standard, WS-1), corresponding to a gas sample size range of ~70–250 nanomoles.

Our experiments to date have revealed a minor non-linearity in the signal vs. sample size relationship when examined over a large range of signal sizes. A strong linear relationship we observe in the >70 nmole sample size range appears not to hold for smaller samples, and additional testing is required for samples in the 10–70 nmole range. We have also observed some very minor non-linearity in isotopic compositions for analyses of N_2 standards spanning a wide range of m/z 28 beam intensities, near the linearity specification for dual-inlet pressure mismatch for the MAT 252. Applying a simple linear correction to analyses of samples of > 70 nmole-size yields corrected isotope ratios with apparent precision approaching 0.06‰ 1σ for multiple silicate standards. Without such a correction, 1σ is at present typically 0.10–0.20‰ and mean isotope values are extremely near (within 0.1–0.2‰ of) accepted values for the standards.

5 ON THE HORIZON

Our ongoing research includes: (1) further work on reducing procedural blanks, (2) investigation of the minor non-linearities among sample size, m/z 28 beam intensity, and $\delta^{15}N$ identified in the early testing of the system, (3) incorporation of laser heating in the extractions, (4) production of one or two silicate N-isotope standards for wider distribution to the community, and (5) collaboration with James Brenan and Natalie Caciagli (Univ. Toronto) in an attempt to experimentally determine mica-fluid N-isotope fractionations over a temperature range appropriate for study of crustal processes and N isotope behavior in subducting slabs. Approaches (1) through (3) are aimed at extending our methods to the critical 10–70 nanomole sample size range (and conceivably to even smaller sample sizes). This research represents an extension of our efforts to apply the underutilized N isotope system to understanding of material transfer in the Earth's crust, or more specifically, to ascertain the degree to which the system can yield unique insight into such processes as large-scale volatile transport, sources of magmas, and ore deposition. Applications we plan over the next five to ten years will in part be aimed at tracing organic signatures into the crust and mantle, and investigating early-Earth N cycling as related to atmosphere evolution.

ACKNOWLEDGEMENT

This research was in part funded by a supplement to NSF grant EAR-0079331 (to GEB).

REFERENCES

Bebout, G.E. 1997. Nitrogen isotope tracers of high-temperature fluid-rock interactions: case study of the Catalina Schist, California. *Ear. Planet. Sci. Lett.* 151: 77–91.

Bebout, G.E. & Fogel, M.L. 1992. Nitrogen-isotope compositions on metasedimentary rocks in the Catalina Schist, California: implications for metamorphic devolatilization history. *Geochim. Cosmochim. Acta* 56: 2139–2149.

Bebout, G.E. & Sadofsky, S.J., in press. $\delta^{15}N$ analyses of ammonium-rich silicates by sealed-tube extractions and dual inlet, viscous-flow mass spectrometry. In P. de Groot (ed.), *Handbook Stable Isotope Techniques*, Elsevier.

Boyd, S.R. 2001. Nitrogen in future biosphere studies. *Chem. Geol.* 176: 1–30.

Brand, W. 1995. High precision isotope ratio monitoring techniques in mass spectrometry. *Jour. Mass Spectrom.* 31, 225–235.

Jia, Y., Kerrich, R., Gupta, A.K. & Fyfe, W.S. 2003. ^{15}N-enriched Gondwana lamproites, eastern India: crustal N in the mantle source. *Ear. Planet. Sci. Lett.* 215: 43–56.

Kyser, K., Klassen, K. & Mitchell, R. 2000. Analysis of N isotopes in micas and amphiboles using EA-CF-IRMS: problems, solutions and implications for the origin of N in hydrous phases from the mantle. *Proc. GeoCanada 2000*, Calgary, Alberta, May, 2000.

Macko, S. 1981. *Stable Nitrogen Isotope Ratios as Tracers of Organic Geochemical Processes*. Unpub. Ph.D. Thesis. Univ. Texas.

Merritt, D.A. & Hayes, J.M. 1994. Factors controlling precision and accuracy in isotope-ratio-monitoring mass spectrometry. *Anal. Chem.* 66: 2336–2347.

Sadofsky, S.J. & Bebout, G.E. 2000. Ammonium partitioning and nitrogen isotope fractionation among coexisting micas during high temperature fluid-rock interactions: Examples from Vermont and Maine, U.S.A. *Geochim. Cosmochim. Acta* 64: 2835–2849.

Young, E.D., Fogel, M.L., Rumble, D., III & Hoering, T.C. 1998. Isotope-ratio-monitoring of O_2 for microanalysis of $^{18}O/^{16}O$ and $^{17}O/^{16}O$ in geological materials. *Geochim. Cosmochim. Acta* 62: 3087–3094.

Water–Rock Interaction, Wanty & Seal II (eds)
© 2004 Taylor & Francis Group, London, ISBN 90 5809 641 6

Mineral chemical records of prograde devolatilization in Alpine HP/UHP metasediments

G. Bebout & R. King
Earth & Environmental Sciences, Lehigh University, Bethlehem, PA, USA

P. Agard
Laboratoire de Tectonique, Universite PM Curie, Paris Cedex, FRANCE

K. Kobayashi & E. Nakamura
The Pheasant Memorial Laboratory, Institute for Study of the Earth's Interior, Okayama University at Misasa, Tottori-ken, Japan

ABSTRACT: Detailed ion- and electron-microprobe study of HP/UHP metasedimentary suites from the Italian Alps reveals mineral major element chemistry and trace element redistribution reflecting prograde (subduction-related) history and overprinting during exhumation. We have focused on trace elements of particular interest in studies of convergent margin cycling (B, Li, Be, Cs, Sr, Ba); for these elements, whole-rock evolution is tied to evolution of the micas (phengite, and in some samples, also paragonite). For white micas, significant scatter in trace element compositions within individual samples (particularly at higher grades) in part reflects exhumation-related overprinting. Subtle decrease in B concentrations and B/Li occurs between units experiencing peak temperatures <450°C and those experiencing higher temperatures; however, for some of these latter rocks, peak temperatures were experienced during decompression. In higher-grade rocks, garnets retain prograde growth zoning in major/trace element compositions, and contain mineral inclusions reflecting prograde growth, potentially affording reconstruction of prograde chemical evolution.

1 INTRODUCTION

Recent attempts to exploit the geochemical record in "paleosubduction suites" have shown great promise toward reconstruction on the geochemical evolution of forearc (and subarc) regions, contributing geochemical information that can be directly compared with that obtained through study of arc lava suites, and also forearc serpentinite seamounts (in the case of the Mariana margin) and accretionary prism fluids. Although the exposures are fragmentary, these rocks provide direct records of forearc-to-subarc metamorphic reaction history, geochemical evolution, and deformation in their textures, mineralogy, and geochemistry (e.g. chemical changes in subducting sediments indicated by study of the Catalina Schist; Bebout et al. 1999, Fig. 1). The results regarding element mobilities from study of metamorphic suites (Fig. 2), combined with the experimental mineral-fluid trace element partitioning database, should ultimately afford attempts to quantitatively assess forearc modification and the compositions of any "fluids" leaving the subducting slab and sediments and interacting with the mantle wedge.

Subduction-zone metamorphic rocks (particularly UHP suites) are notorious for their superposition of

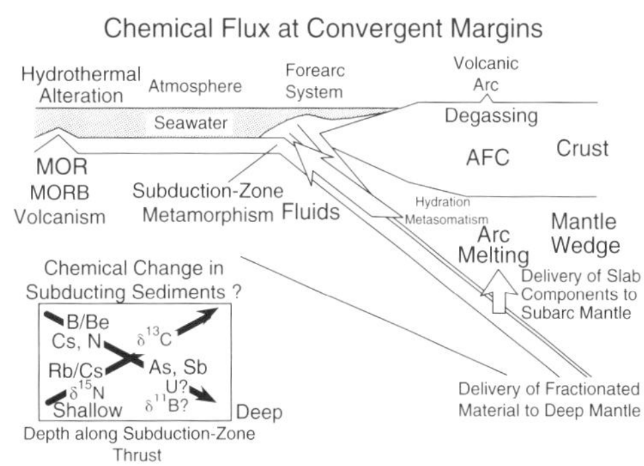

Figure 1. Cartoon illustrating chemical fluxes, and possible change in composition of subducting sediments, in an ocean-continent convergent margin.

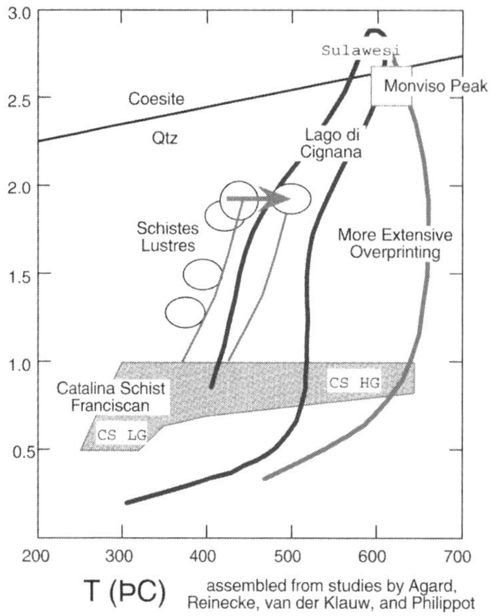

Figure 2. Peak P-T and P-T paths of suites discussed in abstract. Peak P-T of Schistes Lustres units (ovals) are from Agard et al. (2002). Peak P-T for Franciscan overlap those of the low-grade Catalina Schist (CS LG)–high-grade Catalina rocks (CS HG) experienced peak T > 600°C. Depending on the exhumation P-T path of the UHP suites, overprinting can be extreme. For the Schistes Lustres, peak T for at least the higher-grade units is thought to reflect underplating and exhumation, and all SL units have experienced differential overprinting by recrystallization during exhumation (evidenced by neoblastic white-mica with differing Ar age relations; Agard et al. 2002). At LDC, only garnets, tourmalines, apatite and dolomite preserve prograde history in mineral inclusions (including coesite) and mineral chemistry (Reinecke 1998).

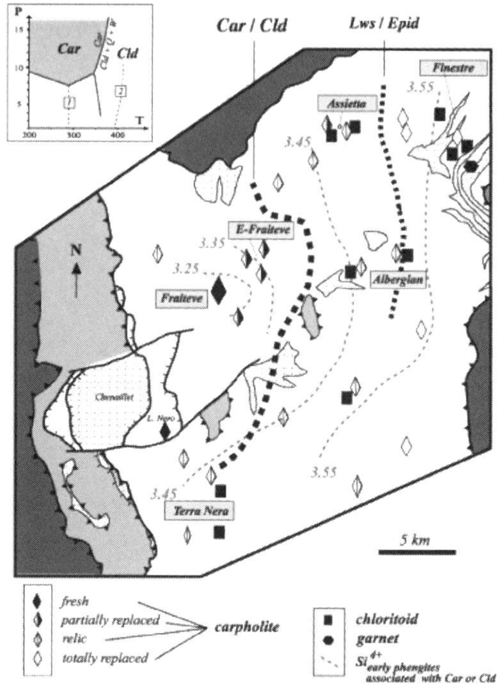

Figure 3. Simplified map of the Cottian Alps exposure of the Schistes Lustres studied in detail by Agard et al. (2002). Grade increases from the SW to the NE (north is up).

prograde mineral assemblages and exhumation-related assemblages. Relationships among the prograde and exhumation-related histories of some HP and UHP suites recently studied for their geochemistry are illustrated in Figure 2. Catalina Schist metasedimentary rocks (Bebout et al. 1999) reflect metamorphism in a newly initiated subduction zone and most of the units were metamorphosed at peak temperatures higher than those at analogous depths in modern subduction zones (only the low-grade units were metamorphosed along prograde paths similar to those in modern systems). Franciscan metasedimentary rocks exposed in the Coast and Diablo Ranges (California) experienced prograde P-T evolution more consistent with paths experienced in modern systems (Sadofsky & Bebout 2003). Intact exposures of metasedimentary rocks subducted to depths greater than those represented by the Catalina and Franciscan suites (max. depths ~50 km) have not been investigated in detail. Because of their greater overprinting during exhumation (relative to mafic rocks), related to more open-system behavior during exhumation-related deformation, metasediments must be carefully selected.

We have undertaken a whole-rock-EPMA-SIMS analysis of records of prograde devolatilization and exhumation in metasedimentary suites in the W. Alps, Italy (Fig. 3; Schistes Lustres, Cottian Alps, SL; coesite-bearing Lago di Cignana, LDC). These suites represent peak conditions of 1.5 to 2.9 GPa, 300–625°C. Prograde histories in the higher-grade SL and at LDC are known, from petrologic and thermochronological study, to be variably to highly obscured by overprinting during underplating and exhumation (Reinecke 1998; Agard et al. 2002). Thus, analytical methods with high spatial resolution were employed (SIMS, EPMA) to separate the geochemical effects of exhumation from those that can be used to ascertain prograde devolatilization history.

2 DISCUSSION OF RESULTS

In the lower-grade SL, heterogeneity in mica trace-element content (B, Li, Be, Cs, Ba, Sr) exists at

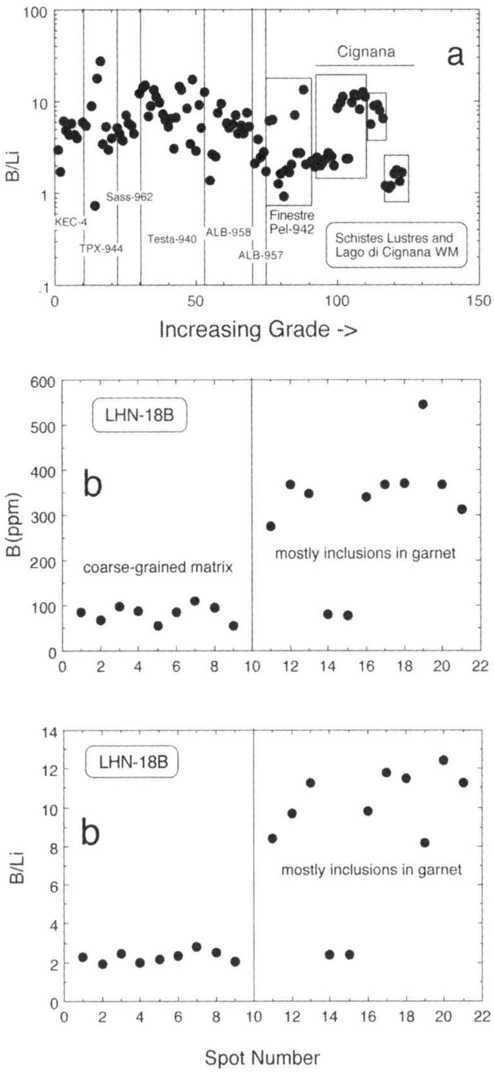

Figure 4. (a) SIMS analyses of white-mica B-Li (phengites verified by EPMA) in metasedimentary rocks of the Schistes Lustres (Cottian Alps) and from Lago di Cignana. Most of the Schistes Lustres samples are those analyzed by Agard et al. (2002) in a detailed laser Ar thermochronology study. Boron and Li contents (and B/Li) are in some samples fairly uniform; however, most samples show significant variation. At Cignana, phengite inclusions in garnet have B contents higher than those of matrix phengites, perhaps related to whole-rock B loss after garnet growth, and thus open-system behavior (b). Also, in high-grade Schistes Lustres and at Cignana, coarse-grained paragonite has higher B and Sr, and lower Cs and Ba, than intergrown phengite in the same samples (Fig. 5). B and Li are of particular interest, as they have been shown to fractionate during prograde metamorphism in high-grade Catalina Schist units (Bebout et al. 1999).

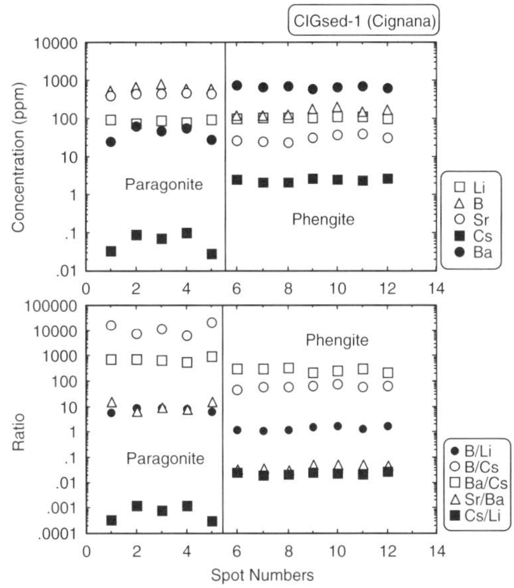

Figure 5. Demonstration of contrasting trace-element compositions of coexisting, coarse-grained phengite and paragonite in one sample of metasedimentary rock from the LDC locality.

sub-mm scales, in some cases obviously related to mica texture or grain size. In the high-grade SL and at LDC, multiple coarse-grained mica generations (some paragonitic) have different trace element contents (Fig. 4a) and, in some LDC samples, phengite inclusions in garnet contrast in trace element chemistry with matrix phengites (Fig. 4b). Other than in B contents, white mica does not obviously vary systematically in trace-element contents across grade, with any prograde variation variably obscured by heterogeneity in part produced by exhumation-related overprinting. B concentrations, and B/Li ratios (Fig. 4a), appear to show some decrease in the higher-grade units which experienced peak temperatures (but not necessarily at peak P) of >450°C, perhaps in part reflecting tourmaline growth. In these higher-grade metasedimentary rocks, scatter in trace element contents is related to the presence of obviously later-stage mica generations (based on textures) and the presence of paragonite, which is very different in its trace element compositions (Fig. 5).

Garnet porphyroblasts show some exciting potential for yielding information regarding earlier, prograde history of the higher-grade SL (and LDC) metasedimentary rocks. In high-grade rocks from both localities, garnets contain cores with major and trace element zoning (and mineral inclusions) consistent with prograde growth (also see Reinecke 1998, for detailed study of garnet zoning at LDC).

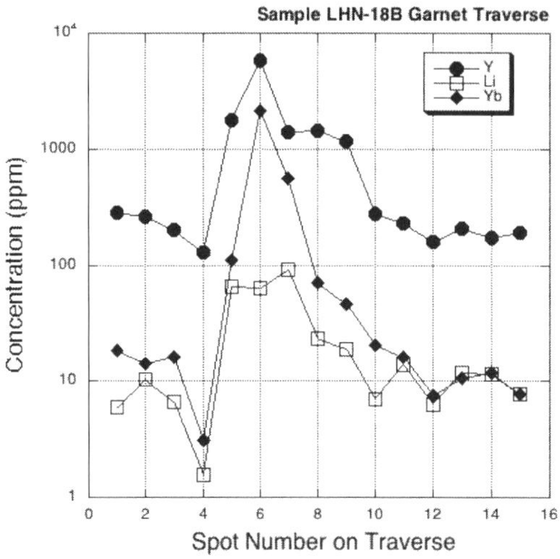

Figure 6. Rim-core-rim traverse of Y, Yb, and Li contents of a garnet in a metasedimentary LDC sample. Regions toward rims with more uniform composition contain inclusions of quartz + clinozoisite; cores are relatively inclusion-free.

Figure 6 shows the zoning in Y, Yb, and Li in one garnet in a LDC sample. In this, and garnet in other samples from this locality and the higher-grade units of the Schistes Lustres, flattening of zoning profiles occurs in a mantle nearer the rim and this likely corresponds to the section of the garnets crystallized during exhumation (mineral inclusions in these outer regions include quartz and clinozoisite). Reinecke (1998) demonstrated prograde garnet cores, mantled first by thin resorption zones (containing coesite), then by broader zones attributable to decompression.

Reinecke (1998) reported that tourmaline cores preserve prograde mineral assemblages (as inclusions, e.g. coesite, rutile, pyropic garnet). We investigated major element and $\delta^{11}B$ of tourmaline in the low- to high-grade SL metasedimentary rocks, and in metasedimentary rocks from the LDC locality (Bebout & Nakamura 2003). Low-grade SL contains very minor fine-grained tourmaline with $\delta^{11}B_{SRM951}$ ranging from −10 to 15‰; some higher-grade SL contains abundant, coarse-grained tourmaline with $\delta^{11}B$ of −12 to −1‰ consistent with B derivation from devolatilizing lower-$\delta^{11}B$ micas, perhaps during initial underplating and exhumation. Tourmaline in siliceous LDC rocks preserves distinct generations of low-$\delta^{11}B$, high-Mg tourmaline attributable to UHP metamorphism, and high-$\delta^{11}B$, lower-Mg tourmaline attributed to exhumation (Bebout & Nakamura 2003). Highly overprinted pelitic LDC rocks contain texturally late-stage, zoned tourmalines with $\delta^{11}B$ of −7 to 1‰ similar to that in high-grade SL, compatible with a late-stage origin of B from devolatilizing white mica (see fractionation model in Bebout & Nakamura 2003).

3 SCHISTES LUSTRES + CIGNANA – A USEFUL REFERENCE SECTION FOR SEDIMENT SUBDUCTION TO 100 KM?

Recent study of the effects of prograde devolatilization on compositions of metasedimentary lithologies has thus far mostly been for paleoaccretionary suites representing subduction to maximum depths of ~50 km (Bebout et al. 1999, Sadofsky & Bebout 2003, Bebout et al., in review). To understand the effects of forearc devolatilization to depths beneath arcs, it is necessary to identify metasedimentary suites that experienced peak, subduction-related metamorphism at 50–100 km depths. The Schistes Lustres (with LDC) is one candidate reference section for use in identifying compositional effects of deep forearc metamorphism.

In a recent study of N, K, Rb, Cs, and H contents of rocks from the same localities, Busigny et al. (2003) concluded that the lack of obvious variation in such ratios as K/N, K/Rb, and K/Cs reflected the efficient retention of the LILE and N, particularly in white-mica, in these metasedimentary rocks. In their study, N contents and isotope compositions were obtained for hand-picked, coarse-grained white micas, and whole rocks were analyzed for major and trace element (K, Rb, Cs) compositions. It is interesting that ratios such as Rb/Cs and $^{15}N/^{14}N$ remain constant (similar to that of seafloor sediments; see Bebout et al., in review) in these units, apparently preserved over a wide range in prograde P-T history and during variable to extensive recrystallization (and possible metasomatism) during exhumation. It is unlikely that reliable measures of the effects of prograde devolatilization can be obtained through such an approach, as it has been demonstrated that the suites experienced appreciable overprinting during exhumation (in some cases, extremely extensive) resulting in some considerable geochemical heterogeneity. This overprinting, known to have redistributed trace elements, is thought to have possibly been accompanied by some open-system behavior (for LDC, Reinecke 1998, Bebout & Nakamura 2003, this study). Open-system behavior would be expected to differentially affect various chemical/isotopic signatures, depending on the compositions of any infiltrating fluids, and the associated fluid-rock mass-balance. Further geochemical research on these rocks is certainly warranted, but additional work should employ high-spatial resolution methods capable of distinguishing geochemical signatures of prograde metamorphism and exhumation-related overprinting.

ACKNOWLEDGEMENTS

This research was funded by NSF grant EAR-0079331 (to GEB) and by the 21st Century Center of Excellence Program, Ministry of Education, Culture, Sports, Science and Technology, Japan.

REFERENCES

Agard, P., Monie, P., Jolivet, L. & Goffe, B. 2002. Exhumation of the Schistes Lustres complex: in situ laser probe $^{40}Ar/^{39}Ar$ constraints and implications for the Western Alps. *J. Meta. Geol.* 20: 599–618.

Bebout, G. E., Bebout, A. E. & Graham, C. M., in review, Cycling of B, Li, and LILE (K, Cs, Rb, Ba, Sr) into convergent margins: SIMS evidence from metasedimentary rocks of the Catalina Schist. *Geochem. Geophys. Geosyst.*

Bebout, G. & Nakamura, E. 2003. Record in metamorphic tourmaline of subduction-zone devolatilization and B cycling. *Geology* 31: 407–410.

Bebout, G. E., Ryan, J. G., Leeman, W. P. & Bebout, A. E., 1999. Fractionation of trace elements during subduction-zone metamorphism: impact of convergent margin thermal evolution. *Ear. Planet. Sci. Lett.* 171: 63–81.

Busigny, V., Cartigny, P., Philippot, P., Ader, M. & Javoy, M., 2003. Massive recycling of nitrogen and other fluid-mobile elements (K, Rb, Cs, H) in a cold slab environment: evidence from HP to UHP oceanic metasediments of the Schistes Lustres nappe (western Alps, Europe). *Earth Planet. Sci. Lett.* 215: 27–42.

Reinecke, T., 1998. Prograde high- to ultrahigh-pressure metamorphism and exhumation of oceanic sediments at Lago di Cignana, Zermatt-Saas Zone, western Alps. *Lithos* 42: 147–189.

Sadofsky, S. J. & Bebout, G. E. 2003. Record of forearc devolatilization in low-T, high-P/T metasedimentary suites: significance for models of convergent margin chemical cycling. *Geochem. Geophys. Geosyst.* 4.

Water–Rock Interaction, Wanty & Seal II (eds)
© 2004 Taylor & Francis Group, London, ISBN 90 5809 641 6

Geochemistry of quartz and calcite veins in a paleoaccretionary complex

Z. Berner & D. Stüben
Institut für Mineralogie und Geochemie, Universität Karlsruhe, Germany

E. Seidel
Institut für Mineralogie und Geochemie, Universiät Köln, Germany

ABSTRACT: Stable isotope (O, C, H) and trace element geochemistry of quartz and calcite veins, located in a subducted and subsequently exhumed sedimentary series [the high pressure-low temperature (HP-LT) metamorphic Phyllite-Quartzite (PQ) unit of Crete] were used to put constraints on the origin and composition of fluids, and the scale and mechanisms of fluid movement in deep levels of accretionary complexes. Results indicate that veins were formed at different stages of the tectonic evolution, basically from progressively heated connate waters/formational brines to which metamorphic fluids were admixed from dehydration reactions.

1 INTRODUCTION

The massive release of fluids from shallow parts of accretionary wedges is well documented and was studied in detail at active accretionary prisms. In contrast, information on fluids in deeper parts of subducted crustal slabs is more limited. Exhumed parts of fossil accretionary complexes proved to yield valuable information on fluids circulating in such tectonic settings.

Situated in the central fore-arc of the Hellenic subduction zone, the tectonic edifice of Crete is considered to be the result of the subduction of a microcontinent located on the African plate beneath the Aegean plate to the north (e.g. Thomson et al. 1998). During collision, part of the sedimentary cover was scraped off and was accreted at shallow crustal levels, while another part, including the Phyllite-Quartzite (PQ) Unit, was buried to about 30 to 35 km depth and metamorphosed under HP-LT conditions during Oligocene to early Miocene time (25–20 Ma) (e.g. Seidel et al. 1982, Jolivet et al. 1996). Peak metamorphic conditions are constrained by minerals and mineral parageneses in metavolcanic and metasedimentary rocks to 10 ± 2 kbar and 400 ± 50°C, with a slightly increasing trend in intensity from the east to the west of the island (e.g. Theye & Seidel 1991). According to a model proposed by Thomson et al. (1998), roll-back associated with slab-pull and tectonic unroofing in an extensional regime favored a buoyant uprise of the HP-LT metamorphic rocks within a few million years to a crustal level of less than 10 km and temperatures below 300°C. The main objective of this study is to gain insight into the fluid history of this subducted and subsequently exhumed crustal segment, using trace element and isotope data on the quartz and calcite veins of the PQ Unit and on fluid inclusions contained therein.

2 THE PHYLLITE-QUARTZITE UNIT

The PQ Unit mainly consists of a succession of metapelites, quartzites and rare metabasites resulted from a HP-LT overprint of a predominantly siliciclastic sequence of late Carboniferous to late Triassic age. The deposits are locally intercalated with conglomerates, greywacke, and carbonate rocks or with evaporitic sequences (mainly gypsum) deposited in a lagoonal environment. Quartz and/or calcite veins up to several dm thick are common throughout the outcrop area of the PQ Unit. Most of the veins are either of quartz or calcite, but locally dolomite and chlorite and high pressure minerals like carpholite, lawsonite, and aragonite may occur. Based on structural features different types of veins can be identified (Stöckhert et al. 1999). Some veins follow joints normal to bedding formed during diagenesis; a major vein formation period was contemporaneous with the HP-LT metamorphism, whereas later formed, undeformed veins crosscut all earlier structures. Quartz and calcite veins and their host rocks were sampled over the

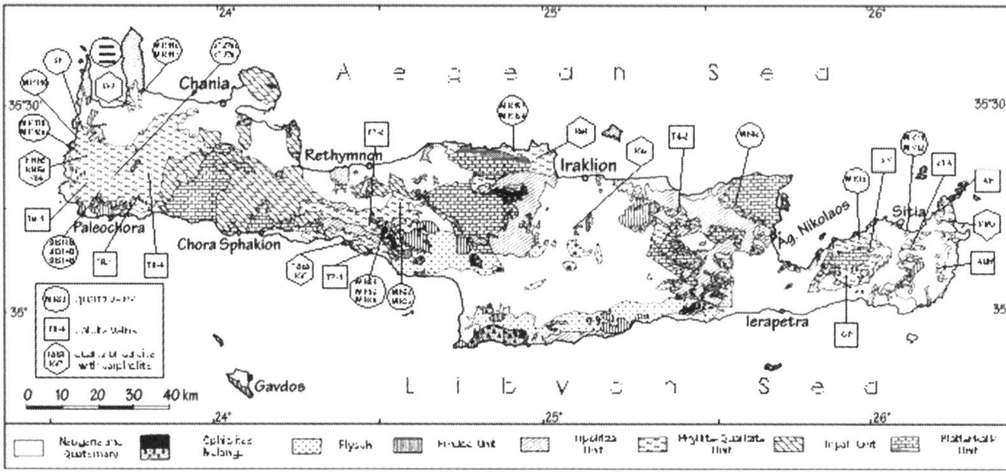

Figure 1. Schematic geologic map of Crete with the indication of the sampled outcrops.

outcrop area of the PQ Unit (Fig. 1), also including samples from the study of Küster (1994) to enable the integration of her fluid inclusion and microstructural data. Some of the outcrops were sampled in more detail.

3 CALCITE VEINS

3.1 *Isotopic composition*

The isotopic composition of the calcite veins is in a range between -8 and 3‰ (V-PDB) for $\delta^{13}C$ and 17–26‰ (V-SMOW) for $\delta^{18}O$. The spread of values is much less at the scale of an outcrop (up to 2‰ for $\delta^{18}O$ and 0.6–1.0‰ for $\delta^{13}C$) and is even less in single veins. No significant difference could be observed between the isotopic composition of veins discordant to or concordant with schistosity nor with clear zoning in respect to the distance from host rock. The isotopic composition of the veins is broadly similar to that of the carbonate in the surrounding rocks (marbles, carbonate-bearing schists, phyllites), although veins are generally lower in $\delta^{18}O$ and slightly higher in $\delta^{13}C$ compared to the rocks in direct contact.

Calcite veins with low $\delta^{13}C$ values are mostly located in initially organic rich lithologies. The isotopic compositions define two end members (Fig. 2): one is low in $\delta^{13}C$ (about -8‰ V-PDB) and the other in $\delta^{18}O$ (19–20‰ V-SMOW) relative to Triassic marine limestones (i.e. the presumed protoliths). Several models were proposed to explain the origin of carbonates with low $\delta^{13}C$. Although the depletion of ^{13}C due to fluid infiltration and metamorphic volatilization generally leads to a positive correlation between $\delta^{13}C$ and $\delta^{18}O$ (e.g. Valley

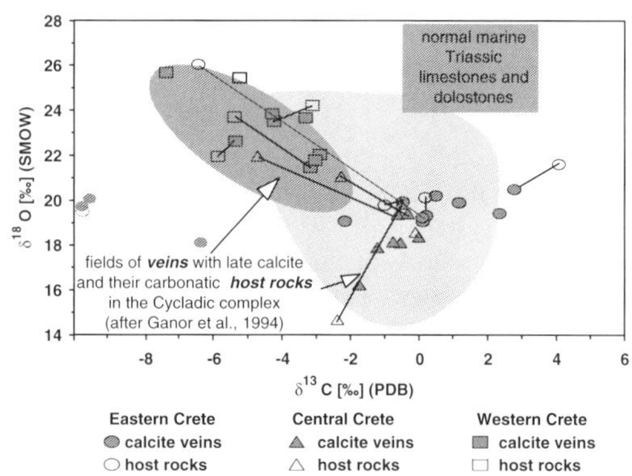

Figure 2. Distribution of the $\delta^{13}C$ and $\delta^{18}O$ values in the calcite veins and in the calcite fraction of the host rocks. Vein and the corresponding host rock samples are connected with a straight line.

1986), a relationship similar to the trend in Figure 2 was also reported in veins from metamorphic terrains. Ganor et al. (1994) explained such a relationship by mixing of a late fluid depleted in ^{13}C with fluids in contact with marbles equilibrated with the intercalated silicate rocks.

3.2 *Trace element contents*

Early HP (aragonite-)marble was found to be free of Mg, but contains a significant amount of Sr, whereas late vein forming calcite has low Sr, but commonly has high contents of Mg. Due to the preferred incorporation of Sr into aragonite and phase equilibrium considerations, early aragonite has the highest Sr contents, but becomes increasingly depleted as

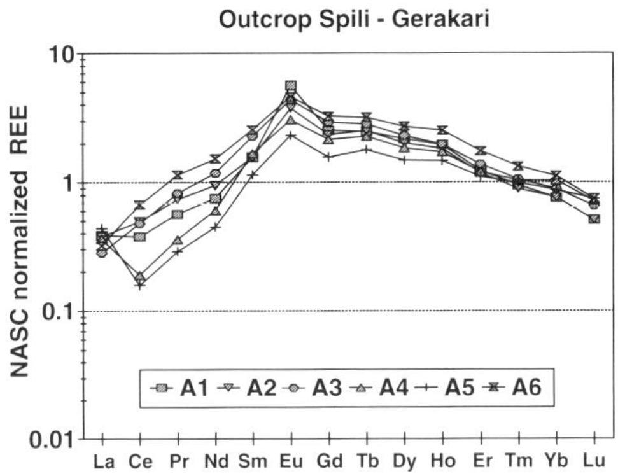

Figure 3. NASC normalized REE patterns in calcite veins (distinct signatures) from an outcrop near to the localities Spili – Gerakari. Schistosity conforming veins (A4, A5) show distinct negative Ce anomalies, suggesting interaction with host rocks of marine origin.

reaction progresses (Theye & Seidel 1993). Later, during the retrograde transformation a Sr-rich calcite will form which is unstable and gradually loses its Sr content, but incorporates some Mg from the adjacent dolomitic rocks. Consequently, in latest stages, a high-Mg calcite with low contents of Sr forms. The distribution of the NASC-normalized REE contents of the calcite veins shows a characteristic roof shaped pattern, typical for carbonate precipitating from crustal fluids interacted with pelitic sediments (Hecht et al. 1999).

4 QUARTZ VEINS

The $\delta^{18}O$ values of the quartz veins are between 13 and 26‰ (V-SMOW), but most of them fall in a range typical for metapelitic rocks (14–18‰). The oxygen isotopic composition of discordant veins scatter over a wider range and tend to be higher (14–24‰) than those of veins concordant with schistosity or bedding (14–18‰). An increasing trend from west (14–16‰) to east (18–22‰) can be also observed. The uniformity of the oxygen isotopic data in veins with different structural positions studied in detail in a single outcrop is in agreement with observations in thin sections that suggest that all the vein types suffered a plastic deformation probably in the framework of a single tectonic episode (B. Stöckhert, pers. comm.).

Bulk $\delta^{18}O$ values of the quartz rich host rocks are generally similar to those of veins concordant with schistosity, but decrease with increasing amount of phyllosilicates (e.g. 10.9‰ in phyllite, 13.9‰ in mica bearing quartzite and 15.1‰ in quartzite), indicating the isotopic equilibration of the host rock minerals. In contrast, the $\delta^{18}O$ of late formed discordant veins is systematically 1–2‰ higher compared to the host rocks.

To estimate the state of isotopic equilibration of the quartz veins with the surrounding rocks, the chemical index "I" according to Garlick (1966) was calculated using rock analyzes recalculated on a carbonate free basis. Considering an equilibration temperature of 390°C, rock samples from western Crete plot along a line that extrapolated to I = 1 (pure quartzite) points to a $\delta^{18}O$ value which is identical with the average of the quartz veins concordant with schistosity (15.5‰), suggesting the equilibration of the rock forming minerals with the vein quartz.

5 OXYGEN ISOTOPE THERMOMETRY

Isotope thermometry on vein minerals (calcite, quartz, carpholite) indicates a broad thermal range of formation. Except for a discordant quartz–calcite vein giving 180°C, temperature estimates based on the equations of Sharp & Kirschner (1994) vary between 245 and 463°C, with most of the values falling between 300 and 350°C. Using the equations of Zheng as given in Henry et al. (1996), mineral pairs involving the HP mineral carpholite yielded temperatures between 286 and 463°C, with consistently higher values indicated by the calcite–carpholite pair (376–463°C) as compared to quartz–carpholite (245–463°C).

6 FLUID INCLUSIONS

6.1 *Isotopic composition*

Fluid inclusions released by thermal decrepitation delivered excessively high δD values (from −37 to 31‰ for inclusions in quartz; and from −37 to −8‰ for those in calcite), which are compatible only with fluids evolved from highly concentrated basinal brines. Nevertheless, except for some of the quartz samples, the fluid-inclusion data plot in a δD vs. $\delta^{18}O$ diagram in the field of metamorphic fluids (Fig. 4). The vague trend of increasing $\delta^{18}O$ values parallel to decreasing δD indicates a local buffering of the saline connate waters by the surrounding rocks and the admixing of water from metamorphic devolatilization of hydrated minerals (clay minerals, gypsum). The influence of the host rocks on the fluid composition is supported by the $\delta^{13}C$ values of the CO_2 in fluid inclusions, which locally is as low as −27.5‰.

6.2 *Chemical composition*

The chemical composition of the fluid inclusions was evaluated by crush leaching. Fluids released from

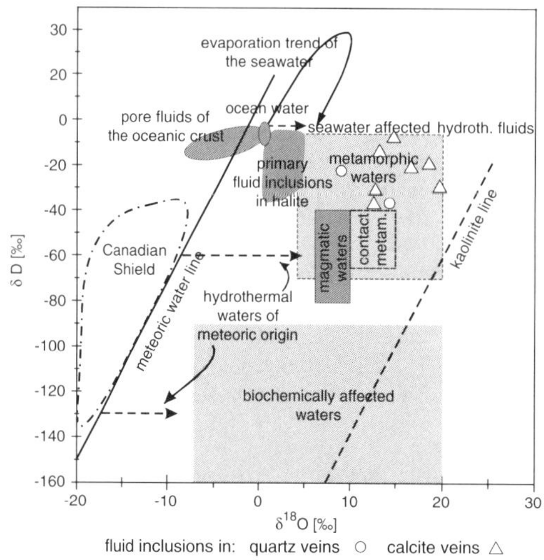

Figure 4. Isotopic composition of the fluid inclusions in quartz and calcite veins (Fields are compiled after Sheppard 1986).

quartz samples containing predominantly primary inclusions are essentially brines of Na-Cl composition (with up to 25 wt. % salinity), depleted in Mg and in sulfate, and enriched in K, Ca and in trace elements like Sr, Li, Rb, and F. High contents (up to several hundred mg/L) of heavy metals like Zn, Pb, Cu and Ni were recorded in the inclusions with high salinity. Fluids with high Na/K ratios generally have higher contents of Sr and Ba, whereas those with low Na/K ratios are enriched in Rb, Cs and U. These compositional features together with the Na-Cl-Br systematics of the fluids suggest diagenetically altered seawater as their origin, the composition of which was gradually changed due to interaction with the surrounding rocks and due to their mixing with fluids derived from metamorphic dehydration reactions, including fluids from a presumably deeper seated slab of continental crust.

7 CONCLUSIONS

Quartz and calcite veins of the PQ Unit were formed during different stages of the tectonic evolution of the area. Stable isotopic and chemical data indicate that the mineralizing fluids evolved from progressively heated connate waters/formational brines buried with the subducted and accreted sediments, which were mixed with metamorphic fluids generated by dehydration reactions (gypsum, clay minerals). Fluids equilibrated during the HP-LT metamorphic stage with the surrounding rocks. Their solute load developed from a mixture between the initial brines and the dissolution or reactions with the minerals of the surrounding rocks. The chemical composition of the fluid inclusions indicate the participation of fluids from the deeper subducted (granitic?) continental crust. During peak metamorphism, fluid movement had chiefly a pervasive character, with low fluid/rock ratios and a short distance of transport, leading to the formation of veins in form of local segregations. Further veins formed along extensional fractures during exhumation. Local disequilibria between discordant veins and host rocks suggest channeled fluid flow allowing longer transport along small fracture systems and higher time-integrated fluid/rock ratios at this stage.

ACKNOWLEDGEMENTS

The project was funded by the Deutsche Forschungsgemeinschaft. T. Vennemann measured the isotopic composition of the fluid inclusions. Some of the quartz samples were analyzed in the stable isotope laboratory of J. Hoefs. B. Stöckhert made valuable comments and suggestions on an extended version of the manuscript. A part of the samples were made available to us by M. Küster and J. Krahl. We express our gratitude to all of them.

REFERENCES

Ganor, J., Matthews, A. & Schliestedt, M. 1994. Post-metamorphic low $\delta^{13}C$ calcite in the Cycladic complex (Greece) and their implications for modeling fluid infiltration processes using carbon isotope compositions. *European Journal of Mineralology* 6: 365–379.

Garlick, G.D. 1966. Oxygen isotope fractionation in igneous rocks. *Earth. Planet. Sci. Lett.* 1: 361–368.

Hecht, L., Freiberger, R., Gilg, A.H., Grundmann, G. & Kostitsyn, Y.A. 1999. Rare earth element and isotope (C, O, Sr) characteristics of hydrothermal carbonates: genetic implications for dolomite-hosted talc mineralization at Göpfersgrün (Fichtelgebirge, Germany). *Chemical Geology* 155: 115–130.

Henry, C., Burkhard, M. & Goffé, B. 1996. Evolution of synmetamorphic veins and their wallrocks through a Western Alp transect: no evidence for large-scale fluid flow. Stable isotope, major- and trace-element systematics. *Chemical Geology* 127: 81–109.

Jolivet, L., Goffé, B., Monié, P., Truffert-Luxey, C., Patriat, M. & Bonneau, M. 1996. Miocene detachment in Crete and exhumation P-T-t paths of high-pressure metamorphic rocks. *Tectonics* 15: 1129–1153.

Küster, M. 1994. *Fluid-Einschlüsse in niedrigtemperierten Hochdruckmetamorphiten und Geothermometrie am Beispiel der Phyllit-Quarzit-Einheit auf Kreta (Ägäis).* Unpub. Ph.D. Thesis. Fakultät für Geowissenschaften an der Ruhr-Universität Bochum.

Seidel, E., Kreuzer, H. & Haare, W. 1982. A late Oligocene / Early Miocene high pressure belt in the External Hellenides. *Geologisches Jahrbuch* 23: 165–206.

Sharp, Z.D. & Kirschner, D.I. 1994. Quartz-calcite oxygen isotope thermometry: A calibration based on natural isotopic variations. *Geochim. Cosmochim. Acta* 58: 4491–4501.

Sheppard, S.M.F. 1986. Characterization and isotopic variations in natural waters. *Reviews in Mineralogy* 16: 165–180.

Stöckhert, B., Wachmann, M., Küster, M. & Bimmermann, S. 1999. Low effective viscosity during high pressure metamorphism due to dissolution precipitation creep: the record of HP-LT metamorphic carbonates and siliciclastic rocks from Crete. *Tectonophysics* 303: 299–319.

Theye, T. & Seidel, E. 1991. Petrology of low-grade high-pressure metapelites from the External Hellenides (Crete, Peloponese). A case study with attention to sodic minerals. *European Journal of Mineralology* 3: 343–366.

Theye, T. & Seidel, E. 1993. Uplift-related retrogression history of aragonite marbles in Western Crete (Greece). *Contributions to Mineralogy and Petrology 114: 349–356.*

Thomson, S.N., Stöckhert, B. & Brix, M.R. 1998. Thermochronology of the high-pressure metamorphic rocks of Crete, Greece: implications for the speed of tectonic processes. *Geology* 26: 259–262.

Valley, J.W. 1986. Stable isotope geochemistry of metamorphic rocks. In J.W. Valley, H.P. Taylor, Jr. & J.R. O'Neil (eds.), *Stable Isotopes in High Temperature Geological Processes. Reviews in Mineralogy* 16: 445–490. Washington, D.C.: Mineralogical Society of America.

Water–Rock Interaction, Wanty & Seal II (eds)
© 2004 Taylor & Francis Group, London, ISBN 90 5809 641 6

Modeling contact metamorphism of siliceous dolomite via kinetic control of overall reactions

E.W. Bolton, D.M. Rye & J.J. Ague
Yale University, Department of Geology and Geophysics, New Haven, CT, USA

A. Lüttge
Rice University, Department of Earth Science, Houston, TX, USA

ABSTRACT: We present results of a two-dimensional model of contact metamorphism of siliceous dolomite. Competing rates of fluid advection, basal heating, and reaction determine the spatial and temporal evolution of the system initially composed of quartz and dolomite. Eight minerals are included in the system composed of the elements Si, Ca, Mg, C, H, and O (CMS), and a supercritical binary fluid composed of CO_2 and H_2O. The kinetic model includes 15 overall reactions with rates based on experimental data. Thermodynamic data of minerals and properties of fluid mixtures allow the calculation of the change of Gibbs free energies of reactions throughout temperature, pressure, and compositional space, even far from equilibrium. Reaction rates based on the deviation from equilibrium and mineral-surface areas combine with a self-consistent calculation of the flow of heat and fluid to yield temperature-composition pathways that can differ substantially from traditional equilibrium-based models. Kinetic effects could be important for determining styles of flow and reaction near plutons.

1 INTRODUCTION

The central role of heat and fluid flow in contact aureoles has long been recognized. Use of thermodynamic equilibrium concepts has provided insight into the spatial distribution of mineral phases in contact metamorphic settings. Both field work and numerical modeling have highlighted important aspects of the flow of heat and fluids, as well as the accompanying mineral reactions (cf. Ferry et al. 2002, Cui et al. 2002, as well as references therein). Several recent papers have focused on the relevance of kinetic control in metamorphic systems (Baxter & DePaolo 2000, 2002a, b, Lasaga & Rye 1993, Lasaga et al. 2000). Of the two-dimensional modeling studies, that of Balashov & Yardley (1998) seems unique in its use of kinetic reaction control. Here we extend the model introduced in Lüttge et al. (2004) to two spatial dimensions.

2 MATHEMATICAL FORMULATION

2.1 *Supercritical binary fluids: conservation and transport*

The supercritical mixture of CO_2 and H_2O is treated as a binary fluid (Bird et al. 1960), with pore velocities related to the barycentric (mass averaged) velocity. A mass–conservation equation includes reaction production and consumption of fluid species. For the fluid composition equation, in addition to the normal advection and diffusive fluxes, full dispersion was included for the enhanced mixing effects of the porous medium. In addition, a simple compaction model was used (see below). Darcy's flux could be partitioned into a buoyancy generated part (stream function) and a reaction produced flow component (velocity potential). Fluid pressure was assumed essentially hydrostatic below a lid of fixed pressure. Through this assumption and an anelastic approximation for the fluid mass conservation equation, sound waves were filtered out.

2.2 *Thermodynamic and kinetic considerations*

We used the thermodynamic database of Berman (1988, and updates) to calculate the Gibbs free energy of the mineral phases at metamorphic conditions. We used the fugacity and equation of state estimates for H_2O-CO_2 mixtures of Kerrick & Jacobs (1981) and Jacobs & Kerrick (1981). Consider:

$$3\,\text{dolomite} + 4\,\text{quartz} + 1\,H_2O \Leftrightarrow 1\,\text{talc} + 3\,\text{calcite} + 3\,CO_2 \quad (1)$$

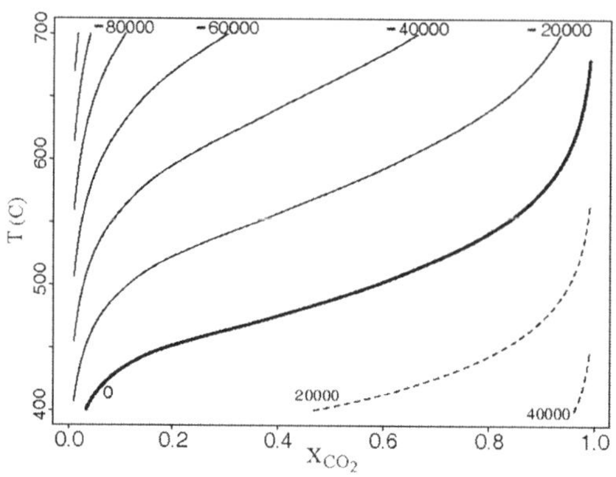

Figure 1. The change in the Gibbs free energy of reaction (ΔG, in J/mol) for reaction 1 as written: (3 dolomite + 4 quartz + 1 $H_2O \Leftrightarrow$ 1 talc + 3 calcite + 3 CO_2) at 3 kbar. ($\Delta G = 0$: bold solid, for equilibrium; $\Delta G < 0$: thin solid, favoring forward reaction; $\Delta G > 0$: dashed, favoring backward reaction). Calculated from Berman (1988) and Kerrick & Jacobs (1981).

The combination of these thermodynamic calculations for overall reactions yields results as shown in Figure 1. We used kinetic rates of overall reactions, normalized to one mole of CO_2 produced, extracted from experimental data (cf. Lasaga et al. 2000). The formulation for the rate dependence on deviation from equilibrium and the evolving mineral surface areas is described in Lüttge et al. (2004). Thermal evolution depended on advection of heat by the fluid, diffusion in the matrix, and heats of reaction (both endothermic and exothermic as required by the reaction rates and enthalpies).

2.3 *Subgridscale grain model for grain size, porosity, permeability and compaction*

Each grid point of the finite difference model contains a host of mineral grains, spaced at a characteristic distance in a cubic array. Fluid was assumed to bathe the entire surfaces of the mineral grains (making effective rates of reaction faster than a model with fluid bathing only grain triple junctions). Grain sizes were initiated on the grid consistent with the assumed modal abundance of each mineral. Minute quantities of each low-abundance mineral were imposed as "seeds" for nucleation and growth. Grain sizes changed in each grid point consistent with the reactions occurring and conservation of mass. From the nucleation densities and the evolving grain sizes, the porosity, mineral surface areas, and permeability (based on the fluid gap between grains) could be updated.

A simple compaction model was used to modify the porosity, which was compatible with pressure and temperature conditions utilized. This procedure yielded a characteristic compaction time of 10 yrs (cf. Balashov & Yardley 1998, Zhang et al. 1994), which contributed some to flow expulsion from the reaction zone during periods with net volume loss of the solid phases.

2.4 *Solution method*

The two-dimensional model is based on the finite difference method. An assessment is made at each local node for which phases are present and which reactions are possible based on the thermodynamics described above. The possible reactions are then allowed to proceed using the kinetic rate equations (provided that mineral reactants are present in sizes larger than the "seeds"), the volume change of the minerals is updated, and the molar changes of CO_2 and H_2O are computed (the only fluid components transported), leading to a new molar volume of the fluid mix. The Darcy velocities are then calculated from a combination of buoyancy effects and the fluids generated by the reactions (through a sparse matrix solver of the coupled Poisson equations). Overstepping of reactions can occur via transport of the fluid phase, even when the solid phases remain in place or slightly compact. Thermal evolution is then calculated from advection effects and heats of reaction.

3 MODELING DESCRIPTION AND RESULTS

3.1 *Initial and boundary conditions*

We present results for two cases that differ only by their initial grain size (and spacing). Both cases presented here used ramped heating on the left portion of the lower boundary, with a 0.1 C/yr maximal rate. The sides were thermally insulating, while the top and right-lower boundaries were constant temperature. All sides were no-flux with respect to fluid, except for the top (where fluid was allowed to escape to maintain constant fluid pressure). The finite difference resolution was 67 grid points in *x* (horizontal) and 61 grid points in *z* (vertical). The initial mineral composition was 85% dolomite and 15% quartz, with a porosity fraction of 0.004.

3.2 *Small grain size case*

This case was initiated with grain centers with 0.4 mm spacing. Figure 2A shows results for fluid velocities and composition after 800 yrs of heating. Note the dramatic expulsive form of the flow (with dominantly down-temperature flow). In Figure 2B the final T-X pairs are displayed for each grid point in the simulation domain, superimposed on the $\Delta G = 0$

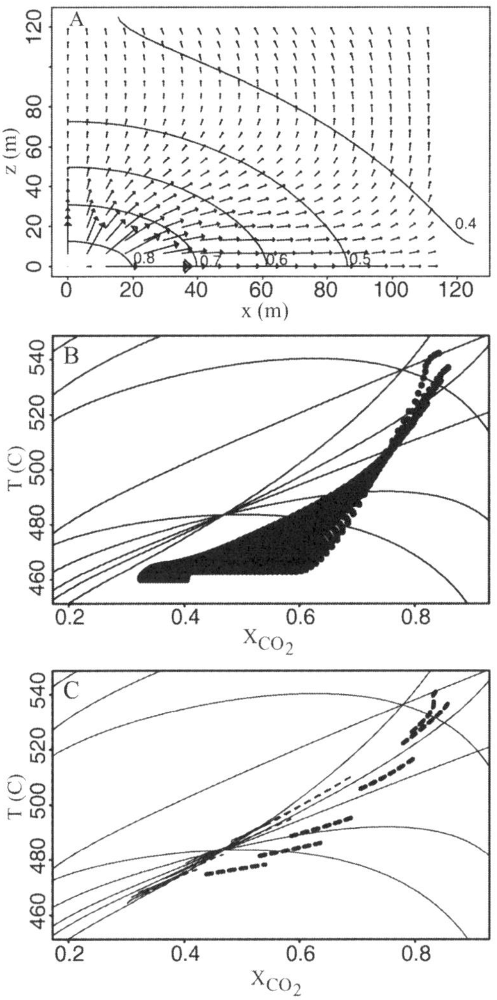

Figure 2. Results from the small grain-size simulation. A. Darcy flux vectors (0.75 m/yr maximum) with CO_2 contours at 800 yrs, note reaction driven flow away from heated zone, B. T-X pairs at 800 yrs, superimposed on the $\Delta G = 0$ curves for 15 reactions, C. T-X pathways for selected point in the lower left quadrant of the domain during the various time intervals (thin solid: 160–320 yr; thin dashed: 320–480 yr; thick dashed: 640–800 yr). See also text.

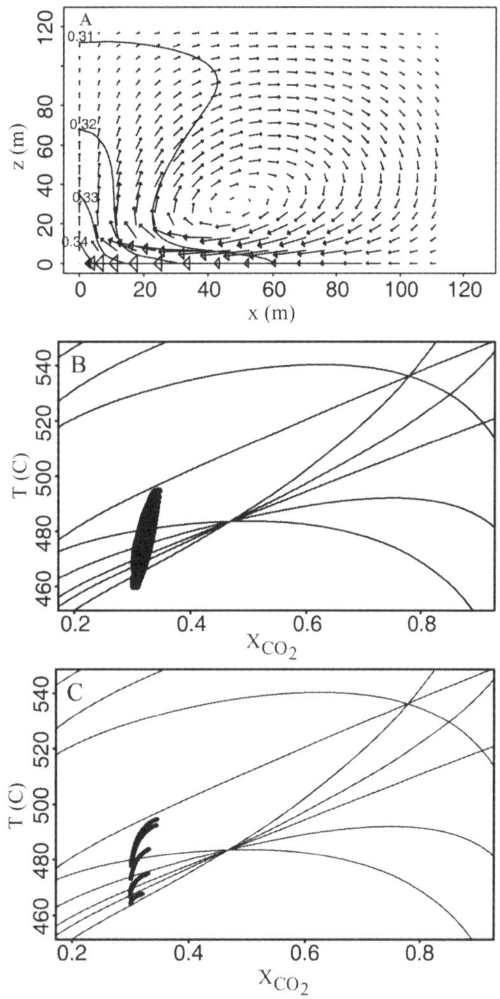

Figure 3. Results from the large grain-size simulation at 320 yrs. A. Darcy flux vectors (4 m/yr maximum) with CO_2 contours: note buoyancy driven flow cell, B. T-X pairs, C. T-X pathways during the 160–320 yr interval. See also Figure 2 caption and text.

reaction curves. The lower left curve for $X_{CO_2} < 0.45$ corresponds to the reaction of Equation 1 above. Figure 2C shows the T-X evolution of selected points in the lower-left quadrant of the domain. In general, the high X_{CO_2} of the expulsed fluid tends to drive much of the domain toward high X_{CO_2} well below the first dolomite–quartz reaction, rather than following univariant reaction curves. In addition to results shown, a thermal wave moved outward from the heat source in a form similar to the fluid composition (Fig. 2A). Waves of mineralogical changes also moved out in pulse-like form, mimicking time-dependent isograds.

3.3 *Large grain size case*

This case was initiated with grain centers with 4.0 mm spacing, resulting in larger permeability and lower mineral surface area to volume ratios. Both of these factors could contribute to the difference in flow style as compared to the small grain size case discussed above. Results in Figure 3A show a buoyancy cell with a weak plume in the fluid composition. This case has fluid flowing toward higher temperature at the base. Figure parts B and C show dramatic overstepping of the univariant curves, with trends toward higher CO_2.

4 CONCLUSIONS

Kinetic modeling of contact metamorphism using realistic heating and reaction rates results in temperature/composition pathways that can deviate strongly from traditionally assumed equilibrium pathways. Not only does this fact highlight the need for additional kinetic data, it also calls into question the reliability of conclusions drawn from models built on equilibrium assumptions (fluid-flux calculations, thermobarometry, and strict use of mineral assemblages in inverting P-T conditions). Metastable reactions are an important part of these calculations, and must be included in kinetic models that use overall reactions. Future modeling will focus on longer simulations, as well as both prograde and retrograde heating cycles.

Our model calculations were based on overall mineral reaction kinetics. It would be more realistic to base the model on individual mineral dissolution–nucleation- and growth-reactions (Lüttge & Metz 1991). Such an extension could be based on recent theoretical developments (Lasaga & Lüttge 2001, 2003), and on experimental data as it becomes available. In addition to the mineral reactions, more data are needed on the solubilities of minerals in supercritical metamorphic fluids. Future models should also move beyond the anelastic approximation, yet include some effective way of filtering out sound waves. Other matrix effects, including fracture dynamics, should be considered.

ACKNOWLEDGEMENT

We received research support from DOE-BES grants DE-FG02-90ER14153 and DE-FG02-01ER15216.

REFERENCES

Balashov, V.N. & Yardley, B.W.D. 1998. Modeling metamorphic fluid flow with reaction-compaction-permeability feedbacks. *Am. J. Sci.* 298: 441–470.

Baxter, E.F. & DePaolo, D.J. 2000. Field measurements of slow metamorphic reaction rates at temperatures of 500° to 600°C. *Science* 288: 1411–1414.

Baxter, E.F. & DePaolo, D.J. 2002a. Field measurement of high temperature bulk reaction rates: I. Theory and technique. *Am. J. Sci.* 302: 442–464.

Baxter, E.F. & DePaolo, D.J. 2002b. Field measurement of high temperature bulk reaction rates: II. Interpretation of results from a field site near Simplon Pass, Switzerland. *Am. J. Sci.* 302: 465–516.

Berman, R.G. 1988. Internally-consistent thermodynamic data for minerals in the system Na_2O-K_2O-CaO-MgO-FeO-Fe_2O_3-Al_2O_3-SiO_2-TiO_2-H_2O-CO_2. *J. Petrol.* 29: 445–522.

Bird, R.B., Stewart, W.E. & Lightfoot, E.N. 1960. *Transport Phenomena*. New York: Wiley.

Cui, X., Nabelek, P.I. & Liu, M. 2002. Numerical modeling of fluid flow and oxygen isotope exchange in the Notch Peak contact-metamorphic aureole, Utah. *Geol. Soc. Am. Bull.* 114: 869–882.

Ferry, J.M., Wing, B.A., Penniston-Dorland, S.C. & Rumble, D. III. 2002. The direction of fluid flow during contact metamorphism of siliceous carbonate rocks: new data for the Monzoni and Predazzo aureoles, northern Italy, and a global review. *Contrib. Mineral. Petrol.* 142: 679–699.

Jacobs, G.K. & Kerrick, D.M. 1981. APL and FORTRAN programs for a new equation of state H_2O, CO_2 and their mixtures at supercritical conditions. *Comp. Geosci.* 7: 131–143.

Kerrick, D.M. & Jacobs, G.K. 1981. A modified Redlich-Kwong equation for H_2O, CO_2 and H_2O-CO_2 mixtures at elevated pressures and temperatures. *Am. J. Sci.* 281: 735–767.

Lasaga, A.C. & Rye, D.M. 1993. Fluid flow and chemical reactions in metamorphic systems. *Am. J. Sci.* 293: 361–404.

Lasaga, A.C. & Lüttge, A. 2001. Variation of crystal dissolution rate based on a dissolution stepwave model. *Science* 291: 2400–2404.

Lasaga, A.C. & Lüttge, A. 2003. A model for crystal dissolution, *Eur. J. Min.* 15: 603–615.

Lasaga, A.C., Lüttge, A., Rye, D.M. & Bolton, E.W. 2000. Dynamic treatment of invariant and univariant reactions in metamorphic systems. *Am. J. Sci.* 300: 173–221.

Lüttge, A., Bolton, E.W. & Rye, D.M. 2004. A kinetic model of metamorphism: an application to siliceous dolomites. *Contrib. Mineral. Petrol.* 146: 546–565.

Lüttge, A. & Metz, P. 1991. Mechanism and kinetics of the reaction: 1 dolomite + 2 quartz $\leftrightarrow$ 1 diopside + 2 CO_2 investigated by powder experiments. *Can. Mineral.* 4: 803–821.

Zhang, S., Paterson, M.S. & Cox, S.F. 1994. Porosity and permeability evolution during hot isostatic pressing of calcite aggregates. *J. Geophys. Res.* 99(B8): 15,741–15,760.

Water–Rock Interaction, Wanty & Seal II (eds)
 ISBN 90 5809 641 6

Sulfur geochemistry of Paleozoic rocks, U.S. mid-continent: implications for large-scale fluid flow and ore genesis

M.B. Goldhaber
U.S. Geological Survey, Denver, Colorado, USA

ABSTRACT: The Paleozoic platform carbonates of the U.S. midcontinent were the aquifers for large-scale hydrothermal fluid flow in the early Permian. One consequence of this fluid flow system was the formation of world class Zn-Pb ore deposits. S isotopic data obtained on regionally distributed epigenetic sulfides formed along the flow paths suggest that H_2S was produced by two mechanisms. Thermochemical (abiologic) reduction generated isotopically heavy H_2S in the upper Bonneterre formation. This heavy sulfide can be traced widely in the Ozark region and was the S source for the Viburnum Trend district of southeast Missouri. A second and quantitatively much larger source was likely introduced by fluids migrating from the adjacent Arkoma Basin. The origin of this H_2S is hypothesized to be reduction of diagenetic pyrite by organic matter at elevated temperature in deeply buried basin sediments.

1 INTRODUCTION

This study addresses the geochemistry of H_2S generation and migration in the Ozark region of the south central United States. The Ozark region hosts two world-class Zn-Pb ore districts and several smaller districts in Paleozoic carbonate-rich platform sedimentary rocks (Fig. 1). These "Mississippi Valley-Type" (MVT) districts formed from saline basin brines at temperatures typically in the range of 100–150°C. Combined they contain hundreds of millions of tons of sulfide ore minerals and pyrite. Transport of the Zn, Pb, and other metals was by a hydrothermal flow system that extended hundreds of kilometers. Flow was driven by late Paleozoic Ouachita tectonism (Garven & Raffensperger 1997; Leach et al. 2001). The source of sulfide in the ores is, however, problematic, as is the site of sulfide generation; both local (district scale) and distant H_2S sources have been proposed (Goldhaber et al. 1995).

A regional, "far-field" approach is emphasized in this investigation, which combines stable isotopic, chemical analytic and petrographic study of samples remote from the ores to constrain MVT sulfur geochemistry. Ore districts reflect the impact of intense local, district scale or smaller processes such as fluid mixing and temperature change to form the immense quantities of metal sulfides localized there. Some of this complexity can potentially be simplified by examining rock–water interactions in aquifers that fed ore fluids to the deposits. A far-field approach

Figure 1. Map of mid-continent of the U.S. showing locations ore districts (gray shading) and location of cores sampled for this study (dots). Dotted line shows the location of the core fence of Figure 3.

implicitly assumes that ore-formation involved large-scale fluid flow over much of the Ozark region and thus, hydrothermal fluid migration was not a phenomenon localized to the immediate vicinity of the ores themselves. A second assumption is that hydrothermal fluids left trace sulfide phases during their migration through this large area.

2 SAMPLING STRATEGY AND ANALYTICS

The predominant sample medium analyzed in this study was 1 N HCl acid insoluble residues of the

Paleozoic platform carbonate rocks. Regional subsurface geochemical data obtained from analyses of tens of thousands of such insoluble residues prepared from drill-hole samples in the midcontinent region have shown that anomalous concentrations of metals, including Zn and Pb, are widespread in association with vuggy carbonate rocks of Cambrian through Mississippian age (Erickson et al. 1981). The vugs contain coarse dolomite formed from highly saline (MVT type) fluids at temperatures in the range 80–160°C. The residence of the anomalous metal concentrations is predominantly epigenetic pyrite/marcasite. This conclusion is based on chemical analyses of handpicked FeS_2 grains separated from the bulk residue (Erickson et al. 1981; Goldhaber et al. 1995) and SEM examination of polished grain mounts (this study). The link between this epigenetic vug-filling FeS_2 and ore fluids for MVT deposits is very clearly documented by comparison of Pb isotopic analyses on Pb-rich insoluble residue FeS_2 samples with Pb isotopic data on ore sulfides (Goldhaber et al. 1995).

The temperature range for formation of both the MVT ores and the far-field FeS_2 samples (as deduced from the paragenetically associated vuggy dolomite) is generally higher than the physiological limits of sulfate-reducing bacteria (Goldhaber 2003). Thus, non-bacterial processes must be considered in sulfide-generating reactions.

The insoluble residues were analyzed by oxidizing the sulfide phases to aqueous SO_4^{2-} using Br_2/aqua-regia. The resulting SO_4^{2-} was precipitated as $BaSO_4$, which was analyzed isotopically either at Global Geochemistry in Canoga Park CA, or in the laboratory of R.O. Rye at the USGS in Denver, CO.

3 ORE DEPOSIT ISOTOPIC COMPOSITIONS

The benchmark for comparison with the regional sulfur-isotope sample data is the sulfur isotope systematics of the ore districts. The range for the world-class Tri-State, and the Viburnum Trend (VT) plus genetically related Old Lead Belt (OLB) districts are shown as vertical bars on Figure 2. These two differ, with the VT/OLB being enriched in isotopically heavy S compared to the Tri-State district.

4 SULFIDE RESULTS

4.1 *Isotopically heavy ores of the VT*

In order to characterize the S isotopic values of fluid with a range of flow vectors towards the VT/OLB districts, $\delta^{34}S$ values on FeS_2 sampled from a core fence were measured (Fig. 1).

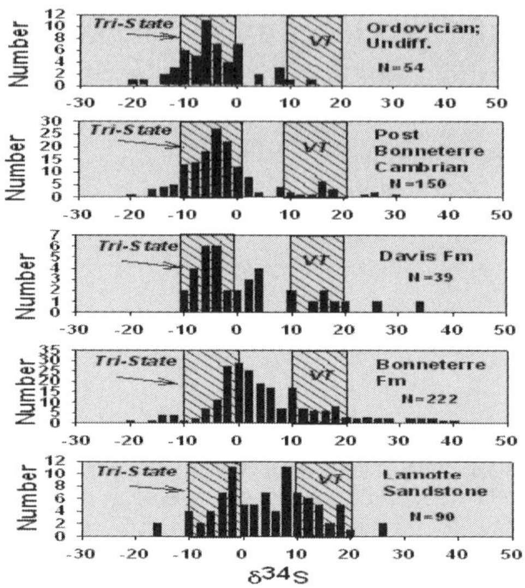

Figure 2. Histogram showing sulfur isotopic data from this study. Vertical hachured bars show range of isotopic values for the Tri-State district (left) and Viburnum Trend/Old Lead Belt (right). Histograms are in stratigraphic order.

Figure 3 shows $\delta^{34}S$ data from the segment of the stratigraphic section hosting the VT ores. This section extends from the top of the Cambrian Lamotte Sandstone through the carbonate dominant Bonneterre Formation, into the shale-rich Davis Formation or dolomite-dominated post Bonneterre Cambrian (undifferentiated) units, which commonly overlie the Bonneterre depending on geographic location. Importantly, VT ore is localized in the upper portion of the Bonneterre Formation.

There is a pronounced maximum in $\delta^{34}S$ values associated with the upper Bonneterre Formation (Sullivan Siltstone Member), and no maximum where this member is absent. This isotopic maximum extends to just west of the VT (Core 2; Fig. 1). A closely spaced eastward extension of this core fence through the VT (data not plotted) shows that the positive (i.e. isotopically heavy) $\delta^{34}S$ anomaly in the upper Bonneterre extends to the ore district but only for a limited distance east of the ore trend.

The isotopically heavy sulfides present in the core fence are reflected in the overall data set (Fig. 2) as a tail to positive values. This tail progressively decreases in prominence up the stratigraphic section and is nearly absent in the Ordovician-hosted samples.

The data indicate the transport of isotopically heavy H_2S in the deeper part of the stratigraphic section, and specifically the upper Bonneterre Formation (the ore host). We attribute these positive sulfur-isotope values (>10‰) to H_2S that was formed by thermochemical sulfate reduction (TSR). TSR is

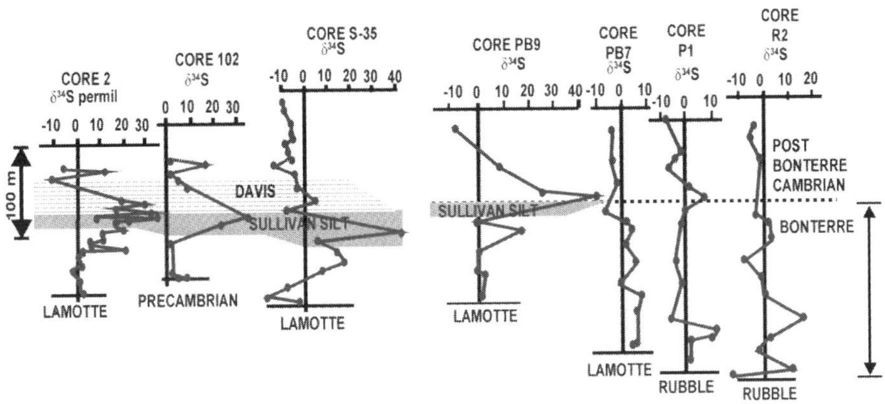

Figure 3. Sulfur isotopic data for core fence shown in Figure 1. East is on the right, west is on the left. Core 2 is adjacent to the Viburnum Trend. Note the heavy values in the upper Bonneterre (Sullivan Siltstone Member).

the abiologic reduction of SO_4^{2-} to H_2S by organic matter (Goldhaber & Orr 1994) and is known to produce H_2S with only a small isotope fractionation compared to starting SO_4^{2-} (Ohmoto & Goldhaber 1997), which is itself typically heavy. A limited number of organic carbon analyses indicate that the Sullivan Siltstone Member is enriched in organic carbon compared to other portions of the Bonneterre Formation consistent with this hypothesis.

4.2 *Sulfide source for the Tri-State ores*

The most conspicuous feature in the overall data set (Fig. 2) is the presence of a well-defined sulfur isotopic peak at about −5‰. Sulfide with this isotopic composition predominates in the Post Bonneterre Cambrian and Ordovician portion of the stratigraphic section, and occurs over the entire study area. Limited sulfur isotopic data from Mississippian rocks (not plotted) shows the same narrow range of values. It is significant that this far field sulfide closely matches the range of values for the Mississippian hosted Tri-State District, and is similar to values for ore and gangue sulfides for several smaller MVT districts not discussed here. Given the links described above between Pb isotopes in ores and far field sulfides, it is unlikely that the similarity between this peak in far field values and Tri-State ores is coincidental. The conclusion drawn from the S isotope data is that ore and gangue sulfides in the Tri-State district were formed from dissolved H_2S transported by the same large-scale flow system as formed the far field sulfides.

If this conclusion is correct, then the "light" far field sulfides and ores have both a narrow isotopic spread and a homogeneous source in common. In addition, the mass of this sulfide was substantial. A rough calculation for the post Bonneterre Cambrian units alone, assuming an average thickness of 160 m and an average sulfide content of 0.08 wt. % (based on data from this study) is about 2.5×10^{10} metric tons of such S in the study area. This is a minimum estimate because Ordovician and Mississippian rocks contain sulfide of this isotopic composition.

It is unlikely that the H_2S that formed the epigenetic sulfide was formed within the platform carbonate area of the midcontinent. The potential platform sources are unlikely to account for either the isotopic composition or mass of H_2S required. First, sulfide with this isotopic composition was not produced by TSR because it is too light and organic C is very low in these platform carbonates, which formed under high energy conditions not conducive to organic matter accumulation. It was not produced bacterially because the temperatures recorded in cogenetic dolomite are too high for bacterial viability. Where did the H_2S come from? The likely candidate for this enormous amount of sulfide is an adjoining sedimentary basin.

Figure 4 shows a cross-section into the Arkoma Basin from the platform edge towards the south. A substantial thickness of sediment was deposited in this basin, much of which is now eroded (Byrnes & Lawyer 1999). Also shown on Figure 4 are estimated temperature contours for the time of ore-formation (late Pennsylvanian-early Permian) by Garven et al. (1993) based on burial and fluid flow modeling.

At temperatures greater than about 200°C, pyrite can be reduced to pyrrhotite and H_2S:

$$FeS_2 + H_2 = FeS + H_2S \quad (1)$$

The source of reducing power was likely organic matter according to reaction 2, where organic matter is abbreviated as CH_2O:

$$2FeS_2 + CH_2O + H_2O = 2FeS + 2H_2S + CO_2 \quad (2)$$

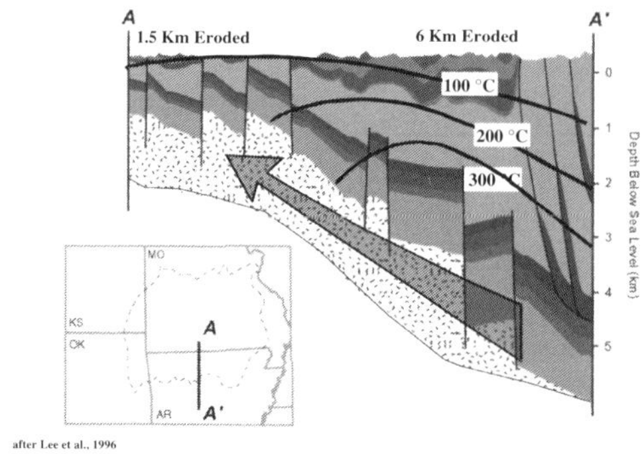

Figure 4. Cross section of Arkoma basin (see inset for location of section) showing temperature contours at the time of ore formation. Note the elevated temperatures in deeper part of basin. Arrow shows direction of hydrothermal fluid flow during late Paleozoic.

Organic matter is abundant in a number of strata in the Arkoma Basin stratigraphic section. Much of the stratigraphic section in this basin contains organic matter-bearing marine shale and sandstone (Byrnes & Lawyer 1999) and was a major oil and gas source. Reaction 2 potentially can provide enormous amounts of H_2S to fluids migrating onto the platform. Furthermore, the isotopic composition of the resulting H_2S may have been in the correct range. The formation of H_2S by reaction 2 does not involve a large isotope fractionation, so the resulting aqueous sulfide inherits its isotopic composition from its precursor pyrite (Ohmoto & Goldhaber 1997). Modern pyrite has a median isotopic composition of about -17‰ (Goldhaber 2003). However, in Cambro-Ordovician time, when a significant portion of the Arkoma Basin sedimentary section was deposited, seawater sulfate was about 10‰ heavier than today, resulting in pyrite heavier by the same amount than modern pyrite and in the range required to explain the observed platform values.

5 CONCLUSIONS

A large database of S isotopic measurements has been compiled on epigenetic sulfide, predominantly pyrite, from hydrothermally altered Paleozoic platform carbonate rocks in the U.S. midcontinent. The samples studied came from the aquifer for fluids responsible in the latest Paleozoic for formation of large Zn/Pb ore districts. Fluid inclusion data record temperatures above physiological limits for sulfate reducing bacteria. Thus, sulfide formed by abiologic processes. The results indicate that the isotopically heavy sulfides of the main stage ore of the Viburnum Trend were formed by thermochemical sulfate reduction in fluids migrating within the upper Bonneterre Formation. Organic matter is required for this reaction and the upper Bonneterre was enriched in organic matter compared to other portions of the section. In contrast, the fluids migrating higher in the section contained H_2S with a rather uniform and slightly negative isotopic composition (~-5‰). There is no evident source for this enormous amount of sulfide within the carbonate platform rocks. An external source of H_2S is proposed from the adjoining sedimentary basin. This aqueous sulfide source is hypothesized to have formed by reduction of pyrite to form pyrrhotite and H_2S.

REFERENCES

Byrnes, A.P. & Lawyer, G. 1999. Burial, maturation, and petroleum generation history of the Arkoma Basin and Ouachita Foldbelt, Oklahoma and Arkansas. *Natural Resources Research* 8(1): 3–26.

Erickson, R.L., Mosier, E.L., Odland, S.K. & Erickson, M.S. 1981. A favorable belt for possible mineral discovery in the subsurface Cambrian rocks in southern Missouri. *Econ. Geol.* 76: 921–933.

Garven, G., Ge, S., Person, M.A. & Sverjensky, D.A. 1993. Genesis of stratabound ore deposits in the Midcontinent basins of North America; 1, The role of regional groundwater flow. *Am. J. Sci.* 293(6): 497–568.

Garven, G. & Raffensperger, J.P. 1997. Hydrogeology and geochemistry of ore genesis in sedimentary basins. In H. Barnes (ed.) *Geochemistry of Hydrothermal Ore Deposits; 3rd Ed.*: 125–189. New York: John Wiley & Sons.

Goldhaber, M.B. 2003. Sulfur Rich Sediments. In F.T. Mackenzie (ed.) *Sediments, Diagenesis, and Sedimentary Rocks, Volume 7; Treatise on Geochemistry*. Elsevier.

Goldhaber, M.B., Church, S.E., Doe, B.R., Aleinikoff, J.N., Brannon, J.C., Podosek, F.A., Mosier, E.L., Taylor, C.D., & Gent, C.A. 1995. Lead and sulfur isotope investigation of Paleozoic sedimentary rocks from the southern Midcontinent of the United States: Implications for paleohydrology and ore genesis of the Southeast Missouri lead belts. *Econ. Geol.* 90: 1875–1910.

Goldhaber, M.B. & Orr, W.L. 1994. Kinetic controls on thermochemical sulfate reduction as a source of sedimentary H_2S. In M.A. Vairavamurthy, M.A.A. Schoonen, T.I. Eglinton, G.W. Luther III & B. Manowitz (eds.) *Geochemical transformations of sedimentary sulfur* 612: 412–425. American Chemical Society.

Leach, D.L., Bradley, D., Lewchuk, M.T., Symons, D.T.A., de Marsily, G. & Brannon, J. 2001. Mississippi Valley-type lead-zinc deposits through geological time: implications from recent age-dating research. *Mineralium Deposita* 36(8): 711–740.

Ohmoto, H. & Goldhaber, M.B. 1997. Sulfur and carbon isotopes. In H.L. Barnes (ed.) *Geochemistry of Hydrothermal Ore Deposits*: 517–612. New York: John Wiley & Sons.

Water–Rock Interaction, Wanty & Seal II (eds)

About the role of water in the metamorphic alteration of ultrabasic rocks

A.A. Grigorian
ARSAT Center, Yerevan, Armenia

S.B. Abovian
Institute of Geological Sciences NAS, Yerevan, Armenia

ABSTRACT: Laboratory studies of a series of rock samples, and geological and geophysical investigations for ophiolite zones in Armenia are compared in this paper. Investigated peridotite and dunite samples, as well as their respective serpentinitic alteration products, show structural, mineralogical, and chemical differences. Hydration reactions are considered for the primary minerals – olivine and pyroxene. The possible role for hydrous minerals in the processes of serpentinization and deserpentinization at high thermobaric conditions is discussed. Fluid-rock interactions are shown to result in the alteration of physical and chemical properties of considered rocks. Serpentinites, being a rock rich in water, are typically related to tectonic discontinuities with relatively easy influx of water. A comparison with deep drilling data demonstrates complexity in metamorphic conditions and geodynamics of the Earth's crust in Armenia. Possible implications of the results are discussed regarding hydrological balance, the Earth's crustal geodynamics, and earthquake sources.

1 INTRODUCTION

It is very difficult to imagine geochemical processes occurring in the Earth's crust and mantle without the influence of water. Crustal water at different depths can affect rocks as a hydraulic pressure medium, as a solvent, and as chemically active reagent. Water-rock interaction plays an essential role in metamorphic processes. These processes were studied by field geological and geophysical investigations as well as by laboratory experiments. As determined by a literature survey, metamorphosed rocks in Armenia can be related to gabbro-peridotite complexes. These rock types include serpentinite, derived from automorphic transformation of ultramafic rock, as well as some other rock types, according to Abovian (1981). The original type of ultramafic rock may be determined by a simple petrologic examination even if the serpentinite has no relict minerals because of the chemical composition. Phase equilibrium involving different metastable phases as well as melting in a complex system like MgO-FeO-CaO-SiO_2 system including water, studied by numerous researchers in a wide range of pressures and temperatures, was reviewed by Herzberg (1995). Modern static high-pressure apparatus make it possible to obtain temperatures and pressures up to 3000 K and 300 Gpa, respectively, and thus simulate conditions at depths up to 500 km (Mueller et al. 2002). Recent publications highlight substantial interest in water transporting mechanisms via subduction and water recycling in mantle by studying high-hydrous materials at high P-T conditions (e.g. Bagdassarov & Dingwell 1993; Schilling et al. 2003). Such studies are also of importance in understanding partial melting of the mantle wedge and brittle faulting by increasing pore pressure in the crust caused by dehydration reactions of hydrous minerals (Peacock 1993; Kirby et al. 1996).

Rheology and microstructures of olivine aggregates deformed in torsion were studied using a high-pressure and high-temperature torsion apparatus at 1473 K and 300 MPa (Bystricky et al. 2000). Serpentinization and deserpentinization are considered to be major processes in metamorphism, especially for the problem of ophiolite-belt evolution (e.g. Coleman 1977; Abovian 1981). Both serpentinization and deserpentinization processes one may consider as a complicated process, involving water as an active agent, although the possibility of water-absent equilibrium at high P-T conditions is also known, e.g. for the talc + 3 brucite = 2 serpentine reaction (Pugin et al. 1969).

This paper presents some recently obtained results on laboratory and field investigations of the processes of serpentinization and deserpentinization in some rocks. Considered results are combined with previous laboratory and field data on geologic and geochemical evolution of ultrabasic rocks from the ophiolite belt of Armenia and are discussed to better understand the mechanism of rock metamorphism.

2 EXPERIMEMENTAL

2.1 *Methods*

Laboratory studies were conducted using a series of optical, spectral, thermal and chemical analytical techniques. An optical microscope was used for surface observations as well as for petrographic and mineral analysis to study bulk textural and mineralogical features. X-ray diffraction (XRD), IR-Spectral analysis and Electron Microscopy were used for surface investigations. Differential-thermal analysis (DTA) was used to study water extraction and to simulate the deserpentinization process at standard pressure in ambient inert gas and temperatures up to 1500 K. Chemical and elemental compositions were determined by standard chemical and spectral analysis, respectively. An apparatus with 0.5 GPa and 1200 K confining pressure and temperature, developed by Veguni & Bdoyan (1987), was used to obtain high P-T data to simulate deep crustal conditions.

Ultrabasic and metamorphosed rock samples were collected in different regions, determined previously as regions with active metamorphic processes and generally associated with ophiolite belt of Armenia on the basis of investigations by Abovian (1981). Peridotite and dunite, as well as their serpentinized equivalents, collected mainly at the Lake Sevan ophiolite zone, were of special interest. Some results on spectroscopic studies in two rock samples: #4247 (partly serpentinized lherzolite) and #4248 (serpentinite), collected at the Ashok's deposit, are considered below.

2.2 *Optical measurements*

The structure of rock samples both before and after metamorphism was investigated by microscope analysis in a visible spectrum. An optically transparent thin section of a respective rock sample was prepared and after polishing was placed between the two thin glass plates for further petrographic analysis.

Rock samples as well as respective thin sections were investigated using an apparatus and measurement technology developed in ARSAT Center. A POLAM-111 optical microscope combined with a photographic camera was used for petrographic investigations both with and without a polarizing filter by transmitted light from an incandescent lamp with a blue filter. A low-sensitivity Kodak ProFoto100 color film was used to obtain higher resolution in zoomed image. After a standard development process and film scanning, further measurements were made on a PC with image processing software.

3 RESULTS AND DISCUSSION

Some results of petrographic analysis are presented below. The same numbers are used for a thin section of a rock sample as described above.

Rock sample #4247 is a fine-grained rock, with greyish-black color and porphyritic inclusions. It consists of olivine (70–77%), orthopyroxene (20–25%), and clinopyroxene (5–10%). Each olivine is surrounded by a rim of thin lizardite veinlets. Orthopyroxene grains are fresh and less than 10 mm in diameter, and are covered by numerous parallel cracks with local lizardite inclusions. Clinopyroxene clusters are fresh and formed small grains up to 2 mm size. Both olivine and pyroxene exhibit a wave-like extinction probably caused by uniaxial deformation. An opaque dusting is observed along with lizardite veinlets. An accessory mineral is represented by ferruginous spinel (pleonaste) with brownish translucency, partly darkened by isolated iron. The rock is partially serpentinizated, estimated up to 15–20%.

Rock sample #4248 is a fine-grained rock, greenish-grey in color, with irregular fractures. The sample has partly interlocking network-ribbon texture because of the alteration of olivine to lizardite and partly to brucite. All of the alteration rims are stretched in one direction, probably due to uniaxial deformation or to preferred direction of plastic flow during the metamorphism under high P-T conditions. The outer part of the rim consists of lizardite aggregate, and the inner part consists of serpentine (serpofite), and is slightly transparent in polarized light. The rock is cut by later serpentine veinlets with the opaque dustings in the middle. The accessory mineral is chromite with idiomorphic grain shapes, surrounded by thin lizardite edges.

Figures 1 and 2 are photomicrographs of these samples.

DTA and XRD analyses confirm, that the serpentinization process that resulted in the formation of rock sample #4248 was almost complete, and lizardite formation is accompanied by a minor amount of brucite. The same measurements made for the rock sample #4247 show moderate and low exothermic behavior at temperatures 955 and 1091 K, respectively. The endothermic behavior of lizardite decomposition and the exothermic behavior of forsterite formation were also estimated by DTA measurements.

Kinetics-based estimation of activation energy of dehydration reactions for several serpentinites shows

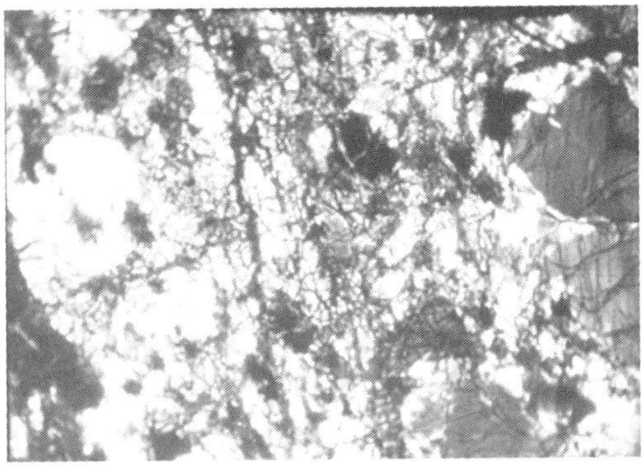

Figure 1. Microscope image of thin-section of partly serpentinized lherzolite. Rock sample #4247 – "fresh" rock.

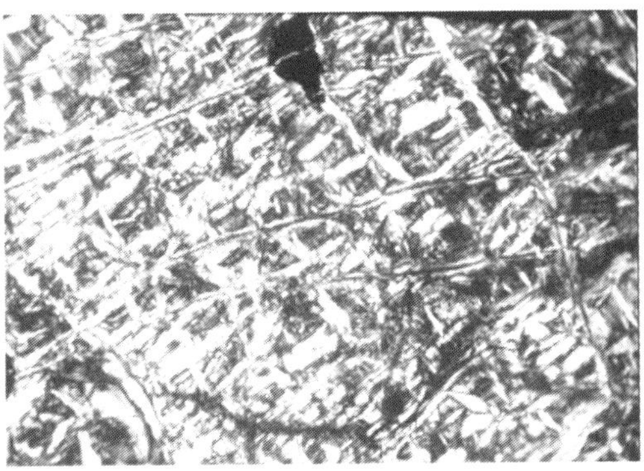

Figure 2. Microscope image of thin-section of serpentinite Rock sample #4248 – "metamorphosed" rock.

that the process is limited by diffusion and has an activation energy in the range of 142–227 kJ/mole (Khachatrian et al. 1981). However, the results are sensitive to the method of estimation, and need further specific investigation.

The results agree with our laboratory study of rocks under high P-T conditions. Deserpentinization of initial serpentinites and formation of secondary olivines, studied at high pressures up to 0.5 GPa, show a dependence on rock type within the temperature interval of 770–970 K (Haroutunian & Grigorian 2000). The metamorphism in some cases was accompanied by loss of a small amount of fluid of unknown composition. DTA analysis performed after cooling determined metamorphic reactions in the stressed rock sample. The low temperature DTA interval indicates water extraction.

Some aspects of the discrepancy in free water, bonded water and hydroxyl groups imbedded in the crystal lattice of a respective mineral, are discussed concerning their influence to the metamorphism process.

Laboratory experiments were performed to study the chemical composition of fluids in gas-and-liquid inclusions from magmatic and hydrothermal quartz and calcite at pressure 0.15 GPa. Homogenization temperatures of fluids were determined as about 720 K in magmatic quartz, 420–720 K in different generations of geothermal quartz, and 560–580 K in calcite (Saghatelyan & Grigorian 2000). These temperatures demonstrate the possibility for moderate to low temperatures of ore formation in calcite and quartz associations.

Some data on results of deep drilling at borehole "Oktemberyan-1", 4550 m confining depth, are outlined below. Extrusive rocks of basaltic composition with alternations of spilites, and sub-volcanic meta gabbro and diabase appear at lower than 3243 m depth, and intrusive inclusions of metagabbro and gabbro-pyroxenite are found from 3664 m depth. Igneous rocks exhibit fracturing, crushing and shear deformation. Textural and chemical analysis shows their association to the Vedi ophiolite zone of Armenia. Igneous rock samples of both basic and ultrabasic compositions with densities of 2.76–2.97 g/cm^3, moisture contents of 0.03–0.17% were separated from cores obtained at depths below 3225 m. A mean value of open porosity of 0.56% displays low near-logarithmic increases at depths between 3225 and 4495 m.

Water-mass balance in hydration-dehydration chemical reactions during the metamorphism processes at deep crustal conditions may lead to a critical alteration in rock volume and thus may have a crucial role in local stress fields. There is much significance for the latter in seismology, especially to better understanding earthquake-source mechanisms, as suggested by Grigorian (2002). One may consider that the crustal water budget is supported by balance among three main factors: both free and bonded water in rocks and hydrous minerals, water migration induced by pressure gradients through pores and cracks, and geodynamically-driven migration along with associated rocks rich in water. In the case of water deficiency, ultrabasic igneous rock may absorb water from ambient sedimentary rocks with higher hydrostatic pressure, as was suggested by Hess (1955), and also it is likely that some dense hydrous silicates could migrate toward greater depths (Peacock 1993).

One may consider the deep crust-mantle water amount in a fixed local volume at a relevant depth as a dynamic equilibrium determined by the two factors: input-output flux, and chemical reactions (hydration-dehydration and other chemical reactions). This perspective seems to be useful in relation to an approach developed by Woith et al. (1998) to study the association of hydrologic systems to earthquakes (Grigorian, unpubl.), as well as to a new approach used in the systematic geophysical study of the Lake Sevan mountain hydrogic system.

Better understanding of these processes is of great importance not only for our knowledge about the Earth's crust and mantle evolution, but also may have a wide range of practical implications in geophysics, ore-deposit exploration and the technology of material production.

4 CONCLUSIONS

1 The serpentinite sample, in contrast to the mildly serpentinized sample, exhibits distinctly visible quasi-ordered textures with macroscopic anisotropy.
2 A comparison of rock samples from the surface and rock samples collected from borehole core confirms a complexity both in metamorphic conditions and geodynamics of the Earth's crust in Armenia.
3 One may consider serpentinites as a probable source of fluids derived at corresponding depths of the Earth's deep crust and mantle during deserpentinization; therefore, a rock rich in serpentinites may be considered as a potential source of water. Water, released during deserpentinization, may cause further stages of metamorphism.
4 Serpentinization, deserpentinization, and metamorphism in general, are not simple one-stage processes, but are complicated multi-stage physical and chemical processes. A reliable model to describe these processes could be possibly designed based on the theory of nonlinear chain reactions with feedback. This needs further studies to obtain additional kinetic data by improved measurement techniques, partially by dynamic thermal measurements, as well as new data on metamorphism at high P-T conditions.
5 Laboratory studies combined with interpretations of geologic and geophysical data make it possible to define more accurately a geological chronology in the serpentinization of ultrabasic and basic rocks of Armenia.

ACKNOWLEDGEMENTS

The authors would like to thank all who collaborated within the research in Armenia and some other persons for assistance. One of the authors (A.A.G.) is especially grateful to: Dr. Nshan Zulumyan, IGIC, Armenia – for providing data (partly unpublished) and for helpful consultations on TGA, XRD techniques and Dr. Yuri Kagramanov, IGS NAS, Armenia – for providing deep drilling data; Dr. Wolfgang Kramer, GFZ-Potsdam, Germany – for fruitful discussions on ophiolite-related geology and also for his support and assistance; Dr. Heiko Woith, GFZ-Potsdam, Germany – for a stimulated discussion on hydrological aspects of water-rock interaction related the earthquake problems; and Dr. Nikolai Bagdassarov, Inst. of Meteorology and Geophysics, Frankfurt Univ., Germany – for helpful discussions and assistance. The authors are thankful to Dr. Robert R. Seal, USGS, who carefully edited the initial manuscript and made some essential improvements.

This work was carried out under the auspices of ARSAT Foundation, Armenia and in the framework of Research Program on Earthquake Source Mechanism, Advanced Research Lab., ARSAT Center.

REFERENCES

Abovian, S.B. 1981. *Mafite-ultramafite intrusive complex of the ophiolite belts of Armenia (in Russian).* Yerevan: Izd. AN Arm.SSR.

Bagdassarov, N.S. & Dingwell, D.B. 1993. Frequency dependent rheology of vesicular rhyolite. *J. Geophys. Res.* 98(4): 6477–6487.

Bystricky M., Kunze, K., Burlini, L. & Burg, J.-P. 2000. High Shear Strain of Olivine Aggregates: Rheological, Textural and Seismic Consequences. *Science* 290: 1564–1567.

Coleman, R.G. 1977. *Ophiolites.* Berlin-Heidelberg-New York: Springer-Verlag.

Grigorian, A.A. 2002. A Multidisciplinary approach for understanding the earthquake phenomenon. *ASC2002 Symposium on Seismology, Earthquake Hazard Assessment and Risk Management, Kathmandu, Nepal, 24–26 November.*

Haroutunian, A. & Grigorian, A. 2000. Formation and migration of deep crustal fluids of the Earth's crust. *EURESCO Conferences, Geochemistry of Crustal Fluids – Fluids in the Crust and Chemical Fluxes at the Earth's Surface, Granada, Spain, 2–7 December.*

Herzberg, C. 1995. Phase equilibria of common rocks in the crust and mantle. In Thomas J. Ahrens (ed.). *A Handbook of Physical Constants: Rock Physics and Phase Relations.* AGU Reference Shelf 3: 166–177.

Kirby, S.H., Engdahl, E.R. & Denlinger, R.P. 1996. Intermediate-depth intraslab earthquakes and arc volcanism as physical expression of crustal and uppermost mantle metamorphism in subducting slabs. In G.E. Bebout, D.W. Scholl, S.H. Kirby & J.P.Platt (eds.). *Subduction – Top to Bottom. Geophy.Monogr.* 96: 195–214.

Khachatrian, L.A., Kostanian, K.A. & Arutyunian, V.V. 1981. Kinetic investigations of the dehydration process of serpentinites of Armenia (in Russian). *Arm.Khim. Zhurn.* 34(1): 7–12.

Mueller, H.J., Schilling, F.R. & Lauterjung, J. 2002. In-situ investigation of physical properties of rocks and minerals at lower crustal and mantle conditions – methods, measurements, challenges. *Z. Geol. Wiss. Berlin* 30(1/2): 49–76.

Peacock, S.M. 1993. Large-scale hydration of the lithosphere above subducting slabs *Chem. Geol.* 108: 49–59.
Pugin, V.A., Khitarov, N.I., Slutsky, A.B., Revin, N.I. & Kuskov, O.L. 1969. Serpentinization and deserpentinization reaction (in Russian). *Geokhimia* 10: 1188–1194.
Schilling, F.R., Sinogeikin, S.V. & Bass, J.D. 2003. Single-crystal elastic properties of lawsonite and their variations with temperature. *Phys. Earth Planet. Interiors* 136: 107–118.
Saghatelian, E. & Grigorian, A. 2000. Fluids in ore-bearing hydrothermal paleosystems (on an example of Hyots-Dzor ore field of Armenia). *EURESCO Conferences, Geochemistry of Crustal Fluids – Fluids in the Crust and Chemical Fluxes at the Earth's Surface, Granada, Spain, 2–7 December.*
Veguni, A.T. & Bdoyan, A.A. 1987. Investigation of rock properties at phase transformation by ultrasonic technique at high pressures and temperatures (in Russian). *Eleventh Int. Conference MARIVD.* Kiev, USSR.
Woith, H., Milkereit, C., Zshau, J., Maiwald, U. & Pekderer, A. 1998. Monitoring of thermal and mineral waters in the frame of READINESS. In G.B.Arehart & J.R.Hulston (eds.). *Water Rock Interaction.* WRI-9: 809–812. Rotterdam: Balkema.

Water–Rock Interaction, Wanty & Seal II (eds)
© 2004 Taylor & Francis Group, London, ISBN 90 5809 641 6

Contribution of fertilization practices to nitrate contamination of groundwater

Huaming Guo, Guanghe Li, Dayi Zhang, Fei Yan & Xu Zhang
Department of Environmental Science and Engineering, Tsinghua University, Beijing, PR China

Changai Lu
Chinese Academy of Agricultural Science, Beijing, PR China

ABSTRACT: Groundwater contamination by nitrate associated with fertilization practices is a ubiquitous environmental issue, and consequently of world-wide concern. NO_3 flux from the unsaturated zone to groundwater in related to fertilization practices was studied, based on in situ experiments. Results show that NO_3-N flux was 47.12 kg/(ha · a) at the HF (High Fertilization plot), 21.37 kg/(ha · a) at the CK (Control plot), 13.46 kg/(ha · a) at the LF (Low Fertilization plot), and 9.96 kg/(ha · a) at the NF (No Fertilization plot). Application of abundant fertilizer would result in severe loss of NO_3-N with increasing loads to groundwater contamination. In order to reduce the potential of groundwater contamination in agricultural areas, the amount of fertilizer applied during agricultural cultivation must effectively be controlled in view of systematic combination of agricultural production and environmental protection.

1 INTRODUCTION

Groundwater contamination by nitrate associated with fertilization practices is a ubiquitous environmental issue, and consequently of world-wide concern. Much knowledge has accumulated on the climate, soil, topography, land use and nitrogen management factors influencing the risk associated with nitrate leaching (Zebarth et al. 1998, Babiker et al. 2004, Hubbard & Sheridan 1994). Large inputs from animal manure and inorganic N fertilizers may result in high nitrate concentrations in the soil. The leakage of the soil can in some cases result in deterioration of groundwater quality. The NO_3 flux from the unsaturated zone to the groundwater system must be minimized in order to significantly reduce the loss of N fertilizers and effectively control groundwater contamination by nitrate.

In recent years, eutrophication of Dianchi Lake, Kunming, China, has become more and more serious. This situation mainly results from non-point source agricultural pollution in spite of the fact that considerable effort is focused on industrial discharge and sludge water. Contamination of groundwater caused by non-point source pollution contributes partly to the lake eutrophication because groundwater recharges the lake. However, the contribution of contaminated groundwater to the lake eutrophication is still unknown. As a primarily step concerning this issue, NO_3 flux from unsaturated zone to groundwater in related to fertilization practices was studied.

2 MATERIALS AND METHODS

Dayu, located in approximately 25 km southwest of Kunming city in southwestern China, lies on the east shore of Dianchi Lake, one of the largest lakes in Yunnan province. Mean temperature at this area is 14.7°C with a maximum of 31°C and minimum of −8.1°C. Mean annual precipitation is 782.5 mm, more than ninety percent of which falls from May to October.

Hydraulic characteristic of soils at different depths are presented in Table 1. Due to cultivation practices, the arable layer has a high porosity with hydraulic conductivity of 10.36 m/d. The soils are compressed more tightly with increasing of depth due to long-term geological stress. Although the soils contain some gravels at depths between 30 and 100 cm, hydraulic conductivity and porosity of those soils were low. Hydraulic conductivity of the soils is 1.5 m/d at depths between 30 and 60 cm, and 2.29 m/d at depths between 60 and 100 cm. Hydraulic conductivities are between 0.02 and 0.04 m/d at depths between 100 and 280 cm (Guo et al. in press).

Table 1. Hydraulic characteristic of soils at different depths.

Depth (cm)	Soil description	Hydraulic conductivity (m/d)	Porosity (%)	Bulk density ($\times 10^3$ kg/m^3)
0~30	Dark brown clay, loose tillage layer	10.36	52.88	1.23
30~60	Brown clay dotted by few gravel	1.40	47.69	1.36
60~100	Brown gravelly clay	2.29	47.85	1.36
100~180	Yellow clay	0.02	48.75	1.33
180~280	Yellow silty clay	0.04	46.83	1.38

Table 2. Fertilization treatments in the field experiment in Dayu, Kunming city.

		Base fertilizer N (kg/ha)			P_2O_5 (kg/ha)			K_2O (kg/ha)		
Top application	Time	CK/NF	LF	HF	CK/NF	LF	HF	CK/NF	LF	HF
	Feb. 24, 2003	0	45	120	0	135	540	120	120	120
		0	405	1079	0	90	360	240	240	240
1	Mar. 15, 2003	0	40	108	0	18	72	30	30	30
2	Apr. 6, 2003	0	81	216	0	36	144	60	60	60
3	May 1, 2003	0	162	431	0	36	144	90	90	90
4	May 26, 2003	0	122	324	0	0	0	60	60	60

In situ experiments were carried out in the greenhouse cultivating celeries with an area of $46.5 \times 4.0\,m^2$. The pilot site occupied an area of $14.0 \times 4.0\,m^2$ within the greenhouse, which was divided into four plots, including three differently fertilized plots (HF, LF and NF) and one uncultivated plot (CK), with an area of $1.8 \times 7.0\,m^2$ for each plot. Fertilizer treatments are summarized in Table 2. Irrigation intensity at CK plot was the same as that at HF, LF and NF plot. The irrigation was generally carried out once every three days, with an essentially fixed value of 28.6 mm. Guo et al. (in press) described the irrigation intensity in more detailed. Celeries were transplanted on February 25, 2003 and harvested on Jun. 10, 2003.

Prior to celery transplantation, TDR (TRIME-T3) and WM-1 tensiometers were installed at depths of 0.3, 0.6, 1.0, 1.5, 2.0, 2.8 m. TDR was used to monitor soil moistures with measured ranges of 0–60% in volumetric moisture content (Musters & Bouten 2000, Brandelik & Hubner 1996), and WM-1 was used to monitor matrix potentials (Jin et al. 1994).

Simultaneously, the soil water samplers were set up at locations corresponding to TDR and WM-1. In order to avoid disturbing soil characteristics, the soils were carefully dug out and replaced at their original locations after installation of experimental equipments. A well to 2 m east of the experimental site penetrating into 5 m depth below land surface was installed to monitor groundwater table during celery cultivation.

3 RESULTS AND DISCUSSIONS

3.1 *Groundwater table*

Groundwater table slightly fluctuated during the study from a minimum of 151 cm, and a mean of 191 cm below land surface (Fig. 1). The water table at the experimental site was controlled by the regional groundwater system, mildly affected by agricultural irrigation (Guo et al. in press). Therefore, soil waters extracted at the depth of 2.0 m were considered as groundwater. Correspondingly, the chemical components in the soil water at the 2.0-m depth generally represented contaminant concentrations in the water entering groundwater from unsaturated zone.

3.2 *Vertical infiltration recharge*

The zero flux-plane method was used to calculate vertical infiltration recharge, which was based on the soil–water hydraulic theory that the water flux did not exist when a gradient of total potential of soil–water was zero. On the zero flux plane, soil water obeyed the following formula:

$$q = -K(\theta)\frac{\partial \phi}{\partial z} = 0 \qquad (1)$$

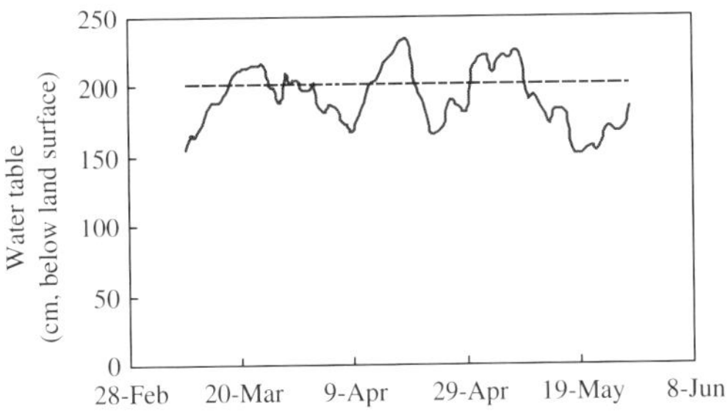

Figure 1. Groundwater table during the period of celery growth.

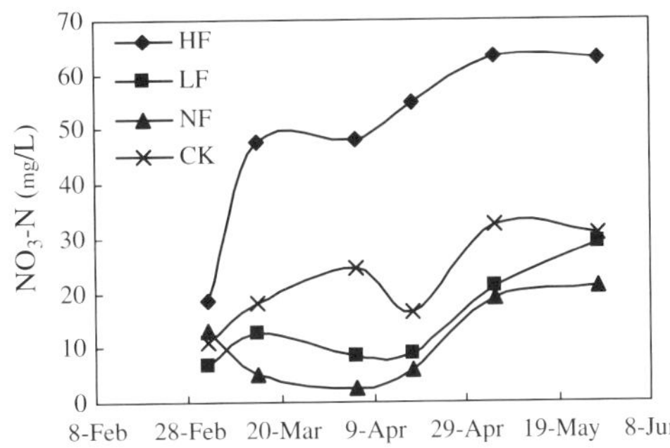

Figure 2. NO_3-N concentrations of soilwaters at depth of 2.0 m below land surface at four plots.

where q represents water flux; θ represents soil moisture content; $K(\theta)$ represents hydraulic conductivity; ϕ represents total potential of soil–water; z represents the depth.

Irrigation was uniform throughout the experimental site and the surface of the site was flat. Therefore, lateral flow of soil water could be considered to be negligible. The formula can be adapted to describe vertical water equilibrium for the quantities of our experiment.

$$q(z) - q(z') = -\int_{z'}^{z} \frac{\partial \theta}{\partial t} dz \qquad (2)$$

$$Q(z) - Q(z') = \int_{z'}^{z} \theta(z, t_1)\, dz - \int_{z'}^{z} \theta(z, t_2)\, dz \qquad (3)$$

where $q(z)$ and $q(z')$ represents the water flux at depths z and z', respectively; $Q(z)$ and $Q(z')$ represents the quantity of water through the plane at depth of z and z', respectively; $\theta(z, t_1)$ and $\theta(z, t_2)$ represents volumetric moisture content of soil at t_1 and t_2, respectively.

When the zero flux plane is at depth of z', $Q(z')$ and $Q(z')$ are equal to zero. Equations 2 and 3 can be simplified as.

$$q(z) = -\int_{z'}^{z} \frac{\partial \theta}{\partial t} dz \qquad (4)$$

$$Q(z) = \int_{z'}^{z} \theta(z, t)\, dz - \int_{z'}^{z} \theta(z, t)\, dz \qquad (5)$$

The zero flux plane can be achieved by using total potential data in soil profiles obtained by WM-1 tensiometers. If the zero flux plane exists, flux and quantity of water through any plane at depth of z can be calculated by using Equations 4 and 5, respectively, based on moisture contents in soil profiles measured by TDR. When z is located near the groundwater table, $Q(z)$ is considered to be the vertical infiltration recharge.

On a basis of total potential data, the zero flux plane was determined at depth of 0.6 m. As mentioned above, groundwater table is about 2.0 m below land surface. Therefore, z' and z are 0.6 and 2.0, respectively. Finally, the vertical infiltration recharge was calculated and evaluated as an average of 0.335 mm/d at four plots.

Table 1 shows that hydraulic conductivity of soils below the arable layer was very low, with a minimum of 0.02 m/d in soils at depths between 100 and 180 cm, which generally disagrees with the water flux calculated by using zero flux-plane method. This disagreement probably results from soil heterogeneity and the existence of joint cracks, which would promote downward movement of soil water (Hendriks et al. 1999).

3.3 *NO_3-N flux to groundwater*

NO_3^- concentrations of soil waters near the water table at different plots were shown in Figure 2. NO_3-N concentrations of soil waters at depth of 2.0 m generally increased during celery cultivation in relation to all plots. NO_3-N increased especially, from 18.52 to 62.68 mg/L from March 4 to May 28 at the highly fertilized plot. Figure 2 indicates that NO_3-N concentrations at the CK plot were higher than those at the LF and NF plots. That was because the background nitrate content of arable soil was rather high, reaching 15.59 g/kg (Guo et al. 2003). The leakage from the high nitrate content soil would reach deeper into the unsaturated zone in the case of regular irrigation. On the other hand, the uncultivated land (CK) favored leakage.

Based on water fluxes from the unsaturated to the saturated zone and NO_3-N concentrations in those waters, NO_3-N flux at each plot during the period of celery cultivation was calculated. The result is shown in Table 3. NO_3-N loss was the highest at the HF plot, followed by the CK, LF and NF plots, which indicated that application of abundant fertilizers would

Table 3. The amount of NO_3-N entering groundwater at each plot during the period of celery cultivation.

Plot	N fertilizer used (g N/plot)	NO_3-N flux (g/plot)	Percentage of N flux to fertilizer N (%)
HF	2016.00	19.30	0.96
LF	756.00	5.52	0.73
NF	0.00	4.08	
CK	0.00	8.76	

result in severe loss of NO_3-N with increasing contamination of groundwater. Therefore, to reduce the potential of groundwater contamination in agricultural areas, the amount of fertilizer applied must effectively be controlled in view of systematic combination of agricultural production and environmental protection.

Although N fertilizers were not applied to the CK, NO_3-N loss at this plot was the second highest among the four plots. This phenomenon generally resulted from the leakage of the high nitrate content of arable soils without cultivation of celery in the case of regular irrigation. Consequently, surface covering would mitigate contamination of groundwater below the contaminated soils. It can be applied to decrease drastically infiltration of surface water and further reduce the downward rate of movement of the contamination zone at contaminated sites. In the case of nitrate contamination of surface soils, agricultural cultivation would remediate contamination of groundwater present below the nitrate contaminated soils without application of fertilizers, which can be seen by the comparison of the NO_3-N flux into groundwater at NF with that at CK.

In the Dianchi catchment, farmers use chemical fertilizers in amounts comparable to those applied at HF. NO_3-N flux from unsaturated zone to groundwater would be 47.12 kg/(ha · a) at the area with a low water table (i.e. 200 cm depth), based on our *in situ* experiments. It was speculated that NO_3-N flux at the area near Dianchi Lake with a high water table (usually about 40 cm depth) was much more than that at the experimental site. Another study was being performed by us near Dianchi Lake to assess the NO_3-N flux in those areas with a high water table. Further results will be reported soon in international journals.

ACKNOWLEDGEMENTS

The research work was financially supported by China's National Key Technologies R & D Program in the 10th Five-Year Plan (No.2001BA610A-04) and China Postdoctoral Science Foundation.

REFERENCES

Babiker, I.S., Mohamed, M.A.A., Terao, H., Kato, K. & Ohta, K. 2004. Assessment of groundwater contamination by nitrate leaching from intensive vegetable cultivation using geographical information system. *Environment International* 29: 1009–1017.

Brandelik, A. & Hubner, C. 1996. Soil moisture determination-accurate, large and deep. *Phys. Chem. Earth.* 21(3): 157–160.

Guo, H. Li, G. Yan, F. Zhang, D. & Zhang, X. 2004. Moisture content and matrix potential of soils under intense irrigation in celery greenhouse, Dianchi catchment. *Geological Science and Technology Information* (to be published).

Guo, H., Li, G., Zhang, X., Yan, F., Zhang, D. & Lu, C. 2003. A spatial study of denitrification and nitrification potential of agricultural soils in relation to fertilization practice. In Y. Wang, A. Zhou & X. Liang (eds), *Proceeding of International Symposium on Water Resources and the Urban Environment*: 179–185.

Hendriks, R.F.A., Oostindie, K. & Hamminga, P. 1999. Simulation of bromide tracer and nitrogen transport in a cracked clay soil with FLOCR/ANIMO model combination. *Journal of Hydrology* 215: 94–115.

Hubbard, R.K. & Sheridan, J.M. 1994. Nitrates in ground water in the southeastern USA. In D.C. Adriano, A.K. Iskander & I.P. Murarka (eds), *Contamination of Groundwaters. Advances in Environmental Science*: 303–345.

Jing, E., Fei, J., Zhang, X., Han, S. & Xu, W. 1994. Experimental study on soilwater flux. Beijing: Seismal Press, 138–148.

Musters, P.A.D. & Bouten, W. 2000. A method for identifying optimum strategies of measuring soil water contents for calibrating a root water uptake model. *Journal of Hydrology* 227: 273–286.

Zebarth, B.J., Hii, B., Liebscher, H., Chipperfield K., Paul, J.W., Grove, G. & Szeto, S.Y. 1998. Agricultural land use practices and nitrate contamination in the Abbotsford Aquifer, British Columbia, Canada. *Agriculture, Ecosystems and Environment* 69: 99–112.

Water–Rock Interaction, Wanty & Seal II (eds)

Mineralogy and mobility of chemical elements at water–basalt interaction, Hole 504B

V.B. Kurnosov, B.P. Zolotarev & A.V. Artamonov
Geological Institute, Moscow, Russia

ABSTRACT: All studied basalts and dolerites from Hole 504B (DSDP/ODP Legs 69, 70, 83, 111, 137, 140, and 148) from the Costa Rica Rift are altered to various degrees in low to high-temperature hydrothermal environments. Temperatures of alteration reached >380°C in the lower dike section (1500–2111 m below sea floor). Basalts from the volcanic section were smectitized. Chlorites occur lower into the sheeted dike complex. In the sheeted dike complex, wide spread amphibolitization of igneous rocks is common. Results on the mobility of elements during alteration of igneous rocks from the drill hole revealed that in the reaction zone (the lower dikes) the most marked loss of elements is found for Si, Ti, Fe, Mn, Mg, Zn, La, Sm, Yb, Eu, and Lu. The elements Al, Ca, Na, P, Cu, Zr, Ce, Nd, and Tb are variably depleted or enriched in the main zone.

1 INTRODUCTION

Mid-ocean ridges stretch for more than 60,000 km in the World's oceans. They are characterized by the presence of convective hydrothermal cells. Water–rock interaction at this cells results in hot springs ("black smokers") on the floor of axial valleys. Our present knowledge of alteration of igneous rocks in these giant structures needs further improvement. We have continued study of alteration of young basalts from mid-ocean ridges in the Pacific Ocean, first in DSDP/ODP Hole 504B.

Site 504 is located in the eastern equatorial Pacific near the Costa Rica Rift axis (Fig. 1). Hole 504B was drilled in seven DSDP/ODP legs: DSDP 69, 70, and 83, and ODP 111, 137, 140, and 148. The recovered basalts have an age of 5.9 Ma (Becker et al. 1988). Hole 504B penetrated 2111.0 m of rock. Beneath 274.5 m of sediment the hole penetrated 571.5 m of volcanic rocks (pillow lavas and flows) – from 274.5 to 846 meters below sea floor (mbsf), a 209 m thick transition zone (846–1055 mbsf), and 1050 m thick sheeted dike complex (upper dikes – 1055–1500 m and lower dikes – 1500–2111 m). The hole stayed in the sheeted dike complex (near or in uppermost third layer of the oceanic crust).

Alteration of igneous rocks (mineralogy, chemical composition, and distribution and zonation of secondary minerals, replacement of phenocrysts, minerals and interstitial glass of groundmass, formation of

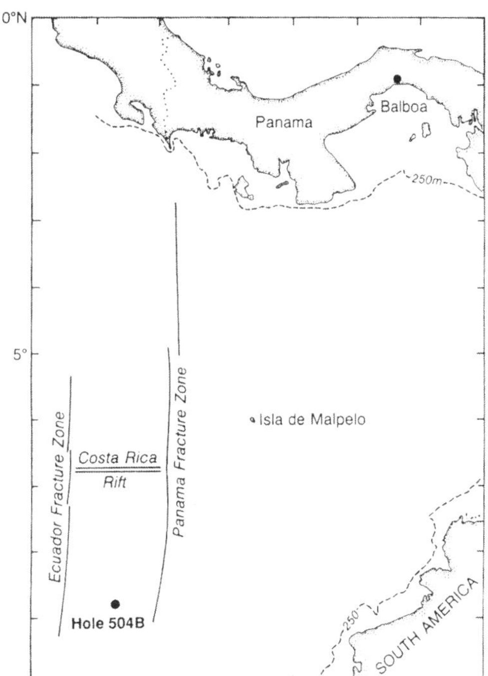

Figure 1. Location map of DSDP/ODP Hole 504B in the eastern equatorial Pacific.

veins and filling of vesicles, conditions and sequence of secondary minerals formation) from Hole 504B was studied in great detail by Adamson (1983), Alt & Emmerman (1985), Alt & Chaussidon (1989), Alt et al. (1983, 1985a, b, 1989, 1995, 1996a, b), Becker et al. (1988), Honnorez et al. (1983, 1985), Ishizuka (1989), Laverne (1983), Noack et al. (1983), Pertsev & Boronikhin (1983a, b, 1996), Sparks (1995), Laverne et al. (1995), Zuleger et al. (1995), and Vanko et al. (1996). We participated in the study of alteration mineralogy of the volcanic section from Hole 504B, Legs 69/70 DSDP (Kurnosov et al. 1983). Synthesis of data on the distribution of secondary minerals and the mobility of elements in these processes was made by Alt et al. (1996a).

We have studied additional samples of basalts and dolerites recovered during DSDP Legs 69, 70, and 83, and ODP Legs 111, 137, 140, and 148, Hole 504B. The sampling was made from slightly, moderately, and highly altered rock, from massive rocks, and from fractured and brecciated zones.

2 IGNEOUS ROCKS CHEMISTRY

Basalts and dolerites are essentially undifferentiated in SiO_2, Al_2O_3, Fe, Mg, and Ti. They belong to the low-alkaline type of the tholeiitic MORB with abundant siderophile and chalcophile minor elements; but somewhat depleted in Nb and Zr, and Rb and Sr. The normalized REEs indicate that igneous rocks from Hole 504B are depleted in large-ion lanthanides and Eu. They were related to a homogenous group of rocks.

3 PETROGRAPHY

Petrographic description of thin sections on selected igneous rocks shows that they are represented by aphyric (from pillow lavas) and phyric varieties (from sheeted dike complex) with various amounts of plagioclase, clinopyroxene, and olivine phenocrysts.

4 ALTERATION

In the igneous rocks we examined, secondary minerals replace phenocrysts, minerals and interstitial glass of the groundmass, and partially occur in veins and vesicles. The total distribution of all secondary minerals was presented by Alt et al. (1996a). Also we studied mobility of elements in relation of alteration of basalts under different conditions: smectitization, chloritization, and amphibolitization.

4.1 *Secondary minerals in igneous rocks*

Basalts from upper 580 meters of Hole 504B (up to deep about 850 mbsf) are smectitized. Trioctahedral smectites (saponite type) are the principal secondary minerals in the basalts from the volcanic section. Beneath 850 mbsf, clay minerals are represented by chloritic minerals. In both the transition zone and dike sections for mixed-layer chlorite-smectite, the content of swelling interlayers decreases with depth. Below 1815 mbsf, minerals with swelling interlayers are not formed.

Amphiboles and chlorite are wide spread in both the transition zone and the sheeted dike complex. However, they are most wide spread in the lower dike section, particularly from 1630 mbsf and downwards to the bottom of the hole. Zones of alternation of mainly chloritization and amphibolitization dolerites are located in this interval. Amphiboles replace of pyroxene. Locally in centers of tremolite grains are small grains of titanite. Titanite was also found in chlorite aggregates. Chlorites are located in interstices between plagioclase laths. They replace olivine, clinopyroxene, and plagioclase. Albite was found from 1000 mbsf and lower. It replaces plagioclase. The most amount of talc was found in the lower dikes (1712–1731 mbsf).

4.2 *Minerals from veins*

Vein minerals from the volcanic section (274.5–846 mbsf) are trioctahedral and dioctahedral smectites, Fe-Mg-chlorite, hematite, phillipsite, natrolite, aragonite, calcite, Mn-mineral, analcime, okenite, gyrolite, thomsonite (thomsonite was identified by Pertsev), anhydrite, quartz, cristobalite, apophylite, actinolite, pyrite, chalcocite, bismuthinite, foujasite, Fe-hydrosilicate, and magnesite (Kurnosov et al. 1983). Below 846 mbsf (in the transition and dike sections) we studied only a few samples with veins.

The study of clay minerals from veins and their comparison with clay minerals from igneous rocks showed that conditions of clay–mineral formation in host rocks and in fractures in the volcanic section and upper part of the transition zone was not identical. Veins often contain chloritic components whereas host rocks only are smectitized. Solutions in the fractures were hotter than those altering the basalts. This phenomenon was not determined below 908 mbsf. The increased depths probably equalized the temperatures between the fractures and country rock. The compositions of the fluids were likely similar as well.

4.3 *Mobility of elements*

In bulk samples of igneous rocks, we analyzed major elements by classical wet-chemistry. To determine minor elements, we used both atomic emission spectroscopy and XRF. REEs were analyzed by neutron activation analysis. Increases and decreases of elements in altered rocks were considered in comparison with the concentrations of elements in fresh analogues. The comparison was made in different groups of rocks: aphyric and glassy basalts, phyric basalts, oxidized basalts, dolerites, and brecciated or fractured rocks.

In the upper volcanic section, a decrease in Si, Fe_{Total}, Ca, and Ce contents for altered basalts from all group rocks was identified, as well for Na, P, V, Zr, La, Sm, Yb, Y, Nd, and Eu contents (except for oxidized rocks for Na, aphyric basalts for P, and dolerites for the rest elements). Ti is depleted in several samples of altered phyric and oxidized basalts. Al (for all group of rocks, except dolerites) is depleted or enriched. Contents of Mn increase in several samples of altered aphyric and phyric basalts, and dolerites. In the uppermost part of the upper volcanic section, Mn contents decrease in altered phyric and oxidized basalts; below Mn contents increase in phyric and aphyric basalts, and in dolerites. Mg is either depleted (altered aphyric and phyric basalts and several samples of fractured rocks) or enriched (altered dolerites and several samples of oxidized rocks). K contents are enriched in altered basalts from all groups of rock. It should be emphasized that K contents have wide range of values in fresh rocks from the upper volcanics. In upper part of the section, Cu, Zn, Tb, and Lu are depleted (altered phyric and oxidized basalts, and fractured rocks). Contents of Ni decrease in altered phyric and aphyric basalts, and in oxidized and fractured rocks. Co, Rb, and Sr are either depleted or enriched. Ba is enriched (altered phyric and aphyric basalts, and oxidized rocks).

In the lower volcanic section, chemical changes due to alteration of rocks were rather similar, but were less intense to nonexistent. In addition, they have other geochemical features. In samples of altered aphyric basalts, Ti, Na, P, Cu, Zr, Nd, Sr, La, Y, Ce, Nd, and Tb are enriched. Contents of Mg, K, Co, and Ba decrease in altered fractured rocks.

In the transition zone, chemical changes of rocks are similar to changes in the volcanic section. There are some differences in behavior of elements. Altered rocks record that alteration leads to a decrease in Mg (fractured rocks and dolerites) or a increase in Mg (phyric basalt). Na and K are enriched, especially Na. It should be emphasized that K contents in fresh rocks of the transition zone and in the dike section are very low. Accordingly, low K contents in the altered rocks from those sections are not associated with alteration of the igneous rocks. There is a marked increase in Cu and Zn (Cu in altered dolerites and aphyric basalts and Zn in fractured rocks and in dolerite and phyric basalt). Nb, La, Ba, Ce, and Nd are enriched (Nd and La in phyric basalt, Ba in fractured rock, and Ce and Nd in fractured and aphyric basalt). Sm, Yb, Eu, and Tb are either depleted or enriched.

The behavior of elements during alteration of igneous rocks from the upper dike section differ slightly from chemical changes of altered rocks from overlying sections, regardless of the rock types located in the zone of strong chloritization. In this section, Mg is depleted or enriched (several samples of dolerite).

Results on the mobility of elements due to alteration of igneous rocks show that in the reaction zone (lower sheeted dike complex) the greatest loss of elements is found (Si, Ti, Fe, Mn, Mg, Zn, La, Sm, Yb, Eu, and Lu), whereas concentration of Cr, Ni, Co, V, and Sr increase. In altered dolerites, K and Rb from lower part of the section are slightly decreased. Al, Ca, Na, P, Cu, Zr, Ce, Nd, and Tb are depleted or enriched.

5 CONCLUSIONS

Results show that amphibolitization, chloritization, and smectitization zones (change of greenschist facies to smectite facies) demonstrate vertical zonation of temperatures in the upper part of the oceanic crust. In interval 560–580 mbsf, smectitization of basalts produced chloritic minerals (mixed-layer chlorite-smectite mineral and mixed-layer chlorite-swelling chlorite mineral). This phenomenon shows that there was a heterogeneous environment of hydrothermal alteration in basalts (sub-horizontal hot water flows).

The temperature of the convective cell probably was not uniform with time. For example, uralitic amphibole was partly replaced by chlorite in samples from 1781.3 and 1992.14 mbsf (retrograde metamorphism); in sample from 1813.16 mbsf, chlorites are replaced by amphiboles (prograde metamorphism).

Absence in the volcanic section of chloritized basalts indicates that Hole 504B never intercepted the discharge zone of hot solutions.

Results on the mobility of elements due to alteration of igneous rocks revealed that in reaction zone (the lower dikes), the most significant mobility of elements occurred.

ACKNOWLEDGMENTS

We thank DSDP/ODP for providing the samples. The study was supported by Russian Foundation for Basic Research: grant N 02-05-64207.

REFERENCES

Adamson, A.C. 1983. Chemistry of alteration minerals from Deep Sea Drilling Project Sites 501, 504, and 505. In J.R. Cann, M.G. Langseth, J. Honnorez, R.P. Von Herzen & S.M. White (eds.), *Init. Repts. DSDP* 69: 551–563. Washington DC: U.S. Govt. Printing Office.

Alt, J.C. & Chaussidon, M. 1989. Ion microprobe analyses of the sulfur isotopic composition of sulfides in hydrothermally altered rocks, DSDP/ODP Hole 504B. In K. Becker & H. Sakai (eds.), *Proc. ODP, Sci. Results* 111: 41–45. College Station, TX: Ocean Drilling Program.

Alt, J.C. & Emmermann, R. 1985. Geochemistry of hydrothermally altered basalts: Deep Sea Drilling Project Hole 504B, Leg 83. In R.N. Anderson, J. Honnorez & K. Becker (eds.), *Init. Repts. DSDP* 83: 249–262. Washington DC: U.S. Govt. Printing Office.

Alt, J., Honnorez, J., Hubberten, H.-W. & Saltzman, E. 1983. Occurence and origin of anhydrite from Deep Sea Drilling Project Leg 70, Hole 504B, Costa Rica Rift. In J.R. Cann, M.G. Langseth, J. Honnorez, R.P. Von Herzen & S.M. White (eds.), *Init. Repts. DSDP* 69: 547–550. Washington DC: U.S. Govt. Printing Office.

Alt, J.C., Laverne, C. & Muehlenbachs, K. 1985a. Alteration of the upper oceanic crust: mineralogy and processes in Deep Sea Drilling Project Hole 504B, Leg 83. In R.N. Anderson, J. Honnorez & K. Becker (eds.), *Init. Repts. DSDP* 83: 217–248. Washington DC: U.S. Govt. Printing Office.

Alt, J.C., Saltzman, E.S. & Price, D.A. 1985b. Anhydrite in hydrothermally alterd basalts: Deep Sea Drilling Project Hole 504B, Leg 83. In R.N. Anderson, J. Honnorez & K. Becker (eds.), *Init. Repts. DSDP* 83: 283–288. Washington DC: U.S. Govt. Printing Office.

Alt, J., Anderson, T.F., Bonnell, L. & Muehlenbachs, K. 1989. Mineralogy, chemistry, and stable isotopic compositions of hydrothermally altered sheeted dikes: ODP Hole 504B, Leg 111. In K. Becker & H. Sakai (eds.), *Proc. ODP, Sci. Results* 111: College Station, TX: Ocean Drilling Program. 27–40.

Alt, J.C., Zuleger, E. & Erzinger, J.A. 1995. Mineralogy and stable isotopic compositions of the hydrothermally altered lower sheeted dike complex, Hole 504B, Leg 140. In J. Erzinger, K. Becker, H.J.B. Dick & L.B. Stokking (eds.), *Proc. ODP, Sci. Results*, 137/140: 155–166. College Station, TX: Ocean Drilling Program.

Alt, J., Laverne, C., Vanko, D.A., Tartarotti, P., Teagle, D.A.H., Bach, W., Zuleger, E., Erzinger, J., Honnorez, J., Pezard, P.A., Becker, K., Salisbury, M.H. & Wilkens, R.H. 1996a. Hydrothermal alteration of a section of upper oceanic crust in the eastern equatorial Pacific: a synthesis of results from Site 504 (DSDP Legs 69, 70, and 83, and ODP Legs 111, 137, 140, and 148). In J.C. Alt, H. Kinoshita, L.B. Stokking & P.J. Michael (eds.), *Proc. ODP, Sci. Results*, 148: 417–434. College Station, TX: Ocean Drilling Program.

Alt, J., Teagle, D.A.H., Laverne, C., Vanko, D.A., Bach, W., Honnorez, J., Becker, K., Ayadi M. & Pezard, P.A. 1996b. Ridge flank alteration of upper ocean crust in the Eastern Pacific: Synthesis of results for volcanic rocks of Holes 504B and 896A. In J.C. Alt, H. Kinoshita, L.B. Stokking & P.J. Michael (eds.), *Proc. ODP, Sci. Results*, 148: 435–450. College Station, TX: Ocean Drilling Program.

Becker, K. & Sakai, H. 1988. In K. Becker & H. Sakai (eds.), *Proc. ODP, Sci. Results* 111: College Station, TX: Ocean Drilling Program.

Honnorez, J., Emmermann, R., Hubberten, H.W., Laverne, C. & Muehlenbachs, K. 1983. Alteration processes in layer 2 basalts from Deep Sea Drilling Project Hole 504B, Costa Rica Rift. In J.R. Cann, M.G. Langseth, J. Honnorez, R.P. Von Herzen & S.M. White (eds.), *Init. Repts. DSDP* 69: 509–546. Washington DC: U.S. Govt. Printing Office.

Honnorez, J., Alt, J.C., Honnorez-Guerstein, B.-M., Laverne, C., Muehlenbachs, K., Ruiz, J. & Saltzman, E. 1985. Stockwork-like sulfide mineralization in young oceanic crust: Deep Sea Drilling Project Hole 504B. In R.N. Anderson, J. Honnorez & K. Becker (eds.), *Init. Repts. DSDP* 83: 263–282. Washington DC: U.S. Govt. Printing Office.

Ishizuka, H. 1989. Mineral paragenesis of altered basalts from Hole 504B, ODP Leg 111. In K. Becker & H. Sakai (eds.), *Proc. ODP, Sci. Results* 111: 61–76. College Station, TX: Ocean Drilling Program.

Kurnosov, V.B., Kholodkevich, I.V., Chubarov, V.M. & Shevchenko, A.Y. 1983. Secondary minerals in basalts from the Costa Rica Rift, Holes 501 and 504B, Deep Sea Drilling Project Legs 68, 69, and 70. In J.R. Cann, M.G. Langseth, J. Honnorez, R.P. Von Herzen & S.M. White (eds.), *Init. Repts. DSDP* 69: 573–583. Washington DC: U.S. Govt. Printing Office.

Laverne, C. 1983. Occurrence of melanite and aegirine-augite in Deep Sea Drilling Project Hole 504B. In J.R. Cann, M.G. Langseth, J. Honnorez, R.P. Von Herzen & S.M. White (eds.), *Init. Repts. DSDP* 69: 393–605. Washington DC: U.S. Govt. Printing Office.

Laverne, Ch., Vanko, D.A., Tartarotti, P. & Alt, J.C. 1995. Chemistry and geothermometry of secondary minerals from the deep sheeted dike complex, Hole 504B. In J. Erzinger, K. Becker, H.J.B. Dick & L.B. Stoking (eds.), *Proc. ODP, Sci. Results*, 137/140: 167–189. College Station, TX: Ocean Drilling Program.

Noack, Y., Emmermann, R. & Hubberten, H.-W. 1983. Alteration in Site 501 (Leg 68) and Site 504 (Leg 69) basalts: preliminary results. In J.R. Cann, M.G. Langseth, J. Honnorez, R.P. Von Herzen & S.M. White (eds.), *Init. Repts. DSDP* 69: 497–508. Washington DC: U.S. Govt. Printing Office.

Pertsev, N.N. & Boronikhin, V.A. 1983a. Alteration zones near veins in basalts from Deep Sea Drilling Project Sites 501/504 and 505, Costa Rica Rift. In J.R. Cann, M.G. Langseth, J. Honnorez, R.P. Von Herzen & S.M. White (eds.), *Init. Repts. DSDP* 69: 565–572. Washington DC: U.S. Govt. Printing Office.

Pertsev, N.N. & Boronikhin, V.A. 1983b. Secondary K-feldspar in basalts at Deep See Drilling Project Hole 504B and the problem of K-feldspathization in oceanic basalts. In J.R. Cann, M.G. Langseth, J. Honnorez, R.P. Von Herzen & S.M. White (eds.), *Init. Repts. DSDP* 69: 89–591. Washington DC: U.S. Govt. Printing Office.

Pertsev, N.N. & Boronikhin, V.A. 1996. Metasomatic and metamorphic peculiarities of the diabase dike alteration in Hole 504B, Leg 140. In J.C. Alt, H. Kinoshita, L.B. Stokking, & P.J. Michael (eds.), *Proc. ODP, Sci.*

Results, 148: 97–110. College Station, TX: Ocean Drilling Program.
Sparks, J.W. 1995. Geochemistry of the lower sheeted dike complex, Hole 504B, Leg 140. In J. Erzinger, K. Becker, H.J.B. Dick & L.B. Stokking (eds.), *Proc. ODP, Sci. Results*, 137/140: 81–98. College Station, TX: Ocean Drilling Program.
Vanko, D.A., Laverne, C., Tartarotti, P. & Alt, J.C. 1996. Chemistry and origin of secondary minerals from the deep sheeted dikes cored during Leg 148 (Hole 504B). In J.C. Alt, H. Kinoshita, L.B. Stokking & P.J. Michael (eds.), *Proc. ODP, Sci. Results*, 148: 71–86. College Station, TX: Ocean Drilling Program.
Zuleger, E., Alt, J.C. & Erzinger, J.A. 1995. Primary and secondary variations in major and trace element geochemistry of the lower sheeted dike complex: Hole 504B, Leg 140. In J. Erzinger, K. Becker, H.J.B. Dick & L.B. Stokking (eds.), *Proc. ODP, Sci. Results*, 137/140: 65–80. College Station, TX: Ocean Drilling Program.

Water–Rock Interaction, Wanty & Seal II (eds)
© 2004 Taylor & Francis Group, London, ISBN 90 5809 641 6

Water–rock interaction responsible for the origin of high pH mineral waters (S. Portugal)

J.M. Marques, M.J. Matias, M.J. Basto, R.C. Graça & L. Aires-Barros
Instituto Superior Técnico, Lisboa, Portugal

M. Andrade & P.M. Carreira
Instituto Tecnológico e Nuclear, Sacavém, Portugal

F. Goff
Los Alamos National Laboratory, New Mexico, USA

L. Rocha
Junta de Freguesia, Cabeço de Vide, Portugal

ABSTRACT: Water–rock interaction studies have greatly contributed to the conceptual hydrogeological circulation model of the Cabeço de Vide mineral waters. The studied mineral waters belong to the Na-Cl/Ca-OH-type waters (pH ≈ 11.5) and issue from the intrusive contact between the mafic/ultramafic rocks and a local carbonate sequence. Mg-HCO_3-type waters constitute most of the shallow groundwaters discharging from local serpentinized ultramafic rocks. Similar δ^2H and $\delta^{18}O$ values found in these two types of waters suggest a common meteoric origin and the evolution of the Mg-HCO_3 waters towards Cabeço de Vide-type mineral waters. The Mg-HCO_3-type waters seem to be generated in an initial step under open CO_2 conditions by water–serpentinite interaction in a shallow environment; the Cabeço de Vide mineral waters could be produced in a second step under closed-system conditions as the result of water–dunite interaction at depth.

1 INTRODUCTION

During the hydrogeological investigations at Cabeço de Vide region (southern Portugal) to refine the conceptual circulation model of the local mineral waters, several questions arose: (i) what causes the unusual high pH values of the mineral waters? (ii) are these mineral waters the result of meteoric water interaction with local ultramafic rocks? and (iii) why do these waters have low Mg and SiO_2 concentrations?

Hydrogeological investigations of mineral, dilute and rain waters from the surroundings of the Cabeço de Vide spas, in order to answer such kind of questions ascribed to Cabeço de Vide mineral waters. Local mafic/ultramafic rocks and mineral phases (from outcrops and drill-cores from boreholes) have also been analysed, as a comparison to the fluids, in order to assess water–rock interaction processes occurring at depth.

2 GEOLOGICAL SETTING

The study region (Fig. 1) is situated at the Ossa Morena Zone of the Iberian Hercynian Belt. The Lower Cambrian carbonate sequence was intruded and metamorphosed by mafic and ultramafic rocks forming a NW-SE cumulate-type structure of Ordovician age, which has been subjected to serpentinization-rodingitization processes (Costa et al. 1993). Cabeço de Vide mineral waters issue at the intrusive contact between the mafic/ultramafic rocks and the carbonate sequence, associated with the main regional NNE-SSW trending fault. The ultramafic rocks range from partially serpentinized dunites to serpentinites.

The resulting serpentinites may contain some relict olivine, but most commonly, they contain only serpentine minerals (lizardite, antigorite and chrysotile), magnetite, and residual chromite, plus brucite and carbonates. The mafic rocks adjacent to the

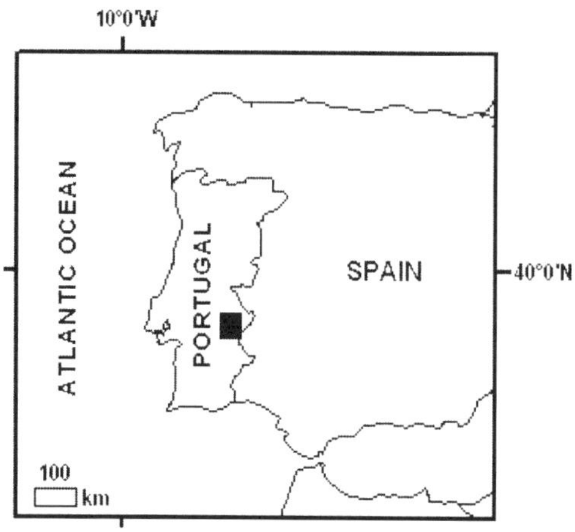

Figure 1. Location map of Cabeço de Vide region.

Table 1. Chemical composition (major elements in wt. %) of (1) gabbros and (2) serpentinites from Cabeço de Vide (CV) area. Analysis by E. Kluk, LANL.

Rock type	1(a)	1(b)	2(a)	2(b)
SiO_2	48.0	42.0	38.6	36.4
TiO_2	0.444	2.61	0.360	0.078
Al_2O_3	13.2	8.64	1.69	1.05
Fe_2O_3	1.42	3.12	10.48	9.49
FeO	2.72	9.02	2.88	3.13
MnO	0.080	0.213	0.250	0.108
MgO	11.9	10.9	32.7	36.4
CaO	18.6	17.5	2.77	0.21
Na_2O	0.62	0.67	0.00	0.00
K_2O	0.48	0.39	0.00	0.00
P_2O_5	0.031	1.47	0.009	0.013
LOI	1.72	2.12	9.15	11.4
Total	99.26	98.66	98.92	98.32
Cr (ppm)	265	271	2970	3230
Ni (ppm)	156	225	1600	2090

(1a) M^{te} das Fontainhas outcrop; (1b) drill core from borehole AC1 (188.5 m depth); (2a) CV railway station outcrop; (2b) drill core from borehole AC2 (105.8 m depth).

ultramafics are mostly coarse- to fine-grained gabbros with Ca-plagioclase, Ti-clinopyroxene, olivine and some Al-hornblende (Costa et al. 1993). Representative chemical analysis of these rock types are presented in Table 1.

3 WATER CHEMISTRY VS WATER–ROCK INTERACTION

The different geologic formations led to the origin of various chemical types of groundwaters in the region

Table 2. Physico-chemical characteristics of representative waters from Cabeço de Vide region, issuing from distinct geological formations.

	(L)	(G)	(Sp)	(CV)
T(°C)	19.4	18.2	19.2	19.6
Cond	724	785	740	646
pH	6.96	7.37	7.54	10.93
Na	14.3	9.53	9.56	49.0
K	0.96	0.35	0.44	5.04
Ca	100.0	41.9	38.7	22.5
Mg	26.5	77.3	72.2	0.16
HCO_3	469.2	540.3	542.9	n.d.
CO_3	0.24	0.70	1.06	6.9
SO_4	20.56	14.72	14.42	5.18
NO_3	19.48	12.33	9.07	6.22
Cl	16.68	14.44	9.08	32.14
SiO_2	24.6	74.3	72.8	5.50
OH	n.d.	n.d.	n.d.	33.02

Waters issuing from (L) limestones (G) gabbros, (Sp) serpentinites, and (CV) Cabeço de Vide mineral waters. Concentrations in mg/L; Electrical conductivity (Cond) in μS/cm; n.d. – below detection limits.

of the Cabeço de Vide spas (Table 2). The Cambrian limestones represent an important aquifer in the area, where the fractured carbonate rocks allow the development of Ca-HCO_3 waters. The Mg concentrations found in this type of waters can be attributed to the presence of dolomite found in the limestone rocks. Mg-HCO_3 type waters constitute most of the surface (stream) and shallow groundwaters discharging from springs, and of dug and drilled wells located in the gabbros and serpentinites.

Barnes et al. (1967) found similar Mg-HCO_3 type groundwaters in the serpentinized ultramafic rocks of the California Coast Ranges. According to those authors, the solution of serpentine tends to remove H^+ as:

$$\text{serpentine} + 6H^+ = 2SiO_2(aq) + 3Mg^{2+} + 5H_2O \quad (1)$$

leading to high Mg and SiO_2 concentrations in waters.

Using WATEQ4F (Ball & Nordstrom 1991), a hydrogeochemical code that calculates the speciation of natural waters, it was possible to observe that the Mg-HCO_3 waters of Cabeço de Vide area are undersaturated with respect to chrysotile (log S.I. < -3), suggesting that water chemistry is strongly dependent of serpentine dissolution, which may explain the high Mg (Fig. 2) and SiO_2 concentrations found in this type of water.

The main characteristics of Cabeço de Vide mineral waters can be summarized as follows: (i) Na-Cl/Ca-OH type, (ii) very alkaline pH values, and (iii) low

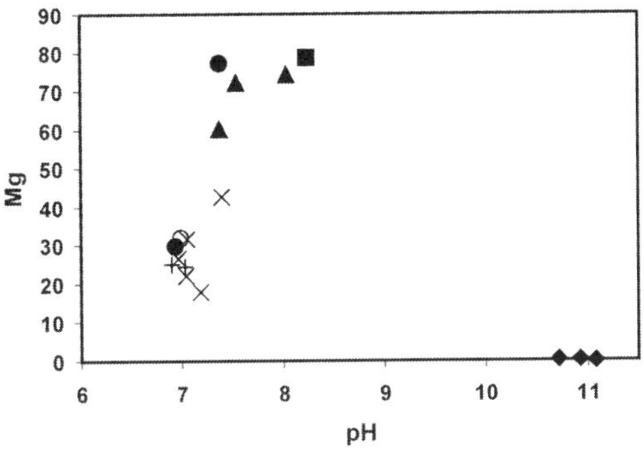

Figure 2. Mg (mg/L) vs pH plot of waters from different geological formations: (X) limestones; (●) gabbros; (▲) serpentinites; (○) contact gabbros/serpentinites; (+) contact gabbros/limestones. (■) Cabeço de Vide stream; (◆) Cabeço de Vide mineral waters.

Mg and SiO_2 concentrations (Table 2). This type of water has been well documented in the literature (e.g. Barnes & O'Neil 1969, Barnes et al. 1967, 1973, 1978, Bruni et al. 2001).

Cabeço de Vide mineral waters are undersaturated with respect to forsterite (log S.I. < -1) and supersaturated with respect to chrysolite (log S.I. $> +4$), suggesting that serpentinization of the ultramafic rocks may be an important process contributing to the groundwater chemistry through the reaction:

$$2Mg_2SiO_4 + 3H_2O = Mg(OH)_2 + Mg_3Si_2O_5(OH)_4 \quad (2)$$

which may explain the low Mg and SiO_2 concentrations found in the Cabeço de Vide mineral waters (Fig. 2).

4 ISOTOPE GEOCHEMISTRY

Cabeço de Vide mineral waters and the local $Mg\text{-}HCO_3$ waters have similar δ^2H_{VSMOW} and $\delta^{18}OV_{SMOW}$ values (around -27 and -4.5‰, respectively), indicating a common meteoric origin for these waters (Marques et al. 2003). The absence of a $\delta^{18}O$ shift from the global meteoric water line indicates that mineral water–rock interaction may have taken place at quite low temperatures. This is consistent with the low temperature ($\sim$19.5°C) of the Cabeço de Vide mineral waters. As stated by Marques et al. (2003), in view of the lower δ^2H and $\delta^{18}O$ values ($\delta^2H = -32.2$‰; $\delta^{18}O = -5.44$‰) of the long-term mean of precipitation (ITN 2002) from the Portalegre meteorological station (597 m a.s.l., approximately 15 km NE of Cabeço de Vide), the major source of recharge of the Cabeço de Vide mineral waters would appear to be ascribed to lower

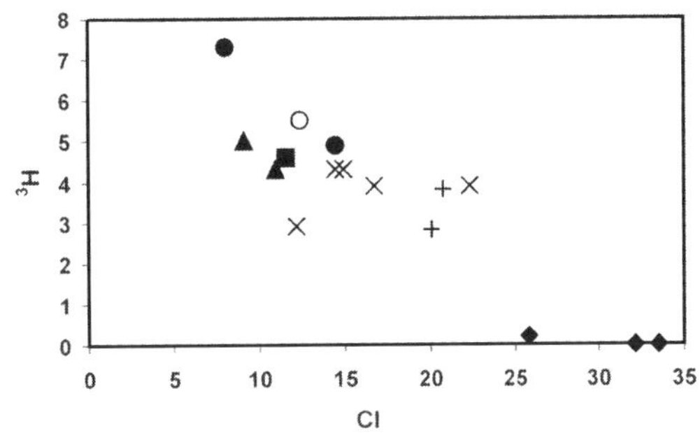

Figure 3. Plot of ^3H (T.U.) vs Cl (mg/L). Symbols as in Figure 2.

elevation sites (250–300 m a.s.l) located at the local mafic/ultramafic outcrop.

The very low ^3H values found in the Cabeço de Vide mineral waters suggest an age older than 60 years (Fig. 3). Two samples for ^{14}C-concentration determinations of Cabeço de Vide mineral waters were collected (AMS determinations; Geochron Laboratories, USA) to support these results. The values obtained range between 69.12 ± 0.28 and 65.24 ± 0.35 percent modern carbon, indicating a reliable apparent groundwater age between 2970 ± 40 and 3430 ± 50 years BP, respectively. Assuming a common origin for the $Mg\text{-}HCO_3$ and the Cabeço de Vide mineral waters, the high Cl concentrations (accompanied by low ^3H values) found in the mineral waters could be attributed to the increased water–rock interaction, as the result of a large flow path and/or residence time in the subsurface rocks (Fig. 3).

The $\delta^{13}C_{PDB}$ values of -22.9‰ and -18.0‰ of Cabeço de Vide mineral waters (boreholes AC3 and AC5, respectively) suggest (i) an organic origin for the carbon in the Cabeço de Vide mineral waters, and (ii) negligible contribution from the local carbonate rocks ($\delta^{13}C$ values $\sim$1.48‰) to the isotopic signatures the mineral waters.

5 CONCLUDING REMARKS

Because both $Mg\text{-}HCO_3$ waters and Cabeço de Vide mineral waters have similar isotopic (δ^2H and $\delta^{18}O$) signatures, we believe that the $Mg\text{-}HCO_3$ waters evolve towards the Cabeço de Vide mineral waters through interactions with the country rocks.

Bruni et al. (2001) studied the progressive evolution from rainwater to neutral $Mg\text{-}HCO_3$ waters and to mature, high-pH, Ca-OH-type waters, due to water–ultramafic rock interaction in the absence of carbon sources. The computed Ca, Mg, and pH values presented by Bruni et al. (2001) are compatible with

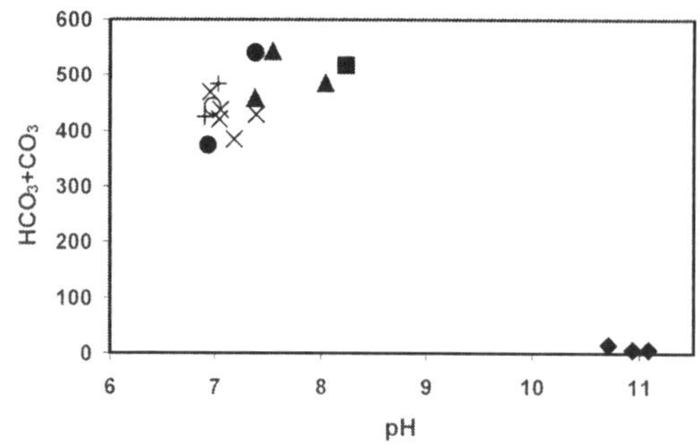

Figure 4. Plot of $HCO_3 + CO_3$ (mg/L) vs pH. Symbols as in Figure 2.

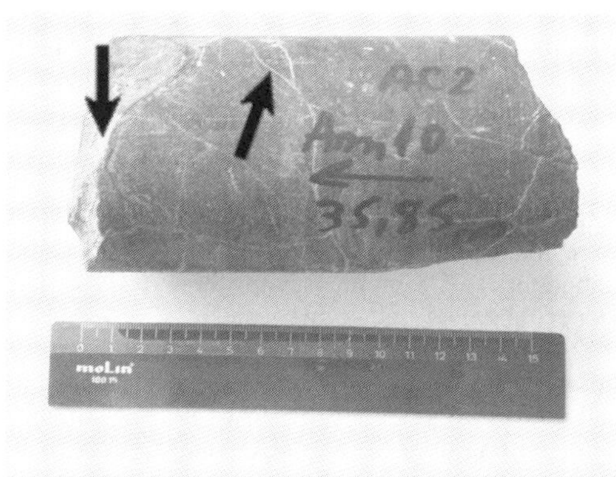

Figure 5. Drillcore from borehole AC2 (Cabeço de Vide Spas). Serpentinized dunite. The arrows indicate the presence of veins filled with calcite.

our data from the Cabeço de Vide mineral waters, indicating that the irreversible water–rock mass transfer could be a realistic model to describe the water–rock interactions occurring at depth in the Cabeço de Vide area.

In such water–rock interaction model, the local Mg-HCO_3-type waters seem to be generated in an initial step, under open CO_2 conditions, due to meteoric water–serpentinite interactions in a shallow environment, whereas the Cabeço de Vide mineral waters will be produced in a subsequent step, under closed CO_2 conditions. In this second step, the increased pH values (due to water–dunite interaction at depth) will favor calcite precipitation, depleting the aqueous solution in C ($HCO_3 + CO_3$) species (Fig. 4). In fact, the carbonate deposits occurring along veins of drill cores from Cabeço de Vide area are mainly calcite (Fig. 5). Because most the magnesium will be retained in serpentines and brucite (Fig. 6), Ca concentrations should increase proportionately in

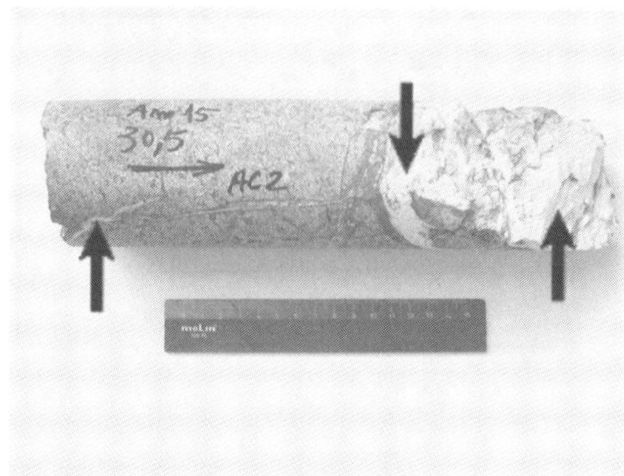

Figure 6. Drillcore from borehole AC2 (Cabeço de Vide Spas). Serpentinized dunite. The arrows indicate the presence of serpentine and brucite along fractured zones.

waters during the evolution of the Mg-HCO_3-type waters towards the Cabeço de Vide mineral-water end member.

The results obtained in this study document a locally recharged hydrologic system, dominated by interactions between meteoric waters and ultramafic rocks. These hydrogeological signatures are being used by the concessionaire of Cabeço de Vide Spas for the redefinition of well-head and aquifer protection areas, drilling strategies, and development plans.

ACKNOWLEDGEMENTS

This work has been supported by the Junta de Freguesia de Cabeço de Vide, under the Research Contract TERMAVIDE/No. 693. The authors thank Robert Seal and an anonymous reviewer for significant improvements of the manuscript.

REFERENCES

Ball, J.W. & Nordstrom, D.K. 1991. User's manual for WATEQ4F with revised thermodynamic data base and test cases for calculation speciation of minor, trace and redox elements in natural waters. *U.S. Geol. Surv. Open File Rep.* 91–183.

Barnes, I. & O'Neil, J.R. 1969. The relationship between fluids in some fresh alpine-type ultramafics and possible modern serpentinization, Western United States. *Geol. Soc. America Bull.* 80: 1947–1960.

Barnes, I., LaMarche, V.C. Jr. & Himmelberg, G.R. 1967. Geochemical evidence of present-day serpentinization. *Science* 56: 830–832.

Barnes, I., O'Neil, J.R., Rapp, J.B. & White, D.E. 1973. Silica-carbonate alteration of serpentine: wall rock alteration in mercury deposits of the California Coast Ranges. *Econ. Geol.* 68: 388–390.

Barnes, I., O'Neil, J.R. & Trescases, J.J. 1978. Present-day serpentinization in New-Caledonia, Oman and Yugoslavia. *Geochim. Cosmochim. Acta* 42: 144–145.

Bruni, J., Canepa, M., Cipolli, F., Marini, L., Ottonello, G., Vetuschi Zuccolini, M., Chiodini, G., Cioni, R. & Longinelli, A. 2001. Reactions governing the chemistry of waters interacting with serpentinites. In R. Cidu (ed.), *Water-Rock Interaction 10* 2: 145–148.

Costa, I.R., Barriga, F., Mata, J. & Munhá, J.M. 1993. Rodingitization and serpentinization processes in Alter-do-Chão Massif (NE Alentejo). In F. Noronha, M. Marques & P. Nogueira (eds.), *Proceedings of the IX Semana de Geoquímica*: 27–31: Universidade do Porto. Faculdade de Ciências. Museu e Laboratório Mineralógico e Geológico.

Gonçalves, F. & Fernandes, A.P. 1973. Carta Geológica de Portugal, na escala 1:50 000, Notícia Explicativa da Folha 32-B Portalegre. Serviços Geológicos de Portugal.

ITN [Instituto Tecnológico e Nuclear] 2002. Precipitation data base. Portalegre meteorological station (1988–2002).

Marques, J.M., Andrade, M., Goff, F., Basto, M.J., Matias, M.J., Graça, R.C., Carreira, P.M., Aires-Barros, L. & Rocha, L. 2003. Origin and evolution of high pH mineral waters traced by isotope geochemistry (S-Portugal). *Book of Extended Synopses of the International Symposium on Isotope Hydrology and Integrated Water Resources Management* – IAEA-Vienna: 180–181.

Water–Rock Interaction, Wanty & Seal II (eds)
© 2004 Taylor & Francis Group, London, ISBN 90 5809 641 6

Role of water–rock interaction in the formation of La Azul fluorite deposit, southern Mexico

T. Pi & J. Solé
Institute of Geology, UNAM, Coyoacan, Mexico D.F., Mexico

Y. Taran
Institute of Geophysics, UNAM, Coyoacan, Mexico D.F., Mexico

P. Alfonso
Facultat de Geologia, Universitat de Barcelona, Barcelona, Spain

ABSTRACT: A comprehensive geochemical study of La Azul, a fluorite deposit in southern Mexico near the city of Taxco, includes petrography, microthermometry of fluid inclusions, stable and radiogenic isotope analyses, geochemistry of REE, and (U-Th)/He geochronology. We differentiated three stages of mineralization using a correlation between REE patterns and their concentrations in fluorites and absolute ages of the ore-forming events estimated from (U-Th)/He ratios in fluorites. The early fluorite precipitated with a high Sr and low REE content by replacement of Cretaceous limestone from a hydrothermal fluid associated with Eocene-Oligocene rhyolite intrusions at T ~150°C. The later stages are characterized by more oxidized fluid at lower temperature, higher concentrations of REE and lower Sr concentrations. The Sr-Nd isotopic systematics of fluorites suggests a mixed source of the ore-forming fluids with no more than 10% of Sr leached from magmatic rocks.

1 INTRODUCTION

Mexico has been one of the main producers of fluorite for a long time. However, until now only deposits of the northern part of country have been characterized (Ruiz 1983). La Azul fluorite deposit is located 13 km to the east of the city of Taxco, center of the major and ancient Ag-Cu-Zn mining district to the south of the Mexican Volcanic Belt (Fig. 1). The fluorite body has a dimension of approximately 200 m × 150 m × 40 m and is associated with or near the contact between Mesozoic metasedimentary rocks (Taxco Schist), Cretaceous limestone (Morelos Formation) and Tertiary (Eocene-Oligocene) volcanic rocks, chiefly of rhyolitic composition. We characterize the genesis of the deposit as a multi-stage fluid-rock interaction at middle to low temperature on the basis of a geochemical study using radiogenic and stable isotopes, abundances and patterns of REE, and a first attempt to estimate the age of mineralization using (U-Th)/He dating in fluorites.

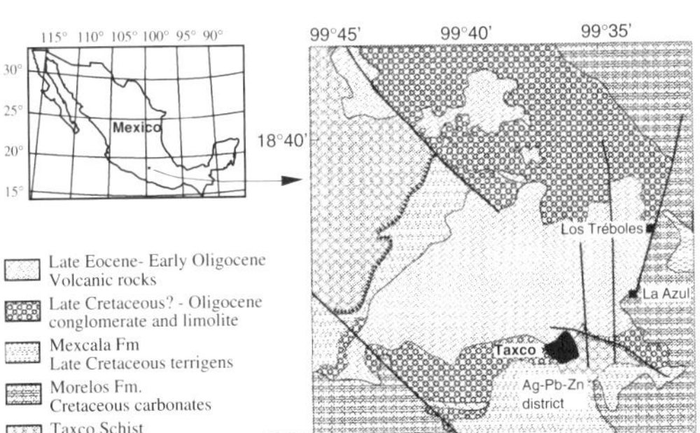

Figure 1. Simplified geological map of the Taxco area and the localization of La Azul deposit, according to Morán-Zenteno et al. (1998), Rivera et al. (1998) and Alaníz-Álvarez et al. (2002).

2 PETROGRAPHY & MINERALOGY

The paragenesis of the deposit is simple. The major mineral assemblage is represented by fluorite – quartz ± calcite ± barite, and among accessory minerals the most abundant are uraninite, realgar, pyrite, marcasite, hematite, dolomite, kaolinite, Mn-oxides, sphalerite and celestite. The color of the fluorite is variable (purple, brown, yellow, colorless) with the darkest samples enriched in uranium and the formation of an opaque, uraninite (UO_2)-enriched phase. There are two main groups of fluorite textures, distinguished as "early" and "late". The early fluorite mainly replaces carbonate and is represented by the massive banded bodies. Late fluorite is represented by nodules, sealing cavities and breccias and is associated with the later hydrothermal activity.

3 FLUID INCLUSION & STABLE ISOTOPE GEOTHERMOMETRY

Two-phase primary fluid inclusions in fluorites were studied. The inclusions of approximately 10 μm in dimension are most common. On the basis of the microthermometry of 121 inclusions, the homogenization temperatures ranged from 130 to 200°C and the salinity of ore-forming solution from 0 to 7.5 wt% eq. NaCl.

The $\delta^{13}C$ of calcite varies from −5 to 0‰ (V-PDB) and the $\delta^{18}O$ of calcite and quartz from 12 to 23‰ (V-SMOW). A difference in $\delta^{18}O$ between calcite and quartz in primary veins corresponds to isotopic equilibrium at ~150°C, in agreement with the fluid inclusion data. Therefore, low- to medium-temperature hydrothermal fluorite deposition is inferred to have occurred from water with a calculated $\delta^{18}O$ of 3 to 6‰.

Table 1. Analytical data from (U-Th)/He experiments on early and late fluorites from "La Azul" deposit, Taxco mining district, Mexico. Analyses were made at the Instituto de Geología, Universidad Nacional Autónoma de México. ^4He was extracted by laser fusion and measured with a MM1200 noble gas mass spectrometer. Sr, U and Th were measured by ICP-MS. Complete procedures are described in http://geologia.igeolcu.unam.mx/GEOQ/SoleJ.html.

Sample	Sr (μg/g)	U (μg/g)	Th (μg/g)	^4He (10^{-9} moles/g)	Age (Ma)
Early fluorite					
A20	300	68.50	0.68	12.20	32.9 ± 1.7
Az5	147	3.63	0.10	0.60	30.6 ± 1.5
Az7a	379	4.83	0.34	0.89	33.7 ± 1.7
Az7b	486	4.83	0.34	0.86	32.3 ± 1.7
Az20	214	21.00	<0.05	3.75	33.1 ± 1.7
Az24	511	7.56	<0.05	1.27	31.1 ± 1.6
Az92a	955	94.40	<0.05	17.20	33.8 ± 1.7
Az92b	409	51.50	<0.05	9.59	34.5 ± 1.8
T3	380	26.90	0.06	4.63	31.9 ± 1.6
Late fluorite					
N1b	47	1.54	0.14	0.18	21.0 ± 1.5
Az45	78	1.62	0.15	0.17	19.2 ± 1.8
N5a	91	11.60	0.23	0.81	12.9 ± 2.3
N5b	99	5.90	0.26	0.34	10.4 ± 0.6
Az54	62	28.10	0.22	0.87	6.1 ± 0.7

4 SOURCES & AGE OF ORE-FORMING FLUIDS (RADIOGENIC ISOTOPES, (U-Th)/He DATING & REE CHEMISTRY)

4.1 *Sr-Nd classification of fluorites and host rock*

Early and late fluorites from La Azul are characterized by rather uniform $^{87}Sr/^{86}Sr$ ratios in the range of 0.7074 to 0.7081 and ε_{Nd} values from −5 to −1, although the early minerals are more enriched in Sr and depleted in Sm and Nd than the late ones. Limestone of the Morelos Formation has a narrow range of $^{87}Sr/^{86}Sr$ close to 0.7075. Metasedimentary rocks of the Taxco Schist are characterized by a typical crustal value near 0.7120. Volcanic rocks in the area are enriched relative to the mantle in radiogenic Sr with $^{87}Sr/^{86}Sr$ in the range of 0.705–0.706 and ε_{Nd} values from −2 to 0. It can be suggested, based only on Sr data, that fluorites were deposited from a mixed fluid with an approximate proportion of Sr from limestone, and magmatic and crustal end-members of 80:10:10.

4.2 *(U-Th)/He dating of fluorites*

The problem with the (U-Th)/He dating is the diffusive loss of He and until now very few minerals have been used to date ore deposits by this method (see Farley (2002) for review). Our optimism for the application of the (U-Th)/He dating in fluorite is based on both unpublished data on He diffusivity in fluorite (P. Reiners, pers. comm.) and the obtained ages (Table 1). The ages for the primary fluorite fall in a narrow range of 35 to 31 Ma, which is in excellent agreement with the age of volcanic events in the area (Moran-Zenteno et al. 1998, Alaníz et al. 2002). The later stages of mineralization are characterized by ages from 20 to 6 Ma. Variations of the amount of Sr in fluorites over the (U-Th)/He time are shown in Figure 2. This graph gives an approximate time scale for the formation of La Azul and such a scale is very difficult to obtain using other methods of absolute dating.

4.3 *REE geochemistry and evolution of the ore-forming fluid*

The patterns and relative abundances of REE in fluorites and host rocks show that the higher concentrations of REE in fluorites have patterns similar to those of local volcanic rocks (Fig. 3). The distribution of REE in fluorites that replaced carbonates is similar to that in the carbonates and these fluorites

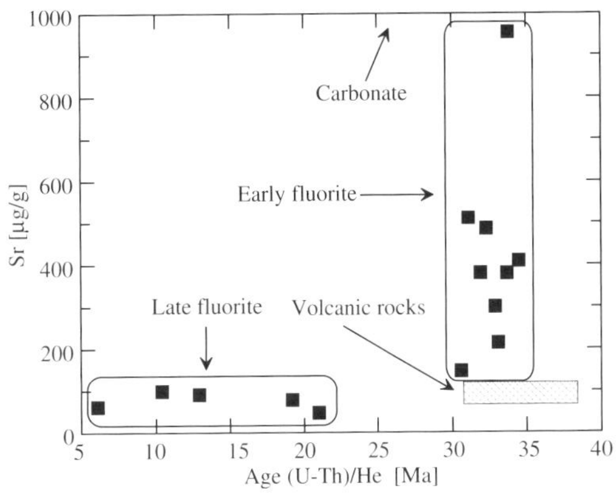

Figure 2. Correlation between (U-Th)/He age and concentration of Sr in early (>30 Ma) and late (<30 Ma) fluorite in "La Azul" fluorspar deposit.

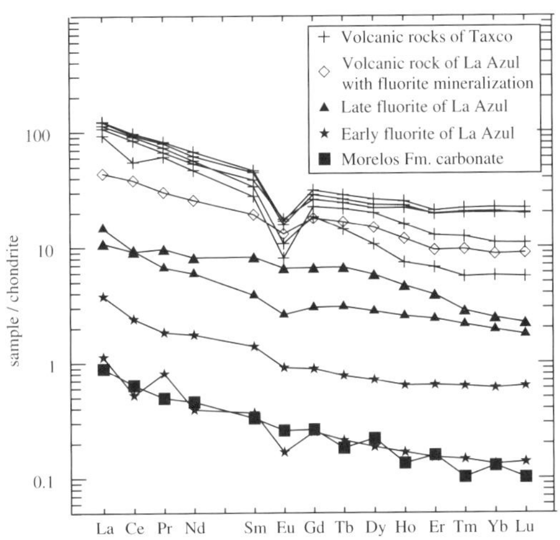

Figure 3. Patterns of lanthanides of early and late fluorite, volcanic rocks and carbonate of La Azul fluorspar deposit.

are characterized by low to very low concentrations of REE. Most of the fluorites with higher REE concentration show a negative Eu anomaly, similar to the host rock. However, some of the dark, organic- and U-enriched fluorites have a strong positive Eu anomaly, most probably due to the highly reduced conditions of the ore formation (Bau 1991). The most characteristic feature of the REE chemistry of fluorites is that the patterns and absolute concentrations of REE are age dependent (in terms of our (U-Th)/He age determinations, Pi et al. 2003). Figure 4 shows that at the early stage of mineralization, fluorite deposited by replacement of calcite incorporated lower amounts of REE with patterns similar to those of the host calcite. Next stages are characterized by

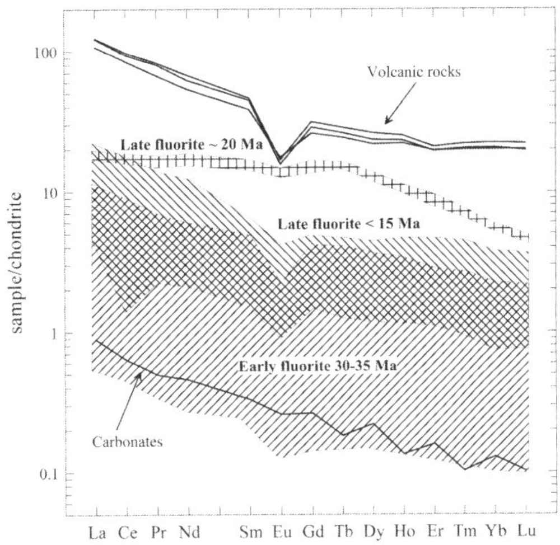

Figure 4. Patterns of lanthanides and age (U-Th)/He of the three phases of mineralization identified in La Azul fluorspar deposit.

higher concentrations of REE. The latest fluorite contains high concentrations of REE, but still one order of magnitude lower than volcanic rocks and with patterns similar to those of the volcanic rocks.

5. CONCLUSIONS

The good correlation between REE concentrations and patterns in fluorite and absolute age of the ore forming events estimated from (U-Th)/He ratios in fluorite help us to characterize three stages of mineralization in the La Azul (Taxco, México) fluorspar deposit. These results, together with petrographic, fluid inclusions, stable and radiogenic isotope data permit us understand the role and processes of water-rock interaction in the genesis of the deposit. We think that the method can be applied with success to other multi-stage deposits with fluorite.

ACKNOWLEDGEMENT

The analytical part of this work has been supported by CONACYT project n° 32475-T. Ma del Sol Hernández, Juan Julio Morales and Gabriela Solís are knowledged by their analytical support for Sr and Nd isotopic measurements. The paper benefited from comments by Greg Arehart and Bob Seal.

REFERENCES

Alaniz-Álvarez, S.A., Nieto-Samaniego, A.F., Morán-Zenteno, D.J. & Alba-Aldave, L. 2002. Rhyolitic

volcanism in extensión zone associated with strike-slip tectonics in the Taxco region, southern México. *J. Volc. Geotherm. Res.* 118:1–14.
Bau, M. 1991. Rare-earth element mobility during hydrothermal and metamorphic fluid-rock interaction and the significance of the oxidation state of europium. *Chem. Geol.* 93:219–230.
Farley, K.A. 2002. (U-Th)/He dating: Techniques, calibrations, and applications. In: P.D. Porcelli, C.J. Ballentine & R. Wieler (eds) *Noble Gas Geochemistry. Rev Mineral Geochem* 47:819–843.
Morán-Zenteno, D.J., Alba-Aldave, L., Martínez-Serrano, R.G., Reyes-Salas, M.A., Corona-Esquivel, R. & Angeles-García, S. 1998. Stratigraphy, geochemistry and tectonic significance of the tertiary volcanic sequences of the Taxco-Quetzalapa region, southern México. *Rev. Mexicana Cienc. Geol.* 15:167–180.
Pi, T., Solé, J. & Taran, Y. 2004. (U-Th)/He dating of fluorite. Application to a Mexican fluorspar deposit. Submitted to *Mineralium Deposita.*
Rivera, E., De la Teja- Segura, M.A., Miranda, A.H., Lemus, O.B., Montolinea, O.G., León, V.A. & Moctezuma, MDS. 1998. *Carta Geológico-Minera Cuernavaca E14–5.* Consejo de Recursos Minerales, México.
Ruiz, J. 1983. *Geology and geochemistry of fluorite ore deposits and associated rocks in Northern México.* Unpub. Ph.D. thesis. Ann Arbor: University of Michigan.

Water–Rock Interaction, Wanty & Seal II (eds)
© 2004 Taylor & Francis Group, London, ISBN 90 5809 641 6

Barite as an indicator of cross-formational fluid flow

D. Savage
Quintessa Limited, West Bridgford, Nottingham, UK

ABSTRACT: Published data on the occurrence of pervasive barite ($BaSO_4$) cement associated with faulting in two sandstone reservoir systems, one containing hydrocarbon gas in the southern North Sea, the other bearing metalliferous minerals in the onshore UK Cheshire Basin, have been used to attempt to constrain the amounts of fluid involved in cross-formational fluid flow. Mass balance calculations using plausible estimates of source fluid compositions suggest fluid volumes of at least 3–40 km^3 were responsible for barite precipitation. These fluid volumes could have transported a substantial proportion of the metals and hydrocarbons observed in the two reservoir systems.

1 INTRODUCTION

Barite ($BaSO_4$) is a sparingly soluble mineral, with a solubility product at 25°C of approximately 10^{-10} (Blount 1977). In natural systems, it may occur associated with heavy metal mineralization or as stratiform deposits in sedimentary rocks. Modern seafloor barite occurrences are thought to be the product of warm Ba-rich fluids from sub-seafloor sediments driven by tectonic activity along fault zones mixing with cold seawater. Barite also forms during the diagenesis of sedimentary rocks, either in relatively small amounts resulting from the dissolution in the subsurface of Ba-bearing detrital mineral grains such as potassium feldspar, or as a more pervasive cement attributed to mixing of fluids accompanying fault movement.

This latter occurrence, coupled with its low solubility, implies mixing of two chemically-distinct fluids, one Ba-rich and SO_4-poor; the other Ba-poor and SO_4-rich. Hanor (1994) noted the inverse relationship between barium and sulfate concentrations in sedimentary brines worldwide, suggesting control by equilibrium with barite. Barite occurrence and abundance has been used previously as an indicator of fluid mixing (e.g. Breit et al. 1990) and in a similar study to that described here, has been used to estimate the volume of fluid necessary to form the mineral deposit at Chamberlain Creek, Arkansas (Hanor & Baria 1977).

Here, evidence from studies of two sandstone reservoir systems, one containing hydrocarbon gas in the southern North Sea (Amethyst Field – Gluyas et al. 1997), the other bearing metalliferous minerals in the onshore Cheshire Basin, UK (Alderley Edge – Rowe & Burley 1997) has been used to attempt to constrain the amounts of fluid passing along their respective fault networks during diagenesis.

2 FIELD EVIDENCE

2.1 *The Amethyst Field*

The Amethyst Field lies in the southern North Sea Basin, approximately 65 km east of Hull. Gas is trapped in a gentle NW-SE trending anticline over an area of 32 by 10 km. The reservoir consists of 15–40 m of Permian Rotliegend sandstones, with a caprock of Permian Zechstein evaporites and carbonates. The top of the reservoir is 2713 m deep. Current reservoir conditions are 90°C and 280 bars. Approximate gas reserves are $2.9 \times 10^{10} m^3$ (92 mol% CH_4). Thermally mature coals in the underlying Carboniferous are believed to be the source of the gas. Analyses of current formation waters are presented in Table 1. These formation fluids are saline (TDS = 250 g/L) with relatively low contents of Ba (<0.01 g/L), with either negligible or relatively low (for southern North Sea reservoir fluids) contents of dissolved sulfate (<1 g/L).

Gluyas et al. (1997) have documented pervasive cementation of a number of the wells in the field with anhydrite and/or barite, linking this mineralization with fluid flow along a number of major east-west trending faults in the field (Fig. 1). The precipitation of anhydrite and barite (and introduction of gas) occurred late in the history of the field during basin inversion in the Tertiary.

Table 1. Formation waters from the Amethyst Field (g/L). From Warren & Smalley (1994).

	47/14a-6	47/14a-8	47/14a-9	47/13a-3
pH	6.03	6.69	5.40	5.56
Na	61.8	60.1	65.6	61.5
K	1.23	1.28	1.33	1.33
Ca	22.0	18.9	21.5	20.9
Mg	2.52	3.06	2.84	3.61
Sr	1.03	1.16	1.20	1.04
Ba	–	0.01	–	0.01
Cl	149.0	140.8	147.2	144.1
SO_4	0.83	–	0.35	0.36
HCO_3	0.16	0.10	0.04	0.06

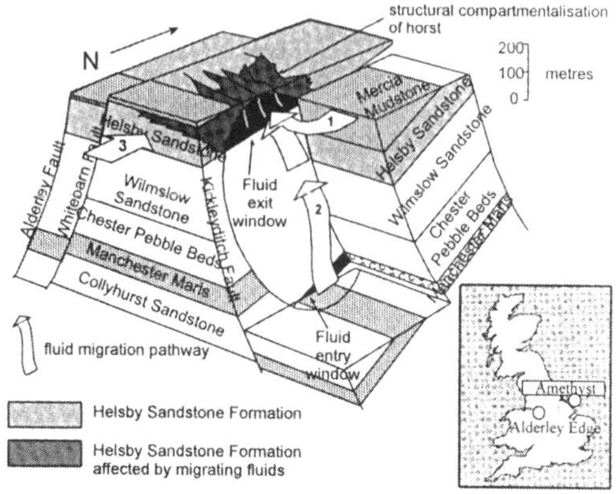

Figure 2. Schematic model of fluid flow and mineralization at Alderley Edge according to Rowe & Burley (1997).

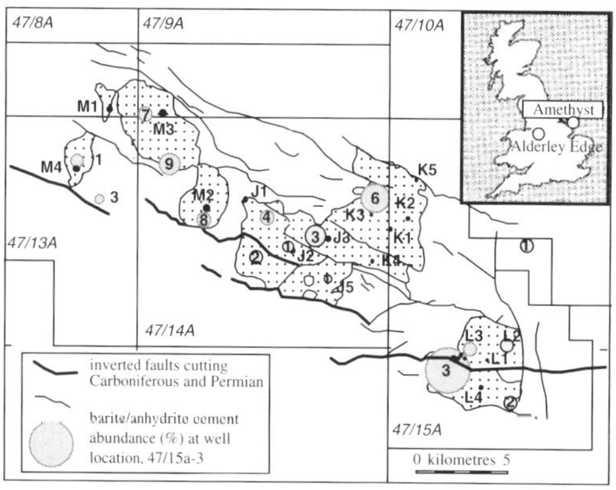

Figure 1. Abundance of sulfate cements associated with E-W faulting in the Amethyst Field. The diameters of the circles are proportional to the abundance of anhydrite and barite cements. From Gluyas et al. (1997).

Barite and anhydrite locally constitute up to 20% of the sandstone volume. Sulfur isotopic data indicate a Zechstein source for the sulfur, whereas the Ba (and possibly Ca) was most likely derived from the underlying Carboniferous (Gluyas et al. 1997). Gluyas et al. (1997) recognized the incompatibility of transport of both Ba and SO_4 in the same fluid and invoked fluid mixing as a deposition mechanism. Gluyas et al. (1997) do not present precise volumes of anhydrite and barite cements throughout the reservoir volume, so a conservative estimate of the proportion of barite in the field was considered to be 5% of the total rock volume across a 100 m distance from the major E-W faults cutting the field (total E-W fault length also conservatively estimated as 50 km). If a mean reservoir thickness of 20 m is assumed, then a total volume of barite is estimated to consist of $5 \times 10^6\,m^3$ ($0.05 \times 50\,000 \times 20 \times 100$).

2.2 *Alderley Edge*

Heavy metal sulfide and barite mineralization at Alderley Edge in the Cheshire Basin, UK has been described by Rowe & Burley (1997). Alderley Edge is a 3 km-wide tilted horst structure, dipping to the SSW, near the eastern margin of the Cheshire Basin (Fig. 2). The horst is bounded by the normal Alderley and Kirkleyditch Faults, with throws of 250 and 330 m, respectively. The horst is crossed by subordinate WNW-ESE dip-slip faults with displacements of less than 10 m. This faulting is associated with hematite reduction, barite cementation and sulfide mineralization. Mineralization is in the Helsby Sandstone Formation of the Sherwood Sandstone Group (Triassic). There is up to 2000 m of Mercia Mudstone overlying this formation with interbedded mudstones, siltstones and evaporites. Mineralization is spatially associated with faulting and there is a primary sulfide assemblage of bravoite, pyrite, chalcopyrite, sphalerite and galena, which has been locally re-mobilized by recent groundwater movement to a mixture of oxides, hydroxides, carbonates and sulfates. Sulfur isotopic studies indicate that the sulfur for the mineralization was derived from evaporites in the overlying Mercia Mudstone formation, with metals and barium being sourced most probably from the underlying Carboniferous. Mineralization occurred where warm (60–150°C) reducing brines carrying metals (and probably hydrocarbons) were transported from the Carboniferous up basin margin faults to mix with sulfate-bearing brines present in the Helsby Sandstone Formation. In situ reduction of the sulfate (microbially-catalysed) by hydrocarbons must be invoked to account for the extensive precipitation of sulfides. The total mineralized volume of rock at Alderley Edge is developed over an area of 2.5 by

1.8 km, through an average sandstone thickness of 100 m, which represents a total volume of 0.45 km^3. Rowe & Burley (1997) liken this to the volume of a moderate-sized hydrocarbon reservoir.

According to Rowe & Burley (1997), 18% of the mineralized sandstone consists of barite, i.e. 8.1×10^7 m^3.

3 SOURCE FLUID COMPOSITIONS AND FLUID VOLUMES

In order to be able to calculate how much water would be needed to transport the amounts of Ba to the Amethyst and Alderley Edge reservoirs, some estimate of the likely amount of Ba in their respective (Carboniferous) source fluids is required. There are only two analyses of formation fluid analyses available for the Carboniferous sections of North Sea oil and gas fields (from the Visund Field) as compiled by Warren & Smalley (1994), but Edmunds (1974) presents 13 analyses of brines in the Coal Measures of the UK Durham coalfield (marginal to the North Sea Basin). The analyses presented by Edmunds (1974) reveal a range of Ba contents from below analytical detection limits (<0.001 g/L), up to a maximum of 4.2 g/L in a brine containing 105 g/L Cl^-. Edmunds (1974) noted that SO_4 and Ba concentrations in these groundwaters are mutually exclusive. Edmunds (1974) also reported a maximum value of Ba in groundwaters worldwide as 5.53 g/L. The dataset of the Durham coalfield brines and the two Carboniferous brines from the North Sea Formation Fluids Atlas (Warren & Smalley 1994) is not large enough to perform a statistically-meaningful analysis, but nine of the 13 fluid analyses have a Ba content of less than 0.5 g/L.

If the Ba content of the Carboniferous groundwaters that produced the cement in the Amethyst Field is assumed to be 0.5 g/L, then the amount of fluid required to precipitate all the barite observed in the Amethyst Field is 2.6×10^{10} m^3 (26 km^3). If a more conservative assumption is adopted and the maximum Ba content for groundwaters worldwide is assumed for the source of the Ba in the Amethyst Field (~5 g/L), then the amount of fluid required to precipitate all the barite observed is correspondingly an order of magnitude less, i.e. 2.6×10^9 m^3 (2.6 km^3). In this context, it is noteworthy that Lange et al. (1983) considered it likely that the barite associated with the Mississippi Valley-type ore deposits of southeast Missouri, USA formed from a fluid containing only 0.025 g/L Ba, emphasizing the conservatism of the calculations presented here.

It is not unreasonable to assume that the groundwater transporting Ba from the Coal Measures underlying the Amethyst Field was saturated with methane under ambient pressure–temperature–salinity conditions. Methane solubility increases with increasing temperature and pressure, but decreases with increasing salinity (Haas 1978). If the P-T conditions of the Carboniferous beneath Amethyst are taken to be 150°C and 500 bars (equivalent to 4–5 km of burial) and the salinity of these pore fluids is analogous to those of the Durham coalfield brines (2.0 M NaCl equivalent average), then data for methane solubility (Haas 1978) indicate a methane concentration of 0.16 moles/L. The fluid volume estimated from the transport of Ba (using the most conservative 5 g/L figure for the Ba content of source fluids) could thus theoretically transport 4.2×10^{11} moles of CH_4. Assuming ideal gas behavior ($PV = nRT$), then the volume of gas under current reservoir conditions (90°C, 280 bars) would be 4.5×10^7 m^3. This volume is almost three orders of magnitude less than actual gas in place of 2.9×10^{10} m^3 (Gluyas et al. 1997), but illustrates that a substantial volume of gas could have been transported in a single-phase solution with the groundwater responsible for the barite precipitation. If at the time of barite and gas emplacement reservoir fluids in the Amethyst Field were halite-saturated (derived from the overlying Zechstein), then effectively all of this gas could have been exsolved from the fluid phase due to the lower methane solubility in halite-equilibrated fluids (<0.03 moles/L at 90°C, 280 bars, 6 M NaCl – Haas 1978).

In an analogous manner to the calculations presented above for the Amethyst Field, assumption of a Carboniferous brine source fluid for the mineralization at Alderley Edge with a Ba content of 5 g/L implies that 4.2×10^{10} m^3 (42 km^3) of fluid could have been responsible for the transport of barium and other heavy metals to the Helsby Sandstone. As a check on this calculation of fluid volume, the concentration of lead in the fluid which would be required to precipitate all the observed galena (PbS) at Alderley Edge was estimated. Rowe & Burley (1997) cited an average total sulfide volume of 2% in the mineralized sandstone at Alderley Edge, of which it is assumed that 0.1% is galena (S. Burley, pers. comm.). If 0.1% of the mineralized sandstone (0.45 km^3) were galena, then 4.5×10^5 m^3 or 3.4×10^{12} g of galena would be present. This is equivalent to 1.4×10^{10} moles of galena or 2.9×10^{12} g of lead. Dividing the mass of lead by the total volume of fluid calculated above results in a concentration of lead in the mineralizing fluid of 0.069 g /L. This concentration is within the range (up to 0.1 g/L Pb) in brines thought to be possible ore-forming fluids for Mississippi Valley-type mineral deposits of the USA. This implies that the volume of fluid estimated from the distribution of barium is consistent with the amount of galena present in the Helsby Sandstone at Alderley Edge using plausible concentrations of lead in ore-forming solutions.

4 CONCLUSIONS

Calculations presented here using abundances of barite in sandstones and likely abundances of barium in source fluids indicate large volumes of groundwater (3–40 km^3) interacting with sandstones during fault-related fluid flow. The insolubility of barite, its presence in significant quantities in these sandstone reservoirs (up to 20%), and the geological setting (mineralization adjacent to faults) indicate an origin due to fluid mixing associated with cross-formational flow along faults. The volumes of fluid estimated could have transported significant amounts of metals and/or hydrocarbons to form economic accumulations of such resources.

Despite the uncertainties concerning the abundances of barite in the reservoirs and the precise Ba content of source fluids, the conservative assumptions made concerning these parameters suggest that the volumes of groundwater calculated are underestimates of actual volumes. Better data for barite abundances in the reservoirs concerned and data for the Ba contents of diagenetic fluids in fluid inclusions would help reduce the uncertainties concerning precise fluid volumes. The lower estimates of the volumes of groundwater calculated here are at least an order of magnitude greater than those released during "single-event" historic large magnitude normal-fault earthquakes documented in the literature (e.g. Muir-Wood & King 1993), whereas the upper estimated volumes are a factor of 3–4 less than that calculated to have been involved with formation of the Chamberlain Creek barite deposit in Arkansas, USA (Hanor & Baria 1977).

The occurrence of large accumulations of barite provides a ready indicator of fluid mixing processes in the sub-surface. Although cementation of sandstones by barite during diagenesis may be relatively rare, it provides evidence for the movement of appreciable quantities of pore fluids in the subsurface (Breit et al. 1990). Given suitable source rocks, it is almost inevitable that economically-significant constituents (heavy metals, oil, gas) will accompany this fluid, either as a single, or multi-phase fluid. Quantifying the amounts and mechanisms of fluid movement in the deep subsurface will lead to a better understanding of the accumulations of economic resources.

REFERENCES

Blount, C.W. 1977. Barite solubilities and thermodynamic quantities up to 300°C and 1400 bars. *Am. Mineral.* 62: 942–957.

Breit, G.N., Goldhaber, M.B., Shawe, D.R. & Simmons, E.C. 1990. Authigenic barite as an indicator of fluid movement through sandstones within the Colorado Plateau *J. Sed. Petrol.* 60: 884–896.

Edmunds, W.M. 1974. Geochemistry of brines in the Coal Measures of northeast England. *Inst. Mining Metall.* 84: 339–352.

Gluyas, J., Jolley, L. & Primmer, T. 1997. Element mobility during diagenesis: sulphate cementation of Rotliegend sandstones, Southern North Sea. *Marine Petrol. Geol.* 14: 1001–1011.

Haas, J.L. Jr. 1978. An empirical equation with tables of smoothed solubilities of methane in water and aqueous sodium chloride solutions up to 25 weight percent, 360°C and 138 MPa. *US Geol. Surv. Open-File Report* 78–1004.

Hanor, J.S. 1994. Origin of saline fluids in sedimentary basins. In J. Parnell (ed.), *Geofluids: Origin, Migration and Evolution of Fluids in Sedimentary Basins Geological Society of London Special Publication* 78: 151–174.

Hanor, J.S. & Baria, L.R. 1977. Controls on the distribution of barite deposits in Arkansas. In C.G. Stone (ed.), *Geology of the Ouachitas* 2: 48–55. Arkansas Geological Commission.

Lange, S., Chauduri, S. & Clauer, N. 1983. Strontium isotopic evidence for the origin of barites and sulfides from the Mississippi Valley-type ore deposits in southeast Missouri. *Econ. Geol.* 78: 1255–1261.

Muir-Wood, R. & King, G.C.P. 1993. Hydrological signatures of earthquake strain. *J. Geophys. Res.* 98: 22035–22068.

Rowe, J. & Burley, S.D. 1997. Faulting and porosity modification in the Sherwood Sandstone at Alderley Edge, northeastern Cheshire: an exhumed example of fault-related diagenesis. *Petroleum Geology of the Irish Sea and Adjacent Areas, Geological Society of London Special Publication* 124: 325–352.

Warren, E.J. & Smalley, P.C. 1994. North Sea Formation Waters Atlas. *Geological Society of London Memoir* 15.

Water–Rock Interaction, Wanty & Seal II (eds)
© 2004 Taylor & Francis Group, London, ISBN 90 5809 641 6

Micropore-diffusion catalytic effect significance in water–rock interaction

L.I. Shabalin
Siberian Research Institute of Geology, Geophysics and Mineral Resources, Novosibirsk, Russia

ABSTRACT: The author has elaborated upon and experimentally verified the concept of micropore-diffusion catalytic effect (MDC-effect). Its essence is as follows: in very small pores having diameters less two mean distances between molecules of any dissolved substance, two processes occur simultaneously (1) the acceleration of these molecules as they exit from the micropores, and (2) the acceleration of chemical interactions among dissolved molecules within the micropores due to more frequent collisions among them. The force of decondensation of the surficial layer of water (FDSL) takes part in its action. This effect produces high chemical activity of solution during water–rock interactions.

1 INTRODUCTION

It is known that in the Earth's crust almost all chemical reactions take place in micropores inside the layer of so called bonded or non-dissolving water. Thus, this water is the medium of all chemical interactions in water–rock systems (Udodov et al. 1983), including petroleum generating processes (Hunt 1982).

Derouan et al. (1987) have suggested that the micropores are traps where catalytic transformations of molecules takes place. They explained this process as a result of surface curvature effects, which in their opinion are due to increases in the physical adsorption of molecules to the walls of micropores at the expense of the van der Waals forces.

2 ESSENCE OF THE MDC-EFFECT

The author has a different explanation increases in the catalysis chemical processes inside micropores and proposed the idea of the micropore-diffusion catalytic effect (MDC-effect). The essence of the MDC-effect is as follows: inside micropores having diameters less two mean distances between molecules of a dissolved substance, two processes occur simultaneously: (1) acceleration of these molecules as they exit from the micropores, (2) the acceleration of chemical interactions among dissolved molecules within the micropores due to more frequent collisions among them. Thus, the smaller the micropores are, the more intensively these processes occur.

This effect is based on the author's hypothesis about the existence in the nature of a yet force unknown called molecular-kinetic "force of decondensation of the surficial layer", abbreviated FDSL, of liquids, solids and gaseous substances and especially water. The cause of the force for water is as follows: during water flow along any surface and as the area of its surficial layer increases, new layers of water come to form parts of this layer and here they are decondensating with the force equal the force of temperature expansion for the water (temperature component – T-FDSL) and with the force of osmotic pressure for dissolved molecules (osmotic component – O-FDSL).

This idea has been discussed in several books and articles (Shabalin 1997, 1998, 1999a,b, 2000, 2001, 2002, 2003); therefore, a full explanation of this force will not be given here. I will only explain the details of the osmotic component of FDSL, which plays the most important role in the MDC effect. It is precisely this component that promotes the acceleration of molecular motion as water exits micropores or moves to broader parts of the pores.

It is known that the concentration of dissolved substances in contact with hydrophilic colloidal particles and minerals is less than that in the volume of water. This fact has been confirmed experimentally by using dissolved substances or tracers (Dumanskii 1960), by pressing water from rocks (Krukov 1971) and from the practice of sugar production. It is considered that a non-dissolving layer of bonded water is the cause of this phenomenon. Hunt (1982) used this fact to explain the mechanism of diffusional migration of hydrocarbon molecules from clay-rich oil-producing sediments to surrounding sandstones. This diffusion, he explained by considering that dissolved hydrocarbons and the surfaces of clay minerals are surrounded by the layer of bonded water. According to existing ideas, their approach and contact will lead to the destruction of the bonded water must occur. For this process, energy is needed, i.e. the layers tend to protect themselves from penetration from each other. Therefore, the motion of hydrocarbons molecules from narrow parts of micropores to wider parts along boundaries with free water will occur.

In my opinion, the suggestion by Hunt about diffusional mechanisms of dissolved components migrating from micropores is correct in principle, because any force promoting the generation of a non-dissolving layer with lower concentrations of molecules must push out molecules of dissolved substances from it and from micropores. However, the problem is that this mechanism will not act inside of clay-rich sediments because here almost 100% of water is represented by bonded water and the boundary between free and bonded water is absent.

The MDC-effect generated by the action of FDSL does not have this disadvantage. Nothing impedes its action in fine-grained microporous rocks of the Earth's crust, because migration of dissolved substances from the narrowest parts of micropores to broader parts will occur independently from presence of boundary between bonded and free water. It is explained by a different principle of the action of the MDC-effect when the pushing of dissolved molecules from the micropore walls takes place at the expense of a peculiar molecular-kinetic mechanism explained by the author at the examples of several models (Shabalin 2000, 2001, 2002, 2003). This mechanism here is not explained due to lack of space.

3 EXPERIMENTS

The simplest verification of MDC-effect is an experiment with the corrosion of metal. A polished surface of iron plate, covered by thin layer of water is needed to draw a furrow by the tip of knife. A symmetrical aureole of brown iron oxides appears around the deformed furrow within 2–5 minutes. Over it at the surface of water ultrasmall colloidal transparent particles of iron oxides appear, which are floating like an umbrella. Here, it is quite obvious that the very quick corrosion is due to the appearance of a great number of micropores and cracks around the furrow.

The next experiments verify the existence of the MDC-effect all the more convincingly. If you take a white china plate with water and put simple iron objects, a bended paper clip and nail, on the bottom, you will see brown iron oxides appear in 4–5 hours only around those parts, where these subjects are touching the bottom. These oxides cover all surfaces surrounding the contact points. If these objects were then moved away, it will be seen that under them in places of contact, the plate is quite white, i.e. free from iron oxides at a distance related to the thickness of the objects. It is quite obvious that corrosion of iron subjects took place most intensively in the confinements of those "micropores", which occurred in the places of contact of the objects with the bottom of the plate. The resulting iron oxides are moving from here and depositing on nearby surfaces. Here, the "micropores" at the contact points were acting as catalysts for the corrosion, and not the china plate. This can be confirmed by similar experiments putting paper clips and nails on glass, wood, plastic or others kinds of surfaces soaked by water.

4 CONCLUSIONS

In detail, these simple experiments are the key for understanding of action of the MDC-effect in rocks of the Earth's crust. It is precisely by its action, as noted by Shvartzev (1985), that the "ability of water everywhere during all geologic history to dissolve endogenetic aluminosilicates" is explained. The MDC-effect acts within micropores of rocks, to give water high chemical aggressiveness at great depths within the Earth's crust, which promotes maintenance of micropores in an open state for metasomatic processes during endogenic ore and magma formation. Under subsurface conditions, it promotes the generation of oxide zones in deposits, the migration of disseminated petroleum from source rocks, and also results in the geochemical composition of ground waters. The effect promotes the action of a mechanism that is called diffusional fluid replacement by the author. It promotes the formation of ore deposits and magmatic melts, and it regulates earthquakes in the Earth's crust (Shabalin 2003). Based on this mechanism, the author has devised and patented a method for the prevention of catastrophic earthquakes called "Method of stabilization of ground massifs in the earthquake-prone regions" (Shabalin 1999b). Also, a new concept was proposed: a molecular-kinetic concept of ore and magma formation (Shabalin 2002).

REFERENCES

Derouane, E.G., Andre, J.M. & Licas, A.A. 1987. A simple van der Waals model for molecular-curved surface interaction in molecular-sized micropores solids. *Chemics Physics Letters* 137(4): 336–340.

Dumanskii, A.V. 1960. *Liophility of Dispersion System.* Kiev: AS USSR.
Hunt, J. 1982. *Geochemistry and Geology of Petroleum and Gas.* Moscow: Mir.
Krukov, P.A. 1971. *Rock, Soil and Mud Solutions.* Novosibirsk: Nauka.
Shabalin, L.I. 1997. Micropore diffusion catalytic effect. Experimental verification and role in formation of magmatic rocks and ore bodies. *Izvestiya VUZov. Ser. Geology and Prospecting.* N1: 79–89.
Shabalin, L.I. 1998. Self-organization system of the nature. Novosibirsk: SNIIGGiMS.
Shabalin, L.I. 1999a. How is self-progressing the alive and non-alive nature. Tomsk: Tomsk State University.
Shabalin, L.I. 1999b. Method of stabilization of ground massif in seismic prone regions: Patent of Russia, No 2140492, priority from 24.06.98. Inventions. No 30. P.393.
Shabalin, L.I. 2000. Force of decondensation of the superficial layer of water as fundamental nature phenomenon, producing exchange of substances (claim for discovery). Novosibirsk: SNIIGGiMS.
Shabalin, L.I. 2001. Force of decondensation of the superficial layer of liquid, solid and gaseous substances. Novosibirsk: SNIIGGiMS.
Shabalin, L.I. 2002. Basis of molecular-kinetic conception of ore-and magma formation. Novosibirsk: SNIIGGiMS.
Shabalin, L.I. 2003. Force of decondensation of the superficial layer if water and its role in geological processes. Novosibirsk: SNIIGGiMS.
Shvartzev, S.L. 1985. Physico-chemical and geological evolution of water–rock system. *Underground Water and Evolution of the Lithosphere* 1: 253–266. Moscow: Nauka.
Udodov, P.A., Korobeinikova, E.S. & Nazarov, A.D. 1983. *Geochemical Properties of Porous Solutions in Rocks.* Moscow: Nedra.

Water–Rock Interaction, Wanty & Seal II (eds)
© 2004 Taylor & Francis Group, London, ISBN 90 5809 641 6

Fluid–rock interactions within Paleoproterozoic Au mineralization of the Sakoli Group, India

A.S. Venkatesh, B. Jose & T. Srinivas
Department of Applied Geology, Indian School of Mines, Dhanbad, India

ABSTRACT: Au mineralization occurs in the Palaeoproterozoic Sakoli Group of rocks in central India associated with quartz-sulfide ± carbonate veins within the metavolcano-sedimentary sequence of the southern block of the Central Indian Tectonic Zone (CITZ). The mineralization is focused predominantly within the E-W trending major shear zone correlated with the regional schistosity (S_2) plane. Gold occurs both as invisible and native forms with other polymetallic associations. The mineralization exhibits both lithological and structural controls and is associated with pervasive alteration resulting from intensive ore-fluid–rock interaction, which produced considerable mass and volume changes near the shear zone. The fluid inclusion study reveals that the fluids responsible for gold mineralization are approximated by the H_2O–NaCl–CO_2 ± CH_4 system. The system appears to have evolved from a carbonate dominated regime to an increasingly aqueous dominated regime through time.

1 INTRODUCTION

The Paleoproterozoic Sakoli Group forms a triangular pattern within the Bastar Craton in the Central Indian Shield constituting part of the Central Indian Tectonic/Suture Zone (CITZ). The CITZ hosts various styles of mineralization within deformed volcano-sedimentary basins in diverse tectonic settings. The ESE-WNW-trending CITZ divides the Indian sub-continent into southern and northern crustal provinces; the Sakoli fold belt (SFB) falls within the wider southern tectonic block. Rogers & Santosh (2002) suggested that the CITZ was an orogenic belt and showed it to have been an extensional belt along with the Western Australian and North American mobile belts, which were part of Columbia supercontinent. Stein et al. (2003) suggested that the CITZ was Late Archean in age on the basis of a Re-Os date from molybdenite from the Malanjkhand porphyry copper–molybdenum deposit. The CITZ has generated considerable interest among geologists world wide because of its typical Archean metallogenic features (Yedekar 1990, Acharyya 2003).

There are two diverse styles of mineralization occur within in the SW portion of the Sakoli basin in Central India. The one bears similarities of a lode type gold–copper mineralization in Pular-Parsori-Tuthanbori areas (Mishra et al. 2001) where as the other one has a VHMS affiliated stratabound Zn–W–Au mineralization in parts of Kolari-Bhaonri areas (Gadhadharan & Jog 1990, Bhoskar 1990) that lies about 7 km south of the Au-Cu belt. Both the mineralizing events lies within the volcano sedimentary sequence on a regional extent comprising quartz–chlorite–sericite schist and chlorite-muscovite schist (with garnet porphyroblasts) is associated with stratabound and stratiform tourmalinites, coticules and BIF that belong to one geological formation (Bhiwapur Formation as given in Bandyopadhyay et al. 1995) within the SW portions of SFB. The total reserves of zinc ore is 3.9 million tones over a strike length of 2.80 km with an average grade of 5.41% Zn estimated up to a maximum vertical depth of 210 m. A resource of 2219 tonnes of 65% WO_3 concentrate over an aggregate strike length of 1,380 m and to a maximum depth of 200 m has been estimated by Gadhadharan & Jog (1990). The polymetallic mineralization within the SFB is analogous to that in the Abitibi and Yilgarn belts (Yeats & Groves 1998).

2 LITHOLOGY

The study area, Pular-Parsori-Tuthanbori belt occurs in the SW margin of the SFB where gold-copper mineralization has been reported. Various workers from the Geological Survey of India (GSI) have also delineated several auriferous zones including over 500 old workings of gold within the shear zone associated with sulfides in the Pular-Parsori and

Tuthanbori area. Saha & Mohan (1994) have estimated reserve of gold that extended to a depth of up to 105 m and for 300 m along strike. The gold content is highly variable, the highest grade reported from the belt being 8.4 g/t over a width of 1.66 m (Saha et al. 1996).

The original basin geometry, superimposed deformations, and three prominent ductile shear zones account for the present triangular outcrop of SFB, which forms a broad synclinorium (Bandyopadhyay et al. 1995). Detailed structural studies indicate the presence of various planar and linear features that are correlated spatially and temporally with mineralization and major deformation events, suggesting strong structural controls on the mineralization in the region. The trend of mineralization follows the regional schistosity (S_2). Felsic and mafic metavolcanic rocks are the predominant hosts for gold mineralization. Structurally, there are three generations of deformation and metamorphism, which has caused mineralizing fluids to be, channelized within the shear fractures marking the quartz – sulfide ± carbonate veins. Thus, the mineralization exhibits both lithologic and structural controls.

3 MINERALIZATION AND PARAGENETIC RELATIONSHIPS

Gold mineralization within the belt is characterized by the presence of both native and lattice-bound invisible gold. The deposits exhibit a wide range of polymetallic associations of Cu, W, Pb, Zn, Ag, Mo and Biin addition to gold. Gold at Pular-Parsori-Thutanbori belt of Sakoli Group, Central India occurs in two forms: (i) as invisible form in sulfide minerals, and (ii) as native state. The presence of both types has been reported from many other mesothermal deposits; for e.g. Mumin et al. 1994, Genkin et al. 1998, Ashley et al. 2000, Saha & Venkatesh, 2002. Both the types of gold are spatially, genetically and compositionally different from each other and form at different stages (Bonnemaison & Marcoux 1990, Mumin et al. 1994, Saha & Venkatesh 2002). The existence of both invisible and native forms of gold in a single deposit involves a complex evolutionary process comprising of successive stages of fluid evolution and concentration of gold. The three-stage model includes progressive increase in the concentration of gold, which can be seen as an early stage containing invisible gold, an intermediate stage with fine grained submicroscopic substrates associated with sulfides, and a late stage with gold of economic value containing gold nuggets in rich lodes (Mishra et al. 2001).

Mineral chemical analysis by electron probe microanalysis (EPMA) and Scanning Electron Microscopy (SEM) reveals the presence of a substantial amount of lattice-bound gold that varies from 440 ppm to 1790 ppm in the arsenopyrite, arsenian pyrite, and chalcopyrite. The substitution of Fe^{3+} by Au^{3+}instead of Au substituting into Assites within the sulfide minerals, adsorption–reduction, and fluid-rock interaction/phase immiscibility may be the plausible mechanisms for the formation of invisible gold, microscopic substrates and visible gold respectively (Mishra et al. 2001, Saha & Venkatesh 2002).

Wood & Sterns (1979) suggested that Au^{3+} might replace the Fe^{3+} in the octahedrally coordinated iron sites in arsenopyrite. Substitution mechanism of 2As = Au (or Sb) + Fe has been proposed by Johan et al. (1989) and shown the existence of Fe^{3+} in arsenopyrite structure which may imply the presence of Au^{3+}in arsenopyrite·Arehart et al. (1993) proposed coupled substitution mechanism by which Au substitutes Fe, As for S, and one Au and As for two Fe. The variability in the content of invisible gold in different sulfides could be due to deposition of gold under fluctuating physiochemical environments during the formation.

4 WATER–ROCK INTERACTION

Water–rock interactions have been documented in the form of predominant wall rock alteration produced by the hydrothermal fluids that are responsible for the ore deposition. The hydrothermal alteration in the area is characterized by strong sericitization and chloritization particularly close to quartz ± carbonate veins within the shear fractures. Mass-balance calculations based on major, trace and rare earth element data, combined with field and petrologic studies reveal the zonation of alteration from proximal to distal zones as sulfidation, silicification, chloritization, sericitization, and carbonatization finally leading to the garnetiferous schistose rock forming part of the variants of host felsic metavolcanic rocks similar to the one reported from the Yilgarn gold belt (Yeats & Groves 1998).

Considerable mass changes up to 122% along with limited volume changes of up to 105% suggesting effective ore bearing fluid–rock interactions along macroscopic and microscopic shear fractures adjacent to mineralized zones involving the sulfidation reactions necessary for the deposition of gold in the form of quartz-sulfide veins close to the main shear system. Two contrasting, yet significant styles of alteration have been observed predominantly typifying the differing phases of mineralizing events, one being characteristic of the lode-gold deposition and the other one representing volcanic-hosted mass sulfide (VHMS) mineralization within the same belt.

The extensive alteration zones surrounding the deep crustal scale shear zone include the aluminosilicate minerals such as chloritoid and andalusite, along with iron carbonate, quartz, sericite, chlorite, and carbonate, which may reflect buffering at a weakly acidic pH by water–rock reactions under relatively reduced conditions. The alteration reflects an intricate interplay among rheological behavior, fluid dynamics, chemical exchange involving the composition of host units and the composition of the hydrothermal fluids, temperature, and pressure.

5 FLUID EVOLUTION

Fluid inclusions were examined in quartz samples from auriferous (qz + sulfide + Au {both native and invisible variety}) and barren veins to characterize the nature of fluids, their temperature and evolution patterns in the belt. Fluid inclusion types at Pular-Parsori-Tuthanbori belt have been classified based on the geometry, phases and phase proportions and composition as type I, type II (IIa and IIb) and type III (IIIa, IIIb and IIIc). The details about the classification with respect to auriferous/barren and other microthermometric details of types and subtypes are given in Table 1.

The type I and type II inclusions form the carbonaceous variety of inclusions. The type I are predominantly CO_2-rich inclusions containing considerable amounts of CH_4 while the type II inclusions also contain variable CO_2 contents with minor to no amounts of CH_4. The presence of dissolved gases, in addition to CO_2, is indicated in this group of fluid inclusions by a depression of T_mCO_2 to –66°C and below, and by an elevation of T_mClath to 11°C and above from 7° to 10°C. The microthermometric data suggest that the additional phase possibly due to CH_4 as the triple point is below that of pure CO_2 (–56.6°C) (Ho 1987). The type III inclusions contain aqueous liquid (H_2O) with or without vapour and solid phases ranging in size from 4 to 15 μm with variable salinities. On the basis of the presence or absence of liquid, vapour and solid and the origin of the inclusions they are further subdivided into three categories (Table 1).

Occurrence of fluid inclusion (carbonic and aqueous) in different generations of quartz veins showing varying homogenization temperatures between 220° and 400°C reflects the progressive cooling and waning stages of the hydrothermal systems. Unmixing of CO_2- and H_2O-rich fluids in the area is indicated by

Table 1. Fluid inclusion characteristics correlated with vein paragenesis at Sakoli Group, Central India.

Types	Type I	Type II		Type III		
	$(CH_4)_V + (H_2O)_V + (H_2O)_L$	$(CO_2 \pm CH_4)_V + (CO_2 \pm H_2O)_L + (H_2O)_L$		(H_2O)	$((H_2O)_V + (H_2O)_L)$	$(H_2O)_{V+L+S}$
Properties	I	IIA	IIB	IIIA	IIIB	IIIC
Geometry of fluid inclusions					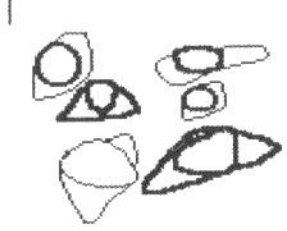	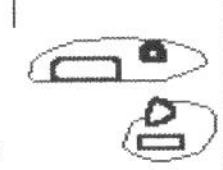
Origin and distribution in vein stages	Pre-shearing laminated veins: Stage I	Syn-shearing and post-shearing Stage II and III	Syn-shearing - post shearing: Stage II and III	Pre-shearing veins: Stage I	Pre-shearing and syn - shearing veins: Stage I and II	Syn-shearing brec ciated veins: Stage II
Abundance	-------------5%-------------	--------------------20%------------------		------------------------------75%------------------------------		
Size (μm)	4-22	5-22	12-30	2-25	8-30	10-25
Degree of fill	60-80	30-55	65-80	100	40-90	45-70
Tm CO_2 (°C)	-66 to -52	-58 to –52	-54 to –56	---		---
Tfm ice (°C)	-52 to -12	-38 to –17	-27 to –12	---	-54 to –22	-26.5 to –27.3
Tm ice (°C)	-11.2 to –1.2	-15.4 to –2.6	-11.2 to –1.2	-23.7	-18.6 to –1.5	-6.4 to –1.3
Tm clath (°C)	5.5 to 7.2	-5.9 to –2.7		-7 to –5	---	---
Th CO_2 (L)(°C)	26 to 30	29.8 to 30	18 to 26	---	---	---
Th total (L/V °C)	350 to 375	275 to 375	250 to 350	---	175 to 273	175 to 300
Density (g/cc)	0.937 to 0.784	0.997 to 0.83	0.9581		0.855	0.995
Salinity (wt% NaCl equiv)	3.5 to 7.5	8.5 to 9.2	9.5 to 12.5		9.5 to 8.4	16.5 to 17.5

the coevally-trapped CO_2- and H_2O-dominant fluid inclusions. The variation of the V/L ratios (20/80 to 70/30) for type II inclusions (a and b) in veins of different generations suggests that gas–liquid immiscibility occurred intermittently throughout the vein history (Table 1).

The presence of carbonic inclusions with the sulfide minerals like arsenopyrite, pyrite and chalcopyrite suggests that the fluids responsible for gold mineralization in Au-bearing quartz veins in the mineralized belt are approximated by the H_2O–$NaCl$–$CO_2 \pm CH_4$ system, whereas those of barren veins, are characterized by the CO_2–H_2O–$NaCl$ systems respectively. The carbonic phases of type I and II inclusions range from pure CH_4, through CH_4 and CO_2, to pure CO_2 phases. Transient P–T–X conditions were such that phase separation occurred and produced carbonic and aqueous fluids. These fluids may have physically separated and subsequently evolved independently with progressive phase separation producing increasingly saline, aqueous fluids and increasingly volatile rich fluids. Overall, the system appears to have evolved from a carbonate dominated regime to an increasingly aqueous dominated regime through time. It appears that syn-mineralization shearing is responsible for the enhancement of the weak zones in to which there has been a further hydrothermal fluid influx leading to the main ore deposition in the form of lodes.

6 CONCLUSIONS

The shear zone hosted gold and base metal mineralization in Sakoli Group envisages a three-stage model for the gold deposition within the metavolcanic host rocks. Ore-bearing hydrothermal fluids that deposited the metals along the shear zone have brought intense water–rock interaction. The ore–fluid system appears to have evolved from a carbonate-dominated regime to an increasingly aqueous-dominated regime through time.

ACKNOWLEDGEMENTS

Dr Greg Arehart and Dr Bob Seal improved the manuscript through their critical and constructive reviews. ASV is thankful to the Department of Science and Technology, India for funding. We thank the Director, ISM and Head-Applied Geology for the facilities.

REFERENCES

Acharyya, S.K. 2003. The Nature of Meso-Proterozoic Central Indian suture zone with exhumed and reworked older granulites. *Gondwana Res.* 6(2): 197–214.

Arehart, G.B., Chryssoulis, S.L. & Kesler, S.E. 1993. Gold and arsenic in iron sulphides from sediment hosted disseminated gold deposits: implications for depositional processes. *Econ. Geol.* 88: 171–185.

Ashley, P.M., Creagh, C.J. & Ryan, C.G. 2000. Invisible gold in ore and mineral concentrates from the Hillgrove gold-antimony deposits, NSW, Australia. *Mineral. Deposita.* 35: 285–301.

Bandyopadhyay, B.K., Roy, A. & Huin, A.K. 1995. Structure and tectonics of a part of the Central Indian Shield. *Mem. Geol. Soc. India* 31: 433–467.

Bhoskar, K.G. 1990. Banded Iron Formations of the Sakoli Group and related metallogeny. *GSI Spl. Pub.* 28: 658–675.

Bonnemaison, M. & Marcoux, E. 1990. Auriferous mineralization in some shear zones: a three-stage model of metallogenesis. *Mineral. Deposita.* 25: 96–104.

Gadhadharan, A. & Jog, R.G. 1990. Stratabound Zinc and Tungsten mineralization in Kolari-Bhaonri area, Maharastra. *GSI. Spl Pub.* 28: 696–713.

Genkin, A.D., Bortnikov, N.S., Cabri, L.J., Wagner, F.E., Stanley, C.J., Safonov, Y.G., McMahon, G., Freidl, J., Kerzin, A.L. & Gamyanin, G.N. 1998. A multidisciplinary study of invisible gold in arsenopyrite from four deposits in Siberia, Russian Federation. *Econ. Geol.* 93: 463–487.

Ho, S.E. 1987. Fluid inclusions: their potiential as an exploration tool for Archaean gold deposts. In S.E. Ho & D.I. Groves (eds.), *Recent Adcvances in Understanding Precambrian Gold Deposits.* Univ. W Australia, Geology Dept. *Univ. Extension Publ.* 11: 239–263.

Johan, Z., Marcoux, E. & Bonnemaison, M. 1989. Arsenopyrite aurifere: Mode de Substitution de Au dans la Structure de Fe As S. *Comptes Rendus* 308(II): 185–191.

Mishra, T., Jose, B., Deepa, K.K., Sreehari, S.M.S. & Venkatesh, A.S. 2001. Auriferous mineralisation in the Sakoli Group, Central India, with particular reference to sulfide- bound gold. *Trans. Instn. Min. Metall.* (Sec. B: Appl. Earth Sci.) 110: B103–B109.

Mumin, A.H., Fleet, M.E. & Chryssoulis, S.L. 1994. Gold mineralization in As-rich mesothermal gold ores of the Bogosu-Prestea mining district of the Ashanti Gold Belt, Ghana: remobilization of "invisible" gold. *Mineral. Deposita.* 29: 445–460.

Rogers, J.J.W. & Santosh, M. 2002. Configuration of Columbia, a Mesoproterozoic Supercontinent. *Gondwana Res.* 5: 5–22.

Saha, A.K. & Mohan M. 1994. Preliminary exploration for gold in Pular-Parsori-Mokhabardi-Khabri-Thuttanbori area, Nagpur and Bhandara Dists. Maharashtra, Central India. *Rec. Geol. Surv. India* 127(6): 34–36.

Saha, A.K., Sasidharan, K., Padhi, R.N. & Bhoskar, K.G. 1996. Gold mineralization in Sakoli Supra crustals of Central Indian shield, Maharashtra. Golden Jubilee symposium, Nagpur University. In N.K. Mohabey et al. (eds.), *Mineral and Ground Water Resourses of Vidharbha,* 13–22.

Saha, I. & Venkatesh, A.S. 2002. Invisible gold within sulfides from Archean Hutti-Maski schist belt, Southern India. *J. Asian Earth Sci.* 20: 449–457.

Stein, H.J., Zimmerman, A., Hannah, J.L. & Markey, R.J. 2003. Late Archean–Early Proterozoic timing for an

Andean style Cu-Mo deposit at Malanjkhand, Central Indian Tectonic Zone: Implications for a late Archean Super continent. *Geophys. Res. Abstracts* 5: 07496.
Yedekar, D.B., Jain, S.C., Nair, K.K.K. & Dutta, K.K. 1990. Central Indian collision suture. *Geol. Soc. India Spec. Pub.* 28: 1–43.
Yeats, C.J. & Grovcs, D.I. 1998. The Archaean Mount Gibson gold deposits,Yilgarn Craton,Western Australia: Products of combined synvolcanic and syntectonic alteration and mineralisation. *Ore Geol. Rev.* 13: 103–129.
Wood, B.H. & Strens, R.G.J. 1979. Diffuse reflectance spectra and optical properties of some sulphides and related minerals. *Mineral. Mag.* 43: 509–518.

Water–Rock Interaction, Wanty & Seal II (eds)

Influence of grain boundary energy on fluid distribution in deep-seated rocks

D.A. Wark & E.B. Watson
Rensselaer Polytechnic Institute, Troy, New York

ABSTRACT: Pore microstructure in fluid-bearing rocks of the deep crust is controlled largely by the ratio of grain-boundary energy (γ_{gb}) to the interfacial energy of the pore wall (γ_{cf}). With the goal of better characterizing pore topology in equilibrated rocks, we have developed a technique to quantify γ_{gb}/γ_{cf} for different degrees of lattice "misfit" at a grain boundary. We first juxtapose two oriented crystal surfaces in the presence of fluid at elevated pressure and temperature. Lenticular-shaped pores that form along that surface are then characterized by atomic force microscopy to allow quantitative measurement of wetting properties, from which the ratio γ_{gb}/γ_{cf} can be extracted. Initial results show promise not only for quantifying grain-boundary energies, but also for understanding the influence of crystal lattice orientation on the shape of pore walls.

1 INTRODUCTION

1.1 *Grain boundaries and pore microstructure*

Few details are known about the influence of grain boundaries on fluid transport in deep-seated, microstructurally equilibrated rocks (i.e. those annealed at elevated pressure and temperature). Even less is known about the influence of "interphase boundaries"; that is, those separating dissimilar solid phases. Each type of boundary, however, almost certainly plays an important role, because the energy of a grain or interphase boundary has a direct influence on fluid "wetting" behavior. This, in turn, controls pore connectivity at the grain scale, with implications for transport rates of fluid by porous flow, and of solutes by diffusion through the fluid medium.

The nature of the relation between grain-boundary energy (γ) and fluid distribution is evident in the well known force–balance equation (Smith 1969),

$$\cos\left[\frac{\theta}{2}\right] = \frac{\gamma_{gb}}{2\gamma_{cf}} \tag{1}$$

where θ is the dihedral angle (Fig. 1), which is a function of the relative magnitudes of crystal-fluid interfacial energy (γ_{cf}) and the energy of the grain boundary (γ_{gb}). The value θ is considered a general indicator of fluid-phase connectivity through a crystalline aggregate: for $\theta \leq 60°$ the fluid forms grain-edge channels that interconnect in three dimensions; for $\theta > 60°$, fluid tends to occupy isolated pores at grain corners. Only when $\gamma_{cb}/\gamma_{cf} \geq 2$ will θ be zero, resulting in the wetting of all surfaces, including grain boundaries.

1.2 *Limitations to existing wetting data*

The dihedral angle is typically estimated by measuring the distribution of angles at the junctions of two pore walls with a grain boundary, on a polished section through a quenched polycrystalline – usually monomineralic – aggregate. Although generally reliable, this statistically based strategy does not

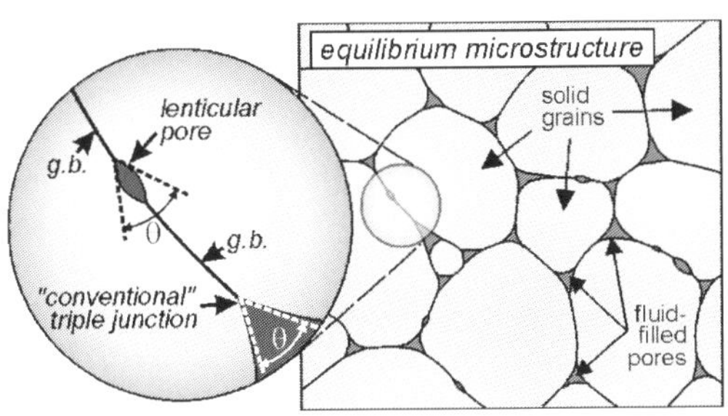

Figure 1. Schematic representation of slice through a microstructurally equilibrated, fluid-bearing rock, showing distribution of pores primarily at junction of three grains. Magnified view shows that some fluid also gets trapped along grain boundaries as lenticular pores. Dihedral angle (θ) is the angle formed between two pore walls, where they meet the grain boundary.

result in a unique measurement of θ. Rather, it yields a distribution of *apparent* angles (a true angle can be measured only if the pore walls are perpendicular to the plane of the polished section), the median value of which is considered the "correct" θ. Previously published θ values have illuminated the general wetting behavior of various mineral-fluid systems at elevated P and T (e.g. Watson & Brenan 1987, Holness 1982). In all cases, grain boundaries are demonstrably "dry", with reported θ values (all >0) interpreted to indicate that fluid is either interconnected or is restricted to isolated "pockets" at the junction of multiple grains, depending on the system and pressure–temperature conditions of interest.

Most reported dihedral angles, while useful for predicting the degree to which fluids are interconnected in a general way, are, in fact, only *averages*. In reality, all polycrystalline materials will exhibit some range of θ values because γ_{gb}, and hence θ (see Equation 1) will vary with extent of lattice "misfit" where two crystals meet. Even γ_{cf}, while probably relatively uniform in most systems, may vary, thereby affecting the shape of pore walls (see our results for calcite, below). The effect of variation in γ_{cf} and/or in γ_{gb} on wetting properties is likely to be particularly strong for anisotropic minerals such as carbonates, pyroxenes, and amphiboles. In addition, variable γ_{gb} and γ_{cf} will probably have the greatest effect on fluid transport for systems with median θ near 60°, because in these systems there should be both isolated and interconnected pores.

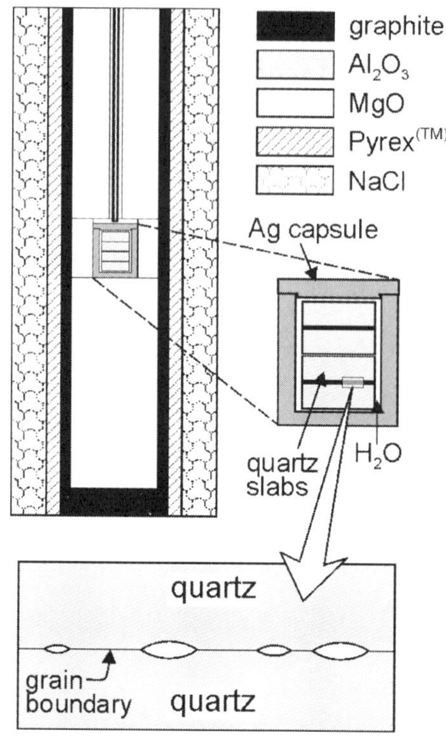

Figure 2. Schematic representation of piston–cylinder assembly and procedure for producing "synthetic" lenticular pores at grain boundaries.

2 A NEW WAY TO CHARACTERIZE WETTING

With the goal of characterizing the true distribution of dihedral angles, which is necessary to accurately model both pore geometry and fluid transport, we are investigating the influence of lattice orientation on the energies of grain and crystal–fluid boundaries in simple, monomineralic systems. Ultimately, we plan to explore the energetics of interfaces between dissimilar phases, which will be useful in modeling fluid distribution in more complex, polymineralic rocks. Our emphasis is not upon direct imaging techniques applied to determination of atomic structure (e.g. TEM), but upon characterization of a property that is observable at a larger scale: the wetting behavior (in the form of the dihedral angle) of synthetic grain boundaries, which provides quantitative information (using Equation 1) about grain-boundary energy.

Our approach to measuring θ circumvents many of the limitations noted above for previous studies of fluid wetting behavior. We prepare two or more oriented and polished slabs of the mineral(s) of interest, stack them in a piston–cylinder pressure cell, and anneal them at high PT conditions in the presence of a fluid (see Figure 2). At run conditions, the slabs weld together to form a grain boundary. Some fluid becomes trapped along this grain boundary, and quickly evolves into lenticular pores. The outer edges of these pores are continuous, nearly circular junctions between the pore walls and the grain boundary (Fig. 3).

Post-experiment separation of the grain boundary, followed by imaging of the lenticular pore halves (which appear as well-formed dimples on the separated surfaces) by atomic force microscopy (AFM), provides digital, 3-D information (Fig. 4a) from which geometric relationships can be extracted (Fig. 4b). The dihedral angle, in fact, can be measured unequivocally with the AFM for a specific crystal pair and degree of lattice misfit.

3 PRELIMINARY DATA

3.1 *Quartz*

The AFM image of Figure 4a illustrates typical pores formed at synthetic grain boundaries produced

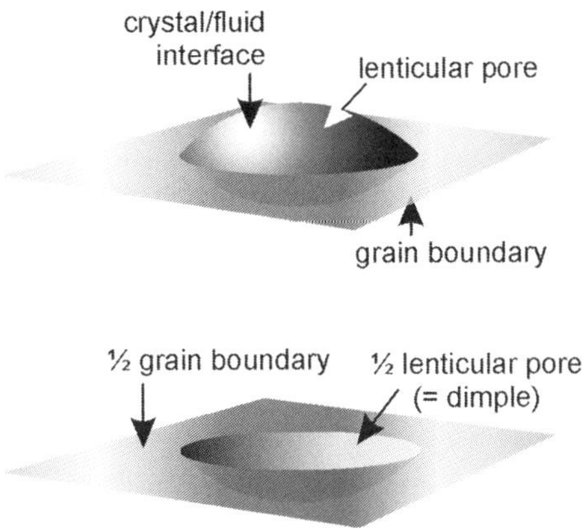

Figure 3. 3-D representation of a singular lenticular pore (top) along with the split pore (dimple) formed after separation of the grain boundary.

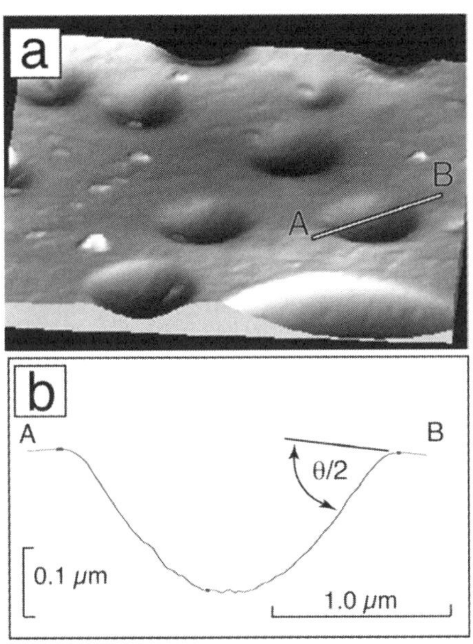

Figure 4. (a) AFM image of one surface of a separated quartz-quartz grain boundary produced as illustrated in Figure 2. The dimples represent halves of lenticular pores formed along the grain boundary at run conditions. (b) Digital AFM profile across a dimple along the line in (a).

between oriented crystals of quartz. The pores appear to conform quite well to the geometry of spherical caps, which implies that the quartz/fluid interfacial energy (γ_{cf}) is essentially independent of lattice orientation. This means that variations in θ with degree of grain boundary misfit can be attributed with confidence to changes in the grain boundary energy (γ_{gb}) itself. This may not always be the case, and we anticipate that systems involving highly anisotropic minerals (e.g. pyroxenes and amphiboles) may show more complex behavior. For the experiment shown in Figure 4, the grain boundary was roughly perpendicular to the c axis of one crystal and parallel to the c axis of the other, although the relative rotational position of the two lattices is unknown. The measured dihedral angle was 46° (±3).

To date, we have successfully measured dihedral angles on AFM images of two synthetic grain boundaries of known degree of lattice misfit, which were produced by annealing for several hours at 850°C and 1 GPa. A low-angle boundary (i.e. one formed with the crystal lattices closely aligned) was created by juxtaposing two {1100} surfaces, differing only by a 2° rotation about [1100]. The other boundary, with stronger lattice mismatch, was produced by juxtaposing a {1100} surface with one that was produced after a 5° rotation about [1000]. This difference in nature of grain boundaries had a strong effect on wetting: measured θ values were 67° (±1) and 35° (±4) for the low and "higher" angle boundaries, respectively. Using equation (1) and assuming uniform γ_{cf}, this decrease in dihedral angle reflects a 15% increase in grain boundary energy with increasing amount of lattice mismatch.

Interestingly, our measured dihedral angles straddle the "average" θ value of ~55° that was reported by Watson & Brenan (1987) for quartzites annealed at similar conditions, but determined using the "bulk sectioning" technique.

3.2 *Calcite*

Although we have not yet successfully measured lenticular pores along grain boundaries with known lattice orientation, examination of calcite grain boundaries in polycrystalline, synthetic rocks shows that pore shapes deviate significantly from spherical caps, sometimes assuming a (partial) negative crystal shape (Fig. 5). This behavior implies that not only γ_{gb}, but also γ_{cf}, varies with orientation of the calcite lattice, and it complicates the simple interpretation described above. However, because the AFM produces a 3-D digital image, the complexity may provide the opportunity to compute relative γ_{cf} values for different lattice orientations, using (when facets are present) the classical Wulff construction (e.g. Choi et al. 1997) as well as directional characterization of the dihedral angle. The dependance of pore microstructure on lattice orientation appears to be quite strong (relative to quartz) in calcite (and probably in other minerals, as well), perhaps indicating the need to refine existing models of pore connectivity in deep-seated rocks.

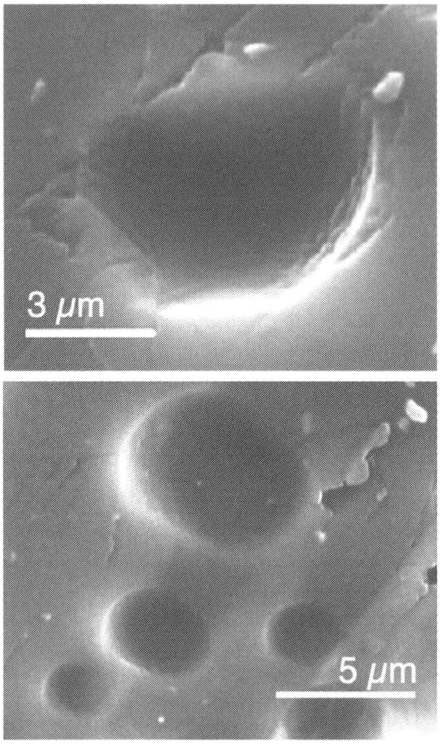

Figure 5. SEM images of separated grain boundaries in a synthetic marble. These are from the same experiment, but they represent different degrees of lattice misfit. Note apparent facets on the lenticular pores in top image. Facets are absent in the bottom image, but the pores are elongate.

4 CONCLUSIONS

The new technique described here promises to provide quantitative information about the influence of grain boundary energy on pore microstructure in deep-seated, microstructurally equilibrated rocks, which will serve as valuable input to models of fluid transport in the deep crust.

ACKNOWLEDGEMENTS

Funding is provided by the Office of Basic Energy Sciences, U.S. Department of Energy, Award #DE-FG02-94ER14432.

REFERENCES

Choi, J.-H., Kim, D.-Y., Hockey, B.J., Wiederhorn, S.M., Handwerker, C.A., Blendell, J.E., Carter, W.C. & Roosen, A.R. 1997. Equilibrium shape of internal cavities in sapphire. *J. Am. Ceram. Soc.* 80: 62–68.

Holness, M.B. 1992. Equilibrium dihedral angles in the system quartz-CO_2-H_2O-NaCl at 800°C and 1–15 kbar: the effects of pressure and fluid composition on the permeability of quartzites. *Earth Planet. Sci. Lett.* 114: 171–184.

Smith, C.S. 1969. Some elementary principles of polycrystalline microstructure. *Metall. Rev.* 9: 1–47.

Watson, E.B. & Brenan, J.M. 1987. Fluids in the lithosphere. 1. Experimentally-determined wetting characteristics of CO_2-H_2O fluids and their implications for fluid transport, host-rock physical properties, and fluid inclusion formation. *Earth Planet. Sci. Lett.* 85: 497–515.

Water–Rock Interaction, Wanty & Seal II (eds)
© 2004 Taylor & Francis Group, London, ISBN 90 5809 641 6

Kinetics study of silicate mineral dissolution in water at high temperatures near the critical point of water

R. Zhang & S. Hu
Laboratory of Geochemical Kinetics, Institute of Mineral Resources, Chinese Academy of Geological Sciences, Beijing, China

X. Zhang
School of Material Science and Technology, China University of Geosciences, Beijing, China

C. Shi & J. Zhang
College of Physics Science, Nankai University, Tianjin, China

ABSTRACT: Kinetic experiments of silicate dissolution for albite (Ab), actinolite (Act), and diopside (Di) were carried out using a flow-through experimental reactor (packed bed reactor) in an open system in the temperature range 25–400°C and at 22 MPa. We found that a wide fluctuation in silicate mineral (Ab, Act and Di) dissolution rates occurred near the critical point of water. The silicate mineral dissolution was always incongruent in water at most temperatures, but became nearly congruent at 300°C (close to the critical point, 374°C). The maximum dissolution rate (release rate of Si) of the minerals occurs at 300°C. The incongruent dissolution shows that Na, Ca, Mg, Fe, and Al dissolve easier than Si in the temperature range from 25–300°C. As temperature increases from 300–400°C, Si more easily dissolves than the other elements.

1 INTRODUCTION

In most cases, water–rock interactions in lithosphere occur at temperatures above 300°C, but limited experiments on silicate mineral dissolution have been performed in the temperature range above 300°C and at pressures near the critical point of water. Therefore, it is necessary to carry out the kinetic experiments of mineral dissolution (albite, actinolite, diopside) at temperatures above 300°C. Our experiments using a flow-through experimental reactor (packed bed reactor) in an open system in the temperature range of 25–400°C and at 22 MPa, provide new information for understanding natural geothermal processes and kinetic behavior near the critical point of water.

2 EXPERIMENTAL APPROACH

A packed bed, flow-through reactor was used to perform the kinetic study of silicate mineral dissolution in the temperature range 25–400°C and 22 MPa. This equipment used in the present experiments was the same as described in previous studies (Zhang et al. 1990, 1998, 2000, 2002, Zhang & Hu 1996) and included a pressure vessel, liquid pump, back pressure regulator, temperature controller, heat source (furnace), pressure gauge, electric conductivity detector and computer. Natural albite crystals were obtained from the Geological Museum, Beijing. Actinolite was collected at Maanshan mine. Diopside is from Lohe iron deposit in a volcanic area. Mineral particles were crushed in a shatter box and sieved to 20–40 mesh. Pure mineral separates were then handpicked with the aid of an optical microscope at 50X magnification. Mineral particles were crushed again in a shatter box and sieved to 20–120 mesh. Mineral particles were cleaned ultrasonically using acetone to remove fine particles, rinsed with distilled water, and dried at 70–80°C. Surface areas of representative samples were measured with a single point BET method, using Ar-He as the absorbate gas. For 20–40 mesh of albite, surface area is 1.1 m^2/g. For fresh actinolite, the surface area is 2.9 m^2/g; after reaction with water, it is 4.22 m^2/g; after reaction with an aqueous HCl solution, it is 3 m^2/g. For fresh diopside, the surface area is 9.876 m^2/g.

The total mass of Ab grains placed into the pressure cell was 13 g and the volume of fluid was approximately 3 mL. The total mass of actinolite grains put in the vessel was 7.8 g with approximately 3 mL of fluid.The total mass of diopside grains put in the vessel was 14.26 g with approximately 3 mL of fluid.

The pressure vessel was mounted vertically, and mineral particles were put in the pressure vessel. De-ionized and degassed water was introduced into the vessel from the bottom at different flow rates (0.5–3 mL/min). Output solutions were sampled and the composition of these solutions was analyzed using the ICP-MS method. The steady state of the experimental system was tested in real time, using a conductivity detector connected to the outlet.

3 EXPERIMENTAL RESULTS OF ALBITE IN WATER

3.1 *Albite in congruent dissolution*

The albite dissolution experiments in this study show that the Al/Si and Na/Si release ratios decrease from a positive value to zero as temperature increases from 25–300°C. Dissolution is congruent at 300°C. In the case of $T < 300°C$, as temperature increases, the Al (or Na) is released faster than Si. As temperature increases from 300–400°C, the Al/Si and Na/Si release ratios are negative, i.e. Si is released faster than Al and Na (Hellmann et al. 1989, Zhang et al. 2000). The Al/Na release is congruent at 300°C. The Al/Na release ratio is negative over the temperature range from 25–400°C except at 300°C. The Na is released faster than Al in most cases.

3.2 *Dissolution rates of albite*

In a flow-through experimental reactor, the dissolution rate of the mineral (mol/min/m^2) derived from the following mass balance expression for the concentration of the ith solute in a reactor cell:

$$\text{Rate} = (C_i - C_0)/t(A/V) \quad (1)$$

where C_i is the total output concentration (mol/L) of species i, C_0 is the initial concentration of species i, A is the total reactive surface area of the mineral (m^2), t is the average fluid residence time, and V is the volume of the pressure vessel (mL). The derivation of this equation is presented by Zhang et al. (1992).

Equation 1 is valid for albite dissolution rates in the case of a flow-through experiment and is similar to that reported for the mineral dissolution in other flow-through experiments (Hellmann et al. 1989, Mogollon et al. 1996).

Maximum release rates for Na, Al and Si in flowing hydrothermal systems under all flow velocities were always observed at 300°C. The release rates increased as the temperature increased from 25–300°C, but decreased as temperature increased from 300–400°C.

The dissolution reaction of albite in water can be expressed as follows:

$$NaAlSi_3O_8 + 8H_2O = Na^+ + Al(OH)_4^- + 3H_4SiO_4 \quad (2)$$

Assuming that the above reaction is the key step of the whole albite dissolution process, then the reaction mechanism in the experimental system can be evaluated theoretically based on the dissolution-rate data. Assuming that this reaction is near equilibrium, or that it has two reaction rates in both positive and negative directions, then the proximity of the reaction to equilibrium or saturation state of the solution is described by the Gibbs free energy (or chemical affinity A) of the reaction.

$$\Delta G_r = RT\ \ln(Q/K) \quad (3)$$

where R is the gas constant, T is temperature (K), Q represents activity product, calculated using experimental data, and K is the equilibrium constant.

The values of Q were calculated from the expressions for the reaction as written in equation 2 with actual activities of aqueous species, where activity of water is assumed to be equal to 1. See Appendix for calculation method of activity of an aqueous species of i.

The ΔG_r function can be obtained readily if the reaction is an elementary reaction. In this case, the dependence of the rate on ΔG_r, as derived from Transition State Theory (TST), as has been demonstrated (Lasaga 1981, Aagaard & Helgeson 1982). Therefore, a general rate law for mineral dissolution and precipitation can be written as

$$R_{net} = -k_+ \Pi a_i^m (1 - \exp(\Delta G_r/RT)) \quad (4)$$

where R_{net} (mol/min/m^2) is the net rate of reaction (forward rate minus reverse rate), k_+ (mol/min/m^2) is the rate constant of the forward reaction, a_i is the activity of species i in the rate determining reaction raised to some power m. Here, Πa_i^m is an activity product from n aqueous species.

4 EXPERIMENTAL RESULTS OF ACTINOLITE IN WATER

The dissolution reaction of actinolite in water can be expressed as follows:

$$Ca_2Mg_5Si_8O_{22}(OH)_2 + 22H_2O = 2Ca^{2+} + 5Mg^{2+} + 8H_4SiO_4 + 14OH^- \quad (5)$$

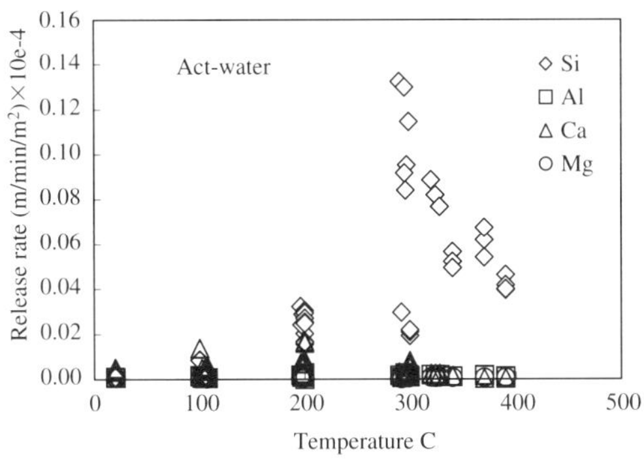

Figure 1. Release rate of Si, Al, Ca and Mg vs. temperature for actinolite dissolution in water in the flow system at 22 MPa.

Chemical analysis of the actinolite indicates that it contains Ca, Mg, Si and trace Fe and Al. It was also found that a wide fluctuation in actinolite dissolution rates occurred near the critical point of water. Actinolite dissolution is incongruent in water at most temperatures, but became congruent at 300°C (near the critical point, 374°C). This incongruent dissolution of actinolite shows that Ca, Mg, Fe, Al dissolve easier and enter to water in the temperature range from 25–300°C than does Si. As temperature increases from 300–400°C, the dissolution behavior indicates that the Si more easily dissolves than other elements.

Dissolution rates (for release rate of Si) for actinolite increased as the temperature rose from 25–300°C and decreased when the temperature rose continuously from 300–400°C. Maximum actinolite dissolution rates in the flowing hydrothermal systems were independent of flow rates, and release rates of Si were reproducible at 300°C and 22 MPa (Fig. 1).

5 EXPERIMENTAL RESULTS OF DIOPSIDE IN WATER

The dissolution reaction of diopside in water can be expressed as follows:

$$CaMgSi_2O_6 + 6H_2O = Ca^{2+} + Mg^{2+} + 2H_4SiO_4 + 4OH^- \quad (6)$$

Chemical analysis of the diopside indicates that it contains Ca, Mg, Si and trace Na, Fe, and Al. It was also found that a wide fluctuation in diopside dissolution rates occurred at 300°C near the critical point of water. Diopside dissolution is incongruent in water at most temperatures, but became congruent at 300°C (near the critical point, 374°C). This incongruent dissolution of diopside shows that Ca, Mg, Fe, Al dissolve easier and enter to water in the

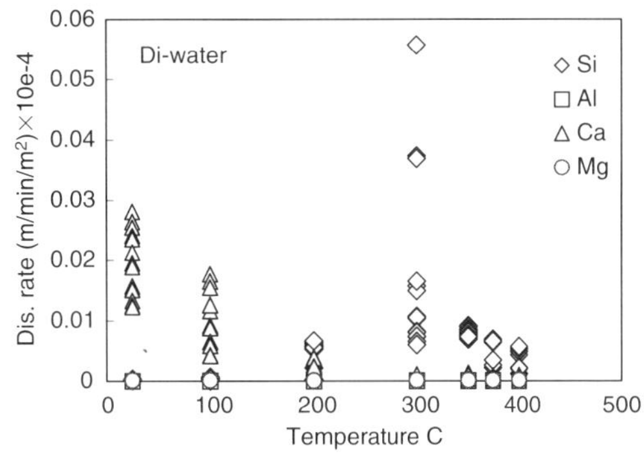

Figure 2. Plot of dissolution rates vs. temperature for diopside dissolution in water in the flow system at 22 MPa.

temperature range from 25–300°C than does Si. As temperature increases from 300–400°C, their dissolution behavior indicates that Si more easily dissolves than other elements (Fig. 2).

Dissolution rates (for release rate of Si) of diopside increased as the temperature rose from 25–300°C and decreased when the temperature rose continuously from 300–400°C.

6 DISCUSSION

Generally, Na and Al in albite (or Ca, Mg, Fe, Al in actinolite and diopside) dissolve easier and enter the water in the temperature range from 25–300°C than does Si. As temperature increases from 300–400°C, Si more easily dissolves than other elements. Also, the molar concentration ratios of Si/Al and Mg/Ca in output solutions for silicate dissolutions increase with increasing temperature.

The kinetic experiments of mineral–aqueous solution interactions at temperatures above 300°C at a pressure of 22 MPa indicate that the kinetic behavior is affected by water-property variations associated with crossing the critical point, such as the density and dielectric constant of water. The reaction process during increasing temperature and at 22 MPa is isobaric expansion.

It is generally considered that the Al–O bridging bond is comparatively long and weak and more easily broken than the Si–O bridging bond. At room temperature, the Al–O and Si–O bridging bonds exhibit ionic bond properties and the polar bond properties, respectively. The Ca–O and Mg–O bridging bonds exhibit the ionic bond properties. As temperatures rise, due to the decrease in the water density and dielectric constant, the water gradually becomes capable of dissolving polar-bonded substances (e.g. breaking of the Si–O bond). Therefore, while temperatures rise, breaking of Ca–O, Mg–O and Al–O bridging bonds becomes increasingly difficult. There is a chance for the breakup of both the Al–O and Si–O bridging bonds by water molecules at the same release rate. This event probably occurs at 300°C and 22 MPa, when congruent dissolution of ablite takes place.

At the critical point of water, the rapid decrease in the dielectric constant and density of water results in the destruction of the hydrogen bond network of the water, which may affect the rates of ionic reactions and silicate framework hydration. Therefore, in the process of crossing the critical point of water (at 300–400°C), both the ionic and hydration reactions are weakening. However, owing to differences in polarizability among the $Ca–O_{br}$, $Mg–O_{br}$, $Al–O_{br}$ and $Si–O_{br}$ bonds, the $Si–O_{br}$ bond is more easily hydrated at 300–400°C. Thus, the release rate of Si is faster.

7 CONCLUSIONS

The silicate mineral (albite, diopside and actinolite) dissolutions were always incongruent in water at most temperatures, but became congruent at 300°C (near the critical point, 374°C). The maximum dissolution rate (for Si) occurs in this case. The incongruent dissolution of albite (or Act, Di) shows that Na and Al (or Ca, Mg, Fe, Al) dissolve easier and enter to water than does Si in the temperature range from 25–300°C. As temperature increases from 300°C–400°C, the Si more easily dissolves than other elements. Molar concentration ratios of Si/Al and Mg/Ca in output solutions increase with increasing temperature. From 300–400°C at 22 MPa, water properties vary significantly and affect the dissolution behavior. For example, fluctuation of silicate dissolution rates occurred in this region.

ACKNOWLEDGEMENTS

We would like to thank the Ministry of Science and Technology and the Ministry of Land and Resources for supporting our project: 20010302, DK2001022, and support of 2001DEA20023, 2001DEA30041, G1999043212, and also Chinese NSF, Grant no. 4003011.

REFERENCES

Aagaard, P. & Helgeson, H.C. 1982. Thermodynamic and kinetic constraints on reaction rates among minerals and aqueous solutions. l. Theoretical considerations. *Amer. J. Sci* 282: 237–285.

Hellmann, R., Crerar, D. & Zhang, R. 1989. Albite feldspar hydrolysis to 300°C. *Reactivity of Solids: Proc. 11th Symp., Solid State Ionics* 32/33: 314–329.

Lasaga, A.C. 1981. Rate laws of chemical reactions. In Kinetics of geochemical processes In A.C. Lasaga & R.J. Kirkpatrick (eds.), *Rev. Mineral.* 8: 1–68.

Mogollon, J.L., Ganor, J., Soler, J.M. & Lasaga, A.C. 1996. Column experiments and the full dissolution rate law of gibbsite. *Am. J. Sci.* 296: 729–765.

Shvarov Yu. W. 1989. A numerical criterion for existence of the equilibrium state in an open chemical system. *Sci. Geol. Bull.* 42(4): 365–369.

Zhang R., Posey-Dowty, J., Hellmann, R., Borcsik, M. Crerar, D. & Hu S. 1990. Kinetics of mineral-water reactions in hydrothermal flow systems at elevated temperatures and pressures. *Science in China (Series B)* 33(9): 1136–1152 (English edition).

Zhang, R. & Hu, S. 1996. Reaction kinetics of fluorite in flow systems and surface chemistry. *Science in China (Series D)* 39(6): 561–575.

Zhang, R., Hu, S., Tong, J. & Jiang, L. 1998. *Mineral-Fluid Reaction Kinetics in Open Systems*. Beijing: Science Publishing House (English abstract).

Zhang. R. & Hu, S. 2000. Kinetics of hydrothermal reactions of minerals in near-critical and supercritical water. *Acta Geologica Sinica* 74(2): 400–405.

Zhang, R., Hu, S. & Su, Y. 2002. Alteration zoning and kinetic rocesses of mineral-water interactions. *Acta Geologica Sinica* 76(3): 351–366.

APPENDIX

Calculation method of the activity of a given aqueous species:

For example of albite reacted with water, the Na, Al and Si concentrations in aqueous solutions, i.e. reaction products under the given physical and chemical condition, can be measured and obtained in the experiments of our open flowing systems. The dominate aqueous species of the dissolved Si, Al and Na and their species activities at run temperatures were calculated from total concentrations of dissolved elements in output solutions of our kinetic experiments using Shvarov software HCh. (Shvarov 1989). The theoretical calculation provides those activity coefficients of aqueous species in output solutions of the experiments. Calculated results indicate that activity coefficients of H_4SiO_4 approximated to 1 in most cases of the experiments at temperature range from 25°C–400°C. Activity coefficients of $Al(OH)_4^-$ varied gradually from 0.91–0.99 at 25°C–0.89–0.92 at 300°C, then changed to 0.9–0.93 until 400°C. Activity coefficients of Na^+ decreased slightly from 0.91–0.99 at 25°C–0.89–0.92 at 300°C, then, it increased to 0.9–0.95 until 400°C. Note that activity coefficients of aqueous species at each temperature varied with concentrations of dissolved elements.

Water–rock interactions in groundwater systems and sedimentary basins

Water–Rock Interaction, Wanty & Seal II (eds)
© 2004 Taylor & Francis Group, London, ISBN 90 5809 641 6

Geothermal potential of thermal waters of Manikaran, Himachal Pradesh, India

M.A. Alam & D. Chandrasekharam
Department of Earth Sciences, Indian Institute of Technology Bombay, Mumbai, India

A. Minissale
Consiglio Nazionale delle Ricerche (CNR), Florence, Italy

ABSTRACT: Based on major ion chemistry, the thermal waters of the Manikaran can be classified as Na-Cl-HCO_3 type. High salinity of the thermal waters is indicating mixing between a saline and groundwater source. The source of this saline component is probably ancient formation waters trapped in the geological formations, or magmatic or metamorphic waters. Presence of saline fluids in older granitic rocks of the Himalayas (based on INDEPTH studies) and Manikaran Quartzite (based on fluid inclusion studies) supports this view. The composition of the thermal waters can be described in terms of the variations in the relative proportions of two end member components: saline water and fresh water. Depletion in the calcium content can be explained in terms of the formation of travertine deposits along the vents of the thermal springs. Excellent prospects exist for the utilisation of the thermal waters for direct and indirect uses.

1 INTRODUCTION

In this work thermal springs in the Manikaran Geothermal Area, in Parbati Valley Geothermal Field, Kullu District, Himachal Pradesh, has been studied, in order to understand the geochemistry of the thermal as well as cold waters, evolutionary history of the thermal waters and to evaluate the prospects of thermal springs as geothermal energy source of the area.

Manikaran (32°02′N, 77°21′E) is a small village on the banks of the Parbati River, a tributary of the Beas River, in the Kullu District of Himachal Pradesh (India), at an elevation of 1760 m in the North-Western Himalayas. Reconnaissance surveys on some of the springs have been carried out earlier by a few workers (Chaturvedi & Raymahashay 1976, Gupta et al. 1976, Jangi et al. 1976, Giggenbach et al. 1983). The present work aims at study of the chemical characteristics of thermal waters and their use for direct and indirect applications.

2 METHODOLOGY

Detailed geological mapping in and around Manikaran has been done (Fig. 1) and chemical analyses on thermal and cold waters around Manikaran have been carried out following the methods prescribed by the American Public Health Association (APHA 1985). The results of chemical analyses are shown in Table 1.

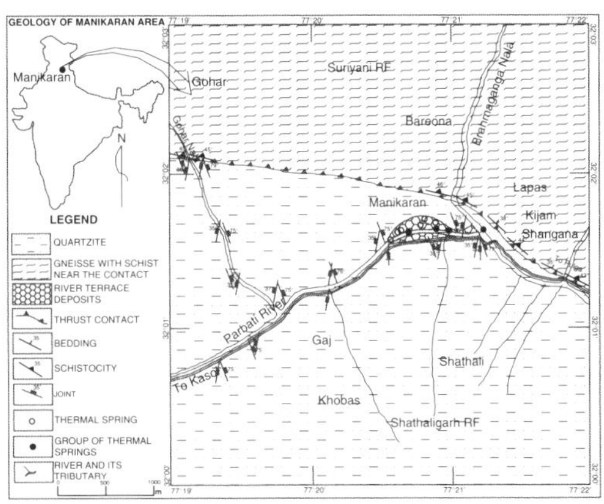

Figure 1. Geology of the Manikaran Geothermal Area.

3 GEOLOGY OF THE AREA

The main formation in the area, exposed around Manikaran village and along the road to Kasol from

Table 1. Results of the chemical analyses of the Manikaran thermal and cold water samples.

S.N.	T (°C)	pH	SiO_2 (ppm)	Major ions (ppm)						
				Ca^{2+}	Mg^{2+}	Na^{+}	K^{+}	SO_4^{2-}	Cl^{-}	HCO_3^{-}
TW1	85	7.6	104.9	41.9	4.6	106.3	20.0	54.6	106.9	130.0
TW2	86	6.4	103.9	38.1	4.6	77.5	17.5	50.7	104.4	140.0
TW3	72	7.2	112.6	38.1	4.6	83.8	20.0	54.3	87.4	120.0
TW4	73	7.3	117.1	41.9	6.9	87.5	21.3	53.3	126.3	160.0
TW5	92	6.7	105.2	47.6	5.8	96.3	25.0	49.7	143.3	150.0
TW6	78	7.6	86.2	70.4	10.4	75.0	15.0	70.0	116.6	200.0
RW7	4	7.4	9.8	9.5	2.3	8.8	6.3	40.4	9.7	30.0
RW8	6	7.6	10.8	13.3	3.5	3.8	1.3	39.0	7.3	30.0

TW – Thermal water samples, RW – River water samples.

Manikaran, is a well jointed, white to greyish, thick sequence of quartzite, which has been named the Manikaran Quartzite (Srikantia & Bhargava 1998), along with minor phyllites and slates. In the western and northern part of the area, near Gohar Village, granitic gneisses are exposed.

The contact between the quartzite–phyllite group and the overlying schistose rocks is a thrust contact. This thrust forms a part of the Central Himalayan Thrust, which is an extensive feature related to Cenozoic folding in the Himalayas (Srikantia & Bhargava 1998). It is marked by brecciated and at places tightly folded schists. The thrust contact is inferred based on the higher grade of metamorphism of the gneissic rocks that overlie the quartzites and it runs in a north-westerly direction (Alam 2002). Due to the inaccessibility of the terrain between Gohar Village and Bareona Village, the contact between the two is inferred on the map (Fig. 1).

Thermal springs issue through the three sets of shear-joints in the quarzites with the average spacing of joints varying from 0.3 to 0.5 m (Alam 2002). Two sets of joints have strike parallel to the strike of the formation (~NW-SE) and one set strikes perpendicular to that of the above two sets of joints (Alam 2002). The most prominent joint is a bedding joint striking N 50° W across the river dipping around 75° NE (upstream).

4 HYDROGEOCHEMISTRY

The analytical data on the water samples shown in Table 1 are plotted in Langelier Ludwig diagram (after Langelier & Ludwig 1942; Fig. 2).

The thermal waters are of Na-HCO_3-Cl type, while the cold waters are of Ca-SO_4 character. More clear distinction between these two groups of waters is seen in the anion variation diagram (Fig. 3).

Depletion in the Ca content in the thermal waters can be explained in terms of the formation of

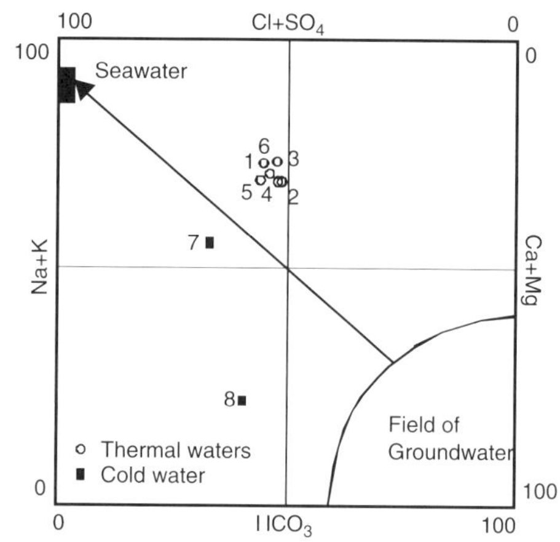

Figure 2. Langelier–Ludwig diagram (after Langelier & Ludwig 1942). Samples 1–8 on this diagram are presented in Table 1.

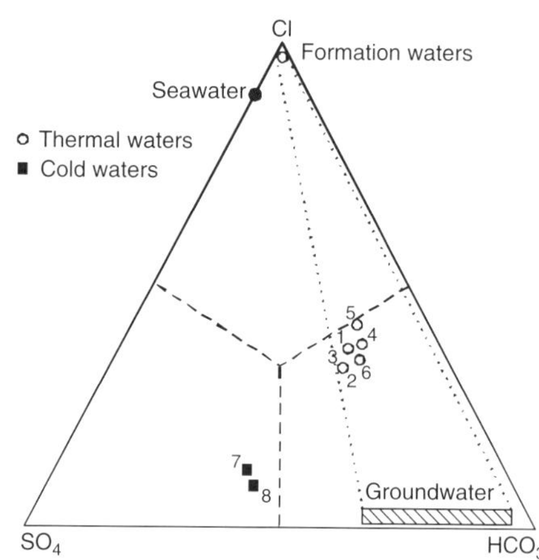

Figure 3. Anion variation diagram (after Giggenbach et al. 1983). Samples 1–8 on this diagram are presented in Table 1.

travertine deposits along the vents of the thermal springs as a result of evaporation and loss of CO_2.

In fact, the thermal waters fall between the formation water and groundwater fields of Giggenbach et al. (1983) indicating mixing between a saline (not yet identified) and freshwater source. It is probable that the saline waters represent ancient formation waters trapped in the geological formations or magmatic or metamorphic waters.

The presence of saline fluids in the older granitic rocks of the Himalayas has been reported based on INDEPTH (International Deep Profiling of Tibet and the Himalayas) studies carried out along the Indus-Tsangpo Suture Zone (Chandrasekharam 2001a). The "*seismic bright spots*" located during the INDEPTH studies in Tibet region, north-east of Manikaran, were attributed to the presence of magmatic melts and/or saline fluids within the crust (Makovsky & Klemperer 1999). Highly saline fluids are also found in Ladakh Granite (~60 Ma) as inclusions, which are attributed to the high volatile content in the granitic melts (Sachan 1996). Although INDEPTH investigation has not been carried out in the Manikaran area, considering the proximity of INDEPTH investigation sites in Tibet, the probability of occurrence of such "*seismic bright spots*" in this area is high. This inference gains strength from the 1 Ma anatexis process recognized in Nanga Parbat (Chichi Granite Massif) in the Pakistan Himalayas (Schneider et al. 1999). Similar processes must be in operation on the eastern side of Nanga Parbat also (including Manikaran area). This evidence confirms that the present day observed high heat flow (>100 mW/m^2) and geothermal gradient (>100°C/km) values are related to a crustal melting process at shallow depth associated with subduction tectonics. These facts support the view that the source of the saline component is magmatic.

Fluid inclusion studies on the Manikaran Quartzite (Sharma & Misra 1998), through which the thermal waters are issuing, suggests the composition of the trapped fluid as H_2O-NaCl-KCl, since the biphase aqueous inclusions show a eutectic temperature varying from −21.1 to −23.7°C (Crawford 1981) and with salinity varying from 2.7 to 10.6 wt% NaCl equivalents (Bodnar 1993). Therefore, magmatic/hydrothermal fluids locked up in the granites and quartzites probably are the main sources of salinity in the geothermal system, due to a high degree of water–rock interaction at elevated temperatures.

Dilution of such saline fluids could have occurred during their ascent to the surface before emerging as thermal springs. This is indicated by large variation in the temperature and chloride content (a relatively non-reactive constituent) of the thermal waters of the area. Thus on hydrogeochemical considerations, it could be concluded that thermal waters of the Manikaran area receive significant contributions from magmatic sources and dilution from the shallow groundwater.

The fraction of shallow groundwater mixing with deep thermal waters can be estimated using a graphical procedure suggested by Fournier & Truesdell (1974) with a set of enthalpy values and solubility of silica (quartz) at different temperatures. By this method, the estimated fraction of shallow groundwater present in thermal waters turns out to be 0.54–0.65, with an average value of 0.60.

Detailed work on the trace elements and isotopic signatures of the thermal and cold waters is being carried out.

5 GEOTHERMOMETRY

Reservoir temperatures estimated based on cations and silica using the equations recommended by Fournier (1983), Fournier & Potter (1982), Fournier & Truesdell (1973), Giggenbach et al. (1983) and Verma (2000) give values between 150 and 250°C.

The reservoir temperature obtained using the values of the cold water fractions from the mixing model, as explained above, varies from 190 to 210°C, with an average value of 202°C.

Considering the reported high heat flow (100 mW/m^2), high geothermal gradient (100°C/km) and presence of younger granites, coupled with ongoing shallow magmatic processes (Chandrasekharam 2001a), it is likely that temperatures over and above those estimated may be encountered in this area.

6 GEOTHERMAL POTENTIAL

Manikaran thermal springs at present are being used for therapeutic purposes but can be utilized for various direct and indirect applications. Manikaran is in a region with high altitudes and rugged mountain topography. Therefore, it is not economical to transmit power by conventional coal or hydropower grid. Although the local government has installed transmission cables to this remote village, the power supply has not been reliable. With the existing geothermal resources and available technology, it is possible to generate power locally, which can provide a regular supply of electricity to this village economically, without much loss of power as is incurred during transmission over long distances.

Besides power generation, direct utilization of geothermal energy will be more beneficial and economical in these regions. For direct use the thermal waters of the area can be channelised and used for space heating of houses, community centres, schools, dispensaries, hotels, tourist complexes and other establishments as they grow with the development. Thermal waters can also be used in the functioning of cold storages, which may give a great boost to the fruit industry by preventing damage to produce and loss of revenue at the time of disruption of road transport during rainy and snowy seasons. It can also be used in setting up plants for drying, processing, preserving and canning of fruits and fruit products at specially created centres near the sites of the thermal springs. It will reduce the manufacturing cost marginally and the food products could be sold at competitive prices.

Additionally, large scale development of sheep breeding farms under controlled conditions, wool processing, carpet shawl garment manufacturing, animal husbandry, poultry and fish farming, small scale industries based on timber and other forest based products are all well within the realm of possible uses for the geothermal resources in the area.

Successful experiments with winter cultivation, using geothermal greenhouses in the Puga-Chhumathang area (Ravi Shanker 1986), north of Manikaran, should encourage large scale greenhouse cultivation in the Manikaran area also. Using greenhouses, seed and plant nurseries can also be developed under controlled conditions for rapid increase in the production of vegetables and other crops. This could become a flourishing and profitable industry and contribute to the development of the area, since fresh vegetables are being transported by air or road at exorbitant cost. The state of Himachal Pradesh cultivates 5000 tonnes of fruits per year and at present has only two fruit processing plants (Chandrasekharam 2001b). The region has to depend on other states for other farm products. This state can become one of the major food producing and processing regions, with Manikaran as a major centre, if available geothermal energy sources are utilized to their full potential.

7 CONCLUSIONS

Based on major ion chemistry the thermal waters of the Manikaran area are characterized as $Na\text{-}Cl\text{-}HCO_3$ type. High salinity of the thermal waters indicates possible mixing between a saline (not yet identified) and groundwater source. The source of this saline component is probably ancient formation waters trapped in the geological formations, magmatic or metamorphic waters. The presence of saline fluids in older granitic rocks of the Himalayas based on the INDEPTH studies supports this view. The composition of the thermal waters can be described in terms of the variations in the relative proportions of two end member components: saline water and fresh water. Depletion in the calcium content can be explained in terms of the formation of travertine deposits along the vents of the thermal springs.

The high heat flow (100 mW/m^2) and the high geothermal gradient (100°C/km) are due to the presence of younger granites and ongoing shallow magmatic processes.

Excellent prospects exist for utilisation of thermal waters for indirect (power generation) as well as direct use (recreation, agriculture, space heating, cold storage, food processing, greenhouse farming, etc).

REFERENCES

Alam, M.A. 2002. Hydrogeochemistry of thermal springs in Manikaran, Kullu district, Himachal Pradesh (India). *M.Sc. Dissertation Report, Department of Earth Sciences, Indian Institute of Technology, Bombay, India (unpublished).*

APHA 1985. Standard methods for examination of water and wastewater. *APHA, AWWA, WPCF; (16th ed.)* 199, 210, 228, 237, 246, 269, 287, 467.

Bodnar, R.J. 1993. Revised equation and table for determining the freezing point depression in H_2O-NaCl solutions. *Geochim. Cosmochim. Acta* 57: 683–684.

Chandrasekharam, D. 2001a. HDR prospects of Himalaya geothermal province. In K. Popovski and B. Sanner (eds) *Proceeding, International Geothermal Days "Germany 2001" – International Summer School on Direct Application of Geothermal Energy, Bad Urach, Germany* 269–273.

Chandrasekharam, D. 2001b. Use of geothermal energy for food processing – Indian status. *Geo-Heat Center Quarterly Bulletin, Klamath Falls, Oregon* 22: 8–12.

Chaturvedi, N.N. & Raymahashay, B.C. 1976. Geological setting and geochemical characteristics of the Parbati Valley Geothermal Field, India. In *Proceedings of the 2nd UN Symposium, Developments and Use of Geothermal Resources, San Francisco, CA, USA* 1: 329–338.

Crawford, M.L. 1981. Phase equilibria in aqueous fluid inclusions. *Mineralogical Association of Canada, Short course handbook* 6: 75–100.

Fournier, R.O. 1983. A method of calculating quartz solubilities in aqueous sodium chloride solutions. *Geochim. Cosmochim. Acta* 47: 579–586.

Fournier, R.O. & Potter, R.W. 1982. A revised and expanded silica (quartz) geothermometer. *Geothermal Resources Council Bulletin* 11(10): 3–12.

Fournier, R.O. & Truesdell, A.H. 1973. An empirical Na-K-Ca geothermometer for natural waters. *Geochim. Cosmochim. Acta* 37: 1255–1275.

Fournier, R.O. & Truesdell, A.H. 1974. Geochemical indicators of subsurface temperature, Part 2: Estimation of temperature and fraction of hot water mixed with cold water. *J. Res. U.S. Geol. Survey* 2: 263–276.

Giggenbach, W.F., Gonfiantini, R., Jangi, B.L. & Truesdell, A.H. 1983. Isotopic and chemical composition of Parbati Valley geothermal discharges, northwest Himalaya, India: *Geothermiocs* 12: 199–222.

Gonfiantini, R., Gallo, G., Payne, B.R. & Taylor, C.B. 1976. Environmental isotopes and hydrochemistry in groundwater of Grain Canaria: In *Interpretation of isotope and hydrogeochemical data in groundwater hydrology, International Atomic Energy Agency* 159–170.

Gupta, M.L., Saxena, V.K. & Sukhija, B.S. 1976. An analysis of the hot spring activity of the Manikaran Area, Himachal Pradesh, India, by geochemical studies and tritium concentrations of springs waters. In *Proceedings of the 2nd UN Symposium, Developments and Use of Geothermal Resources, San Francisco, CA, USA* 1: 741–744.

Jangi, B.L., Gyan Prakash, Dua, K.J.S., Thussu, J.L., Dimri, D.B. & Pathak, C.S. 1976. Geothermal exploration of the Parbati Valley Geothermal Field, Kulu District, Himachal Pradesh, India. In: *Proceedings of the 2nd UN Symposium, Development and Use of Geothermal Resources, San Francisco, USA* 2: 1085–1094.

Langelier, W.F. & Ludwig, H.F. 1942. Graphical method for indicating the mineral character of natural waters. *J. Am. Water Works Assoc* 34: 335–352.

Makovsky, Y. & Klemperer, S.L. 1999. Measuring the seismic properties of Tibetan bright spots: Evidence of free aqueous fluids in the Tibetan middle crust. *J. Geophys. Res.* 104: 795–825.

Ravi Shanker, 1986. Scope and utilisation of geothermal energy for area development in backwards, hilly and tribal regions in India. *Indian Minerals* 40(2): 49–61.

Sachan, H.K. 1996. Cooling history of subduction related granite from the Indus Suture zone, Ladakh, India: evidence from fluid inclusions. *Lithosphere* 38: 81–92.

Schneider, D.A., Edwards, M.A., Kidd, W.S.F., Zeitler, P.K. & Coath, C.D. 1999. Early Miocene anatexis identified in the western syntaxis, Pakistan Himalaya. *Earth. Planet. Sci. Lett.* 167: 121–129.

Sharma, R. & Misra, D.K. 1998. Evolution and resetting of the fluids in Manikaran Quartzite, Himachal Pradesh: applications to burial and recrystallization. *J. Geol. Soc. India* 51: 785–792.

Srikantia, S.V. & Bhargava, O.N. 1998. Geology of Himachal Pradesh. *Geological Society of India, Bangalore, India* 51–52, 202–205.

Verma, M.P. 2000. Chemical thermodynamics of silica: a critique on its thermometer. *Geothermics* 29: 323–346.

Water–Rock Interaction, Wanty & Seal II (eds)
 ISBN 90 5809 641 6

Geochemical and isotope features of brines of the Siberian platform

S.V. Alexeev & L.P. Alexeeva
Institute of the Earth's Crust SB RAS, Irkutsk, Russia

O. Shouakar-Stash & S.K. Frape
Department of Earth Science, University of Waterloo, Waterloo, Ontario, Canada

ABSTRACT: New stable isotope data (^{2}H, ^{18}O, ^{37}Cl) of the brines occurring down to depths of 3500 m of the Siberian platform have been obtained. Most $\delta^{2}H/\delta^{18}O$ values of groundwater lie below the Global Meteoric Water Line (GMWL). Isotopic composition correlates with the salinity and occurrence depth of the brines: the more salinity and occurrence depth of the brines the more $\delta^{2}H$ and $\delta^{18}O$ values. The $\delta^{37}Cl$ values are controlled mainly by the various cation compositions (Na or Ca) of the saline groundwaters. The research results allow determine possible the genetic brine types of the Siberian platform.

1 INTRODUCTION

The problem of genesis and formation of the concentrated chloride Ca and Mg brines in the sedimentary rocks of the Siberian platform has always attracted researchers' attention and now it is widely discussed among world scientists (Kraynov & Ryzhenko 1997; Shvartsev 1998, 2000; Shoukar-Stash et al. 2002). Understanding the origin of groundwater is best achieved by a combination of hydrochemistry and stable isotopes. The brines of the sedimentary cover of the Siberian platform have complex origins and geochemical evolution. The objective of this article is to analyze the vast amount of data on regional hydrogeological and hydrogeochemical features of the Siberian platform (Fig. 1), and to use hydrochemistry and stable isotopes to determine the origins of the brines.

The saline water and the brines under consideration were sampled from the boreholes and the springs from five sites of the Siberian platform: the Daldyn-Alakit (I) and the Malo-Botuobiya (II) regions, including in the Yakutian artesian basin; the Nepa basin (III) and the Irkutsk basins (V), making part of the Angara-Lena artesian basin; and the Tungusska artesian basin (IV). The drilling data from the numerous boreholes show that acid to weakly alkaline (pH = 3.2–7.5) Ca chloride brines are the major geochemical type of groundwater in the sedimentary aquifers. They occur in the central part of the Tungusska and Angara-Lena artesian basins and in the west part of the Yakutian artesian basin.

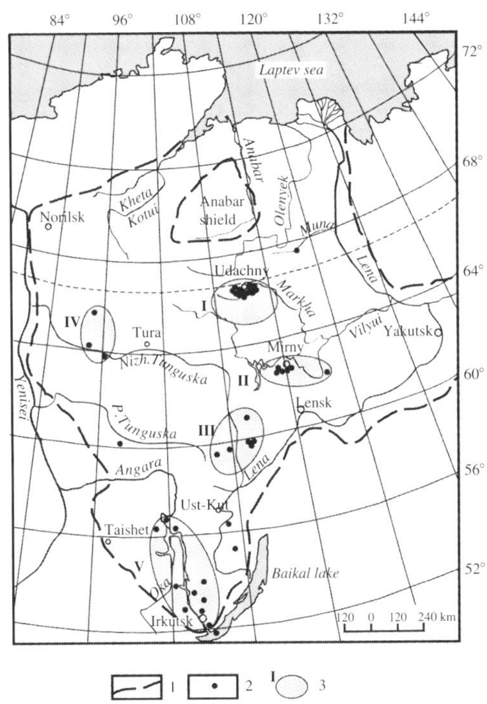

Figure 1. Map of the Siberian platform. 1 – the boundary of the Siberian platform; 2 – the points of water sampling; 3 – the studied sites.

Na chloride brines are mainly confined to halite-bearing formations.

It is important to note that in the Daldyn-Alakit region there are no salt-bearing deposits in the geological section, and in the Tungusska artesian basin there are a few halite-bearing layers. The brines under consideration have high TDS contents (up to 626 g/L), Cl/Br weight ratios of 30 to 80 (Ca-brines) and 500–1000 (Na-brines), and high concentrations of Sr, Br, and Li. The high TDS of Na-brines in the halite-bearing layers is obviously caused by the leaching process, and the high salt concentration of Ca-brines is the subject of discussion among hydrogeologists. This question is closely connected with the problem of brine genesis. Groundwater samples were taken from depths between 120 and 3395 m.

Table 1. Major chemical and isotope components in the brines of the Siberian platform.

No.	Studied site	Depth, m	pH	Density	Na, mg/L	Ca, mg/L	Cl, mg/L	Br, mg/L	TDS, g/L	Average δ^2H	Average $\delta^{18}O$	$\delta^2H/\delta^{18}O$	Average $\delta^{37}Cl$
Na-chloride brines													
1	II	600–900	6.5	1.06	9569	3707	49895	749	85.65	−108.81	−14.06	−0.787	1.302
3	IV	300	5.6	1.22	70728	26899	165215	1243	318	−88.46	−12.00	−3.595	−0.105
4	V	1630–1660			9921	986	165345	2578		−90.73	−8.83	0.000	−0.285
7	III	120	7.4	1.05	27076	792	38423	36	77	−152.08	−20.01	−0.506	−0.255
8	III	419–481			13427	1094	21857	17	44	−147.46	−18.47	−0.298	−0.131
9	V	600			2218	283	3500	4	7	−113.24	−15.71	−0.062	0.102
10	V	455–500			14758	1275	22088	20	64	−133.97	−16.72	−0.478	0.081
11	V	700–1000			98388	1114	132866	92	296		−15.25		−0.063
12	V	733–760			7667	2656	17538	219	32	−96.44	−12.20	−0.332	0.564
13	II	500	8.4	1.06	27996	1028	42159	51	87.4	−113.17	−9.57	−0.772	1.536
14	III				3685	270	639	2	11.2	−124.04	−17.65	−0.090	0.349
15	II	410							70.2	−160	−19.7	−0.439	
17	II	430							84.9	−163	−19.2	−0.521	
18	II	470							91.1	−166	−18.4	−0.549	
19	II								51.9	−171	−21.4	−0.304	
20	II	410							36.6	−169	−19.4	−0.217	
Ca-chloride brines													
21	I	500–550	6.4	1.24	16451	48518	160931	3648	323	−66.38	−3.04	−4.866	−0.403
22	I	260–270	7.4	1.08	4843	12737	59535	1167	111.3	−123.00	−13.10	−0.905	0.009
23	V	1818–1826	<4	1.41	1988	95351	220840	8587	589.7	−47.28	−9.91	−12.47	−0.527
24	I	805–1150	4.9	1.26	10171	37157	155949	3621	353.2	−61.68	−2.13	−5.727	−0.346
25	I	1390–1567	4.9	1.28	12098	61202	179098	4910	395.8	−45.33	1.73	−8.732	−0.337
26	II	600	5.5	1.31	4025	46417	173338	4712	434.9	−39.30	−1.92	−11.06	−0.318
27	I	210	7.8	1.03	4352	6225	26305	412	39.92	−139.59	−16.45	−0.286	0.524
28	I	110–120	7.1	1.08	8545	15479	76242	1458	98.34	−109.36	−8.92	−0.899	0.208
29	IV	2100	7.5	1.26	40000	78830	239000	4950	386.3	−121.12	−15.22	−3.189	0.579
30	V	3–400			208	222		0		−131.89	−16.76	0.000	−0.126
31	V	2756–2782	3.2	1.28	22266	75330	179793	4828	366.2	−35.86	−2.47	−10.21	−0.156
32	V	3067			8764	96021	161699	4252	446.9	−49.41	−0.20	−9.043	−0.315
33	V	3382–3395			37177	42733	169649	4188	358.9	−38.44	−0.30	−9.338	−0.405
34	II	884–1024	6	1.25	11294	86041	158918	3816	362.9	−51.76	−5.23	−7.011	−0.151
35	II	600	5.5	1.27	19191	73761	163375	3421	388	−50.47	−4.94	−7.687	0.044
36	I	449–800	5	1.24	15872	74691	192444	4440	317	−67.30	−5.52	−4.710	−0.215
37	I	313	6.8	1.07	19172	14461	62810	849	91.3	−95.90	−10.22	−0.952	−0.398
38	I	400–650	3.5	1.23	21600	63465	151202	3229	315.2	−69.93	−4.37	−4.507	−0.230
39	I	504–750	4.5	1.21	19870	70923	139435	3094	316.1	−70.52	−4.96	−4.482	−0.202
40	I	834–1475	4.2	1.27	5066	94723	212280	5637	381.4	−73.32	−8.20	−5.201	−0.269
41	I	553–952	4.1	1.26	19720	69117	199151	4354	344.6		−2.00		−0.238
42	V	1830–1888			1151	110952	231899	6317	626	−69.64	−10.93	−8.990	−0.082
43	III	1674–1667	5.4	1.29	8228	100808	184261	5206	384	−44.89	−1.37	−8.554	−0.208
44	III	1623–1633	4.2	1.22	18559	76395	171839	3928	358	−78.26	−7.21	−4.574	−0.238
45	III	1380–1408			9229	77988	164859	2026	399	−194.59	−12.30	−2.050	−0.389
46	II	2777–2784			10406	108747	187668	4725	396	−30.62	−6.15	−12.93	−0.666
47	V	2258–2296			2775	119885	205786	5135	511	−50.69	−6.25	−10.08	−0.143
48	II	890							315	−91.00	−10.40	−3.462	
49	II	900							331	−89.00	−8.60	−3.719	

2 GEOCHEMICAL AND ISOTOPIC FEATURES OF GROUND BRINES OF THE SIBERIAN PLATFORM

The chemistry of the brines is summarized in Table 1. The brines are divided into Na-chloride and Ca-chloride types. For each group of brines, the concentration of major cations and anions, and stable isotope (H, O, Cl) ratios are shown.

The pattern of $\delta^{18}O$ and $\delta^{2}H$ values are similar to that shown by Pinneker et al. (1987). Figure 2 shows that isotopic composition is below the Global Meteoric Water Line, that is Na and Ca chloride brines are depleted in ^{2}H and enriched in ^{18}O compared to the meteoric and the surface waters. The trends in the brine samples lie at angle from the GMWL, however, some groundwater samples have $\delta^{18}O$ values close to sea water (SMOW). This could mean that the deep brines were formed as a result of burying sea water together with syngenetic deposits and subsequent metamorphosis during water-rock interaction.

The $\delta^{18}O$ and $\delta^{2}H$ values allow the brines to be divided into two groups. The Na chloride brines have lower $\delta^{18}O$ values (–20.01 to –9.57‰) and $\delta^{2}H$ values (–152.08 to –88.46‰), which are similar to the surface waters. The Ca chloride brines have higher $\delta^{18}O$ values (–16.76 to +1.73‰) and $\delta^{2}H$ composition (from –194.59 to –30.62‰) that are much higher than the surface waters. The correlation also extends to TDS as the Ca chloride brines have TDS contents of >300 g/L.

Figure 3 shows the relation between $\delta^{18}O$, $\delta^{2}H$ and depth and the salinity of the brine on the Siberian platform: as the salinity and depth of its occurrence increase, so do $\delta^{18}O$ and $\delta^{2}H$ values. The brines from a depth of <500 m have low values of oxygen-18 and deuterium. The lowest values of $\delta^{18}O$ (–17 to –20‰) are typical of the groundwater samples taken from the Nepa potash-bearing basin.

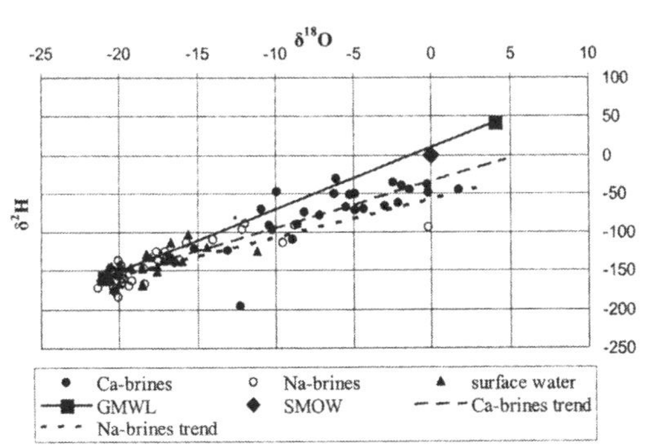

Figure 2. $\delta^{18}O$ and $\delta^{2}H$ values of the surface water and in the brines from sedimentary rocks on the Siberian platform.

Salt precipitation may be the method that depletes the brines in ^{2}H and ^{18}O (Bases of hydrogeology, 1982; McCaffray et al. 1987). On the contrary, the brines from the deep rock strata are enriched in the heavy isotope of oxygen ($\delta^{18}O$ increases up to +1.73 ‰ at 1500–3400 m) and of deuterium (–30.62‰ at 2780 m). Thus, even though direct connection between the isotopic composition and depth is not apparent, two groups of the brines can be identified. Their formation may have to do with the degree of hydrogeological isolation of the brines and with their possible mixing with meteoric water. The shallow groundwater depleted in $\delta^{18}O$ and $\delta^{2}H$ belongs to one group, the other group includes deep brines that are much more enriched in $\delta^{18}O$ and $\delta^{2}H$.

The relationship between $\delta^{37}Cl$ and Cl-concentration in the brines of the area is shown in figure 4. The value of $\delta^{37}Cl$ varies from –0.4 to +1.5‰.

Again the brines are clearly divided into two groups:

- the first group includes mainly Ca chloride deep brines with high TDS contents (from 300 to 600 g/L). In this group there are only three samples of Na chloride composition, which have high TDS (about 300 g/L). High Cl-content (>130 g/L) and a comparatively small range of $\delta^{37}Cl$ (from –0.666 to +0.579‰).
- the second group includes mainly Na chloride brines with TDS contents of <100 g/L, and four

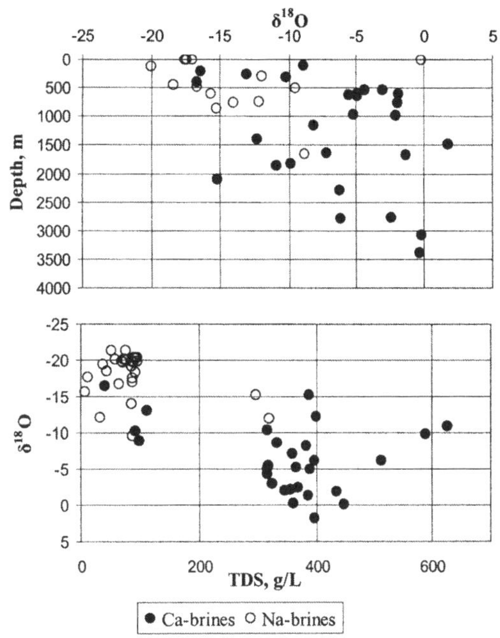

Figure 3. $\delta^{18}O$ and $\delta^{2}H$ values of *vs.* depth of the brines of the Siberian platform.

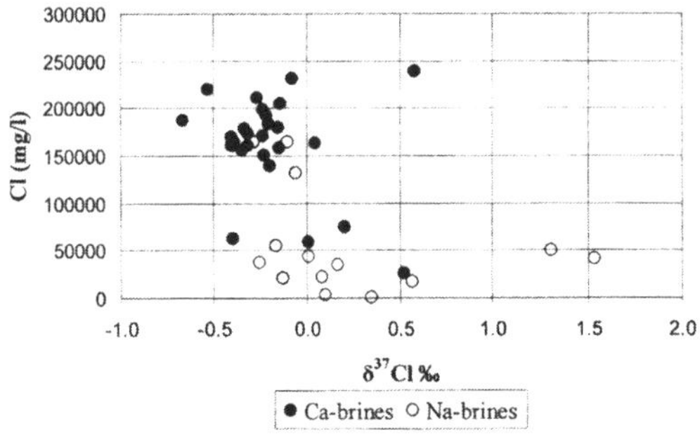

Figure 4. δ^{37}Cl *vs.* chloride concentration for the brines on the Siberian platform.

samples of Ca-rich brines with low TDS concentration (40–98 g/L). The brines in this group have low Cl concentrations (<77 g/L) and a larger range of δ^{37}Cl values (−0.398 to +1.536‰).

3 SUMMARY

The brines from the Siberian platform have fairly uniform anion composition (Cl^-) and either Na or Ca cation composition. Chemistry and stable isotopes demonstrate the distinct division of the brines into two groups, and lead to the following possible conclusions about their origins.

1. The brines in the sedimentary rocks on the Siberian platform are of different origin: Na chloride brines are of meteoric origin and Ca chloride brines are of metamorphic origin.
2. The brines in the sedimentary rocks on the Siberian platform have the same origin (the mixing of the ancient infiltrative water or sea water). During their evolution, the brines were subjected to different processes of transformation (metamorphism, leaching, and other interaction with enclosed rocks).
3. The brines in the sedimentary rocks on the Siberian platform have the same genesis, but the degree of dilution caused by infiltrative waters is different. Moreover, the degree of dilution for Ca-chloride brines is much smaller than Na-chloride brines, because the former were in a more passive water-exchange system.

These possible conclusions are not a solution of the brine genesis problem but the stable isotopic data of the brines in the Siberian platform may be helpful for the explanation of the saline water origin.

ACKNOWLEDGEMENT

The authors wish to acknowledge to the Russian Fund for Basic Research (Project 01-05-64012)

REFERENCES

Bases of hydrogeology, 1982. Geological action and water history in the Earth interior. Novosibirsk, Nauka. 239 p., (in Russian).

Frape, S.K., Fritz, P. & McNutt, R.H. 1984. Water-rock interaction and chemistry of groundwaters from the Canadian Shield. *Geochim. Cosmochim. Acta* 48(8): 1617–1629.

McCaffray, M.A., Lasar, B. & Holland, H.D. 1987. The evaporation path of seawater and the coprecipitation of Br^- and K^+ with halite. *J. Sed. Petrol.* 57(5): 928–937.

Kraynov, S.R. & Ryzhenko, B.N. 1997. Origin of chloride groundwater and brines in crystalline massifs: evidence from thermodynamic modeling of geochemical processes in water-granite systems. *Geochemistry* 10: 1035–1057 (in Russian).

Shoukar-Stash, O., Alexeev, S.V., Frape, S., Alexeeva, L.P. & Pinneker, E.V. 2002. Geochemistry and stable isotopic signatures of deep groundwaters and brine from the permafrost zone of the Siberian platform, Russia. GSA Annual Meeting. Denver. Colorado. USA, p. 218–224.

Pinneker, E.V., Borisov, V.N., Kustov, U.I., Brandt, S.B. & Dneprovskaya, L.V. 1987. New data about isotope composition of oxygen and hydrogen of brines of the Siberian platform. *Water Resources* 3: 105–115 (in Russian).

Shvartsev, S.L. & Bukaty, M.B. 1995. The role of rocks in formation of calcium chloride brines. *Reports of Academy of Sciences* 342(4): 530–533 (in Russian).

Shvartsev, S.L. 1998. Brines in Siberian Platform: Geochemical and isotopic evidence for water-rock interaction. In G. Arehart & Hulston, (eds.): 357–360. *Proc. 9th Intern. Symp. WRI.* Rotterdam: Balkema.

Shvartsev, S.L. 2000. Chemical composition and strontium isotopes in brines of Tungussky basin connecting with its formation. *Geochemistry* 11: 1170–1184 (in Russian).

Water–Rock Interaction, Wanty & Seal II (eds)

Modeling of tracer behaviour and dominant reactions during the Pore Water Chemistry (PC) experiment in the Opalinus Clay, Switzerland

D. Arcos & L. Duro
Enviros Spain, Barcelona, Spain

T. Gimmi
RWI, Institute of Geological Sciences, University of Bern, and Paul Scherrer Institut, Villigen, Switzerland

H.N. Waber
RWI, Institute of Geological Sciences, University of Bern, Switzerland

ABSTRACT: The Pore Water Chemistry (PC) experiment, conducted at the Mont Terri Rock Laboratory (Switzerland) is designed to equilibrate an artificial water injected into a packed-off interval with the Opalinus Clay formation pore water. Observed changes in added tracers and the total chemistry were simulated with a 2-D diffusion model and a 1-D reactive transport model. The results show that diffusion is the main transport process with diffusion coefficients similar to those previously obtained in the laboratory. The main geochemical process is organic matter oxidation due to sulphate-reducing bacteria.

1 INTRODUCTION

The knowledge of pore water chemistry and solute transport behaviour in low permeability argillaceous rocks is important for the design and safety assessment of radioactive and toxic waste disposals. Within this framework an international and multidisciplinary project on the Opalinus Clay formation has been carried out at the Mont Terri Rock Laboratory (Switzerland) since 1996.

The application of different techniques including long-term borehole water sampling, squeezing, leaching, extraction, and geochemical modelling resulted in a considerable understanding of the pore water chemistry and solute transport in the Opalinus Clay (Pearson et al. 2003). Nevertheless, relevant uncertainties with regard to pH and the carbonate and redox systems under *in-situ* conditions remain.

The Pore Water Chemistry (PC) experiment aimed to gain more insight into this problem based on a long-term diffusive exchange experiment (Wersin et al. 2004).

2 PC EXPERIMENT DATA

To minimise induced oxidation, borehole BPC-1 was drilled with N_2 and filled with Ar prior to equipment installation. On March 22, 2002, artificial water was filled into a packed-off interval and allowed to equilibrate with the formation pore water. The artificial water was designed by geochemical modelling to be in equilibrium with the mineralogy and cation exchange properties of the rock (Pearson et al. 2003) and traced with Cl, Br, SO_4, ^2H, ^{18}O, and ^{13}C. The concentrations of interest of the expected formation water, the artificial water injected in the packer interval, and samples PC-2 to PC-5 taken after 12, 116, 264, and 529 days from the installed closed sampling system are given in Table 1. The number of samples withdrawn was small to minimise disturbances. Plotting relative concentrations in the borehole as $c = (C - C_{FW})/(C_{AW} - C_{FW})$, where C is the actual and C_{AW} the initial borehole concentration, and C_{FW} the background concentration of the formation water, allows direct comparisons between all tracers.

3 TRACER MODELLING

Diffusion in Opalinus Clay is anisotropic, with a diffusion coefficient parallel to bedding about a factor 5 larger than the one perpendicular to bedding (Van Loon et al. 2003). Diffusion was expected to occur in an anisotropic way also in this experiment, even though the test interval was located in the zone

Table 1. Anionic and isotopic concentrations of formation water (FW), injected artificial water (AW), and water samples (PC-1 to PC-5) taken from the interval; full analyses are given in Wersin et al. 2004 (Eh in mV vs. SHE, concentrations in mol/kgH_2O, Alk. = titrated total alkalinity, and isotopes in permil relative to V-SMOW ($\delta^{18}O$, δ^2H) and PDB ($\delta^{13}C$); n.m. = not measured).

	FW[1]	AW	PC-2	PC-3	PC-4	PC-5
pH	7.96	7.74	7.77	7.00	6.74	6.68
Eh	−10	+90	+127	−126	−171	−220
Fe_{tot}	n.m.	2.3E-6	n.m.	<3.6E-6	4.5E-6	2.3E-6
Cl	287	262	267	283	285	286
Br	0.4	29.4	26.9	12.7	8.2	7.3
SO_4	13.7	14.7	14.6	13.3	10.5	6.6
HS^-	n.m.	n.m.	n.m.	1.2	0.3	0.6
Alk.	2.5	1.0	1.1	6.5	7.0	11.6
δ^2H	−54.7	370.0	251.1	38.8	4.3	−32.6
$\delta^{18}O$	−8.5	−10.46	−10.08	−9.02	−8.43	−8.62
$\delta^{13}C$	−13.1	−29.1	−16.3	−11.4	−3.2	−15.9

[1] Sample taken on April 28, 1999, from BWS-A1 borehole (Pearson et al. 2003).

of the main fault, where the bedding is partly disrupted. The borehole BPC-1 is oriented at an angle of roughly 45° with respect to the bedding, as is schematically shown in Figure 1. The concentration changes in the borehole will be dominated by diffusion along the bedding planes, because initially diffusion perpendicular to the bedding is only relevant at the upper and lower end of the interval, and at larger times the concentration gradients will be low anyway. Thus, we analysed the tracer data with a 2D, and not with a 3D solution.

3.1 *Model set-up*

The plane parallel to the bedding intersects the borehole in form of an ellipse, as indicated in Figure 1. We simplified this geometry slightly by assuming a radial symmetry, where the now circular borehole area (radius $a' = 0.032$ m) has the same circumference as the ellipse.

The diffusion equation for this radial system, including the borehole area, was solved in Laplace space and inverted numerically with the Talbot algorithm, similar as described in Gimmi & Waber (2003). The solution in Laplace space is:

$$\hat{c}(r;s) = \frac{a'K_0(rq)}{qD_p[a'qK_0(a'q) + \alpha K_1(a'q)]} \quad (1)$$

where r is the radial distance from the center, s the Laplace transform variable, α two times the rock capacity per rock area divided by the borehole capacity per borehole area, D_p the pore diffusion coefficient, $q = (s/D_p)^{1/2}$, and K_0 and K_1 modified Bessel functions of order zero and one.

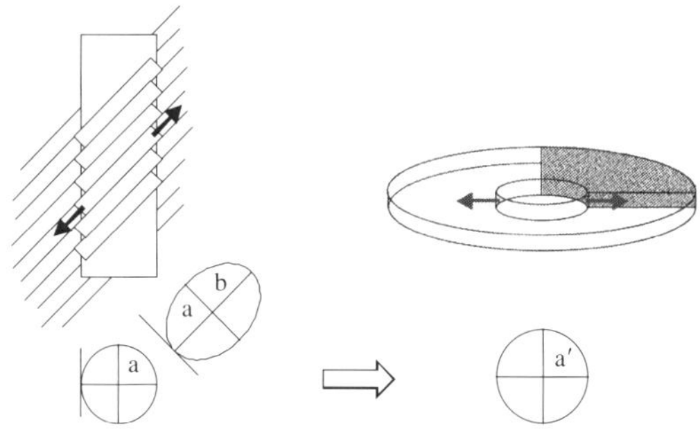

Figure 1. Schematic drawing of the geometry of the borehole and of the modelling approach.

3.2 *Parameter estimation*

Diffusion coefficients for Opalinus Clay samples from Mont Terri were determined in lab experiments by Van Loon et al. (2003). As a first estimate, we used the value of $D_e = 4.3 \cdot 10^{-11}\,m^2s^{-1}$ for HTO parallel to the bedding (all values were adapted to a temperature of 15°C) for HDO and for $H_2^{18}O$. Because the test interval lies in the zone of the main fault, the anisotropy may be less pronounced than in the undisturbed matrix. Thus, we performed also simulations with a value of $D_e = 8.7 \cdot 10^{-12}\,m^2s^{-1}$ (perpendicular to the bedding), as well as with the geometrical mean of the two values ($D_e = 1.9 \cdot 10^{-11}\,m^2s^{-1}$). For Cl and Br, the laboratory value for ^{36}Cl of $1.3 \cdot 10^{-11}\,m^2s^{-1}$ (parallel), and values of $2.5 \cdot 10^{-12}\,m^2s^{-1}$ (perpendicular), and $5.7 \cdot 10^{-12}\,m^2s^{-1}$ (geometric mean) were used. A porosity ε of $0.17\,m^3m^{-3}$ was assumed for the water tracers, and of $0.082\,m^3m^{-3}$ for the anions (Van Loon et al. 2003). Note that $D_p = D_e/\varepsilon$. The capacity ratio α was 9.2ε.

3.3 *Modelling results*

Figure 2 shows the measured and calculated tracer data. Both water tracers follow more or less the simulation for a D_e of about $2 \cdot 10^{-11}\,m^2s^{-1}$ and a porosity of $0.17\,m^3m^{-3}$. The D_e value is somewhat lower than typical values reported for diffusion along the bedding for Mont Terri samples. Because the experiment was performed in the fault zone, where the bedding is disturbed, a value intermediate between those along and perpendicular to bedding seems reasonable. The deviations of the measurements 12 days after the injection of the tracer from the simulations may be due to the installation of the surface equipment, which occurred on day 11.

The anionic tracers Cl and Br show generally the same behaviour. The concentration changes can be described approximately by using a D_e of about $6 \cdot 10^{-12}\,m^2s^{-1}$ and a porosity of $0.082\,m^3m^{-3}$. Again, the D_e value lies between those expected for

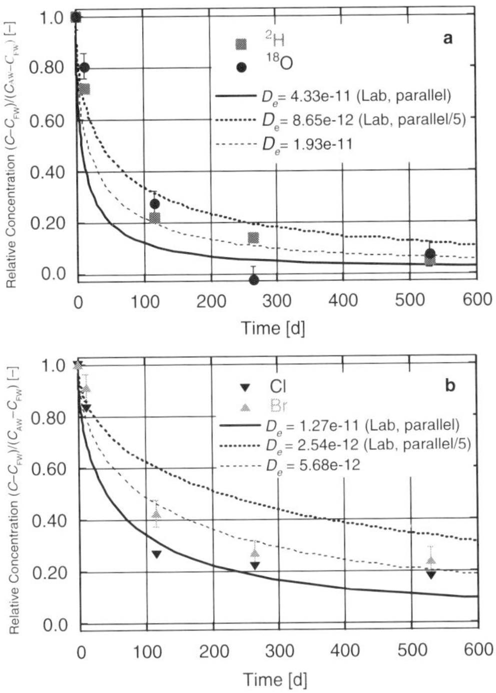

Figure 2. Measured and simulated relative concentrations c for, (a), the water tracers and, (b), the anionic tracers Cl and Br. The relative measurement uncertainties of ± 0.005 (^2H), ± 0.05 (^{18}O), and ± 0.05 (Br) are indicated by vertical bars. For Cl, they equal about ± 1.0.

diffusion parallel and perpendicular to the bedding. The measurement uncertainties are large for Cl, thus those data should be interpreted with care.

4 MODELLING OF DOMINANT REACTIONS

4.1 *Geometric domain and flow conditions*

Modelling of the dominant reactions in the system was performed utilising a one dimensional reactive transport model PHREEQC. The modelling domain has been divided into 30 cells, 29 out of which correspond to the clay formation and the remaining cell corresponds to the test borehole. The length of each cell is 0.03 m.

Diffusion is the only transport process accounted for in the system under consideration. The right boundary, in contact with the cell representing the experimental borehole is assumed a no flow boundary, whereas diffusion is allowed in the left boundary assuming a constant initial pore water composition in the near vicinity within the Opalinus Clay.

Effective diffusion coefficients (D_e) of $1-3\times 10^{-11}\,m^2s^{-1}$ (Pearson et al. 2003; Van Loon et al. 2003) were used. The model can accurately describe the experimental data for the conservative elements Cl and Br (Fig. 3a).

4.2 *Geochemical system*

The chemical composition of the initial pore water within the Opalinus Clay (FW in Tab. 1) is that corresponding to water sampled in long-term experiments from borehole BWS-A1 (Pearson et al. 2003). The initial water composition in the borehole is the in-filled artificial water (AW in Tab. 1).

The water–rock reactions modelled within the formation are mineral precipitation and dissolution and cation exchange reactions. The solid phases present in the Opalinus Clay include calcite (6–39 wt.%), dolomite (0.6–2.7 wt.%), quartz (6–30 wt.%), celestite (0.02–0.06 wt.%) and pyrite (0.7–3.2 wt.%) beside the dominating clay minerals (45–76 wt.%) (Pearson et al. 2003). Aragonite, dolomite and gypsum are allowed to form as secondary phases as soon as they oversaturate in the system, while the precipitation of amorphous FeS is permitted according to the kinetic rate law determined by Van Cappellen & Wang (1996).

The initial carbonate concentration in the formation pore water is fixed to be in equilibrium with a pCO_2 of $10^{-2.85}$ bars according to the experimental values obtained by Lassin et al. (2003).

The degradation of dissolved organic matter is included in a kinetic Monod-like fashion. Sulphate is assumed to be the only electron acceptor species. Organic matter is conceptualised by the generic formula CH_2O. According to this process, the reaction included in the code is

$$2CH_2O + SO_4^{2-} \leftrightarrow H_2S + 2HCO_3^-$$

with the following rate law

$$R = k_{max}\frac{ED}{K_{ED}+ED}\frac{EA}{K_{EA}+EA}$$

where k_{max} stands for the maximum rate of the metabolisable organic fraction and K_{EA} and K_{ED} are the half-saturation concentration of the electron acceptor (SO_4^{2-} in this case) and of the electron donor (dissolved organic matter), respectively.

In the absence of site-specific data on the parameters included in the rate law, we have conducted a sensitivity analysis to obtain the values of those three parameters that better reproduce the on-line measured data, obtaining the following values:

- $k_{max} = 3\times 10^{-11}\,mol\cdot L^{-1}\cdot s^{-1}$, which is in the range of values reported in the literature (from 3×10^{-5} to $1.26\times 10^{-12}\,mol\cdot L^{-1}\cdot s^{-1}$)
- $K_{ED} = 10^{-3}\,mol\cdot L^{-1}$, which is on the order of literature data ($1.1\times 10^{-4}\,mol\cdot L^{-1}$)

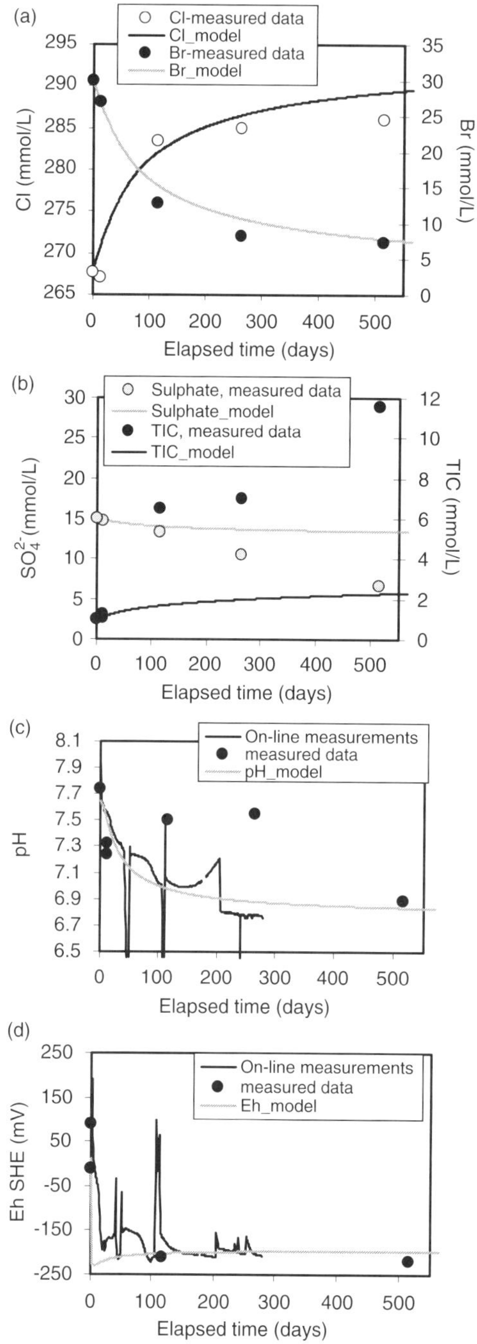

Figure 3. Comparison between measured values and calculated results. (a) Conservative elements; (b) Sulphate and total inorganic carbon (TIC); (c) on-line, measured and model results for pH; and (d) on-line, measured and model results for redox potential.

- $K_{EA} = 1.62 \cdot 10^{-3}\,\mathrm{mol \cdot L^{-1}}$, corresponding to the maximum of the values reported in the literature (from 3×10^{-5} to $1.62 \times 10^{-3}\,\mathrm{mol \cdot L^{-1}}$), given the high sulphate concentration in the system.

4.3 *Results*

The comparison between the model results and the measured data is shown in Figure 3. The master variables of the water (pH and redox potential) are fairly well represented by the model. One must notice the different pH values obtained in laboratory determinations in the three first water samples in comparison with the less alkaline values obtained on-line. It is likely that a degassing process during sample manipulation occurred in this case, given that the value of the fourth sample is in close agreement with the on-line measurements and the modelling results. The model is also able to describe the redox evolution of the system, indicating the adequacy of considering the role of bacterially mediated organic matter oxidation.

An apparent discrepancy between data and calculation appears when dealing with sulphate and TIC. Although the model describes the patterns of these two components, the extent of sulphate decrease and TIC increase is much lower in the model compared to the actual observations. This can be a direct effect of the fact that the parameterisation of the kinetics of organic matter decomposition has been conducted without *in situ* specific data. Indeed, from the comparison of the results in Figure 3b, it is clear that a faster kinetics of the OM oxidation will induce a faster decrease in the SO_4^{2-} concentration as well as a faster increase in the TIC, which would be in closer agreement with the experimental results.

5 CONCLUSIONS

- Diffusion seems to dominate the exchange of conservative tracers between the injected artificial pore water and the formation water. No influence of fissures was detected.
- The conservative tracers did behave according to present-day understanding: differences between anionic and water tracers were as expected from laboratory observations.
- The process controlling the geochemical behaviour of the system is biodegradation of organic matter via sulphate reduction.
- Differences between measured and modelled values are related to the simplification of the organic matter degradation reaction, the selection of the type of dissolved organic matter and the parameters used for the kinetic degradation of organic matter.

ACKNOWLEDGEMENTS

This work is part of the International Mont Terri Project and supported by ANDRA, BGR, CRIEPI, ENRESA, JNC and NAGRA.

REFERENCES

Gimmi T. & Waber H.N. 2003. Modelling of profiles of stable water isotopes, chloride, and chlorine isotopes of pore water in argillaceous rocks in the Benken borehole. *Nagra Technical Report 03-xx, in print.*

Lassin A., Gaucher E. & Crouzet C. 2003. Dissolved Carbon Dioxide and Hydrocarbon Extraction. Annex 6 *in* Pearson et al. 2003. *Reports of the Swiss Federal Office for Water and Geology (FOWG), Geology Series, Nr. 5*, Bern, Switzerland.

Pearson F.J., Arcos D., Bath A., Boisson J.-Y., Fernández A.M., Gaebler H.E., Gaucher E., Gautschi A., Griffault L., Hernan P. & Waber H.N. 2003. Geochemistry of water in the Opalinus Clay formation at the Mont Terri Rock Laboratory – Synthesis Report. *Reports of the Swiss Federal Office for Water and Geology (FOWG), Geology Series, Nr. 5*, Bern, Switzerland.

Van Cappellen P. & Wang Y. 1996. Cycling of iron and manganese in surface sediments: A general theory for the coupled transport and reaction of carbon, oxygen, nitrogen, sulphur, iron, and manganese. *Am. J. Sci.* 296, 197–243.

Van Loon L.R., Soler J.M. & Bradbury M.H. 2003. Diffusion of HTO, $^{36}C^{-}$, and $^{125}I^{-}$ in Opalinus Clay samples from Mont Terri and Benken: Diffusion parallel and perpendicular to bedding under different confining pressures. *PSI/Nagra Technical Report*, in preparation, Paul Scherrer Institut, Villigen, Switzerland.

Wersin P., Gautschi A., Vinsot A., de Canniè-re P., Hernán P., Gäbler H.-E., Hama K., Mahara Y. & Gaucher E. 2004. Results from the Porewater Chemistry (PC) Experiment at Mont Terri, Switzerland. *This volume.*

Water–Rock Interaction, Wanty & Seal II (eds)
© 2004 Taylor & Francis Group, London, ISBN 90 5809 641 6

Aspects of groundwater geochemistry from Middle Atlas and Saiss basin (Northern Morocco)

T. Bahaj, M. Wartiti & M. Zahraoui
University Med V, Faculty of Sciences Rabat-Agdal, Earth Sciences Department, Morocco

A. Essahlaoui
University Mly Ismail Faculty of Sciences Meknès, Morocco

R. Caboi
Università degli Studi di Cagliari, Dipartimento di Scienze della Terra, Italy

ABSTRACT: Both the Middle Atlas and Saiss basins have important water resources. Because of their meteoric origin and the nature of the reservoirs in the areas, which is made by calcareous and detrital rocks, these groundwaters are alkaline earth to Ca-HCO_3 waters. They generally have good quality. Although during the last years the region has experienced an important increase in population and a development of agricultural industries. This paper aims to identify the natural and anthropogenic geochemical processes.

1 INTRODUCTION

The Middle Atlas is a subtabular highland. Its elevation is around 1100 m and it receives a significant quantity of precipitation (up to 1200 mm/year). The nature of its geology, made by karstified limestone and dolomites, gives it the characteristics of a voluminous water reserve. In addition, the landscape in the area is covered by protected forests, which limit industrial activities.

The Saiss basin has economic and historic significance; it hosts two major Moroccan cities, Fès and Meknès. It is a very populated region. The weather is semi arid and the population exploits the shallow aquifer for drinking and irrigation purposes.

2 GEOLOGY AND HYDROLOGY

The Middle Atlas region is underlain by karstified Liasic dolomites and limestones. Paleozoic formations occur locally and consist on schists and quartzites. The deformations are due to tectonic activity and consist on normal faults and lineaments. Thus, individual terraces give the Middle Atlas important hydrological characteristics.

The lithology of the Saiss basin is related to its paleogeography. After the sedimentation of marls during the Miocene, the Saiss basin was isolated. The sediments during the early Pliocene are sandstones with carbonate cement; during the Middle Pliocene, the sediments are a complex of yellow sands. During the late Pliocene, the basin evolved as a large lake and the sedimentation is dominated by lacustrine limestones and travertine with local clayey intercalations to the south, west and center of the basin. The northeastern sediments were supplied by the Pliocene Sebou River as conglomerates, sands, and clays (Fig. 1) (Ahmamou 1987, Benaabidate 2000, Gourari 2001).

Due to it's geographic location (1000 m elevation), the Middle Atlas receives significant quantities of precipitation and snow; thus contributing to surficial waters and also to the recharge of the karstic aquifer, which is drained by springs and artesian wells.

The Saiss basin aquifer is recharged by meteoric waters and by waters coming from the Middle Atlas. It is hosted generally by lacustrine limestones, and locally by the complex of yellow sands. In the northwestern side of the basin, the aquifer consists of conglomerates, clays, and marls (Fig. 2) (Martin 1981, Essahlaoui 2000).

3 GROUNDWATER GEOCHEMISTRY

3.1 *Materials and methods*

The distribution of the water samples was as strategic as possible. Different physical parameters were

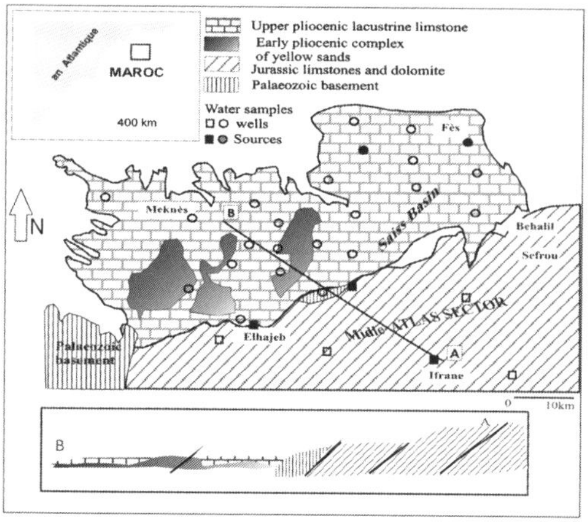

Figure 1. Simplified geological map of the study area.

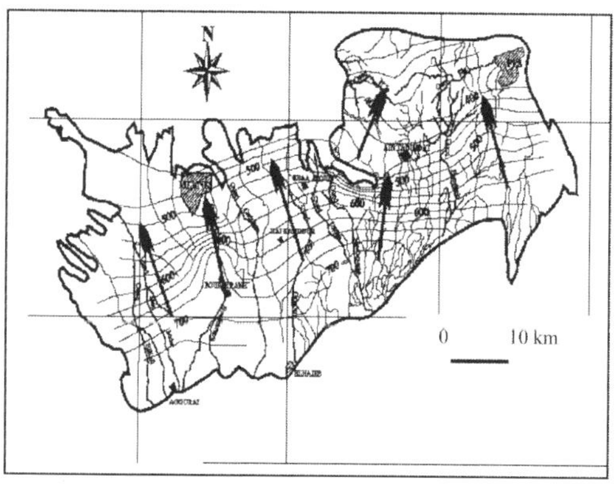

Figure 2. Piezometric map of the Saiss aquifer. State of 2000 (Essahlaoui 2000).

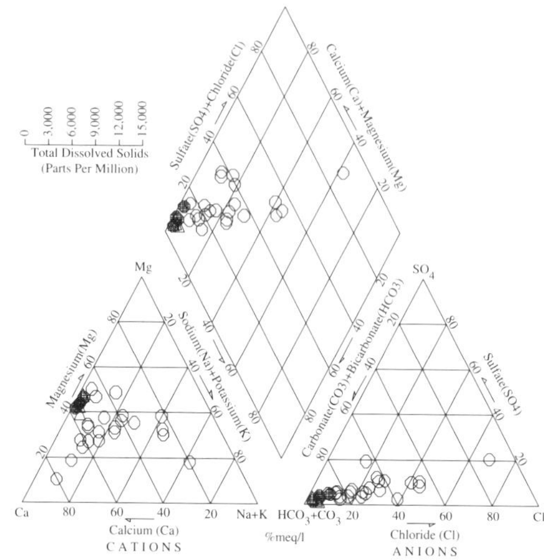

Figure 3. Piper diagram showing the different water families. Black symbols: Middle Atlas sector, white circles: Saiss basin.

measured during the sampling: pH, electrical conductivity and the redox potential.

The anions Cl, SO_4, and NO_3 were measured by ionic chromatography (HPLC); analysis of Na, Ca, Mg, and Si were made by ICP-OES; K was analysed by AAS; heavy element concentrations were analysed by ICP-MS. The quality of the analyses was monitored by analysis of international standard samples (SRM 1643c) of the National Institute of Standards and Technology (USA).

3.2 *Water families*

Waters from the Middle Atlas region appear to be very homogenous. They are dominated by alkaline earths and bicarbonate. The Ca/Mg ratio equals unity and shows that waters are hosted dominantly by dolomites or this rock dissolves easier than the limestone. HCO_3 represents up to 60% of the total dissolved solids.

During the migration of water to subsurface from the Middle Atlas region to Saiss basin, it interacts with different lithologies. Thus, alkaline earth waters ($Ca/Mg = 0.9$) were identified in the southeastern side of the basin. In the center of the basin, a family of mixed waters (HCO_3-Cl-Ca, $Ca/Mg = 0.5$) is identified. Finally, Na-Cl and SO_4 waters circulate in the northwestern part of the basin. In the basin HCO_3, Cl, Ca, and Na are the most abundant elements and contribute largely to the water salinity.

3.3 *Water–rock interaction aspects*

Beginning with the fact that all these waters are meteoric and the aquifer is made by carbonates, aluminosilicates, and silicates, and with HCO_3 being the dominant anion, mineral dissolution depends on their different reaction rates.

To determine the natural dissolution-precipitation processes, some water samples were used for geochemical calculations using PHREEQCI (v. 2.4.2). Model calculations indicate that the dissolution of carbonate minerals is essential, whereas that of the silicates is less important. Waters are supersaturated with respect to calcite and dolomite (Fig. 4). The weathering of carbonates and silicates is

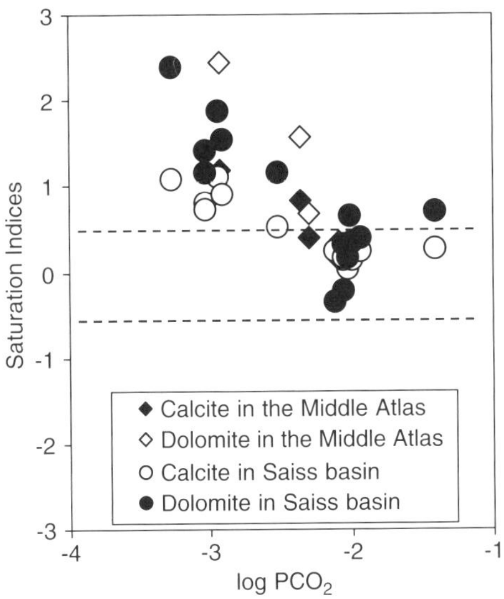

Figure 4. Plot of saturation index of water relative to calcite and dolomite in the studied waters versus log PCO_2.

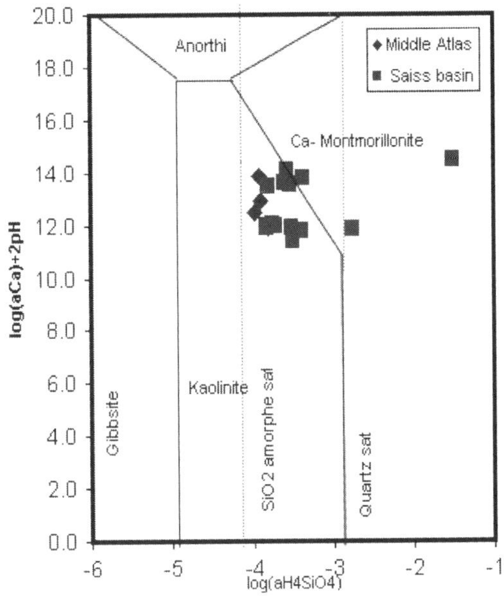

Figure 6. Activity diagram ($\log(aH_4SiO_4)$ versus $\log(aCa^{2+})+2pH$) in the system Na_2O-Al_2O_3-SiO_2-H_2O.

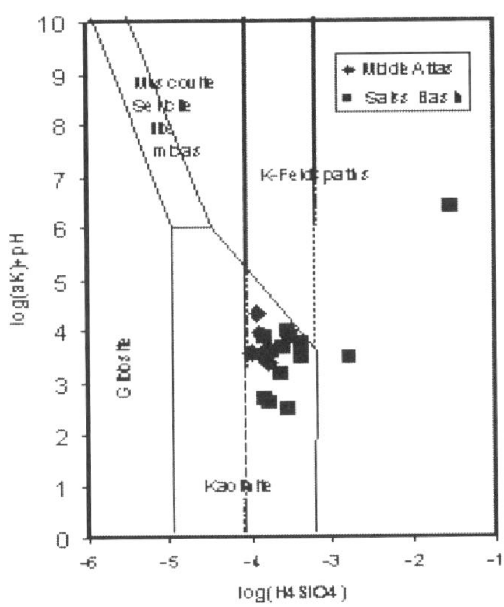

Figure 5. Activity diagram ($\log(aH_4SiO_4)$ versus $\log(aK^+)+pH$) in the system K_2O-Al_2O_3-SiO_2-H_2O (Drever 1982).

accompanied by precipitation of secondary minerals, which are generally clays. Activity diagrams in Figures 5 & 6 show that in the study area, kaolinite is likely to precipitate as secondary mineral (Mathess 1982, Appelo 1993).

4 WATER QUALITY

Natural sources of water contamination were not found in the two study areas. All the waters are fresh, near neutral, and have low electrical conductivity and total salinity (TDS). However, water contamination from anthropogenic sources was identified. In fact, the intense agricultural activity has caused increases in the levels of nitrates (up to 180 ppm) in the Saiss basin. The waters from the Middle Atlas have good quality. Nevertheless the intense pastoral and tourist activities have caused some wells to show contamination of nitrates up to 90 mg/L (Fig. 7).

In the Middle Atlas the concentration of heavy metals and trace elements is usually below the limit detection of the ICP-MS. In the vicinity and downstream of Fès (also downstream of the aquifer), the waters show some abnormal levels of Ni, Cr, and Hg, which indicate water contamination from the Sebou river which drains untreated urban and industrial waste waters (Fig. 8).

5 CONCLUSIONS

The Middle Atlas region has important fresh groundwater reserves of a good quality, because of their meteoric origins and the simple lithological nature of the aquifer. However, anthropogenic activity has

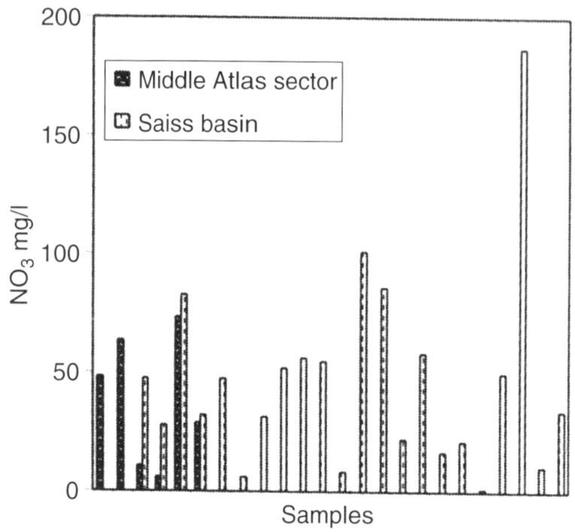

Figure 7. Plot of nitrate versus TDS.

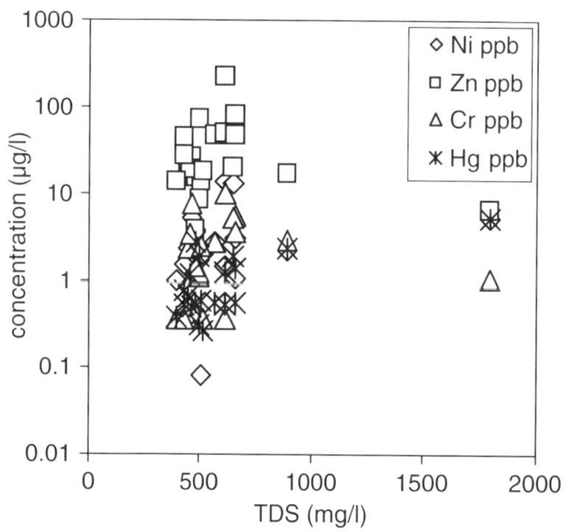

Figure 8. Plot of some heavy metals and trace elements versus the TDS.

begun to contaminate the area. It must be protected soon. The Saiss basin has very important water resources, but hosts significant agricultural and industrial development, which causes the inhabitants to use the accessible phreatic levels of the water resource for drinking and irrigation purposes. The waters in the area have an acceptable quality. However, downstream of Fès, water can be a cause of serious health problems.

REFERENCES

Ahmamou, M. 1987. *Etude sédimentologique des calcaires lacustres saissiens (Pilo-Quaternaires) du bassin de Fès-Meknès.* Unpub. Thesis. Aix-Marseilles. FST St Jérome.

Appelo, C.A.J. & Postma, D. 1993. *Geochemistry, Groundwater and pollution.* Rotterdam: Balkema.

Bahaj, T. 2002. *Hydrogéochimie et qualité des eaux des nappes phréatiques du Causse Moyen Atlasique, du Bassin de Saiss, de la Plaine du Gharb, de la plaine de la Maâmora et de la région de Zemmour.* Unpub. Ph.D. Thesis. Rabat: Faculté des Sciences Rabat.

Benaabidate, L. 2000. *Caractérisation du bassin versant de Sebou: Hydrogéologie, qualité de l'eau et géochimie des eaux thermales.* Unpub. Thesis. Fès: Fac. Sci. Et techniques Fès- Saiss.

Drever J. 1982. *The Geochemistry of Natural Waters.* New Jersey: Prentice-Hall. Inc.

Essahlaoui, A. 2000. *Contribution a la reconnaissance des formations aquifères dans le bassin de Meknès-Fès: Prospection géoélectrique, étude hydrogéologique et inventaire des ressources en eau.* Unpub. Thesis. Ecole Mohammdia des ingénieurs.

Gourari, L. 2001. *Etude Hydrochimique, morphologique, lithostratigraphique, sédimentologique et pétrographique des dépots travetino-détritiques actuels et plio-quaternaires du bassin karstique de l'Oued Aggai (Causse de Sefrou, Moyen-Atlas, Maroc).* Unpub. Ph.D. Thesis. Fès: Univ. Med Ben Abdellah.

Martin, J. 1981. Etude géomorphologique du Moyen Atlas. *Notes et mém. Sérv. Géol.* 258.

Mathess, G. 1982. *The Properties of Groundwater.* New York: J. Wiley & Sons.

Water–Rock Interaction, Wanty & Seal II (eds)

Processes affecting groundwater quality in a basalt aquifer system in southern Australia

D.A. Bennetts & J.A. Webb
Department of Earth Sciences, La Trobe University, Bundoora, Victoria, Australia

ABSTRACT: ^{36}Cl and major element chemistry indicate that solutes from a basalt aquifer system located near Hamilton, southern Australia, are predominantly of cyclic origin. ^{36}Cl signatures reflect that of local precipitation, showing no evidence of Cl^- derivation from connate sources or mineral weathering. Mass balance modeling shows that in the freshest groundwaters (<300 mg/L Cl^-), up to 55% of all solutes are derived from the weathering of silicate minerals. These low salinity waters are recharged rapidly through eruption points and rocky outcrops, allowing only small amounts of evapotranspiration to occur. Higher salinity groundwaters (>1000 mg/L Cl^-) are recharged through thick clay soils, where significant evapotranspiration takes place. These groundwaters show no evidence of mineral weathering, the effects of which are masked by clay mineral reactions and evapotranspiration. This saline water is progressively added to the groundwater system downflow, causing an increase in salinity along flowpath.

1 INTRODUCTION

Poor quality, saline surface and groundwater commonly occur across the Plio-Pleistocene Newer Volcanics Province of southwestern Victoria, Australia, threatening agricultural land, infrastructure and water resources. This study investigates the controls on groundwater quality in the province, in particular in a fractured basalt aquifer system near Hamilton, using hydrogeological, hydrochemical and isotopic methods.

1.1 *Geological and hydrogeological framework*

The Newer Volcanics Province is dominated by basalt lavas, which within the study area are tholeiitic to transitional in composition (Price et al. 1997), and typically consist of 50–60% plagioclase ($Na_{0.49}Ca_{0.50}Al_{1.34}Si_{2.62}O_8$), 20–25% clinopyroxene (augite; $Ca_{0.85}Mg_{0.82}Al_{0.18}Fe_{0.25}Si_{1.89}O_6$), and 15–20% olivine ($Mg_{1.6}Fe_{0.4}SiO_4$), as determined using electron microprobe analysis during the present study. Accessory minerals include oxides such as ilmenite ($Fe_{0.92}Ti_{0.32}O_{3.12}$). The basalts are typically 30–40 m in total thickness, although at eruption points they may be 140 m thick (Mannet al. 1992), and are underlain by Tertiary marine sediments throughout much of the study area.

There were three major phases of volcanicity, at ~4 Ma, ~2 Ma and <0.3 Ma (Bennetts et al. 2003). Each phase forms an individual aquifer separated by palaeo-weathering profiles, which behave as semi-permeable aquitards. The 4 Ma basalts form a regional aquifer up to 50 km in length beneath most of the study area (Fig. 1), and are veneered by a clay-rich weathering profile up to 15 m thick, except on eruption points where rocky outcrop is common. The 2 Ma basalts confine the 4 Ma basalts in the east of the study area, and have a flow length of up to 25 km toward the west and at least 40 km to the south (Fig. 1); these basalts have thinner clay-rich soil profiles (up to 5 m thick), with rocky outcrop being abundant on eruption points. The surface of the recent (<0.3 Ma) basalts is dominated by rocky outcrop with little soil development; these basalts form a local aquifer within the lava aprons associated with Mounts Napier and Rouse.

Eruption points form groundwater divides, and groundwaters flow from these areas in several directions (Fig. 1). Near Hamilton, groundwaters flow towards a central plain, which is occupied by a disrupted drainage network and several salt lakes.

1.2 *Methods*

Major element analyses of 31 groundwater samples extracted from government bores during this study were combined with 164 incomplete analyses (missing Si data) contained within the Victorian State Groundwater Database. Five Groundwater samples

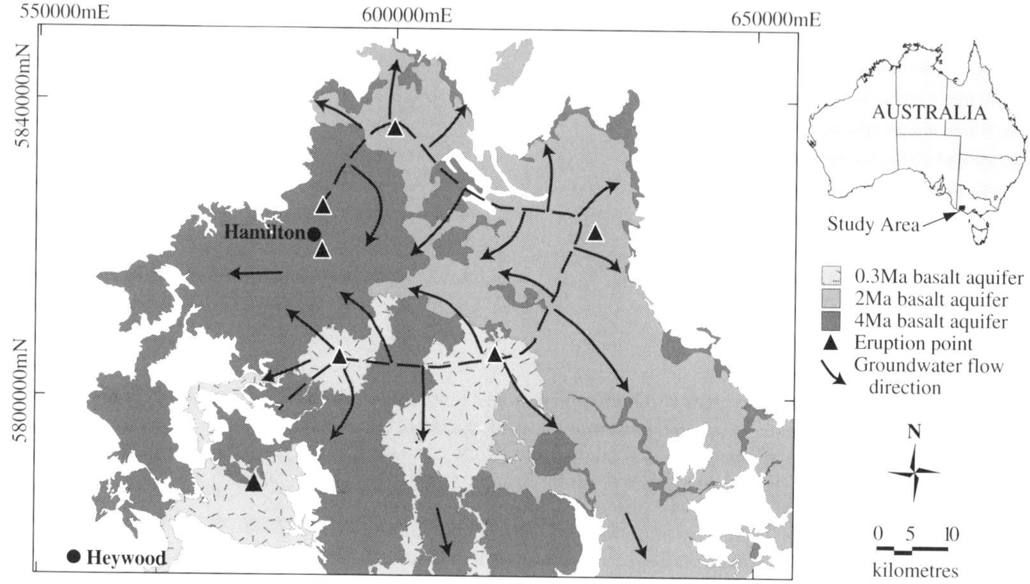

Figure 1. Aquifer distribution and groundwater flow directions. Non-basaltic rocks are shown in white.

obtained during this study were analyzed for their $^{36}Cl/Cl^-$ ratio.

2 RESULTS AND DISCUSSION

2.1 *Source of solutes*

Groundwaters within the basalts are Na^+ and Cl^- dominated, accounting for ~70% of the total dissolved solids (by weight). ^{36}Cl analysis of 5 representative groundwaters gave a median of 19×10^{-15} $^{36}Cl/Cl^-$, consistent with atmospheric precipitation in southwest Victoria (Davieet al. 1989). Thus the groundwater Cl^- is probably sourced from cyclic sources (rainfall and/or windblown dust); contributions from connate water and/or basalt weathering are unlikely. Given the 301,000 year half-life of ^{36}Cl, the $^{36}Cl/Cl$ ratio of connate water in the underlying Tertiary marine sediments would have decayed to zero, or approached secular equilibrium with the aquifer sediments (~4×10^{-15} for sandstone aquifers; Bentley et al. 1986). In any case, carbon-14 and hydraulic estimates of groundwater residence time are much younger (in the order of thousands of years) than the Tertiary sediments (>4 Ma), indicating that any connate water would have long been removed. The $^{36}Cl/Cl^-$ ratio of the 2 Ma and 4 Ma basalts would have approached secular equilibrium, likely to be close to zero given the low Th and U content of basalts in the area (Price et al. 1997). In the more recent basalts, the $^{36}Cl/Cl^-$ ratio could potentially be higher, but the groundwater $^{36}Cl/Cl^-$ ratios in both older and younger basalts are almost the same.

Most solutes in the groundwaters (Na^+, Ca^{2+}, Mg^{2+}, K^+, HCO_3^-, H_4SiO_4, SO_4^{2-}) are also of cyclic origin, with some additions from mineral weathering (see discussion below). Since Cl^- behaves conservatively, it is used here as a tracer to determine the relative influence of mineral weathering and evapotranspiration on the groundwater composition.

2.2 *Influence of silicate weathering*

Three classes of water can be identified within the basalt aquifer based on their Cl^- composition: <300 mg/L, 300–1000 mg/L and >1000 mg/L. Waters with <300 mg/L Cl^- show strong evidence of silicate mineral weathering, exhibiting ratios of Na^+/Cl^-, Ca^{2+}/Cl^-, Mg^{2+}/Cl^-, H_4SiO_4/Cl^- and HCO_3^-/Cl^- well above those in local rainfall (Hutton & Leslie 1958). The pH of recharge rainfall is sufficiently low (~5.6; Drever 1997) to attack silicate minerals; as the acidity is neutralized, the waters become less aggressive and mineral weathering all but ceases, as is evident in the ion ratios of higher Cl^- waters (see later discussion).

Following the original method of Garrels & MacKenzie (1967), it is possible to calculate the contributions of weathering reactions to the groundwater composition (Table 1). The composition of the freshest groundwater sample (bore 111527) was used, so as to exclude the influence of clay mineral reactions (see later discussion). The Cl^- content in local

Table 1. Silicate weathering mass balance calculations (na – not analysed).

	Na^+ $M \times 10^{-5}$	Ca^{2+} $M \times 10^{-5}$	Mg^{2+} $M \times 10^{-5}$	K^+ $M \times 10^{-5}$	H_4SiO_4 $M \times 10^{-5}$	Cl^- $M \times 10^{-5}$	SO_4^{2-} $M \times 10^{-5}$	HCO_3^- $M \times 10^{-5}$
Composition of freshest groundwater sample	169.6	31.9	50.9	11.8	59.3	169.3	5.2	181.5
Standardized composition of local rainfall	162.8	10.1	20.1	4.9	na	169.3	na	33.3
Total contribution from silicate weathering	6.8	21.8	30.8	6.9	59.3	0.0	0.0	148.2
Weathering of 13.9 plagioclase to 9.3 kaolinite	6.8	7.0	0.0	0.0	17.9	0.0	0.0	21.0
Weathering of 17.4 augite to 1.6 kaolinite	0.0	14.8	14.3	0.0	29.8	0.0	0.0	83.7
Dissolution of 10.3 olivine	0.0	0.0	16.5	0.0	10.3	0.0	0.0	41.4
Deficit	0.0	0.0	0.0	6.9	1.3	0.0	5.2	2.1

rainfall (8.85 mg/L, 18-month average at Heywood; Fig. 1; Hutton and Leslie 1958) was normalized to that of this groundwater, and the resultant composition subtracted from the groundwater analysis to give the ions derived from silicate weathering. Alteration of plagioclase and augite to kaolinite and dissolution of olivine can account for all ions (Table 1) except for K^+ and a small amount of silica probably from volcanic glass (Eggleton et al. 1984), and SO_4^{2-}. Hutton and Leslie (1958) did not analyze the local rainfall for SO_4^{2-}, but rainfall samples acquired during this study and from Mt. Gambier, 95 km to the west (Blackburn & Mcleod 1983), contain 0.1–3.2 mg/L SO_4^{2-}, accounting for the apparent discrepancy. The weathering reactions also cause a rise in pH, from that of rainwater (~5.6; Drever 1997), to 6.2–8.36.

Augite weathering dominates water-rock interaction, accounting for 54% (by molarity) of the total dissolved ions added by silicate weathering; plagioclase and olivine supply 20% and 26% respectively. Augite supplies most of the Ca^{2+}, H_4SiO_4 and HCO_3^-, whereas plagioclase contributes all additions of Na^+ and olivine adds most of the Mg^{2+}. The ratio in which plagioclase, augite and olivine contribute to the water composition is 1:1.25:0.74 (Table 1), whereas the ratio in which they occur in the basalt is 1:0.3–0.5: 0.25–0.4. This indicates that on average augite is weathering three times as fast as plagioclase, whereas olivine is weathering only twice as fast as plagioclase. Augite typically weathers more rapidly than plagioclase (Eggleton et al. 1984), but it is interesting to note that augite is weathering more rapidly than olivine in these basalts.

In the freshest groundwater sample, mineral weathering accounts for 40% (by molarity; 55% by weight) of the total solute content (Table 1). In more saline groundwaters the influence of mineral weathering is greatly reduced by clay mineral reactions and by concentration due to evapotranspiration (see below), and contributes little to the overall water composition.

2.3 *Influence of clay mineral reactions*

Waters with >1000 mg/L Cl^- show little evidence of silicate weathering, with ion/Cl ratios consistently below those of local rainfall (Hutton & Leslie 1958). Comparing the median compositions of groundwaters with <300 mg/L and >1000 mg/L Cl, the former standardized to the Cl^- concentration of the latter, shows losses of Na^+, Ca^{2+}, Mg^{2+}, K^+, H_4SiO_4 and HCO_3^- with increasing salinity. Waters with <300 mg/L Cl^- plot in the kaolinite field on clay mineral stability diagrams (Fig. 2), whereas those with >1000 mg/L Cl^- plot in the smectite field. Thus the losses in many ions with increasing Cl^- concentration are probably due to the conversion of kaolinite to smectite. This reaction occurs in aquifers in the nearby Otway Basin (Blake 1989) and also in the Great Artesian Basin in central Australia (Herczeg et al. 1991), and consumes Ca^{2+}, Mg^{2+}, Na^+ and H_4SiO_4, matching the observed changes with increasing Cl^-. The reaction also produces acid, causing a progressive decline in the HCO_3^-/Cl^- ratios. K^+ is probably lost by conversion from kaolinite to illite; similar studies elsewhere in Australia have identified illite formation as a major sink for K^+ (e.g. McArthur et al. 1989).

Ca^{2+} and HCO_3^- concentrations are also influenced by calcite precipitation. Calcite saturation indices for groundwater with >300 mg/L Cl^- are typically >0.5, the level at which precipitation usually commences (Herman & Lorah 1987). Calcite is commonly found coating fractures and vesicles within the basalts (Mannet al. 1992).

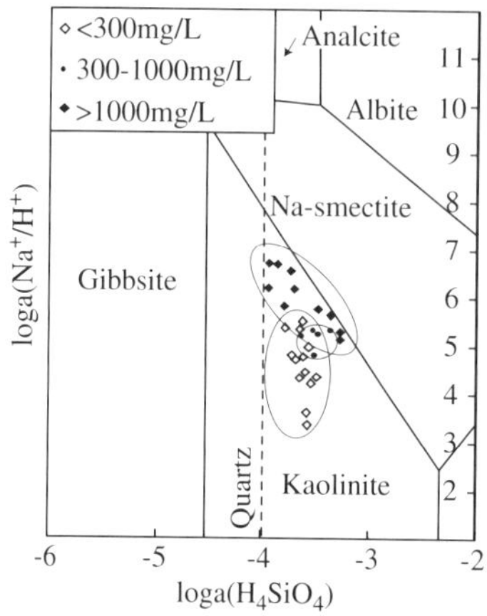

Figure 2. Stability diagram for minerals in Na_2O-Al_2O_3-SiO_2-H_2O system (after Drever 1997). Samples are grouped according to Cl^- content.

2.4 *Influence of evapotranspiration*

Evapotranspiration is the overriding influence on the quality of groundwaters in the Hamilton basalts. In even the fresher samples with <300 mg/L Cl^-, <55% of the total solute content is due to mineral weathering. The less saline groundwaters are located in close proximity to eruption points and within <0.3 Ma basalts, where soils are thin and rocky outcrop is common. In these areas recharge is rapid and residence times within the unsaturated zone are short, allowing relatively little time for evapotranspiration to occur. However, the lowest Cl^- concentration in these waters (47 mg/L) still represents concentration by a factor of 5.3 relative to local rainfall (8.85 mg/L; Hutton & Leslie 1958). Waters in these areas are also characterized by high Eh values (up to 168 mV) and moderate to high concentrations of nitrate (up to 23 mg/L NO_3^- as N), probably from nitrate-bearing fertilizers.

Waters recharged through thick clay soils across the older basalts away from the eruption points are typically much more saline, owing to their greater residence time within the unsaturated zone. These waters are also characterised by low to intermediate Eh values (as low as −226 mV) and trace amounts of nitrate, presumably due to denitrification within the soil zone.

2.5 *Change in water quality along flow path*

There is a clear trend of increasing salinity and Cl^- content as waters flow away from recharge areas (eruption points) within the 2 Ma and 4 Ma basalts (Bennetts et al. 2003). Since there are no sources of Cl^- within the aquifer, this must be due to the influence of evapotranspiration. Fresh waters enter the ground in rocky areas (eruption points and <0.3 Ma basalts). Additional recharge is progressively added along flowpaths, recharged through thick clay soils. This progressive leakage continually adds saline water, making the groundwater increase in salinity downflow. For example, waters flowing from a prominent 4 Ma eruption point (Mt. Bainbridge) triple in salinity along a 2–5 km flowpath.

3 CONCLUSIONS

Groundwater quality is dominantly influenced by evapotranspiration, the degree of influence being dependant on the rate of recharge. Mineral weathering plays a significant role but is only evident in the freshest waters; clay reactions remove its influence as salinity increases. This information may be used to delineate flowpaths and determine the origin of recharge waters.

ACKOWLEDGEMENTS

We thank Dr Jake Turin for assistance during ^{36}Cl extraction and Mr. Ian McCabe and Mr. Danny McDonald for assistance with major element analysis. The project was supported by the Glenelg-Hopkins Catchment Management Authority.

REFERENCES

Bennetts, D.A., Webb, J.A. & Gray, C.M. 2003. Distribution of Plio-Pleistocene basalts and regolith around Hamilton, western Victoria, and their relationship to groundwater recharge and discharge. In: I.C. Roach (ed.) *Advances in Regolith*: 11–15: CRC LEME.

Bentley, H.W., Phillips, F.M., Davis, S.N., Habermehl, M.A., Airey, P.L., Calf, G.E., Elmore, D., Gove, H.E. & Torgersen, T. 1986. Chlorine-36 dating of very old groundwater; 1, The Great Artesian Basin, Australia. *Water Resources Research* 22: 1991–2001.

Blackburn, G. & Mcleod, S. 1983. Salinity of atmospheric precipitation in the Murray-Darling drainage division, Australia. *Austral. J. Soil Res.* 21: 411–434.

Blake, R. 1989. The origin of high sodium bicarbonate waters in the Otway Basin, Victoria, Australia. In: D.L. Miles (ed.) *Proc. International Symposium on Water-Rock Interaction 6*, pp. 83–85: Balkema: Rotterdam.

Davie, R.F., Kellett, J.R., Fifield, L.K., Evans, W.R., Calf, G.E., Bird, J.R., Topham, S. & Ophel, T.R. 1989. Chlorine-36 measurements in the Murray Basin; preliminary results from the Victorian and South Australian Mallee region. *BMR J. Austral. Geol. Geophys.* 11: 261–272.

Drever, J.I. 1997. *The geochemistry of natural waters: surface and groundwater environments*. Upper Saddle River, N.J.: Prentice Hall.
Eggleton, R.A., Smith, K.L. & Binns, R.A. 1984. Basalt weathering in eastern Australia. *Geol. Soc. Austral. Abstracts* 12: 159.
Garrels, R.M. & Mackenzie, F.T. 1967. Origin of the chemical compositions of some springs and lakes. In: R.F. Gould (ed.) Equilibrium Concepts in Natural Water Systems, Washington D.C.: American Chemical Society.
Herczeg, A.L., Torgersen, T., Chivas, A.R. & Habermehl, M.A. 1991. Geochemistry of ground waters from the Great Artesian Basin, Australia. *J. Hydrol.* 126: 225–245.
Herman, J.S. & Lorah, M.M. 1987. CO2 outgassing and calcite precipitation in Falling Spring Creek, Virginia, U.S.A. *Chem. Geol.* 62: 251–262.
Hutton, J.T. & Leslie, T.I. 1958. Accession of non-nitrogenous ions dissolved in rainwater to soils in Victoria. *Austral. J. Agricultural Res.* 9: 59–84.
Mann, B.S., Stanley, D.R. & Bolger, P.F. 1992. Basalt plains hydrogeological salinity investigation. Progress Report no. 2 – Hamilton-Dunkeld Sub-Region. Rural Water Commission Report, 1992/1.
McArthur, J.M., Turner, J., Lyons, W.B. & Thirlwall, M.F. 1989. Salt sources and water-rock interaction on the Yilgarn Block, Australia; isotopic and major element tracers. *Appl. Geochem.* 4: 79–92.
Price, R.C., Gray, C.M. & Frey, F.A. 1997. Strontium isotopic and trace element heterogeneity in the plains basalts of the Newer Volcanic Province, Victoria, Australia. *Geochim. Cosmochim. Acta* 61: 171–192.

Water–Rock Interaction, Wanty & Seal II (eds)

Rapid growth of meter-scale calcite speleothems in the Mission Tunnel, Santa Barbara, CA

J.R. Boles
Department of Geological Sciences, University of California, Santa Barbara, California, USA

ABSTRACT: Groundwater seeping from the unlined arkosic sandstone sections in the Mission Tunnel, completed in 1912, has resulted in large calcite speleothems including stalactites and flowstone (1.5 m), and thick deposits on wooden beams and cables. Oxygen and carbon isotope values for the calcite speleothems range from −4.5 to −20 and −4 to −33‰ (PDB), respectively and they show increasing values with decreasing age. The oxygen isotopic composition in the most recent calcite is about 2 per mil higher than expected for isotopic equilibrium with the present 26°C seep water. Oxygen and carbon isotopes show a systematic covariation trend with a slope of $\delta^{13}C/\delta^{18}O = 1.7$, which is intermediate to that calculated for the model of Bigeleisen & Wolfsberg (1958) and Slater (1953) for kinetic non-equilibrium fractionation. The calcite forms during rapid degassing of CO_2 charged ground water entering the tunnel.

1 INTRODUCTION

Carbonate precipitation in man-made tunnels is of interest because of their well documented short temporal history of fluid chemistry and physical conditions in the tunnel, which is usually much better constrained than in natural caves (e.g. Genty 1992). In such environments it may be easier to identify calcite precipitation processes and the extent to which equilibrium is achieved between the speleothem chemistry and the seep waters than in cave environments. In addition, sampling in man-made tunnels is usually not constrained by conservation issues, as is the case in natural caves.

In this paper I describe some very large calcite speleothems in the Mission Tunnel and discuss their origin based on oxygen and carbon isotopic results (PDB scale). These results have an important bearing on the interpretation of the isotopic geochemistry of speleothems, and how such a system might evolve. The main conclusion here is that these speleothems have formed rapidly and their isotopic composition is an excellent example of kinetic non-equilibrium effects. They are not useful for estimating paeloclimatic conditions.

2 THE MISSION TUNNEL

The Mission Tunnel was dug between 1906 and 1911 through the Santa Ynez Mountains for the purpose of bringing water to Santa Barbara from Gibralter Reservoir on the Santa Ynez River, Santa Barbara County, California. The tunnel is 5974 m long, essentially horizontal, and approximately 2 m high by a meter and a half in width. The tunnel at its deepest point is 716 m beneath the mountain crest. The bed rock is steeply dipping arkosic sandstone and shales ranging from Cretaceous to Tertiary age. The ends of the tunnel are closed with a grated door through out the year except for the bi-annual inspection. The reservoir water typically flows throughout the year in the tunnel and depth of flow is less than 30 cm. The total fracture seepage amounts to about 1000ACF per year (Rademacher et al. 2002). Even in the driest periods, the tunnel walls are moist due to water seeping from fractures, indicating that humidity is always high and there are no periods of significant evaporation. Continuous flowing reservoir water along the bottom of the tunnel also contributes to the humid environment.

The seep water is a mixture of pre-atomic bomb groundwater (i.e. 1960's)and more modern meteoric water. The spectral record of seep rate compared with a 30 year-old rainfall record shows that on average the seepage lags precipitation by 3 months, which demonstrates that flow path is unsaturated and that rainy periods saturate the system and force old trapped water into the tunnel (Rademacher et al. 2002).

3 SPELEOTHEMS

3.1 *Morphology*

The speleothems in the Mission Tunnel have grown from seepage out of fractured arkosic sandstone. The speleothems are remarkably large and well developed considering their age. They exhibit a variety of shapes including stalactites hanging from the ceiling, flowstone occurring at major seeps from fractures on the sides of the tunnel, thick coatings on beams, cables and other anthropogenic objects, and cave bacon or drapes extending from fracture controlled seeps. These features exceed a meter in size at several places in the tunnel and commonly are on a scale of tens of centimeters (Fig. 1). Also occurring, but volumetrically less important, are thin (2–10 mm) straw stalactites with lengths up to 30 mm. The straw stalactites occur in both the unlined sandstone and shale intervals. It is the only type of speleothem found in the shale sections. A few stalagmites can also be seen, the biggest of which is about 80 mm across by 130 mm high. Stalagmites may be relatively rare due to the lack of a stable substrate on the bottom of the tunnel. Cut samples of stalagmites and cored flowstone show that the most recent growth is commonly finer grain and less porous than the early growth period.

3.2 *Isotopic composition*

The oxygen and carbon isotopic values of the calcite speleothems (PDB scale) exhibit a striking covariation, which is not systematic with distance into the tunnel. All of the data taken together form a trend line of $Y = 1.67X + 2.17$, where X = the oxygen and Y = the carbon isotopic value (Fig. 2). The R^2 value of the data is 0.85. Some speleothem types, such as straw stalactites have a $\delta^{13}C/\delta^{18}O$ slope of 1.8 with $R^2 = 0.95$. Considering that the speleothems have formed within the past 100 years in a system where water isotopic composition and temperature has changed very little, this trend and the magnitude of the variation is unexpected if the speleothems are forming by inorganic equilibrium crystallization.

Some of the straw stalactites have much lower isotope values than other speleothem types, in some cases in the −10 to −20‰ range for carbon and oxygen. One sample had an extremely low carbon and oxygen value of −32.96 and −19.76‰, respectively.

Sequential samples in the direction of growth of speleothems typically have carbon and oxygen trends which become increasingly heavier up to the youngest (modern) surface. The flowstone samples exhibit the clearest systematic change with growth direction, although stalactites and rare stalagmites also show the same general trend. Both the stalactites and the stalagmites show a more systematic decrease in the carbon value than the oxygen value.

3.3 *Evidence for disequilibrium calcite oxygen isotopic composition*

The youngest calcite appears to be out of isotopic equilibrium with the present day temperature and water composition. The oxygen isotopic composition of the present seep water ranges from about −6.9 to −7.6‰ (SMOW) for distances into the tunnel from between 248 to 4000 m (Rademacher et al. 2002). These values are mostly averages of three values taken at different years, either in the spring or fall. As shown by Rademacher et al. (2002), these waters fall close to the meteoric water line indicating they are largely of meteoric origin. There is no systematic isotopic composition changes with distance into the tunnel although some values towards the center of the tunnel tend to be several per mil higher than the outer parts. Temperature of the modern seep water ranges from 19 to 28°C and there is a trend of increasing

Fig. 1. Meter-scale calcite flowstone enclosing power cable, Mission Tunnel.

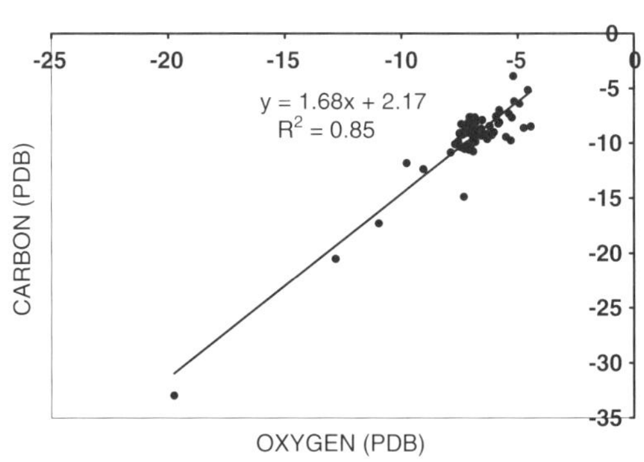

Fig. 2. Oxygen and carbon isotopic composition (PDB) of Mission Tunnel speleothems.

temperature with distance to the surface (or towards the tunnel center).

A comparison between the measured oxygen isotopic values *of the latest calcite in the growth trends* (i.e. outer most calcite on speleothem) and the predicted calcite isotopic composition for the water composition and temperature at that locality, indicates that, except for one sample, the speleothem calcite has a significantly heavier isotopic composition than predicted. For example at 3139 m distance into the tunnel, the latest calcite on the flowstone is −6.6‰ whereas the predicted value at this locality is −9‰. Similarly, the stalagmite at 2874 m has an oxygen value of −6.8‰ whereas the predicted value is approximately −9.5‰. Overall, all of the most recent speleothem calcite layers are about 2‰ heavier than expected for inorganic equilibrium crystallization.

3.4 *Evidence for kinetic isotopic fractionation effects*

Fantidis & Ehhalt (1970) and Hendy (1971) discuss the effect of *rapid* CO_2 loss and the resultant kinetic fractionation between HCO_3^- and CO_2. Calcite precipitation in such cases will show simultaneous enrichment of carbon and oxygen isotopes. Criteria that demonstrate kinetic isotopic fractionation effects usually are based on samples from a single lamina of a speleothem in order to ensure that temperature and parent water composition are constant at the time of calcite crystallization (Fantidis & Ehhalt 1970; Hendy 1971; Schwarcz 1986). If isotopic covariation is found in such a case, an argument can be made for kinetic disequilibrium.

In the case of the Mission Tunnel, we know that seep temperature and water composition, within the short hundred year growth time of the speleothems, is relatively constant, yet the speleothems show a marked covariation of carbon and oxygen and a relatively wide range of carbon (25‰ variation) and oxygen (15‰ variation) isotope values. Therefore the suite of values from the tunnel is interpreted as evidence of kinetic isotopic fractionation, even though the samples are not from individual laminae. The fact that the oxygen isotopic values of the most recent speleothem surfaces are higher than predicted for equilibrium crystallization from the modern seep water is consistent with a kinetic non-equilibrium interpretation. Considering the relatively rapid growth rates of the Mission Tunnel speleothems, isotopic disequilibrium is not surprising.

Hendy (1971) calculated the $\delta^{13}C/\delta^{18}O$ covariation slope for kinetic non-equilibrium fractionation from high temperature limiting isotopic reaction rates, for the reaction:

$$H^+ + HCO_3^- = CO_2 + H_2O$$

based on two competing models: 1) the Slater model (Slater 1953) and 2) the Bigeleisen & Wolfsberg model (Bigeleisen & Wolfsberg 1958). Hendy calculated a $\delta^{13}C/\delta^{18}O$ slope for the Slater model of 8.3 and for the Bigeleisen & Wolfsberg model of 0.6 and concludes that it is not possible from published experiments to determine which model is preferable. The observed covariation between oxygen and carbon speleothems for all data from the Mission Tunnel ($\delta^{13}C/\delta^{18}O$) is about 1.7–1.8 (e.g. Fig. 2), which is intermediate to that calculated for the model of Bigeleisen & Wolfsberg and Slater. In conclusion, the P_{CO_2} in the seep water must be considerably higher than in the tunnel, causing rapid loss of CO_2. Unfortunately it is not easy to measure the pre-loss P_{CO_2} in the seep water values because by the time the water exits into the tunnel it has already been exposed to the tunnel atmosphere in the open cracks that extend deep into the overlying country rock.

4 DISCUSSION

The stable isotopic data clearly indicate that the Mission Tunnel speleothems have formed as a result of rapid CO_2 degassing. What is not clear is why there is a trend of increasingly higher isotopic values and increasingly finer grain and less porous calcite in the most recent growth. The latter is interpreted as a slowing of the growth rate, although there are no hard numbers to support this interpretation. The trend of increasing oxygen and carbon isotope composition in the speleothems, from a kinetic non-equilibrium interpretation of their growth mechanism, suggests that there is an increasing rate of CO_2 loss from the seep water relative to the tunnel atmosphere because of greater partial pressure of dissolved CO_2 or decrease in the P_{CO_2} in the tunnel atmosphere. Presumably a greater amount of degassing would result in increasingly higher isotope values in the seep water. Alternatively, the isotopic composition of the dissolved CO_2 in the seep water may have been changed to higher values (change in source composition), independently of the rate of gas loss. One possible mechanism for this is that the growth of microbial communities in the tunnel has systematically consumed increasing quantities of isotopically light dissolved CO_2 which leaves increasingly heavier CO_2 for degassing.

5 CONCLUSIONS

Speleothems of the type found in the Mission Tunnel cannot be used for paleo temperature estimates although they are excellent examples of CO_2 degassing processes and evolution of a water–rock system.

They are among the best examples in the literature of the isotopic effects of rapid calcite crystallization resulting from CO_2 degassing. We are currently studying the biological processes in this system and their resultant effects on the isotopic signatures.

REFERENCES

Bigeleisen, J. & Wolfsberg, M. 1958. Theoretical and experimental aspects of isotopic effects in chemical kinetics. *Advan. Chem. Phys.* 1: 15–76.

Fantidis, J.F. & Ehhalt, D.H. 1970. Variations in the carbon and oxygen isotopic composition in stalagmites and stalactites: evidence of non-equilibrium isotopic fractionation. *Earth Plan. Sci. Letters* 10: 136–144.

Genty, D. 1992. Les Speleothems du tunnel de godarville (Belgique)-un example exceptionnel de concretionnement moderne-interet pour l'etude de la cinetique de la precipitation de la calcite et de sa relation avec les variations d'environnement. *Speleochronos* 4: 3–29.

Hendy, C.H. 1971. The isotopic geochemistry of speleothems I. The calculation of the effects of different modes of formation on the isotopic composition of and their applicability as paleoclimate indicators. *Geochim Cosmochim Acta* 35: 801–824.

Hill, C.A. & Forti, P. 1997. *Cave minerals of the world, 2nd ed.* Huntsville , AL: National Speleological Society.

Rademacher, L., Clark, J.F. & Boles, J.R. 2002. Groundwater residence times and flow paths in fractured rock determined using environmental tracers in the Mission Tunnel; Santa Barbara County, California, USA. *Environ. Geol.* 43: 557–567.

Schwarcz, H.P. 1986. Geochronology and isotopic geochemistry of speleothems. In P. Fritz & J.Ch. Fontes (eds.), *Handbook of Environmental Isotope Geochemistry* 2: 271–303.

Slater, N.B. 1953. Gaseous unimolecular reactions: theory of the effects of pressure and vibrational degeneracy. *Trans. Roy. Soc. London* A246: 57–80.

Water–Rock Interaction, Wanty & Seal II (eds)

U transport at Itataia U-deposit, Brazil

D.M. Bonotto & R.C.S. Kesselring
Instituto de Geociências e Ciências Exatas, UNESP, Rio Claro, São Paulo, Brazil

J.R.L.V. Leal
Indústrias Nucleares do Brasil S.A., INB, Fortaleza, Ceará, Brazil

ABSTRACT: This paper describes the results of a study on ground waters from the Itataia U deposit, Ceará State, Brazil. The purpose of this study was to use chemical and U-isotope data to evaluate the migration of U under pre-mining existing conditions. Values between 2 and 190 μg/l for dissolved U content were found, which allowed a comparison of the inventory element in the water phase from the U occurrence at Poços de Caldas plateau, Brazil.

1 INTRODUCTION

The Itataia U deposit is located about 170 km southwest of Fortaleza city, capital of the state of Ceará, Brazil (Fig. 1). It is situated in Precambrian metamorphic terrains of the Caicó complex lithostratigraphic unit (Mendonça et al. 1980). The main country rocks surrounding the Itataia deposit are paragneisses and large carbonate lenses that are cut by several granitic and pegmatitic apophyses. U mineralization occurs in uniform masses of collophane (collophanite), as well in collophane veinlet and stockwork ores in marbles, gneisses and episyenites (Mendonça et al. 1985). U occurs in the cryptocrystalline hydroxy-apatite, with the phosphatic rocks containing about 80% of collophane, so that the ore reserve of Itataia deposit is estimated to be 142,500 tons of U_3O_8 and about 18 million tons of P_2O_5 (Mendonça et al. 1985).

No mining activities are presently being performed at Itataia area, since most of the U-mining in Brazil is being realized at the U-district of Lagoa Real, which is located in the southern-central part of Bahia State (Carvalho et al. 2001). Determining the background conditions of U migration under pre-mining conditions in the area is necessary for planning of future activities. Furthermore, the scarcity of water resources appropriate for drinking purposes in the region indicates that much attention must be paid to the U-transfer among different phases, since the maximum allowed limits may be exceeded, even under strict natural (geogenic) water–rock interactions. This paper describes a hydrogeochemical survey of the Itataia U-deposit focusing mainly on the migration of the U isotopes ^{238}U and ^{234}U.

2 GENERAL ASPECTS OF THE STUDY AREA

The Itataia U deposit was discovered in 1975 by a carborne radiometric survey of 35,000 km^2 (Mendonça et al. 1985). It is located 45 km southeast of Santa Quitéria municipality and 15 km WSW of Itatira municipality at Ceará State. The climate of the region is the type Bsh (Köppen classification), characterized by well-defined rainy and dry seasons. The rainy period is between January and June, with maximum rainfall occurring in March–April (Fig. 2) and annual rainfall rarely exceeding 700 mm; the average annual air temperature is 27°C, whereas the wind preferential directions are SE-NW and E-W. *Caatinga* species dominate the vegetation of the area, with noncalcic plansols and cambisols being the dominant soils (Mendonça et al. 1985). The Acaraú river basin is the most important surface drainage system of the region, more specifically the Groaíras river risings that exhibit a dense and sinuous drainage pattern. The *sertão* depression and residual plateaux comprise the main geomorphological units in the Itataia area, largely occurring deeply eroded hills supported by precambrian rocks (Mendonça et al. 1985). The rocks are linearly folded regionally as a result of compressive tectonic. Wrench faults of Itatira and Groaíras river form shearing couples in which the drag folds climax are thrusting faults with axial plane dipping north. The highest grade ore is in a very fractured zone,

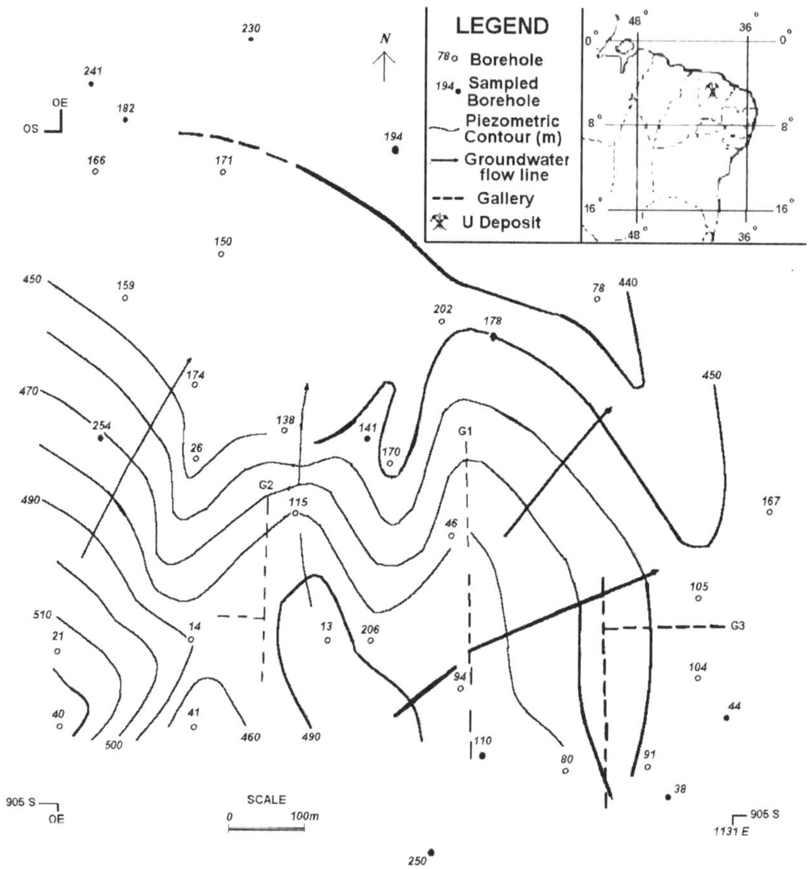

Figure 1. Location of the Itataia U-deposit at northeast Brazil, and piezometric map of the area (IPT 1981) showing the location of the sampling points for water analysis.

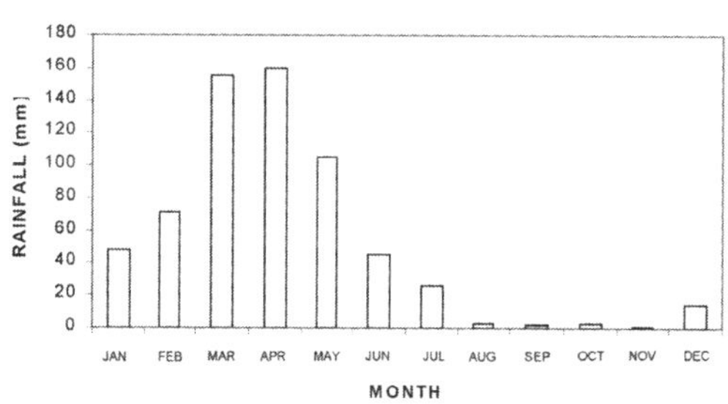

Figure 2. Average monthly rainfall (mm) in the period 1973–1993 at Itatira municipality, Ceará State. Data are from Brasil (1990) and FUNCEME.

associated with marble (Mendonça et al. 1980). The hydrogeology at the Itataia area is similar to that of the crystalline region covering most of the Brazilian Northeast formed by exposed Precambrian rocks: the groundwater is stored in cracks and faults, as well in alluvium which serves as the basis for intermittent rivers draining the region. Boreholes (BH) drilled over the prospecting activities in the area provided some information about the characteristics of the subsurface flow (Fig. 1).

3 SAMPLING AND ANALYTICAL METHODS

Two sampling campaigns to collect water were performed in 1993 during rainy (February) and dry (July) periods. Because of the scarcity of ground water in the area, it was only possible to sample the boreholes shown in Figure 1. One water sample from Quixaba dam (*açude*) and two samples from tubular wells used for water supply where no mining has occurred; however, these sampling points are not shown in Figure 1.

Some parameters (pH, Eh, conductivity, alkalinity, and temperature) were measured *in situ*. Water samples (~5 kg) were stored in polyethylene bottles

Table 1. Results of groundwater analysis for borehole samples collected in Feb 1993 at Itataia U-deposit.

Parameter	Unit	BH141	BH182	BH230	BH241
Depth	m	260.6	100.1	150	180
Water table depth	m	36.2	10.3	12.5	8.3
Temperature	°C	29	28	28	28
Conductivity	μS/cm	3300	4800	3500	3100
pH	–	6.53	6.53	6.62	6.53
Eh	mV	412	414	423	287
Sodium	mg/l	144.9	276.0	131.1	259.9
Potassium	mg/l	14.6	11.3	11.0	11.7
Calcium	mg/l	313.8	275.6	148.4	84.8
Magnesium	mg/l	86.1	78.0	97.5	64.0
Bicarbonate	mg/l	155	185	65	175
Chloride	mg/l	600	630	690	670
Sulfate	mg/l	153.7	103.1	72.7	4.3
Nitrate	mg/l	0.16	0.03	0.07	<0.10
Fluoride	mg/l	2.35	1.90	0.65	0.58
Phosphate	mg/l	0.06	0.13	0.33	0.05
Silica	mg/l	6.3	8.8	7.9	0.5
Load	mg/l	744.0	54.5	284.0	11.7
TDS	mg/l	681	287	4261	1783

and sent to LABIDRO-Isotopes and Hydrochemistry Laboratory-UNESP-Rio Claro, where they were filtered through 0.45 μm membranes, separated into different aliquots, and preserved with appropriate acids, depending on the requirements of the analyses.

Standard analytical techniques used to determine the composition of the major and minor elements in the waters include titration, colorimetry, flame photometry, and ion selective electrodes. The total dissolved solids (TDS) content was evaluated by evaporating the filtrate in a weighed flask that achieved constant weight at 180°C. The increase in flask weight represented the TDS (APHA 1989).

The U content and $^{234}U/^{238}U$ activity ratio (AR) were measured by alpha spectrometry using the radiochemical procedure described by Bonotto (1996). The α-activities were determined with Si(Au) surface barrier detectors (Osmond & Cowart 1976), and the spectra for natural U and ^{232}U tracer added were recorded on an EG&G ORTEC 919 Spectrum Master Multichannel Buffer. 0.5 g of rock/mineral samples (collophanite, marble, and calcite), as well particle solids (>0.45 μm) from some boreholes were also analyzed for U-isotopes after their digestion/dissolution with proper acids and application of the same chemical procedures utilized for the waters.

Tables 1 and 2 report all obtained data, which allowed estimate association ratios for U expressed by ml/g and defined as the ratio of the elemental concentration in suspended particles to the groundwater concentration converted into units per ml (Miekeley et al. 1991).

Table 2. Results of U-isotopes analysis for samples collected at Itataia U-deposit.

Identification	U content*	$^{234}U/^{238}U$ A.R.*
Water sample collected in February 1993		
Well 2	10.22 μg/l	1.40
Quixaba *açude*	5.22 μg/l	1.02
BH141	186.38 μg/l	1.42
BH182	117.23 μg/l	1.22
BH230	6.65 μg/l	1.14
BH241	1.72 μg/l	1.30
Water sample collected in July 1993		
Well 1	2.78 μg/l	1.46
BH44	4.64 μg/l	1.34
BH110	16.27 μg/l	1.50
BH178	1.91 μg/l	1.47
BH182	7.93 μg/l	1.02
BH194	4.12 μg/l	1.45
BH230	27.01 μg/l	1.39
BH241	8.00 μg/l	1.77
BH250	42.73 μg/l	1.10
BH254	2.49 μg/l	1.04
Suspended particles – Samples collected in Feb 1993		
BH141	21.46 μg/g	0.95
BH182	66.56 μg/g	1.12
BH230	7.80 μg/g	0.98
Rock/mineral		
Collophanite	280.50 μg/g	0.95
Marble	35.16 μg/g	0.99
Calcite	3.51 μg/g	0.98

*Analytical uncertainty 10% at 1σ standard deviation.

4 DISCUSSION AND CONCLUSIONS

Uncertainty of the water analysis is reported as twice the charge imbalance (Ball & Nordstrom 1992) and ranged from 4–34%. The temperature of the waters is practically constant (28°C), and odor is absent. TDS ranged from 287 to 4261 mg/l, typical values for waters from semi-arid region in Brazil (Oliveira 1997). The higher conductivity value was found at BH182, like dissolved sodium. All available data were plotted in an Eh-pH diagram (Fig. 3), which shows that they fall into the oxidizing field. When the chemical data are plotted on a partial Piper diagram (Fig. 4), it is evident that the waters are Cl-dominated, with non-dominance of dissolved cations.

U-accumulation in collophanite is evident from data in Table 2, as well its relatively low abundance in calcite. Quite variable dissolved U content was found at the studied area (values between 2 and 190 μg/l), and no unique U-transport pattern related to seasonal variability was verified, since it was possible to identify an increase during rainy conditions (BH182), as well dry (BH230 and BH241) conditions (Table 2). Additionally, no unique dissolved

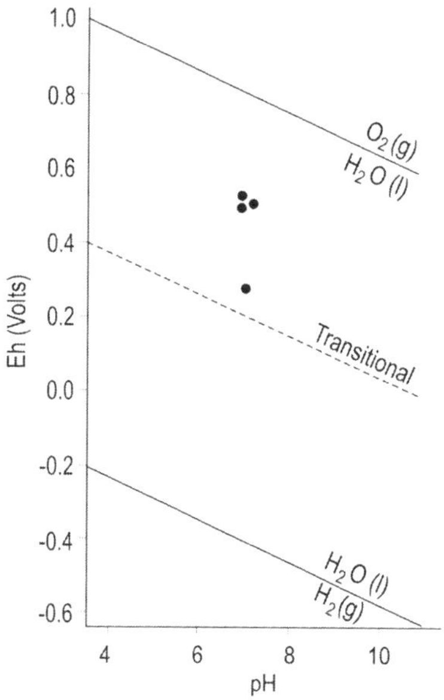

Figure 3. Data for groundwaters from the studied area plotted on an Eh–pH diagram.

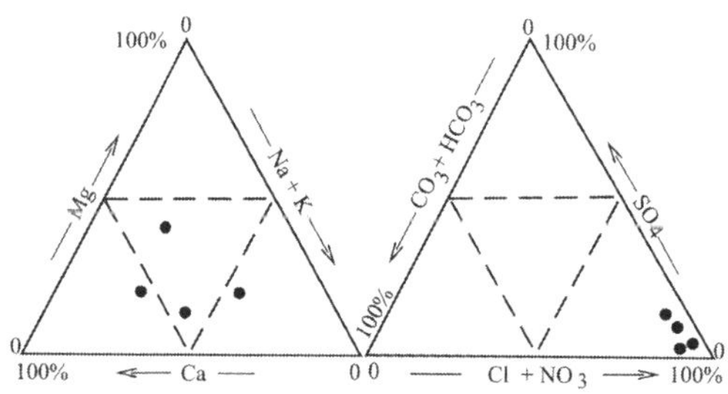

Figure 4. Chemical data for groundwaters from Tubarão aquifer plotted on a partial Piper diagram.

U content distribution pattern relative to the groundwater flow direction was observed, because it was identified a trend of increasing at north, and decreasing at south (Fig. 1). Therefore, unequal U-presence in the mineral/rock matrices, topography, different orientation of faults and joint sets certainly are, among others, factors responsible by the inhomogeneous U-distribution in the liquid phase.

Preferential removal of ^{234}U to the liquid phase occurred for most of the analyzed waters as a consequence of water–rock interactions (Rosholt et al. 1963; Kigoshi 1971; Fleischer 1975), implying on the generation of ARs >1. $^{234}U/^{238}U$ ratio in suspended solids is close to 1, within experimental errors, suggesting mobilization without fractionation of both U-isotopes (^{238}U and ^{234}U), in accordance with the AR of the rock matrices (collophanite/marble). Neither ^{238}U nor ^{234}U exceeded the allowed limit for drinkable water of 4 Bq/l established by WHO (2002), even for samples of BH141 and BH182 with high dissolved U content collected during the rainy period.

Speciation calculations using the obtained chemical data as inputs in the WATEQ4F software (Ball & Nordstrom 1992) indicate that U is transported mainly as $UO_2(HPO_4)_2^{2-}$ (BH230) and $UO_2\ (HPO_4)_2^{2-} + UO_2(CO_3)_2^{2-}$ (BH141, BH182, and BH241), in agreement with major minerals occurring in the area.

U exhibits the tendency to be associated with suspended particles (> 0.45 μm) relative to the liquid phase (<0.45 μm) in the studied area, as indicated by the association ratios between 115 and 1173 ml/g. Association ratios measured in this study are lower than those typically ranging from 10^4 to 10^5 ml/g for waters from the Osamu Utsumi U mine, Poços de Caldas plateau (Miekeley et al. 1991), which is attributed to the higher concentrations and TDS in waters from Itataia U-deposit. Therefore, the chemistry of major dissolved elements (ionic strength) in water plays an important role on the U mobilization in Itataia area.

REFERENCES

APHA (American Public Health Association) 1989. *Standard Methods for the Examination of Water and Wastewater*. Baltimore: Port City Press.

Ball, J.W. & Nordstrom, D.K. 1992. *WATEQ4F: a program for the calculating speciation of major, trace, and redox elements in natural waters*. International Ground water modeling center, Golden, Colorado.

Bonotto, D.M. 1996. *Comportamento hidrogeoquímico do ^{222}Rn e isótopos de urânio ^{238}U e ^{234}U sob condições controladas de laboratório e em sistemas naturais*. Post Ph.D. thesis: Rio Claro. (in Portuguese).

Brasil. 1990. *Dados pluviométricos mensais do nordeste-Ceará*, 1: 300–304. Recife: SUDENE.

Carvalho, I.G., Perin, H.M., Pitsch, H., Beaucaire, C., Fanfani, L., Cidu, R., Lorrai, M., Ferreyra, R.E., Lopez, L.E. & Zuddas, P. 2001. Hydrogeochemistry in the U-district of Lagoa Real, Brazil. In R. Cidu (ed.) *Proc. WRI – 10*: 1193–1196. Rotterdam: Balkema.

Fleischer, R.L. 1975. On the dissolution of respirable PuO_2 particles. *Health Phys.* 29: 69–73.

IPT (Instituto de Pesquisas Tecnológicas do Estado de São Paulo S.A.) 1981. Estudos geológico-geotécnicos para projetos de engenharia na Mina de Itataia – Município de Santa Quitéria-CE (Technical Report 15523), 1. São Paulo.

Kigoshi, K. 1971. Alpha-recoil ^{234}Th: dissolution into water and the $^{234}U/^{238}U$ disequilibrium in nature. *Science* 173: 47–48.

Mendonça, J.C.G.S., Braga, A.P.G. & Campos, M. 1980. *Proc. XXXI Congr. Brás. Geol.* (SBG ed.), 4: 2472–2486.

Mendonça, J.C.G.S., Campos, M., Braga, A.P.G., Souza, E.M., Favali, J.C. & Leal, J.R.L.V. 1985. Jazida de urânio de Itataia-Ceará. In DNPM (ed.), *Principais depósitos minerais do Brasil*, 1: 121–131. Brasília: CPRM.

Miekeley, N., Jesus, H.C., Silveira, C.L.P. & Degueldre, C. 1991. Chemical and physical characterization of suspended particles and colloids in waters from the Osamu Utsumi mine and Morro do Ferro analogue study sites, Poços de Caldas, Brazil. *Poços de Caldas Technical Report Series* 9, 78 p.

Oliveira, S.P.B. 1997. *Estudos dos processos de salinização das águas subterrâneas dos poços do cristalino do nordeste.* Ph.D. thesis: Rio Claro. (in Portuguese).

Osmond, J.K. & Cowart, J.B. 1976. The theory and uses of natural uranium isotopic variations in hydrology. *At. Energy Rev.* 14: 621–679.

Rosholt, J.N., Shields, W.R. & Garner, E.L. 1963. Isotope fractionation of uranium in sandstone. *Science*. 139: 224–226.

WHO (World Health Organization) 2002. *Guidelines for drinking water quality: radiological aspects.* http:www.who.int/water_sanitation_health/GDWQ/Radio/radioasp.htm

Water–Rock Interaction, Wanty & Seal II (eds)
© 2004 Taylor & Francis Group, London, ISBN 90 5809 641 6

Smectite-to-illite reaction vs. quartz kinetics in shale from the Haltenbanken area

M. Bruvoll, J.S. Jahren & P. Aagaard
Department of Geosciences, University of Oslo, Oslo, Norway

ABSTRACT: Clay mineralogy from the Late Jurassic Melke and Spekk black shale Formations shows that smectite transformation is linked to precipitation of autigenic quartz in the sediment. X-ray diffraction (XRD) quantitative analysis demonstrates that the early stages of smectite dissolution coincides with incipient quartz precipitation at a temperature around 65°C. The silica concentration in the pore water is closely related to the effective precipitation of quartz, and the coupled nature of quartz formation and the smectite to illite transformation also producing quartz, along with the sensitivity of the smectite dissolution, makes mudstones very well suited for quartz kinetic studies.

1 INTRODUCTION

Clay minerals are very reactive components in a wide range of geologic systems. Clay mineral studies are important in understanding coupled dissolution precipitation reactions in mineral transformations in these systems. In this study the smectite-to-illite transformation is studied in detail.

The shale samples used in this study were taken from the Haltenbanken area. The Haltenbanken area is situated offshore Mid-Norway and is the host of several large petroleum fields. For this reason the area is well explored and a large amount of sample material is available. Numerous publications, the majority concerning sandstones, in different disciplines are available.

The Kimmeridgian tectonic episode (Whitley 1992), with regional extensional tectonics, led to an uplift of the footwalls, erosion and formation of large scale rotated fault blocks in the area. At the end of the Mid Jurassic and to the beginning of the Cretaceous, a major transgression caused deposition of marine shales with a moderate to high organic content (Melke and Spekk Fm.). Both the Melke and the Spekk Formations are laterally extensive and are found at a wide depth range. Several episodes of rifting have brought the fault blocks to progressively deeper positions towards the north-west. This makes the Haltenbanken area a natural laboratory for diagenetic studies.

2 MATERIALS AND METHODS

2.1 *Sample material*

37 samples from 19 wells in the Haltenbanken area were studied. The samples are sidewall cores and cuttings and range from a depth of 1377 meters below seafloor (mbsf) to 3891 mbsf. They consist of dark grey to black marine mudstone with a moderate to high organic content. Deposition took place in a marine environment with restricted water circulation (Johannesen 1995; Whitley 1992).

2.2 *XRD sample preparation*

The samples were wet ground in a McCrone micronizing mill. This produces a narrow grain size distribution (O'Connor & Chang 1986) and is recognized as the most effective way of assuring required reproduction of intensities without excessive damage to the crystalline structure and of mixing samples with internal standards (Moore & Reynolds 1997; Hillier 2000; Środoń et al. 2001). Two parallel samples were prepared for each depth and the samples were saturated with $MgCl_2$ and KCl respectively before they were washed with deionized water to rinse off the excess chloride ions. Oriented smear slides were prepared for analysis using the method described by Moore & Reynolds (1997).

The samples were analyzed at the Department of Geology and Geophysics at Louisiana State

University, USA, on a Siemens diffractometer. Traces were recorded from 2 to 36° 2θ by using nickel-filtered copper radiation generated at 40 kV and 30 mA, 1° beam slit and a solid state detector, scanning 0.02° per 2 seconds.

2.3 *Quantification procedure*

In order to identify overlapping X-ray diffraction reflections a decomposition routine to separate the contributions from different mineral phases was run using MacDiff 4.2.5 (Petschick 2002) profile-fitting peak decomposition program. This was carried out on the Mg-saturated and ethylene glycol-solvated sample patterns to determine the precise positions and intensities of the individual peaks within the broad diffraction bands. Based on this the layer types in mixed-layer clay minerals were identified qualitatively following the methods of Moore & Reynolds (1997) and Środoń (1984).

A quantitative distribution of the clay mineral phases present in the samples was achieved by simulation and whole pattern matching. One-dimensional patterns for pure clay minerals and for mixed-layer clays were generated using MULCALC (Le & Ferrell 1996). The parameters: mixed-layer order (R), chemical composition of each layer, exchangeable cation, and other parameters related to crystallite size distribution and the diffractometer can be changed. The simulations produced a XRD library of potential phases with peaks that matched the positions determined by the peak decomposition.

CLAY++ (Huang et al. 1993; Huang & Ferrell 1998) was used for the whole-pattern matching. The program calculates clay mineral percentages from the XRD data and determines the fractional contribution of patterns in the library database to the whole observed pattern by fitting with a least-squares minimization procedure. The method is thoroughly described in Aparicio & Ferrell (2001).

3 RESULTS

Quantitative results from XRD analysis of the clay fraction of the samples are presented in Table 1.

A significant change in the mineral composition of the samples occurs between 2749 mbsf and 2830 mbsf. This leads to a natural subdivision of the results into two groups, 1377 to 2749 mbsf and 2830 to 3891 mbsf. There is no sample coverage between 1458 and 1979 mbsf.

3.1 *1377 to 2749 mbsf*

This interval covers 16 samples and the mineral composition is dominated by discrete clay mineral phases with smectite being the most abundant.

Discrete smectite is the dominating phase in the interval and shows a decreasing trend from 59 wt% in the upper part of the interval to 12 wt% for the deepest samples in the interval. The major depletion occurs between 1458 and 2165 mbsf (~50 wt%). The smectite 001 reflection is sharp in the shallower samples but the peak height/width ratio becomes increasingly smaller as the peak becomes progressively broader and weaker with depth. From 2288 to 2749 mbsf, smectite-illite with 90% smectite layers is identified by CLAY++. It does not show any clear internal trend.

Illite is identified as discrete illite and as illite(0.9)/smectite (statistical approach to maximize the curve fitting in CLAY++) and the mineral content varies between 6 and 19 wt%. Below 1458 mbsf the total amount of illite (ideal layers) starts to increase. The discrete illite peak is well defined in the samples but grows a shoulder on the high angle side from 1458 mbsf and deeper, resulting in a broadening of the peak width from ~0.5 to ~2.5° 2θ.

From 1458 mbsf there is an increasing amount of chlorite in the system. The chlorite reflections at 6.2 and 18.8° 2θ are often weak. This is most likely an effect of either a low chlorite concentration or a high Fe content in the chlorite (Moore & Reynolds 1997). In the 2314 to 2749 mbsf interval the chlorite 001 reflection is recognizable on the high angle shoulder of the ~16 Å peak. The width of the peak is shorter than the height and the chlorite reflection becomes more detached as the smectite content decreases.

Mixed-layer illite-smectite is present in amounts ranging from 0–19 wt% from 2165 mbsf to the end of the interval.

A slight increase in the amount of quartz is observed from 1979 mbsf. Caution should be made with respect to the amount of detrital versus authigenic quartz in the samples since there is no control on the initial quartz content in the sediment.

3.2 *2830 to 3891 mbsf*

Between 2830 and 3891 mbsf the dominating mineral phase changes from discrete smectite to mixed-layer illite-smectite phases. For the purpose of this paper there will not be given an individual examination of the identified mixed-layer phases, but they will be treated as a group.

Minor amounts (0–5 wt%) of discrete smectite are observed in scattered samples throughout the interval.

The discrete illite content ranges between 8 wt% and 30 wt%, with the average being 20 wt%. There is no trend with increasing depth but compared to the shallower interval (1377–2749 mbsf) the illite content is higher here.

The amount of chlorite varies from 0–30 wt% and shows no development within the interval, but as for

Table 1. Mineral composition and quantitative representation (QR) (wt%) obtained with CLAY++.

Phases present Depth (mbsf)	Smectite	Illite	Mixed-layer illite/smectite	Chlorite	S(.9)/I	I(.9)/S	Quartz	Kaolinite
1377	59					7	27	7
1411	88					7	2	3
1420	68					12	3	17
1453	100							~
1454	100							~
1458	84					10	2	4
1979	51	14		7			14	14
2165	16	19	19	34			3	9
2288	54	10	8		12		8	8
2314	26	15	9	9	26		6	9
2321	40	6	6		30		11	8
2321	26	10			27		27	10
2493	18	9	15	15	24		12	6
2525	37	12	3	15	12		12	9
2722	4	14	14	24	34		2	8
2749	12	19	15	15	19		8	11
2830		26	34	20			7	13
3254	5	17	51	5			13	9
3340		20	33	30			3	13
3424	5	14	53	5			9	14
3479	5	20	40	5			15	15
3496		19	30				38	13
3501		29	36	14			14	7
3529		25	37	19			6	13
3600	5	24	39	14			9	9
3610	3	20	42	24			3	7
3666		19	30	19			25	6
3753		15	32	7			38	8
3756	3	17	38	26			8	8
3762		24	29	19			14	14
3777		22	35	14			22	7
3786		8	41	4			42	4
3816		30	24	16			22	8
3846		19	44	6			19	12
3882	5	17	30	19			23	5
3883		26	35	14			13	13
3891	2	12	56	21			4	5

illite, the mineral content is higher here than in the first interval.

The average quartz content is 33 wt%, this is slightly higher than in the shallow interval.

4 DISCUSSION

The samples encompass a depth range from 1377 to 3891 mbsf and during this interval the temperature increases from ~55°C at 1377 mbsf to ~155°C at 3891 mbsf. The calculated geothermal gradientis 40°C/km.

Depletion of smectite as a discrete phase occurs over a distance of ~1000 meters, through which the temperature increases ~40°C. After 2749 mbsf, most of the smectite is transformed to mixed-layer illite-smectite phases, but also to discrete illite, discrete chlorite and authigenic quartz. There is a restricted range of coexistence between discrete smectite and mixed-layer illite-smectite.

Smectite(0.9)/illite identified in the 2288 to 2749 mbsf interval may represent a "high charge" smectite where hydrated interlayers are collapsed and potassium ions are fixed in the interlayer space. This would decrease the repeat distance between the individual clay particle layers, and can lead to a peak shift towards higher angles in X-ray diffractograms if it occurs in sufficient XRD scattering domains. The statistical approach of the curve-fitting procedure in CLAY++ may record the occurring reflection as a mixed-layer smectite/illite phase, while it is interpreted to represent an effect of the initial smectite dissolution stage.

The temperature and depth of the zone where smectite starts to dissipate coincides with the temperature of initial quartz precipitation. This is in accordance with the silica concentration in the pore water being the controlling factor for smectite dissolution. The silica concentration in the pore water is controlled by the precipitation of quartz above a certain threshold temperature. Walderhaug (1994) sets the first detection of quartz cementation at 75°C based on homogenization temperatures for fluid inclusions trapped within quartz overgrowths in sandstones from the North Sea, whereas work by Saigal et al. (1992) and Ehrenberg (1990) suggests the formation temperature for quartz cementation to start at 60°C. The homogenization temperature for quartz in mudstones may differ from this. Once the precipitation of quartz in the sediment is fast enough to keep the silica concentration at quartz saturation, smectite becomes unstable and dissolution commences.

No correction has been done to account for the temperature variance associated with differential subsidence. Based on the Arrhenius function, Walderhaug (1996) came up with a relationship for the temperature dependence of the quartz precipitation rate at diagenetic temperatures:

$$r = a10^{bT} \tag{1}$$

where a is a constant with units of mol cm^{-2} s^{-1}, b is a constant with units of °C^{-1}, and T is temperature in°C. The amount of quartz precipitated will be a function of the time spent at different temperatures (time-temperature integration). The Haltenbanken area experienced a subsidence of ~1000 m during the last 2 Ma. This represents a temperature increase of 40°C. The rapid subsidence changed the time-temperature integral and as a result of this the transition of smectite was extended over a longer temperature interval. The detected onset of smectite depletion would therefore occur at higher temperatures with rapid subsidence than with a slower burial rate.

5 CONCLUSIONS

The clay mineralogy in the black shale samples from Haltenbanken shows a classic diagenetic development going from smectite domination in the shallow samples to illite domination in the deeper samples. Initiation of smectite dissolution starts at 65°C/1458 mbsf (note the absence of samples between 1458 and 1979 mbsf), and is controlled by the silica concentration in the pore water. When the pore water approaches quartz saturation, smectite starts to dissolve. The silica concentration in the pore water is closely related to the effective precipitation of quartz and the coupled nature of these reactions and the sensitivity of the smectite dissolution reaction makes mudstones very well suited for studies of quartz kinetics.

ACKNOWLEDGEMENTS

The authors wish to thank Professor Ray E. Ferrell and Wanda LeBlanc for help with X-ray diffraction at Louisiana State University.

REFERENCES

Aparicio, P. & Ferrell, R.E. 2001. An application of profile fitting and CLAY++ for the quantitative representation (QR) of mixed-layer clay minerals. *Clay Mins.* 36, 501–514.

Ehrenberg, S.N. 1990. Relationship between diagenesis and reservoir quality in sandstones of the Garn Formation, Haltenbanken, mid-Norwegian continental shelf. *AAPG Bull.* 74, 1538–1558.

Hillier, S. 2000. Accurate quantitative analysis of clay and other minerals in sandstones by XRD: comparison of a Rietveld and a reference intensity ratio (RIR) method and the importance of sample preparation. *Clay Mins.* 35, 291–302.

Huang, K. & Ferrell, R.E. 1998. CLAY++: A computer program for quantitative clay mineral analysis. Department of Geology and Geophysics, E235, Howe/Russell Geoscience Complex, Baton Rouge, LA 70803, USA.

Huang, K., Ferrell, R.E. & LeBlanc, W.S. 1993. Computer assisted interpretation of clay mineral diffraction patterns. *30th Annual Meeting of the Clay Minerals society, San Diego*, p. 51.

Johannesen, J. 1995. *Source Rock Characterization of the Spekk Formation, Haltenbanken (offshore Mod-Norway).* Cand. Scient. thesis, University of Bergen, Norway.

Le, T.H. & Ferrell, R.E. 1996. MULCALC. Department of Geology and Geophysics, E235, Howe/Russell Geoscience Complex, Baton Rouge, LA 70803, USA.

Moore, D.M. & Reynolds, R.C.Jr. 1997. *X-ray Diffraction and the Identification and Analysis of Clay Minerals.*2nd edition. New York: Oxford University Press.

O'Connor, B.H. & Chang, W.J. 1986. The amorphous character and particle size distribution of powders produced with the micronizing mill for quantitative X-ray powder diffractometry. *X-ray Spectrometry* 15, 267–270.

Petschick, R. 2002. MacDiff 4.2.5. Powder diffraction software. Available from the author at www.geol.uni-erlangen.de/html/software/Macdiff.html.

Saigal, G.C., Bjørlykke, K. & Larter, S. 1992. The effects of oil emplacement on diagenetic processes – examples

from the Fulmar reservoir sandstones, central North Sea. *J. Petrol. Geol.* 8, 163–186.
Środoń, J., Drits, V.A., McCarty, D.K., Hsieh, J.C.C. & Eberl, D.D. 2001. Quantitative XRD analysis of clay rich rocks from random preparations. *Clay Mins.* 48, 514–528.
Środoń, J. 1984. X-ray powder diffraction identification of illittic materials. *Clay Mins.* 32, 337–349.
Walderhaug, O. 1996. Kinetic modeling of quartz cementation and porosity loss in deeply buried sandstone reservoirs. *AAPG Bull.* 80, 731–745.
Walderhaug, O. 1994. Temperatures of quartz cementation in Jurassic sandstones from the Norwegian continental shelf – Evidence from fluid inclusions. *J. Sed. Res.*A64, 311–323.
Whitley, P.K. 1992. The geology of Heidrun, a giant oil and gas field on the mid-Norwegian Shelf. In: Halbouty, M.T. (ed.). *Giant Oil and Gas Fields of the Decade 1978–1988. AAPG Mem.* 54, 383–406.

Water–Rock Interaction, Wanty & Seal II (eds)

Origin of brines from the Bachu Uplift, Tarim Basin, China: chemistry and $\delta^{11}B$ evidence

Chunfang Cai
Institute of Geology and Geophysics, CAS, Beijing P. R. China

Hong He
Exploration & Production Research Institute, SINOPEC, Beijing P. R. China

Bowen Mei
Changjiang University, Jingzhou, Hubei P. R. China

Yingkai Xiao
Qinghai Institute of Salt Lake, CAS, Xining P. R. China

ABSTRACT: Major ion chemistry, $\delta^{11}B$, $^{87}Sr/^{86}Sr$, δD and $\delta^{18}O$ of brines together with clay mineral XRD were analyzed from the Carboniferous, Ordovician and Cambrian basinal brines in the Bachu Uplift, Tarim Basin, NW China. The brines are thought to originate from seawater evaporation and have subsequently mixed with meteoric water. The brines show B, Li and Sr enrichment in the Carboniferous and Ordovician and K enrichment in part of the Ordovician relative to seawater, and have $\delta^{11}B$ from +19.7‰ to +32.2‰. With increasing depth, B concentrations increase, $\delta^{11}B$ values decrease and illite percentages increase among clay minerals, thus B with low $\delta^{11}B$ values is interpreted to have been added from thermal degradation of kerogen (Williams et al. 2001). However, those most K-enriched brines have the highest B and Li and lowest $\delta^{11}B$, suggesting that the additional B, Li and K may have the same source, for which the basalts in the Cambrian and Lower Ordovician cannot be ruled out.

1 INTRODUCTION

Oilfield brine chemistry, δD, $\delta^{18}O$ and $^{87}Sr/^{86}Sr$ have been intensively used to account for the origin and migration of brines (e.g. Cai et al. 2001) while very limited $\delta^{11}B$ data have been published (e.g. Land and Macpherson 1992; Moldovanyi et al. 1993; Vengosh et al. 1998; Xiao et al. 1997). However, $\delta^{11}B$ may be a valuable parameter to reflect water–rock interaction due to large variations in $\delta^{11}B$ values in sedimentary basins.

The study area is located in the Bachu Uplift of the Tarim Basin, NW China with gas pools in the Hetianhe Field, south of the Bachu Uplift. The previous studies show that the gases and condensate in Carboniferous and Ordovician reservoirs were derived from Cambrian and Carboniferous source rocks, respectively, indicating a complicated history of fluid migration in the area. Cai et al. (2002) showed that meteoric water mixing along the C/O unconformity has led to secondary alteration of the natural gases. This manuscript presents Carboniferous, Ordovician and Cambrian brine chemistry, boron, strontium, oxygen and hydrogen isotopes, fluid inclusion salinity and homogenization temperature, and clay mineral XRD data to address the origin and evolution of the brines and fluid flow history in this area.

2 GEOLOGICAL SETTINGS AND ANALYSIS METHODS

The study area, the Bachu Uplift, is located in west of the center of the Tarim basin. The geological setting has been described elsewhere (Cai et al. 2002). Briefly, the Cambrian and Lower Ordovician strata are composed of platform carbonates, bedded anhydrite and salt with dark grey gabbro, coarse-grained basalt and diabase. The Upper and Middle Ordovician strata are composed of limestone, wackestone and interlayered limestone with mudstone. The

Carboniferous is generally 800–1000 m thick and composed of platform limestone, beach facies coarse clastics and lake and swamp facies carbonaceous mudstone and sandstones. It was divided into eight units: Upper Carboniferous Xiaohaizi limestone (C_2^1) and sandstone and mudstone (C_2^2) members; Lower Carboniferous upper mudstone (C_1^3), standard limestone (C_1^4), middle mudstone (C_1^5), bioclastic limestone (C_1^6), lower mudstone (C_1^7) and sandstone and conglomerate (C_1^8) members. The Permian strata consist of terrestrial alluvial plain coarse clastic red beds and basalt. It is postulated that this area was subaerially exposed from the end of the Permian to the early Tertiary and then subsided as a result of marine transgression.

Water samples were obtained through drill-stem testing (DST) from the C_2^2 sandstone, C_1^4, C_1^6, Ordovician and Cambrian carbonates and were filtered though a 0.45 μm filter. The analytical methods for water chemistry and δD and $\delta^{18}O$ have been published in Cai et al. (2001). Boron isotope analysis used the procedure reported by Xiao et al.(1988) based on $Cs_2BO_2^+$ with graphite loading. The average measured $^{11}B/^{10}B$ of NIST SRM 951 standard was 4.0505 ± 0.0065 (2σ) and the isotopic composition of boron in samples was expressed as $\delta^{11}B = [(^{11}B/^{10}B)_{sample}/(^{11}B/^{10}B)_{standard} - 1] \times 1000$ (Xiao et al. 1997).

Mud-bearing limestones (<5% mud) were analyzed for clay minerals by X-ray diffractometry (XRD) and chemical composition using the method of atomic absorption spectrometry with an analytical precision of 5%.

3 RESULTS AND DISCUSSION

3.1 *General chemistry and $\delta^{18}O$*

The pH of the basinal brines ranges from 6.3 to 8.0. Three different water types can be identified in terms of TDS, Cl concentrations and molar Na/Cl ratios. Type I brinesare characterized by the highest TDS in the range of 183 to 229 g/L, Cl concentrations >80 g/L and molar Na/Cl ratios <0.86. These brines were sampled from the deeper Cambrian and Ordovician dolomite reservoirs, and are considered to be evaporated seawater as supported by the relationship between Na and Cl (Fig. 1) and the occurrence of the evaporites. Type II brines are characterized by TDS from 9.5 to 49.5 g/L, Cl concentrations from 4.1 to 28.2 g/L and molar Na/Cl ratios from 0.9 to 2.6. These brines were found in reservoirs close to the C/O unconformity. They show a trend of increasing TDS and less negative $\delta^{18}O$ values from the west to the east in the Hetianhe Field with the lowest TDS values near the C/O unconformity, indicating that meteoric water recharged in the west and flowed eastward along the unconformity.

Type III brines have TDS, Cl and molar Na/Cl ratios between the type I and type II brines. These brines may have mixed with type I and/or type II waters. They occur in the Ordovician and Carboniferous carbonates and Carboniferous sandstones. These reservoir rocks had numerous microfractures. Fluid inclusions within fracture-filling calcite have salinities (up to 14.3 wt %) significantly higher than the TDS of the present brines in the reservoirs (unpublished data), supporting significant cross-formation fluid flow and mixing with more saline type I brines.

3.2 *B, Li and K concentrations, $\delta^{11}B$ and $^{87}Sr/^{86}Sr$*

The brines have boron concentrations from 0.3 to 121 mg/L, and show significant B, Li and Sr enrichment relative to evaporated seawater in the C and O brines and K enrichment in part of the Ordovician brines (Fig. 1). The B, Li, K and Sr concentrations are shown to increase with depth (Fig. 2). It is impossible for elevated K and Sr of the brines to have been derived from leaching of limestones since no decrease for K and Sr of limestones with depth is observed (Fig. 2), and sedimentary microfacies for the Ordovician reservoir limestones are similar.

In sedimentary basins, the boron isotopic composition shows large variations: seawater +39.5‰, kerogen ~−2‰, adsorbed-B on marine sediment

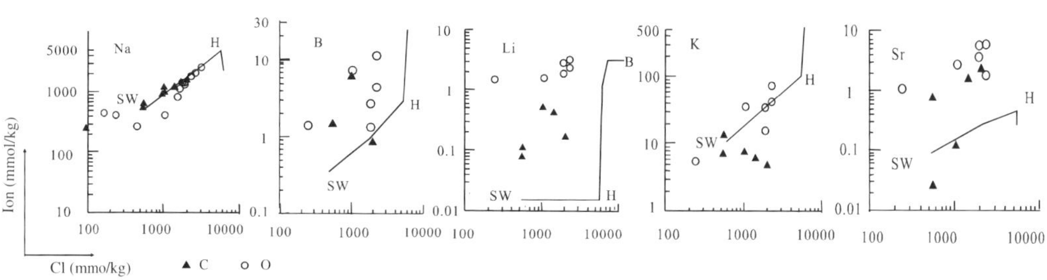

Figure 1. Relationships of Na, B, Li, K and Sr concentrations to Cl relative to evaporated seawater.

~+15‰, fixed B in authigenic illite of sandstones ~−2‰, B-containing marine evaporites +20 to +31‰ and nonmarine evaporites −31 to +7‰ and basalts from −3.7 to +6.2‰ (Coplen et al. 2002; Ishikawa et al. 2001; Kloppmann et al. 2001; Spivack et al. 1987; Vengosh et al. 1998; Williams et al. 2001; You et al. 1995). Boron isotopes may be used to monitor fluid/rock interactions during diagenesis. Studies of B in marine sediments have shown that preferential adsorption of ^{10}B on clay minerals leaves pore waters enriched in ^{11}B (Spivack et al. 1987). During diagenesis, clay minerals recrystallize and incorporate ^{10}B preferentially into the mineral structure (Williams et al. 2001). On the other hand, desorption of B from clay mineral surfaces is considered to contribute B with higher ^{10}B content to pore water below 100°C, thus decreasing pore water $\delta^{11}B$ (Vengosh et al. 1998). Above the temperature adsorbed coefficient (KD) equals 0 (You et al. 1995), thus there should no adsorbed B on the sediments.

The brines in the study area have $\delta^{11}B$ from +19.7 to +32.2‰ and show a decreasing $\delta^{11}B$ and increasing B with an increase in illite content of clay minerals, illite percentage in mixed layer I/S and depth (Fig. 2). Under normal conditions of burial diagenesis, illitization of smectite leads to preferential incorporation of ^{10}B into the lattice. Thus pore water boron concentrations are expected to decrease when temperatures >100°C. The rise of B in brines and clay minerals suggests a source of B from unstable detrital minerals such as micas, OM, or basalts in the deep formations. Kerogen has been measured to have boron of about 140 ppm with $\delta^{11}B$ about −2‰ and is considered to the source of B for deeper formation waters in the Gulf Coast Basin (Williams et al. 2001) at temperatures >120°C to 150°C. For the Bachu Uplift, the Cambrian kerogen, the source of gas in the C and O reservoirs, experienced temperatures >130°C to 180°C during the Permian (Cai et al. 2002). The homogenization temperatures of aqueous two-phase inclusions in healed microfractures in the C and O reservoirs is up to 133°C. These temperatures are high enough for the kerogen in the Cambrian to supply B to upward-migrating brines.

That there are positive correlations between K and B and K and Li, and negative correlations of $\delta^{11}B$ to B and K (Fig. 3), suggests a common source for K, Li and B with low $\delta^{11}B$ values. Increased Li concentrations and higher $\delta^{6}Li$ values have been suggested to be derived from water-sediment interactions in the Heletz-Kokhav field, Israel (Chan et al. 2002). Basalts were shown to have B from 13 to 48 ppm, $\delta^{11}B$ from −3.7 to +6.2‰ and are enriched in K, Li and Sr (Ishikawa et al. 2001). Thus, it cannot be ruled out that K, Li and B with low $\delta^{11}B$ values were derived from basalt-water interaction in the Lower Ordovician and Cambrian. Similarly, Molodovanyi et al. (1993) suggested that consistently increasing K, Li, B, Sr and Rb of Smackover brines with depth are derived from upward-migrating brines through faults in the Gulf Coast Basin.

$^{87}Sr/^{86}Sr$ values of six water samples from C and O limestones range from 0.7090 to 0.7110. Most of the values are significantly higher than those of Ordovician carbonates in Central Tarim (0.7064 to 0.7094; Cai et al. 2001) and seawater in the periods of Ordovician and Carboniferous from 0.7074 to 0.7090, indicating a significant ^{87}Sr contribution from rock-water interaction.

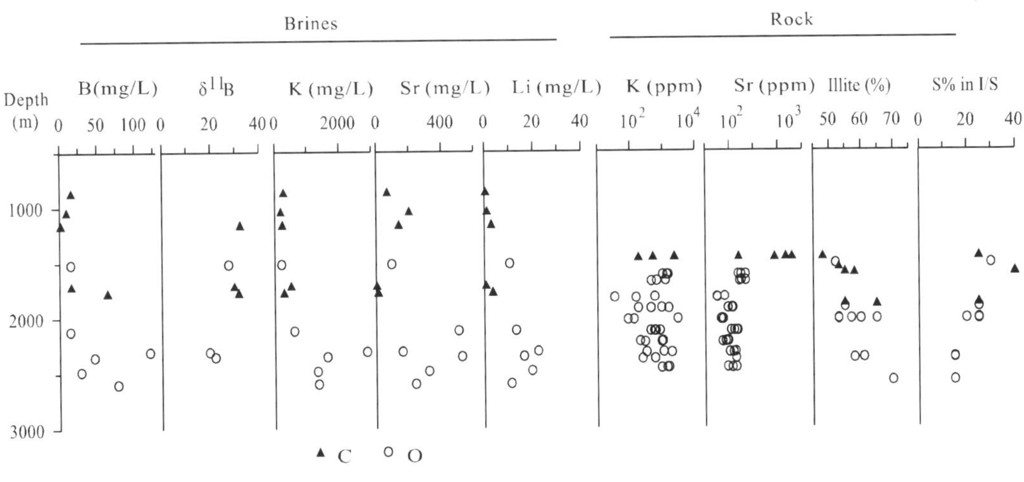

Figure 2. Variation of brine B, K, Li, Sr and $\delta^{11}B$ and rock K, Sr, illite content among clay minerals and smectite percentage in I/S interlayer with depth.

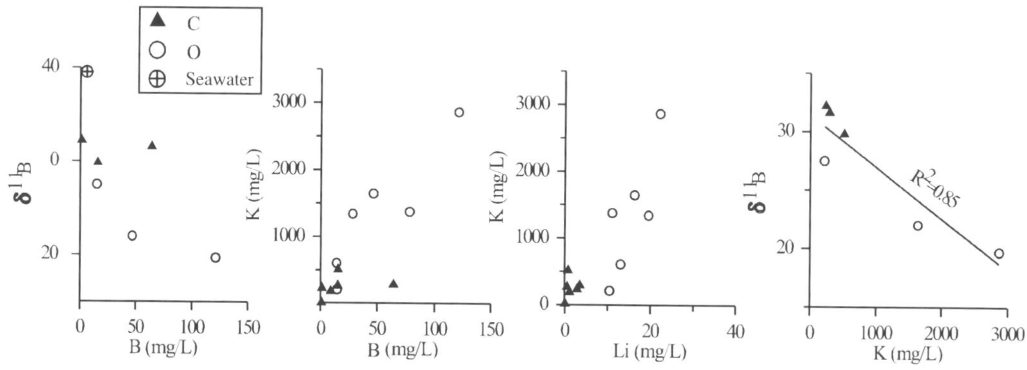

Figure 3. Relationships between $\delta^{11}B$, B, K and Li concentrations showing negative correlations of $\delta^{11}B$ to B and K.

4 CONCLUSIONS

The brines from Cambrian, Ordovician and Carboniferous reservoirs in the Bachu Uplift originated from seawater evaporation, cross-formation fluid flow and mixing, water-rock interaction and meteoric water infiltration. B, Li, K and Sr concentrations show enrichment relative to evaporated seawater and increase with depth while $\delta^{11}B$ values decrease. Organic matter and/or basalts in the Cambrian and Lower Ordovician are potential sources for the additional ^{10}B, K and Li of the brines.

ACKNOWLEDGEMENTS

The research is financially supported by the National Natural Science Foundation of China (Grant No. 40173023).

REFERENCES

Cai, C.F., Franks, S. & Aagaard, P. 2001. Origin and migration of brines from Paleozoic strata in Central Tarim, China: constraints from $^{87}Sr/^{86}Sr$, δD, $\delta^{18}O$ and water chemistry. *Appl. Geochem.* 16: 1269–1284.

Cai, C.F., Worden, R.H., Wang, Q.H., Xiang, T.S., Zhu, J.Q. & Chu, X.L. 2002. Chemical and isotopic evidence for secondary alteration of natural gases in the Hetianhe Field, Bachu Uplift of the Tarim Basin. *Org. Geochem.* 43: 1415–1427.

Chan, L.H., Starinsky, A. & Katz, A. 2002. The behavior of lithium and its isotopes in oilfield brines: Evidence from the Heletz-Kokhav field, Israel. *Geochim. Cosmochim. Acta* 66: 615–623.

Coplen, T.B., Bölke, J.K., Bièvre, P.De, Ding, T., Holden, N.E., Hopple, J.A., Krouse, H.R., Lamberty, A., Peiser, H.S., Révész, K., Rieder, S.E., Rosman, K.J.R., Roth, E., Taylor, P.D.P., Vocke, R.D. & Xiao, Y.K. 2002. Isotope-Aboundance Variations of Selected Elements. *Pure Appl. Chem,* 74: 1987 – 2017.

Kloppmann, W., Négrel, Ph., Casanova, J., Klinge, H., Schelkes, K., & Guerrot, C. 2001. Halite dissolution derived brines in the vicinity of a Permian salt dome (N German Basin). Evidence from boron, strontium, oxygen, and hydrogen isotopes. *Geochim. Cosmochim. Acta* 65: 4087–4101.

Ishikawa, T., Tera, F. & Nakazawa, T. 2001. Boron isotope and trace element systematics of the three volcanic zones in the Kamchatka arc. *Geochim. Cosmochim. Acta* 65: 4523–4537.

Land, L.S. & Macpherson, G.L. 1992. Origin of saline formation waters. Cenozoic Section, Gulf of Mexico Sedimentary Basin. *Am. Assoc. Petrol. Geol. Bull.* 76: 1344–1362.

Moldovanyi, E.P., Walter, L.M. & Land, L.S. 1993. Strontium, boron, oxygen, and hydrogen isotope geochemistry of brine from basal strata of the Gulf – Coast sedimentary basin, USA. *Geochim. Cosmochim. Acta* 57: 2083–2099.

Spivack, A.J., Palmer, M.R. & Edmond, J.M. 1987. The sedimentary cycle of the boron isotopes. *Geochim. Cosmochim. Acta* 51: 1033–1043.

Vengosh, A., De Lange, G.J. & Starinsky, A. 1998. Boron isotope and geochemical evidence for the origin of Urania and Bannock brines at the eastern Mediterranean: Effect of water-rock interactions. *Geochim. Cosmochim. Acta* 62: 3221–3228.

Williams, L.B., Hervig, R.L, Wieser, M.E. & Hutcheon, I. 2001. The influence of organic matter on the boron isotope geochemistry of the Gulf Coast sedimentary basin, USA. *Chem. Geol.* 174: 445–461.

Xiao, Y.K., Beary, E.S. & Fassett J.D. 1988. An improved method for the high-precision isotopic measurement of boron by thermal ionization mass spectrometry. *Int. J. Mass Spectrom. Ion Proc.* 85: 203–213.

Xiao, Y.K., Vocke, R.D., Swihart, G.H. & Xiao, Y.K. 1997. Boron volatilization and its isotope fractionation during evaporation of boron solution. *Anal. Chem.* 69: 5203–5207.

Xiao Y.K., Shirodkar, P.V., Liu, W.G., Wang, Y.H. Jin, L. 1999. The investigation of isotopic compositions of boron in salt lake brine, Qaidam Basin. *Progress in Natural Sciences* 9: 612–618 (in Chinese).

You, C.F., Spivack, A.J., Gieskes, J.M., Rosenbauer, R. & Bischoff, J.L.1995. Experimental study of boron geochemistry: Implications for fluid processes in subduction zones. *Geochim. Cosmochim. Acta* 59: 2435–2442.

Water–Rock Interaction, Wanty & Seal II (eds)
© 2004 Taylor & Francis Group, London, ISBN 90 5809 641 6

Use of a gel probe sampling device to examine porewater profiles and the effects of porewater composition on As sorption in sediments

K.M. Campbell, S. Dixit & J.G. Hering
Environmental Science & Engineering, California Institute of Technology, Pasadena, CA, USA

ABSTRACT: The effect of sediment porewater chemistry on As sorption to Fe oxyhydroxide minerals is being investigated using a gel probe sampling device. The gels are comprised of a polyacrylamide gel matrix, and can be used to measure porewater concentrations of As and other constituents (clear gels) or sorption behavior by doping the gel matrix with hydrous ferric oxide (HFO). The gels can be placed into a probe, inserted into the sediments, and allowed to reach equilibrium with the sediment porewater. The results of laboratory validation and a preliminary field study are presented. The field site is a reservoir in which an Fe-As floc is deposited as a consequence of treatment to remove As from drinking water supplied to the City of Los Angeles. The sediments are Fe- and As-rich, and provide a unique test site to investigate the mechanisms of As mobilization in sediment porewaters.

1 INTRODUCTION

Arsenic has long been known to have severe adverse health effects for humans. The greatest exposure to As is through ingestion, thus measurement, regulation and treatment of As in drinking water are extremely important. As in drinking water affects millions of people around the world. In cases of chronic As exposure, skin lesions, melanosis (skin discoloration), keratosis, cancer and possible reproductive effects have been observed in affected populations (Bhattacharyya et al. 2003; Ng et al. 2003). An estimated 41 million people are potentially exposed to elevated levels of As, from 10 to 10,000 μg/L (Nordstrom 2002). The current drinking water standard in the United States is 10 μg/L, recently decreased from 50 μg/L in 2002 by the US EPA. However, there is a greater challenge to find acceptable treatment options in less developed regions such as Bangladesh and West Bengal where the affected population is primarily rural and large scale treatment is prohibitively expensive.

Since As contamination is a problem with widespread and serious implications, it is necessary to understand the specific biogeochemical controls of As behavior in subsurface environments. By understanding the processes that govern mobilization, it may be possible to predict future problems of As contamination and design effective treatment methods. Arsenic mobilization in Bangladesh has been attributed to the reductive dissolution of Fe oxyhydroxides by microorganisms (Nickson et al. 1998; Acharyya et al. 1999; Akai et al. 2004). In order to study this phenomenon in more detail, a specific test site has been selected in California that may provide information on the general processes that influence As release to groundwater.

2 FIELD SITE: HAIWEE RESERVOIR, OLANCHA, CA, USA

The City of Los Angeles (LA) imports water from several different sources, including the Sierra Nevada Mountains in central California. The water is transported through the Los Angeles Aqueduct (LAA), which feeds several reservoirs, including North and South Haiwee Reservoirs. Natural geothermal springs in Hot Creek contribute high concentrations of As to the LAA source waters. Because of this, the LA Department of Water and Power (LADWP) has developed an interim management plan to remove As from the LAA (Stolarik & Christie 1999). The treatment method is the addition of ferric chloride as a coagulant directly into the LAA at the Cottonwood Treatment Plant, about 27 km upstream of Haiwee reservoir. Cationic polymer is added as a flocculent aid. The ferric chloride forms an amorphous Fe oxide floc in the aqueduct channel. Dissolved As readily adsorbs and/or co-precipitates with the floc, and is subsequently deposited in the inlet channel to Haiwee reservoir. The treatment goal is to decrease the dissolved As concentration to the new standard of <10 ppb, and has been quite successful. However, the result is substantial deposition of Fe- and As-rich sediments in the inlet channel. These sediments are a reasonable test site for investigating biogeochemical controls on As mobilization.

The reddish-brown floc can be seen flowing into the inlet channel from the aqueduct. The sediment is visibly red at the edge of the channel where the floc is deposited. The upper portion (<5 cm) of the sediment is red-brown in color, fine, and porous, while the lower sediment is fine, black and thicker than the upper sediment. The flow is higher and faster in the middle of the channel, although the deposition in that region has not been sampled.

3 PREVIOUS STUDY RESULTS: HAIWEE RESERVOIR

In a collaborative study previously done on Haiwee sediments, As, Mn and Fe were found to be mobilized at depth in the sediments. Mn was mobilized closer to the sediment/water interface below a depth of 5 cm, while As and Fe in the porewater mobilized below 12 cm and their concentrations in the porewater were highly correlated. The peak As concentration was 1.3 mg/L, but it was estimated that less than 2.5% of total As content of the sediment was mobilized to yield this concentration. Similarly, less than an estimated 1% of the total Fe was mobilized into the porewater (Kneebone 2000, Kneebone et al. 2002).

Previous studies by Kneebone et al. (2002) suggest that As is rapidly reduced from As^{V} to As^{III} in the sediment. In the overlying reservoir waters, dissolved As is present predominantly as As^{V} and it is probable that the As associated with Fe floc deposited in the inlet channel is also As^{V}. However, As^{III} is detected by X-ray absorption spectroscopy (XAS) even in the surface sediments and only As^{III} is present associated with the sediments at depth.

Porewater composition may play an important role the As mobilization at depth in Haiwee sediments. Specifically, resorption of As may be inhibited by porewater chemistry, such as elevated levels of phosphate, dissolved Fe, and organic matter. In this study, the porewater composition and As sorption behavior at depth will be investigated using a gel probe sampling device.

4 GEL PROBES

4.1 *Description of gel probe*

Gel probes are sampling devices that allow porewater profiles to be obtained with higher resolution and shorter equilibration times than other methods, such as core squeezing and dialysis membranes known as "peepers". In addition, they can be deployed with minimal sediment disturbance. Gel probes originally constructed by Davison et al. (1994) were made of an inert polyacrylamide gel matrix, designed to come to equilibrium with the surrounding solution. The device is known as DET (diffusive equilibrium in thin films; Davison et al. 1994). The gel was cast into a sheet that was placed into a Plexiglas probe, and inserted into sediments. The porewater was allowed to equilibrate with the gel matrix, the probe was retrieved, and the gels sliced for analysis of particular analytes. Kneebone modified the gel probe design by cutting the gels into slabs (2.5 cm × 0.5 cm × 0.2 cm) and placing them in a Plexiglas vertical "ladder" such that each gel was separated. The gels were covered by a 0.45 μm cellulose nitrate filter and secured with a cover plate. While some vertical resolution was lost compared to Davison's probe, it was sufficient for obtaining porewater profiles in Haiwee sediments, and was easier to manipulate in the field.

In this work, porewater profiles were obtained for Haiwee sediments using a gel probe modified from Kneebone's work. P, As, Mn, and Fe in the porewater were measured; pure polyacrylamide gels (referred to as clear gels), were used for this purpose.

In addition, a second type of gel was prepared to investigate the sorption behavior of As. These gels were polyacrylamide doped with synthetic hydrous ferric oxide (HFO). HFO is similar to the Fe floc precipitated in the LAA. By doping the gels with HFO, fresh sorption sites are introduced into the subsurface environment when the probe is deployed. The extent of As uptake onto fresh HFO should provide insight into the effects of porewater composition on As sorption. A profile of sorption behavior onto fresh HFO can be obtained, along with porewater composition profiles (clear gels) to investigate whether porewater chemistry is inhibiting As sorption, and thus affecting As mobility.

4.2 *Laboratory validation*

The gels were cast using a modified method of Krom et al. (1994). Once cast, the gels were placed in 18 MΩ Milli-Q water for hydration. After hydration, the gels were 2 mm thick. The gels were cut into slabs 2.5 × 0.5 cm using a thin plastic cutter, and stored in Milli-Q water. The gels were found to be 92% water by volume, as determined from the difference in the measured weights of fully hydrated and dried clear gels. In addition, the variation of doped HFO per gel slab was less then 10%.

The performance of the gels was tested in the laboratory prior to field deployment. As^{III} recovery was determined for both clear and HFO gels by placing the gels in a solution of 10 μM arsenite with ionic strength of 0.01 M and pH 8 (HEPES buffer). The loss of As from solution was measured. The gels were re-equilibrated with 1% HNO_3 (clear gels) or 35% HNO_3 (HFO gels), and As in the re-equilibration solution was measured. The 35% acid ensured that the HFO completely dissolved in the doped gels and the polyacrylamide gel matrix was unaffected by acid. As was measured by inductively coupled mass spectrometry (ICP-MS), and Fe was measured by a colorimetric method. Recovery of As from both clear and HFO gels was better than 95%.

Total As sorption as a function of the amount of Fe per gel slab was measured. The amount of Fe in the HFO gels ranged from 1×10^{-6} to 8×10^{-6} mol/gel slab and placed in a solution of 10 μM As^{III} with an ionic strength of 0.01 M and a buffered pH of 8. The amount of Fe used in further laboratory studies and field deployment was 4×10^{-6} mol Fe/gel slab. Equilibration time was determined for HFO gels by taking a series of samples over time and measuring the loss of As from solution. Based on these results, an equilibration time of >6 hours was used.

In order to fully understand As sorption behavior, it was necessary to determine experimentally the total site density of HFO in the polyacrylamide gel matrix. This was determined by a sorption isotherm in which As sorption was measured as a function of dissolved As concentration. The total site density as a function of Fe^{III} concentration was found to be $[\equiv FeOH]_T = [Fe^{III}]$ (0.12 mol sites/mol Fe).

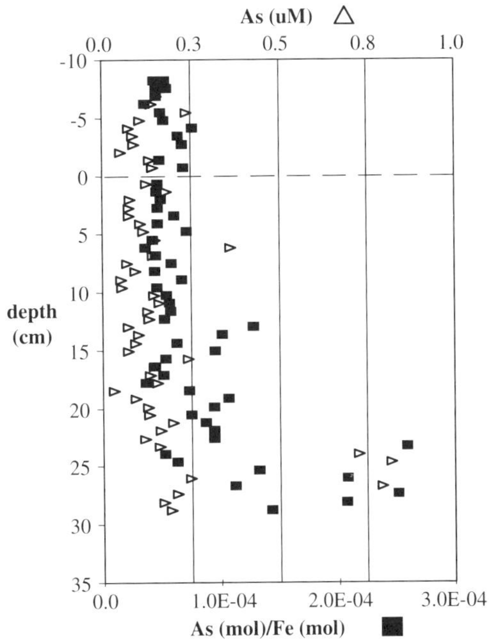

Figure 1. Profile from July 2003 deployment of gel probe sampler. The sediment/water interface is at depth = 0. Open triangles are porewater concentrations (clear gels) and closed boxes are amount of As sorbed in HFO gels normalized to the amount of Fe per gel. The deployment time for this probe was 7.5 hours. The clear gels and the HFO gels were deployed simultaneously, and were placed within 3 cm of each other.

4.3 *Field deployment*

Gel probes were taken into the field on two occasions, July 2003 and October 2003. In July, a combination of clear and HFO gels were inserted into the probe in order to compare the sorption behavior in HFO gels to a porewater profile in close proximity. In October, only clear gels were deployed in several locations to estimate spatial variability.

There is evidence of As mobilization at depth from the field deployment. However, the concentrations of As in the porewater (clear gels) are substantially less when placed with HFO gels than in previous studies and in the October sampling. As the porewater comes into equilibrium with the clear gels, minimal perturbation is expected due to the small volume of each gel. However, the HFO gels may perturb the system significantly by concentrating As and other constituents onto the HFO through sorption, thus depleting the surrounding porewater. This is possibly reflected in the low concentrations in the clear gel profile when placed next to HFO gels. The extent of perturbation is determined by diffusion of constituents from surrounding porewaters, ability of the sediment to resupply As to the porewater, and kinetic controls on As sorption to HFO. To assess the effect of chemistry on sorption onto the HFO gels, the sediment characteristics of resupply, diffusion, and sorption kinetics must be determined, as well as an appropriate equilibration time for deployment in the field.

4.4 *Sediment microcosms*

In order to address the issue of sediment resupply behavior and equilibration time, a laboratory sediment microcosm can be employed to investigate natural sediments in a controlled experiment. Preliminary results indicate that equilibration times on the order of 24 hours may be necessary to allow the sediment to re-equilibrate with the porewater following the perturbation by the HFO gels.

5 FURTHER WORK

Further research is in progress to more fully characterize the sediment-porewater-HFO gel system through sediment microcosms. Once the behavior of the gel system is understood, the gel probes can be used to observe the effects of *in situ* biogeochemical processes in the sediment environment.

REFERENCES

Acharyya, S.K., Chakraborty, P., Lahiri, S., Raymahashay, B.C., Guha, S. & Bhowmik, A. 1999. Arsenic poisoning in the Ganges Delta. *Nature* 401: 545–546.

Akai, J., Izumi, K., Fukuhara, H., Masuda, H., Nakano, S., Yoshimura, T., Ohfuji, H., Anawar, H.M. & Akai, K. 2004. Mineralogical and geomicrobiological investigations on groundwater arsenic enrichment in Bangladesh. *Appl. Geochem.* 19: 215–230.

Bhattacharyya, R. & Charrerjee, D. 2003. High arsenic in groundwater: Mobilization, metabolism, and mitigation – an overview of the Bengal Delta Plain. *Molecular Cellular Biochem.* 253: 347–355.

Davison, W.H., Zhang, H. & Grime, G.W. 1994. Performance characteristics of gel probes used for measuring the chemistry of porewaters. *Environ. Sci. Technol.* 28(9): 1623–1632.

Kneebone, P.E. 2000. Arsenic geochemistry in a geothermally impacted system: The Los Angeles Aqueduct. *Environmental Engineering Science.* Pasadena: California Institute of Technology.

Kneebone, P.E., O'Day, P.A., Jones, N. & Hering, J.G. 2002. Deposition and fate of arsenic in iron- and arsenic-enriched reservoir sediments. *Environ. Sci. Technol.* 36(3): 381–386.

Krom, M.D., Davison, P., Zhang, H. & Davison, W. 1994. High resolution porewater sampling with a gel sampler. *Limnol. Oceanogr.* 39(8): 1967–1972.

Ng, J.C., Wang, J. & Shraim, A. 2003. A global heath problem caused by arsenic from natural sources. *Chemosphere* 52: 1353–1359.

Nickson, R., McArthur, J., Burgess, W., Ahmed, K.M., Ravenscroft, P. & Rahman, M. 1998. Arsenic poisoning of Bangladesh groundwater. *Nature* 395: 338.

Nordstrom, D.K. 2002. Worldwide occurrences of arsenic in groundwater. *Science*: 296.

Stolarik, G. & Christie, J.D. 1999. Interim Arsenic Management plan for Los Angeles. *Proc. 1999 Am. Water Works Assoc. Ann. Conf.e* Chicago, 20–24 June, 1999.

Water–Rock Interaction, Wanty & Seal II (eds)

^{14}C and ^{36}Cl as indicators of groundwater flow, Bland Catchment, NSW

E.A. Carrara & T.R. Weaver
Hydrogeology & Environment Research Group, School of Earth Sciences, University of Melbourne, Melbourne, Victoria, Australia

I. Cartwright
Hydrogeology & Environment Research Group, School of Geosciences, Monash University, Clayton, Victoria, Australia

R.G. Cresswell
Bureau of Rural Sciences, Canberra, ACT, Australia

ABSTRACT: Multi-component geochemical tracers together with physical parameters were used to identify the processes controlling groundwater chemistry and flow in the Bland Catchment, NSW Australia. These results are here combined with ^{36}Cl and ^{14}C data to provide a better understanding of the Bland and Lachlan hydrogeological systems for improved groundwater management. In the Lake Cowal region ^{14}C corrected ages indicate that groundwater flow in the Lachlan Formation has not always been consistent with current groundwater flow directions. ^{14}C data also show that the reversal of groundwater flow from the Lachlan Valley to the Bland Catchment has persisted for thousands of years and is not a direct effect of current irrigation practices. ^{36}Cl results support Cl/Br ratios and $\delta^{18}O$ and $\delta^{2}H$ data, showing evaporation in groundwater in the Cowra Formation aquifer in the north center of the catchment.

1 INTRODUCTION

The study area encompasses the Bland Catchment (940,000 ha) and is located within the Lachlan River Catchment of NSW. It is situated immediately outside the northeastern boundary of the sedimentary Murray Basin, but within the broader Murray-Darling River Basin as shown in Figure 1.

Radiometric carbon is used to ascertain groundwater age in the study area. Age data can be used to investigate groundwater flow directions and velocities, hydraulic conductivities, and recharge to groundwater (Zhu 2000; Iwatsuki et al. 2001; Edmunds et al. 2003). Dissolved inorganic carbon (DIC) from various sources can interfere with radiocarbon decay from the atmospheric source. Hence, a thorough investigation on various sources of DIC is essential to transform ^{14}C activity to groundwater age (Edmunds et al. 2003). In this study a $\delta^{13}C$-mixing model is used (Clark & Fritz 1997). ^{36}Cl is a useful tool for understanding environmental conditions such as evaporation (Cresswell et al. 1998), for identifying the contribution of different Cl sources (Bird et al. 1989; Yechieli et al. 1996; Davis et al. 2001; Edmunds et al. 2003), and for identifying groundwater flow paths (Lehmann et al. 2003) and recharge or discharge mechanisms (Davie et al. 1989).

The main objectives of this study are to investigate: (a) groundwater residence times, flow directions and recharge areas in the parts of the Bland Catchment, (b) the degree of connectivity or separation between the Cowra and the underlying Lachlan Formation, and (c) to substantiate the environmental processes (e.g. evaporation) recognised in the stable isotope and major ion data.

2 HYDROGEOLOGY AND GEOCHEMISTRY

The study area is situated within the Palaeozoic Lachlan Fold Belt that extends across eastern Australia. The dominant hydrogeological units in the catchment are Tertiary alluvial deposits infilling the buried Lachlan and Bland palaeochannels that incised the Palaeozoic bedrock (Williamson 1986). These are the most important sources of groundwater. The Miocene to Pliocene Lachlan Formation occupies the base of the palaeovalleys, whereas the less

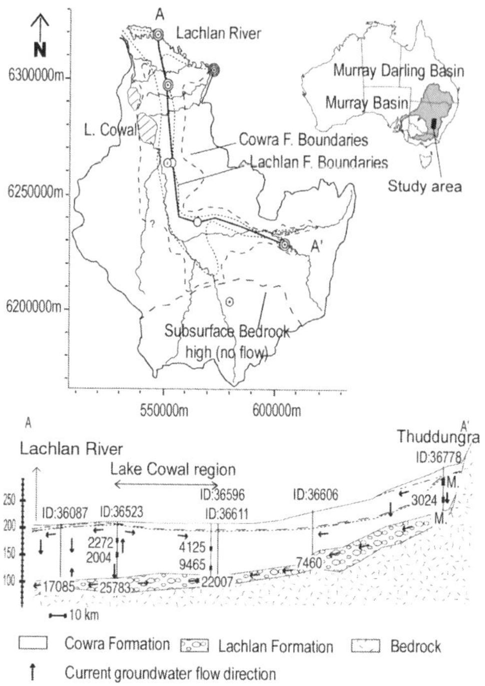

Figure 1. Location of major aquifers and spatial distribution of ^{14}C corrected groundwater ages along Bland palaeochannel. Ages are in order of depth from shallower to deeper screen interval. (M. = Modern-age groundwater).

permeable Pleistocene Cowra Formation occupies the upper, wider portion of the palaeovalleys. The Bland palaeochannel intersects the Lachlan palaeochannel at its outlet, and Lake Cowal is located at the lower end of the Bland Catchment before it enters the Lachlan Valley (Fig. 1).

Groundwater in the Lachlan Fm. in the Bland C has relatively low electrical conductivity values (EC < 5000 μS/cm) whereas groundwater in the more heterogeneous Cowra Fm. can be more saline (EC 741–48,400 μS/cm). Generally EC values increase along the flow path towards Lake Cowal in both aquifers in the Bland C. Groundwater in the Lachlan Valley has much lower EC values (max 1573 μS/cm). Major ions in the Lachlan and Cowra Fms. present a spatial trend of increasing Cl and Na + Mg consistent with both EC increase and groundwater flow toward Lake Cowal. Na/Cl ratios in groundwater range from 0.5 to 8.14. The variable Na content indicates minor water–rock interaction, mainly localised in the Lachlan palaeochannel and in the southeastern area of the Bland palaeochannel. Most Cl/Br molar ratios throughout the study area range from 400 to 750, thereby excluding significant halite dissolution. $\delta^{18}O$ and $\delta^{2}H$ values of groundwater from the Cowra Fm. aquifer have a slope of 3.8 in the Lake Cowal region which indicates evaporation. Elsewhere samples lie on the meteoric water line.

3 METHODOLOGY AND RESULTS

In November 2002, 13 groundwater samples were collected for ^{14}C, $\delta^{13}C$ and ^{36}Cl analysis from the observation bores shown in Figure 1. Where possible nested piezometers screened in both the Cowra and Lachlan Formations and located along the main groundwater flow path in the Bland and Lachlan palaeovalleys were sampled.

^{14}C and $\delta^{13}C$ analyses were performed through CSIRO Land and Water, Adelaide. ^{14}C contents were determined by AMS on the 14UD tandem electrostatic accelerator at the Australian National University. The ^{14}C concentrations are expressed as percent modern carbon (pMC).

Samples for ^{36}Cl analyses were precipitated at Bureau of Rural Sciences lab facilities following the procedure described by Cresswell et al. (1998). Subsamples of the precipitate were then analysed for ^{36}Cl by the AMS system at ANU, using the 14UD accelerator (Fifield et al. 1987). ^{36}Cl concentrations are expressed as $^{36}Cl/Cl \times 10^{-15}$, and as total $^{36}Cl \times 10^{7}$ at/L. Table 1 lists the results for groundwater samples in the study area.

$\delta^{13}C$ and pH values were used to correct the radiocarbon data and determine net groundwater age. Groundwater ages were derived using the variation of the Pearson model described in Clark & Fritz (1997) with a correction parameter (q) calculated as:

$$q = (\varepsilon^{13}C_{DIC} - \delta^{13}C_{carb})/(\delta^{13}C_{rech} - \delta^{13}C_{carb}) \quad (1)$$

where $\delta^{13}C_{rech} = \delta^{13}C_{soil} + \varepsilon^{13}C_{DIC-CO_2}(soil)$; $\delta^{13}C_{carb} \sim 0‰$; $\delta^{13}C_{soil} \sim 23‰$; $\varepsilon^{13}C_{DIC-CO_2}(soil)$ = pH dependent enrichment between soil CO_2 and aqueous carbon at a given temperature, in this case 15°C. Consequently, the decay equation corrected for $\delta^{13}C$ is:

$$t = -8267 \ln[a_t\,^{14}C/(q a_o^{14}C)] \quad (2)$$

where $a_t^{14}C$ = radiocarbon activities pMC; $a_o^{14}C$ = 100 pMC; 8267 = 1/lλ; $\lambda = (\ln 2)/t_{1/2}$; ^{14}C $t_{1/2}$ = 5730 yrs.

Groundwater ages are reported in Table 1 and range from modern (younger than 1950s) to a maximum of 25,783 yr. $^{36}Cl/Cl$ ratios in groundwater range from 122×10^{-15} to 216×10^{-15}. ^{36}Cl content vary from 3×10^{7} at/L to 2423×10^{7} at/L. The relative short residence time of groundwater (determined with ^{14}C) in the study area suggests that meaningful ages cannot be determined using ^{36}Cl (half-life 301,000 yr).

Table 1. ^{14}C and ^{36}Cl results.

Samples	$\delta^{13}C$	^{14}C pMC	σ	^{14}C* Age yrs	σ	$^{36}Cl/Cl$ 10^{-15}	σ	Cl mg/l	^{36}Cl 10^7 at/L
36079p4 L/LF	−15.9	56.8	1.3	5266	184	122.4	6.9	20.95	4.36
36087p3 L/LF	−14.4	11.3	0.5	17085	367	138	9	279.68	65.56
36523p1 B/CF	−17.2	70.6	1.6	2272	178	145.7	7.5	12.81	3.17
36523p2 B/CF	−17.1	72.2	1.6	2004	178	156	10	13.5	3.58
36523p3 B/LF	−13.7	3.7	0.2	25783	375	179	11	360.76	109.7
36596p1 B/CF	−13.1	34	1.4	4125	319	159	11	8972	2423.32
36596p2 B/CF	−14.3	25.1	0.6	9465	204	141	9	9648	2310.9
36606 B/LF	−12.6	33.3	0.9	7460	208	B.P.	10	1020.2	284.22
36611 B/LF	−11.4	5	0.2	22007	347	164	11	169.48	62.19
36738p2 B/LF	−13.4	14.2	0.4	14668	234	159.7	8.2	509.21	137.97
36778p1 B/CF	−11	95.4	2.1	M.	173	159.5	10	35.5	10.49
36778p2 B/CF	−16	68.7	1.6	3024	181	174	10	46.5	12.88
36778p3 B/LF	−13.6	83.9	1.8	M.	172	163	8.6	1463	396.89

L = Lachlan Valley, B = Bland Catchment, M = Modern, LF = Lachlan Formation, CF = Cowra Formation, B.P. = Bomb Pulse, * Correction = variation Pearson model (Clark & Fritz 1997).

4 DISCUSSION

4.1 *^{14}C dating*

Figure 1 shows that groundwater in the Lachlan Fm. gets older along the flow path from Thuddungra (modern) to the north of Lake Cowal (~26,000 yr) and then younger again (~17,000 yr) towards the Lachlan Valley. Similarly, groundwater in the Cowra Fm. gets older from Thuddungra to the south of Lake Cowal (~9500 yr), but then gets younger (2000 yr) towards the Lachlan Valley. No data are available for Cowra Fm. groundwater in the Lachlan Valley.

In this case a shift in the location of the greatest groundwater age in the two aquifers most likely indicates a change in groundwater flow direction. The ^{14}C ages in the Lachlan Fm. indicate long-term groundwater flow from Thuddungra to the north of Lake Cowal in the Bland palaeochannel. By comparison the oldest groundwater in the Cowra Fm. (~9500 yr) is located to the south of Lake Cowal in accord with modern groundwater flow directions (Fig. 1). These data indicate that the modern groundwater flow reversal from the Lachlan Valley to south of Lake Cowal has persisted for thousands of years and is not solely a result of modern irrigation in the north area of the Bland C and Lachlan Valley. Consequently a zone of low flow or stagnation may exist in the Lake Cowal area, due to the influence of flooding from the Lachlan River.

Two ^{14}C groundwater ages calculated for the Lachlan Fm. in Lachlan Valley indicate that groundwater increases in age along the main flow path from Jemalong Gap (5266 yr) to the west (17,085 yr). This significant increase is inconsistent with the fresh groundwater quality and the high hydraulic conductivity (>10 m/day) of the Lachlan Fm. However the increase in EC from east (437 μS/cm) to west (1573 μS/cm) and the significant increase in age most likely indicate a limited input of older and saltier groundwater to the Lachlan Valley from the Lachlan Fm. in the Bland Catchment.

In the Thuddungra region, groundwater age increases downward in the Cowra Fm. from modern to 3024 yrs old, and then decreases in the Lachlan Fm. to modern. The presence of modern-age groundwater confirms this as a recharge area for both aquifers.

^{14}C values in pMC show a general linear correlation with the depth of sampling (27–125 m), with younger groundwater present closer to the ground surface. Radiocarbon dating also shows a generalised preferential lateral flow throughout the deeper Lachlan F. In fact, groundwater ages do not indicate any significant vertical recharge to the Lachlan Fm. except in the Thuddungra region. By contrast, groundwater ^{14}C data in the shallower Cowra Fm. show a system in which vertical infiltration dominates and lateral flow is limited and local. The limited lateral flow probably reflects the occurrence of sand and gravel lenses.

Figure 2 shows that ^{14}C values (pMC) and the $^{36}Cl/Cl$ ratios are not correlated. There are two main populations of samples. Group A has high ^{14}C values (>56 pMC) whereas group B has less than 35 pMC implying older groundwater ages. All samples except 36606 have similar $^{36}Cl/Cl$ ratios ($122 - 179 \times 10^{-15}$). Sample 36606 has a high $^{36}Cl/Cl$ ratio that indicates bomb impulse input, but also has a low ^{14}C content. The bomb impulse is visible in the chloride isotopic ratio but not in the ^{14}C activity value probably because of dilution of the ^{14}C bomb input.

4.2 *^{36}Cl as an indicator of the processes affecting groundwater chemistry*

A range of environmental processes influence ^{36}Cl and the total chloride contents of groundwater samples differently (Mazor 1992; Yechieli et al. 1996). As well as indicating some very recent groundwater recharge, as at 36606, ^{36}Cl values are here used to assist in the interpretation of environmental conditions

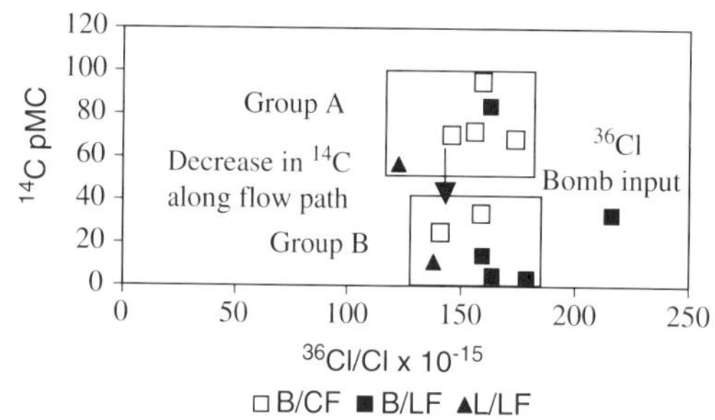

Figure 2. ^{14}C vs. $^{36}Cl/Cl$. (Legend as per Table 1).

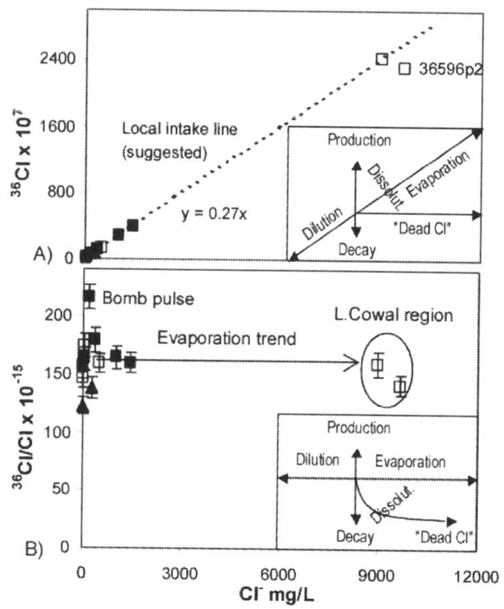

Figure 3. A) ^{36}Cl against Cl. The data lie on or below the suggested local intake line, following Mazor (1992). Sample 36596p2 shows halite dissolution as well as evaporation. Environmental processes affecting ^{36}Cl following Bird et al. (1989). B) ^{36}Cl against Cl. Diagram showing evaporation as dominant process in groundwaters in the Cowra Formation from Lake Cowal region. (Legend as per Figure 2).

such as evaporation (e.g. Edmunds & Smedley 2000). Figure 3a indicates that all samples in the study area fall on a dilution-evaporation line (Bird et al. 1989).This line (slope 0.27) is well in accord with the intake line proposed by Mazor (1992) for the Lachlan Fan area (slope ~0.25) located 150 km east of the study area. The positive correlation between ^{36}Cl and Cl content, in combination with $\delta^{18}O$ and $\delta^{2}H$ values clustering along a slope of 3.8 and Cl/Br molar ratios generally lower than 650, all suggest that evaporation is an important environmental process for the buildup of salinity in the study area. This is also supported by the data in Figure 3b, where an increase in Cl content together with a relatively constant $^{36}Cl/Cl$ ratio indicates evaporation. Both $^{36}Cl/Cl$ and stable isotopes data indicate that evaporation is most apparent in the Cowra Fm. groundwater in the Lake Cowal region.

There is no significant dissolution of halite, as this would displace samples away from the proposed intake line on Figure 3a. Only sample 36596p2 is slightly displaced to the right of the intake line. This result, together with a Cl/Br ratio of 714, which is higher than the average for the area, could indicate minor dissolution of halite affecting deep groundwater in the Cowra Fm. around Lake Cowal.

5 CONCLUSIONS

The use of ^{14}C dating improved the understanding of the groundwater flow paths in the Bland C, clarifying the timing of the reversal of groundwater flow. This finding will influence future groundwater management of the area. ^{36}Cl results supported the previous geochemical findings showing the importance of evaporation in the past hydrogeological regime.

ACKNOWLEDGMENTS

This research is supported by a SPIRT grant from the Bureau of Rural Sciences and the Australian Research Council. The authors also thank the Department of Land and Water Conservation of Forbes, NSW for allowing access to bores and data. In particular we are grateful to Martin O'Rourke for his assistance.

REFERENCES

Bird, J.R., Calf, G.E., Davie, R.F., Fifield, L.K., Ophel, T.R., Evans, W.R., Kellett, J.R. & Habermehl, M.A. 1989. The role of ^{36}Cl and ^{14}C measurements in Australian groundwater studies. *Radiocarbon* 31(3): 877–883.

Clark, I. & Fritz, P. 1997. *Environmental Isotopes in Hydrogeology*. New York: Lewis Publishers.

Cresswell, R., Fifield, K., Wischusen, J. & Jacobson, G. 1998. Groundwater sustainability in Central Australia studied using Chlorine-36. *IAH Intern GW conf 1998*. Groundwater: sustainable solutions, Australia.

Davie, R.F., Kellett, J.R., Fifield, L.K., Evans, W.R., Calf, G.E., Bird, J.R., Topham, S. & Ophel, T.R. 1989. Chlorine-36 measurements in the Murray Basin: preliminary results from the Victorians South Australian Malee Region. *BMR Australia Geol. Geophys.* 11: 261–272.

Davis, S.N., Cecil, L.D., Zreda, M. & Moysey, S. 2001. Chlorine-36, bromide, and the origin of spring water. *Chem. Geol.*179(1–4): 3–16.

Edmunds, W.M., Guendouz, A.H., Mamou, A., Moulla, A., Shand, P. & Zouari, K. 2003. Groundwater evolution in the Continental Intercalaire aquifer of southern Algeria and Tunisia: trace element and isotopic indicators. *App. Geochem.* 18(6): 805–822.

Edmunds, W.M. & Smedley, P.L. 2000. Residence time indicators in groundwater: the East Midlands Triassic sandstone aquifer. *App. Geochem.* 15(6): 737–752.

Fifield, K., Ophel, T.R., Bird, J.R., Calf, G.E., Allison, G.B. & Chivas, A.R. 1987. The chlorine-36 measurements program at the Australian National University. *Nucl. Instr. Methods Phys Res.* B29: 114–119.

Iwatsuki, T., Xu, S., Mizutani, Y., Hama, K., Saegusa, H. & Nakano, K. 2001. Carbon-14 study of groundwater in the sedimentary rocks at the Tono study site, central Japan. *App. Geochem.* 16(7–8): 849–859.

Lehmann, B.E., Love, A., Purtschert, R., Collon, P., Loosli, H.H., Kutschera, W., Beyerle, U., Aeschbach-Hertig, W., Kipfer, R. & Frape, S.K. 2003. A comparison of groundwater dating with ^{81}Kr, ^{36}Cl and ^{4}He in four wells of the Great Artesian Basin, Australia. *Earth Planet. Sci. Lett.* 211(3–4): 237–250.

Mazor, E. 1992. Reinterpretation of ^{36}Cl data: physical processes, hydraulic interconnections and age estimates in groundwater systems. *App. Geochem.* 7(4): 351–360.

Williamson, W.H. 1986. Investigation of the Lachlan Valley alluvium. Part 1: Cowra to Jemalong Weir. Water Resources Commission, NSW.

Yechieli, Y., Ronen, D. & Kaufman, A. 1996. The source and age of groundwater brines in the Dead Sea area, as deduced from ^{36}Cl and ^{14}C. *Geochim. Cosmochim. Acta* 60(11): 1909–1916.

Zhu, C. 2000. Estimate of recharge from radiocarbon dating of groundwater and numerical flow and transport modeling. *Water Resources Res.* 36(9): 2067–2620.

Water–Rock Interaction, Wanty & Seal II (eds)

Hydrogeology of the Goulburn Valley, Murray Basin, Australia

I. Cartwright
Hydrogeology & Environment Research Group, School of Geosciences, Monash University, Clayton, Vic, Australia

T.R. Weaver
Hydrogeology & Environment Research Group, School of Earth Sciences, University of Melbourne, Parkville, Vic, Australia

ABSTRACT: Major ion geochemistry, stable isotopes, $^{87}Sr/^{86}Sr$ ratios and $R^{36}Cl$ values show that groundwater in the Goulburn Valley of the Murray Basin contains solutes derived mainly from evaporation of rainfall and minor silicate dissolution. Shepparton Formation groundwater recharges vertically across the region, while Calivil-Renmark groundwater shows a greater component of lateral flow. Groundwater in both formations yields ^{14}C ages of up to 20 ka, implying that recharge rates are low and that the resource is vulnerable to over abstraction. Many tracers indicate that, locally, vertical recharge of groundwater into the Calivil-Renmark Formation has occurred, indicating that this resource is also vulnerable to contamination.

1 INTRODUCTION

Groundwater in Australia is increasingly used for irrigation, stock, and domestic supply, especially during drought. Sustainable groundwater use requires the hydrogeology and hydrogeochemistry to be understood. Here we examine the hydrogeology of the Goulburn Valley of the Murray Basin, which is an area of increased groundwater use that has hitherto undergone little detailed study.

The Murray Basin is a shallow (generally <400 m thick) saucer-shaped basin (Fig. 1). In the Riverine Province, the lowermost Calivil-Renmark Formation consists of fluvial clays, silts, sands, and gravels that form a confined aquifer. In the west of the province, these units are intercalated with lower permeability clays and marls; however, in much of the province such low permeability units are absent. In ancestral drainage channels of present day rivers, the Calivil-Renmark Formation comprises coarse sands and gravels that are channels of preferential groundwater flow (Brown 1989; Cartwright & Weaver 2003). The overlying Shepparton Formation comprises laterally-discontinuous fluvio-lacustrine clays, sands, and silts that form a highly heterogeneous aquifer system. Groundwater recharge to the Shepparton Formation occurs across the Riverine Province and the heterogeneity may inhibit lateral flow, promoting downward leakage into the underlying Calivil-Renmark Formation.

The Goulburn Valley deep lead of Calivil-Renmark sediments is up to 200 m thick in the north (Fig. 1). The Shepparton Formation overlies both the Calivil-Renmark sediments and Palaeozoic basement rocks (indurated turbidites and granites). The depth to the Palaeozoic basement below the sediments of the Riverine Plain increases northwards to as much as 250 m. Groundwater quality in the Goulburn region is highly variable with total dissolved solids (TDS) of 53 to 28,500 mg/L. However, there is little systematic pattern to the salinity variation although, regions of relatively uniform fresh water in the north and southeast of the region exist. Additionally, the deeper groundwater in the Calivil-Renmark deep lead system tends to have lower TDS values than that in the shallower Shepparton Formation.

2 FLOW SYSTEMS

The distribution of hydraulic heads implies that groundwater flow is dominantly northwards. A groundwater mound exists in the north of the area. Horizontal potentiometric gradients are $\sim 10^{-4}$ m/m. At a few localities, dominantly in the Goulburn River floodplain, head gradients are upward; however, at most localities head gradients are typically <0.2 m/m and downwards. The highest vertical gradients are in the vicinity of the groundwater mound.

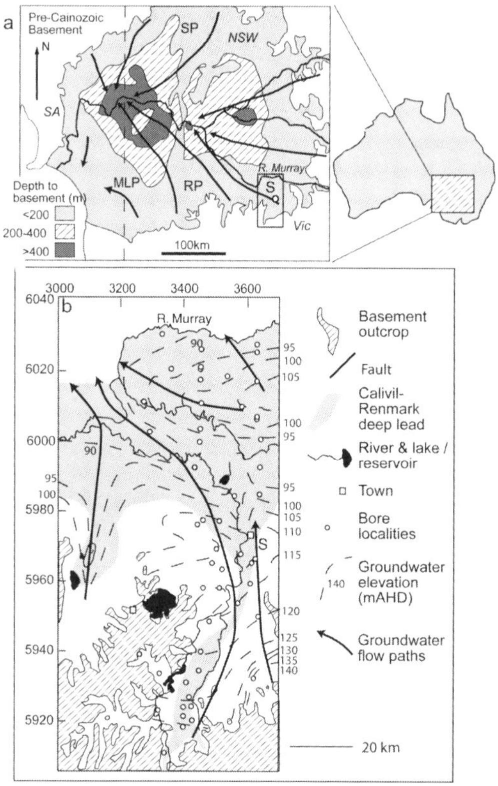

Figure 1a. Murray Basin. RP, SP, MLP are Riverine, Scotia, and Mallee-Limestone Provinces; S = Shepparton. 1b. Goulburn Valley showing the deep lead and groundwater flow paths. Coordinates are from the Australian Map Grid.

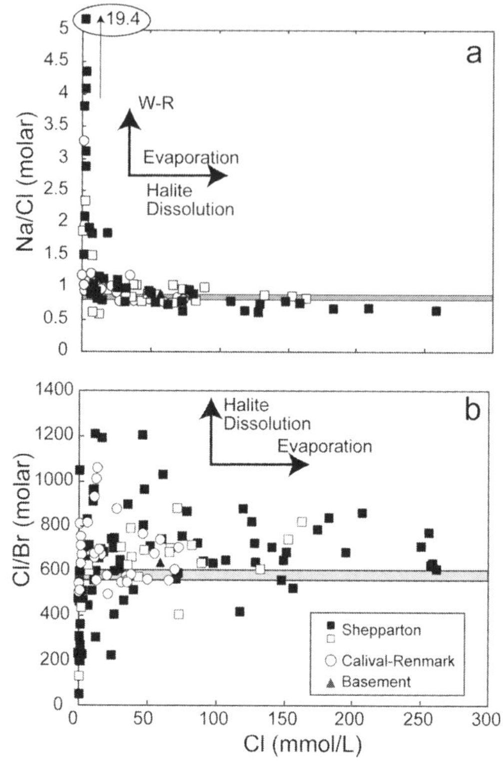

Figure 2. Molar Na/Cl (2a) and Cl/Br (2b) ratios vs Cl for Goulburn Valley groundwater. Open squares are data from Watkins et al. (1999).

Recharge to the unconfined Shepparton Formation occurs across most of the Riverine Plain. The Calivil-Renmark Formation does not crop out and is, therefore, recharged through the Shepparton Formation. This is most likely to occur in the south of the region where depth to the Calivil-Renmark Formation is locally only 20 m and in the north where high vertical head gradients exist in the vicinity of the groundwater mound.

3 HYDROCHEMISTRY

3.1 *Major ions*

Many solutes (Na, Ca, Mg, K, Si, Sr, Mn, Li, Al, Zn) are derived partly from dissolution of silicate minerals, with some control by ion exchange and reactions between clay minerals (Fig. 2a). High Na/Cl ratios in the freshest groundwater are due to alteration of albitic feldspar to kaolinite, which is observed in the aquifer matrix.

The groundwater is undersaturated with respect to calcite, dolomite, and gypsum, and dissolution of these minerals is not significant in controlling groundwater chemistry. Most dissolved inorganic carbon (DIC) is derived from the soil zone. Locally high NO_3 and F concentrations are probably due to recent surface contamination from agricultural land. There are no regular lateral trends in groundwater chemistry or TDS contents, suggesting that there is a major component of vertical flow superimposed on the lateral flow that is implied by the head data.

Br, Cl, and some Na are derived from evaporated rainfall (Fig. 2). Halite dissolution provides some additional Na and Cl. In this area as elsewhere in the southeastern Murray Basin, Cl/Br ratios become less variable with increasing Cl contents (Fig. 2b). This is the result of mixing of groundwater with initially heterogeneous Cl/Br ratios (specifically, inland semi-arid recharge with low Cl/Br ratios and low Cl contents mixing with groundwater that had dissolved minor halite which consequently has higher Cl/Br ratios but still low Cl contents) combined with evaporation to increase the overall salinity.

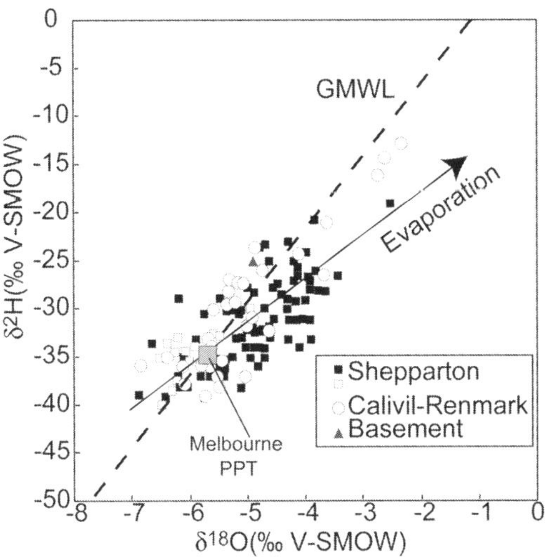

Figure 3. Goulburn Valley groundwater $\delta^{18}O$ and $\delta^{2}H$ values. GMWL = global meteoric water line, Melbourne PPT = average composition of Melbourne rainfall. Open squares are data from Watkins et al. (1999).

3.2 *Stable isotopes*

$\delta^{18}O$ vs $\delta^{2}H$ values of the Goulburn groundwater cluster near the average isotopic composition of modern precipitation in the Melbourne region (Fig. 3). There are no areas where groundwater has distinctly different $\delta^{18}O$ and $\delta^{2}H$ values, implying that climatic conditions during recharge were similar to those of today. The best-fit line through the samples has a slope of ~4.5, confirming evaporation as a major process. As with Cl/Br ratios, $\delta^{18}O$ values become less variable with increasing Cl content as a result of groundwater mixing along flow paths DIC $\delta^{13}C$ values vary from −21.3 to −7.7‰. These values span those typically associated with groundwater where the majority of the carbon is derived from mixtures of C3 and C4 plants (e.g. Clarke and Fritz 1997). There is no correlation of $\delta^{13}C$ values with DIC contents, pH, depth, Cl/Br ratios, or position in the flow system, suggesting that the carbon source has heterogeneous $\delta^{13}C$ values.

3.3 *Spatial variations*

Cl/Br and $\delta^{18}O$ values are more variable in shallow (<20 m) Shepparton groundwater throughout the area in deeper Shepparton and Calivil-Renmark groundwater from the regional recharge area in the south and the area of the groundwater mound in the north. NO_3, and F concentrations are higher in the groundwater with variable Cl/Br and oxygen isotope ratios. These parameters map out areas of recharge and define

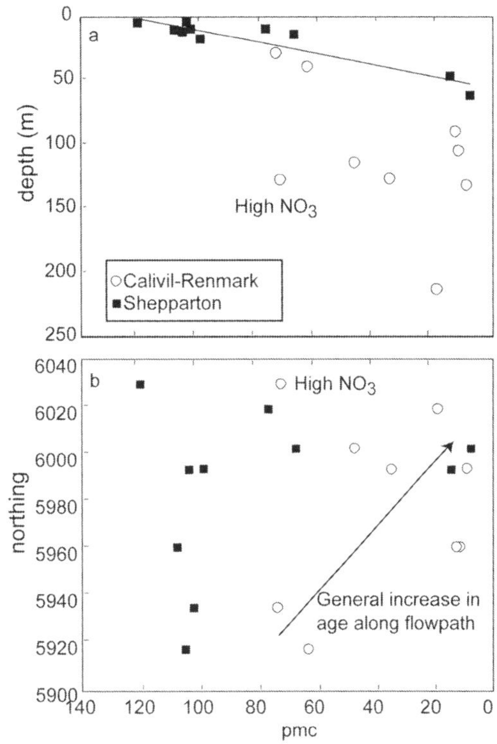

Figure 4. pmc vs depth and distance (northing) for Goulburn Valley groundwater.

where leakage of groundwater through the Shepparton Formation into the underlying Calivil-Renmark Formation occurs.

3.4 *^{14}C*

Percent modern carbon (pmc) values are 7.7 to 120.6 in Shepparton groundwater and 8.8 to 74.2 in Calivil-Renmark groundwater (Fig. 4). A correlation between pmc and depth in the Shepparton Formation groundwater implies that this formation experiences vertical recharge. Calivil-Renmark groundwater shows a broad decrease in pmc values northwards, suggesting that there is considerable lateral flow in this formation.

Samples with anomalously high pmc values given their position in the flow system commonly have high NO_3 concentrations, confirming that nitrate traces the influx of surface water. Overall these data indicate the dual lateral and vertical groundwater flow in the Calivil-Renmark Formation.

Hydraulic parameters may be derived from the ^{14}C and head data. Oldest groundwater residence times from the Shepparton and Calivil-Renmark groundwater are 18–20 ka. For the Calivil-Renmark

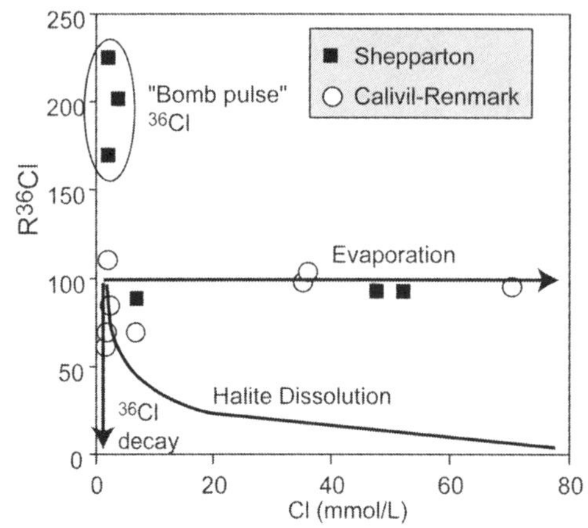

Figure 5. $R^{36}Cl$ vs Cl concentrations for Goulburn Valley groundwater.

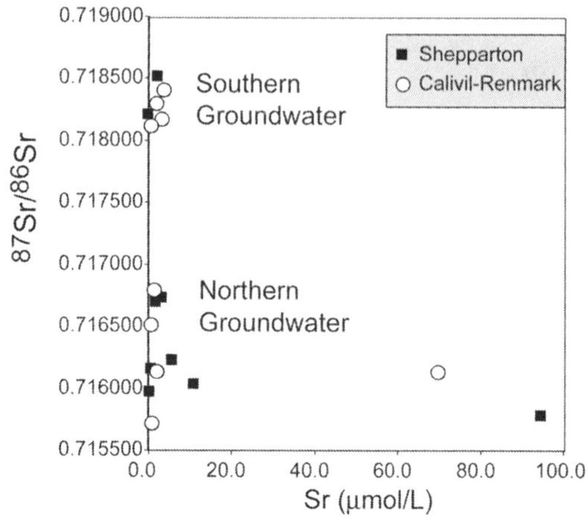

Figure 6. $^{87}Sr/^{86}Sr$ ratios vs Sr concentrations for Goulburn Valley groundwater.

Formation hydraulic conductivities are 2700–6700 m/year. For the Shepparton Formation, vertical recharge rates are ~0.25 cm/year and hydraulic conductivities are 3×10^{-3} to $3\ 10^{-2}$ m/year. These low values may be due to the deeper groundwater in the Shepparton Formation being trapped in sand lenses surrounded by clay. However, the evidence of leakage of groundwater through the Shepparton Formation into the underlying Calivil-Renmark Formation suggests the presence of interconnected sand lenses in the Shepparton Formation that form pathways of relatively rapid flow. Thus, the Shepparton Formation cannot be relied on everywhere to protect the underlying Calivil-Renmark Formation from surface or near-surface contamination.

3.5 *$R^{36}Cl$*

Groundwater from several of the bores screened in both formations has uniform $R^{36}Cl$ values of ~100 regardless of Cl content that we interpret as the $R^{36}Cl$ value of groundwater that recharged the system. The uniform $R^{36}Cl$ values again indicate evaporation rather than halite dissolution. Three groundwater samples with lower $R^{36}Cl$ values have dissolved halite. These are from a range of depths and have a range of pmc values, suggesting that halite dissolution in the aquifer is not solely related to modern dryland or irrigation salinity in this area.

3.6 *$^{87}Sr/^{86}Sr$*

Sr concentrations of Calivil-Renmark and Shepparton groundwater are generally <20 μmol/L and $^{87}Sr/^{86}Sr$ ratios fall into two groups (Fig. 6). Groundwater from the north of the area has $^{87}Sr/^{86}Sr$ ratios of 0.7157 to 0.7168, while groundwater from the south has higher $^{87}Sr/^{86}Sr$ ratios of 0.7181 to 0.7185. The high $^{87}Sr/^{86}Sr$ ratios and low total Sr are similar to those recorded in groundwater from granitic environments (e.g. Harrington & Herczeg 2003), reflecting the predominance of clastic rocks over carbonates in the aquifers of the Goulburn Valley. The northward increase in $^{87}Sr/^{86}Sr$ ratios is probably due to changes in the aquifer matrix.

With one exception, groundwater from the two formations at each locality has closely similar $^{87}Sr/^{86}Sr$ ratios. Since the Shepparton Formation is dominated by vertical flow and the Calivil-Renmark Formation has a greater component of horizontal flow, this is surprising. We interpret these data to indicate that the source of Sr is dominantly in the Shepparton Formation and the $^{87}Sr/^{86}Sr$ ratios are an indication of vertical leakage.

4 CONCLUSIONS

Evaporation is the dominant process in increasing salinity in the Goulburn Valley groundwater, with halite dissolution and weathering playing minor roles. Combined ^{14}C and ^{36}Cl data show that the minor halite dissolution is a long-term process reflecting a continuing semiarid climate. Several geochemical tracers track local recent influx of surface waters into the Calivil-Renmark Formation; however, the Sr data imply that vertical flow between the Shepparton and Calivil-Renmark Formations is a long-term feature of this aquifer system. Major ion and stable isotope data do not show systematic lateral trends implying that the aquifer system has a significant component of vertical recharge that makes it vulnerable to surface contamination.

REFERENCES

Brown, C.M. 1989. Structural and stratigraphic framework of groundwater occurrence and surface discharge in the Murray Basin. Bureau Mineral Resources. *J. Australian Geol. Geophys.* 11: 127–146.

Cartwright, I. & Weaver, T.R. 2003. Modern recharge, water-rock interaction, and groundwater ages in the Goulburn Valley region of the Murray Basin, Australia: implications for resource vulnerability. *Hydrogeol. J.: in press.*

Clarke, I.D. & Fritz, P. 1997. *Environmental Isotopes in Hydrogeology.* Lewis, New York, USA, 328p

Harrington, G.A. & Herczeg, A.L. 2003. The importance of silicate weathering of a sedimentary aquifer in arid Central Australia indicated by very high $^{87}Sr/^{86}Sr$ ratios. *Chem. Geol.* 199: 281–292.

Watkins, K.L., Ivkovic, K.M. & Bauld, J. 1999. Bureau of Rural Sciences Australian Groundwater Quality Assessment Project Report 7. Canberra, 104p.

Water–Rock Interaction, Wanty & Seal II (eds)

Mobilization of F^- from rocks and soils in to the groundwater, Morel River basin, Rajasthan, India

Hema T. Chaturvedi & D. Chandrasekharam
Department of Earth Sciences, IIT-Bombay, Mumbai, Maharashtra, India

ABSTRACT: The drinking water of about 260 million people in 30 countries worldwide contains more than 1 ppm of F^-. In India alone, approximately 62 million people suffer from dental and skeletal fluorosis. We present here results of our investigation from the eastern part of Rajasthan, one of the three states in India that are highly endemic for F^-. Groundwater was sampled from 69 villages, covering approximately 2200 km^2 of the Morel River basin, and F^- concentrations of 0.1 to 13.8 ppm were measured. In the absence of known anthropogenic activity, thick alluvial deposits and the underlying bedrock appear to be the main source of F^- in the water. Local units include F-rich rocks such as mica schist, granite gneiss and conglomerate that in most areas are overlain by 50–100 m thick alluvial cover. Alluvium of the area is rich in micaceous minerals and produced the highest F concentrations in leaching experiments. Although the conglomerates have high F content, the amount of F^- leached from conglomerates was low (on the order of 0.75 ppm after 105 days of contact time). However, larger amounts of F^- (1.2–1.5 ppm) were leached from granite gneiss and mica schist over a similar amount of contact time. The present study indicates that prolonged water–rock and water–alluvium interaction due to deficient rainfall and lack of sufficient flushing of aquifers are factors causing poor water quality in the Morel basin.

1 INTRODUCTION

India is one of 23 countries in the world in which some people depend on groundwater with a F^- content above the limit of 1.5 ppm recommended by the World Health Organization (WHO 1993). About 62 million people in India are suffering from dental, skeletal or non-skeletal fluorosis.

We present here results of an investigation from the Morel River basin in a semi-arid portion of the Indian state of Rajasthan. The F^- content in groundwater is typically above the recommended limit. Due to the dry, hot conditions, individual water consumption is estimated to be around 3–4 L/day, substantially higher than the 2 L/adult/day that is used for calculations of endemic F^- by WHO; hence a much higher population is at risk here. Nearly 0.15 million rural populations depend on groundwater for both drinking and irrigation purposes, and >90% of them suffer from dental and skeletal fluorosis. In order to understand the influence of local geology on this problem, we conducted groundwater analyses and a water/rock–alluvium interaction experimental investigation.

2 STUDY AREA

The Morel River basin covers an area of about 5491 km^2 in eastern Rajasthan. The present study covers an area of about 2200 km^2 and is confined to the part of this basin falling within the districts of Jaipur and Dausa (Fig. 1).

Most of the study area is covered by alluvial deposits that are 50–100 m thick, consisting mainly of sand, silt, unconsolidated clay, and caliche. There are also some large exposures of Mesoproterozoic metamorphic rocks, mainly in the northwestern and eastern parts of the study area (Fig. 1), in addition to an exposure of conglomerate. Most of the metamorphic rocks are quartzite, which form ridges. Other rock types which form the basement of the aquifers include Pre-Cambrian granitic-gneisses and schists. The rocks are highly jointed and fractured. Calcite veins are commonly seen traversing the gneisses. Muscovite and biotite are the major F-bearing minerals in the schist and gneisses, and hornblende is the major F-bearing mineral in the conglomerates.

The alluvium of the study area is characterized by good drainage and aeration, rapid infiltration, good

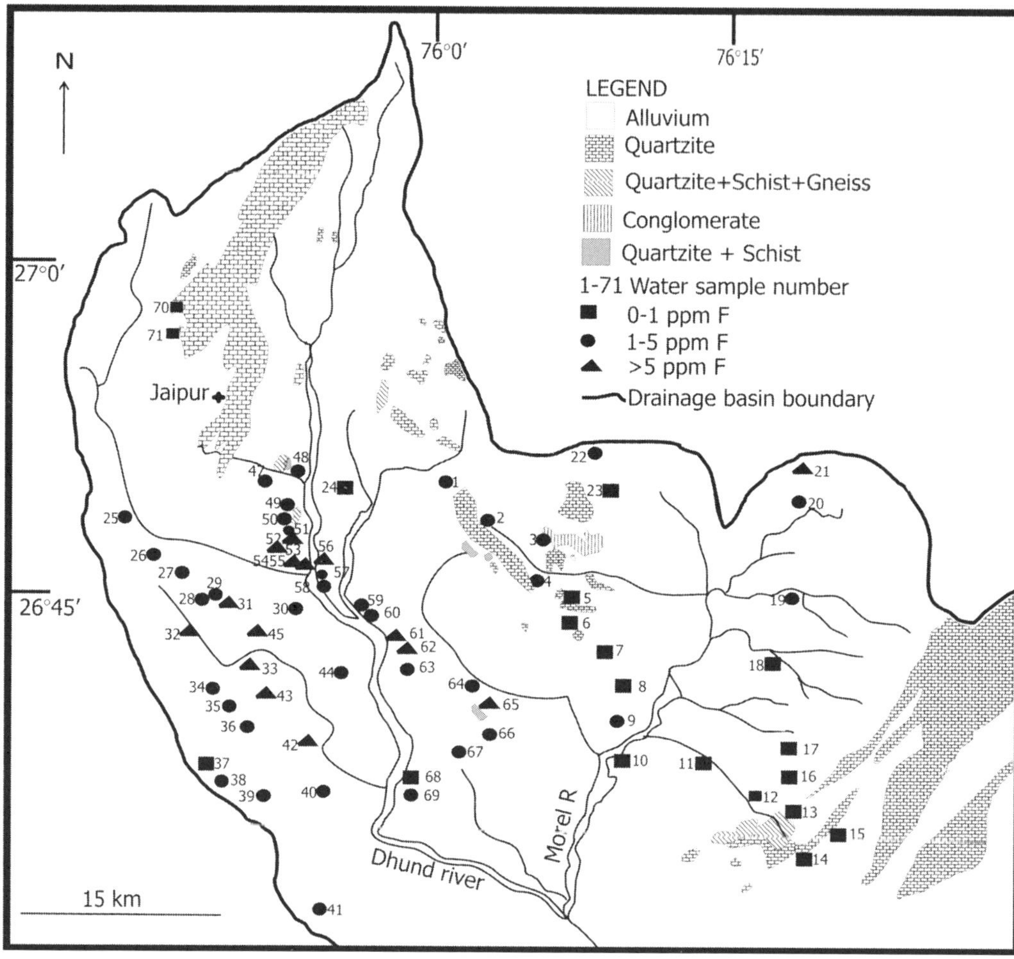

Figure 1. Geological and water sample location map of the study area.

permeability, low moisture retention and holding capacity. The alluvium of the semi-arid parts of Rajasthan are known to be rich in illite and montmorillonite. These minerals in the alluvium have been inherited partly from the parent materials (brought up by the tributaries of the Yamuna River) rich in micaceous minerals (Saxena & Singh 1983) and also partly from the aeolian dust from the western Rajasthan Thar desert.

3 MATERIALS AND METHODS

Water samples were collected from 69 drilled wells and 2 dug wells during the pre-monsoon (May 2002), monsoon (July–August 2002) and post-monsoon (Dec–Jan 2003) seasons (Fig. 1). Six additional drilled wells were sampled during the monsoon and the post-monsoon seasons. Depth of the wells ranges from 13.5 m to 70 m. The majority of the wells tap the alluvial aquifers, however few wells also tap the hard rock aquifers. Water samples were collected in clean polythene bottles washed with distilled water and rinsed with respective water samples. Temperature and pH were measured in the field. The concentrations of the major ions in the water samples were measured following the methods recommended by APHA (1981). The F^- concentration was measured using ion analyzer model EA940 with a F^- ion-selective electrode (Orion model 96-09).

F^- concentrations were determined by the ion-selective electrode method (Bodkin 1977) in eleven rock samples and seven alluvium samples. Mineral identification of the alluvium samples was done using XRD (Rigaku Diffractometer Gigaflex, Right angle goniometer). Seven of these rock samples were used for additional leaching experiments. These samples include quartzite, conglomerate, gneiss, mica-schist, weathered gneiss and weathered mica-schist, and a sample of gneiss with a quartz-calcite vein. For each rock, 100 gm of sample was crushed; sieved and only coarse sand-sized particles with size varying from 1/16 mm to 2 mm were taken. These were placed in covered glass jars and allowed to interact with 1 liter of de-ionized water for a period of 200 days. Similar leaching experiments were carried out on the five analyzed alluvium samples that had the highest F^- content (Table 2). In the alluvium-water interaction experiments, 5 gm of dried, powdered soil sample was stirred for 1 hour with 100 ml of de-ionized water. Immediately after stirring the mixture was centrifuged and the supernatant was analyzed for F^-.

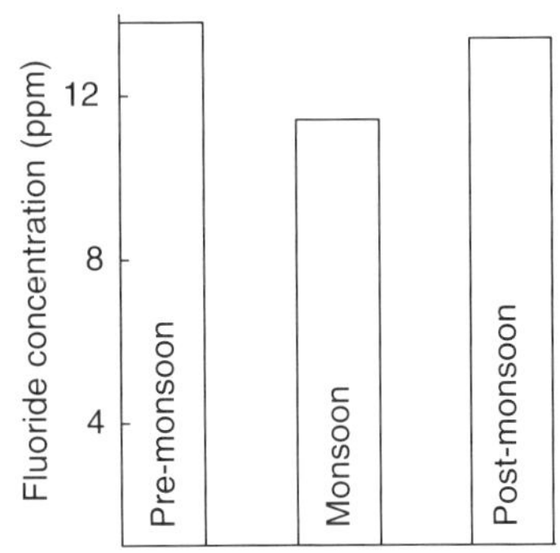

Figure 2. Temporal variation of F^- content.

4 RESULTS

F^- concentrations in the wells varied from 0.02 to 14 ppm. More than half of the wells had F^- levels greater than 1.5 ppm. Although variation in F^- content between wells was evident, the temporal variation in individual wells was not appreciable (Fig. 2). This lack of variation can be attributed, at least in part, to negligible amounts (16.79 cm) of rainfall during the timeframe of the study. Water samples collected from a drilled-well located in quartzite (well 70) had the lowest F^- concentration (0.02 ppm). High F^- values were found in the alluvial region, maximum being 13.8 ppm (well 33). Several wells tapping the alluvium aquifer adjacent to the river beds had low F^- contents (0.2–0.6 ppm). Wells near the quartzite ridges (e.g., wells 1–24) had lower F^- levels (<5 ppm) than most wells in the thick alluvium deposits (>5 ppm) in the western part of the basin. One of the shallow dug-wells (depth-21 m; well 10) had a low F^- content (0.6 ppm) whereas the other (well 42; depth-19 m) was very high (10.2 ppm), attesting to the variation in water quality within the shallow aquifer. All the water samples belong to Na-HCO_3 type. Na concentrations varied from 20–1100 ppm, while Ca concentrations varied from 8–260 ppm. HCO_3 concentration varied from 110–1400 ppm while Cl varied from 4.9–670 ppm. Sulfate concentrations varied from 17–530 ppm. The well water samples were deficient in Ca as compared to Na.

All of the eleven rock samples analyzed contained F-bearing minerals with F content varying from 310–820 ppm (Table 1). The conglomerate (1LG) recorded maximum F (820 ppm), followed by mica schist (4DD) with 740 ppm. F contents in the gneisses varied from 540–610 ppm.

Alluvium samples had much higher levels of F compared to the rock samples (Table 2). Three alluvium samples from a single locality spanned the range of measured F^- concentrations, with values ranging from 2200 ppm in alluvium near a brick kiln to 350 ppm in alluvium near a quartzite quarry. The anomalously high level of F in the alluvium near the brick kiln could possibly be due to certain chemicals which are added during the process of brick making. Maximum F content was found in the alluvium from the mica-schist region at Khori while minimum was also found in the alluvium sample taken from the quartzite quarry at Khori. In the alluvium samples

Table 1. F content of rocks.

Sample no.	Rock type	F (ppm)
1LG*	Conglomerate	820
7KY	Gneiss	544
5DD*	Gneiss (weathered)	580
6SD*	Gneiss	600
8KH*	Gneiss + Qtz-Calc. vein	608
9KH*	Quartzite	368
2LG	Quartzite	384
3LS	Quartzite	308
10KH*	Mica-schist (weathered)	668
4DD*	Mica-schist	740
18TH	Pegmatite	532

* Samples taken for leaching experiments.

Table 2. F content of alluvium samples.

Sample no.	Location	F (ppm)
2s*	Khori brick kiln	2228
5s*	Khori	1338
4s	Khori quartzite quarry	354
17s*	Thalli	866
15s*	Thalli	731
3s*	Alovera	777
1s	Morel River bed	452

* Soil samples taken for leaching experiment.

Table 3. Rock–water interaction results. Results are shown as F^- concentration in ppm as a function of time.

	Days						
Sample	0	45	75	105	135	170	200
Conglomerate 1LG	0.1	0.5	0.6	0.8	0.7	0.8	0.8
Mica schist 4DD	0.1	1.0	1.1	1.1	1.0	1.1	1.0
Gneiss (weathered) 5DD	0.1	1.1	1.4	1.5	1.3	1.4	1.6
Gneiss 6SD	0.1	0.7	0.9	0.9	0.9	1.1	1.1
Gneiss + Qtz-calc. vein 8KH	0.1	0.7	0.7	1.0	1.0	1.0	1.0
Quartzite 2LG	0.1	0.4	0.6	0.7	0.7	0.7	0.9
Mica-schist (weathered) 10KH	0.1	0.7	0.8	0.7	0.7	0.8	0.9

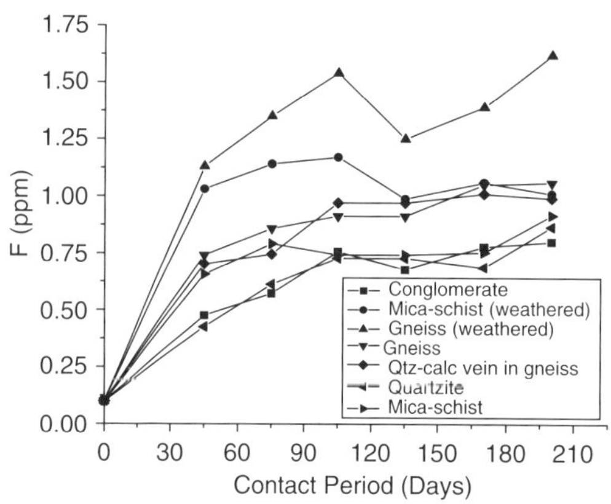

Figure 3. Variation of F^- with contact period.

taken from depths of 1 and 5 m from a drilled-well at Thalli, F increased with depth from 731 to 866 ppm (Table 2).

The results of the rock–water interaction experiments indicate that a considerable amount of F is leached from the rocks (Table 3). The greatest amount of F^- was leached from the weathered gneiss and mica-schist (Fig. 3). This may be due to their weathered nature and/or due to the release of interstitial F (Wanty et al. 1992). Leaching of F^- from the other samples was broadly similar over time.

As indicated in Table 4, the alluvium samples, which were exposed to water for 1 hour as opposed to 200 days, released F^- very rapidly compared to the rock samples. Maximum F^- was leached by the alluvium sample 15s (1.2 ppm) which had lower F^- content (730 ppm) as compared to sample 2s (1.2 ppm) which had maximum (2200 ppm).

Table 4. Alluvium–water interaction results.

Sample no.	Location	Depth (m)	Leached (ppm)
2s	Khori brick kiln	1	1.2
5s	Khori	1	0.5
3s	Alovera	1	0.4
15s	Thalli	1	1.2
17s	Thalli	5	0.9

5 DISCUSSION

The leaching experiments suggest that the bedrock lithology is the probable source of F^- in the groundwater. Compared to the rock samples, the alluvium samples of the area are also capable of contributing substantial quantities of F^- to the aquifer system. XRD analysis of these samples revealed the presence of quartz, anorthite, lepidolite, calcite, halloysite and hendricksite. Rapid leaching of F^- is due to the F^- being adsorbed on the clay mineral surface. Presence of calcite in the alluvium indicates that Ca is not easily leached into the water and thus waters are low in Ca and this enhances F^- concentration in the water.

6 CONCLUSIONS

The present study clearly points towards a geogenic source of F^- in the groundwater of this region. Prolonged rock–water interaction due to low recharge and high evapotranspiration is the main cause for high F^- concentration in the groundwater in the study area. The alluvium also acts as both a source and sink for F^-. Caliche precipitation coupled with low Ca^{++} content in groundwater is responsible for high F^- levels in groundwater.

REFERENCES

APHA 1981. *Standard methods for the examination of water and waste waters*. Washington, DC, American Public Health Association.

Bodkin, J.B. 1977. Method for analysis of with ion-selective electrode for rocks and minerals. *Analyst* 102: 409.

Saxena, S.C. & Singh, K.S. 1983. Clay mineralogy of Semi-arid region soils of Rajasthan. *Indian Soc. Soil Sci.* 31: 85–93.

Wanty, R.B., Folger, P.F., Frishman, D., Bricggs, P.H., Day, W.C. and Poeter, E. 1992. Weathering of Pikes Peak Granite: Field, experimental, and modeling observations. In Y.K. Kharaka & A.S. Maest (eds), *WRI 7:* 599–602. Rotterdam: Balkema.

WHO 1993. Guidelines for drinking water quality. *Revision of the 1984 guidelines*, Final task group meeting, 21–25 September 1992, 188p.

Water–Rock Interaction, Wanty & Seal II (eds)
© 2004 Taylor & Francis Group, London, ISBN 90 5809 641 6

Geochemical baseline as basis for the European Groundwater Directive

W.M. Edmunds & P. Shand
British Geological Survey, Wallingford, UK

ABSTRACT: Defining the natural geochemical baseline conditions is a necessary pre-requisite for aquifer characterization and management and these principles being adopted in the new European Groundwater Directive. The basis for defining baseline is derived using a data set from 25 reference aquifers from 12 countries. Dating tools necessary for recognizing pristine and modern groundwater are discussed as well as the information derived from time series. The importance of 3-D variations in the hydrochemistry of groundwater bodies and changes with time are emphasized as important for characterization.

1 INTRODUCTION

For the purposes of aquifer characterisation and groundwater management there is a need for an improved understanding of the geochemical processes controlling natural water quality. Entirely natural processes may breach existing drinking water quality limits and it is important to be able to assess whether or not anthropogenic pollution is taking place against the natural baseline. A geochemical approach is necessary to define the overall controls and reactions giving rise to the water quality and its diversity, the age distribution of groundwater, and the rates at which natural processes and changes are occurring. The extent to which pristine waters (unaffected by anthropogenic influences) are being depleted or modified by contamination must also be addressed. In this paper, a geochemical basis for a standardised Europe-wide approach is proposed, forming the basis to be used in the new Water Framework Directive and the related Groundwater Directive.

Results from 12 EU member states have been used to illustrate the baseline concept. A series of 25 reference aquifers have been studied which represent aquifer types across Europe. These have been supplemented by in-depth studies looking at groundwater residence times, water quality trends, applications of geochemical modelling, monitoring of quality and transfer to policy (Edmunds et al. 2003).

2 BASELINE DEFINITION

The baseline concentration of a substance in groundwater is here defined as:

"The range of concentrations of a given element, species or chemical substance present in solution being derived from natural geological, biological, or atmospheric sources."

Pristine baseline quality is established by geochemical reactions on a geological timescale within groundwater flow paths, having also been influenced by climate and land use factors. Recent borehole construction and groundwater abstraction induce rapid changes in natural quality disturbing the natural quality stratification including saline intrusion or up-coning from depth. Thus, the baseline will be represented by a range of values (Fig. 1).

The quality of natural groundwater is typically variable and can be described as a distribution around a mean or median value. Natural hydrochemical distributions are typically neither normal nor log-normal. They are often skewed, have outliers and originate from more than one process. Cumulative frequency graphs (log probability plots) are used as the main representation of the baseline range. These have been used extensively in the mining

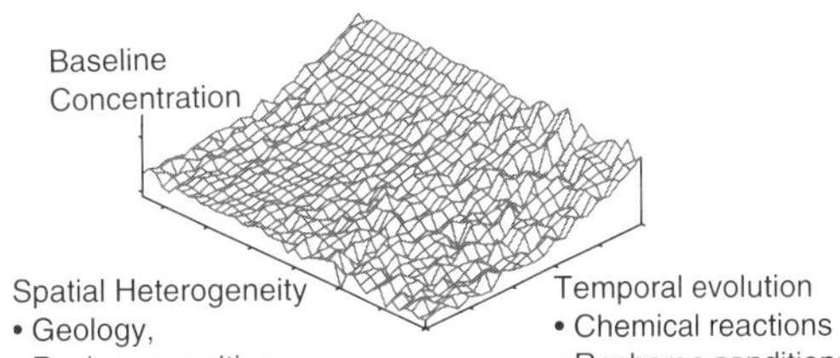

Figure 1. Graphical representation of the groundwater baseline.

industry (Runnells 1998) but less frequently in hydrogeological studies (Langmuir 1997; Edmunds et al. 2003). The operational definition of the baseline concentration is the median value, and the range is from 2.3% to 97.7% of sample population. However, care is needed in applying statistical treatment of data. For instance, when the number of observations is lower than 37, the minimum and maximum values measured will define the baseline range. In defining baseline aquifer properties this statistical approach is only an adjunct to a field understanding of processes.

3 BASELINE WATER QUALITY IN EUROPE

3.1 *Overview of data*

The European data from 25 aquifers (Table 1) have been summarised (for most elements and where numbers of data permit) using cumulative probability plots, which highlight the ranges of data as well as helping to discriminate different populations. The baseline studies cover an extensive range of inorganic species (over 80 elements or species) as well as a detailed evaluation of natural organic carbon (DOC, TOC). Representative plots for 4 elements/species are shown in Figure 2 (NH_4, Fe, F). It can be seen that the concentration ranges typically vary over several orders of magnitude within individual aquifers and there is usually an overlap between different aquifers regardless of rock type. This highlights the major control on water chemistry of a range of hydrochemical processes such as mixing, redox reactions and ion-exchange reactions, in addition to lithology. These, in combination with residence time, dominate the evolution of groundwater chemistry in aquifers.

A number of solutes found naturally in groundwater (including Fe, Mn, F, Cl and SO_4) exceed existing regulatory standards for drinking water, but also noteworthy are the number of aquifers with relatively high concentrations of the trace elements As, Ba and Ni.

The statistical data and plots are useful for comparison between areas and are of strategic value since they include data for elements for which there are as yet no existing guidelines, but which may still be harmful (e.g. Be, Se, Sb, Tl, U). The baseline ranges for a system can be used to recognise contamination incidents by uncommon elements. Comparison of the TOC (based on 439 analyses) and for the DOC (based on 246 analyses) show that both populations resemble a lognormal distribution.

Overall, the results show that the baseline values for TOC and DOC are comparable. The mean and median concentrations were around $2\,mg\,C\,L^{-1}$, while minimum and maximum values ranged from $<0.5\,mg\,C\,L^{-1}$ to $>50\,mg\,C\,L^{-1}$.

Table 1. European Reference aquifers showing location, aquifer age/lithology and numbers of samples (n).

Country	Code	n	Aquifer
United Kingdom	UK 1	43	Triassic Sandstone (Yorks)
	UK 2	56	Triassic Sst. (East Midlands)
	UK 3	59	Wessex Basin Chalk
	UK 4	86	Lower Greensand
Spain	E 2	98	Donana Sands (Pliocene)
	E 1	0	Madrid Basin (Tertiary)
Portugal	P 1	72	Aveiro (Quarternary)
	P 2	90	Aveiro (Cretaceous)
Belgium	B 1	256	Ledo-Paniselian (Eocene)
	B 2	84	Neogene Sands
France	F 2	42	Fontainbleau Sands (Oligocene)
	F 3	14	Valreas (Miocene)
	F 1	18	Lorraine (Lower Triassic)
Denmark	DK 2	64	Miocene Sands
	DK 3	?	Quaternary Sands
	DK 1	35	Tertiary Limestone/ Chalk
Poland	PL 1	40	Kedzieryn Tertiary Sands
	PL 2	33	Bogucice Tertiary Sands
Estonia	EE 2	2394	Devonian Limestone
	EE 1	1551	Cambrian-Vendian Sands
Czech Republic	CZ 1	43	Cenomanian (mixed)
	CZ 2	61	Turonian (mixed)
Bulgaria	BL 1	19	Thracian Plio-Quat. Sandstone
	BL 2	20	Pre-Cambrian Razlog Marble
Malta	M 1	7	Mean Sea-Level

3.2 *Controls*

It is clear that within any groundwater body there are likely to be *geochemical gradients* (sequential changes in chemical properties with time and distance from recharge areas), and also *boundaries* which may give rise to abrupt changes in quality (solution limits such as decalcification; mineral saturation boundaries; redox boundaries; exchangeable cation boundaries). There is commonly a zonation in redox processes as groundwaters evolve along flow lines within aquifers. A redox boundary often exists close to the junction between unconfined and confined parts of the aquifer and this typically coincides a change in residence time caused by the juxtaposition of different flow systems. This may give rise to very different baseline concentrations.

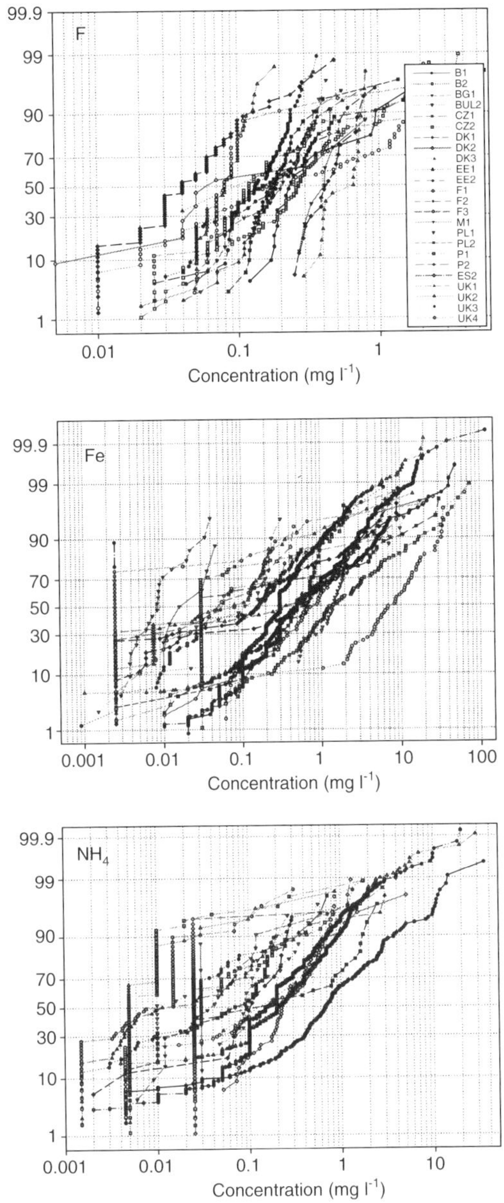

Figure 2. European summary data for NH_4, F, Fe in 25 aquifers shown in Table 1.

Ion-exchange reactions are important controls on baseline where the adsorbed "exchangeable cations" reach equilibrium within a given water–rock system.

This may lead to complex spatial variations, which may continue to evolve over the scale of thousands of years. Mixing with saline formation water is also a dominant process in most of the reference aquifers. This is an important process which may derogate the water in terms of drinking water or ecosystem balance.

It is often difficult to distinguish between the groundwater chemistry in different aquifer types (e.g. carbonate vs. non-carbonate). This is due to the fact that many non-carbonate aquifers contain trace calcite as cement: carbonate minerals dissolve easily and dominate the groundwater chemistry. Only in cases where the carbonate cement has been dissolved (e.g. in recharge zones) will silicate mineral dissolution predominate.

3.3 *Timescales*

In order to define the baseline water quality, some knowledge of groundwater ages (timescales) is required. For this purpose, both inert and reactive chemical and isotopic tracers are used. By investigating the evolution of water quality along flow lines, relative timescales may also be established.

Groundwater residence times in the investigated reference aquifers range from a few days up to many thousands of years. Therefore, a series of tracers have been used to establish timescales in the different European aquifers as the foundation for evaluating time evolution of hydrochemistry and in detecting any modern inputs. Different processes act as "clocks" and can be used to estimate groundwater residence times; accumulation processes (e.g. ^{4}He and chemical parameters), radioactive decay (e.g. ^{39}Ar, ^{14}C, ^{3}H, ^{85}Kr), variable inputs with known chronology (e.g. ^{3}H, ^{85}Kr, CFCs, SF_6,$\delta^{2}H$, $\delta^{18}O$) or combinations of these processes (e.g. ^{85}Kr, ^{3}H). As an example, *palaeowaters*, waters recharged under other climatic conditions than those prevailing during the Holocene and at the present day, can be identified using the isotopic composition of water molecules and the concentrations of stable noble gases in water. Furthermore, ^{39}Ar has been used successfully to date waters of intermediate age (<1000 years), which are between the age ranges obtainable by routinely used tracer methods.

The identification (or absence) of marker species related to activities of the industrial era, such as tritium (^{3}H), dissolved anthropogenic gases (chlorofluorocarbons, SF_6, ^{85}Kr) and certain micro-organic pollutants clearly provide evidence of a *recent* component in groundwater. For the purpose of baseline definition, four age groups of groundwater are recognised and defined:

i. *palaeowater*, water recharged during or before the last glacial era,
ii. *pre-industrial Holocene water* (free of any anthropogenic components);
iii. *pre-thermonuclear era Holocene water* (water free of tritium), and

iv. *modern water*, i.e. water identified by presence of tritium or CFCs or by presence of other key anthropogenic components.

An ideal starting point to establish baseline concentrations is to locate waters where there are no traces of human impact (essentially those from the pre-industrial era), although this is not always easy for several reasons indicated above. Unconfined aquifers and confined aquifers in recharge areas usually contain, especially at shallow depths, anthropogenic components from diffuse sources (e.g. urban inputs, industrial emissions and agrochemicals). As a result of porosity and permeability differences, groundwater is often a mixture of water flowing rapidly through fractures and older, slower-moving water held in the matrix.

The UK East Midlands Triassic Sandstone aquifer has been the subject of inter-comparison studies where the age structure helps to explain the observed chemistry. The combination of careful study of chemical trends along flow lines coupled with age determination and followed by modelling presents a powerful approach for understanding the baseline conditions.

3.4 *Hydrochemical time series*

The use of historical data provides one of best and most direct methods to determine the natural baseline. Unfortunately, there are few good quality records that go back further than a few decades. Even where good time series exist, interpretation may be difficult as a result of changing analytical methods and staff, as well as variable abstraction rates. Moreover, for trace constituents, reliable records are invariably limited to one or two decades. Two examples are shown (Fig. 3).

Grove is a large abstraction site in the East Midlands Triassic Aquifer. It is situated in the confined aquifer and (aerobic) groundwater is shown to be in excess of 10,000 years old. The available records from the mid-1950s to the present day show that nitrate is at or around detection limit and pollution is absent. Chloride *decreases* over the 60-year period from 15 to around 6 mg l^{-1}. This aquifer contains pristine water up to 30,000 years old. The trends therefore indicate that purer (older low Cl) water is being pumped with time and this is consistent with *younger* Holocene water replacing *older* water with slightly higher Cl. Clearly there is no influence of modern (anthropogenically influenced high Cl) water and shows that very small changes in Cl can be used to monitor baseline changes.

In the second example (Everton) in the same aquifer, closer to the outcrop, time series exist from the 1930s. Initial nitrate concentrations are in the range 0–1 mg l^{-1} NO_3-N and represent the pre-development baseline, together with chloride, which initially lay in the range 10–15 mg l^{-1}, consistent with values found in the pristine section of the aquifer (see Grove). Clear and roughly parallel trends in both Cl and NO_3 however indicate the advance of water of the modern era. At Everton for example this trend was marked first by an increase in Cl in the late 1950s and around two decades later by an increase in nitrate.

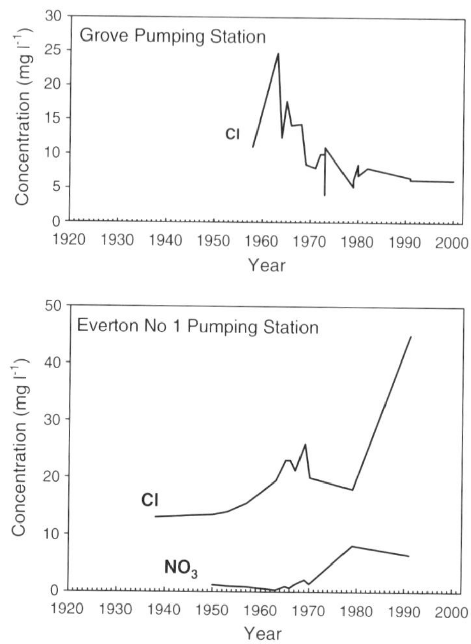

Figure 3. Time series for Cl and NO_3 from the East Midlands aquifer, UK from 1930–2000.

4 CONCLUSIONS

Geochemical criteria form the basis for the characterisation, assessment and management of groundwater quality. Statistical treatment is only an aid and not an end; controls on water quality are three-dimensional, resulting from geological factors and groundwater residence times must also be fully taken into account in understanding water quality.

ACKNOWLEDGMENTS

This paper is a summary of some results of the EU funded BaSeLiNe project (EVK1-CT-1999-0006). The work is that of many authors (Anon 2003) and we acknowledge their permission to present this summary.

REFERENCES

Edmunds, W.M. et al. 2003. Natural baseline quality in European aquifers: a basis for aquifer management. *Final Report to EU Contract EVK-CT1999-0006*. Brussels.

Runnells, D.D., Dupon, D.P., Jones, R.L. & Cline, D.J. 1998. Determination of background chemistry of water at mining and milling sites, Salt Lake Valley, Utah, USA. In G.Arehart & J.Hulston (eds.), *Proc. WRI-9*: 997–1000. *Rotterdam: Balkema*.

Langmuir, D. 1997. *Aqueous Environmental Chemistry*. New Jersey: Prentice Hall.

Water–Rock Interaction, Wanty & Seal II (eds)
© 2004 Taylor & Francis Group, London, ISBN 90 5809 641 6

Geochemistry of groundwater in the Gharb plain aquifers (Northern-Morocco)

N. El Mahmouhi, M. El Wartiti & M. Zahraoui
Département de Science de la Terre – Université Mohammed V, Rabat – Morocco

R. Caboi
Dipartimento di Scienze della Terra, Università di Cagliari, Italy

ABSTRACT: The Gharb plain (Northern Morocco) benefits from a privileged geographical situation, very fertile soils, and relevant water resources. Geochemical characterisations of the groundwater and of the water-rock interaction processes are suggested. A shallow aquifer and a deep multi-layered aquifer have been distinguished. The shallow aquifer is hosted in silty clay formations, it supplies an alkaline-chloride type of water with the total dissolved solids (TDS) typically in the 0.7–1.8 g/l range. The dissolved contents depend on the dissolution of sea salts, calcite, gypsum, and dolomite. Leaching of salt domes and cationic exchange or inverse exchange processes may occur locally, increasing the TDS up to 4 g/l. Less salty water (TDS: 0.2–0.9 g/l) comes from a multi-layered aquifer, and shows a calcium bicarbonate composition related mainly to the interaction with carbonate minerals which prevail in the sediments. All waters are over-saturated with respect to calcite, and under-saturated with respect to gypsum.

1 INTRODUCTION

The Gharb plain is located in the Atlantic coastal zone at the junction of two structural domains of northwest Morocco. The plain extends across 4000 km^2, and represents about 50% of the total surface of the Gharb – Maamora hydrogeological basin (Fig. 1). During the past decades it has witnessed a population explosion (1,500,000 inhabitants with an average annual growth evaluated at 3%). The Gharb plain benefits from a privileged geographical situation in the Kingdom of Morocco, very fertile soils, and important water resources. The renewable water resource in the area is supposedly 150 million m^3/y, and the reservoir/renewable resource ratio amounts to 130 years. These data show the importance of this resource for a semi-arid country like Morocco in a period of increasing drought. This study aims at a preliminary geochemical characterisation of the groundwater in order to set out the water-rock interaction processes as a premise to distinguish natural phenomena from others linked to the anthropogenic contamination of the water. A campaign to measure the physico-chemical parameters and to sample the water has been carried out. The data from the field and laboratory investigations have been integrated with data on the lithology of different drillings, gathered from the General Management of Hydraulics (GDH) and from the Management of Water Research and Planning (DRPE) of the Kingdom of Morocco.

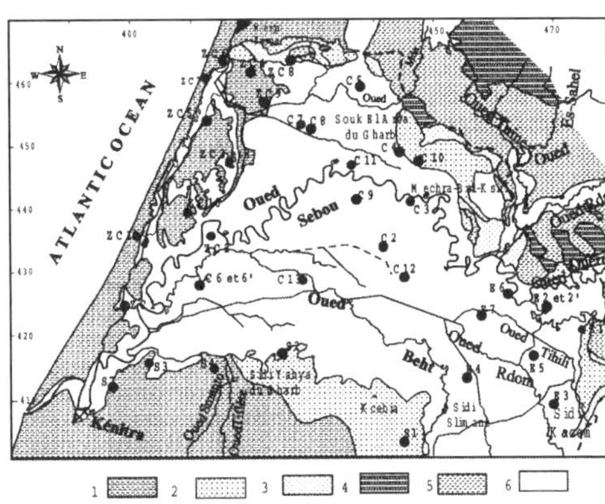

Figure 1. Schematic map of the lithology outcropping in the Gharb-Maamora hydrogeological basin (Kacimi 1992, modified). legend 1: dune, sandstone and sands; 2: sandstone and gravels; 3: sand and sandstone; 4: marls and marly limestone; 5: blue marls; 6 clay and silty clay, thin silt and silty argillaceous sands.

1.1 *Previous studies*

The hydrogeochemical characteristics of surface and shallow waters of the Gharb plain are reported in previous studies on the geochemistry of water carried out on the entire basin of the Sebou River (Caboi et al. 2000; Saddiki et al. 2001; Bahaj et al. 2001) and in a hydrogeological study with Fe and Mn data from the water of Gharb (Kacimi 1992). This paper presents a more detailed study on the geochemistry of shallow and deep water in the Gharb plain.

2 GEOLOGY

Two important geomorphological units are distinguished in the Gharb basin: i) the plain, which represents 90% of the basin, has a triangular form and is crossed by the Oued Sebou and its tributaries (Ouergha, Beht); ii) the coast, formed by a system of sub-parallel dunes which separate the plain from the Atlantic Ocean. The Gharb is limited to the north by Mesozoic and Tertiary formations of the Rif, folded during the Alpine Orogeny, and to the south-west by the Moroccan Meseta, made up of Hercynian basement. The subsidence of the geological basin hosting the Gharb aquifers is related to the same tectonic phenomena that created the Rif Chain (Michard 1976). The bottom of the basin, consisting of a tec tonic nappe dating back to the Miocene, is made of detrital clay, turbidites, and blue gypsum-marl with interbedded Triassic evaporites. The Middle Miocene is represented by 150–200 m thick limestone (Cirac 1985). In the upper Pliocene, the sedimentation is characterised by coastal type deposits: calcarenite sand and sandstone in the southeast, and conglomerates in the north. During the Quaternary, the first transgression is marked by marine deposits essentially made up of calcarenites (Aberkan 1989). This coarse and permeable material makes up the water reservoir of the deep aquifer hosted in the coastal dune complex. Later Quaternary transgressions formed other dunes of essentially marine origin along a parallel strip on the beach.

3 HYDROGEOLOGY

Two aquifer units have been distinguished: i) a shallow aquifer that spreads over the entire plain, except at the borders of the basin where a deep formation outcrops; ii) a deep multi-layered aquifer.

Shallow water is contained in superficial silty clay with low permeability. The upper part of the reservoir is made of clay with a hydraulic conductivity between 1.10^{-7} and 5.10^{-7}, or of sandy silts with a hydraulic conductivity between 1.10^{-7} and 4.10^{-6} m/s. In the coastal zone of the plain, the flow direction is mainly towards the Atlantic Ocean.

The system of deep multi-layered aquifers consists of coarse levels of gravels, and sands whose hydraulic conductivity varies between 1.10^{-4} and 6.10^{-3} m/s, and in the west, part the aquifer sediments are finer grained, essentially made up of sands, silt, and sandy clays whose hydraulic conductivity is estimated between 1.10^{-5} and 5.10^{-5} m/s. The multi-layer character is related to the presence of lenses of permeable coarse sandy material including pebbles, limestones, calcarenites, and sandstones, at times separated by layers of clay. The flow in the deep aquifer is prevalently directed towards the centre of the plain, because of its pan shape, and towards the Atlantic Ocean only on the west side. Except for the borders of the basin where deep formations outcrop, the deep aquifer is covered by an impermeable high layer of clay. Therefore the recharge of the aquifer, which is estimated up to 150 million m^3/y, mainly occurs at the basin borders.

4 HYDROGEOCHEMISTRY

4.1 *Sampling and analytical methods*

Forty-one wells, used for irrigation and the supply of drinking water, randomly distributed on the plain, were sampled in September 2001 and January 2002. Temperature, pH, redox potential, conductivity, alkalinity, NH_4^+, and NO_2 species were measured at the sampling sites. In the laboratory, anions were determined by ion chromatography on filtered, non-acidified samples. Cations were analysed by ICP-OES on filtered samples acidified by 1% HNO_3.

4.2 *Groundwater classification*

Generally, the type of water of the different aquifers of the Gharb varies between bicarbonate and chloride for anions, and between calcium and sodium for cations. Most waters of the deep aquifer can be classified as calcium bicarbonate water. The shallow water can be classified as the alkaline-chloride type with a variable alkaline-earth ion contribution between 20 and 30%. The variation in the concentration of the major ions in relation to total salinity (TDS: total dissolved solid) in the Gharb aquifer (Fig. 2) shows that the water in the deep aquifer is less salty (TDS: 0.24–1.2 g/l). Bicarbonate for anions and calcium for cations give the greatest contribution to salinity.

Comparison of the two aquifers shows the water of the shallow aquifer has greater salinity (most samples in the range: 0.9–1.8 g/l). Sodium and chloride predominate, while sulphate and magnesium are always

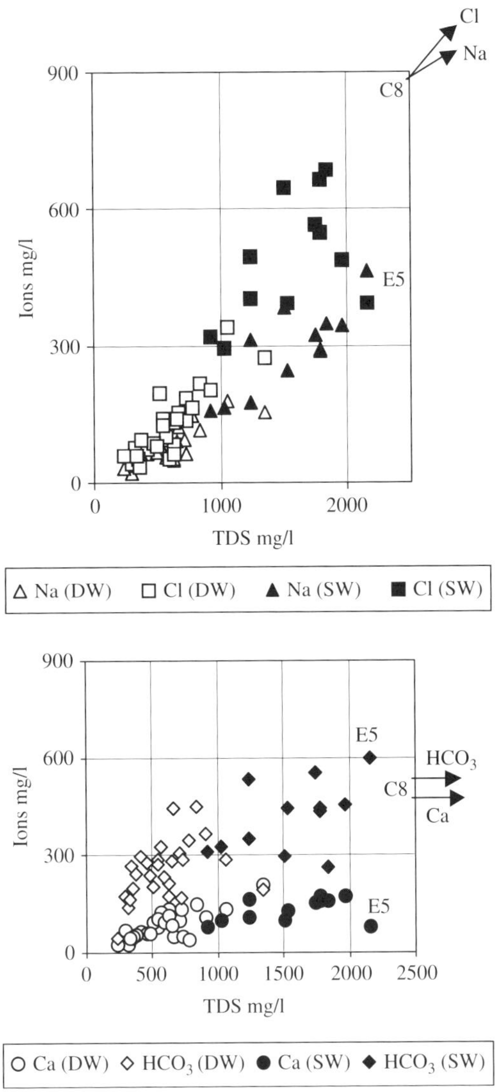

Figure 2. Concentration of main cation and anion vs TDS in the shallow (SW black symbols) and deep water (DW open symbols) samples of the Gharb aquifers.

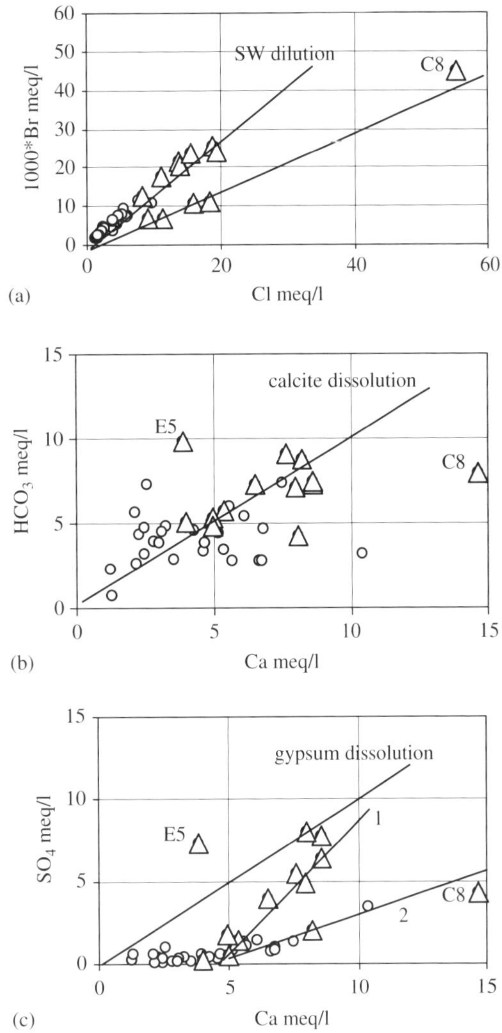

Figure 3. Shallow and deep water evolution lines (see text; circles: deep water; triangle: shallow water).

subordinated to other ions. Two samples, E5 and C8, with respectively TDS = 2.2 and 4.5 g/l reveal a few peculiarities, which will be discussed in the next section.

4.3 *Water geochemistry*

The diagrams in Figure 3 help distinguish the different mineral–water interaction processes in the soils and the sedimentary aquifers. Most samples, both in deep and shallow water, lie on a line with a Br/Cl ratio that can be referred to the sea water dilution line (Fig. 3a). Another trend with a lower Br/Cl ratio than that of sea water results from halite dissolution from salts domes (Hounslow 1995) that are widely spread in the area.

4.4 *Mineral–water interaction*

Calcite dissolution is among the main processes of mineral–water interaction (Fig. 3b). The calcium deficit (about –5 meq/l with respect to other shallow waters lying on the calcite dissolution line) in the "E5" type water can be attributed to the cationic exchange reaction:

$$5Na_2\text{-montm} + 5Ca^{2+} \rightarrow 5Ca\text{-montm} + 10Na^{+} \quad (1)$$

concordant with a +10 meq/l Na excess with respect to most shallow waters. Some other deep watersalso underwent a similar process. The different trends observed in the SO_4 vs Ca diagram (Fig. 3c) show that: i) gypsum dissolution is not the predominant mineral-water process; ii) the two evolution lines of the shallow water suggest a gypsum plus calcite (and dolomite) dissolution for trend 1 and the overlapping of this process with a reaction of inverse cation exchange for trend 2 as:

$$Ca\text{-montm} + 2Na^{+} \rightarrow Na_2\text{-montm} + Ca^{2+} \quad (2)$$

Trend 2 also includes some deep water samples, especially sample C8 which is the saltiest, whose Br/Cl ratio points to interaction with evaporites causing an increase in Na concentration due to halite dissolution.

In Figure 4 SI_{gypsum} vs $SI_{calcite}$ are shown (gypsum and calcite saturation indices, $SI = \log(IAP/K_s)$). Most shallow and deep waters are oversaturated with respect to calcite and all the samples are undersaturated with respect to gypsum. The saturation indices of gypsum for the shallow water were always less than $SI_{gypsum} = -0.5$. This could indicate a control on the calcium concentration is caused by calcite precipitation.

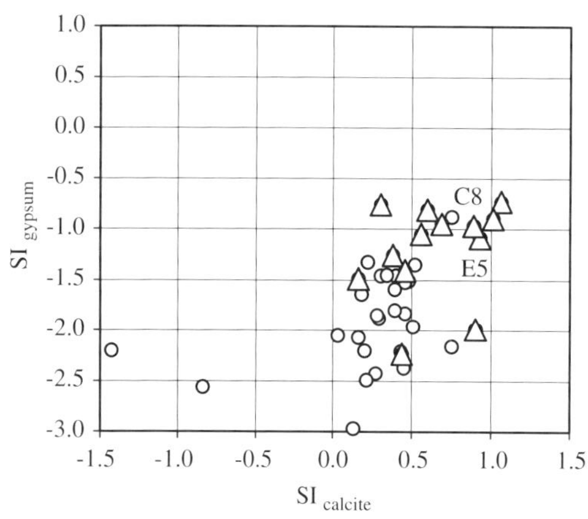

Figure 4. Calcite and gypsum saturation indices in the shallow and deep water of Gharb plain; (symbols as in fig. 3).

5 CONCLUSIONS

The geochemistry of the shallow and deep waters of the Gharb plain depends on overlapping different interaction processes of water and the minerals of the sedimentary marine deposits of the Lower Miocene and the Pliocene.

5.1 *Shallow aquifer*

This aquifer, contained in a superficial silty clay formation with low permeability, can be classified as the alkaline-chloride type with a variable alkaline-earth ion contributions between 20 and 30%. TDS is generally in the 0.9–1.8 g/l range, depending on the dissolution of sea salts, calcite, gypsum and dolomite. At times, cationic exchange or inverse exchange processes which raise the TDS to 4 g/l, can occur.

5.2 *Deep groundwater*

The deep multi-layer aquifer supplies less saline water (TDS: 0.2–1.2 g/l) of the calcium bicarbonate type. Most of these waters are oversaturated in calcite and are linked to interaction with limestones, calcarenites, and conglomerates in the eastern and northern parts of the plain. Interaction with layers of clay minerals interbedded in the permeable levels locally causes Ca/Na or a Na/Ca exchange.

ACKNOWLEDGEMENTS

This study was supported by funds from the Sardinian Regional Government (RAS).

REFERENCES

Aberkan, M. 1989. *Etudes des formations quaternaires des marges du bassin du Gharb (Maroc nord-occidental).* Thèse de doctorat d'état, Université de Bordeaux 1, no. 964, 290 p.

Bahaj, T., El Wartiti, M., Zahraoui, M., Caboi. & Cidu, R. 2001. Shallow groundwater in the Sebou basin (Northern Morocco). In *Proceedings of the 10th Internat. Symp. on Water-Rock Interaction – WR-10, Villasimius (Italy).* 445–447 Rotterdam: Balkema.

Caboi, R., Cidu, R., Carcangiu, M., El Wartiti, M., Zaharaoui M., Bahaj T., Saddiki N. & Bahaj S. 2000. New hydrogeochemical data in the Oued Sebou basin (Morocco). *Proceed. of the 5th International Conference on the Geology of the Arab World,* 226–227. Cairo University, Egypt, February 2000.

Cirac, P. 1985. Le bassin sud-rifain occidental au néogène supérieur. Evolution de la dynamique sédimentaire et de la paléogéographie au cours d'une phase de comblement. *Mem. Inst. Géol. Bassin d'Aquitaine*, Université de Bordeaux 1, no. 21, 287 p.

Hounslow, A.W. 1995. *Water Quality Data: Analysis and Interpretation*. Boca Raton: Lewis Publishers.

Kacimi, I. 1992. *Quelques Aspects sur l'hydrogéologie et l'hydrochimie des nappes aquifères du Gharb - Maroc. Le problème particulier du fer et du Manganèse*. Maitrise en Sciences de la Terre, Université du Québec, Montréal, 154 p.

Michard, A. 1976. Eléments de géologie marocaine, *Notes et Mém. Serv. Géol. Maroc*, Rabat, no. 252, 408 p.

Saddiki, N., El Wartiti, M., Zahraoui, M., Caboi. & Cidu, R. 2001. Anthropogenic impact on the quality of surface waters in the Fès-Meknès plateau (Morocco). In *Proceedings of the 10th Internat. Symp. on Water-Rock Interaction – WR-10, Villasimius (Italy)*. 1135–1338 Rotterdam: Balkema.

Water–Rock Interaction, Wanty & Seal II (eds)
© 2004 Taylor & Francis Group, London, ISBN 90 5809 641 6

Hydrogeochemistry of the Lake Cooper region, Murray Basin

K. Hannam & I. Cartwright
Hydrogeology & Environment Research Group, School of Geosciences, Monash University, Clayton, Vic, Australia

T.R. Weaver
Hydrogeology & Environment Research Group, School of Earth Sciences, University of Melbourne, Parkville, Vic, Australia

ABSTRACT: Groundwater in the Lake Cooper region at the southern margin of the Murray Basin is saline (TDS up to 40,000 mg/L). Cl/Br and stable isotope ratios indicate that evapotranspiration combined with minor halite dissolution is the dominant process in increasing salinity in the area. ^{14}C data indicate that the deeper Calivil-Renmark groundwater is up to 14.3 ka and significantly older than groundwater in the adjacent deep leads. The distribution of ^{14}C ages and Cl/Br ratios suggest that groundwater flow has a large vertical component. Differences in the degree of evaporation relative to transpiration in the shallow and deep groundwater probably reflect rising water tables resulting from land clearing in the last 200 years. The predominance of vertical recharge limits the system's ability to respond to rising water tables.

1 INTRODUCTION

Most research into the hydrogeology of the Murray Basin, southeast Australia, has focused on low-salinity groundwater that represents a potential domestic or agricultural resource. However, much of the Murray Basin groundwater has total dissolved solids of >5000 mg/L. The saline groundwater exists in regional recharge and discharge areas, suggesting that it is not the result of progressive water–rock interaction, and in temperate to semi-arid climatic zones, suggesting that is not simply due to evaporation. Understanding the processes that produce high salinities, especially in recharge areas, are critical to understanding basin dynamics.

1.1 *Hydrogeological setting*

The Murray Basin (Fig. 1) is a shallow ~300,000 km^2 in SE Australia (e.g. Brown 1989). In the Riverine Province, the lowermost Calivil-Renmark fluvial clays, silts, sands, and gravels form a confined aquifer. In the west of the province, these units are intercalated with lower permeability clays and marls; however, elsewhere low permeability units are absent. In palaeoriver channels ("deep leads"), the Calivil-Renmark Formation comprises coarse sands and gravels that are channels of preferential groundwater flow. The Calivil-Renmark Formation between the deep leads, such as at Lake Cooper, has a higher proportion of silts and lower hydraulic conductivities. The uppermost Shepparton Formation comprises fluvio-lacustrine sediments clays, sands,

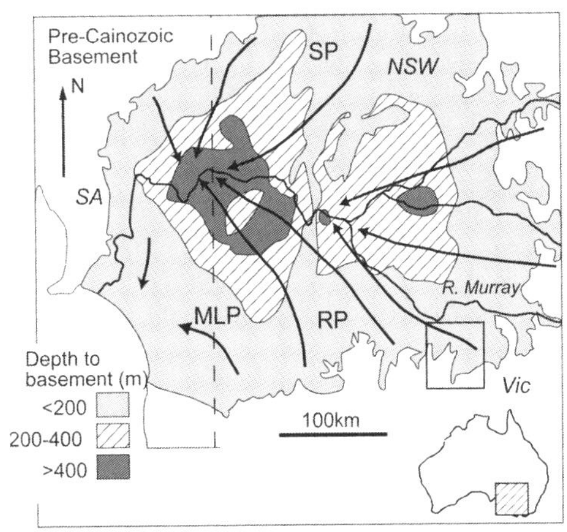

Figure 1. Map of Murray Basin: RP, MLP, SP = Riverine, Mallee-Limestone, and Scotia Provinces. Box shows the location of Lake Cooper (Fig. 2a).

and silts that are laterally discontinuous resulting in a highly heterogeneous aquifer system.

Groundwater recharge to the Shepparton Formation occurs across the Riverine Province and the heterogeneity may inhibit lateral flow, promoting downward leakage into the underlying Calivil-Renmark Formation. Groundwater from both Shepparton and Calivil-Renmark Formations at Lake Cooper is generally saline (Total Dissolved Solids, TDS, up to 40,000 mg/L), and high salinities are recorded up to the margin of the Murray Basin. Groundwater flows to the north and at most localities vertical head gradients are downward.

2 ^{14}C AND FLOW PATHS

Shallow (<25 m) Shepparton groundwater has high percent modern carbon (pmc) values (>94) regardless of location (Fig. 2a). Deeper Shepparton and Calivil-Renmark groundwater has pmc values of 3.9 to 59.1 that show systematic variation. Pmc values generally decline with depth. However, groundwater from the Campaspe and Goulburn deep leads has relatively high pmc values of 39.9 and 59.1 given the depth of these samples (106 and 97 m, respectively). Groundwater from 30–50 m at the southern and western margins of the Lake Cooper embayment has relatively low pmc values (3.9–31.6) given their depths.

Groundwater residence times were calculated from the pmc values using an alkalinity model. All groundwater from <20 m yielded negative "ages", indicating recharge since atmospheric nuclear testing in the 1960's. This reflects the rising water tables in the region and the progressive saturation of the pre-land clearing unsaturated zone. Residence times of groundwater in the Campaspe and Goulburn deep leads are 3.8 and 2.0 ka, respectively. Calf et al. (1986) reported ^{14}C ages of 3.2 to 5.5 ka for Calivil-Renmark groundwater in the Campaspe deep lead, similar to the one age from this study. Calivil-Renmark groundwater from the Goulburn deep lead is younger than that elsewhere in that system (15–19 ka: Calf et al. 1986; Cartwright & Weaver 2003). However, groundwater in the north of the Goulburn deep lead locally has young ages (1–5 ka) due to vertical recharge, and the age from this study is within that range.

Groundwater from Lake Cooper embayment has residence times of up to 25 ka that are significantly older than that in the deep leads, which is in keeping with the higher hydraulic conductivity of the leads. That pmc values do not decrease regularly northward along the groundwater flow paths but increase broadly with depth implies that vertical flow is important. This is due to the Shepparton Formation being an unconfined aquifer system.

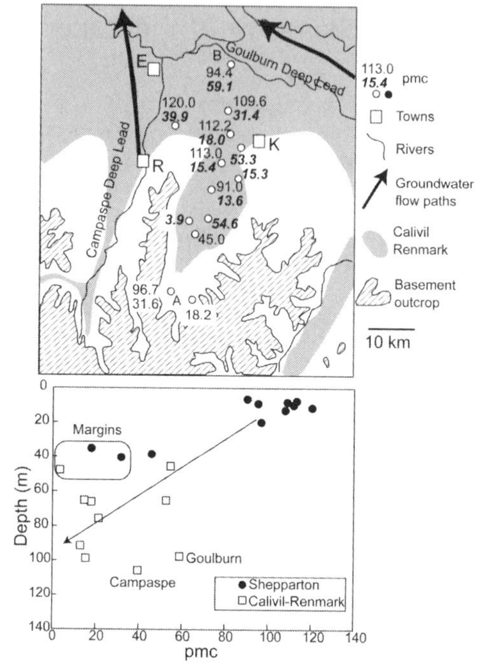

Figure 2a. Map of the Lake Cooper region showing pmc values of Shepparton (normal) and Calivil-Renmark (italic) groundwater. 2b. Variation of pmc values with depth.

3 MAJOR IONS

3.1 *Cl/Br ratios*

With increasing salinity, Cl/Br ratios increase slightly and then remain relatively uniform. However, unlike groundwater from elsewhere in the southern Murray Basin (Cartwright & Weaver 2003; Cartwright et al. 2003), there are two trends in the Lake Cooper groundwater (Fig. 3). Calivil-Renmark and some Shepparton groundwater has Cl/Br ratios of 800–1100, while another group of Shepparton groundwater has somewhat higher Cl/Br ratios (1300–1600).

As halite probably has Cl/Br ratios of 10^4–10^5 (Cartwright et al. 2003), the Cl/Br trends are most constant with minor halite dissolution occurring, but evapotranspiration being the major process in increasing overall salinity. Cl/Br ratios in the different Murray Basin sub-catchments vary proportional to the amount of halite locally present. Compared to the adjacent Campaspe and Goulburn deep leads, Lake Cooper groundwater tends towards higher Cl/Br ratios implying larger stores of halite in this region, which is consistent with the higher volumes of clay-rich sediments.

Aside from the samples from the Goulburn deep lead, the decreasing Cl/Br ratios northward in the

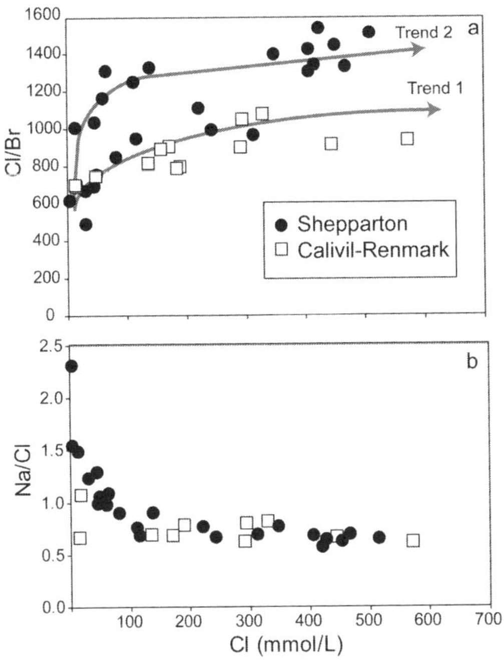

Figure 3. Molar Cl/Br (3a) and Na/Cl (3b) ratios of Lake Cooper groundwater.

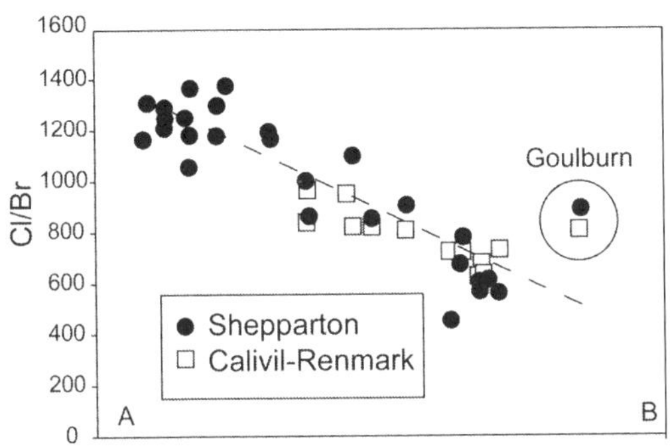

Figure 4. Molar Cl/Br ratios vs distance north. Points A & B are shown in Fig. 2a.

Lake Cooper area (Fig. 4) probably reflects the distribution of halite in the unsaturated zone, with higher concentrations in the south of the area. Groundwater flow is to the north and, given that the groundwater is below halite saturation and water–rock interaction (halite or silicate dissolution) is a potential source of Cl but not Br, it is difficult to interpret the trend in Fig. 4 as being due to groundwater evolution. Paired bores in the Sheparton and Calivil-Renmark Formations also have similar Cl/Br ratios. These data are most consistent with a high degree of vertical flow in the system, as implied by the ^{14}C data.

3.2 *Na/Cl ratios*

Na/Cl ratios of most Shepparton and Calivil-Renmark groundwater are 0.6 to 0.8 (Fig. 3b). Only the least saline samples (Cl <50 mmol/L) have significantly higher Na/Cl ratios (up to 2.4), which reflects weathering of albitic feldspar. However, the relative contribution of weathering to the solute load of the Lake Cooper groundwater is minor. Overall, Na/Cl ratios are again most consistent with the geochemistry of the groundwater being controlled by evaporation. Modern rainfall in this region has a Na/Cl ratio of ~0.8 (Arad & Evans 1987). Groundwater with the lower Na/Cl ratios probably reflects loss of Na by ion exchange with clays.

3.3 *Carbonate and sulfate system*

There is no correlation between Ca or Ca+Mg and total carbon. However, $\delta^{13}C$ values of DIC in groundwater in both formations are +1 to −15‰. This implies that calcite (probably from lacustrine sediments in the Shepparton Formation) contributes to the solutes, but is not a major control on groundwater chemistry. That the high $\delta^{13}C$ carbonate is in the Shepparton Formation again suggests that vertical flow occurs across the area.

Molar Ca/S ratios in the Calivil-Renmark Formation groundwater vary from as high as 6.5 in the less saline water to <1 in the more saline. This reflects a high relative input of Ca from silicate weathering (probably plagioclase) in the least saline samples and a loss of Ca from the groundwater due to carbonate precipitation in the more saline samples. By contrast, molar Ca/S ratios of the Shepparton Formation groundwater generally increase to ~1 with increasing salinity. However, Na/Cl ratios suggest that feldspar weathering contributes to the geochemistry of these waters. Unless the feldspars are dominantly albitic, Ca may be lost onto clays.

Saturation indices with respect to gypsum in all groundwater samples increase as salinity increases, which is the trend expected if evaporation were the dominant process in controlling the chemistry of the groundwater. Log saturation indices with respect to gypsum in the most saline groundwater are close to 0 indicating that gypsum precipitation from these samples may have occurred.

3.4 *Other ions*

The other cations and silica are primarily derived from rock weathering. For many cations, their ratio

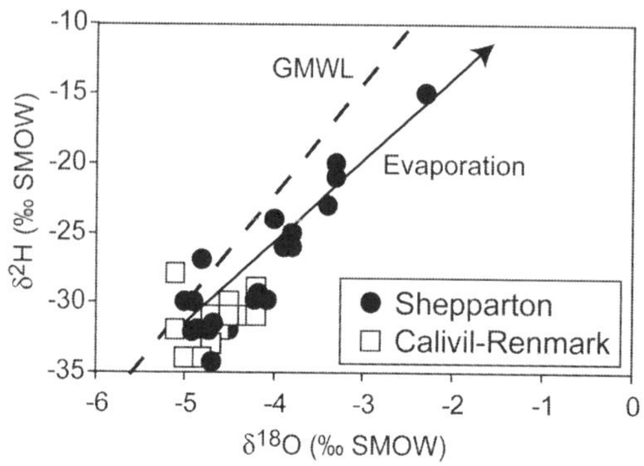

Figure 5. Isotopic composition of Lake Cooper groundwater. GMWL = global meteoric water line.

to Cl decreases with increasing salinity (similar to Na/Cl: Fig. 3b), consistent with rock weathering dominating the chemistry of the least saline groundwater. Nitrate concentrations are locally up to 100 mg/L, especially in the Shepparton Formation groundwater. Such high nitrate concentrations probably reflect surface agricultural sources and provide further evidence of vertical flow.

4 STABLE ISOTOPES

Shepparton groundwater defines a $\delta^{18}O$ vs δ^2H array with a slope of ~5.8 that intersects the local (Melbourne) meteoric water line at approximately the composition of modern precipitation for Melbourne (Fig. 5).

Calivil-Renmark groundwater defines an array with a similar slope and intercept, but with much less variation. In agreement with the Cl/Br data, we interpret the stable isotope data as indicating that evapotranspiration has occurred.

The slopes of these arrays imply that evaporation occurred with a relative humidity of 70–80%, which is appropriate for the climate of northern Victoria. The difference in the relative amount of displacement from the meteoric water line may reflect the relative importance of evaporation in the shallower groundwater vs transpiration in deeper older groundwater. If that is so, the differences may reflect landuse and water table changes.

5 DISCUSSION

The combined use of major ion geochemistry, stable isotopes and ^{14}C has allowed the origins and evolution of groundwater and solutes in the Lake Cooper area to be constrained. The system is dominated by evaporation with silicate and halite dissolution playing relatively minor roles.

The distribution of ^{14}C ages and Cl/Br ratios together with the similarity of geochemistry of groundwater from paired bores in the Shepparton and Calivil-Renmark Formations suggest that the system has a large component of vertical flow. Given this, it should be possible to use ^{14}C ages to constrain flow rates and hydraulic conductivities. Using measured vertical head gradients, ^{14}C ages and an assumed porosity of 0.2 yields vertical flow rates of 10^{-4} to 10^{-3} m/year and hydraulic conductivities of 10^{-10} to 10^{-8} m/s. That these values are low may reflect the clay-rich nature of the Shepparton Formation. Alternatively, modern vertical head gradients are almost certainly higher than those that would have existed prior to land clearing and the subsequent rise in the water table (locally by several metres), while groundwater chemistry reflects longer-term processes. Thus, vertical flow rates over the last 10–20,000 years may have been much lower than those implied by the present day transient head gradients.

Rising water tables may explain differences in the $\delta^{18}O$ and δ^2H values of the groundwater from the two formations. The Calivil-Renmark groundwater would have recharged prior to land clearing when water tables were lower than at present Transpiration, rather than evaporation, of this groundwater by deep-rooted native vegetation may have been the dominant process in increasing salinity. Currently water tables in this region are <5 m, the area is covered dominantly by shallow-rooted grasses, and flood irrigation is or has been practiced. These conditions would potentially lead to a higher degree of evaporation. Some of the increase in Cl/Br ratios in the south of the area might also relate to rising water tables intercepting halite in the previously unsaturated zone.

The ^{14}C data implies that much of the shallow Shepparton Formation groundwater was recently recharged. This also may reflect the recent rise of the water table as the saturated zones may have been below the level of those bores pre land clearing.

While this study has emphasised vertical groundwater flow, clearly there must be transport of groundwater northward out of the region. This may be via zones of locally higher hydraulic conductivity (such as palaeoriver channels). Understanding the fate of saline groundwater from this area is important to assess whether it may impact fresher groundwater in the adjacent deep lead systems (especially as lateral head gradients have increased post land clearing) or contribute salts to the major rivers. The dominance of vertical flow also indicates that the rise in water tables due to land clearing will not be readily relieved by increased lateral flow even if land use practices change rapidly.

REFERENCES

Arad, A. & Evans, R. 1987. The hydrogeology, hydrochemistry and environmental isotopes of the Campaspe River aquifer system, north-central Victoria, Australia. *J. Hydrol.* 95: 63–86

Brown, C.M. 1989. Structural and stratigraphic framework of groundwater occurrence and surface discharge in the Murray Basin. Bureau Mineral Resources. *J. Australian Geol. Geophys.* 11: 127–146.

Cartwright, I. & Weaver, T.R. 2003. Modern recharge, water-rock interaction, and groundwater ages in the Goulburn Valley region of the Murray Basin, Australia: implications for resource vulnerability. *Hydrogeol. J.*: in press.

Cartwright, I., Weaver, T.R., Fulton S., Nichol, C., Reid, M. & Cheng, X. 2003. Hydrogeochemical and isotopic constraints on the origins of dryland salinity, Murray Basin, Victoria, Australia. *Appl. Geochem.*, in press.

Calf, G.E., Ife, D., Tickell, S. & Smith, L.W. 1986. Hydrogeology and isotope hydrology of Upper Tertiary and Quaternary aquifers in northern Victoria. *Australian J. Earth Sci.* 33: 27–42.

Water–Rock Interaction, Wanty & Seal II (eds)
© 2004 Taylor & Francis Group, London, ISBN 90 5809 641 6

Origin of dolomites in the phosphatic Upper Cretaceous Duwi Formation, Eastern Desert, Egypt

H. Holail, M. Shaaban, M. El-Askary & M. Rashed
Geology Department, Faculty of Science, Alexandria University, Alexandria, Egypt

ABSTRACT: The Upper Cretaceous phosphatic Duwi Formation in the Eastern Desert, Egypt, is characterized by a hemipelagic facies at its base overlain by reef sediments. Detailed petrographic and geochemical investigations (trace elements and C and O stable isotopes) reveal the presence of six dolomite petrotypes. These include: 1) authigenic dolomite within phosphorites (type A), 2) dolomitized reefs, which involve selective dolomitization (type B), pervasive mimic and non-mimic dolomite replacement (type C) and void-filling dolomite (type D), 3) dolomite lenses (type E), and 4) phosphatic marly dolomite (type F). Although each of the recorded dolomite petrotypes possesses a distinct and characteristic crystal-size distribution, crystal boundary shape, zoning, and luminescence, they are all ordered and near stoichiometric. Dolomitization is closely associated with the development of secondary porosity; dolomitization pre- and post-date dissolution and corrosion and no secondary porosity generation is present in the associated upper Cretaceous phosphatic-limestone sequence. The most common porosity types are non-fabric selective moldic and vuggy porosity and intercrystalline porosity. The absence of evidence for hypersaline or freshwater conditions during deposition, the marine $\delta^{18}O$ values (+2.0 to −4.7‰ PDB) of the dolomites, and the $\delta^{13}C$ values (−2.5 to −6.8‰ PDB) and high organic carbon content (0.3 to 2.1%) supports the formation of dolomite types A, B, C, E, and F from marine-derived pore waters within the sulphate-reduction zone. The clear limpid euhedral form and limited distribution of dolomite type D suggest formation from mixed meteoric-marine water derived as a consequence of uplift during post-Eocene time.

1 INTRODUCTION

The essence of the dolomite problem is based upon two facts; the first is the presence of giant bodies of ancient massive dolomites with no present-day analogues while the second is the failure to synthesize dolomite under normal sedimentary conditions. Ancient dolomites, however, exhibit complex geological histories documenting the effect of the variable conditions involved in their formation as well as the multiple stabilization events they pass through.

In spite of extensive studies there are still many problems remaining in our understanding of dolomites. In this study, we present integrated field, petrographic, and geochemical investigations of dolomite from the Upper Cretaceous phosphatic Duwi Formation in the Eastern Desert of Egypt. Petrographic characteristics and quantitative modeling of compositional variations are used to interpret the origin of the dolomite and to place constraints on the origin, composition and modification of Upper Cretaceous fluids in the phosphate basin.

2 GEOLOGICAL SETTING

Klitzsch & Squyres (1990) demonstrated how marine transgressions have covered large areas of Egypt concurrent with the time at which the huge Pangaea continent began to break up and the African plate returned to its original slightly northward dip. During this period, the uplift of Egypt came to an end and the depositional basin was shifted further southward. The onset of the marine deposition in the Eastern Desert of Egypt took place during Senonian time.

The Upper Cretaceous comprises four successive formations; from base to top: Nubia Formation, Quseir Formation, Duwi Formation and the lower portions of the Dakhla Formation. The Duwi Formation, which contains marine deposits of the first marine transgression, is composed of a wide variety of rock types that belong to two main depositional realms (Glenn & Arthur 1990). The first is hemipelagic deposition and is represented by organic-rich black shale, cherts and siliceous claystones and is confined to the base of the formation. The second is characterized by reef limestones. Phosphorite beds are intercalated with the hemipelagic and reef sediments. The examined dolomites are recorded as void fillings and replacements, which occur as bed or lenses.

3 PETROGRAPHY

Dolomitization, as observed in the Duwi Formation, is a complex diagenetic process. The present study has revealed the presence of various forms and types of dolomite including: 1) Authigenic dolomite within phosphorites (type A), 2) dolomitic oyster reefs, which involve partial (selective) dolomitization (type B), pervasive dolomitization (type C) for both the grains and the matrix, and void-filling dolomites (type D), 3) dolomite lenses intercalated with dolomitized oyster reefs (type E), and 4) phosphatic marly dolomites (type F). None of these petrographic types of dolomites exhibit any evidence of an association with pre-existing evaporite deposits. SEM study reveals that porosity associated with type A dolomite is generally represented by some minute intracrystalline and intercrystalline pores. Some of these pores invaded by biogenic silica. In general, intercrystalline porosity is associated with dolomitization of the study upper Cretaceous sequence. Moldic porosity is quite common within the study sequence. It is evident that some of the original limestone grains are partly or completely dissolved contributing to the development of porosity.

4 GEOCHEMISTRY

4.1 *Degree of non-stoichiometry*

All of the investigated dolomite types, except for a few samples, possess a uniform and constant degree of non-stoichiometry. XRD analysis reveals that they generally display sharp and well-developed (211) and (111) superstructure reflections, indicating their highly ordered structures. Few of the analyzed samples fall within the ankerite range (more than 2% $FeCO_3$).

4.2 *Trace elements*

Trace element analyses for Sr, Fe and Mn were undertaken for all dolomite types, except type D which is difficult to sample using a micro-driller. Sr values in general exhibit a noticeable variability among the investigated dolomite types. Dolomite types A and B possess similar Sr values (122 to 357 ppm). Dolomite types C and E on the other hand exhibit lower Sr value (96 to 237 ppm). Dolomite type E possesses the lowest Sr values (62 to 174 ppm). Fe contents are relatively high but few samples possess an ankerite composition. Mn contents, however, show consistent values although some variability in the concentration of this element is noticed among the different study areas. This is probably attributable to variability in the terrestrial supply to these areas.

4.3 *Oxygen and carbon stable isotopes*

The isotopic composition of Upper Cretaceous dolomites has been previously studied in Egypt by several workers (e.g. Holail et al. 1988, 1994). In this study, no major variations are recorded among the $\delta^{18}O$ values for all dolomite types (from +2.0 to −4.7‰ PDB). The results are consistent with the precipitation from slightly modified marine water. Plotting the average values of the investigated dolomites in a temperature/$\delta^{18}O$ dolomite diagram contoured in $\delta^{18}O$ water, and considering that Cretaceous seawater possessed a $\delta^{18}O$ value around −2‰ (Holail et al. 1988), it can be estimated that these dolomites were probably formed within a temperature range between 25° and 28°C.

The negative $\delta^{13}C$ values reported for all studied dolomite types is evidence for the extensive role of sulfate-reducing bacteria as well as the role of organic matter diagenesis in dolomite precipitation through the massive production of ^{13}C-depleted HCO_3^- ions as a result of organic matter degradation.

5 DISCUSSION

5.1 *Stoichiometry of the dolomitization reaction*

It is evident from the negative $\delta^{13}C$ values in all the examined dolomite types, the presence of some scattered pyrite cubes and the high C_{org} content that sulfate-reducing bacteria must have played a role in supplying the HCO_3^- ions required for dolomitization. The precipitation of dolomite type A probably occurred within the upper few centimeters below the sediment/water interface in the sulfate reduction zone according to the reaction:

$$Ca^{2+} + Mg^{2+} + 2HCO_3^- = CaMg(CO_3)_2 + 2H^+$$

where Ca^{2+} and Mg^{2+} ions were derived by diffusion from the overlying seawater.

Fe released from some detrital terrestrial particles probably reacted with the present sulfides to form Fe sulfide minerals:

$$Fe^{2+} + 2H_2S = FeS_2 + 2H^+$$

For dolomite type B, dolomitization started to work on the fine grained-matrix which provides possible enhancement due to the availability of maximum nucleation sites per volume. Meanwhile, dolomite mimic-replacement of the grains started to take place till it was finally interrupted by silicification. The dedolomitization within some of the dolomite rhombs of type B may be enhanced through the production of excess Ca ions. Slower rates of precipitation of type B dolomites are evidenced from the perfect euhedral crystal forms of these dolomites and Fe enrichment within their lattices.

The noticeable similarity in both trace element and isotopic composition for dolomite types C, E, and F suggest that they may have formed as a result of the following mechanism:

$$(2 - x)CaCO_3 + Mg^{2+} + xCO_3^{2-} = CaMg(CO_3)_2 + (1 - x)Ca^{2+}$$

The excess Ca^{2+} released to the solution according to this mechanism may react with some of the SO_4^{2-} present and account for the formation of minute gypsum inclusions observed within some of the dolomite crystals. The relatively low Sr contents and the more stoichiometric nature of these dolomites suggest some further stabilization events. Dolomite type D was probably formed during late-stage of diagenesis from mixed meteoric-marine water derived as a consequence of uplift during post-Eocene time. The clear limpid euhedral forms of these dolomites and their limited distribution support this origin (Folk & Land 1975).

5.2 *Kinetic controls on dolomitization*

A noticeable variation in dolomite supersaturation state can be deduced for all the studied dolomite types. The replacement of both fossil particles and matrix in dolomite type C and the complete dolomitization for the original components in dolomite types E and F, as evidenced from the presence of dolomite fossil ghosts, indicate higher degrees of supersaturation at the time of formation of these dolomites. This results usually in a random addition of ions to the growing dolomite crystals causing formation of an early Ca-rich dolomite phase. The formed dolomites possess non-planar crystal boundaries as a consequence of this sort of crystal growth (Sibley & Gregg 1987).

On the other hand, the relatively lower degrees of supersaturation prevailing during the formation of dolomite type B led to selective dolomitization. This selectivity results from the variable availability of nucleation sites where the fine-grained matrix provides a more favorable substrate for dolomite nucleation while the relative coarseness of the calcite crystals composing the fossil fragments does not enhance complete dolomitization at this saturation state. The association of both dolomite type B and dolomite type C in some localities is caused primarily by temporal changes in the degree of dolomite supersaturation via the variation of the ion activity product in pore water.

The precipitation of the void-filling dolomite types A and D possibly took place from a lower supersaturation state as indicated from their euhedral crystal forms and the planar crystal boundaries (Sibley & Gregg 1987). The relative coarseness of crystals comprising these types of dolomite compared to replacement dolomites may be attributed to the fewer nucleation sites allowed for the dolomite cements, in general causing faster growth rate (Sibley 1982).

All in all, it is obvious that the major kinetic constraints that might have influenced the formation of the dolomite types under study are: sulfate inhibition, effect of the present organic material, activity of CO_3^{2-} and HCO_3^- ions in pore water, Mg diffusion rate, ionic strength of the dolomitizing fluid, difference in the hydration behavior of Ca and Mg ions, the presence of some defects in the crystal lattice of the precursor mineral, degree of super saturation of the precipitating fluid with respect to dolomite and time available for the dolomitization reaction.

5.3 *Nature of the carbonate precursor*

It is obvious that variations among chemical, mineralogical and texture of the host carbonate rock may leave some impacts on the replacement dolomite rock. The possible water-rock interaction involved during dolomitization may be then determined by the original host rock composition, chemistry of the potential dolomitizing fluids and the ambient conditions (e.g. pH and Eh). The relative decrease in $\delta^{13}C$ signatures of replacement dolomites under study rather than those obtained for the calcitic skeletal particles, that preserve their original structure, is attributed to the derivation of part of carbon of the dolomitized particles from HCO_3^- ions supplied to the pore water via organic matter degradation. It is evident that sedimentary dolomites are enriched in their $\delta^{18}O$ by 3–4‰ with respect to their co-existing calcites (Land 1982). The average $\delta^{18}O$ signatures of the studied dolomites, which are enriched by 2–3‰ from values obtained for the preserved calcitic bioclasts, are consistent with Land's conclusion and with the precipitation under low-temperature, shallow-burial conditions.

5.4 *Water chemistry and nature of dolomitizing fluid*

Seawater is the most common Mg-rich fluid in the Earth's crust and it is considered as the major agent of massive dolomitization (Land 1985). However modifications of seawater either by increasing salinity (Carpenter 1980) or decreasing salinity (Folk & Land 1975) or removing sulfate ions (Baker & Kastner 1981) may also produce solutions of high dolomitizing potential. The investigated dolomite types are differentiated into two groups on the Sr/ $\delta^{18}O$ relationship. Group one includes dolomite types A and B which possess the highest Sr values with the latter type exhibiting relatively heavier $\delta^{18}O$ content. Group two comprises dolomite type F, which possesses the lowest Sr values and the narrowest range of $\delta^{18}O$ values. Dolomite types C and E are distributed between the two groups.

The anoxic nature of the dolomitizing water may reduce the Fe and Mn oxide coatings on the detrital terrigenous sediments and increase the activity of the Fe^{2+} and Mn^{2+} ions in solution. The high contents of these two elements in the well preserved calcitic bioclasts (Shaaban et al. 1995) suggest that the investigated dolomites might have inherited some of the Fe and Mn from the $CaCO_3$ precursor. It is only after the dissolved sulfide concentrations have been lowered, which might occur via the transport of gaseous H_2S to the more oxic waters and subsequent oxidation. Thus the Fe sulfide precipitation would have ceased, the reduced Fe would be concentrated in solution and incorporated within the dolomite crystal lattice. In turn, this may cause the precipitation of Fe-rich dolomites.

6 CONCLUSION

Six dolomite types, which exist either as cement materials or replacement products, are recorded within the Duwi Formation based on integrated petrographic and geochemical investigation.

The absence of hypersaline or freshwater indicators – evidenced from field and petrographic studies – together with marine signatures and negative $\delta^{13}C$ values may support the precipitation of five of the studied dolomite types A, B, C, E and F from marine-derived pore waters in the sulfate reduction zone. Dolomite type D was probably formed during a later stage of diagenesis from mixed meteoric-marine water composition.

REFERENCES

Baker, P.A. & Kastner, M. 1981. Constraints on the formation of sedimentary dolomite. *Science* 213: 214–216.

Carpenter, A.B. 1980. The chemistry of dolomite formation. I. The stability of dolomite. In D.H. Zenger, J.B. Dunham & R.L.Ethington (eds.), *Concepts and Models of Dolomitization:* 111–121. *SEPM Special Pub.* 28.

Folk, R.L. & Land, L.S. 1975. Mg/Ca ratio and salinity: two controls over crystallization of dolomite. *Am. Assoc. Petrol. Geol. Bull.* 59: 60–68.

Glenn, C.R. & Arthur, M.A. 1990. Anatomy and origin of a Cretaceous phosphorite-greensand giant, Egypt. *Sedimentol.* 37: 123–154.

Holail, H., Lohmann, K.C., & Sanderson, I. 1988. Dolomitization and dedolomitization of Upper Cretaceous carbonates: Bahariya Oasis, Egypt. In V. Shukla & P.A. Baker (eds.), *Sedimentology and Geochemistry of Dolostones:* 191–207. *SEPM Special Pub.* 43.

Holail, H., Shaaban, M. & Khalifa, M. 1994. Petrographic microheterogeneity in dolomite, Bahariya Oasis, Egypt, *Qatar University Sci. J.* 14: 373–380.

Klitzsch, E. & Squyres, C.H. 1990. Paleozoic and Mesozoic geological history of north eastern Africa based upon new interpretation of Nubian strata. *Am. Assoc. Petrol. Geol. Bull.* 74: 1203–1211.

Land, L.S. 1982. Dolomitization. *Am. Assoc. Petrol. Geol. Ed. Course Series, 24*: 20pp.

Land, L.S. 1985. The origin of massive dolomite. *J. Geol.* 33: 112–125.

Shaaban, M., Holail, H., El-Askary, M. & Rashed, M. 1995. Environment and diagenesis of an Upper Cretaceous bioclastic oyster limestone bed, Red Sea coast, Egypt. *Facies* 33: 121–128.

Sibley, D.F. 1982. The origin of common dolomite fabrics: Clues from the Pliocene. *J. Sed. Petrol.* 52: 1087–1100.

Sibley, D.F. & Gregg, J.M. 1987. Classification of dolomite rock textures. *J. Sed. Petrol.* 57: 967–975.

Water–Rock Interaction, Wanty & Seal II (eds)
© 2004 Taylor & Francis Group, London, ISBN 90 5809 641 6

Groundwater age, chemistry and evolution in fractured granite, Uveghuta, Hungary

I. Horvath, T. Szocs & Gy. Toth
Geological Institute of Hungary, Budapest, Hungary

ABSTRACT: The research was conducted in the context ofexploration for the final disposal of low- and intermediate-level radioactive waste. The study area is approximately 25 km^2 in SW Hungary. The shallow and the deep aquifers were studied to understand the flow system and the hydrogeochemical evolution of waters in the study area, and to provide data for flow and transport model calibration. The upper zone, in which shallow groundwater can be found, consists of Pleistocene loess and weathered granite, whereas the deep flow system comprises fractured granite. Water samples were collected for major elements, trace elements, stable isotopes, and radiocarbon age determinations from precipitation, springs, surface waters, shallow dug wells, shallow (less than 50 m deep) and deep boreholes. Good sample coverage was available for both shallow and deep flow systems.

1 INTRODUCTION

The average annual precipitation is between 630 and 721 mm, from which the net infiltration is around 20 to 25 mm/year. The water table ranges between 20 and 70 m under the ridges of hills and between 0.5 to 2 m under the valleys. The conductivity of the weathered granite is much higher (with 2–3 order of magnitude) than the fresh granite, such that approximately 95% of the infiltrating water flows laterally in this zone and comes to the surface as springs or in the surface waters. The amount of water entering the deep granite zone is only 1 to 1.2 mm/year. From this fact we can conclude that the largest mass transfers take place in the shallow groundwater zone.

2 GENERAL WATER COMPOSITION

2.1 *Main components*

The comparison of medians of main water constituents shows that waters originating from different sources have similar compositions (Fig. 1). The waters have moderate (less then 1000 mg/L) TDS, and except for the deep boreholes, they are Ca-Mg-HCO_3 type waters. In the deep flow system, the amount of Ca and HCO_3 decreases whereas the Na content increases and gives rise of Na-Ca-Mg-HCO_3 type waters. The nitrate and sulfate content also decreases in the deep flow system.

The Si and the Fe (the latter not shown on Figure 1) have higher concentrations in the deep compared to the shallow aquifers.

2.2 *$\delta^{18}O$ & δD distribution*

The $\delta^{18}O$ and δD analyses offer the opportunity to distinguish the Holocene from the Pleistocene waters. The boundary is continuous, partly because of the climate change and partly because of the mixing of

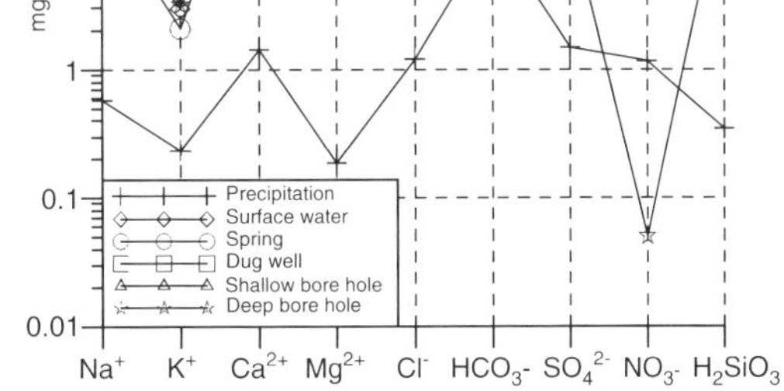

Figure 1. The medians of main water constituents of different origins.

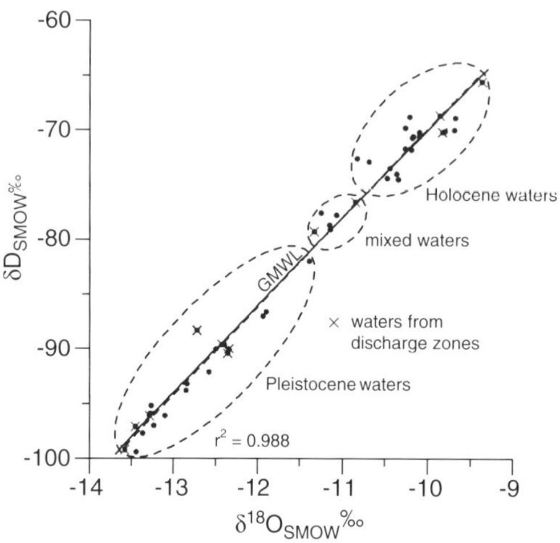

Figure 2. The relation of δD and $\delta^{18}O$ in the deep groundwaters (GMWL – Global Meteoric Water Line; Craig 1961).

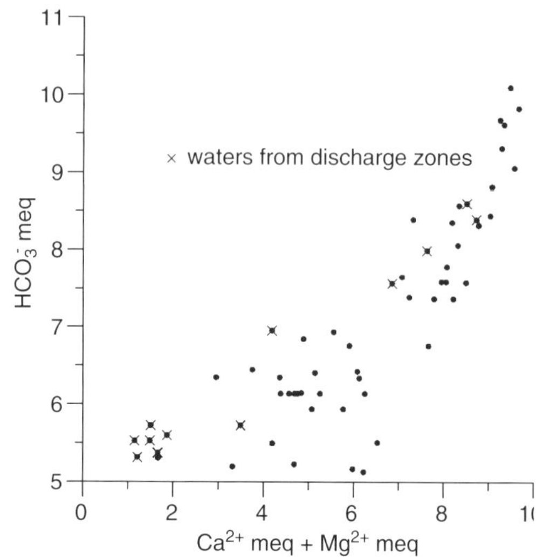

Figure 3. The changes in calcium and magnesium content in the groundwaters.

different waters. Figure 2 illustrates that both the shallow and the deep groundwaters are of meteoric origins. The differences in the isotope values are due to changes in climate. There is no indication of isotopic disturbances that would shift our data from the Global Meteoric Water Line. The points are clustered in three well defined groups: young or Holocene waters, mixed waters, and old (Pleistocene) waters. The mean $\delta^{18}O$ value of Holocene waters is −9.7‰.

The $\delta^{18}O$ values of shallow groundwater from the boreholes situated at the ridges of hills are typically less then −10‰. The values fall between −10.6 and −10.0‰. This isotopic shift cannot be explained only by the known (i.e. altitude, continental effect) influences. The explanation is that the precipitation that recharges the groundwater at the ridges of hills infiltrated 150–300 years earlier in the so called Little Ice Age, when the average temperature was 1°C less than during the rest of the Holocene. The infiltration time was calculated on the basis of tritium measurements from pore waters (Appelo & Postma 1993). The corresponding temperature difference is consistent ~0.5‰ decrease in $\delta^{18}O$.

We have δD and $\delta^{18}O$ data that show mixing only from four shallow boreholes. For the shallow boreholes situated in valleys near discharge areas, we have old (Pleistocene) water only from one sample ($\delta^{18}O \sim -12.5$‰).

3 WATER–ROCK INTERACTIONS

Two main processes characterize the water-rock interactions in the study area: the calcite-dolomite precipitation and the ion exchange.

3.1 *Carbonate precipitation*

The infiltrating groundwaters are oversaturated for calcite and dolomite. The bicarbonate and Ca + Mg concentrations change simultaneously (Fig. 3). The correlation between them is $r^2 = 0.719$. The amount of Ca and Mg decrease defines the amount of carbonate precipitation. Almost all of the young (Holocene) waters have $SI_{calcite}$ between 0.0 and 1.5. Except three samples, the old (Pleistocene) waters have $SI_{calcite}$ between −0.9 to 0.0. The HCO_3^- concentration for $SI_{calcite} > 0$ falls between 400 to 625 mg/l and the saturation index is almost all the time less than 0 for HCO_3^- concentrations between 300 and 400 mg/l. The precipitating carbonate is mostly calcite.

The Mg concentration of the shallow groundwater is around 3.5 meq, whereas the Ca concentration decreases from 5.5 meq to 2 meq. Thus, the shallow groundwater can be considered as the main calcite precipitation zone. Below this area, both values decrease.

3.2 *Ion exchange*

The infiltrating Ca-Mg-HCO_3 type waters gradually change to Na-Ca-Mg-HCO_3 type. Whereas the calcite and dolomite precipitation takes the Ca and Mg out of the water, the weathering of plagioclase and biotite releases Na and K into the water. This process can be seen in Figure 4. Longer the flow paths provide the longer times for water-rock interactions. The relation is neither linear nor logarithmic but looks similar to

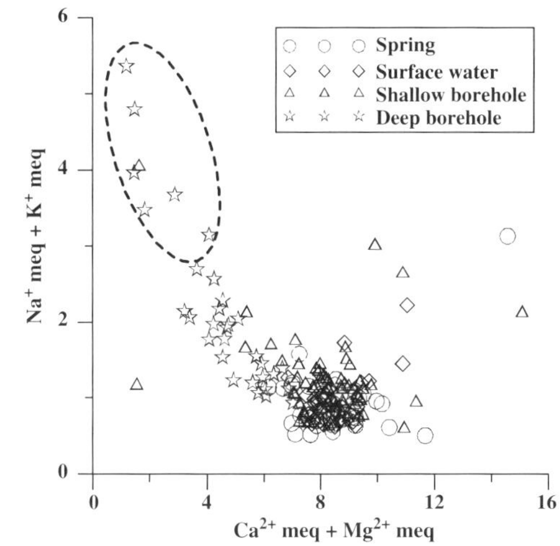

Figure 4. The relation between the Ca + Mg and Na + K content based on shallow and deep groundwater data.

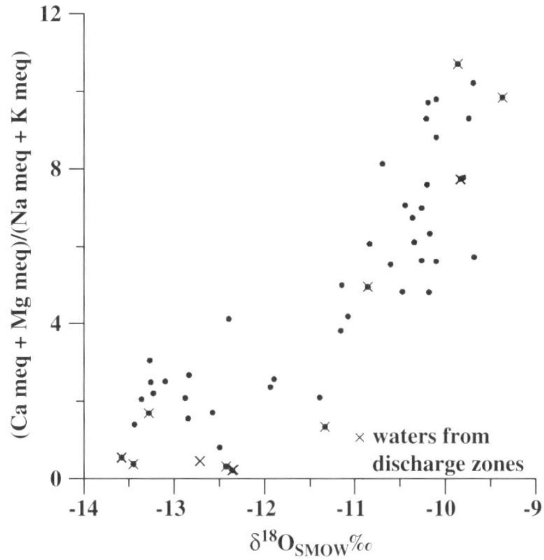

Figure 5. The relation between the (Ca + Mg)/(Na + K) ratio and the $\delta^{18}O$ data.

the latter. These data correlate very well with the results of $\delta^{18}O$, δD, $\delta^{13}C$, and ^{14}C measurements. The samples circled by broken line are the ascending "old" waters of the studied area.

The (Ca + Mg)/(Na + K) ratio and the $\delta^{18}O$ values can be used to follow the evolution of ion exchange (Fig. 5). The inferred $\delta^{18}O$ of meteoric waters at the Holocene and the Pleistocene boundary (~10,000 years) waters is about −11.5‰. The intensive ion exchange takes places in the Holocene. The flow path through which this water goes consists of loess and weathered granite. The weathered granite consists of clay zones that explain the strong ion exchange capacity. The formation of clay minerals in the deeper zone is due to local hydrothermal alteration zones.

4 GROUNDWATER AGE & EVOLUTION

To determine the age of the sampled waters, $\delta^{13}C$ and ^{14}C measurements were used. The current ^{14}C is measured, but the initial ^{14}C is estimated. The most uncertain part in ^{14}C water age calculations is the estimation of initial ^{14}C activity. There are several factors that may affect it such as carbon introduced from soil zone, weathering and precipitation/dissolution of minerals, isotopic exchange, sulfate reduction, etc.

The initial ^{14}C activity was calculated using a modified equation from Ingerson & Pearson (1964):

$$A_{in} = \frac{\delta^{13}C - \delta^{13}C_c}{\delta^{13}C_{HCO_3} - \delta^{13}C_c} \times (A_{HCO_3} - A_c) + A_c \qquad (1)$$

where A_{in} is the initial ^{14}C activity (%); $\delta^{13}C$ is the measured $\delta^{13}C$ of water sample (‰); $\delta^{13}C_c$ is the measured $\delta^{13}C$ of parent carbonate mineral (‰); $\delta^{13}C_{HCO3}$ is the mean of measured $\delta^{13}C$ contents of shallow groundwater (as recharge water;‰); A_{HCO_3} is the initial specific ^{14}C activity (%); and A_c is the ^{14}C activity of parent rock (%).

We used two different initial specific activity values, one for the young and one for the old waters. For the young waters, we used the mean value of 92%, and for the old waters, we used 77% as initial A_{HCO_3}.

To improve the age calculation, we determined the $\delta^{13}C$ values of the parent rock, which gave an average value of –4.2‰. We used this value instead of 0‰ proposed by Ingerson & Pearson (1964). Because there is carbonate precipitation along the flowpaths, we introduced another correction factor (q) called "carbonate quotient". The precipitating carbonates have a decreasing effect on the ^{14}C content of DIC. To account for this, we multiplied the measured ^{14}C activity with the quotient of mean initial bicarbonate concentration (490 mg/L) and the bicarbonate concentration measured in the sample.

The age calculation was done according to:

$$Age(year) = \frac{5730}{\ln 2} * \ln\left(\frac{A_{in}}{A * q}\right) \qquad (2)$$

The calculations show that the waters in the study area are younger than 25,000 years. We used the Netpath geochemical model to account for these

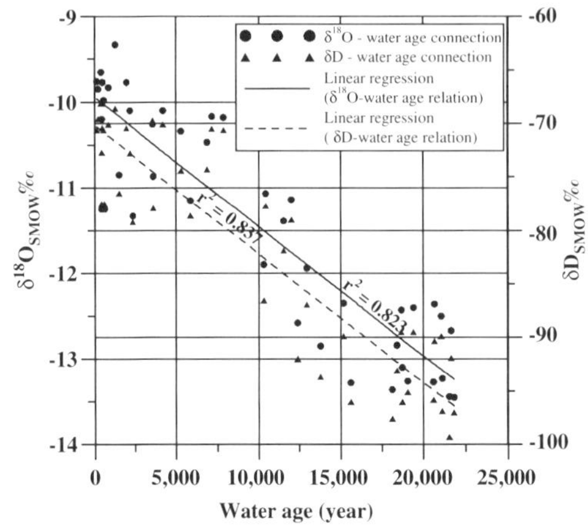

Figure 6. The relation between the water ages and the δD and $\delta^{18}O$ data.

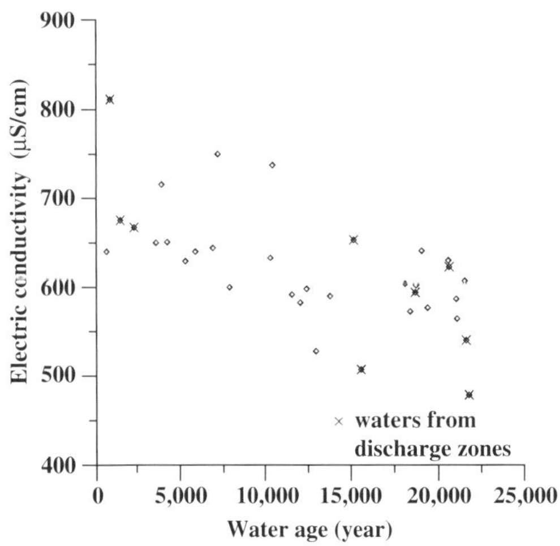

Figure 7. The relation between the electric conductivity and water age.

processes. The models strengthened our statement; calcite precipitates near slight dolomite and CO_2 gas dissolution. The computed radiocarbon ages are in agreement with our age calculations.

To check the age indicating capability of $\delta^{18}O$ and δD values, they were studied in comparison with water ages obtained from $\delta^{13}C$ and ^{14}C data (Fig. 6).

The $\delta D < -80‰$ and $\delta^{18}O < -11.5‰$ values indicate with more than 90% probability that the age of the samples is older than 10,000 years. Two δD and three $\delta^{18}O$ measurements have higher values and ages older than 10,000 years. For the waters older than 10,000 years, the age indicating capability of

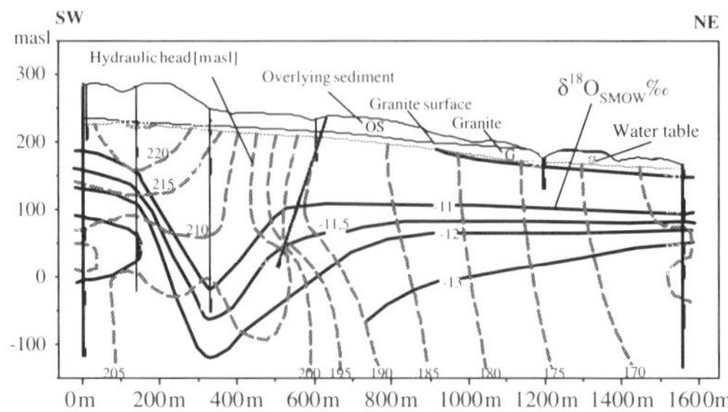

Figure 8. Hydraulic head and $\delta^{18}O$ isopleths along a SW-NW profile.

$\delta^{18}O$ and δD is less reasonable because the Würm glaciation started around 65,000 years ago, and the oxygen isotope values of the infiltrating precipitation between 10,000 and 65,000 years ago could range between −13.5 and −11.5‰.

The total dissolved solid content is relatively low and decreases with age, due mostly to precipitation and partly to reduction processes. These processes are shown in Figure 7.

The distribution of the hydraulic heads describes the complete flow system (Fig. 8). The main recharge zone with downward groundwater movement is at the SW part of the section. The $\delta^{18}O$ values of the waters clearly demonstrate the low velocity of the groundwater movement. The position of the −11.5 $\delta^{18}O$ isopleth is at 100 to 300 m meter depth. This isopleth represents an isochron of 10,000 years age. Thus, the groundwater velocity varies between 0.01 and 0.03 m/year at the recharge zone. The other components show similar spatial distribution indicating their evolution in time.

5 CONCLUSIONS

Several conclusions can be made by the study: 1. water–rock interactions are characterized by two main processes: calcite-dolomite precipitation and ion exchange; 2. in the study area, the waters are younger than 25,000 years; 3. the total dissolved solid content of the waters decreases with the water ages; 4. ascending deep waters, like surface waters or springs, could not be detected at the discharge areas; and 5. we found deep waters from only one of the shallow boreholes, and we detected week deep water mixing in four shallow boreholes.

ACKNOWLEDGEMENTS

The laboratory analyses were done at the Geological Institute of Hungary, Water Resources Research

Centre, Hungarian Academy of Sciences Laboratory of Geochemical Research and Institute of Nuclear Research.

REFERENCES

Appelo, C.A.J. & Postma, D. 1993. *Geochemistry, Ground-Water and Pollution*. Rotterdam: Balkema.

Craig, H. 1961. Isotopic variations in meteoric waters. *Science* 133: 1702–1703.

Ingerson, E. & Pearson, F.J. Jr. 1964. Estimation of age and rate of motion of groundwater by ^{14}C-method. In *Recent Researches in the Fields of Atmosphera, Hydrosphera, and Nuclear Geochemistry, Sugawara Festival Volume*: 263–283. Tokyo: Maruzen Co.

Water–Rock Interaction, Wanty & Seal II (eds)

Geochemical characteristics in the coastal aquifer of the Mamora plain

I. Kacimi, T. Bahaj, N. Kassou & M. Wartiti
Applied Geology Laboratory, Faculty of Sciences Rabat, Morocco

R. Caboi
Laboratorio di Geochimica delle Acque, Univ. Cagliari, Italy

ABSTRACT: The study focuses on the hydrogeochemical characteristics of the coastal aquifer in the Mamora plain. The area has a great hydraulic potential and it supplies drinking water for many cities. The Mamora plain is alluvial. It has shifted about during the Quaternary due to a subsidence. The groundwater aquifers include carbonate and detrital material (sands, clays, and muds). The groundwater flows from the east to the west (towards the Atlantic Ocean), the water is dominated by bicarbonate, chloride, and calcium. All the dissolved elements increase with the total dissolved solids; nitrates are present in elevated concentrations.

1 INTRODUCTION

The coastal zone of Mamora plain is a narrow band of about 20 km wide and 80 km long. The plain is parallel to the coast line between Rabat in the south and Larache in the north. The climate is subhumid (Kacimi 1992). The annual rainfall varies from 350 up to 750 mm. The zone is a center for population growth, which has caused important intensification of agriculture and industry. Groundwater is the main water resource. However, previous work has highlighted increasing regional impacts on water quality (Bahaj 2002).

2 GEOLOGICAL FRAMEWORK

The base of the sedimentary sequence consists of a deposit of a marly sediments several hundred meters thick of Miocene to lower Pliocene age, which are overlain by a complex of sands with carbonate matrix (Fig. 1a). The smectite and the illite are the most abundant clay minerals.

From middle Pliocene to Quaternary part of the section consists of a bioclastic limestone, sandstones with carbonate matrix, and lenses of gravels. The dominant clay minerals are kaolinite and chlorite. The dunal deposits are aligned parallel to the shore and cover important surfaces of this zone (Cirac 1985).

3 HYDROGEOLOGY

Drill holes in the region have identified an aquifer principally constituted by sands, clays, calcarenites, and bioclastic and karstic limestone, the impermeable marly substratum forms the aquitard (Fig. 1). The water reservoir show hydraulic characteristics quite erased. The electric resistivity is from 100 up to 1600 Ωm (El Mansouri 1999), $T = 5 \times 10^{-3}\,m^2/s$, $K = 4 \times 10^{-4}$ m/s and $S = 3.6 \times 10^{-2}$.

The aquifer is dominantly a single layer, commonly unconfined, forming a large band of tens of kilometers wide parallel to the Atlantic coast. The piezometric map shows a southeast to northwest flowing axis towards the Atlantic ocean related to the topographic gradient. The dune ridges are assures the groundwater recharge because of their hydraulic characteristics that are very eraised, $T = 10^{-2}$ m/s, $S = 1.2 \times 10^{-3}$(Fig. 1b) and the sediment nature meuble sands (Combe 1969).

4 GROUNDWATER GEOCHEMISTRY

4.1 *Methods & analysis*

In addition to twelve water analyses made by the ONEP (National Office of Drinking Water) in 1994 (Fig. 1), twelve other points were sampled. The choice of the samples was made on the basis of location in the recharge areas of the ONEP designated urban water supply. Different physical parameters

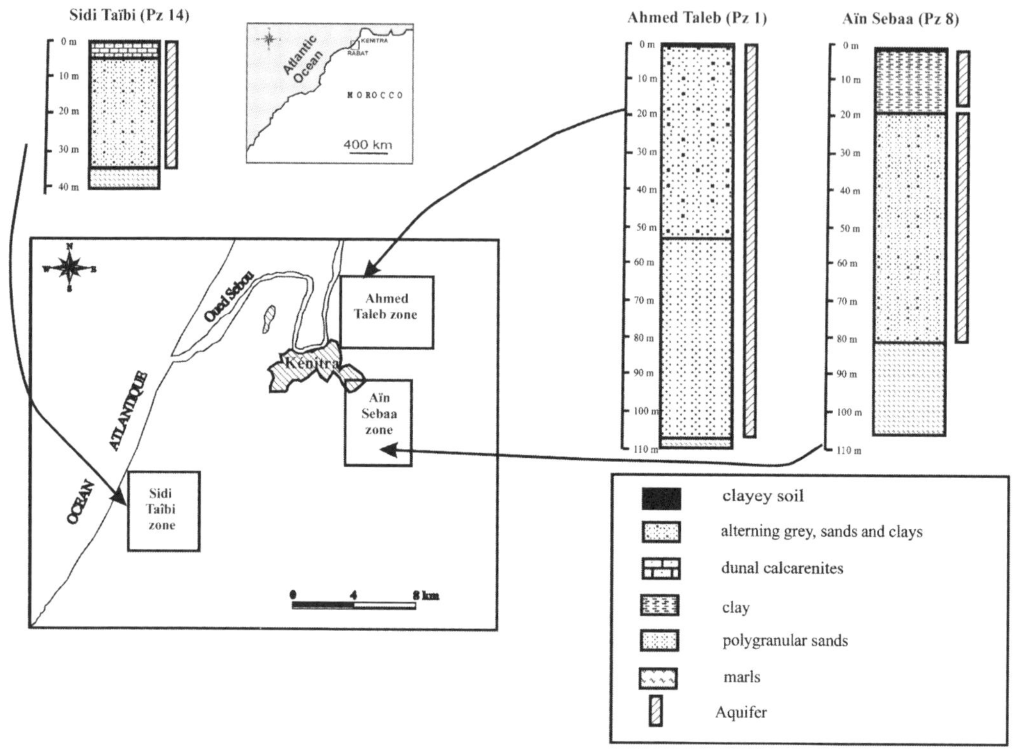

Figure 1. Locations of sampled wells and stratigraphic logs of the different recharge areas of the study area.

were measured during the sampling, including pH, electrical conductivity, and redox potential. Anions (Cl, SO_4, and NO_3) measured by ionic chromatography (HPLC). Analyses of Na, Ca, Mg, and Si were made by ICP-OES. Potassium was measured by AAS. Heavy element concentrations were measured by ICP-MS. The quality of the analyses was monitored by analysis of international standard samples (SRM 1643c) of the National Institute of Standards and Technology (USA).

4.2 *Hydrochemical features*

The Piper diagram of water samples shows that waters belong to mixed chemical features of a bicarbonate-chloride and calcium facies (Fig. 2).

Samples show a gradual evolution of their concentrations of sodium-potassium, because of water–rock interactions. In fact, waters flowing through dune ridges and in the limestone show lower concentrations of sodium and in potassium compared to those flowing through muddy and clayey horizons. A single water samples from the intertidal zone shows concentration quite low in bicarbonates due to the

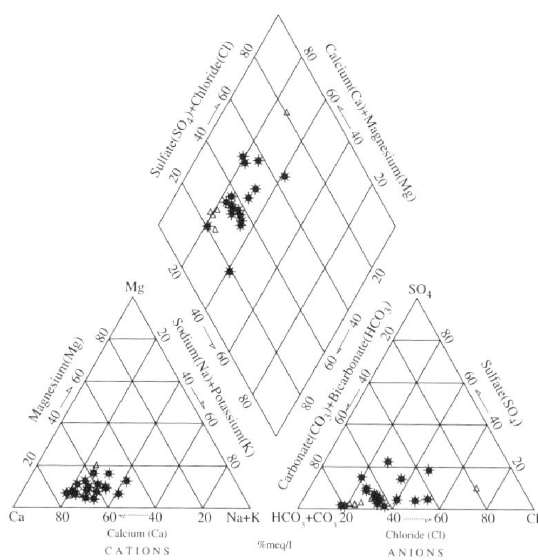

Figure 2. Piper diagram of the groundwater △ Dunal cordons ✱ Interne zone.

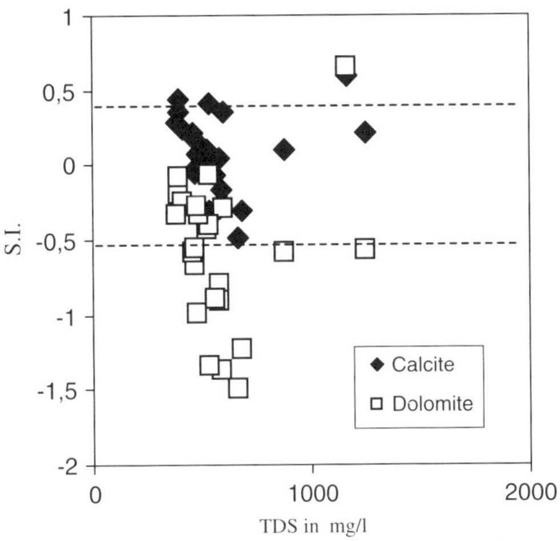

Figure 3. Saturation state of waters relative to calcite and dolomite.

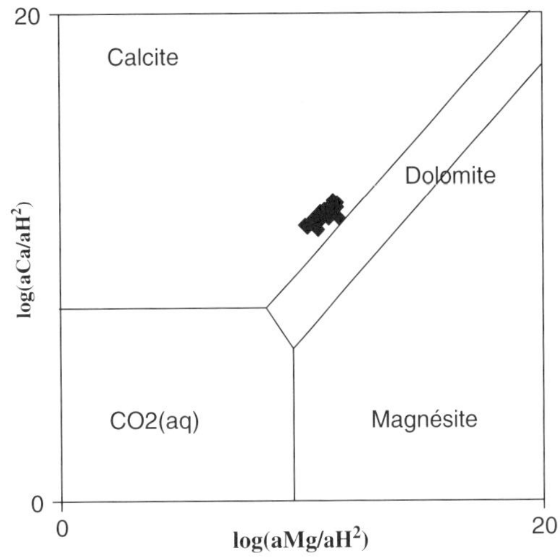

Figure 4. Diagram of log activity (aMg/aH^{+2}) versus log(aCa/aH^{+2}) Stability fields are drawn for 20°C and 1 atm conditions (Kretzschamar et al. 2001).

precipitation of $CaCO_3$ in limestones (Fig. 2) during the dry season where small pellets of limestone are present. Waters have moderate total dissolved solids (TDS) between 600 and 750 mg/L, with the exception of two samples from mud- and clay-rich impermeable levels, which have 1200 mg/l TDS.

On the basis of the water analyses, saturation indices for several minerals were calculated with the geochemical speciation program PHREEQCI (v. 2.15). All water samples were saturated with respect to calcite.

Many samples show undersaturation with respect to dolomite (Fig. 3). In fact, previous studies (Snoussi 1986) noted low concentration of magnesium and high concentrations of calcium in surficial sediments of the Maamora plain. Activity-activity diagrams indicate that calcite should be the stable carbonate phase (Fig. 4).

Figure 5 indicates that kaolinite should be the stable clay mineral in the K_2O-Al_2O_3-SiO_2-H_2O systemat is in equilibrium with the water samples. These two diagrams show a direct relationship with the lithology of the carbonate and siliciclastic aquifers, respectively.

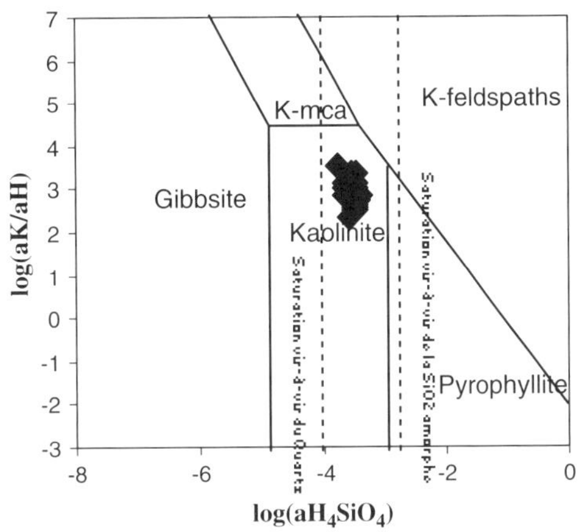

Figure 5. log activity(aH_4SiO_4) versus log(aK^+/aH^+) in the system K_2O-Al_2O_3-SiO_2-H_2O (Drever 1982).

4.3 *Heavy & trace elements in ground water*

In general, and from the water analyses carried out during summer 2002 period in the study area, we noticed that the heavy and trace elements concentrations vary with the TDS (Fig. 6) and are the result of water–rock interaction processes. Two samples taken in the perimeter of Kenitra city show abnormal concentrations for all studied elements; the origin of these elements can only be attributed to anthropogenic activity, because they are situated close to several industrial sites and are in mud- and clay-rich sediments having low permeability.

5 QUALITY OF GROUNDWATERS

Elevated concentrations of dissolved ions were not detected in the study area, except for the two wells showing anomalies with heavy and traces metals. The salinity does not exceed recommended values for agricultural use or for human water supply.

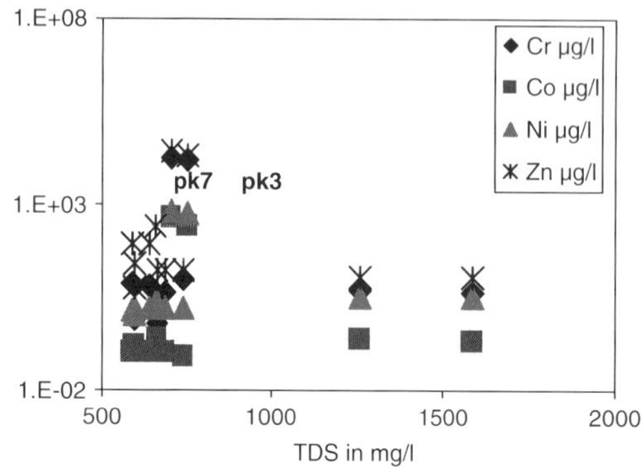

Figure 6a. Variation of heavy metals with TDS.

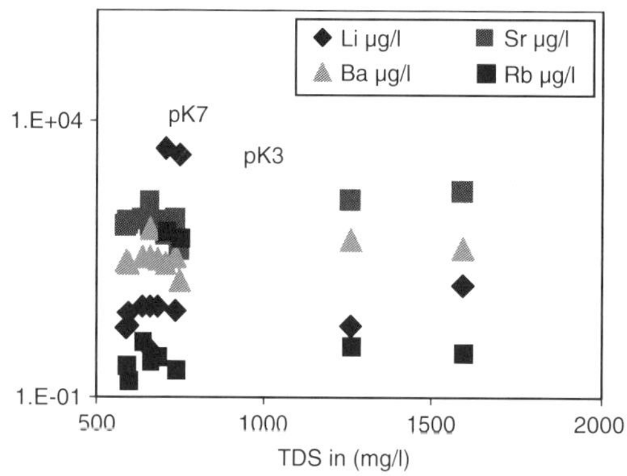

Figure 6b. Variation of trace elements with TDS.

The nitrate (NO_3^-) in the study area shows levels from 50 to 410 mg/L. These levels are due to the use of nitrogenous fertilizers and of manure in which the migration towards the groundwater increases because of the lithological nature of the aquifer being very permeable ($k = 1.8 \times 10^{-3}$ m/s) (Mouaddine 1997) and consequently very vulnerable at the agricultural contamination. Two samples of water show some elevated concentrations of heavy and trace metals (Zn, Cu, Co, Cr, Cd, and As) greater than 500 ppb with zinc being the highest, largely exceeding permissible maximal standards, which could be a serious water-quality problem and may have repercussions on the consumers' health.

6 CONCLUSIONS

The geochemical study shows that the groundwater in the coastal aquifer from the Mamora plain that water belongs to the HCO_3-Cl-Ca facies; it has a low salinity. The water is saturated with respect to calcite and the clay mineral kaolinite. It is undersaturated with respect to dolomite. Heavy and trace elements have a low concentration in the aquifer with the exception of contamination in the industrial zone of Kenitra city. The inorganic water quality is becoming degraded because of recent contamination of groundwater by nitrates from agricultural sources.

REFERENCES

Bahaj, T. 2002. *Hydrogéochimie et qualité des eaux des nappes phréatiques du Causse Moyen Atlasique, du Bassin de Saiss, de la Plaine du Gharb, de la plaine de la Maâmora et de la région de Zemmour*. Unpub. Ph.D. Thesis. Rabat: Faculté des Sciences.

Cirac, P. 1985. Le bassin sud-rifain occidental au Néogène supérieur. Evolution de la dynamique sédimentaire et de la paléogéographie au cours d'une phase de comblement. *Mem. Inst. Géol. Bassin d'aquitaine, Université de Bordeaux*1(21).

Combe, M. 1969. Cartes hydrogéologiques de la plaine du Rharb au 1/100000. *Notes et mem. Serv. Géol.*, Maroc 221.

Drever, J. 1982. *The Geochemistry of Natural Waters*. Prentice-Hall. Inc.

El Mansouri, B. 1999. *Développement d'outils et concepts pour la gestion des eaux souterraines. Applications à l'aquifère côtier du Gharb*. Unpub. Ph.D. Thesis. Kénitra: Faculté des Sciences de Kénitra.

Kacimi, I. 1992. *Quelques aspects sur l'hydrogéologie et l'hydrochimie des nappes aquifères du Gharb. Le problème particulier du fer et du manganese*. Unpub. M.Sc. Thesis. UQAM.

Kretzschmar, T.G., Schulze-Makuch, D. & Torres-Alvarado, I.S. 2001. Chemical evolution of groundwater in the Tularosa Basin in southern New Mexico. In R. Cidu (ed.), *Water–Rock Interaction 10*: 545–548.

Mouaddine, L. 1997. Utilisation des cartes de vulnérabilité pour la délimitation des zones de protection des captages d'eau potable: Cas du champ captant de Fouarat. DES Fac. Sci. Semlalia Marrakech.

Snoussi, M. 1986. Nature, estimation et comparaison des flux de matières issus des bassins versants de l'Adour (France), du Sebou, de l'Oum-Er- Rabia et du Souss (Maroc). Impact du climat sur les apports fluviatiles à l'océan. *Mém. Inst.Géol. du Bassin d'Aquiitaine. Univ. Bordeaux* 1(22).

Water–Rock Interaction, Wanty & Seal II (eds)
© 2004 Taylor & Francis Group, London, ISBN 90 5809 641 6

Earthquakes influencing geothermal water–rock interaction in Iceland

H. Kristmannsdóttir
University of Akureyri, Faculty of Natural Resource Sciences, Solborg, Akureyri, Iceland

V. Hardardóttir
ÍSOR, Iceland Geosurvey, Orkugarður, Grensásvegur, Reykjavík, Iceland

ABSTRACT: It has been demonstrated in Iceland and elsewhere that earthquakes may be preceded by a Rn anomaly in the groundwater and the monitoring of Rn concentration in groundwater/geothermal water may therefore be useful as an alarm system to predict earthquakes. By the study of groundwater chemistry and chemical monitoring of geothermal fields in Iceland some other possible chemical precursors of earthquakes and effects on the groundwater chemistry have been detected. The concentration of F^- in geothermal water has been observed to increase temporarily due to seismic activity. In the wake of a major earthquake occurring in the Southern Lowlands of Iceland in the year 2000 the concentration of Mg in geothermal water was dramatically increased for several weeks inducing Mg-silicate scaling in production wells. Further studies and enhanced monitoring of groundwater chemistry are needed to explain the mechanism of these phenomena, as well as their magnitude and possible use to predict earthquakes.

1 INTRODUCTION

Some twenty years ago studies were carried out on the radon (Rn) concentration in geothermal water in several geothermal wells in Iceland and several anomalous changes preceding local earthquake activity were observed (Hauksson & Goddard 1981).

The program was concentrated in two different areas of transform faulting, the Southern Lowlands and the Tjörnes Fracture Zone. Since this program was carried out there has been no regular long-term monitoring of Rn in groundwater in Iceland until ~6 months before the start of a major period of earthquake activity (of magnitude 6–6.4) in Southern Iceland in the year 2000 (Stefánsson et al. 2000) when a new program was started (Theodórsson et al. 2002).

Chemical monitoring of production water from geothermal wells is conducted in most producing geothermal fields in Iceland to manage the field and to optimize the use of the resources economically and environmentally (Kristmannsdóttir & Ármannsson 1996). Chemical monitoring of other species than Rn has not been conducted in connection with seismic activity in Iceland. Results from the reservoir monitoring programs of geothermal fields have in some cases revealed effects induced by seismic activity. In a few cases water samples collected within the programs aimed for reservoir monitoring have later been analysed for other species to search for possible effects of seismic activity.

2 RADON MONITORING

A monitoring program was conducted to demonstrate the possible connection between Rn concentration in groundwater and seismic activity. Samples were collected from a network of nine stations for analysis of Rn (Hauksson & Goddard 1981). During the years the monitoring was conducted several anomalous changes in Rn content were observed to precede local earthquakes. The anomalies appeared to occur farther away for larger earthquake magnitude. There were a few false alarms, and several incidences where no anomaly was observed. The probability of observing Rn anomalies before earthquakes was found to be about 65%. Unfortunately this program was terminated abruptly due to lack of funds. During the last few years a new program has begun (Theodórsson et al. 2002). In the new program the Rn concentration is measured continuously in discharged water from several production wells within geothermal fields in the Southern Lowlands where there is a potential for rather big earthquakes to occur. During the large earthquakes in the year 2000 the data revealed that Rn anomalies did in fact precede seismic activity at some of the observation sites (Theodórsson et al. 2002).

The Rn anomalies appear to be due to an opening of fractures prior to an earthquake allowing more Rn to be dissolved from the rocks. As Rn is a conservative element, its dissolution is in direct relation to its availability in the rock. The short half-life of the element restricts the distance and time for its role as an active precursor for earthquakes. Still it is a very interesting option to develop further this technique and possibly combine it with other geochemical monitoring methods.

Data of Rn concentration in Icelandic waters have been scarce, but currently a research program is being conducted to study trace elements in Icelandic groundwaters. The program entails studies of over 30 trace elements, including Rn, in groundwaters of varying temperatures and chemical nature from all over the country.

3 CHANGES IN FLUORIDE CONTENT

One of the geothermal fields in Southwest Iceland at Bakki, Ölfus (Fig. 1, 2) has been utilized for a central heating system for the small town of Thorlákshöfn some 13 km away for over twenty years (Kristmannsdóttir & Sveinbjörnsdóttir 1992). The reservoir temperature of the field exceeds 120°C, the wells are free flowing, and water is separated at the well heads at atmospheric pressure. The field is located on the border of the Western active zone of rifting and volcanism in an area where three tectonically active regions cross (Rögnvaldsson et al. 1998).

It is therefore an area of intense crustal deformation and since 1994 it has been very active. Several earthquake swarms have been recorded since 1994, probably linked with an uplift in the Hengill volcanic center (Sigmundsson et al. 1997). In June 1998 a large earthquake swarm occurred with the strongest single earthquake of magnitude 5.1, with an origin less than 10 km away from the Bakki geothermal field (Fig. 2). Subsequently after the largest earthquake the pressure of the geothermal field was observed to increase and the yield of the production wells was improved. Later the same year, in November, there was another rather strong earthquake (magnitude 5.0) originating just by the geothermal field. That earthquake also induced major changes in the geothermal area, but this time the reservoir pressure was greatly reduced. One of the drillholes even stopped flowing and could not be activated by pressure to flow freely until after several days of activation (Thorhallsson & Björnsson 1998).

The tectonic activity in June did increase the stress at the site of the Bakki geothermal field, inducing increased pressure at the well heads and increased flow rate. In the November seismic activity, however, coseismic stress changes induced a deflation of the field (Rögnvaldsson et al. 1998) and subsequent disruption or decrease of the flow rate (Kristmannsdóttir et al. 1998).

Coincidently with the tectonic activity in June and increase in flow rate, a distinct geochemical anomaly was detected in the geothermal water in one production well (Kristmannsdóttir et al. 1998). A sharp increase in the concentration of F^- was observed, from 0.4–0.5 mg/l up to 0.7–0.9 mg/l. The samples were taken at different flow rates (5–20 l/s) and the

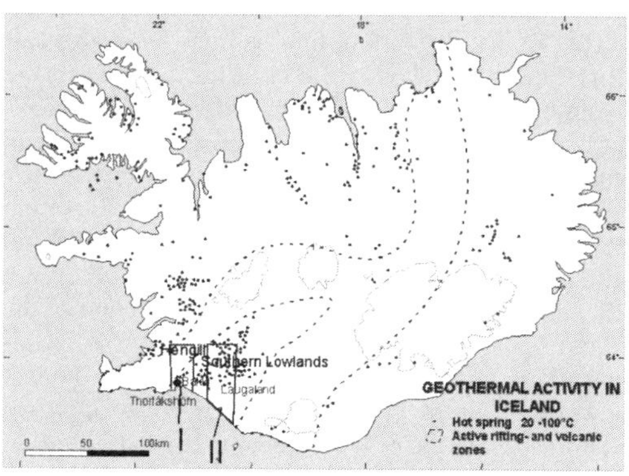

Figure 1. Location of the Bakki geothermal field (I) and the Laugaland-Kaldárholt geothermal fields (II). Also shown on the figure is the location of low temperature geothermal activity and the active zones of volcanism and rifting.

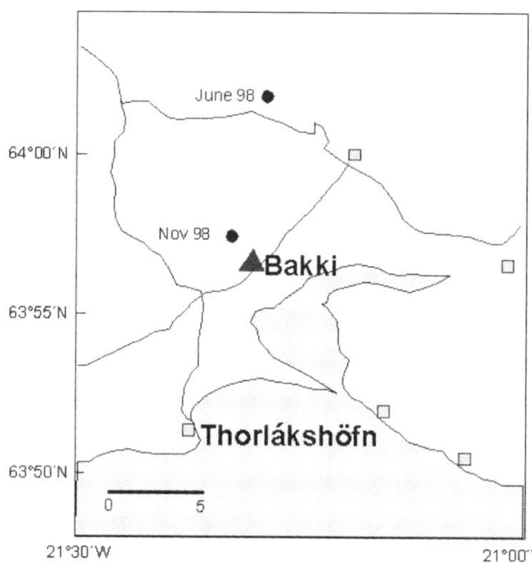

Figure 2. The location of the two earthquakes occurring in the Ölfus area in 1998 (•June 98 and •Nov 98). The location of the Bakki geothermal field and the town of Thorlákshöfn is also shown. The map corresponds to section I in Fig. 1.

F^- anomaly was found to occur in all the samples taken at well head. The well was also sampled by a downhole sampler at 400 and 500 meters depth and no F^- anomaly was detected at any flow rate in any of those samples. The anomaly is therefore found to be introduced through a big aquifer at 300 m depth, but not observed in the deeper aquifers. In a few days the concentration of F^- was back to normal in the geothermal water.

During the November tectonic event, no F^- anomaly nor any other changes in the chemistry of the geothermal water were observed. In that event free flow from one of the drillholes was disrupted for several days.

F^- has not been monitored frequently in many Icelandic geothermal fields as it is normally not very useful as an indicator for temperature changes in geothermal systems or inflow of cold water. In the basaltic reservoir rocks the concentration of F^- seems to rely mostly on the concentration in the rocks and equilibrium with F^--bearing minerals will not be attained except in acidic reservoir rocks. In the big earthquake events in June 2000 F^- was monitored in a few of the geothermal fields which were hit the hardest by tectonic activity, but no major change or a connection between the activity and changes in F^- content could be demonstrated neither in the areas of increased nor reduced pressure. One possible difference between that event and the one in the Ölfus area could be because the rock formations involved are a lot older and their available concentrations of F^- are probably lower.

4 INCREASE IN MAGNESIUM CONTENT

After the big earthquakes that occurred in June 2000 in the Southern Lowlands (Fig. 3), some mechanical damage to the geothermal production wells in the area was observed. By later inspection some unfamiliar scaling was found in several wells, which was later identified as a Mg-silicate precipitate of the same type as commonly found in surface installations of heating systems utilizing heated fresh water (Kristmannsdóttir et al. 2002). Similar scales are found to form when geothermal water is mixed with fresh water (Kristmannsdóttir et al. 2001).

In one of the larger heating systems in the Southern Lowlands, Hitaveita Rangæinga, two geothermal production fields were greatly affected by the earthquakes. In one of the fields, Laugaland (Fig. 3), the pressure in the reservoir was reduced and a sudden drawdown in the water level occurred, whereas in the other, Kaldárholt, the reservoir pressure was greatly increased and many wells that previously had water levels some meters below the ground surface began to flow freely.

The concentration of Mg in production water from the Laugaland field (Fig. 4) was found to increase sharply just after the seismic incident, when pumping could be resumed. No changes in chemistry of the production water from the other production field, Kaldárholt, were detected. The Mg concentration of the production water dropped quickly, but not to the original levels until several months after the earthquakes.

Inflow of cold groundwater into geothermal fields is to be expected due to pressure draw down or in the wake of seismic activity decreasing pressure in some parts of the field. In the case of Laugaland the pressure of the rock formations was reduced by the earthquakes so an inflow of cold water was to be expected. The inflow appears to have happened solely after the big earthquakes as no increase in Mg concentration has been observed before the events. However, there was no continuous monitoring of Mg – only grab samples taken approximately

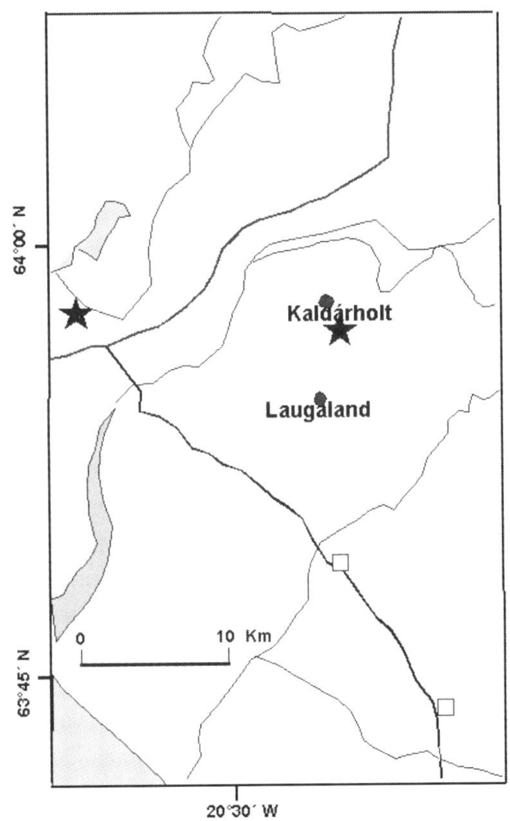

Figure 3. The location of the two earthquakes (★) occurring in the Southern Lowlands area in June 2000. The location of the Laugaland and Kaldárholt geothermal fields are also shown. The map corresponds to section II in Fig. 1.

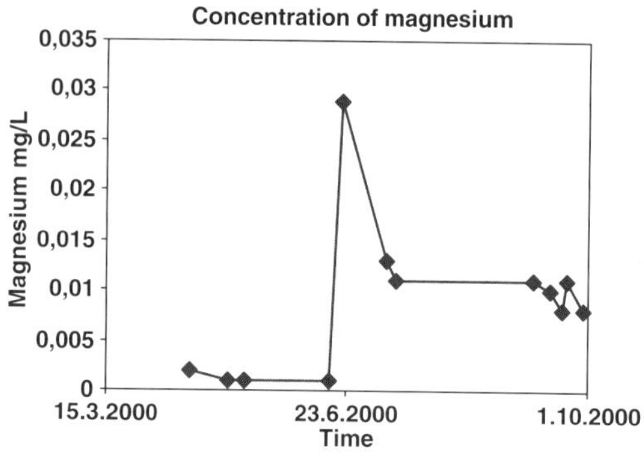

Figure 4. Changes in the concentration of Mg in samples of geothermal water from the Laugaland geothermal field from March to October year 2000.

twice a week – so a preceding anomaly could have been missed. Mg-silicate have a very low solubility at elevated temperatures and it would be expected to precipitate rather quickly in the affected areas (Kristmannsdóttir et al. 2001). In the case of Laugaland a considerable inflow of cold water appears to have been maintained, which reached the producing aquifer quickly enough for the Mg to remain in solution at high levels of supersaturation with respect to Mg silicates like saponite.

5 CONCLUSION

Rn concentrations have been monitored occasionally in a few areas in Iceland, with the goal of forecasting earthquakes. In some cases this monitoring has yielded positive results. Other changes in the geochemistry of groundwater due to earthquake activity, mainly increase in F^- and Mg concentration, have been detected in a few incidences by reservoir monitoring of geothermal fields. The F^- anomaly, as well as the Rn anomalies, are probably due to the opening of new fractures during seismic activity and, being mainly conservative substances, their concentration is in direct correlation to availability and contact with rock surfaces. Opening of fractures in the beginning phases of seismic activity could thus induce anomalous concentrations in the groundwater. The Mg concentration, on the other hand, is due to the inflow of cold Mg-rich water into Mg depleted geothermal water. For an anomaly to be detected there needs to be a continuous inflow near the sampled aquifer or else it will precipitate as Mg-silicates.

Changes of other parameters in the groundwater like Cl^-, B, Ca and K, as well as the stable isotopes, could be expected during seismic events and subsequent inflow of cold water. Unfortunately the groundwater and geothermal reservoir monitoring has been both irregular and not primarily aimed at the parameters which would be expected to show the main changes due to seismic activity. There appears to be possibilities to obtain information about seismic activity by the monitoring of certain geochemical parameters in groundwater and the interpretation of their changes.

ACKNOWLEDGMENTS

The authors thank Páll Theodórsson for providing information about the new Rn project for seismic monitoring. The directors of the Heating companies in Thorlákshöfn and Rangæinga, Guðlaugur Sveinsson and Ingvar Baldursson are duly thanked for cooperation and support.

REFERENCES

Hauksson, E. and Goddard, J.G., Radon Earthquake Precursor Studies in Iceland. J. Geophys. Res., 86, 7037–7054.

Kristmannsdóttir, H. and Ármannsson, H., 1996. Chemical Monitoring of Icelandic Geothermal Fields during Production. Geothermics, 25, 349–364.

Kristmannsdóttir, H. and Sveinbjörnsdóttir, Á.E., 1992. Changes of stable isotopes and chemistry of fluids in the low-temperature geothermal field at Bakki-Thóroddsstadir, Ölfus, SW-Iceland. Proc. of the 7th International Symposium on Water-Rock Interaction, 13–18 July 1992, Park City, Utha, USA 1992, 951–954.

Kristmannsdóttir, H., Hauksdóttir, S., Thorhallsson, S., Saemundsson, K., Axelsson, G. and Thordarson, S., 1998. "Thorlákshöfn Heating System. The results of downhole temperature logging and sampling of water from well HJ-01 and motions for mitigation measures to maintain the production capacity of the heating system". Orkustofnun report (in Icelandic). OS-98048, 17 p.

Kristmannsdóttir, H., Thórhallsson, S., Ólafsson, M., Hauksdóttir, S. and Ármannsson, H., 2001. "Magnesium silicate scales in geothermal production in Iceland" (in Icelandic) Orkuþing 2001 (ed. Gunnarsdóttir, M.J.) Reykjavík 2001, 404–407.

Kristmannsdóttir, H., Axelsson, G. and Harðardóttir, V., 2002 "The Rangæinga Heating Company – Monitoring of the geothermal production in the production fields at Laugaland, Holt and Kaldárholt during the year 2001" Orkustofnun report (in Icelandic), OS-2002/009, 21p.

Rögnvaldsson, S.Th., Árnadóttir, Th., Ágústsson, K., Skaftadóttir, Th., Guðmundsson, G.B., Björnsson, G., Vogfjörð, K., Stefánsson, R., Böðvarsson, R., Slunga, R., Jakobsdóttir, S.S., Thorbjarnardóttir, B., Erlendsson, P., Bergsson, B.H., Ragnarsson, S., Halldórsson, P., Thorkelsson, B. and Ásgeirsdóttir, M., 1998. "An earthquake swarm in Ölfus in November 1998". The Icelandic Meteorological Institute Report (in Icelandic). VÍ-G98046-JA09, 19 p.

Sigmundsson, F., Einarsson, P., Rögnvaldsson, S.Th., Foulger, G., Hodkinsson, K. and Thorbergsson, G.,

1997. 1994–1995 seismicity and deformation at the Hengill triple junction, IcelND: Triggering of Earthquakes by an Inflating Magma Chamber in a Zone of Horizontal Shear Stress. J Geophys. Res. 102, 15151–15161.

Stefánsson, R., Gudmundsson, G.B. and Halldórsson, P., 2000. The two large earthquakes in the South Iceland seismic zone on June 17 and June 21, 2000. The Icelandic Meteorological Institute Report (in Icelandic), VÍ-G00010-JA04. (Also at http://hraun.vedur.is/ja/skyrslur/June17and21_2000 /index.html

Theodórsson, P., Einarsson, P. and Gudjónsson. G.I., 2002. "New radon systems and the earthquakes in South Iceland year 2000" (in Icelandic). Physics in Iceland, Publ. of The Physical Society of Iceland, X, June 2002.

Thorhallsson, S. and Björnsson, G., 1998. "The effects on an earthquake swarm on the production wells of Thorlákshöfn heating system and motions for improvement". Orkustofnun Report (in Icelandic). SÞ/GrB-1998/01, 3p.

Water–Rock Interaction, Wanty & Seal II (eds)
© 2004 Taylor & Francis Group, London, ISBN 90 5809 641 6

Fracture-related geochemical controls on As concentrations in ground-water

Gail Lipfert & Andrew Reeve
University of Maine, Orono, Maine, USA

ABSTRACT: A small watershed in coastal Maine, USA underlain by sulfidic pelitic schist intruded by small granitic and dioritic plutons is investigated geochemically to understand the controls of As distribution and transport. Chemical analysis of ground-water from domestic wells and boreholes within the watershed show a positive correlation between As and Si and Mg. There is also a positive correlation between As and Si leached from fracture surfaces using NaOH and buffered oxalic acid solutions. Spatial distribution of As in the ground water and these As/Si relationships indicates a probable correlation of As ground-water concentrations and fracture-related flow conditions. Results from the NaOH leach tests suggest that adsorption/desorption reactions of As with amorphous ferric oxyhydroxides play a role in As solute transport.

1 INTRODUCTION

A small watershed in Northport, Maine, USA contains wells with As ground-water concentrations of up to 2300 μg/L, the highest concentrations in the state (Fig. 1). The maximum contaminant level set by the US Environmental Protection Agency for As is now 10 μg/L, so there is great interest in understanding the occurrence of sites like these. This area is being studied in detail to understand the controls on the distribution and transport of As in fractured bedrock aquifers.

Located along the coast of central Maine, this watershed is composed of till-covered sulfidic pelitic schist that has been intruded by numerous small granitic to dioritic plutons. The till is thin to nonexistent in the upland areas, but is locally thick (12 to 24 m) in other areas of the watershed. The till is well compacted and poorly sorted with occasional lenses of sand and gravel.

Four bedrock cores were drilled in the watershed (Fig. 1). Arsenopyrite in these cores is commonly associated with thin tourmaline-quartz-carbonate veins. These veins are plentiful, especially in the upland, recharge area and are a source of As within the bedrock.

Other studies have suggested that ferric oxyhydroxides control the concentration of As in ground-water systems (e.g. Pierce & Moore 1982; Welch et al. 2000). These studies suggest two primary mechanisms control As levels: adsorption/desorption and reductive dissolution. Under neutral or acidic

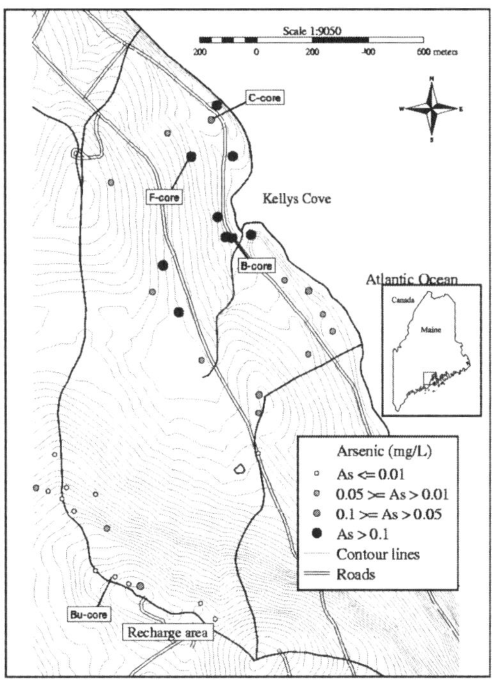

Figure 1. As concentrations and borehole locations showing a cluster of high As in down-gradient area.

conditions, As has a strong affinity for ferric oxyhydroxides and adsorbs onto the aquifer materials. Under high pH conditions As desorbs is more mobile. Under reducing conditions, ferric oxyhydroxides dissolve or As reduces from $H_2AsO_4^-$ or $HAsO_4^{2-}$ to $H_3AsO_3^0$ and As is released. Our study uses sequential leach tests of fracture surfaces to evaluate the impact of these mechanisms on As concentrations in this watershed. The fracture surfaces are leached sequentially with a NaOH solution that should quantify adsorbed As, and an oxalic acid buffer solution that should show how much As is co-precipitated in fracture coatings.

2 METHODS

2.1 *Water sampling*

Ground-water samples are taken from 32 bedrock wells, 4 open boreholes, and 7 shallow wells within the watershed. Samples are taken before any treatment systems using a flow-through apparatus after allowing pH, Eh, temperature, and specific conductance to stabilize. Major cations are analyzed with an ICP and the major anions are analyzed with an ion chromatograph. Total Fe and Fe^{2+} concentrations are measured using a spectrophotometer in the field. As is speciated in the field using strong anion exchange resin (Edwards et al. 1998). Total As and As^{3+} are analyzed with the ICP and As^{5+} is calculated by difference.

2.2 *Leach tests*

Sequential leach tests are performed on 15 fracture surfaces that have been cut from the four bedrock cores. The sides and the bottoms of the fracture samples are coated with paraffin wax leaving the fracture surfaces exposed. Two leaching solutions recommended by Daus et al. (1998) are used for the leach tests, a NaOH solution and an oxalate buffer solution. Each fracture is first leached in 100 ml of 1 mol/L NaOH in zip-lock bags for 1 h. The fractures and the bags are rinsed three times with deionized water and each fracture sample is leached in 100 ml of an oxalate buffer solution containing 0.2 mol/L oxalic acid and 0.2 mol/L ammonium oxalate for 24 hours. The leachates are then analyzed in the same way as the ground-water samples.

3 RESULTS AND DISCUSSION

3.1 *Ground-water chemistry*

As concentrations in the bedrock wells are shown in Figure 1. Important observations are that the As concentrations are low to undetectable in the upland areas and the highest As concentrations are clustered in a narrow region of the discharge area suggesting fracture flow control of As concentrations. The upland area has ample arsenopyrite in the bedrock yet very low As in the ground water. Obviously, there are other controls on the of As in ground water than the presence of As in the bedrock.

In general, there are poor correlations between As and the other analytes. There does, however, appear to be a relationship between As and Si (Fig. 2), As and Mg, and Si and Mg. All high As concentrations occur at high Si and high Mg concentrations. Si and Mg are positively related due to weathering of chlorite, a common mineral in the schist. The association of high As with high Si and Mg indicates that As may be controlled by residence time. Either the ground-water flow rates are slow or the flow paths are long.

As concentrations tend to increase with decreasing Eh values, and a graph of As concentrations vs. Fe^{3+}/Fe^{2+} shows that As decreases with increasing Fe^{3+}/Fe^{2+}, indicating a redox control on As concentrations (Fig. 3).

There are strong inverse relationships between As and nitrate, Cl, and trace metals such as Pb, Cu, and Zn. The inverse relationship of As and the cationic

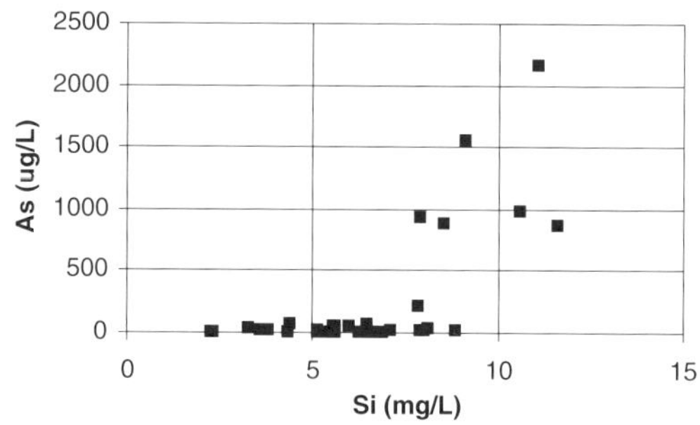

Figure 2. As versus Si concentrations in ground-water samples of the bedrock wells and boreholes showing a positive correlation of As with Si.

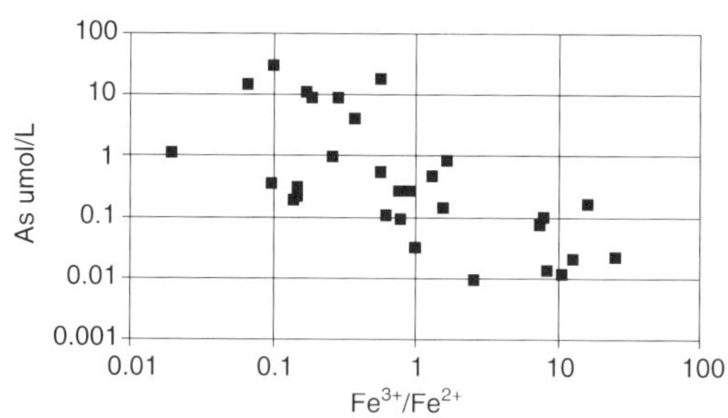

Figure 3. As concentration vs. the ratio of oxidized Fe to reduced Fe.

trace metals is due to the tendency of cationic trace metals to desorb at low pH and As to adsorb at low pH. The inverse relationship of As to nitrate may be related to redox conditions, but it may be simply that the high nitrate concentrations are found in the upland areas where the thin overburden is unable to mitigate the nitrate. The inverse As/Cl relationship may be due to greater interception of marine aerosols by the upland area and recharge of the down gradient area by rain with lower Cl concentrations.

Unlike some studies (e.g. Pierce & Moore 1982; Welch et al. 2000) that show increasing As with increasing pH, we note that the highest As concentrations are centered near neutral pH in our wells (Fig. 4). A similar relationship is also found between pH and Si, Fe, and Mn.

3.2 *Fracture surface descriptions*

The surfaces of the fractures in the four bedrock cores vary greatly, physically and chemically. Many fractures occur along veins and the surface reactions are controlled by the vein mineralogy instead of the primary rock mineralogy. So far, XRD analysis shows no crystalline forms of Fe oxides, but many of the surfaces appear to be coated with amorphous Fe oxides, particularly in the upper parts of the cores. The B-core in the discharge area has framboidal pyrite below 50 m that appear to have grown on the fracture surfaces. Many of the fractures within the schist show alteration or precipitation products, although those in granitic rock typically show little alteration except for Fe oxide precipitates in the upper sections. The alteration materials on the fractures can vary significantly in appearance even within a single core. Using XRD, electron microprobe analysis, and visual inspection, some of the materials identified so far are amorphous Fe oxides, chlorite, and some clays.

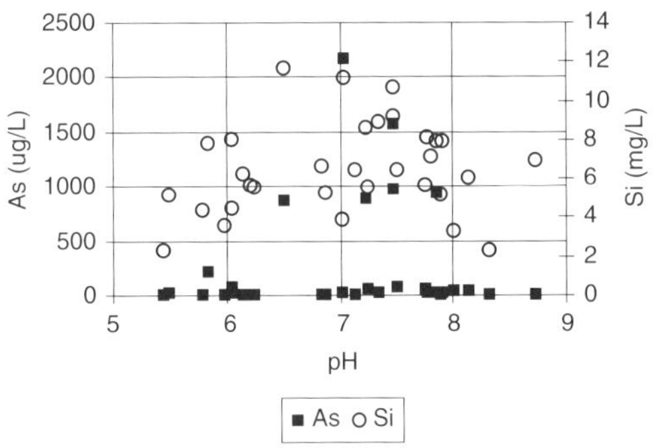

Figure 4. As and Si have high values at circum-neutral pH.

3.3 *Leach tests*

The purpose of the leach tests is to determine how much As is adsorbed versus how much incorporated in the minerals lining the fracture surfaces. Association of analytes should give us some idea of what is happening on the fracture surfaces. The NaOH leaching solution increases the pH, which releases adsorbed As to the leachate. The oxalate buffer solution should dissolve amorphous and weakly crystalline materials that may have taken in As.

Arsenic is not present on all the fracture surfaces even in those cores that have high As concentrations in the ground water (Fig. 5). This absence of As on some fracture indicates that these fractures are either not in hydraulic communication with the primary water-supplying fractures in the core or they are fresh, unweathered fractures.

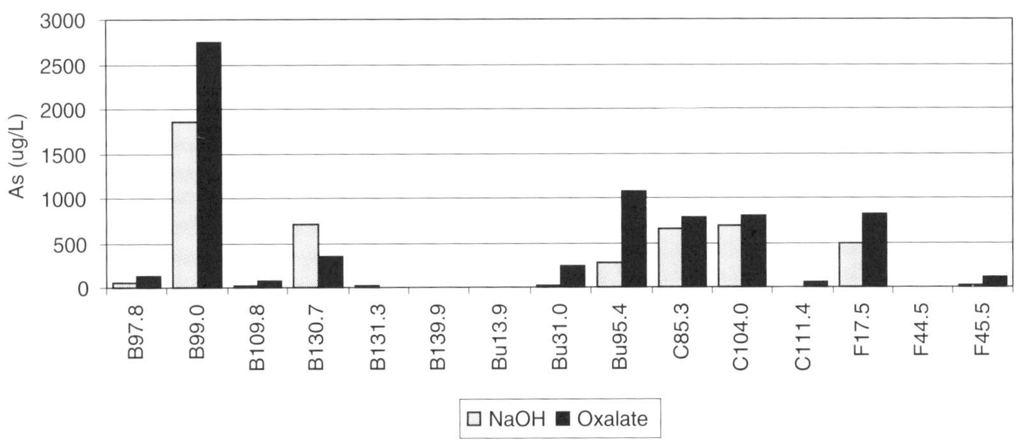

Figure 5. As concentrations in NaOH and oxalate leachates from 15 fracture leach tests.

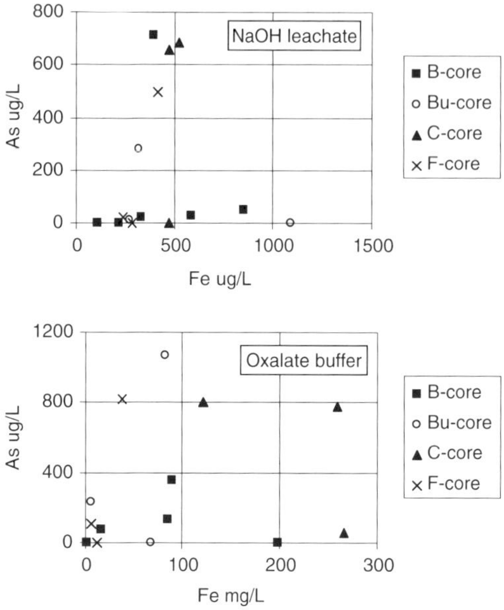

Figure 6. As versus Fe concentrations in the NaOH leachates (above) and the oxalate leachates (below) showing a correlation of As with Fe in the NaOH leachates.

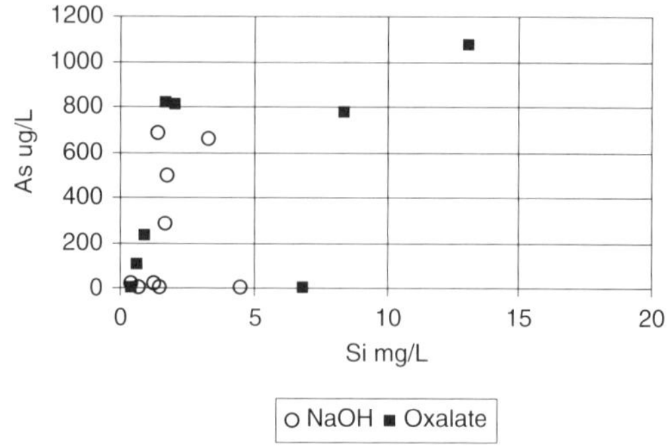

Figure 7. As versus Si concentrations in the NaOH and oxalate leachates showing increasing As with increasing.

Except for two fractures in the B-cores, the concentration of As is greater in the oxalate buffer leachate than in the NaOH leachate. Visual inspection indicates that only a small percentage of the fracture surface materials dissolved in the 24-hour leach test, suggesting that these materials represent a significant source of relatively accessible As.

There is a correlation of As with P in both the NaOH and oxalate leachates (not shown) that can be explained by the competition of As and P, which are both anions, for the same surface sites.

There is a correlation between As and Fe in the NaOH leachate, but not in the oxalate leachate (Fig. 6). Figure 6A shows two different As/Fe trends, a steeper, high As trend and a shallower, low As trend, that do not depend on the core the fractures were taken from (Fig. 5A), but may result from the fractures having hydraulic connection to different sources of As. These correlations indicate that adsorption-desorption reactions may play a part in controlling As concentrations in the ground water although the As/pH graph indicates that it is not a significant control.

As with the ground-water analysis, there is a positive correlation between As and Si in the NaOH and oxalate leachates (Fig. 7). The As/Si correlation in the NaOH leachates suggests that As may be associated with amorphous Si on the fracture surfaces. The oxalate buffer leach tests also appear to dissolve a Si-rich weathering product. These As/Si relationships further support the hypothesis that As concentrations are related to weathering processes and are probably controlled by flow conditions.

4 CONCLUSIONS

Arsenic concentrations in ground water occur in a tight cluster in the discharge zone of a small bedrock watershed. As concentrations in ground-water samples are positively correlated with Si and Mg concentrations. In leach tests of fracture surfaces from cores within this watershed, As again correlates with Si in both the NaOH and the oxalate leachates. The relationship of high As concentrations with high Si and high Mg concentrations indicates significant water–rock interaction from slow ground-water movement, or long flow paths, or both. The fracture surface chemistry, the ground-water relationships, and the spatial distribution of ground-water chemistry all support a flow-related control on the concentrations of As in the ground water of this watershed.

REFERENCES

Daus, B., Weiβ, H., & Wennrich, R. 1998. Arsenic speciation in iron hydroxide precipitates. *Talanta* 46: 867–873.

Edwards, M., Patel, S., McNeill, L., Chen, H., Frey, M., Eaton, A.D., Antweiler, R.C., & Taylor, H.E. 1998. Consideration in As analyses and speciation. *J. Am. Water Works Assoc.* 90(3): 103–113.

Pierce, M.L. & Moore, C.B. 1982. Adsorption of arenite and arsenate on amorphous iron hydroxide. *Water Research* 16: 1247–1253.

Welch, A.H., Westjohn, D.B., Helsel, D.R. & Wanty, R.B. Arsenic in ground water of the United States: occurrence and geochemistry. *Ground Water* 38(4): 589–604.

Water–Rock Interaction, Wanty & Seal II (eds)

 ISBN 90 5809 641 6

Renewal of Fe used to remove TCE in simulated high alkalinity groundwater by ultrasound

Fei Liu
Department of Earth and Space Sciences, Beijing University, PRC Department of Water Resources and Environment, China University of Beijing, Beijing PRC

Robert Gillham & Lai Gui
Department of Earth Sciences, University of Waterloo, Waterloo, Canada

ABSTRACT: In some field tests, dechlorination rates and flow velocities through permeable reactive barriers (PRBs) of granular Fe were observed to decline. Laboratory column tests were conducted to study the phenomena because these reductions are important to limit the application of PRBs. The results of experimental simulations of groundwater with high alkalinity showed that carbonate precipitates eventually coat the Fe surface and reduce the reactivity of the Fe. Ultrasound was used to remove these precipitates, but this study indicated that ultrasonic treatment is not practical or cost-effective to use in PRBs which were modeled experimentally using Fe-packed columns.

1 INTRODUCTION

During the past decade, PRBs of granular Fe have shown great promise for *in-situ* remediation of chlorinated solvents in groundwater. However, reductions in dechlorination rates and flow velocity have been observed in some field tests (Mackenzie et al. 1996). Reductions may be related with mineral precipitation on the Fe surface. Studies have shown that these precipitates are complex mixtures of Fe oxides, carbonates and other minerals (Agrawal & Tratnyek 1996). In highly alkaline groundwater, carbonate minerals are the main precipitates that eventually coat the Fe surface and reduce the reactivity of Fe (Farrell et al. 2000). A technique that can maintain or restore PRB performance is needed. The use of ultrasound (US) for this purpose was first tested by Geiger et al. (2002) at the University of Central Florida. Though batch experiments and surface analysis in their study showed the first-order rate constants improved as much as 78% with 2h US (20 KHz, 330 Wh) pretreatment, the study didn't examine the precipitates that were removed by US (Geiger et al. 2002).

US refers to periodic stress waves that occur at frequencies in excess of 20 kHz. There are two modes of cavitational bubble collapse that can affect the surface of Fe. First, cavitational bubbles collapse directly on the surface, which may cause defects by shock waves. Second, cavitational collapse near the solid surface in the liquid phase can cause microjets to hit the surface and produce a nonsymmetrical shock wave. This mode is the well-known cleaning action of US. As a result, more reactive surface area can be formed to enhance surface reaction (Gotpagar et al. 1999).

The objectives of this study were to (1) examine the effectiveness of US energy to rejuvenate Fe surfaces by conducting column experiments with the key goal of restoring the rate of TCE degradation, and (2) to explore the type and extent of mineral precipitates that were removed.

2 EXPERIMENTS

2.1 *Column design and packing*

Five columns were constructed for this study. The columns consisted of 10 cm Plexiglas tubes, with an internal diameter of 3.81 cm (Gillham & O'Hannesin 1994). Each column had three equidistant sampling ports for taking aqueous samples and three other equidistant ports for taking solid Fe samples for surface analysis.

Connelly Fe (UW 230) was added to the column. The dry bulk densities in the columns, porosity, pore volume, specific surface area and ratio of surface area to volume of solution are listed in Table 1.

2.2 *Influent solution, sample collection and analysis*

Reagent-grade $NaNO_3$, $CaCO_3$ and distilled water were used to prepare the various influent solutions. 500 mg/L $CaCO_3$ solution was obtained by adding 9.5 g $CaCO_3$ solid to 19 L distilled water and purging CO_2 gas to dissolve $CaCO_3$. 500 mg/L $CaCO_3$ solution and distilled water were purged using N_2 gas to keep dissolved oxygen <0.3 mg/L. The initial TCE influent solution was obtained by spiking a stock solution of 203,282 ppm TCE dissolved in methanol to 30 L Teflon bags. The influent solution was pumped from the bottom of the column using an Ismatec^{TM} multiple channel peristaltic pump. A flow rate of 0.5 mL/min was maintained throughout the experiments. Samples were collected from the sampling ports, and from the influent and effluent lines using a glass syringe. 4 ml solution for chlorinated byproducts were taken by glass syringes, put in 10 ml glass vials, and sealed with crimp-caps with Teflon-lined septa. For TCE analysis, a 2.0 ml sample was taken and added to 2 ml patene which contained 500 μg/L of 1,2-dibromoethane in a 5 ml glass vial capped with a Teflon-faced septum. These samples were shaken on an orbital shaker for 15 minutes at 300 RPM. The patene phase was removed to a 2 ml glass crimp-top GC vial and put on the instrument.

Analyses for chlorinated chemicals were conducted using an HP 5890 Series II GCs. One GC equipped with an electron capture detector (ECD) was used to analyze TCE and another GC equipped with a PID (photoionization detector) was used to analyze the less chlorinated products. The GC with HNU PID has a bulb ionization potential of 10.2 eV.

Table 1. Physical parameters of the columns.

Column No.	Influent	Material density (g/cm^3)	Porosity	Pore volume (ml)
509	TCE	3.148	0.679	77.4
510	TCE + $CaCO_3$	3.200	0.665	75.8
511	TCE	3.082	0.648	73.9
512	TCE + $CaCO_3$	2.838	0.596	67.9

Notes: Specific surface area of Fe used is 1.52 cm^2/g. Columns 509 and 510 had no US treatment but columns 511 and 512 were treated with US for 1 h once every two weeks.

The GC was equipped with a fused silica capillary NSW-plot column (15 × 0.53 mm ID). Using an HP 7694 headspace auto-sampler, with a 1 ml stainless steel sample loop, the samples were injected into the GC. At 75°C, the samples were kept in the analyzer for 2 minutes and subsequently injected into the GC. The GC oven temperature increased from 50°C to 200°C at 20°C/min, and was held for 7 min. The injector and detector temperatures were 100°C and 120°C, respectively. MDLs (method detection limits) were 2.5 ppb, 3.3 ppb, 4.0 ppb, and 2.0 ppb for cis-DCE, trans-DCE, 1,1-DCE and VC respectively. The MDLs were determined using the EPA procedure for MDL (USEPA 1982).

TCE was analyzed by GC with a ^{63}Ni ECD. Using an HP 7673 liquid auto sampler, a 1 μl sample was injected into a DB-624 capillary column. The detector temperature ramp is 50°C to 150°C at a rate of 15°C/min and then held for 1 min. The detector and injector temperatures were 300°C and 200°C, respectively. The carrier gas is pre-purified He with a total flow rate of 25 ml/min and the make-up gas is 5% CH_4, 95% Ar. The MDL for TCE is 1.0 μg/L.

Ca and total dissolved Fe were analyzed by Inductively coupled Plasma-Atomic emission spectroscopy (ICP-AES). Eh and pH were checked using a Markson Model NO. 95 Digital pH/mv/temperature Meter. Alkalinity was titrated using 0.08 M H_2SO_4 using a Hach Digital Titrator.

3 RESULTS AND DISCUSSION

3.1 *Effect of US treatment on the reactivity of the Fe*

Columns 509 and 510 were constructed to study the effectiveness of US treatment for rejuvenating granular Fe in the absence and presence of $CaCO_3$, respectively. In column 509, the influent solution consisted of 1 mg/L TCE.

Figure1 shows pairs of TCE degradation profiles taken right before and several pore volumes after US in this column. There is no apparent difference in

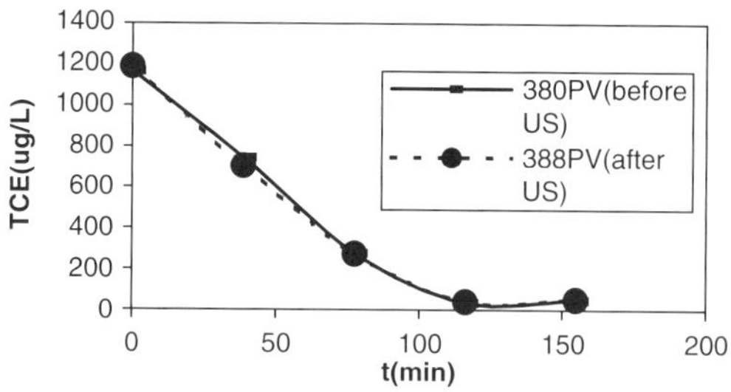

Figure 1. Comparison of TCE profiles before and after US in column 509 containing TCE only.

TCE disappearance profiles before and after US treatment, indicating a limited effect of US in regenerating granular Fe reactivity at the US intensity used. This result might be due to two possibilities. One, the US wave is too weak to penetrate the column wall or to remove the surface oxides. And two, no passivation happened on the Fe surface. If the second case applies, no differences should be observed in Fe reactivity, regardless the application of US treatment.

To make sure there is obvious mineral precipitation, column 510 was constructed with the influent consisting of 1 mg/L TCE and 500 mg/L $CaCO_3$. In groundwater containing high concentrations of carbonate, the Fe can lose its reactivity faster compared to groundwaters with low $CaCO_3$ concentration due to carbonate-mineral precipitation on the Fe surface. The high concentration of $CaCO_3$ in the influent solution quickly resulted in an obvious white-grey precipitation on the Fe surface.

However, comparison of TCE profiles taken before and after US treatment shown in Figure 2, there is still no observable effect of the US treatment. Therefore, it is possible that the US wave was not able to penetrate the column wall to remove the carbonate precipitates.

The data in Figure 3 demonstrate that the TCE profiles have shifted upwards with increasing amount of solution passing through the column. The rate of removal becomes slower in both columns 510 and 512. Though this indicated that passivation of the Fe has happened during the operation period of the columns, comparison of TCE profiles taken at similar PV in the column that received regular US treatment and in the column which did not receive treatment indicated that there is no dramatic difference (Fig. 4). This result implies that US might have no effect on the surface of reactive materials.

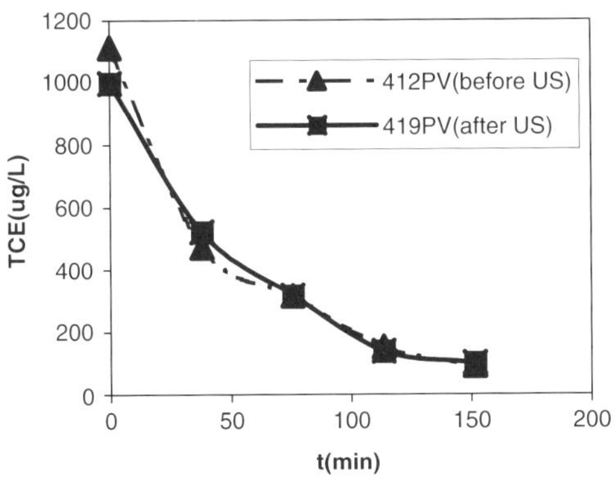

Figure 2. Comparison of TCE profiles before and after US in column 510 containing TCE + $CaCO_3$.

3.2 *Inorganic constituents*

It is assumed that the concentration of Ca in the effluent would rise after US treatment if the US could remove the deposits on the Fe. But the results do not support this assumption. There are two explanations, one of which is that the deposits on the Fe surface were not removed. Another possibility is that the precipitates removed from the Fe surface of Fe were not carried by the flow but settled down in the column. So far, the results are ambiguous.

3.3 *The penetration ability of the US produced by a US bath*

On the basis of the above discussion, it is unclear why the US bath had no effect on the reactivity of the Fe. The intensity of US was detected using Al foil to get explanation.

A sheet of Al foil was put in a column filled with water inside a big beaker with water. Some holes can be observed on the Al foil treated in US bath for 0.5 h. A sheet of Al foil was put in a column filled with

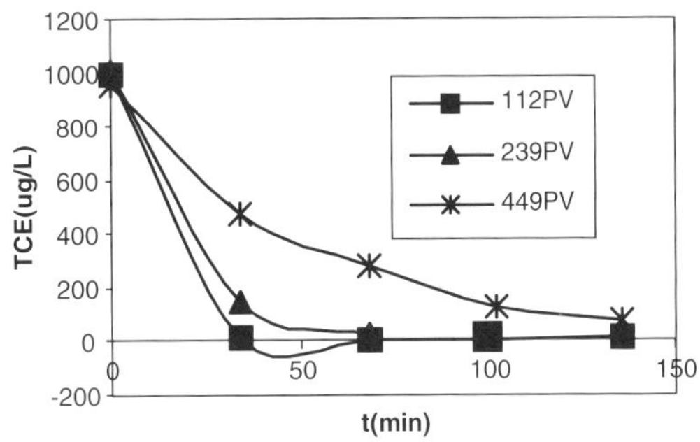

Figure 3. Change in TCE profiles of column 512 (without US) in $CaCO_3$ solution.

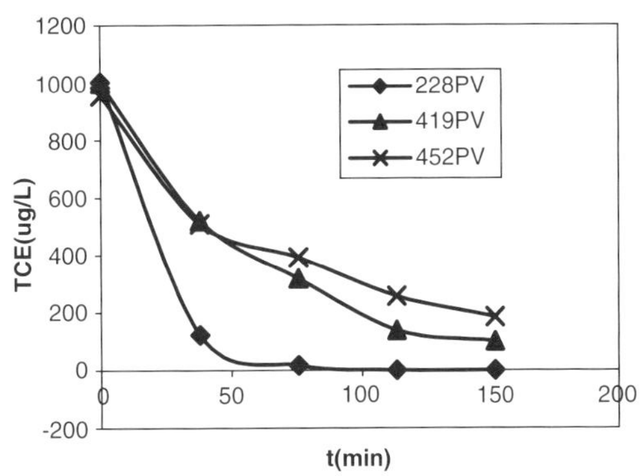

Figure 4. TCE profile shifts for column 510 (with US) in $CaCO_3$ solution.

water in a large beaker, when wet Fe was packed between the column wall and beaker wall, nothing happened to the Al foil that was treated in the US bath for up to 3 hours. The thickness of the Fe layer was about 2.7 cm. The US wave produced by a 400 w, 20 KHZ-US bath is not strong enough to penetrate a 2.7-cm Fe layer. So it may be deduced that the US wave passing through column wall just acts on the small fraction of the Fe that is close to the column wall. This action is too weak to improve the reactivity of the Fe in the entire column. In other words, packed Fe attenuates the intensity of US wave dramatically.

4 CONCLUSIONS

The reaction of TCE with Fe has been proven to occur on the surface of Fe. As precipitates of $CaCO_3$ build up on the surface, the reaction rate tends to decrease. Although other researchers have demonstrated that US can remove mineral deposits from the Fe surface and restore reactivity, the effect is limited when granular Fe is packed in columns. There are no obvious differences between columns with and without US. The Al foil experiment showed that a packed Fe layer attenuated a US wave. Although US may rejuvenate the Fe by removing the surface deposits in batch systems, this study showed that it is not practical and cost-effective to use in packed columns, and, by extension, probably will be ineffective in rejuvenating PRBs. Packed Fe attenuates US waves rapidly, so US technology is not promising because it will require very closely spaced US wells, which would be expensive and impractical.

REFERENCES

Agrawal, A. & Tratnyek, P.G. 1996. Reduction of nitro aromatic compounds by zero-valent iron metal. *Environ. Sci. Technol.* 30: 153–160.

Farrell, J., Kason, M., Melitas, N. & Li, T. 2000. Investigation of the long-term performance of zero-valent iron for reductive dechlorination of trichloroethylene. *Environ. Sci. Technol.* 34: 514–521.

Geiger, C.L., Ruiz, N.E., Clausen, C.A., Reinhart, D.R. & Quinn, J.W. 2002. Ultrasound pretreatment of elemental iron-kinetic studies of dehalogenation reaction enhancement and surface effects. *Water Research* 36: 1342–1350.

Gillham, R.W. & O'Hannesin, S.F. 1994. Enhanced degradation of halogenated aliphatics by zero-valent iron. *Groundwater* 32: 958–967.

Gotpagar, J., Lyuksyutovm, S., Cohn, R., Grulke, E. & Bhattacharyya, D. 1999. Reductive dehalogenation of trichloroethylene with zero-valent iron: surface profiling microscopy and rate enhancement studies. *Langmuir* 15: 8412–8420.

Mackenzie, P.D., Horney, D.P. & Sivavec, T.M. 1996. Mineral precipitation and porosity losses in granular iron columns. *J. Hazard. Mater.* 68: 1–17.

Water–Rock Interaction, Wanty & Seal II (eds)
© 2004 Taylor & Francis Group, London, ISBN 90 5809 641 6

Processes controlling groundwater chemistry of a coastal area in SE Sardinia (Italy)

M. Lorrai, L. Fanfani & P. Lattanzi
Dipartimento Scienze della Terra, Università di Cagliari, Italy

R.B. Wanty
U.S. Geological Survey, Denver CO, USA

ABSTRACT: The groundwaters of a small coastal area in south-eastern Sardinia, circulating mainly in granitic aquifers, show a wide range of salinity. This feature is difficult to explain only on the basis of water-rock interaction processes. Hydrogeological considerations exclude direct seawater intrusion for almost all aquifers sampled. On the basis of Cl/Br ratios, the important role played by sea spray is identified. Assuming all the Cl^- is derived from sea spray, the approximate amount derived from seawater and from water-rock interaction processes can be calculated for the other chemical species. With this approach, ion-exchange processes were singled out, which explains Ca, Na, and K patterns. The same process or microbial sulfate reduction is suggested to explain some anomalous dissolved Ba and sulfate concentrations.

1 INTRODUCTION

A hydrogeochemical investigation has been performed in a coastal area close to the village of Tertenia, with the goal of explaining the relatively high U concentration of some groundwaters (Lorrai et al. 2003). Since a wide range of salinity was found in these groundwaters, attention has been focused also on the processes responsible for the observed concentrations and patterns of major ions.

The study area is characterized by a low-grade metamorphic basement of Cambrian to lower Carboniferous age intruded at the end of the Hercynian orogeny by calc-alkaline magmas mainly of granodioritic composition (Secchi et al. 2001). Post Hercynian (Permian) rhyolitic volcanic rocks are present in the northern part of the study area. Mineral deposits in the area are mainly represented by several barite veins.

2 FIELD & ANALYTICAL METHODS

Water samples were collected from wells and springs. Conductivity, pH, Eh, temperature, and alkalinity were measured in the field. Water samples were filtered (0.45 μm). Samples for major cations as well as minor and trace-element analyses were filtered and collected in acid-washed polyethylene bottles, and then acidified to 1% v/v with superpure HNO_3. All the samples were analyzed by IC (major anions), ICP-OES (major and some minor cations) and ICP-MS (minor and trace elements).

3 GROUNDWATER CHEMISTRY & SOURCE OF CHLORIDE

The groundwaters of the area show a wide range of salinity, from 130 to 3800 mg/L, as well as variable main chemical features (Fig. 2). To identify the source of salinity we can look closely at the behavior of Cl^- because its conservative nature may be used for comparison with the concentration trends of other ions. The Br vs. Cl plot (Fig. 3) shows a good correlation ($R = 0.99$) between these elements. Their pattern matches the seawater dilution line.

Hydrogeological considerations do not support a direct seawater intrusion for all aquifers sampled, if we exclude a few wells close to the coastline. Moreover, appreciable contributions of Cl^- and Br^- from water–rock interaction processes can be excluded as these would explain neither the wide range of concentrations observed nor Cl/Br ratios so close to seawater in all the aquifers of different lithologies.

The source of Cl and Br is identifiable in sea spray because Cl and Br are both conservative elements and the formation and atmospheric transport of sea spray

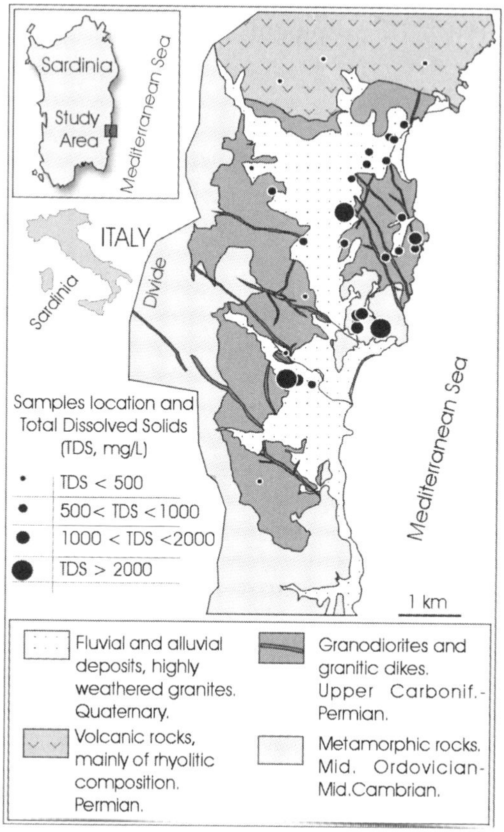

Figure 1. Schematic geological map and location of samples, showing the range of groundwater salinity.

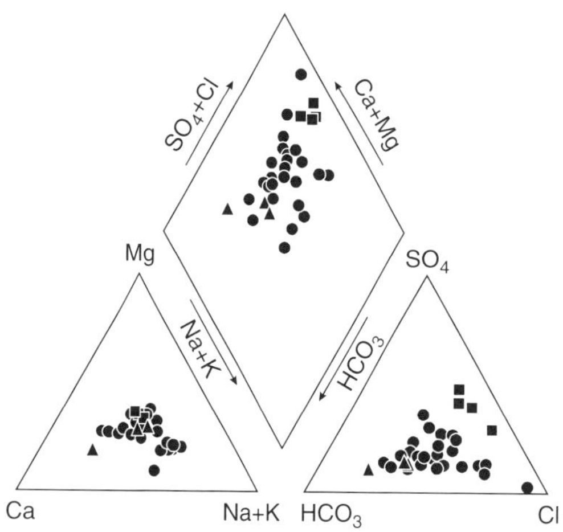

Figure 2. Piper diagram showing the main chemical features of the groundwater samples. Triangles: from volcanic rock aquifers; Squares: from metamorphic rock aquifers; Circles: from granitic rock aquifers.

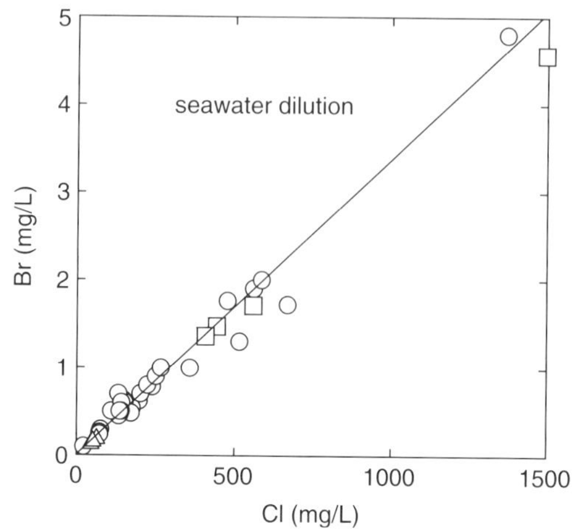

Figure 3. Br vs. Cl graph showing the good correlation between these elements. Their ratios fit that of seawater dilution. Symbols as in Figure 2.

is not accompanied by ion fractionation (Mazor 1997). This is a well known phenomenon in coastal areas where large amounts of water droplets are produced by breaking waves in surf zones. The droplets and evaporative particulates are transported by wind and deposited inland.

Experimental measurements (Gustafsson et al. 2000) have shown that the concentration and flux of salt decrease with downwind distance from the coast as an inverse power function. The salt deposition rate decreases rapidly over the first 2–5 km from the coast followed by a slower decreasing rate further inland. Salt accumulation in the soil depends on the topography and exposure of the land surface to prevailing winds. After deposition of sea salts in the soil, cyclic dissolution/evaporation processes could be responsible for ion concentration in the unsaturated zone as well as in shallow aquifers.

The wide range of Cl^- concentrations in groundwaters of this area could indicate different degrees of exposure as well as different stages of evolution of these processes prior to infiltration in the saturated zone and deeper aquifers.

4 BEHAVIOR OF CA, NA & K: ION-EXCHANGE PROCESSES

Figure 4 shows a good correlation between Cl^- and Na^+, with Cl/Na ratios close to seawater for most samples. However many samples have Cl/Na ratios substantially higher than that of seawater. The same pattern is observed for K (Fig. 5).

The behavior of Ca is mainly controlled by calcite dissolution in the initial stage of evolution, while at

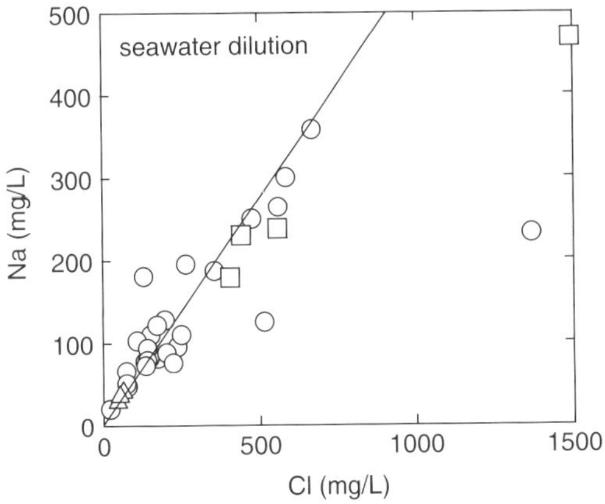

Figure 4. Na vs. Cl graph showing that in many samples Na is significantly depleted with respect to seawater ratios.

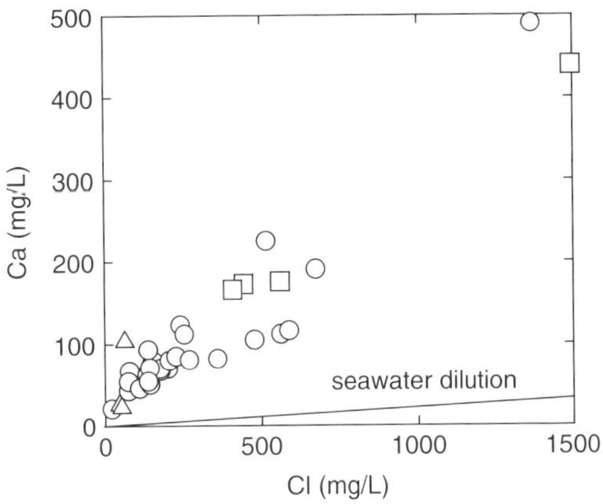

Figure 6. Ca vs. Cl graph showing that most calcium is derived from WRI processes.

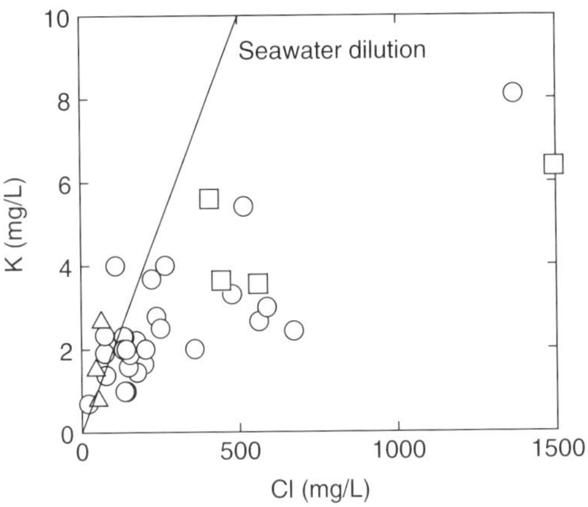

Figure 5. K vs. Cl graph showing that in many samples K is depleted with respect to seawater ratios.

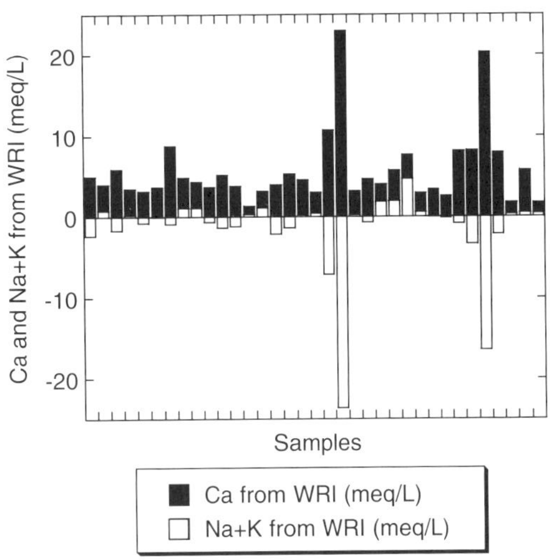

Figure 7. Calculated amounts of Ca and Na+K derived from WRI processes. The greatest enrichment in Ca is linked with the greatest depletion in Na+K.

higher salinity the waters are saturated with respect to calcite. As opposed to Na and K, the Ca vs. Cl graph (Fig. 6) shows an enrichment of Ca with respect to seawater ratios. These patterns can be explained by ion exchange processes between clay minerals and groundwaters. Na and K are taken up by the clays, and Ca is released. A simple mass-balance calculation helps in explaining this phenomenon.

Assuming that all the Cl is derived from atmospheric deposition, it is possible to calculate for each sample the amount of each component ultimately derived from seawater. For example, the amount of Na derived from seawater in a sample ($Na_{SEA\text{-}DERIVED}$) is:

$$Na_{SEA\text{-}DERIVED} = Cl_{SAMPLE} \cdot (Na_{SEAWATER}/Cl_{SEAWATER})$$

$Cl_{SEAWATER}$ and $Na_{SEAWATER}$ are the concentrations in average seawater.

The difference between the observed concentration of sodium (Na_{SAMPLE}) and the amount derived from seawater ($Na_{SEA\text{-}DERIVED}$) is the amount derived from water–rock interaction processes (Na_{WRI}):

$$Na_{WRI} = Na_{SAMPLE} - Na_{SEA\text{-}DERIVED}$$

Since K and Na patterns (relative to Cl) are very similar, we can consider their sum $(Na+K)_{WRI}$ and compare with Ca_{WRI}. Figure 7 shows that the greatest depletion in $(Na+K)_{WRI}$ is accompanied by the greatest enrichment in Ca_{WRI}, in good agreement with the hypothesized ion-exchange processes.

A few samples show a markedly positive $(Na+K)_{WRI}$. These samples are collected from wells located very close to the coastline, and these are the ones where a direct encroachment of seawater into the aquifer could be assumed. One of these wells, sampled before and after the period of maximum pumping (August), showed a marked increase in salinity (TDS from 450 to 830 mg/L). After the period of maximum pumping this water showed a Na-Cl character, whereas before this period it was of the Na-HCO_3 type. This could be attributed to removal of Ca from recharging waters and release of Na. In this respect, Na-HCO_3 type waters are indicated as typical of situations where salt water is flushed by fresh water (Appelo & Postma 1993). This result also means that possible seasonality of the ion-exchange process must be taken into account.

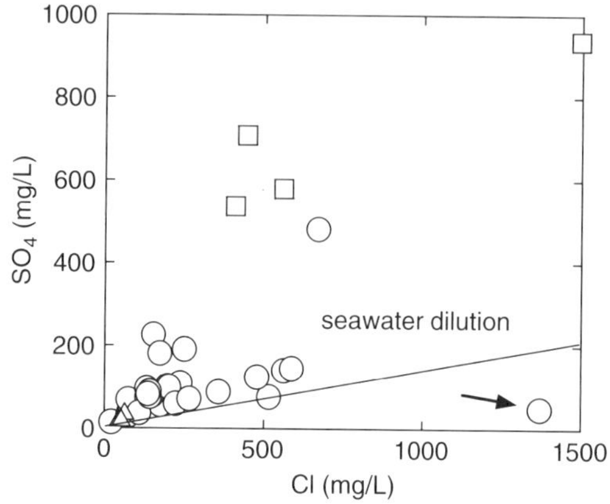

Figure 8. SO_4 vs. Cl graph showing the depletion in SO_4 with respect to seawater ratios in the sample indicated by the arrow.

5 BEHAVIOR OF SULFATE

The sulfate vs. chloride graph (Fig. 8) shows that almost all the samples are enriched in sulfate compared to seawater. In particular, groundwaters from metamorphic aquifers (squares in the graph) have the highest sulfate content, probably linked with the oxidation of pyrite veins present in that aquifer.

One sample falls below the seawater dilution line. This sample has less sulfate than expected on the basis of Cl concentration. Precipitation of gypsum, barite, and microbiological reduction of sulfate could be responsible for this sulfate removal. The following observations pertain to this sample: it is from a shallow well; its Eh is positive (0.5 V); it is undersaturated with respect to gypsum ($S.I = -1.6$); it has the highest dissolved Ba (Fig. 9); and it is slightly supersaturated with respect to barite ($SI = 0.4$).

Several studies have shown that sulfate-reducing bacteria are able to use barite as a source of sulfate (Bolze et al. 1974; Phillips et al. 2001). In these experimental studies, the sulfate-reduction process converted sulfate derived from barite to sulfide, resulting in the release of Ba into solution. Breit et al. (1996) found that sulfate-reducing bacteria were responsible for high dissolved Ba associated with low sulfate in groundwaters from a shallow alluvial aquifer. As described previously, several barite veins constitute the main mineralization in the area.

Other processes that could explain the high Ba in solution are the supply of some Ba from dissolution of feldspars or ion-exchange processes exactly as suggested for Ca. Figure 9 shows that Ba enrichment seems to be linked to Na + K depletion. The supply of Ba to groundwater by ion-exchange could precipitate barite and remove sulfate from solution.

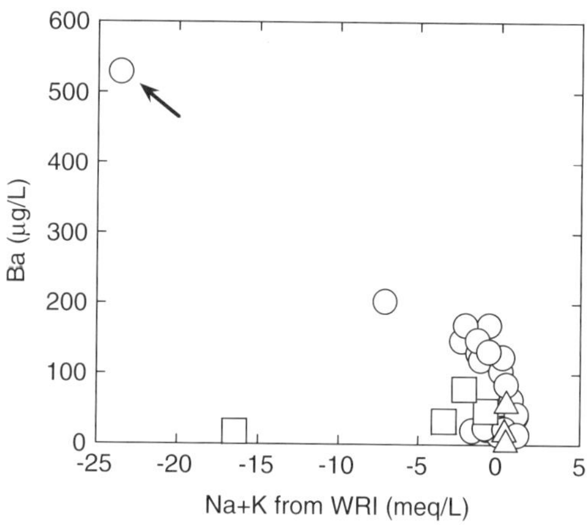

Figure 9. Relationship between Ba and Na + K deficit in solution.

6 CONCLUSIONS

Cl/Br ratios in groundwaters of this coastal area indicate the important role played by sea spray as a source of salts. The wide range of Cl concentrations is indicative of different degrees of sea spray contribution.

Ion-exchange processes are proposed to explain Na, K and Ca patterns, whereas the same process or biological processes involving sulfate-reducing bacteria are proposed to explain the high Ba and low sulfate observed in one groundwater sample.

ACKNOWLEDGEMENT

The authors thank an anonymous reviewer for providing insightful comments.

REFERENCES

Appelo, C.A.J. & Postma, D. 1993. *Geochemistry, Groundwater and Pollution*. Rotterdam: Balkema.

Bolze, C.E., Malone, P.G. & Smith, M.J. 1974. Microbial mobilization of barite. *Chem. Geol.* 13: 141–143.

Breit, G.N., Cozzarelli, I.M., Johnson, R.D. & Norvell, J.L.S. 1996. Interaction of alluvial sediments and a leachate plume from a landfill near Norman, Oklahoma. *Geological Society of America Abstracts with Programs* 28(7): A-258.

Gustafsson, M.E.R. & Franzén, L.G. 2000. Inland transport of marine aerosols in southern Sweden. *Atmospheric Environ.* 34: 313–325.

Lorrai, M., Fanfani, L., Lattanzi, P. & Wanty, R.B. 2003. Uranium mobilization in groundwaters and water-rock interaction modeling: the example of a coastal area of eastern Sardinia (Italy). *Proc. Intern. Conf. Uranium Geochemistry. Nancy, France, April 13–16,* 231–234.

Mazor, I. 1997. *Chemical and Isotopic Groundwater Hydrology: the Applied Approach*. New York: Dekker.

Phillips, E.J.P., Landa, E.R., Kraemer, T. & Zielinski, R. 2001. Sulfate-reducing bacteria release barium and radium from naturally occurring radioactive material in oil-field barite. *Geomicrobiol. J.* 18: 167–182.

Secchi, F., Cincotti, F., Cherchi, G.P. & Sarria, E. 2001. Geological and petrographical aspects of late Hercynian intrusive sequences from southern Ogliastra area (SE Sardinia, Italy). *Periodico di Mineralogia* 70: 34–74.

Water–Rock Interaction, Wanty & Seal II (eds)
© 2004 Taylor & Francis Group, London, ISBN 90 5809 641 6

New method for porewater extraction from claystone and determination of transport properties with results for Opalinus Clay (Switzerland)

U.K. Mäder, H.N. Waber
Rock-Water Interaction, Geological Sciences, University of Bern, Switzerland

A. Gautschi
Nagra, Wettingen, Switzerland

ABSTRACT: A new technique to extract porewater from claystone employs advective displacement of the in situ porewater by a traced artificial porewater. Monitoring of tracer breakthrough yields species-specific transport properties. Results for Opalinus Clay from the Mont Terri Research Laboratory indicate that the chemical disturbances due to the method are minimal, and that the observed significant difference in transport properties for Br^- and ^2H are in agreement with existing data. Sampling times are 2–4 months, and observation of tracer breakthrough takes 12–24 months at a hydraulic conductivity of $\sim 10^{-13}$ m/s.

1 INTRODUCTION

Knowledge of the in situ porewater composition and transport properties in very low permeability rocks such as claystone is important. The quantification of transport and water–rock interaction and its rates are used to interpret gradients across formations in isotopic, chemical, and gas composition that have evolved over thousands to millions of years (Rübel et al. 2002; Pearson et al. 2003). This quantitative understanding allows evaluation of a clay formation as a barrier for deep disposal of radioactive waste (e.g. Nagra 2002), or the behavior of pollutants in clay barriers. Also, the quantification of diagenetic processes and the evolution and migration of oil and gas hinge on the knowledge of in-situ chemical conditions and transport properties.

Difficulties in sampling porewater from claystone are due to very low hydraulic conductivities ($\sim 10^{-13}$ m/s), the presence of reactive components such as pyrite, carbonate, organics, and in-gassing or degassing of CO_2 during sampling. Methods in use for extracting porewater from claystone include long-term sampling in boreholes, porewater squeezing of rock samples, leaching of rock samples, in-situ diffusive exchange methods, and modeling of water-rock interaction processes (Pearson et al. 2003). All methods have their merits but are associated with poorly understood artifacts that modify the collected sample composition.

The aim of our new method is to minimize artifacts and to obtain both chemical and transport information from a single experiment. This contribution is associated with an ongoing in situ experiment at the Mont Terri Research Laboratory (Swiss Jura Mountains) that monitors and interprets long-term diffusion and reaction processes in Opalinus Clay (Wersin et al. 2004; Arcos et al. 2004).

2 METHOD

The method is forcing displacement of the original porewater contained in a drill core sample by advection of a traced artificial porewater. The feasibility of the method hinges on the well connected porosity and homogeneity of most claystones that result in approximate one dimensional advective-dispersive flow when applying a large hydraulic gradient across the length of a cylindrical sample. Also, the original moisture content has to be physically and chemically preserved during drilling, recovery of the core and sample preparation. The tracers in the infiltrating fluid allow for quantification of mixing with the *in situ* porewater, and to observe tracer breakthrough behavior when monitoring the infiltration experiment over a sufficiently long time period. The induced fluid fluxes are low (0.1–0.5 ml/day) and the dispersion is large and dominated by diffusion. Our experience indicates that 10–30 ml of little disturbed original porewater can be extracted from claystones with 10–20% porosity. Orientation of forced flow may either be parallel or perpendicular to bedding.

The method is called "porewater extraction by advective displacement" (weeks to few months), and "core infiltration experiment" when monitoring over extended time periods (months to few years) to obtain information on species-specific transport properties.

The method was developed from experience gained during a long-term experiment with hyperalkaline fluid and Opalinus Clay (Adler et al. 2001). The unique prototype apparatus was designed at the Paul Scherrer Institut and subsequently optimized at the University of Bern.

3 EXPERIMENTAL DETAILS

The heart of the apparatus is a pressure vessel that subjects the rock core to a hydraulic confining pressure (Fig. 1). The core is sealed from the confining medium by a sleeve (Teflon–rubber–shrink tubing) and two coupling pieces made of titanium. Porous disks (Teflon, Ti, PEEK) separate the top and base of the core from the couplings to allow for distribution or collection of infiltrating or extracted fluid. The spindle allows adjustment to the length of a rock core.

Fluid infiltration is driven from a He-pressurized tank. Fluid outflow was directed through a small-volume electrical conductivity cell and collected in a sampling device. We used an acrylic sample container for this experiment that was subjected to a small protective He pressure, allowing for periodic sampling (few days to weeks).

Measurement of pH and Eh was performed from time to time in-line with micro flow-through cells. Measurements of pH were performed in a cell that also subjected the combination electrode to a He confining pressure, and that minimized dead volume to 20 μl. The Eh cell was equipped with an Au electrode and an Ag/AgCl reference.

A claystone core (Opalinus Clay, mid Jurassic) of 120 mm length and 79 mm diameter was sealed with Teflon and rubber (outer layer), and fitted with porous Teflon discs at the top and base. The sample was subjected to 65–75 bar confining pressure, and a 40–60 bar infiltration pressure. Temperature was kept at 15°C corresponding to in situ conditions at Mont Terri. The core was drilled with nitrogen at the Rock Laboratory (borehole BPC-A1, from 10 m depth) and prepared and mounted on-site to minimize drying and exposure to atmosphere. An artificial porewater of similar salinity traced with bromide and deuterium was used (Table 1). The experiment started on March 27, 2002, and is still on-going as of October 2003.

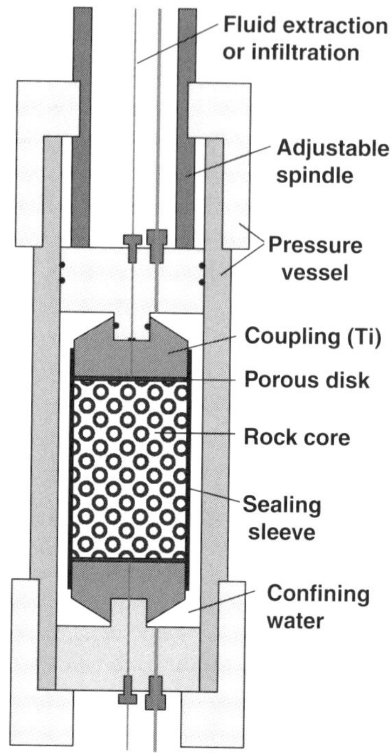

Figure 1. Pressure vessel containing the rock core assembly with tubing for confining pressure (thick lines) and infiltration (thin lines).

Table 1. Composition (mg/l) of early porewater extracts.

	#02	#05	#06	#07	APW[a]
pH (lab)	7.41	7.38	7.53	7.44	7.74
pH (in-line)	7.5 ± 0.2				n.a.
Ca	522	580	587	610	638
Mg	320	361	366	380	438
Na	5230	5410	5400	5560	5870
K	64	217	69	67	57
Alk. (meq)	<0.72	1.77	1.90	2.08	1.00
Cl	8300	9080	9000	9050	9300
SO_4	1360	1415	1410	1410	1410
Br	29.2	36.7	55.9	108	2350
TIC	<20	25.3	20.4	24.8	n.a.
TOC	84.5	58.2	59.3	40.4	<0.2
Si	2.1	2.43	2.41	2.54	0
Al	<0.1	<0.1	<0.1	<0.1	0
Sr	31.6	35.0	35.4	36.7	39.0
Li	0.36	0.41	0.43	0.43	0
Mn	0.40	0.35	0.27	0.24	0.22
Fe (tot)	0.23	0.24	0.24	0.25	0.13
$\delta^{18}O$ (‰)	−8.88	−8.98	−8.93	−9.04	−10.46
$\delta\,^{2}H$ (‰)	−54.8	−51.3	−52.2	−47.9	+370

[a] artificial porewater used for displacement.

4 APPLICATIONS

The apparatus is unique and versatile and can be used for porewater extraction as well as transport experiments with focus on chemical rock-water interaction and/or tracer transport and retardation. Past work included infiltration of hyperalkaline fluid into claystone (Adler 2001, Adler et al. 2001), infiltration of high-pH fluid in a granitic shear zone (Mder et al. 2003), and porewater extraction from the Callovo-Oxfordian clay formation from the URL under construction by ANDRA at Bure, France.

The strength of the method is to transfer approximate in-situ conditions to the laboratory and minimize unwanted effects due to stress relief, redox processes (pyrite, organics), and reactions within the carbonate system. A single carefully prepared long-term experiment can thus yield a comprehensive data set to constrain rock-water interaction processes as well as transport and retardation properties.

5 RESULTS FOR OPALINUS CLAY

Six small porewater samples totaling 23 mL were collected during the first 7 weeks of porewater extraction. Two samples were discarded for reasons of initial contamination by distilled water, or electrode contamination. Four samples were fully analyzed by the laboratories of the British Geological Survey (Table 1) and are shown in comparison to the artificial porewater used as input containing bromide (NaBr substituting for NaCl) and deuterium as tracers.

Long-term monitoring and regular sampling provided information on the evolution of the hydraulic conductivity calculated from integrated fluxes, hydraulic gradient and sample dimension (Fig. 2), and the breakthrough of tracers Br and δ^2H (Fig. 3). Sulfate concentrations remained constant over time indicating negligible disturbance from potential pyrite oxidation and/or microbial activity (in contrast to the in-situ experiment, Wersin et al. 2004). Breakthrough of Ca, Mg, Na, and K (not shown) displayed minor deviations from pure mixing trends due to limited interaction with the exchanger population in the claystone (~120 meq CEC per kg rock, Pearson et al. 2003).

The pH was measured four times during the first six months and converged to 7.5 ± 0.2 during periods of a few days each. Eh was measured two times in a micro flow-through cell. Redox potentials gradually dropped to approximately −130 mV (SHE) over the course of a few days. The Eh device was not yet refined, but did indicate that reducing conditions prevailed during the experiment – in agreement with the observed absence of effects that would be associated with pyrite oxidation.

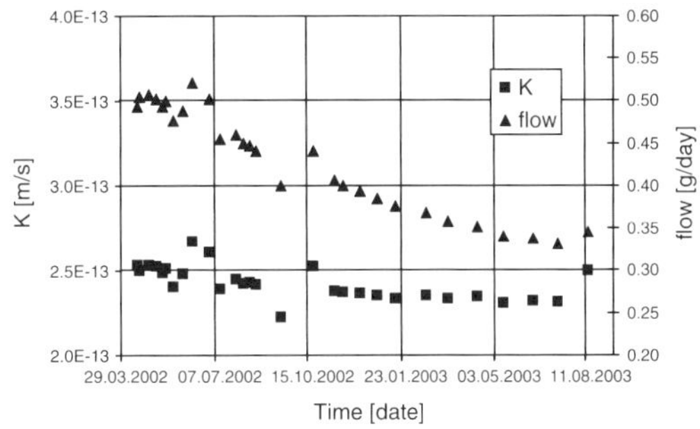

Figure 2. Evolution of hydraulic conductivity (K) and flow.

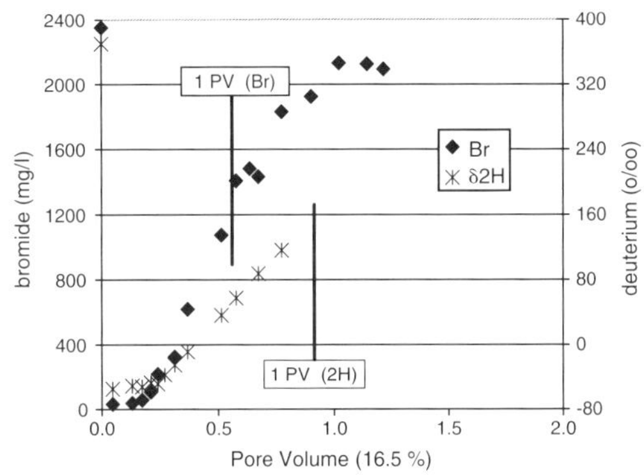

Figure 3. Breakthrough of Br and δ^2H.

The initial porewater content was determined on adjacent samples by weight loss on drying at 105°C. Four samples averaged 6.8 wt% that can be transformed into 16.5% porosity using a grain density of 2.72 g/cm^3. This is consistent with numerous comparable samples measured by different techniques (Pearson et al. 2003). The total porewater content of the sample is approximately 97 ml, a value used to convert the time axis into numbers of pore volumes percolated through the sample core (Fig. 3).

6 INTERPRETATION

Samples #05–#07 are interpreted to be the best representatives of the in-situ porewater extracted by this method (Table 1). Sample #05 was slightly contaminated by traces of electrode reference solution (3 M KCl) as indicated by elevated K^+ concentrations (and Cl^-). The mixing indicators Br^- and δ^2H show progressive admixing contributions from the artificial porewater of no more than 1–3%

Table 2. Modeled parameters using PHREEQC.

Parameter/Sample	#05	#06	#07
SI (calcite)[a]	−0.06	0.24	0.21
log $P(CO_2)$[a]	−2.55	−2.67	−2.54
SI (calcite)[b]	0.27	0.34	0.35
log $p(CO_2)$[b]	−2.64	−2.88	−2.72
SI (calcite)[c]	0.24	0.15	0.25
log $P(CO_2)$[c]	−2.61	−2.70	−2.62
SI (calcite)[d]	0.0	0.0	0.0
log $P(CO_2)$[d]	−2.61	−2.92	−2.75
Alk. (meq)[d]	1.57	1.11	1.31
SI (dolomite)[d]	0.13	0.13	0.13
SI (celestite)[d]	0.00	0.00	0.00

Modeling Assumptions: [a] measured total alkalinity, lab pH; [b] measured TIC, lab pH; [c] measured TIC, in-line pH; [d] TIC adjusted for calcite saturation, lab pH. Partial pressure in bars.

for sample #07 compared to #05. Sample #02 appears to have been diluted by pure water during a small accidental ingression of confining water into the extraction system at the start of the experiment. The elevated TOC content in #02 may be due to contamination by organic lubricants used in the pressure vessel in contact with the confining water. Compositions of major components and tracers are compatible with a dilution of 5–10%.

Sample #05 represents the most undisturbed composition (except for K^+) and suggests a distinctly lower salinity (Cl+Na) compared to the artificial porewater (Cl+Br+Na) designed for a location closer to the maximum observed salinity across the Opalinus Clay formation (Pearson et al. 2003).

Samples #05–#07 were interpreted using PHREEQC and the Nagra-PSI thermodynamic data base to model saturation constraints and processes that may have affected the carbonate system (Table 2). Samples recalculated on the basis of laboratory pH and measured total alkalinity or TIC resulted in slight supersaturation with respect to calcite (options a, b, c). Our preferred interpretation is option d whereby calcite saturation was imposed and TIC adjusted at laboratory pH. This interpretation (and others) would imply partial pressures of CO_2 of $10^{-2.6}$–$10^{-2.9}$ bar at the time of extraction. In situ pH cannot be rigorously reconstructed. The in-line measured pH of 7.5 ± 0.2 is plausible, and in combination with the associated $p(CO_2)$ is in agreement with multiple lines of evidence for this stratigraphic location (Pearson et al. 2003).

The breakthrough behavior of Br^- and δ^2H (Fig. 3) displayed a distinctly species-specific character. The time axis was converted to numbers of pore volumes percolated through the core based on a measured water-loss porosity of 16.5%. The shapes of the breakthrough curves are compatible with one-dimensional advective-dispersive flow. The half-height in breakthrough concentration should occur exactly at one pore volume percolated if chemical retardation of the tracers can be excluded. Using this half-height as marker (vertical lines in Fig. 3) one can derive species-specific effective pore volumes (or porosities) relevant for tracer transport. The interpretation is that the effective porosity for ^2H (as H_2O) is equal or slightly smaller than the water-loss porosity, and that for Br^- amounts to only 60% of the water-loss porosity. The ratio of the effective transport porosity for $Br^-/^2H$ is about 2/3. This is in excellent agreement with data for effective diffusion porosity derived from in situ and laboratory experiments (Pearson et al. 2003). The effect is described as anion-exclusion due to the large amount of charged surface in claystones which decreases the amount of connected pore space that can be used by anions for diffusion or advection.

7 CONCLUSIONS

The new method of porewater extraction offers advantages compared to both in situ techniques and alternate laboratory methods. It appears to minimize artifacts that disturb the porewater composition during sampling and extraction. The small but continuous flow of extracted sample would allow for even more sophisticated in-line measurement techniques, of which so far only electrochemical methods were employed.

Species-specific transport properties and processes of rock–water interaction can be quantified when extending the experimental duration to months or years in order to monitor the breakthrough behavior of tracers and major components.

REFERENCES

Adler, M. 2001. Interaction of claystone and hyperalkaline solutions at 30°C: A combined experimental and modeling study. Ph.D. Thesis, University of Bern.

Adler, M., Mder, U.K. & Waber, H.N. 2001. Core infiltration experiment investigating high-pH alteration of low-permeability argillaceous rock at 30°C. *Proceedings of the 10th International Symposium on Water-Rock Interaction*, 1299–1302, Rotterdam: Balkema.

Arcos D., Gimmi T., Duro L. & Waber, H.N. 2004. Modelling of Tracer Behaviour and Dominant Reactions during the Pore Water Chemistry (PC) Experiment in the Opalinus Clay, Switzerland. *This volume*.

Mder, U., Frieg, B., Puigdomenech, I., Decombarieu, M. & Yui, M. in press. Hyperalkaline cement leachate-rock interaction and radionuclide transport in a fractured host rock (HPF Project). *Sci. Basis Nucl. Waste Manag. XXVII, Mat. Res. Soc. Symp. Proc.*,

Nagra 2002. Projekt Opalinuston – Synthese der geowissenschaftlichen Untersuchungsergebnisse. *Nagra Technischer Bericht NTB 02–03*, Wettingen, Switzerland.

Pearson, F.J., Arcos, D., Bath, A., Boisson, J.-Y., Fernández A.M., Gaebler, H.E., Gaucher, E., Gautschi, A., Griffault, L., Hernan, P. & Waber, H.N. 2003. Geochemistry of water in the Opalinus Clay formation at the Mont Terri Rock Laboratory – Synthesis Report. *Reports of the Swiss Federal Office for Water and Geology (FOWG), Geology Series, Nr. 5*, Bern, Switzerland.

Rübel, A., Sonntag, C., Lippmann J., Gautschi, A. & Pearson, F.J. 2002. Solute transport in formations of very low permeability: Profiles of stable isotope and dissolved gas contents of the pore water in the Opalinus Clay, Mont Terri, Switzerland. *Geochim. Cosmochim. Acta* 66: 1311–1321.

Wersin, P., Gautschi, A., Vinsot, A., de Canniere, P., Hernan, P., Gbler, H.-E., Hama, K., Mahara, Y. & Gaucher, E. 2004. Results from the Porewater Chemistry (PC) experiment at Mont Terri, Switzerland. *This volume.*

Water–Rock Interaction, Wanty & Seal II (eds)
© 2004 Taylor & Francis Group, London, ISBN 90 5809 641 6

^{13}C enrichments in periglacial biogenic carbonates may be due to acetogenic processes instead of methanogenesis

M.E. Marschner & I.D. Clark
University of Ottawa

ABSTRACT: Biomineralized secondary carbonate deposits in periglacial limestone fissures typically show enrichments in heavy carbon (^{13}C) compared to the host rock. Previous investigations have implicated the role of methanogenic bacteria in creating these enrichment signals. Acetogenic bacteria may play an equal or greater role in cycling carbon. Our research proposes a new "mixotrophic" acetogenic reaction where CO_2 and a partially-reduced organic compound (such as methanol) combine to reduce the dependence of acetic acid production on molecular hydrogen (H_2). Furthermore, acetogenesis appears to enhance limestone dissolution and hence reduce the depletion of residual DIC by biogenic CO_2. Radiogenic age dating using biogenic carbonates needs to account for the different mixing ratios of new and old carbon-bearing materials.

1 INTRODUCTION

Secondary carbonate within limestone fissures in Canadian periglacial terrains is a newly identified material that holds interesting paleoclimatic information. These deposits, referred to as *endostromatolites* for their growth within the rock mass and hidden from light, precipitated in equilibrium with meteoric groundwater (Fig. 1). Analyses of the ^{18}O isotopic signal from these calcretes suggest that past climatic conditions in the Canadian arctic were warmer than at present (Clark et al. 2004).

Isotopic analyses of microsamples taken from longitudinal profiles of endostromatolite growth columns have shown a progressive enrichment of ^{13}C increasing away from the contact surface of the host carbonate rock (Lauriol & Clark 1999). Similar enrichments in heavy carbon seen in other carbonate materials are ubiquitously assumed to be the product methanogenic activity which preferentially releases isotopically lighter $^{12}CH_4$. Other evidence, such as banded microstructures, proves that the isotopic enrichment and growth of columns could not be due to purely inorganic processes (Fig. 2).

Arctic wetlands provide a significant flux of methane to the global atmospheric budget (cf. Martens et al. 1992). Our investigations in and around a small-scale limestone terrain near Inuvik, N.W.T., have shown a positive correlation between the occurrence of limestone outcrops containing fissure

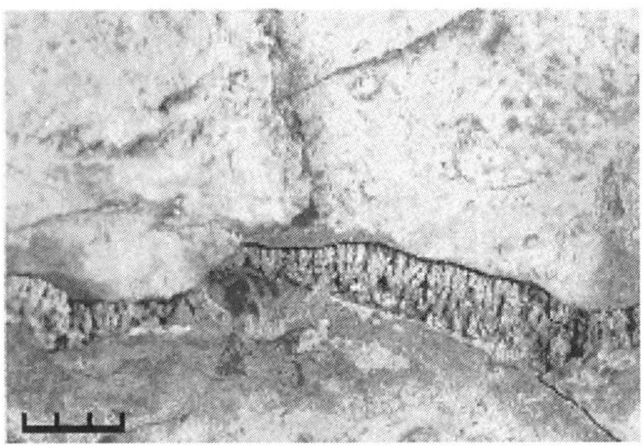

Figure 1. Biomineralized "endostromatolites" growing within a limestone fissure. Columns are attached to one side of the fissure and grow toward the other side. Scale bar shown in gradations of 1 mm. Photo courtesy I. Clark.

calcretes and dissolved methane in thawed active layer waters (Marschner, unpublished data). Samples of natural materials from the study area have been used to investigate the in-situ microbial communities and the geochemistry of incubated growth systems inoculated with these materials. Experimental incubations performed seek to determine the mechanisms of carbon cycling that led to the growth of biomineralized columns. This information is critical to understanding the geochemical conditions in the environment of deposition and the role various

Figure 2. Thin section of endostromatolite mounted on a normal microscope slide (field of view is about 1 cm across). Courtesy I. Clark.

carbon-bearing organic and inorganic compounds played. Radiocarbon modeling using fissure calcrete also relies on results from this investigation as ^{14}C age dating requires knowledge of the inputs of "dead" carbon from limestone sources.

2 EXPERIMENTAL AND ANALYTICAL PROCEDURES

Closed-system incubation experiments were performed to monitor the geochemical evolution of methanogenic systems. Incubation vessels were partially filled with growth media before being inoculated and sealed with a nitrogen headspace. The inoculum for various experiments included arctic soil (preserved by refrigeration) and bacterial cultures grown from frozen arctic soil. Early experiments used enriched growth media. Subsequent experiments made use of natural materials which more closely resemble in-situ growth conditions. Important variables monitored include pH, IC (inorganic carbon), OC (organic carbon), CH_4 (methane) and CH_3CO_2H (acetic acid). "IC" refers to the combined total of dissolved inorganic carbon (DIC) in the aqueous phase and gaseous CO_2 in the headspace.

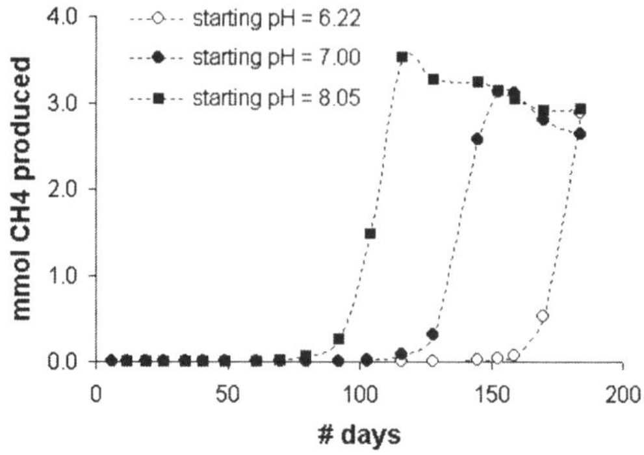

Figure 3. Methane production over time in systems with different starting pH conditions.

3 RESULTS AND DISCUSSION

3.1 *Incubation systems*

The inoculated incubation systems which showed the most methanogenic activity (i.e. methane production) contained one of two preferred growth stimulants: methanol or trimethylamine. By contrast, systems provided with formate or acetate produced very little methane. Previous investigations of non-pathogenic methanogens in natural consortiums have focused on acetic acid and CO_2 as two precursor compounds to methane formation. Published data concerning the levels of other methanogenic precursor compounds in natural systems is surprisingly lacking. The H_2/CO_2 methanogenic reaction pathway is widely accepted as a viable process for autotrophic and mixotrophic organisms despite the fact that ambient levels of H_2 in most natural systems are maintained below threshold detection levels. It may be possible that other methanogenic precursors are likewise maintained at imperceptible levels. We have chosen to use methanol as a growth stimulant and substrate since short-chain alcohols are natural fermentation products.

Our methanogenic cultures showed a preference for growth at higher pH conditions. Figure 3 shows the amount of methane produced by three identical inoculated systems which were started at different pH values. The most alkaline system underwent accelerated methane production earlier than the more acidic systems. Additional systems containing solid limestone produced results similar to the alkaline system in Figure 4 (data not shown). Limestone dissolution is a natural pH buffering reaction. Hence the affinity of methanogens for limestone is at least

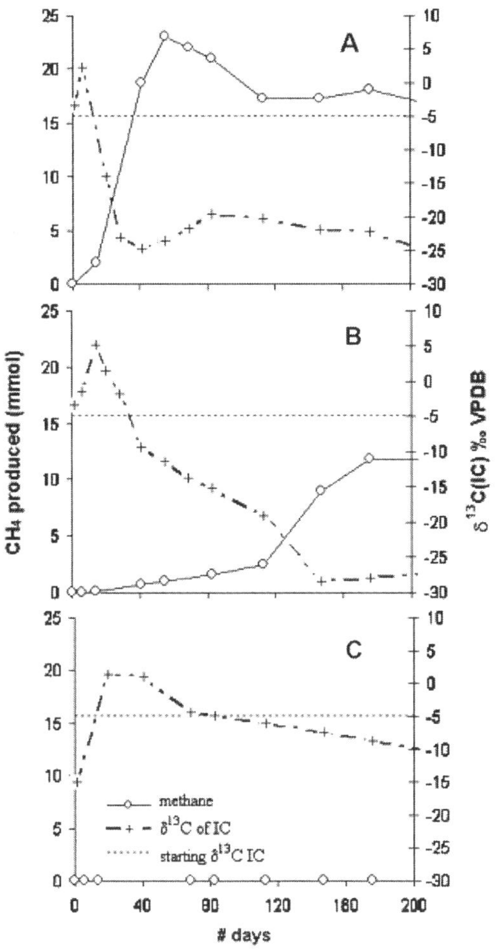

Figure 4. Methane production and carbon-isotopic composition of IC versus time in three distinct incubation systems where methanol is the main carbon substrate. Both Systems A and C were not given any source of IC initially. System B was prepared with $NaHCO_3$ to a concentration of 0.048 M.

partially related to suitable geochemical conditions created by limestone buffering.

Our systems containing methanol ubiquitously became acidic over time. By contrast, systems containing trimethylamine remained circumneutral and did not show much variance over time. Selective analysis of fatty acids showed that systems containing methanol yielded significant quantities of acetic acid, a stronger acid than CO_2. Hence the drop in pH can be attributed to acetic acid production and deprotonation.

Acetic acid production was highest in systems containing both methanol and high levels of IC. Clearly, acetogenesis is enhanced by both of these compounds. Homoacetogenesis proceeds via the following reaction:

$$2CO_2 + 4H_2 \rightarrow CH_3CO_2H + 2H_2O \qquad (1)$$

If one equivalent of methanol can substitute for one equivalent of CO_2, a new "heterotrophic" reaction pathway is created as follows:

$$CO_2 + CH_3OH + H_2 \rightarrow CH_3CO_2H + H_2O \qquad (2)$$

At standard conditions, the autotrophic reaction (Eq. 1) pathway is thermodynamically favorable over the heterotrophic pathway (Eq. 2). However, limited availability of H_2 in natural systems favours the substitution of CO_2 with a partially-reduced species like methanol, and hence the heterotrophic reaction becomes favorable.

Both autotrophic acetogenesis and methanogenesis create a strong demand for IC. As well, both groups of microorganisms preferentially consume isotopically lighter carbon (^{12}C) with similar fractionation factors between substrate and product (Gelwicks et al. 1989). The uptake of ^{12}C-enriched IC both lowers aqueous DIC concentration and enriches the residual DIC fraction in ^{13}C. Previously, only methanogenic activity was invoked to explain the isotopic enrichment profiles seen in endostromatolites. Isotopic data from our incubation systems show that DIC-enrichment in heavy carbon can be generated by acetogenic activity.

Three different systems prepared with methanol were compared for methane production and $\delta^{13}C$ of IC over time (Fig. 4). System A was not given any IC. System B was prepared with sodium bicarbonate to a concentration of 0.048 M. System C was selectively inhibited for methanogens with 2-bromoethanesulfonic acid (BES) to a concentration of 0.014 M. The starting $\delta^{13}C$ of IC was −4.8‰ VPDB. The IC of all three vessels underwent a similar ^{13}C enrichment followed by rapid depletion. In system B, the enrichment preceded any appreciable methane production, so the increase in $\delta^{13}C(IC)$ from −4.8 to +5.1‰ VPDB must have been due to acetogenic activity. The decrease in pH in these systems as a result of acetic acid production induced the release of CO_2 to the headspace of each vessel, but numeric modeling and calculations have proven that the isotopic enrichment caused by this reallocation of DIC is less than 2‰ (data not shown). Most of the IC present at the start of incubation in System B was eventually consumed. The rapid turnover of IC created by fermentative respirers consuming organic carbon (OC) (−26‰ VPDB) is responsible for erasing the initial ^{13}C enrichments and keeping $\delta^{13}C(IC)$ values in equilibrium with OC and residual methanol (initially −40‰ VPDB).

All three systems produced acetic acid, though production in system B was several times greater than either of the other two "unbuffered" systems (data not shown).

3.2 *Implications on endostromatolite formation*

In the anaerobic depositional setting of limestone fissures, the availability of organic substrates is more limited than IC. Biogenic CO_2 and dissolved organics are imported from overlying soil horizons by groundwater flowing through zones of talik. Previous research has shown that dissolved CO_2 is the preferred species of DIC used by methanogens (Vorholt & Thauer 1997). However, our results have shown that methanogenic production proceeds more quickly at higher pH conditions (Fig. 3). Limestone dissolution increases pH by consuming CO_2 as follows:

$$CO_2 + H_2O + CaCO_3 \leftrightarrow 2HCO^-_{3(aq)} + Ca^{2+}_{(aq)} \quad (3)$$

The mixing of biogenic CO_2 and dissolved limestone ($\delta^{13}C_{CaCO3} = 0$ to $-3‰$ VPDB) at a 1:1 ratio creates a DIC pool with an intermediate isotopic composition around -13 to $-15‰$ VPDB. To precipitate $CaCO_3$ that is enriched in ^{13}C compared to the limestone host rock, $^{12}CO_2$ has to be selectively consumed and removed such that it cannot remix with the DIC pool. Both acetogens and methanogens have the ability to do this. From our experimental results, these two groups of microorganisms use DIC competitively. Methane generated in our incubation systems above (Systems A, B, and C; see Fig. 4) had $\delta^{13}CCH_4$ values typically around -80 to $-85‰$ VPDB, indicating that the hydrogenotrophic pathway was being used by methanogens (Whiticar et al. 1986). So competitive use of DIC was likely the result of H_2 availability. Hydrogenotrophic CO_2-reduction requires 4 equivalents of H_2 per equivalent of CO_2 used by methanogens:

$$CO_2 + 4H_2 \rightarrow CH_4 + 2H_2O \quad (4)$$

By contrast, homoacetogenesis requires only 2 equivalents of H_2 (see Eq. 1). Heterotrophic acetogenesis (Eq. 2) requires only 1 equivalent of H_2. While our experimental evidence confirms that the acetogenic reaction pathways are favoured when both methanol and DIC are present under limited H_2 availability, there are likely other unknown parameters which control the prevalence of one bacterial group over another.

The interaction of acetogenic and methanogenic processes in limestone fissures is complex and not well understood. The growth of endostromatolite columns on one side of the fissure toward the opposite side would seem to suggest that separate processes are operating on each side of the fissure. The microlaminated internal structures of growth columns would instead suggest that cyclical processes are occurring over time, with alternating periods of high and low organic content. Our experimental evidence points to acetogenic activity being responsible for the ^{13}C-enrichments of IC observed in our incubation systems. Acetic acid production, like methane production, creates a product that is strongly depleted in heavy carbon and which does not seem to participate in further processes. One advantage acetic acid has in enhancing ^{13}C-enrichment of DIC is that it releases protons which react with limestone to liberate more DIC to solution without diluting the DIC pool in lighter carbon:

$$CH_3CO_2H + CaCO_3 \rightarrow HCO_3^- + Ca^{2+} + CH_3CO_2^- \quad (5)$$

So acetogenesis favours adding proportionately more limestone DIC into solution than does inorganic limestone dissolution using biogenic CO_2 alone. If acetogenic processes are significant in the environment of endostromatolite formation, models of radiocarbon age dating using endostromatolites need to be reevaluated. Marine limestone lacks cosmogenic ^{14}C. Age dating using ^{14}C requires knowledge of $\delta^{13}C$ of the source material so that a correction for dilution with a "dead" source and for isotopic fractionation during formation of the material. As we have seen above, both acetogenesis and methanogenesis produce similar isotopic fractionations for DIC, but the mixing ratios of old and modern carbon sources by these processes are potentially quite different. More investigation needs to be accomplished to determine the significance of acetogenic and methanogenic processes in endostromatolite formation to determine the validity of using these biogenic carbonates for paleoenvironmental interpretation.

4 CONCLUSIONS

Secondary carbonate deposits within limestone fissures are created indirectly by biogenic processes which selectively remove light carbon from the DIC pool. Competitive substrate usage appears to favour acetogenesis under natural conditions. Acid production by acetogenesis releases proportionally more radiogenically "dead" DIC from the dissolution of limestone than does methanogenesis, and hence the significance of the former process in creating ^{13}C-enriched growth columns appears to be underrated. Reassessment of ^{14}C age-dating models may need to take into account the higher inputs of ^{13}C-enriched limestone to the DIC pool in fissure groundwaters by acetogenic processes.

REFERENCES

Clark, I.D., Lauriol, B. & Sabourin, N. 2004. Biomineralized fissures in the Yukon: orbital climate forcing and arctic hydrology. *Can. J. Earth Sci.* (in review).

Gelwicks, J.T., Risatti, J.R. & Hayes, J.M. 1989. Carbon isotope effects associated with autotrophic acetogenesis. *Org. Geochem.* 14: 441–446.

Lauriol, B. & Clark, I. 1999. Endostromatolites in the arctic: a paleohydrologic indicator. *Appl. Geochem.* 14: 775–785.

Martens, C.S., Kelley, C.A., Chanton, J.P. & Showers, W.J. 1992. Carbon and hydrogen isotopic characterization of methane from wetlands and lakes of the Yukon-Kuskokwim delta, Western Alaska. *J. Geophys. Res.* 97: 16,689–16,701.

Vorholt, J.A. & Thauer, R.K. 1997. The active speices of "CO_2" utilized by formylmethanofuran dehydrogenase from methanogenic Archaea. *European J. Biochem.* 248: 919–924.

Whiticar, M.J., Faber, E. & Schoell, M. 1986. Biogenic methane formation in marine and freshwater environments: CO_2 reduction vs. acetate fermentation – Isotope evidence. *Geochim. Cosmochim. Acta* 50: 693–709.

Water–Rock Interaction, Wanty & Seal II (eds)
 ISBN 90 5809 641 6

Why are recent groundwaters sea-tagged and fresh, whereas formation waters are brine-tagged and saline?

Emanuel Mazor
Weizmann Institute of Science, Rehovot, Israel

ABSTRACT: The active groundwater cycle operates within the zone of hydraulic differences, existing at the landscape relief that protrudes above the terminal base of drainage – the sea surface. Sea-spray is efficiently driven far inland and washed into the groundwater system, tagging the latter by Cl, Br, Na and a list of additional ions, in relative abundances as in seawater. The cycling water washes salts from the continent into the sea and by the constant evaporation and formation of new precipitation fresh water is common. At greater depth extends the zone of zero hydraulic difference, at which formation water is trapped in hydraulically isolated rock-compartments, as is born out by their high ages and spatial composition variability. The isotopic composition discloses that formation waters are of meteoric origin. They are tagged by brine-spray, reflecting the evaporitic paleo-landscapes at which they have been formed. These were vast flat lowlands frequented by alternations of sea invasions and retreats. Thus, formation water properties disclose that the host rocks were formed in vast subsiding basins, i.e. at landscapes formed during phases of orogenic tranquility.

1 INTRODUCTION

The number of different groundwater types is uncountable. Looking at the values of pressure, chemical and isotopic compositions, age and additional parameters, every water sample seems to be a distinct item. Can we see the forest beyond the trees and recognize general patterns in the domain of groundwater?

Scanning data reported in many publications it turns out that there are two groundwater mega-systems (Fig. 1): (a) the shallow cycling groundwater that flows within the landscape relief that protrudes above sea-level, and (b) formation waters statically entrapped in hydraulically isolated rock-compartments at depths deeper than sea-level.

Waters in the shallow active system have a light isotopic composition that signifies meteoric (continental) groundwaters, and they are (a) recent-decades to a-few-thousand years old; (b) commonly fresh; and (c) tagged by sea-spray, even in sites thousands of kilometers inland (Fig. 2). In contrast, formation waters have a meteoric (continental) isotope composition as well, and are: (a) old, i.e. 10^4 to 10^8 years old; (b) saline; and (c) tagged by brine-spray (Fig. 3).

Why do the shallow cycling groundwater system and the deep static groundwater system differ systematically in their salinity and ionic-tagging? What can be learned from this key observation?

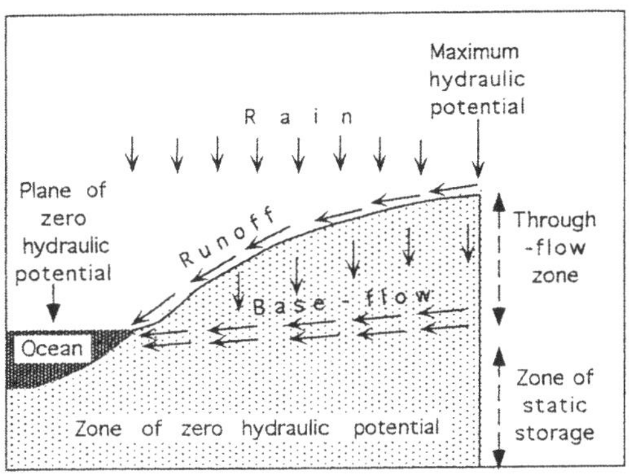

Figure 1. The two main groundwater zones or mega-systems. The shallow system flows from locations of high hydraulic potential to the plane of zero potential at the ocean surface. Fossil formation waters are entrapped in isolated rock-compartments, situated within the zone of zero hydraulic potential (Mazor 1997).

2 SEDIMENTARY ROCKS ACCUMULATED IN MARGINAL OCEANS

Observation: Our Earth experienced stages of orogenic mountain building that alternated with intermediate stages of orogenic tranquility. During the

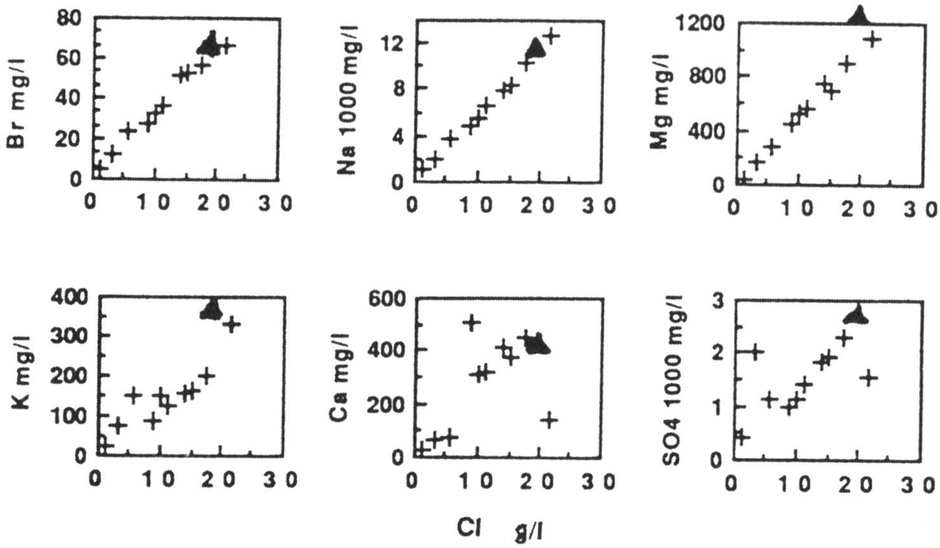

Figure 2. Ionic concentrations as a function of Cl in groundwaters encountered in the Wheatbelt, Australia (Mazor, 1997). Linear correlations with Br, Na and Mg are well seen, and the sea values (solid triangle) fall on these lines. Thus, tagging by sea-spray is obvious.

latter, erosion lowered the landscape, resulting in large-scale flat lowlands.

Second observation: Sea sediments get quantitatively accumulated at the ocean part that is close to a continent, because the raw material, from which marine sediments are formed, is mainly of continental origin – the products of erosion.

Third observation: Continental sediments, e.g. soils, calcretes or conglomerates, are washed into the ocean – they rarely accumulate on the land.

Conclusion: The thousands-of-meters of sedimentary rocks, in which formation waters are encountered, were accumulated in a marginal marine environment of subsiding basins.

Discussion: The meteoric composition discloses that formation waters were formed as continental groundwaters within marine rocks that were exposed on the continent during frequent sea regressions.

3 THICK SEDIMENTARY ROCK ASSEMBLAGES DISCLOSE EVAPORITIC ENVIRONMENTS, FREQUENTED BY SEA INVASIONS AND RETREATS OVER LARGE PALEO-REGIONS

Observation: The rocks filling the sedimentary basins are rich in clay, marl, sandstone, dolomite, as well as gypsum and halite (Mazor, in print).

Conclusion: These rocks were formed on flat lands that hosted sabkhas and lagoons, and were frequented by sea transgressions and regressions.

Observation: The named different rock types are stacked on each other, disclosing frequent and rather abrupt changes in the environment of the sediment accumulating system, i.e. frequent changes in the depth of the sea, and common alternations between sub-aerial exposure and a sub-marine environment.

Observation: Many of the sedimentary basins are of lateral extent of hundreds to thousands of kilometers.

Conclusion: Sedimentary rocks accumulated in sedimentary basins, which evolved in paleo-landscapes of large flat lowlands of evaporitic facies. Hence, practically all the preserved fossil groundwaters are brine-tagged.

4 THE ROCKS RECORDED THE SEA TRANSGRESSIONS, THE FORMATION WATERS RECORDED THE REGRESSIONS

The geological evidence provided by rocks is biased towards the marine facies, as marine rocks are well preserved. Phases of sea retreat are poorly recorded, because hiatuses, or unconformities, are easily overlooked. At this point the meteoric brine-tagged

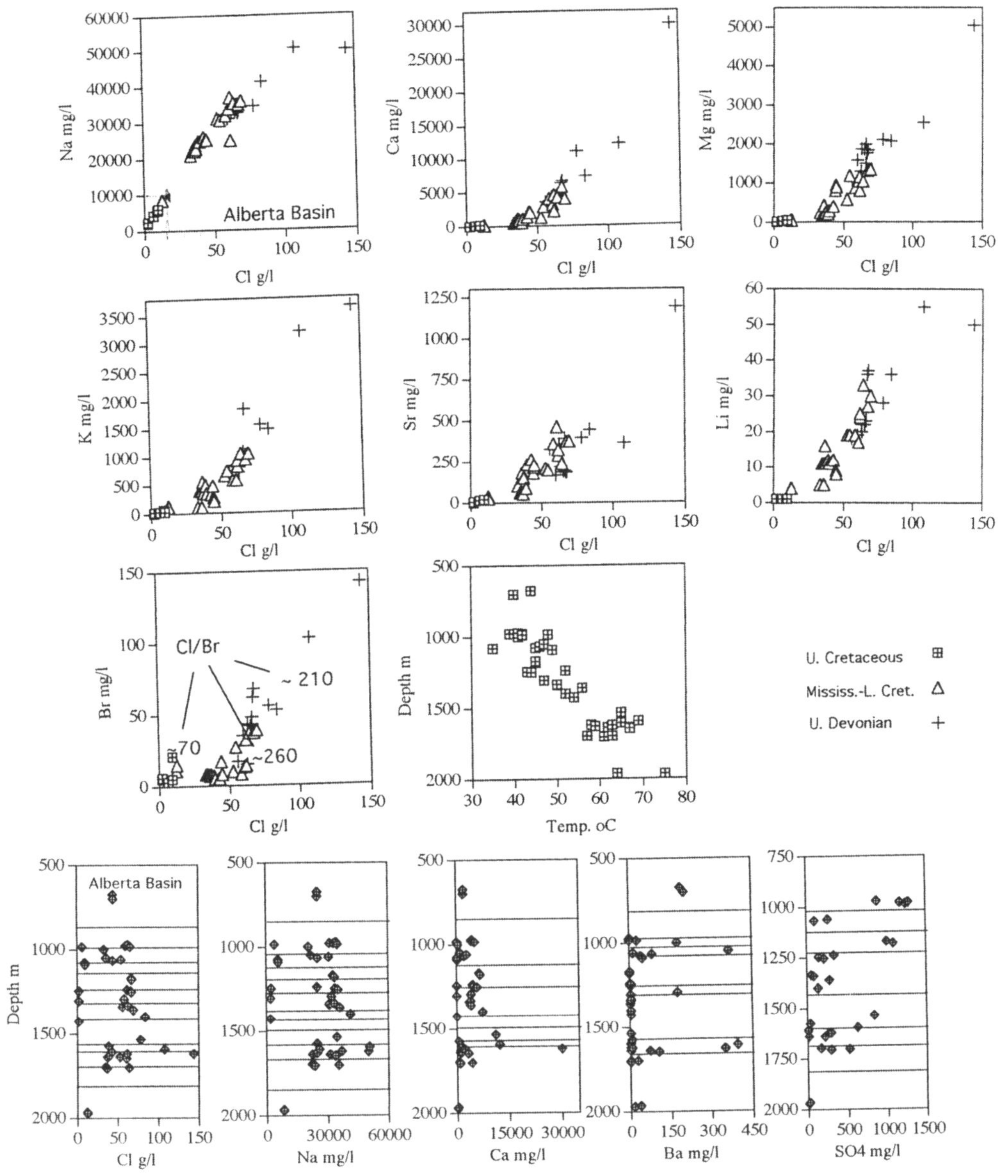

Figure 3. Formation waters encountered within the Alberta Basin, Canada, reveal linear correlations between the concentrations of Cl, Na, Ca, Mg, K, Sr and Li, indicating an common external origin of these ions (data from Connolly et al. 1990). The Cl/Br weight ratio is in the range typical to evaporitic brines, and the importance of Ca besides Na is another marker of tagging by evaporitic brines. The depth profiles reveal abrupt changes in ion concentrations (marked by horizontal lines), revealing that these formation waters are stored in distinct, hydraulically isolated, rock-compartments.

formation waters are of great importance – they clearly disclose frequent phases of exposed land. In other words, the brine-tagged formation waters complement the geological record provided by their host rocks.

5 THE ANSWER TO THE PUZZLE

At present Earth's landscape is rather mountainous – an echo of young orogenic activity, and flat evaporitic lowlands are scarce. In such mountainous landscapes the open ocean provides the spray, transported by the atmosphere inland, and the result is the commonly observed sea-tagged recent groundwaters. In elevated landscapes all the groundwater is drained and recycled into the oceans and, hence, it is in most cases fresh. Such regimes were common during previous mountain building periods, as well, but practically no terrestrial sediments were preserved of these periods, and hence, also no related sea-tagged waters were confined and preserved.

In contrast, the thick sedimentary rock assemblages were accumulated and preserved along with their contained paleo-groundwaters, formed in low flatlands of evaporitic environments, frequented by sea invasions and retreats, over large paleo-regions, as already shown. Hence, the preserved formation waters are as a rule saline and tagged by brine-spray.

REFERENCES

Connolly, C.A., Walter, L.M., Baadsgaard, H. & Longstaffe, F.J. (1990a): Origin and evolution of formation waters, Alberta Basin, Western Canada Sedimentary Basin. I. Chemistry. *Appl. Geochem.* 5, 375–395.

Mazor, E. 1997. *Chemical and Isotopic Groundwater Hydrology.* Marcel Dekker, Inc.; New York (3rd edition in print).

Mazor, E. In print. *Global Water Dynamics.* Marcel Dekker, Inc.; New York.

Water–Rock Interaction, Wanty & Seal II (eds)
© 2004 Taylor & Francis Group, London, ISBN 90 5809 641 6

Impact of Pleistocene glaciation on hydrogeochemistry of Devonian carbonate groundwaters, northern margin Michigan Basin

J.C. McIntosh & L.M. Walter
Department of Geological Sciences, University of Michigan, Ann Arbor, Michigan, USA

ABSTRACT: The hydrogeochemistry of the Devonian carbonate regional aquifer system was investigated along the northern margin of the Michigan Basin to study the impact of Pleistocene glaciation on deep groundwater flow systems. Groundwaters in glacial drift aquifers in upland areas are predominately Ca^{2+}–Mg^{2+}–HCO_3^- waters with high Ca/Mg molar ratios (>2), low Sr^{2+} concentrations (<10 μM), and $\delta^{13}C$ values for dissolved inorganic carbon (DIC) typical of open system carbonate mineral dissolution (about −12‰ VPDB). Confined groundwaters in bedrock aquifers, beneath impermeable shale units or lake-bed clays, have been significantly altered by incongruent dissolution of carbonate minerals, cation-exchange with Na^+-rich clays, and anhydrite dissolution. These fluids have high Sr^{2+} concentrations (10–50 μM), low Ca/Mg ratios (<2), and $\delta^{13}C$ values for DIC up to −3‰ (VPDB). The $\delta^{18}O$ and δD values for these dilute waters, relative to modern precipitation, indicate they were recharged during Pleistocene glaciation.

1 INTRODUCTION

Multiple episodes of continental glaciation, most recently during the Pleistocene, profoundly altered groundwater flow systems in the mid-continent region of the United States (Siegel & Mandle 1984; Grasby et al. 2000). Subglacial meltwater preferentially discharged into the Silurian-Devonian carbonate regional aquifer system along the margins of the Michigan, Illinois, and Appalachian basins (Gupta & Bair 1997; McIntosh et al. 2002; Fig. 1). Increased hydraulic head from the continental ice sheets forced fresh water basinward along the dip of the carbonate strata to great depths, displacing saline fluids and significantly reorganizing salinity structures (McIntosh et al. 2002). Much of this glacial-age water has been flushed from shallow flow systems by modern meteoric recharge. However, remnant glacial meltwater remains at depth along the basin margins trapped by overlying impermeable deposits, such as lake-bed clays and Devonian shales (Martini et al. 1998; Eberts & George 2000).

This paper focuses on the hydrogeochemistry of the Devonian carbonate aquifers along the northern margin of the Michigan Basin. This study is part of a larger investigation into the impact of Pleistocene glaciation on the hydrogeochemistry of deep groundwater flow systems along the margins of the Michigan and Illinois basins (McIntosh & Walter, in prep).

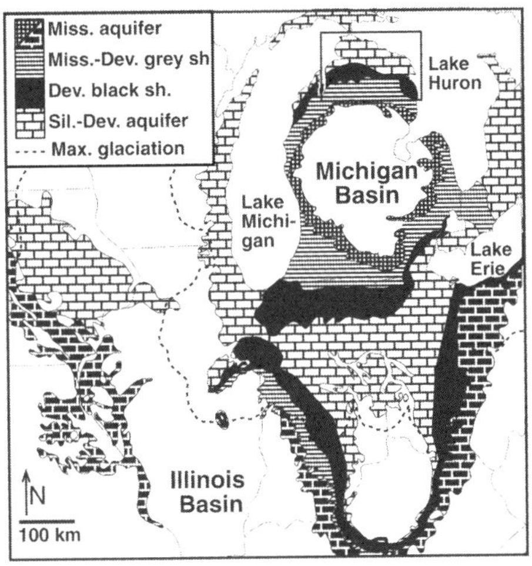

Figure 1. Hydrogeologic framework of the mid-continent region, USA, with the southern extent of Pleistocene glaciation highlighted. The hydrogeochemistry of Devonian carbonate groundwaters and overlying glacial drift were investigated along the northern margin of the Michigan Basin (study area outlined by black box).

2 HYDROGEOLOGY

The subcrop of the Silurian-Devonian carbonate aquifer system is expansive in the mid-continent region, US (Fig. 1). Along the margins of the Michigan, Illinois and Appalachian basins, the carbonate subcrop is overlain by varying thicknesses and permeabilities of glacial deposits. Within the sedimentary basins, the Silurian-Devonian aquifer system is confined by overlying Mississippian-Devonian shales, and underlying Ordovician shales and bedded evaporite deposits in the Silurian section of the Michigan Basin. In the Michigan Basin, the Devonian carbonates contain limestone, dolomite, and localized beds of anhydrite and halite (Dorr & Eschman 1970).

Along the northern margin of the Michigan Basin, most of the topographic relief (up to ~500 m) is related to thickness of glacial deposits (up to 350 m), and shallow groundwater flow is controlled by permeability of glacial sediments. Regional groundwater flow is from topographic highs, composed of permeable glacial till and outwash, to low-lying lakes and streams, underlain by impermeable clays, and to Lakes Huron and Michigan. During glaciation, the regional groundwater flow system was reversed, with discharge of dilute waters into the Michigan Basin, along the dip of the Devonian carbonate strata.

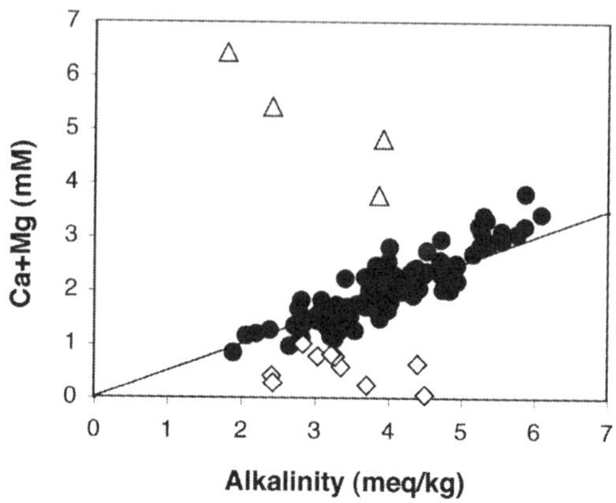

Figure 2. Ca + Mg vs. alkalinity for groundwaters in glacial drift and bedrock aquifers. Majority of samples (shown in black dots) plot along the 1:2 Ca + Mg: Alkalinity line, representative of calcite and dolomite dissolution. Triangles are fluids that have dissolved anhydrite, while diamonds are waters that have been altered by cation exchange with Na-rich clays.

3 RESULTS AND DISCUSSION

Ninety groundwater samples were collected from household wells screened in glacial drift and Devonian carbonate bedrock along the northern margin of the Michigan Basin. Twenty-nine additional groundwaters from shallow glacial drift aquifers and one bedrock well in the study area, published by Ku (2001), were included in the data set. Well locations were chosen to achieve a diverse spatial sampling both with depth and laterally across the study area. In addition, fluid chemistry was interpreted in relation to detailed hydrogeologic mapping and cross-sections.

Temperature, pH, dissolved oxygen and conductivity were measured at the well head. Filtered (0.45 μm) samples were collected for elemental chemistry and isotope analyses, and kept refrigerated. Sample aliquots for cation chemistry were acidified to $pH < 2$ with nitric acid. Aliquots for dissolved inorganic carbon (DIC) and $\delta^{13}C$ DIC were preserved with $CuCl_2$. Total alkalinity was measured by the Gran titration method, DIC by coulometry, anions by IC, and cations by ICP-AES. Stable isotope compositions were measured by mass spectrometry. Complete methods and analyses, and location of the groundwaters are presented in McIntosh & Walter (in prep) along with an extended suite of data including S and Sr isotope chemistry, tritium, and ^{14}C activity.

3.1 *Element geochemistry*

The majority of groundwaters in glacial drift and underlying Devonian carbonate aquifers, along the northern margin of the Michigan Basin, are Ca^{2+}–Mg^{2+}–HCO_3^- waters. These fluids have Ca + Mg vs. alkalinity molar ratios of 1:2 (Fig. 2), indicative of carbonate mineral dissolution, according to the stoichiometry of the following reactions:

$$CaCO_3 + CO_2 + H_2O = Ca^{2+} + 2HCO_3^- \qquad (1)$$

$$CaMg(CO_3)_2 + 2CO_2 + 2H_2O = Ca^{2+} + Mg^{2+} + 4HCO_3^- \qquad (2)$$

Several groundwaters plot below the calcite and dolomite dissolution line, depleted in Ca^{2+} and Mg^{2+}, relative to bicarbonate, and enriched in Na^+. These Na^+–HCO_3^- waters are located in discharge areas near the Lake Huron shoreline, confined by lake-bed clays. They likely evolved from cation exchange of Ca^{2+} and Mg^{2+}, derived from carbonate mineral dissolution, for Na^+ in smectite-rich clay sequences (Hendry & Wassenaar 2000).

A few of the groundwaters are highly enriched in Ca^{2+} and Mg^{2+} relative to HCO_3^-, and charge balanced by SO_4^{2-}. These waters are deep carbonate bedrock wells, overlain by impermeable glacial drift. Although they are supersaturated with respect to calcite and dolomite, they are still undersaturated with respect to gypsum/anhydrite. Dissolution of anhydrite by dilute waters at depth in evaporite-bearing carbonate units increases Ca^{2+} and SO_4^{2-} concentrations. As $[Ca^{2+}]$ increases calcite may precipitate, decreasing Ca^{2+} and HCO_3^- concentrations. The decrease in $[HCO_3^-]$ may cause dolomite to become undersaturated and dissolve, adding Mg^{2+} to

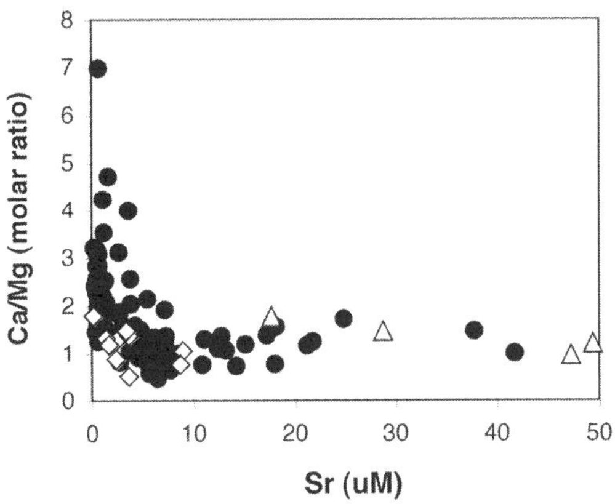

Figure 3. Ca/Mg molar ratios versus Sr^{2+} concentration.

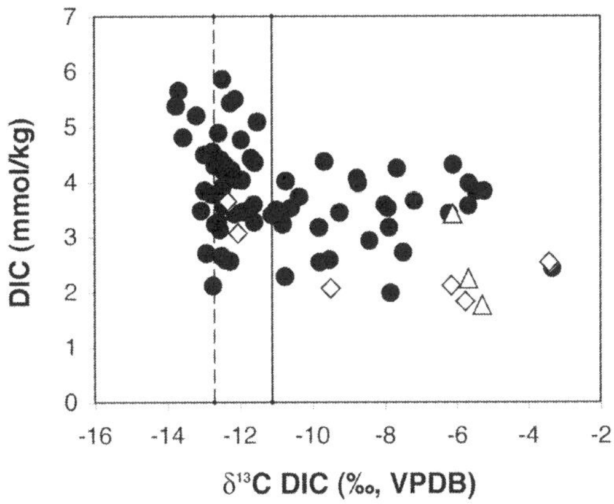

Figure 4. Dissolved inorganic carbon (DIC) vs. $\delta^{13}C$ of DIC. $\delta^{13}C$ values for carbonate mineral dissolution under open (dashed line) and closed system (solid line) conditions are shown for reference (Ku 2001).

the water and decreasing the Ca/Mg ratio. The net reaction for this process is:

$$CaMg(CO_3)_2 + CaSO_4 * 2H_2O = 2CaCO_3 + Mg^{2+} + SO_4^{2-} + 2H_2O \quad (3)$$

Pyrite oxidation and subsequent carbonate mineral dissolution produces the same products as reaction (3). However, very positive $\delta^{34}S$ values of SO_4^{2-} in the groundwaters show that the sulfate-rich waters are derived from anhydrite dissolution, rather than pyrite oxidation (McIntosh & Walter, in prep).

Incongruent dissolution of anhydrite, calcite, and dolomite also enriches the fluids in Sr^{2+}, in addition to Mg^{2+} (Plummer & Back 1980). Groundwaters in northern Michigan show increasing $[Sr^{2+}]$ with decreasing Ca/Mg ratios (Fig. 3), evidence of progressive water-rock interaction.

3.2 *$\delta^{13}C$ DIC evolution of groundwaters*

Carbon system parameters and $\delta^{13}C$ values of dissolved inorganic carbon (DIC) are also important in determining sources of carbon and tracing water-rock interactions. Within the observed pH range of groundwaters in northern Michigan, the DIC is equivalent to the alkalinity. Carbon isotope values appear to increase with decreasing DIC, and range from −13.8‰ to −3.4‰ (VPDB; Fig. 4). The $\delta^{13}C$ value for soil gases in northern Michigan is approximately −22‰ (Ku 2001). Carbonate minerals in the glacial drift and bedrock have $\delta^{13}C$ values ranging from −4.1 to +0.5‰ (Ku 2001; Budai et al. 2002). Congruent dissolution of carbonate minerals with soil $CO_{2(g)}$ would produce $\delta^{13}C$ DIC values of approximately −12.5 and −11.1‰ for open versus closed system conditions, respectively.

Groundwaters in the recharge areas with relatively permeable glacial drift have C isotope values consistent with open system carbonate mineral dissolution. Fluids with lower DIC (<4.5 mmol/kg) and high $\delta^{13}C$ values (>−10‰) likely evolved via incongruent dissolution of carbonate minerals under closed system conditions. Groundwaters affected by anhydrite dissolution, dedolomitization, and cation exchange reactions also have high $\delta^{13}C$ and low DIC values, reflecting their isolation from freshwater recharge and longer residence times. Carbonate dissolution likely occurs first under open system conditions in the recharge areas, near the water table, and then under closed system conditions in the discharge areas and along the regional groundwater flow paths.

3.3 *Source of waters*

O and H isotope chemistry helps to constrain fluid migration, water source, and relative age of meteoric recharge. The $\delta^{18}O$ and δD values for groundwaters in glacial drift and bedrock in northern Michigan range from −15.0 to −10.4‰, and −103.4 to −71.8‰, respectively (Fig. 5). The lowest $\delta^{18}O$ and δD values were found in confined groundwaters in the discharge zone along the Lake Huron shoreline. These waters also have been affected by cation-exchange with overlying lake-bed clays.

Modern precipitation in northern Michigan ranges from −11 to −8‰ $\delta^{18}O$ and −78 to −54‰ δD (IAEA/WMO 2001), while estimated glacial meltwater values range from −25 to −11‰ $\delta^{18}O$ and −190 to −78‰ δD (McNutt et al. 1987; Remenda et al. 1994). Groundwaters show mixing of modern precipitation with glacial meltwater at depth. The presence of thick aquitards in low-lying discharge areas helped to prevent flushing of glacial meltwater in underlying aquifers by more modern recharge. Results of tritium

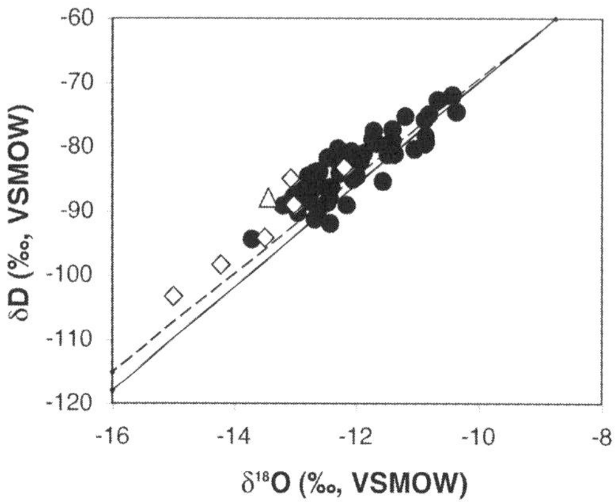

Figure 5. O and H isotope chemistry of groundwaters. Global meteoric water line (Craig, 1961) is shown in the solid black line, while the Ottawa meteoric water line (IAEA 2001) is shown by the dashed line.

and ^{14}C activity on a subset of these groundwaters are presented in McIntosh & Walter (in prep).

4 CONCLUSIONS

Elemental and stable isotope geochemistry of groundwaters in glacial drift and bedrock aquifers, along the northern margin of the Michigan Basin, are closely coupled with hydrogeology, glacial history, water-rock interactions, and fluid residence times. Groundwaters in glacial drift aquifers in upland areas are predominately Ca^{2+}–Mg^{2+}–HCO_3^- waters with high Ca/Mg ratios and low $[Sr^{2+}]$. C isotope values of DIC reflect open system dissolution of carbonates, and $\delta^{18}O$ and δD values show mixing of modern precipitation with glacial meltwater at depth.

Confined groundwaters, on the other hand, are much more evolved. These fluids reside in deep bedrock aquifers beneath lake-bed clays or impermeable surficial deposits, and in discharge zones for groundwaters to the Great Lakes and/or into the deeper flow system of the Michigan Basin. O and H isotopes indicate these waters were likely recharged during Pleistocene glaciation and are out of contact with more recent recharge within overlying shallow flow systems. Many of these waters have dissolved anhydrite, been altered by cation-exchange reactions with Na^+-rich clays, and have incongruently dissolved carbonates under closed system conditions, as shown by high $\delta^{13}C$ values for DIC, high $[Sr^{2+}]$, and low Ca/Mg ratios.

Impermeable surficial deposits and confining bedrock units have played an important role in preserving glacial meltwater at depth along the margins of sedimentary basins.

REFERENCES

Budai, J.M., Martini, A.M., Walter, L.M. & Ku, T.C.W. 2002. Fracture-fill calcite as a record of microbial methanogenesis and fluid migration: a case study from the Devonian Antrim Shale, Michigan Basin. *Geofluids* 2: 163–183.

Craig, H. 1961. Isotopic variations in meteoric waters. *Science* 133: 1702–1703.

Dorr, J.A. & Eschman, D.F. 1970. *Geology of Michigan.* The University of Michigan Press.

Eberts, S.M. & George, L.L. 2000. Regional ground-water flow and geochemistry in the Midwestern basins and arches aquifer system in parts of Indiana, Ohio, Michigan, and Illinois. *U.S. Geological Survey Professional Paper 1423-C.*

Grasby, S., Osadetz, K., Betcher, R. & Render, F. 2000. Reversal of the regional-scale flow system of the Williston basin in response to Pleistocene glaciation. *Geol.* 28: 635–638.

Gupta, N. & Bair, E.S. 1997. Variable-density flow in the Midcontinent basins and arches region of the United States. *Water Resources Research* 33: 1785–1802.

Hendry, M.J. & Wassenaar, L.I. 2000. Controls on the distribution of major ions in pore waters of a thick surficial aquitard. *Water Resources Research* 36: 503–513.

IAEA/WMO 2001. Global network for isotopes in precipitation. The GNIP Database. 2002: URL: http:www.iaea.org/programs/ri/gnip/gnipmain.htm.

Ku, T.C.W. 2001. *Organic carbon-mineral interactions in terrestrial and shallow marine environments.* Ph.D. thesis, University of Michigan.

Martini, A.M., Walter, L.M., Budai, J.M., Ku, T.C.W., Kaiser, C.J. & Schoell, M. 1998. Genetic and temporal relations between formation waters and biogenic methane: Upper Devonian Antrim Shale, Michigan Basin, USA. *Geochim. Cosmochim. Acta* 62: 1699–1720.

McIntosh, J.C. & Walter, L.M., in prep. Hydrogeochemistry of Silurian-Devonian carbonate groundwaters, mid-continent US.

McIntosh, J.C., Walter, L.M. & Martini, A.M. 2002. Pleistocene recharge to mid-continent basins: effects on salinity structure and microbial gas generation. *Geochim. Cosmochim. Acta* 66: 1681–1700.

McNutt, R.H., Frape, S.K. & Dollar, P. 1987. A strontium, oxygen and hydrogen isotopic composition of brines, Michigan and Appalachian Basins, Ontario and Michigan. *Appl. Geochem.* 2: 495–505.

Plummer, L.N. 1980. The mass balance approach: Application to interpreting the chemical evolution of hydrologic systems. *Am. J. Sci.* 280: 130–142.

Remenda, V.H., Cherry, J.A. & Edwards, T.W.D. 1994. Isotopic composition of old ground water from Lake Agassiz: implications for Late Pleistocene climate. *Science* 266: 1975–1978.

Siegel, D.I. & Mandle, R.J. 1984. Isotopic evidence for glacial meltwater recharge to the Cambrian-Ordovician Aquifer, North-Central United States. *Quaternary Research* 22: 328–335.

Water–Rock Interaction, Wanty & Seal II (eds)
© 2004 Taylor & Francis Group, London, ISBN 90 5809 641 6

Incipient alteration of biotite by groundwater in fractured rock

J. McMurry
Atomic Energy of Canada Ltd. (AECL), Whiteshell Laboratories, Pinawa, Manitoba, Canada

ABSTRACT: In a paleohydrogeological investigation of mineral indicators of past changes in groundwater chemistry in the Canadian Shield, biotite in core samples of fractured granitic rock was examined for evidence of low-temperature reaction with groundwater. Compositional variation of biotite in shallow fractures, where present-day groundwaters are advective and oxidizing, was contrasted with biotite exposed in fractures to depths of up to ~1000 m below the surface. Minor incipient alteration of biotite (loss of K, Ti, Si) was observed in fractures at all depths. Oxidation of Fe^{II} to Fe^{III} by reaction with groundwater was noted only in the most shallow fractures (<10 m deep), where present-day groundwaters are oxygenated. The lack of reactivity in biotite in deeper fractures supports other paleohydrogeological evidence that saline, reducing, and largely stagnant groundwaters at depth in the study area remained decoupled from the more active shallow part of the flow system during repeated episodes of continental-scale glaciation in the past million years.

1 INTRODUCTION

The hydrogeology of the Canadian Shield is characterized in general by relatively shallow and advective flow systems in near-surface fractures in crystalline rock and by stagnant, more isolated sets of groundwaters at depth. Determining the history of stability of such flow systems in crystalline rocks, as well as the hydrogeological and chemical impact of perturbations by glacial meltwaters, is of interest with respect to the isolation of nuclear waste in a deep geological repository (e.g., McMurry et al. 2003). For example, groundwaters at depth typically are reducing, a desirable characteristic for a repository because it would limit the dissolution and transport of many radionuclides.

The work reported here is part of a paleohydrogeological investigation of fractured granitic rock in the Canadian Shield (Gascoyne et al. 2003; McMurry & Ejeckam 2002). One focus of the investigation was to examine biotite in the subsurface for evidence of redox-buffering reactions with oxygenated groundwater as a possible indicator of past changes in groundwater redox potential, such as may have been associated with changes in flow conditions during glaciation.

The study area is located in the Lac du Bonnet batholith, a Proterozoic intrusion in southeastern Manitoba (Brown et al. 1989). Pink, hematite-bearing granite, thought to be part of the original roof zone of the intrusion, predominates to depths of about 200 m, below which the main phase is a grey, magnetite-bearing granite chemically similar to the pink phase except for a lower Fe_2O_3/FeO ratio. Dark green biotite is the most common ferromagnesian mineral, typically accounting for about 1 to 10 percent of the mode of the rock. Most fracture development as well as high-temperature mineral alteration took place during cooling and uplift of the intrusion more than 2 billion years ago. The batholith has been at or near the present-day surface several times in the last 500 million years, having experienced several cycles of crustal downwarping, sedimentary deposition, uplift, and erosion in that time (Everitt & Osadetz 2000). More recently (in the past 1 million years), the batholith also has been subjected to nine or ten cycles of continental-scale glaciation, which may have resulted in episodic changes in regional groundwater movement and chemistry.

The hydrogeological model for the groundwater flow system is structured by two main fracture types – (1) subhorizontal fractures that are part of several large-scale, low-dipping fracture zones at various depths in the batholith, and (2) near-vertical fractures (joints) that are only abundant in the upper 200 m. The subhorizontal fracture zones are the main discharge pathways for deeper groundwaters but under special conditions in the past – such as the accumulation and melting of thick glacial ice – they may have served as conduits for recharge. In the upper ~200 m of the batholith, near-modern levels of tritium and ^{14}C and a warm-climate $^{2}H/^{18}O$ groundwater signature indicate that interconnected vertical and subhorizontal fractures host an active shallow groundwater circulation system (Gascoyne 2000). Shallow groundwater is dilute (TDS < 0.3 g/L) and oxygenated, although chemical and biological reactions reduce the O_2 content to <2 μg/L by a depth of about 60 m. Groundwaters at moderate depths (~150–400 m) are more saline (TDS ~1 g/L) and have lower redox potential. Fewer fractures are present, most of which are associated with subhorizontal fracture zones. Isotopic data indicate a cool-climate signature, suggesting a groundwater residence time of at least 10^3 to 10^5 years. The moderate-depth groundwaters form a transition zone between the upper, advective flow regime and a deeper regime in sparsely fractured or unfractured rock, where groundwaters are typically very saline (TDS > 35 g/L), reducing, and have warm-climate, pre-glacial isotopic signatures indicative of residence times in excess of 10^6 years. Biotite or chlorite is probably the dominant mineral control on redox potential, particularly at depth (Gascoyne 2000).

Fracture characterization in the Lac du Bonnet batholith is based in part on more than 130 boreholes, many of them cored, that have been drilled from surface to depths of up to 1200 m over the past 20 years. Additional cores and other detailed information about subsurface characteristics have been obtained from the Underground Research Laboratory, which was excavated to a depth of 443 m to conduct *in situ* experiments in crystalline rock.

2 METHODS

For sampling purposes, fracture types in the core samples were divided into five groupings on the basis of depth and orientation of the fracture:

Group 1: shallow (0–150 m), depth, vertical dip
Group 2: shallow (0–150 m), subhorizontal dip
Group 3: moderate (150–300 m), vertical dip
Group 4: moderate (150–300 m), subhorizontal dip
Group 5: deep (>600 m).

Most deep fractures, although scarce, are subhorizontal.

The composition of biotite was determined by electron microprobe analysis, with a special focus on biotite that was likely to have been in direct contact with groundwater in rock fractures – e.g., grains exposed at fracture surfaces or in fracture-filling microbreccia (Fig. 1). Analyses were performed on carbon-coated polished thin sections using a Cameca SX-50 electron microprobe at the University of Manitoba. All compositions were determined by wavelength-dispersive analysis using natural and synthetic mineral standards. More than 600 separate

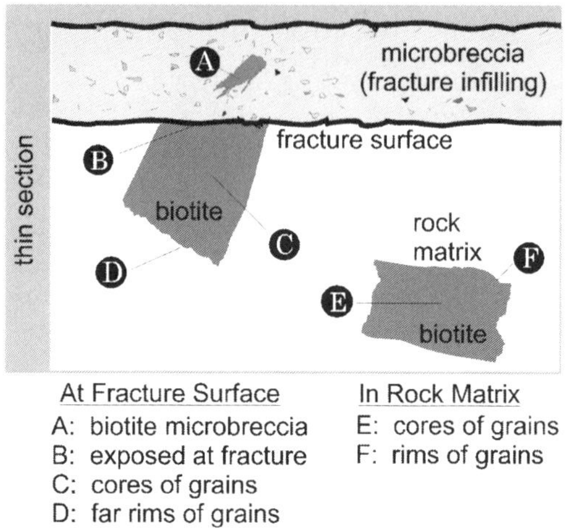

Figure 1. Sampling schematic for microprobe analyses.

point locations on more than 100 individual grains of biotite were analyzed, with representation among all five groupings of fracture types. The analytical strategy for individual grains was (1) to determine if biotite grains adjacent to fracture surfaces were more altered than biotite grains in the rock matrix, and, on a finer scale, (2) to determine if the portion of an individual biotite grain exposed at a fracture surface was more altered than the centre (core) or opposite rim of the same grain (Fig. 1). The criteria used to identify incipient alteration included a decrease in K and Ti concentrations and a decrease in the relative proportion of Si to Al (Dong et al. 1998). Variations in Fe and Mg also are associated with biotite alteration, but their behaviour is complicated by additional processes such as oxidation of Fe^{II} to Fe^{III}, substitution of Mg cations for Fe in the altered biotite crystal structure, and in situ formation of very fine-grained Fe- or Mg-bearing solid phases among cleavage planes in the biotite (Fordham 1990; Gilkes et al. 1972).

3 RESULTS

3.1 *Variation with depth*

In petrographic terms, biotite in all of the samples was present as discrete, subhedral flakes, typically smaller than 1 mm in diameter, with dark green – olive green pleochroism. Reddish-brown biotite was noted only in a few samples from the shallowest fractures (<10 m deep), and even in these cases the biotite was green except for the portion of the grain that was exposed at the fracture surface itself. In hand samples of these shallow cores, the biotite at the fracture surfaces had a bronze-coloured, weathered

Table 1. Average primary (unaltered) biotite compositions in fracture-type groupings.

Fracture depth & orientation	n	SiO_2	TiO_2	Al_2O_3	FeO*	MnO	MgO	Na_2O	K_2O	F	Cl	Total
(1) Shallow vertical	108	36.3	2.2	15.8	20.9	0.5	9.4	0.1	9.6	0.60	0.03	95.6
(2) Shallow subhorizontal	65	36.9	2.1	15.6	20.7	0.4	9.4	0.1	9.6	0.79	0.05	95.7
(3) Moderate vertical	88	36.5	2.1	15.5	21.2	0.4	9.8	0.1	9.3	0.78	0.04	95.8
(4) Moderate subhorizontal	152	36.9	1.8	15.6	19.9	0.5	10.6	0.1	9.3	0.78	0.03	95.5
(5) Deep	95	35.9	2.1	16.1	2.0	0.4	9.7	0.1	9.4	0.60	0.03	95.3

* Total Fe reported as FeO.

appearance. The only other examples of discoloured biotite were tiny shredded flakes of brown biotite entrained as microbreccia which were noted in fracture infillings from all five fracture-type groupings. In the most altered examples, tiny granules of iron oxide were visible as inclusions in wavy exfoliations of the biotite in the microbreccia.

Primary (unaltered) biotite compositions are summarized in Table 1. Individual biotite grains were unzoned (little or no variation in core-to-rim composition), so core and rim data points were combined for all analyzed grains from each fracture grouping to provide the average values. Total iron is reported as FeO because microprobe analysis cannot distinguish between the different valence states of Fe. A minimum estimate of Fe^{III} was made on the basis of excess Fe in calculated structural formulae, which indicated that only a few of the most altered portions of the analyzed samples contained Fe^{III}. However, previously published biotite analyses for the Lac du Bonnet batholith (Taylor & Owen 1997), in which Fe^{III} was measured by titration, suggest that approximately one-third of the iron reported in Table 1 is likely to be Fe^{III}.

Little variation in primary biotite composition was noted in the analyzed samples, indicating that the biotite had formed under relatively uniform magmatic or post-magmatic conditions regardless of location in the crystallizing magma. There was no significant difference, for example, between biotite in the upper 200 m of the batholith (pink granite) and in the lower portion (grey granite). One exception is biotite from the moderate-depth subhorizontal fractures (Group 4), which has higher Mg and lower Fe than biotite from the other groups.

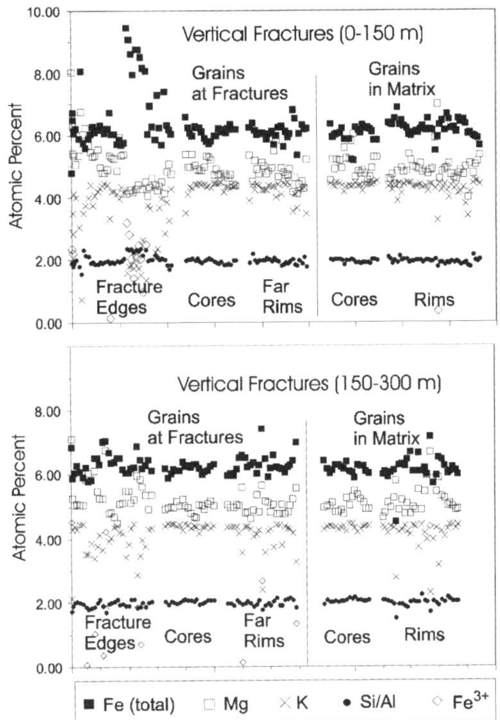

Figure 2. Comparison of variation in biotite composition for grains exposed at fracture surfaces and in the intact rock matrix. Incipient alteration is indicated by decrease in K and Si/Al. Total Fe values greater than ~6.5 (at.%) included Fe^{III} (plotted separately). Data are shown for two of the fracture-type groupings (shallow vertical and moderate-depth vertical).

3.2 *Localised alteration*

Biotite grains in the matrix of all rocks had less incipient alteration than biotite grains that had been dissected by fractures, as illustrated by Figure 2. Moreover, the portion of any single grain exposed at the fracture surface typically was more altered than the opposite rim of the same grain, suggesting that interaction with water at the fracture surface had promoted the alteration of the biotite on an extremely localized scale (micrometers). In general, the most altered grains were those entrained as small flakes in microbreccia.

4 DISCUSSION

Biotite is more resistant to alteration than most ferromagnesian minerals in closed, low-volume systems, but it is susceptible in open systems (i.e., those that contain a cation sink or in which the solution is replenished), even at low temperatures (Taylor & Owen 1997; Newman & Brown 1966). For example, biotite flakes isolated from soils and weathered rocks consistently contain more ferric iron than flakes in unaltered parent material (Gilkes et al. 1972).

Overall, the amount of biotite alteration at rock-water interfaces in fractures in the analyzed samples is minimal. It may be that in fairly stagnant groundwaters that have long residence times, closed-system conditions are approached that would limit the dissolution of biotite. Thermodynamic properties indicate that biotite generally is reactive with the groundwaters in the study area (Gascoyne 2000), although the dissolution rate of biotite is slow (Nagy 1995). Even so, the relatively minor amount of incipient alteration is surprising, given that the individual biotite grains exposed in fracture surfaces presumably have been in contact with water for millions of years or longer. Biotite exposed to oxidizing groundwater in shallow fractures has been subjected to more aggressive alteration than biotite in deeper fractures. In all cases, however, the most extensive alteration of individual grains was limited to a narrow region (typically micrometers) exposed at the fracture surface itself. This suggests that reaction times for biotite oxidation are slow under low-temperature conditions.

5 CONCLUSIONS

Conspicuous low-temperature alteration of biotite was noted petrographically only in a few of the most shallow fractures (<10 m deep), where present-day groundwaters are known to be oxygenated. In these cases, the outermost portions of biotite grains exposed at the fracture surface were a bright reddish-brown in thin section, whereas the remainder of the grain remained unaltered (dark green). Microprobe analysis of the reddish-brown regions indicated excess Fe in the biotite structural formula, suggesting that Fe^{II} had been converted to Fe^{III} in the altered portion of the grains. Otherwise, at all depths in the batholith only minor alteration of biotite was noted at any fracture surface, despite a long history of exposure to groundwater in the fractures. These observations suggest that flow system stability has been maintained during past episodes of glaciation, at least with respect to a lack of deep recharge by oxygenated glacial meltwaters.

ACKNOWLEDGMENTS

The work reported here is part of a broader paleohydrogeological study of fracture mineralization in the Lac du Bonnet batholith, in which Reg Ejeckam and Mel Gascoyne are collaborators. Richard Everitt and Dave Woodcock provided core samples and original fracture log summaries from the geological data collection at the Underground Research Laboratory. Ron Chapman assisted with microprobe analyses. Mark Jensen, Mel Gascoyne, Reg Ejeckam, Richard Everitt, and Ian Cartwright are thanked for helpful discussions and comments. Funding for the study was provided by Ontario Power Generation.

REFERENCES

Brown, A., Soonawala, N.M., Everitt, R.A. & Kamineni, D.C. 1989. Geology and geophysics of the Underground Research Laboratory site, Lac du Bonnet batholith, Manitoba. *Can. J. Earth Sci.* 26: 404–425.

Dong, H., Peacor, D.R. & Murphy, S.F. 1998. TEM study of progressive alteration of igneous biotite to kaolinite throughout a weathered soil profile. *Geochim. Cosmochim. Acta* 62: 1881–1887.

Everitt, R. & Osadetz, K. 2000. Application of apatite fission track thermochronology to dating basement reactivation. Ontario Power Generation, Nuclear Waste Management Division Report 06819-REP-01200-10040-R00. Toronto, Ontario.

Fordham, A.W. 1990. Formation of trioctahedral illite from biotite in a soil profile over granite gneiss. *Clays Clay Mins.* 38: 187–195.

Gascoyne, M. 2000. Hydrogeochemistry of the Whiteshell Research Area. Ontario Power Generation, Nuclear Waste Management Division Report 06819-REP-01200-10033-R00. Toronto, Ontario.

Gascoyne, M., McMurry, J. & Ejeckam, R.B. 2003. Paleohydrogeologic interpretation of fracture infill mineralogy, geochemistry and isotope systematics. In *Proc., 4th Joint IAH-CNC/CGS Conference, Winnipeg, Canada, Sept. 29-Oct.1 2003.*

Gilkes, R.J., Young, R.C. & Quirk, J.P. 1972. The oxidation of octahedral iron in biotite. *Clays Clay Mins.* 20: 303–315.

McMurry, J. & Ejeckam, R.B. 2002. Paleohydrogeological study of fracture mineralogy in the Whiteshell Research Area. Ontario Power Generation, Nuclear Waste Management Division Report 06819-REP-01200-10082-R00. Toronto, Ontario.

McMurry, J., Dixon, D.A., Garroni, J.D., Ikeda, B.M., Stroes-Gascoyne, S., Baumgartner, P. & Melnyk T.W. 2003. Evolution of a Canadian deep geologic repository: base scenario. Ontario Power Generation, Nuclear Waste Management Division Report 06819-REP-01200-10092-R00. Toronto.

Nagy, K.L. 1995. Dissolution and precipitation of sheet silicates. In A.F. White & S.L. Brantley (eds.), *Chemical weathering rates of silicate minerals, Reviews in Mineralogy* 31: 173–233. Washington, DC: Mineralogical Society of America.

Newman, A.C.D. & Brown, G. 1966. Chemical changes during the alteration of micas. *Clay Minerals* 6: 297–310.

Taylor, P. & Owen, D.G. 1997. Biotite dissolution and oxygen consumption in aqueous media at 100°C. Atomic Energy of Canada Limited Report, AECL-11571, COG-96-157-I. Chalk River, Ontario.

Water–Rock Interaction, Wanty & Seal II (eds)

Hydrochemistry of groundwater on both sides of the Jordan – Dead Sea – Arava transform fault

P. Möller & P. Dulski
Geo-Research Centre Potsdam, Germany

S. Geyer
Centre for Environmental Research Leipzig-Halle, Germany

E. Rosenthal & Y. Guttman
Tel Aviv University, Israel

E. Salameh
University of Amman, Jordan

ABSTRACT: Major and minor elements, rare earth elements and yttrium, and O and H stable isotopes yield different insights into the source of groundwaters, their aquifers and their salinisation. In comparable aquifers on both sides of the Jordan – Dead Sea – Arava Fault, groundwaters are chemically and isotopically different. Interaquifer exchange of water is common on both sides. Salinisation is due to dissolution of dispersed evaporites associated with limestones and sandstones, and mixing with Ca-Mg-Cl brines generated by leaching of post-halite evaporites.

1 INTRODUCTION

In the years 1999 to 2002 about 70 wells and springs of the Jordan and Arava Valleys, Middle East, were analysed for major and minor element hydrochemistry and H, O, & S isotope distribution. The valleys form a unique oasis, which includes the Dead Sea as the lowest-elevation terrestrial point on earth and Lake Kinneret as the most important fresh water reservoir of Israel. Several water samples from the Jordanian Highland are included. The goals of the study were to identify (i) the recharge areas of individual wells and springs, (ii) the types of rocks and sediments the groundwaters interacted with, (iii) the extent of groundwater flow between formations, (iv) the sources of salinisation in general and of Ca-Mg-Cl brines in particular, and (v) if groundwater flow crosses the Jordan – Dead Sea – Arava transform fault.

2 RESULTS AND DISCUSSION

A rough characterization of the waters with respect to locality, aquifer rocks and geologic formation of their origin is given in Table 1. The chemical analyses of waters are presented in spider diagrams (Fig. 1), rare earth plus yttrium distribution patterns (henceforth referred to as REY; Fig. 2), and a plot of δ^2H vs. $\delta^{18}O$ (Fig. 3). In the spider diagrams the concentrations of all species in water are normalized to those in seawater. These diagrams allow an easy visual comparison of the chemical composition of waters. According to their trends spider diagrams are obtained such as shown in Figs. 1a and b, in which mainly waters from sandstones and limestones are grouped, respectively. They differ in normalized Ca and Sr and Ca-Sr relationship. Those of the thermosaline waters (Fig. 1c) show patterns with very high normalized contents of nearly all species (except HCO_3, Y and U). Patterns of waters from basaltic aquifers are highly variable (Fig. 1d).

Based on similarities in REY distribution patterns, groups of groundwaters are obtained in which mainly waters from sandstones (Fig. 2a), limestones (Fig. 2b, c) and basalts (Fig. 2d) are compiled. This REY grouping is also used to characterize the waters in Figure 3.

According to the spider diagrams groundwaters from mainly Mesozoic sandstones (Fig. 1a) are

Table 1. Compilation of discussed water and their origin.

Locality	Rock Formation	Aquifer rock
Israel: Lake Kinneret and its surroundings		
Hammei Tveria, spring Sp. main spring	Upper Cretaceous	Limestone
Hammei Tveria, Kikar, well	Upper Cretaceous	Limestone
Sartan spring, saline	Eocene	Chalk
Sartan spring, fresh	Eocene	Chalk
Tabgha, Ein Sheva	Eocene	Chalk
Kinneret 7, well	Eocene	Chalk
Druzi spring	Upper Cretaceous	Limestone
Haon beach, well	Quaternary	Basalt
Hamat Gader, Ein Reach	Eocene	Chalk
Hamat Gader, Ein Makla	Eocene	Chalk
Israel: Upper Jordan Valley and Yarmouk Valley		
Bira 2, well	Tertiary	Basalt
Ein Moda, spring	Eocene	Chalk
Ein Amal, spring	Eocene	Chalk
Meizar 3, well	Upper Cretaceous	Limestone
Israel: Arava Valley		
Paran 18, well	Lower Cretaceous	Limestone
Eilat 103, well	Quaternary	Sediments
Samar 2, well	Quaternary	Sediments
Yotvata 13, well	Quaternary	Sediments
Yaalon 117, well	Lower Cretaceous	Sandstone
Bet Yosef 5, well	Tertiary	Chalk, basalt
Jordan: Jordan Valley		
JICA well at Sweima	Lower Cretaceous/Triassic/Jurassic	Sandstone
Manshiya 2, artesian well	Upper Cretaceous	Limestone
Shuna-N, artesian well	Upper Cretaceous	Limestone
Mukheiba 8, artesian well	Upper Cretaceous	Limestone
Jordan: Dead Sea Area		
Zara Princess Basma Farm,	Lower Creataceous	Sandstone
Zarqa Main: waterfall;	Lower Cretaceous	Sandstone
Ibn Hammad 2, spring	Lower Cretaceous	Sandstone
Araba Valley		
Wadi Araba 5, artesian well	Paleozoic Ishrin Fm	Sandstone
Disi 6, well	Paleozoic Disi Fm	Sandstone
Disi 1, well	Paleozoic Disi Fm	Sandstone
Jordan: Jordanian Highland		
Hammat Jerash, spring	Lower Cretaceous	Sandstone
Azraq, well 8	Eocene	Chalk
Azraq, well 15	Eocene	Chalk
Ruwayshed 1 (old well)	Eocene/Up. Cretaceous	Limestone
Ruwayshed 2 (new well)	Eocene/Up. Cretaceous	Limestone
Bir Sihan, spring	Upper Cretaceous	Limestone

dominantly mineralised by dissolution of gypsum ($Log(Ca/SO_4)_{sw} = 0.44$), calcite ($Log(Ca/HCO_3)_{sw} = -11.4$) and halite ($Log(Na/Cl_{sw} = 0.07$). Besides the dissolved evaporitic marine sulfate, $\delta^{34}S$ values indicate variable fractions of sulfate derived from igneous sulfur pointing to interaction of waters with intrusive rocks. Most of the waters from Lower Cretaceous and Triassic/Jurassic sandstones show characteristic REY patterns (Fig. 2a). The δ^2H vs. $\delta^{18}O$ diagram shows that this type of groundwater represents a mixture of residual Pleistocene water with more recent precipitation. Different from Israel, in Jordan these waters are additionally shifted to greater $\delta^{18}O$ values indicating oxygen isotope exchange in hydrothermal environments. The isotopically most extreme sandstone-sourced groundwater in Jordan is from Triassic/Jurassic sandstones. In contrast, the waters from the Paleozoic sandstone in Jordan do not fit into the above discussed patterns because their spider diagrams and REY patterns

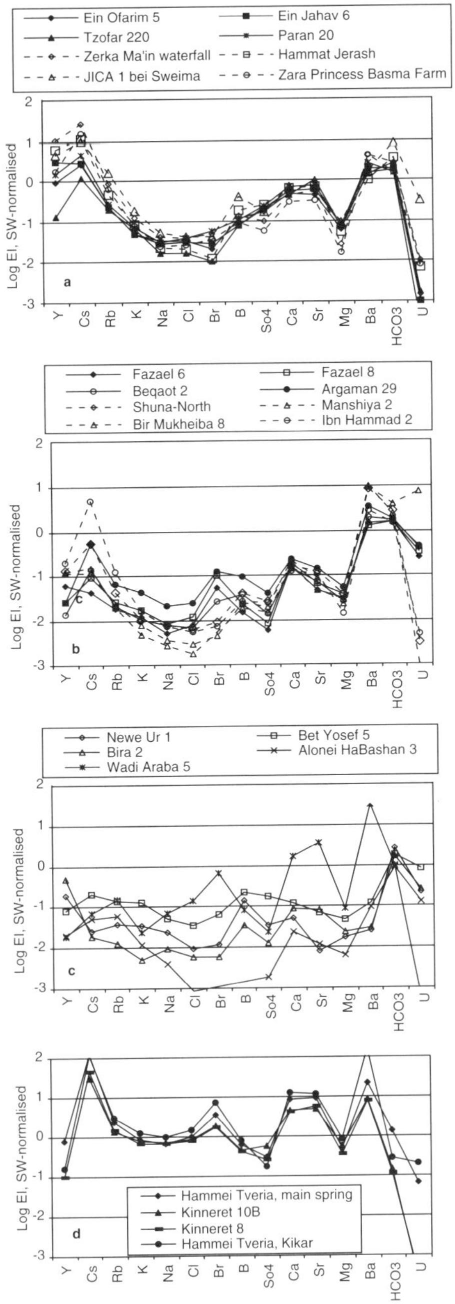

Figure 1. Spider diagrams of selected groundwaters from Lower Cretaceous and Triassic/Jurassic sandstones (a), Upper Cretaceous to Eocene limestones (b), and Neogene basalts (c), Termomineral waters (d) are from Upper Cretaceous limestones. Dashed and solid lines indicate groundwaters from Jordan and Israel (including adjoining territories), respectively.

clearly mark them as water derived from Upper Cretaceous limestones.

The groundwaters from mainly Upper Cretaceous to Eocene limestones are dominated by dissolution of limestone with variable fractions of gypsum and halite (Fig. 1b). Locally, sulfate derived from igneous sulfur is present. Locally, the waters are affected by silicified limestones (Fig. 2e), which indicates strong hydrothermal overprinting of aquifer rocks by Neogene mafic in- and extrusions. The Eocene limestones are distinguishable from the diagenetic Upper Cretaceous by their negative Ce anomaly at least in Israel (Figs. 2b and c). On Jordan territory this Ce anomaly mostly vanished due to REY exchange with abundant clay minerals from marls and weathering of the basaltic cover. The Upper Cretaceous and Avdat waters cluster near the eastern meteoric water line (EMWL; Fig. 3), but some mixing with Pleistocene water is evident from stable isotopes. The waters from Eocene limestones in Israel show lower stable isotope ratios than most of the waters from the Upper Cretaceous limestone. In Jordan, half of these groundwaters are shifted to greater $\delta^{18}O$ values according to hydrothermal overprinting by Neogene basalts. Groundwaters and brines that are from rocks older than the Pleistocene were not sampled in this study.

The thermosaline waters of Israel result from mixing of groundwater from Upper Cretaceous aquifers with a Ca-Mg-Cl brine that evolves from locally infiltrating waters dissolving underlying evaporites. In the surroundings of Lake Kinneret, the thermosaline waters show a trend of increasing TDS with increasing $\delta^{18}O$ values, which is related to infiltrating water from Lake Kinneret leaching post-halite evaporites. The most saline water contains 2.8 T.U., i.e., thereby indicating that the leaching of evaporites is going on today. Locally, brines are also generated in southern Israel and in Jordan.

Cl/Br mass ratios vary from 80 to 1000 (seawater: 298). The high ratios are common in the saline waters, and indicate halite dissolution, whereas the low ratios point to leaching of Br from a bituminous-rich layer and/or phosphorites occurring near the Cretaceous/Tertiary boundary (Mller et al. 2003). Halite dissolution plays an important role in Jordan because Na/Cl molar ratios are nearly 1 (Salameh & Rimawi 1988), whereas in Israel Cl is always greater than Na. In Jordan, meteoric waters contain dissolved calcite, silica, gypsum, and halite, and clayey particles. The most saline waters in Israel show high B/Cl weight ratios that are in the range of evaporitic halite composition. Waters from sandstones of both sides of the transform differ significantly in B, Br, Na/Cl ratio (Fig. 1a), and stable isotopes (Fig. 3). Waters from limestones are similar in stable isotope ratios but different in Br contents and Na/Cl ratios (Fig. 1b).

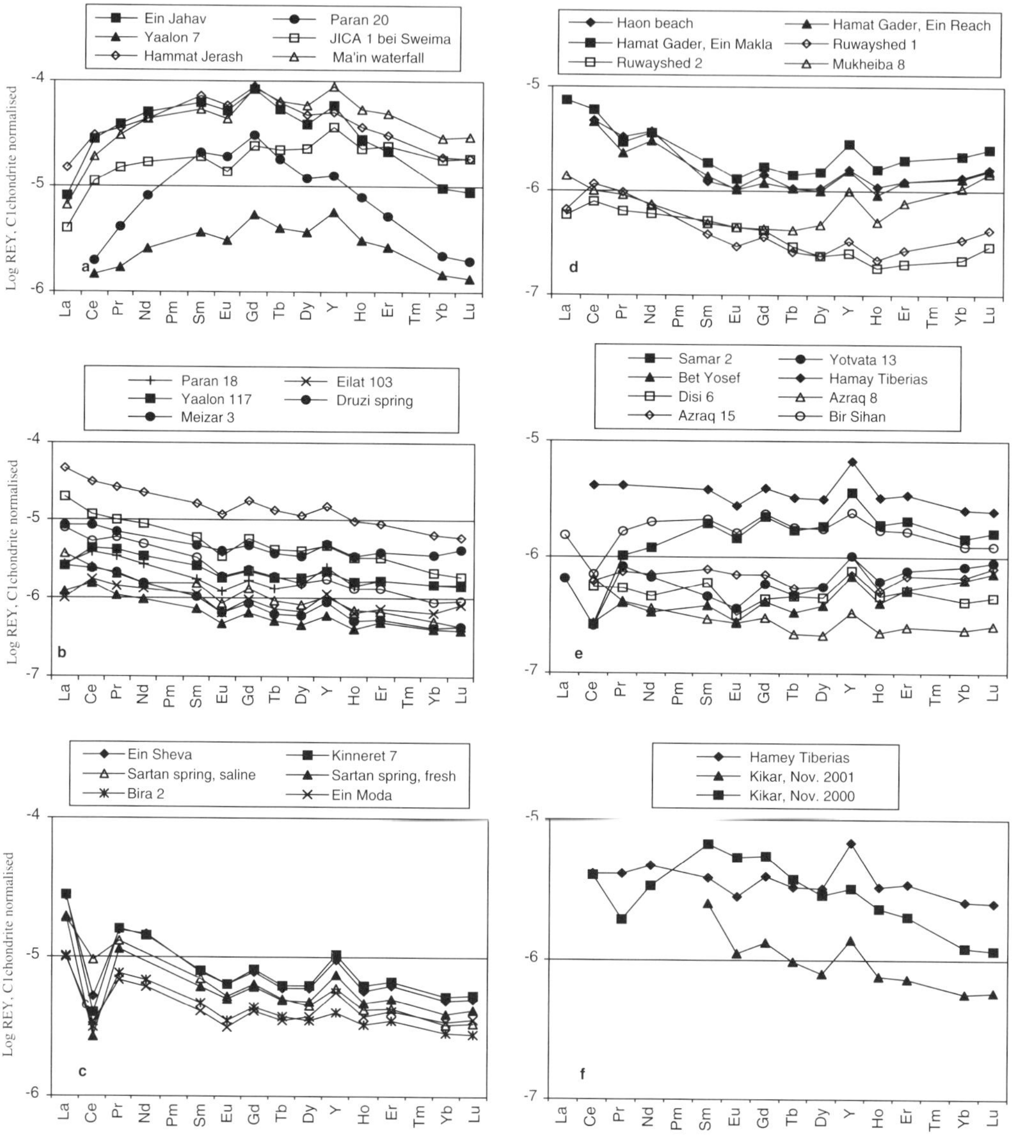

Figure 2. Grouping of REY patterns of groundwaters according to their trends and anomalies. They represent waters mainly derived from sandstones (a), diagenetic limestone (b), Eocene limestone (c), basalts (d), silicified limestone (e), and thermosaline waters (f).

The presence of Ca-Mg-Cl brines is defined by the excess of Cl over alkalis and excess of Ca + Mg over SO_4^{2-} and HCO_3^-. Both excess values are equal in quantity. These brines are possibly caused by dissolution of post-halite evaporitic Mg salts. Such Mg brines are either present as such or they are subsequently altered by dolomitization of limestone or albitization of mafic rocks. These brines constitute the high salinity thermosaline waters (Figs. 1d and 2f).

Waters from basaltic aquifers have highly variable chemical composition (Fig. 1c), REY patterns (Fig. 2f), and isotopic compositions (Fig. 3).

Very often the interpretation of REY signatures does not correspond with the lithology of origin of waters which proves that flow between different aquifers occurs. For instance, the REY patterns waters from Paleozoic Disi sandstones in Jordan represent waters which have been recharged in areas

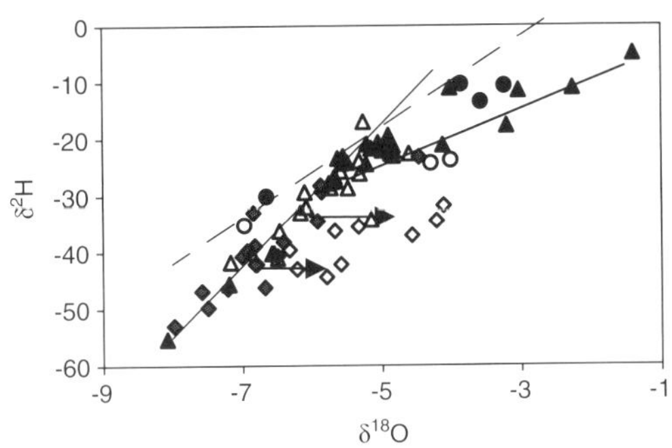

Figure 3. Cross plot of δ^2H vs. δ^{18}O (‰VSMOW) in groundwaters from Jordan (open symbols) and Israel (closed symbols). Diamonds, triangles and circles represent waters with REY signatures inherited from sandstones, limestones and basalts, respectively, according to the grouping in Fig. 2. Dashed line = eastern Mediterranean meteoric water line; Thin solid line = Pleistocene to recent groundwaters; Bold solid lines = thermosaline waters of Israel. Arrows indicate high-temperature interaction of Jordan water samples. Diamonds denote water from sandstones; Triangles indicate waters from limestone aquifers.

covered by Upper Cretaceous limestones. In other localities water drawn from basaltic aquifers has spider diagrams typical for such aquifers but the REY patterns indicate that the waters are recharged over limestone areas. Thus, different from the major element hydrochemistry, REY data yield reliable information on water–rock interaction and, in many cases, on the recharge area.

3 CONCLUSIONS

The chemical composition of groundwater is the result of water–rock interaction which comprises (i) dissolution of major and minor minerals, (ii) ion exchange, (iii) alteration of minerals, (iv) co-precipitation of minor elements with alteration minerals, and (v) ad- and desorption processes. Major and minor element hydrochemistry is controlled by soluble minerals in aquifer rocks, whereas REY are mainly controlled by rocks with the largest fraction of most soluble accessory minerals. If this is the case in the recharge area then the groundwater carries this REY signature along through all subsequent aquifers.

Stable H and O isotopes yield information on climatic conditions of meteoric precipitation. Sulfur isotopes indicate the presence of mostly unknown igneous intrusions.

Salinisation is either due to dissolution of evaporites from the rock sequence through which infiltration occurred, or by mixture with Ca-Mg-Cl brines that are locally ascending due to groundwater exploitation.

Israel, low δ^{18}O and δ^2H values indicate the presence of the groundwater of Pleistocene age. In Jordan, however, no such low δ^2H and δ^{18}O ratios are found (Fig. 3). Groundwaters with contributions of Pleistocene precipitates (given examples of precipitates here) are common in Lower Cretaceous and Triassic/Jurassic sandstones. Where Paleozoic rocks are near to the eroded surface, they are recharged by water from Cretaceous limestones, as in the case of the Paleozoic Disi sandstones in Jordan. The groundwaters from Cretaceous aquifers still have minor contributions of Pleistocene water, whereas the waters from Eocene limestones cluster along EMWL, at least in Israel. In Jordan the Eocene is much more affected by basaltic rocks than in Israel due to which significant isotopic and chemical changes occur.

In general, the chemical analyses and their interpretation, visualized in the spider diagrams, do not reveal any groundwater exchange across the Jordan – Dead Sea – Arava Fault. There may be one exception in the Arava Valley, where probably flood waters from Jordan pass the sedimentary cover of the fault. Having in mind that the transform was activated in the Oligocene one has to assume that at least from this time onwards groundwater flow across the transform was interrupted.

The combination of major and trace element- and of stable isotope distribution yield important information for the development of both conceptual and numerical models. The results are vital for all water management in these arid areas.

REFERENCES

Mller, P., Rosenthal, E., Dulski, P., Geyer, S. & Guttman, Y. 2003. Rare earths and yttrium hydrostratigraphy along the Lake Kinneret – Dead Sea – Arava transform fault, Israel and adjoining territories. *Appl. Geochem.* 18: 1613–1628.

Salameh, E. & Rimawi, O. 1988. Hydrochemistry of precipitation of northern Jordan. *Intern. J. Environmental Studies*, 32, 2003–216.

Water–Rock Interaction, Wanty & Seal II (eds)
© 2004 Taylor & Francis Group, London, ISBN 90 5809 641 6

U-series evidence of water–rock interaction at Yucca Mountain, Nevada, USA

James B. Paces
U.S. Geological Survey, Denver Federal Center, Denver

Leonid A. Neymark
S.M. Stoller Corp, c/o U.S. Geological Survey, Denver Federal Center, Denver

ABSTRACT: Radioactive disequilibrium among ^{238}U, ^{234}U, and ^{230}Th in rock, mineral, and water samples has been used to investigate water–rock interaction in the unsaturated zone at Yucca Mountain, Nevada, USA, site of the nation's proposed high-level radioactive waste repository. Cumulative results indicate that ^{238}U-^{234}U-^{230}Th disequilibrium is dominated throughout most of the thick unsaturated zone by alpha-recoil-related processes rather than bulk U dissolution. Greater amounts of U dissolution are present in a near-surface fault zone, suggesting greater water fluxes and focused flow. Results illustrate the utility of U-series isotopes for testing models of unsaturated flow and transport.

1 INTRODUCTION

The proposed repository for high-level radioactive waste is sited in Miocene-age felsic tuffs at Yucca Mountain, Nevada, USA (Fig. 1). Repository performance has been predicted by numerical models of flow and transport through a thick (about 400 to 750 m) unsaturated zone (UZ). Although these models are based on measured hydraulic properties, direct measurement of natural flow has not been possible. Therefore, alternative means of testing both conceptual and numerical flow models are desirable.

Water percolating through the continuum of matrix and fracture networks causes changes in U-Th isotope concentrations over time in both tuffs and water. In unmodified tuffs, ^{234}U and ^{230}Th will reach secular radioactive equilibrium with the parent ^{238}U such that $^{234}U/^{238}U$ and $^{230}Th/^{238}U$ activity ratios (AR) equal 1.0. However, greater solubility of U relative to Th, and ^{234}U relative to ^{238}U (due to alpha-recoil-related decay processes, including radiation damage and oxidation, Gascoyne 1992, sec. 2.5.1), results in $^{234}U/^{238}U$ AR and $^{230}Th/^{238}U$ AR that deviate from equilibrium values in rocks. $^{234}U/^{238}U$ AR in percolating water also are impacted by these processes.

Previous work showed that U-series radioactive disequilibrium is common in samples of tuff from the Yucca Mountain UZ (Gascoyne et al. 2002). In this

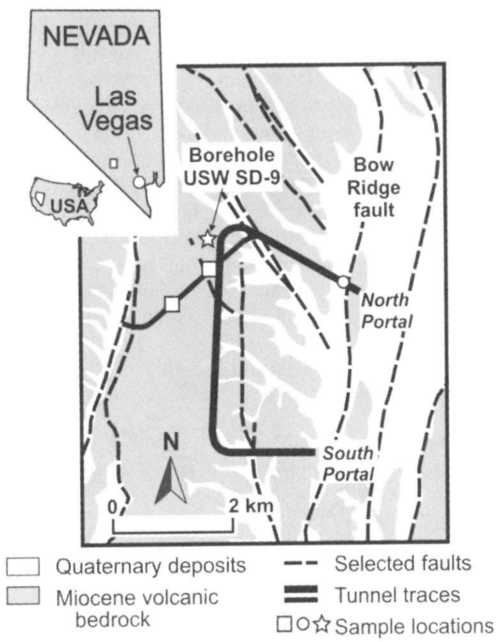

Figure 1. Location map of Yucca Mountain showing selected faults, traces of tunnels, and sample locations (basemap modified from Day et al. 1998).

paper, U-series analyses of tuff in contrasting UZ hydrologic environments are presented along with depth-$^{234}U/^{238}U$ AR relations for percolating water. These data are useful for investigating water–rock interaction under natural flow conditions over the past 10^3 to 10^6 years, including evidence for flow focusing along structural pathways and relative differences in percolation fluxes in different parts of the flow system. This work was conducted under an Interagency Agreement between the U.S. Geological Survey and U.S. Department of Energy.

2 DESCRIPTION OF WORK

2.1 *Felsic tuffs*

Whole-rock grab samples of felsic tuffs were collected from tunnel walls in 2 areas at Yucca Mountain. Three samples of welded tuff are from the proposed repository horizon at a depth of 200 to 300 m below land surface (squares in Fig. 1), including 2 samples immediately underlying cavities containing secondary mineral deposits (dominantly calcite and opal) indicative of fracture flow. Eleven samples of welded and nonwelded tuffs are from a profile across the Bow Ridge fault zone at depths of only 20 to 30 m below land surface (circle in Fig. 1). Five of these samples are welded tuff from the footwall block, with 1 from the footwall surface; 3 are brecciated tuff from the 2-m-wide fault zone; and 3 are nonwelded tuff from the hangingwall block. Samples across the Bow Ridge fault were selected to investigate the effects of structural focusing of flow. To evaluate whether water–rock interaction is greater on fracture surfaces relative to rock interiors (preferential fracture versus matrix flow), profiles of 2 to 4 subsamples each were taken across 8 of the individual welded tuff samples described above. Subsamples were either milled from natural fracture surfaces or wafered from 1.2-cm-diameter core that extended as much as 25 mm into rock interiors.

2.2 *Percolating water*

$^{234}U/^{238}U$ AR in UZ water were determined from modern pore water and from secondary fracture minerals. Pore water was extracted from samples of core from a surface-based borehole (USW SD-9; star in Fig. 1) and from underground boreholes. Depth relations for pore-water samples were determined relative to the base of a series of nonwelded tuffs overlying the proposed repository horizon. Although active fracture flow has not been observed at Yucca Mountain, secondary minerals formed from past fracture flow are well exposed on tunnel walls (Paces et al. 2001, Whelan et al. 2002). These minerals have U-series and U-Pb ages ranging from several thousand to more than 10 million years (Paces et al. 2001, Neymark et al. 2002, Paces et al. 2004). U isotopic compositions of the fracture water from which these minerals formed was estimated by calculating initial $^{234}U/^{238}U$ AR for minerals with the youngest $^{230}Th/U$ ages at a given sample location.

2.3 *Analytical methods*

$^{234}U/^{238}U$ and $^{230}Th/^{238}U$ AR were analyzed by thermal ionization mass spectrometry on 0.02- to 0.05-g aliquots of powdered whole-rock samples or 0.2 to 4 mm of pore water. Analytical methods are described elsewhere (Paces et al. 2002, Paces & Neymark 2003). Isotope standards for U (SRM 4321) and Th (IRMM-035 and IRMM-036), as well as an in-house secular equilibrium standard, were analyzed throughout the data acquisition period. Estimates for analytical precision and accuracy typically range from 0.2 to 1.5 percent for $^{234}U/^{238}U$ AR and 0.5 to 2.5 percent for $^{230}Th/^{238}U$ AR.

3 ANALYTICAL RESULTS

3.1 *Felsic tuffs*

Analyses of welded tuff samples from the proposed repository horizon show small degrees of ^{238}U-^{234}U-^{230}Th disequilibrium (Fig. 2). Both $^{234}U/^{238}U$ and $^{230}Th/^{238}U$ AR are typically less than 1.0 and plot near the $^{230}Th/^{234}U$ equiline (locus of points with $^{230}Th/^{234}U$ AR = 1.0), indicating that ^{230}Th is in radioactive equilibrium with its immediate parent, ^{234}U, but not with its grandparent, ^{238}U. The one sample not visibly associated with any secondary mineral deposits (solid ellipses in Fig. 2) has

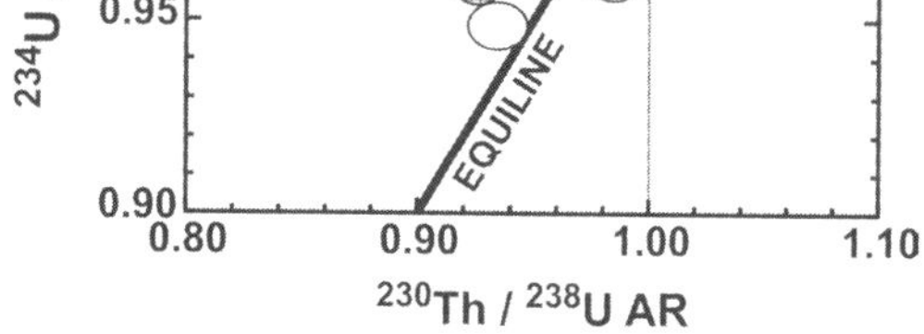

Figure 2. $^{234}U/^{238}U$ and $^{230}Th/^{238}U$ activity ratios (AR) for 3 welded tuffs from the proposed repository horizon. Fill patterns for 2σ error ellipses indicate subsamples from the same sample. The dark equiline represents points with equal ^{230}Th and ^{234}U activities ($^{230}Th/^{234}U$ AR =1).

$^{234}U/^{238}U$ and $^{230}Th/^{238}U$ AR closer to 1.0 than other samples with nearby fracture minerals.

Analyses of tuff from the Bow Ridge fault zone have greater ^{238}U-^{234}U-^{230}Th disequilibrium than those from the proposed repository horizon, with $^{234}U/^{238}U$ AR as low as 0.76 and $^{230}Th/^{238}U$ AR between 0.82 and 1.16 (Fig. 3). Although some analyses plot near the equiline, many have lower or higher $^{230}Th/^{234}U$ AR. In particular, welded tuff from the footwall surface of the fault has $^{230}Th/^{234}U$ AR >1, whereas tuff farther into the footwall block tends to have $^{230}Th/^{234}U$ AR <1.

Subsample profiles for both shallow and deep tuffs do not show obvious differences in U-series disequilibrium between fracture surfaces and rock interiors. $^{234}U/^{238}U$ AR show small variations across individual profiles (typically <1.5 percent), whereas $^{230}Th/^{238}U$ AR show larger variations (typically <6 percent) and can have either higher or lower values for subsamples from fracture surfaces relative to rock interiors. Therefore, systematic variations in the degree of $^{234}U/^{238}U$ or $^{230}Th/^{238}U$ disequilibrium indicating greater amounts of water–rock interaction on fracture surfaces are not observed.

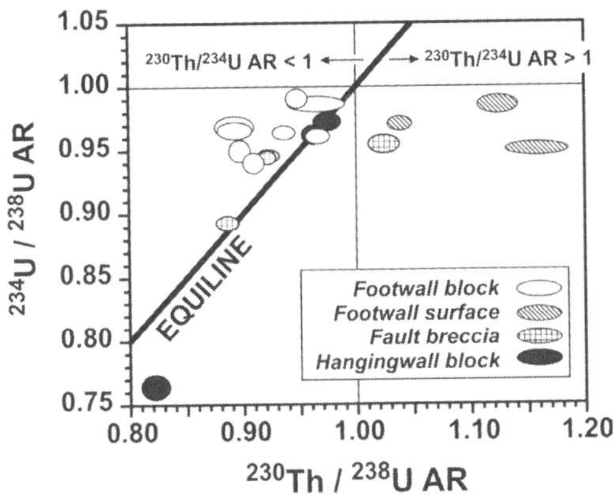

Figure 3. $^{234}U/^{238}U$ and $^{230}Th/^{238}U$ activity ratios (AR) for subsamples of 11 welded and nonwelded tuffs (shown as 2σ error ellipses) associated with the Bow Ridge fault zone.

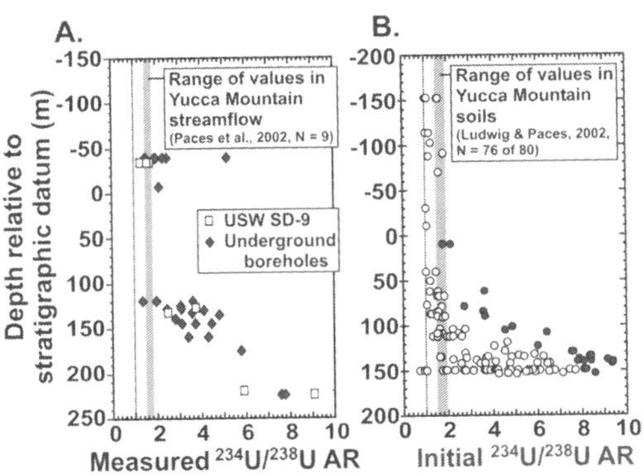

Figure 4. Depth profiles showing variations of (A) measured $^{234}U/^{238}U$ activity ratios (AR) in modern pore water, and (B) initial $^{234}U/^{238}U$ AR in secondary fracture minerals. Initial $^{234}U/^{238}U$ AR in minerals with $^{230}Th/U$ ages $<10^5$ years (filled circles) best reflect $^{234}U/^{238}U$ AR values in fracture water.

3.2 *Percolating water*

Values of $^{234}U/^{238}U$ AR measured for pore water and inferred from fracture minerals increase with depth through the UZ (Fig. 4). Streamflow and soil water (inferred from analyses of calcrete) have $^{234}U/^{238}U$ AR between about 1.4 and 2.0 (Ludwig & Paces 2002, Paces et al. 2002). As water percolates into the subsurface, it becomes enriched in ^{234}U relative to ^{238}U. Pore water extracted from nonwelded tuffs in the shallow UZ typically has $^{234}U/^{238}U$ AR of 1.5 to 2.5 (Fig. 4A). In deeper tuffs, $^{234}U/^{238}U$ AR in pore water are as high as 9. Similar values and trends in initial $^{234}U/^{238}U$ AR are observed for secondary minerals precipitated from fracture water (Fig. 4B). $^{234}U/^{238}U$ AR of 8 to 9 for percolating water in the deep UZ are similar to values for perched water bodies and saturated zone ground water beneath Yucca Mountain (Paces et al. 2002).

4 DISCUSSION

^{238}U-^{234}U-^{230}Th disequilibrium in whole-rock samples from the Yucca Mountain UZ shows the preferential loss of ^{234}U relative to ^{238}U over the last 10^3 to 10^6 years. ^{234}U lost from rocks (either fracture surfaces or matrix pore walls) contributes to the elevated $^{234}U/^{238}U$ AR in water percolating through the UZ. Insignificant differences in whole-rock $^{234}U/^{238}U$ or $^{230}Th/^{238}U$ AR between subsamples of fracture surfaces and rock interiors imply that water–rock interaction is not restricted to fracture surfaces and that matrix percolation may be equally important. A similar conclusion was reached using U-series results from whole-rock samples in areas of greater and lesser overall fracturing (Gascoyne et al. 2002). Roughly equivalent fracture and matrix fluxes also are implied by similar $^{234}U/^{238}U$ AR in fracture and pore water (Fig. 4), which integrate larger parts of the flow network.

The small amounts of ^{238}U-^{234}U-^{230}Th disequilibrium in most rocks and the large $^{234}U/^{238}U$ AR in water in the deep UZ imply minimal U mobility during water-rock interaction. Significant losses (or gains) of soluble U relative to insoluble Th would result in a wider dispersion of data away from the $^{230}Th/^{234}U$ equiline (Fig. 2). Liberated ^{238}U also

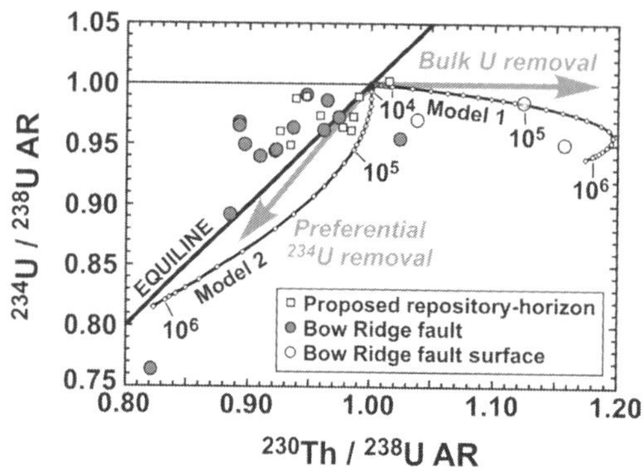

Figure 5. $^{234}U/^{238}U$ and $^{230}Th/^{238}U$ activity ratios (AR) for whole-rock samples of tuffs along with examples of the continuous U leaching model described in the text. Model 1 is dominated by bulk U removal with relatively high leach constants for ^{238}U (c_8) and ^{234}U (c_4) that are approximately equal ($c_8 = 0.2 \times \lambda_{230}$ and $c_4 = 1.1 \times c_8$). Model 2 simulates lower bulk U removal rates and preferential ^{234}U removal ($c_8 = 0.01 \times \lambda_{230}$ and $c_4 = 8 \times c_8$). Tick marks on model paths represent time in log increments of years from 10^3 to 10^6. Leach constants are given as fractions of the decay constant of ^{230}Th (λ_{230}).

would buffer $^{234}U/^{238}U$ AR so that deeper UZ water could not attain elevated values. However, elevated $^{230}Th/^{234}U$ AR for tuffs along the Bow Ridge fault surface are consistent with greater degrees of water–rock interaction caused by flow focusing down the fault plane. Some of the U removed from tuffs forming the footwall surface may have been transported to tuffs farther into the footwall block, resulting in $^{230}Th/^{234}U$ AR <1.

Water–rock interaction in the UZ is likely produced by numerous or even continuous percolation events. Initial efforts to simulate Yucca Mountain U-series disequilibrium assumed continuous U removal at constant rates using the leach model of Latham & Schwarcz (1987). $^{234}U/^{238}U$ and $^{230}Th/^{238}U$ AR calculated using this model depend only on leach constants for ^{234}U and ^{238}U (c_4 and c_8, respectively). Both $^{234}U/^{238}U$ and $^{230}Th/^{238}U$ AR reach steady-state values after approximately 3 million years; however, the pattern of isotope evolution (Fig. 5) depends on bulk U removal (absolute value of c_8) and the degree of fractionation between ^{234}U and ^{238}U (c_4/c_8). Aggressive leaching (larger c_8 values and $c_4/c_8 \approx 1$) results in elevated $^{230}Th/^{238}U$ AR and minor changes in $^{234}U/^{238}U$ AR (Model 1 in Fig. 5), similar to results from the Bow Ridge fault-surface sample (open circles in Fig. 5). In contrast, less aggressive leaching (smaller c_8 values and $c_4/c_8 > 1$) results in whole-rock $^{230}Th/^{234}U$ AR that remain near the equiline (Model 2 in Fig. 5) and moderate to high $^{234}U/^{238}U$ disequilibrium in both tuffs and percolating water. Analyses plotting to the left of the equiline require gains of U relative to Th and cannot be explained by a simple leaching model.

Steady-state leaching conditions result in constant ^{238}U-^{234}U-^{230}Th disequilibrium. At Yucca Mountain, slow, uniform secondary mineral growth rates imply stable hydrologic conditions in the deeper parts of the UZ on million-year time scales (Neymark et al. 2002, Paces et al. 2004). Therefore, whole-rock disequilibrium likely reflects long-term rather than transient hydrologic conditions.

5 CONCLUSIONS

^{238}U-^{234}U-^{230}Th disequilibrium in tuffs and elevated $^{234}U/^{238}U$ AR in percolating water are indications of water–rock interaction in the UZ at Yucca Mountain. Data imply that both matrix and fracture flow result in similar degrees of U-series disequilibrium, and that the amount of bulk U dissolution at repository depths (200 to 300 m) is very low. U-series characteristics in both rock and water are controlled dominantly by alpha-recoil-related processes rather than aggressive leaching of U. Greater amounts of $^{230}Th/^{234}U$ disequilibrium in samples from the shallow Bow Ridge fault zone (20 to 30 m) imply that focused flow causes greater U mobility. Results from this work show that zones of lesser and greater UZ percolation flux at Yucca Mountain can be identified using U-series data. Ultimately, these data may be useful for testing numerical models of UZ flow and transport that are fundamental to evaluating long-term repository performance.

REFERENCES

Day, W.C., Potter, C.J., Sweetkind, D.S., Dickerson, R.P. & San Juan, C.A. 1998. Bedrock geologic map of the central block area, Yucca Mountain, Nye County, Nevada: *U.S. Geol. Surv. Misc. Invest. Series Map I-2601*: 2 sheets.

Gascoyne, M. 1992. Geochemistry of the actinides and their daughters, in M.Ivanovich & R.S.Harmon (eds.), *Uranium-series disequilibrium: Applications to Earth, Marine, and Environmental Sciences (2d ed.)*: 34–61. Oxford, UK: Clarendon Press.

Gascoyne, M., Miller, N.H. & Neymark, L.A. 2002. Uranium-series disequilibrium in tuffs from Yucca Mountain, Nevada, as evidence of pore-fluid flow over the last million years. *Appl. Geochem.* 17: 781–792.

Latham, A.G. & Schwarcz, H.P. 1987. On the possibility of determining rates of removal of uranium from crystalline igneous rocks using U-series disequilibria-1: a U-leach model, and its applicability to whole-rock data. *Appl. Geochem.* 2: 55–65.

Ludwig, K.R. & Paces, J.B. 2002. Uranium-series dating of pedogenic silica and carbonate, Crater Flat, Nevada. *Geochim. Cosmochim. Acta* 66: 487–506.
Neymark, L.A., Amelin, Y., Paces, J.B. & Peterman, Z.E. 2002. U-Pb ages of secondary silica at Yucca Mountain, Nevada: implications for the paleohydrology of the unsaturated zone. *Appl. Geochem.* 17: 709–734.
Paces, J.B., Ludwig, K.R., Peterman, Z.E. & Neymark, L.A. 2002. $^{234}U/^{238}U$ evidence for local recharge and patterns of ground-water flow in the vicinity of Yucca Mountain, Nevada, USA. *Appl. Geochem.* 17: 751–779.
Paces, J.B. & Neymark, L.A. 2003. U-series disequilibrium as a test for unsaturated-zone hydrologic models at Yucca Mountain, Nevada. *Proceedings, 10th International High-Level Radioactive Waste Management Conference, Las Vegas, NV, March 30–April 2, 2003.* American Nuclear Society, LaGrange Park, IL.: 27–38.
Paces, J.B., Neymark, L.A., Marshall, B.D., Whelan, J.F. & Peterman, Z.E. 2001. Ages and origins of calcite and opal in the Exploratory Studies Facility tunnel, Yucca Mountain, Nevada. *U.S. Geol. Surv. Water-Resources Invest. Rep. 01–4049.*
Paces, J.B., Neymark, L.A., Wooden, J.L. & Persing, H.M. 2004. Improved spatial resolution for U-series dating of opal at Yucca Mountain, Nevada, USA, using ion-microprobe and microdigestion methods. *Geochim. Cosmochim. Acta*, in print.
Whelan, J.F., Paces, J.B. & Peterman, Z.E. 2002. Physical and stable isotope evidence for formation of secondary calcite and silica in the unsaturated zone, Yucca Mountain, Nevada. *Appl. Geochem.* 17: 735–750.

Water–Rock Interaction, Wanty & Seal II (eds)

The evolution of groundwater in the Tyrrell region, south-central Murray Basin, Victoria, Australia

B. Petrides & I. Cartwright
Hydrogeology and Environmental Research Group, School of Geosciences, Monash University, Clayton, Vic, Australia

T.R. Weaver
Hydrogeology and Environmental Research Group, School of Earth Sciences, The University of Melbourne, Parkville, Vic, Australia

ABSTRACT: The southern Murray Basin is one of Australia's most important agricultural areas. This research is being conducted in the Mallee region of the Murray Basin to determine the source of primary salinity, and its long residence times within the basin. Groundwater from the Parilla Sands aquifer and the underlying Renmark Group aquifer is relatively saline (TDS up to 200,000 mg/L), and is Na-Cl type. There is no regular change in salinity along flow paths implying that vertical flow is important. Stable isotopes and Br/Cl ratios indicate that groundwater is entirely of meteoric origin and has undergone evaporative enrichment with only minor halite dissolution. The groundwater chemistry also shows the effects of minor rock weathering and ion-exchange and sorption reactions with clays. Similarity in chemistry of all groundwater in the region implies relative consistency in climatic conditions.

1 INTRODUCTION

Nowhere on the Australian continent is the problem of determining the origin of saline waters of more practical importance than in the Murray Basin, which contains some of the most important agricultural land on the continent. The clearing of land in the groundwater recharge zones, accompanied by irrigation development, and the evaporative concentration of salts, has resulted in the development of secondary salinity (Brown et al. 1988). However, it is clear that much of the basin was always saline. The source of this primary salinity may be evaporation (Bonython 1956), remnant sea-water entrapped in the sedimentary basin at the time of their deposition (Wopfner & Twidale 1967, Johnson 1979), airborne oceanic aerosols (cyclic salts) transported inland into basins by rainfall (Jack 1921, Evans et al. 1990, Mazor & George 1992) or a combination of rock weathering and marine aerosol accession (Chivas et al. 1991, Herczeg & Lyons 1991).

Here, we use a combination of major ion and stable isotope geochemistry to examine the origin of groundwater and solutes in groundwater from the Parilla Sands and Renmark Group aquifers in the Tyrrell area of the Murray Basin (Fig. 1). We also assess whether these have varied over time and the overall flow paths in the region. Most studies of Murray Basin hydrogeology have focused on the regions of fresher groundwater that form a potential resource. However, areas dominated by saline groundwater (such as Tyrrell) are more abundant and understanding them is critical to our overall knowledge of basin hydrogeology.

2 GEOLOGICAL & HYDROGEOLOGICAL SETTING

The Murray Basin is a roughly circular basin covering approximately 300,000 km^2 of south-eastern Australia (Fig. 1). The geology and hydrogeology have been described in detail by Brown (1989) and Evans & Kellet (1989). The study area (Fig. 1) is located in the south-central Murray Basin. This is an area of $\sim$5000 km^2 of flat sandy plains that is underlain by up to 600 m of Cainozoic sediments. These sediments are subdivided into two main aquifer systems. The deepest aquifer, the Renmark Group, comprises continental fluvial or lacustrine deposits of predominantly quartz with minor clay and carbonate minerals that locally attain a maximum thickness of

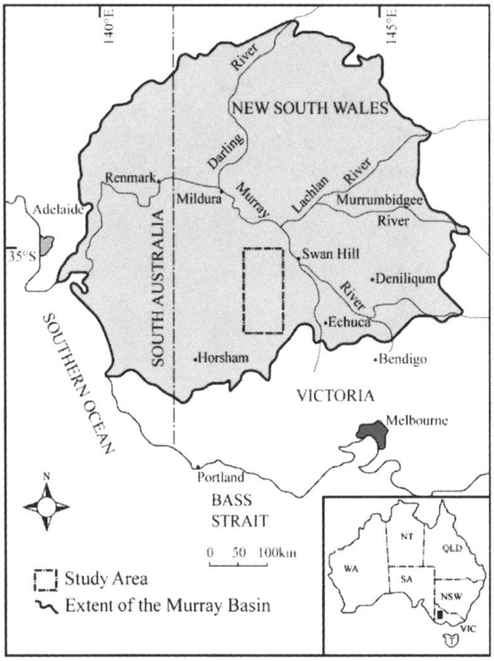

Figure 1. Location map of the Murray Basin and the Tyrrell region study area.

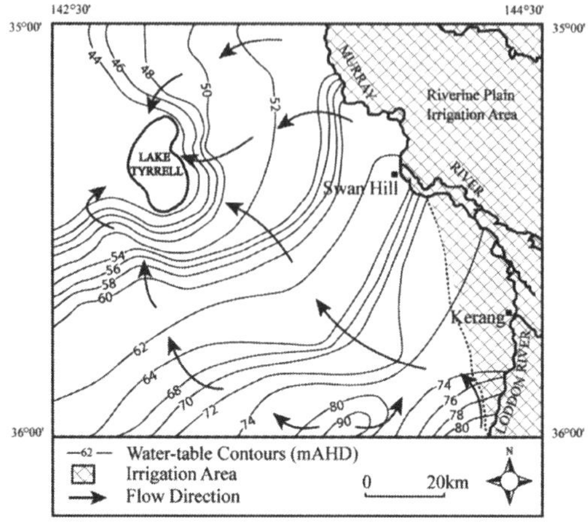

Figure 2. Water-table contours (mAHD) of the Parilla Sands aquifer. Contours are density corrected.

300 m. This aquifer is overlain in the western region of the study area by the mid-Tertiary low permeability barrier, the Geera Clay. The youngest sediments in this region are the Parilla Sands, which directly overly the Renmark Group except in the west. They are composed of semi-consolidated sands and clays deposited in marginal marine or fluvio-lacustrine environments (Herczeg et al. 2001). Both the Renmark Group and Parilla Sands overlie a basement of Palaeozoic rocks.

The major surface-water system in the Tyrrell region is the Murray River and its tributaries. Groundwater flow converges on the Murray River from the north and south of the basin. In the Tyrrell region, the Parilla Sands aquifer is unconfined and recharged by rainfall over most of the region. The deeper Renmark Group receives water by downwards leakage of infiltrated rainfall through the overlying Parilla Sands aquifer, especially in the south and south-east where the Parilla Sands are thinnest.

Groundwater flow is to the north or north-west (Fig. 2). Typical hydraulic gradients in both aquifers are approximately 0.0025 m/m. Horizontal hydraulic conductivity is $\sim$0.5 to 5 m/day in the Parilla Sands and $\sim$1 to 7 m/day in the Renmark Group (Lawrence 1975).

Vertical hydraulic gradients indicate a potential for upward flow in the central and northern parts of the region. Here groundwater flows from the Renmark Group into the overlying Parilla Sands. Discharge from the Parilla Sands is into the Murray River, which incises into the Parilla Sands from the eastern margin of the study area. Discharge from the Parilla Sands also occurs in playas. These salt lakes occur where the water-table comes within a few meters of the ground surface and groundwater is able to evaporate. They are characterized by the precipitation of salts at the ground surface and highly saline groundwater (TDS >100,000 mg/L). Lake Tyrrell is the largest of these and it is underlain by extremely saline groundwater (over 150,000 mg/L TDS) (Macumber 1992).

3 GROUNDWATER CHEMISTRY

52 observation bores were sampled during May 2003 and analysed for major ion chemistry and stable isotopes (^{2}H, ^{18}O and ^{13}C). The groundwater chemistry shows little spatial variation. However, the TDS of groundwater from the unconfined Parilla Sands (TDS up to 200,000 mg/L) are generally higher than those of groundwater in the underlying semi-confined Renmark Group. Groundwater from both the aquifers is Na-Cl type, with Na comprising more than 80% of the cations and Cl more than 85% of the anions.

3.1 *Origin of primary salinity*

Relationships between Na, Cl and Br allow origins of the salts to be constrained. Figure 3a shows Cl/Br ratios versus Total Cl concentrations. Most groundwater has a Cl/Br ratio similar to the oceans, although there is a slight increase in Cl/Br ratios as the salinity

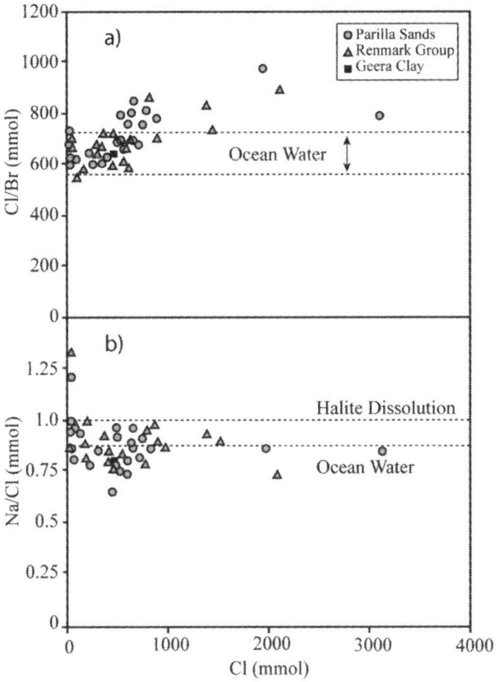

Figure 3. a) Hydrochemical plot of Cl/Br ratios vs Total Cl; b) Na/Cl ratios vs Total Cl.

increases. These data imply that evaporation rather than halite dissolution is the major process controlling the salinity of groundwater.

Na/Cl ratios decrease from 1.28 in the least saline groundwater to as low as 0.61 as salinity increases (Fig. 3b). The higher Na/Cl ratios imply that weathering of minerals such as albite influence Na concentrations in the fresher groundwater. However, overall, albite weathering does not contribute significant quantities of solutes to the groundwater. The chemistry of the more saline groundwater is consistent with evaporation being the more dominant process. The lowest Na/Cl ratios are probably due to ion-exchange reactions with clays that remove Na from groundwater and add divalent cations such as Ca and Mg.

3.2 *Stable isotopes*

The δ^2H and $\delta^{18}O$ values of groundwater from the Renmark Group and Parilla Sands in the Tyrrell region range from −42‰ to 12‰ and −5.8‰ to 3.8‰ (relative to SMOW), respectively. Most of the groundwaters fall to the right of the Global and Local Meteoric Water Lines (Fig. 4). The groundwater defines a single trend with a slope of 4.88 which is consistent with evaporation in a semi-arid climate.

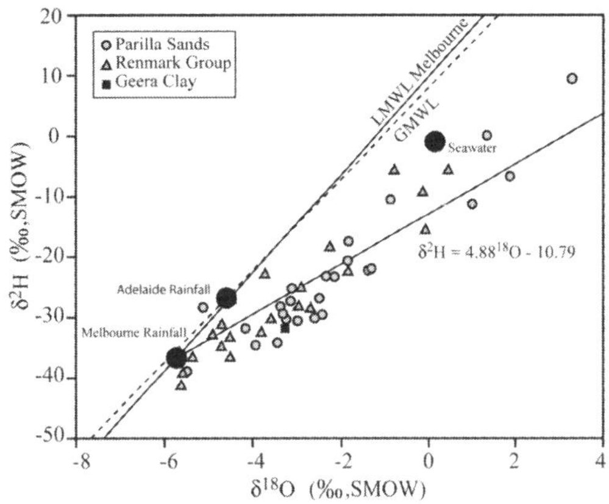

Figure 4. Stable isotope ratios. The groundwater defines a single trend with a slope of 4.88.

The data intercept the Meteoric Water Line at a value between the isotopic composition of rainfall in Melbourne and Adelaide. The stable isotopic data therefore provide compelling evidence that all water molecules are derived from rainfall, and that none of the groundwaters are remnant seawater. Although groundwater in the Murray Basin is known to vary in age from modern to >20,000 years there is no suggestion of different groups of water in the Tyrrell region. This suggests that climatic conditions over the last several thousand years were similar to those of today.

$\delta^{13}C$ values of DIC range from −21.3‰ to 0.73‰ (relative to PDB). The lower values indicate carbon sourced from organic matter in the soil zone. Given the lack of carbonate units in the area, the higher values are most likely to reflect carbon sourced from carbonate cements.

3.3 *Other major ions*

The low saturation indices for gypsum imply that gypsum dissolution does not control Ca concentrations, although gypsum dissolution is the most likely source of sulphate. Ca/SO_4 ratios (Fig. 5) are <1 implying a loss of calcium. Many samples are oversaturated with respect to calcite implying that calcite precipitation may have occurred, removing calcium from the groundwater. The loss of calcium can also be attributed to sorption onto kaolinite at low TDS, and the formation of Ca-beidellite clays. This is consistent with groundwater plotting in the stability field of kaolinite and Ca-beidellite.

A lack of dolomite in the region indicates low Mg concentrations within the system thereby suggesting the bulk of the major ions are derived from silicate weathering. Reactions between clays also control the

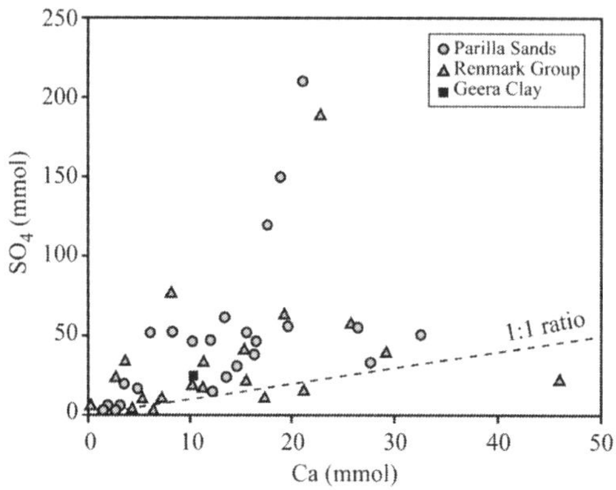

Figure 5. Hydrochemical plot of Ca vs SO_4.

concentrations of the major ions in the groundwater. However the degree of water–rock interaction in this region is low as saturation with amorphous silica is yet to be reached.

3.4 *Spatial trends*

There are no regular spatial trends in groundwater chemistry. In particular salinity does not increase systematically along the hydraulic gradient. At individual sites the groundwater from the Parilla Sands is generally more saline than that from the underlying Renmark Group; however, the chemistry is generally similar. These data imply that vertical groundwater flow in this area is important. This is also confirmed by the high salinity groundwater underlying Lake Tyrrell that results from brine reflux.

4 CONCLUSIONS

The solutes in the groundwater are dominantly the result of evapotranspiration with minor halite dissolution and silicate weathering. Sorption and ion exchange reactions involving clays are also important in modifying groundwater chemistry. The similarity of the chemistry of the groundwater in the Renmark Group and Parilla Sands aquifers together with the lack of systematic chemical trends along flow paths suggests that vertical flow is important. The groundwater in this system will have recharged at various times, however, the stable isotope and major ion chemistry of all groundwater is similar implying that climate change of this period is not recorded in the groundwater. This may indicate a high degree of mixing that has minimised any chemical fluctuations over time. Alternatively rainfall patterns and temperatures in this area may not have varied as much as in temperate coastal zones.

REFERENCES

Bonython, C.W. 1956. The salt of Lake Eyre – its occurrence in Madigan Gulf and its possible origin. *Royal Society of South Australian Transactions*. 79: 66–90.

Brown, C.M. 1989. Structural and stratigraphic framework of groundwater occurrence and surface discharge in the Murray Basin, southeastern Australia. *BMR J. Geol. Geophys*. 11: 127–146.

Brown, C.M., Kellet, R.J. & Evans, W.R. 1988. Summary report on Murray Basin 88 Conference, geology, groundwater, and salinity management. *Bureau of Mineral Resources*. Rec. 1988/43. 32pp.

Chivas, A.R., Andrew, A.S., Lyons, W.B., Bird, M.I. & Donnelly, T.H. 1991. Isotopic constraints on the origin of salt in Australian playas. 1. Sulphur. *Palaeogeography Palaeoclimatology Palaeoecology*. 84: 309–331.

Evans, W.R., Brown, C.M., & Kellet, R.J. 1990. Geology and Groundwater. In N.Mackay & D.Eastburn (eds). *The Murray*. 77–94.

Evans, W.R. & Kellet, R.J. 1989. The hydrology of the Murray Basin, southeastern Australia. *BMR J. Geol. Geophys*. 11: 147–166.

Herczeg, A.L., Dogramaci, S.S. & Leaney, F.W.J. 2001. Origin of dissolved salts in a large, semi-arid groundwater system: Murray Basin, Australia. *Marine Freshwater Research*. 52: 41–52.

Herczeg, A.L. & Lyons, W.B. 1991. A chemical model for the evolution of Australian sodium chloride lake brines. *Palaeogeography Palaeoclimatology Palaeoecology*. 84: 43–53.

Jack, R.L. 1921. The salt and gypsum resources of South Australia. *Geological Survey of South Australia Bulletin No. 8*. 118pp.

Johnson, M. 1979. The origin of Australia's salt lakes. *NSW Geological Survey Technical Report* No. 19(2): 221–266.

Lawrence, C.R. 1975. Geology, Hydrodynamics and Hydrochemistry of the Southern Murray Basin. *Geological Survey of Victoria*.

Macumber, P.G. 1992. Hydrological processes in the Tyrell Basin, southeastern Australia. *Chem. Geol*. 96: 1–18.

Mazor, E. & George, R. 1992. Marine airborne salts applied to trace evaporation, local recharge and lateral groundwater flow in Western Australia. *J. Hydrol*. 136: 63–77.

Wopfner, H. & Twidale, C.R. 1967. The chemical characteristics of lentic surface waters in Australia. In A.H. Weatherby (ed.) *Australian Inland Waters and their Fauna: Eleven Studies*.

Water–Rock Interaction, Wanty & Seal II (eds)
© 2004 Taylor & Francis Group, London, ISBN 90 5809 641 6

Geochemical and isotopic study of groundwaters in Cerro Prieto, BC, México

E. Portugal & G. Izquierdo
Instituto de Investigaciones Eléctricas, Cuernavaca, Mor., México

J. Álvarez
Residencia de Estudios, Cerro Prieto, BC, México

ABSTRACT: Results of a geochemical and isotopic study of the groundwaters from the phreatic aquifer on and surrounding the Cerro Prieto geothermal field are presented. Based on their chemical composition three types of water were recognized: chloride, sulfate and bicarbonate. Average temperature is 25°C; with few exceptions in the south where temperature can be as high as 47°C. Stable isotope compositions from some waters plot close to the global meteoric water line, corresponding to the ancient waters of the region. The hydrogeochemistry varies in relation to three principal processes: evaporation, infiltration of water used in agriculture, and interactions with evaporitic deposits.

1 INTRODUCTION

The study area is located in the plain of the Mexicali Valley, about 30 km south of the United States-Mexico border. It has an area of approximately 400 km^2, is located between 655000–685000 west longitude and 3605000–3576000 north latitude relative to UTM coordinates (Fig. 1), and includes the Cerro Prieto geothermal field (CPGF).

This paper reports the results of a systematic study on an extended area that includes the Cerro Prieto geothermal field. The principal objectives of this contribution are to present the geochemical and isotopic results of the groundwaters from the phreatic system in and surrounding the Cerro Prieto geothermal field in order to improve the knowledge of the system and to identify the processes that control the chemical changes in the waters.

2 HYDROGEOLOGY

The Mexicali Valley is located in the Salton Depression, which was formed in the Miocene period. New limit definition between the North American and Pacific plates, and the incorporation of the Baja California Peninsula into the Pacific Plate is postulated to be cause of its formation. The activation of lateral NW-SE trending shear systems produced the extensional rift system and pull-apart basin of the Extensional Gulf province (Stock & Hodges 1989; Lee et al. 1996).

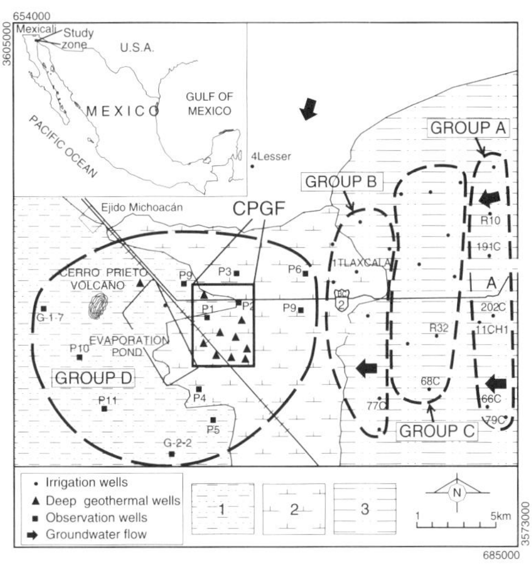

Figure 1. Hydrogeological sketch of the study area showing the sampled sites (agricultural and observation wells), the groundwater flow (indicated by arrows), geothermal area and the main sedimentation environment; 1 alluvial, 2 estuarine and 3 fluvial.

The phreatic aquifer of the zone is an extension of the hydraulic system of the Mexicali Valley. A lithological study carried out in the area by CICESE (1998) from the surface down to 200 m revealed that the quaternary sediments were alluvial to the W, estuarine at the center and deltaic to the E. It is worth pointing out the sharp change between them. Facies in the East part are formed by clay, sand and coarse gravel (fine, medium and coarse grain). Gravel with high transmisivity outcrops to the NE. To the W and SW the thicknesses of these materials decrease as depth increases. The central part consists of lagoonal or estuarine facies with clays and silts the main components along with sands. The extent of these facies to the North and South are not known due to lack of observation wells. In the west, alluvial distal fans are distinguished from the western limit of the mud zone to Sierra Cucapa (not shown in the Fig. 1). Such fans are formed by coarse sand, gravel at depth and by rounded granules. In the CPGF area, earlier studies have shown that no-consolidated sediments occurs in the NNW-SSW direction from 350 to 400 m depth; while at the NE are up to 2250 m and at 2000 m in the NE (Puente & de la Peña 1979). Figure 1 presents a simplified map of the area where sampled sites are shown, as well as the localization of the sedimentation environments, the paths of the groundwater flow (indicated by arrows) and the geothermal area.

The Mexicali Valley is characterized by arid climate. In summer the average temperature is 42°C and 6°C in winter. Rainfall and evaporation for a period of 25 and 14 years are 55.4 mm and 2205 mm, respectively (ERIC 1999). The regional aquifer is recharged by water infiltration from the Colorado River (Payne et al. 1979, Makdisi et al. 1982). Water table measurements in the study area indicate that groundwater flows from NE and E to SW (Fig. 1).

3 RESULTS

3.1 *Geochemical characteristics*

Three main groundwater types have been recognized:

- Type 1. Chloride type. Most of the waters are of the sodium-chloride type. These waters are extensive around the study area.
- Type 2. Sulfate type. Samples in the east are Ca- and Na-sulfate waters including the sites 191C, 202-C, R32, I-Tlaxcala, 54CH and 42CH.
- Type 3. Bicarbonate-type. Samples in the north-east and east part of the area are Ca- and Na-bicarbonate waters.

By means of semi logarithmic diagram of Schoeller–Berkaloff, which is based on the chemical composition, the waters were classified in four groups defined as A, B, C and D. Same grouping was found when the statistical multi-variable method of cluster analysis was applied. A strong enhance of salinity in the behavior patterns is observed as follows A < B < C < D. The Figure 2 (Piper diagram) shows the symbols that identify each one of the groups and also illustrates the major ion geochemistry in the study zone. Arrows indicate chemical evolution towards saline water located in the silts zone. In the Figure 1 broken lines enclose areas indicating the water groups identified. According to the path of the groundwater flow (E to W), a sequence of A → C → B → D is found; which under mineralization criteria for an aquifer does not fit at all (Hem 1992).

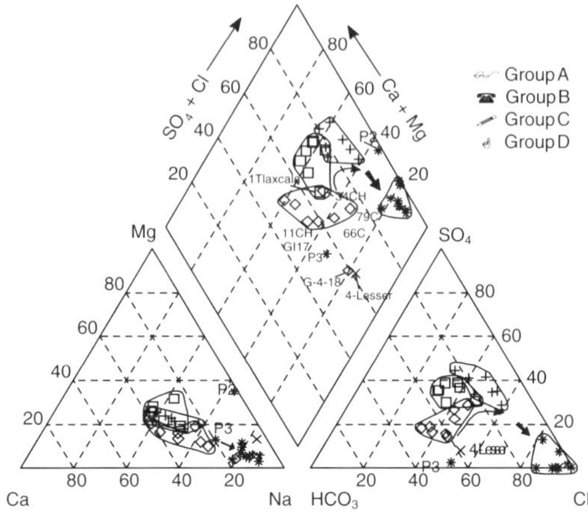

Figure 2. Piper diagram illustrates the major ion geochemistry of the study area.

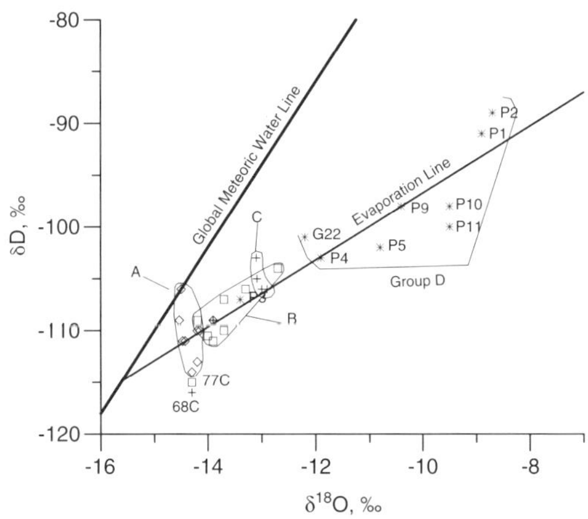

Figure 3. Stable isotopic composition of groundwater in the zone.

3.2 *Stable isotopes*

The stable isotopic data for the waters show dispersion of 18 units in deuterium and 4.7 units in oxygen-18.

In the δD vs $\delta^{18}O$ diagram (Fig. 3), Group A plots near the global meteoric water line as defined by Craig (1961). Other waters plot on an evaporation trend, whose behavior is described by the following relation:

$$\delta D = 3.7\delta^{18}O - 58.6 \quad (1)$$

The evaporation line intercepts the world meteoric line at −116.4‰ δD and −15.8‰ $\delta^{18}O$, indicating the recharge water composition. The lighter values indicate recharge at a higher elevation. Groundwater samples R10, 34C, 4Lesser, G-1-17 (group A), 77C (group B) and 68C (group C) are the most depleted in heavy isotopes. The average stable isotopic composition of this group is $\delta D = -112 \pm 3‰$ and $\delta^{18}O = -14.4 \pm 0.09‰$, which is very similar to those found by Payne et al. (1979) who defined it as old Colorado River water.

Evaporation increases the concentration of dissolved salts in water as a function of its initial composition, the intensity of evaporation, atmospheric humidity etc. In arid zones this phenomena occurs at the surface and during water infiltration to deep saturated zones (Clark & Fritz 1997). Despite the scarcity of rainfall and near lack of runoff, this process is quite likely in the Mexicali Valley due to the intense recycling of groundwater used for agriculture purposes and because of the installed complex irrigation and drainage network.

The importance of evaporation and dissolution of evaporates can be seen in the Cl- $\delta^{18}O$ diagram (Fig. 4); which includes data for groundwater collected in this study and for surface samples reported by Payne et al. (1979) and Ramírez (1997). In order to present the wide range in chloride concentration a semi-logarithmic diagram was used. Despite the apparent scatter of data, Figure 4 shows a correlation between increase in chloride concentration and increase in oxygen-18 for some water samples of groups A, B, C and D. From this data, mixing between surface water and groundwater does not appear to be an important process for all the groundwater samples. Instead is noted that for some waters from A and B groups, high chloride salinity is mainly due to the dissolution of evaporitic deposits (e.g. 79C → 77C sites) and that for other sites from the same groups evaporation seems to be the best explanation. This is the case of the 191C → 1Tlaxcala sites. While chemical and isotopic compositions of some water samples are due to the occurrence of both processes as for site 79C → 77C → P4.

The isotopic composition of A and B groups is very similar, indicating a strong relationship among fluids, and suggesting that waters from A and B groups are fed by the same aquifer.

The heavier isotopic composition, higher salinity and separate location suggest that group C are fed by

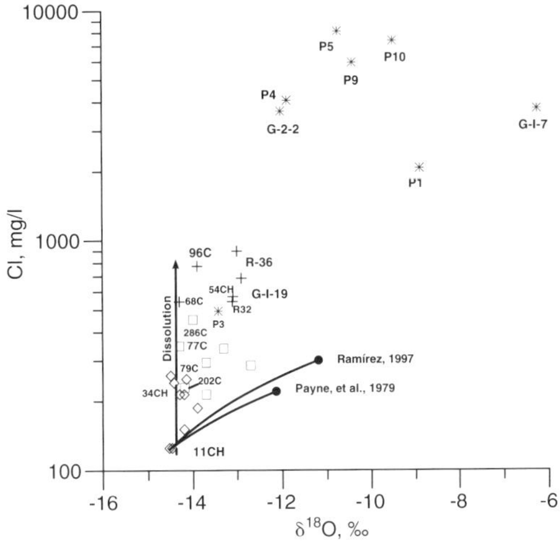

Figure 4. The Cl-$\delta^{18}O$ relationship of groundwater in the Mexicali Valley.

a different aquifer, whose circulation is through a shallower strata. Surface evaporation and infiltration of water used in agriculture can explain high salinity; but partial dissolution of evaporitic deposits could also be responsible for elevated SO_4, Cl and Na concentrations. However, the latter explanation is more likely since the isotopic composition data from group C are only slightly heavier.

Sites 77C and 68C (classified in B and C groups respectively) are isotopically lighter than other waters from the same groups. This can be taken as an indication that these waters are deeper with salinity due to the dissolution of evaporates.

3.3 *Chemical and isotopic trends*

Figure 5 presents a map showing of $\delta^{18}O$ (‰) trend. As it can be seen, the contour lines indicate to values isotopically heavier from northeast to southwest and from east to west. Similar trend are exhibited by contour of concentration of Cl and δD. The distribution of iso-concentration of these conservative species agree perfectly with main directions of groundwater circulation in the zone deduced of the equipotential lines. In Figure 5 there is a dark fringe showing an area where well produce water with elevated sulfate concentration (this waters corresponds to those of group C). The changes in sulfate concentration that occurs during water circulation in the study zone are as follows: Group A has an average values of 120 mg/L, while groups C and B have average values of 754 mg/L and 404 mg/L respectively. The average concentration for group D falls down to 11 mg/L (the minimum values for group D waters were 1.1 mg/L and 0.7 mg/L, P1 and P9 wells).

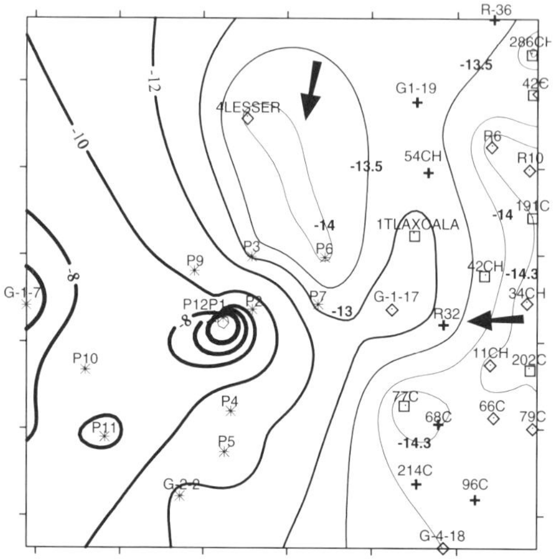

Figure 5. Contours $\delta^{18}O$ for groundwater in the study area.

The sulfate removal from the aqueous phase in sedimentary environments can be explained by several mechanisms: i.e. H_2S formation by reduction of sulfate as described by the reaction:

$$SO_4^{2-} + 8e^- + 10H^+ \leftrightarrow H_2S + 4H_2O \qquad (2)$$

H_2S may be removed by degassing or by precipitation as a metal sulfide, which has low solubility at low temperature (Poreda & Cox 2000).

Sulfate reduction by organic matter catalyzed by bacteria is a favorable process in sedimentary environments. However, metal sulfides were not found in the limited sediment samples analyzed from drilling at 95 m depth from some observation wells (P1, P3, P4, P7 and P12); the only sulfate phase detected was gypsum. However, chemical equilibrium modeling showed that waters are undersaturated in gypsum; whereas all other samples are saturated with respect to calcite, except for water from the sites P2, P11, G-2-2 and G-1-7.

X-ray diffraction analysis revealed that the mineralogy of the sediments is composed by quartz, calcite and plagioclase as main phases and smectite, kaolinite, halite, sylvite and gypsum as minor phases. The clarification of the processes that control the concentration of sulfate is an ongoing topic.

4 CONCLUSIONS

The study area is located in an arid climate, with high temperatures in summer and scarce annual precipitation. Because Mexicali Valley is an agriculture zone, the area is subjected to intensive pumping; most of the water comes from the Colorado River. The groundwater in the zone comes from the north and east from the regional Mexicali Valley aquifer system.

Based on their chemical composition, three groundwater types were recognized: sodium-chloride, calcium- and sodium-sulfate, and bicarbonate. Sodium-chloride type waters dominate the area.

The hydrogeochemistry varies in relation to three principal processes: evaporation, infiltration of water used in agriculture, and interactions with evaporitic deposit. The groundwater evaporation effect is more important for groundwater in lacustrine areas, where a stagnant aquifer should exist.

REFERENCES

CICESE 1998. Estudio hidrogeológico del campo geotérmico de Cerro Prieto: *Informe técnico* RE 05/98.

Clark, I. & Fritz, P. 1997. *Environmental isotopes in hydrology*. Lewis Publishers: 328 p.

Craig, H. 1961. Isotopic variations in meteoric waters. *Science* 133: 1702–1703.

ERIC 1999. Extractor Rápido de Información Climatológica. IMTA, México.

Hem, J. 1992. Study and interpretation of the chemical characteristics of natural water. *U.S. Geological Survey Water-Supply paper* 2254. Third edition: 263 p.

Lee, J., Miller, M.M., Cripper, R., Hacker, B. & Ledesma-Vazquez, J. 1996. Middle Miocene extension in the Gulf Extensional Province, Baja California: Evidence from the southern Sierra Juárez. Geol.l *Soc. Am. Bull.* 108: 505–525.

Makdisi, R.S., Truesdell, A.H., Thompson, J.M., Coplen, T.B. & Sanchez, J. 1982. Chemical evolution of Mexicali Valley groundwater. *Proc 4° Symp. Cerro Prieto Geothermal Field, Guadalajara Jalisco, México*: 551–562.

Payne, B.R., Quijano, L. & La Torre, D. C. 1979. Environmental isotopes in a study of the origin of salinity of groundwater in Mexicali Valley. *J. Hydrol.* 41: 201–215.

Poreda, M. & Cox, M.E. 2000. Sediment-water interaction, acidity and other waters quality parameters in a subtropical setting, Pimpama River, southeast Queensland. *Environ . Geol.* 39: 319–329.

Puente, C.I. & de la Peña, A. 1979. Geology of the Cerro Prieto geothermal field. *Geothermics* 8: 155–175.

Ramírez, J. 1997. *Estudio de las relaciones geohidrológicas del acuífero superior del Valle de Mexicali con las aguas geotérmicas superficiales*. Unpub. PhD thesis. Spain: Universidad de Alcalá.

Stock, J.M. & Hodges, K.V. 1989. Pre-pliocene extension around the Gulf of California and the transfer of Baja California to the Pacific Plate. *Tectonics* 8(1): 99–115.

Water–Rock Interaction, Wanty & Seal II (eds)
© 2004 Taylor & Francis Group, London, ISBN 90 5809 641 6

Chemical and isotopic characteristics and origin of high-pH thermal springs near Acapulco, State of Guerrero, México

A. Ramírez-Guzman & Y. Taran
Instituto de Geofisica, UNAM, México

ABSTRACT: Several groups of warm (39–43°C) springs are located within the southern part of the state of Guerrero, Mexico, close to the city of Acapulco within an area of approximately $100 \times 50\,km^2$ along the Pacific coast. They discharge low-salinity (TDS < 0.5 g/kg) and highly alkaline (pH = 9.5–10) waters from fractures in granites and gneisses. We report here our data on chemical composition of waters and bubbling gases, $^3He/^4He$, D/H and $^{13}C/^{12}C$ isotopic ratios and results of the multi-component chemical equilibrium calculations of rock-water system supporting our suggestion that the high pH of waters can be caused by water-granite interaction at 25–100°C and very high (>50) water/rock ratio.

1 DESCRIPTION OF THE STUDY AREA

The coastal part of the state of Guerrero, Mexico is a classical continental margin related to the modern subduction of the Cocos oceanic plate beneath the North America continental plate. The area belongs to the so-called Xolapa complex or terrain. According to Herrmann et al. (1994) it consists of high-grade metamorphic to migmatitic orthogenesis and para-genesis with metamorphism and migmatization occurred from 66 to 46 Ma. Oligocene (35–27 Ma) granitic plutons crop out parallel to the coast. There are 6 groups of thermal springs within an area of about $50 \times 100\,km^2$ (Fig. 1). Each group issues thermal water with temperatures close to 40°C on contacts between plutonic (granites, diorites, mon-zonites) and metamorphic (mostly gneisses) rocks. The flow rate of the springs varies between about 2 kg/s for the Paso Real group, to more than 20 kg/s for the total discharge of all Agua Caliente springs.

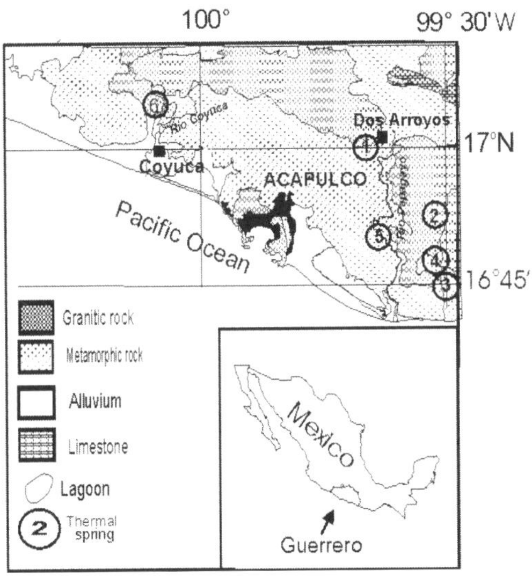

Figure 1. A simplified geological and location map of the Guerrero State near Acapulco. 1 – Dos Arroyos; 2 – Tamarindo; 3 – Cocoyul; 4 – Rio Cortes; 5 – Agua Caliente; 6 – Paso Real.

2 FIELD RESULTS

Water compositions are shown in Figure 2. All waters are very dilute, with TDS ~ 500 mg/l. Waters from El Tamarindo and Rio Cortes are similar in anionic composition while water from the Dos Arroyos springs is more of the SO_4-HCO_3 type. All waters are relatively high in F with the maximum of 11 mg/kg in the Cocoyul spring.

The common feature for all waters is their extremely high pH and high negative Eh. Major cation composition of springs is plotted on the log Na/K vs log K^2/Mg diagram (Fig. 2). According to Giggenbach's (1988) classifications, all waters are partially equilibrated with hydrothermally altered rock at 80–100°C.

The gases are relatively rich in He (up to 900 ppm). The N_2/Ar ratios in all springs are close to the atmospheric air value. All gases are extremely low in

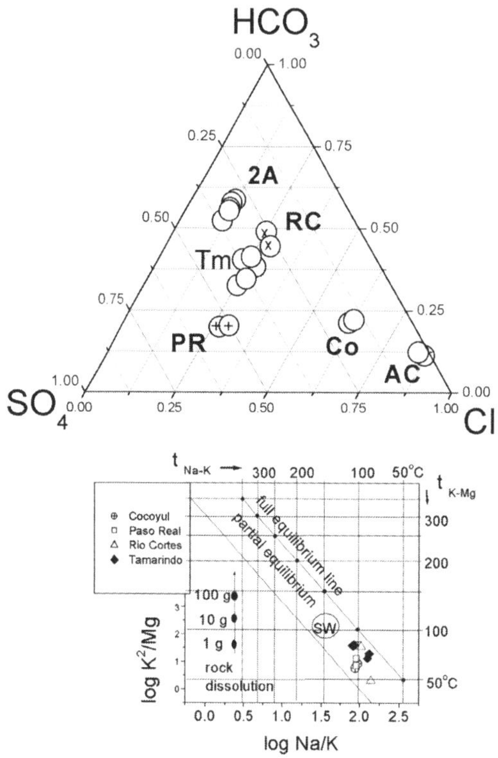

Figure 2. Major anion (triangle) and cation composition in the Guerrero thermal waters: 2A – Dos Arroyos, RC – Rio Cortes, Co – Cocoyul, Tm – Tamarindo, PR – Paso Real, SW – seawater.

CO_2; usually its concentration in bubbles is lower than that in the air (0.034 vol%).

The isotopic composition of waters ($\delta D \sim -60‰$ and $\delta^{18}O \sim -8.5‰$) corresponds to meteoric waters of some higher elevations than the springs are located. Isotopic ratios $^3He/^4He$ are about 5 times lower than in the atmosphere. These are probably the lowest values of $^3He/^4He$ measured in thermal waters of Mexico (Taran et al. 2002 and references therein). Isotopic composition of carbon in CH_4 in the Cocoyul spring ($\delta^{13}C = -25‰$) corresponds to "thermogenic" CH_4.

3 THERMOCHEMICAL MODELING AND ORIGIN OF HIGH-pH WATERS IN GRANITIC AQUIFERS

3.1 *Processes responsible for the formation of high-pH waters*

Several analyses of high-pH waters together with water from the Cocoyul spring are shown for comparison in Table 1. There are two main types of high-pH ($\sim$>10) waters in nature: waters of "soda" - lakes and springs, and waters discharging from ultramafic rocks. "Soda-waters" contain up to 85 g/L of total carbonate and are characterized by pH up to 10.1. (e.g. Cioni et al. 1992). High-pH waters associated with ultramafic rocks are, in contrast, usually diluted and carbonate-poor (Barnes et al. 1978; Bruni et al. 2002), although their pH can reach values as high as 12. Another known type of alkaline waters is high-temperature waters from so-called "weir-box" – waters from deep geothermal wells or boiling springs, which lost steam at atmospheric pressure. These waters can approach pH values up to 9.6. However, thermal waters from the granite-gneiss regional aquifer of the Guerrero state are neither associated with carbonatites nor ultrabasic rocks, nor are they of high-temperature hydrothermal systems. There are not too many examples of such waters. Hem (1985) showed one analysis of a 57°C spring in Oregon discharging a very dilute water from a quartz monzonite aquifer with pH = 9.4. Nordstrom et al. (1989) published data on chemical composition of cold dilute waters from a Stripa well (Sweden) with pH up to 10.0.

Table 1. Chemical composition of the Cocoyul thermal water and other high-pH waters of different origin.

	1	2	3	4	5	6
t°C	41	57	27.1	15.6	8.6	12
pH	9.98	9.4	10.10	11.78	10.06	11.86
Na	88	72	39300	40	218	23.7
K	0.84	2.4	547	1.1	0.44	3.1
Ca	0.5	1.3	1.82	48	94	47.4
Mg	0.02	0.3	1.11	0.4	0.03	0.02
Cl	61.3	6.5	7460	32	460	21.2
HCO_3	37	20	85400	0.0	18	1.3
SO_4	20	32	144	1.4	57	0.14
F	10.4	16	1350	0.00	5.3	–
SiO_2	57	75	378	5.2	18	1.5

1 – Cocoyul spring, State of Guerrero, Mexico. Water-bearing formation – granite (?) (this work); 2 – Spring in Idaho, USA, quartz monzonite (Hem 1985); 3 – Lake Bogoria, Kenya. Alkaline volcanic rocks (Cioni et al. 1992); 4 – Spring in Red Mountain, California. Ultrabasic rock (Hem 1985); 5 – Stripa well, 960 m, Sweden. Granite (Nordstrom et al. 1989); 6 – Cold spring, Genova province, Italy. Serpentinites (Bruni et al. 2002).

Waters from Guerrero are similar to the dilute hot and cold waters from crystalline (igneous and metamorphic) rocks. Probably, the mechanism controlling pH during the interaction of water with granites or metamorphic crystalline rocks critically depends on the water/rock ratio and carbonate C content in the primary rock High pH can be attained at high water/rock ratios and low partial pressure of CO_2 at the initial stage of rock dissolution, when concentrations of mineral-forming ions (mainly Na^+) are still low enough for the re-crystallization of secondary minerals (low albite, paragonite, Na-clays) and at low CO_2 concentration the excess of Na^+ is compensated by the formation of OH^-.

3.2 *Thermodynamic modeling of water-rock interaction: "thermo-gradient" waters in fractured granitic bodies*

Waters from Guerrero are so-called "thermo-gradient" waters, heated at considerable depth to a given temperature corresponding to the regional thermal gradient. An estimation of this maximum temperature can be the temperature calculated from the Na/K ratio of water samples. It can be seen from Figure 2 that for all Guerrero springs Na/K temperatures are close to 100°C. Therefore, we can imagine a recharge-discharge "convective cell" with the rain water infiltrating and descending through the fractured granites up to ≥100°C level (the depth depends on the local thermal gradient) and then rising up to the discharge (spring) site with temperature after conductive cooling to about 40°C. This model is similar to the Bowers and Taylor (1985) or Grichuk (2000) models of submarine hot hydrothermal systems of spreading zones (flow-through multi-step reactor). Using the CHILLER computer code (Reed and Spycher 1984) we "titrated" granitic rock on the way down and up calculating the equilibrium composition of water and equilibrium set of secondary minerals at each step at given temperature and water-to-reacted rock (w/r) ratio. Chemical composition of local granites was taken from Schaaf (1990). Descending flow is divided into 6 steps in the range of 25°C to 100°C. The ascending branch also consists of 6 steps covering the range of 90 to 40°C. The amount of reacted granite at each step was calculated using methods described by Grichuk (2000).

In our model of the "thermo-gradient" water we calculated as the first approximation relative amounts of secondary minerals and the composition of resulting water and gas phases in equilibrium after passing one kg of water with a given initial composition through 13 isothermal reactors filled with fresh granite. The water batch (1 kg) is "pressed" through a reactor at given temperature with dissolution of a portion of granite. That mass is proportional to the mass of the total dissolved rock (15 g) and to the average relative reaction rate. Water with a composition changed after the dissolution of granite and deposition of secondary minerals, reacts with another portion of fresh granite in the next reactor and so on.

3.3 *Results of numerical modeling and comparison with the Guerrero thermal waters*

Chemical composition of water, its pH and relative amounts of secondary minerals deposited on the way down and up are shown in Figures 3 & 4 as a function of the amount of rock reacted with 1 kg of meteoric oxygen-free water along the "convective cell". The resulting water (modeled "spring water") is essentially a solution of Na_2CO_3 with pH > 10 and high F

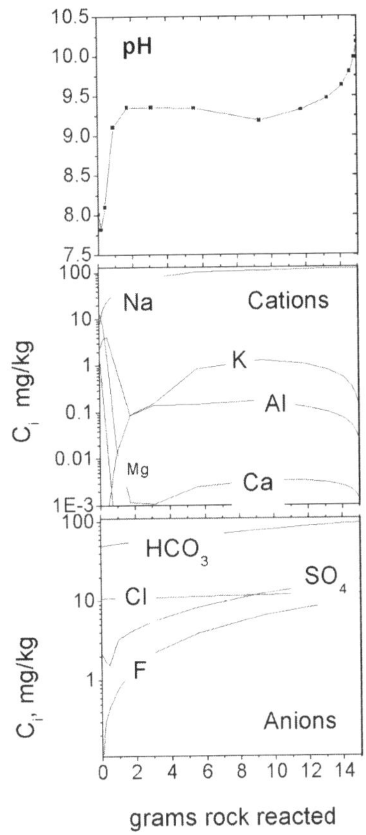

Figure 3. pH and chemical composition of the modeled water as a function of the amount of granitic rock reacted.

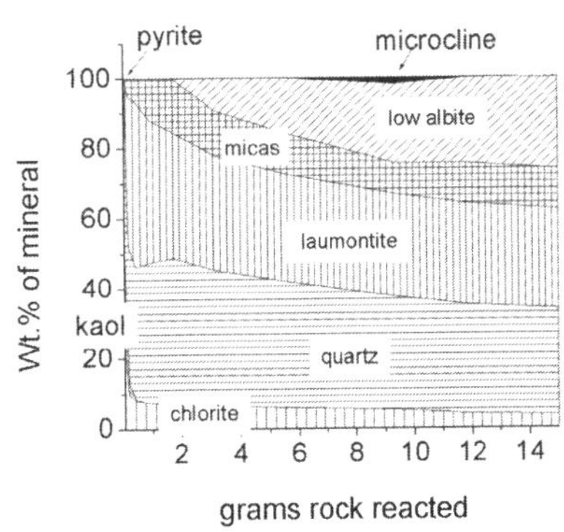

Figure 4. Mineral deposition along a "convective cell".

content because of a very low concentration of Ca which prevents precipitation of fluorite. Extremely low Ca and Mg concentrations in the modeled water are probably a consequence of the thermodynamic data used for low-temperature clay minerals.

ACKNOWLEDGEMENTS

This work was supported by grant from CONACyT project # 38578-T and partially from the CONACyT project # G35442-T.

REFERENCES

Barnes, I., O'Neil & Trescases, J.J. 1978. Present day serpentinization in New Caledonia, Oman and Yugoslavia. *Geochim. Cosmochim. Acta* 42: 144–145.

Bowers, T.S. & Taylor, H.P. 1985. An integrated chemical and stable-isotope model of the origin of mid-ocean ridge hot spring systems. *J. Geophys. Res.* 90: 12583–12606.

Bruni, J., Canepa, M., Chiodini, G., Cioni, R., Cipolli, F., Longinelli, A., Marini, L., Ottonello, G. & Zuccolini, M.V. 2002. Irreversible water-rock mass transfer accompanying the generation of the neutral, Mg-HCO_3 and high-pH Ca-OH spring waters of the Genova province, Italy. *App. Geochem.* 17: 455–474.

Cioni, R., Farelli, G., Guidi, M., Kinyariro, J.K. & Marini, L. 1992. Lake Bogoria hot springs (Kenya): geochemical features and geothermal implications. *J. Volcanol. Geotherm. Res.* 50: 231–246.

Giggenbach, W.F. 1988. Geothermal solute equilibria. Derivation of Na-K-Mg-Ca geoindicators. *Geochim. Cosmochim. Acta* 52: 2749–2765.

Grichuk, D.V. 2000. *Thermodynamic Models of Submarine Hydrothermal Systems.* Moscow, Scientific World, 304 p.

Hem, J.D. 1985. Study and interpretation of the chemical characteristics of natural waters. *US Geological Survey Water-Supply Paper* 2254, 264 p.

Herrman, U.R., Nelson, B.K. & Ratschbacher, L. 1994. The origin of a terrain: U/Pb zircon geochronology and tectonic evolution of the Xolapa complex (southern Mexico). *Tectonics* 13: 455–474.

Nordstrom, D.K., Ball, J.W., Donahoe, R.J. & Whittemore, D. 1989. Groundwater chemistry and water-rock interaction at Stripa. *Geochim. Cosmochim. Acta* 53: 1727–1740.

Reed, M. & Spycher, N. 1984. Calculation of pH and mineral equilibria in hydrothermal waters with application to geothermometry and studies of boiling and dilution. *Geochim. Cosmochim. Acta* 48: 1479–1492.

Schaaf, P. 1990. Isotopengeochemische untersuchungen an granitoiden gesteinen eines activen Kontinentaerandes. *PhD Thesis*, Munchen, 202 p.

Taran, Y.A., Inguaggiato, S., Varley, N., Capasso, G. & Favara, R. 2002. Helium and carbon isotopes in thermal waters of the Jalisco Block, Mexico. *Geofisica Int.* 41: 459–566.

Venturelli, G., Boschetti, T. & Duchi, V. 2003. Na-carbonate waters of extreme composition: Possible origin and evolution. *Geochemical J.* 37: 351–366.

Water–Rock Interaction, Wanty & Seal II (eds)
© 2004 Taylor & Francis Group, London, ISBN 90 5809 641 6

Two stages of cation exchange in the geochemical evolution of seawater intrusion into the coastal aquifer of Israel

O. Sivan
Institute of Earth Sciences, Hebrew University, Jerusalem, Israel
Geological Survey of Israel, Jerusalem Israel
Current address: Department of Earth and Planetary Sciences, Harvard University, Cambridge, MA, USA

Y. Yechieli
Geological Survey of Israel, Jerusalem, Israel

B. Herut
Israel Oceanographic and Limnological Research, National Institute of Oceanography, Haifa, Israel

B. Lazar
Institute of Earth Sciences, Hebrew University, Jerusalem, Israel

ABSTRACT: Inspection of chemical and isotopic data from saline and brackish groundwater samples at the coastal aquifer of Israel indicates that there are three main stages in the geochemical evolution of seawater intrusion toward the coastal aquifer. In the first stage (SWS water type), which is within the marine bottom sediments, anaerobic oxidation of relatively old organic carbon might occur. Later on, but still far from the fresh-saline water interface (FSI) in the aquifer, the intrusion is slightly diluted with about 15% fresh groundwater (SDS water type). During this dilution cation exchange reactions yield the sediments sites to be close to 90% from their saturation with seawater composition. At the FSI a conservative behavior or additional ion exchange were observed.

1 INTRODUCTION

The rise of sea level since the end of the last glacial period (18,000 y) caused worldwide intrusions of seawater into coastal aquifers, replacing fresh groundwater with seawater (*e.g.* Jones et al. 1999). Most studies on seawater intrusion have dealt with the water-rock interaction processes at the fresh-saline water interface (FSI, *e.g.* Magaritz & Luzier 1985; Price & Herman 1991; Appelo & Postma 1996). These studies suggest that cation exchange is a major process in the FSI, considering the cation exchange capacity (CEC) of the aquifer and the composition of the fresh and the saline water (*e.g.* Appelo & Postma 1996). In the latter study CEC was calculated according to the empirical formula developed by Breeuwsma et al. (1986):

$$\text{CEC (meq/100 g)} = 0.7 \cdot (\%\,\text{clay}) + 3.5 \cdot (\%\,\text{organic C}) \quad (1)$$

The present research describes the geochemical processes, including cation exchange, associated with the progressive intrusion of seawater into marine sediments along the coast of Israel.

The coastal aquifer of Israel consists of inter-layered sand, calcareous sandstone (Kurkar), red loam (Hamra) and marine clays of the Pleistocene age (Issar 1968), which overly impervious marine clays of the Pliocene age (the Saqiye Group).

2 METHODS

Profiles through saline, brackish and fresh groundwater horizons were taken seasonally from 1998 to 2001 from Poleg 1 research well, located in the coastal aquifer of Israel (Netanya, 134.59E; 187.08N), 70 m from the shoreline, 3.1 m above sea level. The samples were collected by a stainless steel bailer. In addition, a Multi Layer Sampler (MLS,

Ronen et al. 1987) was deployed several times to obtain very detailed chemical profiles (resolution of 10 cm) of the interface for 1–2 months each time (summer 2000, autumn 2000 and spring 2001). Additional saline samples were obtained from producing wells and observation wells along the coastal aquifer of Israel using a Grundfos submersible pump.

3 RESULTS AND DISCUSSION

Chemical compositions of the non-conservative solute species in Poleg 1 and other wells (Fig. 1) suggest that there are three distinct water types (end members) in the coastal aquifer: SWS which includes samples from the Mediterranean Sea, marine pore-waters and saline groundwaters (with at least 95% of seawater salinity); SDS which includes slightly diluted saline-groundwater with salinity of about 85% seawater salinity and Fresh Groundwater (FGW).

These compositions represent different stages in the groundwater chemical evolution. During the earliest stage of penetration (SWS water type), within marine bottom sediments, the concentrations of dissolved inorganic carbon (DIC), total alkalinity, ammonia and phosphate already increased in some cases and the $\delta^{13}C_{DIC}$ and $^{14}C_{DIC}$ decreased due to anaerobic oxidation of relatively old organic carbon.

The next stage was dilution with about 15% fresh groundwater far from the FSI (SDS water type). During this dilution the SWS was chemically modified to SDS that was enriched with Ca^{2+} (excess of 10 meq L^{-1}) and Sr^{2+} and depleted by ~50% in K^{+} and slightly with the Na^{+} as compared to 15% conservative dilution of seawater (Fig. 1). These changes were caused by a "chromatography-like" cation exchange mechanism that yielded sediments to be close to 90% from their saturation with seawater composition.

The Kurkar country rock in the Poleg 1 area contained up to 0.3–1% clays and 0.8% organic carbon. This means that the minimum CEC was 3 meq per 100 g of dry sediment (ds) according to Eq. 1. This CEC, expressed in units of meq L^{-1} groundwater (gw), was found to be 185 meq L^{-1}, assuming a specific density of 2.65 g cm^{-3} and a porosity (ϕ) of 0.3 (100 g ds associated $37.7 \cdot 10^{-3}$ L ds and $16.2 \cdot 10^{-3}$ L gw ($37.7 \cdot 10^{-3} \cdot \phi/(1-\phi)$), and therefore $3/16.2 \cdot 10^{-3} = 185$ meq L^{-1} gw).

We calculated that that CEC could easily provide the excess of 10 meq L^{-1} of Ca^{2+}. The calculation used the parameter β_I, which is the equivalent fraction of the exchangeable sites for ion I^{i+} $(I - Xi)$ from the total capacity of the exchangeable sites. An additional parameter is $K_{Na\backslash I}$, the exchange coefficient of I with respect to Na^{+}. Based on our analyses

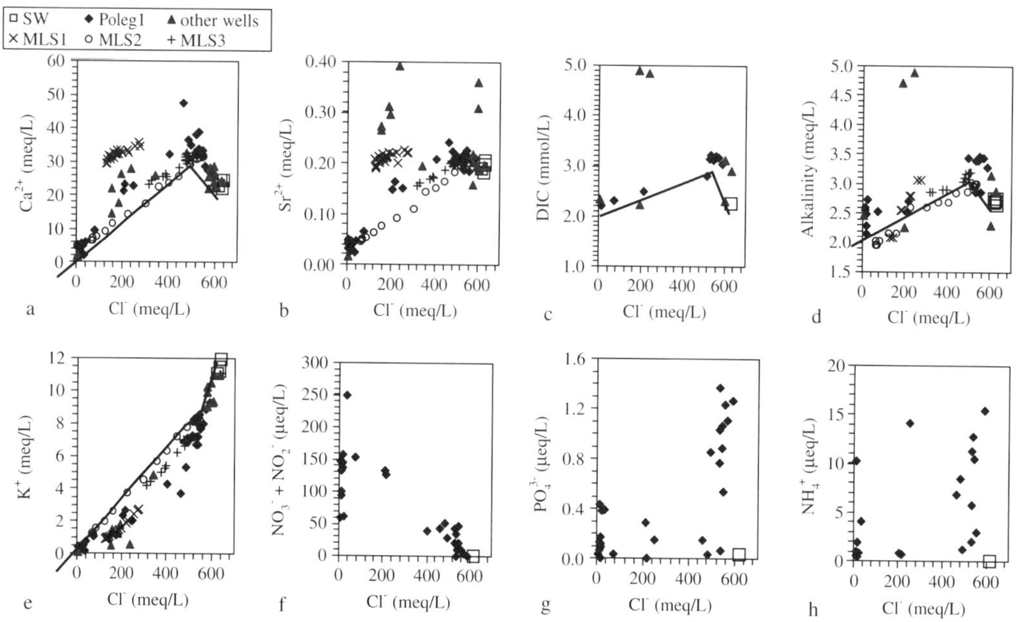

Figure 1. Chemical composition of the non-conservatives parameters *vs* Cl^- in Poleg 1 (by bailer or by MLS) and other wells (a) Ca^{2+}; (b) Sr^{2+}; (c) DIC (dissolved inorganic carbon); (d) total alkalinity; (e) K^{+}; (f) NO_3^-, (g) PO_4^{3-}; and (h) NH_4^+. The vertical and horizontal bars are smaller than the symbols. The curves are schematic mixing lines between the three end-members: SWS, SDS and FGW.

we assumed that the main cations were Na^+, Ca^{2+}, K^+ and Mg^{2+}. Therefore the sum of their β was equal to 1, and each β could be calculated according to Appelo & Postma (1996):

$$\beta_{Mg} = \frac{\beta_{Na}^2 \cdot [Mg^{2+}]}{K_{Na\backslash Mg}^2 \cdot [Na^+]^2} \quad (2)$$

$$\beta_{Ca} = \frac{\beta_{Na}^2 \cdot [Ca^{2+}]}{K_{Na\backslash Ca}^2 \cdot [Na^+]^2} \quad (3)$$

$$\beta_k = \frac{\beta_{Na} \cdot [K^+]}{K_{Na\backslash K} \cdot [Na^+]} \quad (4)$$

$$\beta_{Na}^2 \cdot \left(\frac{[Mg^{2+}]}{K_{Na\backslash Mg}^2 \cdot [Na^+]^2} + \frac{[Ca^{2+}]}{K_{Na\backslash Ca}^2 \cdot [Na^+]^2} \right) + \frac{\beta_{Na} \cdot [K^+]}{K_{Na\backslash K} \cdot [Na^+]} + \beta_{Na} - 1 = 0 \quad (5)$$

Equations 2–5 were used to calculate β values for FGW and seawater using $K_{Na/Ca}$ of 0.4, $K_{Na/Mg}$ of 0.5 and $K_{Na/K}$ of 0.2 (after Gaines & Thomas 1953; Appelo & Postma 1996), and the molar concentrations of the relevant cations in FGW and seawater. The calculated β values for FGW were: $\beta_{Na} = 0.093$, $\beta_{Mg} = 0.353$, $\beta_{Ca} = 0.554$ and $\beta_K = 0$; and for seawater (SW) the values were: $\beta_{Na} = 0.571$, $\beta_{Mg} = 0.281$, $\beta_{Ca} = 0.084$ and $\beta_K = 0.062$. The calculated $\Delta\beta_i$, for the above cations explained the depletion in Na^+ and K^+ (having $\Delta\beta_{Na} = 0.478$, $\Delta\beta_K = 0.062$) and the gain in Ca^{2+} and Mg^{2+} (having $\Delta\beta_{Ca} = -0.470$, $\Delta\beta_{Mg} = -0.072$). The magnitude of the change in concentration for each of the cation can be estimated from the product $\Delta\beta_i \cdot CEC$. According to this calculation, the maximum possible increase in Ca^{2+} is 87 meq L^{-1}. Therefore, the observed 10 meq L^{-1} Ca^{2+} excess in Poleg and the other wells across the coastal aquifer fits the condition in which almost 90% of the exchange sites are already released in the early steps (at the end of the "washing" the cations show conservative behavior; Appelo & Postma 1996).

At the FSI, the major ions generally exhibited conservative behavior (Fig. 1). This indicates that the "column" (the sediments) was already saturated with the "flushing" water when the samples were taken. One profile (September 2000), however, did show enrichment in Ca^{2+} and Sr^{2+} and depletion in K^+. In this case the profiler was deployed at the beginning of the summer rise of the FSI when saline groundwater started to replace the fresh groundwater releasing Ca^{2+} and Sr^{2+} due to ion exchange. Later on, during that summer ("MLS 2") and after the continued flushing of the aquifer caused to all the possible exchangeable adsorbed Ca^{2+} to be released, the chemical composition of the FSI water returned to conservative behavior. Lowering again the FSI due to recharging the aquifer recharge with meteoric waters started a new cycle of Ca^{2+} and Sr^{2+} "loading" on the sediments that would be released again during the next FSI rise. In general, samples that show more intensive ion exchange represent the early stage of the system, where the column is far from saturation, or might be connected to higher content of organic matter and clays in the aquifer.

Inspection of chemical data from coastal aquifers around the world indicates that these main paths of intrusion, and intensive ion exchange with slight dilution with fresh groundwater, are global phenomena.

REFERENCES

Appelo, C.A.J. & Postma, D. 1996. Ion exchange and sorption. In *Geochemistry, groundwater and pollution*: 142–201. Balkema: Rotterdam.

Breeuwsma, A., Wosten, J.H.M., Vleeshouwer, J.J., Van Slobbe, A.M. & Bouma, J. 1986. Derivation of land qualities to assess environmental problems from soil surveys. *Soil Sci. Soc. Am. J.* 50: 186–190.

Gaines, G.L. & Thomas, H.C. 1953. Adsorption studies on clay minerals. II. A formulation of the thermodynamics of exchange adsorption. *J. Chem. Phys.* 21: 714–718.

Issar, A. 1968. Geology of the central coastal plain of Israel. *Israel J. Earth Sci.* 17: 16–29.

Jones, B.F., Vengosh, A., Rosenthal, E. & Yechieli, Y. 1999. Geochemical investigations. In: Seawater intrusion in coastal aquifers – concepts, methods and practices. Kluwer Academic Publishers, The Netherlands, pp 51–72.

Magaritz, M. & Luzier, J. 1985. Water-rock interactions and seawater-freshwater mixing effects in the coastal dunes aquifer, Coos Bay, Oregon. *Geochim. Cosmochim. Acta.* 49: 2515–2525.

Price, R.M. & Herman, J.S. 1991. Geochemical investigation of salt-water intrusion into a coastal carbonate aquifer: Mallorca, Spain. *Geol. Soc. Am. Bull.* 103: 1270–1279.

Ronen, D. Magaritz, M. & Levy, I. 1987. An *in situ* multilevel sampler for preventive monitoring and study of hydrochemical profiles in aquifers. *Groundwater Monitoring Review* 7: 69–74.

Water–Rock Interaction, Wanty & Seal II (eds)
© 2004 Taylor & Francis Group, London, ISBN 90 5809 641 6

Experimental and numerical simulation of carbonate dissolution and porosity development by organic acids

H.-Y. Tseng & G.A. Otten
ExxonMobil Upstream Research Company, Houston, Texas, USA

V. De Choudens-Sánchez
University of Kansas, Lawrence, Kansas, USA

A.B. Herhold
ExxonMobil Research and Engineering Company, Annandale, New Jersey, USA

ABSTRACT: A series of flow-through experiments were conducted to identify and quantify the effects of acetic acid on secondary porosity generation in carbonate reservoirs during deep burial. For the acetic acid concentrations studied (typical of oil field waters), dissolution occurred as a sharp reaction front on a core scale. Along the reaction front, acetic acid dissolved the material by attacking calcite cleavage planes. Calcium concentrations measured from effluent samples indicated that the pore fluid reached its equilibrium concentration rapidly. Dissolution resulted in a bulk porosity increase of less than 0.5% after 14 days, contributed mostly by the dissolution at the core inlet. The experimental results validated predictions of reactive transport models and the theoretical derivation of dissolution patterns. Mass balance calculations suggest that the porosity increases by organic acid dissolution are not volumetrically important.

1 INTRODUCTION

Generation of secondary porosity in carbonate reservoir rocks is a complex problem. Surdam et al. (1984) and Mazzullo & Harris (1992) have suggested that, in the deep subsurface, chemically aggressive fluids evolved during organic diagenesis (organic matter maturation and subsequent hydrocarbon degradation, including thermochemical sulfate reduction) may increase porosity by enhancing pre-existing pores or by creating new pores in carbonate rocks. These fluids are charged with organic acids, carbon dioxide and hydrogen sulfide. The concentrations of organic acids, specifically acetic acid, could range up to 10,000 ppm, dominating the formation water alkalinity and controlling carbonate mineral stability. Such diagenetic changes can be related temporally and spatially to the burial depth-temperature-hydrologic history of sedimentary basins and to their hydrocarbon maturation-migration-destruction history. The generated late-stage pores are generally rounded and cut all textural elements. They can modify significantly or even obscure the relationship between reservoir properties and the original sediment texture. Failure to consider the effects of aggressive pore fluids on late-stage dissolution could result in underestimation of reservoir quality and errors in volume calculations and reservoir performance prediction.

At this point, we do not have enough data to fully evaluate the impact of late organic diagenetic processes on porosity and permeability evolution in carbonate rocks. Some have argued that the porosity enhancement may be limited by (1) insufficient quantity of organic acids generated (Lundegard and Land 1986) and (2) neutralization reactions attributed to long-distance migration pathway and slow fluid flow rates (Giles et al. 1994). When considering influx of carbonic acid, Morse & Mackenzie (1993) argued that reaction conditions are likely to generate sharp reaction fronts on a reservoir scale. Fredd & Fogler (1998) have conducted reactive flow experiments with various acids in limestone cores, including acetic acid, but mostly under flow rates much higher than subsurface systems. Experimental and numerical studies to quantify the effect of organic acids on the magnitude of carbonate dissolution, subsequent porosity/permeability development, and textural changes under conditions close to subsurface environments will help to constrain the geologic problem.

The main objective of this study was to evaluate quantitatively the potential impact of organic acids on the evolution of pore systems in mineralogically stabilized limestone under simulated deep burial conditions. We conducted laboratory experiments and numerical reactive transport modeling on calcium carbonate reacting with a range of organic acid concentrations at constant temperature, pressure, flow rates and fluid composition. Acetic acid, a monocarboxylic acid, was chosen because it is the most abundant organic acid in oilfield waters. Experimental results were used to validate predictions from computational models. The models were then used to define the boundary conditions under which such interactions may be significant in subsurface environments. Textural observations of experimental products identify dissolution features by organic acids in the rock record. Results discussed here confirm the theoretical models describing transport and reaction of reactive fluids in porous media (Daccord et al. 1993) and can be applied to predict carbonate dissolution patterns under various burial conditions.

2 METHODS

2.1 *Experimental procedure*

The experimental apparatus consisted of a pressure vessel mounted inside an oven, an influent reservoir and an outlet sampling port. The externally heated pressure vessel contained a reaction cell with independently controlled confining pressure and pore-fluid pressure. A manually operated back-pressure regulator was used to extract reacted pore fluid at constant pressure from the outlet sampling port. Unreacted fluid in the influent reservoir was simultaneously added at the vessel inlet by an ISCO syringe pump at a controlled flow rate. Aqueous solutions flowed through a cylindrical (1 cm diameter × 2.5 cm length) limestone core placed in the reaction cell at a constant flow rate of 0.001–0.002 mL/min. The experiment was performed at 100°C and 2000 psi pore pressure to simulate deep burial temperature-pressure regimes.

A series of experiments was conducted on specimens of the same rock type: isolated oomoldic grainstones with an average porosity of 13% and permeability (N_2) of 0.3–1.1 mD. The size of the isolated molds ranged from 50 to 500 μm, where the size of the microporosity viewed with the SEM was less than about 20 μm. X-ray diffraction analysis indicated the presence of a single major phase, low-Mg calcite, and a trace amount of dolomite.

Aqueous solutions were prepared with deionized water, 500–1000 ppm acetic acid, and a sodium-acetate pH buffer. The concentration of total acetate in the solution was chosen to be around 1000 ppm, which is typically of oilfield waters (Surdam et al. 1984). The initial pH values at 100°C (3.3–4.7) were close to the range of calculated values (4.0–5.5) for reservoir waters at subsurface temperatures corrected for known amounts of degassing of CO_2 and H_2S (Kharaka et al. 1986).

Prior to initiation of flow, the pore space was pressurized with N_2. The first solution showed at the outlet after 1–2 days, and the effluent sample was collected each day for 11–15 days. Both the influent and the effluent samples were analyzed for their calcium concentration using a pH/ion meter with a calcium ion selective electrode and Inductively Coupled Plasma-Atomic Emission Spectroscopy (ICP-AES). The changes in the pore structure of samples prior to and after experiments were examined using scanning electron microscopy (SEM) and X-ray computer tomography.

2.2 *Reactive transport modeling (RTM)*

The reactive transport modeling performed for this study used GWB/Xt (Bethke 2001; 2002) to calculate the amount of calcite dissolution expected under experimental conditions, to assess the effects of dissolution on the rock matrix, and to determine whether consequent changes in rock properties can, in turn, affect the flow system. Figure 1 presents a flow chart summarizing the modeling approach that simulated the experimental procedure. The model employed the Debye-Hückel method to calculate activity coefficients and the saturation states of calcite in aqueous solutions sampled from the experiments. A simple rate law, $R = (A/V)(k)(1 - Q/K)$ where A is the mineral's surface area, V is the volume of solution, k is the rate constant, and Q and K are the activity product and equilibrium constant, was applied to simulate kinetics of calcite dissolution (Morse & Arvidson 2002). We assumed that calcite dissolution rates for conditions far from equilibrium

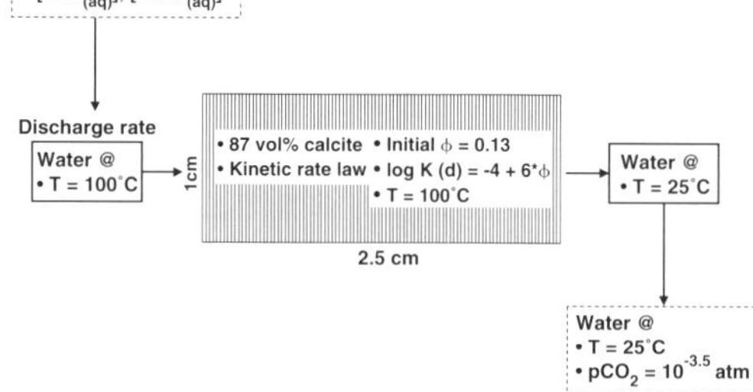

Figure 1. Flow chart showing reactive transport modeling approach.

ranged from 10^{-8} to 10^{-9} mol/cm²/s at pH = 3.3–4.7 (Arvidson et al. 2003). The influence of temperature on calcite reaction kinetics in the pH range of this study is relatively small (Plummer et al. 1979) and, thus, was neglected in our calculation. Model results calculated from short-term simulations were calibrated by experimental results.

3 RESULTS AND DISCUSSION

3.1 *Experimental and RTM results*

Experimental results indicated a complete dissolution of the sample at the fluid–rock reaction front, consistent with the RTM predictions. The sharp front arises because the reaction rate is very fast compared to the flow rate, hence the fluid equilibrates within each pore and the mass removal is limited solely by convection. The result is qualitatively similar to the low-flow-rate results of Fredd & Fogler (1998). Because of the relatively short testing time (<15 days) and the small amount of total fluid (<25 cc.), only the entry face of the sample was affected. Before the experiments, calcite crystals at the reaction front exhibited a rhombohedral crystal habit with no evidence of dissolution along surface planes. After the experiments, the sample surface was covered by needle-shaped fabric, and the original morphology of the calcite crystals could not be distinguished (Fig. 2). A close inspection of the needle orientation suggested they were the remnants of calcite crystal dissolution along cleavage planes. The thickness of the reaction front where such dissolution textures occurred was less than 50 μm, too small to be clearly identified under the optical microscope. Some roughness to the reaction front was also observed, implying that heterogeneity in the pore sizes and shapes influenced the dissolution.

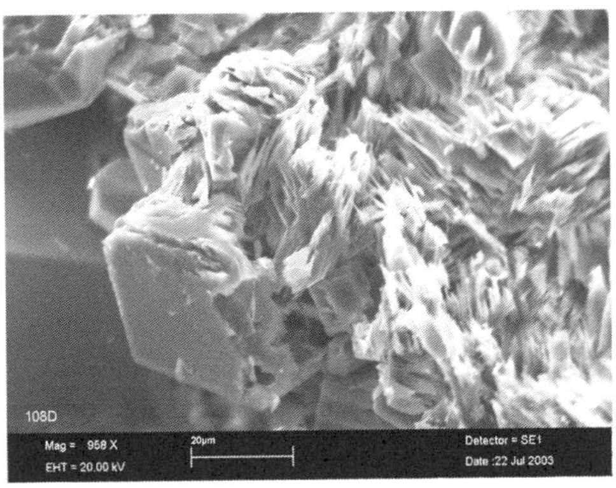

Figure 2. SEM microphotograph of reaction front.

Calcium concentrations measured from effluent samples were relatively constant with time. Because of the high ratios of mineral surface area to fluid volume and low flow rates used in this experimental system, the rock sample dissolved rapidly, and the pore fluid reached its equilibrium concentration in less than one day. The steady-state effluent concentrations of calcium averaged 360 ppm for the 1000 ppm acetic acid influent (pH 3.3) and 240 ppm for the 500 ppm acetic acid influent with 500 ppm sodium acetate buffer (pH 4.7).

The volume of dissolved calcite can be directly related to the bulk porosity increase. The porosity change as inferred from mass loss was calculated using the measured calcium concentration, the volume of effluent, and the calcite density of 2.71 g/cc. Results indicated a porosity increase of less than 0.5% after 14 days, contributed mostly by the dissolution at the core inlet (Table 1). The experimental results validated the estimations of the RTM simulations.

3.2 *Geologic implications*

3.2.1 *Dissolution patterns*

The occurrence of a sharp dissolution front observed under our experimental conditions can be described by a theoretical phase diagram of dissolution patterns that includes the combined effects of fluid convection, mass transport and surface reaction processes in porous media (Daccord et al. 1993). The phase diagram is defined by two dimensionless parameters: the Peclet number, $N_{pe} = vL/D$, where v is the interstitial velocity, L is the characteristic pore size, and D is the diffusion constant; and the product of the Peclet and Damköhler numbers, $N_{ki} = K_s L/D$, where K_s is the reaction rate (Fig. 3). Three different reactive-flow regimes (convection-limited, mass transport-limited and surface-reaction-limited) correspond to three types of dissolution patterns observed (compact or front reaction, wormhole and uniform dissolution). Given a set of initial experimental conditions (i.e. flow rate, diffusivity, pore size, dissolution rate

Table 1. Comparison of porosity increase% due to organic acid dissolution.

Experiment	Lab measurement	GWB/X1t Model
Exp. 1. (pH 3.3) 1000 ppm HAc 0.002 ml/min	0.42% (11 days)	0.44% (11 days)
Exp. 2. (pH 3.3) 1000 ppm HAc 0.001 ml/min	0.29% (14 days)	0.29% (14 days)
Exp. 3. (pH 4.7) 1000 ppm HAc + NaAc 0.001 ml/min	0.17% (14 days)	0.18% (14 days)

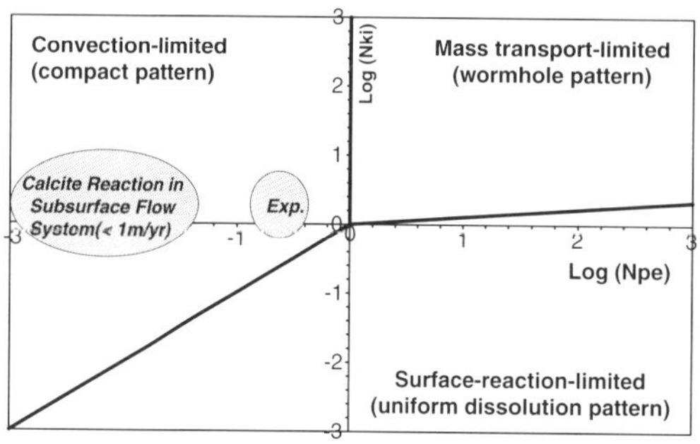

Figure 3. Convection vs. reaction behavior diagram.

and solubility), the calculated corresponding points $\{N_{pe}, N_{ki}\}$ for our experiments lie in the zone of convection-limited regime, which yields compact dissolution patterns. Since the geometry of the pore system changes with time due to the reaction, the pore size varies, and, consequently, the trajectory of the experimental points in the plane will also change. Keeping all the parameters constant except flow rates, we found that, on the matrix or pore scale, carbonate dissolution kinetics in deep subsurface flow systems (<1 m/yr) are limited by convection, resulting in compact dissolution patterns as simulated experimentally and numerically (Fig. 3).

3.2.2 *Mass balance calculation for a geologic scenario*

A mass balance calculation was performed to determine the impact that acetic acid production in a source rock could have on the porosity evolution in a nearby carbonate reservoir. We based these calculations on the assumption that 500 L of water flow through 740,000 cc of source rock with 5% kerogen in one million years and that each gram of kerogen produces 5 mg of acetic acid. In one million years, the source rock will produce 500 L of 1000 ppm acetic acid. Based on the stoichiometry of the dissolution reaction (and validated by the experiments described in this paper), 1 mL of 1000 ppm HAc will dissolve 3.4×10^{-4} cc of calcite, so 500 L of 1000 ppm acetic acid will dissolve approximately 170 cc of calcite. For a reservoir with volume equal to the source rock (740,000 cc) and a porosity of 20%, organic acid dissolution would cause a porosity increase of 0.03%. If we assume the source rock is a hundred times the size of the reservoir rock, the increase in reservoir porosity is about 3%.

These estimated porosity increases are not very important in a volumetric sense. In nature, however, the rapid organic acid dissolution rates and heterogeneous flow profiles due to the broad pore size distribution of carbonate rocks (i.e. fractures, stylolites, pore structures created in the shallow diagenetic environment) could lead to the formation of highly conductive flow channels. The dissolution may become important when it is concentrated along such high-permeability pathways. Enhanced permeability along such conduits may facilitate oil migration.

4 CONCLUSION

From this study, we conclude that

- In natural systems, organic acids may contribute to the enhancement of permeability in carbonates by dissolving material along high-permeability pathways. Any increase in matrix porosity is not volumetrically important.
- The dissolution textures observed in these experiments may be used to identify dissolution by organic acids in the rock record.
- For carbonates with homogeneous composition and pore structure, rapid dissolution by organic acids at low flow rates in the subsurface results in compact dissolution patterns on a core scale. However, an analytical reactive flow phase diagram suggests that increased heterogeneity of the rock matrix could promote the development of wormhole structures (highly conductive flow channels) under certain conditions. As a result, the porosity and permeability of reservoirs could be enhanced by aggressive dissolution along such high permeability pathways.

ACKNOWLEDGEMENTS

The authors thank their colleagues Roy Brown for SEM examination, S.D. Cameron for ICP measurements and J.H. Dunsmuir for tomography analysis. They also thank M.M. Thomas and D. Ertas for their valuable discussions and suggestions.

REFERENCES

Arvidson, R.S., Ertan, I.E., Amonette, J.E. & Luttge, A. 2003. Variation in calcite dissolution rates: A fundamental problem? *Geochim. Cosmochim. Acta* 67(9): 1623–1634.

Bethke, C.M. 2001. The Xt model of transport in reacting geochemical systems: Reference manual for X1t, X2t, and Xtplot. Hydrogeology Program, University of Illinois.

Bethke, C.M. 2002. The Geochemist's Workbench release 4.0: A user's guide to Rxn, Act2, Tact, React, and Gtplot. Hydrogeology Program, University of Illinois.

Daccord, G., Liétard, O. & Lenormand, R. 1993. Chemical dissolution of a porous medium by a reactive fluid – II. convection vs reaction behavior diagram. *Chem. Eng. Sci.* 48(1): 179–186.

Fredd, C.N. & Fogler, H.S. 1998. Influence of transport and reaction on wormhole formation in porous media. *Am. Inst. Chem. Eng. J.* 44(9): 1933–1949.

Giles, M.R., de Boer, R.B. & Marshall, J.D. 1994. How important are organic acids in generating secondary porosity in the subsurface? In E.D. Pittman & M.D. Lewan (eds), *Organic acids in geological processes:* 449–470. Springer Verlag.

Kharaka, Y.K., Law, L.M., Carothers, W.W. & Goerilitz, D.F. 1986. Role of organic species dissolved in formation waters from sedimentary basins in mineral diagenesis. In D.L. Gautier (ed.), *Roles of organic matter in sediment diagenesis:* 111–122. Society of Economic Paleontologists and Mineralogists Special Publication No. 38.

Lundergard, P.D. & Land, L.S. 1986. Carbon dioxide and organic acids: Their role in porosity enhancement and cementation, Paleogene of the Texas Gulf Coast. In D.C. Gautier (ed.), *Roles of organic matter in sedimentary diagenesis*: 129–146. Society of Economic Paleontologists and Mineralogists Special Publication No. 38.

Mazzullo, S.J. & Harris, P.M. 1992. Mesogenetic dissolution: Its role in porosity development in carbonate reservoirs. *Am. Assoc. Petrol. Geol. Bull.* 76(5): 607–620.

Morse, J.W. & Mackenzie, F.T. 1993. Geochemical constraints on $CaCO_3$ transport in subsurface sedimentary environments. *Chem. Geol.* 105: 181–196.

Morse, J.W. & Arvidson, R.S. 2002. The dissolution kinetics of major sedimentary carbonate minerals. *Earth Sci. Rev.* 58: 51–84.

Plummer, L.N., Parkhurst, D.L. & Wigley, T.M.L. 1979. Critical review of the kinetics of calcite dissolution and precipitation. In E. Jenne (ed.), *Chemical modeling – speciation, sorption, solubility and kinetics in aqueous systems*: 537–573. American Chemical Society, Washington, D.C.

Surdam, R.C., Boese, S.W., & Crossey, L.J. 1984. The chemistry of secondary porosity. In D.A. McDonald & R.C. Surdam (eds), *Clastic diagenesis. American Association of Petroleum Geologists Memoir* 37: 127–149.

Water–Rock Interaction, Wanty & Seal II (eds)
© 2004 Taylor & Francis Group, London, ISBN 90 5809 641 6

Controls on REE during groundwater flow and mixing in fractured rock aquifers

S.O. Tweed[1,2] & T.R. Weaver[1]
[1]*Hydrogeology and Environment Research Group, School of Earth Sciences, University of Melbourne, Victoria, Australia*

I. Cartwright[2]
[2]*Hydrogeology and Environment Research Group, Department of Earth Sciences, Monash University, Clayton, Victoria, Australia*

ABSTRACT: When used in conjunction with major ion chemistry, rare earth elements (REE) can refine understanding of groundwater flow in and between different aquifers in complex fractured-rock systems. Within the Dandenong Ranges, SE Australia, groundwater REE patterns in local flow systems reflect dissolution from basalt or sedimentary host rocks during recharge. Where groundwater pH is ≤ 6, REE concentrations are higher (ΣREE: 29 to 119 ng/L), reflect local mineralogy, and sorption is lower. Where pH is >6 REE concentrations are lower (ΣREE: 14 to 28 ng/L) and are derived from mineral dissolution that occurred up-gradient in lower pH environments. Groundwater flow from recharge areas to the regional sedimentary aquifer (higher pH environment), results in progressive sorption of REE and groundwater mixing affecting REE patterns. Here, REE concentrations reflect early-stage mineral dissolution so they are useful tracers of mixing that could not be distinguished from the accumulated major ion concentrations.

1 INTRODUCTION

This study addresses the issues of time-integrated processes controlling REE concentrations along groundwater flow paths, and the effectiveness of using REE as regional groundwater tracers. Previous research into REE behavior in groundwater during low-temperature water–rock interaction has presented important links between source rock inheritance and subsequent solution and surface complexation that affect groundwater REE signatures (e.g. Smedley 1991; Johannesson et al. 1999; Biddau et al. 2002). In addition, understanding whether REE signatures represent accumulative, most-recent, or early-stage water rock interaction along groundwater flow paths has important consequences for using REE to trace groundwater in multi-aquifer systems.

The area investigated is the Dandenong Ranges sub-catchment, Victoria, Australia (Fig. 1). Major aquifers include Silurian-Devonian marine sediments, Devonian acidic volcanics, and Tertiary Older Volcanic basalts (Fig. 1). The groundwater system is characterized by fracture-hosted flow, is relatively fresh (TDS = 59–595 mg/L), and has a pH range of 4.8–8.6 (Tweed et al. 2003a). To determine REE

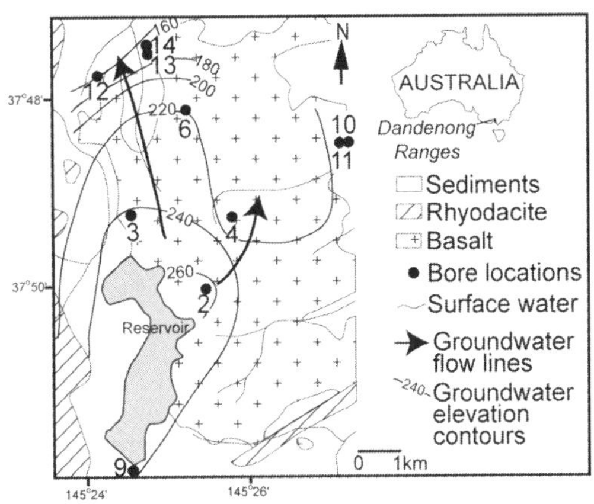

Figure 1. Geology of the Dandenong Ranges sub-catchment and location of groundwater samples for REE analysis (after Tweed et al. 2003b). Groundwater elevation m AHD (Australian height datum) contours shown for summer (December–February) 2000.

systematics along defined flow paths, REE chemistry is compared to major ion chemistry, and the effects of host rock lithology, speciation, sorption, and mixing on REE patterns and yttrium (Y) are identified.

2 METHOD

Groundwater samples from 10 boreholes located throughout the Dandenong Ranges (Fig. 1) were collected using a micropurge technique with a submersible pneumatic bladder pump after pH, Eh and electrical conductivity (EC) had stabilized. Samples were filtered in the field through 0.45 μm cellulose-acetate filters and were then acidified to pH < 2 using ultrapure 6 M HCl. Groundwater samples were pre-concentrated, the method adapted for this research was originally developed by Shabani et al. (1992) and is described in detail in Tweed et al. (2003b). Samples were analysed at the School of Geosciences, Monash University, using a Finnigan Mat Element HR-ICP-MS.

3 RESULTS AND DISCUSSION

REE_{CN} (chondrite normalized, e.g. Nakamura 1974) patterns are presented in Figure 2. Dominant complexes for groundwater samples, and whether LREE (e.g. Nd) or HREE (e.g. Yb) abundances are greater, are shown in Figure 3. Relative abundances of REE complexes were calculated using PHREEQC (Parkhurst & Appelo 1999). PHREEQC modeling indicates that REE are mostly present as free ions or carbonate complexes.

3.1 *REE variability in groundwater from the basalt aquifer*

All groundwater sampled from the basalt aquifer shows similar REE patterns. These patterns are similar to REE signatures of basalt rocks throughout Victoria (Fig. 2a). Both water and rock show enrichment in LREE, and $(Yb/Nd)_{CN}$ ratios in groundwater (0.32 to 0.42) are within the range for the basalt rocks ($(Yb/Nd)_{CN}$: 0.19 to 0.64).

The major difference between groundwater and the basalt rocks is the relative enrichment of yttrium (Y), Gd and Lu in groundwater (Fig. 4). REE speciation modeling indicates that, within the pH range, sorption of REE has the potential to occur in groundwater from samples BHs 2, 3 and 6 to kaolinite, montmorillonite, and FeO(OH) (e.g. Drever 1997; Fig. 3). This could produce the observed Y, Gd and Lu enrichment (e.g. Möller et al. 1998; Bau 1999). In BHs 2, 3 and 6, both LREE and HREE are present as positively charged carbonate complexes and free ions. Therefore, during sorption these REE will

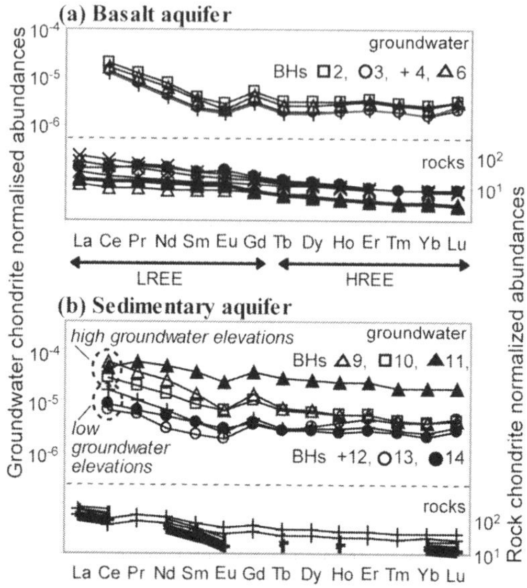

Figure 2. Chondrite normalized REE abundances for groundwater samples in (a) basalt and (b) sedimentary aquifers. Rock REE data is from similar lithologies (from Tweed et al. 2003b).

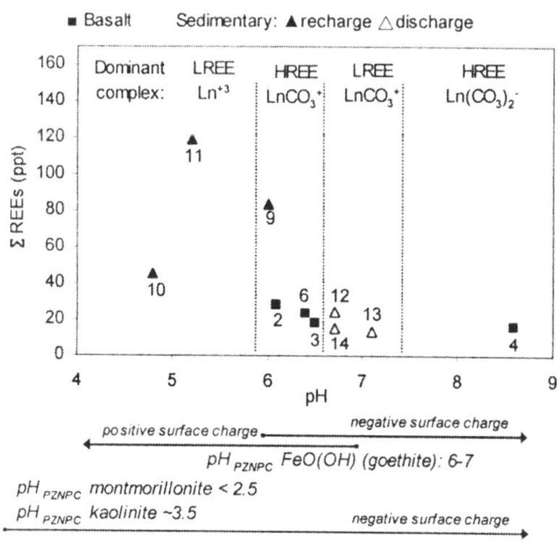

Figure 3. Decreasing REE concentrations with increasing pH. Dominant REE complexes and whether there is a greater percentage of LREE or HREE are shown. LREE are represented by % Nd complex and HREEs are represented by % Yb complex. The points of zero net proton charge (PZNPC; Drever 1997) for goethite (FeO(OH)), montmorillonite and kaolinite also are shown.

behave similarly, resulting in the limited fractionation observed between LREE and HREE. Groundwater at BH 4 is more basic (pH = 8.6), and speciation modeling indicates that REE occur

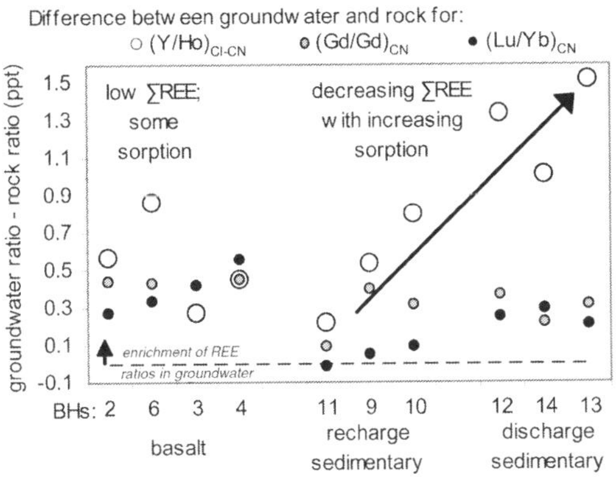

Figure 4. Degree of sorption of REE in groundwater relative to similar host rocks is represented by the differences in $(Y/Ho)_{Cl\text{-}CN}$, $(Gd/Gd)_{CN}$ which is $Gd_{CN}/[0.33Sm_{CN} + 0.6Tb_{CN}]$ (Bau & Dulski 1996), and $(Lu/Yb)_{CN}$ ratios in groundwater to the maximum ratio values for the rocks (rock data from references in Fig. 2).

predominantly as $Ln(CO_3)_2^-$ (Fig. 3). Under these pH conditions, the net surface sites will be negatively charged, although there may be specific sorption onto positive sites. So the REE are expected to behave relatively conservatively. However, compared with groundwater from other areas of the basalt aquifer, groundwater from BH 4 has a similar fractionation of Y and Gd, and only slightly higher fractionation of Lu (Fig. 4). This similar REE fractionation indicates input of groundwater from up-gradient in the basalt aquifer.

Groundwater within the basalt aquifer flows from BHs 2 and 3, to 4 and 6 (Fig. 1). From BHs 2 and 3 to 6 the fractionation of $(Y/Ho)_{Cl\text{-}CN}$ (Cl chondrite normalised, e.g. Anders & Grevesse 1989) in groundwater increases slightly, and from BH 2 to 4, fractionation of $(Lu/Yb)_{CN}$ increases with increasing pH values (Fig. 4). This is probably due to progressive sorption along the flow path, although not enough to significantly alter REE patterns. The overall consistency of REE patterns and concentrations (Fig. 2a) as groundwater interacts with the basalt aquifer indicates fairly conservative REE behavior within this aquifer. Therefore, REE in solution may reflect early dissolution or desorption during infiltration of slightly acidic rainwater into the weathered basalt.

Groundwater from the basalt aquifer has low total REE concentrations (18 to 28 ng/L) compared with groundwater from recharge areas of the sedimentary aquifer (46 to 119 ng/L). However, fractionation of REE during sorption is not significantly higher in groundwater from local flow paths in the basalt aquifer (BHs 2 to 6) compared with the sedimentary aquifer (BHs 9 to 11; Fig. 4). Therefore lower REE concentrations in groundwater from the basalt aquifer may reflect lower REE concentrations typical of basalts (Fig. 2a; ΣREE: 32 to 168 ppm) compared to those of sediments (Fig. 2b; ΣREE: 119 to 338 ppm).

In contrast to REE, groundwater interaction with the basalt aquifer results in the ongoing accumulation of major ions, as reflected in increasing TDS contents (132–288 mg/L). Also, REE signatures of groundwater interacting with the basalt aquifer are maintained in areas where specific major ion concentrations indicate mixing with groundwater from the deeper sedimentary aquifer (e.g. BHs 2 and 4) (Tweed et al. 2003a). This suggests limited input of REE from the regional sedimentary aquifer.

3.2 *REE variability in groundwater from the sedimentary aquifer*

REE patterns in groundwater from the sedimentary aquifer show the most variability (Fig. 2). The highest total REE concentrations (46 to 119 ng/L: BHs 9 to 11 occur in groundwater from recharge areas, which has the lowest TDS contents (59 to 82 mg/L) and lowest pH values (4.8 to 6.0). These total REE concentrations are significantly greater than those in groundwater from the regional sedimentary aquifer in discharge areas (14 to 25 ng/L: BHs 12–14; Fig. 2b), where TDS contents (267 to 595 mg/L) and pH (6.7 to 7.1) are higher.

3.2.1 *Recharge areas*

In areas of higher groundwater elevation in the sedimentary aquifer, one sample (BH 9) is relatively enriched in LREE ($(Yb/Nd)_{CN}$: 0.15) compared with groundwater from BHs 10 and 11 ($(Yb/Nd)_{CN}$: 0.27 to 0.34) and rocks similar to the host aquifer ($(Yb/Nd)_{CN}$: 0.20 to 0.38; Fig. 2b). All of these samples show fractionation of Y and Gd compared to similar host rocks (Fig. 4), indicating that some sorption occurs. In the slightly acidic groundwater at BHs 10 and 11, REE are present predominantly as free ions, and sorption would be restricted to clay sites (Fig. 3). Groundwater from BH 9 has a slightly higher pH (6.0), and LREE are mostly present as free ions although HREE are mostly present as carbonate ions. Under these pH conditions, REE have the potential to sorb to both clays and iron oxyhydroxides (Fig. 3) and sorption will result in little fractionation between LREE and HREE.

3.2.2 *Discharge areas*

Flow from recharge to discharge areas in the sedimentary aquifer results in increased pH values and TDS contents due to increased water-rock interaction, but the REE concentrations decrease (Fig. 2b). Groundwater from the discharge areas of the sedimentary aquifer (BHs 12 to 14) is characterised by low REE concentrations and variable REE patterns

(Fig. 2b). These groundwater samples have higher positive Y, Gd and Lu anomalies than similar host rocks and groundwater from the recharge area, which is accompanied by a decrease in total REE concentrations (Fig. 4). This may be due to increased sorption of REE carbonate complexes onto iron oxyhydroxides and clays (Fig. 3). Increased sorption may be facilitated by the higher sorption potential of kaolinite in higher pH environments (Coppin et al. 2002). Also, increased residence for groundwater in the regional sedimentary aquifer would allow for increased sorption.

Groundwater from BHs 12 to 14 is down-gradient from recharge areas in both the sedimentary and basalt flow systems. Consequently, in addition to sorption processes, inter-aquifer mixing is likely to have significant effects on REE concentrations. Groundwater from BH 14 is screened from 18 to 300 m and has a REE pattern similar to groundwater from recharge areas of the sedimentary aquifer (Fig. 2b). This indicates input from the shallow sedimentary aquifer to the deeper sedimentary flow system, which will not be homogeneous with depth. More likely, REE represent the depth from where the sample came from along the 282 m screened interval of this artesian bore. REE concentrations in groundwater from BH 14 are lower than those in groundwater from the shallow weathered sediments ($\Sigma REE = 16$ vs. 46–119 ng/L respectively). Therefore, as groundwater flows through the regional sedimentary aquifer, the contribution of REE is small relative to the amount of sorption occurring as pH values increase (Fig. 4). Groundwater from BH 13 has lower total REE concentrations, increased Y fractionation (Fig. 4), and greater HREE enrichment than that from BH 14. This is probably due to increased sorption of REE onto more prevalent clays and iron oxyhydroxides in the shallower weathered sediments at BH 13 (screened: 20–44 m).

Effects of groundwater mixing from different aquifers are also observed in groundwater from BH 12. Groundwater from BH 12 and the basalt aquifer (e.g. BH 6) has similar REE patterns, concentrations (24 and 25 ng/L), and ratios (e.g. $(Yb/Nd)_{CN} = 0.37$ in both; $(Lu/Yb)_{CN} = 1.33$ and 1.27; $(Gd/Gd)_{CN} = 1.63$ and 1.56). Therefore REE at BH 12 reflect local flow from the basalt aquifer into the regional sedimentary aquifer. In comparison, major ion concentrations at BH 12 reflect cumulative water-rock interaction predominantly within the regional sedimentary aquifer (Tweed et al. 2003a).

4 CONCLUSIONS

In the Dandenong Ranges flow system, REE reflect an open flow system, and highlight more interaction between local groundwater flow and the underlying regional sedimentary aquifer than was previously suspected. This has implications for water resource management, where protection of local groundwater in both the basalt and sedimentary aquifers is required to maintain a regional groundwater resource in the sedimentary aquifer.

In using REE as tracers of mixing in multi-aquifer systems, this research indicates that REE concentrations are dominated by early-stage water–rock interaction. REE concentrations represent dissolution from weathered host rocks in acidic groundwater environments (i.e. in recharge areas). Progressive sorption (amount depends on the groundwater pH environment and residence times) and groundwater mixing between aquifers affect the subsequent fate of these REE.

As pH is a controlling mechanism of REE behaviour along flow paths, and since pH values are influenced by water–rock interaction, REE concentrations and fractionation reflect their position in the flow system. This has implications for using REE as a sole tracer in a complex system. Without an understanding of major ion chemistry and pH values, it would be challenging to ascertain mechanisms controlling REE. However, REE are effective tracers of groundwater, as they refine flow paths and may clarify regions where mixing with groundwater from other aquifers occurs. In this case study, REE were able to provide an additional level of understanding of groundwater flow patterns to a system previously defined using major ions chemistry.

REFERENCES

Anders, E. & Grevesse, N. 1989. Abundances of the elements: meteoric and solar. *Geochim. Cosmochim. Acta*53: 197–214.

Bau, M. 1999. Scavenging of dissolved yttrium and rare earths by precipitating iron oxyhydroxide: experimental evidence for Ce oxidation, Y-Ho fractionation, and lanthanide tetrad effect. *Geochim. Cosmochim. Acta* 63(1): 67–77.

Bau, M. & Dulski, P. 1996. Anthropogenic origin of positive gadolinium anomolies in river waters. *Earth Plan. Sci. Lett.* 143: 245–255.

Biddau, R., Cidu, R. & Frau, F. 2002. Rare earth elements in waters from albite-bearing granodiorites of Central Sardinia, Italy. *Chem. Geol.* 182: 1–14.

Coppin, F., Berger, G., Bauer, A., Castet, S. & Loubet, M. 2002. Sorption of lanthanides on smectite and kaolinite. *Chem. Geol.* 182: 57–68.

Drever, J.I. 1997. *The Geochemistry of Natural Waters, Surface and Groundwater Environments,*3rd ed, (ed R.A. McConnin), Englewood Cliffs, NJ: Prentice-Hall.

Johannesson, K.H., Farnham, I.M., Guo, C. & Stetzenbach, K.J. 1999. Rare earth element fractionation and concentration variations along a groundwater flow path within a

shallow, basin-fill aquifer, southern Nevada, USA. *Geochim. Cosmochim. Acta* 63(18): 2697–2708.
Möller, P., Dulski, P. & Gerstenberger, H. 1998. Rare earth elements, yttrium and H, O, C, Sr, Nd and Pb isotope studies in mineral waters and corresponding rocks from NW-Bohemia, Czech Republic. *Appl. Geochem.* 13(8): 975–994.
Nakamura, N. 1974. Determination of REE, Ba, Fe, Mg, Na and K in carbonaceous and ordinary chondrites. *Geochim. Cosmochim. Acta* 38: 757–775.
Parkhurst, D.L. & Appelo, C.A.J. 1999. User's guide to PHREEQC (Version 2) – a computer program for speciation, batch-reaction, one-dimensional transport, and inverse geochemical calculations. *U.S. Geological Survey Water-Resources Investigations Report* 99–4259.
Shabani, M.B., Akagi, T. & Masuda, A. 1992. Preconcentration of trace rare-earth elements in seawater by complexation with Bis(2-ethylhexyl) hydrogen phosphate and 2-ethylhexyl dihydrogen phosphate adsorbed on a C18 cartridge and determination by inductively coupled plasma mass spectrometry. *Anal. Chem.* 64: 737–743.
Smedley, P.L. 1991. The geochemistry of rare earth elements in groundwater from the Carnmenellis area, south west England. *Geochim. Cosmochim. Acta* 55: 2767–2779.
Tweed, S., Weaver, T.R. & Cartwright, I. 2003a. Distinguishing flow in different fractured-rock aquifers using groundwater chemistry: Dandenong Ranges, southeast Australia. *submitted to Hydrogeology Journal.*
Tweed, S., Weaver, T.R. & Cartwright, I. 2003b. Behavior of rare earth elements in groundwater during flow and mixing in fractured rock aquifers. An example from the Dandenong Ranges, southeast Australia. *submitted to Geochim. Cosmochim. Acta.*

Water–Rock Interaction, Wanty & Seal II (eds)

The origin and age of groundwater in the Nubian Sandstone aquifer in the Negev, Israel

A. Vengosh, S. Hening & J. Ganor
Department of Geological and Environmental Sciences, Ben Gurion University, Beer Sheva, Israel

T.D. Bullen
US Geological Survey, Menlo Park, CA, USA

C.E. Weyhenmeyer
Lawrence Livermore National Laboratory, Livermore, CA, USA

N.C. Sturchio
Department of Earth and Environmental Sciences, University of Illinois at Chicago, Chicago, IL, USA

A. Paytan
Department of Geological and Environmental Sciences, Stanford University, Stanford, CA, USA

ABSTRACT: Future prospects of utilization of the Lower Cretaceous Nubian Sandstone aquifer in many basins in the Middle East are overshadowed by the lack of modern replenishment and groundwater salinity. Here we examined the chemical and isotopic ($^{87}Sr/^{86}Sr$, $\delta^{11}B$, $\delta^{18}O$, $\delta^{2}H$, ^{14}C, ^{36}Cl) compositions of brackish to saline groundwater from the Nubian Sandstone aquifer in the Negev, Israel. The data reveal two major recharge sources, one with low D-excess (~10), originated from recharge of southern wet air masses in southern Sinai and a second with high D-Excess (~16) derived from local and dry Mediterranean air masses recharged in the Negev. Groundwater salinity is derived from both flushing of residual entrapped brines and dissolution of gypsum and halite. Groundwater from both sources have a low $^{36}Cl/Cl$ ($5–20 \times 10^{-15}$) signal, but measurable ^{14}C values (0.9–3.9 pmc) suggest dissolution of old (>1 Ma) salts by Late Pleistocene (23 to 35 ka) meteoric water.

1 INTRODUCTION

The scarcity of water resources in arid zones often requires exploitation of "fossil" groundwater (i.e., water that is not participating in the present hydrological cycle) that was recharged into the aquifers during the Late Pleistocene. Groundwater in many basins across the Sahara and Sahel regions in northern Africa and in the arid zones of the Middle East have deuterium-excess values (i.e., the intercept of $\delta^{18}O$-$\delta^{2}H$) close to that of the Global Meteoric Line (~10) and lower than modern precipitation (e.g., eastern Mediterranean = 22; Gat 1981). The low D-excess values typically reflect paleorecharge during periods of higher humidity (Gat 1981). The quality of the fossil groundwater varies from fresh water (e.g., Disi aquifer, Jordan) to brackish and saline groundwater (e.g., Zarq'a Maiin, Jordan and the Nubian Sandstone aquifer in the Negev, Israel). Tracing the sources of salinity is crucial for future exploitation of these basins. Groundwater salinity can be derived from residual formation brines, dissolution of evaporites, and active intrusion of brines.

Here we investigate the chemical and isotopic compositions of brackish to saline groundwater from the Nubian Sandstone aquifer in the Negev and Arava Valley, Israel (Figs. 1 and 2). We measured the major inorganic dissolved salts, the isotopic composition of the water ($\delta^{18}O$-$\delta^{2}H$), and the isotopic composition of dissolved constituents ($^{87}Sr/^{86}Sr$, $\delta^{11}B$, ^{14}C, ^{36}Cl).

2 RESULTS AND DISCUSSION

2.1 *Sources and timing of recharge*

Our study reveals several water types separated geographically (Fig. 2) and distinguished by their

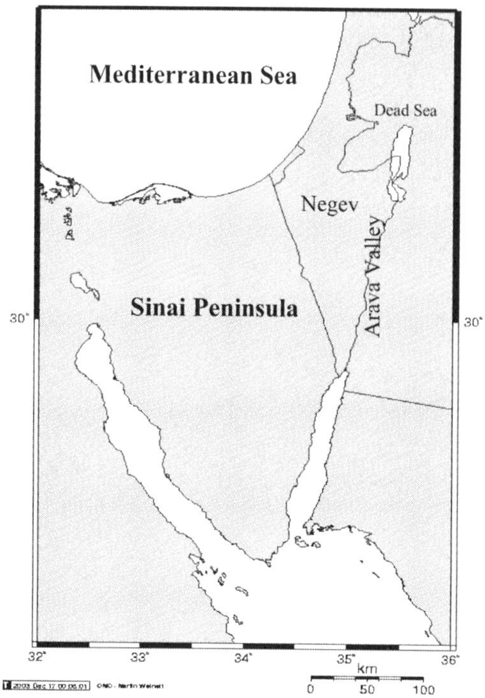

Figure 1. Location map of the study area.

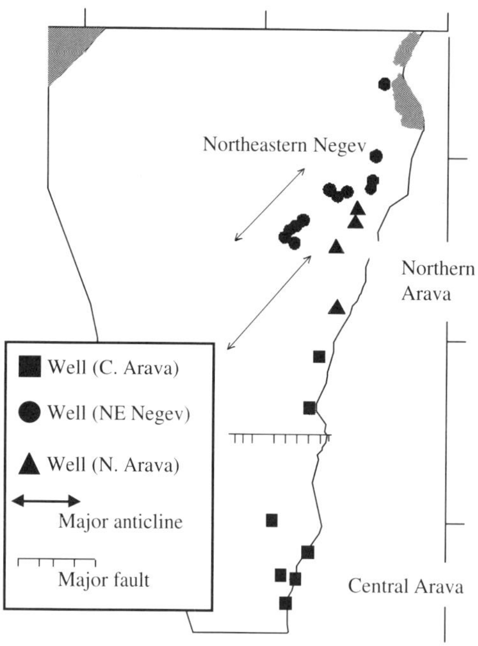

Figure 2. Location map of the wells from the Nubian Sandstone aquifer. Note the geographical distribution of the different water groups.

Table 1. Average geochemical and isotopic characteristics of the major water types in the Nubian Sandstone aquifer.

Water type	Cl mg/l	Na/Cl molar	D-Ex* ‰	$^{87}Sr/^{86}Sr$	δ^{11}B** ‰
NE Negev	740	0.93	16	0.70776	30
Fresh (NW Negev)	280	1.15	13	0.7087	25
Saline (NE Negev)	2000	0.93	10	0.70762	33
Northern Arava	654	0.76	14	0.70774	32
Southern Arava	626	0.67	10	0.70743	34

* Deuterium excess.
** Relative to NBS SRM–951.

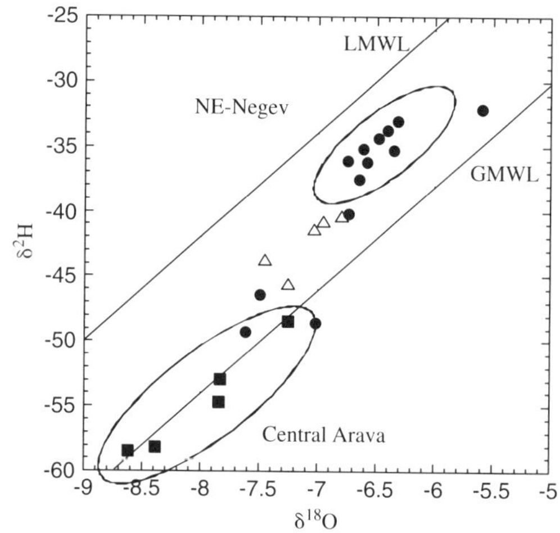

Figure 3. δ^2H versus $\delta^{18}O$ of groundwater from the Nubian sandstone aquifer in the Negev. Closed circles, triangles, and squares mark groundwater from the NE-Negev, Northern Arava, and Southern Arava, respectively. Note that groundwater samples from the central Arava valley have D-excess values close to that of Global Meteoric Water Line (GMWL) whereas in the northeastern Negev the D-excess values are higher, in between Local Meteoric Water Line (LMWL) and GMWL.

salinity, chemical composition (e.g., Na/Cl and B/Cl ratios), and oxygen, deuterium, strontium and boron isotopic compositions (Table 1).

The stable isotope data (Fig. 3) indicate that groundwater in the northeastern Negev (Fig. 2) has a high D-excess value (16), higher than that of the Global Meteoric Water Line. In contrast, groundwater from Sinai (Abed El Samie & Sadek 2001) and central Arava has a D-excess of ~10, which is identical to the Global Meteoric Water Line and reflects recharge in humid climatic conditions relative to modern precipitation. This isotopic distinction suggests recharge from two different sources: southern humid and northern arid (Mediterranean) air masses. This infers that groundwater in the Nubian sandstone aquifer was derived from two recharge regimes: (1) recharge in southern Sinai from which the groundwater flows to the central Arava; and (2) local recharge in northeastern Negev, probably in areas where the Nubian sandstone is not confined along the Hatira Anticline ("Machtesh Gadol"). The higher D-excess values (16) infer relatively arid conditions, closer to modern Mediterranean air masses (22; Gat 1981). However, preliminary AMS ^{14}C data (0.9 to 3.9 pmc) of groundwater with high D-excess values suggest an apparent age of 23 to 35 ka (correction factor of 0.7), indicating that the groundwater is not recent but rather old. Although previous ^{14}C data from the central Arava are scarce and less accurate, it seems that the reported values of 1 to 7 pmc are consistent with Late Pleistocene recharge. We therefore suggest that during the Late Pleistocene the rain in Sinai originated mainly from humid air masses whereas the rain in the Negev originated for the most part from dryer (Mediterranean) air masses. We conclude that a large fraction of the groundwater in the northeastern Negev was formed from direct recharge sources. This conclusion contradicts previous models of long groundwater flowpaths from the Sinai Peninsula (e.g., Issar et al. 1972; Issar 1981; Rosenthal et al. 1998).

2.2 *Sources and ages of dissolved salts*

The southern water group in the central Arava valley has geochemical features that resemble diluted brine with relatively low Na/Cl. high Br/Cl, and high $\delta^{11}B$ values. Similar low Na/Cl ratios were observed in saline groundwater from the Nubian aquifer in Sinai (data from Abed El Samie & Sadek 2001). Hence both the geochemical and stable isotope data (low D-excess) strongly suggest that the groundwater from the central Arava originated from flowpaths beginning in the Sinai Peninsula. The original "brine" composition is buffered by gypsum dissolution as evident from the Ca/SO_4 ratios of $\sim$1 found in both the central Arava and Sinai groundwater. The low $^{87}Sr/^{86}Sr$ (0.70730) ratio of the groundwater from central Arava Valley is distinctly different from the Dead Sea Rift brines (0.7080; Stein et al. 1997), and thus suggests that the gypsum (presumably the primary source for Ca and Sr) originated from Lower Cretaceous seawater (i.e., the Sr-isotope record in seawater shows similar low $^{87}Sr/^{86}Sr$ ratios during the Early Cretaceous; Jones & Jenkyns 2001).

In contrast, groundwater in the northeastern Negev has higher Na/Cl ($\sim$1) and low Br/Cl ratios (<seawater), which suggests that halite dissolution is the primary source of the salts. The most diluted groundwater found in the northeastern Negev,

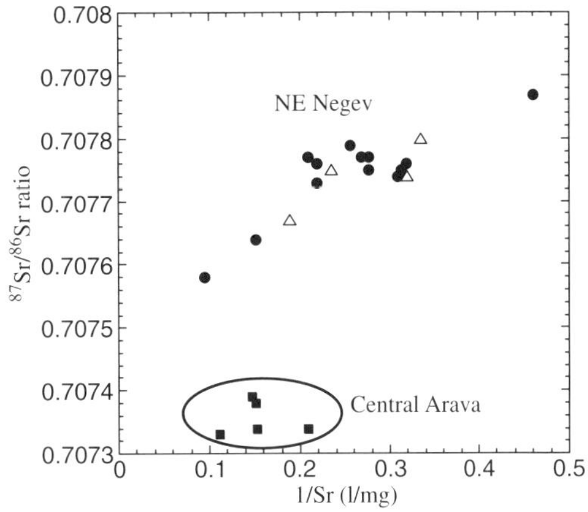

Figure 4. $^{87}Sr/^{86}Sr$ ratios versus reciprocal of strontium in groundwater from the Nubian sandstone aquifer in the Negev. Symbols are explained in Figure 2. Note the linear correlation found in groundwater from the NE-Negev, suggesting mixing of saline source with a low $^{87}Sr/^{86}Sr$ ratio and freshwater with a high $^{87}Sr/^{86}Sr$ signature.

underlying the anticline (Machtesh 1 well), has the highest $^{87}Sr/^{86}Sr$ (0.70787) and the lowest $\delta^{11}B$ (25‰) values (Table 1). We attribute this composition to water-rock interactions that occurred during recharge of groundwater in the unsaturated zone. Since elemental Sr concentration is correlated with chloride, the linear correlation between $^{87}Sr/^{86}Sr$ ratio and reciprocal of Sr observed in groundwater from the NE-Negev (Fig. 4) suggests that this groundwater is derived from two primary sources: (1) saline groundwater with a low $^{87}Sr/^{86}Sr$ signal, similar to the composition of groundwater from the Central Arava, and (2) fresh groundwater in which Sr is derived from a significantly higher $^{87}Sr/^{86}Sr$ source, originated during recharge in the unsaturated zone. This is consistent with the stable isotope data since the higher $^{87}Sr/^{86}Sr$ groundwater has also higher D-excess values. The Sr mixing relationships suggest that the saline component reflects residual saline groundwater flowing from Sinai that was mixed with fresh groundwater. We suggest that the local recharge in the northeastern Negev, which is characterized by higher D-excess and $^{87}Sr/^{86}Sr$ values, was mixed with subsurface flow from the Sinai Peninsula.

The $^{36}Cl/Cl$ results (5 to 20×10^{-15}) indicate that the age of the (chlorine) salts is considerably old. We assume that meteoric water during the Late Pleistocene had $^{36}Cl/Cl$ ratios of 100 to 200×10^{-15} and thus the significantly lower values obtained in the groundwater samples only reflect subsurface nucleogenic production of ^{36}Cl. Hence the "age" of the salts, both the "evaporite" source in the Northeastern Negev and the "brine" source in central Arava is older than 1 Ma (i.e., several half lives of ^{36}Cl decay).

3 CONCLUSIONS

The combination of several geochemical and isotopic tracers enabled us to reconstruct the groundwater recharge regime, delineate possible flowpaths, and identify the sources and ages of the dissolved salts. We were able to distinguish two major recharge sources that originated from two distinctive air masses during the Late Pleistocene, southern humid and northern arid types. Hence the common assumption that paleo-waters are always depleted in ^{18}O with low D-excess is not valid for this aquifer as we obtained high D-excess and low ^{14}C values. We show that salts dissolved in groundwater may be derived from residual brines that were preserved in the aquifer, as well as dissolution of evaporites (gypsum and halite). The Sr-isotope ratios enabled us to distinguish between "original" (Lower Cretaceous) and "external" (i.e., formed in the unsaturated zone above the aquifer) gypsum sources. We argue that the age of the salts dissolved in groundwater may be significantly different from than the recharge age.

REFERENCES

Abed El Samie, S.G & Sadek, M.A. 2001. Groundwater recharge and flow in the Lower Cretaceous Nubian Sandstone aquifer in the Sinai Peninsula, using isotopic techniques and hydrochemistry. *Hydrogeol. J.* 378–389.

Gat, J.R. Groundwater. In *Technical Report Series No. 210, Stable isotope Hydrology: Deuterium and oxygen-18 in the Water Cycle.* International Atomic Energy Agency.

Issar, A., Bein, A. & Michaeli, A. 1972. On the ancient water of the Upper Nubian Sandstone aquifer in central Sinai and southern Israel. *J. Hydrol.* 17: 353–374.

Issar, A. 1981. The rate of flushing as a major factor in determining the chemistry of water in fossil aquifers in southern Israel. *J. Hydrol.* 54: 285–296.

Jones, C.E. & Jenkyns, H.C. 2001. Seawater strontium isotopes, oceanic anoxic events, and seafloor hydrothermal activity in the Jurassic and Cretaceous. *Am. J. Sci.* 301: 112–149.

Rosenthal, E., Jones, BF. & Weinberger, G. 1998. The chemical evolution of Kurnub Group paleowater in the Sinai-Negev province– a mass balance approach. *Appl. Geochem.* 13(5): 553–569.

Stein, A., Starinsky, A., Katz, A., Goldstein, S.L., Machlus, M. & Schramm, A. 1997. Strontium isotopic, chemical, and sedimentological evidence for the evolution of Lake Lisan and the Dead Sea. *Geochim. Cosmochim. Acta* 61: 3975–3992.

Water–Rock Interaction, Wanty & Seal II (eds)

Geochemical characterization of a volcanic-sedimentary aquifer in Central Italy

R. Vivona, E. Preziosi, G. Giuliano, D. Mastroianni & F. Falconi
Water Research Institute, IRSA-CNR, Rome, Italy

B. Madé
Ecole des Mines de Paris, Fontainebleau, France

ABSTRACT: A hydrogeochemical study of the groundwater system was carried out in an area north of Rome (central Italy) where volcanic products of the Quaternary alkaline-potassic magmatism outcrop. Recently widespread high concentrations of toxic elements (F, As, Mn) have been observed, thus the importance of gaining insight into the geochemical and hydrogeological processes leading to these anomalous scenarios. Water samples were collected from private wells, rivers and springs for field parameter determination, chemical analysis and discharge measurements. A geochemical modeling tool was used to determine the speciation of the elements and to simulate the equilibrium state of the aquatic system. Preliminary results show the existence of a basal regional aquifer, flowing from west to east and circulating either in the volcanic rocks and in the underlying Pleistocene sedimentary units (in the eastern part). The relationship between As, F and Ca suggests a possible enrichment of the waters flowing from the volcanic to the sedimentary units in Ca and HCO_3^- and a related depletion of the cited toxic elements.

1 INTRODUCTION

The aim of this study is the characterization of chemical composition and hydrodynamic features of a groundwater system in an area north of Rome (Central Italy), where anomalous concentrations of minor toxic elements such as As, F, Mn of natural origin have been detected (Brondi et al. 1986; Dall'Aglio et al. 1994; Dall'Aglio et al. 2001). Nevertheless, in this area groundwaters are largely exploited mainly for agriculture and drinking purposes. Thus it seems important to get a better knowledge of the geochemical processes leading to anomalous concentrations of these elements, especially in relation to As, which is becoming recognized as a major health risk. Arsenic concentrations greater than the guideline value in drinking water have recently been observed also in many other countries in the world (WHO 2001).

A number of samples were taken from 42 wells, 16 river sites and 8 springs over an area of about 100 km^2 (Fig. 1), and they were analyzed for chemical components, such as major and trace elements, by means of ICP-MS and IC. The piezometric heads in wells and the discharge of rivers were also measured.

A chemical modeling analysis has been conducted to investigate the processes of water rock interaction, simulating the equilibrium state of the aquatic system and studying the behaviour of the main geochemical species.

2 GEOLOGICAL AND HYDROGEOLOGICAL SETTING

The study was conducted in an area located 100 km north of Rome, between Vico Lake and the Tiber valley. In this area alkaline-potassic volcanic rocks of Quaternary age largely outcrop; in the eastern edge they overlay the Pleistocene alluvial deposits of Paleo-Tiber represented by carbonate gravels and sands.

The hydrogeological setting of the area is characterized by a regional mainly unconfined aquifer; its flow is in general eastward, moving from the highest part of the volcanic complex to the Tiber valley (Boni et al. 1986). The aquifer feeds the base flow of several small rivers (up to a few hundreds of liters per second) and a number of small springs.

Geothermal activity related to a deep carbonate reservoir of Mesozoic age is testified by the diffuse heat flow (Barbier et al. 1998; CNR-PFE-RF15 1982) and by the presence of thermal springs (Minissale

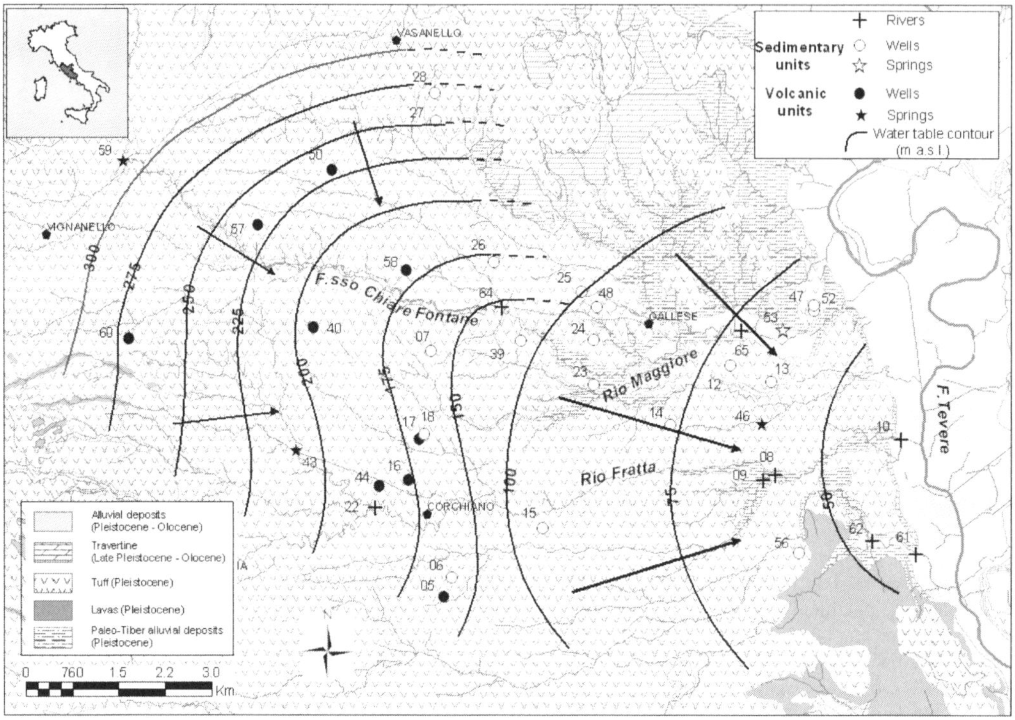

Figure 1. Simplified geological map of the study area with sample location and number.

1991). At present the reservoir is exploited by several wells for energy production. In the study area Pliocene clays separate the regional unconfined aquifer from the geothermal reservoir.

3 METHODS

Field activity was conducted in drought periods (between July and September of 2002 and 2003). In these conditions the discharge of the rivers is represented entirely by the contribution of groundwaters.

Field parameters such as temperature, pH, Eh, electrical conductivity, dissolved O_2 and alkalinity were determined at sampling points. The samples were collected in polyethylene bottles, after having been rinsed, and transported in refrigerated containers for chemical analysis. Major anions (F, Cl, PO_4, NO_3, SO_4) were determined by ion chromatography (IC); major cations (Na, K, Ca, Mg) and trace metals were analysed by inductively coupled plasma mass spectrometry (ICP-MS). These samples were acidified in the field with HNO_3 to $pH<2$ to avoid metal precipitation or metal adsorption on the bottle surface.

River discharge measures were performed in order to evaluate the base flow variation where the Paleo-Tiber deposits outcrop. Water samples were taken at each measured site.

Through the geochemical modelling tool of JCHESS (Van der Lee 1998), the speciation of the chemical elements and the calculation of the saturation index of the minerals at equilibrium with the solution were performed. Dissolution-precipitation processes of the main solid phases were also considered.

4 RESULTS AND DISCUSSION

Preliminary results confirm the existence of a regional aquifer laying on the Pliocene clays, flowing from the highest altitude areas towards the Tiber valley; the groundwater circulates in the volcanic units in the western part and in the sedimentary Paleo-Tiber formation to the east. The central part of the basal aquifer is characterised by a high hydraulic gradient west to east (Fig. 1); moreover, a local perched aquifer has been observed in the volcanic units overlaying the sedimentary complex. The perched aquifer is going to be abandoned because of its overexploitation during the last tens of years. Nowadays, the deeper aquifer represents the most

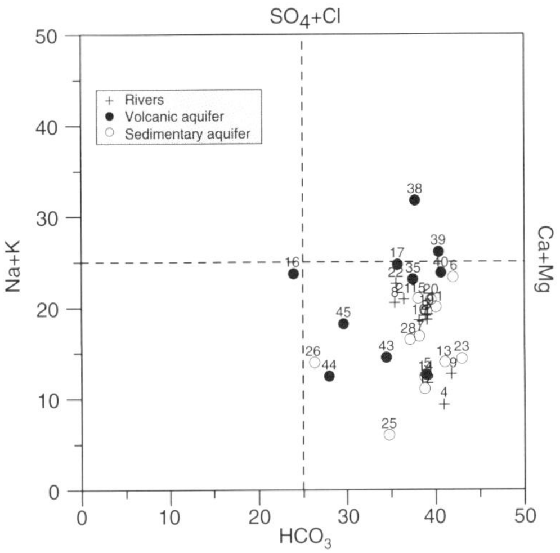

Figure 2. Chebotarev diagram.

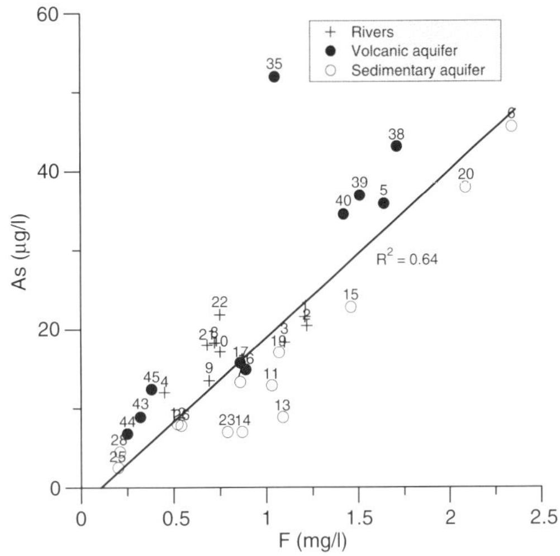

Figure 3. As vs F concentration.

interesting reservoir of natural groundwater resources in the area, used mainly for agricultural and drinking purposes.

Chemical analysis of sampled waters shows that they belong to the alkaline-earth bicarbonate hydrofacies (Chebotarev 1955; Fig. 2).

Relatively high concentrations of minor toxic elements (As and F) have been found. Values range from 4 up to 52 µg/l of As and from 0.2 up to 2.3 mg/l of F, thus exceeding in several cases the maximum concentration limit (10 µg/l for As and 1.5 mg/l for F) fixed by the European Council Directive 98/83/EC.

A good linear correlation between As and F was observed ($R^2 = 0.64$; Fig. 3): it is likely related to the strong chemical affinity of F for the As in ground waters (Levy et al. 1999). There is no evidence of the presence of As-bearing minerals because of the lack of suitable pH and redox conditions, and their SI are always <0. The speciation calculation performed with equilibrium constants of the CHESS data base (Wolery 1992) shows the presence of As and F aqueous complexes, mostly $(AsO_3F)^{2-}$ (80%) and $(HAsO_3F)^-$ (20%; Fig. 4), shown also by an As stability diagram computed for specific As and F activities (Fig. 5).

In general, the samples collected in the sedimentary aquifer show an enrichment in Ca and HCO_3 together with concentrations in As and F as high as those collected in the volcanic aquifer.

The relationships As vs Ca (Fig. 6) and F vs Ca show a similar trend because of the described linear correlation of As to F (Fig. 3). They suggest a possible evolution of the chemical facies in the waters flowing from the volcanic units to the sedimentary ones towards an enrichment of Ca and HCO_3 as a

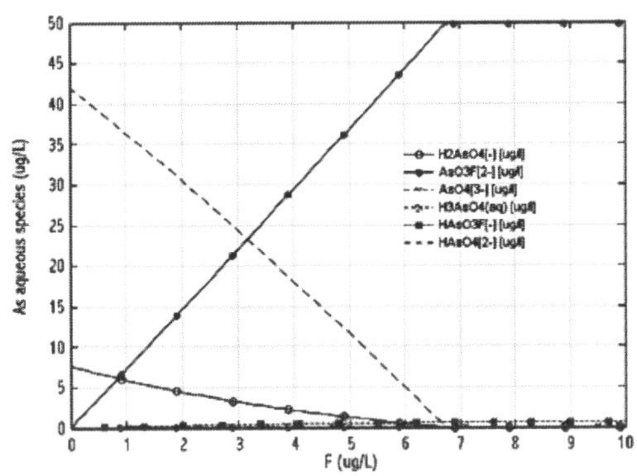

Figure 4. Speciation diagram of As aqueous species (pH = 7.5, F ranging from 0 to 10 µg/l, As = 50 µg/l).

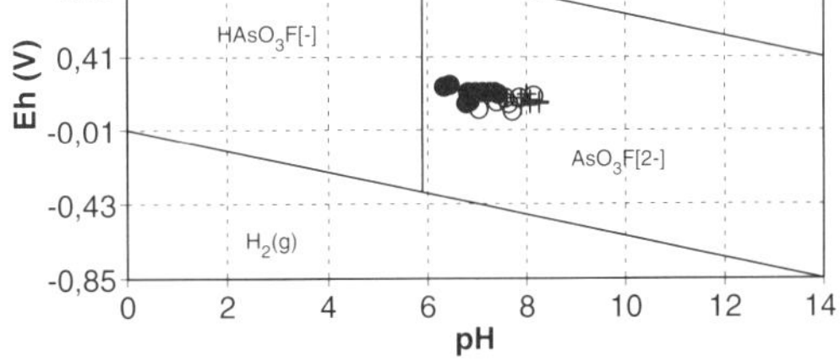

Figure 5. As specific activity stability diagram ($F = 1.25 \times 10^{-4}$, $As = 3.5 \times 10^{-7}$). Water sample symbols same as in the other figures.

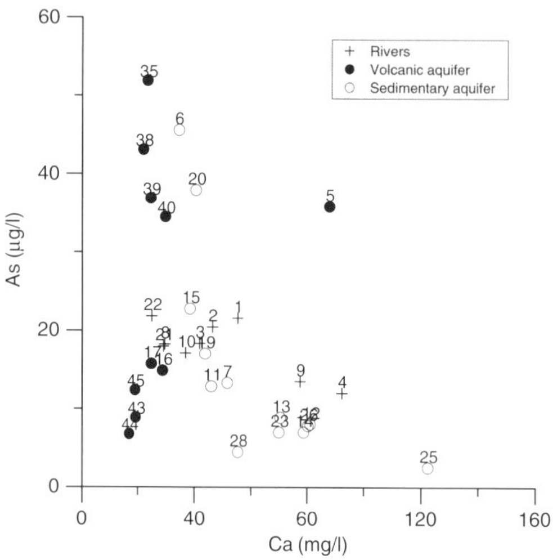

Figure 6. As vs Ca concentration.

consequence of leaching processes, with a related progressive reduction of the minor elements.

The equilibrium condition of fluorite (CaF_2) and the control of F in relation to high Ca concentration are well known in the literature (Appelo & Postma 1993). The results obtained through geochemical modelling show undersaturation with respect to fluorite while calcite is in equilibrium with the aqueous solution for most of the samples. It is likely that calcite controls Ca in solution, thus the disequilibrium of fluorite. Nevertheless, fluorite is known to be present in the volcanic rocks related to the alkaline-potassic magmatism of central Italy; therefore, fluorite can make part of the mineral paragenesis of the rocks that have been leached during water–rock interaction processes. Moreover, fluorapatite is known to be a typical volcanic rock mineral bearing a certain percentage of fluorine; therefore, it can be considered another source of fluorine. Its saturation index is always >0 in the modeling calculations.

The deep fluids coming from the geothermal reservoir showing high concentrations of As and F (Calamai et al. 1975 report values as high as 8.3 ppm for As and 100 ppm for F) could contribute to the enrichment of the phreatic waters in the above cited elements. Moreover, the detected high contents of acid gas (CO_2, CO, H_2S; Chiodini 1994; Chiodini et al. 1995; Duchi et al. 2003) could improve their mobilization.

These observations and the hydrodynamic setting suggest that the sedimentary aquifer is fed by vertical percolation through the volcanic units overlaying the Paleo-Tiber gravels and/or by lateral flow from the volcanic units of the western side.

The newly collected data will allow a further evaluation of the factors controlling the concentrations of the minor elements considered up to now, in particular fluorine and arsenic, and the role of carbonate gravels and sands in the chemical evolution pathways of the groundwater.

ACKNOWLEDGEMENTS

Research funded by the National Group for the Defense from Hydrogeological Hazards, MIMA Project (Integrated Methodologies of Monitoring Aquifer), U.O. 4/13-IRSA.

REFERENCES

Appelo, C.A.J. & Postma, D. 1993. *Geochemistry, groundwater and pollution*. Rotterdam: Balkema.

Boni, C. Bono, P. & Capelli, G. 1986. Schema idrogeologico dell'Italia centrale. *Mem. Soc. Geol. It.* 35: 991–1012.

Barbier, E. Musmeci, F. & Saracco, L. 1998. Banca Nazionale Dati Geotermici. CNR-Istituto Internazionale per le Ricerche Geotermiche: Pisa.

Brondi, M., Dall'Aglio, M., Ghiara, E. & Gragnani, R. 1986. Distribuzione degli elementi minori ed in traccia di interesse tossicologico e nutrizionale nelle acque italiane. *Acqua-Aria* 10: 1043–1061.

Calamai, A., Castaldi, R., Dall'Aglio, M. & Ferrara, G.C. 1975. Preliminary report on the Cesano hot brine deposit (Northern Latium, Italy). *2nd UN Symposium on geothermal energy 1, San Francisco, USA, 20–90 May 1975*: 305–313.

Chebotarev, C. 1955. Metamorphism of natural water in the crust of weathering. *Geochim. Cosmochim. Acta* 8: 22–48: 198–212.

Chiodini, G. 1994. Temperature, pressure and redox conditions governing the composition of the cold CO_2 gases discharged in north Latium (Central Italy). *Appl. Geochem.* 9: 287–295.

Chiodini, G., Frondini, F. & Ponziani, F. 1995. Deep structures and carbon dioxide degassing in Central Italy. *Geothermics* 24: 81–94.

Council Directive 98/83/EC on the quality of water intended for human consumption. *Official Journal of the European Communities* 5/12/98.

CNR-PFE-RF 15 1982. Contributo alla conoscenza delle potenzialità geotermiche della Toscana e del Lazio.

Dall'Aglio, M., Duchi, V., Minissale, A., Guerrini, A. & Tremori, M. 1994. Hydrogeochemistry of the volcanic district in the Tolfa and Sabatini Mountains in Central Italy. *J. Hydrol.* 154: 195–217.

Dall'Aglio, M., Giuliano, G., Amicizia, D., Andrenelli, M.C., Cicioni, G.B., Mastroianni, D., Sepicacchi, L. & Tersigni, S. 2001. Assessing drinking water quality in Northern Latium by trace elements analysis. In R.Cidu (ed.), *Proc. International symposium on Water Rock Interaction (WRI-10:* 1063–1066. Rotterdam: Balkema.

Duchi, V., Matassoni, L., Tassi, F. & Nisi, B. 2003. Studio geochimico dei fluidi (acque e gas) circolanti nella

regione vulcanica dei M.ti Vulsini (Italia Centrale). *Boll. Soc. Geol. It.* 122: 47–61.
Minissale, A. 1991. Thermal springs in Italy: their relation to recent tectonics. *Appl. Geochem.* 6: 201–212.
Levy, D.B., Schramke, J.A., Esposito, K.J., Erickson, T.A. & Moore, J.C. 1999. The shallow ground water chemistry of arsenic, fluorine, and major elements: Eastern Owens Lake, California. *Appl. Geochem.* 14: 53–65.
Van der Lee, J. 1998. Thermodynamic and mathematical concepts of CHESS. Technical report LHM/RD/98/39. CIG Ecole des Mines de Paris, Fontainebleau: France.
WHO 2001. Fact sheet No 210. http://www.who.int/inf-fs/en/fact210.html
Wolery, T. 1992. EQ3/6: A software package for geochemical modeling of aqueous systems: package overview and installation guide (version 7.0). LLNL report, CA, USA.

Water–Rock Interaction, Wanty & Seal II (eds)

Models for development of dryland salinity, Murray Basin, Australia

T.R. Weaver, A. Bush & T. Singh
Hydrogeology & Environment Research Group, School of Earth Sciences, University of Melbourne, Parkville, Victoria, Australia

I. Cartwright & E. Irvine
Hydrogeology & Environment Research Group, School of Geosciences, Monash University, Clayton, Victoria, Australia

ABSTRACT: Dryland salinization is a major problem in Australia. Current models propose that recharge occurs mainly on upland areas and that the saline groundwater results from mobilization of salt in the unsaturated zone. By contrast, Cl/Br and stable isotope values demonstrate that evaporitic concentration, rather than halite dissolution, is the dominant cause of salinization in four areas of the southeastern Murray Basin. Recharge is shown to occur across the areas, and reversal of hydraulic gradients during wet winters in areas where groundwater discharge otherwise occurs allows saline groundwater to re-enter the flow systems, further contributing to salt load. Existing models imply that land and water salinization will cease to expand once systems reach new hydraulic equilibria; however, our data imply that evaporation at discharge areas and on the surrounding plains will continue to increase groundwater salinity while water tables remain shallow. These studies indicate that management strategies must focus on the entire catchments, not just the uplands.

1 INTRODUCTION

Dryland salinity is one of Australia's most pressing environmental problems, adversely impacting agricultural productivity, ecosystems, and water quality (George et al. 1997). The expansion of dryland salinity is particularly problematic in semi-arid regions of the southeast Murray Basin. Current models (e.g. Ghassemi et al. 1995; Coram et al. 2000) propose that dryland salinity results primarily from mobilization of salts concentrated in the unsaturated zone of hills flanking the basin. Land clearing following European settlement increased recharge in these areas causing the water table to rise and mobilizing previously concentrated salts in the unsaturated zone (e.g. Cook et al. 1994). Evaporation further concentrates these salts at discharge areas where the water table approaches or intersects the ground surface (Fig. 1). This model predicts that once a new hydraulic equilibrium has been reached, areas affected by dryland salinity should stop expanding. Here we refine this model using data from four sites in the southeastern Murray Basin.

1.1 *Hydrogeological setting*

The four sites investigated are Sheep Pen Creek and Honeysuckle Creek in the Goulburn-Broken Catchment, and Springhurst and Everton in the Ovens Catchment (Fig. 2). Each area is used primarily for stock grazing on dryland and is located on the flanks of the catchments where Tertiary basin sediments onlap bedrock hills comprised of granite (Everton), volcanics (Honeysuckle Creek), conglomerate (Springhurst), or metasediments (Sheep Pen Creek,

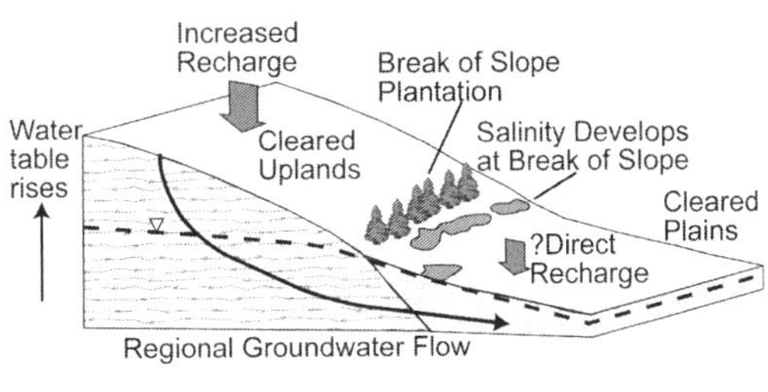

Figure 1. Cross section showing current model of dryland salinity development in southeastern Australia.

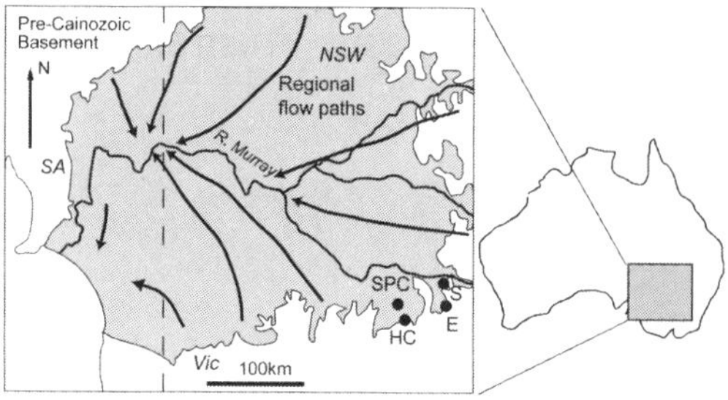

Figure 2. Location of the four sites in the southeast Murray Basin (shaded). E: Everton; HC: Honeysuckle Creek; SPC: Sheep Pen Creek; S: Springhurst. *SA*: South Australia, *Vic*: Victoria.

Everton, Springhurst). All four areas show increasing dryland salinity, although the extent of salinization in the Goulburn-Broken Catchment is significantly greater than in the Ovens because of the more arid nature of that region. Water levels and electrical conductivity (EC) values have been collected from these sites for the last 20 years by the Department of Natural Resources and Environment.

Hydraulic head data show that the bedrock and adjacent sediments form a single aquifer at each site, although hydraulic gradients between the hills and the plain have increased over the last 20 years at Sheep Pen Creek, Honeysuckle Creek, and Everton as the water table on the hills has risen due to higher recharge after land clearing over the last 100 years.

At Springhurst, water levels have remained constant over the same monitoring period in spite of land clearing having occurred at a similar time. At each site, groundwater discharge sites have developed at the break of slope and it is here that the effects of salinization are most apparent.

2 GROUNDWATER CHEMISTRY

EC values of groundwater in each of the four areas are highly variable, ranging from 0.513–29.0 mS/cm (Sheep Pen Creek), 0.245–24.5 mS/cm (Honeysuckle Creek), 0.160–10.0 mS/cm (Everton), and 1.89–19.8 mS/cm (Springhurst). Although salinity varies, the major-ion composition of groundwater in all four areas is similar. All samples are dominated by Na and Cl. SO_4 is an important anion, particularly in the more saline samples, and HCO_3/Cl ratios increase in the fresher samples from Everton and Honeysuckle Creek. At Honeysuckle Creek and Sheep Pen Creek, NO_3 concentrations are typically <0.16 mmol/L, but concentrations may increase to about 1.13 mmol/L in some samples. NO_3 concentrations are significantly greater in groundwater from Springhurst and Everton; up to 2.10 mmol/L in areas of intense grazing.

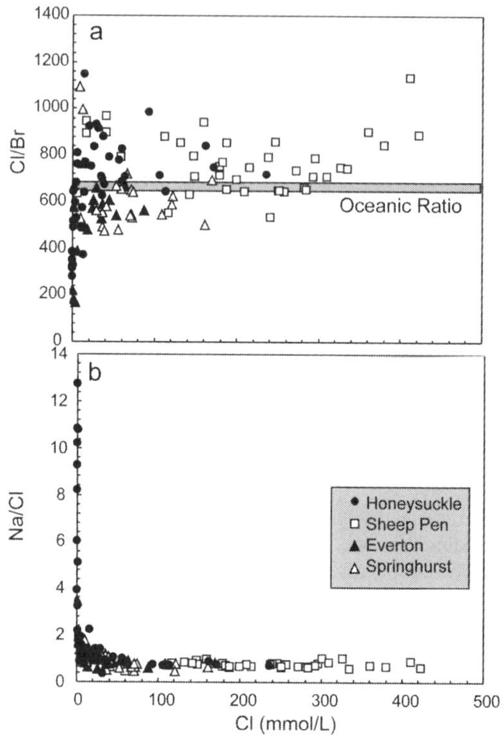

Figure 3. Molar Cl/Br (3a) and Na/Cl (3b) ratios vs Cl concentrations for the four sites.

2.1 *Na/Cl and Cl/Br ratios*

A comparison of Na/Cl molar ratios shows that the bedrock contributes Na from the dissolution of plagioclase feldspar at each site (Fig. 3b). Higher Na/Cl ratios (3–13) are indicative of higher amounts of water-rock interaction in the granite and volcanic uplands of Everton and Honeysuckle Creek and contrast with Na/Cl ratios of 0.7–1.2 in groundwater from Sheep Pen Creek, Springhurst, and the plains areas of Everton and Honeysuckle Creek. These values are more consistent with evaporation, during or after recharge, of oceanic salts derived from rainfall (cyclic salts). The higher Na/Cl ratios also correspond with higher Si concentrations and occur only in the least saline groundwater (Cl < 20–50 mmol/L) at each site.

Across the four sites, Cl/Br molar ratios vary from 170–1146. As with Na/Cl ratios, the variability of Cl/Br ratios at each site declines as Cl concentrations increase above 20 mmol/L (Fig. 3a). The low Cl/Br ratios evident in fresh groundwater at all but Sheep Pen Creek most likely indicate recharge occurring in a semi-arid climate, inland from coastal areas (e.g. Davis et al. 1998). The highest Cl/Br ratios that occur

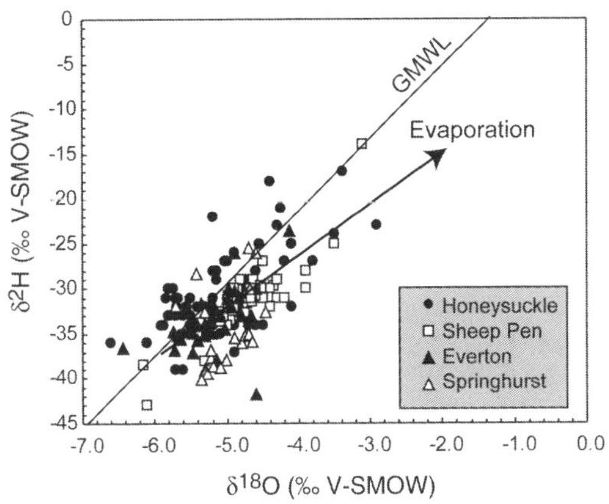

Figure 4. Stable isotope ratios of groundwater from the four sites.

in the lower salinity groundwater result from minor halite dissolution. Halite has Cl/Br ratios of 10^4–10^5 (e.g. Cartwright et al. 2003). Since Cl/Br ratios do not increase with increasing Cl additional halite dissolution cannot be the cause of the high Cl concentrations. Instead, evaporation must be the major process responsible for increasing groundwater salinity in these dryland salinity affected areas. The progressive homogenization of Cl/Br ratios results from mixing of groundwater with initially heterogeneous Cl/Br ratios.

2.2 *Stable isotopes*

A plot of $\delta^{18}O$ and δ^2H data further indicates the importance of evaporation in contributing to salinity at these sites. Most fresh groundwater lies on or near the meteoric water line (GMWL), while groundwater from the discharge areas and some parts of the plains lies off the meteoric water line along trend lines with slopes of 3.8–4.3 that indicate evaporation (Fig. 4). Some groundwater samples from the hills also lie off the meteoric water line, indicating that in some areas, evaporation also occurs during recharge.

Seasonal variability in $\delta^{18}O$ and δ^2H values highlights the relatively rapid groundwater movement that occurs in some parts of the flow systems. This is particularly evident in the uplands at Everton where $\delta^{18}O$ and δ^2H values from the same bore change from −4.1‰ and −24‰ in late summer to −6.4‰ and −37‰ in mid winter. This change is accompanied by a decline in EC values and increased head, indicating recent recharge of fresher water after winter rains. Samples from discharge areas were also displaced towards the meteoric water line from the summer, when evaporation dominated, to the winter, when recharge was occurring.

3 COMPARISON WITH GROUNDWATER FLOW PATHS

Comparison of groundwater quality with position in the flow system at each site shows clear controls on salt input and salinity development. Hydraulic gradients are consistently downward in the uplands where previous salinity models have implied that most groundwater recharge occurs. However, groundwater chemistry in the recharge areas is highly variable, reflecting different processes that occur during recharge in the uplands. Most groundwater samples in the igneous-rock uplands of Honeysuckle Creek and Everton are characterized by low Cl/Br ratios (170–390) and low salinity (Cl < 4 mmol/L). However, some groundwater samples at Everton contain higher salt concentrations, with near-average Cl/Br ratios (e.g. Cl/Br = 564; Cl = 88 mmol/L) and stable isotope ratios that plot to the right of the meteoric water line, indicating that evaporation must have occurred during recharge in the uplands. Locally, halite dissolution also occurs during recharge on some of the hills. At Springhurst, minor halite in the unsaturated zone has been redissolved and introduced to groundwater, producing some of the highest Cl/Br ratios (1093, 998) of the four sites. Although Cl concentrations increase in the associated discharge areas (8–12 vs 169 mmol/L), Cl/Br ratios decrease, indicating that halite dissolution during recharge is not the primary control on salinization in these areas.

At each of the four sites, the most saline groundwater occurs at discharge areas where upward hydraulic gradients dominate throughout the summer and autumn. Upward flow continually provides water for evaporation, increasing the salt concentration, but Cl/Br ratios remain nearly constant. At all sites, these discharge areas are located at, or near, the break of slope between the hills and the plains (e.g. Fig. 1). Examples of Cl concentration due to evaporation are seen at each of the four sites (Honeysuckle Creek: Cl/Br = 840, Cl = 162 mmol/L; Everton: Cl/Br = 541, Cl = 53 mmol/L; Sheep Pen Creek: Cl/Br = 737, Cl = 272 mmol/L; Springhurst: Cl/Br = 697, Cl = 169 mmol/L). Salt efflorescence at some discharge areas in summer indicates that evaporation can occur to halite and gypsum precipitation.

These groundwater discharge areas do not always behave as such. Hydrograph data over the last 20 years at all four sites show that, in wet winters, the strongly upward hydraulic gradients that persist over summer and autumn reverse with the onset of higher rainfall (e.g. Fig. 5). This gradient reversal allows shallow saline groundwater and re-dissolved precipitated salts to re-enter the groundwater flow system, and contributes saline water to the regional flow system down gradient of the discharge areas.

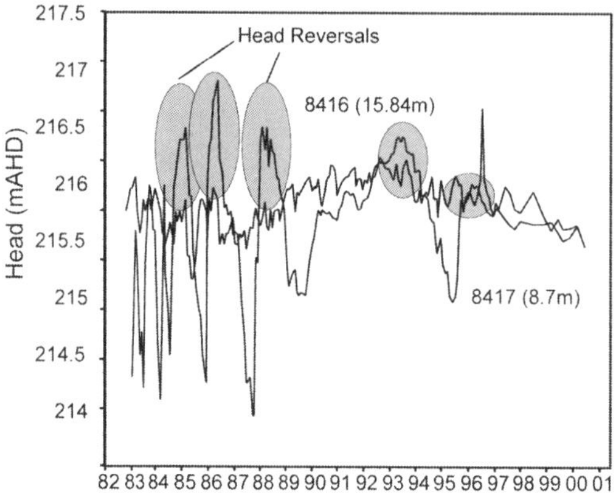

Figure 5. Hydrographs of nested bores from HC discharge areas showing downward hydraulic gradients in wetter winters.

In some areas, groundwater farther downgradient from the discharge areas shows a slight improvement in quality. This is due to recharge on the plains that dilutes the salt introduced from the discharge areas during gradient reversals. However, the water table across the plains is sufficiently shallow that evaporation and evapotranspiration still occur during recharge and directly from the water table (e.g. Honeysuckle Creek plains have Cl/Br = 718 and 747, Cl = 236 and 172 mmol/L and Honeysuckle Creek discharge areas have Cl/Br = 841, Cl = 162 mmol/L). Consequently, as long as the water table remains high, salinity will continue to rise and water quality will continue to degrade in the plains areas.

4 CONCLUSIONS AND IMPLICATIONS

This research compares groundwater salinization across four areas of the southeastern Murray Basin, in some of the region's most important recharge areas. The sites are consistent with the current model of dryland salinity development in that increased recharge on the hills after land clearing has led to rising water tables and the development of new discharge areas at the break of slope between the uplands and the plains. However, our research clearly shows the importance of recharge across the entire flow system and not only highland areas. The discharge areas that develop at the break of slope are critical in the concentration of salt through evaporation but hydraulic gradient reversals that allow recharge to occur from these areas during wet periods reintroduce significant concentrations of salt to the groundwater flow system down gradient of the discharge areas. Recharge also occurs across the plains, contributing to the water table rise and raising the potential for further groundwater evaporation.

An important prediction based on this improved understanding of groundwater movement is that 1) as long as local discharge areas occur and the water table is shallow, evaporation will continue to increase salt concentrations, and 2) these salts will be reintroduced to the regional flow system during periodic gradient reversals at discharge areas.

This prediction has significant implications for management of dryland salinity areas. To date, most focus has been on minimizing recharge on the hills, a strategy that will, eventually, reduce the rate of water level rise and thus reduce the time for a new hydraulic equilibrium to be reached. Other strategies have included fencing discharge areas to keep stock away and planting salt tolerant vegetation around discharge areas to minimize erosion. However, these strategies will do little to reduce evaporation in the discharge areas, and will not prevent gradient reversals during wet periods. Reducing recharge on the plains by revegetation or improved drainage will help to keep the water table lower; however, it will also reduce the amount of dilution occurring in these regions.

ACKNOWLEDGEMENTS

This research was funded by ARC-SPIRT to Weaver and Cartwright, and by the Centre for Land Protection Research and Dept Natural Resources and Environment, Victoria.

REFERENCES

Cartwright, I. Weaver, T.R., Fulton S., Nichol, C., Reid, M. & Cheng, X. 2003. Hydrogeochemical and isotopic constraints on the origins of dryland salinity, Murray Basin, Victoria, Australia. *App. Geochem.*, in press.

Cook, P.G., Jolly, I.D., Leaney, F.W., Walker, G.R., Allan, G.L., Fifield, L.K. & Allison, G.B. 1994. Unsaturated zone tritium and chlorine 36 profiles from southern Australia: their use as tracers of soil water movement. *Water Resources Res.* 30: 1709–1719.

Coram, J.E., Dyson, P.R., Houlder, P.A. & Evans, W.R. 2000. Australian groundwater flow systems contributing to dryland salinity. Bureau of Rural Sciences, Canberra, 74 p.

Davis, S.N., Whittemore, D.O. & Fabryka-Martin, J. 1998. Uses of chloride/bromide ratios in studies of potable water. *Ground Water* 36: 338–351.

George, R.J., McFarlane, D. & Nulsen, R. 1997. Salinity threatens the viability of agriculture and ecosystems in Western Australia. *Hydrogeol. J.* 5: 6–21.

Ghassemi, F., Jakeman, A.J., Nix, H.A. 1995. Salinisation of land and water resources: Human causes, extent, management, and case studies. Univ. New South Wales Press, Sydney, Australia, 526 p.

Water–Rock Interaction, Wanty & Seal II (eds)
© 2004 Taylor & Francis Group, London, ISBN 90 5809 641 6

Results from the Porewater Chemistry Experiment in Opalinus Clay at Mont Terri, Switzerland

P. Wersin & A.Gautschi
NAGRA, Wettingen, Switzerland

A. Vinsot
ANDRA, Bure, France

P. DeCannière
SCK CEN, Mol, Belgium

P. Hernán
ENRESA, Madrid, Spain

H.-E. Gäbler
BGR, Hannover, Germany

K. Hama
JNC, Horonobe, Japan

Y. Mahara
CRIEPI, Abiko, Japan

E. Gaucher
BRGM, Orélans, France

F.J. Pearson
New Bern, USA

ABSTRACT: A long-term *in situ* porewater experiment is being carried out in the Opalinus Clay formation under anoxic conditions at Mont Terri using the diffusive equilibration method. In addition to the expected evolution of conservative tracers, microbially- mediated sulfate reduction leading to high pCO_2 and alkalinity, a decrease in sulfate, and an increase in sulfide, is observed. The high pCO_2 does not agree with laboratory studies on core samples. The source of degrading organic carbon remains to be identified.

1 INTRODUCTION

Low permeability argillaceous rocks are considered potential host rocks for radioactive waste disposal in a number of countries. Characterization of the porewater chemistry in these rocks is important for prediction of radionuclide release, but is technically challenging because of the intimate clay-water association (Sacchi et al. 2000; Pearson et al. 2003).

An international multidisciplinary project on the Opalinus Clay formation was begun at the Mont Terri Rock Laboratory (Switzerland) in 1996. Considerable understanding of the porewater chemistry has been gained through the application of different techniques including *in situ* extraction, laboratory squeezing, leaching and geochemical modeling (Pearson et al. 2003). However, significant uncertainties with regard to *in situ* pH/pCO_2 and Eh values remain. To reduce these uncertainties and to gain deeper insight into geochemical processes regulating pH and Eh, the Porewater Chemistry (PC) experiment was started in 2002 which is based on the diffusive equilibration method described below. This paper presents and discusses initial results obtained from this experiment.

2 METHODS

2.1 *Set up and monitoring of the* in situ *experiment*

Solute transport in the Opalinus Clay has been shown to occur predominantly by diffusion. Artificial porewater with a composition similar to the formation water is circulated in a packed-off borehole and equilibrates with the formation water of the clay via diffusion. Because of the excess of formation water compared to the circulating water the equilibration will occur relatively fast. The diffusive exchange process is monitored by conservative tracers added to the artificial porewater.

The layout is illustrated in Figure 1. The borehole BPC-1 was drilled with N_2 and filled with Ar prior to equipment installation to minimize disturbance by oxygen. Diffusive equilibration started on March 22,

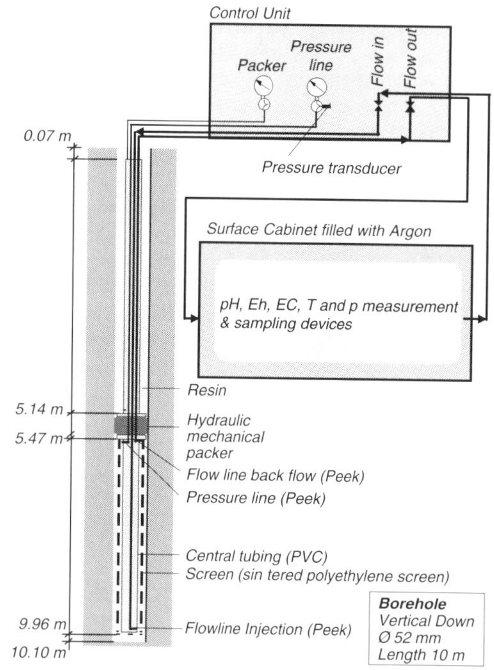

Figure 1. Layout of *in situ* experiment (from Solexperts

2002. The chemical and isotopic composition of the circulating water containing the tracers Br^-, HDO, and ^{13}C is shown in Table 1. The sum of Cl^- and Br^- concentration in the artificial water was prepared to correspond to that expected for chloride in the formation water. The artificial water was designed by geochemical modeling to be in equilibrium with the mineralogy and cation exchange properties of the rock.

All the parts contacting the circulating water are non-metallic (lines, screen, and pump) to avoid any influence on Eh (Fig. 1). The parameters pH, Eh, electrical conductivity, temperature and hydraulic pressure were continuously recorded. Water samples were taken periodically and analyzed for inorganic and organic compounds as given in Table 1.

In addition, one sample was taken for microbial analysis on December 11, 2002. These analyses included (i) direct counting of the fresh sample using a Thoma cell, (ii) enumeration of growing bacteria by the Most Probable Number (MPN) method.

2.2 *CO_2 measurements of core samples*

CO_2 measurements on two core samples from the BPC-1 borehole were performed by the outgassing method described in Pearson et al. (2003). In addition

Table 1. Selected chemical and isotopic parameters. Concentrations in mmol/L.

	Test water	PC-2	PC-3	PC-4	PC-5	Formation water (A1)[a]
Sampling date		3 Apr. 02	16 Jul. 02	11 Dec. 02	2 Sep. 03	28 Apr. 99
pH	77.4	7.77	7.00	6.74	6.68	7.96
Eh (mV)	n.m.	+127	−126	−171	−220	n.m.
Sample name						
Sodium	255	258	253	252	247	245
Potassium	1.45	1.52	1.8	1.8	1.9	1.1
Calcium	15.9	16.0	17.4	15.8	16.0	15.2
Magnesium	18.0	18.4	19.6	19.0	19.5	17.1
Ammonium	0.02	0.03	0.40	0.47	0.57	0.57
Chloride	262	267	283	285	286	287
Sulfate	14.7	14.6	13.3	10.5	6.6	13.7
Bromide	29.4	26.9	12.7	8.2	7.3	0.4
Alkalinity	1.0	1.1	6.5	7.0	11.6	2.5
HCO_3^-	0.82	0.88	4.93	5.78	7.00	0.8
TOC	0.02	0.13	0.65	14.0	9.9	1.1[b]
Sulfide (tot)	n.m.	n.m.	1.2	0.3	0.6	n.m.
δ^2H (‰)	+370	251.5	38.8	4.3	−32.6	−54.7
$\delta^{18}O$ (‰)	−10.46	−10.08	−9.02	−8.43	−8.62	−8.53
$\delta^{13}C$-DIC (‰)	−29.1	−16.3	−11.4	−3.2	−15.9	−13.1[b]

[a] Pearson et al. (2003), sample taken on April 28, 1999 in borehole A1 (close to borehold BPC-1).
[b] Sample taken on Feb. 18, 1998.

$\delta^{13}C$ and $\delta^{18}O$ analysis of the CO_2 gas and the carbonate minerals in the sample were carried out.

3 RESULTS AND DISCUSSION

3.1 *Tracer evolution*

The temporal evolution of the tracers is shown in Figure 2. Br^- and HDO showed a steady decrease with time. After 530 days of circulation Br^- decreased to 25% and HDO to about 5% relative to the difference between initial and background concentrations. These changes can be modeled assuming a diffusion-controlled process (Arcos et al. 2004). The slower decrease of Br^- is consistent with the lower diffusivity of anion relative to the self-diffusion of water in Opalinus Clay. Chloride concentrations showed the expected increase of concentration with time. This indicates that diffusive equilibration with the formation water in fact occurs with no notable contribution of advective transport.

The behavior of $\delta^{13}C$ in the dissolved inorganic carbon (DIC) cannot be explained solely by diffusive exchange with the surrounding formation water, but appears to be influenced by redox processes occurring in the borehole as discussed below.

3.2 *Evolution of chemical composition*

Table 1 shows the chemical and isotopic composition of the artificial test water, the circulating water in the BPC-1 borehole at different times and the formation water analyzed in a previous porewater sampling study (Pearson et al. 2003) in a borehole termed A1.

The composition of major cations Na^+, Ca^{2+}, $Mg^2$and K^+ remained relatively constant with time and was found to be similar to that obtained for the formation water. This indicates that only minor reactions occurred by ion exchange with the clay.

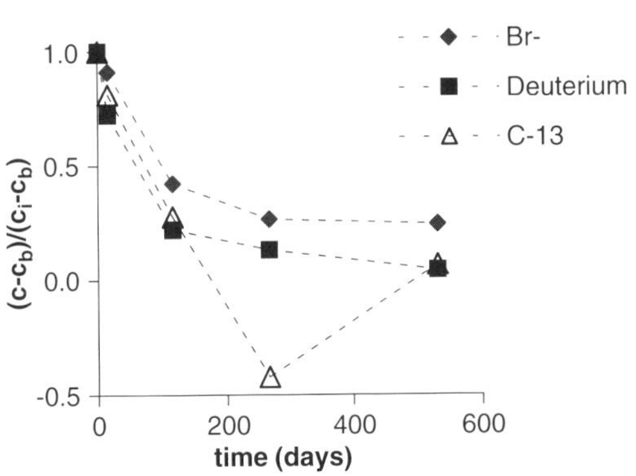

Figure 2. Evolution of tracers in borehole. c_b: expected background conc., c_i: initial conc.

On the other hand, significant trends for the major anions Cl^-, SO_4^{2-} and HCO_3^- were noted. Whereas the increase of Cl^- in the borehole results from diffusive equilibration with porewater (last section), the evolution of the other two anions cannot be explained by this process. Bicarbonate concentrations (Table 2) showed a seven fold increase to 7 mmol/L after 550 days which is much higher than expected according to previous data. It is important to note that a significant fraction of the measured alkalinity is made up of weak organic acids and sulfide (Table 1). Concomitant with the bicarbonate increase, pH values decreased from 7.8 to 6.7, which is much lower than any obtained in the porewater sampling study of Pearson et al. (2003).

The Eh values showed an initial rapid decrease from 0 to about −140 mV (SHE) within the first 4 months of circulation followed by a slow steady decrease to −220 mV (data not shown). On various occasions the circulation system was exposed to air due to sampling, electrode calibration and technical problems. After restarting the circulation system the former reducing conditions were rapidly re-established (within 1–2 days). Thus, the added oxygen was effectively depleted by redox reactions. Concomitant with decreasing Eh, NH_4^+ and sulfide generally increased whereas sulfate decreased to concentrations well below those expected in the formation water (Table 1).

3.3 *Comparison of pCO_2 with laboratory studies*

The CO_2 partial pressure was calculated from equilibrium modeling using the PHREEQC code. The pCO_2 in the *in situ* experiment showed a 100 fold increase yielding values of $10^{-1.2}$ bar (Table 2). As illustrated in Table 2 this is in contrast with CO_2 measurements on fresh core samples which indicate significantly lower CO_2 partial pressures of about $10^{-2.8}$ bar. The CO_2 in those samples showed a $\delta^{13}C$ and $\delta^{18}O$ composition indicating equilibrium with the carbonate minerals in Opalinus Clay. Moreover, data on porewater extracted by advective displacement which was carried out on a core sample in the vicinity

Table 2. pH/pCO_2 conditions determined in different studies.

Study/method	pH	HCO_3^- mM	log pCO_2 bar
This (water sample PC-5)	6.7	7.0	−1.2
This (core sample)			−2.8
Porewater extraction[a]	7.5	1.4	−2.7
Seepage water from A1[b]	8.0	0.8	−3.0

[a] Mäder et al. (2004), [b] Pearson et al. (2003).

of the PC experiment (Mäder et al. 2004), also indicated significantly lower pCO_2 and higher pH (Table 2). Finally, in the water sampling study by Pearson et al. (2003) pH/pCO_2 conditions were noted that are at odds with those observed in our *in situ* experiment, but are in rather good agreement with the core studies.

Overall these results indicate that CO_2-producing reactions occurred in the borehole, which are not indicative of pH/pCO_2 conditions of the undisturbed porewater.

3.4 *Evidence of microbially-mediated sulfate reduction*

There are several lines of evidence pointing to significant sulfate reduction induced by microbial activity. These include:

- Decrease of S(VI), increase of S(-II): a novelty at Mont Terri, where seepage waters from boreholes always showed increase in sulfate with time due to pyrite oxidation. This effect was avoided in the PC – 1 borehole because of the care taken to minimize oxygen contamination (e.g. drilling with N_2).
- Presence of sulfate reducing bacteria ($2 \cdot 10^3$ bacteria/mL) in the circulating water was shown by microbial analysis. The total number of facultative and strict anaerobes was $>7 \cdot 10^4$ per mL.
- The high pCO_2 and alkalinity could be explained by a sulfate reduction process using coupled diffusion-reaction modeling (Arcos et al. 2004). In that study sulfide was assumed to be controlled by mackinawite precipitation.

It must be pointed out that the redox processes taking place are not yet fully understood. Because of limited amount and availability of Fe(III) this electron acceptor plays only a minor role. On the other hand, the relevance of methanogenic reactions needs to be further investigated.

3.5 *Source of degrading organic carbon*

The dissolved organic carbon (DOC) in the borehole showed a moderate increase to values of 1 mmol/L during the first four months. Subsequently, however, a strong increase to about 10 mmol/L was noted. Most of this DOC was identified as acetic acid. The levels of dissolved methane (data not shown) remained low throughout in the range of about 0.04 mmol/L.

The source of degrading organic carbon is not understood at present because of the poorly known nature and reactivity of organic carbon in the Opalinus Clay. Moreover, degradation of the polyethylene filter screen material or of the PVC and PEEK tubing cannot be ruled out, although it is not considered probable.

4 CONCLUSIONS

The main original goal of the PC experiment, namely the more precise determination of *in situ* pH and Eh conditions in the porewater, could not be achieved because of bacterially-mediated disturbance of the sulfur and carbon system. Whether the active bacteria are autochthonous or brought in by the drilling or installation procedure is not known.

The finding is of interest for safety assessments because redox processes enhanced by bacteria significantly modify chemical conditions of the porewater in the disturbed zone around a repository where microbial activity is favored by increased pore space. However, it is important to identify unambiguously the source of reactive organic carbon in this system (e.g. to exclude degradation of equipment material) to be able to draw conclusions on the relevance of sulfate reduction in the Opalinus Clay under repository conditions. This has become the main goal of the continuing PC experiment.

REFERENCES

Arcos, D., Duro, L., Gimmi, T. & Waber H. N. 2004. Modeling the Porewater Chemistry experiment at Mont Terri. *This volume.*

Mäder, U., Waber, H.N. & Gautschi, A. 2004. A new method for porewater extraction and ionic transport characterization from clay stone with an application to Opalinus Clay (Switzerland). *This volume.*

Pearson, F.J., Arcos, D., Bath, A., Boisson, J.Y., Fernández, A.M., Gäbler, H.-E., Gaucher, E., Gautschi, A., Griffault L., Hernán, P. & Waber, H.N. 2003. Geochemistry of water in the Opalinus Clay formation at the Mont Terri Rock Laboratory. – Synthesis Report. Reports of the Swiss Federal Office for Water and Geology (FOWG), Geology Series, Nr. 5, Bern, Switzerland.

Sacchi, E. & Michelot, J.-L. 2000. Porewater extraction from argillaceous rocks for geochemical characterisation. OECD/NEA, Paris.

Water–Rock Interaction, Wanty & Seal II (eds)

Hydrochemistry of groundwaters in Quaternary aquifers: a case study at Taiyuan, China

S. Yan, Q. Guo & Y. Wang
School of Environmental Studies, China University of Geosciences, Wuhan, P.R. China

R. Wang
Geoenvironment Monitoring Center, Bureau of Land and Resource of Shanxi Province, Taiyuan, P.R. China

ABSTRACT: Hydrochemcial zones were delineated for groundwaters in Quaternary aquifers in Taiyuan city: SO_4-HCO_3 water, HCO_3-SO_4 water, and HCO_3 water occur sequentially from mountainous areas to basinal areas. Geochemical modeling indicates that mixing of karstic water, shallow groundwater and river water is a major control on hydrochemistry of the groundwaters.

1 INTRODUCTION

As the most important water source for local agriculture and industry, groundwater is vitally important for the development of industry and agriculture in Taiyuan City, one of the largest cities in northern China. Taiyuan City lies in the north of the Jinzhong faulted basin. Mountains surround the city from the north, east and west. The main surface water bodies are the Fen River and Jinyang Lake (an artificial lake). Regional groundwater flow direction is generally from the mountains to the basin. The lithology of the main aquifers in mountain areas include karstified-fissured Cambrian-Ordovician carbonate rocks and Carboniferous-Permian sandstones. In basin areas, three subsystems of Quaternary aquifers can be roughly divided: the shallow aquifers with bottom depths down to 50 m; intermediate aquifers with bottom depths from 50 to 200 m, and deep aquifers with bottoms deeper than 200 m. The groundwaters in the basin areas are recharged in three ways: groundwaters from overlying or underlain aquifers, seepage of surface water (Fen River) and lateral permeation of karst-fissure water from mountain areas.

Zonation is a natural occurrence found in many geological phenomena, such as geologic structure, magmatic activity, and geochemical distributions (Naito et al. 1995, Robion et al. 1997, Wogelius et al. 1997). This paper reports our work on the hydrochemial zones and the impact of mixing processes on hydrogeochemical features of intermediate aquifers at Taiyuan.

2 HYDROCHEMISTRY

Twenty-five groundwater samples were collected at Taiyuan in May of 2002. Sampling locations are shown in Figure 1 and the hydrochemical analysis results are summarized in Table 1.

The groundwaters have pH values between 7.1 and 7.8, and the main cations are Ca^{2+}, Mg^{2+}, K^+, and Na^+ and the main anions are HCO_3^-, SO_4^{2-}, and Cl^-. The TDS contour map (Fig. 2) and the contents of all the main ions of water samples for this study show that their spatial variations are similar. In other words, the general tendency is for ion con-concentrations of groundwaters to increase from north to south.

The hydrochemical types of groundwaters classified according to the predominant anion(s) follow a spatially zoned pattern: from SO_4-HCO_3 water, HCO_3-SO_4 water to HCO_3 water sequentially from the boundary to the middle of the basin (Fig. 1). The increase in chloride concentration around Jinyang Lake is probably related with the anthropogenic contamination.

The hydrochemical type of groundwater in the northern basin is HCO_3-Ca-Mg. In addition, the seepage from the Fen River water dilutes groundwater, resulting in the distribution of HCO_3 water from both sides of Fen River. However, affected by other factors, TDS and the contents of HCO_3^-, SO_4^{2-}, Cl^-, Ca^{2+}, Mg^{2+}, and Na^+ increase from north to south along Fen River (Fig. 3).

In the eastern and western mountains, the karstic water types are HCO_3-SO_4-Ca-Mg and

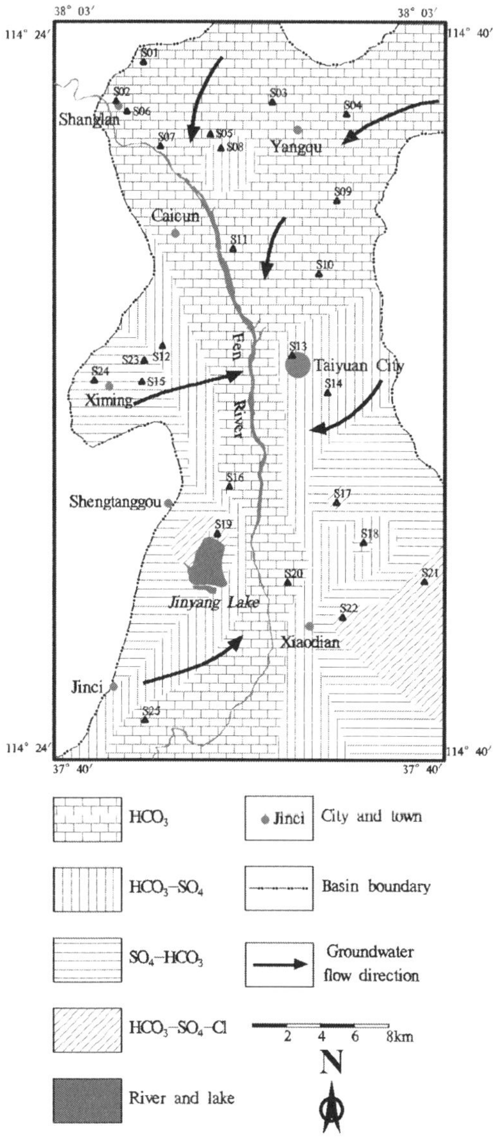

Figure 1. Hydrochemical zones of groundwater at Taiyuan.

SO_4-HCO_3-Ca-Mg, which are related to dissolution of sulfate minerals and leakage of mine drainage. Recharged by the karst water, the groundwater at the eastern and western boundary of the basin is SO_4-HCO_3 water, and becomes HCO_3-SO_4 water in the center of the basin (Fig. 1).

Around Jinyang Lake and in the southeastern part of the basin, the groundwater is HCO_3-SO_4-Cl water, probably resulting from human activity of these areas.

Table 1. Range of hydrochemical properties of Groundwater samples from Taiyuan (in mg/L except pH). Samples were collected in September 2000.

Parameters	Minimum	Maximum
pH	7.1	7.8
Ca^{2+}	37.1	468.0
Mg^{2+}	21.3	241.0
K^+	0.9	2.0
Na^+	10.0	214.0
Cl^-	10.6	840.0
SO_4^{2-}	2.4	941.0
HCO_3^-	220.0	510.0
NO_3^-	1.0	70.0
NO_2^-	0.0	16.0
F^-	0.3	4.6
CO_2	0.0	25.3
TDS	252.5	2847.3

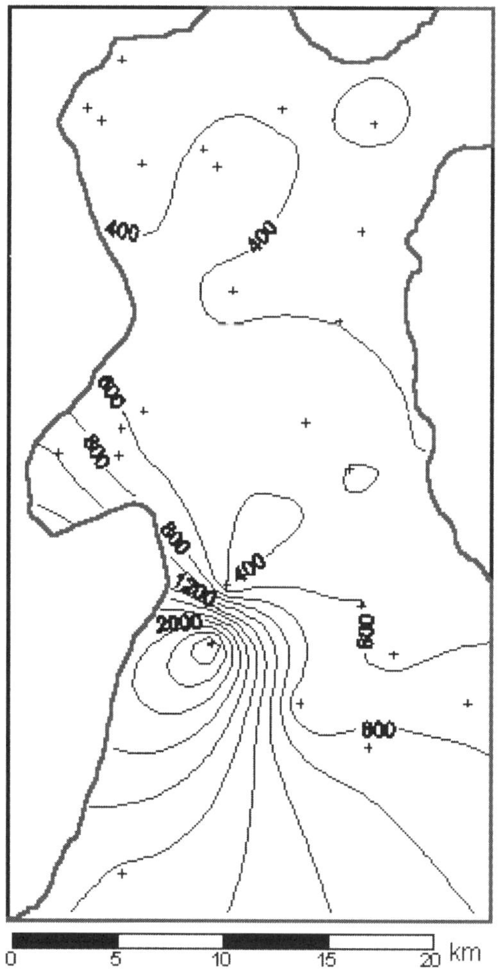

Figure 2. Contour map of TDS of groundwater.

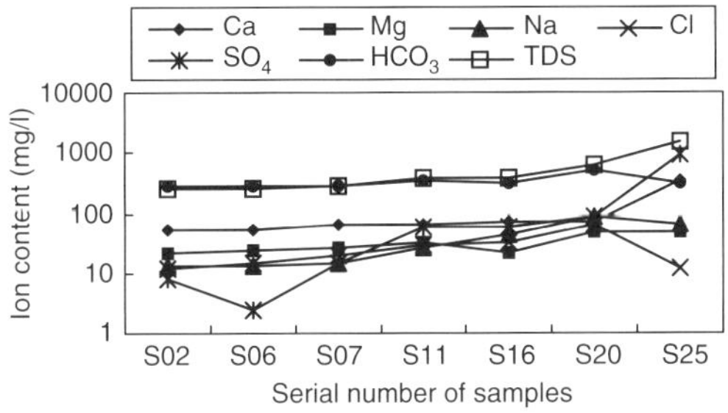

Figure 3. Variation of major ion contents of the HCO_3 water.

3 GEOCHEMICAL MODELING OF MIXING PROCESSES

To understand the impact of mixing processes on the hydrochemistry of groundwaters at Taiyuan, a series of groundwater mixing models were set up using PHREEQC. The hydrochemistry of shallow groundwaters from the northern and western basin, the karstic water in the northern and western mountains, and Fen River water (Table 2) were used in our modeling.

As mentioned above, sources of recharge for HCO_3 water of the northern basin include karstic water of the northern mountain area, the shallow groundwaters of the northern basin, and the Fen River water. Therefore, they are used as end members in the mixing model. The model indicates that when the mixing ratio of the three end memebers is 70:1:20, the saturation indices of the four main minerals (aragonite, calcite, dolomite, and gypsum) of the resulting solution are closest to those of groundwaters of the northern basin (Table 3). Thus, it can be inferred that karstic water of the northern mountains is the most important source of recharge for HCO_3 water in the northern basin.

The hydrochemical types for groundwater of the western basin include SO_4-HCO_3 water, HCO_3-SO_4 water and HCO_3-SO_4-Cl water, and the sources of recharge are karstic water of the western mountains and the shallow groundwater of the western basin. When the mixing ratio of two sources is 50:50, the saturation indices of the four main minerals of the mixing solution best match those of groundwater of the western basin (Table 3). Thus, the contribution of two recharge sources to the groundwater of the western basin are identical.

The groundwater level in the southern basin is the lowest, so the HCO_3 water of the southern basin should be the result of mixing of groundwaters of the northern and western basins. In comparison with the HCO_3 water of the northern basin, the HCO_3 water of the southern basin has high TDS and ion contents.

Table 2. Hydrogeochemistry of karst water, shallow groundwater and Fen River water (in mg/L except pH).

	pH	Ca^{2+}	Mg^{2+}	K^+
I	7.6	70.9	46.2	1.02
II	7.4	360.3	102.1	9.92
III	7.8	60.4	28.6	0.86
IV	7.4	139.0	42.3	2.40
V	7.0	118.0	35.9	13.00
	Na^+	Cl^-	SO_4^{2-}	HCO_3^-
I	82.1	62.6	127.4	370.0
II	173.7	421.6	811.3	333.7
III	12.0	15.4	51.1	268.7
IV	19.4	23.1	325.5	232.0
V	129.0	206.0	198.0	168.0
	NO_3^-	F^-	TDS	
I	21.3	0.6	597.1	
II	24.0	0.8	2070.6	
III	20.0	0.3	323.0	
IV	1.0	1.0	669.6	
V	60.0	8.0	851.9	

Note: I: Shallow groundwater in the northern basin; II: Shallow groundwater in the western basin; III: Karst water in the northern mountain area; IV: Karst water in the western mountain area; V: Fen River water; VI: Groundwater in the northern basin; VII: Groundwater in the western basin; VIII: Groundwater in the southern basin; solution1: III: I: V = 70%: 10%: 20% solution2: IV: II = 50%: 50%; solution3: VI: VII = 50%: 50%.

Table 3. Saturation indices of aragonite, calcite, dolomite, and gypsum in the mixed solution and groundwater.

Mineral	Aragonite	Calcite	Dolomite	Gypsum
Solution1	0.18	0.32	0.63	−1.72
VI	0.19	0.33	0.70	−2.37
Solution2	0.40	0.54	0.89	−0.66
VII	0.43	0.58	1.04	−0.97
Solution3	0.42	0.56	1.00	−1.19
VIII	0.51	0.66	1.17	−1.36

The mixing model shows that when the mixing ratio of groundwaters of the northern and the western basins is 50:50, the saturation indices of the four main minerals of the mixing solution best fit with those of groundwater of the western basin (Table 3). It proves that the HCO_3 water of the southern basin is recharged by groundwater of the northern and western basin.

Geochemical modeling of the groundwater of the eastern basin was not made due to lack of data for the hydrochemistry of karstic water in the eastern mountains. Moreover, besides the mixing action, other hydrochemical processes such as ion exchange and evaporation should have effects on the hydrochemistry of groundwater of the eastern and western basin. However, the mixing process is the most important among all the hydrogeochemical processes for groundwaters from Quaternary aquifers at Taiyuan.

4 CONCLUSIONS

1. Hydrochemical zones have been identified at Taiyuan in this study. From the boundary to the center of the basin, the hydrochemical type of groundwater in Quatenary aquifers sequentially changes from SO_4-HCO_3 through HCO_3-SO_4 to HCO_3 water.
2. Mixing is the most important process in forming the hydrochemistry of groundwaters at Taiyuan. Groundwater of the northern basin was recharged by karstic water of the northern mountains, the shallow groundwater of the northern basin, and Fen River water, with karstic water of the northern mountains being the most important source of recharge. Groundwater of the western basin is recharged by karstic water from the western mountains and shallow groundwater from the western basin. Hydrochemistry of groundwater in the southern basin is affected by the mixing of groundwater from the western basin with that from the northern basin.

ACKNOWLEDGEMENTS

The research work was financially supported by National Natural Science Foundation of China (NSFC-49832005), the Doctoral Thesis Funding Program and the Teaching and Research Award Program for Outstanding Young Teachers in Higher Education Institutions of Ministry of Education, P. R. China.

REFERENCES

Naito, K, Fukahori, Y. & Peiming, H. 1995. Oxygen and carbon isotope zonations of wall rocks around the Kamioka Pb-Zn skarn deposits, central Japan: application to prospecting. *J. Geochem. Explor.* 54: 199–211.

Robion, P., Kissel, C. & Lamotte, D.F. 1997. Magnetic mineralogy and metamorphic zonation in the Ardennes Massif (France-Belgium). *Tectonophys.* 271: 231–248.

Wogelius, R.A., Fraser, D.G. & Wall, R.T. 1997. Trace element and isotopic zonation in vein calcite from the Mendip Hills, UK, with spatial-process correlation analysis. *Geochim. Cosmochim. Acta* 61: 2037–2051.

Water–Rock Interaction, Wanty & Seal II (eds)
© 2004 Taylor & Francis Group, London, ISBN 90 5809 641 6

Geochemistry and fluxes of strontium and barium in subsurface water in the central part of the Russian plate

V.P. Zverev & I.A. Kostikova
Institute of Environmental Geoscience RAS, Moscow, Russia

N.N. Bogdanova
Center region monitoring, Moscow, Russia

ABSTRACT: Strontium and barium fluxes in subsurface water are estimated and mapped for the first time in subsurface water using the central part of the Moscow artesian basin as an example. It is quantitatively shown that transport of strontium and barium by subsurface water depends both on the content of these elements in Carboniferous and Devonian limestone-dolomite rocks and on the water-exchange intensity in the upper parts of the Earth's Crust.

1 INTRODUCTION

Studying fluxes of chemical elements in subsurface water caused by its interaction with rocks, which controls chemical weathering and many other exogenous geological processes, is one of the major problems in modern fundamental hydrogeology. The scales of transfer of major dissolved components by subsurface water were estimated by Zverev (1982, 1999) both on regional and global levels. To gain a more comprehensive idea of water-rock interaction processes, it is essential to study the flux in subsurface water of trace components derived from rock.

2 HYDROGEOCHEMISTRY

Hydrogeochemical features were investigated and fluxes of strontium and barium (only stable isotopes) were estimated, the latter being the chemical elements of the second group of the Periodic system, which are syngenetic to limestone-dolomite and sulfate deposits of the upper Devonian and Carboniferous age in the central part of the Russian plate. This area represents a watershed of the Baltic, Black, Azov and Caspian seas, where the largest rivers of European Russia, i.e., Volga, Oka, Don, Dnepr and Zapadnaj Dvina Rivers originate. The investigated area represents the central part of the Moscow artesian basin (Hydrogeology 1966), where hydrodynamic features and groundwater chemical composition are mainly determined by the geological structure of the Moscow syncline.

In the Moscow artesian basin, groundwater chemical composition follows general hydrogeochemical characteristics with zones of elevated concentrations of strontium and barium. These zones are a result of abnormally high content of strontium and barium sulfate minerals in rocks (Bogdanova 2003).

Limestones and dolomitic carbonate rocks, with interspersed layers of lagoonal sulfate-dolomitic rocks, are the principal aquifers in the central part of the Moscow artesian basin.

Syngenetic celestite is found in dolomites of the Ozersk horizon in the peripheral part of the sulfate-dolomitic zones. The thickest and most extensive layers of celestite were formed by a sharp change of depositional conditions at the gypsum-dolomite contact. These formations were undoubtedly the source of widespread epigenetic celestite, which is traced in a border zone of gypsum dissolution. The maximum content of strontium (more than 4000 g/t) is observed in the Devonian aquifers in the Don River basin (Evseeva 1999).

In the studied area, barite in Carboniferous deposits is the most common of the barium minerals. According to spectral analysis data, the maximum content of barium is registered in the terrigenic carbonate sediments of the Kashira horizon.

Chemical elements are leached out of rock in the study area by subsurface water predominantly in the zone of active water exchange mainly encompassing the upper two water-bearing horizons. This water is relatively low-TDS bicarbonate-calcium (more rarely, magnesium-calcium) water (200–750 mg/l), with strontium and barium concentrations fluctuating within a relatively wide range (1–21 and 0.02–0.25 mg/l, respectively) (Table 1).

Thermodynamic modeling of element speciation in subsurface water of the upper Devonian and Carboniferous limestone-dolomitic formation in the central part of the Moscow artesian basin was performed using SOLMINEQ–88 (Perkins et al. 1990), and showed that strontium and barium migrate predominantly in the form of Sr^{2+} and Ba^{2+} ions, with the concentration of neutral species $SrSO_4^{\circ}$ and $BaSO_4^{\circ}$ being an order of magnitude lower. This modeling also allowed us to assess the equilibrium state of subsurface water with minerals of the limestone-dolomite aquifer (Table 2).

Table 1. Saturation index (log Q/K_0 10°C) of minerals in the water-rock system.

Mineral	C_2	C_2	C_1	C_1	D_3	D_3
Anhydrite	−2.734	−2.113	−1.929	−1.796	−0.712	−2.009
Aragonite	−1.166	−0.299	−0.364	−0.387	0.053	−0.095
Barite	−0.478	0.288	0.944	1.008	–	–
Calcite	−1.012	−0.146	−0.211	−0.234	0.207	0.058
Celestite	−2.310	−0.937	−0.491	−1.316	−0.272	−0.793
Dolomite	−1.299	0.732	0.588	0.261	1.241	0.885
Gypsum	−2.314	−1.694	−1.510	−1.377	−0.293	−1.590
Magnesite	−1.938	−0.773	−0.852	−1.156	−0.617	−0.824
Quartz	0.034	0.636	0.432	0.638	–	–
Strontianite	−1.975	−0.357	−7.468	−1.141	−0.740	−0.113
Witherite	−4.479	−3.467	−3.060	−3.152	–	–

Table 2. Chemical composition of subsurface waters of Carboniferous and Devonian limestone-dolomite sediments, mg/l.

Components	C_2	C_2	C_1	C_1	D_3	D_3
pH	6.95	7.3	7.35	6.85	7.06	7.37
M	230	588	588	763	2072	558.5
Cl	15	38	25	38	10.2	15.3
SO_4^{2-}	26	95	156	124	1229.9	124.7
HCO_3^-	126	295	246	397	408.8	427.1
Na^+	6.5	20.5	34.2	14.3	65.75	96.1
K^+	2.5	12.8	13	9.2	–	–
Ca^{2+}	38	67	64	123	352.7	66.1
Mg^{2+}	12	42	39	39	143.4	23.1
Sr^{2+}	1.2	11.9	20.6	4.4	11.0	12.8
Ba^{2+}	0.021	0.053	0.15	0.249	0.0	0.0
SiO_2	4	16	10	16	–	–

It has been established that ground water does not attain equilibrium with the principle rock-forming mineral (i.e., calcite) and, in most cases, it occurs in near-equilibrium with dolomite. Almost all water samples showed nonequilibrium with strontium-containing minerals (celestite and strontianite). As for the barium-containing minerals, barite turned out to be, for the most part, near equilibrium with subsurface water and witherite greatly undersaturated.

3 FLUXES

The equilibrium-nonequilibrium conditions of subsurface water with strontium and barium minerals shows favorable conditions for leaching strontium as well as redistribution and leaching of barium in the upper aquifers in the central part of the Moscow artesian basin.

The fluxes of groundwater to the surface were estimated according to the ground water input to rivers. These fluxes were calculated as a half-sum of water-flow discharges in winter and summer low-water 30-day-long periods from 1981 until 1990, published in annual State water summaries. The data on 310 hydrological posts in basins of the Volga, Oka, Don, Dnepr, and Zapadnaj Dvina Rivers, as well as their numerous tributaries, were used. These data formed the basis for mapping groundwater flux modules in $l/s \cdot km^2$ for the central part of the Moscow artesian basin to a scale of 1:500,000. As a result (Table 3), it was found that the groundwater flux is

Table 3. Fluxes of chemical elements in subsurface waters of the central part of the Moscow artesian basin.

See basin	River	Area, 10^3 km²	Normalized subsurface water fluxes, $l/s \cdot km^2$	Subsurface water fluxes, 10^3 l/s	Fluxes, g/s Σ	Ca^{2+}	SO_4^{2-}	Cl^-	Sr^{2+}	Ba^{2+}
Baltic	Z.Dvina	13.54	1.16	15.66	5275.62	1340.81	558.91	379.54	57.58	0.21
Black	Dnepr	33.26	1.75	58.26	21327.76	4230.48	2109.49	394.41	154.85	1.18
Azov	Don	15.07	1.79	27.04	10447.79	1618.38	1866.80	427.19	156.41	0.00
Caspian	Volga	31.21	1.72	53.62	20984.24	4262.87	2505.18	372.42	87.45	4.33
	Oka	90.21	1.76	158.57	72508.52	12643.01	7105.05	1184.71	512.97	10.33
All region		183.29	1.71	313.15	130543.93	24095.55	14145.43	2758.27	969.26	16.05

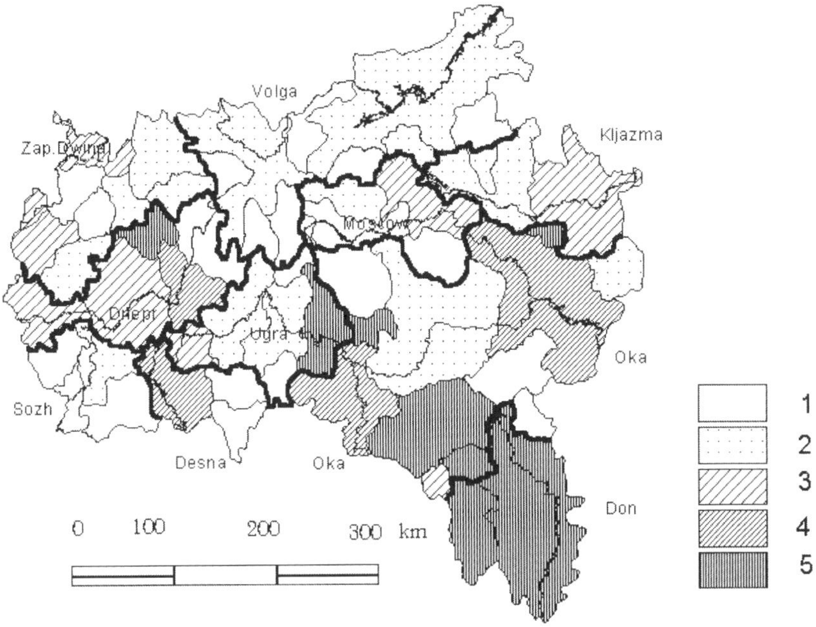

Figure 1. Strontium fluxes in subsurface water of the central part of the Moscow artesian basin. The normalized fluxes of strontium in subsurface water of mg/l · km^2: 1 – <2.5; 2 – 2.5–5; 3 – 5–7.5; 4 – 7.5–10; 5 – >10.

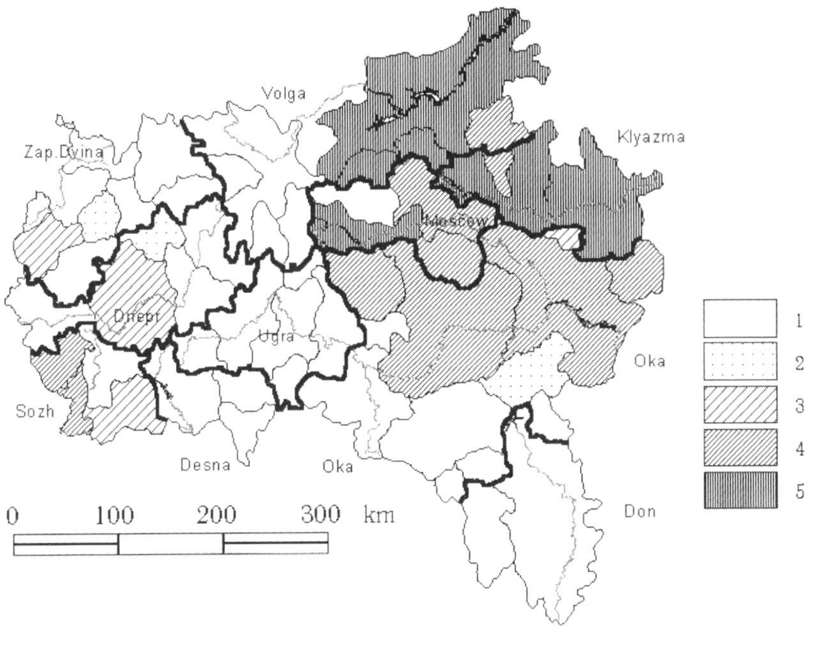

Figure 2. Barium fluxes in subsurface water of the central part of the Moscow artesian basin. The normalized fluxes of barium in subsurface water of mg/l · km^2: 1 – <0.025; 2 – 0.025–0.05; 3 – 0.05–0.02; 4 – 0.01–0.02; 5 – >0.2.

Table 4. Strontium fluxes in subsurface waters of the central part of the Moscow artesian basin.

Region of discharge	Area, $10^3\,km^2$	Subsurface water fluxes, 10^3 l/s	Normalized fluxes of subsurface waters, l/s · km^2	Average concentration of strontium, mg/l	Strontium fluxes, 10^{-3} g/s	Normalized fluxes of strontium, 10^{-3} g/s · km^2
Devonian and Carboniferous rocks on the South of Region	26.36	38.01	1.44	3.93	152.39	5.78
Devonian and Carboniferous rocks on the West of Region	33.61	52.23	1.55	6.06	364.68	10.85
The rest part of Region	123.32	222.91	1.80	2.02	452.19	3.67
The Center of the Moscow artesian basin	183.29	313.15	1.71	3.09	949.26	5.29

Table 5. Barium fluxes in subsurface waters of the central part of Moscow artesian basin.

Region of discharge	Area, $10^3\,km^2$	Fluxes of subsurface waters, 10^3 l/s	Normalized fluxes of subsurface waters, l/s · km^2	Average concentration of strontium, mg/l	Barium fluxes, 10^{-3} g/s	Normalized fluxes of barium, 10^{-3} g/s · km^2
Upper and middle Carboniferous rocks on the Nord-East of Region	75.59	138.75	1.83	0.099	13.81	0.183
The rest part of Region	107.70	174.4	1.62	0.013	2.25	0.021
The Center of the Moscow artesian basin	183.29	313.15	1.71	0.051	16.05	0.087

equal to 313×10^3 l/s in the area of $183. \times 10^3\,km^2$, with the average module being equal to 1,71 l/s · km^2.

By using several thousand chemical analyses of subsurface water discharging to rivers, the total fluxes of major chemical components, as well as Ca^{2+}, SO_4^{2-}, Cl^-, Sr^{2+} and Ba^{2+} ions, have been evaluated (Table 3). As seen in the table, strontium and barium fluxes vary from 0.58 to 20.54 and from 0.006 to 0.364 mg/s · km^2, respectively. On the basis of these analyses, strontium and barium fluxes were mapped across the basin for the first time. The normalized fluxes of strontium and barium by subsurface water in mg/s · km^2 within the limits of minor river basins were used for mapping at a scale of 1:500,000. The schematic maps of Sr and Ba fluxes in subsurface water are given in Figures 1 and 2.

Mapping of strontium and barium fluxes showed significant heterogeneity of these parameters in the study area, with the following patterns. The greatest leaching of strontium occurs in the south and southwest, where Carboniferous and Devonian sediments with elevated strontium contents are found (Fig. 1). Barium is mainly leached in the northwestern part of the study area, where the upper and middle Carboniferous deposits with high concentrations of this element in limestone and dolomite are found (Fig. 2). These patterns are apparent in Tables 4 and 5, showing that normalized fluxes of strontium and barium in these areas greatly exceed their average values both for the entire investigated area and, in particular, for the areas in which rocks have much lower amount of strontium and barium.

4 CONCLUSIONS

A close relationship is established between the content of such elements as Sr and Ba in the Devonian and Carboniferous limestone-dolomite deposits and the fluxes of these elements in subsurface water. Assuming celestite and barite to be the major sources of strontium and barium, respectively, we determined that their leaching intensity for the central part of the Russian plate, is equal to 6.29×10^7 and 0.89×10^7 g/year, respectively, in the area of about $183 \times 10^3\,km^2$. The research preformed permitted us for the first time to assess quantitatively and to map strontium and barium fluxes, the minerals of which are syngenetic to limestone and dolomite deposits in the central part of the Russian plate. It is established that their fluxes depend both on the content of these elements in rocks and subsurface water and on the water-exchange intensity in the upper horizons of the Moscow artesian basin.

ACKNOWLEDGMENTS

This work was financially supported by the Russian Foundation for Basic Research, project no. 02-05-64098a.

REFERENCES

Bogdanova, N.N. 2003. Natural hydrogeochemical anomalies of strontium and barium in the central region. *Proceedings of the Conference "Sergeevskie chtenia"* 5: 313–317. Moscow: GEOS.

Evseeva, V.I. 1999. Prospects in hydrogeochemical research of strontium-bearing deposits in the southwestern slope of the Moscow syncline. *Geologicheskii vestnik tsentral'nykh raionov Rossii* 4: 16–23.

Hydrogeology of the USSR, v.1. 1966. Moscow and Central districts. Moscow: Nedra.

Perkins, E.H., Kharaka, Y.K., Gunter, W.D. & Debraal, J.D. 1990. Geochemical modeling of water–rock interaction using SOLMINEQ 88. *ACS Symposium series* 416: 117–127.

Zverev, V.P. 1982. Role of subsurface water in migration of chemical elements. Moscow: Nedra.

Zverev, V.P. 1999. Massflows of the subsurface hydroshere. Moscow: Nauka.

Carbon dioxide and hydrogen sulfide sequestration

Water–Rock Interaction, Wanty & Seal II (eds)
 ISBN 90 5809 641 6

Investigations of dawsonite-forming reactions for geological CO_2 sequestration

P. Bénézeth, D.A. Palmer, L.M. Anovitz, D.J. Wesolowski & J.G. Blencoe
Oak Ridge National Laboratory, Oak Ridge TN, USA

ABSTRACT: Identification and characterization of mineral assemblages that will influence mineral trapping and caprock integrity are essential to determine the economic, technical and environmental feasibility of sequestering large amounts of CO_2 in deep saline aquifers. Numerous modeling studies indicate that injection of CO_2 into these formations will stabilize various carbonate minerals, including calcite ($CaCO_3$), dolomite [$CaMg(CO_3)_2$], magnesite ($MgCO_3$), siderite ($FeCO_3$), and dawsonite [$NaAlCO_3(OH)_2$]. However, the equilibria and kinetics of carbonate formation, which determine the extent and rate of formation of stable carbonate minerals, are poorly understood or lacking at high CO_2 pressures. Very few experiments are being performed to validate and improve computer codes for modeling CO_2 sequestration in deep saline formations. In this study, we propose to measure, *in situ* and in real time, the mineralization equilibria and kinetics pertinent to this critically important technology for CO_2 disposal.

1 INTRODUCTION

Global warming, observed since the beginning of the industrial revolution (Marchetti 1977), is being attributed largely to the accumulation of anthropogenic CO_2 in the Earth's atmosphere. Among the many solutions posed to reduce further emissions and to remediate the existing problem, capture and storage of CO_2 in geologic formations is one of the most promising approaches. CO_2 can be sequestered in geologic media: (1) through injection of CO_2 associated with enhanced oil recovery; (2) by storage in depleted oil and gas reservoirs; (3) through replacement of absorbed methane by CO_2 in deep coalbeds; (4) by injection into deep saline aquifers, as demonstrated at the Sleipner gas field in the Norwegian North Sea, (5) through storage in salt caverns; and (6) by replacement of CH_4 in seafloor clathrate hydrates (Yezdimer et al. 2002).

This study performed experiments to constrain conditions suitable for CO_2 sequestration in deep saline aquifers, which represent one of the most promising geological options. They have the largest potential storage capacity for CO_2 (estimates are on the order of several 100 y of anthropogenic CO_2 emissions) because they are widespread (Bergman & Winter 1995). They are close to CO_2 emission sources thereby reducing the cost of CO_2 transportation. Suitable aquifers also exist at depths well below freshwater aquifers (3000 to 12,000 feet). Since the carbonates (e.g. calcite) formed already exist in vast quantities in nature, mineral sequestration provides a permanent and environmentally benign solution. Major issues for CO_2 sequestration in deep saline aquifers, in addition to minimizing the costs of CO_2 injection, and gaining public acceptance, are: injecting the CO_2 into isolated, subterranean rock formations ("hydrodynamic trapping"); determining the permeability of the host formation; reducing the risk of CO_2 leakage through fault zones, and further ensuring long-term CO_2 retention by studying the impacts of mineral-brine-CO_2 reactions on the integrity of the overlying seal ("caprock"). Several projects in the USA and foreign countries are examining the feasibility of injecting and storing CO_2 in deep saline aquifers. Only one commercial deployment exists in Norway, where approximately 10^6 metric tons of CO_2 are being injected annually since 1996 into the saline Utsira Formation at the Sliepner natural gas field in the North Sea.

2 BACKGROUND

The sequestration reactions of prime interest are: "solubility trapping" (dissolution and long-term retention of CO_2 in saline groundwater), and "mineral trapping" – the reaction of CO_2 with non-carbonate minerals to form stable carbonates (Dove et al. 1998, Johnson et al. 2001, 2002). When high-pressure CO_2 is injected into deep aquifers, it acidifies the water through the reaction:

$$CO_{2(aq)} + H_2O \Leftrightarrow H_2CO_{3(aq)} \Leftrightarrow H^+ + HCO^-_{3(aq)} \quad (1)$$

This increases the dissolution rate of minerals in place in the host rock (e.g. silicates), releasing cations that will in turn react with HCO_3^- (and possibly $CO_{2(aq)}$) to form new minerals, including various carbonates. Computer simulations (e.g. GEO-SEQ, Johnson et al. 2001, 2002) indicate that reactions between silicate minerals, water, and injected CO_2 will produce expected calcite-group carbonates (e.g. calcite, siderite, dolomite, and magnesite) according to the reactions:

$$Ca^{2+} + HCO_3^- \Leftrightarrow CaCO_3 + H^+ \quad (2)$$

$$Fe^{2+} + HCO_3^- \Leftrightarrow FeCO_3 + H^+ \quad (3)$$

$$Ca^{2+} + Mg^{2+} + 2HCO_3^- \Leftrightarrow CaMg(CO_3)_2 + 2H^+ \quad (4)$$

$$Mg^{2+} + HCO_3^- \Leftrightarrow MgCO_3 + H^+ \quad (5)$$

Perhaps more importantly, formation of substantial quantities of dawsonite and quartz is predicted in the "intra-aquifer plume" (Johnson et al. 2001, 2002) by reaction with potassic feldspars:

$$KAlSi_3O_8 + Na^+ + CO_{2(aq)} + H_2O \Leftrightarrow NaAlCO_3(OH)_2 + 3SiO_2 + K^+ \quad (6)$$

which shows that dawsonite precipitation is favored by: high Na^+ concentrations in the brine, elevated solubility of CO_2, and acid-induced K-feldspar (orthoclase) dissolution. In fact, there are natural analogs for this process in a variety of settings, for example, the presence of pervasive dawsonite cement in the Bowen-Gunnedah-Sydney Basin, Eastern Australia, which has been interpreted to reflect magmatic CO_2 seepage on a continental scale (Baker et al. 1995).

Simulations performed by Zerai et al. (2002) using Geochemist Workbench™ of CO_2 sequestration in Ohio's deep saline aquifers (The Rose Run sandstone aquifer) indicate that under certain conditions, dissolution of albite ($NaAlSi_3O_8$, an abundant feldspar in that aquifer) can lead to mineral trapping of CO_2 in dawsonite through a reaction analogous to reaction (6). This reaction can be broken down into two separate processes: (i) dissolution of feldspar (reactions involving Na- or Ca-rich feldspars can be substituted) by exposure to CO_2-induced acidic conditions:

$$KAlSi_3O_8 + 4H^+ \Leftrightarrow K^+ + Al^3 + 3SiO_2 + 2H_2O \quad (7)$$

followed by (ii) precipitation of dawsonite by interaction of the released Al with the plume fluids, high in Na^+ and $CO_{2(aq)}$:

$$Na^+ + Al^{3+} + CO_{2(aq)} + 3H_2O \Leftrightarrow NaAlCO_3(OH)_2 + 4H^+ \quad (8)$$

The importance of the above reactions for (i) CO_2 sequestration in mineral phases, (ii) changes in aquifer porosity and permeability, and (iii) caprock integrity (e.g. through fracture sealing), and (iv) validation of reservoir processes and parameters predicted by computer models, underscores the need for extensive and reliable experimental data on silicate-carbonate reaction equilibria and kinetics under conditions similar to those expected in saline aquifers during and after CO_2 injection (a partial pressure of CO_2 as high as 100 bars, brine salinities between 10 and 30 weight percent NaCl, and temperatures from 30 to 80°C). It is noteworthy that there are currently no data available to adequately assess either the equilibrium solubility of dawsonite or the rates of relevant reactions controlling precipitation of dawsonite via reaction (6). In most reactive geochemical transport simulations, the rate constant for the dawsonite has been set to either K-feldspar (Xu et al. 2003 from Blum & Stillings 1995) or estimated as intermediate to calcite and magnesite (Johnson et al. 2001, 2002).

The objective of this study is to investigate *in situ* the effect of pCO_2, pH and salinity on solubility and precipitation rates of dawsonite through reactions (6) & (8), and the processes controlling them. The actual speciation of Al^{III} in the simulated brines varies with pH (Palmer et al. 2001, Bénézeth et al. 2001a); it is not known which aqueous Al, carbonate, and other brine species will actually be the rate-determining species for reactions such as (6) and (8) above.

3 APPROACH

3.1 *Experimental procedure*

Coupling of reactions (1) and (3) through mass action equilibria dictates that if the pH of the solution is controlled by reaction (1), then any reaction that consumes CO_2 (e.g. reactions 6 and 8) must also perturb solution pH. In our previous studies we have demonstrated the ability to accurately measure *in situ* pH during homogeneous and heterogeneous aqueous reactions over long periods of time (days to weeks) at

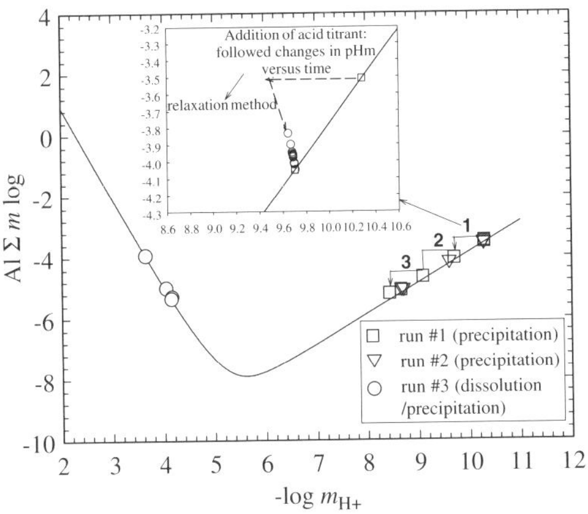

Figure 1. Kinetics of gibbsite precipitation at 50°C in 0.1 molal ionic strength, NaCl.

temperatures from 0 to 300°C and salinities ranging up to NaCl saturation, using our hydrogen-electrode concentration cell (HECC) described in numerous publications (e.g. Palmer et al. 2001). We also have investigated mineral dissolution and precipitation reactions of boehmite, AlOOH, and gibbsite, $Al(OH)_3$, in basic NaCl solutions at 50–250°C (Bénézeth et al. 2001b), and demonstrated that the ability to perturb pH isothermally by addition of acidic or basic titrant allows studies of the kinetics of dissolution/precipitation at near-equilibrium conditions, even for relatively fast reactions. An example is given in Figure 1 for gibbsite precipitation runs, obtained in a single experiment at 50°C and 0.1 molal NaCl.

After letting the solid equilibrate with a basic solution (point 1, square symbol in Figure 1), an acid titrant is added (step 1), the solution composition moves off the curve (see inset in Fig. 1), then relaxes back to the equilibrium state with pH and total aluminum varying according to the stoichiometry of the net reaction. Solution pH can then be changed again by adding more titrant (steps 2, 3,...), and precipitation/dissolution cycles can be performed and repeated in a single experiment. Therefore, in such experiments, pH is a direct and sensitive indicator of reaction progress, which can be continuously and precisely monitored using this potentiometric method. The resolution of our instrumentation permits observation of pH changes of <0.001 log units, or 0.25% in the concentration of H^+. We suspect, however, that monitoring pH alone, even coupled with solution sampling to determine the changes in dissolved metal concentrations with time, will be insufficient to quantify the rates of reactions (6) or (8).

A powerful constraint on the reaction mechanism would be the ability to track changes in the concentrations of dissolved carbonate species throughout the course of each experiment. Theoretically, this could be done very accurately by monitoring the system pressure (generated by the coexisting vapor phase consisting of gaseous H_2O, CO_2 and H_2 (the latter gas being needed to allow the HECC to function as a pH-measurement system). This, coupled with (i) *in situ* pH monitoring, together with sampling and analysis for total carbonate and metal contents, (ii) data on the solubility of CO_2 gas in NaCl brines (i.e. the Henry's Law Constant, available for 0–6 molal NaCl brines from Drummond 1981; Corti et al. 1990), and (iii) the dissociation constants of carbonic acid and bicarbonate (Patterson et al. (1982, 1984), should allow accurate tracking of mass balance. Under such conditions, there is no reason why both equilibrium and kinetic experiments cannot be performed under significant CO_2 partial pressures. However, reaction (6) is problematic in that "protolytic side reactions," which consume H^+, such as the dissolution of feldspar via reaction (7), will complicate the simple relationship between the progress of reaction (6) and the change in pH of the solution. Some information is available on the dissolution kinetics of orthoclase, but virtually nothing is known about the solubility and dissolution/precipitation rates of dawsonite (reaction 8), which are discussed below.

3.2 *Dawsonite*

As already mentioned dawsonite is a naturally occurring crystalline material that is believed to be formed by a natural CO_2 sequestration process at depth in the Earth's crust. It was first discovered in 1862 at the Old Francon quarry on the island of Montreal city. Since then, dawsonite has been found in widely distributed localities worldwide. In the USA, dawsonite occurs in relative abundance in the Eocene Green River Formation oil shales at the Piceance Creek Basin of northwestern Colorado (Smith & Milton 1966), which has been estimated to contain 26 billion metric tons of dawsonite (e.g. Hite & Dyni 1967). Synthetic orthorhombic dawsonite was first obtained by Bader (1938) and later, by Besson et al. (1973). Patents have also been issued on dawsonite synthesis as it is widely used as an antacid and as a flame retardant material.

We have synthesized dawsonite by homogeneous nucleation (e.g. reacting an aqueous sodium aluminate solution with sodium bicarbonate urea, and hydroxycarboxylic acid) following various patented procedures (e.g. Herold 1992). We are still refining the procedures, but we have already produced dawsonite in highly pure form as confirmed by XRD. Figure 2 is an SEM photograph of dawsonite crystals we have synthesized that will be used for measuring

Figure 2. SEM photograph of synthesized dawsonite.

solubility and dissolution/precipitation rates in aqueous carbonate solutions. Following that, we will perform experiments to characterize reaction (6) at fixed pH, as the kinetics of this reaction are likely to be pH dependent.

4 CONCLUDING REMARKS

The kinetic and solubility data we plan to obtain will allow us to precisely determine which species control dawsonite dissolution and precipitation rates. Careful characterization of K-feldspar and products of reactions (6) and (8) will be incorporated into this effort. It will be important to determine the crystallinity and chemical composition of the surface layers of the mineral powders before and after dissolution/precipitation cycles. An array of analytical methods will be utilized in this work, including FTIR, XRD, SEM, Raman and BET.

ACKNOWLEDGMENTS

Research sponsored by the Division of Chemical Sciences, Geosciences, and Biosciences, Office of Basic Energy Sciences, U.S. Department of Energy, under contract DE-AC05-00OR22725 with Oak Ridge National Laboratory, managed and operated by UT-Battelle, LLC.

REFERENCES

Bader, E. 1938. Formation, constitution and synthetic preparation of dawsonite. *Neues Jahrb. Miner. Geol. Paleont.* B 74A: 449–465.

Baker, J.C., Bai, G.P., Hamilton, P.J., Golding, S.D. & Keene, J.B. 1995. Continental-scale magmatic carbon dioxide seepage recorded by dawsonite in the Bowen-Gunnedah-Sydney Basin system, eastern Australia. *J. Sed. Res.* 65: 22–530.

Bénézeth, P., Palmer, D.A. & Wesolowski, D.J. 2001a. Aqueous high temperature solubility studies. II. The solubility of boehmite at 0.03 M ionic strength as a function of temperature and pH as determined by "in situ" measurements. *Geochim. Cosmochim. Acta* 65: 2097–2111.

Bénézeth, P., Palmer, D.A., Wesolowski, D.J., Xiao, C. & Wood, S.A. 2001b. Solubility and reaction rates of oxides and hydroxides to high temperature with *in situ* pH measurement. In R. Cidu (ed.), *Proc. Tenth Intl. Symp. Water–Rock Interaction*: 235–238. Rotterdam: Balkema.

Bergman, P.D. & Winter, E.M. 1995. Disposal of CO_2 in aquifers in the U.S. *Energy Convers. Mgmt.* 36: 523–526.

Besson, H., Caillere, S., Henin, S. & Prost, R. 1973. Formation expérimentale et conditions de gisement de la dawsonite. *C.R. Acad. Sci. Ser.D* 277: 261.

Blum, A.E. & Stillings, L.L. 1995. Feldspar dissolution kinetics: In A.F. White & S.L. Brantley (eds.), *Chemical Weathering Rates of Silicate Minerals: Rev. Min.*31: 291–351. Washington, DC: Min. Soc. Amer.

Corti, H.R., De Pablo, J.J. & Prausnitz, J.M. 1990. Phase equilibria for aqueous systems containing salts and carbon dioxide: application of Pitzer's theory for electrolyte solutions. *J. Phys. Chem.*94: 7876–7880.

Dove, P., Harris, J., Logan, J.M., Pyrak-Nolte, L., Orr, F.M. Jr., Ortoleva, P.J., Richter, F., Rudnicki, J.W., Warpinski, N.R., Wawersik, W.R., Wilson, J.L., & Wong, T.F. 1998. Terrestrial Sequestration of CO_2 – Assessment of Research Needs. In W.R. Wawersik & J.W. Rudnicki (eds.). *Report to the Office of Basic Energy Sciences (OBES), US Dept. Energy.*

Drummond, S.E. 1981. *Boiling and Mixing of Hydrothermal Fluids: Chemical Effects on Mineral Precipitation*. Ph.D. Thesis, The Pennsylvania State University.

Herold, G. 1992. Process for producing dawsonite. US patent 5,078,983.

Hite, R.J. & Dyni, J.R. 1967. Potential resources of dawsonite and Nahcolite in the Piceance Creek Basin, Northwest Colorado. *Quarterly of the Colorado School of Mines* 62: 25–38.

Johnson, J.W., Nitao, J.J., Steefel, C.I. & Knauss, K.G. 2001. Reactive transport modeling of geologic CO_2 sequestration in saline aquifers: the influence of intra-aquifer shales and the relative effectiveness of structural, solubility, and mineral trapping during prograde and retrograde sequestration. *NETL Publications 2001, 1st Natl. Conf. Carbon Sequestration.*

Johnson, J.W., Nitao, J.J. & Steefel, C.I. 2002. Fundamental elements of geologic CO_2 sequestration in saline aquifers. *ACS Fuel Chemistry Division Symposia Preprints* 47: 41–42.

Marchetti, C. 1977. On geoengineering and the CO_2 problem. *Climate Change* 1: 59–68.

Palmer, D.A., Bénézeth, P. & Wesolowski, D.J. 2001. Aqueous high temperature solubility studies. I. The solubility of boehmite at 150°C as a function of ionic strength and pH as determined by "in situ" measurements. *Geochim. Cosmochim. Acta* 65: 2081–2095.

Patterson, C.S., Slocum, G.H., Busey, R.H. & Mesmer, R.E. 1982. Carbonate equilibria in hydrothermal systems: First

ionization of carbonic acid in NaCl media to 300°C. *Geochim. Cosmochim. Acta* 46: 1653–1663.
Patterson, C.S. Busey, R.H. & Mesmer, R.E. 1984. Second ionization of carbonic acid in NaCl media to 250°C. *J. Solution Chem.* 13: 647–661.
Smith, J.W. & Milton, C. 1966. Dawsonite in the Green River formation of Colorado. *Econ. Geol.* 61: 1029–1042.
Xu, T., Apps, J.A. & Pruess, K. 2003. Reactive geochemical transport simulation to study mineral trapping for CO_2 disposal in deep arenaceous formations. *J. Geophys. Res.* 108 (B2): 2071.
Yezdimer, E.M., Cummings, P.T. & Chialvo, A.A. 2002. Determination of the Gibbs free energy of gas replacement in sI clathrate hydrates by molecular simulation. *J. Phys. Chem. A* 106: 7982–7987.
Zerai, B., Matisoff, G. & Saylor, B.Z. 2002. Modeling of CO_2 – Brine – Rock Reaction in a mixed carbonate – siliciclastic aquifer. *Geol. Soc. Am.* 34, 6: 194–195.

Water–Rock Interaction, Wanty & Seal II (eds)
© 2004 Taylor & Francis Group, London, ISBN 90 5809 641 6

The effect of $CO_{2(aq)}$, $Al_{(aq)}$ and temperature on feldspar dissolution

S.A. Carroll & K.G. Knauss
Lawrence Livermore National Laboratory, Livermore, California, USA

ABSTRACT: We measured labradorite ($Ca_{0.6}Na_{0.4}Al_{1.6}Si_{2.4}O_8$) dissolution rates using a mixed flow reactor from 30 to 130°C as a function of CO_2 (3×10^{-3} and 0.6 M), and Al (10^{-6} to 10^{-3} M) at pH 3.2. Over these conditions, labradorite dissolution can be described with a single rate expression that accounts for observed increases in dissolution rate with temperature and decreases in dissolution rate with dissolved Al:

$$\mathrm{Rate_{Si}}(\mathrm{mol\ cm^{-2}\ s^{-1}}) = k'' \times 10^{-Ea/2.303RT} \left[\frac{\left(\frac{a_{H+}^{3n}}{a_{Al3+}^{n}}\right) K_T}{\left(1 + K_T\left(\frac{a_{H+}^{3n}}{a_{Al3+}^{n}}\right)\right)} \right]. \tag{A1}$$

where the apparent dissolution rate constant, $k'' = 10^{-5.69}$ (mol Labradorite $cm^{-2}s^{-1}$); the net activation energy, $E_a = 42.1$ (kJ mol^{-1}); H^+-Al^{3+} exchange coefficient, $n = 0.31$; and silica-rich surface complex formation constant, $K_T = 4.5$ to 5.6 from 30 to 130°C. The effect of $CO_{2(aq)}$ on mineral dissolution is accounted for by changes in solution pH. At temperatures below 60°C, labradorite dissolves incongruently with preferential dissolution of Na, Ca and Al over Si.

1 INTRODUCTION

Geologic sequestration of CO_2 in sedimentary basins or in saline aquifers appears to be a promising method for disposal, because the reaction of CO_2, water, and primary silicate minerals produces more stable secondary silicates, hydroxides, and carbonates. Measurement of mineral dissolution and precipitation rates is fundamental to the successful disposal of greenhouse CO_2-rich gases in aquifers, because reservoir storage capacity is related to the conversion of $CO_{2(g)}$ to carbonate minerals, and because secondary mineral precipitation will alter reservoir porosity and permeability. Our interest in quantifying labradorite ($Ca_{0.6}Na_{0.4}Al_{1.6}Si_{2.4}O_8$) dissolution rates with a single rate equation as a function of pCO_2, dissolved Al, and temperature was motivated by the need to understand the source term for secondary mineral precipitation caused by the neutralization of injected acid CO_2 by rock-water interactions. Labradorite dissolution rates were measured in acidic waters saturated with supercritical CO_2 or with atmospheric CO_2 as a function of dissolved Al (10^{-6} to 10^{-3} M) and temperature (30 to 130°C). The resulting rate equation can be directly incorporated into reactive transport codes to better evaluate geologic storage of CO_2.

2 EXPERIMENTAL METHODS

Ground labradorite ($Ca_{0.6}Na_{0.4}Al_{1.6}Si_{2.4}O_8$) phenocrysts from a weathered basalt flow near Pueblo Park was used in this study (Casey et al. 1988). Waters saturated with supercritical CO_2 were made by reacting pressurized liquid CO_2 at 165°C. All other solutions were reagent grade HCl and $Al(Cl)_3nH_2O$. We used a titanium mixed flow reactor to measure the effects of dissolved CO_2 and Al on labradorite dissolution kinetics in several stacked experiments. The net rate of reaction was calculated from changes in inflow and outflow solution composition, flow rate, and mineral surface area (0.03 m^2/g as determined by nitrogen adsorption BET method).

3 RESULTS AND DISCUSSION

3.1 *Incongruent labradorite dissolution*

In our study we observed incongruent dissolution at 60°C and below, with enhanced release of Na, Ca and Al over silica in solutions equilibrated with either laboratory air, $CO_{2(aq)} \approx 3 \times 10^{-3}$ M, or equilibrated with supercritical CO_2, ≈ 0.6 M (Fig. 1). Labradorite dissolution is congruent above 100°C. At 130°C we

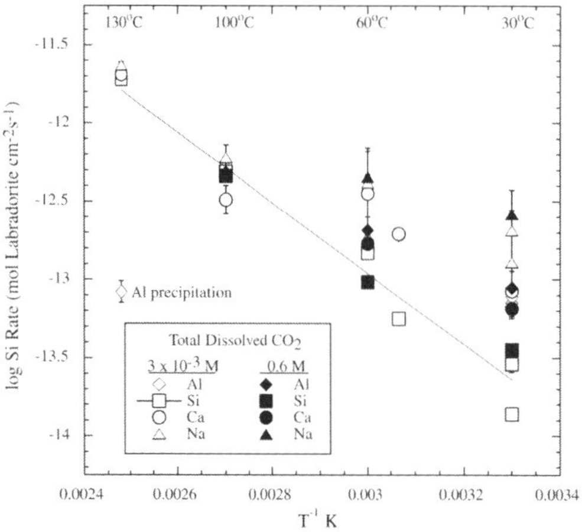

Figure 1. Comparison of labradorite dissolution rates at pH 3.2 in waters equilibrated with atmospheric CO_2 (open symbols) and equilibrated with supercritical CO_2 (filled symbols, 0.6 M CO_2) as function of temperature and release of Na, Ca, Al, and Si. All rates are normalized to their mole fraction concentrations in labradorite at a constant flow rate to normalize effect of dissolved Al on dissolution rate.

clearly see evidence for precipitation of an Al-bearing mineral from the Al-normalized dissolution rate, which is more than 25 times lower than the silica-normalized dissolution rate. This suggests that the secondary precipitate is an Al-hydroxide, which exhibits retrograde solubility in acid solutions, due to the temperature dependence of Al hydrolysis (Bourcier et al. 1993). Although these waters were also supersaturated with respect to kaolinite ($Al_2Si_2O_5(OH)_4$), congruent Ca, Na, and silica release as well as systematic increase of the silica dissolution rates with temperature (log Rate vs. T^{-1}) shown in Figure 1 suggest no precipitation of a silica-bearing solid.

Several other studies have shown that anorthitic feldspars similar in composition to labradorite used in this study dissolve incongruently for long periods of time (as much as 3000 hours) at 25°C (Stillings & Brantley 1995, Casey et al. 1989a, b, Nesbitt et al. 1991, Shotyk & Nesbitt 1990, Taylor et al. 2000) and one study at 45°C (Casey et al. 1988). The temperature dependence suggests that congruent dissolution occurs when the hydrolysis rate of the Al-O-Si bridging bonds exceeds the effect of H^+ exchange for Na^+, Ca^{2+} and some Al^{3+} on dissolution. This interpretation is consistent with other studies. Nesbitt et al. (1991) concluded that incongruent dissolution was a product of ion exchange, because they observed congruent labradorite dissolution in electrolyte solutions and incongruent dissolution in pure HCl-H_2O solutions. Similarly, Brantley & Stillings (1996) and Stillings & Brantley (1995) found that in dilute solutions ion exchange enhances dissolution.

3.2 *Effect of $CO_{2(aq)}$*

We observed no direct effect of dissolved CO_2 on labradorite dissolution from 30 to 130°C (Fig. 1). Labradorite dissolution rates measured at pH 3.2 with much lower dissolved CO_2 ($\sim 3 \times 10^{-3}$ M in dilute HCl solutions saturated with laboratory air) are within experimental uncertainty of those reacted in waters equilibrated with supercritical CO_2 containing about 0.6 M dissolved CO_2 and the same pH. These results agree with labradorite dissolution at much lower concentrations of CO_2 (Brady & Carroll 1994). Any effect of high levels of CO_2 in sequestration schemes on mineral dissolution rates can be accounted for by CO_2 solubility and the resulting acidity in the formation waters, greatly simplifying the dissolution rate equation. As the formation waters are neutralized downstream from the injection well, CO_2 may have a slight impact on silicate dissolution kinetics. Berg & Banwart (2000) show that Ca-feldspar dissolution rates are slightly enhanced at near neutral pH by dissolved CO_3^{2-} (Rate = k_{diss} $[CO_3^{2-}]^{0.24}$).

3.3 *Effect of dissolved aluminum and temperature*

Oelkers and coworkers (Gautier et al. 1994, Oelkers et al. 1994, Oelkers & Schott 1995, Devidal et al. 1997) account for the decrease in silicate dissolution with increasing dissolved Al by the formation of rate-limiting silica rich surface complexes. In this model, the concentration of these complexes is dependent on the dissolved Al concentration according to the following mass balance equation:

$$\text{M-AlSiO} + 3nH^+ = \text{M-SiO} + nAl^{3+} \quad (1)$$

where M-AlSiO represents an Al-filled mineral surface site, n is a stoichiometric exchange coefficient for H^+ and Al^{3+}, and M-SiO is the silica rich complex at the mineral surface. At conditions far from equilibrium this rate expression is independent of reaction affinity.

We extend this rate model to account for the effect of temperature as well as the effect of Al on labradorite dissolution using the following equation:

$$\text{Rate}_{Si}(\text{mol Labradorite cm}^{-2}\,\text{s}^{-1}) = k'' \times 10^{-Ea/2.303RT}\left[\frac{\left(\frac{a_{H+}^{3n}}{a_{Al3+}^{n}}\right)K_T}{\left(1 + K_T\left(\frac{a_{H+}^{3n}}{a_{Al3+}^{n}}\right)\right)}\right] \quad (2)$$

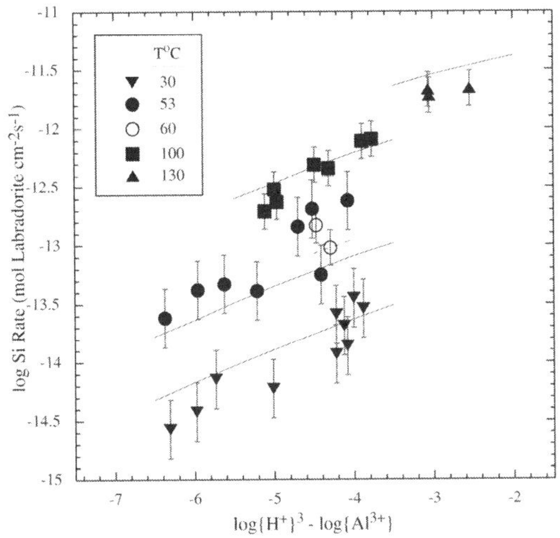

Figure 2. Temperature and aluminum dependence of labradorite dissolution plotted as log Si rate (mol Labradorite cm^{-2} s^{-1}) versus $\log\{H^+\}^3/\{Al^{3+}\}$.

where k'' is the apparent dissolution rate constant, E_a is the activation energy. In Equation 2, the temperature dependence of the formation constant, K_T, can be estimated from the van't Hoff equation:

$$\log\left(\frac{K_1}{K_2}\right) = \frac{\Delta H}{2.303R(T_2^{-1} - T_1^{-1})} \qquad (3)$$

Using Equations 2 and 3 we derive k'', E_a, and n, K_T, and ΔH, the respective stoichiometric exchange coefficient, formation constant at temperature, and the enthalpy of reaction for $H^+ - Al^{3+}$ exchange to form the silica-rich surface complex from a non-linear multiple regression of rate data from 30 to 130°C. The fit to the data is shown in Figure 2 and corresponds to $\log k'' = -5.69$ (mol Labradorite cm^{-2} s^{-1}), $E_a = 42.1$ (kJ/mol), $\Delta H = 2.3$ (kJ/mol) and $n = 0.31$. Regressed values for K_T are 4.5, 4.8, 4.9, 5.3, and 5.6 for 30, 53, 60, 100, and 130°C, respectively. We normalize the rate data to the stoichiometric release of dissolved silica, because decomposition of the Al-O-Si bridging bonds is believed to control feldspar dissolution. Equation 2 does not account for the incongruent dissolution observed at lower temperatures.

4 SUMMARY

The focus of this work was to measure mineral dissolution rates needed to predict changes in reservoir storage capacity due to injection of supercritical CO_2 in the subsurface where it will react with aquifer water and rock. We describe labradorite ($Ca_{0.6}Na_{0.4}Al_{1.6}Si_{2.4}O_8$) dissolution with a single rate equation as a function of pCO_2 (3×10^{-3} and 0.6 M), dissolved Al (10^{-6} to 10^{-3} M), and temperature (30 to 130°C). The resulting temperature dependent rate equation is an extension of the model proposed by Oelkers and coworkers in which the dependence of silicate dissolution rates on dissolved Al is attributed to the formation of rate-limiting silica-rich surface complexes. This study adds to a growing body of literature that shows increasing concentrations of dissolved Al inhibit alumino-silicate dissolution rates. Although the effect of dissolved Al on dissolution rates is real, its effect on rates is secondary compared to that of temperature. Any enhanced dissolution rates in solutions with high dissolved CO_2 are due to the increase in acidity from the solubility of CO_2 and can be modeled with a rate equation that accounts for solution pH and dissolved Al. At 60°C and below, labradorite dissolution is incongruent with calcium rates as high as three times the silica rates. These rates appear to persist for long periods of time, and may need to be incorporated in CO_2 sequestration simulations to accurately predict the storage of CO_2 in carbonate minerals.

REFERENCES

Berg, A. & Banwart, S.A. 2000. Carbon dioxide mediated dissolution of Ca-feldspar: Implications for silicate weathering. *Chem. Geol.* 163: 25–42.

Bourcier, W.L., Knauss, K.G. & Jackson, K.J. 1993. Aluminum hydrolysis constants to 250°C from boehmite solubility measurements. *Geochim. Cosmochim. Acta* 57: 747–762.

Brady, P.V. & Carroll, S.A. 1994. Direct effects of CO_2 and temperature on silicate weathering: Possible implications for climate control. *Geochim. Cosmochim. Acta* 58: 1853–1856.

Brantley, S.L. & Stillings, L.L. 1996. Feldspar dissolution at 25°C and low pH. *Am. J. Sci.* 296: 101–127.

Casey, W.H., Westrich, H.R. & Arnold, G.W. 1988. Surface chemistry of labradorite feldspar reacted with aqueous solutions at pH 2, 3, and 12. *Geochim. Cosmochim. Acta* 52: 2795–2807.

Casey, W.H., Westrich, H.R., Arnold, G.W. & Banfield, J.F. 1989a. The surface chemistry of dissolving labradorite feldspar. *Geochim. Cosmochim. Acta* 53: 821–832.

Casey, W.H., Westrich, H.R., Massis, T., Banfield, J.F. & Arnold, G.W. 1989b. The surface of labradorite feldspar after acid hydrolysis. *Chem. Geol.* 78: 205–218.

Devidal, J., Schott, J. & Dandurand, J. 1997. An experimental study of kaolinite dissolution and precipitation kinetics as a function of chemical affinity and solution composition at 150°C, 40 bars, and pH 2. 6.8, and 7.8. *Geochim. Cosmochim. Acta* 61: 5165–5186.

Gautier, J., Oelkers, E.H., & Schott, J. 1994. Experimental study of K-feldspar as a function of chemical affinity at 150°C and pH 9. *Geochim. Cosmochim. Acta* 58: 4549–4560.

Nesbett, H.W., MacRae, N.D. & Shotyk, W. 1991. Congruent and Incongruent dissolution of labradorite in dilute, acidic salt solutions. *J. Geol.* 99: 429–442.
Oelkers, E.H., Schott, J. & Devidal, J. 1994. The effect of aluminum, pH, and chemical affinity on the rates of aluminosilicate dissolution reactions. *Geochim. Cosmochim. Acta* 58: 2011–2024.
Oelkers, E.H. & Schott, J. 1995. Experimental study of anorthite dissolution and the relative mechanisms of feldspar hydrolysis. *Geochim. Cosmochim. Acta* 59: 5039–5053.
Shotyk, W. & Nesbitt, H.W. 1990. A ligand-promoted dissolution of plagioclase feldspar: A comparison of the surface chemistry of dissolving labradorite and bytownite using SIMS. *Chem. Geol.* 84: 320–321.
Stillings, L.L. & Brantley, S.L. 1995. Feldspar dissolution at 25°C and pH 3: Reaction stoichiometry and the effect of cations. *Geochim. Cosmochim. Acta* 59: 1483–1496.
Taylor, A.S., Blum, J.D. & Lasaga, A.C. 2000. The dependence of labradorite dissolution and Sr isotope release rates on solution saturation state. *Geochim. Cosmochim. Acta* 64: 2389–2400.

Water–Rock Interaction, Wanty & Seal II (eds)
© 2004 Taylor & Francis Group, London, ISBN 90 5809 641 6

Geochemical monitoring of the Weyburn CO_2-injection EOR site, Saskatchewan, Canada

K. Durocher, I. Hutcheon, M. Shevalier & B. Mayer
Applied Geochemistry Group, University of Calgary, Calgary, Alberta, Canada

W. Gunter & E. Perkins
Alberta Research Council, Edmonton, Alberta, Canada

ABSTRACT: The Weyburn Oil Field is the site of EnCana's CO_2-injection enhanced oil recovery (EOR) project in a carbonate reservoir in southern Saskatchewan. An IEA-sponsored program was initiated to evaluate the geological storage of greenhouse gases. Geochemical monitoring and modeling plays an important role in assessing fluid–rock interactions. Pre- and post-injection samples (10 monitoring trips, each sampling ~50 wells) of produced fluids have been analyzed for chemical and isotopic parameters since August 2000. Two primary processes are observed after CO_2 injection:

(1) *Carbonate mineral dissolution*: Increases in [HCO_3], [Ca], and [Mg] point toward limited dissolution of Mississippian carbonate minerals ($\delta^{13}C = 4‰$) in the reservoir; and
(2) *Dissolution of injected CO_2*: Injected CO_2, with a $\delta^{13}C$ value of –34‰, has resulted in decreasing $\delta^{13}C$ values of produced HCO_3^-.

These measurable processes indicate that it is possible to trace the injected anthropogenic CO_2 through a combination of geochemical and isotopic techniques.

1 INTRODUCTION

1.1 *Geological setting*

The Weyburn field is one of a number of large oilfields that lie along the Mississippian subcrop belt on the northern extent of the Williston Basin. The Weyburn field is located approximately 130 kilometers southeast of Regina, Saskatchewan, Canada (Fig. 1) where medium gravity crude oil is produced from the Midale beds of the Mississippian Charles Formation (Burrowes & Gilboy 2000).

The Midale beds of the Weyburn field were discovered in 1954 and produced by primary depletion until 1964, when an inverted nine spot water flood was implemented (El Sayed et al. 1993). Secondary oil recovery via the water flood continued for approximately 36 years. By 1996, cumulative production reached 328 million barrels of oil or 23% of the 1.4 billion barrels of the original oil-in-place. In 1997, PanCanadian (now EnCana) announced the Weyburn miscible flood project, which is forecasted to add incremental oil recovery of 120 to 140 million barrels. The CO_2 used in the miscible flood originates from the Great Plains Synfuel coal gasification plant in Beulah, North Dakota, USA by way of a 320-kilometer pipeline (Fig. 1).

1.2 *Miscible CO_2 flooding*

Under the correct conditions of pressure, temperature and oil composition, the CO_2 becomes miscible with the oil and causes the residual oil to swell, lowering its viscosity. CO_2 is injected with water to retard CO_2 mobility. A factor thought to increase oil recovery during CO_2 injection is the increase in permeability due to the acidity of carbon dioxide-water mixtures. Acid gases have been suggested to dissolve carbonate minerals in the formation matrix resulting in a permeability increase (Sayegh et al. 1990).

1.3 *IEA monitoring program*

Before the EOR program had begun, it was recognized that this process was a short-term analogue to the geological storage of CO_2, a major greenhouse gas. Under the guidance of the International Energy

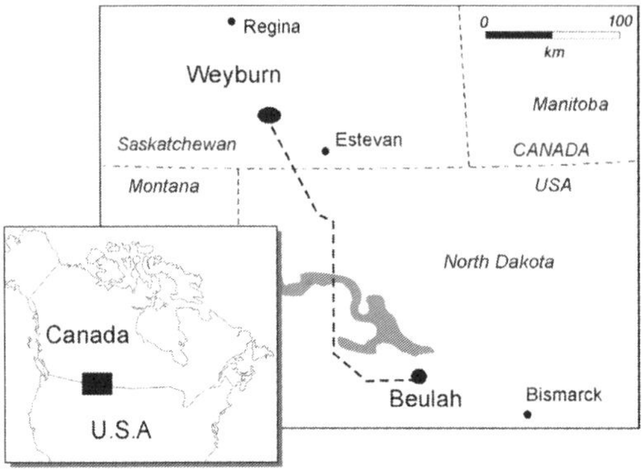

Figure 1. Location of the Weyburn field and CO_2 source in Beulah, North Dakota. The dashed line represents the approximate trace of the CO_2 pipeline (after IEA Weyburn website diagram).

Agency (IEA), the Petroleum Technology Research Centre (PTRC, Regina, Saskatchewan) is heading a research program to study greenhouse gas storage, using the EOR process as a modern analogue. The Weyburn monitoring program, slated for completion in mid-2004, has a budget of approximately \$21 million from a number of international government and industry sponsors. The program, managed by Waleed Jazrawi of the PTRC, employs approximately 25 research groups and organizations that provide analytical data and interpretations. These research providers are based throughout Canada, the United States, and Europe.

1.4 *Geochemical modeling of CO_2 storage*

Carbonate reservoirs typically containing Fe-, Mg- and Ca-bearing carbonates have limited trapping potential for CO_2 as can be shown by the reactions:

(1) CO_2 dissolution:

$$H_2O + CO_2 \Rightarrow H_2CO_3 \Rightarrow H^+ + HCO_3^-$$

(2) Calcite dissolution:

$$H^+ + CaCO_3 \Rightarrow Ca^{2+} + HCO_3^-$$

(3) Both CO_2 and calcite dissolution:

$$H_2O + CO_2 + CaCO_3 \Rightarrow Ca^{2+} + 2HCO_3^-$$

Reaction 1 represents CO_2 dissolution, and results in the modified fluid having a lower pH and higher [HCO_3^-]. Reaction 2 represents reservoir calcite dissolution. This reaction leads to higher carbonate alkalinity and higher cation concentrations. If both reactions take place (3), then for each mole of injected CO_2 that reacts, we can expect one mole of cation, and 2 moles of HCO_3^- dissolved in the *in situ* fluid.

Although the reservoir rocks are dominantly carbonate, even small amounts of potentially reactive silicate minerals can lead to the precipitation of injected and dissolved CO_2 as newly formed carbonate minerals over the long term (100's to 1000's of years; Gunter et al. 2000). Silicate mineralogy and CO_2 trapping potential is the subject of current research by the University of Calgary, the Alberta Research Council, and BRGM (France).

2 ANALYTICAL RESULTS

Fluids and gases from forty-one wells distributed across the Phase 1A CO_2 injection pilot area were collected for the Baseline study of August 2000. Nine post-injection monitoring trips have taken place since the Baseline study. The Baseline to Monitor 1 (after approximately 6 months of injection) geochemical observations were presented at the Kyoto GHGT-6 conference (Emberley et al. 2002). This abstract details geochemical results after approximately 3 years of CO_2 injection.

After 930 days of CO_2 injection (Monitor 7 sampling trip), approximately 1.9 million standard m^3 of CO_2 had been injected with the highest volumes of CO_2 having been injected in the central and south-central portion of the pilot area. The same wells from the Baseline study and additional horizontal producing wells were sampled for the Monitor-7 study of March 2001.

Geochemical and isotope data from the baseline study were plotted and contoured over the Phase 1A injection area. The most obvious trend in the baseline data is the NW-SE trend in the concentration gradient of many cations, anions, and isotopic compositions (Fig. 2). This is likely due to the mixing of fluids from two aquifers: One having a shallow origin and flowing from the NW along the Mississippian unconformity, and the second, a basinal brine, flowing from the SE. These two aquifers meet and mix across the Weyburn field (Bailey et al. 1973, Emberley 2003).

Table 1 shows selected production-weighted mean geochemical values for Baseline (pre-CO_2 injection) and Monitor-7 (930 days post-CO_2 injection) samples.

After 930 days of CO_2 injection, several geochemical parameters have dramatically changed (Table 1). As expected from the reactions described above, [HCO_3], [Ca], and [Mg] have increased across the field (Fig. 2). Of note is the appearance of a "bullseye" pattern in the lower central portion of the injection pilot area. The observed high concentrations coincide with the highest injected CO_2 volumes and may reflect a zone wherein calcite and dolomite dissolution has been accelerated relative to elsewhere in the pilot area.

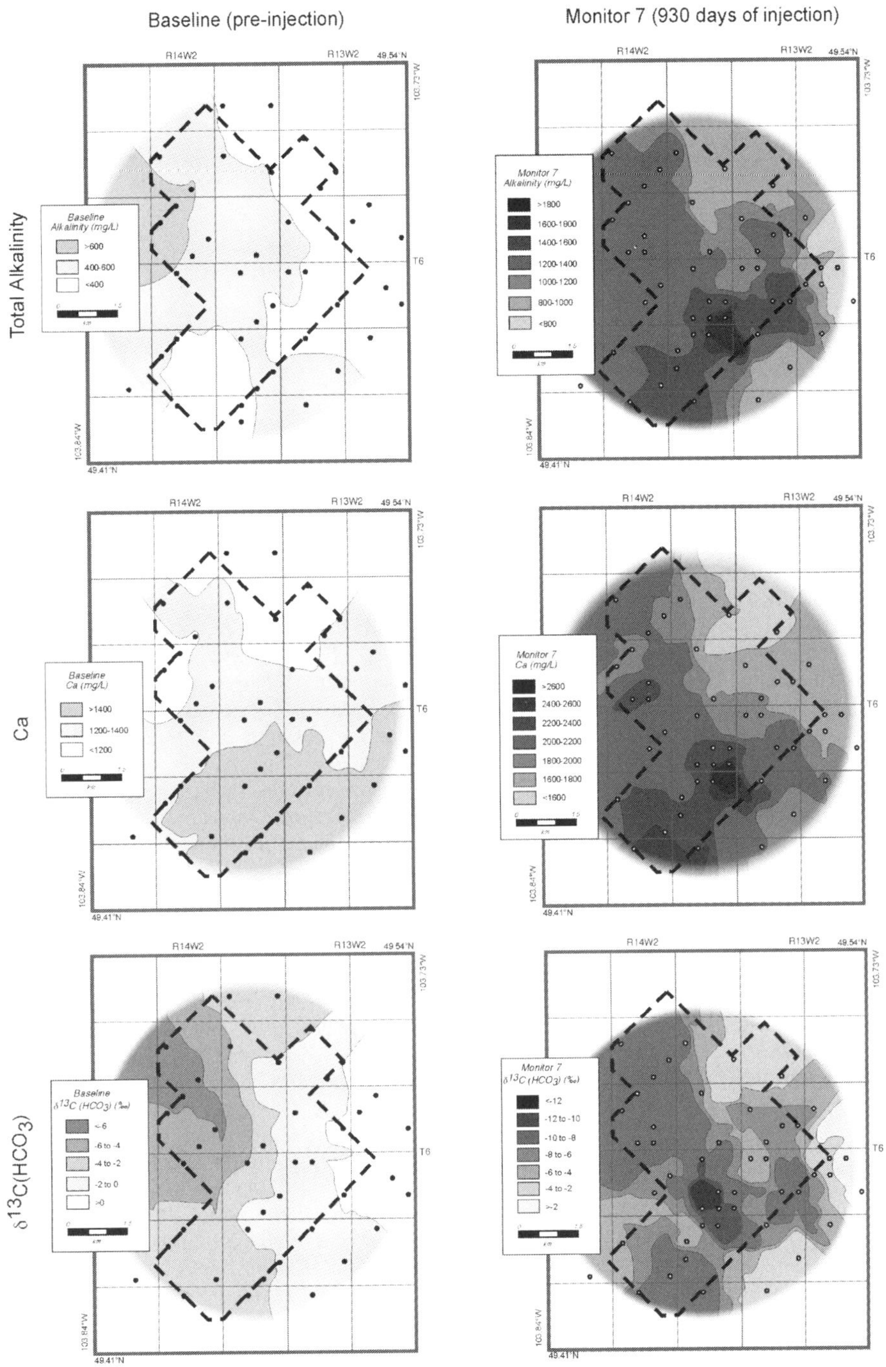

Figure 2. Contour maps showing total alkalinity, [Ca], and $\delta^{13}CHCO_3$ across the Phase 1A injection pilot area (shown by dashed line). Pre-injection contoured measurements are shown on the left whereas post-CO_2 injection values (after 930 days of continuous injection) are contoured on the right. Dots represent well locations sampled during each trip.

Table 1. Average values of selected geochemical parameters for the Weyburn pilot injection area.

	Pre-injection (n = 41*)		930d of Injection (n = 53**)		
	avg.	s.d.	avg.	s.d.	% change
Measured pH	6.7	0.2	6.7	0.3	−1
Total Alk. (mg/L)	428.8	103.9	1168.1	418.3	172
Ca (mg/L)	1371.9	186.0	1979.6	336.1	44
Mg (mg/L)	373.8	27.2	608.1	71.9	63
Resist. (ohm m)	0.149	0.023	0.103	0.012	−45
$\delta^{13}C$-HCO_3 (‰, PDB)	−2.8	2.7	−6.5	5.1	n/a
Cl (mg/L)	38443.0	7929.8	40055.2	6838.2	4
SO_4 (mg/L)	3464.5	206.7	3563.7	256.7	3

* = 41 vertical producing well data points.
** = 41 vertical and 12 horizontal producing well data points.
Abbreviations: Alk. = alkalinity, resist. = resistivity.

The carbon isotope compositions of produced fluids and gases are important variables in tracing the dissolution of reservoir minerals and injected CO_2. Reservoir calcite and dolomite have $\delta^{13}C$ values between 3 and 5‰. Pure injected CO_2 had a $\delta^{13}C$ value of approximately −34‰. Pre-injection (Baseline) HCO_3^- has a value of approximately −3‰ (Fig. 2; Table 1). Dissolution of injected CO_2 drives carbonate mineral dissolution (Reaction 3) in the reservoir. The net effect is decreasing $\delta^{13}C$ values for produced HCO_3^-. As shown in Figure 2 and Table 1, a slight drop in the average $\delta^{13}C$ value of HCO_3^- is apparent, with the same "bullseye" pattern exhibited in the [Ca] data.

The average measured pH has essentially remained constant from pre- to post-injection, however "bullseyes" are observable (not shown here) and spatially correlative with alkalinity and dissolved cation anomalies.

Fluid resistivity has decreased as TDS increased after CO_2 injection. Resistivity contour maps (not shown) show progressive field-wide decreases with continued CO_2 injection.

3 DISCUSSION

3.1 *Carbonate mineral and injected CO_2 dissolution*

[HCO_3] has more than doubled after CO_2 injection. Combined with the negligible change in pH, and an increase in [Ca], this suggests that injected CO_2 has dissolved into the reservoir fluid *and* that CO_2 injection has resulted in some carbonate mineral dissolution in the reservoir.

CO_2 and carbonate dissolution have an effect on the carbon isotope composition of produced HCO_3. With injected CO_2 dissolution, $\delta^{13}C$ of produced HCO_3^- decreases markedly. CO_2 dissolution drives carbonate mineral dissolution in the reservoir. Reservoir calcite and dolomite have positive $\delta^{13}C$ values. The net effect of dissolution of these isotopically distinct C sources is a decrease in $\delta^{13}C$ of produced HCO_3^-. The average $\delta^{13}C$-HCO_3values decreased by 3‰ and locally by as much as 10‰ after injection. Ongoing mass balance calculations will quantify the degree of carbonate dissolution since the onset of CO_2 injection approximately 3 years ago.

Increased TDS and decreased resistivity may be attributed not only to carbonate dissolution, but also to CO_2-enhanced "sweeping" of previously inaccessible (and more saline) fluid from less permeable reservoir rocks.

A first-order estimate of the reservoir brine's capacity to store CO_2 can be calculated, given an average baseline water composition, the reservoir temperature and pressure (approximately 60°C and 1 kbar, respectively), and the gas solubility equations of Barta & Bradley (1985). These data suggest that the average baseline fluid sample is capable of storing approximately 1.2 moles (53 grams) of dissolved CO_2 per liter of brine. An injection area of 20 km by 20 km by 20 m, a reservoir porosity of 20% and an oil:water ratio of 0.3 will contain roughly 2 trillion liters of brine which could potentially store up to 60 million metric tones of dissolved CO_2, approximately 6 times the amount of dissolved CO_2 in the average baseline water composition (~0.2 g/L).

4 CONCLUSIONS

The trends in chemical composition and the isotope data suggest that injected CO_2 has been dissolved in the formation brine and has lead to the limited dissolution of reservoir carbonate minerals. The areas where these processes are most pronounced coincide with the highest CO_2 injection volumes. Basinal brine chemistry suggests that significant amounts of injected CO_2 and CO_2 produced via mineral dissolution may be stored as bicarbonate ion.

ACKNOWLEDGEMENTS

The authors greatly acknowledge the efforts of many individuals who aided in the collection and interpretation of the geochemical data presented here, including: Geoff Burrowes, Sandy Graham, Dave Stachniak, Ken Ferguson, Kelly Vandesype, and Trevor Westman at EnCana Corporation, Bernice Kadatz at the Alberta Research Council, and Steve Taylor of the Isotope Science Laboratory at the U of Calgary. Dr. D. Hayba is thanked for his review of the manuscript.

REFERENCES

Bailey, N.J.L., Krouse, H.R., Evans, C.R. & Rogers, M.A. 1973. Alteration of crude oil by waters and bacteria – Evidence from geochemical and isotope studies. *AAPG Bulletin* 57: 1276–1290.

Barta, L & Bradley, D.J. 1985. Extension of the specific interaction model to include gas solubilities in high temperature brines. *Geochim. Cosmochim. Acta* 49: 195–203.

Burrowes, G. & Gilboy, C. 2000. *Investigating Sequestration Potential of Carbonate Rocks During Tertiary Recovery from a Billion Barrel Oil Field, Weyburn, Saskatchewan: The Geoscience Framework.* IEA Weyburn CO_2 Monitoring and Storage Project: Report.

El Sayed, S.A., Baker, R., Churcher, P.L. & Emunds, A.C. 1993. Multidisciplinary reservoir characterization and simulation of the Weyburn Unit. *J. Petrol. Technol.* 45: 930–974.

Emberley, S. 2003. *Geochemical Monitoring of Fluid-rock Interaction and CO_2 storage at the Weyburn CO_2 Injection Enhanced Oil Recovery site, Saskatchewan, Canada.* Unpub. M.Sc. Thesis. Calgary: University of Calgary.

Emberley, S., Hutcheon, I., Shevalier, M., Durocher, K., Gunter, W.D. & Perkins, E.H. 2002. Geochemical monitoring of fluid-rock interaction and CO_2 storage at the Weyburn CO_2-injection enhanced oil recovery site, Saskatchewan, Canada; GHGT-6. *Sixth International Conference on Greenhouse Gas Control Technologies.* Kyoto, Japan.

Gunter, W.G., Perkins, E.H. & Hutcheon, I. 2000. Aquifer disposal of acid gases: Modeling of water-rock reactions for trapping of acid waters. *J. Appl. Geochem.* 15: 1086–1096.

Sayegh, S. G., Krause, F. F., Girard, M., & DeBree, C. 1990. Rock/fluid interactions of carbonated brines in a sandstone reservoir: Pembina Cardium, Alberta, Canada. *J. Soc. Petrol. Eng.* 30: 399–405.

Water–Rock Interaction, Wanty & Seal II (eds)

Coupled carbonate mineral dissolution and growth: calcite and strontianite

K.G. Knauss
Earth Sciences Division, Lawrence Livermore National Laboratory, Livermore, CA

C.M. Eggleston & B. Greer
Department of Geology and Geophysics, University of Wyoming, Laramie, WY

S.R. Higgins
Department of Chemistry, Wright State University, Dayton, OH

ABSTRACT: As part of a larger project aimed at reconciling results from microscopic and macroscopic reactive transport experiments and modeling, we studied the coupled dissolution of calcite and growth of strontianite. Although our analyses are ongoing, preliminary results suggest that the experiments designed to examine the two scales yielded similar results in terms of processes. Quantification of rates extracted from experimental results at each scale continues. The comparison of macroscopic experimental results to *a priori* reactive transport simulation presented here shows that, in general, the model over-predicts both dissolution and growth rates to some extent. The model does represent processes well and captures solution concentrations in time and mineral distribution in space for the macroscopic experiment reasonably well.

1 INTRODUCTION

The ability to accurately scale-up from microscopic reactive transport processes (e.g. studied using the Hydrothermal Atomic Force Microscope – HAFM (Higgins et al. 1998)) to the corresponding macroscopic ones (e.g. studied using the Plug Flow Reactor – PFR) and having them both coincide with *a priori* geochemical modeling simulation is a holy grail of aqueous geochemistry. This work is directed towards that goal.

Because carbonate mineral dissolution and growth rates are reasonably rapid, they are amenable to study at both microscopic and macroscopic scales. Furthermore, carbonate minerals are important players in mineral-trapping of carbon in CO_2 sequestration, in metal contaminant transport in general, and in soil development. For these reasons we chose to use the carbonate minerals calcite and strontianite as vehicles for trying to scale-up our prior HAFM molecular-scale studies of coupled dissolution and growth processes (Greer et al. 2002) to the macroscale represented by corresponding PFR experiments. These PFR experiments were also amenable to simulation using reactive transport codes.

In this paper we present the preliminary results of the PFR experiment and the reactive transport simulation of coupled calcite dissolution and strontianite growth.

2 MATERIALS AND METHODS

2.1 *Solutions and minerals*

The PFR was packed with crushed, optically clear spar calcite obtained from Ward's Minerals. The grains were sieved to between 425 and 850 μm and cleaned repeatedly in alcohol using an ultrasonic bath. The estimated geometric specific surface area is 52 cm^2/g. The inflowing solution was prepared from reagent grade 0.002 m $Sr(NO_3)_2$ in a 0.01 m $NaNO_3$ background electrolyte made using 18 MΩ-cm deionized water, assumed to be equilibrated with the atmosphere. The measured pH was 5.5, while the pH calculated for this solution using EQ3/6 (Wolery et al. 1990) and assuming equilibration with the atmosphere is 5.6, supporting this assumption.

2.2 *Experimental and analytical apparati*

The design of the PFR is described in detail in Knauss et al. (2003). It uses a recoverable Au liner that permits easy sub-sampling of the reacted core to map out mineral distributions in space. Flow is precisely controlled using a Quizix pump and a back pressure regulator maintains system pressure. The high aspect ratio (length:diameter = 50:1) makes the PFR an ideal 1D physical model.

The aqueous solutions were analyzed for cations using ICP-ES. Solids were analyzed using AFM, SEM/EDS and XRD techniques. The solution pH was measured using a solid state ISFET pH probe, following NIST protocol.

3 EXPERIMENTAL DESIGN

We had earlier studied coupled calcite dissolution and strontianite growth at 40°C using the HAFM (Greer et al. 2002). However, preliminary reactive transport modeling suggested that the amount of strontianite produced at 40°C in our PFR using the same mass flow rate as in the HAFM would be difficult to quantify using macroscopic techniques, e.g. XRD. Simulations suggested that at 100°C easily measurable quantities of strontianite should be produced, permitting us to map out its distribution in space. Consequently, the $Sr(NO_3)_2/NaNO_3$ solution was passed through the 30 cm core, held isothermally (i.e. without a thermal gradient along its length) at 100°C for 2 weeks. A back pressure of 5 MPa was maintained to prevent atmospheric gases from exsolving upon entering the furnace, thus, a single liquid phase existed in the core during the run.

Based on the *a priori* modeling, and the chosen mass flow rate (500 g/d, equivalent to a Darcy flux of 6812 m^3/m^2-y in the PFR), we collected samples closely spaced in time over the first 2 days and then, when modeling predicted achievement of steady-state, we sampled less frequently. As a consequence, pH was not measured during the first day of the experiment. The solution pH was measured on later samples that had cooled to near room temperature and at least partially degassed, i.e. they had not re-equilibrated with the atmosphere at room temperature. Sampling for cations, however, was performed even during the core heat-up phase, which lasted 60–90 minutes.

At the conclusion of the run the gold core liner was removed and cut into 1 cm sections. The solids recovered from each section were then examined using SEM/EDS, AFM and XRD methods.

4 REACTIVE TRANSPORT MODELING DESIGN

4.1 *Thermodynamic and kinetic data*

The thermodynamic data used in model calculations were taken from the EQ3/6 database (Wolery et al. 1990). Although preliminary simulations were run with the entire suite of minerals in the database, based on their stability under these experimental conditions, only the minerals calcite and strontianite were active in the problem. The rate law used in the reactive transport model (see below) allows for parallel rate laws for each mineral, spanning the pH range. For the acid catalyzed and neutral pH rate laws we have extracted kinetic parameters from Sonderegger (1976) for strontianite and from Alkattan et al. (1998) and Chou et al. (1989) for calcite.

4.2 *Reactive transport simulator*

We used the reactive transport code CRUNCH (Steefel 2001), which coupled a chemical model with simplified fluid flow, to simulate the results of injection of the $Sr(NO_3)_2/NaNO_3$ solution through the calcite core and have accounted for mineralogical changes along the flow path. The rate law that we used in CRUNCH had a transition-state-theory-like (Laidler 1987) formulation that accounts for kinetic inhibition or catalysis by any species in solution. In this formulation the same rate law was used to describe both dissolution and growth according to the principle of microscopic reversibility and this is, most certainly, an empirical approximation anywhere in free energy space removed from equilibrium.

CRUNCH was used to help design the experiment, and then the results from the PFR experiment (an ideal 1D physical model) were used to evaluate the performance of the model. This direct simulator validation is a critical, but rarely taken, step in the development and acceptance of reactive transport simulators.

5 RESULTS AND DISCUSSION

5.1 *PFR experimental data*

5.1.1 *Aqueous phase*

In Figure 1 we plot the aqueous concentration data and pH for samples from the PFR experiment.

We have plotted all the data from the 2-week experiment, although it is clear that a steady-state was achieved early in the run, as predicted (see below). In Section 5.2 we will replot the measurements for the first few hours of the experiment in order to compare them more clearly to the simulated results.

5.1.2 *Solid phase*

The SEM examination of the reacted grains showed that at the front of the core, the solution was clearly undersaturated with respect to calcite evidenced by the presence of etch pits. It also showed that a secondary mineral was growing in abundance on the surface, coating the calcite grains and in places

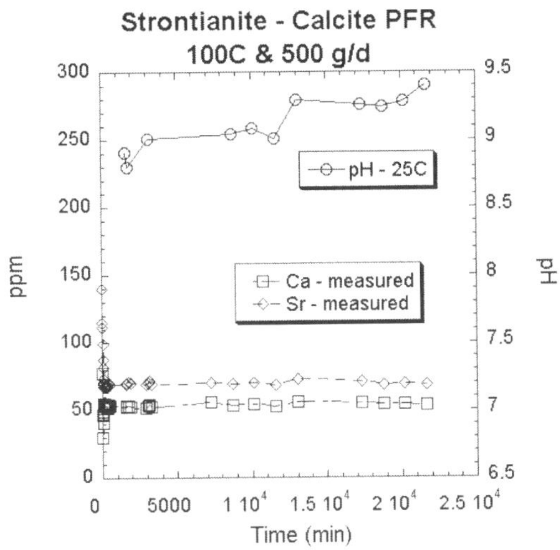

Figure 1. PFR measured concentrations and pH.

Figure 2. SEM photo showing etch pit on dissolving calcite at core entrance sample. The mineral crust is at the lower left.

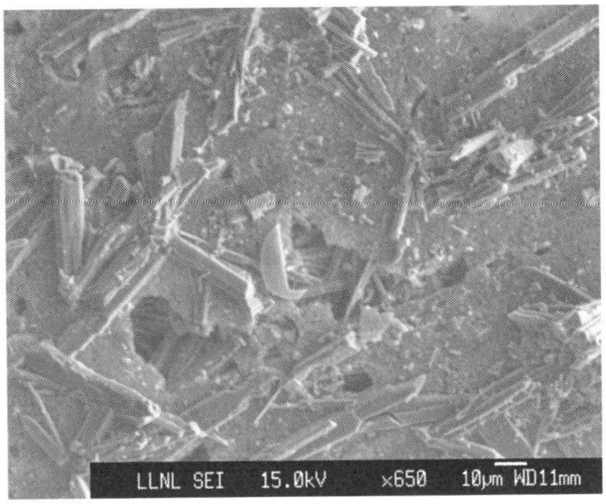

Figure 3. SEM photo of strontianite crust at core entrance.

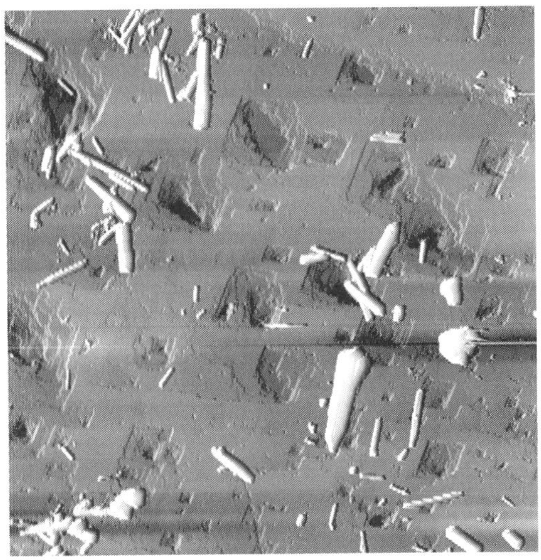

Figure 4. Blended height and deflection contact mode AFM images of core exit sample. Scan length is 20 μm.

loosely cementing them to each other. Where the coating was pulled away from the underlying calcite grain, the coupled nature of the processes can be clearly seen.

In Figure 2 we have marked two sides of a large etch pit to illustrate that this sample from the entrance to the core is clearly undersaturated at this location underneath the overlying crust of secondary mineral grains. In Figure 3 it can be seen that, although the crust is quite thick, it is not a complete barrier to the bulk solution, hence calcite can remain undersaturated. EDS analyses of the crust show it to be strontianite.

Toward the exit of the core the calcite grains still show evidence of dissolution, although etch pits in AFM images (Fig. 4) also show evidence of either backfilling with calcite (or high-Sr calcite) or, more likely, dissolution under near equilibrium conditions. A few isolated grains of strontianite can also be found.

The XRD analyses done on each section of core show clearly that strontianite abundance is high (>5%) over the first 10 cm, but then decreases further along the core. It is most abundant a few cm into the core (Fig. 5).

5.2 *CRUNCH reactive transport simulation*

As seen in Figure 6, the reactive transport simulation predicts that steady-state is achieved quite quickly.

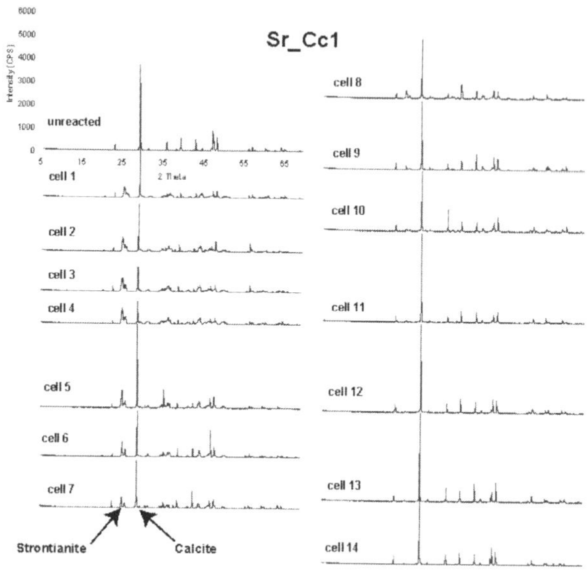

Figure 5. XRD analyses of samples from front half of core.

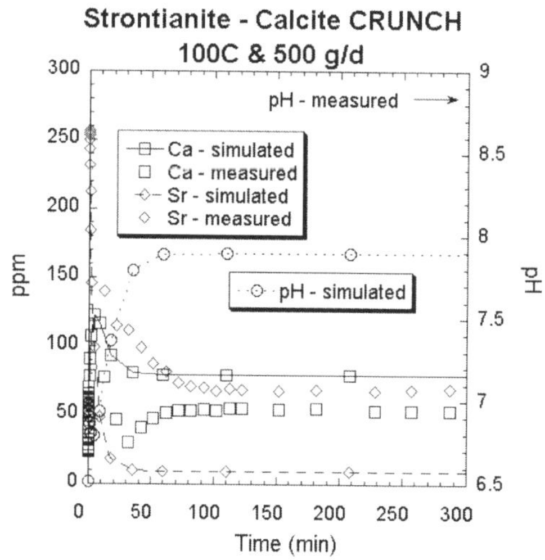

Figure 6. Plot of simulator and PFR experiment results.

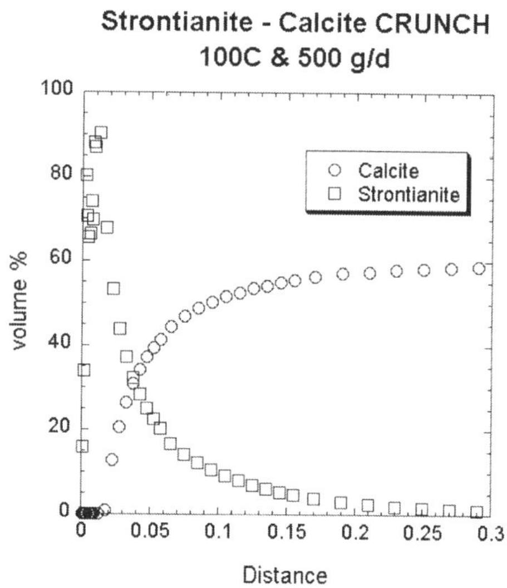

Figure 7. Plot of simulator predicted mineral distribution after 14 days.

This steady-state in the simulation results from rapid equilibration with calcite and rapid and abundant precipitation of strontianite.

Part of the lag in measured vs. predicted concentrations results from the time required for the furnace to heat up to run temperature. This offsets the measured values to longer time. Although the trends are in good agreement, the simulation over-predicts the Ca concentration slightly (and consequently under-predicts the pH), while it more significantly under-predicts the Sr concentration (probably from over-predicting the strontianite growth rate).

The simulation also predicts (Fig. 7) extensive calcite dissolution in the front of the core and abundant strontianite growth throughout much of the core, peaking at about 1 cm. Comparing the simulation to the XRD data (Fig. 5), the model has over-predicted both calcite dissolution and strontianite growth, especially at the front of the core.

Reconciling the physical and computational models can be accomplished in several ways. The physical model should include direct measurement of in situ carbonate concentration. This simply requires that samples be collected in gas-tight syringes preloaded with NaOH to prevent CO_2 evolution. We elected not to do this in the first experiment to gather more frequent cation samples. The 25°C measured pH will always be subject to error caused by partial equilibration with the atmosphere as it cools, so the requirement to calculate in situ pH will remain. The model must be able to account for partially occluded surface area, when thick crusts form, and the present growth formalism (running the dissolution rate law backwards) clearly needs to be expanded to accommodate more realistic mineral growth rate laws.

6 CONCLUSION

Changes to rate law formalisms in our simulators can bring model predictions in line with macroscopic ideal physical model results. In this way we validate the reactive transport simulators, a critical step in their confident use.

REFERENCES

Alkattan, M., Oelkers, E. H., Dandurand, J.-L. & Schott, J. 1998. An experimental study of calcite and limestone dissolution rates as a function of pH from −1 to 3 and temperature from 25 to 80C. *Chem. Geol.*151: 199–214.

Chou, L., Garrels, R. M. & Wollast, R. 1989. Comparative study of the kinetics and mechanisms of dissolution of carbonate minerals. *Chem. Geol.*78: 269–282.

Greer, B., Higgins, S. R., Eggleston, C. M. & Knauss, K. G. 2002. Study of heterogeneous kinetics at calcite surfaces in the presence of strontium using atomic force microscopy. ACS National Meeting, Orlando, FL.

Higgins, S.R., Eggleston, C.M., Knauss, K.G. & Boro, C.O. 1998. A hydrothermal atomic force microscope for imaging in aqueous solution up to 150°C. *Rev. Sci. Instrum.* 69: 2994–2998.

Knauss, K.G., Johnson, J.W. & Steefel, C.I. 2004. Impact of CO_2, contaminant gas, aqueous fluid, and reservoir rock interactions on the geologic sequestration of CO_2. *Chem. Geol.* In review.

Laidler, K. J. 1987. *Chemical Kinetics.* Harper & Row, N.Y.

Sonderegger J. L., Brower K. R., & Lefebre V. G. 1976. A preliminary investigation of strontianite dissolution kinetics. *Am. J. Sci.* 276: 997–1022.

Steefel, C. I. 2001. CRUNCH. Lawrence Livermore National Laboratory. Internal Report.

Wolery T. J., Jackson K. J., Bourcier W. L., Bruton C. J., Viani B. E., Knauss K. G. & Delany J. M. 1990. The EQ3/6 software package for geochemical modeling: Current Status. In D. C. Melchior & R. L. Bassett (eds.), *Chemical Modeling in Aqueous Systems II*, ACS Symp. Ser. 416: 104–116.

Water–Rock Interaction, Wanty & Seal II (eds)
© 2004 Taylor & Francis Group, London, ISBN 90 5809 641 6

Geochemical impact assessment of CO_2 storage in the North German Basin

F. May
Bundesanstalt für Geowissenschaften und Rohstoffe, Hannover, Germany

ABSTRACT: Most of the storage capacity for CO_2 in Germany is within the saline aquifers of the North German Basin. Fast reactions involving carbonates and evaporites may cause geotechnical problems during the injection phase, while the slow alteration of silicates may contribute to long-term storage safety and capacity. Ideally, geochemical modeling should be based on natural analogues and on injection experiments in situ and in labs. Although not all of these prerequisites are possible for feasibility studies of CO_2 injection in the North German Basin, reaction progress and reactive transport processes have been simulated for generic models. Predictive long-term simulations do require additional information though, e.g. from pilot projects.

1 INTRODUCTION

The underground storage of CO_2, as the main greenhouse gas is one of the options considered to meet the challenge of climate change mitigation. Germany is the largest CO_2 emitting country in Europe, releasing about 850 Tg of CO_2 annually. The storage capacity of several European countries has been assessed in the GESTCO project (Christensen & Holloway 2003). About three quarters of the estimated storage capacity in Germany is within the North German Basin (NGB), which is a part of the southern Permian Basin extending from England to Poland. Most of this capacity is in deep saline aquifers (33 ± 10 Tg of CO_2) and less in depleted gas reservoirs (2.7 Tg) (May et al. 2003). The main aquifers suitable for storage are listed in Table 1.

Formation waters within the aquifers of the North German Basin are usually highly concentrated brines due to their contact with various evaporites, especially the wide spread, thick Upper Permian "Zechstein" salt that has penetrated younger strata in numerous salt domes. The deep saline aquifers are separated from shallow fresh water aquifers by the thick Lower Tertiary "Rupelton" clay formation.

2 POSSIBLE REACTIONS

Geochemical reactions between CO_2, saline formation water and reservoir rocks may cause geotechnical problems, they can cause economic losses, and they even may endanger public safety when corrosion leads to the escape of CO_2 to the surface. Thus, the

Table 1. Lithology of the main aquifers in the North German Basin.

Aquifer	Lithology
Lower Cretaceous "Wealden Sandstein"	Coal bearing fluviatile sandstones
Middle Jurassic, "Dogger" Aalenium – Bathonium	Ceakly cemented sandstones
Lower Jurassic, "Lias" Hettangium, Sinemurium	Barely cemented sandstones with shale layers
Upper Triassic, "Keuper" "Rhät-Sandstein"	Mature quartz sandstones, partly anhydrite cement
Upper Triassic, "Keuper" "Schilfsandstein"	Coal bearing fluviatile sandstones
Lower Triassic "Mittlerer Buntsandstein"	Iron oxide coated sandstones with clay and carbonate in the basin centre
Upper Permian "Oberrotliegend"	Sandstones – greywacke with shale and clay stone Intercalations, carbonate cements

prediction of the possible impact of geochemical reactions is a very important part in the planning of sequestration projects. In principle a distinction can be made between fast geochemical reactions important during the injection period and slow reactions that influence long-term safety.

2.1 *Fast reactions*

Some of the chloride, sulfate, and carbonate minerals dissolve and precipitate relatively fast. Anhydrite, rock salt, and gypsum form cap rocks of aquifers at different stratigraphic levels. They are also common cement minerals within the aquifers. Injection rates may benefit from fast calcite dissolution reactions close to the injection wells in carbonate bearing sandstones (acid stimulation). As CO_2 injection fronts expand, the displaced water may soon get saturated with respect to calcite, so that little further dissolution is expected farther away from injection wells. Fast dissolution of carbonates may also be a threat to the stability of wells e.g. in Upper Cretaceous Chalk in the overburden, or chalk reservoirs in the Northern Permian Basin. Precipitation of salt may result from the evaporation of water into the dry CO_2 phase injected into salt saturated brines. Reactive transport simulations indicate however, that halite precipitation is probably a near-well problem that can be treated with fresh water flooding, as practiced in some of the gas fields in the NGB.

2.2 *Slow reactions*

Reaction rates of complex sheet and framework silicates are usually rather slow and some minerals may be metastable, withstanding dissolution over geological periods of time even though the surrounding formation water is under-saturated with this mineral phase. On the other hand, minerals that could precipitate from formation water, according to their calculated saturation indices, e.g. K-feldspar cannot be detected in the aquifers. One of the benefits of silicate alteration could be the in situ mineralization, and thus long-term fixation of CO_2. Chlorites, that occur within "Rotliegend" sandstones could react to kaolinite, silica and Fe-Mg-carbonates. This type of alteration has been observed in chlorite bearing aquifers containing CO_2-rich waters in the Rhenohercynian basement that underlies parts of the NGB (May 1998). The bulk reaction equation is given below (1):

$$Fe_5Al_2Si_3O_{10}(OH)_8 + 5CO_2 \rightarrow 5FeCO_3 + Al_2Si_2O_5(OH)_4 + SiO_2 + 2H_2O \quad (1)$$

Heavy metals such as Ni that can be released from chlorites may not be incorporated in secondary minerals and accumulate in the aqueous phase. This would be an additional environmental problem in the case of CO_2 leakage into shallow fresh water aquifers. Feldspar forms another group of minerals that can contribute to the natural "mineral sequestration" of CO_2. Intensive silicate alteration, in "Rotliegend" greywacke can reduce porosity and create an additional internal barrier for dissolved C-species and supercritical CO_2 remaining within the aquifers. Barium or calcium released from feldspar dissolution could lead to the precipitation of gypsum or barite from formation waters rich in sulfate.

3 PREDICTIVE MODELING

Quantitative predictions of possible geochemical reactions are essential to understand their impact on CO_2 storage operation and safety. Predictions contribute to the assessment of geotechnical, economic, and environmental risks. The comparison of various simulations can be used to reduce these risks and to optimize injection strategies.

3.1 *Geochemical modeling*

Geochemical modeling is often done in a rather straightforward manner. However predictive geochemical modeling of complex alteration reactions is a rather challenging task. The quantification of mass transfer between various fluid and solid phases and the temporal and spatial resolution of reactions in dynamic systems require careful consideration of each modeling step in order to avoid irrelevant results and misleading interpretations. A controlled, stepwise approach is suggested. The ideal sequence of the principal steps of this procedure is discussed below and illustrated with examples from the NGB.

3.1.1 *Natural analogues*

Information about possible reactions between CO_2 and aquifer rocks can be obtained from the analysis of alteration products in natural CO_2-bearing aquifers. Such waters are produced from aquifers at the southern margin of the basin – the Buntsandstein or Rotliegend aquifers in the Vorderröhn. The latter has been studied in the EU supported natural analogue study NASCENT. However, most of the aquifer structures in the NGB do not contain CO_2-bearing waters. The altered wall rocks from mineral water or natural gas-production wells may reveal valuable information especially about slow reactions. The mineral composition of the main aquifer rocks has been analysed from cores drilled for natural gas exploration. Ample analyses of deep formation waters provide information for geochemical modeling.

However, most of these analyses are restricted to the major ions only. Elements like Si or Al that would be required to calculate saturation indices of silicates rarely have been determined. Salinities generally increase with depth, ranging from about 100 kg/m^3 at 1 km depth up to about 350 kg/m^3 at more than 2 km depth (Bandlowa et al. 1997).

3.1.2 *In-situ and lab experiments*

Natural analogue studies should be supplemented with geochemical studies of alteration reactions from field pilot projects and lab experiments that can be conducted under known and controllable conditions. The latter investigations would aim at rather fast reactions. The injection of CO_2 into the "Schilf-sandstein" aquifer structure near Ketzin, west of Berlin, is currently being prepared in the European experiment CO_2SINK (Zemke et al. 2003). This field pilot project is accompanied with core flooding experiments (G. Borm, pers. comm.). Fluid fluxes, which are often hard to determine in natural systems, can well be measured in lab experiments. Together with geochemical and mineralogical analyses of fresh and altered rocks, analyses of the primary and altered pore water chemistry, and known gas compositional data these allow the quantification of mass transfer among fluid and solid phases.

3.1.3 *Reaction progress simulations*

Numerical reaction progress simulations can be used to predict water evolution, rock alteration, mineral precipitation, and volume changes resulting from the reactions inferred from the lab or analogue studies. The validation of the models has to be done by comparison of modelled and observed compositions, fluxes, and masses of reactants. Further parameter variations are needed to calibrate the geochemical models and to assess the uncertainties of predictions.

A reaction progress simulation for the injection of CO_2 into the "Rotliegend-Hauptsandstein" aquifer has been attempted as a case study within the GESTCO project. One of the questions addressed by this simulation was about the possible contribution of alteration reactions to the long-term safe storage of CO_2 in solid carbonates and dissolved bicarbonate. The result of a simulation for reservoir rocks containing chlorite and calcite (no feldspar) is given below (Fig. 1).

Sandstone with a typical chlorite content of about 1 mass % has been assumed for this simulation. Thermodynamic data for Fe-free chlorite have been used. Therefore no Fe-carbonate precipitates. The simulated dolomitization of calcite by Mg released from chlorite is used as a proxy for reactions involving solid solutions of carbonates according to the simplified reaction equation 2.

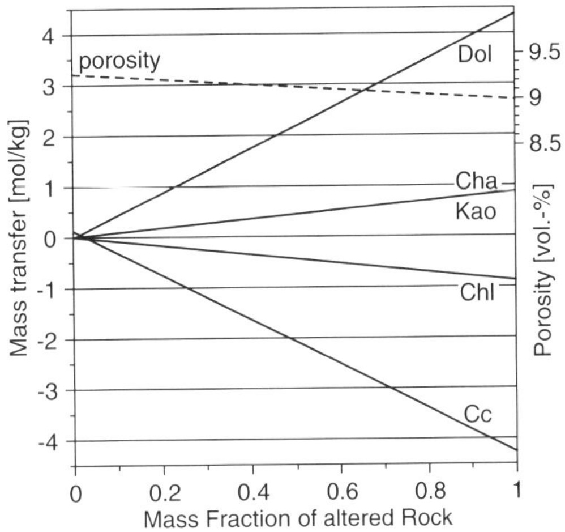

Figure 1. Porosity change resulting from dissolution of primary phases (negative) and precipitation of solid reaction products (positive): kaolinite (Kao), chalcedony (Cha), chlorite (Chl), dolomite (Dol), halite (Ha), calcite (Cc).

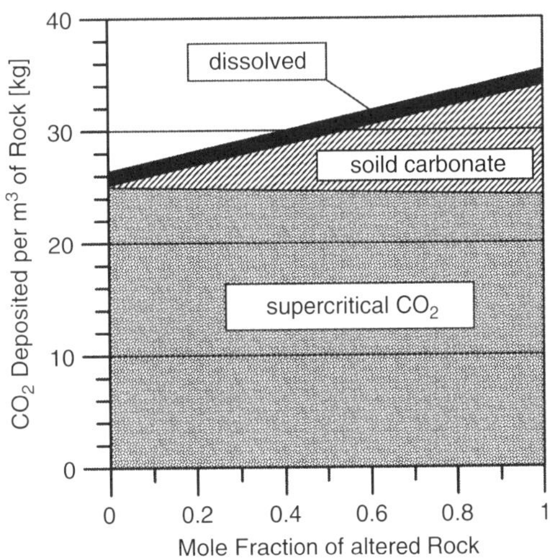

Figure 2. Mass of CO_2 that can be deposited in one cubic meter of reservoir rock at various alteration degrees in supercritical (hatched), liquid (solid) and solid phases (stippled).

$$\text{Chlorite} + 5\ \text{Calcite} \rightarrow 5\ \text{Dolomite} + \text{Kaolinite} + SiO_2 \quad (2)$$

The porosity could decrease from 9.3 initially to 9.0 vol. % when all of the chlorite is altered (Fig. 1). Rock alteration will transform some of the injected CO_2 into carbonates (Fig. 2). The main fraction is

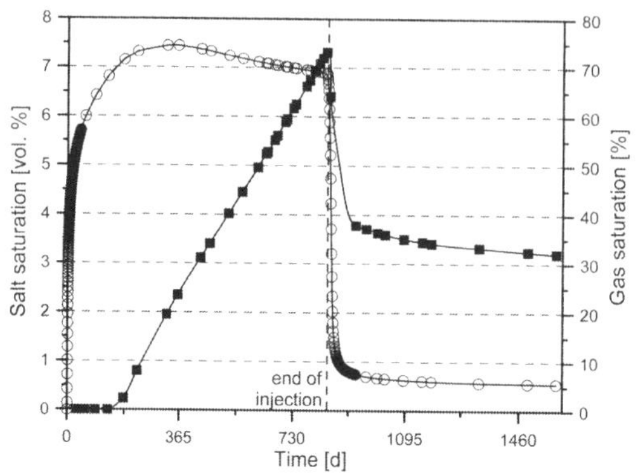

Figure 3. Simulated temporal evolution of gas saturation (circles) and salt saturation (squares) of the pore space within the element of CO_2 injection.

expected to be fluid CO_2 though. Dissolved species probably do not contribute much to the storage capacity in chlorite bearing "Hauptsandstein" beds.

3.1.4 *Reactive transport models*

The impact of reactions close to injection wells or at a propagating gas/water displacement fronts may be misjudged by simulations that consider aquifers as a whole and integrate over the entire aquifer volume. Thus reactions have to be modeled in space and time using reactive transport simulators.

One of the reactions that may be problematic during the CO_2 injection phase is the precipitation of salt from highly concentrated brines. The effects of this reaction have been simulated for a test of CO_2 injection into the "Mittlerer Buntsandstein" aquifer in the north of the NGB. According to the dimensions of the 3-D grid elements used in the simulation, salt precipitation from the highly concentrated brines will start early, after about six months of CO_2 injection. Salt precipitation is limited to the grid element of CO_2 injection however (affected rock volume: 35,100 m^3). According to the constant injection rates (20 kg/s), the total amount of salt precipitated increases linearly, until about 7.3 vol. % of the original pore space is filled with halite (Fig. 3). The absolute porosity decrease from 25 to 23.2 vol. % has no marked effect on the pressure necessary to maintain constant injection rates. After the end of injection, CO_2 rises rapidly within the aquifer and formation water flows back into the injection element and dissolves precipitated salt again. However the salt concentrations of the water flushing back from adjacent elements soon reach halite saturation so that not all of the salt can be dissolved again. In practice this means that salt precipitation from the formation water is irreversible and could require freshwater flushing if the permeability would drop too much.

3.2 *Limitations*

Necessary information to take all of the steps of the ideal modeling procedure is rarely available. All modeling steps involve specific uncertainties which make predictive geochemical modeling such a challenging issue. Natural occurrences of CO_2-rich waters do not exist in all aquifers. Reaction conditions of natural analogues are not known precisely, e.g. the composition of paleo-fluids. CO_2 injection experiments are generally missing. One of the main limitations of numerical modeling is the applicability of thermodynamic data to natural minerals, solid solutions, and organic matter in the coal and hydrocarbon bearing sandstones. The prediction of reactions involving clay minerals is especially difficult. They may be metastable and precipitate as mixed layer minerals with various lattice structures and morphology. The determination of appropriate thermodynamic data is hardly possible, so that parameter variations are the only way to constrain modeling uncertainties. In situ fluid conditions often have to be "reconstructed" from formation water analyses altered by sampling conditions, e.g. degassing, and sensitive redox equilibria. An additional limitation of reactive transport models is the spatial resolution of the generally quite simple geologic aquifer models. The resolution of geophysical exploration methods is not sufficient to generate detailed sedimentological models, which would be required to calculate fluid transport in heterogeneous aquifers with alternating bedding in cm to dm dimensions. Nevertheless, such a detailed model would exceed the computational capabilities of non-linear coupled reactive multiphase-flow simulators.

4 CONCLUSIONS

Predictive numerical modeling requires extremely well studied cases to provide all of the necessary input data. For reservoirs containing abundant reactive minerals, modeling cannot stand alone but has to be supported by lab experiments and field tests. Such experiments and investigations are planned for the EU supported pilot project CO_2SINK. Improved reactive multiphase fluid flow simulators are needed for the complex geochemical simulations of reactions resulting from CO_2 injection, taking into account reservoir architecture and injection strategy.

Fast geochemical reactions, such as salt precipitation, will probably be a practical issue for the sequestration field operators in the NGB. Quantitative predictions about fast reactions during the phase of CO_2 injection, based on experimental studies and simulations, seem to be possible within an acceptable range of uncertainty.

Geochemical long-term modeling will not be able to make predictions within the level of uncertainties requested by the financial planners of sequestration projects. However, the results of numerical process modeling may be interpreted in a qualitative manner, i.e. for the formulation of principal processes, the establishment of a hierarchy of processes, the recognition of likely consequences, the estimation of the magnitude of processes in order to understand and to appraise the chances and risks of geochemical reactions for a specific sequestration project.

REFERENCES

Bandlowa, T., Fischer, M., Krull, P., Schulz, P. & Stiewe, H. 1997. *Tiefversenkung von Abwässern und flssigen Abfällen in den östlichen Bundesländern.* Berlin: BGR.

Christensen, N.P., Holloway, S. 2003. *The GESTCO project summary report.* Kopenhagen: GEUS.

May, F. 1998. Thermodynamic modelling of hydrothermal alteration and geoindicators for CO_2-rich waters. *Zeitschrift der Deutschen Geologischen Gesellschaft* 149: 449–464.

May, F., Brune, S., Gerling, P. & Krull, P. 2003. Möglichkeiten zur untertägigen Speicherung von CO_2 in Deutschland – eine Bestandsaufnahme. *Geotechnik* 2003(3): 162–172.

Zemke, J., Stöwer, M., Arnold, C., Becker, W., May, F., Gerling, P. & Krull, P. 2003 CO_2 sequestration in Germany – General conditions and first field studies. *Proceedings 22nd world Gas Conference, Tokyo 2003.* Tokyo: International Gas Union.

Water–Rock Interaction, Wanty & Seal II (eds)
© 2004 Taylor & Francis Group, London, ISBN 90 5809 641 6

Sequestration of CO_2 in redbeds: siderite precipitation from CO_2 and SO_2 waste gas

J.L. Palandri & Y.K. Kharaka
US Geological Survey, Menlo Park, California, USA

ABSTRACT: We present a method for geologic sequestration of CO_2 as siderite, using Fe^{III} in sediments. Fe is reduced using SO_2 that may be a component of CO_2-dominated flue gas, or by adding SO_2 or H_2S derived from other industrial processes. Geochemical simulations show that Fe and dissolved sulfide derived from SO_2 or H_2S can be converted almost entirely to siderite and dissolved sulfate. In sediments containing only Fe^{III} as a trap for CO_2, the stoichiometry of the reaction requires 0.5 moles of SO_2 or 0.125 moles of H_2S per mole of CO_2. For a synthetic redbed, a quartz arenite with an initial porosity of 21% and containing 0.25 wt% Fe_2O_3, approximately 2 g of CO_2 is sequestered per kg of rock, and the porosity decrease is less than 0.2%. Sequestration of CO_2 using Fe^{III} also has the advantage of disposing of SO_2 that may already be present in flue gas.

1 INTRODUCTION

Geologic sequestration of anthropogenic CO_2 is a potential means of reducing the amount of greenhouse gases released to the atmosphere. In geologic sequestration, CO_2 captured from fossil fuel-fired power plants may be stored in deep saline aquifers or depleted petroleum or gas reservoirs, as (1) supercritical fluid, (2) as a dissolved constituent in formation water, or (3) in carbonate minerals. Most studies of carbonate mineral trapping (e.g. Gunter et al. 2000) consider the use of Ca- and Mg-bearing or glauconitic sediments, to trap CO_2 in calcite, magnesite, siderite, or ankerite. Glauconitic beds, which contain the desired Fe^{II}, are generally of limited thickness and geographical occurrence. However, Fe^{III}-bearing sediments, including redbeds, have the advantages of widespread geographic distribution, generally greater thickness, and higher porosity and permeability.

Fe must be in its ferrous oxidation state for it to precipitate in carbonate minerals. Fe^{III} requires a reductant to be reduced to Fe^{II}, including SO_2 or H_2S. We show using thermodynamic calculations that SO_2, which may be a component of flue gas from fossil fuel-fired power plants, or SO_2 or H_2S derived from other industrial processes, can reduce Fe^{III} to Fe^{II}, while maintaining fugacities of H_2S, a product of SO_2 disproportionation in water, and CO_2 so that precipitation of Fe-carbonate is favored over Fe-sulfide. Simultaneously, reduced S in these gases can be oxidized almost entirely to dissolved sulfate.

2 METHODS

Thermodynamic reaction path calculations are performed with the computer programs CHILLER (Reed 1998) and GAMSPATH (Perkins et al. 1997), which compute the distribution of thermodynamic components among aqueous species, minerals. System variables may be changed incrementally, e.g. temperature (T), pressure (P) or composition, with equilibrium re-computed at each step.

GAMSPATH computes reaction paths constrained by a general rate equation for mineral dissolution and precipitation (modified from Lasaga 1995)

$$\frac{dm}{dt} = -S\sum_i [A_i e^{-E_i/RT} \prod_j a_{i,j}^{n_{i,j}} (1- = \Omega^{p_i})^{q_i}] \qquad (1)$$

where each term in the sum represents a single reaction mechanism. The rate is dm/dt (mol s^{-1}), S is the mineral surface area (m^2), A is the Arrhenius pre-exponential factor (mol m^{-2} s^{-1}), E is the activation energy (J mol^{-1}), T is the temperature (K), and R is the gas constant. The dimensionless product of $a_{i,j}$ accounts for the activity of the ith species in the jth reaction mechanism, raised to reaction order $n_{i,j}$. The dimensionless function of the chemical affinity $(1 - \Omega^p)^q$ where Ω is the mineral saturation index, accounts for slowing of reactions as equilibrium is approached. Precipitation rates are computed by dividing the dissolution rate constant by the equilibrium constant. GAMSPATH simulates an infinitely well-stirred batch reactor, and thus reaction kinetics are strictly surface controlled. The simulations therefore estimate a minimum time for equilibration.

The time-dependent simulation considered is restricted to low pH and $f(O_2)$ conditions, and therefore only one term in Equation 1, depending on the activities of H^+, Fe^{3+} and carbonate species, affects the reaction rates. Thus a much simpler equation, Equation 6 below, yields rates nearly identical to the full equation. Nucleation is simulated by precipitating 10^6 spherical shells with outer radii of 1 μm, and inner radii controlled by the amount of mineral that precipitates in each time step. Once the shells are filled, precipitation proceeds on the outer surface; a saturation index of 1.0 must be attained for mineral precipitation to be initiated. To facilitate comparison among simulations, all shown in figures herein are constructed with P = 200 bar, T = 100°C, an initial 1.0 kg of H_2O at 1.0 NaCl molality with pH adjusted to 5.1, and 10 kg of synthetic redbed composed of 99.75 wt% quartz and 0.25 wt% hematite.

3 RESULTS AND DISCUSSION

Below, we consider geochemical reactions in order of increasing complexity. First, we examine the reaction of SO_2 gas with water, which produces sulfide, sulfate, and low pH. Next, we consider the reaction of SO_2 gas with hematite that produces solid Fe-sulfide and dissolved sulfate. These are considered because they are essential to the desired overall reaction with CO_2. We then investigate reaction of gas mixtures consisting of mostly CO_2 and lesser SO_2, with a synthetic redbed, that produces siderite and dissolved sulfate.

It has been observed in volcanic systems, and confirmed with thermodynamic calculations (Symonds et al. 2001), that SO_2 gas reacts with water to produce H_2S, H_2SO_4, and very low fluid pH:

$$4SO_2 + 4H_2O = H_2S + 3H_2SO_4 \tag{2}$$

which is confirmed in simulations from 25 to 300°C, producing sulfate dominantly as SO_4^{2-} or HSO_4^- depending on pH, and sulfide mostly as dissolved H_2S with minor HS^-.

Within the last decade, research has been conducted on in situ sequestration of H_2S (Gunter et al. 2000), a natural contaminant in natural gas, by precipitation of Fe-sulfide minerals in Fe-bearing sediments. Reduction of Fe^{III} is accomplished with a portion of the added sulfide, which oxidizes to sulfate. One possible reaction for H_2S sequestration is

$$\begin{aligned}4Fe_2O_3(\text{hematite}) + 15H_2S + H_2SO_4 \\ = 8FeS_2(\text{pyrite}) + 16H_2O\end{aligned} \tag{3}$$

Owing to relative precipitation rates, both pyrite and pyrrhotite become supersaturated, and the product observed in experiments (Palandri 2000, and data of others therein) is less-stable amorphous FeS:

$$\begin{aligned}4Fe_2O_3(\text{hematite}) + 9H_2S \\ = 8FeS(\text{amorphous}) + H_2SO_4 + 8H_2O\end{aligned} \tag{4}$$

The time for the reaction to attain 90% completion ranges from days to weeks, depending primarily on the experimental T of 28° to 71°C. There is a steady increase in dissolved Fe^{II}, despite its removal by Fe-sulfide precipitation, which is available for precipitation in carbonate minerals.

Minerals may be suppressed from simulations, where a less stable phase is expected; pyrite is not suppressed to show that the most stable of the Fe-sulfide minerals is a less stable repository for Fe than are Fe-bearing carbonates. Simulated reaction of SO_2 with hematite in water from 25 to 300°C yields products as in Equations 3 and 4, depending on suppression of pyrite and pyrrhotite.

The reaction of interest is that of CO_2 and SO_2 mixtures with Fe^{III} in hematite, and water

$$\begin{aligned}Fe_2O_3(\text{hematite}) + 2CO_2(g) + SO_2(g) + H_2O \\ = 2FeCO_3(\text{siderite}) + H_2SO_4\end{aligned} \tag{5}$$

but the gases must react first with water to produce carbonate- and sulfide-bearing species. The gas composition required, beyond the amount that dissolves, is 66.7/33.3 mol% CO_2/SO_2 for sediments containing only Fe^{III}. Where other divalent metals (e.g. Ca, Mg) capable of trapping CO_2 are present, the amount of SO_2 may be lower.

3.1 *Equilibrium simulations*

The simulations were constructed by first equilibrating brine with rock, yielding a porosity of 21%, and then adding gas. The first simulation uses 99/1 wt% CO_2/SO_2, which approximates a composition that might be encountered from combustion of sulfur-bearing fossil fuel. The desired product, siderite (Fig. 1A), is stable over a wide range of total gas added, from −2.3 to 0.34 log kg of gas added (kga), i.e. 5.0 to 2190 g. Pyrite replaces siderite, but only after hematite has dissolved completely. A gas phase saturates near −1.5 log kga (32 g), and its H_2S content is <1.0 ppm until siderite dissolves completely.

At small values of kga, where siderite is absent, pH decreases imperceptibly (Fig. 1B) as hematite dissolution consumes copious H^+; total concentrations of Fe^{II}, sulfide, sulfate and carbonate increase steadily. Where siderite precipitates, H^+ is produced, and the pH decreases more sharply.

Total sulfide decreases as decreasing pH drives further dissolution of hematite and iron reduction. A significant result is that the total amount of sulfide remains less than one part per trillion until hematite disappears. At the siderite maximum the porosity decrease is small, from 21% to 20.9%., and approximately 2 g of CO_2 is trapped per kg of rock. Pyrite eventually replaces siderite in response to increasing sulfide.

A second simulation was constructed using $T = 50°C$ (not shown), confirming that the reaction proceeds at lower T. Results show that: (1) gas saturation is delayed to a slightly larger amount of gas added (−1.3 log kga) because of increased gas solubility; (2) siderite precipitation is also delayed (−2.1 log kga) because siderite is less stable; and (3) the amount of sulfide is lower where pyrite is present, because pyrite is more stable.

A third simulation (Fig. 1C) tests the effect of increasing SO_2 to the stoichiometry indicated in Equation 5, 66.7/33.3 mol% (58.2/41.8 wt%) CO_2/SO_2. A small amount of pyrite precipitates before siderite because of greater dissolved sulfide relative to carbonate; carbonate eventually reaches a concentration where siderite replaces pyrite. Siderite is replaced by pyrite, and pyrite by native S in response to increasing sulfide. If the gas contains excess SO_2, some Fe that could trap CO_2 in siderite will precipitate in pyrite. If the gas contains insufficient SO_2 to reduce enough Fe to fix all of the CO_2 in siderite, e.g. flue gas, targeted rocks must contain other divalent metals, e.g. Ca and Mg, capable of precipitating as carbonate minerals. Conversely, targeted rocks need contain only a small amount of Fe^{III} to effectively scrub almost all SO_2 from the injected gas.

If H_2S is used instead of SO_2, approximately the same results are obtained with an H_2S concentration one quarter that of SO_2 as indicated by Equation 2, but yielding much less dissolved sulfate.

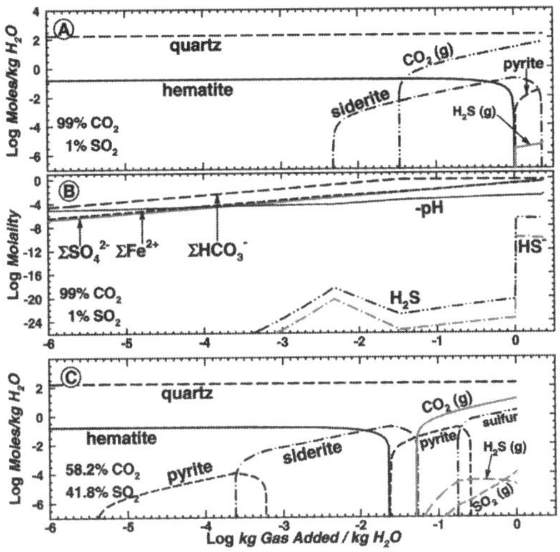

Figure 1. Simulated equilibrium reactions of CO_2/SO_2 with synthetic redbed: A) 99/1 wt% CO_2/SO_2, minerals; B) 99/1 wt% CO_2/SO_2, aqueous species; C) 58.2/41.8 wt% CO_2/SO_2, minerals.

3.2 *A time-dependent simulation*

The time-dependent simulation of the CO_2-SO_2-hematite-H_2O reaction uses a rate expression that is simplified from Equation 1

$$\frac{dm}{dt} = -S \cdot Ae^{\frac{-E}{RT}} a_{H^+}^{n_1} a_{HCO_3^-}^{n_2} a_{H_2CO_3}^{n_3} a_{Fe^{3+}}^{n_4} (1 - \Omega^p)^q \qquad (6)$$

Mineral dissolution rate parameters are summarized in Table 1. Chemical affinity parameters p and q are assumed to be unity because data are absent.

Because of software limitations, the fluid must be charged with gas before adding rock, which leads to native S saturation. Rate data for S are absent, and therefore S is suppressed from the simulation. In the

Table 1. Rate equation parameters.

Mineral	Size/morphology	Rate constant[a] log moles m^{-2} s	Activation energy KJ $mole^{-1}$	Reaction order[b] w.r.t. activity of H^+	Reaction order[b] w.r.t. given species
Quartz[c]	0.1 mm spheres	−13.34	90.1	0.00	–
Hematite[d]	0.01 mm spheres	−9.39	66.2	1.00	0.23, HCO_3^-
Pyrite[e]	0.1 mm spheres	−7.52	56.9	−0.50	0.50, Fe^{3+}
Siderite[f]	0.1 mm spheres	−0.30	14.4	1.00	1.00, H_2CO_3

[a] At 25°C, and $a(H^+) = a(Fe^{3+}) = a(HCO_3^-) = a(H_2CO_3) = 1.0$.
[b] Default reaction order is 0.0 for species not in the reaction mechanism, or data are not available.
[c] Data regressed from Tester et al., 1994.
[d] Data regressed from Bruno et al., 1992; reaction order w.r.t. HCO_3^- from Runa and Gilkes, 1995.
[e] Data from McKibben and Barnes, 1986.
[f] Rate set equal to calcite, data regressed from Talman et al., 1990.

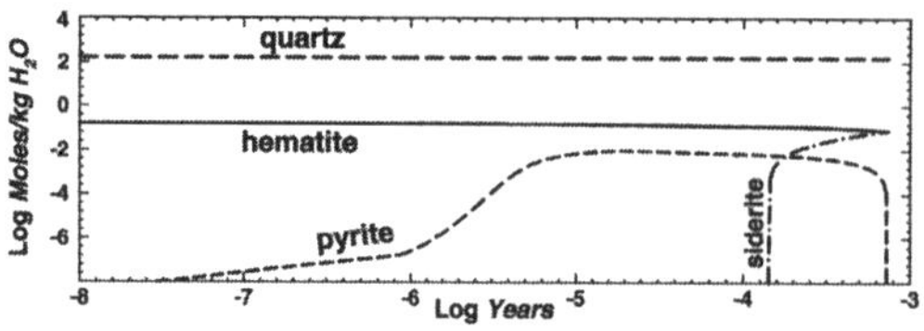

Figure 2. Simulated time dependent reaction of CO_2/SO_2 with synthetic redbed, minerals only.

context of subsurface injection, it is important to recognize that native S may precipitate if its precipitation is fast relative to sulfide oxidation by Fe^{III}, or if Fe^{III} is absent or has been depleted as would occur proximal to an injection well. This may decrease formation porosity and permeability.

We used 10 kg of synthetic redbed of the same composition as in the equilibrium simulations. The gas composition is 90/10 wt% CO_2/SO_2 and totals 100 g. The SO_2 is slightly less than required to reduce all of the Fe, and the CO_2 is in excess of the amount that could precipitate in siderite.

Hematite controls the time for the system to equilibrate, because pyrite dissolves and siderite precipitates relatively quickly (Fig. 2). Siderite precipitates after ~1.2 hours, and complete equilibration occurs in ~6.6 hours. Equilibration rates in aquifers are expected to be slower, primarily because CO_2 sequestration in aquifers involves flow through porous media, which allows the development of concentration gradients near mineral surfaces, resulting in decreased chemical affinity and slower rates; other differences may occur because of variation in grain size, aquifer inhomogeneity, minerals coatings, and preferred fluid flow paths.

Chemical reactions in this simulation are similar to those discussed in section 3.1; differences occur because the fluid is charged with gas before reacting with rock instead of initially equilibrating the fluid with the rock. Published rates for dissolution of a single mineral typically show uncertainties of up to one order of magnitude within a single work, and up to several orders of magnitude among different workers. Therefore, uncertainties of similar magnitude should be assumed in this simulation. This uncertainty should be addressed in experiments specifically designed to examine the reaction of CO_2-SO_2 gas mixtures with Fe-bearing rocks.

4 CONCLUSIONS

Fe^{III} in sediments can be used to trap CO_2 in Fe^{II}-bearing carbonate minerals, by including SO_2 or H_2S in the waste gas to reduce Fe. Fe can be made to reside almost entirely in siderite, with S existing predominantly as dissolved sulfate. In rocks that contain only Fe^{III} as a potential trap for CO_2, the appropriate gas composition is approximately two moles of CO_2 to one mole of SO_2. Given that the waste gas from fossil fuel-fired power plants likely has a higher ratio of CO_2 to SO_2, targeted sediments must contain other divalent metals such as Ca or Mg, or sulfur-bearing gas must be added to the waste gas stream, to trap all of the CO_2 in siderite. Conversely, sediments need contain only a small amount of Fe^{III} to convert reduced S to dissolved sulfate; most sediments contain sufficient Fe^{III} to effectively scrub SO_2 from waste gas typical from the combustion of fossil fuel.

These conclusions are based on idealized computer simulations. Although the thermodynamic part of the calculations are built upon a solid framework, there exist considerable uncertainties in the kinetics of mineral precipitation and dissolution, and of aqueous speciation. Therefore, rigorous experiments to refine rate parameters should be conducted in the future, both for modeling studies in general, and for the process considered herein in particular.

The use of Fe-rich sediments, e.g. redbeds, introduces a CO_2 mineral trap with relatively high porosity and permeability, widespread geographic distribution, and great thickness. The benefits are twofold: Fe^{III} can be used as a trap for CO_2, and SO_2 that may already be part of the flue gas can be injected with the CO_2, so that it does not have to be scrubbed at high cost from the waste gas stream.

REFERENCES

Bruno J., Stumm, W., Wersin, P. & Brandberg, F. 1992. On the influence of carbonate in mineral dissolution; 1, The thermodynamics and kinetics of hematite dissolution in bicarbonate solutions at T = 25°C. *Geochim. Cosmochim. Acta* 56: 1139–1147.

Gunter, W.D., Perkins, E.H. & Hutcheon, I. 2000. Aquifer disposal of acid gases: modelling of water-rock reactions for trapping of acid wastes. *Appl. Geochem.* 15: 1085–1095.

Lasaga, A.C. 1995. Fundamental approaches to describing mineral dissolution and precipitation rates. In A.F. White & S.L. Brantley (eds.) *Chemical Weathering Rates of Silicate Minerals, Reviews in Mineralogy* 31: 23–86. Washington, D.C.: Mineralogical Society of America.

McKibben, M.A. & Barnes, H.L. 1986. Oxidation of pyrite in low temperature acidic solutions: Rate laws and surface textures. *Geochim. Cosmochim. Acta* 50: 1509–1520.

Palandri, J.L. 2000. Applications of computed chemical equilibria: I. Reconstruction of *in situ* composition of sedimentary formation waters; II. Disposal of H_2S from natural gas; III. Serpentinization, related rodingitization, and silica-carbonate alteration of ultramafic rock. Unpub. Ph.D. Thesis. Eugene: University of Oregon.

Perkins, E.H., Gunter, W.D., Nesbitt, H.W. & St-Arnaud, L.C. 1997. Critical review of classes of geochemical computer models adaptable for prediction of acidic drainage from mine waste rock. In H.W. Perkins (ed.), *Proceedings: Fourth international conference on acid rock drainage*: 587–601. Ottawa: MEND-CANMET.

Reed, M.H. 1998. Calculation of simultaneous chemical equilibria in aqueous-mineral-gas systems and its application to modeling hydrothermal processes. In J. Richards and P. Larson (eds.), *Techniques in Hydrothermal Ore Deposits Geology. Reviews in Econ. Geol.* 10: 109–124. Littleton, Colorado: Soc. Econ. Geol.

Ruan, H.D. & Gilkes, R.J. 1995. Acid dissolution of synthetic aluminous goethite before and after transformation to hematite by heating. *Clay Mins.* 30: 55–65.

Symonds, R., Gerlach, T. & Reed, M. 2001. Magmatic gas scrubbing: Implications for volcano monitoring. *J. Volcanol. Geotherm. Res.* 108: 303–341.

Talman, S.J., Wiwchar, B., & Gunter, W.D. 1990. Dissolution kinetics of calcite in the H_2O-CO_2 system along the steam saturation curve to 210°C. In R.J. Spenser & I.-M. Chou(eds.), *Fluid-Mineral Interactions: A Tribute to H. P. Eugster:* 41–55. San Antonio, Texas: Lancaster Press, Inc.

Tester, J.W., Worley, W.G., Robinson, B.A., Grigsby, C.O. & Feerer, J.L. 1994. Correlating quartz dissolution kinetics in pure water from 25 to 625°C. *Geochim. Cosmochim. Acta* 58: 2407–2420.

Water–Rock Interaction, Wanty & Seal II (eds)
© 2004 Taylor & Francis Group, London, ISBN 90 5809 641 6

CO_2 geological sequestration in Italy: state of the art and potential sites

F. Quattrocchi, R. Bencini, D. Cinti, G. Galli, L. Pizzino, N. Voltattorni & A. Navarra
INGV, Istituto Nazionale di Geofisica e Vulcanologia, Laboratory of Fluid Geochemistry, Rome, Italy

ABSTRACT: On the basis of the experience from ongoing geological sequestration of CO_2 in the Weyburn field (Canada) by EOR techniques and from the study of "CO_2 analogues", a multidisciplinary study was undertaken to assess the possibility of geological sequestration of CO_2 in Italy. This paper discusses several possible cases: 1) a deep saline aquifer (San Gavino Basin, Sardinia); 2) a coal region where the ECBM technique may be applied (Sulcis, Sardinia); and 3) oil fields where EOR technique may be exploited (Southern Apennines, Adriatic foredeep, Po Valley, Sicily). The San Gavino Basin (Campidano Graben), where geologic and fluid geochemical information where revisited and geochemical surveys were performed, is discussed in details. We stress the importance, especially for geodynamically active areas, of performing both seismic risk assessments and the gas hazard/CO_2 micro-seepage evaluations before applying each technique. Public acceptance for these greenhouse gases sequestration techniques in the densely populated regions of Italy is essential.

1 TEST SITE SELECTION CRITERIA

1.1 *Perspectives on the global scale and in Italy*

It is generally accepted that anthropogenic emissions of greenhouse gases are causing changes in the climate (IPCC 1592 projection). The major contributor is carbon dioxide, which comes mainly from burning fossil fuels. As a result, the concentration of CO_2 in the atmosphere, which was 280 ppm in pre-industrial times and is 370 ppm today, is predicted to reach about 700 ppm by 2100 in the absence of remedial measures, while the $\delta^{13}C$ data (NOA-CMDL and University of Colorado-INSTAAR) clearly record the carbon evolution in atmosphere. Developed countries agreed to reduce their emissions to 5.2% below 1990 levels over the period 2008–2012 through the Kyoto Protocol, within the context of the UNFCCC, signed in December, 1997, but the agreement is not still ratified as a whole. In September 2003, in Sacramento, California, a bilateral agreement between Italy and USA was signed, concerning CO_2 geological sequestration and clean energy technologies. Broadly speaking, all ways of converting fossil fuels, including the "*hydrogen carrier*", produce a waste stream containing variable quantities of CO_2. Using complex technologies as EOR and ECBM (Emberley et al. 2003, Riding et al. 2003, NETL, NOVEN and TNO reports), it is possible to capture this stream, inject it underground and effectively store it. The potential storage of CO_2 in Italy has never been fully evaluated. Few experiences exist worldwide, and virtually none are applied in geodynamically active areas and throughout densely populated regions such as Italy. Storage capacity in geologic reservoirs in northwest Europe alone could be as much as 8000 G tonnes of CO_2 (most of this in deep saline aquifers and reservoirs). Gas and oil fields have similar potential (an average of 10 G tonnes of CO_2 for each of them). The EOR (*Enhanced Oil Recovery*) Weyburn field is foreseen to inject 20 G tonnes of CO_2 in 15 years. In Italy, CO_2 geological sequestration may be considered, initially by EOR techniques of about 5 G tonnes for a first stage (10 years). At the present time, with the exception of EOR techniques applied to pre-existing oil wells, the geological sequestration of CO_2 is uneconomic for industrial applications. After the first papers dealing with many EOR early experiences (Quattrocchi et al., 2004 and references herein), both geochemical and geophysical monitoring methods adopted recently in the context of EOR projects (Riding et al. 2003, Quattrocchi et al. 2004) are specifically focused on environmental and "public acceptance" aspects.

Potential sites for the geological sequestration of CO_2 in Italy are:

1. gas and oil fields, depleted or abandoned, where the reservoirs are located in proven traps, known to have held liquids and gases for hundred thousands

or millions of years, the geology is well known, and there is the potential of re-using some existing equipment to transport and inject the CO_2;
2. deep saline reservoirs (more than 800 m deep) that have a relatively impermeable cap rock and good permeability inside the reservoir;
3. suitable unminable coal beds, where the ECBM (*Enhanced Coal Bed Methane*) technique may provide net storage of CO_2.

Test sites for the geological sequestration of CO_2 should be selected on the basis of *i)* their proximity to the CO_2 sources and *ii)* their structural and geochemical characteristics (i.e. WRI potential, pH buffering capacity). Regarding the first characteristic, CO_2 sources in Italy have already been partially inventoried; further work is being planned on the basis of the experience gathered by both DOE in USA and NOVEM-TNO. Evaluation of the annual volume that may be captured from chemical plants and transported to sequestration sites has begun (for comparison, 3.4 M tonnes/year in The Netherlands, NOVEM data). This evaluation was not fully accomplished in Italy yet, but a national project is ongoing. The CO_2 sources/streams have been divided into three groups of "end-of-the-chimney" types: concentrated CO_2 streams (>99%; e.g. chemical industry), dilute CO_2 streams (4–21%, e.g. flue gas from most of plants, where capture is needed), and diffuse CO_2 streams.

The main sources of CO_2 are hydrogen plants, the steel industry, the cement industry (900 kg of CO_2/1000 kg of cement, *Greenhouse Issue*, November, 2002), the industrial production of iron and other metals, and power plants (a typical gas-fired power plant emits about 1.1 million tonnes of CO_2/year). Considering the economics of the geological sequestration of CO_2 as a whole, the IEA GHG R&D Program has estimated that storage in deep saline reservoirs, as well as in depleted oil and gas fields costs US$ 1–3/tonne of CO_2, excluding the cost of CO_2 transportation. The costs are expected to fall as the technology matures and the scale of the application increases. The cost of CO_2 capture and storage may correspond approximately to an increase in the price of electricity from 1.5 to 3 US cents/kWh (Bocciarelli unpublished thesis by Univ. Roma Tre-INGV; www.iea.org). The economic evaluations may lead to an European carbon tax at or below 25 Euro/tonne of CO_2.

1.2 *Geochemical CO_2 storage potential*

The most important prerequisite for geological storage of CO_2 is the pH buffering capacity of the host rocks: water–rock interaction and risk-assessment approaches may use the experience gathered in monitoring "CO_2 analogues". The concept of buffering has to be accepted by society as a key factor of the "public acceptance of the geological sequestration of CO_2". The buffer capacity of a reservoir is the ability of the water–rock system to resist pH changes. If pH is buffered, additional CO_2 may cause the precipitation of carbonates, and thus, permanent storage of CO_2. Gunter et al. (1997, 2000, see complete references in Quattrocchi et al., 2004) have grouped mineral trapping into different types. The silicate-carbonate mineral assemblage has the highest buffer capacity, trapping CO_2 as dissolved CO_2 and HCO_3^- ion. It was effectively verified in the field, i.e. 130% increasing of bicarbonate ions both at Rangely Colorado and at Weyburn, two year after injection. Particular care should be given to monitoring dissolved gases (Quattrocchi et al. 2004). Dissolved gases may be more indicative than the mixed stale gas phase collected from the casing head for drawing quantitative conclusions.

The geochemical modeling must evaluate if silicates (i.e. siliciclastic Tertiary basins in Italy) may be present in sufficient quantities to buffer pH, resulting in the eventual precipitation of injected CO_2 as new minerals. Silicate hydrolysis may buffer the pH of the solution through reactions as:

$$2\text{Albite} + 2H_2O + 2CO_2 \leftrightarrow$$
$$2HCO_3^- + 2\text{Na} + \text{Paragon.} + 6\text{Quartz}$$

The silicate minerals, such as kaolinite, smectite, and possibly illite, influence water composition and may result in enough quantity to buffer pH, leading to precipitation of injected CO_2, and allowing more carbon dioxide to be taken up in the aqueous phase than that predicted from the solubility of CO_2 alone.

2 THE SALINE AQUIFER OF THE SAN GAVINO BASIN

Sardinia represents a typical seismically quiescent site, displaying an almost complete lack of both historical earthquakes and of instrumentally recorded seismicity: low geodynamic activity is a good prerequisite for geological sequestration of CO_2 from a public acceptance perspective. Around sixty groundwater sites were surveyed to evaluate the best CO_2 storage sites, sampling thermal-mineralized and cold springs or wells specifically in the *Campidano Graben*, *Tirso Valley*, *Logudoro* and *Casteldoria* provinces. Fluid geochemistry surveys were carried out including physical and chemical parameters, major and minor elements within the liquid phase, and selected isotope ratios (H, O, C, He). ^{222}Rn concentrations and other dissolved gases have been investigated to discriminate faults and convection at depth.

Based on water chemistry and dissolved gases, two different fluids throughout the island have been recognized: *i*) CO_2-rich gases, poor in He and Rn, with a relatively high $^3He/^4He$ ratios (up to $R/R_a = 2.32$), associated with Na-HCO_3-(Cl) cold groundwater; and *ii*) gases rich in N_2 and He, poor in CO_2 and Rn, with low $^3He/^4He$ ratio, associated with alkaline thermal waters, related to CO_2-impermeable fault segments.

The distribution of these two groups of fluids characterizes the Sardinian tectonic systems: CO_2 "baseline" fluxes are not homogeneous, being mainly related both to the fault segments discriminated in this study and to the areas where Quaternary basaltic rocks crop out. The *San Gavino* Miocene-Pliocene sedimentary basin is located along the NNW-SSE elongated *Campidano Graben* (Fig. 1). The basin is characterized by a gravity minimum at the depocenter of the graben (Angelone et al. 2004). Understanding the role of inactive faults as permeability barriers or as pathways for deep fluids is a prerequisite to evaluate the potential for the geological sequestration of CO_2. The remarkable correspondence between high gravimetric and magnetometric gradients and the location of sampled thermal springs (*Sa Guardia* and *Su Concali* sites and *Sardara* site, SAR 17, 18, 19 in Fig. 1) can be useful in the analysis. The *S'Acquacotta* spring (SAR 33, Fig. 1) is associated with the main fault along the western side of the graben, specifically the *Serramanna-Serrenti Basin*, which is the extreme extension of the *S. Gavino Basin*. In this sector of the graben, with the exception of the *Santa Maria Neapolis* sample, all thermal waters show high dissolved CO_2 and radon. A previous isotopic study (Minissale et al. 1999) on dissolved carbonates ($\delta^{13}C \approx -3‰$) is in agreement with our data, suggesting that CO_2 derives from the metamorphism of marine carbonates. Water and dissolved gas "baseline" chemistry highlighted the CO_2-impermeable fault segments and the typical water–rock interaction in the siliciclastic lithology in the center of the basin, where injection, deep in the sedimentary basin of San Gavino Basin, has the largest potential for CO_2 disposal. Siliciclastic aquifers are predicted to have the best potential for trapping CO_2 by precipitating carbonate minerals, when they contained an assemblage of basic alumino-silicate minerals, such as feldspars, zeolites, illites, chlorites, and smectites. When reacting with CO_2, these minerals break down to form kaolinite, and the CO_2 is neutralized. Depending on whether the basic alumino-silicates mineral contains alkali or alkaline earth cations, the formation water will evolve in one or more ways. In the San Gavino

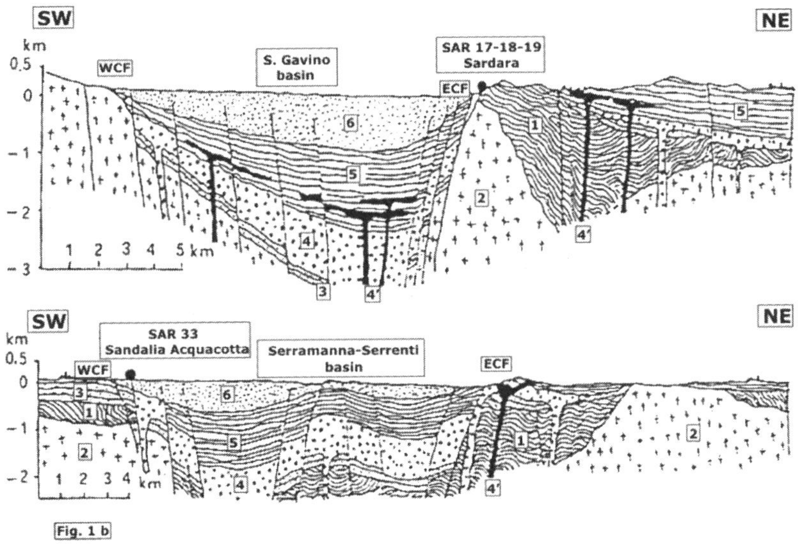

Figure 1. Geological-stratigraphical cross-sections of Campidano along the S. Gavino basin (upper), S'Acquacotta and Serramanna-Serrenti basins (lower) (after Balia et al., 1991): (1) Schists (Paleozoic), (2) Granitic rocks (Paleozoic), (3) Sandstones (Eocene), (4) Andesitic lavas and breccias (Oligocene-Lower Miocene), (4') Andesitic dikes (Miocene), (5) Prevalently marine arenaceous-marly or calcareous sediments (Miocene), (6) Alluvial and lacustrine deposits (Pliocene). WCF = Western Campidano Fault; ECF = Eastern Campidano Fault. The centre of the sedimentary basin is the possible location of CO_2 geological sequestration.

area, metals, and in particular the Al and SiO_2 monitoring is strategic (Quattrocchi et al. 2004).

The expected reactions are similar to those in the *Glauconite Sandatone* aquifer (Gunter et al. 1997 see complete reference in Quattrocchi et al., 2004). The Utsira sandstone formation, for which a huge CO_2 storage potential has been anticipated (SACS Project, 200 m thick, 2800 tonne/day, 600 10^9 tonnes storage capacity in 600 years, as to store the entire CO_2 emissions from all power stations in Europe) has a mineralogy similar to that of the *San Gavino Basin*. Geochemical modeling with the NUFZ code (Johnson J.W., LLNL, California, GHGT-6 presentation, Kyoto, 2002) demonstrates that acidic conditions trigger silicate dissolution, the critical forerunners of mineral trapping. In the vicinity of the injection wells, the reaction of intra-plume dawsonite cementation is anticipated as follows:

$$KAlSi_3O_8 + Na^+ + CO_{2(aq)} + H_2O \leftrightarrow NaAlCO_3(OH)_2 + 3SiO_2 + K^+$$

To reach the maximum performance both in the sandstones and andesitic lithology as host-rock (see Fig. 1), it is important i) the integrity of the cap-rocks, ii) shales inside the aquifer, iii) salinity of the formation waters, iv) suitable mineralogy of the aquifer as well as of the cap-rock, v) sufficient reservoir permeability. Moreover, the basin is characterized by an industrial pole, possibly expanding in future years, where CO_2 sequestration may became necessary. A feasibility study is strongly recommended.

3 THE ECBM SITE IN SULCIS (SARDINIA)

NOVEM (The Netherlands) coordinates the European strategy of ECBM, assisted by TNO. In Europe, the most important project is RECOPOL (coordinator: TNO-NITG) aimed at the reduction of CO_2 emission by means of CO_2 storage in coal seams of the Silesian Basin, Poland. Coal has enormous advantages over all the other potential repositories because it is present worldwide (there are reserves in almost one hundred countries), it maintains stable prices, and it is a resource to be exploited in order to avoid the dependence from other fuels. In Italy, significant coal reserves are not present with the exception of the *Sulcis Coal Province* (Sardinia), which is not exhausted yet. The Sulcis region occupies the southwestern end of Sardinia, from the valley of the Cixerri River to the Gulf of Palmas in front of which the isles of San Pietro and Sant-Antioco are located. In the last months of 1936, the Italian Coal Firm (A.Ca.I) started to exploit the Sirai-Serbariu coal field. After an initial period of intense extractive activity, as the second World War continued, production slowed remarkably. In the 1950's and 1960's, there was renewed activity, as the Sulcis Tertiary coal represented the only coal available in Italy for the renewal of the national industry. The *Sotacarbo* gasification plant project, applied to the Sulcis site, is supported by the Government of Italy and the EU together with the ATI Sulcis consortium. A feasibility study named "ECBM Sulcis" is ongoing by a Sotacarbo-INGV-ENEA consortium.

4 EOR POTENTIAL SITES IN ITALY

The creation of joint ventures between industrial partners and research institutions is indispensable to implement and monitor EOR test-sites for the geological sequestration of CO_2 in Italy and abroad. Typically, the researchers undertake the monitoring work of EOR projects, such as: *i*) the study of possible gas leakage at surface (soil gas survey), *ii*) the environmental hazards assessment, *iii*) the geochemical modelling of the reservoir to assess the CO_2 sequestration potential and the buffer capacity; *iv*) geophysical modelling of the reservoir by 4D seismic survey both to quantify the CO_2 at depth and to monitor its movement and micro-seismicity.

In Italy, several small and large oil fields have been discovered in the past. They contain variable quantities of crude oil of variable quality. Oil quality is important, because certain types of heavy oil are less suitable for CO_2-based EOR techniques. The oil field size is also important because the larger is the original oil accumulation, the larger is the potential for CO_2 injection and associated additional oil recovery. Some fields are already depleted, some are under production and others have not started their productive life yet. These impacts on the timing can be forecasted for the start of the EOR operations, and hence on the timing of the CO_2 injection. Italian oil reservoirs are at various depths, between few hundred and approximately 6000 meters. CO_2 EOR operations are successful only if the reservoir pressure is such that CO_2 is able to be miscible within the oil phase ($>$depth 1000 meters). On the other hand, if the oil reservoir is very deep, approximately more than approximately 3000 meters, the cost of compressing CO_2 for injection may become significant. Italian oil fields are located in the Po Valley, in the Southern Apennines, Sicily and Adriatic cost. Here, the "*EOR Sibilla*" project was launched recently by INGV.

Some oil fields are located in the immediate vicinity of large sources of anthropogenic CO_2, whereas others are located at significant distance from any CO_2 source. It is important to note that oil fields may offer significant additional geologic CO_2 sequestration potential even after the end of EOR operation, i.e. after the end of their productive life. This is possible when the oil accumulation is located in a large structure that can contain CO_2 in excess of the CO_2 that is necessary for EOR operations. In practice, after the economic production of crude oil is terminated, the same injection facilities could be used (at reduced cost relative to new sites) to inject extra CO_2 in the deep saline aquifer below the original oil-water contact, provided that such an aquifer and the structural trap are large enough to contain additional CO_2. On the basis of technical and economic selection criteria listed above, the first "*EOR Sibilla*" demonstration site in Italy is a small, depleted, light oil field deeper than 1000 meters, located near a CO_2 source, almost "pure". The oil field is object of a feasibility study for a possible additional deep saline aquifer sequestration potential too.

5 CONCLUSIONS

Italy offers sites for all principal types of geological sequestration of CO_2: injection in deep saline aquifers; ECBM; and EOR techniques. Considering the complex geodynamic settings of Italy for each potential CO_2 geological storage/sequestration site, a number of uncertainties require detailed studies, such as: 1. the minimum duration that CO_2 can remain effectively stored underground and the effects of well drilling on the integrity of natural sealing capacity; and 2. the effects of possible slow or sudden release of CO_2, particularly for sites in seismic areas and near active faults, as catalogued by INGV. The compilation of the national inventory of potential sites, including information about the proximity of each site to nearby anthropogenic CO_2 sources and the potential CO_2 storage capacity of each site, is a national priority, together with the starting of test-site studies.

REFERENCES

Quattrocchi, F. , Barbieri, M., Bencini, R., Cinti, D., Galli, G., Pizzino, L. & Voltattorni, N. 2004. Water chemistry, isotopic chemistry ($^{87}Sr/^{86}Sr$) and dissolved gases variations at the Weyburn oil-field (Canada). Possible water-rock processes during the CO_2 injection. *Oil and Gas Science and Technology, IFP*, submitted.

Riding, J.B., Czernichowsky-Lauriol, L. S., Quattrocchi, F., Rochelle, C., Savage, D. & Springer, N. 2003. The International IEA Weyburn CO_2 Monitoring Project – The European Dimension. "Greenhouse gas control technologies" Eds. Pergamon (Gale & Kaya), 1879 pp., 1629–163.

Water–Rock Interaction, Wanty & Seal II (eds)
 ISBN 90 5809 641 6

An experiment frozen in time – cryo-SEM imaging of rapidly-formed CO_2 hydrate

C.A. Rochelle, A.E. Milodowski, K. Bateman, D.A. Gunn, P.D. Jackson, L.M. Nelder & J.G. Rees
British Geological Survey, Kingsley Dunham Centre, Keyworth, Nottingham, UK

M.A. Lovell
Department of Geology, University of Leicester, Leicester, UK

ABSTRACT: Cryogenic SEM techniques have been used to obtain high-resolution images of very rapidly formed, laboratory-grown CO_2 hydrate. Careful etching and ablation of the sample inside the SEM revealed that hydrate crystals were formed in a matter of seconds during a rapid depressurization event. This event resulted in a very dynamic, fizzing crystal "mush", in which water, gaseous CO_2, liquid CO_2 and CO_2 hydrate coexisted. This short-lived phenomenon was captured through quenching the experiment by rapid freezing.

1 INTRODUCTION

A potential method of reducing anthropogenic emissions of greenhouse gases to the atmosphere, without drastic changes to our energy-producing technologies, is through their sequestration in other domains of the planet. The most important anthropogenic greenhouse gas is CO_2, and much international effort has been expended to investigate its storage within deep geological formations (e.g. Holloway 1996), within deep oceans (e.g. Marchetti 1977, Austvik & Løken 1992, Golomb et al. 1992, Wilson 1992) and in deep-water sediments (Koide et al. 1997). The latter two approaches rely heavily on the formation of CO_2 hydrate to immobilize the CO_2.

Natural CO_2 hydrate is relatively rare on the Earth (Sakai et al. 1990) because, like many other gas hydrates, it is only stable under high pressures and low temperatures (Fig. 1, and see Sloan, 1998). However, it is likely to be more common elsewhere in the solar system (e.g. Duxbury et al. 2001). Consequently, study of CO_2 hydrate systems is usually performed in laboratory settings, although a few in-situ experiments have been undertaken in the deep ocean (e.g. Brewer et al. 1999).

Detailed observations of gas hydrates are hampered by their high pressure/low temperature stability range. This necessitates pressure vessels and highly penetrating analytical methods (e.g. X-rays, NMR, neutron diffraction), or transparent pressure windows. This limits fine-scale imaging, which is particularly important when considering the form of hydrate within the water column or within sediments. As a potentially useful step in overcoming some of

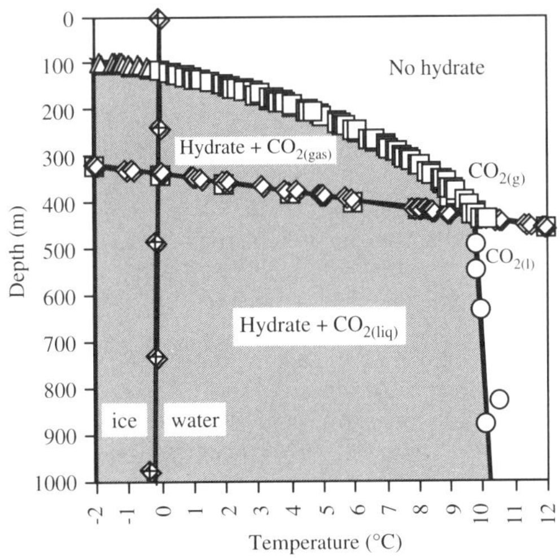

Figure 1. Stability range of CO_2 hydrate over a range of temperatures and water depths.

these problems of resolution, this study utilises cryogenic scanning electron microscopy (cryo-SEM) techniques to make high-resolution observations of deeply frozen samples of CO_2 hydrate.

Although gas hydrate stability conditions seem at odds with those normally required for SEM observations (usually warm and at high vacuum), samples can be stored at relatively low pressures for prolonged time periods at low temperatures (Stern et al. 2000). As a consequence of this, instruments fitted with cryogenic sample chambers can be used to undertake high resolution imaging (e.g. Kuhs et al. 2000).

2 METHODS

2.1 *Experimental procedures*

CO_2 hydrate was made in a stainless steel pressure vessel by adding approximately 20 mL of distilled deionized water, and applying approximately 130 mL of CO_2 at a constant 250 bar using an ISCO 260D syringe pump. The vessel was then set to cool slowly to about 1°C. However, after approximately 12–16 hours, one of the high-pressure gas pipes failed, resulting in the rapid and acute loss of CO_2 from the system. It is estimated (based upon previous "failures") that the temperature in the vessel fell to at least c. −70°C during depressurization. This froze all the water in the experiment, and a solid white "puck"-shaped mass was recovered. This was rapidly cooled to liquid nitrogen temperature, broken into smaller pieces, and stored in aluminium capsules over liquid nitrogen.

2.2 *Analytical procedures*

SEM samples were prepared by cleaving the hydrate under liquid nitrogen using a chisel to produce flakes that were typically 5–10 mm across and up to 5 mm thick. These were placed in a pre-cooled sample holder under a stream of helium gas, and transferred into the cryo-transfer chamber of the SEM, thus preventing atmospheric moisture from depositing frost on the sample.

Observations were made with a Leo 435VP variable pressure SEM with an Oxford CT1500 cryogenic sample handling and cold stage facility. Backscattered electron (BSEM) imaging was used to enhance hydrate/ice phase contrast. The cold stage could maintain a minimum temperature of around −170°C, although this was raised to approximately −60°C to "develop" or "etch" the sample by initiating hydrate breakdown. Typical N_2 pressures within the chamber were ~0.45 Torr (~6×10^{-4} bar). Samples were not coated, and the instrument operated in a low vacuum mode to prevent sample charging.

3 RESULTS

3.1 *Initial observations of CO_2 hydrate and ice*

Visual inspection of the sample revealed a homogeneous white solid. The sample degassed slowly when warmed, suggesting that some, although probably not all, of it was CO_2 hydrate.

Initial backscattered SEM imaging was conducted at approximately −160°C, and fine-scale features did not change over this time. The surface of the sample was found to be covered (to a variable degree) with small ice crystals in the form of "hoar frost" (Fig. 2). These crystals resulted from contact of the very cold sample with atmospheric moisture during the sample cleaving and handling procedure. Although they partially obstructed the sample surface, they did not totally preclude detailed observations.

For the contrast settings used on the SEM, there was relatively little difference in appearance between CO_2 hydrate and water (Fig. 2). CO_2 hydrate was identified as a layer some 10–50 μm wide surrounding voids in the sample. It is believed that during the sudden depressurization and cooling event, degassing of the water created many CO_2-filled bubbles. These bubbles then acted as a source of CO_2 for very rapid hydrate formation, prior to the (slightly less rapid) freezing of the remaining water. Hydrate formation could have only lasted a few 10s of seconds at most, and this concurs with observations of almost instant hydrate formation in in-situ experiments in the deep ocean (Brewer et al. 1999).

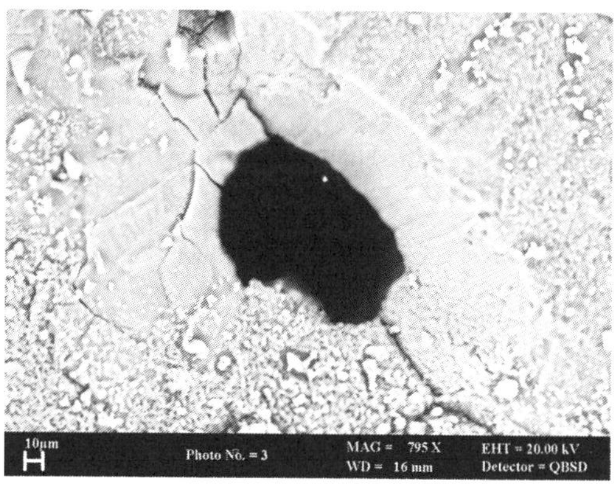

Figure 2. BSEM image showing a "halo" of hydrate c. 20–50 μm thick surrounding a former CO_2 bubble. Note the "powdery-looking" hoar frost on the surface of the sample. 10 μm scale bar.

3.2 *Sample "development" through hydrate destabilization*

To improve imaging of the sample, its temperature was raised to approximately –60°C to cause the hoar frost to ablate. This was achieved successfully. However, this procedure also destabilized the CO_2 hydrate within the sample, resulting in loss of CO_2 and conversion of CO_2 hydrate to ice (see also Milodowski & Rochelle this volume).

When viewed for a second time, the sample appeared very much different to what it had before (Fig. 3). Many features were now visible that before had been hidden, or at least were difficult to observe. In part this was due to the removal of the hoar frost. However, the main factor aiding observations was the conversion of CO_2 hydrate to ice. Ice has a density of approximately 0.9, and CO_2 hydrate has a density of approximately 1.1 (Sloan 1998). Assuming loss of CO_2 only, this would lead to a volume reduction (or porosity generation) on the order of 18%. That this secondary ice pseudomorphing initial CO_2 hydrate has a smaller volume, leads to a good morphological difference between it and the primary ice formed directly from freezing free water. The calculated degree of volume change is consistent with the cryo-SEM observations. What is noteworthy was that all the solid phases within sample were now ice (hence the back-scattered detector would not reveal any contrast/density differences), and interpretations are thus based on morphological and textural differences alone. This has two benefits in terms of sample handling and observations:

- The sample needs only to be held at low temperatures rather than cryogenic temperatures.
- It is easier to distinguish between primary ice and secondary "porous" ice, than it is between primary ice and CO_2 hydrate.

Backscattered SEM imaging of regions free of bubbles revealed that they too were originally filled with CO_2 hydrate. This occurred as pseudo-hexagonal crystals around 100–200 μm across (Fig. 4). It is estimated that some 50% of the sample was originally CO_2 hydrate. The relatively large size and well-formed shape of these crystals suggests that they may have started to form prior to the depressurization event.

In detail, many of the secondary ice pseudomorphs showed several characteristic features. These included:

- Smaller volumes compared to the original hydrate.
- Commonly, a slightly elongated form to individual crystals of the secondary ice. This gave a collection of them a serrated or toothlike appearance.
- Small, micrometer-sized holes/channels within, and between, the secondary ice crystals.

The slightly elongated form to the secondary ice may be a function of how the hydrate breakdown reaction front moved through the individual crystals of CO_2 hydrate (see Milodowski & Rochelle this volume). The small holes and channels are thought to be gas escape structures created through the venting of CO_2 as hydrate broke down deeper within the sample.

The primary ice matrix showed a typical "scalloped" surface texture resulting from ice ablation, but is otherwise homogeneous.

Backscattered SEM imaging of regions containing bubbles (Fig. 5) revealed that the position of former hydrate was much easier to resolve than studying the

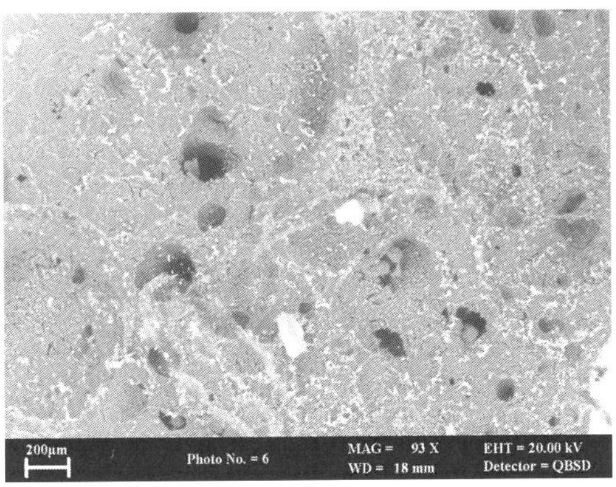

Figure 3. Low resolution BSEM image of the sample after "development". Many former gas bubbles can be seen, together with "porous ice". 200 μm scale bar.

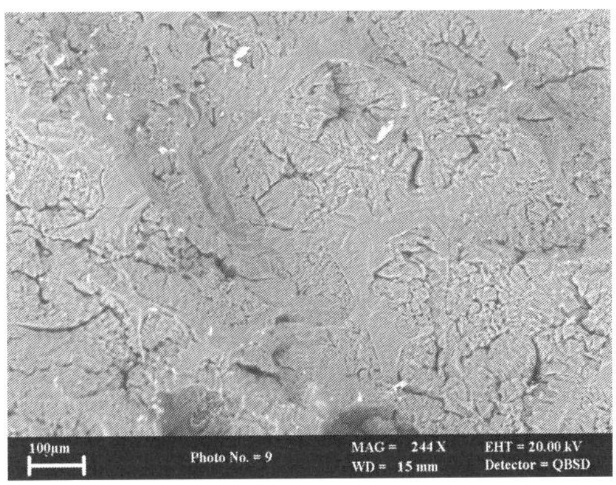

Figure 4. BSEM image showing a bubble-free region of the sample with pseudo-hexagonal pseudomorphs of secondary ice after CO_2 hydrate within a matrix of primary ice. 100 μm scale bar.

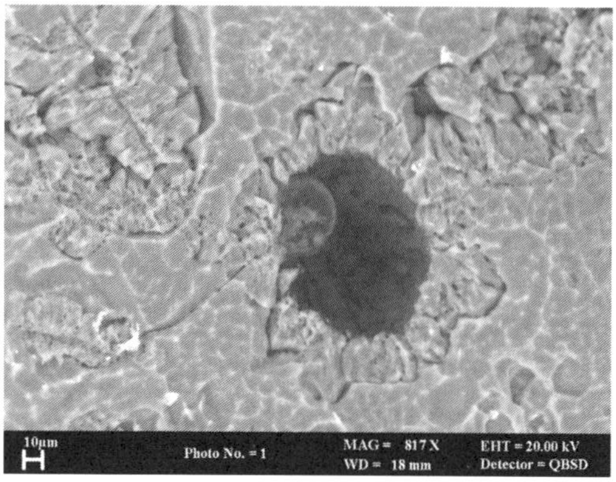

Figure 5. Close up BSEM image of a former bubble surrounded by ice pseudomorphs after CO_2 hydrate within a matrix of primary ice. Note gas escape structures (holes) within the secondary ice, together with a former droplet of liquid CO_2 within the bubble itself. 10 μm scale bar.

hydrate directly (cf Fig. 2). The "halo" of secondary ice representing former small hydrate crystals around the bubbles showed morphologies (teeth-like features and small holes) identical to those of the larger pseudo-hexagonal crystals. It is estimated that approximately 5% of the sample may have been made up of bubbles.

However, about 10% of the gas bubbles contained other features, and in particular, small, spherical, droplet-like objects (Fig. 5) some 20 50 μm in diameter. These structures appear to be made up of a very thin (~1–2 μm) crust of ice pseudomorphs, and are interpreted originally to have been a CO_2 hydrate veneer around a droplet of liquid CO_2. The bubbles therefore, record a very transitory moment during this somewhat serendipitous "failed" experiment, and they have close analogies with gas-rich fluid inclusions in minerals. They represent a very dynamic period when water was rapidly bubbling, CO_2 degassing was occurring, and when three phases of CO_2 (gas, liquid and hydrate) were in coexistence.

4 SUMMARY

The results from this on-going study show that cryo-SEM can be a useful technique for studying gas hydrates. The sample described here was very rapidly frozen during a very dynamic degassing event, and this has preserved very short-lived features of the rapidly-changing system. The use of cryo-SEM imaging allows us to construct a reaction history for this particular experiment:

1. As the experiment cooled, small crystals of CO_2 hydrate *may* have started to nucleate.
2. On rupture of the CO_2 inlet pipe, there was rapid and acute loss of CO_2 from the system. Rapidly decreasing pressure caused degassing of the water and bubble formation. At the same time, vaporization of liquid CO_2 caused intense local cooling.
3. Further nucleation and rapid (less than a few 10s of seconds) growth of CO_2 hydrate crystals throughout the water. In the bulk solution these had hexagonal crystal shape and reached c. 100–200 μm across. A "halo" of smaller CO_2 hydrate crystals formed around gas bubbles. Some bubbles also contained a droplet of liquid CO_2. The water might have had the appearance of a fizzing crystal "mush".
4. Rapid cooling continued past the point of ice formation, trapping bubbles as they rose through the water, and sealing the entire experimental charge in a block of ice. A block of solid CO_2 probably formed above the frozen experimental charge.

Although textural details of the CO_2 hydrate were lost when it was converted to ice, accurate crystal shapes and thin veneers were preserved. Consequently, "developing" or "etching" the sample through hydrate destabilization was a useful method in aiding identification of regions of former CO_2 hydrate. Once this had been done, the entire sample was composed of ice, but there were readily identifiable morphological difference between primary ice and secondary ice pseudomorphing CO_2 hydrate.

The use of electron-beam imaging to investigate gas hydrates has received relatively little attention in the past. However, given the worldwide availability of such equipment, there is much potential for their greatly expanded use in understanding gas hydrate systems. The ability to easily discriminate between ice and gas hydrate could also have important implications in terms of identifying natural samples of CH_4 hydrate. Core samples recovered from deep drilling activities are often deep frozen at liquid nitrogen temperatures to preserve any CH_4 hydrate for later study. Controlled etching and ablation of deeply frozen samples using cryo-SEM could be a valuable tool for studying CH_4 hydrate (especially disseminated hydrate) and its relationship between sediment grains and pore water.

ACKNOWLEDGEMENT

This paper is published with the permission of the Director of the British Geological Survey, NERC.

REFERENCES

Austvik, T. & Løken, K.P. 1992. Deposition of CO_2 on the seabed in the form of hydrates. *Energy Conversion and Management* 33: 659–666.

Brewer, P.G., Friederich, G., Peltzer, E.T. & Orr, F.M.Jr 1999. Direct experiments on the ocean disposal of fossil fuel CO_2. *Science* 284: 943–945.
Duxbury, N.S., Nealson, K.H. & Romanovsky, V.E. 2001. On the possibility of clathrate hydrates on the moon. *J. Geophys. Res.* 106 (E11): 27811–27813.
Golomb, D.S., Zemba, S.G., Dacey, J.W.H. & Michaels, A.F. 1992. The fate of CO_2 sequestered in the deep ocean. *Energy Conversion and Management* 33: 675–683.
Holloway, S. (ed.) 1996. The underground disposal of carbon dioxide. Final report for the CEC of contract number JOU2-CT92–0031. *British Geological Survey.*
Koide, H., Takahashi, M., Shindo, Y., Tazaki, Y., Iijima, M., Ito, K., Kimura, N. & Omata, K. 1997. Hydrate formation in sediments in the sub-seabed disposal of CO_2. *Energy* 22: 279–283.
Kuhs, W.F., Klapproth, A., Gotthard, F., Techmer, K. & Heinrichs, T. 2000. The formation of meso- and macrporous gas hydrates. *Geophys. Res. Lett.* 27: 2929–2932.
Marchetti, C. 1977. On geoengineering and the CO_2 problem. *Climate Change* 1: 59–68.
Sakai, H., Gamo, T., Kim, E.-S., Tsutsumi,M., Tanaka, T., Ishibashi, J., Wakita, H., Yamano, M. & Oomori, T. 1990. Venting of carbon dioxide-rich fluid and hydrate formation in mid-Okinawa trough backarc basin. *Science* 248: 1093–1096.
Sloan, E.D. Jr 1998. *Clathrate Hydrates of Natural Gases.* New York: Marcel Dekker Inc.
Stern, L.A., Kirby, S.H., Durham, W.B., Circone, S. & Waite, W.F. 2000. Laboratory synthesis of pure methane hydrate suitable for measurement of physical properties and decomposition behaviour. In M.D. Max (ed.) *Natural Gas Hydrate in Oceanic and Permafrost Environments*, p. 323–348. Amsterdam: Kluwer Academic Publishers.
Wilson, T.R.S. 1992. The deep ocean disposal of carbon dioxide. *Energy Conversion and Management* 33: 627–633.

Water–Rock Interaction, Wanty & Seal II (eds)
© 2004 Taylor & Francis Group, London, ISBN 90 5809 641 6

Processed kimberlite–water interactions, Ekati diamond mine, NWT, Canada

H.A. Rollo[†] & H.E. Jamieson
Department of Geological Sciences and Geological Engineering, Queen's University, Kingston, Ontario, Canada
[†]*Current address: Lorax Environmental Services Ltd., Vancouver, British Columbia, Canada*

ABSTRACT: The interaction of processed kimberlite fines and associated waters from the Ekati diamond mine that have been stored in a containment facility in the Canadian Arctic for three years was investigated. The kimberlite is dominated by serpentine and olivine, with small amounts of gypsum and pyrite restricted to mud xenoclasts. All waters that have come in contact with processed kimberlite are enriched in SO_4, Mg, and Ca relative to natural watercourses in the region. As well, all waters are approximately pH neutral with total dissolved Fe concentrations that do not exceed 3 mg/L. Reaction and inverse modeling suggest that the most important processes controlling the water composition are evapoconcentration, dissolution of chrysotile, and precipitation of silica and magnesium carbonate. The dissolution-precipitation reaction, if it occurs spontaneously, would also involve dissolution of CO_2 and may represent an opportunity to sequester atmospheric CO_2.

1 INTRODUCTION

Since the 1991 discovery of the first economically viable diamondiferous kimberlites in the Canadian Arctic, diamond mining has become one of Canada's newest mining sectors. Operations began at Canada's first diamond mine, Ekati, in August 1998, whereas the second diamond mine, Diavik, opened in January 2003. By the end of 2003, the combined production at these two mines will account for approximately 10% to 15% of the world diamond production by value (Baker et al. 2003).

The Ekati diamond mine is located 300 km northeast of Yellowknife, NWT, within the Lac de Gras drainage basin, forming the headwaters of the Coppermine River that flows north to the Arctic Ocean. The mine is in the region of continuous permafrost, which experiences annual temperatures ranging from as low as −35°C up to 25°C. Waste kimberlite and process water are stored in an engineered containment facility, formerly two lake basins; Nancy Lake in the north and Long Lake in the south (Fig. 1). Natural lakes in the area are characterized by extremely low total dissolved solids and little buffering capacity. Water from the processed kimberlite containment facility is discharged to the receiving environment over the ice-core dam, provided it meets regulatory requirements stipulated in the Ekati Water License.

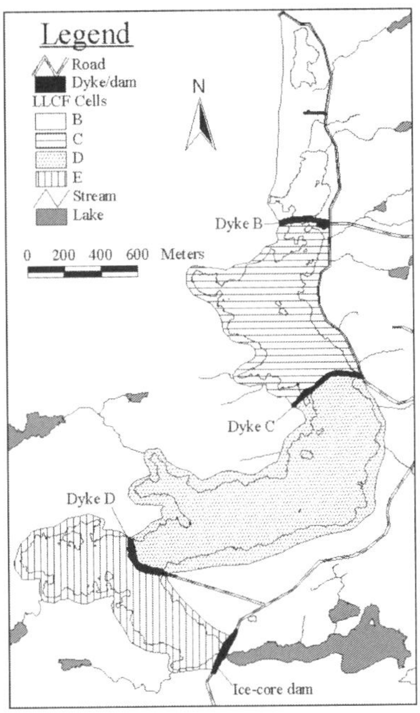

Figure 1. Long Lake Containment Facility (LLCF). Shaded regions represent the approximate maximum extent of each cell when filled with processed kimberlite fines.

Processing of the kimberlite ore involves crushing, washing, and sorting without the use of chemical reagents. All material <1 mm in diameter is hydraulically pumped to the containment facility, at a solids content of approximately 45%, with each cell being filled in order, from north to south. At the time of this investigation, only Cell B and the northern portion of Cell C contained processed kimberlite fines.

The goals of this project were to define chemically and mineralogically the components of the mineral–water system within the processed kimberlite containment facility, and to use major element water chemistry to determine the processes controlling the composition of the water in the facility.

2 WATER COMPOSITION

Water samples were collected each summer for three years (2000 to 2002). These included (1) processed kimberlite fines pore-water, (2) process plant discharge water, and (3) surface waters from the north side of Dyke B and Dyke C (Fig. 1). Pore waters were collected from the active (seasonally unfrozen) layer of the processed fines using both suction lysimeters and a core squeezer.

Major dissolved constituents in containment facility water include SO_4, Mg, Ca, HCO_3, Na, and K with the first three dominating the system. Water pH is approximately neutral and total Fe concentrations are less than 3 mg/L. Each of the three water types has its own characteristic chemical signature (Fig. 2). Pore-water samples, representing the most concentrated waters, are characterized by high SO_4 (up to 4080 mg/L), Mg (up to 870 mg/L), and Ca (up to 473 mg/L) equivalent percent. Discharge water samples, which are less concentrated than pore-water samples, are characterized by the highest Na (22 mg/L) and K (79 mg/L) equivalent percent, and the lowest Ca equivalent percent of all waters sampled. Finally, the two characteristic features of the surface waters, which are the most dilute samples, are that they contain the highest Ca and lowest Mg equivalent percent of all samples collected.

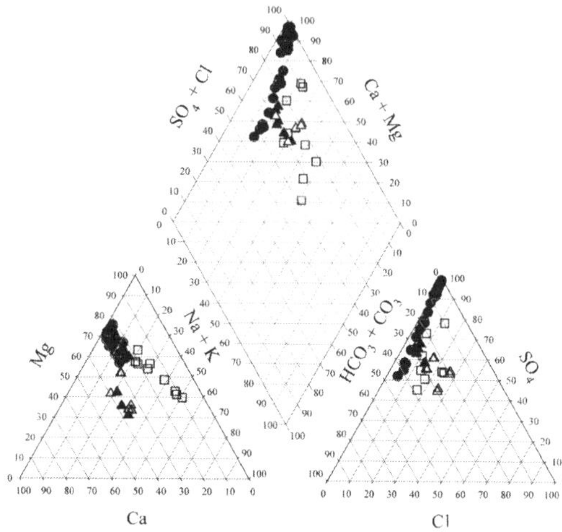

Figure 2. Piper diagram for water samples collected from 2000 to 2002. Solid black circles represent processed kimberlite fines pore-water, open squares represent process plant discharge water, and solid and open triangles represent surface water samples from Dyke B and Dyke C, respectively.

Table 1. Inorganic carbonate, sulfate, and sulfide content of solid materials studied.

	Inorganic CO_2 (wt. %)		
Sample (number)	P_5*	Mean	P_{95}*
Mud xenoclast (6)	0.40	0.50	0.68
Kimberlite ore (3)	1.10	1.40	1.91
Kimberlite fines (17)	1.38	2.38	3.04
Sulfate-S			
Mud xenoclast (6)	0.18	0.27	0.38
Kimberlite ore (3)	0.12	0.14	0.16
Kimberlite fines (17)	0.01	0.04	0.09
Sulfide-S			
Mud xenoclast (6)	1.82	2.54	3.19
Kimberlite ore (3)	0.20	0.21	0.22
Kimberlite fines(17)	0.17	0.25	0.35

* P_5 and P_{95} represent the 5th and 95th percentile values calculated from the analytical data.

3 KIMBERLITE MINERALOGY

Kimberlite ore and processed fines examined for this study originated from the upper portions of the Panda pipe. The ore is dominated by volcaniclastic kimberlite with lenses and blocks of epiclastic sediments throughout the pipe. The mineralogy is predominantly serpentine and forsteritic olivine, with lesser calcite, spinel, pyroxene, garnet, phlogopite, pyrite, and smectite. Minor sulphate and sulphide minerals appear to be restricted to the mud xenoclasts found throughout the kimberlite. The processsed kimberlite, with similar mineralogy to the kimberlite ore, is medium green-grey with no observable reaction fronts or iron oxide staining. Saponite, chrysotile, and phlogopite are present in the clay fraction. Sulfides are dominated by pyrite, which appears to be unoxidized.

As seen in Table 1, inorganic-C is higher in the fines relative to the ore, sulfate-S is concentrated in the mud, and almost absent from the fines, and

Table 2. Mean saturation indices calculated for pore, discharge, and surface water samples from the processed kimberlite containment facility. SIs reported for montmorillonite and saponite represent the range of values calculated for the Ca, Mg, Na and K end-members.

Mineral	Pore	Discharge	Surface
Chrysotile	−1.4	−0.6	−6.0
Forsterite	−7.3	−6.5	−10.1
Diopside	−3.9	−3.6	−7.4
Gypsum	−0.8	−1.9	−2.3
Montmorillonite	4.5–5.1	2.2–2.8	1.1–1.6
Saponite	3.4–4.0	3.1–3.7	2.1–2.8
Calcite	−0.2	−0.1	−1.0
Magnesite	−0.4	−0.1	−1.3
Am SiO_2	−0.8	−1.0	−1.4
$CO_{2(g)}$	−3.3	−3.6	−3.2

sulfide-S is concentrated in the mud, and does not change from ore to fines.

This is consistent with the observed fine-grained unoxidized pyrite and gypsum in the mud, and suggests that dissolved sulfate mostly originates from sulfate minerals present in the mud.

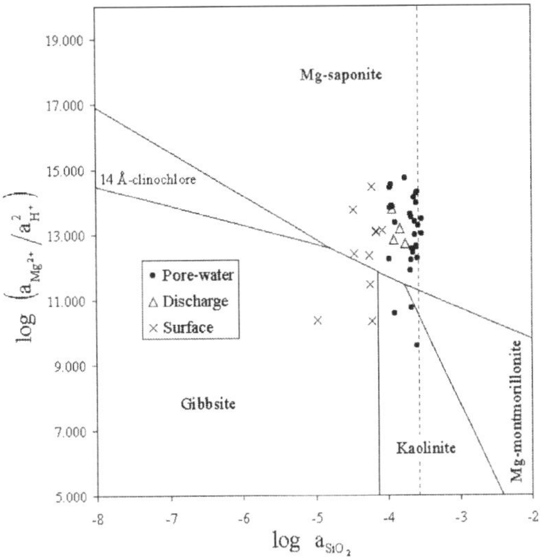

Figure 3. Mineral stability diagrams for the $MgO-SiO_2-Al_2O_3-H_2O$ system constructed at 12°C. Note that Al is assumed to be conservative in the solid phase in these diagrams. The vertical dashed line represents silica solubility.

4 CHEMICAL EVOLUTION OF DISCHARGE & PORE WATERS

Saturation indices (SIs) calculated with PHREEQC (Parkhurst & Appelo 1999) indicate that chrysotile, forsterite, diopside, and gypsum are undersaturated in all waters from the containment facility, whereas montmorillonite and saponite are consistently oversaturated (Table 2). Calcite, magnesite, and amorphous silica are close to saturation for all processed kimberlite pore- and process plant waters.

During prolonged periods without rainfall an efflorescent crust composed of hexahydrite ($MgSO_4 \cdot 6H_2O$) forms on the surface of the processed kimberlite fines, suggesting that evaporation is affecting pore-water composition. Analysis of data from the suite of pore-water samples collected for this study indicates that SO_4, Ca, and Mg, the most abundant solutes, remain concentrated in solution as evaporation progresses. This suggests that no minerals containing these elements are precipitating from the pore-waters sampled (with the exception of the Mg sulfate forming locally from extreme evaporation of near-surface solutions). Silica and HCO_3 concentrations are constant at all degrees of evaporation. Aqueous mineral stability diagrams (Fig. 3) and mineral saturation calculations (Table 2) suggest that amorphous silica and ambient CO_2 are controlling the concentration of these constituents.

Inverse modeling has been used, together with representative water analyses and kimberlite mineralogy, to calculate phase transfers that could be responsible for the water chemistry being discharged from the process plant. The majority of water that enters the containment facility has been used in the processing of kimberlite ore. Therefore, the reactions that occur during the mineral processing circuit directly affect the composition of waters in the containment facility because process plant discharge water is the starting point from which all other waters evolve.

Although numerous models of calculated mass transfers are mathematically possible, geochemically reasonable scenarios can be selected based on knowledge of the minerals involved and their stability under the conditions present in the processing plant. Table 3 shows the most reasonable calculated model. Rejected models include those that involve forsterite precipitation or very large mole transfers. The proposed reaction involves the dissolution of chrysotile and CO_2 in the process plant water and the precipitation of magnesite and amorphous silica. In addition, a small amount of gypsum and an even smaller amount of pyrite dissolve. The mole transfers involving various saponite minerals can be interpreted as cation exchange reactions.

These inverse calculations suggest that the gypsum present in the mud xenoclasts, although volumetrically minor, is one of the most significant contributors to water composition. Furthermore, pyrite dissolution, which is not apparent in thin section, is

Table 3. Mass transfer model calculated using PHREEQC. Positive mole transfers indicate dissolution and negative mole transfers indicate precipitation.

Mineral	Formula	Amount of material transferred (moles/kg solution)
Magnesite	$MgCO_3$	−17.63
Chrysotile	$Mg_3Si_2O_5(OH)_4$	5.875
CO_2		17.63
Forsterite	Mg_2SiO_4	–*
Gypsum	$CaSO_4 \cdot 2H_2O$	0.0004164
Halite	NaCl	0.0005577
Saponite-Ca	$Ca_{0.165}Mg_3Al_{0.33}Si_{3.67}O_{10}(OH)_2$	−0.01917
Saponite-K	$K_{0.33}Mg_3Al_{0.33}Si_{3.67}O_{10}(OH)_2$	0.004239
Saponite-Mg	$Mg_{3.165}Al_{0.33}Si_{3.67}O_{10}(OH)_2$	0.01493
Saponite-Na	$Na_{0.33}Mg_3Al_{0.33}Si_{3.67}O_{10}(OH)_2$	–
Silica	SiO_2	−11.75
Pyrite	FeS_2	0.000000252

* Dashes indicate that the phase was not indicated to dissolve or precipitate by the model.

relatively unimportant at this stage of the chemical evolution of these waters.

5 POTENTIAL FOR CO_2 SEQUESTRATION

Mineral carbonation has recently been proposed as a method to permanently store industrially produced CO_2 because the products of this process are geologically stable, common in nature, and easily transported and stored (Lackner 2003). Mg-silicate, particularly serpentine, is suggested as the most suitable reactant mineral for this process. The results of our inverse modeling (Table 3) suggest that the following reaction is occurring:

$$Mg_3Si_2O_5(OH)_{4(s)} + 3CO_{2(g)} \longrightarrow 3MgCO_{3(s)} + 2SiO_{2(s)} + 2H_2O_{(l)}$$

This is the same reaction proposed for most mineral carbonation schemes. This suggests that CO_2 sequestration is actively occurring in the kimberlite processing plant. As kimberlite ore is crushed and washed, CO_2 from the atmosphere appears to dissolve in the process water producing protons. These protons, together with dissolved carbonate, then react with chrysotile to form magnesite and silica. These results suggest that the potential exists to use serpentine-rich kimberlite to reduce CO_2 emissions at the diamond mine.

6 CONCLUSIONS

The most significant controls on the chemical composition of the containment facility waters during the first five years of water-rock interaction are evaporation and reactions that occur within the processing plant. In particular, evaporation during the short, dry Arctic summer increases the SO_4, Ca, and Mg composition of the pore waters. Vigorous crushing and washing of kimberlite ore within the processing plant appears to result in the dissolution of chrysotile and precipitation of amorphous silica and magnesite. This reaction also involves significant dissolution of CO_2 gas.

Sulfide oxidation and dissolution does not seem very important in controlling the major element water composition at this time. This is indicated by the fact that there is no significant decrease in the amount of sulfide-S from ore to fines as well as the observation that pyrite exhibits no petrographic evidence of oxidation. Instead, the aqueous sulfate appears to have originated from dissolution of gypsum in the mudstone, which is shown by the significant decrease in sulfate content from the ore to the fines. Inverse modeling indicates that almost all of the dissolved sulfate in the discharge water can be explained by the dissolution of sulfate rather than the oxidation of sulfide. Evaporation also increases the concentration of sulfate in the pore water. Calculated SIs show that gypsum is slightly undersaturated in all waters. In extreme cases of evaporation, hexahydrate forms on the fines as a surface crust.

Evapoconcentration calculations suggest that the concentration of dissolved silica is controlled by the precipitation of a silica phase. This is consistent with the calculated saturation indices, and the inverse modeling results, which predict the precipitation of amorphous silica.

Magnesite is also expected to precipitate according to the inverse modeling results. This is consistent with the fact that the inorganic carbon content increases from ore to fines, and the calculated saturation index is near saturation.

This study is one of the first field-based investigations of water-rock interactions in kimberlite waste in the Canadian Arctic. Further work will focus on the longer-term stability of the fines, the geochemical controls on dissolved trace elements, and the potential for kimberlite waste to sequester anthropogenic CO_2 (Rollo 2003).

REFERENCES

Baker, M.J., Blowes, D.W., Logsdon, M.J. & Jambor, J.L. 2003. Environmental Geochemistry of Kimberlite Materials: Diavik Diamonds Project, Lac de Gras, Northwest Territories, Canada. *Exploration and Mining Geology* 10(3): 155–163.

Lackner, K.S. 2002. Carbonate Chemistry for Sequestering Fossil Carbon. *Annual Reviews of Energy and the Environment* 27: 193–232.

Parkhurst, D.L. & Appelo, C.A.J. 1999. User's guide to PHREEQC (Version 2) – A Computer Program for Speciation, Batch-Reaction, One Dimensional Transport, and Inverse Geochemical Calculations. U.S. Geological Survey Water-Resources Investigations Report 99–4259.

Rollo, H.A. 2003. *Processed Kimberlite-Water Interaction in Diamond Mine Waste, Ekati Diamond Mine, NWT, Canada.* Unpub. MSc thesis. Kingston, Ontario: Queen s University.

Water–Rock Interaction, Wanty & Seal II (eds)

The kinetics of hydrogen sulfide sequestration in sandstones

M. Schoonen & Y. Xu
Department of Geosciences, Stony Brook University, Stony Brook, NY, USA

A. Carpenter
Chevron Petroleum Technology Co., La Habra, CA, USA (rtd.)
Present Address: Mineralogical Museum, Harvard University, Cambridge, MA

ABSTRACT: Oil and gas exploitation often produces a considerable amount of H_2S. The standard method of disposal is conversion to elemental sulfur. An alternative is to inject the produced H_2S into a reactive sandstone. The objective is to sequester the hydrogen sulfide in the subsurface as pyrite via a rapid reaction with iron-bearing minerals, preventing H_2S migration back to the surface. To further evaluate this alternative, it is important to understand the kinetics of pyrite formation in sandstones. Batch experiments and flow-through experiments were conducted at temperatures up to 160°C. Mössbauer spectroscopy was used to determine changes in the iron mineralogy as the reaction progressed. The activation energy of the process is 15 kJ mole^{-1}, which indicates that the process is transport limited. The rate is first order in H_2S concentration. Total iron content and presence of Fe^{III}-oxides or hydroxides are poor predictors of total sequestration capacity. Gray sandstones with Fe present in clays appear to be excellent candidates for H_2S sequestration.

1 INTRODUCTION

1.1 *Magnitude of problem*

Production of hydrogen sulfide (sour gas) is a considerable problem at many oil and gas fields (Chesnoy 1997). Most of the hydrogen sulfide produced in North America is converted to elemental sulfur and shipped elsewhere for the manufacture of sulfuric acid and other products. At some locations, there is enough hydrogen sulfide to be a problem but not enough to support the cost of conversion to elemental sulfur. In these cases, hydrogen sulfide has been injected into the subsurface as in Western Canada (Wichert & Royan 1997) and Abu Dhabi but there has been no attempt to sequester the hydrogen sulfide in a form that cannot be remobilized. Unlike radioactive isotopes, the toxicity of hydrogen sulfide does not decrease over time and it is only prudent to dispose of the hydrogen sulfide in a form that cannot migrate through fractures or be accidentally released by drilling. In relatively remote areas such as the oil fields in the vicinity of the Caspian Sea there is little demand for elemental sulfur and it is allowed to accumulate at the surface. Permanent sequestration of hydrogen sulfide in the subsurface is in principle an attractive alternative in such cases.

The research presented here addresses the characterization of sandstones in terms of hydrogen sulfide sequestration capacity and the kinetics and mechanism of sequestration. This effort is in support of two long-term goals. The primary objective is permanent disposal of hydrogen sulfide in the subsurface by selecting host rocks that will convert hydrogen sulfide to pyrite. A secondary, more ambitious and challenging goal is removal of hydrogen sulfide from produced gas in the subsurface by passing it through a reactive interval before the gas reaches the wellhead.

1.2 *Scope of study and research strategy*

Several types of experiments were conducted to determine the rate of the process, the capacity of various sandstones, and the changes in iron mineralogy. A large number of batch experiments were conducted to determine the activation energy of the process. The activation energy is determined so that the rate of reaction can be calculated for a range of operational temperatures. Batch experiments are a simple and convenient approach to provide detailed kinetic data for the sequestration reaction. The conditions in a batch experiment are, however, significantly different from those encountered in sandstone formations. For example, the water/rock ratio in the batch experiments is very high to ensure complete mixing; the sandstone formations are, on the other hand, a rock-dominated environment with a solution passing through a network of connected pores. Therefore, the sequestration reaction was further studied under conditions close to the natural environments of sandstones. Two types of flow-through systems were developed in this study: one using a chromatographic column packed with crushed sandstone material or sands, the other using sandstone cores. Both red and gray sandstones were evaluated.

Whereas this study was conducted in the framework of the possible development of subsurface sour gas sequestration methods, the results are of wider interest. Although the formation pyrite under early diagenetic and hydrothermal conditions has received considerable attention, there has been little or no experimental work on formation of pyrite in sandstones under deep burial conditions. The results of this investigation will assist in the modeling and interpretation of pyrite occurrences in this type of setting.

2 METHODS

2.1 *Batch experiments*

The rate of H_2S consumption by sandstone powders was studied in batch experiment as a function of initial H_2S concentration, solution/rock ratio, and temperature. The reactivity of different types of red and gray sandstone were determined on samples selected from non reservoir rocks in the Tengiz and Provara oil fields, Kazahkastan, reservoir rocks in the Elk Hill and South Timbalier oil fields, USA, and an outcrop in Flagstaff, USA. The experiments were conducted at temperatures between 50 and 160°C. The experiments were conducted in a 100 mL, glass-lined 316-alloy stainless steel autoclave. The autoclave was magnetically stirred. The rock specimens were crushed and sieved through an 80-mesh sieve (180 μm). Sandstone powder was weighed and loaded into the autoclave. After the autoclave was evacuated, 0.1 M boric acid solution was added to the autoclave. After the solution was heated to the desired temperature, 10 mL of a 0.1 M Na_2S solution was injected into the autoclave. All starting solutions were prepared from 18 MegaΩ, deionized water purged with N_2 for 2 hours to remove dissolved molecular oxygen. The reaction progress was determined by analyzing the concentration of H_2S, SO_4^{2-} and $S_2O_3^{2-}$ periodically withdrawn from the autoclave.

2.2 *Flow-through experiments with crushed materials*

Crushed starting materials were loaded into a chromatographic column made out of PEEK, a polymer inert to H_2S. Experiments were usually run at 90°C, because PEEK cannot be used at temperatures above 100°C. A water bath was used to heat the reaction column. As discussed later this is not a serious limitation because the reaction has a weak temperature dependence. Because the rate data obtained in a flow-through experiment is based on the difference in inflow H_2S concentration and its concentration in the effluent, it was important to determine both inflow and outflow concentrations accurately. We found that it is necessary to determine the inflow concentration after the solution had been heated and passed through a heated chromatographic column packed with glass spheres.

In experiments with this system, it is typical to see no H_2S exit the column for several hours to days, followed by a slow rise in H_2S in the effluent until it reaches a steady state. The uptake of H_2S by sandstone material was calculated from the difference in H_2S concentration in the effluent from the reaction column and that from the effluent from the control column. Hence, H_2S uptake data obtained from the system was corrected for H_2S loss unrelated to the sequestration reaction.

2.3 *Flow-through experiments with cores*

In the flow-through experiments with sandstone cores, a H_2S solution was advected through a core (1.9 cm diameter) fitted with two custom-made Teflon "column" adapters placed at the top and bottom of the core (Fig. 1). The sandstone core and

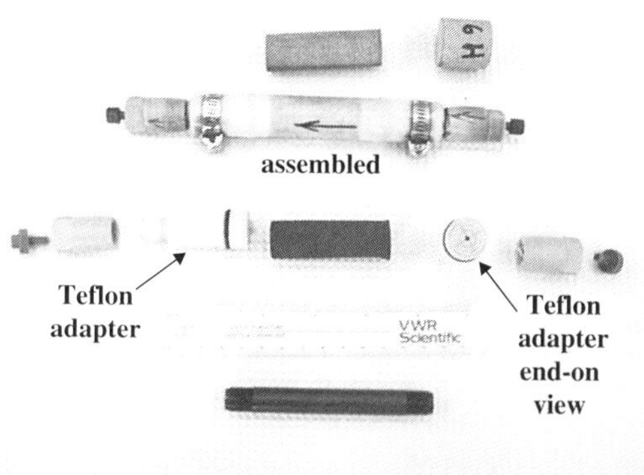

Figure 1. Sandstone core "column" setup. Black 10 cm PEEK chromatographic column shown for scale in addition to scale bar. Note grooves machined into the Teflon adapter, see end-on view.

the two TEFLON adapters were encapsulated by two layers of TEFLON shrink tube, separated by a layer of Al-foil. Metal Al is impermeable and chemically inert to H_2S under the experimental conditions. The Al-foil eliminates diffusive loss of H_2S through the permeable Teflon layers. A H_2S solution flow line was hooked up to the inlet adapter; the effluent was hooked up to an in-line analysis system (described below). The encapsulated core and adapters was placed in an autoclave filled with H_2O as pressure media under a confining pressure of 1000 to 2000 psi. The difference in H_2S concentration between inflow and effluent of the core was used to calculate the uptake of H_2S as a function of time and flow rate. This system is essentially an adaptation of a system used to measure permeability of sandstones.

2.4 *Hydrogen sulfide analyses and Mössbauer spectroscopy*

To evaluate the uptake of hydrogen sulfide it is important to be able to analyze for this compound accurately. The experiments presented a challenge because most ran for several hundred hours and the H_2S concentration varied between micromolar levels to millimolar levels. In addition, H_2S is highly volatile and reactive to air or dissolved molecular oxygen. Hence, the H_2S determination method had to meet the following requirements: (1) detection range must span several orders of magnitude combined with a low detection limit; (2) method must show little drift and be highly reproducible for hundreds of hours, and (3) contamination or exposure to molecular oxygen must be avoided.

To meet these requirements, we developed an automated in-line ion chromatography system. In this system, a H_2S-containing solution was pumped through either a rock core or a column packed with crushed sandstone. The effluent of the core or column was directed through an automated HPLC injection valve equipped with a 25-µL sampling loop. The injection valve is controlled by a two-position actuator, which in turn was connected to a PC equipped with a data acquisition and control interface. At specified time intervals, the actuator injects a 25 µL aliquot of solution to an on-line ion chromatography unit, which uses a Dionex AS4A column and a AG4A column for species separation. Concentration of S(-II) was determined with a variable-wavelength UV-detector set at 215 nm. A 25 mM $Na_2B_4O_7$ solution was used as eluant. The detection limit of this system is about 10 µmol/L. This system can collect data unattended for at least 24 hours. In practice, we recalibrated the system by analyzing H_2S standards once a day.

Mössbauer spectra were collected on samples cooled down to −197°C, using a LN_2 cryogenic sample holder/chamber (a few of the early experiments were analyzed at 25°C). Spectra collected at LN_2 are often sharper, which facilitates the interpretation of the data. An Fe foil was used to calibrate the instrument (Elscint) and sample spectra were compared to spectra collected on pure Fe phases (oxides and sulfides).

3 RESULTS AND DISCUSSION

3.1 *Batch experiments*

The reaction is first order with respect to H_2S. Because the batch experiments are conducted with an excess of sandstone, a pseudo-first-order rate constant can be extracted. The reaction has activation energy of 15 kJ $mole^{-1}$. This value suggests a transport-controlled reaction rate.

3.2 *Flow-through with crushed material*

All flow-through experiments with crushed sandstone showed an initial period of complete H_2S uptake, followed by a transitional period in which the reactivity of the material diminished and the effluent concentration reached a steady state concentration. An example of a typical experiment is shown in Figure 2. The time of the first occurrence of H_2S in the effluent (t_i), the time to reach an effluent concentration equal to 50% of the input concentration (t_{50}), and the total sequestration capacity varied considerably among the sandstone samples (Table 1). Interestingly, two of the red sandstones tested in this study have the lowest sequestration capacity along with the lowest t_i and t_{50} values. This indicates that gray sandstones can be as, or more, efficient in sequestering H_2S than red sandstones.

3.3 *Experiments with sandstone cores*

Only sandstone cores from the Provra and Elk Hills fields were studied. The other sandstones were either too impermeable or too friable for this work. The

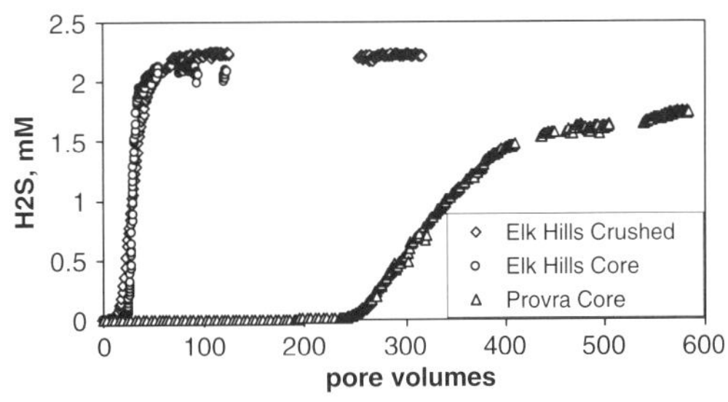

Figure 2. H_2S uptake by sandstone cores and crushed sandstone. Flow rate is 0.05 mL/min in all experiments.

Table 1. Flow-through experiments with crushed sandstones*.

Sample	Color	Fe** (wt%)	S_{tot}# (mg S/g)	t_i (hr)	t_{50} (hr)
Tengiz	Red	1.0	0.01	1.0	2.0
Elk Hills	Red	1.0	2.74	5.0	9.8
Provra	Gray	0.51	6.68	20.0	25.5
S. Timbalier-2	Gray	0.73	2.2	3.0	5.0
S. Timbalier-17	Gray	0.68	2.51	9.0	13.5

* All experiments conducted with 10 cm column filled with crushed sandstone of same size fraction.
** Total Fe in the rock.
# S_{tot} mg of S sequestered per g of material.

change in permeability as a result of the sulfidation reaction was negligible in for both cores (Elk Hills permeability did not change from 30 mDarcy; Provra changed from 0.04 mDarcy to 0.039 mDarcy). Importantly, the sequestration capacity of the Provra rock was a factor of 2 to 3 lower that the capacity determined for crushed Provra sandstone. This is expected as only material lining the fluid path is accessible for reaction. The experiments with the Elk Hills core illustrate this further. Normalized on the basis of porevolumes of fluid advected (Fig. 2.), the capacity of crushed and intact Elk Hills material is about the same. This suggests that a combination of a H_2S flow-through experiment on crushed material and a determination of pore volume can be used to estimate sequestration capacity.

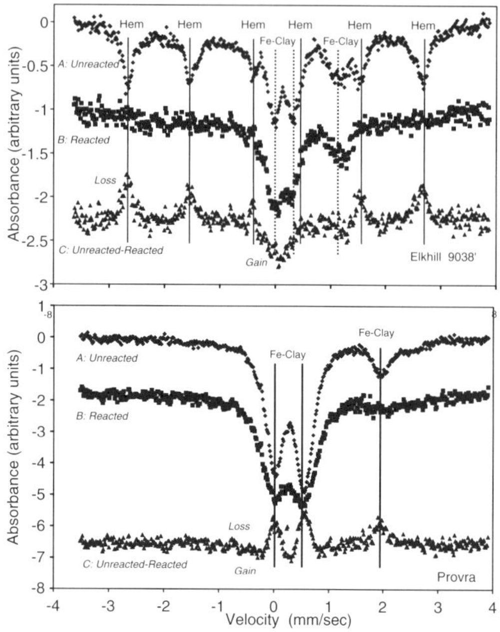

Figure 3. Mossbauer spectra for unreacted and reacted sandstones. The bottom spectrum in each panel is the difference between a spectrum of the reacted sandstone minus one of the reacted sandstone. The top panel is for an experiment with a red sandstone (Elk Hills). The bottom panel is for a gray sandstone (Provra). Note that hematite is lost in the red sandstone, while pyrite is formed. In the gray sandstone, iron associated with mixed-layer clays is converted to pyrite.

3.4 *Changes in iron mineralogy*

By comparing the Mössbauer spectra of reacted and unreacted materials it is possible to determine how the mineralogy of the sandstones was changed by exposure of the sandstone to hydrogen sulfide. Hematite is present in all red sandstones. Exposure of the red sandstone from the Elk Hills field leads to a nearly complete loss in hematite and a gain consistent with the formation of pyrite (doublet with peaks at 0 mm/sec and 0.6 mm/sec; Fig. 3, top panel). Most of the reactive iron in the gray sandstone from the Provra field is associated with the mixed layer clay fraction, Figure 3, bottom panel. Upon exposure, there is significant loss in spectral weight in the region at 2 mm/sec. A similar loss is expected for the doublet between 0 and 1 mm/sec, but this loss may be obscured by an offsetting gain in absorbance due to the formation of pyrite. Spectra for a Tengiz sample, not shown, show that the hematite present in this material is essentially unreactive.

4 CONCLUSIONS

Sulfidation of sandstone has a low activation energy. Hence, even shallow, permeable, Fe-containing rocks are candidates for the disposal of H_2S. One important result is that red sandstones are not necessarily the best hosts. Equally important is the fact that the reaction did not result in a loss of permeability. The gray sandstones studied here appear to have a reactive pool of iron associated with mixed layer clay minerals. The relative inertness of hematite in some of the red sandstones tested in this study may explain why red sandstone can retain their color even upon prolonged burial in sedimentary basins. It also suggests that it is possible to have sulfidic ore solutions migrate through a sedimentary sequence that includes red beds without significant loss of H_2S and metal-sulfide complexes. The lack of equilibration between a sulfide solution and a red bed may have been of importance in the migration of MVT ore fluids.

ACKNOWLEDGMENTS

This work was supported by the Chevron Petroleum Technology Company. Teng-fong Wong, Stony Brook University, is thanked for his help in the design of experiments with cores and for access to his rock-coring facilities. Brent Lindquist, Stony Brook University, is thanked for comments on the manuscript.

REFERENCES

Chesnoy, A.-B. 1997. S_8 threatens natural gas operations, environment: *Oil & Gas Journal*: 74–79.

Wichert, E. & Royan, T. 1997. Acid gas injection eliminates sulfur recovery expense: *Oil & Gas Journal*: 67–72.

Water–Rock Interaction, Wanty & Seal II (eds)
© 2004 Taylor & Francis Group, London, ISBN 90 5809 641 6

Preliminary experimental results of CO_2 sequestration with brine

Y. Soong, D.E. Allen, J.R. McCarthy-Jones, D.K. Harrison, S.H. Hedges & J.P. Baltrus
U.S. Department of Energy, National Energy Technology Laboratory, Pittsburgh, PA, USA

C. Zhu
Department of Geological Science, Indiana University, Bloomington, IN, USA

ABSTRACT: To evaluate mineral trapping of CO_2 above ground as one of the possible carbon sequestration options, a natural brine solution was carbonated under a controlled temperature and P_{CO_2} environment in an autoclave reactor. Oriskany Formation brine was obtained from Indiana County, Pennsylvania, and was used as the reaction medium and source of cations for carbonate precipitation. Experimental conditions varied with respect to pH (from 3.9 to 11.0), temperature (from 25°C to 170°C), and P_{CO_2} (from 0.34 to 7.55 MPa). Preliminary results show that the amount of carbonate formed was a strong function of pH while increases in temperature and P_{CO_2} had an insignificant impact on the amount of carbonate precipitated. Regulation of pH is more easily achieved above ground indicating that carbonation of oil- and gas-field wastewater brines may represent an alternative method of CO_2 sequestration.

1 INTRODUCTION

Fossil fuels, which provide more than eighty-five percent of the world's energy needs, are a major contributor of CO_2 to the atmosphere and, therefore, to the "greenhouse effect" (DOE 1999). Various studies on oceanic, terrestrial, geological, biological, and advanced chemical approaches on sequestering CO_2 are currently being conducted in order to help mitigate the impact of increasing levels of CO_2 in the atmosphere (Smith 1999). Geological sequestration is an immediately available and technologically feasible option (Smith 1999). Carbon dioxide may be sequestered in various geological formations, including depleted oil and gas reservoirs, unmineable coal seams, basalt formations, and deep saline aquifers (DOE 1999). Among these geologic CO_2 sequestration options, injection into saline aquifer formations is one of the most promising. This option is attractive because there are regionally extensive aquifers capable of accepting large volumes of CO_2 from power plants without the need for long transport pipelines.

Upon injection of CO_2 into saline aquifers, CO_2 may be stored by hydrodynamic, solubility, or mineral trapping effects. In hydrodynamic and solubility trapping, CO_2 is stored in saline aquifers as a gas ($CO_{2(g)}$) or aqueous component ($CO_{2(aq)}$, H_2CO_3, HCO_3^-, CO_3^{2-}). The most critical concern regarding hydrodynamic trapping is the potential for CO_2 leakage due to imperfect confinement. The main issue affecting solubility trapping is the limited solubility of gasses in brines due to salting out effects (Cramer 1980). In mineral trapping, CO_2 is converted into carbonate minerals by a series of reactions with aqueous ions in the saline aquifers. Ions in the aqueous fluid exist as dissolved constituents and are derived from interactions between the aquifer fluid and host rock. Carbonates such as calcite, magnesite, dolomite, and siderite can be formed in the brine aquifers and can potentially be stored for millions of years (Bachu et al. 1994, DOE 1999, Smith 1999, Bruent et al. 2002). However, conversion of CO_2 to stable carbonate minerals is expected to be slow. For example, the Alberta Research Council in Canada conducted a computer simulation of the mineral trapping process under typical reservoir conditions and concluded that the time required for precipitation of the various carbonates is on the order of hundreds of years (Hitchon et al. 1999, Gunter et al. 2000). Moreover, effective carbonate precipitation requires favorable conditions that may not be available or controllable during under-ground sequestration efforts.

Currently 20–30 billion barrels of saline wastewater are produced annually with the production of oil and gas in the USA (Kharaka et al. 1998). About 65% of this wastewater is re-injected into reservoirs for pressure maintenance, and the remaining wastewater is treated and discharged as surface water. In Pennsylvania, the typical treatment cost for the wastewater is approximately \$3.00/barrel (Stefanik, pers. comm.). Many of the oil-field brines have high concentrations of dissolved Ca, Mg, and Fe in addition to the dominant Na and Cl ions. This suggests the possibility of using Ca, Mg, and Fe ions from saline wastewater for CO_2 sequestration via the formation of carbonate minerals in above-ground processes.

The process of sequestering CO_2 in brines above ground can be summarized as follows: (1) pump the brine out of the aquifer or use the brine produced during oil and gas production; (2) add CO_2 to the brine under conditions that favor the formation of carbonate minerals: (3) separate the carbonate minerals from the reacted brine; (4) dispose of the carbonate minerals through landfill or use for other possible applications (dolomite and some carbonates can be used for agricultural purposes); (5) dispose of the reacted brine back into the aquifer for further trapping of CO_2. This concept may offer some advantages versus that of hydrodynamic/mineral trapping within the aquifer itself. The above-ground process may reduce the time required for mineral trapping from a geological time scale of hundreds of years to hours. The above-ground process also allows for control of critical variables such as pH, P_{CO_2}, and temperature. Furthermore, this process can fully utilize the saline wastewater already available from oil and gas production.

This study attempts to evaluate the feasibility of using Oriskany gas-field brine alone to serve as a mineral-trapping medium. Specifically, the effects of pH, CO_2 pressure, and temperature on the reaction between CO_2 and brine to form carbonate minerals have been studied.

2 PROCEDURES

Brine samples were collected at a depth of 2,800 meters from the Oriskany Sandstone aquifer in Indiana County, Pennsylvania. Samples were collected after purging from the well operated by Cabot Oil & Gas Corporation. No effort was made to prevent oxygen from contacting the brine so the effectiveness of siderite precipitation cannot be evaluated due to oxidation and precipitation of Fe^{++} as iron oxides and/or hydroxides. The brine was tested without filtration. Brine carbonation experiments were carried out in a half-liter autoclave (Hastelloy C-276) manufactured by Progressive Equipment Corp. The details of experimental set-up are reported elsewhere (Soong et al. 2003).

In a representative experiment, the reactor was charged with 180 mL of brine, and then purged/evacuated with carbon dioxide three times to remove any residual oxygen and nitrogen. A predetermined amount of CO_2 was charged into the reactor to achieve the desired testing pressure. The brine-CO_2 mixture was agitated at 400 or 800 rpm to prevent settling of precipitate during the experiments. Upon completion of each experiment, the slurry was cooled to room temperature. The remaining CO_2 was vented and the slurry was removed from the reactor and filtered to separate the solids from the aqueous solution. A digital pH meter (Sentron – 1001 pH) was used to determine the pH of the brine before and after reaction.

3 EXPERMENTAL RESULTS

3.1 *Effect of pH*

The ionic composition of selected elements in the Oriskany brine is listed in Table 1 under the column labeled "Raw". The effects of pH on the reaction of CO_2 with brine were systematically evaluated as a function of increasing pH by addition of KOH at constant temperature (155°C) and CO_2 pressure (6.87 MPa) (Table 1). Due to the configuration of the current reactor, however, in-situ pH was not measured during the reaction. When the pH of the raw brine was increased from 3.6 to 11.0, the aqueous concentrations of Ca and Mg decreased by 25% and 17%, respectively, indicating increased formation of precipitates at high pH. This result is consistent with the total mass of precipitate isolated after reaction, which increased from 0.020 to 1.355 grams. The concentration of Ca in the precipitate increased by about one order of magnitude after the pH was adjusted to 9.0 and higher. XRD analyses of the precipitate identified calcite as the primary carbonate present in samples collected from the reaction at all pH values above 9.0. After reaction, however, the pH of the solution was measured and found to be lower than that of the starting pH.

Table 1. Results of varying pH at constant temperature (155°C) and P_{CO_2} (6.87 MPa).

Liquid		Raw			
pH	Initial	3.6	7.2	9.0	11.0
pH	Final		4.4	4.6	4.1
Ca	mg/L	35000	34300	30400	26450
Mg	mg/L	2250	2190	1885	1860
Fe	mg/L	230	100	40	30
Solid	grams (total)	0.020	0.068	0.222	1.355
Ca	mg/kg	34900	87900	325000	346000
Mg	mg/kg	2010	1480	2210	4250
Fe	mg/kg	311000	220000	49200	24600

1.3.2 *Effect of* P_{CO_2}

The effect of CO_2 pressure on brine carbonation was investigated with a series of experiments at constant initial pH (pH = 11.0) and temperature (155°C) (Table 2). The CO_2 pressure was varied from 0.34 to 7.55 MPa. The pH after reaction, concentration of the aqueous components and grams of Ca per total gram of precipitate did not change significantly as CO_2 pressure was increased. It is important to note, however, that measured pH values will be affected by CO_2 degassing at the conclusion of the run. These results suggest that pressure does not play a major role in the amount of carbonate minerals formed in Oriskany brines. Minor increases in P_{CO_2} did not have a large effect on the total amount of carbonate precipitated. This implies that strong controls on pressure regulation are not required especially if pH is maintained at relatively high values.

3.3 *Effect of temperature*

The effect of temperature on brine carbonation was investigated at 0.34 MPa of CO_2 and an initial pH of 11.0 (Table 3). The temperature was varied from 25°C to 170°C. The aqueous concentration of Ca decreased very slightly as the temperature increased. However, the concentration of Ca remained relatively constant in the solid phase. Overall, the effect of temperature on the amount of precipitate formed was negligible. The precipitates collected after reaction at 25 and 170°C were analyzed by XRD and the major crystalline phase was identified as calcite.

3.4 *Effect of reaction time*

The effect of reaction time on brine carbonation was also investigated. For this study, reaction time was varied from one to six hours. Changes in the brine composition before and after the reaction were observed. The most noticeable change was found in the concentration of Ca in the brine after 4 hours of reaction that did not vary appreciably thereafter from 4 to 6 hours. Results suggest that carbonation is nearly complete within 4 hours of reaction time at all temperatures considered.

3.5 *Effect of cooling time*

An experiment was conducted to test the hypothesis that the precipitates formed during reaction re-dissolve during slow quenching of the reactor. A quick quenching of the reactor to ambient temperature indicates no significant difference between quick quenching and slow quenching.

4 DISCUSSION

The dissolution of CO_2 in water results in the formation of carbonic acid (H_2CO_3) that dissociates to HCO_3^- and CO_3^{2-} ions, releasing H^+ to the fluid. Therefore, addition of CO_2 to the fluid results in a decrease in pH. A decrease in pH is further enhanced when carbonate minerals precipitate from ions in solution via a reaction such as:

$$Ca^{2+} + CO_2 + H_2O = CaCO_{3(s)} + 2H^+ \quad (1)$$

Thus, calcite precipitation in the absence of buffered pH will be limited by the generation of H^+ which acts to consume alkalinity resulting in increased solubility of calcite and other carbonate minerals. Effective sequestration of CO_2 is enhanced by fixing the pH at relatively high values in order to counteract the loss of alkalinity (production of H^+) during mineral precipitation and CO_2 dissolution. The ineffectiveness of P_{CO_2} and temperature to impact the amount of carbonate formed observed in the experiments conducted in this study is linked to the pH lowering processes described above. Above-ground carbon sequestration may be more feasible

Table 2. Results of varying P_{CO_2} at constant temperature (155°C) and pH (11.0).

Liquid		Raw			
P_{CO_2}	MPa	-	0.34	4.05	7.55
pH	Final	11.0	6.3	4.4	4.8
Ca	mg/L	34000	31800	32400	32900
Mg	mg/L	20	1740	2040	2110
Fe	mg/L	10	10	10	10
Solid	grams (total)	0.701	1.532	1.586	1.522
Ca	mg/kg	70300	343000	370000	372000
Mg	mg/kg	298000	3290	2930	2840
Fe	mg/kg	44400	27500	27200	30500

Table 3. Results of varying temperature at constant P_{CO_2}(0.34 MPa) and pH (11.0).

Liquid		Raw			
Temp.	°C	25	25	100	170
pH	Final	11.0	4.9	4.9	4.7
Ca	mg/L	37400	36000	32900	32300
Mg	mg/L	10	2380	2290	2210
Fe	mg/L	–	10	–	–
Solid	grams (total)	0.752	1.946	1.866	1.957
Ca	mg/kg	55700	373000	366000	371000
Mg	mg/kg	289000	2800	2870	3000
Fe	mg/kg	35500	23500	24100	21200

because of the necessity to buffer pH at relatively high values.

5 CONCLUSIONS

Mineral trapping experiments using Oriskany Sandstone brines were evaluated with a half-liter autoclave under variable pH, temperature, and P_{CO_2} conditions. Preliminary findings suggest that it is feasible to use brines as a medium to sequester CO_2 above ground within hours of reaction time. Results indicate that the amount of calcite precipitated is strongly dependent upon the pH of the brine and much less dependent upon reaction temperature and P_{CO_2}. The following conditions favor carbonate formation from reaction of CO_2 with brine solutions: pH above 9.0, temperatures at or above 25°C, and a constant supply of CO_2. The P_{CO_2} had little impact on the formation of carbonate in this study. With further modification, it is possible to develop a process that can sequester CO_2 above ground via brine solutions that generate valuable by-products.

ACKNOWLEDGMENTS

The authors appreciate the assistance of Mr. A. Stefanik from Cabot Oil & Gas Corp. in providing the brine samples. The authors appreciate the analyses of the samples provided by R. L. Thompson from Parson Project Services, Inc.

REFERENCES

Bachu, S., Gunter, W.D. & Perkins, E.H. 1994. Aquifer disposal of CO_2: hydrodynamics and mineral trapping. *Ener. Conv. Man.* 35: 269–279.

Bruant, R.G., Guswa, A.J., Celia, M.A. & Peters, C.A. 2002. Safe storage of CO_2 in deep saline aquifers. *Environ. Sci. Technol.* 36(11): 241–245.

Cramer, S.D. 1980. The solubility of methane, carbon dioxide, and oxygen in brines from 0° to 300°C. Report of Investigations 8706, Bureau of Mines, U.S. Department of Interior.

DOE 1999. Carbon Sequestration Research and Development. DOE/SC/FE – 1.

Gunter, W.D., Perkins, E.H. & Hutcheon, I. 2000. Aquifer disposal of acid gases: modeling of water-rock reactions for trapping of acid wastes. *Appl. Geochem.* 15: 1085–1095.

Hitchon, B., Gunter, W.D., Gentzis, T. & Bailey, R.T. 1999. Sedimentary basins and greenhouse gases: a serendipitous association. *Ener. Conv. & Man,* 40: 825–843.

Kharaka Y.K., Leong, L.Y.C., Doran, G. & Breit, G.N. 1998. Can produced water be reclaimed: experience with Placeritra Oil Field, California. Proc. 5th Int. Petrol. Environ. Conf.: 1–23. Oct 20–23, Albuquerque, NM.

Smith, I.M. 1999. CO_2 reduction – prospects for coal. IEA Coal Research, CC/26, December 1999, ISBN 92–9029–336–5.

Soong, Y., Jones, J.R., Harrison, D.K., Goodman, A.L. & Baltrus, J.P. 2003. Mineral trapping of CO_2 with Oriskany Brine. *Proceedings, 2nd Annual Conference on Carbon Sequestration, May 5–8, Alexandria, VA.*

Stefanik, A. 2002. Personal communication. Cabot Oil & Gas Corp.

Water–Rock Interaction, Wanty & Seal II (eds)
© 2004 Taylor & Francis Group, London, ISBN 90 5809 641 6

Mineral alteration due to injection of CO_2, H_2S and SO_2 in deep arkosic formations

T. Xu, J.A. Apps & K. Pruess
Earth Sciences Division, Lawrence Berkeley National Laboratory University of California, Berkeley, CA, USA

ABSTRACT: Numerical modeling of geochemical processes is necessary to investigate long-term CO_2 injection in deep saline formations. We have developed a numerical model to investigate the fate of CO_2, H_2S and SO_2 in arkosic formations using hydrogeologic properties and mineral compositions characteristic of Texas Gulf Coast sediments. The simulations were performed using the reactive fluid flow and geochemical transport code, TOUGHREACT, to analyze mineral alteration and acid-gas sequestration.

1 INTRODUCTION

Numerical modeling of geochemical processes is necessary to investigate long-term CO_2 injection in deep saline formations, because aluminosilicate mineral alteration is very slow under ambient deep-aquifer conditions. Xu et al. (2003) performed reactive transport simulations of a 1-D radial well region under CO_2 injection conditions in order to analyze CO_2 immobilization through carbonate precipitation in Gulf Coast sandstones of the Frio formation of Texas. Sequestering less-pure CO_2containing of H_2S and/or SO_2 requires less energy to separate from flue gas or a coal gasification process. Knauss et al. (2002) investigated the long-term impact of dissolved CO_2, H_2S and SO_2 using reactive transport simulations. Here we present simulations on mineral alteration, and consequent sequestration of CO_2, H_2S and SO_2 in a quartzose lithic arkose from the Gulf Coast Frio formation using a similar 1-D radial well region.

2 MODELING APPROACH

The non-isothermal reactive geochemical transport code TOUGHREACT (Xu & Pruess 2001) was developed by introducing reactive chemistry into the framework of the existing multi-phase fluid and heat flow code TOUGH2 (Pruess et al. 1999). Our modeling of flow and transport in geologic media is based on space discretization by means of integral finite differences. Full details on the numerical methods are given in Xu & Pruess (2001).

The simulator can be applied to one-, two-, or three-dimensional porous and fractured media with physical and chemical heterogeneity, and can accommodate any number of chemical species present in liquid, gas and solid phases. A wide range of subsurface thermo-physical-chemical processes can be taken into account. Transport of aqueous and gaseous species by advection and molecular diffusion can be considered in both liquid and gas phases. Aqueous chemical speciation and gas dissolution and exsolution are assumed to be locally at equilibrium. Mineral dissolution and precipitation can be modeled subject to either local equilibrium or kinetic conditions. Changes in porosity and permeability are monitored from changes in mineral volume fractions without feedback to the fluid flow.

3 PROBLEM SETUP

The response of deep formations to CO_2, H_2S and SO_2 injection will depend on many factors, including formation permeability and porosity, and mineral composition. Before considering site-specific investigations, the general features and issues relating to the injection of acid gases into saline formation should be explored. The arkosic formation is assumed to be infinite and homogeneous with a thickness of 10 m, containing a 1 M saline fluid at a constant temperature of 75°C (about 2 km depth). The well field is modeled as a circular region of 8000 m radius into which acid gases and water are injected. The injection scenarios and rates for each gas and water are given in Table 1. A total of four injection cases are considered. A 1-D radial grid was used with a spacing gradually increasing away from the well. Parameter specifications were chosen to be representative of conditions that may be encountered in saline aquifers at a depth of about 2 km with a porosity of 0.3 and a permeability of 10^{-13} m^2. The injection was assumed to continue for a period of 100 years. The geochemical transport simulation was run for a period of 10,000 years.

The initial mineral composition (Table 2) was taken from the previous modeling study by Xu et al. (2003). The composition is a proxy for a quartzose lithic arkose from the Gulf Coast Frio formation, which is similar to those commonly encountered in sedimentary basins. Calcite and anhydrite were assumed to react at equilibrium. Other minerals were set to react under kinetic constraints. The kinetic parameters used are given in Xu et al. (2004). Prior to the reactive transport simulations, batch geochemical modeling of water–rock interaction was performed to obtain a nearly equilibrated water chemistry, using a pure 1.0 M NaCl solution reacting with the primary minerals listed in Table 2 at a temperature of 75°C. The resulting water chemistry was used for the initial condition of reactive geochemical transport simulations.

Table 1. Injection rate (kg/s) of acid gases and water for each case.

Case	CO_2	Water	H_2S	SO_2
CO_2 only	1.0	0.5		
CO_2+H_2S	1.0	0.49	0.01	
CO_2+SO_2	1.0	0.49		0.01
$CO_2+H_2S+SO_2$	1.0	0.49	0.005	0.005

Table 2. List of initial mineral volume fractions and possible secondary mineral phases.

Mineral	Chemical composition	Vol.%
Primary:		
quartz	SiO_2	57.888
kaolinite	$Al_2Si_2O_5(OH)_4$	2.015
calcite	$CaCO_3$	1.929
illite	$K_{0.6}Mg_{0.25}Al_{1.8}(Al_{0.5}Si_{3.5}O_{10})(OH)_2$	0.954
kerogen-OS	$C_{64}H_{102}O_{40}S_{10}$	0.0
oligoclase	$CaNa_4Al_6Si_{14}O_{40}$	19.795
K-feldspar	$KAlSi_3O_8$	8.179
Na-smectite	$Na_{0.290}Mg_{0.26}Al_{1.77}Si_{3.97}O_{10}(OH)_2$	3.897
chlorite	$Mg_{2.5}\ Fe_{2.5}Al_2Si_3O_{10}(OH)_8$	4.556
hematite	Fe_2O_3	0.497
Secondary:		
anhydrite	$CaSO_4$	
magnesite	$MgCO_3$	
low-albite	$NaAlSi_3O_8$	
dolomite	$CaMg(CO_3)_2$	
siderite	$FeCO_3$	
Ca-smectite	$Ca_{0.145}Mg_{0.26}Al_{1.77}Si_{3.97}O_{10}(OH)$	
pyrite	FeS_2	
ankerite	$CaMg_{0.3}Fe_{0.7}(CO_3)_2$	
dawsonite	$NaAlCO_3\ (OH)_2$	
alunite	$KAl_3\ (OH)_6(SO_4)_2$	

4 RESULTS

The fluid flow pattern is very similar for the four cases. Figure 1 shows water saturations versus radial distance. The reactive chemical modeling results obtained from the CO_2 only case are similar to those from CO_2+H_2S. The results from CO_2+SO_2 case are similar to those from CO_2+SO_2 + H_2S. The results of CO_2 only and CO_2+SO_2 injection cases will be presented here. A much lower pH (Fig. 2, about 0.6 close to the well region) was obtained from CO_2+SO_2 injection. Corrosion and well abandonment is a significant issue. Anhydrite and alunite precipitation occur with SO_2 injection (Fig. 3). Minor pyrite forms with co-injection of H_2S or SO_2. Chlorite and oligoclase dissolution occurs in all four cases (Fig. 4). Significant CO_2 is trapped in ankerite (Fig. 5), dawsonite (Fig. 6), and some siderite. Total CO_2 sequestered in minerals is presented in Figure 7, and can reach 76 kg per cubic meter medium. CO_2+SO_2 injection results in significant increases in porosity close to the well region (Fig. 8). More results on mineral alteration are reported in Xu et al. (2004).

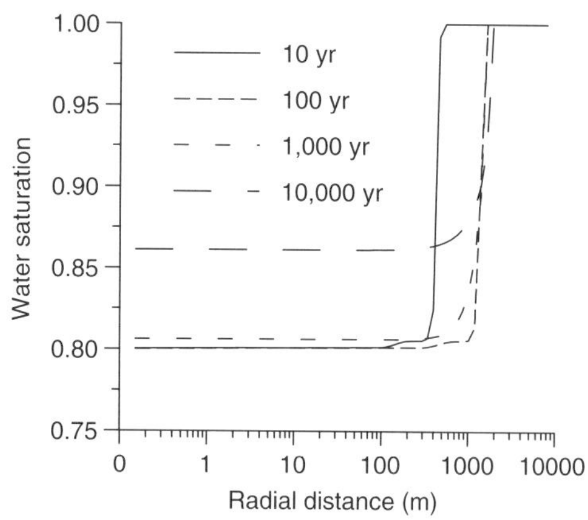

Figure 1. Water saturations at different times.

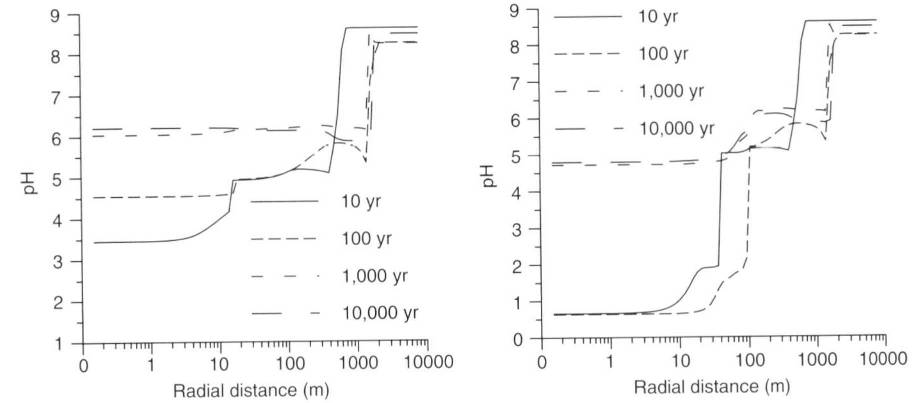

(a) CO_2 (b) $CO_2 + SO_2$

Figure 2. pH at different times for the 1-D radial flow problem.

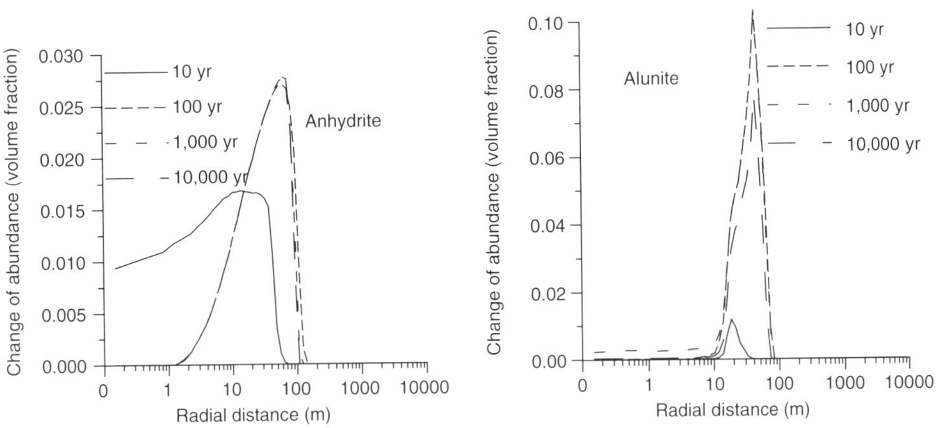

Figure 3. Change of anhydrite and pyrite abundance obtained from $CO_2 + SO_2$ case.

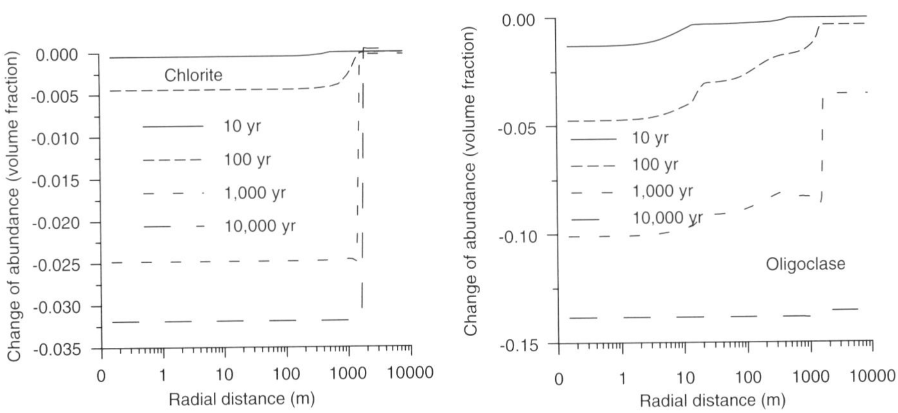

Figure 4. Change of chlorite and oligoclase abundance.

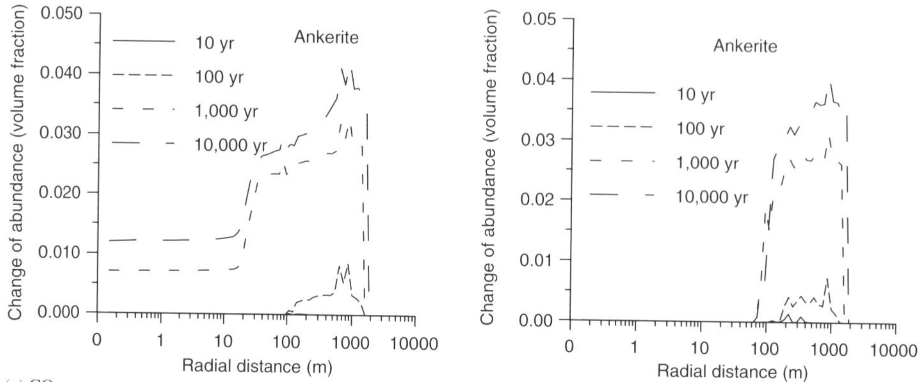

(a) CO_2 (b) $CO_2 + SO_2$

Figure 5. Change of ankerite abundance.

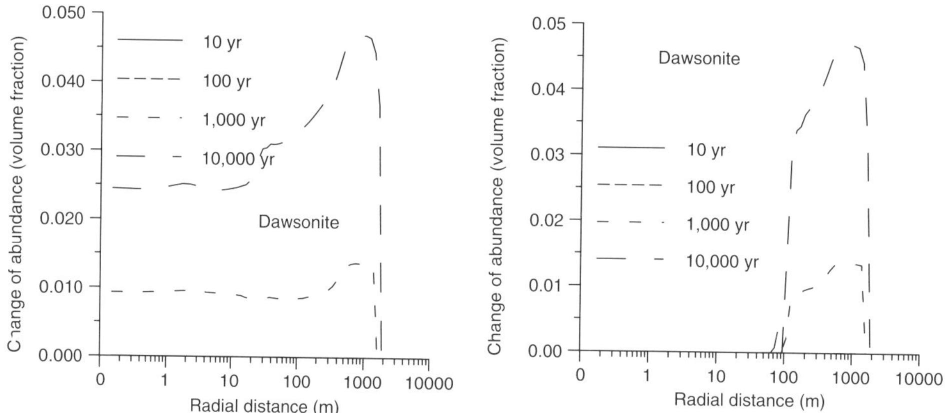

(a) CO_2 (b) $CO_2 + SO_2$

Figure 6. Change of dawsonite abundance.

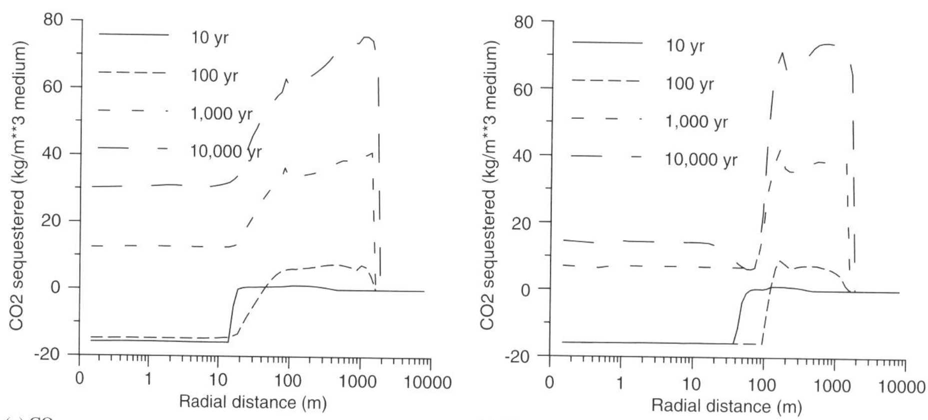

(a) CO_2 (b) $CO_2 + SO_2$

Figure 7. Total CO_2 sequestered in mineral phases.

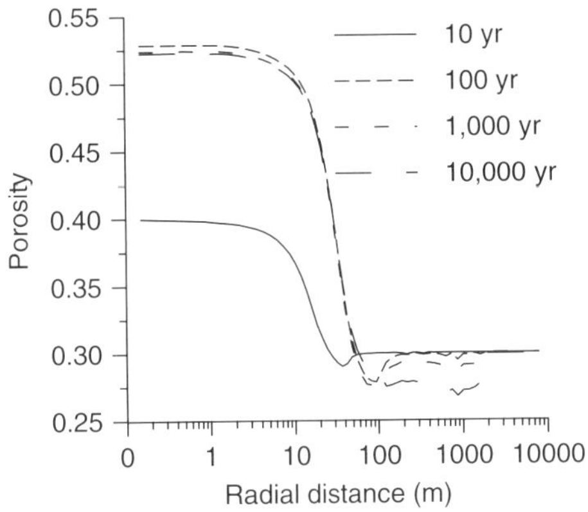

Figure 8. Porosity distribution from $CO_2 + SO_2$ case.

5 CONCLUSIONS

Co-injection of SO_2 results in a larger and stronger acidic zone close to the well. Precipitation of CO_2 trapping minerals occurs in the higher pH ranges beyond acidic zones, whereas, SO_2 trapping minerals are stable in the low pH ranges (below 5) in the front of acidic zone. Corrosion and well abandonment caused by co-injection of SO_2 is a very significant issue. Significant CO_2 is sequestered in ankerite and dawsonite, and some in siderite. CO_2 mineral trapping capability can reach 76 kg per cubic meter medium. Most of SO_2 is trapped by alunite precipitation. Some SO_2 is trapped by anhydrite precipitation, and minor amount is pyrite. Addition of the acid gases and induced mineral alteration result in changes in porosity.

ACKNOWLEDGEMENT

This work was supported by the Director, Office of Science, Office of Basic Energy Sciences, of the U.S. Department of Energy, under Contract No. DE-AC03-76SF00098 with Lawrence Berkeley National Laboratory.

REFERENCES

Knauss, K.G., Steefel, C.I., Johnson, J.W. & Boram, L.H., 2002. Impact of CO2, contaminant gas, aqueous fluid, and reservoir rock interactions on the geologic sequestration of CO2. Geological Society of America Annual Meeting, Session 135 (13), October 27–30, Denver, Colorado.

Pruess, K., Oldenburg, C. & Moridis, G. 1999. TOUGH2 user's guide, Version 2.0. *Lawrence Berkeley Laboratory Report* LBL-43134, Berkeley, California.

Reed, M.H. 1982. Calculation of multicomponent chemical equilibria and reaction processes in systems involving minerals, gases and aqueous phase. *Geochim. Cosmochim. Acta* 46: 513–528.

Steefel, C.I. & Lasaga, A.C. 1994. A coupled model for transport of multiple chemical species and kinetic precipitation/dissolution reactions with applications to reactive flow in single phase hydrothermal system. *Am. J. Sci.* 294, 529–592.

Wolery, T.J. 1992. EQ3/6: Software package for geochemical modeling of aqueous systems: Package overview and installation guide (version 7.0). *Lawrence Livermore National Laboratory Report* UCRL-MA-110662 PT I, Livermore, California.

Xu, T. & Pruess, K. 2001. Modeling multiphase non-isothermal fluid flow and reactive geochemical transport in variably saturated fractured rocks: 1. Methodology. *Am. J. Sci.* 301: 16–33.

Xu, T, Apps, J. A. & Pruess, K. 2003. Reactive geochemical transport simulation to study mineral trapping for CO_2 disposal in deep arenaceous formations. *J. Geophys. Res.*108 (B2): 2071,.

Xu, T, Apps, J.A. & Pruess, K. 2004. Geological sequestration of CO_2 and acid gases in a sandstone-shale system. *Lawrence Berkeley National Laboratory Report*, Berkeley, California.

Advances in spectroscopic and microscopic techniques for the study of water–rock interaction

Water–Rock Interaction, Wanty & Seal II (eds)
© 2004 Taylor & Francis Group, London, ISBN 90 5809 641 6

The role of synchrotron radiation in advancing the frontiers of water–rock interactions

D.L. Sparks
Department of Plant and Soil Sciences, University of Delaware, Newark, Delaware, USA

ABSTRACT: This overview paper discusses the important role that synchrotron-based techniques play in addressing frontiers in water-rock interactions. Over the past decade, with the advent of cutting-edge molecular scale *in-situ* analytical techniques, significant advances have occurred in determining the speciation and sorption/release mechanisms of metals, oxyanions, and other contaminants in soil and water environments. These frontiers in molecular environmental science have major impacts on remediation efforts, development of predictive models, and bioavailability assessments. This paper will emphasize the use of synchrotron-based techniques, including micro-focused X-ray absorption fine structure (XAFS) and X-ray fluorescence (XRF) spectroscopies and micro-tomography to speciate and determine associations of metal(loid)s in soils and plants, and real-time Fourier transform infrared (FTIR) spectroscopy to investigate microbially mediated transformations of contaminants.

1 THE CRITICAL ZONE

The critical zone is "the heterogeneous, near surface environment in which complex interactions involving rock, soil, water, air and living organisms regulate the natural habitat and determine the availability of life sustaining resources" (NRC 2001). The critical zone is an interfacial region of mass and energy flux comprising terrestrial, lacustrine, and marine components of the continental crust and is one of two primary loci of life on Earth and for most of human activity. The other major locus of life is the sun (NRC 2001). The critical zone is comprised of an array of spatial scales, ranging from the atomic to the global, and temporal scales, ranging from seconds to eons. The physical, chemical, and biological processes within the critical zone mediate exchange of mass and energy which is required for biomass productivity, chemical recycling, and water storage. The critical zone is the most heterogeneous and abstruse portion of the entire Earth (NRC 2001). However, we must understand the physical, chemical, and biological processes and reactions in the critical zone over a range of spatial and temporal scales if we are going to sustain the planet for human habitation (Hochella 2002). Processes within the critical zone can be divided into four categories. Each of these affects the other and includes: biologic activity, weathering, fluid transport, and near-surface tectonics.

2 MOLECULAR ENVIRONMENTAL GEOSCIENCE

The research challenges and questions we face in the geosciences occur over multiple scales (Fig. 1) and at interfaces where complex reactions occur involving inorganic, organic and microbial components. While we ultimately want to make accurate predictions at the field/landscape scale, fundamental mechanistic information at smaller scales, i.e. microscopic, molecular, and atomistic is required. Small scale

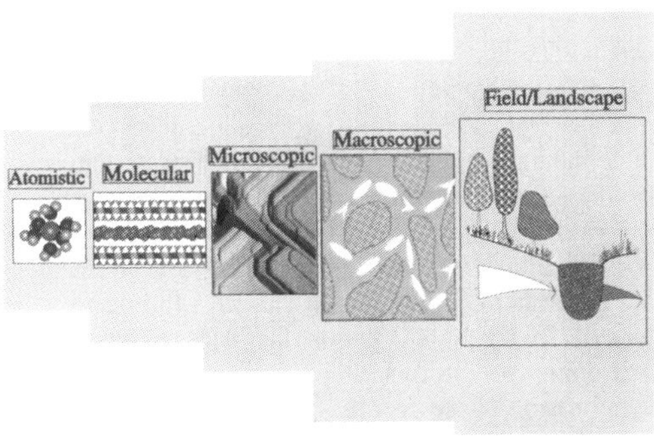

Figure 1. Illustration of the various spatial scales that geoscientists are interested in. From Bertsch & Hunter (1998).

investigations, combined with macroscopic measurements, provides the level of chemical, biological, and physical detail to comprehend contaminant reactivity and bioavailability over a range of environmental conditions. This information then needs to be integrated into field scale hydrologic and geochemical models (O'Day 1999).

With the advent of state-of-the-art analytical techniques, some of which are synchrotron-based, one can elucidate reaction mechanisms at a small scale. This has been one of the major advances in the environmental sciences over the past decade. Undoubtedly, the molecular characterization of microenvironments and interfacial reactions will become increasingly significant in understanding the interactions between chemistry, physics, and biology in the environment. The use of small-scale techniques in environmental research has resulted in a new multidisciplinary field of study in which geoscientists are actively involved – molecular environmental science. Molecular environmental science can be defined as the study of the chemical and physical forms and distribution of contaminants in soils, sediments, waste materials, natural waters, and the atmosphere at the molecular level.

There are a number of areas in the geosciences and environmental sciences where the application of molecular environmental science is resulting in major frontiers. These include: speciation of contaminants which is essential for understanding release mechanisms, spatial resolution, chemical transformations, toxicity, bioavailability, and ultimate impacts on human health; mechanisms of microbial transformations; development of predictive models; effective remediation and waste management strategies; and risk assessment. The application of molecular environmental science will be illustrated throughout the remainder of this overview paper.

3 SYNCHROTRON RADIATION

The use of intense light to understand mechanisms of geochemical reactions and processes has revolutionized the geosciences. Intense light can be produced at a synchrotron facility. Synchrotron radiation is produced over a wide range of energies from the infrared region with energies <1 eV, to the hard X-ray region with energies of 100 keV or more. There are a number of synchrotron facilities throughout the world. In the USA major facilities are found at National Laboratories. One of the most widely used synchrotron-based spectroscopic techniques is X-ray absorption spectroscopy (XAS). Its use has and will continue to revolutionize our understanding of important geochemical properties and processes. XAS can be used to study most elements in crystalline or non-crystalline solid, liquid or gaseous states over a concentration range of a few mg L^{-1} to the pure element. XAS can be used to probe structural sites ranging from those in crystals and glasses to those at interfaces such as the mineral/water interface. With second- and third-generation synchrotron light sources that provide X-ray energies ranging from the infrared to the hard X-ray regions, higher flux and higher brightness, beamline optics that produce micro-focused beams for spectromicroscopy and imaging studies, and state-of-the-art X-ray detectors, important advances in the geosciences will surely occur over the next decade (Sparks 2002).

XAS is also an *in-situ* technique, which means that one can study reactions in the presence of water. This is a major advantage over many molecular scale techniques, which are *ex-situ*, often requiring drying of the sample material, placing it in an ultra-high vacuum (UHV), heating the sample or employing particle bombardment. Such conditions can alter the sample, creating artifacts, and do not simulate most natural conditions. It is important to study geochemical reactions in water as it is the principal medium of transport of inorganic and organic species and biochemical reactions take place in aqueous media and across biological membranes that are water-based (O'Day 1999; Sparks 2002).

XAS is an element specific, bulk method that yields information about the local structural and compositional environment of an absorbing atom. It "sees" only the 2 or 3 closest shells of neighbors around an absorbing atom (0.6 nm) due to the short electron mean free path in most substances. Using XAS, one can ascertain important soil chemical information such as the oxidation state, information on next nearest neighbors, bond distances (accurate to ±0.002 nm), and coordination numbers (accurate to ±15–20%; Brown et al. 1995).

4 SORPTION PROCESSES AND IN SITU MOLECULAR SCALE TECHNIQUES

Over the past 15 years, major advances have occurred in elucidating sorption mechanisms at the water/rock interface. Based on pioneering studies by Stumm and Schindler and their coworkers, it was assumed that major adsorption complexes were outer-sphere (a water molecule is present between the surface functional group and the bound ion or molecule) and inner-sphere (no water molecule is present between the ion or molecule and the surface functional group to which it is bound). However, until molecular scale studies were performed, conclusions about sorption mechanisms were purely speculative. With the advent of *in situ* molecular scale techniques such as Fourier-transform infrared (FTIR), nuclear magnetic resonance (NMR), and XAS spectroscopies, definitive information on sorption products and mechanisms have been gleaned. Numerous studies have appeared in the soil and geochemistry literature over the past 10 years employing XAS and other *in situ* spectroscopic techniques to elucidate metal/metalloid adsorption complexation mechanisms at the mineral/water interface. Many of the XAS studies are reported in Brown & Parks (2001), Brown & Sturchio (2002), and Sparks (2002). From these studies, one can conclude that sorption of alkali and alkaline earth elements primarily occurs via outer-sphere complexes while for most other cations and anions, such as most heavy metals and oxyanions (exceptions being sulfate and selenate in some cases), inner-sphere sorption complexes result. However, as others have recently shown, the type of sorption mechanism for a particular ion is greatly affected by environmental factors such as ionic strength and pH (Strawn & Sparks 1999).

A number of soil chemists and geochemists have shown that sorption of metals such as Co, Cr, Mn, Ni, and Zn on mineral components (Scheidegger et al. 1997; Thompson et al. 1999) and soils results in the formation of metal hydroxide precipitate phases. In the case of Al-bearing soil mineral sorbents, the precipitates are metal-Al hydroxides (hydrotalcite type, layered double hydroxide [LDH] structure) while with non-Al bearing soil minerals, the precipitate phases are metal hydroxides (Scheinost et al. 1999). The surface precipitates occur at metal loadings far below a theoretical monolayer coverage, in a pH-range well below the pH where the formation of metal hydroxide precipitates would be expected according to the thermodynamic solubility product, and at time scales as fast as 15 min (Scheidegger et al. 1998). Using XAS, in combination with diffuse reflectance spectroscopy (DRS) and high resolution thermogravimetric analysis (HRTGA), Ford et al. (1999) found that mixed Ni-Al (LDH) precipitates transform over time to a precursor Ni-Al phyllosilicate phase, which greatly enhances the stability of the metal. For example, Scheckel & Sparks (2001) investigated Ni sorption on several phyllosilicates, gibbsite, amorphous silica, and a mixture of gibbsite and amorphous silica for residence times of 1 h to 2 years and subsequent Ni release using an array of dissolution agents. As residence time increased from 1 h to 2 years, the amount of Ni released from the precipitates decreased from 98% to ~0%, indicating an increase in stability with aging time regardless of sorbent and dissolution agent. This enhanced stability is due to a combination of substitution of Al for Ni in the octahedral layers of the LDH, Si incorporation into the interlayers of the precipitates due to weathering of the sorbent, and perhaps Ostwald ripening (Scheckel et al. 2000). The formation of metal hydroxide precipitates could be an important mechanism for sequestering metals in soil and water environments such that they are less mobile and bioavailable. It is also apparent that to accurately model metal retention/release in soil environments, one must employ sorption models that can describe both adsorption and precipitation phenomena since sorption is often a continuum. At low sorption densities, mononuclear, monodentate and bidentate inner-sphere complexes occur and at higher sorption densities, multinuclear clusters and surface precipitates can form.

5 NEW FRONTIERS IN WATER–ROCK INTERACTIONS

In the remaining section of this review, I want to focus on what I believe are some of the frontiers in the use of molecular scale spectroscopic techniques to investigate natural, heterogeneous materials such as soils and plants. In my view, the combination of advanced technologies and interdisciplinary research will dominate the geosciences field and lead to major discoveries. Molecular-scale investigations have and will open new frontiers concerning water–rock interactions. These include: speciation of contaminants in soils and other heterogeneous systems; cycling of trace elements and nutrients such as C, N, and S and impacts on global climate change; development of models that will accurately predict the rate, fate, and transport of contaminants in the subsurface environment; elucidation of mechanisms for microbial transformations of contaminants; unraveling the precise structure of humic substances; and understanding plant/soil interfacial reactions such as effects of plants on dissolution reactions and the chemistry of the rhizosphere. For the purposes of this review, I wish to focus on speciation of contaminated soils, mineral/bacterial and plant/soil interfacial processes.

5.1 *Speciation of metals in soils*

Standard XAS techniques probe an area of several square mm. However, the most reactive sites in soils have particle sizes in the micrometer range and metal speciation may vary over regions of a few $100\,\mu m^2$. Therefore, the derived structural information may be only an average of several contaminant species. With the advent of third generation synchrotron radiation facilities, which afford enhanced spatial resolution ($<5\,\mu m$) and sensitivity, one can speciate metals in soils at the micro-scale. Thus, one can determine all potential species within a small area in the soil matrix. An example of this capability is shown in Fig. 2 where Zn and other metals were mapped (micro-synchrotron X-ray fluorescence spectroscopy, μ-SXRF) and speciated (micro-extended X-ray absorption fine structure spectroscopy, μ-EXAFS) in a contaminated subsoil sample collected around the Palmerton, Pennsylvania, USA smelting facility near that town (Roberts et al. 2002).

Zn concentrations were several thousand ppm in the soil. From the μ-SXRF maps one sees that Zn is strongly associated with Mn in the center of the samples, with Fe in other portions of the sample, and with neither Mn or Fe in some portions of the sample (Fig. 2a). This micro-scale heterogeneity in elemental associations suggests that Zn could be present in different phases over a small sample area. Bulk XAFS analyses suggested that Zn was bound as an inner-sphere complex to Al, Fe, and Mn oxides but the data interpretation was difficult (Fig. 2b).

μ-EXAFS data were collected on the three regions of the sample (labeled 1, 2, & 3 on the Zn map, Fig. 2a). Spot 1 on the map showed that Zn was octahedrally coordinated and sorbed to an Al-oxide phase (Fig. 2c). Spots 2 & 3, with Zn in a tetrahedral coordination, revealed that Zn was primarily sorbed to Fe-oxide (spot 2) and to Mn-oxide (Spot 3; Fig. 2c). These results suggest that Zn^{2+} could have been released from Zn-bearing mineral phases in the topsoil (containing mainly sphalerite, ZnS and franklinite, $ZnFe_2O_4$) and transported to the subsoil where the Zn^{2+} was partially readsorbed to both Al, Fe, and Mn oxides.

Nachtegaal et al. (2004) investigated the speciation and its relation to the mobility and bioavailability of Zn in a non-remediated and remediated smelter contaminated soil from Belgium. State-of-the-art electron and X-ray microscopic, in combination with μ-EXAFS spectroscopic, studies were conducted on the soils. No significant differences in Zn speciation were found between samples of the remediated and non-remediated soils. Thirty percent of Zn was in smelter related minerals, mainly willemite, 10% of Zn was chemisorbed to Fe- and Mn-oxides and 60%

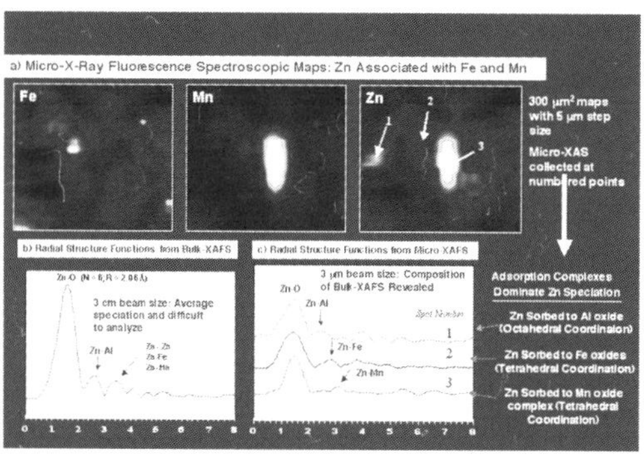

Figure 2. Synchrotron-based speciation of Zn in a smelter-contaminated subsurface soil using a) μ-Synchroton X-ray fluorescence (μ-SXRF) spectroscopy, b) bulk XAFS, and c) μ-XAFS techniques. From Roberts (2001).

of Zn was incorporated into newly formed mixed Zn-Al layered double hydroxides (LDH) in the non-remediated and a Zn-phyllosilicate in the remediated soil. Macroscopic desorption and biosensor studies were conducted on the soils to assess bioavailability of Zn. At the pH of the soils (6.4–6.7), these studies showed that the presence of the LDH phases greatly diminished the bioavailability of Zn in the soils. Desorption of Zn, and consequently bioavailability, was enhanced at lower pHs. The results of this investigation indicate that at higher pHs the formation of (mixed) metal hydroxide precipitates could be an important means of sequestering toxic metals such as Zn in the soil environment.

5.2 *Environmental molecular biology*

The field of environmental molecular biology could significantly advance efforts to enhance environmental quality. The uptake and biochemical transformations of trace elements by plants are important in the production of food and fiber and in soil remediation. One area that will continue to be of great interest to soil and environmental scientists is the study of phytoremediation processes in hyperaccumulating plants. These plants may transform a toxic species into a less toxic form or produce a volatile species of the toxic element. However, we do not understand the chemical composition of metal binding sites within the plant materials. The speciation of the trace elements within the plants and the chemical structure of metal binding proteins are needed. These are areas where soil biochemists, soil chemists, and geochemists, using synchrotron-based techniques, and plant physiologists can work closely with plant molecular biologists, who are employing state-of-the-art genomic, proteomic, and bioinformatic approaches, to produce better metal accumulating plants, enhancing remediation efforts (Tonner & Traina 1998). Such collaborations can also advance our understanding of biogeochemical interactions and cycling among organisms, ecosystems, and the environment.

A fine example of using *in-situ* synchrotron based techniques to study metal speciation on plant roots is the study by Hansel et al. (2002). They investigated spatial and temporal associations of As and Fe species on aquatic plant roots. The soil/root interface is very complex, containing both reduced and oxidized mineral species and organic-metal complexes. It is known that Fe (hydr)oxide precipitates, or plaques, appear on the surfaces of aquatic plants. These probably result from radial oxygen diffusion and subsequent oxidation of Fe^{II}. The Fe-plaques could serve as a substrate for metalloid retention and sequestration. Associated with the plaques are Fe-oxidizing bacteria. However, the role of bacteria in Fe oxidation within the rhizosphere is not well understood. The nature of the plaque mineralogy should indicate the biochemical conditions within the rhizosphere which may depend on plant species and/or site specific soil and aqueous chemistry (Hansel et al. 2002).

Hansel et al. (2002) used X-ray fluorescence microprobe and microtomography to study Fe plaque mineralogy and As speciation on the roots of two aquatic plant species, *P. arundinacea* (reed canarygrass) and *T. latifolia* (cattail). The microprobe analyses provides a vertical average of As and Fe through the plaque, and microtomography gives a cross-sectional slice through the plant root – from the interior of intact roots through the epidermis and plaque. Figures 3a and b shows microtomographic images of reed canarygrass roots.

As exists as isolated hot spots on the root surface and interior. High levels of As do not seem to correspond to higher levels of Fe (Figs 3a and b). X-ray microprobe images of the plaque show heterogeneous precipitation of As and Fe on the surface of the plant root and elevated levels of As correspond with higher Fe concentrations (Fig. 3b). Using linear combination analyses, the Fe plaque on reed canarygrass was 63% ferrihydrite, 32% goethite, and 5% siderite (Hansel et al. 2002). XANES analyses of As in the plaque of reed canarygrass showed that As^{V} predominated with

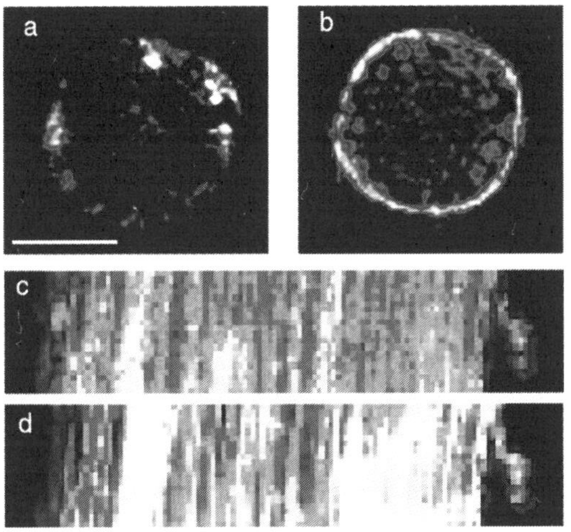

Figure 3. X-ray fluorescence microtomographic images of (a) As and (b) Fe distributions within *P. arundinacea* root cross-sections (slices). X-ray microtomography images were obtained by translating an unaltered (intact) root through an X-ray beam and collecting the fluorescence in order to reconstruct the root cross-section. X-ray microprobe image of (c) As and (d) Fe distributions on the surface of *P. arundinacea* roots. (X-ray microprobe analysis was conducted by physically slicing off a thin film of the Fe plaque from the surface of the root and obtaining a vertical average of the As and Fe concentrations within the plaque.) Scale bar represents 300 μm. From Hansel et al. (2002).

some As^{III}. The As appeared to be a sorbed species on Fe^{III} hydroxides.

Other soil chemical-biological research areas that will be prominent in the future include: understanding the transport and sorption kinetics and mechanisms of contaminants on nanoparticles; mechanisms of microbial-mineral interactions using *in-situ* nanoscale techniques such as biologic force microscopy (BFM; Lower et al. 2002); elucidating microbially mediated redox reactions; effect of biofilms on soil minerals, and metal and oxyanion speciation and retention; the role of biogenically derived minerals on contaminant transport and solubility; redox transformations of C, N, P, and S over different redox boundaries; and real-time biodegradation mechanisms of contaminants.

A recent example of the use of synchrotron radiation to follow the real-time biodegradation of organic chemicals is the research of Holman et al. (2002). It was hypothesized that humic acid accelerates the degradation of polyaromatic hydrocarbons (PAHs) by enhancing the solubility of the PAH, thus increasing PAH bioavailability to microorganisms. However, there was no direct evidence that this indeed occurred until the research of Holman et al. (2002). In this study, the researchers employed synchrotron radiation-based Fourier transform infrared (SR-FTIR) spectromicroscopy to study the effect of HA (Elliott soil HA; abbreviated ESHA) on the degradation kinetics of pyrene by *Mycobacterium* sp. JLS on a magnetite surface.

With the bacteria present there was a reduction in the amount of pyrene on the surface. For samples without ESHA, the degradation of pyrene was slow at first and about 168 hours passed before changes in the spectra were noted. Then, biodegradation increased rapidly and all the pyrene was degraded in nearly 35 hours. Biodegradation of pyrene on samples with ESHA began immediately (about 1 hour) after the bacteria were added and was complete in 4 hours. During the latter part of the pyrene degradation, an increase in biomass absorption was noted, again indicating that biomass formation occurs at the same time as pyrene consumption. At the end of the time resolved studies (about 460 hours), the spatial distribution of pyrene, bacteria, and the ESHA were determined by taking spectra every 5 μm across the center of the bacterial colony containing ESHA. The center of the maps showed a region that contained a high population density of the bacteria and a high concentration of ESHA. Also, at the initiation of the experiment there was a high concentration of pyrene in this central region (the reason this region was studied). At the end of the study, pyrene in the central region was entirely degraded. Where pyrene was present without the bacteria, degradation was insignificant.

6 NEED FOR MULTIDISIPLINARY, INTERDISCIPLINARY AND MULTIFAETE APPROACHES

The future of basic research in the geosciences is exceedingly bright. However, to successfully address and fund the major research needs that have been alluded to, and others that were not mentioned, multidisciplinary and interdisciplinary and multifaceted approaches must be carried out. Geoscientists, soil scientists, chemists, physicists, biologists, material scientists and engineers must and will increasingly collaborate. I predict that one of the major leitmotifs of the field in the 21st century will be the study of biological effects on chemical and physical reactions and processes. This provides geoscientists with a unique opportunity to combine their expertise with that of molecular biologists. We must also employ an array of multiple, molecular-scale techniques over a range of temporal scales in combination with macroscopic approaches and computational modeling to solve complex questions concerning geochemical properties and processes.

REFERENCES

Bertsch, P.M., & Hunter, D.B. 1998. Elucidating fundamental mechanisms in soil and environmental chemistry. The role of advanced analytical, spectroscopic and microscopic methods. In P.M. Huang, D.L. Sparks & S.A. Boyd (eds.), *Future prospects for soil chemistry*: 103–122. Madison, WI: Soil Sci. Soc. Amer. Spec. Publ.

Brown, G.E., Jr., & Sturchio, N.C. 2002. Application of synchrotron radiation in low-temperature geochemistry and environmental science. In P.A. Fenter, M.L. Rivers, N.C. Sturchio, & S.R. Sutton (eds.). *Applications of Synchrotron Radiation in Low-Temperatue Geochemistry and Environmental Science. Rev. in Mineralogy and Geochemistry, Vol. 49*: 1–115. Washington, D.C.: The Mineralogical Society of America.

Brown, G.E., Jr., & Parks, G.A. 2001. Sorption of trace elements from aqueous media. Modern perspectives from spectroscopic studies and comments on adsorption in the marine environment. *Int. Geol. Rev.* 43: 963–1073.

Brown, G.E., Parks, G.A. & O'Day, P.A 1995. Sorption at mineral-water interfaces: Macroscopic and microscopic perspectives. In D.J.Vaughan & R.A.D.Pattrick (eds.), *Mineral Surfaces*: 129–183. London: Chapman & Hall.

Ford, R.G., Scheinost, A.C., Scheckel, K.G. & Sparks, D.L. 1999. The link between clay mineral weathering and the stabilization of Ni surface precipitates. *Environ. Sci. Technol.* 33: 3140–3144.

Hansel, C., LaForce, M.J., Fendorf, S.E. & Sutton, S. 2002. Spatial and temporal association of As and Fe species an aquatic plant roots. *Environ. Sci. Technol.* 36: 1988–1994.

Hochella, M.F.J. 2002. There's plenty of room at the bottom. *Geochim. Cosmochim. Acta* 66: 735–743.

Holman, H.N., Nieman, K., Sorensen, D.L., Miller, C.D., Martin, M.C., Borch, T., McKinney, W.R. & Sims, R.L. 2002. Catalysis of PAH biodegradation by humic acid shown in synchrotron infrared studies. *Environ. Sci. Technol.* 36: 1276–1280.

Lower, S.K., Hochella, M.F.,Jr., Banfield, J.F. & Rosso, K.M. 2002. Nanogeoscience: From movement of electrons to lithosphere plates. *EOS Transactions. American Geophys. Union* pp. 53–56.

Nachtegaal, M.J., Sonke, E., Marcus, M.A., Vangronsveld, J., Livi, K., van der Lelie, D. & Sparks, D.L. 2004. Effects of *in-situ* remediation on the speciation and bioavailability of zinc in a smelter contaminated soil. *Geochim. Cosmochim. Acta* In review.

National Research Council (NRC) 2001. *Basic research opportunities in earth science.* Washington, DC: National Academy Press.

O'Day, P.A. 1999. Molecular environmental geochemistry. *Rev. Geophys.* 37: 249–274.

Roberts, D.R., Scheinost, A.C. & Sparks, D.L. 2002. Zinc speciation in a smelter-contaminated soil profile using bulk and microspectroscopic techniques. *Environ. Sci. Technol.* 36: 1742–1750.

Roberts, D.R. 2001. *Speciation and sorption mechanisms of metals in soils using bulk and micro-focused spectroscopic and microscopic techniques.* Ph.D. Dissertation, University of Delaware, Newark, DE.

Scheckel, K.G. & Sparks, D.L. 2001. Dissolution kinetics of nickel surface precipitates on clay mineral and oxide surfaces. *Soil Sci. Soc. Am. J.* 65: 719–728

Scheckel, K.G., Scheinost, A.C., Ford, R.G. & Sparks, D.L. 2000. Stability of layered Ni hydroxide surface precipitates – A dissolution kinetics study. *Geochim. Cosmochim. Acta* 64(16): 2727–2735.

Scheidegger, A.M., Strawn, D.G. Lamble, G.M. & Sparks, D.L. 1998. The kinetics of mixed Ni-Al hydroxide formation on clay and aluminum oxide minerals: A time-resolved XAFS study. *Geochimica Cosmochimica Acta* 62: 2233–2245.

Scheidegger, A.M., Lamble, G.M. & Sparks, D.L. 1997. Spectroscopic evidence for the formation of mixed-cation, hydroxide phases upon metal sorption on clays and aluminum oxides. *J. Colloid Interf. Sci.* 186: 118–128.

Scheinost, A.C., Ford, R.G. & Sparks, D.L. 1999. The role of Al in the formation of secondary Ni precipitates on pyrophyllite, gibbsite, talc, and amorphous silica: A DRS Study. *Geochim. Cosmochim. Acta* 63: 3193–3203.

Sparks, D.L. 2002. *Environmental Soil Chemistry 2nd edition.*San Diego: Academic Press.

Strawn, D.G. & Sparks, D.L. 1999. The use of XAFS to distinguish between inner- and outer-sphere lead adsorption complexes on montmorillonite. *J. Colloid Interf. Sci.* 216: 257–269.

Thompson, H.A., Parks, G.A. & Brown, G.E., Jr. 1999. Dynamic interactions of dissolution, surface adsorption, and precipitation in an aging cobalt(II)-clay-water system. *Geochim. Cosmochim. Acta* 63: 1767–1779.

Tonner, B. & Traina, S.J. 1998. Microorganisms, organic contaminants, and plant-metalinteractions. *Workshop on scientific directions at the Advanced Light Source.* Berkeley, CA, pp. 170–172.

Water–Rock Interaction, Wanty & Seal II (eds)
© 2004 Taylor & Francis Group, London, ISBN 90 5809 641 6

SEM-CL imaging of hydrothermal quartz – lessons and applications

G. Bignall, N. Hirano, B. Batkhishig & N. Tsuchiya
Graduate School of Environmental Studies, Tohoku University, Sendai, Japan

B. Rusk
Department of Geological Sciences, University of Oregon – Bend Program, USA

K. Sekine
Institute of Fluid Science, Tohoku University, Sendai, Japan

ABSTRACT: Scanning electron microscopy-cathodoluminescence (SEM-CL) is an effective tool for deciphering quartz deposition processes in ancient (fossil) hydrothermal systems, explored for their ore-mineral potential, and in active geothermal systems, which are potential energy resources. SEM-CL imaging may reveal chronological relationships between secondary (hydrothermal) quartz and primary rock-forming minerals, or aid the differentiation of unique hydrothermal events (e.g. at Te Kopia (New Zealand), where CL-dark quartz grew into open space in a 215 ± 15°C reservoir, was dissolved and overprinted by CL-bright quartz). SEM-CL integrated with fluid-inclusion data may enable quartz from specific alteration events to be "fingerprinted", constraining the timing of fluid-inclusion trapping and nature of fluid migration. Flow-reactor autoclave experiments, and SEM-CL examination of laboratory "grown" quartz, are used to model reservoir dynamics in enhanced geothermal systems that depend on fracture surfaces as areas for heat exchange, because quartz precipitation can "plug" these systems.

1 INTRODUCTION

Geothermal reservoirs are utilised worldwide as a renewable or "green" energy resource, with active hydrothermal systems in USA, New Zealand, Japan, and elsewhere having been well studied. Recent investigations, however, have focussed on the development of engineered reservoirs for deep-seated heat extraction, which may complement "conventional" geothermal energy (Tsuchiya et al. 2001). Fluid-rock interactions and the character of fracture systems in Deep-Seated Geothermal Reservoirs (DSGR's) remain poorly understood, yet quartz deposition is important to assess because it closes interconnected pore space and fractures, thus restricting fluid circulation.

In recent years, cathodoluminescence (CL) has been used to reveal microtextural features (cryptic alteration, recrystallisation, and fracturing) in quartz and other minerals, which cannot be seen using other means of observation. Researchers have shown that mineral nonstoichiometry, lattice defects, poor crystallographic ordering, or the incorporation of trace elements into a crystal structure can cause CL of varying wavelengths and intensities (e.g. Goetz et al. 2001).

Textures in quartz, revealed by SEM-CL, have been applied to a wide range of geologic problems, including the interpretation of sedimentary provenance (Seyedolali et al. 1997), diagenetic and pressure solution processes (Schieber et al. 2000), and to the study of sandstone fracturing (Milliken & Laubach 2000). In hydrothermal ore deposits, SEM-CL textures have been used to relate electrum precipitation to specific fluid compositions and quartz-precipitation events (Wilkinson et al. 1999), to interpret the role of temperature – pressure change in (quartz) vein formation (Rusk & Reed 2002), and to relate quartz microtextures and fluid inclusions to distinct structural and hydrothermal events in active geothermal systems (Bignall et al., accepted).

Application of SEM-CL techniques and the interpretation of CL images are not straightforward, however, and require a careful, "hands-on" approach by the SEM operator and hydrothermal specialists. Here, we will address some of our experiences, successes, and recommendations for the effective use of SEM-CL techniques in hydrothermal studies.

2 EXPERIMENTAL PROGRAMMES

Fluid-rock interactions have an important effect on the nature of quartz dissolution and precipitation in magmatic-hydrothermal systems, and are the focus of laboratory experiments undertaken by researchers at the Graduate School of Environmental Studies Tohoku University, where an AKICO-manufactured, high temperature-pressure flow autoclave system has been assembled (Fig. 1). The flow system allows fluid-rock interaction experiments to be conducted at conditions up to 600°C and 60 MPa, using 0–15 cm^3/m fluid flow, in three 1-m Hastelloy-C reactor units (A, B, C) Bignall et al. (2001).

Iidate Granite is sized to 1 cm × 1 cm × 20 cm blocks, polished, and placed into the flow-reactor. Distilled water is heated and injected into the pressurized system, with each reactor temperature controlled by an independent heating system. Experiments are undertaken to induce quartz precipitation (and/or dissolution) along an "artificial fracture" at the interface between two granite blocks. On completion of each experiment, the granite is removed from the flow reactor and examined by optical microscopy, XRD and SEM-CL.

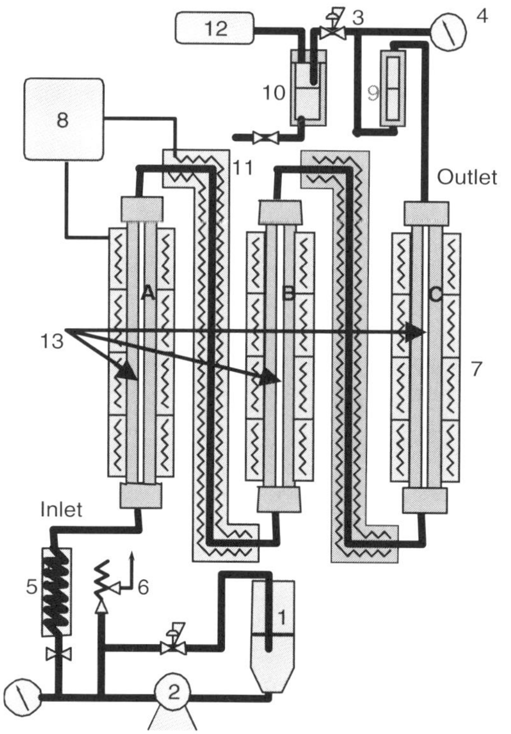

Figure 1. Schematic representation of flow autoclave system. Key: 1 water tank; 2 pump; 3 pressure control valve; 4 pressure gauge; 5 pre-heater; 6 pressure release valve; 7 heater; 8 temperature control unit; 9 separator; 10 sampling chamber; 11 gas chromatography; 12 sampling valve; 13 granite.

3 CATHODOLUMINESCENCE

Mineral CL is caused by the emission of photons in response to electron bombardment, and is observed through a scanning electron microscope equipped with a CL-detector, and photomultiplier. The SEM-CL technique offers high spatial resolution and a wide range of beam currents and acceleration voltages that makes it advantageous for observing minerals with weak luminescence, such as quartz.

Carbon coated, 100 µm-thick, double polished sections of quartz crystals grown in the flow reactor were examined by SEM-CL, using the Graduate School of Environmental Study's Hitachi-S2460N SEM-EDX instrument, which is equipped with an Oxford Mini-CL detector and photomultiplier. The SEM-CL imaging reveals quartz precipitation microtextures resulting from pressure and temperature fluctuations in the artificial, flow-reactor hydrothermal system (Fig. 2).

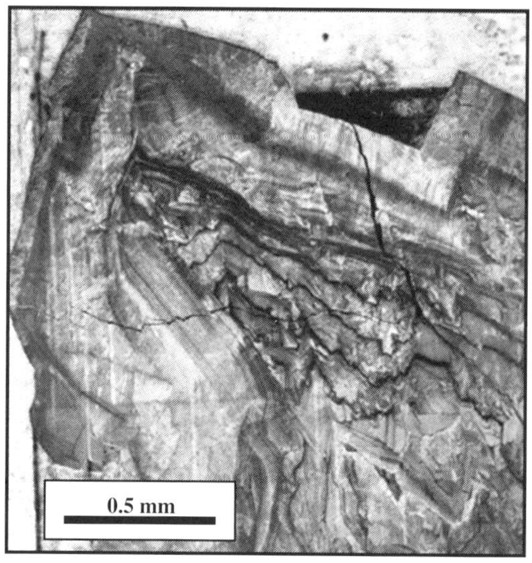

Figure 2. SEM-CL image of optically continuous quartz, from 350–450°C/30 MPa flow-reactor experiment. Dissolution of Iidate Granite occurs near the "inlet" of Reactor A (where temperatures were subcritical: i.e. 350°C). In Reactor B, the internal temperature was set at 350°C (inlet), 450°C (middle) and 425°C (outlet). A temperature gradient of 425°C to 350°C occured in Reactor C. The occurrence of zoned quartz crystals (shown), together with lesser albite and sericite in Reactor B, is attributed to retrograde quartz solubility (at about 410–390°C, as highlighted by Fournier 1985).

It is important to recognize that the apparent intensity of luminescence observed in SEM-CL imaging depends on numerous operating conditions, including subtle adjustment of beam current, accelerating voltage, contrast and brightness adjustments of the photomultiplier, carbon-coat thickness, distance of the sample from the detector, intensity of luminescence of a mineral and sample temperature, all of which require the careful attention of an experienced operator. Here, quartz thick-sections were analyzed at 15 kv with a beam current from 10–30 nA (to obtain optimal contrast in observed luminosity). As the absolute intensity cannot easily be quantified, luminescence intensity is referred to as CL-dark, CL-grey, and CL-bright.

4 FIELD STUDIES

4.1 *Te Kopia geothermal field (New Zealand)*

SEM-CL imaging of (2–5 cm long) hydrothermal quartz crystals from the Te Kopia geothermal area (North Island, New Zealand), complemented by detailed fluid-inclusion examination, reveals a complex history of crystal growth (Bignall et al., accepted). The elongate quartz crystals have a euhedral CL-dark form, consistent with growth into fluid filled space (at least 300–320 m below the present ground surface in a 215 ± 15°C, neutral pH water reservoir), and also with evidence of later dissolution (evident in SEM-CL imaging by crystal rounding) and overprinting.

Progressive movement on the Paeroa Fault, over the last 75,000–80,000 years, has provided pathways for fluids to move through the geothermal system. This has resulted in the precipitation of alternating bands of CL-bright/-dark quartz (growth zones) at the edge of the crystals (Fig. 3), and a wide range of homogenisation (T_h) temperatures for trapped fluids (210 ± 50°C) – consistent with our inference that "pulses" of hot reservoir fluids mixed with cool condensate fluids, which encroached from the margin of the field.

Continued uplift (to present day) has resulted in the quartz, and rocks containing relict alteration assemblage of adularia, illite (sericite) and trace albite (Bignall & Browne 1994) being exposed at the ground surface, and a coating of kaolinite clay and poorly luminescent Cl-dark amorphous silica, characteristic of acidic-style alteration.

4.2 *Shuteen Mineralised Complex (Mongolia)*

The Shuteen Mineralised Complex (SMC, 480 km south of Ulaanbaatar, Mongolia) consists of an extensive, 20 km^2, N-S trending silica-clay lithocap, with anomalous copper and molybdenum values. PIMA analysis, mapping and geochemical studies undertaken by Ivanhoe Mines Ltd. show that the SMC is characterised by multiphase, strong hematitic alteration in the matrix to clast-supported breccias, which form a series of knobby outcrops and may represent a cluster of relict breccia pipes. Numerous, high temperature quartz (± tourmaline) veins occur towards the periphery of the "main alteration zone", that reportedly contain anomalous gold values of up to 57 g/t (Bignall et al. 2003).

Petrographic examination of vein quartz reveals a mosaic of optically continuous, interlocking crystals, but otherwise few distinctive features. In contrast, SEM-CL imaging of the quartz veins (Fig. 4), has provided new insights into the nature of quartz precipitation and Au-bearing quartz vein generation in the SMC, intrusion-related mineral system.

The central part of the quartz veins appears to contain sub- to euhedral CL-bright quartz, which are enveloped by CL-dark bands, with interstitial space filled by CL-grey/bright quartz. The crystals are cut by CL-dark fractures, which are themselves cut by irregular, wispy CL-bright quartz-filled fractures. Outstanding questions remain, however, concerning the nature of vein generation, and relationships between mineralisation (e.g. anomalous Au) and CL luminescence.

Figure 3. SEM-CL imaging of the Te Kopia quartz reveals alternating CL-bright/-dark crystal edge (with fluid inclusion homogenization (T_h) values ranging from 168–248°C (mean 210°C), and a CL-dark core (with T_h values of 209–247°C, mean 213°C), unseen using other observational tools.

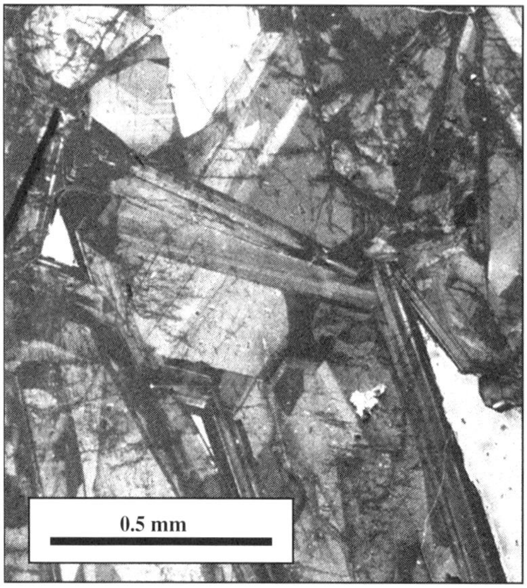

Figure 4. Intergrowth of prismatic quartz crystals, revealed only by SEM-CL, from a quartz vein at Shuteen (Mongolia). Some crystals have CL-bright/dark growth zones, containing fluid inclusions with homogenization (T_h) temperature values ~30–50°C cooler than T_h values for inclusions in crystal cores.

5 CONCLUSIONS

DSGR's have potential to provide renewable energy resources from rock masses hotter and/or deeper than geothermal systems. A key factor for utilisation of a DSGR is the creation, and maintenance of a large surface area for heat exchange, at conditions where weak water-rock interaction preserve the heat exchange surface without plugging by quartz and other secondary minerals.

SEM-CL is an important tool in mineral and geothermal exploration programmes. It may reveal chronological and physico-chemical processes in active/fossil hydrothermal systems, which are not recognised using other tools (transmitted or reflected light microscopy, backscattered electron imaging or secondary electron imaging) – and the nature of quartz vein formation.

Our CL work combines experimental studies and field surveys (in New Zealand and Mongolia) to resolve the process of quartz vein formation in high-temperature (supercritical) magmatic-hydrothermal systems, with special attention to the following:

- Experimental programmes, utilising our high temperature-pressure flow-reactor (autoclave) system, allows for the study and modelling of quartz precipitation processes in an "artificial" fracture, whilst simulating DSGR conditions.
- Quartz crystals at Te Kopia (New Zealand) record processes, which have grown in an active geothermal system. SEM-CL imaging, combined with fluid inclusion microthermometry, reveals multistage crystal growth, with CL-dark luminescence quartz enveloped by concentric CL-dark/-bright growth zones of variable intensity. SEM-CL shows the crystals are fractured and that quartz deposition occurred in open space, underlining a history of repeated microfracturing and resealing through quartz deposition during crystal growth.
- Epithermal, gold-bearing quartz veins on the periphery of the Shuteen Mineralised Complex (Mongolia), are part of a fossil hydrothermal system. Here, SEM-CL imaging, complemented by fluid inclusion data, reveals quartz vein formation involving complex crystal growth, dissolution and fracturing, which is used to resolve the physical and chemical evolution of the SMC, and support ongoing mineral exploration.

In summary, SEM-CL is effective to "fingerprint" hydrothermal quartz. It is a useful tool in relating fluid inclusions to a specific alteration event, which could not otherwise be resolved, and invaluable for deciphering the temperature-pressure and chemical evolution of a hydrothermal system, as trace element abundances (Sekine 2003) or mineral/ fluid stable isotope abundances for parts of individual crystals can be sampled and analysed.

ACKNOWLEDGEMENTS

This study was supported by Grant-in-Aid to GB by Japan Society for Promotion of Science, (Scientific Research B – 14404015). Prof. K. Nakatsuka (Tohoku Univ.) and Prof. O. Gerel (Mongolian Univ. Science and Technology) are acknowledged for their support and logistic assistance. Ivanhoe Mines Ltd is thanked for allowing us to visit the Shuteen prospect.

REFERENCES

Bignall, G. & Browne, P.R.L. 1994. Surface hydrothermal alteration and evolution of the Te Kopia thermal area, New Zealand. *Geothermics* 26: 645–658.

Bignall, G., Hirano, N., Yamasaki, N., Tsuchiya, N. & Nakatsuka, K. 2001. Experimental approach to understanding water-rock interactions in enhanced geothermal systems. Proc. 23rd NZ Geoth. Workshop, 15–20.

Bignall, G., Batkhishig, B. & Tsuchiya, N. 2003. The Shuteen Cu-Au Porphyry Deposit. In O. Gerel, D. Kirwin & R. Seltmann (editors). *Geodynamics and Metallogeny of Mongolia with a special emphasis on Cu-Au porphyry systems. Field excursion to the Oyu Tolgoi District, South Gobi: July 30–August 7, 2003, Excursion Guidebook, IAGOD Guidebook Series* 10.

Bignall, G., Sekine, K. & Tsuchiya, N. (*accepted*). Fluid-rock interaction processes in the Te Kopia Geothermal Field (New Zealand) revealed by SEM-CL Imaging. *Geothermics*

Fournier, R.O. 1985. The behaviour of silica in hydrothermal solutions. Geology and geochemistry of epithermal systems. *Rev. Econ. Geol.* 2: 45–61.

Götze, J., Plotze, M. & Habermann, D. 2001. Origin, spectral characteristics and practical applications of the cathodoluminescence of quartz- a review. *Mineral. Petrol.* 71: 225–250.

Milliken, K. & Laubach, S. 2000. Brittle deformation in sandstone diagenesis as revealed by scanned cathodoluminescence imaging with application to characterization of fractured reservoirs. In M. Pagel, V. Barbin, P. Blanc & D.Ohmenstetter, (eds.) *Cathodoluminescence in Geosciences*. Berlin: Springer.

Rusk, B.G. & Reed, M.H. 2002. Scanning electron microscope-cathodoluminescence of quartz reveals complex growth histories in veins from the Butte porphyry copper deposit, Montana. *Geol.* 30: 727–730.

Sekine, K. 2003. *Development of fracture and fluid migration in granite during uplift and emplacement.* Unpub. Ph.D. thesis. Sendai: Tohoku University. 121p.

Seyedolali, A., Krinsley, D.H., Boggs, S., O'Hara, P., Dypvik, H. & Goles, G.G. 1997. Provenance interpretation of quartz by scanning electron microscope- cathodoluminescence fabric analysis. *Geol.* 25: 787–790.

Schieber, J., Krinsley, D. & Riciputi, L. 2000. Diagenetic origin of quartz silt in mudstones and implications for silica cycling. *Nature* 406: 981–985.

Tsuchiya, N., Hirano, N., Bignall, G. & Nakatsuka, K. 2001. Supercritical Water-Rock Interaction for Development of Deep-Seated HDR Reservoirs. *Proc. 10th Intl. Symp. Water-Rock Interaction, WRI-10*: 209–212.

Wilkinson, J.J., Boyce, A.J., Earls, G. & Fallick, A.E. 1999. Gold remobilization by low-temperature brines: evidence from the Curraghinalt gold deposit, northern Ireland. *Econ. Geol.* 94: 289–296.

Raman spectroscopic method for quantitative analysis of binary aqueous organic solutions

I-Ming Chou & R.C. Burruss
National Center, U.S. Geological Survey, Reston, VA, USA

ABSTRACT: A Raman spectroscopic method was used to determine the composition of three binary aqueous organic solutions containing individually methane, carbon dioxide, and acetic acid at room temperature (22 to 23°C). The results show the linear relation between mole fraction of the dissolved species, i, and the Raman peak area ratio between i and H_2O. The Raman quantification factors (F), a combination of Raman cross-section and instrumental factor, for methane and acetic acid are 1.050 and 0.656, respectively, assuming $F = 1.000$ for H_2O. For carbon dioxide, $F = 0.322$ and 0.084 for the Raman peaks near 1276 and 1383 cm^{-1}, respectively. This *in-situ* nondestructive method is especially useful for the study of diffusion of gases in aqueous solutions. The results also show that modern Raman microprobe systems can detect low concentrations of dissolved methane and carbon dioxide, and can be applied to studies of gases in aqueous fluid inclusions.

1 INTRODUCTION

For quantitative analysis of fluid inclusions in minerals by a laser Raman microprobe (LRM) Wopenka & Pasteris (1986) suggested a practical approach based on the Placzek's ratio method. For Raman active species a and b in the same phase of a fluid inclusion, the relative concentration, C (e.g. mole or mol%), is related to their Raman peak area, A, by the formula:

$$A_a/A_b = (C_a/C_b)(\sigma_a/\sigma_b)(\eta_a/\eta_b) = (C_a/C_b)(F_a/F_b) \quad (1)$$

where σ, η, and F are the Raman scattering cross-section, instrumental efficiency, and Raman quantification factor, respectively. This approach has been successfully applied to the gas mixtures of CH_4 and N_2 (Seitz et al. 1993) and CO_2-CH_4 (Seitz et al. 1996). Sum et al. (1997) used the similar approach for analysis of clathrate hydrates. This study extends this approach to the analysis of three binary aqueous organic solutions containing individually methane, carbon dioxide, and acetic acid. Preliminary results for the CH_4-H_2O system were presented previously (Chou & Burruss 2002).

2 CH_4-H_2O SYSTEM

2.1 *Experimental method*

Distilled deionized water was loaded in a fused silica glass tube (3 mm OD, 1 mm ID, and 25 mm long), and pressurized by methane gas. The glass tube is a part of the optical cell (Chou et al. 1990), which allows direct observation of both the aqueous and vapor phases. The gas pressure was monitored by two Heise gauges; one for pressures below 3.4 MPa, and the other for higher pressures. Measurements were made at pressurization and depressurization steps (i.e. as reversals; see Fig. 1). A JY/Horiba LabRam HR Raman system was used in this study, using 532.06 nm (frequency doubled Nd:YAG) laser excitation, 10x Olympus objective with 0.25 numerical aperture, ~20 mW laser power at sample surface, 50 seconds analysis time with three accumulations per spectrum. Spectra were acquired with a 600-groove/mm grating with a spectral resolution of about 2 cm^{-1}. Each spectrum for the aqueous phase was collected between 2700 and 4000 cm^{-1}, which covers the symmetric stretching mode of aqueous CH_4 (near 2910 cm^{-1}) and the OH stretch of water between 3000 and 3700 cm^{-1}.

Table 1. Experimental results for the CH_4-H_2O system.

P (MPa)	$X(CH_4)$ ×1000	$[A(CH_4)/A(H_2O)]$ ×1000	P (MPa)	$X(CH_4)$ ×1000	$[A(CH_4)/A(H_2O)]$ ×1000
P increase			P decrease		
0.97	0.24	0.24	0.46	0.11	0.13
1.67	0.42	0.42	0.92	0.23	0.38
2.70	0.66	0.62	1.00	0.25	0.34
3.51	0.84	0.80	1.69	0.42	0.45
3.51	0.84	0.88	1.69	0.42	0.45
4.08	0.95	1.15	2.07	0.51	0.58
7.82	1.58	1.69	2.65	0.65	0.66
10.88	2.00	2.12	3.39	0.81	0.92
16.16	2.56	2.33	3.39	0.81	0.91
16.33	2.60	2.75	3.39	0.83	0.97
16.33	2.60	2.57	6.80	1.42	1.96
16.50	2.62	2.81	6.80	1.42	1.87
18.37	2.84	2.89	10.20	1.92	2.28
20.41	2.93	2.92	13.61	2.34	2.54
20.41	2.93	3.18	27.28	3.39	3.62
20.41	2.93	3.11	37.59	3.95	4.32
20.41	2.93	3.13	51.02	4.45	4.41
20.41	2.93	3.15	51.02	4.45	4.56
27.21	3.39	3.58			
27.21	3.39	3.51			
27.21	3.39	3.44			
27.21	3.39	3.37			
27.21	3.39	3.37			
27.21	3.39	3.63			
34.01	3.78	3.63			
40.82	4.09	3.88			
47.62	4.36	4.69			
47.62	4.36	4.36			
54.42	4.54	5.00			
61.22	4.85	5.33			
68.03	5.17	5.62			

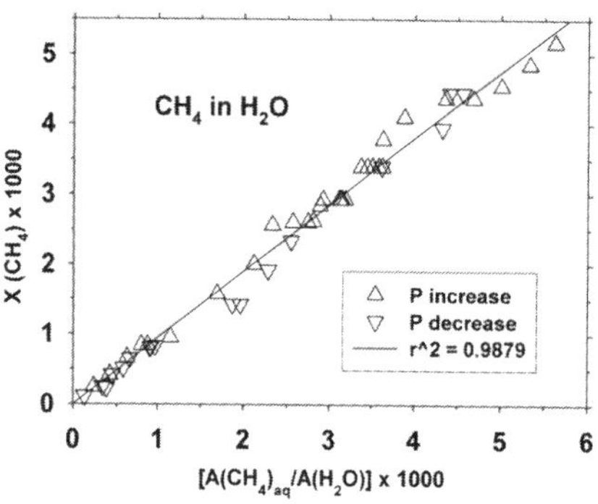

Figure 1. Experimental results showing the relation between dissolved methane concentrations in water (in mole fraction, $X(CH_4)$) and the peak area ratios for aqueous methane and water at room temperature and pressures ranging from 0.5 to 68 MPa. Plotted are the data from Table 1. The straight line represents linear regression of all data. The system is metastable with respect to methane hydrate at $X(CH_4) > 0.0037$ (i.e. $P > 30$ MPa).

2.2 *Experimental results*

The peak areas were calculated for CH_4 from 2890 to 2930 cm^{-1} and for H_2O from 2750 to 3850 cm^{-1} using GRAMS32/AI software (Galactic Industries). For each spectrum, the peak area ratio $[A(CH_4)_{aq}/A(H_2O)]$ was calculated (Table 1) and plotted against the mole fraction of CH_4 in water $[X(CH_4)]$ (Fig. 1). The data can be represented by a linear relation

$$[X(CH_4)] = 0.9521\left[A(CH_4)_{aq}/A(H_2O)\right] \quad (2)$$

Values for $[X(CH_4)]$ at experimental methane pressures (from 0.5 to 68 MPa) were calculated from the solubility data of Culberson & McKetta, Jr. (1951).

3 CO_2-H_2O SYSTEM

3.1 *Experimental method*

The method used is the same as described above for the CH_4-H_2O system, except that the system was pressurized by liquid CO_2 and that each spectrum for the aqueous phase was collected between 700 and 4000 cm^{-1}, which covers the Fermi resonance signals for the dissolved CO_2 at about 1383 cm^{-1} (P1) and 1276 cm^{-1}(P2) and the OH stretch of water between 3000 and 3700 cm^{-1}.

3.2 *Experimental results*

The peak areas were calculated for CO_2 from 1350 to 1425 cm^{-1} for P1 and from 1255 to 1315 cm^{-1} for P2, and for H_2O from 2750 to 3850 cm^{-1}. For each spectrum, the peak area ratio $[A(CO_2)_{aq}/A(H_2O)]$ was calculated (Table 2) and plotted against the mole fraction of CO_2 in water $[X(CO_2)]$ (Fig. 2). The data can be represented by linear relations

$$[X(CO_2)]_{P1} = 3.1013\left[A(CO_2)_{aq}/A(H_2O)\right] \quad (3)$$

$$[X(CO_2)]_{P2} = 11.9267\left[A(CO_2)_{aq}/A(H_2O)\right] \quad (4)$$

Values for $[X(CO_2)]$ at experimental CO_2 pressures (from 0.1 to 20 MPa) were calculated from the equation of state reported by Duan & Sun (2003).

4 ACETIC ACID SOLUTIONS

4.1 *Experimental method*

The sample solutions were prepared from the concentrate solution from Baker (99.9% CH_3COOH; Lot No. 346817). They were contained in either the silica tubes (having the same dimensions as the one

Table 2. Experimental results for the CO_2-H_2O system. $AR(1) = A(CO_2)_{P1}/A(H_2O)$, and $AR(2) = A(CO_2)_{P2}/A(H_2O)$

P (MPa)	$X(CO_2)$ ×1000	AR(1) ×1000	AR(2) ×1000	P (MPa)	$X(CO_2)$ ×1000	AR(1) ×1000	AR(2) ×1000
P increase				P decrease			
0.27	1.60	0.52	0.14	0.10	0.63	0.38	0.26
0.51	2.94	0.82	0.17	0.10	0.63	0.12	0.19
0.51	2.94	0.62	0.13	0.10	0.63	0.48	0.07
0.73	4.16	1.65	0.30	0.10	0.63	0.37	0.13
0.73	4.16	1.04	0.24	0.16	0.95	0.56	0.11
0.73	4.16	1.35	0.38	0.16	0.95	0.52	0.13
0.73	4.16	1.51	0.44	0.18	1.07	0.64	0.19
1.61	9.12	2.83	0.56	0.33	1.91	0.82	0.28
2.50	13.34	3.29	0.68	0.45	2.59	1.20	0.31
3.37	16.93	5.57	1.62	0.45	2.59	1.13	0.33
3.37	16.93	5.16	1.26	0.45	2.59	1.27	0.24
5.10	24.02	7.82	1.99	1.31	7.44	2.79	0.69
5.10	24.02	7.79	2.04	4.08	19.84	6.56	1.81
5.10	22.37	7.39	1.61	6.80	24.98	7.94	2.13
10.20	25.95	8.47	2.17	6.80	24.98	8.00	2.04
13.61	26.72	8.64	2.21	10.20	25.95	8.33	2.22
17.01	27.45	8.73	2.27	13.95	26.79	8.72	2.40
20.41	28.18	8.82	2.38	17.35	27.54	9.01	2.24

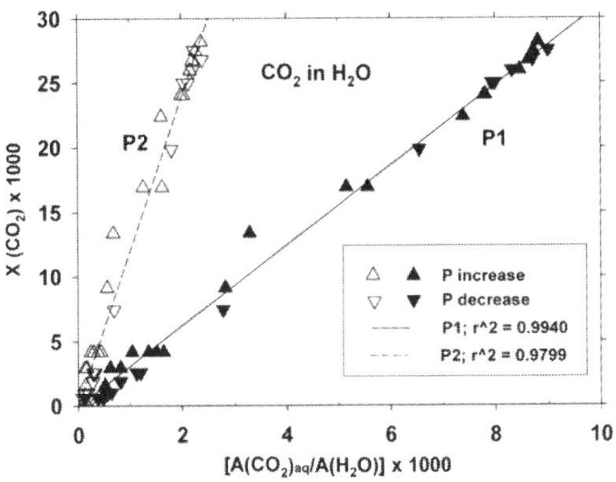

Figure 2. Experimental results showing the relation between dissolved CO_2 concentrations in water (in mole fraction, $X(CO_2)$) and the peak area ratios for aqueous carbon dioxide (P1 ≈ 1383 cm^{-1}and P2 ≈ 1276 cm^{-1}) and water at room temperature and pressures ranging from 0.1 to 20 MPa. Plotted are the data from Table 2. The straight lines represent linear regression of all data.

used in the optical cell described above) or in beakers, such that, during collection of Raman spectra, the sample solutions were exposed to the laser beams either through a wall of the silica tube or directly. Each spectrum for the aqueous phase was collected between 2700 and 4000 cm^{-1}, which covers the peak near 2930 cm^{-1} for CH_3COOH and the OH stretch of water between 3000 and 3700 cm^{-1}.

Table 3. Experimental results for acetic acid (a.a.) solutions.

m/1000 g of H_2O	X (a.a.) ×1000	[A(a.a.)/ $A(H_2O)$] × 1000	m/1000 g of H_2O	X (a.a.) ×1000	[A(a.a.)/ $A(H_2O)$] ×1000
In silica tube			Direct		
0.0145	0.26	0.56	0.0309	0.56	0.40
0.0309	0.56	0.29	0.0339	0.61	0.38
0.0748	1.35	0.54	0.0748	1.35	0.56
0.1966	3.53	2.02	0.0748	1.35	0.52
0.1966	3.53	2.24	0.1966	3.53	1.73
0.1966	3.53	2.11	0.2556	4.58	2.54
0.2556	4.58	2.79	0.2556	4.58	2.53
0.2556	4.58	2.76	0.6114	10.89	6.54
0.2556	4.58	2.69	0.9305	16.49	10.35
0.2556	4.58	2.75	1.2638	22.26	14.69
0.6114	10.89	7.15	1.9585	34.08	22.31
0.6114	10.89	7.39	1.9585	34.08	22.11
0.9305	16.49	10.73			
0.9305	16.49	10.92			
1.2638	22.27	14.65			
1.2638	22.27	14.76			
1.9585	34.08	22.94			
1.9585	34.08	22.59			
1.9585	34.08	22.35			

4.2 *Experimental results*

The peak areas were calculated for CH_3COOH from 2910 to 2980 cm^{-1} and for H_2O from 2750 to 3850 cm^{-1}. For each spectrum, the peak area ratio $[A(CH_3COOH)_{aq}/A(H_2O)]$ was calculated (Table 3) and plotted against the mole fraction of CH_3COOH in water $[X(CH_3COOH)]$ (Fig. 3). The data can be represented by a linear relation

$$[X(CH_3COOH)] = 1.5300\left[A(CH_3COOH)_{aq}/A(H_2O)\right] \quad (5)$$

5 DISCUSSION

The Raman quantification factor, F_i, for the dissolved species i can be calculated from the following equation based on eq. (1):

$$(C_i/CH_2O) = (m_i/mH_2O) \approx X_i = (A_i/AH_2O)(FH_2O/F_i) \quad (6)$$

where m_i is number of moles of i/1000 g of H_2O. Assuming $FH_2O = 1.000$, then F_i values can be calculated from equations 2, 3, 4 and 5, and they are 1.050 and 0.656 for methane and acetic acid, respectively, and 0.322 and 0.084 for CO_2 (P1) and CO_2 (P2), respectively.

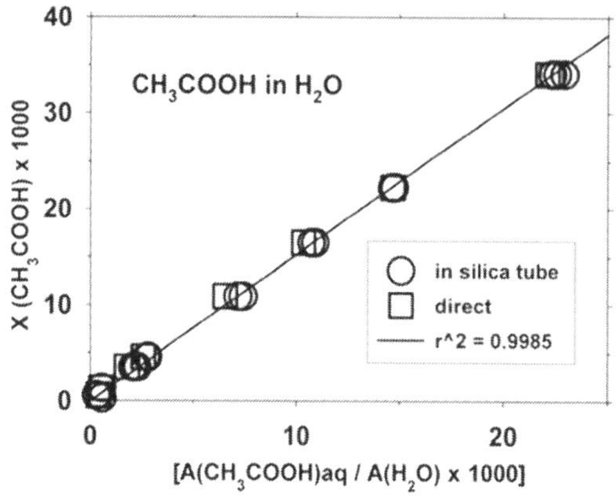

Figure 3. Experimental results showing the relation between acetic acid concentrations in water (in mole fraction, $X(CH_3COOH)$) and the peak area ratios for acetic acid and water at room temperature and 0.1 MPa. The samples were exposed to the laser beams either through a wall of silica tube or directly. Plotted are the data from Table 3. The straight line represents linear regression of all data.

The *in-situ* nondestructive analytical method for aqueous solution described in this study is especially useful for study of diffusion of gases in aqueous solutions. Preliminary results were obtained for diffusion of CH_4 in H_2O at room temperature, in which concentrations of dissolved methane in water were monitored, as a function of time and distance from the vapor-water boundary, immediately after the perturbation of an equilibrium state induced by a sudden change in methane pressure. This type of experiment will also be conducted for CO_2-H_2O and other systems as well as at different temperatures and salt contents in water.

The *in-situ* analysis for acetic acid has been applied to monitor the progress of decarboxylation at 504 and 553°C in sealed fused-silica tubes. This study also established the detection limits for dissolved methane and carbon dioxide applicable to the studies of natural fluid inclusions.

6 CONCLUDING REMARKS

This study shows that modern Raman microprobe systems can detect low concentrations of dissolved methane and carbon dioxide in water at the levels equivalent to the saturated values at 0.5 and 0.1 MPa, respectively. These Raman microprobe systems can also provide quantitative analysis based on Placzek's ratio method. The *in-situ* nondestructive analytical method is suitable for the studies of diffusion kinetics as well as monitoring the progress of chemical reactions.

REFERENCES

Chou, I-M. & Burruss, R.C. 2002. Raman spectroscopic method for the determination of solubility of methane in water at room temperature and elevated pressures. *18th PACROFI Program with Abstracts*: 21–22.

Chou, I-M., Pasteris, J.D. & Seitz, J.C. 1990. High-density volatiles in the system C-O-H-N for the calibration of a laser Raman microprobe: *Geochim. Cosmochim. Acta* 54: 535–543.

Culberson, O.L. & McKetta, J.J., Jr. 1951. Phase equilibria in hydrocarbon-water systems III – The solubility of methane in water at pressures to 10,000 psia. *Petro. Trans. AIME* 192: 223–226.

Duan, Z. & Sun, R. 2003. An improved model calculating CO_2 solubility in pure water and aqueous NaCl solutions from 273 to 533 K and from 0 to 2000 bar. *Chem. Geol.* 193: 257–271.

Seitz, J.C., Pasteris, J.D. & Chou, I-M. 1993. Raman spectroscopic characterization of gas mixtures. I. Quantitative composition and pressure determinations of CH_4, N_2, and their mixtures. *Amer. J. Sci.* 293: 297–321.

Seitz, J.C., Pasteris, J.D. & Chou, I-M. 1996. Raman spectroscopic characterization of gas mixtures. I. Quantitative composition and pressure determination of the CO_2-CH_4 system. *Amer. J. Sci.* 296: 577–600.

Sum, A.K., Burruss, R.C. & Sloan, E.D. 1997. Measurement of clathrate hydrates via Raman spectroscopy. *J. Phys. Chem. B* 101: 7371–7377.

Wopenka, B. & Pasteris, J.D. 1986. Limitations to quantitative analysis of fluid inclusions in geological samples by laser Raman microprobe spectroscopy. *Appl. Spectrosc.* 40: 144–151.

Note: Use of trade names is for information purposes only and does not imply endorsement by the U.S. Geological Survey.

Water–Rock Interaction, Wanty & Seal II (eds)

Synchrotron micro-tomography of porous media for modeling fluid flow

F. Enzmann & M. Kersten
Geoscience Institute, Gutenberg-University, Mainz, Germany

M. Stampanoni
Swiss Light Source, Paul-Scherrer-Institute, Villigen, Switzerland

ABSTRACT: Synchrotron-based X-ray tomographic microscopy (XTM) is suitable to provide data on the real pore microstructure of rocks. We demonstrate that such data are useful for *ab initio* simulations of the fluid velocity field at a sub-μm lateral resolution.

1 INTRODUCTION

The pore space of geologic materials not only controls its hydrodynamic behavior, but is also the site of many relevant chemical reactions involving element partitioning between fluid and solid phases. Most (reactive) advective-dispersive transport models consider the pore space as a black box and try to fit its retarding behavior by gross (empirical) parameters. On the other hand, *ab initio* simulations of fluid transport in real pore space provides an insight into the complexity of hydrodynamic behavior on a microscopic scale (Ferreol & Rothman 1995, Martys & Chen 1996). Our contribution will expand on the latter promising but not yet well-known approach.

2 EXPERIMENTS

The porous geologic material used in this study consisted of different subsamples of a typical sandstone rock from SW Germany ("Pfälzer Buntsandstein"). Micro borehole samples with diameters around 1.5 mm and various lengths up to 15 mm were prepared both with and without epoxy resin impregnation. The porosity of the sandstones varies between 1 and 25% as a result of different geologic formation conditions and degree of pore cementation.

XTM measurements were performed at the SLS beamline MS-Tomo at an energy of 13.5 keV, a sample rotation around the x-axis at 1000 steps each 3.5 s for subsample S1a, and 2.2 s for subsample S1b. Due to the limited vertical beam size, sample S1b was thereby stacked in 200-μm subframes along the whole length x of 1.6 mm. Sinograms were generated on the fly on a SLS-based Linux cluster, together with the back reconstruction into slices from the digital camera data. Figure 1 shows a thus reconstructed slice through sample S1a. The resolution of this example is 500 × 500 pixels of 1.7 μm each. The computer tomography (CT) numbers, converted from the raw intensity data at the sinograms, are scaled in an 8-bit range (0–255 possible values). Figure 2 gives an example for a stacked slices sequence.

For identifying the pore space, the CT-numbers representing the pore voxels are to be selected. In Figure 3, all CT numbers are displayed in a frequency scale. For each phase component in the voxel system (quartz, feldspar, resin, and air), a gaussian distribution of the CT numbers has been identified, and all CT numbers greater than 82 have been allocated to "pore voxels". With the pore identification process thus accomplished, the total pore space can now be

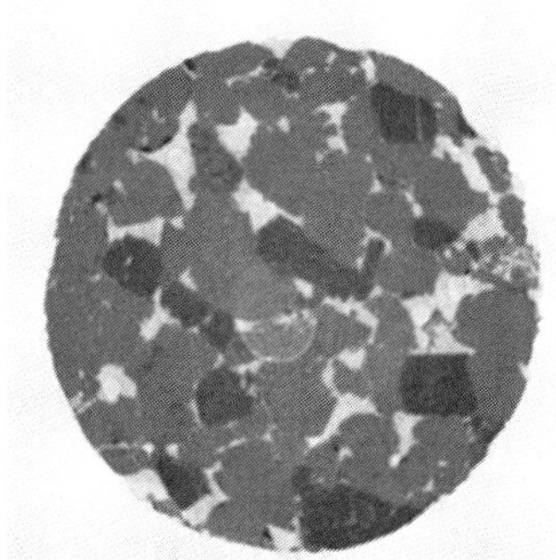

Figure 1. Reconstructed 2D slice through sample S1a, with various phases differentiated by their gray scale.

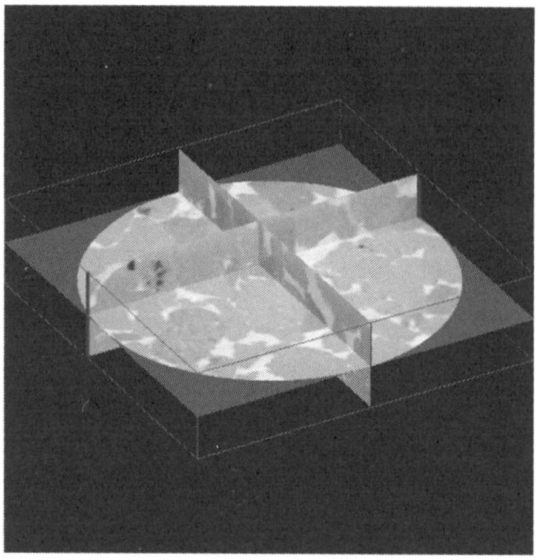

Figure 2. Scheme of a 3D sequence of S1a slice planes.

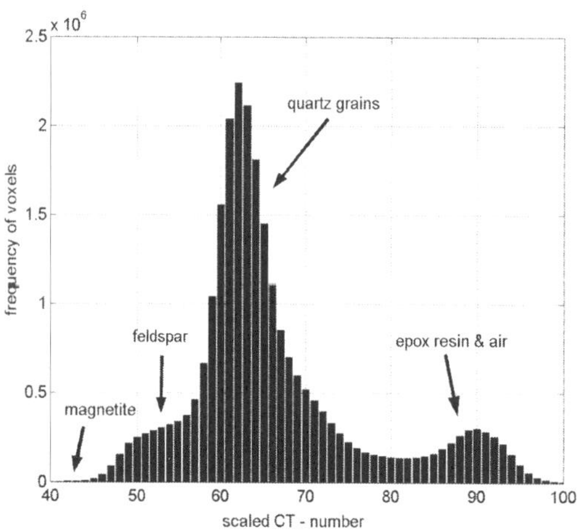

Figure 3. Distribution of scaled CT – numbers for S1a.

displayed as isosurfaces (Fig. 4). For *ab initio* simulations of a migration experiment, only the connected (open) pore structure between two potential boundaries is of interest. Given a digital image or tomogram of a porous material, it is simple to access the connectivity within any component using the so-called "burning algorithm" (Stauffer 1975). This algorithm identifies all cluster members of connected phase voxels with equal or quasi-equal CT numbers in a tomogram.

Figure 5 shows the clustered pore structure, and Figure 6 ultimately the open pore structure as potential fluid migration pathways from bottom to top in sample S1a.

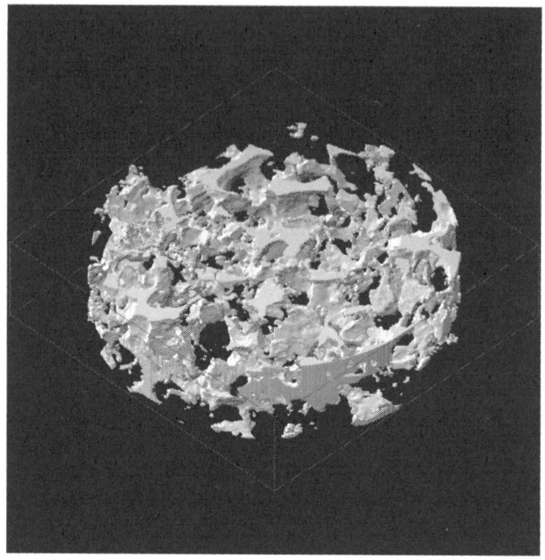

Figure 4. Reconstructed 3D total pore structure S1a.

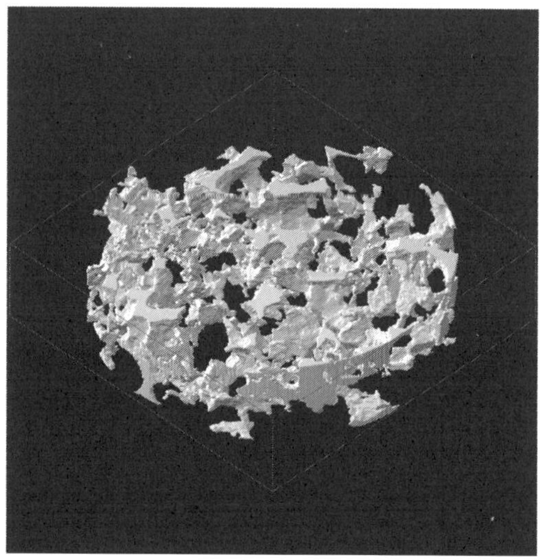

Figure 5. Reconstructed clustered pore structure of sample S1a.

3 MODELLING

For modeling fluid flow, the lattice Boltzmann equation (LBM) method has been used, because it relies on the real pore space structure as a boundary condition input (Ferreol & Rothman 1995). Our code "PoreFlow" uses a "D3Q19" geometry (3 dimensional cell spanned by 19 lattice vectors, Qian et al. 1992). The given formalism and special conditions

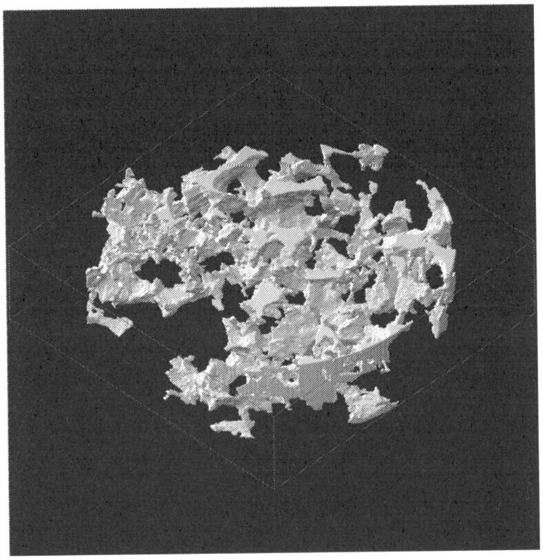

Figure 6. Reconstructed open pore structure of sample S1a.

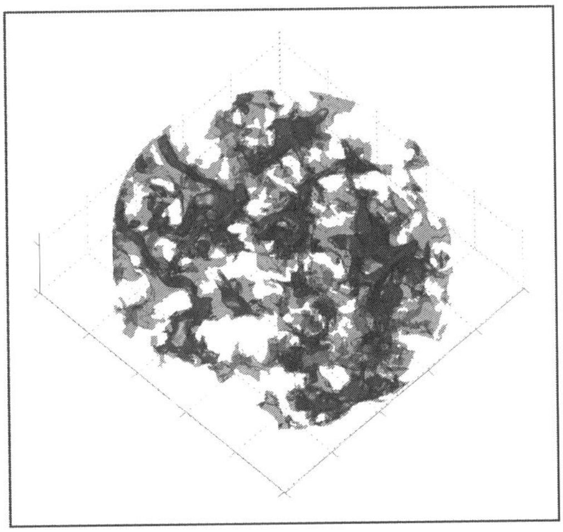

Figure 7. Isosurfaces of simulated fluid velocity in percolated pore space of sample S1a.

(e.g. low numerical Mach and Knudson numbers) lead to a velocity field as a solution of the Navier-Stokes equation at a given kinematic viscosity (Marthys & Chen 1996, Nourgaliev et al. 2003). A preliminary test run was performed with a stationary inflow boundary condition. The result is an inhomogeneous flow field in the connected pore structure, with a high variability in local fluid velocity and direction leading to a disperse transport of conservative tracer particles. Isosurfaces of defined fluid velocities are shown in Figure 7. Note that this

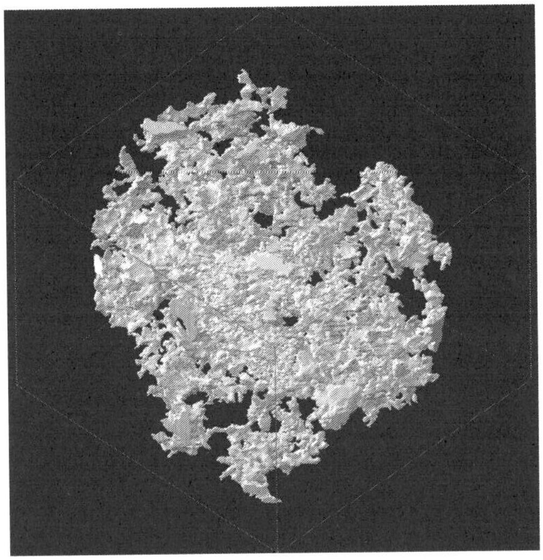

Figure 8. Constructed potential percolated pore structure of sample S1b.

Table 1. Geometric characterization of sample S1a.

500 × 500 × 125 at 1.7 μm Voxels	Total	Cluster	Percolate
Porosity	11.8	11.5	10.1
Surface [m^2]	4.0E-6	3.7E-6	3.2E-6
Specific surface [m^2/g]	1.4E-2	1.3E-2	1.1E-2
Permeability [mD]	360.0	386.2	347.7

velocity structure is based on about 2E6 active nodes for the LBM simulations, which is currently the limit for our parallelized computer code on a 4-processor PC. Figure 8 shows a pore structure with nearly 25 Mio. active nodes ready for fluid flow modeling. An enhanced capacity code for a Linux cluster is in development in order to process this amount of data.

The thus elaborated open pore structure data can be used for direct validation of average hydro-dynamical parameters. Porosity and absolute and specific surface can be calculated directly from the structure. The permeability can be calculated using a relationship between porosity and specific surface (Carman 1937). The only gross parameter that can be estimated by other techniques for porous materials is the pore size distribution, such as derived by mercury porosimetry. Table 1 documents the data for sample S1a, and Table 2 from sample S1b. Note that this is the same sandstone with (a) and without (b) epoxy resin filling of the open porosity.

The thus derived gross parameters represent a typical range for sandstones. However, using the tomography data, we were able to show the

Table 2. Geometric characterization of sample S1b.

$700 \times 700 \times 500$ at 1.7 µm Voxels	Total	Cluster	Percolate
Porosity	9.3	7.9	6.1
Surface [m^2]	3.1E-5	2.2E-5	1.2E-5
Specific surface [m^2/g]	1.4E-2	9.4E-3	6.6E-3
Permeability [mD]	188.3	233.2	187.4

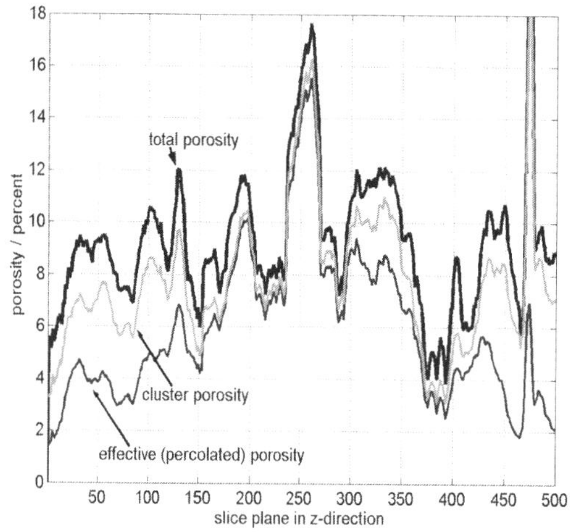

Figure 9. Distribution of local porosity through slice planes percolated from bottom to top of sample S1b.

heterogeneity of these parameters at a microscale, and, moreover, even some anisotropic dispersion effects along with textural inhomogeneities. Figure 9 shows a locally enhanced porosity along the slice planes in the z-direction.

4 CONCLUSIONS

The XTM lateral resolution of 1.7 µm per voxel achieved for the samples S1a and S1b has yet never been reported for a geologic sample. However, it is still not sufficient as demonstrated in Figure 10. The open pore size distribution thus derived from XTM and CT reconstruction still deviates from that derived directly by mercury porosimetry. This deviation results mainly from the larger sample size for the latter method (ca. 1 cm^3), which shows both an open porosity down to 0.7 µm, but also a secondary peak around 100 µm pore size. For the XTM data to cover the full porosity field, we need both to enhance the

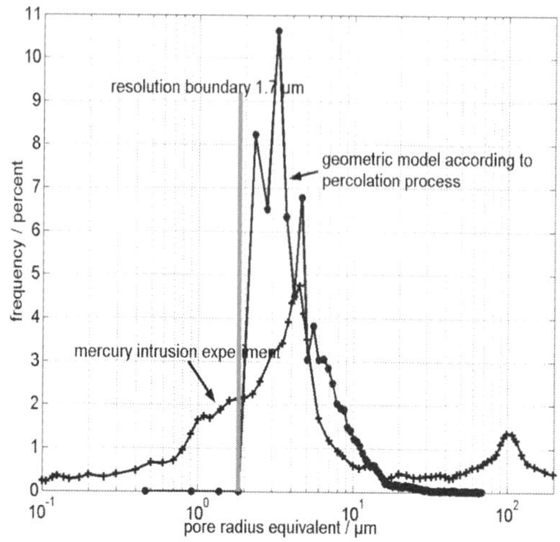

Figure 10. Comparison for pore-size distribution derived from mercury porosimetry data and for a percolation simulation run.

resolution down to the sub-µm size, and to enhance the sample volume by appropriate subframe stacking. However, the main limitation is not the XTM technique (0.7 µm is now possible at SLS), but the computer capacity for both data reduction and modeling. We hope to have soon the software ready and access to a 64 processor Linux cluster.

REFERENCES

Ferreol, B. & Rothman, D.H. 1995. Lattice-Boltzmann simulations of flow through Fontainebleau sandstone. *Trans. Porous Media* 20: 3–20.

Carman, P.C. 1937. Fluid flow through granular beds. *Trans. Inst. Chem. Eng.* 15: 150–166.

Martys, N.S. & Chen, H. 1996. Simulation of multicomponent fluids in complex three-dimensional geometries by the lattice Boltzmann method, *Phys. Rev. E*, 53: 743–750.

Nourgaliev, R.R., Dinh, T.N., Theofanous, T.G. & Joseph D. 2003. The lattice Boltzmann equation method: theoretical interpretation, numerics and implications. *Int. J. Multiphase Flow* 29: 117–169.

Qian, Y.H., D'Humières, D. & Lallemand, P. 1992. Lattice BGK models for Navier-Stokes equation. *Europhys. Lett.* 17: 479–484.

Stauffer, D. & Aharony, A. 1992. *Introduction to Percolation Theory*. London: Taylor & Francis.

Verberg, R. & Ladd, A.J.C. 2002. Simulations of erosion in rough fractures. *Phys. Rev. E* 65: 056311.

Water–Rock Interaction, Wanty & Seal II (eds)
© 2004 Taylor & Francis Group, London, ISBN 90 5809 641 6

^{19}F-NMR study Al-oxyhydroxide surfaces and reactions

B.L. Phillips
Dept. of Geosciences, Center for Environmental Molecular Science, SUNY Stony Brook, Stony Brook, NY

A. Lee & W.H. Casey
Dept. of Land, Air, and Water Resources, University of California, Davis, CA

ABSTRACT: ^{19}F-NMR spectroscopy is a potentially useful technique for characterizing the surface structure of diamagnetic materials. ^{19}F is a very sensitive, 100% naturally abundant NMR-active nucleus, which allows spectra to be obtained at relatively low concentrations, such as on mineral surfaces of moderate specific surface areas and modest loadings. Dissolved fluoride substitutes for both terminal water and bridging hydroxyl groups on Al-oxyhydroxide surfaces and could serve as a probe of reactive sites. A main problem is uncertainty in NMR peak assignments, and we show here that $^{19}F\{^{27}Al\}$ double-resonance techniques could help distinguish fluoride signals arising from terminal vs. bridging configurations. Application to formation of Al-oxyhydroxide gels from the "$GaAl_{12}$" polyoxocation in the presence of fluoride suggests that most terminal fluorides are converted to bridging sites by water-eliminating condensation reactions during precipitation.

1 INTRODUCTION

Nuclear magnetic resonance (NMR) spectroscopy is a useful technique for determining molecular-level structure in crystalline, disordered, and amorphous solids. NMR spectroscopy can provide structural characteristics of mineral/fluid interactions and of sorption complexes in terms of element specific coordination environments, with experiments easily done on multi-phase (e.g., solid/fluid) samples. The major difficulty is sensitivity, which limits application of NMR to materials with moderate to large specific surface area and to sensitive, abundant, NMR-active isotopes. Another problem is that the principal NMR observable, the chemical shift, has no unique relationship to molecular structure. Peak assignments are often made by analogies with data obtained for crystalline compounds, which might not be strictly relevant to asymmetrical mineral surface environments. A related problem is that NMR is a bulk technique, yielding signal from all sites containing the observed isotope. These last two aspects can be addressed by using double-resonance techniques to establish spatial proximity between sorbed or surface constituents and atoms present only in the bulk. Here we show how these techniques can be applied to study surface interactions between dissolved fluoride and Al-oxyhydroxides, and the condensation reactions that occur during precipitation.

The aqueous $GaO_4Al_{12}(OH)_{24}(H_2O)_{12}^{7+}$ molecule ("$GaAl_{12}$"; Fig. 1) can serve as an ordered molecular model for Al-oxyhydroxide surfaces. It contains oxygen sites exposed to the solution that resemble those expected to occur on the surface of Al-oxyhydroxides, including terminal water groups

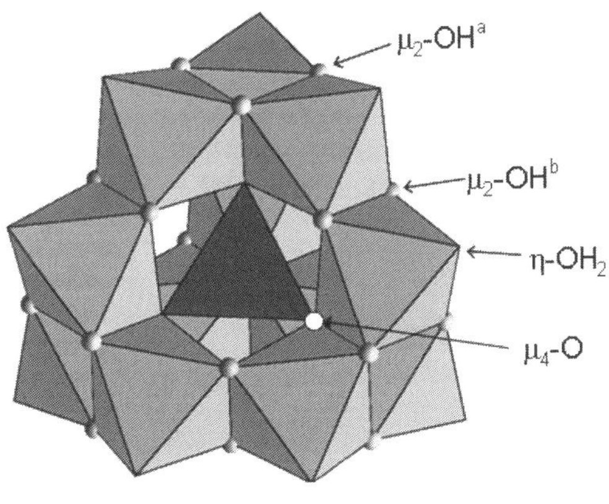

Figure 1. Polyhedral representation of the $GaAl_{12}$ molecule with distinct oxygen sites labeled. The central tetrahedral site is occupied by Ga, the octahedra contain Al.

(η-OH_2) and two distinct types of bridging hydroxyl (μ_2-$OH^{a,b}$). This molecule is metastable in solution near pH 5, and the rates and activation parameters for oxygen exchange between these sites and bulk water have been measured under these conditions (Casey & Phillips 2001). The interior μ_4-O groups are inert to steady exchange with solvent waters.

A solution-state ^{19}F-NMR study of the substitution kinetics of dissolved fluoride on the $GaAl_{12}$molecule showed that fluoride substitutes for both the bridging hydroxyls and terminal water molecules (Yu et al. 2003). At 278 K and a $F^-/GaAl_{12}$ ratio near unity, time constants for fluoride substitution of approximately 4 h and 200 h were observed for the two distinct bridging sites, which are of the order 100 times shorter than the lifetimes for oxygen on these sites in the absence of F^-. Comparison of NMR and potentiometric data show that virtually all the dissolved F^- is taken up by the terminal sites within a few minutes, consistent with the short lifetime of water molecules on these terminal sites (ca. 25 ms at 278 K). These results suggest that F^- might serve as a probe of reactive surface sites on mineral surfaces, which could be studied by ^{19}F-NMR spectroscopy, because ^{19}F is a very sensitive 100% naturally abundant NMR-active nuclide.

2 EXPERIMENTAL

2.1 *Samples*

An amorphous gel was prepared from a solution that contained approximately 15 mM $GaAl_{12}$ in D_2O at pH 4.7, by adding dropwise a 0.1 M NaF/HF solution, also at pH 4.7, until a ratio $F^-/GaAl_{12}$ of 5 was reached. A precipitate gradually formed over the course of about an hour. This precipitate was separated by centrifugation, rinsed thoroughly in D_2O and allowed to dry in air. A crystalline sample was obtained from a similar monospecific $GaAl_{12}$ solution by adding NaF/HF to produce a F/Ga ratio of about 1, aging the resulting solution for about 0.5 h at room temperature, then adding a solution containing an excess of Na_2SeO_4. Crystals subsequently formed over a period of several days, which were washed repeatedly and air-dried. For both samples, the initial $GaAl_{12}$ solutions were prepared by metathetic dissolution of a high-purity crystalline selenate salt of $GaAl_{12}$ in water (or D_2O) containing a slight excess of $BaCl_2$, and then filtering through a 0.2 μm filter to remove the resulting $BaSeO_4$.

2.2 *NMR spectroscopy*

The ^{19}F-NMR spectra of the amorphous gel were taken with a Chemagnetics CMX 360 spectrometer operating at 338.7 MHz and a 42 kHz spinning rate. The $^{19}F\{^{27}Al\}$ TRAPDOR data were obtained as the difference between a ^{19}F spin echo control and an identical experiment except for irradiation at the ^{27}Al frequency, 93.8 MHz, during the interpulse delay. Each spectrum represents between 192 and 2000 acquisitions, which were acquired in groups of 16 transients alternating between TRAPDOR and control. The interpulse delay was incremented by integral numbers of rotor cycles. A ^{71}Ga NMR spectrum was taken under similar conditions at 109.8 MHz. ^{19}F NMR spectra of the crystalline sample were taken with a Bruker Avance 700 spectrometer operating at 658.7 MHz, using a probe assembly configured for 2.5 mm rotors and spinning at 20 kHz. 8000 acquisitions were acquired with a relaxation delay of 5 s. All chemical shifts are reported relative to $CFCl_3$, via a secondary standard of C_6F_6.

3 RESULTS AND DISCUSSION

The ^{19}F-NMR spectra of various phases of fluoride-substituted $GaAl_{12}$ are shown in Figure 2, along with spectra for fluoride sorbed on crystalline $Al(OH)_3$ (bayerite) surfaces from an earlier study (Nordin et al. 1999). The ^{19}F-NMR spectrum of the amorphous gel (Fig. 2a) contains two broad peaks, centered at −136 and −148 ppm and full width at half-maximum of about 10 ppm. The ^{71}Ga NMR spectrum of this sample contains a peak near +130 ppm, indicating the presence of 4-coordinated Ga and suggesting that some aspects of the $GaAl_{12}$ molecular structure is retained in the gel phase (Bradley et al. 1992).

The ^{19}F-NMR chemical shifts observed for the $GaAl_{12}/F^-$ gel are very similar to those observed for peaks assigned to the μ_2-F (−137.3 to −138 ppm) and η-F (−147 ppm) sites of aqueous F-substituted $GaAl_{12}$ (Fig. 2b; Yu et al. 2003). Assignment of the peaks for the dissolved species is based primarily on comparison of the time-dependence of the peak intensities (the peak at −147 ppm dominates at short equilibration times) with kinetic considerations. The other major peaks in Figure 2b (near −157 ppm) arise from monomeric Al-F complexes (Martinez et al. 1996; Bodor et al. 2000) resulting from decomposition of some of the $GaAl_{12}$ molecules during the experiment.

The peak positions and separations for the $GaAl_{12}/F^-$ gel are also similar to those observed for surface-sorbed fluoride on bayerite, −131 and −141 ppm (Fig. 2 d,e). The signals from bayerite were assigned to fluoride in bridging (μ_2-F; −131 ppm) and terminal (η-F; −141 ppm) sites based on analogy with data from F-containing dioctahedral clays, which give a peak near −132 ppm (Huve et al. 1992; Labouriau et al. 1995) and the observation that the relative

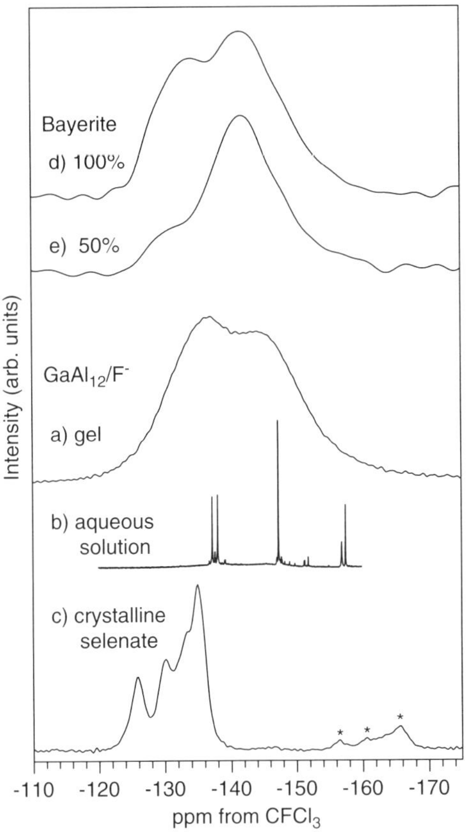

Figure 2. ^{19}F-NMR spectra of various phases of fluoride-substituted $GaAl_{12}$. a) Amorphous gel produced at a ratio $F^-/GaAl_{12} = 5$. b) Aqueous solution with a ratio $F^-/GaAl_{12} = 1$, taken at 278 K, after 94 h of equilibration (after Yu et al. 2003). c) Crystalline selenate salt ($F^-/GaAl_{12} = 1$). Spectra d) and e) are of bayerite with surface-sorbed F^- at 100% (d) and 50% (e) maximum fluoride sorption density at pH 5 (after Nordin et al. 1999). Asterisks denote spinning sideband artifacts.

intensity of the −131 ppm peak increases markedly with fluoride sorption density.

In contrast, the peak positions for the F-containing crystalline $GaAl_{12}$ selenate salt occur at significantly higher frequencies (Fig. 2c). This spectrum can be fit with four symmetrical peaks of subequal intensities at −125.7 ppm (18%, including spinning sidebands), −129 ppm (22%), −132.9 ppm (24%), and −135 ppm (36%). No firm assignment can be made for these peaks at this time, because only three major peaks would be expected in the simplest case: one for each of the exchangeable sites on the molecule (Fig. 1). It is likely that the peak at −135 ppm is due to η-F sites, since these are expected to be most abundant even if F^- had equilibrated with the $GaAl_{12}$ prior to crystallization (Yu et al. 2003) and peaks for η-F sites generally occur upfield from those for μ_2-F (e.g. Taulelle et al. 1999). In solution, the steady-state F^- distribution on the molecule is about 42% η-F and 34% and 24% for the two distinct μ_2-F. It is possible for additional η-F peaks to occur corresponding to those that are adjacent to μ_2-F or μ_2-OH sites, as was found for a crystalline aluminofluorophosphate by Taulelle et al. (1999). Additional experiments are underway, but the main observation is that the ^{19}F-NMR shifts for similar sites appear to differ significantly between the crystal and solution phases, consistent with the idea that in aqueous systems the ^{19}F chemical shifts are very sensitive to hydrogen bonding interactions (Liu & Tossell 2003).

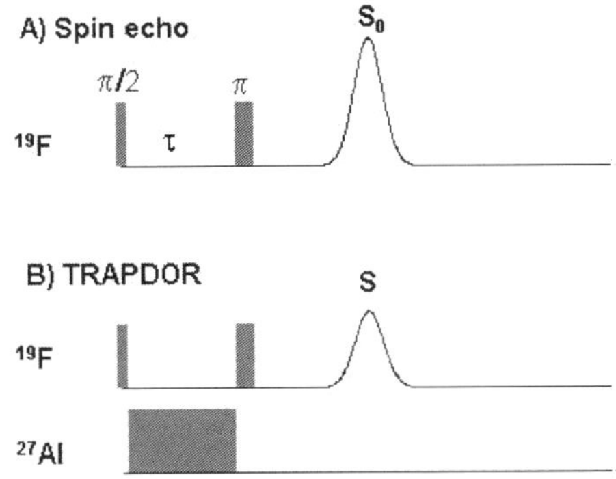

Figure 3. Timing diagram for a $^{19}F\{^{27}Al\}$ TRAPDOR experiment. The ratio S/S_0 is related to the number of Al near the F. The interpulse delay, τ, corresponds to an integral number of rotor periods.

Comparison of the ^{19}F-NMR spectrum of the gel with that of the dissolved F^--substituted $GaAl_{12}/F^-$ (Figs 2a, 2b) would suggest a similar assignment scheme: μ_2-F at −136 ppm and η-F at −148 ppm for the gel. Such an assignment would also be consistent with the relative frequencies assigned to the surface F^- on bayerite (Fig. 2 d,e). However results of the $^{19}F\{^{27}Al\}$ TRAPDOR experiment suggests that both signals from the gel arise from sites having the same number of adjacent Al-atoms.

The TRAPDOR experiment (Fig. 3; Grey & Vega 1995) consists of measuring ^{19}F NMR spin-echo intensity without (S_0; Fig. 3A) and with ^{27}Al irradiation (S; Fig. 3B) during the interpulse delay, τ. Dipolar coupling between ^{19}F and ^{27}Al nuclei reduces the ratio S/S_0 by an amount related to τ and the second moment of the ^{19}F-^{27}Al dipolar coupling, which is proportional to the sum of $[d(\text{Al-F})]^{-6}$, where d(Al-F) is the Al-F internuclear distance. Thus, a significantly faster decrease in S_0 with τ is expected for bridging configurations (two Al at about 0.18 nm) than for terminal sites having one Al-F bond.

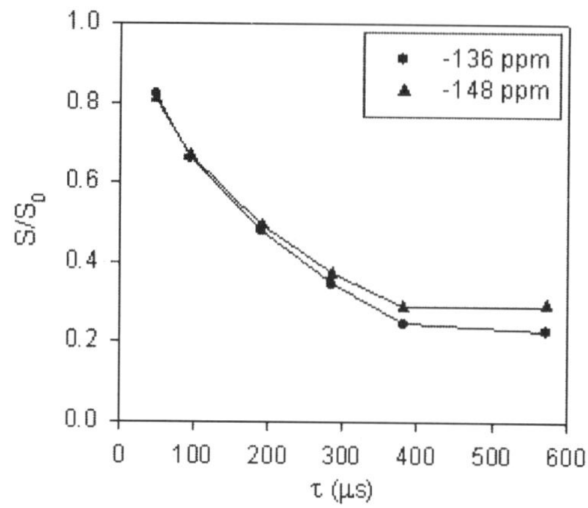

Figure 4. $^{19}F\{^{27}Al\}$ TRAPDOR dephasing curves for the $GaAl_{12}$/F gel. The two sets of data correspond to the peaks in Figure 2a. The variable τ corresponds to the interpulse delay and S and S_0 are the ^{19}F-NMR intensities observed, respectively, with and without ^{27}Al irradiation (Fig. 3).

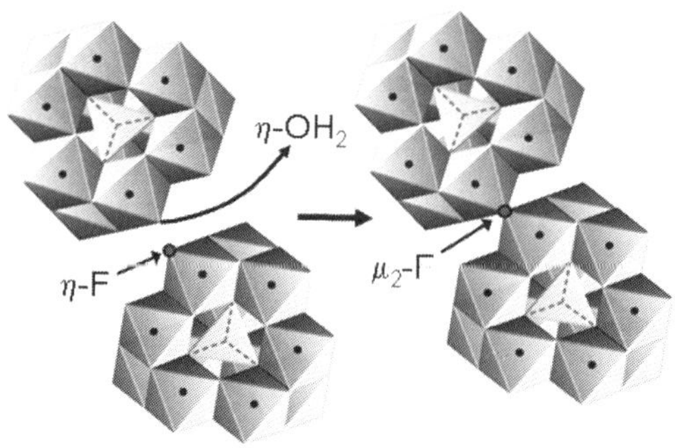

Figure 5. Schematic representation for a possible mechanism for η-F to μ_2-F conversion during gel formation from aqueous F^--substituted $GaAl_{12}$.

The TRAPDOR data for the $GaAl_{12}/F^-$ gel (Fig. 4) show very similar dephasing curves for the signals at −136 and −148 ppm, which suggests that these peaks arise from F^- bonded to the same number of Al. There is a small difference in the TRAPDOR fraction at very long dephasing times that could arise from differences in the number of Al in more distant coordination shells and/or different relative orientations between the ^{19}F shielding and ^{27}Al quadrupole coupling tensors. However, much larger differences, of the order of a factor of two in dephasing rate, are expected for terminal vs. bridging fluoride environments.

This observation suggests that most of the η-F sites on the $GaAl_{12}$ in solution become bridging sites during gel formation. That both ^{19}F-NMR peaks correspond to μ_2-F seems reasonable considering that the F^- substitution reactions on bridging sites should be fast enough at 298 K to yield significant concentrations of μ_2-F prior to precipitation. Furthermore, the presence of 4-coordinated Ga suggests that some of the molecular structure is maintained in the solid phase. Conversion of η-F to μ_2-F can occur via a simple water elimination reaction during condensation of individual F^--substituted $GaAl_{12}$ molecules to produce the gel (Fig. 5). Retention of F^- during precipitation seems likely based on kinetic considerations. The η- F should be much less labile than the η-OH_2, which has a lifetime less than 1 ms at 298 K (Casey & Phillips 2001). Yu et al. 2003 observed only very slow migration of F^- from η-F to μ_2-F (hours), consistent with the results of Bodor et al. (2000) who observed no exchange of F^- on the monomeric AlF^{2+}(aq) complex on the T_1 time scale (seconds).

These results show that ^{19}F-NMR is a promising technique for studying surface reactions and reactive surface sites. However, peak assignments cannot be made solely by comparison of chemical shifts with model compounds; double-resonance techniques appear to be necessary for peak assignments in hydrous, disordered materials because of the sensitivity of the ^{19}F chemical shift on hydrogen bonding interactions.

ACKNOWLEDGEMENTS

We thank Peter Chupas and Clare Grey for help with the TRAPDOR experiments, Martine Ziliox and the Keck NMR Center for Structural Biology at Stony Brook for access to the 700 MHz instrument. This research was supported by a grant from the U.S. NSF, EAR 0101246 (WHC and BLP).

REFERENCES

Bodor, A., Tóth, I., Bányai, I., Szabó & Hefter, G.T. 2000. ^{19}F NMR Study of the equilibria and dynamics of the Al^{3+}/F system. *Inorg. Chem.* 39: 2530–2537.

Bradley, S., Kydd, R.A. & Fyfe, C.A. 1992. Characterization of the $GaO_4Al_{12}(OH)_{24}(H_2O)_{12}^{7+}$ polyoxocation by MAS NMR and infrared spectroscopies and powder X-ray diffraction. *Inorg. Chem.* 31:1181–1185.

Casey, W.H. & Phillips, B.L. 2001. The kinetics of oxygen exchange between sites in the $GaO_4Al_{12}(OH)_{24}(H_2O)_{12}^{7+}$ *(aq)* molecule and aqueous solution. *Geochim. Cosmochim. Acta* 65: 705–714.

Grey, C.P. & Vega, A.J. 1995. Determination of the quadrupole coupling constant of the invisible aluminum spins in Zeolite HY with $^1H/^{27}Al$ TRAPDOR NMR. *J. Am. Chem. Soc.* 117: 8232–8242.

Huve, L., Delmotte, L., Martin, P., LeDred, R., Baron, J. & Saehr, D. 1992. ^{19}F MAS NMR study of structural fluorine in some natural and synthetic 2:1 layer silicates. *Clays Clay Mins.* 40: 186–191.

Labouriau, A., Kim, Y.W., Chipera, S., Bish, D.L. & Earl, W.L. 1995 A F-19 nuclear magnetic resonance study of natural clays. *Clays Clay Mins.* 43: 697–704.

Liu, Y. & Tossell, J.A. 2003. Possible Al-F bonding environment in fluorine-bearing sodium aluminosilicate glasses: from calculation of ^{19}F NMR shifts. *J. Phys. Chem. B*107: 11280–11289.

Martinez, E.J., Girardet, J.-L. & Morat, C. 1996. Multinuclear NMR study of fluoroaluminate complexes in aqueous solution. *Inorg. Chem.* 35: 706–710.

Nordin, J.P., Sullivan, D.J., Phillips, B.L. & Casey, W.H. 1999. Mechanisms for fluoride-promoted dissolution of bayerite [β-$Al(OH)_3(s)$] and boehmite [γ-AlOOH]: F-19-NMR spectroscopy and aqueous surface chemistry. *Geochim. Cosmochim. Acta* 63: 3513–3524.

Taulelle, F., Pruski, M., Amoureux, J.P., Lang, D., Bailly, A., Huguenard, C., Haouas, M., Gerardin, C., Loiseau, T. & Ferey, G. 1999. Isomerization of the prenucleation building unit during crystallization of $ALPO_4$-CJ2: An MQMAS, CP-MQMAS, and HETCOR NMR study. *J. Am. Chem. Soc.* 121: 12148–12153.

Yu, P., Lee, A., Phillips, B.L. & Casey, W.H. 2003. Potentiometric and ^{19}F-NMR spectroscopic study of fluoride substitution in the $GaAl_{12}$ polyoxocation: Implications for Aluminum (hydr)oxide mineral surfaces. *Geochim. Cosmochim. Acta* 67: 1065–1080.

Water–Rock Interaction, Wanty & Seal II (eds)
© 2004 Taylor & Francis Group, London, ISBN 90 5809 641 6

X-ray spectroscopic techniques to probe the microbial dissolution of basaltic glasses

A.S. Templeton & B.M. Tebo
Marine Biology Research Division, Scripps Institution of Oceanography, University of California San Diego, La Jolla CA, USA

H. Staudigel & B.E. Bailey
Institute for Geophysics and Planetary, Scripps Institution of Oceanography, University of California San Diego, La Jolla CA, USA

T.P. Trainor & P.J. Eng
Consortium for Advanced Radiation Sources, University of Chicago, Chicago Il, USA

ABSTRACT: Coupled synchrotron-based X-ray reflectivity and spectroscopy measurements provide a highly surface-sensitive approach to probe the dissolution and redox transformation of minerals and glasses at an angstrom- to nanometer-scale resolution. Using a series of parallel experiments (21 to 270 days), we compare the initial rates of oxidative dissolution of volcanic glasses reacted with cold artificial seawater (10°C) under abiotic conditions to the weathering of glass surfaces by biofilms of Fe- and Mn-oxidizing bacteria, which use the glasses as a primary energy source for growth. Specular reflectivity measurements are used to determine the thickness of the surface "leached" layers, fluorescence-yield profiles quantitatively measure the redistribution of Fe, Mn, Ca, and Ti among the bulk glass, solution, and secondary minerals, and angle-dependent spectroscopic data (e.g. Fe and Mn K-edge XANES) provide depth-resolved measurements of changes in Fe and Mn oxidation state induced by microbial activity vs. diffusion-controlled abiotic reactions.

1 INTRODUCTION

The initial oxidative weathering reactions of oceanic crust are driven by the large disequilibrium between highly reduced volcanic rocks and glasses, and cold, oxygenated seawater. Considerable attention has been directed towards the geochemical characterization of weathering products within rinds and/or fractures developed on basalt surfaces, and historically, these reactions have been considered to be abiotic in nature. The degree of hydration and oxidation of basaltic glasses, as well as the degree of secondary mineral formation, is highly variable, and results in large chemical exchange of constituents such as Ca and Mg and redox transformation of Fe and Mn, leading to the formation of a poorly crystalline mixture of iron oxyhydroxides, clays, and other Fe-bearing minerals such as celadonite.

Generally, the ocean crust is considered to be highly oligotrophic and relatively unsuitable as a host for microbial organisms due to the lack of organic carbon or high concentrations of soluble energy sources. However, volcanic glasses recovered from the oceanic crust commonly show characteristic corrosion features (e.g. Thorseth et al. 1995, Fisk et al. 1998, Furnes & Staudigel 1999) that are considered to be microbial in origin. The pervasive distribution of these corrosion features in the upper 500 m of the oceanic crust (Furnes & Staudigel 1999) suggest that microbial activity may have a direct and significant effect on the chemical exchange between basaltic rocks and seawater. Furthermore, basaltic rocks contain significant amounts of reduced iron (Fe^{II}) and manganese (Mn^{II}), representing potential energy sources for bacterial growth.

Although abiotic oxidation of Fe^{II} commonly occurs, there is also a strong possibility that Fe^{II}-oxidizing microorganisms significantly increase the rate of Fe^{II} oxidation. Recent work has shown that novel Fe^{II}-oxidizing bacteria isolated from iron sulfide surfaces are also able to grow with basaltic glass as a source of Fe^{II} (Edwards et al. 2003). In addition, Mn oxides are commonly intermixed with Fe^{III} phases in the basalt weathering products, an observation which suggests that Mn^{II}-oxidizing bacteria, in addition to Fe^{II}-oxidizing bacteria, may be involved in the oxidative alteration of basalt glasses. Geochemically, the biocorrosion of basaltic glasses may have a significant impact on the rates of dissolution and the chemical composition and stability of secondary phases that form, particularly if microbial activity enhances differential rates of ion exchange and reprecipitation.

At present, there are no quantitative data on chemical transformations of basaltic glass mediated by putative marine "glass-eating" biofilms, nor appropriate methods for probing the initial stages of dissolution over short time-scales. Synchrotron X-ray scattering and spectroscopic measurements are particularly well-suited to probing ion distributions at complex-interfaces and monitoring changes in redox state of transition elements such as Fe and Mn. Detailed chemical information at glass-biofilm-seawater interfaces would provide much needed data for a critical comparison of biologically mediated vs. abiotic weathering of basalt surfaces and provide some of the first information regarding the mechanisms of the initial dissolution process and potentially the mechanisms whereby microorganisms use solid substrates as energy sources for growth.

2 METHODS

Fe-oxidizing bacteria were isolated from basaltic glass surfaces collected from Loihi Seamount using a serial-dilution liquid enrichment technique adapted from Emerson & Moyer (2002) with an anaerobically-prepared artificial seawater medium amended with $FeCl_2$ and O_2. Mn-oxidizers were isolated by using a variety of heterotrophic plates amended with 100 μM $MnCl_2$ and repeated restreaking of colonies that tested positive using a leucoberbelin blue assay sensitive to $Mn^{III/IV}$. A diverse collection of isolates within the α- and γ-Proteobacteria were obtained, a subset of which were used for the glass colonization and dissolution/oxidation experiments.

Basaltic glasses were prepared from fresh natural tholeiitic glass by quenching lava from active Pu'U O'o surface vents (Kilauea, HI). The glass was finely ground, homogenized and remelted in order to remove vesicles and resorb phenocrysts in the melt. The oxidation state of Fe (~10 wt%) and Mn (~0.2 wt%) in the natural and remelted basaltic glasses was determined using X-ray Absorption Near-Edge Spectroscopy (XANES) measurements at the Advanced Photon Source, as well as by Mössbauer spectroscopy (for Fe) at UCSD.

The homogenized glasses were sectioned into 1 cm x 5 cm wafers and epi-polished to produce a surface with a 5 Å roughness. The glass surfaces were acid washed and immersed overnight in 70% ETOH prior to repeated washings with sterile deionized water. Multiple glass wafers were placed in individual sterile polycarbonate tubes and reacted with a regularly-exchanged volume of 10 mL of artificial seawater media with no carbon source. A subset of these wafers was inoculated with either Fe or Mn-oxidizing isolates obtained as described above. Colonization and biofilm development on the glass surfaces was monitored by DAPI-staining and visualizing the bacterial distribution via epifluorescence microscopy at each time point. The remaining crystals served as abiotic controls, and no bacterial colonization was detected on these surfaces. All experiments were conducted at 10°C.

Several combined X-ray scattering and spectroscopic measurements were made for each reacted glass in a grazing-angle geometry using the Kappa diffractometer at the GSECARS beamline 13-IDC at the Advanced Photon Source (Argonne, IL). First, reacted slabs were aligned parallel to the incident beam (30 μm vertical height, 6.5 KeV) and reflectivity profiles were collected from 0 to 1.2 degrees by conducting rocking scans at each incident angle (θ). The incident angle was stepped forward in 0.025 degree increments with the detector positioned at 2θ. Each rocking scan was background subtracted and the integrated specular component was calculated to build the full reflectivity profile. Above the critical angle, rocking scans were obtained over a wide enough range to collect both the specular and diffuse components. Second, fluorescence spectra were collected at each incident angle from 0 to 1.2 degrees, at 7.5 KeV and 7 KeV, with a Ge-detector positioned perpendicular to the incident X-rays. The peak areas for Fe Kα (generated at 7.5 KeV) and Mn Kα, Ca Kα, and Ti Kα (collected at 7 KeV, below the Fe K-edge) were obtained by integration of the peaks from each background subtracted spectrum and compiled into FY profiles. Third, Fe K-edge XANES data were collected at a series of incident angles (e.g. 0.15°, 0.225°, 0.25°, 0.3°, 0.35°, 0.4°) above and below the critical angle to obtain varying depth sensitivity to redox changes. Fourth, Mn K-edge XANES data were collected at a subset of angles (0.15°, 0.25°, 0.4°) representative of the surface, interfacial layer and bulk glass, respectively.

3 RESULTS

Our goal is to compare directly the rate of dissolution and the distribution and speciation/oxidation state of Fe and Mn in bulk basaltic glasses to that within surface leached layers for glasses subjected to abiotic vs. biologically-mediated dissolution.

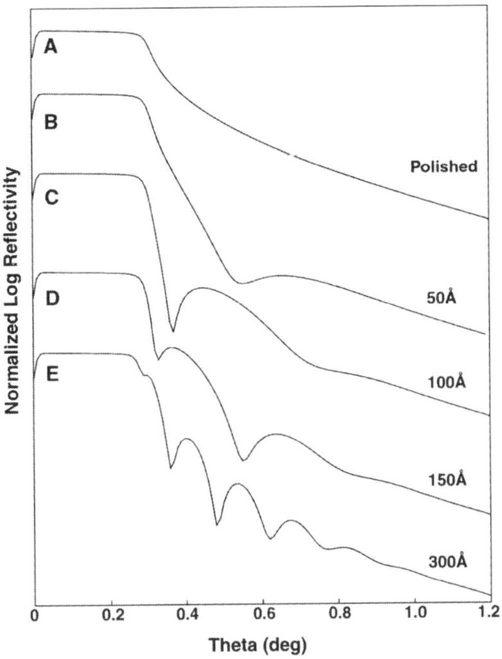

Figure 1. Calculated reflectivity profiles for basalt surfaces with leached layers that range in thickness from 50 Å to 300 Å.

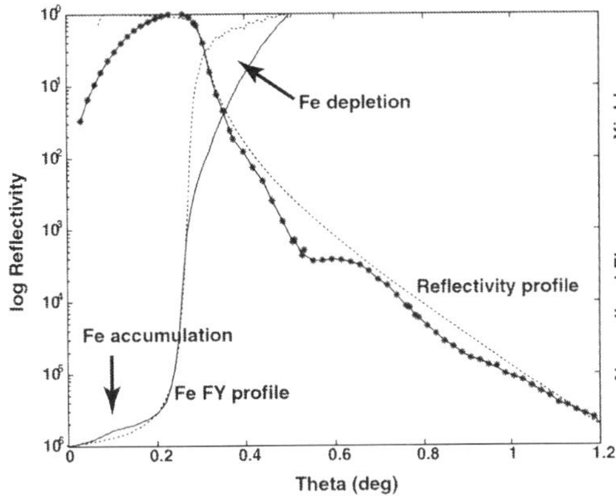

Figure 2. Reflectivity of a glass surface biologically-altered for 4 weeks at 10°C (dotted) vs. reflectivity of the unreacted surface (dashed). Also shown is the normalized Fe FY (solid) for the biologically-altered surface vs. unreacted surface (dashed) as a function of incident angle (at 7.5 KeV).

We first characterize the thickness and density of the hydrated layer that forms on the glass surface using the X-ray reflectivity data. Figure 1 shows predicted reflectivity oscillations for leached layer thicknesses from 50 Å to 300 Å on basalt substrates with an initial density of approximately 3 g/cm^3. The large decrease in the wavelength of the oscillations for calculations A through E shows that systematic increases in the reacted layer thickness should be easily resolved by this technique. The amplitude of the oscillations provide information on the electron density contrast between the bulk glass and surface layer and the slope of the reflectivity curves will also be highly sensitive to changes in surface roughness (calculations not shown).

An example of measured reflectivity oscillations obtained for a glass surface colonized by Fe-oxidizing bacteria (40% surface coverage, epifluorescence image of bacterial distribution not shown) over the course of 4 weeks is shown in Figure 2. Modeling of the reflectivity profile yields a measurement of 145 Å of surface reaction, whereas modeling of the reflectivity profiles obtained for the glass surface reacted abiotically over the same time-period shows the formation of a leached layer of only 45 Å (data not shown). To date, reflectivity profiles have been collected for the Fe-oxidizer vs. abiotic surfaces at 4, 8, 12, and 24 weeks. On the abiotic surface, a linear increase in the thickness of the leached layer has been observed (e.g. 85 Å of leached layer development over 8 weeks vs. 45 Å in 4 weeks). For the biologically colonized glasses, the quality of the reflectivity profiles has significantly diminished over time as the roughness of the surface has increased. However, although the reflectivity of the biologically-altered surfaces significantly diminishes at long reaction times due to increases in surface roughness, angle-dependent XANES data provides an alternate quantitative measure of the depth of propagation of a Fe-oxidation front (see below).

Fe (Fig. 2) and Mn, Ca, and Ti (not shown) fluorescence yield (FY) profiles were collected as function of incident angle in a grazing-incidence geometry to determine the vertical distribution of these elements from the bulk glass through the reacted surface layer (e.g. Fig. 2). When the incident angle is below the critical angle for total external reflection, the X-ray penetration depth within the substrate varies between 30 and 150 Å. Past the critical angle the penetration depth increases dramatically, and the resulting FY is representative of the bulk composition (Fig. 3). Model fits to these FY profiles provide information about the distribution of Mn and Fe in the interfacial region, in particular the removal of Fe and Mn from the surface layer and their potential reaccumulation in overlying phases. Qualitatively, in Figure 2, the reduced Fe FY between 0.25 and 0.5 degrees is indicative of a decreased Fe concentration in the surface layer relative to the bulk concentration. FY profiles for Mn show that Mn (present as Mn^{II} in the bulk glass) is highly soluble

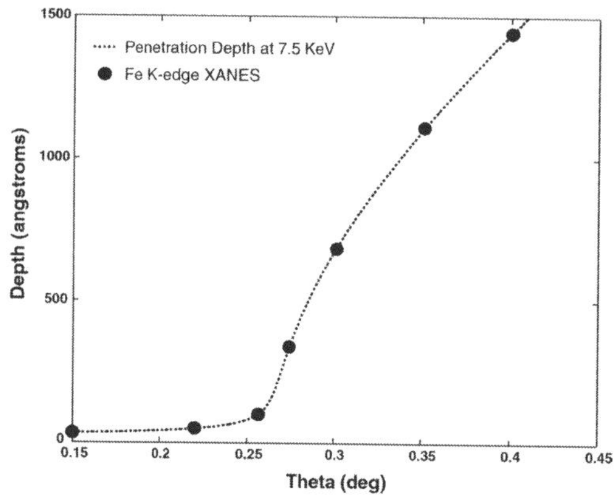

Figure 3. Calculated X-ray penetration depth as a function of angle, which can be controlled with a resolution of ~10 Å. Low angle (<0.25°) measurements are highly surface-sensitive and above the critical angle, the penetration depth rapidly increases. By collecting XANES at a variety of angles, we are able to differentiate between the speciation of Fe and Mn at the glass surface, within the leached layers, and in the bulk.

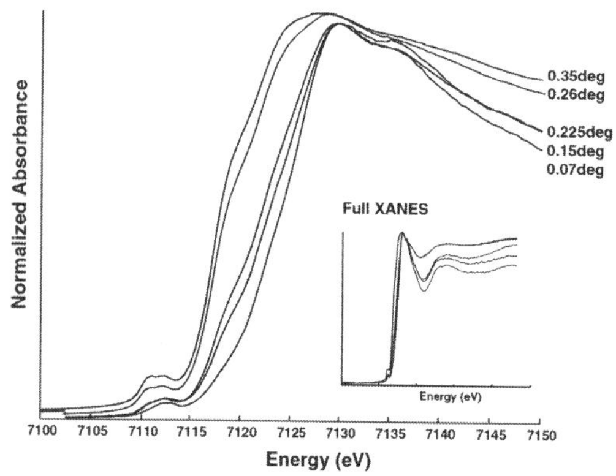

Figure 4. Fe K-edge XANES data collected at several angles. The bulk glass is Fe^{II} dominated (e.g. 0.35°), whereas at surface-sensitive angles (e.g. 0.15°), the XANES spectra are shifted to higher energy because a large fraction of Fe in the leached layer & above the surface has been oxidized to Fe^{III}.

and almost entirely removed from the reacted layer (data not shown).

Fe and Mn K-edge XANES measurements were performed at selected incident angles to determine the oxidation state of these species with varying depth sensitivity. Figure 4 shows an example of the angle-dependent Fe XANES for a Fe-oxidizer colonized glass surface. At high angles (e.g. 0.4°), the XANES is dominated by the bulk glass and the low energy of the edge position demonstrates that the bulk glass is

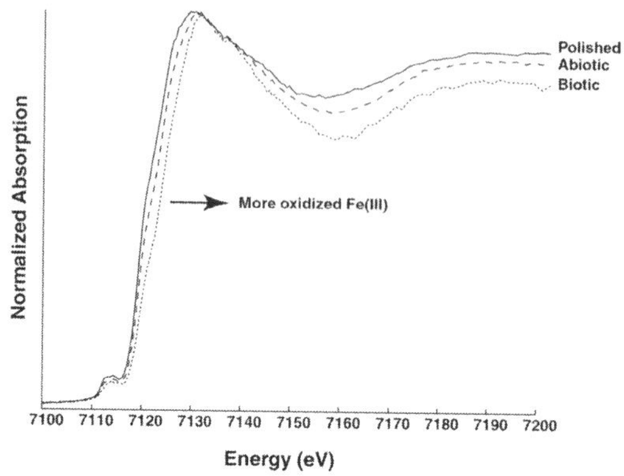

Figure 5. Fe K-edge XANES for unreacted glass vs. abiotic- and biologically-altered glasses reacted with artificial seawater for 4 weeks. All spectra were collected at 0.25 degrees, which is sensitive to Fe speciation in the leached layer.

95% Fe^{II}. The Fe^{II}/Fe^{III} ratio has been calibrated by comparison to XANES data for Fe^{II} model compounds and Mossbauer spectroscopy. In contrast, the Fe XANES data are shifted to higher energies at low angles due to oxidation in the leached layer. A surface layer that is fully oxidized has formed, and the XANES data for the entire leached layer can be quantitatively fit by linear-combinations of the spectra of surface Fe^{III} vs. bulk Fe^{II}.

For all time points, the Fe^{III} fraction is greater at each angle (until reaching bulk at 0.4°) for the biologically vs. abiotically weathered surfaces. An example of this is shown in Figure 5, where Fe XANES data collected at 0.25 degrees for both the biotic vs. abiotic sample (4-week reaction) show significantly greater oxidation (i.e. shift of the edge position to higher energy) for the biological surface, consistent with the formation of a leached layer that is 3-fold deeper (e.g. 145 vs. 45 Å). Mn K-edge XANES are dominated by Mn^{II} for the bulk glasses, and cannot be obtained for the leached layer (biotic and abiotic samples) due to full loss of Mn to solution during dissolution. However, the low angle XANES data are sensitive to the accumulation of oxidized Mn-phases on the surface over time.

4 DISCUSSION

The combined scattering and spectroscopic data provide quantitative information on the element distribution (changes in composition, thickness and density of reacted layers) and metal ion speciation at a complex interface at a scale that cannot be obtained through other geochemical techniques, leading to new insights regarding the mechanisms of abiotic vs. biological oxidative glass dissolution as a function of time. The experiments require high X-ray flux in a tightly collimated, low angular divergence beam, preferably an undulator source, and require precise sample orientation (milli-degree angular resolution) with the ability to simultaneously collect high resolution X-ray reflectivity data. Fortunately, the measurements can be made at ambient environmental conditions in a humid atmosphere, or potentially in the presence of bulk water, rather than in UHV conditions which could induce alteration of the hydrated surface layers. Presently, the reflectivity data and XANES spectra can be robustly modeled and the methods for interpreting the element distribution profiles are currently in development. Systematic comparison of the glass dissolution induced by Fe-oxidizer biofilms vs. abiotic reaction with artificial seawater shows significantly enhanced rates of leaching, Mn remobilization, Fe oxidization and secondary mineral formation during biologically-mediated dissolution.

REFERENCES

Edwards, K.J., Rogers, D.R., Wirsen, C.O. & McCollom, T.M. 2003. Isolation and characterization of novel pychrophillic, neutrophillic α- and γ-Proteobacteria isolated from the deep sea. *Appl. Env. Micro.* 69: 2906.

Emerson, D. & Moyer, C. 2002. Neutrophillic Fe-oxidizing bacteria are abundant at Loihi Seamount hydrothermal vents and play a major role in Fe-oxide deposition. *Appl. Env. Micro.* 68: 3085.

Fisk, M.R., Giovannoni, S.J. & Thorseth, I.H. 1998. Alteration of oceanic volcanic glass: textural evidence of microbial activity. *Science* 281: 978.

Furnes, H. & Staudigel, H. 1999. Biological mediation in ocean crust alteration: how deep is the deep biosphere? *Earth Plan. Sci. Lett.* 166: 97.

Thorseth, I., Torsvik, T., Furnes, H. & Muehlenbachs, K. 1995. Microbes play an important role in the alteration of oceanic crust. *Chem. Geol.* 126: 139.

Water–Rock Interaction, Wanty & Seal II (eds)
© 2004 Taylor & Francis Group, London, ISBN 90 5809 641 6

Solid-phase As speciation in roaster-derived Au mine tailings

S.R. Walker, H.E. Jamieson & C.F. Andrade
Queen's University, Kingston, Ontario, Canada

A. Lanzirotti
Consortium for Advanced Radiation Sources, Brookhaven National Lab, Upton, NY, USA

ABSTRACT: Solid-phase speciation of As in soils, sediments, and mine waste is an important factor that influences the concentration and speciation of dissolved As. Anthropogenic As-bearing Fe oxides can be released into the environment through roasting and smelting of sulfide ores. There is presently little understanding of how As is bound in such Fe oxides. Individual grain analyses of As-bearing Fe oxides from a metallurgical Au roasting operation were conducted to elucidate the solid-phase As speciation. Petrographic examination and micro-analytical techniques, including EMPA, μXANES and μXRD, has confirmed that roasting the sulfide ore produces relic grains of maghemite, hematite, and in some cases a mixture of maghemite and hematite. The grains are nanocrystalline, typically less than 50 μm in diameter, and exhibit complex morphologies ranging from concentric to microporous textures. As is present in mixed oxidation states (As^{III} and As^{V}) in both fresh mill calcine residue and 50-year-old subaerial shoreline tailings.

1 INTRODUCTION

The mobility and toxicity of As depend on its speciation. In the dissolved phase, arsenic is most often present as arsenate ($H_2AsO_4^-$ or $HAsO_4^{2-}$) or arsenite ($H_3AsO_3^0$), with arsenate being the more oxidized and thermodynamically stable form at ambient conditions. However, nonequilibrium behavior of this redox couple is well documented (Inskeep et al. 2002) and both reduced and oxidized forms commonly occur together. This is important because arsenite is generally regarded as more mobile and more toxic (Naqvi et al. 1994).

The concentrations and specific dissolved species of As present in porewater and groundwater are a measure of As mobility. However, it is the solid components of the sediments, soils, and mine waste that provide the biogeochemical substrate that controls As partitioning to the aqueous phase. Therefore, understanding how As is bound to or within the solids (i.e. solid-phase speciation) is critical for understanding the mechanisms that attenuate or release As. In addition, prediction of the stability and mobility of As under changing biogeochemical conditions (e.g. shifts in pH and/or Eh) may be possible with adequate characterization of the solid phases.

Previous studies and this study confirm the presence of anthropogenic metallurgical As-bearing iron oxides (Fig. 1) in the environment (soils, sediments, tailings, and house dust) adjacent to sulfide smelting and roasting operations (Walker et al. in prep., Davis et al. 1996, McCreadie et al. 2000). However, there is

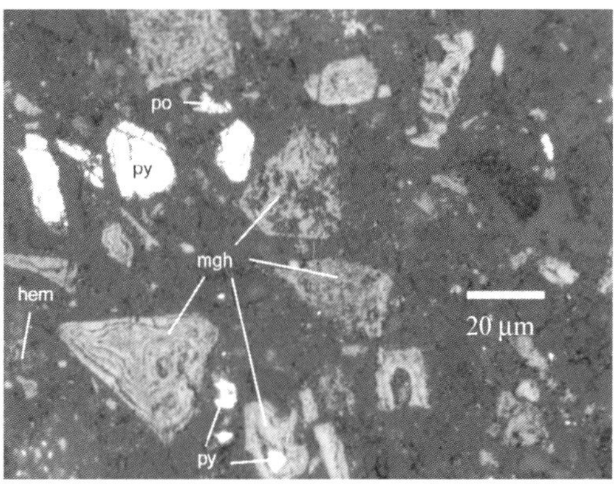

Figure 1. Reflected light photomicrograph of Giant mine calcine residue (roasted sulfide concentrate). Grey grains are arsenic-bearing roaster iron oxides exhibiting microporous and concentric textures. Roaster-derived iron oxides are mostly maghemite (mgh) with some hematite (hem). White grains are relic sulfides including pyrite (py) and pyrrhotite (po).

presently a very limited understanding of the solid-phase As speciation in these iron oxides.

The present study investigates the solid-phase speciation of As associated with roaster-derived Fe oxides produced at the former Giant Mine mill in Yellowknife, NWT, Canada. We have combined powder X-ray diffraction (XRD) and detailed petrographic observation with discrete grain analysis using several micro-beam analytical techniques.

As-bearing minerals exhibit a wide range in oxidation state; nominally – I for sulfides (e.g. arsenopyrite), 0 for As metal, III for arsenites (e.g. As trioxide), and V for arsenates (e.g. scorodite). As of different oxidation states can also be incorporated in mineral phases (e.g. As^{-I} in pyrite (FeS_2) and As^{V} in jarosite ($KFe_3(SO_4)_2(OH)_6$), or sorbed to mineral surfaces (e.g. clay minerals, oxyhydroxides of Fe, Mn and Al, and possibly sulfides).

2 METHODS

2.1 *Samples*

Materials studied include cyanided roaster calcine residue, and a calcine rich layer from shoreline tailings deposited at Yellowknife Bay adjacent to the mine around 1950. Polished thin sections were prepared as a grain mount (mill calcine, Fig. 1), or as intact specimens (shoreline tailings, Fig. 2) by impregnating air-dried samples with room temperature set epoxy. Care was taken to avoid the use of heat or water, and oil was used for all grinding and polishing. Epoxy impregnated samples were cut, ground flat, polished, and fixed to a glass slide using Krazy® glue, before final polishing. Prior to synchrotron analysis, thin sections were immersed in a shallow acetone bath until complete thin section/slide separation had occurred (1 to 2 hours). The detached sample was dried, oriented, and carefully pressed onto Kapton® tape and placed into a 35 mm cardboard slide holder for analysis.

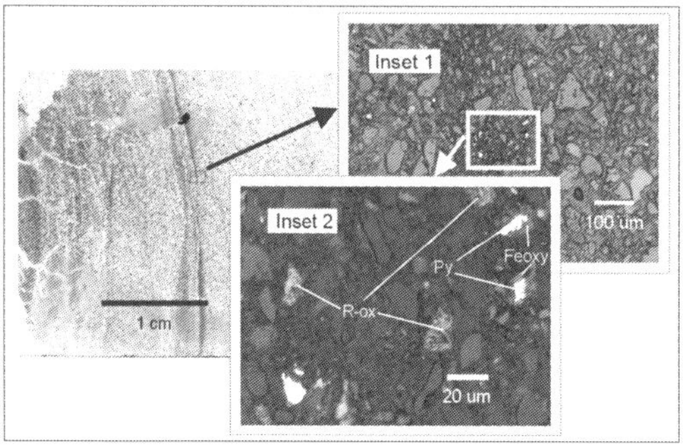

Figure 2. Thin section scan of 50 year old stratified shoreline tailings. Insets 1 and 2 are reflected light photomicrographs, which provide progressive magnification of one of the darker silty layers. Darker layers are calcine-rich and contain As-bearing roaster Fe oxides (R-ox, Inset 2). Two weathered pyrite (py) grains with As-bearing Fe oxyhydroxide rims (Feoxy) are identified for comparison.

2.2 *Analysis*

Thin sections were inspected and photographed under transmitted and reflected light to identify potential arsenic-bearing phases. Target locations were catalogued using coordinates from a thin section scanned image, and petrographic photographs at multiple scales (Fig. 2), to ensure that the small grains (generally <50 μm diameter) could be relocated.

Synchrotron micro-X-ray absorption near edge structure (μXANES) was used to obtain the oxidation state of arsenic in target grains. Synchrotron X-ray micro-diffraction (μXRD) was used to identify finely crystalline phases associated with each μXANES target. Experiments were conducted at beamline X26A, National Synchrotron Light Source. A 350 μm collimated monochromatic beam is focused to 10 μm in diameter using Kirkpatrick-Baez mirrors. The monochromatic beam is achieved using a Si(111) channel-cut monochromator. All μXANES experiments are collected in fluorescence mode by scanning through the As K edge using a Si(Li) detector, and μXRD experiments are imaged in 2D, in transmission mode using a Bruker SMART 1500 CCD diffractometer.

XANES analysis is conducted by scanning across the edge region (11,830 to 11,970 eV) in segments with a 2- to 4-second dwell time by increment, and the finest step size of 0.4 eV over the As K edge (11,850 to 11,890 eV). Standards included: arsenopyrite (FeAsS), As metal (As), reagent grade As trioxide (As_2O_3), and scorodite ($FeAsO_4 \cdot 2H_2O$). To manage self-absorption effects, finely ground standard material was mixed with boron nitride to obtain a total diluted As concentration of approximately 4% (w/w). Each mixed standard and boron nitride was dispersed as a thin layer onto Kapton® tape for analysis. Linear combination fits of mixed oxidation state spectra were completed using standard spectra and WinXAS™ software.

For μXRD, the incident X-ray beam was tuned to 11,950 eV (1.0375 Å). The distance between the camera and the thin section was fixed at approximately 17 cm yielding a calibrated d-spacing range of between 2 and 10 Å. Detector calibration was done against SRM 674a diffraction standard α-Al_2O_3. Calibrations and corrections for detector distortions (camera-sample distance, the camera tilt and rotation, and the beam center on the camera plane) were made using Fit2D™ software.

Total As in grains was determined using electron microprobe analysis (EMPA) in quantitative energy dispersive mode with suitable standards. Bulk XRD was also completed using conventional techniques.

3 RESULTS & DISCUSSION

Petrographic characterization of both the mill calcine and the shoreline tailings confirm the presence of roaster-derived Fe oxides with concentric and micro-porous textures (Fig. 1). Some grains exhibit bright red internal reflections characteristic of hematite, but definitive identification of maghemite or hematite is problematic.

3.1 *Bulk XANES & XRD*

Mixed arsenic oxidation states including As^{-1}, As^{III} and As^{V} (9%, 35%, and 56%, respectively, by linear combination fits of standard spectra) were identified by XANES analysis of bulk powder material of the mill calcine (data not shown).

Powder XRD of samples of the Giant mill calcine and the calcine layer in historic mine tailings confirm the presence of the roaster-derived maghemite. However, no As minerals were detected even though the As content can be relatively high (e.g. 2.5 wt.% in mill calcine). The identification of minor phases in bulk XRD is hindered by the multiplicity of peaks from major phases including: quartz, chlorite, muscovite, and dolomite and minor-phases including calcite, and plagioclase.

The As^{-1} detected by XANES can be attributed to relic arsenical pyrite and arsenopyrite in the calcine. However, based on petrographic observation, bulk XRD, and bulk XANES, the As^{III} and As^{V} cannot be attributed to a specific phase or phases.

3.2 *EMPA*

As content of the roaster Fe oxide grains by EMPA range from <0.5 to 7.6 wt.% As (Table 1) with higher concentrations of As in the recent mill calcine than the shoreline tailings.

Table 1. As content of roaster iron oxides by EMPA.

	Calcine residue wt.% As (n_g/n_a)*	Shoreline tailings wt.% As (n_g/n_a)*
Average	2.9 (27/63)	1.3 (15/32)
Maximum	7.6	2.9
Minimum	<0.5	<0.5
Standard Deviation	1.7	0.6

* n_g – number of grains, n_a – total number of analyses.

3.3 *μXANES and μXRD*

As in mixed oxidation states is also detected within individual grain analyses of roaster Fe oxides (Figs. 3a & b). As^{III} and As^{V} with or without a small (<10%) As As^{-1} component is consistently observed within individual μXANES analyses (e.g. Fig. 3b). The As^{III} and As^{V} mixture is consistently evident in grains with concentric and micro-porous textures and with both high and low arsenic content. This applies to both fresh calcine and 50-year-old subaerial

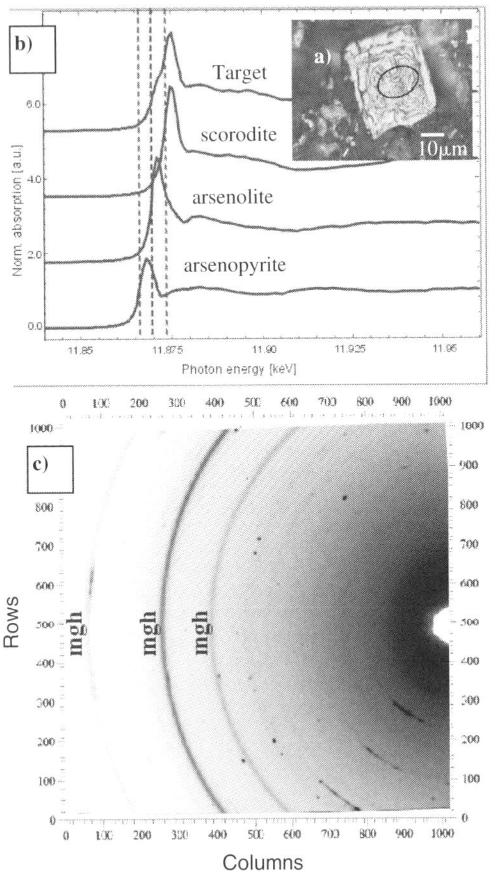

Figure 3. μXANES and μXRD results of grain in roaster calcine residue. (a) Photomicrograph of target grain. Ellipse indicates target beam location on grain. (b) XANES spectrum for grain in a) showing presence of at least two oxidation states (As^{III} and As^{V}). Three standard spectra are included for reference. Spectra are shifted vertically for presentation. Vertical dashed lines indicate first derivative edge position for each standard. (c) μXRD pattern for target in a) showing Debye-Scherrer rings that identify nanocrystalline maghemite (mgh). As content of this grain by EMPA is 4 wt.% As (rim) to 6 wt.% As (core).

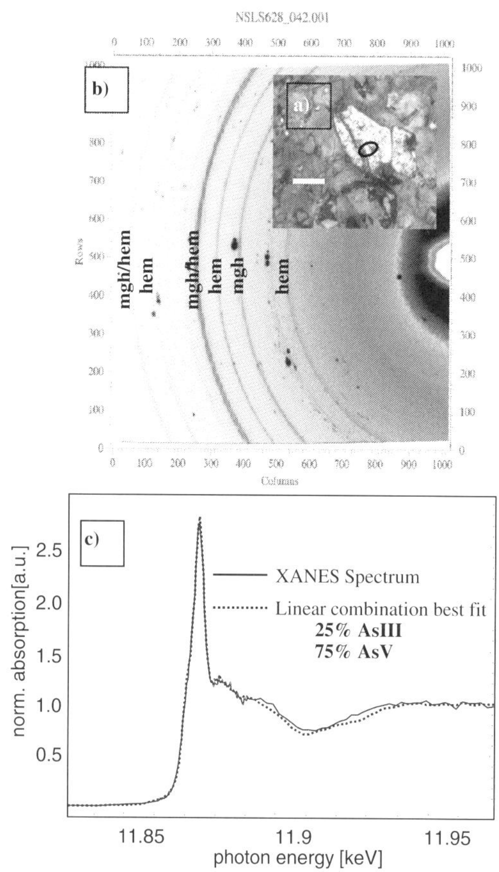

Figure 4. Analysis of roaster Fe oxide grain in 50-year-old shoreline tailings. (a) Two roaster Fe oxide grains. Ellipse indicates the target beam location on the larger grain. (b) μXRD pattern of target in (a) which identifies nanocrystalline maghemite (mgh) and hematite (hem). (c) μXANES spectrum of target in a). A best fit linear combination of standard spectra has determined 25% As^{III} and 75% As^{V}.

shoreline tailings. Variation in the ratio of As^{III} and As^{V} within the range of samples and grains analyzed is the subject of on-going work. Preliminary results suggest that the As^{III}/As^{V} ratio is lower in the shoreline tailings than fresh tailings. This could be explained by 50 years of exposure in the shoreline environment or by the different roasting conditions in the 1950s and 1990s.

Well-defined Debye-Scherrer (powder) patterns are observed for all μXRD images of roaster Fe oxides (e.g. Fig. 3c) indicating the average size of mineral crystallites is small relative to the incident beam diameter. Roaster Fe oxide grains are therefore nanometric polycrystalline particles. Individual crystallites at the beam (micron) scale under monochromatic X-rays would show only portions of reciprocal lattice nets (i.e. individual points, arcs or "spotty" rings representing discrete diffracted beams).

Maghemite is by far the most common phase identified by μXRD of roaster Fe oxide grains in both the calcine residue and shoreline tailings. Hematite is infrequently observed in both calcine residue and the shoreline tailings, but in a few grains μXRD indicates a mixture of hematite and maghemite (Fig. 4) at the micrometer scale.

No discrete As phases were identified by μXRD of roaster Fe oxide grains. The fact that only Fe oxides (maghemite and hematite) are detected by μXRD for grains that have relatively high As content (e.g. 4 to 6 wt. % As, Fig. 3a), suggests that the As is intimately associated with the Fe oxide, and not present as a discrete crystalline As-bearing phase. However, the possibility of a crystalline or amorphous nanometric phase or phases dispersed within the nanocrystalline maghemite (and/or hematite) remains a possibility. Highly soluble phases such as arsenolite are unlikely because the calcine at the Giant mill was subject to a wet regrind prior to cyanidation.

4 CONCLUSIONS

Grain by grain analysis using micro-analytical techniques has provided a better understanding of As speciation in roaster-derived mine waste. It appears that the As^{III} and As^{V} observed in bulk calcine material is largely related to the mixed oxidation state of As in discrete nanocrystalline roaster Fe oxide grains, and not due to separate As^{III} and As^{V} phases at the grain scale. Additional work is necessary to confirm the specific binding mechanisms of the As^{III} and As^{V} in roaster-derived Fe oxides.

Maghemite is the most common roaster Fe oxide found in both young and old calcine rich tailings examined in this study. Hematite is infrequent, but it can be present mixed with maghemite within individual grains in both fresh and 50-year old tailings.

As occurs in a mixed oxidation state (As^{III} and As^{V}) in all roaster Fe oxides examined. Conditions that destabilize these phases (e.g. biologically mediated reductive dissolution) are likely to lead to the direct release of both arsenite and arsenate to the aqueous phase.

Maghemite is generally regarded as thermodynamically metastable, and As^{III} phases tend to have high solubility and will be prone to oxidation. However, the persistence of As^{III} in roaster-derived Fe oxides in shallow subaerial shoreline tailings for over 50 years, suggests that the As-bearing nanocrystalline maghemite is relatively stable under these conditions.

REFERENCES

Davis, A., Ruby, M.V., Bloom, M., Schoof, R., Freeman, G. & Bergstrom, P.D. 1996. Mineralogic constraints on the bioavailability of arsenic in smelter-impacted soils. *Environ. Sci. Technol.* 30: 392–399.

Inskeep, W.P., McDermott, T.R. & Fendorf, S. 2002. Arsenic (V)/(III) cycling in soils and natural waters: chemical and microbiological processes. In William T. Frankenberger, Jr. (ed.), *Environmental Chemistry of Arsenic*: 183–215. New York: Marcel Dekker.

McCreadie, H., Blowes, D.W., Ptacek, C.J. & Jambor, J.L. 2000. Influence of reduction reactions and solid-phase composition on porewater concentrations of arsenic. *Environ. Sci. Technol.* 34: 3159–3166.

Naqvi, S.M., Vaishnavi C. & Singh H. 1994. Toxicity and metabolism of arsenic in vertebrates. In J.O. Nriagu (ed.) *Arsenic in the Environment, Part II: Human Health and Ecosystem Effects*: 55–91. New York: John Wiley & Sons, Inc.

Walker, S.R., Jamieson, H.E., Andrade, C.A. & Lanzirotti, A. In prep. Determining arsenic speciation in iron oxides: Application of synchrotron micro-XRD and micro-XANES at the grain scale.

Water–Rock Interaction, Wanty & Seal II (eds)

Detection of a mass transport front in rocks by thermoluminescence technique

M. Yamamoto & N. Tsuchiya
Department of Environmental Studies, Tohoku University, Sendai, Japan

ABSTRACT: Two-dimensional thermoluminescence and spectroscopy of rock samples were measured, and applied a mathematical model for Fick's diffusions law. Samples were selected from the Kamaishi Mine including the hydrothermal veins and surrounding alteration zone. Results showed some differences between the altered and unaltered zones. Minerals in the altered zone showed stronger thermoluminescence than those in the unaltered zone, and then different phenomena for wavelength could be observed between them. Results of the mathematical model showed that the difference of dimensionless effective diffusion coefficient between the altered zone and unaltered zone is about 100 times. Therefore the possibility could be shown that thermoluminescence could be a chemical sensor that detects mass change (transport front in rocks) by using these thermoluminescence methods.

1 INTRODUCTION

1.1 *Background*

The intention of many nations to dispose of radioactive wastes in low-permeability formations has led to an increased interest in the behavior of contaminants in fractures (Steefel & Lichtner 1998b). In this kind of study, it is important to evaluate the multicomponent reactive transport not only in fractures but also into the rock matrix because the bulk rock matrix porosity is much larger than the fracture porosity. However, little research has been done on the transport of material in the rock matrix (Steefel & Lichtner 1998a). In an assessment of safety of the accumulation of toxic substances such as trichloromethane, diffusion into a rock matrix is also extremely important.

Hareyama et al. (1998) and Hareyama et al. (2000) described a relatively high natural radioactivity in an alteration halo compared to the unaltered rock matrix using the two-dimensional radiation imaging plate detector. These facts indicate enrichment of radionuclides, such as potassium, associated with the secondary minerals and diffusion of hydrothermal fluids containing various chemical species from the altered rock and through the rock matrix.

1.2 *Using thermoluminescence*

Thermoluminescence (TL) is a phenomenon that the mineral emits light when a rock is heated. This method has been applied to date archaeological and geological minerals and recently has been used in geothermal exploration by measuring TL distribution in a field (Tsuchiya et al. 2000).

TL now can be thought of as a geochemical sensor to evaluate mass transport in the rocks because structural changes by the heating or degassing cause distortion and increases the number of electron capture centers in a hydrothermaly altered rock. The TL phenomena of rock-forming minerals reflect their lattice defects and chemical impurities, so the origin and formation processes of such minerals could be estimated by the change in their TL characteristics (Tsuchiya & Nakatsuka 2002).

In addition, different spectra have been observed depending on the various types of structures and compositions of materials. Therefore, spectroscopic analysis of TL can play an important role in revealing the material transport as a physical or chemical sensor.

2 SAMPLE

The sample is the Kurihashi granodiorite from the Kamaishi Mine, Iwate Prefecture (Fig. 1) that shows both altered and unaltered zones around the vein. Figure 1 also shows the vein and its surroundings. XRD shows that the vein is mainly composed of zeolite, and the altered zone is mainly composed of quartz, K-feldspar and plagioclase (anorthite). The sample was cut out as a plate (vertical to the targeted vein); the size is $30 \times 30 \times 1$ mm.

3 TWO-DIMENTIONAL TL MEASUREMENT

3.1 *Equipment and conditions of measurement*

Figure 2 shows the outline of the equipment. Samples are heated from room temperature to 300°C (20°C/min) and the control images are recorded by the CCD camera linked to a personal computer.

3.2 *Results*

A sample image pictured in natural light and pictured in luminescence is shown on Figure 3. Sample A showed strong luminescence around the vein and sample B shows unsystematic quantity of luminescence thought to derive from some small veins besides the targeted vein.

3.3 *Discussion*

3.3.1 *Evaluation of TL distribution*

Tens of feldspar crystals are selected in one sample and for each the quantity of luminescence is plotted versus the distance from vein edge (Fig. 4). The distance from the left end of the sample image is the distance from the vein because the sample was cut vertical to the vein.

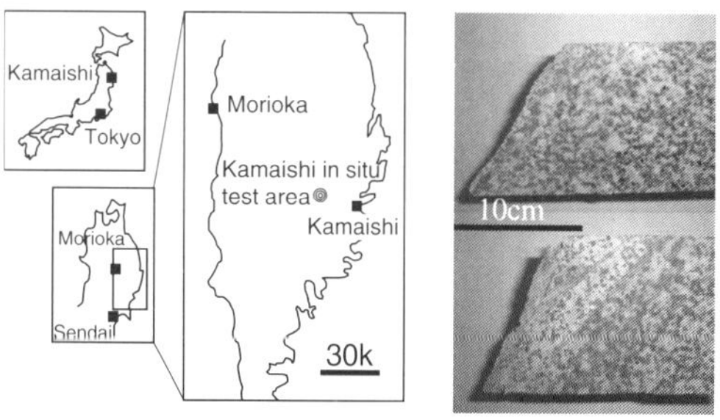

Figure 1. Location of the Kamaishi Mine and samples from the Kamaishi Mine. One main vein and altered zone are relatively clear in sample A (Top), whereas many small veins occur in sample B (Bottom).

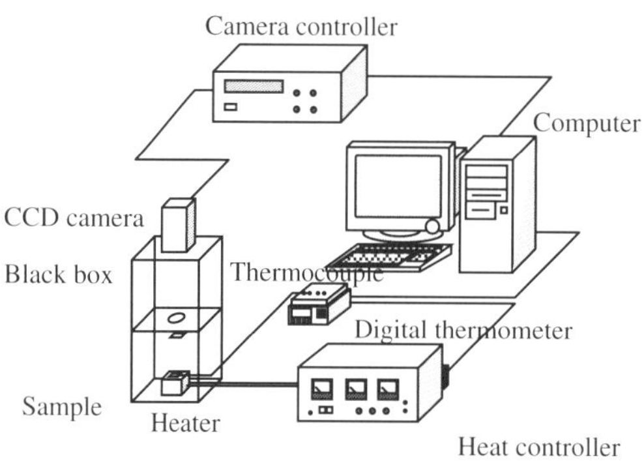

Figure 2. Two-dimensional TL measurement system using a highly sensitive CCD camera with an image intensifier and computer system. Distribution of TL from rock slices can be measured (30 × 30 mm area).

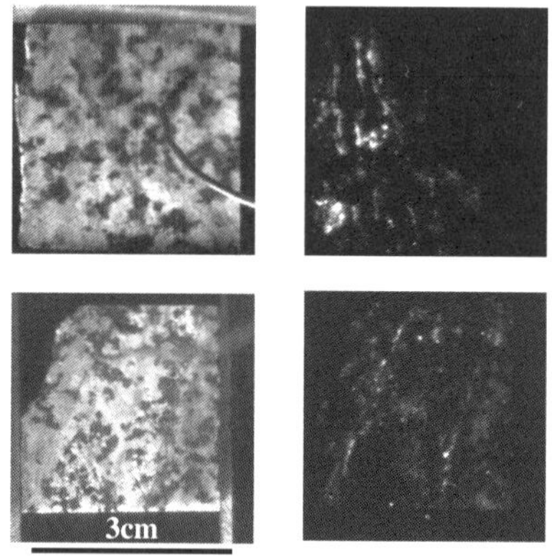

Figure 3. Results of two-dimensional TL measurement for sample A (Top) and sample B (Bottom). Images on left side are samples in natural light and images on right side are natural TL. Vein locates at the left end of the image and sample size is 30 × 30 × 1 mm.

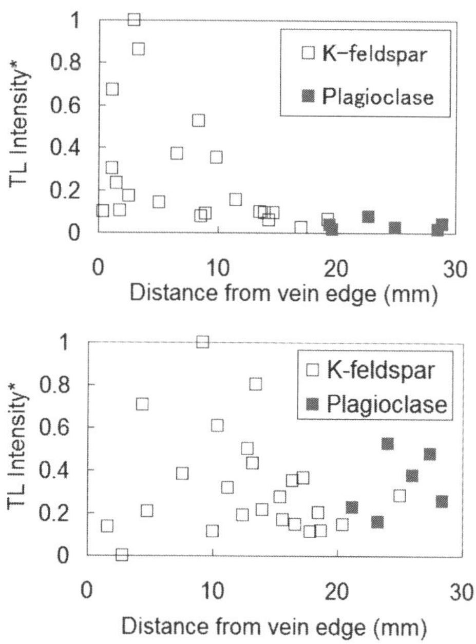

Figure 4. Relationship between TL intensity and distance from vein edge for sample A (top) and sample B (bottom). *TL intensity is normalized intensity by the maximum intensity.

In sample A, there are many points of feldspar that show strong TL around the vein, and TL decreases with increasing distance from the targeted vein. On the other hand, the intensity of TL tends to be unsystematic with distance from vein in the sample B. The reason may be that there are many small veins in addition to the targeted vein in sample B and material movement in this system was complicated compared to the single vein in sample A. It is difficult to evaluate the variation of TL intensity with distance from one targeted vein in sample B.

However, looking from another angle, the results of sample B proved that minerals with strong luminescence occur around the vein so this result supports that of sample A. Some phenomena influence the intensity of TL in the fractures and movement of material.

Moreover, the reason why the intensity of TL is low around the targeted vein in sample B may be that the degree of alteration is too much and crystal structures are disintegrated. Similar tendencies for K-feldspar were observed by Pinnioja et al. (1999).

3.3.2 *Fractures and rock matrix*

On Figure 5, one model can explain the process of the reactions by the fracture appearance and thermal water in the rock.

1 When a fracture appears in the rock, then small fractures or pores appear by the impact of fracture appearance. As a result, thermal water can easily infiltrate into rock matrix around the fracture, that is, the movement of mass is promoted.
2 When thermal water reacts with the rock matrix in a fracture, the number of pores increases at the point as dissolution reactions, then the rate of material transport becomes faster.

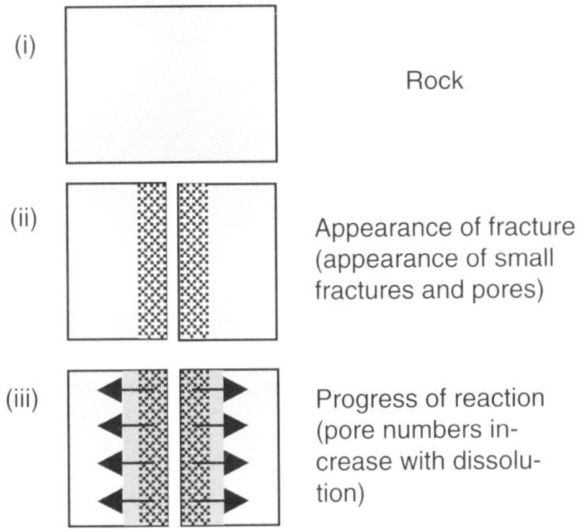

Figure 5. Fracture-rock matrix model. (ii) Small fractures and pores appear from the impact of forming big fracture. (iii) Pore numbers increase in the process of dissolution reactions. As a result, materials transport easily in the rock matrix.

These two reasons explain why material transports in the alteration zone become faster in the altered zone.

To confirm this model, $KMnO_{4aq}$ was dripped onto the sample heated in a vacuum to visualize infiltration into the rock and the results are shown in Figure 6. There are many more reddish-brown precipitates in the altered zone than in the unaltered zone as seen in photomicrographs.

3.3.3 *Mathematical model*

The mathematical formulation for multicomponent reaction in discrete fractures is described in detail in Steefel & Lichtner (1998a), who describe the conservation of solute mass in a single fracture. One equation can be used to describe the relationship between "mass transport" and "TL intensity" derived from Fick's diffusion law.

$$\frac{\partial c}{\partial t} = D_e \frac{\partial^2 c}{\partial x^2} \tag{1}$$

TL intensity is proportional to the number of unpaired electrons (Mckeever 1985) and it can be rationalized that the number of unpaired electrons is in proportion to the concentration of a material. Actually, there are reports that TL intensity should be greatly dependent on the impurity contents, including aluminum and rare earth elements (Hashimoto et al. 1993). Therefore it is assumed that TL intensity ($\equiv I_{TL}(x,t)$) is in proportion to the concentration of a material ($\equiv c$).

$$I_{TL} \propto c \tag{2}$$

The following equation is derived from equations 1 and 2:

$$\frac{\partial I_{TL}}{\partial t} = D'_e \frac{\partial^2 I_{TL}}{\partial x^2} \tag{3}$$

The initial condition is defined as:

$$I_{TL}(i,0) = 0 \tag{4}$$

And boundary conditions are:

$$I_{TL}(0,t) = 1 \tag{5}$$

$$I_{TL}(\infty,t) = 0 \tag{6}$$

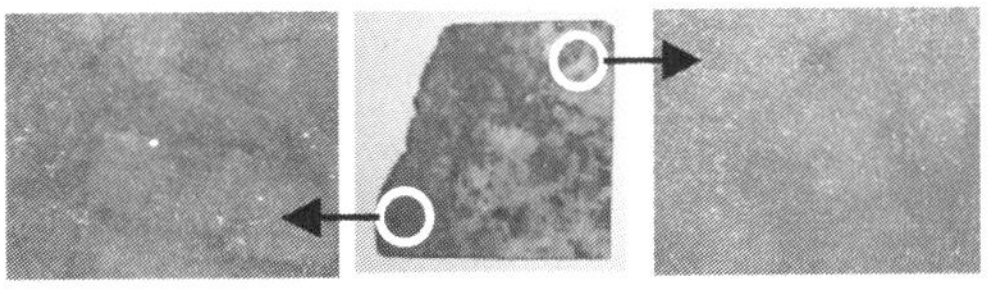

Figure 6. Observation of micro fractures and pores for altered zone (left) and fresh zone (right).

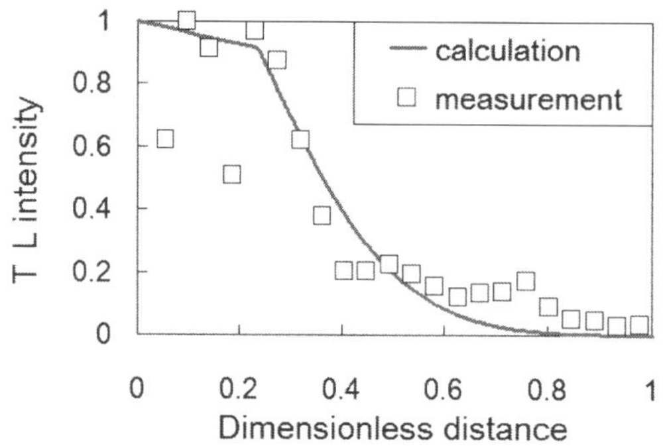

Figure 7. Calculation with diffusion model. The solid line shows the result of these calculations made using equation 3.

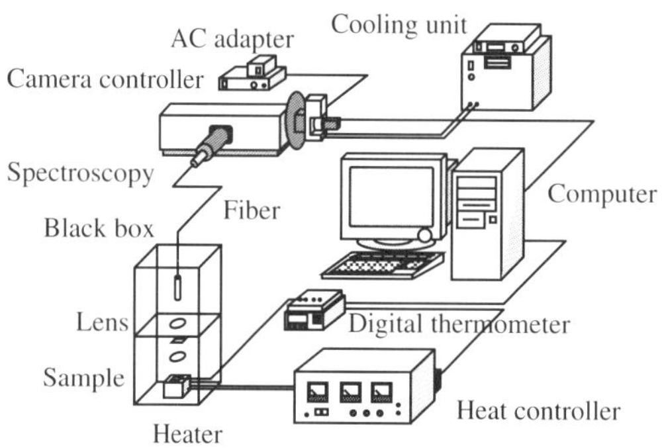

Figure 8. Spectral TL measurement system. Using a spectrometer coupled with a highly sensitive CCD camera. TL from one grain of a mineral can be measured by using two lenses and fiber.

All minerals were selected at the same distance from the vein in sample A for the calculations, and were evaluated by the integrated value of each TL intensity. Considering the differences for material moving quickly between altered and unaltered zones, the dimensionless position of the alteration front and dimensionless effective diffusion coefficient are varied as a parameter. The values of these parameters were deemed to be suitable, based upon the results of calculation on most of the graphs obtained from experiment. The result is shown in Figure 7.

From the result, the difference of dimensionless effective diffusion coefficients between the altered and unaltered zones is about 100 times.

4 SPECTROSCOPY MEASUREMENT OF TL

4.1 *Equipment and conditions of measurement*

Figure 8 outlines the equipment. Samples irradiated by ^{60}Co are heated from room temperature to 400°C

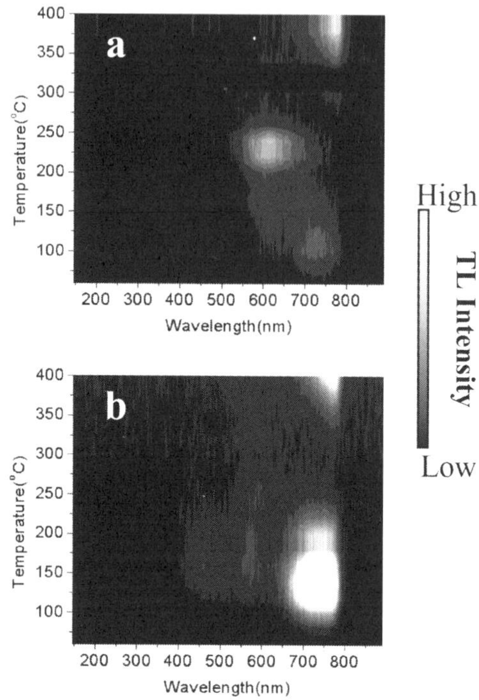

Figure 9. Results of spectral TL measurement for (a) altered feldspar, and (b) unaltered feldspar.

(40°C/min) and the images are recorded by the spectroscope linked to a personal computer.

4.2 *Results*

Results are shown in Figure 9 by three dimensions (temperature, wavelength and intensity). The values of wavelength in the altered and unaltered zones are shown for both sample A and sample B.

In the altered feldspar, the main luminescence is at 250°C, 600 nm (red TL). In the unaltered feldspar, the main luminescence is at 150°C, 730 nm (far red TL), also blue TL could be observed (175°C, 450 nm). The peak at around 400°C in both results is the black body radiation of the heater.

5 CONCLUSIONS

Two-dimensional TL and spectroscopy of TL were measured and evaluated around a vein that is a vestige of past hydrothermal activity.

The fracture resulted from the reaction of thermal water, in an alteration zone forming veins. In this process, an increase of the lattice defects and impurities around the vein affects the TL intensity with the impact of fracture appearance and material dissolution or addition according to reactions. Stronger TL occurred in the altered zone around veins and differences of spectra between the altered and unaltered zones were confirmed by spectroscopy measurement.

Therefore, it is possible that thermoluminescence could be a chemical sensor that detects mass change. TL intensity, the wavelength of TL detect the alteration front more certainly. However it is still necessary to do the some experiments for various samples with veins to understand TL behavior.

REFERENCES

Pinnojia, S. Siitari-Kauppi, M. & Lindberg, A. 1999. Effect of feldspar composition of thermoluminescence in minerals separated from food: *Rad. Phys. Chem.* 54: 505–516.

Steefel, C.I. & Lichtner, P.C. 1998a. Multicomponent reactive transport in discrete fractures: I. Controls on reaction front geometry. *J. Hydrol.* 209: 186–199.

Steefel, C.I. & Lichtner, P.C. 1998b. Multicomponent reactive transport in discrete fractures II: Infiltration of hyperalkaline groundwater at Maquarin, Jordan, a natural analogue site. *J. Hydrol.* 209: 200–224.

Hareyama, M. Tsuchiya, N. & Takebe, M. 1998. Two-dimensional measurement of natural radioactivity of rocks by photostimlated luminescence. *Proceedings Water-Rock Interaction-9*: 835–838.

Hareyama, M. Tsuchiya, N. Takebe, M. & Chida, T. 2000. Two-dimentional measurement of natural radioactivity of granitic rocks by photostimulated luminescence technique. *Geochem. J.* 34: 1–9.

Hashimoto, T. Kojima, N. Shirai, N. & Ichino, M. 1993. Activation energies from blue-and red-thermoluminescence (TL) of quartz grains and mean lives of trapped electrons related to natural red-TL. *Nucl. Tracks Radiat. Meas.* 21(2): 217–223.

Mckeever, S.W.S. 1985. *Thermoluminescence of Solids*. Cambridge University Press.

Tsuchiya, N. Suzuki, T. & Nakatsuka, K. 2000. Thermoluminescence as a new research tool for the evaluation of geothermal activity of the Kakkonnda geothermal system, northeast Japan. *Geothermics* 29: 27–50.

Tsuchiya, N. & Nakatsuka, K. 2002. Application of thermoluminescence technique to geothermal exploration. *Proceedings 24th NZ Geothermal Workshop 2002*: 85–90.

Water–Rock Interaction, Wanty & Seal II (eds)
© 2004 Taylor & Francis Group, London, ISBN 90 5809 641 6

Study of phosphate layers in Lake Baikal sediments by complex of local methods

S.M. Zhmodik, N.V. Verkhovtseva, N.A. Nemirovskaya & A.T. Titov
Institute of Geology of the SB RAS, Novosibirsk, Russia

E.L. Goldberg, M.A. Fedorin & O.M. Khlystov
Limnological Institute of the SB RAS, Irkutsk, Russia

A.G. Mironov & N.S. Karmanov
Geological Institute of the SB RAS, UlanUde, Russia

K.V. Zolotarev
The Budker Institute of Nuclear Physics of the SB RAS, Novosibirsk, Russia

ABSTRACT: The distribution of elements in sediment cores from Lake Baikal reflects the conditions of sediment formation, and thus can be used for palaeoclimatic reconstructions. The ability to identify cyclicity in sedimentation depends on analytical methods, especially the ability to determine the speciation of elements in sediments. Application of autoradiography in combination with SR-XFA, EMPA, SEM, and TEM provides new information on micro- and nanoscale particles, crystals, and layers in the sediments. A detailed study of elemental distributions in Lake Baikal (St 8, 11) sediment cores containing U-bearing phosphates was conducted using activation autoradiography (P, U) and SR-XFA. Results of this approach showed that phosphate formations characterized by anomalous concentrations of P, U, As, Fe, Ca, Sr that should be considered in interpretations of geochemical data.

1 INTRODUCTION

Interest in the study of deep-water sediments of Lake Baikal has grown considerably during last ten years because of the availability of data from lake sediments for palaeoclimatic reconstructions in Holocene-Pleistocene time and earlier periods. Recent investigations show that variations in concentrations of biogenic silica, organic carbon, uranium, thorium and other elements and their ratios provide some of the most sensitive indicators of climate change (Colman et al. 1995, Grachev et al. 1997). Estimation of the periodicity and duration of events by studies of cores of lake sediments depends on the availability of suitable analytical methods for resolving small-scale fluctuations of sediment composition. For example, analysis of a core with a step interval of 100 μm deposited under sedimentation rate of 2.63 cm per 1000 years provides resolution of elemental concentrations on an average scale of every 3.8 years. A critical component of such an analysis is consideration of the influence of terrigenous, diagenetic, and authigenic components. Therefore, it is necessary to use analytical methods that combine high sensitivity and spatial resolution and analyze large areas without destruction of samples.

Results of detailed study of the local distribution of elements in sediments of Baikal Lake containing phosphorite formations are reported. U and P were studied by the activation autoradiography method; the distribution of U, As, Fe, Ca, Sr, Y, Ti, Mn and other elements were studied by SR-XFA (synchrotron radiation X-ray fluorescence analysis) and scanning electron microscopy (SEM). Geochemical features of phosphates in sediments of Baikal Lake were revealed. This study shows that the autoradiography method, in combination with SR-XFA and SEM, is an efficient, non-destructive technique for studying element concentrations and distributions in thin layers (100 μm).

2 SAMPLES AND METHODS

2.1 *Samples*

The sections of sediment cores from Academician Ridge of Baikal Lake at station 8: 53°32′15″E; 107°56′25″N (interval 181.8–234.5 cm) and station 11: 53°33′51″E; 108°00′05″N (interval 88–125 cm) were analyzed. The cores were obtained using a method that allowed extraction of samples with minimal disturbance so that primary structures are preserved. A 50 × 30 × 10 mm slab was cut and then cemented by epoxy. Polished thin sections were made from the central part of the slab. The cores from stations 8 and 11 have good correlations with other sediments of Academician Ridge in terms of primary textures and diatomaceous contents. The intervals studied record a gradual transition from glacial clays to diatomaceous silts. Approximately 30–40% of the diatomaceous horizons consist of diatomaceous algae related to oceanic stage 5a (maximum ~80,000 years; Colman et al. 1995) or to the Kazancevsky interglacial period (Gavshin et al. 1998). In the cores from depths of 219–221 cm and 111 cm, layers and brown intervals enriched in phosphatic Fe oxyhydroxide of material. These intervals were studied by the autoradiography method and SR-XFA.

2.2 *Methods*

Study of the spatial distribution of elements in lake sediments was conducted using the method of activation autoradiography. The distribution of U was studied by neutron-fission ^{235}U (n,f)-autoradiography; P was studied by (n,beta)-autoradiography. Samples were prepared as thin (60 μm) polished plates fixed on a quartz glass support. The samples were irradiated with a neutron flux of $1 \cdot 10^{16}$ neutrons/cm^2 (the irradiation neutron flux gradient in the channel was less than 1–3%). Etching of the detector was performed in 20% HF at 20°C. The track detector used to record fission fragments was synthetic fluorphlogopite. The distribution of the fission tracks was studied using optical microscopy (JENAVAL, Carl Zeiss and Olympus BH-2 with digital Olympus DP12). Natural glasses of basic and andesitic composition with uranium content from 0.2 ± 0.01 ppm U; 1.96 ± 0.1 ppm U and 2.55 ± 0.13 ppm U were used as standards.

The distribution of P was studied by neutron activation autoradiography via $^{31}P(n, \beta)^{32}P$. ^{32}P is a β emitter (100%) with a half-life of 14.5 days and a maximum energy of the beta particles of 1.707 MeV. The β autoradiograms were recorded on Kodak BioMax MS film with exposure times of 1.5 to 14 hours. A series of autoradiograms was obtained at different periods of time after irradiation of the samples. During neutron irradiation a large number of elements (such as Fe, Co, Mn, Cr, Sc and P) were activated. However, their low specific activities in combination with the low concentrations of these elements allowed the assumption that the autoradiograph represented only ^{32}P β rays, from which the areal distribution of phosphorus in the sediments could be obtained.

SR-XFA was conducted at scanning station "Baikal" in The Budker Institute of Nuclear Physics, Novosibirsk, Russia. For SR-XFA scans, the sample was moved vertically across the beam in 0.2 mm steps; step accuracy is on the order of 0.01 mm. Collimation of the beam in these experiments was accomplished by monochromator slits that were 0.2 mm high and 2–5 mm wide. Thus, in the entire sample was exposed to elemental analysis.

3 RESULTS AND DISCUSSION

Figure 1 shows the distribution of a set of elements in two sections of core from St 8 and St 11. Abnormally high concentrations of uranium and phosphorus are observed in these intervals by (n,f) and (n,beta) autoradiography. The anomalous zones occur as thin layers and layered concretions. On autoradiographic images, the detailed structure of uranium-bearing phosphates can be seen as thin layers due to variations of U and P concentrations. Optical and scanning electron microscopy, electronography, and microprobe analysis showed that phosphatic material is Ca-deficient hydroxylapatite, which in most cases replaces internal parts of diatoms. Application of SR-XFA allowed not only confirmation of the existence of high uranium concentrations (up to 100 ppm) in phosphorite formations, but also identified the group of elements accumulating in phosphate (Fig. 1). Cluster and factor analyses indicate an association of elements such as U, As, Y, Ca, Fe, and Sr (Figs. 2 & 3). Concentrations of these elements in phosphate exceed typical background concentrations for argillaceous parts of sediment by a factor of 3 to 30. In contrast, phosphorite concentrations of K, Th, Ti and other elements in are similar to typical diatomaceous sediment due to the presence of clay particles.

The SEM data show that surfaces of diatoms consist of SiO_2 with insignificant admixtures of Na, Mg, Al, K, Ca, Fe (Fig. 4). Phosphatic material is tightly connected with diatoms such that CaO/P_2O_5 ratios vary from 0.66 to 1.23 with CaO concentrations varying from 8.86 to 19–20 wt% with a maximum of 42.06 wt% P_2O_5. Electron-diffraction patterns are characterized by diffuse rings with d/n close to 3.4:2.7:2.3. Experiments and thermodynamic calculations indicate that hydroxylapatite cannot be formed in seawater and fresh water of Baikal

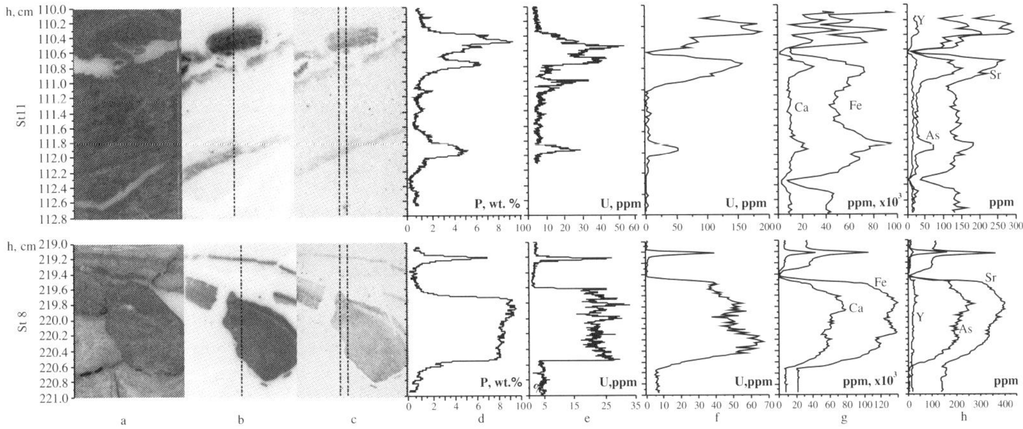

Figure 1. Autoradiographs and distribution profiles of elements in sediment cores from the Academician Ridge of Lake Baikal: (a) thin section; (b) (n, β) – autoradiograph; (c) (n, f) – autoradiograph; (d) phosphorus distribution by the autoradiography method; (e) uranium distribution by the autoradiography method; (f) U distribution by SR-XFA; (g) Ca, Fe distribution by SR-XFA; (h) Y, Sr, As distribution by SR-XFA. Sections of sediment cores consist of clay-diatomaceous material with phosphate layers (upper row) and predominantly diatomaceous material with layers and concretions of phosphate; h, cm – depth below sediment surface.

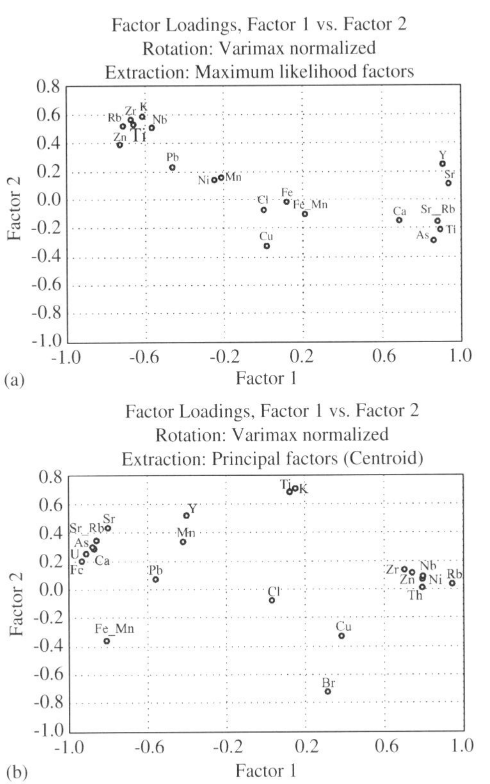

Figure 2. Factor analysis for the cores of lake sediments of (a) Station 11, (b) Station 8.

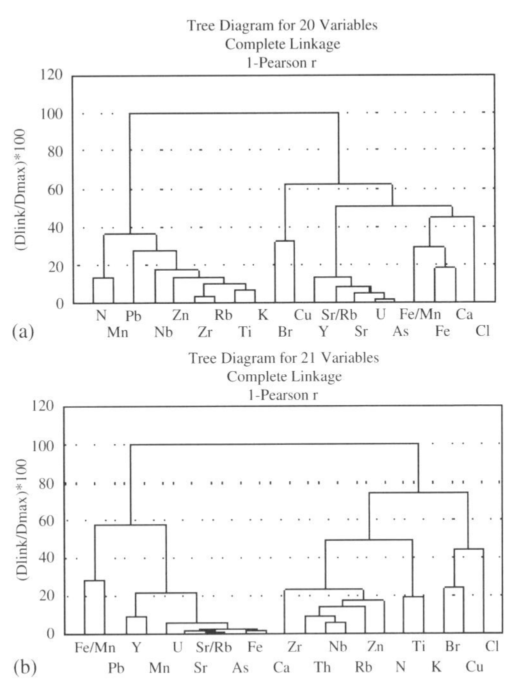

Figure 3. Cluster analysis for the cores of lake sediments of (a) Station 11, (b) Station 8.

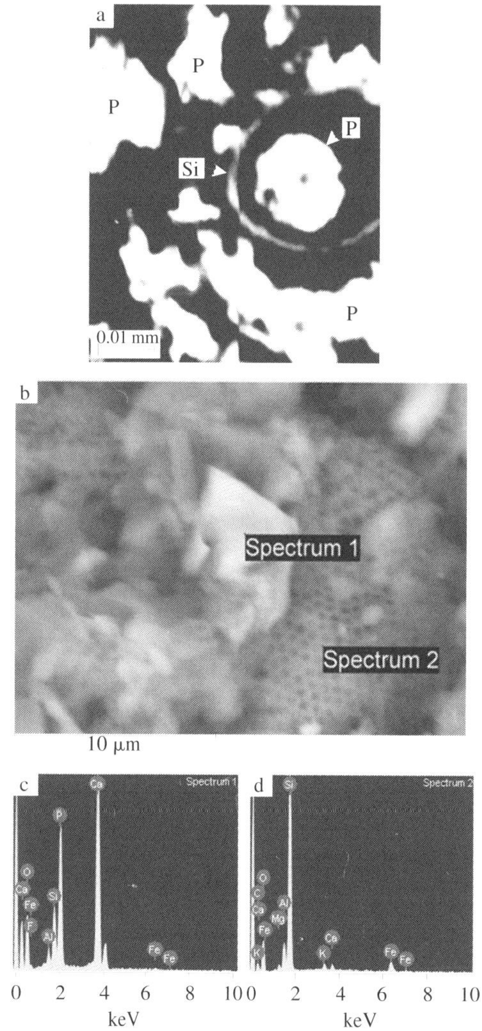

Figure 4. Morphology of phosphatized diatomaceous algae and phosphate grains in a phosphate layer from sediments of Lake Baikal. a – perpendicular section of diatom and clastogene phosphate grains. Covering of diatoms is siliceous (Si); internal composition is calcium phosphate (P). There is a cavity (black) between internal part and external covering of diatom; b – destroyed fragments of phosphatized diatoms with power-dispersion spectra (c, d); SEM "Jeol" and SEM LEO 1430VP EDX Oxford.

(Golubev 2001, Kashik et al. 1996). However, the presence of organic material (albumin) in solution results in precipitation of hydroxylapatite (data of A.T. Titov). In our case, the association of calcium-deficient hydroxylapatite with diatomaceous algae confirms the essential role of organic material for formation of hydroxylapatite from low-temperature solutions. Localization of phosphate inclusions inside silicate coatings allows us to speculate about the initial formation of microcrystalline nuclei of calcium phosphate due to cellular growth of diatoms. The calcium-deficient hydroxylapatite represents an initial stage of crystallization of apatite with well developed surfaces and vacancies (deficient of calcium) that characterizes it as a strong sorbent Authigenic phosphorites in sediments of Lake Baikal concentrator uranium and other attendant elements, which, obviously, form layers distributed over large areas.

4 CONCLUSIONS

Autoradiography and SR-XFA methods allow the study of elemental distributions in cores of lake sediments over intervals with steps of 100 μm and less. This fine-scale resolution makes it possible to obtain compositional information on time scales on the order of a few years, taking into account that the minimal analytical step for (n,f)-autoradiography is close to 10 μm (Zhmodik et al. 2003). Detailed study of lake sediments by autoradiography and SR-XFA methods allow us to determine the nature of abnormal elemental concentrations in sediments.

ACKNOWLEDGEMENTS

This work was supported by RFBR, grants 01–05-97258, 03-05-64266, 02-05-64756 and grants of Presidium of the SB RAS 6.4.1; 121; 170. We would like to express our thanks to Professor M.A. Grachev for his consultations and support to this work.

REFERENCES

Colman, S.M., Peck, J.A., Karabanov, E.B., Carter, S.J., Bradbery, J.B., King, J.W. & Williams, D.F. 1995. Continental climate response to orbital forcing from biogenic silica records in Lake Baikal. *Nature* 378: 769–771.

Gavshin, V.M., Arkhipov, S.A., Bobrov, V.A., Melgunov, M. S., Makarova, I.V., Sukhorukov, F.V., Stepin, A.S., Khlystov, O.M., Kireev, A.D. & Budashkina, V.V. 1998. Distribution of natural radioactive elements in Holocene – Pleistocene deep-water sediments of Lake Baikal and paleoclimate reconstruction. *Geol. Geophys* 39: 1045–1058.

Grachev, M.A., Likhoshvai, E.V., Vorobyova, S.S., Khlystov, O.M., Bezrukova, E.V., Veinberg, E.V. & Goldberg, E.L. 1997. Signals of upper Pleistocene paleoclimates in Lake Baikal sediments. *Geol. Geophys* 38: 957–980.

Golubev, S.V. 2001 Experimental study of dissolubility of magnesium phosphates and magnesium ammonium

phosphate in connection with a mechanism of processes of oceanic phosphorite formation. *XIV Russian Meeting on Experimental Mineralogy*. Chernogolovka, IEM RAS: 161.
Kashik, S.A., Lomonosova, T.K. & Mazilov V.N. 1996. Processes of mineral formation and chemical balance material in Lake Baikal. *Lithosphere of Central Asia*. Novosibirsk: Nauka: 92–98.
Zolotarev, K.V., Goldberg, E.L., Kondratyev, V.I., Kulipanov, G.N., Meginsky, E.G., Tsukanov, V.M., Phedorin, M.A. & Kolmogorov, Yu P. 2001 Scanning SR-XRF beamline for analysis of bottom sediments. *NIMA* 470: 382–387.
Zhmodik, S.M., Verkhovtseva, N.V., Mironov, A.G., Ilić, Nemirovskaya, N.A., Khlystov, O.M., Klerkx, J. & Zhmodik, A.S. 2003 Mapping of uranium and phosphorus in sediments of Lakes Baikal and Issyk-Kul by neutron-induced autoradiography. *Radiation Measurements* 36: 567–579.

Radionuclide interactions with minerals and microbes

Water–Rock Interaction, Wanty & Seal II (eds)

Depleted U in sediments of Patroon Reservoir, Albany, NY, USA

J.G. Arnason & B.A. Fletcher
Department of Earth and Atmospheric Sciences, University at Albany, SUNY, Albany, NY, USA

ABSTRACT: Patroon Reservoir, located 1 km downstream of a former depleted uranium (DU) manufacturing plant, contains sediments with elevated levels of U and other heavy metals. Activities of U and Th series radionuclides were measured by gamma and alpha spectroscopy on selected sediment core samples and were compared to total U concentration profiles. U in the lowermost sediments (>1.9 m) occurs at background concentrations (<3 ppm) and is in secular equilibrium, indicating a natural source. In contrast, U in the overlying sediments (1.9–1.1 m) occurs at elevated concentrations (3–320 mg/kg) and is in secular disequilibrium, indicating that it contains DU derived from the plant. Using a simple mixing model, we show that U in the overlying sediments contains between 25 and 97 weight percent DU.

1 INTRODUCTION

Depleted uranium (DU) is a by-product of the uranium enrichment process used in the manufacture of nuclear fuels. In comparison to naturally occurring uranium, DU contains a lower proportion of the isotopes ^{234}U and ^{235}U and a higher proportion of ^{238}U (Table 1). It also contains trace amounts of ^{236}U that are not present in natural uranium.

DU has been used for manufacturing armor piercing projectiles for the military and for tail counterweights in large civilian aircraft such as the Boeing 747. During the 1991 Gulf War, approximately 340 tons of DU were used in munitions, and an estimated 11 tons were used in the Balkans in the late 1990s (Royal Society 2002). DU is both chemically toxic and radioactive (Royal Society 2002). Its use on the battlefield results in its being pulverized and spread over large areas, leading to potentially hazardous health and environmental consequences.

From 1958 to 1984, projectiles and aircraft counterweights containing DU were manufactured by National Lead Industries (NLI) at their site in Albany, New York, under license from the U.S. Atomic Energy Commission (DOE, 1995). In 1999, we extracted a sediment core from a small reservoir just downstream of the NLI site and discovered elevated concentrations of Cd, Hg, Pb, U and other heavy metals in the reservoir bottom sediments (Arnason & Fletcher 2003). Heavy metal concentration profiles in the core suggest that U, Pb, and Cd were derived from a single anthropogenic source, most probably the NLI site. In the present study, we measured activities of U and Th series radionuclides in order to determine the source of U in the sediments and to further elucidate the deposition history.

Table 1. Abundances of uranium isotopes in natural and depleted uranium.

	^{234}U	^{235}U	^{236}U	^{238}U
Natural U	0.0055	0.720	0.0	99.2745
Depleted U	0.0008	0.2015	0.0030	99.7947

2 MATERIALS AND METHODS

2.1 *Description of the study area*

The 4.5-ha NLI site lies across an unnamed branch of Patroon Creek that flows SSE into the main creek (Fig. 1). Between 1958 and 1984, radioactive U and Th were released into the soil, groundwater, and atmosphere on and adjacent to the site (DOE 1995). Micron-size DU particles from the NLI site were detected in air filters more than 40 km NW of the site, indicating that contamination was widespread (Dietz 1980). In 1984, the site was closed and industrial activity ceased.

The analyzed sediments are from a 3-m-long core extracted from Patroon Reservoir, a small (1.3 ha) reservoir on Patroon Creek approximately 1 km downstream of the NLI site. The core was collected along the central axis of the reservoir near the

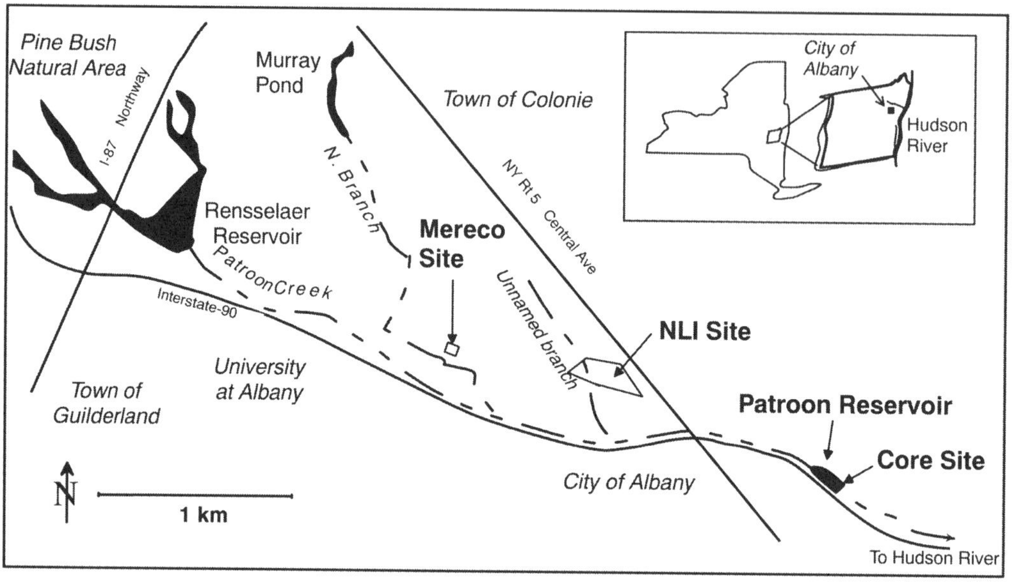

Figure 1. Sketch map of Patroon Creek showing location of the NLI site and the sediment core collected from Patroon Reservoir.

downstream end (42°41′05″N, 73°47′21″W) in approximately 2.5-m water depth.

2.2 *Description of the core*

The physical properties, mineralogy, and heavy metal concentrations profiles of the core have been described previously (Arnason & Fletcher 2003). For the present study, fifteen samples were collected for activity analysis across the depth interval 1.06–2.27 m. This interval includes sediments containing both low and high total U concentrations. By correlating historical industrial activity with metal concentration profiles and assuming a constant deposition rate, Arnason & Fletcher (2003) estimated that this interval corresponds approximately to the time period 1944 to 1973.

2.3 *Analytical methods*

Total U and Th concentrations in sediment were determined previously by ICP-MS as described in Arnason and Fletcher (2003). Radioactivity measurements on sediment core samples were performed at the US EPA National Air and Radiation Environmental Laboratory. The activity and 2-sigma uncertainty for the radionuclides ^{40}K, ^{60}Co, ^{131}I, ^{137}Cs, ^{140}Ba, ^{208}Tl, ^{212}Bi, ^{212}Pb, ^{214}Bi, ^{214}Pb, ^{226}Ra, ^{227}Th, ^{228}Ra, ^{228}Th, ^{230}Th, and ^{232}Th were measured by gamma spectroscopy. The activity and 2-sigma uncertainty for the radionuclides ^{234}U, ^{235}U, and ^{238}U were determined by alpha spectroscopy. All activities are reported as picocuries per gram dry weight. In all samples, the activities of ^{140}Ba, ^{60}Co, and ^{131}I were below detection levels of 110, 0.18, and 140 pCi/g, respectively.

Table 2. Nuclear decay/sec (Bq) for ^{234}U and ^{238}U in 1 g depleted U and in 1 g natural U.

	^{234}U Bq	^{238}U Bq
Natural U	1,815	12,410
Depleted U	12,342	12,342

Assuming a binary mixture of natural and depleted U, the fraction of depleted uranium (X) in each sample was calculated from the measured activities using the equation

$$X = \frac{[12,342 \cdot (1 - R)]}{[64 \cdot R + 10,527]} \tag{1}$$

where R is the ratio of total observed alpha counts of ^{234}U to total observed alpha counts of ^{238}U in a mixture of natural and depleted uranium; 12,342 is the specific activity (Bq) of natural ^{238}U; 64 is the difference between specific activities of ^{238}U in depleted and natural U; and 10,527 is the difference between specific activities of ^{234}U in natural and depleted U (Table 2; L. Dietz, personal comm.). The uncertainty associated with X was calculated using

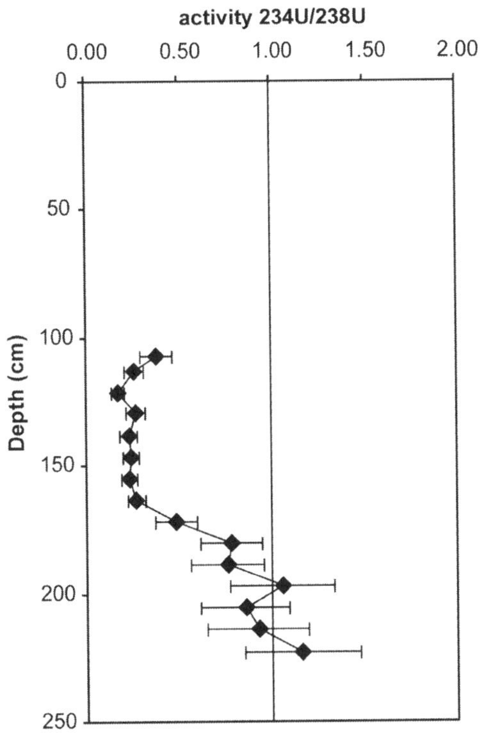

Figure 2. 234U/238U activity ratio vs. core depth in sediments of Patroon Reservoir. An activity ratio of 1.0 (vertical line) indicates secular equilibrium and a natural U source. See text.

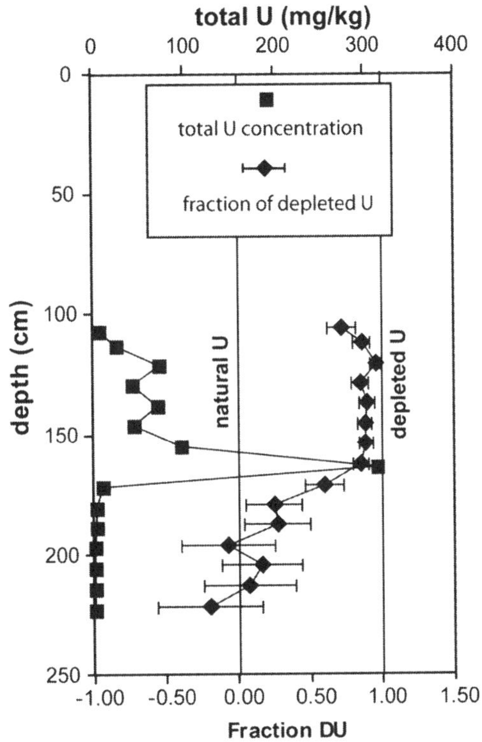

Figure 3. Total U concentrations (squares) and Fraction DU (diamonds) vs. core depth in sediments of Patroon Reservoir. Total U is from Arnason & Fletcher (2003). Vertical lines denote Fraction DU in natural (0.0) and depleted (1.0) uranium. See text for explanation.

standard error propagation methods applied to the 2-sigma uncertainty of the activity measurements.

3 RESULTS AND DISCUSSION

3.1 *U activities*

In Figure 2, the ^{234}U / ^{238}U activity ratio is plotted as a function of depth in the core. Below 1.9 m, ^{234}U activity/^{238}U activity ≈ 1 (i.e., ^{234}U and ^{238}U activities are approximately equal), indicating secular equilibrium and a source dominated by natural U. In contrast, above 1.9 m, the $^{234}U/^{238}U$ activity ratio decreases to a minimum value of 0.18, indicating secular disequilibrium and the presence of a DU component in this portion of the core.

3.2 *Total U concentrations and DU fraction*

Total U concentration and the fraction of DU are plotted as functions of core depth in Figure 3. Fraction of DU is calculated from the activity ratios using equation (1). Below a depth of 1.9 m, total U concentrations are uniformly <3 mg/kg. Above 1.9 m, U concentrations increase gradually to 1.68 m, and then abruptly to 318 mg/kg at 1.6 m before declining gradually towards the sediment-water interface. The U concentrations below 1.9 m are reasonable for background or uncontaminated values, whereas the abrupt increase strongly suggests anthropogenic contamination. Below 1.9 m, the DU fraction is essentially zero; above this level DU fraction ranges from 0.25 to 0.95, i.e., nearly pure DU. The total U and fraction of DU profiles are well correlated, indicating that the increase in total U was accompanied by the introduction of increasing proportions of a DU component.

3.3 *Total Th concentrations and activities*

In contrast to U, total Th concentrations remain within a narrow range (6–11 mg/kg) throughout the entire core interval, showing no stratigraphic variation (not shown). Although historical records indicate that Th was used in various manufacturing processes

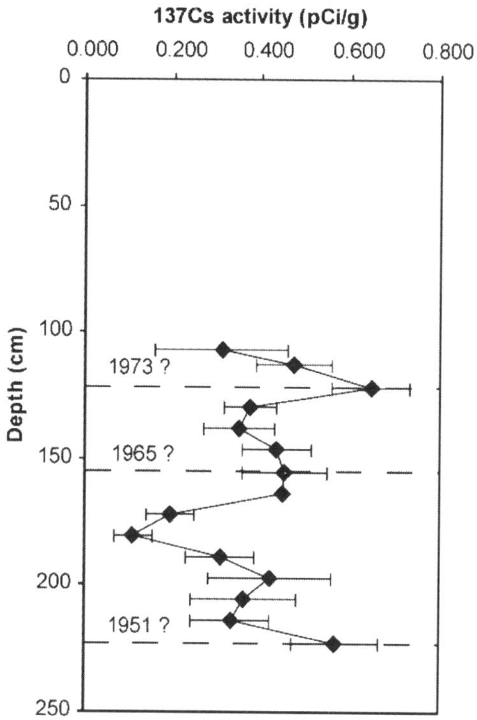

Figure 4. Activity of ^{137}Cs vs. core depth in sediments of Patroon Reservoir. Dashed lines represent inferred dates (see text).

at the NLI site, our results indicate that Th is present at concentrations comparable to typical crustal abundances. This suggests that Th was not used in as significant a quantity as U, or that it was not transported downstream in the watershed to the same extent as U.

Within error, $^{228}Th/^{232}Th$ activity ratios are equal to unity throughout the core, indicating secular equilibrium for the Th series isotopes (not shown). $^{230}Th/^{238}U$ activity ratios, however, indicate that these two isotopes are in secular equilibrium below 1.9 m, and in secular disequilibrium above. This occurs because, during the U enrichment process, ^{230}Th and other daughter isotopes are chemically separated from U (Ketterer et al. 2000).

3.4 ^{137}*Cs activities*

In order to provide more constraints on the age and timing of sediment deposition, we measured ^{137}Cs activity. The activity of this isotope should peak during the period of above ground nuclear testing in the mid 1960's. The ^{137}Cs activities range from 0.11 to 0.65 pCi/g, and show maxima at 2.23, 1.55, and 1.22 m, corresponding approximately to the dates 1951, 1965, and 1973 (Fig. 4). This chronology is based two assumptions: (1) the earliest occurrence of DU in the core (189 cm depth) corresponds to the year 1958; and (2) the sedimentation rate was constant. The 1965 maxima correspond roughly to the time period of peak nuclear testing. The other two maxima, however, do not correspond to any known event. More data from above and below the measured interval are required to accurately constrain the age and timing of deposition using ^{137}Cs.

4 CONCLUSIONS

In Patroon Reservoir, U in the lowermost sediments (>1.9 m) occurs at background concentrations (<3 ppm) and is in secular equilibrium, indicating a natural source. In contrast, U in the overlying sediments (1.9–1.1 m) occurs at elevated concentrations (3–320 mg/kg) and is in secular disequilibrium, indicating that it contains DU derived from the NLI plant. Using a simple mixing model, we show that U in the overlying sediments contains between 25 and 97 weight percent DU.

ACKNOWLEDGEMENTS

The authors wish to thank Barbara Youngberg, John Mitchell, and John Abunaw of the Bureau of Radiation, NYSDEC for providing alpha and gamma spectroscopic data. We also wish to thank Len Dietz, formerly of the Knolls Atomic Power Laboratory, for assistance with calculations and for providing valuable historical information.

REFERENCES

Arnason, J.G. & Fletcher, B.A. 2003. A 40+ year record of Cd, Hg, Pb, and U deposition in sediments of Patroon Reservoir, Albany County, NY, USA. *Environmental Pollution* 123: 383–391.

Dietz, L. 1980. Investigation of excess alpha activity observed in recent air filter collections and other environmental samples. Knolls Atomic Power Laboratory, General Electric Company, January 24, 1980. CHEM-434-LAD.

DOE (United States Department of Energy) 1995. Engineering Evaluation and Cost Analysis (EE/CA) for the Colonie Site. DOE/OR/21950–1008, September, 1995.

Ketterer, M.E., Bonniwell, E.C., Wetzel, W.C., Layman, R.R. & Matisoff, G. 2000. Isotopic studies of sources of uranium in sediments of the Ashtabula River, Ohio, U.S.A. *Environ. Sci. Technol.* 34(6): 966–972.

The Royal Society 2002. The health effects of depleted uranium munitions, Summary. Document 6/02, March 2002, ISBN 0854035753.

Water–Rock Interaction, Wanty & Seal II (eds)
 ISBN 90 5809 641 6

EXAFS study of uranyl adsorption on Wyoming montmorillonite

J.G. Catalano
Department of Geological and Environmental Sciences, Stanford University, Stanford, CA

G.E. Brown, Jr.
Department of Geological and Environmental Sciences, Stanford University, Stanford, CA
Stanford Synchrotron Radiation Laboratory, SLAC, Menlo Park, CA

ABSTRACT: Uranyl adsorbs to montmorillonite through both cation exchange and specific adsorption. EXAFS analysis of uranyl adsorption mode as a function of pH and ionic strength reveals that cation exchange is more significant at circumneutral pH than previously believed. The disagreement between EXAFS and previous surface complexation modeling (SCM) studies may be due either to kinetic effects or to incorrect SCM parameters. EXAFS results indicate that uranyl binds to SiOH, AlOH, and FeOH edge sites as inner-sphere surface complexes. Comparison of uranyl adsorbed to montmorillonites with different octahedral sheet Fe contents suggests that uranyl may preferentially bind to FeOH sites over AlOH sites. The identity of the "indifferent" background electrolyte affects the structure of uranyl adsorption complexes on SWy-2 edge sites. This is likely due either to electrolyte-induced changes in the distribution of uranyl among the three general types of edge sites, or to the formation of ternary surface complexes.

1 INTRODUCTION

Uranium contamination of soils and sediments is widespread at numerous sites across the United States, especially former US Department of Energy (DOE) weapons complexes (Riley & Zachara 1992). In many environments, U occurs as U^{VI}, often as part of the uranyl moiety, UO_2^{2+}. Predicting the future fate and transport of uranyl in contaminated environments requires an understanding of the processes affecting the partitioning of uranyl between aqueous and solid phases. In most systems, this partitioning may occur through either precipitation or adsorption processes.

The solubility of uranyl is strongly pH dependent in systems equilibrated with the atmosphere, with high solubility at low and high pH and low solubility in circumneutral solutions. Uranyl undergoes a number of hydrolysis and complexation reactions in aqueous solutions, especially with dissolved carbonate. The high affinity of uranyl for carbonate anions results in the formation of the highly soluble complex $UO_2(CO_3)_3^{4-}$, accounting for the observed pH-dependent solubility. Precipitation in simple systems results in the formation of schoepite; systems containing dissolved species (e.g. silica and phosphate) may precipitate more complex uranyl phases.

The pH dependence of uranyl adsorption depends on the binding site considered. Fixed-charge cation exchange sites, such as interlayer sites in phyllosilicates resulting from isomorphous substitution, show little pH-dependence. Partitioning of uranyl to these sites depends primarily on the identity and quantity of competing cations. Variable-charge SiOH, AlOH, and FeOH sites occur at metal-oxide surfaces and the exposed edges of phyllosilicates. Cation binding to these sites shows a strong pH-dependence due to variations in the protonation of these sites and the overall surface charge with pH.

A potentially important sorbent of uranyl in soils and sediments is montmorillonite. Montmorillonite occurs in the soils and sediments of a number of US DOE sites, and is the primary clay mineral at the Hanford, WA site. The high cation-exchange capacity and surface area of montmorillonite make it highly reactive towards metal adsorption. Montmorillonite has the ability to bind metals at both fixed- and variable-charge sites.

In this study, we examined the structure and binding sites of uranyl adsorption complexes on montmorillonite using extended X-ray absorption fine structure (EXAFS) spectroscopy. We have investigated the distribution of uranyl between fixed-charge interlayer sites and variable-charge edge sites as a function of pH and background electrolyte concentration. The likely edge sites involved in binding were evaluated, including FeOH edge sites. The effects of various "indifferent" background electrolytes on uranyl surface complexation were probed as well.

2 MATERIALS AND METHODS

Wyoming montmorillonite sample SWy-2 was obtained from the Source Clay Repository of the Clay Mineral Society. The bulk sample was Na-saturated and size fractionated to <2 μm. An X-ray diffraction pattern of the processed sample showed no sign of any crystalline contaminant phases. Cheto-type montmorillonite SAz-1 and ferruginous smectite SWa-1 were prepared using identical methods.

The processed SWy-2 sample was used for all experiments except where specified. A clay concentration of $1\,gL^{-1}$ and an initial U concentration of 5 μM were used in all experiments. Details of the concentrations of other components are specified in the results below. In all experiments, reaction times were 24 hours, after which the samples were centrifuged, and the clay in the form of a wet paste was loaded into a 2 mm Teflon sample holder and sealed with Kapton® tape in preparation for EXAFS analysis. U L_{III}-edge EXAFS spectra were collected at the Stanford Synchrotron Radiation Laboratory (SSRL) on beam lines 11–2 and 4–3 using Si (220) and (111) double-crystal monochromators. All spectra were collected in fluorescence-yield using a 30-element (11–2) or 13-element (4–3) Ge detector. X-ray energy was calibrated using a yttrium foil; the first inflection point in the Y K-edge was set to 17038 eV. Spectra were processed and analyzed using the SixPACK interface (Webb 2003) to IFEFFIT (Newville 2001). The k^3-weighted EXAFS spectra were fit using phase and amplitude functions generated using FEFF 8.20 (Ankudinov et al. 1998). The three multiple scattering paths associated with the axial oxygen atoms of uranyl were included in all fits as specified in Hudson et al. (1996).

3 RESULTS AND DISCUSSION

3.1 *Interlayer versus edge site binding*

As uranyl may bind to both interlayer and edge sites on montmorillonite, an understanding of the relative amount of each type of binding under varying solution conditions is desired. Previous surface complexation modeling (SCM) studies (e.g. McKinley et al. 1995) have generally predicted, at low ionic strength (1–10 mM), that uranyl binds only to interlayer sites at low pH (<4), and only to edge sites at circumneutral pH. At higher ionic strengths (0.1–1 M), uranyl does not bind to montmorillonite at low pH because only cation exchange may occur at these pH values, and that binding mechanism is suppressed by the high cation concentrations. At circumneutral pH and high ionic strength, uranyl was predicted to bind only to edge sites.

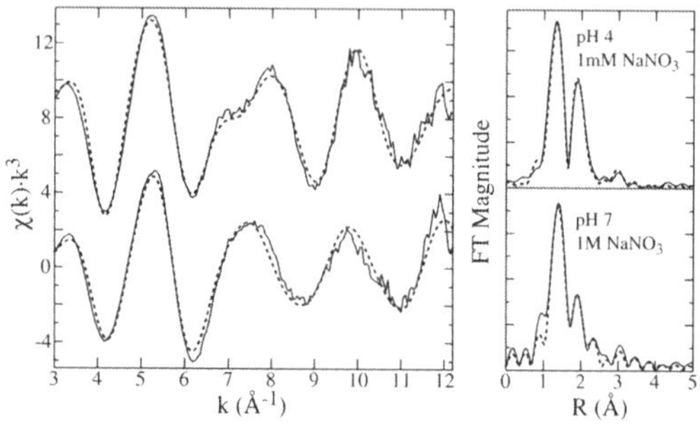

Figure 1. EXAFS (left) and Fourier transform (right) spectra (solid) and fit (dashed) of the different end-member uranyl surface complexes on SWy-2.

Table 1. EXAFS fitting results for end-member species.

		O_{ax}	O_{eq1}	O_{eq2}
pH 4	N	2*	4.9(9)	
1 mM	R(Å)	1.759(5)	2.42(1)	
$NaNO_3$	σ^2(Å^2)	0.0013(4)	0.008(2)	
pH 7	N	2*	3.3(2)	2.7*
1 M	R(Å)	1.794(7)	2.31(2)	2.47(2)
$NaNO_3$	σ^2(Å^2)	0.0020(3)	0.004(1)	0.004*

Note: The estimated standard deviations are listed in parentheses, representing the uncertainties in the last digit.
* Parameter fixed or constrained during fitting.

The relative proportions of uranyl bound to edge and interlayer sites can be determined using EXAFS spectroscopy. The EXAFS spectra of uranyl adsorbed to montmorillonite in systems expected to only contain end-member edge site (pH 7, 1 M $NaNO_3$) and interlayer site (pH 4, 1 mM $NaNO_3$) surface complexes were collected (Fig. 1). Analyses of the spectra (Table 1) are consistent with previous studies of uranyl binding to montmorillonite (e.g. Sylwester et al. 2000) with a single axial oxygen shell (O_{ax}) for both, a single equatorial oxygen shell (O_{eq1}) for the interlayer site and a split equatorial oxygen shell (O_{eq2}) for the edge site.

To evaluate the accuracy of previous SCM studies, EXAFS spectra of samples with compositions intermediate between these end-members were collected (Fig. 2). These spectra were fit with linear

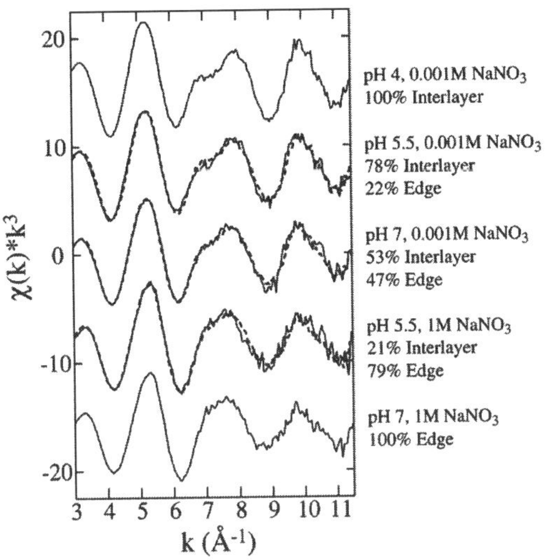

Figure 2. Linear combination fitting of the EXAFS spectra of uranyl adsorbed to SWy-2 at intermediate compositions.

combinations of the spectra of the two end-member species. While the spectra of uranyl adsorbed from high ionic strength solutions generally agree with the predictions of SCM studies, the spectra of samples from low ionic strength solutions show that a greater proportion of uranyl is bound to interlayer sites than predicted by SCM. This finding suggests that under low ionic strength conditions, interlayer sites can effectively compete with edge sites for uranyl. We attribute the discrepancies between the EXAFS and SCM studies either to errors in the SCM parameters due to the use of gibbsite and silica surfaces as model edge sites or to the possibility that the kinetics of cation exchange were faster than the kinetics of edge-site adsorption.

3.2 *Potential edge binding sites*

While inner-sphere surface complexation of uranyl on edge sites of phyllosilicates has often been inferred from a splitting in the O_{eq} shell (e.g. Sylwester et al. 2000), identification of the specific binding site has been identified in only one study (Hennig et al. 2002). Previous SCM studies have suggested that uranyl binds primarily to AlOH surface sites on phyllosilicate edges, although FeOH sites were not considered (McKinley et al. 1995).

We have analyzed the U L_{III}-EXAFS spectrum of uranyl adsorbed to SWy-2 in a CO_2-free, 1 M $NaNO_3$ solution at pH 7 to determine which potential binding sites (SiOH, AlOH, and/or FeOH) may be involved (Fig. 3). The best fit to the EXAFS data suggests that uranyl binds as a bidentate, inner-sphere complex on both SiOH and FeOH sites, as indicated by a Si

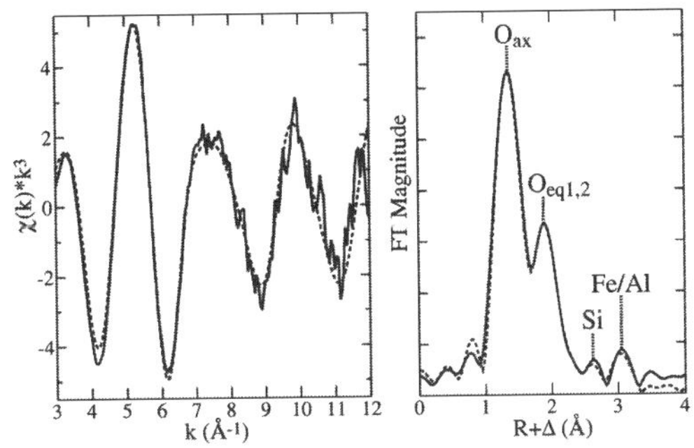

Figure 3. EXAFS (left) and Fourier transform (right) spectrum (solid) and fits (dashed), including binding sites, of uranyl adsorbed to SWy-2 in 1 M $NaNO_3$ background at pH 7.

Table 2. EXAFS fitting results for inner-sphere uranyl surface complex on SWy-2.

Shell	O_{ax}	O_{eq1}	O_{eq2}	Si	Fe
N	2*	3.0(2)	3.0*	0.3(3)	0.2(3)
R(Å)	1.775(8)	2.28(2)	2.44(2)	3.06(7)	3.45(9)
σ^2($Å^2$)	0.0020(5)	0.002(1)	0.002*	0.005*	0.005*

Note: The estimated standard deviations are listed in parentheses, representing the uncertainties in the last digit. * Parameter fixed or constrained during fitting.

neighbor at 3.06 Å and an Fe neighbor at 3.45 Å (Table 2). Alternatively, an Al neighbor at 3.31 Å can be fit, but does not reproduce the Fourier transform feature at 3.0 Å as well as Fe. Fitting with a combination of both Al and Fe neighbors reproduces the data well, but exceeds the number of degrees of freedom. While this prevents us from quantifying the relative proportions of each, we feel both Al and Fe signals are significant components. These data suggest that uranyl may bind to SiOH, AlOH, and FeOH sites, in disagreement with SCM studies. However, the surface coverages required for this study are likely higher than those used in SCM studies; SiOH sites may become less important at lower coverages.

The above results suggests a possible preferential binding of uranyl to FeOH sites over AlOH sites, as Fe only makes up about 10% of the octahedral cation sites in SWy-2 (Mermut & Cano 2001). To further explore this preference, the spectrum of uranyl bound to SWy-2 was compared to uranyl bound to SAz-1, having Fe in 3% of octahedral cation sites, and SWa-1, having Fe in 77% of octahedral cation sites (Fig. 4). The Fourier transform feature at ~3.0 Å of SWy-2 more closely resembles that of SWa-1 than SAz-1, which would be expected if uranyl preferentially binds to FeOH sites. This study suggests that, at a

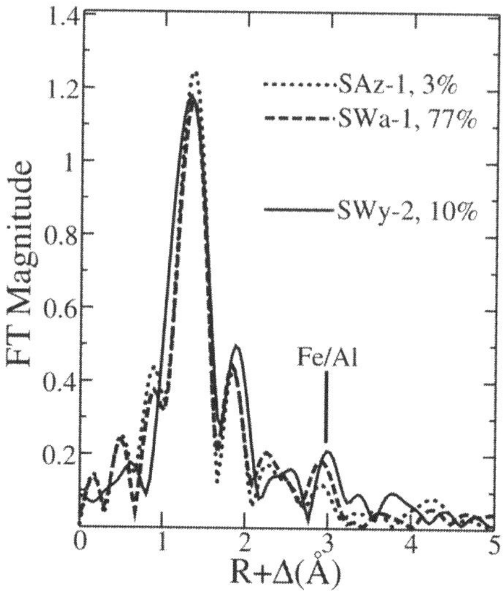

Figure 4. Comparison of the Fourier transformed EXAFS spectra of uranyl bound to the edges of SWy-2, SAz-1, and SWa-1.

minimum, FeOH sites should be considered when constructing future surface complexation models of uranyl adsorption to phyllosilictes with moderate Fe contents (i.e., Fe in>~5% of octahedral sites).

3.3 *Effects of "indifferent" electrolytes*

Recent surface complexation modeling work (Criscenti & Sverjensky 1999) suggests that electrolyte anions traditionally considered "indifferent" (NO_3^-, Cl^-, and ClO_4^-) might actually interact with cations adsorbing onto metal oxide surfaces, forming ternary surface complexes. We have examined the possibility of electrolyte effects on uranyl adsorption to montmorillonite by examining the EXAFS spectra of uranyl adsorbed to SWy-2 from pH 7, 0.1 M $NaNO_3$, NaCl, and $NaClO_4$ solutions. Minor variations in the EXAFS spectra are observed with a change in electrolyte (Fig. 5). The differences primarily lie in the structure of the equatorial oxygen shell.

These variations may be attributed to changes in the distribution of uranyl among SiOH, AlOH, and FeOH edge sites, and/or to the formation of ternary surface complexes. Changes in the distribution of uranyl among the different edge sites might occur if different electrolytes affect the protonation behavior of surface functional groups or if certain electrolyte anions block SiOH, AlOH, and/or FeOH sites. The formation of ternary surface complexes with NO_3^- or Cl^- might be expected as uranyl forms aqueous complexes with these anions (Grenthe et al. 1992); no ternary complexes with ClO_4^- are expected, as uranyl does not form aqueous complexes with this anion

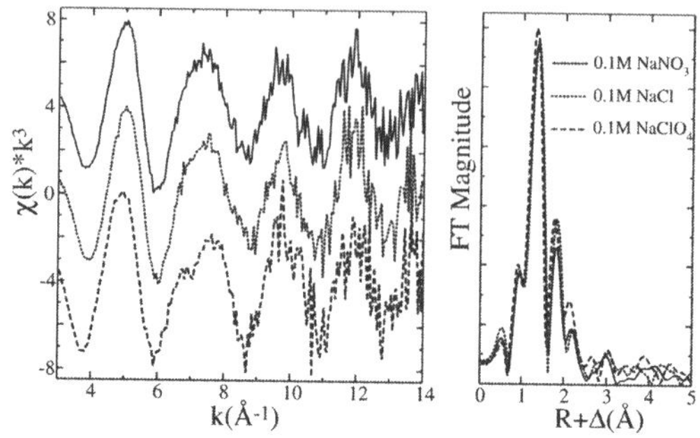

Figure 5. Comparison of the EXAFS (left) and Fourier transform (right) spectra of uranyl adsorbed to SWy-2 in 0.1 M $NaNO_3$, NaCl, and $NaClO_4$ electrolytes.

(Grenthe et al. 1992). The formation of uranyl-nitrato and -chloro ternary surface complexes should be included in future SCM studies.

4 CONCLUSIONS

Uranyl, a widespread soil and groundwater contaminant at US DOE sites, may adsorb to montmorillonite through both cation exchange to interlayer sites or through specific adsorption to hydroxylated edge sites. The binding to each type of site is dependent on solution pH and ionic strength. EXAFS analyses of uranyl adsorption mechanisms as a function of pH and ionic strength reveal that at low ionic strength and circumneutral pH cation exchange is significant, contrary to previous SCM studies. This disagreement may be due either to kinetic effects or to incorrect SCM parameters. The EXAFS results agree with SCM predictions for high ionic strength conditions. Uranyl may bind to SiOH, AlOH, and FeOH edge sites as an inner-sphere surface complex. Comparison of the spectrum of uranyl adsorbed to SWy-2 to the spectra of uranyl on SAz-1 and SWa-1 suggests that uranyl may preferentially bind to FeOH sites over AlOH sites. The identity of the "indifferent" background electrolyte affects the structure of uranyl adsorption complexes on SWy-2 edge sites. This is likely due to electrolyte-induced changes in the distribution of uranyl among the three general types of edge sites, and/or to the formation of ternary surface complexes. Future surface complexation modeling studies of uranyl adsorption to phyllosilicates should consider these binding site and electrolyte effects.

ACKNOWLEDGEMENTS

This research was supported by DOE-EMSP grant FG07-ER0263495 and Battelle contract 409089-A9E-P3623. We thank S.M. Webb (SSRL) for insights on using SixPACK, and J.R. Bargar and J. Rogers (SSRL) for helpful suggestions on the collection and analysis of U EXAFS spectra. We thank J.M. Bigham (Ohio State University) for providing the SWa-1 sample. SSRL is supported by the U.S. Department of Energy, Office of Basic Energy Sciences and Office of Biological and Environmental Research, and the National Institutes of Health, National Center for Research Resources, Biomedical Technology Program.

REFERENCES

Ankudinov, A.L., Ravel, B., Rehr, J.J. & Conradson, S.D. 1998. Real-space multiple-scattering calculation and interpretation of X-ray absorption near-edge structure. *Phys. Rev. B* 15: 7565–7576.

Criscenti, L.J. & Sverjensky, D.A. 1999. The role of electrolyte anions (ClO_4^-, NO_3^-, and Cl^-) in divalent metal (M^{2+}) adsorption on oxide and hydroxide surfaces in salt solutions. *Am. J. Sci.* 299: 828–899.

Grenthe, I., Fuger, J., Konings, R.J.M., Lemire, R.J., Muller, A.B., Nguyen-Trung, C. & Wanner, H. 1992. *Chemical Thermodynamics of Uranium*. Amsterdam: North-Holland.

Hennig, C., Reich, T., Dahn, R. & Scheidegger, A.M. 2002. Structure of uranium sorption complexes at montmorillonite edge sites. *Radiochim. Acta* 90: 9–11.

Hudson, E.A., Allen, P.G., Terminello, L.J., Denecke, M.A. & Reich, T. 1996. Polarized x-ray absorption spectroscopy of the uranyl ion: Comparison of experiment and theory. *Phys. Rev. B* 54: 156–165.

McKinley, J.P., Zachara, J.M., Smith, S.C. & Turner, G.D. 1995. The influence of uranyl hydrolysis and multiple site-binding reactions on adsorption of U(VI) to montmorillonite. *Clays Clay Mins.* 43: 586–598.

Mermut, A.R. & Cano, A.F. 2001. Baseline studies of The Clay Minerals Society Source Clays: Chemical analyses of major elements. *Clays Clay Mins.* 49: 381–386.

Newville, M. 2001. IFEFFIT: interactive XAFS analysis and FEFF fitting. *J. Synchrotron Radiat.* 8: 322–324.

Riley, R. G. & Zachara, J. M. 1992. *Chemical Contaminants on DOE Lands and Selection of Contaminant Mixtures for Subsurface Science Research.* U.S. Department of Energy, Office of Energy Research, Washington, D.C.

Sylwester, E.R., Hudson, E.A. & Allen, P.G. 2000. The structure of uranium (VI) sorption complexes on silica, alumina, and montmorillonite. *Geochim. Cosmochim. Acta* 64: 2431–2438.

Webb, S.M. 2003. *SixPACK.* Stanford Synchrotron Radiation Laboratory, Menlo Park, CA.

Water–Rock Interaction, Wanty & Seal II (eds)

An experimental study of the dissolution rates of apatite and britholite as a function of solution composition and pH from 1 to 12

C. Chaïrat, E. H. Oelkers, S. Köhler, N. Harouiya & J.-E. Lartigue
Géochimie, CNRS/URM 5563, Toulouse, France
CEA Valrhô, Bagnols-sur-Céze, France

ABSTRACT: Long-lived radioactive actinides can be incorporated in the crystalline structure of apatite potentially providing an effective waste storage host. A first step towards evaluating this waste storage strategy is characterizing the stability and reactivity of non-radioactive analogues. Towards this goal steady-state dissolution rates of fluorapatite and britholite ($(Ca_9Nd)(PO_4)_5SiO_2F_2$) have been measured as a function of aqueous solution composition and temperature. Apatite dissolution rates, based on Ca release rates, decrease monotonically with increasing pH; logarithms of measured 25° C rates decrease from −7 to −11.5 mol/m^2/s with increasing pH from 2 to 11.8. Rates at acidic pH are in close agreement with literature values. Stoichiometric dissolution of Ca, P, and F are observed in most experiments. Britholite dissolution rates, based on Ca release rates, are similar to those of apatite, though Nd release rates are found to be nearly an order of magnitude slower than that expected from stoichiometric dissolution in many experiments due to $NdPO_4$ precipitation.

1 INTRODUCTION

The major motivation for this study is the assessment of minerals of the apatite structure to serve as host for the storage of long-lived radionuclides. Apatite is believed to be well suited to the storage of radioactive actinides for a variety of reasons. First, apatite, with the generic formula $M_{10}(XO_4)_6Y_2$, where $M = Ca^{2+}$, Pb^{2+}, Ba^{2+}, etc., $XO_4 = PO_4^{3-}$, VO_4^{3-}, SiO_4^{4-}, etc., and $Y = F^-$, Cl^-, OH^-, Br^-, I^-, etc., can readily incorporate tri and tetravalent actinides (Ac) through a double substitution of $Ac^{(III,IV)}$ for Ca^{II} and PO_4 for SiO_4 leading to the electrically neutral structures $Ca_{10-x}Ac^{III}_x(PO_4)_{6-x}(SiO_4)_xF_2$ and $Ca_{10-x}Ac^{IV}_x(PO_4)_{6-2x}(SiO_4)_{2x}F_2$. Secondly, the apatite structure is believed to be relatively resistant to self-irradiation. Thirdly, 2-billion year old apatite containing tri- and tetravalent actinides have been found in the Oklo, Gabon, natural nuclear reactor, attesting to the potential long term stability of this mineral.

A number of studies have focussed on apatite dissolution kinetics. The majority of these studies were aimed at compositional analogues of human bone or skeletal apatite reacting under conditions appropriate to the human biology (37°C, model body fluids). Comparatively little effort has focused on the dissolution behaviour of apatite in geological systems. One of the first kinetic studies of apatite dissolution to geologic applications was reported by Valsami-Jones et al. (1998). The authors studied apatite dissolution in batch reactor in the pH range from 2 to 7. More recently, Guidry et al. (2003) studied natural fluorapatite dissolution at pH 2–8 and temperature 25–45°C. Building upon these studies and extending this work to the potential long term storage of radioactive waste, the dissolution rates of natural fluorapatite from Paraïba, Brazil, and synthetic britholite have been measured as a function of solution composition over the pH range 3–12. The purpose of this paper is to present the results of these experiments and assess the potential of apatite as a waste storage host.

2 EXPERIMENTAL METHODS

The britholite used in this study was synthesized using the method of Guy et al. (2002) according to the reaction:

$$\frac{1}{2}Nd_2O_3 + CaF_2 + 3CaCO_3 + SiO_2 + \frac{5}{2}Ca_2P_2O_7 = Ca_9Nd(PO_4)_5(SiO_4)F_2 + 3CO_2$$

This britholite and natural pegmatitic apatite from Paraïba, Brazil, was crushed and then ground with an agate mortar and pestle. The size fraction between 250–500 μm for britholite and 113–200 μm for apatite were obtained by sieving; these samples were then ultrasonically cleaned using acetone. The specific surface area of the cleaned britholite and apatite powders are 341 and 1213 m^2/g, respectively, as determined by krypton absorption using the B.E.T. method. The chemical composition of the apatite, as determined by microprobe analysis, is consistent with $Ca_9Nd_{0.9}(PO_4)_{5.0}(SiO_4)_{1.1}F_{1.1}OH_{0.9}$.

Apatite and britholite were dissolved at 25°C as a function of pH and as a function of chemical affinity and aqueous Ca, PO_4, and F concentrations at pH 10. All dissolution experiments were performed in mixed flow reactors using covered Azalon plastic beakers equipped with a 0.45 μm outlet filter. Fluid was injected into the reactors using Gilson peristaltic pumps, which allowed constant fluid flow rates from 0.01 to 10 gm/min. Steady-state dissolution rates, as indicated by constant outlet Ca and P concentration, were obtained after elapsed times ranging from 0.2 to 30 days, depending on flow rate and temperature. Dissolution experiments were performed in fluids comprised of demineralised H_2O and Merck reagent grade HCl, NaOH, NaCl, NH_4Cl, and/or NH_4OH. All inlet fluids had stoichiometric ionic strengths of 0.01 mol/kg. Ca concentrations of the outlet fluids were determined using atomic absorption spectroscopy (Perkin Elmer Zeeman 5000), aqueous F and P were determined using High Pressure Liquid Chromatography, and aqueous Nd concentrations were determined using ICP mass spectroscopy. The reproducibility of chemical analyses was ~±4 percent. Outlet fluid pH was measured at 25°C immediately after sampling.

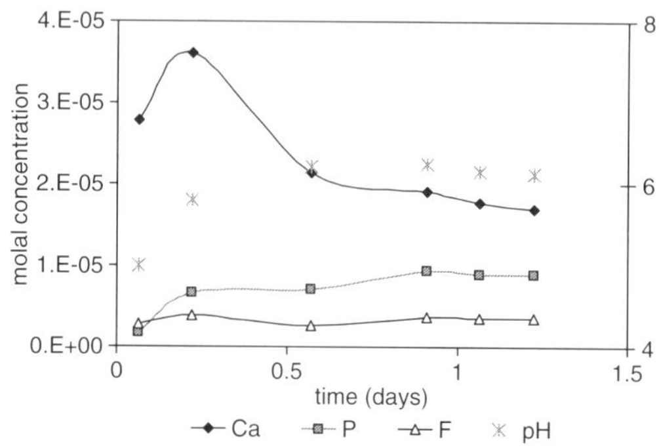

Figure 1. Temporal evolution of outlet solution Ca, P and F concentrations as a function of time during fluorapatite dissolution at pH = 6.2 and 25°C.

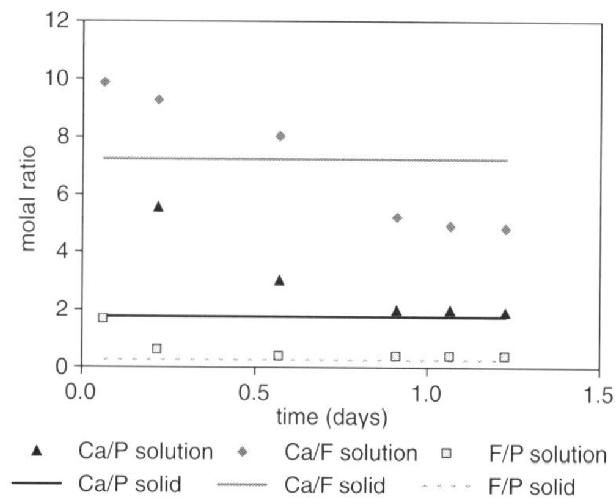

Figure 2. Elemental ratios as a function of time during fluorapatite dissolution at pH = 6.2 and 25°C.

3 RESULTS AND DISCUSSION

3.1 *Dissolution stoichiometry*

A representative example of the temporal variation of outlet solution concentrations during an apatite dissolution experiment is shown in Figure 1. Calcium is liberated preferentially at the onset of dissolution, whereas P and F increase monotonically with time until steady-state is attained.

A representative example of the temporal variation of elemental ratios in the outlet solutions is illustrated in Figure 2. Molar Ca/P and Ca/F ratios of the outlet fluids exceed those of the dissolving apatite before steady-state is attained. The outlet solution Ca/F ratio at steady-state is ~5, which is consistent with a stoichiometric fluorapatite ratio of 10:2 rather than the 10.1:1.4 ratio of the natural apatite considered in the present study.

A typical relative release rate of Nd during the dissolution of apatite and britholite can be assessed with the aid of Figure 3, which illustrates mineral dissolution rates computed from the outlet solution composition of the individual elements. In both cases rates based on Nd outlet solution concentrations are significantly lower than those computed from the outlet solution concentrations of the other elements, suggesting Nd retention in the solid phase. This retention of Nd by the solid phase is interpreted to stem from the precipitation of $NdPO_4$ during these dissolution experiments. This interpretation is supported by the low solubility of $NdPO_4$ over most of the pH range considered in this study and SEM analysis of post-experiment solids. An example of a post-experiment SEM photomicrograph is illustrated in Figure 4.

3.2 *Dissolution rates as a function of pH*

A summary of apatite dissolution rates measured as a function of pH in the present study is summarized in

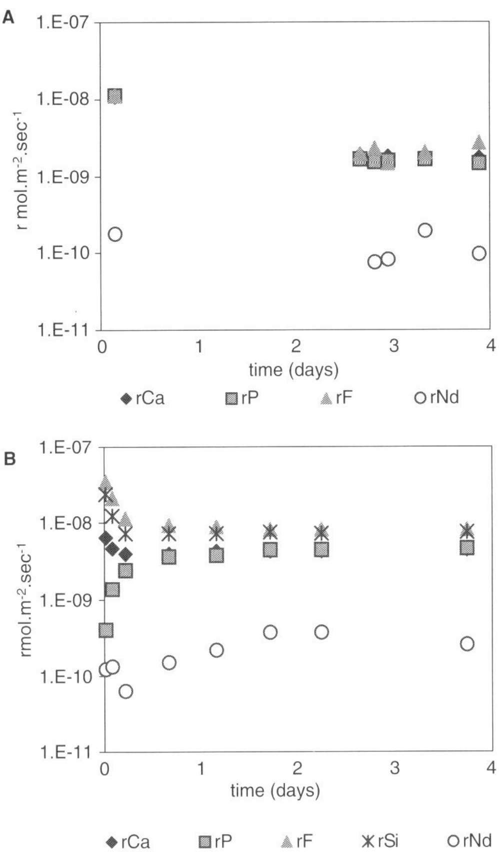

Figure 3. Dissolution rates calculated from outlet solution concentrations of various elements as a function of time. A: Fluorapatite dissolution at 25°C and pH = 4. B: Britholite experiment at 200°C and pH = 4 (Lartigue unpublished results).

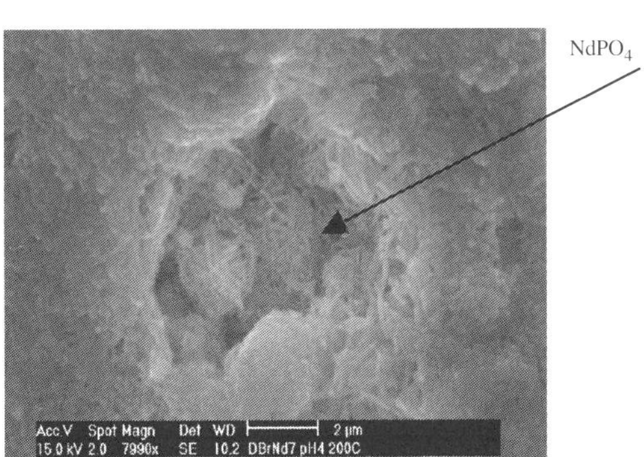

Figure 4. SEM photomicrograph of britholite following its dissolution at 200°C and pH = 4.

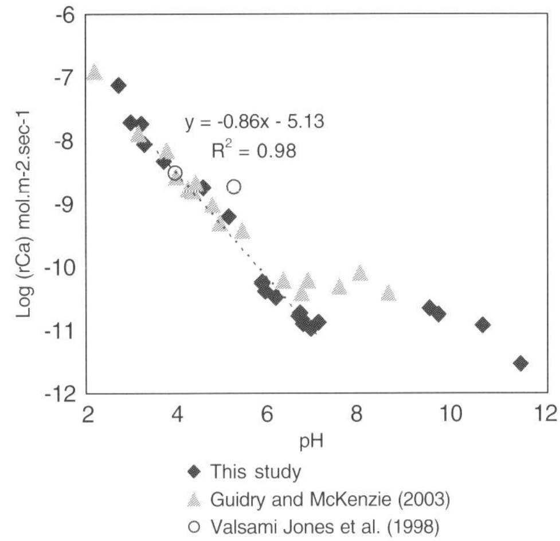

Figure 5. Measured steady-state dissolution rates of natural fluorapatite at 25°C as a function of pH.

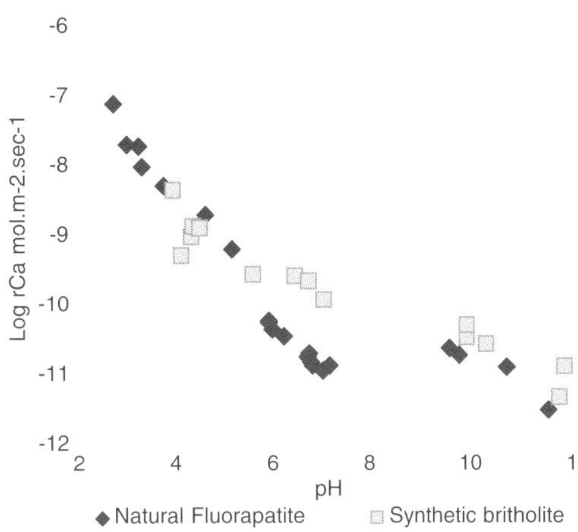

Figure 6. Comparison of apatite and britholite steady-state dissolution rates at 25°C as a function of pH.

Figure 5. Rates measured in the present study are closely consistent with other values reported in the literature. Apatite dissolution rates apparently decrease monotonically with increasing pH. The plot of log rate versus pH exhibits a slope of ~0.86 at pH < 6. Preliminary data suggest that the effect of pH on rates decreases at higher pH.

A comparison of apatite dissolution rates with those of britholite is shown in Figure 6. Britholite dissolution rates are found to be similar to those of apatite; these rates also decrease monotonically with increasing pH to at least pH 12.

4 CONCLUSIONS

There are two major conclusions of this study supporting the use of radioactive actinide-bearing minerals having the apatite structure as waste storage hosts. First, measured apatite and britholite dissolution rates decrease monotonically with increasing pH. This behavior is in contrast to that of borosilicate glasses, which have minimum dissolution rates at intermediate pH and increasing rates with increasing pH at basic conditions. As concrete is among the materials often proposed to be present at long-term radioactive waste sites, strongly basic pH solutions may be expected to be in contact with the waste storage hosts. The variation of apatite and britholite dissolution rates with pH suggests that this mineral will be most resistant to dissolution in these basic solutions. Secondly, the tendency of Nd phosphates to form as a secondary product during apatite and britholite dissolution suggests that actinide elements could also be immobilized by similar precipitates.

REFERENCES

Guy, C., Audubert, F., Lartigue, J.-E., Latrille, C., Advocat, T. & Fillet, C. 2002. New conditionings for separated long-lived radionuclides. *C.R. Physique* 3: 827–837.

Guidry, M.W. & Mackenzie, F. 2003. Experimental study of igneous and sedimentary apatite dissolution: control of pH, distance from equilibrium, and temperature on dissolution rates. *Geochim. Cosmochim. Acta* 67: 2949–2963.

Valsami-Jones, E., Ragnardottir, K.V., Putnis, A., Bosbach, D., Kemp, A.J. & Cressey, G. 1998. The dissolution of apatite in the presence of aqueous metal cations at pH 2–7. *Chem. Geol.* 151: 215–233.

Water–Rock Interaction, Wanty & Seal II (eds)

Mineral formation and radionuclide sorption in waste-impacted Hanford sediments

J. Chorover
University of Arizona, Tucson, AZ, USA

P. Rotenberg
Professional Service Industries, Pittsburgh, PA, USA

R.J. Serne
Pacific Northwest National Laboratory, Richland, WA, USA

ABSTRACT: Leakage of caustic solutions from high level radioactive waste (HLRW) storage tanks drives sediment weathering reactions that may affect radionuclide sorption and precipitation. Batch experiments were conducted with three major sediment types from the Hanford Site (WA) to assess the effects of a synthetic waste solution on mineral transformation and uptake of Cs and Sr. Dissolution of sediment Si and subsequent precipitation with waste Al led to the formation of zeolite and feldspathoid solids coupled to uptake of Cs and Sr over 375 d reaction time.

1 INTRODUCTION

The contamination of vadose zone sediments by leaking radioactive waste tanks at Department of Energy (DOE) sites across the U.S. has prompted the need for a better understanding of contaminant-soil interactions. The Hanford Site (WA) has been a DOE storage and disposal site for radioactive waste resulting from nuclear reactor operations and weapons production since 1944 (Routson et al. 1981). The DOE stores over 60% of its high level radioactive waste (HLRW) at the Hanford facility in 177 buried tanks (Ahearne 1997). Of these tanks, 67 have leaked, releasing over 3800 m^3 of highly radioactive liquid waste containing more than 10^6 Curies of ^{137}Cs into the underlying sediment (Zachara et al. 2002). The leaking radioactive tank waste from the solvent extraction (REDOX) process is characterized by highly elevated pH, ionic strength and aluminate concentrations (Serne et al. 1998). The bulk of the radioactivity is present as ^{137}Cs and ^{90}Sr, which are present in concentrations up to parts per million. Historically, the sorption capacity of natural soils and sediments has been employed to decontaminate waste streams via geochemical attenuation processes.

Contaminant sorption behavior under these caustic waste conditions is likely to be affected by hydroxide-promoted mineral transformation reactions and participation of dissolved waste Al in solid phase product formation. For example, weathering of kaolinite in caustic solutions representative of leaking tank waste resulted in Si release followed by its subsequent precipitation with leachate Al and the uptake of Cs and Sr into neo-formed secondary solid phases (Chorover et al. 2003). Others have also examined the effects of high pH, ionic strength and aluminum concentrations on sorption to Hanford sediments (Serne et al. 1998; McKinley et al. 2001; Zachara et al. 2002); however, few studies have examined how concurrent mineral dissolution and precipitation processes affect contaminant uptake under the extreme geochemical conditions.

The objective of this research was to assess the coupling of Cs and Sr sorption to weathering-induced transformations of the Hanford Sediments, with a particular focus on the potential role of secondary mineral formation in contaminant sequestration. Our approach involves conjunctive use of macroscopic (wet chemistry) techniques, electron microscopy and spectroscopy.

2 MATERIALS AND METHODS

2.1 *Hanford sediments*

Three uncontaminated Hanford sediments (from the same formations as those beneath the storage tanks) were collected: Hanford formation gravel and coarse sand ("Hanford Coarse", HC), Hanford fine sand ("Hanford Fine", HF), and silt-dominated sediment from the Upper Ringold Formation ("Ringold Silt", RS). The three sediments were air dried, sieved <2 mm and used without further modification to simulate natural conditions. Detailed sediment characterization results are reported in Serne et al. (2001) and Rotenberg (2002). Selected data are shown in Table 1. Semiquantitative XRD analyses of the bulk sediments and clay fractions indicate that quartz is the most prevalent mineral in the whole sediments (*ca.* 55, 80 and 30% by mass in HC, HF and RS, respectively), whereas smectite is the most prevalent clay mineral (*ca.* 30% of the clay fraction in all cases). Trace amounts of apatite, Fe-oxides and sepiolite were detected in the clay fractions.

2.2 *Kinetic batch experiments*

All experiments and stock solutions were prepared in acid washed Nalgene™ polypropylene co-polymer (PPCO) vessels using ultrapure water (Millipore MilliQ UV Plus). Reagent grade $NaNO_3$, NaOH, CsCl, $SrCl_2 \cdot 6H_2O$ (J.T. Baker) and $NaAlO_2 \cdot xH_2O$ powder (98% pure, EM Science) were used as obtained.

Solutions were pre-filtered through 0.45 µm Pall-Gelman hydrophilic polypropylene filters prior to reaction with sediments as synthetic tank waste leachate (STWL) composed of 0.05 mol kg^{-1} $NaAlO_2$, 2 mol kg^{-1} Na^+, 1 mol kg^{-1} NO_3^-, and 1 mol kg^{-1} OH^- (pH~13.7). CsCl and $SrCl_2 \cdot 6H_2O$ were both added to the STWL stock (as co-contaminants) to give three initial Cs^+ and Sr^{2+} concentrations: 10^{-5}, 10^{-4} and 10^{-3} mol kg^{-1}, where both cations were present at the given concentrations. Each STWL solution was then stirred for approximately 5 min and used immediately.

Batch reactions were conducted in triplicate in 50 mL PPCO centrifuge tubes containing 0.8 g of sediment in 40.0 g STWL solution. Two types of sediment-free controls were used; the first employed STWL only, and the second was an Al-free STWL solution. Experiments were conducted in a dark constant-temperature room at 25°C on an end-over-end shaker at 1 rpm for time intervals of 1, 7, 29, 92, 183 and 374 d, at which time suspensions were permitted to settle and a subsample of colloidal plus dissolved (CD) fraction (<2 µm) was obtained by aspiration of the supernatant suspension. The remaining suspensions were centrifuged (20 min at 25,261*g*) and supernatant solutions were filtered through 0.2 µm Gelman Acrodisc filters to obtain the dissolved (D) fraction. All solutions were acidified to pH ~ 2 with trace metal grade HNO_3 and preserved at 4° for further analysis by ICP-AES and AA.

Table 1. Selected physical/mineralogical data on Hanford sediments.

Sample	Sand g kg^{-1}	Silt g kg^{-1}	Clay g kg^{-1}	BET SSA[1] (m^2 g^{-1})	Mineralogy[2]
HC	930	29	41	7.57	Q, P, KF, S, I, V, C, K
HF	728	206	66	4.94	Q, P, KF, S, I, C, V, K
RS	18	839	143	21.05	Q, KF, P, S, I, C, V, K

[1] Specific surface area measured by N_2 BET method
[2] Mineralogy listed in order of abundance: Q-quartz, P-plagioclase, KF-K-feldspar, S-smectite, I-illite, C-chlorite, V-vermiculite, K-kaolinite.

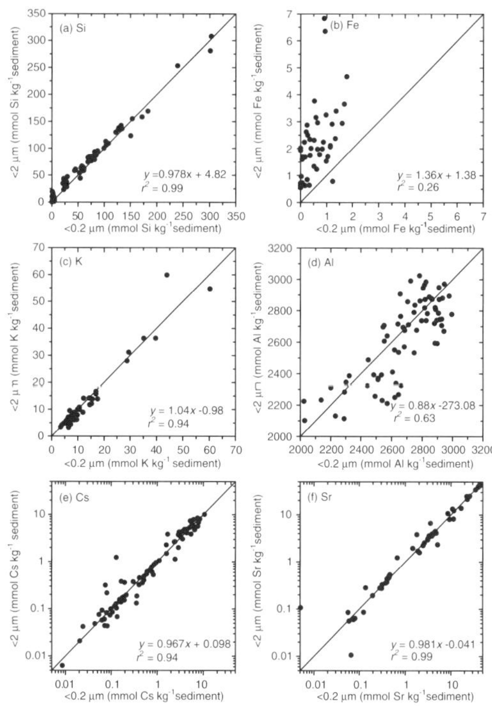

Figure 1. Colloidal plus dissolved [CD < 2 µm] versus dissolved [D < 0.2] concentrations (a) Si, (b) Fe, (c) K, (d) Al, (e) Cs and (f) Sr in reacted suspensions.

2.3 *Solid phase analyses*

Centrifuged sediment pellets were resuspended in 95% ethanol adjusted to pH 10.0 ± 0.1 to remove

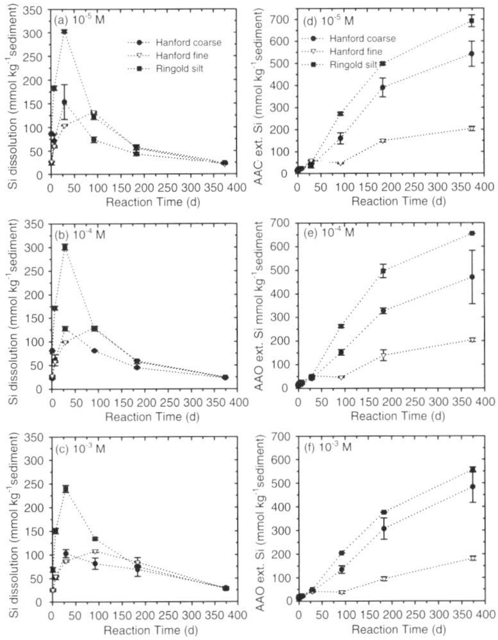

Figure 2. Dissolution (a–c) and re-precipitation of "poorly-crystalline" (d–f) silicon in STWL-sediment systems. Concentrations of Cs & Sr increase from top to bottom of the figure.

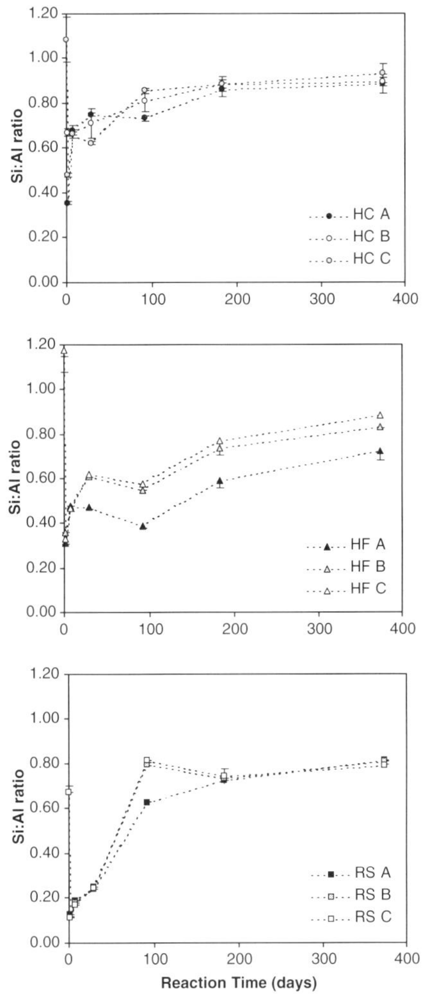

Figure 3. Time dependent change in molar Si/Al ratio of AAO extractable solids.

entrained STWL. Two additional ethanol washes followed, without pH adjustment. One of the three replicates was set aside for mineralogical analysis (DRIFT, XRD, SEM-EDS), and the remaining duplicate samples were reacted with $0.1\,mol\,kg^{-1}$ $Mg(NO_3)_2$ adjusted to pH 7.00 ± 0.05 for 1 h to extract Mg^{2+} "exchangeable" Cs and Sr. Samples were washed three times with ethanol (no pH adjustment), then extracted a second time using $0.2\,mol\,kg^{-1}$ $NH_4C_2H_2O_4$ (pH 3.00 ± 0.05) for 4 h in the dark to extract Cs and Sr bound by "poorly-crystalline" solids. In preparation for solids analysis, sediments were washed successively with 0.1 and $0.01\,mol\,kg^{-1}$ NaCl for 1 h, followed by three ethanol washes. Sediment pellets were resuspended in small amounts of MilliQ water, frozen at $-80°C$ and freeze-dried using a Multi-Dry$_{TM}$, (FTS Systems Inc).

3 RESULTS AND DISCUSSION

Solution pH values were not significantly affected by reaction time, ranging from 13.4 to 13.8 over all samples. The neoformation of colloidal solids (from dissolution of native sediment minerals) was anticipated to result in a significant suspended colloidal fraction of lithogenic elements with possible incorporation of contaminant Cs and Sr, but this was not observed. Figure 1 shows the <2 µm colloidal plus dissolved (CD) fraction plotted against the <0.2 µm dissolved fraction (D fraction) for each element. The close agreement between concentrations in the CD and D pools (with the exception of Fe) indicates that the colloidal pool was negligible and the dissolved pool was dominant. This is despite the fact that colloidal-sized precipitates were clearly observed using SEM. Evidently, these precipitates were either associated with the surfaces of underlying sediment grains (i.e., surface precipitates or colloidal deposits) or the colloids flocculated at the high ionic

Figure 4. Scanning electron micrographs of precipitates formed after 6 months of Hanford sediment weathering in STWL: (a–b) Hanford Coarse, (c–d) Hanford Fine, (e-f) Ringold Silt. Images on left and right for each sediment correspond to 10^{-5} and 10^{-3} M initial Cs and Sr concentrations, respectively.

strength of STWL and these floccules sedimented along with the silt and sand sized particles. Since the colloidal pool was insignificant, further discussion will be limited to the dissolved (D) fraction data.

Dissolution of Si gave maximum aqueous phase concentrations at 30 d reaction time for RS and HC, and at 92 d for HF (Fig. 2). The mass of Si extractable in acid ammonium oxalate (AAO) solution increased with time, with the largest increases occurring after the peak in soluble Si. Evidently, the drop in dissolved Si contributes to the formation of AAO-extractable Si containing solids. Dissolution and reprecipitation of Si show a strong dependence on sediment type (RS > HC > HF) and moderate sensitivity to contaminant concentration; higher co-contaminant concentration leads to decreased Si release and a slower rate of AAO-extractable Si solid formation (Fig. 2). Larger effects of co-contaminant concentration on mineral transformation rate were observed for specimen kaolinite (Chorover et al. 2003) and 2:1 layer type silicate suspensions (Choi et al. in prep.) subjected to the same STWL reaction conditions, but the trends were the same as those reported here. After 374 d, the fraction of total sediment Si that is extractable in AAO ranged from 2% for HF to 7% for RS. Product solids incorporate Al from the STWL as indicated by decreasing soluble Al (not shown), despite increasing molar Si/Al ratio of AAO-extractable solid phase products (Fig. 3).

Data from XRD, FTIR (not shown) and SEM (Fig. 4) are consistent with neoformation of zeolite and feldspathoid solids that have been observed to form from weathering of quartz (Bickmore et al. 2001) and specimen clays (Chorover et al. 2003) under similar conditions (e.g., nitrate cancrinite, sodalite, zeolite A). Effects of mineral transformation on sorption were evident for both Cs and Sr, with the latter showing a larger response (manuscript in prep.). Chemical extraction and differential IR spectroscopy data indicate co-precipitation of Sr with neoformed aluminosilicate phases.

REFERENCES

Ahearne, J.F. 1997. Radioactive waste: The size of the problem. *Physics Today* 50: 24–29.

Bickmore, B.R., Nagy, K.L., Young, J.S. & Drexler, J.W. 2001. Nitrate cancrinite precipitation on quartz sand in simulated Hanford tank solutions. *Environ. Sci. Technol.* 35: 4481–4486.

Chorover, J., Choi, S.K., Karthikeyan, K.G., Amistadi, M.K., Crosson, G. & Mueller, K.T. 2003. Linking cesium and strontium uptake to kaolinite weathering in simulated tank waste leachate. *Environ. Sci. Technol.* 37: 2200–2208.

McKinley, J.P., Zeissler, C.J., Zachara, J.M., Serne, R.J., Lindstrom, R.M., Schaef, H.T. & Orr, R.D. 2001. Distribution and retention of ^{137}Cs in sediments at the Hanford site, Washington. *Environ. Sci. Technol.* 35: 3433–3441.

Rotenberg, P. 2002. Cesium and strontium uptake and mineral weathering of Hanford sediments in the presence of a highly alkaline tank waste. *M.S. Thesis in Soil Science*, The Pennsylvania State University, University Park, PA 154 p.

Routson, R.C., Barney, G.S. & Smith, R.M. 1981. Hanford site sorption studies for the control of radioactive wastes: A review. *Nucl. Technol.* 54: 100–106.

Serne, R.J., Zachara, J.M. & Burke, D.S. 1998. *Chemical Cs Migration from Past Tank Leaks*. PNNL-11495 UC-510. Richland, Washington: Pacific Northwest National Laboratory.

Serne, R.J., Schaef, H.T., Bjornstad, B.N., Williams, B.A., Lanigan, D.C., Horton, D.G., Clayton, R.E., LeGore, V.L., O Hara, M.J., Brown, C.F., Parker, K.F., Kutnyakov, I.V., Serne, J.N., Mitroshkov, A.V., Last, G.V., Smith, S.C., Lindenmeier, C.W., Zachara, J.M. & Burke, D.B. 2001. *Characterization of Uncontaminated Sediments from the Hanford Reservation – RCRA Borehole Core Samples and Composite Samples*. PNNL-2001-1. Richland, Washington: Pacific Northwest National Laboratory.

Zachara, J.M., Smith, S.C., Serne, R.J., McKinley, J.P. & Gassman, P.L. 2002. Sorption of Cs^{+} to micaceous subsurface sediments from the Hanford site, USA. *Geochim. Cosmochim. Acta* 66: 193–211.

Water–Rock Interaction, Wanty & Seal II (eds)

Remobilization of precipitated and cell-associated uranium in brines

C.J. Dodge, J.B. Gillow & A.J. Francis
Environmental Sciences Department, Brookhaven National Laboratory, Upton, NY

ABSTRACT: We investigated the transformations of uranyl-citrate by a denitrifying halophilic bacterium, *Halomonas* sp. (WIPP1A), isolated from the Waste Isolation Pilot Plant (WIPP) repository. The addition of uranyl-citrate to the bacterial growth medium resulted in the precipitation of U. Extended X-ray absorption fine structure (EXAFS) analysis of the precipitate formed in the growth medium was identified as uranyl hydroxophosphato species $[K(UO_2)_5(PO_4)_3(OH)_2 \cdot nH_2O]$. In the presence of growing culture of *Halomonas* sp. the precipitate underwent dissolution to form the uranyl dicarbonate complex $[UO_2(CO_3)_2]^{2-}$. Addition of U to resting cells of the bacterium resulted in intra- and extra-cellular uptake as uranyl phosphate species. The U was removed from the cell surface by washing with bicarbonate.

1 INTRODUCTION

Under appropriate conditions, microorganisms can affect the mobilization or immobilization of U from a waste repository by direct enzymatic or indirect non-enzymatic actions (Francis 1998). Microbes also can affect the transport of U by formation of biocolloids (Francis et al. 1998). Fundamental information on the mechanisms of microbial transformations of actinides under various environmental conditions is useful in predicting the fate and long-term transport of actinides in waste repositories, as well as developing strategies for managing waste. In this paper, we describe the biotransformation of uranyl nitrate and citrate under denitrifying conditions by a halophilic bacterium *Halomonas* sp. isolated from the Waste Isolation Pilot Plant (WIPP, New Mexico, USA) repository.

2 MATERIALS AND METHODS

2.1 *Bacterial culture*

Halomonas sp. (WIPP1A) was isolated from a sediment slurry sample from the WIPP site and grows anaerobically using nitrate as an electron acceptor but does not reduce Fe or U (Francis et al. 1997). The bacterium was grown in a medium at pH 6.5 containing 5.0 g sodium succinate, 1.0 g KNO_3, 0.25 g K_2HPO_4, 0.5 g yeast extract, 200 g WIPP salt (halite), and 1000 ml deionized water.

2.2 *U-citrate complex*

An equimolar 1:1 U:citric acid complex (50 mM) was prepared by slowly combining appropriate amounts of anhydrous citric acid with a uranyl nitrate stock solution while constantly mixing it. The pH of the solution was adjusted to 6.5. The complex was stored in the dark to prevent photodecomposition (Dodge & Francis 1994).

2.3 *U precipitate characterization*

We determined the influence of growth medium components on U solubility. The U precipitate recovered from adding U-citrate to the brine solution and the growth medium was analyzed by extended X-ray absorption fine structure (EXAFS). The precipitate was placed in a heat-sealed polyethylene bag (0.2 mil) and analyzed on the X-18B beamline at the National Synchrotron Light Source (NSLS) at the U L_{III} absorption edge (17.166 keV) using a fluorescence detector with an Ar filled ion chamber. Uranyl hydrogen phosphate standard was synthesized by slowly adding 0.42 M uranyl nitrate solution to 0.84 M phosphoric acid. A multi-step data analysis of the k^3-weighted EXAFS spectra in the k range 3.7–12.5 $Å^{-1}$ included Fourier transformation of the normalized spectra. The theoretical EXAFS modeling code FEFF7 was used to calculate the backscattering phase and amplitude information for individual neighboring atoms. Fitting parameters for the sample then were obtained by comparison with the uranyl hydrogen phosphate standard.

2.4 *Biotransformation of U by growing cells*

One hundred ml of 0.45μm (Nalgene) filter sterilized and N_2-purged bacterial growth medium was added to sterile 160 ml glass serum bottles inside a N_2-filled glove box. Uranyl-citrate stock was added to the growth medium so that the final concentration of U was 10.52×10^{-3} M. Triplicate samples of each medium were inoculated with 2.5 ml of 24-h old culture of *Halomonas* sp. and incubated at $30 \pm 2°C$ under static conditions. Uninoculated samples in duplicate were included for each treatment as abiotic controls. All samples for U analysis were filtered through a 0.22 μm Millex filter (Millipore, MA) and analyzed for total U with a Varian Liberty 150 inductively coupled plasma-atomic emission spectrometer (ICP-AES).

2.5 *Interactions of U with bacterial cells*

For each sample the bacteria were washed and the pellet was resuspended in 40 ml of the electrolyte to an optical density of ~0.6 (at 600 nm). The uptake of uranyl ion by whole cells was determined by adding 5 ml of uranyl nitrate (pH 5) to obtain a final concentration of 0.125 mM. The U and cells were reacted for 2-h under a N_2 purge to exclude air. The filtrate was acidified and analyzed for U by ICP-AES. The amount of U retained on the filter (minus the abiotic control) was reported as associated with the bacteria. Whole cell mounts for TEM and EDS analysis were prepared to determine the intracellular association of U with the bacteria, as described by Krueger et al. 1993.

2.6 *Carbon dioxide production*

The total amount of CO_2 produced by the bacteria incubated for 120 h was determined in separate samples after adding 1 ml of concentrated perchloric acid. Carbon dioxide in the head space was analyzed by a Varian 3400 gas chromatograph fitted with a Porapak QS (Alltech, IL) column (2 m × 3.2 mm) and a thermal conductivity detector with He as a carrier gas.

3 RESULTS

3.1 *Characterization of the U precipitate*

Analysis of the EXAFS spectral data for the precipitate from addition of U-citrate complex to the growth medium was obtained by comparison with the uranyl hydrogen phosphate standard. The atomic parameters for the standard and this precipitate were obtained by fitting k^3-weighted EXAFS spectra.

Figure 1 depicts the Fourier-transformed spectra for uranyl hydrogen phosphate and the U precipitate

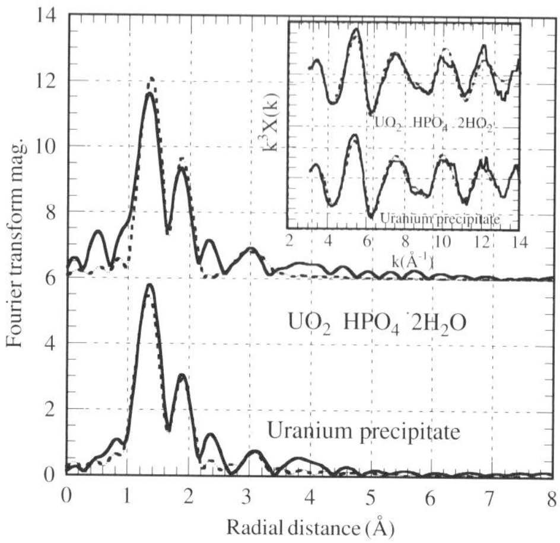

Figure 1. Raw EXAFS data and corresponding Fourier transforms for uranyl hydrogen phosphate and precipitate resulting from addition of citrate to growth medium.

over the k range 3.6–12.5 $Å^{-1}$. Uranyl hydrogen phosphate is composed of a uranyl moiety (2 axial O's at 1.75 Å) surrounded by four equatorial oxygens at 2.30 Å which in turn occupy the corners of a phosphate tetrahedron. The four phosphorus atoms are 3.68 Å from the U. The precipitate is similar in structure to the U phosphate standard and has axial oxygens at 1.80 Å, 4 equatorial oxygens at 2.30 Å, and 4 phosphorus atoms at 3.65 Å from the U.

3.2 *Uranyl-citrate*

Figure 2A depicts the growth of the bacteria in the presence of 0.52 mM uranyl-citrate, as their number increased from 5.3×10^6 to 1.0×10^8. The pH increased from 6.3 to 7.7. The addition of uranyl-citrate to the growth medium caused the dissociation of the complex with the precipitation of U (Fig. 2B). Complete precipitation of U however, was not observed in comparison to uranyl nitrate. The citric acid released from the complex was completely metabolized by the bacteria at the rate of 1.2 mmol h^{-1}. Carbon dioxide production was 30.7 μmol ml^{-1} (data not shown). The concentration of succinic acid decreased from 18.9 mM to 16.6 mM; citric acid appears to be preferentially metabolized by the bacteria. The UV-vis spectra of the filtered culture solution during bacterial growth showed the appearance of peaks at 434, 448, and 464 nm, which is diagnostic for uranyl dicarbonate species (Fig. 2C).

3.3 *Association of U with resting cells*

U accumulated extracellularly and/or intracellularly with the cells. TEM and EDS spectra showed the

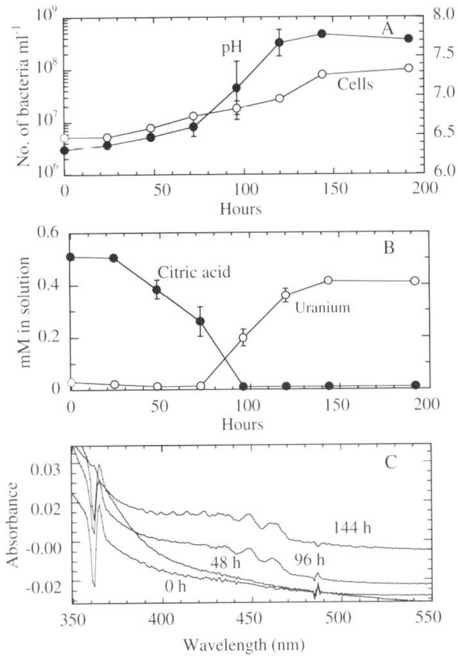

Figure 2. Growth of *Halomonas* sp. and biotransformation of U added as U-citrate complex. (A) increase in pH was concomitant with bacterial growth. (B) U was precipitated from solution at start of experiment; and (C) supernatant shows formation of uranyl dicarbonate.

U was bound to the cell surface (areas of high contrast), while electron-dense intracellular granules were also observed (Fig. 3A, 3B) (Francis et al. 2000).

3.4 *Carbonate extraction of U associated with bacterial cells*

Addition of bicarbonate to the whole or lysed cell fragments exposed to uranyl nitrate and washed in appropriate electrolyte caused the dissolution of U with a concurrent increase in phosphate in solution (data not shown). About 90% of U associated with the whole cells, and about 70% from the lysed cell fragments was solubilized by bicarbonate.

4 DISCUSSION

Although much is known of the behavior of uranyl compounds in very acidic and low ionic strength solutions, little is known of the stability, solubility, and the chemical speciation of the various inorganic- and organic-complexes of U in high ionic strength solutions at physiological pH's (6 to 8).

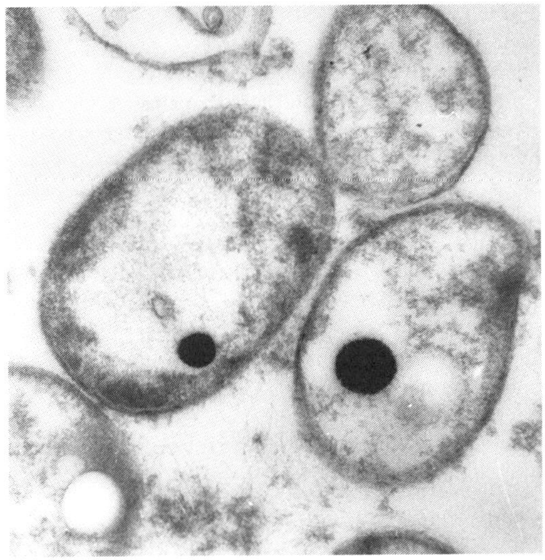

Figure 3A. *Halomonas* sp. accumulated U intracellularly as discrete spherical granules (Francis et al., submitted).

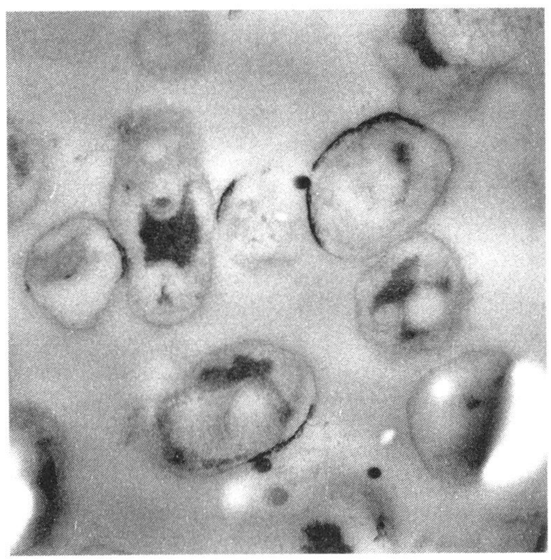

Figure 3B. TEM of lysed cells shows U accumulation on the cell walls (Francis et al., submitted).

The formation of 1:1 and 1:2 uranyl-citrate complexes in 3.4 M brine at pH 5.0 (uncorrected for ionic strength) was observed and the stability constants (log k) were calculated to be 6.04 and 10.98, respectively (Borkowski et al. 1996). These values are orders of magnitude greater than those of U-acetate, U-lactate, U-EDTA, and U-ascorbate complexes. These studies also indicated that the concentration of citrate needed to compete with hydrolysis reactions at pcH 6.0 to obtain 10% hydrolysis in brine was the lowest of all organics tested, thereby contributing significantly to the dissolution of U. Citric acid has been shown to form a binuclear complex with U which may prevent its hydrolysis (Rajan & Martell 1965).

Adding uranyl-citrate to the bacterial growth medium at near neutral pH (6.5) completely precipitated U from solution and was observed only in the presence of phosphate. Elemental analysis and EXAFS analysis confirm the U associates predominantly with the phosphate ligand $K(UO_2)_5(PO_4)_3$, and most likely consists of units of uranyl orthophosphate $(UO_2)_3(PO_4)_2$ due to the interaction with phosphate present in the growth medium. The formation of the precipitate is presented in equation 1.

$$5UO_2^{2+} + 3PO_4^{3-} + 4H_2O + K^+ \rightarrow [K(UO_2)_5(PO_4)_3(OH)_2 \cdot 2H_2O] + 2H^+ \quad (1)$$

The U precipitate was solubilized by adding bicarbonate to the precipitate, or alternatively by the production of CO_2 from the bacterial metabolism of the carbon source succinate, with the formation of uranyl dicarbonate (equation 2):

$$[K(UO_2)_5(PO_4)_3(OH)_2 \cdot 2H_2O] + 10HCO_3^- + 2K^+ \rightarrow 5UO_2(CO_3)_2^{2-} + 3KH_2PO_4 + 2H^+ + 2H_2O \quad (2)$$

The carbonate ion is one of the most environmentally important inorganic complexing agents of U. The speciation and solubility of uranyl carbonate in aqueous solutions has been extensively studied (Clark et al. 1995). Uranyl carbonate can be present as mono-, di-, and tri-carbonato species depending on the pH as well as carbonate concentration in solution. The monomeric and dimeric uranyl carbonate species predominate above pH 5; at uranyl concentrations above 1 mM the trimeric species is favored. Spectroscopic analysis of the carbonate species indicate the presence of U in the dicarbonate form $[UO_2(CO_3)_2]^{2-}$ (log K 16.2). We have demonstrated that an increase in microbial activity can enhance U dissolution due to formation of the soluble uranyl carbonate species. Additional studies are needed to determine the extent of microbial dissolution of U associated with various phosphate minerals.

ACKNOWLEDGEMENTS

This work was supported in part by a subcontract from Sandia National Laboratories (SNL), Albuquerque, New Mexico, AT-8739; Brookhaven National Laboratory, Laboratory Directed Research Development (LDRD); and U.S. Department of Energy's Office of Science Biological and Environmental Research, Natural Accelerated Bioremediation Research (NABIR) program, under Contract DE-AC02-98CH10886.

REFERENCES

Borkowski, M., Lis, S. & Choppin, G.R. 1996. Complexation study of NpO^{+2} and UO_2^{+2} ions with several organic ligands in aqueous solutions of high ionic strength. *Radiochim. Acta* 74: 117–122.

Clark, D.L., Hobart, D.E. & Neu, M.P. 1995. Actinide carbonate complexes and their importance in actinide environmental chemistry. *Chem. Rev.* 95: 25–48.

Dodge, C.J. & Francis, A.J. 1994. Photodegradation of uranium citrate complex with uranium recovery. *Environ. Sci. Technol.* 28: 1300–1306.

Francis, A.J. 1998. Biotransformation of uranium and other actinides in radioactive wastes. *J. Alloys Compd.* 271: 78–84.

Francis, A.J., Gillow, J.B. & Giles, M.R. 1997. Microbial gas generation under expected WIPP conditions. Sandia National Laboratories, Albuquerque, NM, SAND 96–2582.

Francis, A.J., Gillow, J.B., Dodge, C.J., Dunn, M., Mantione, K., Strietelmeier, B.A., Pansoy-Hjelvik, M.E. & Papenguth, H.W. 1998. Role of microbes as biocolloids in the transport of actinides at WIPP. *Radiochim. Acta* 84: 347–354.

Krueger, S., Olsen, G.J., Johnsonbaugh, D. & Beveridge, T.J. 1993. Characterization of the binding of gallium, platinum, and uranium to *Pseudomonas fluorescens* by small-angle X-ray scattering and transmission electron microscopy. *Appl. Environ. Microbiol.* 59: 4056–4064.

Rajan, K.S. & Martell, A.E. 1965. Equilibrium studies of uranyl complexes. III. Interaction of uranyl ion with citric acid. *Inorg. Chem.* 4: 462–469.

Francis, A.J., Dodge, C.J., Gillow, J.B. & Papenguth, H.W. 2000. Biotransformation of uranium compounds in high ionic strength brine by a halophilic bacterium under denitrifying conditions. Environ. Sci. Technol. 34: 2311–2317.

Francis, A.J., Gillow, J.B., Dodge, C.J., Harris, R., Beveridge, T.J. & Papenguth, H.W. Association of uranium with halophilic and non-halophilic bacteria and archaea. Manuscript submitted to Radiochim. Acta.

Water–Rock Interaction, Wanty & Seal II (eds)

U adsorption onto *Bacillus subtilis* bacterial cell walls

D. Gorman-Lewis & J.B. Fein
University of Notre Dame, Civil Engineering and Geological Sciences, Notre Dame, Indiana, USA

ABSTRACT: In this study, we measured aqueous U adsorption onto a common Gram-positive soil bacterium, *Bacillus subtilis*. We performed batch adsorption experiments with a $10^{-5.36}$ M U solution in contact with bacteria as a function of pH, from pH 4 to 9, and as a function of solid:solute ratio. Ionic strength was buffered in all experiments using 0.1 M $NaClO_4$ as the background electrolyte. Control experiments conducted using the same experimental set-up, but without the bacteria in the system, demonstrated that there was negligible U loss from solution during the experiments due to homogeneous precipitation or adsorption onto the experimental apparatus. U adsorption reaches a maximum at pH 6 and decreases with increasing pH under more alkaline conditions, likely due to aqueous carbonate complexation under these higher pH conditions. We model the adsorption reactions using a surface complexation approach, determining the thermodynamic stability of the important uranyl-bacterial surface complexes. The calculated stability constants from this study may be integrated into geochemical models to quantitatively assess the effect of bacterial surface adsorption on the mobility and fate of U in bacteria–water–rock systems.

1 INTRODUCTION

The release of radionuclides into the environment because of mining activities, weapons production, and storage of radioactive waste has focused research on the development of accurate and adaptable tools to predict the fate and transport of the contaminants. The high solubility of uranyl minerals, the extreme toxicity, and complex speciation of aqueous U make the prediction of its behavior in the subsurface environment difficult. Bacteria are ubiquitous in near-surface environments and can affect the mobility of metals (Fein 2000). Therefore, it is crucial to account for U-bacteria adsorption in order to accurately predict contaminant transport through most groundwater systems.

Under oxidizing conditions, U is present in aqueous solutions as the highly soluble uranyl ion. Because the uranyl ion forms a number of stable hydroxide and carbonate complexes in the pH range 5–10, the interpretation of circumneutral U adsorption data can be difficult. A number of experimental studies of bacterial surface uranyl adsorption have avoided this difficulty by studying adsorption over a limited pH range, and by invoking a bulk partitioning approach to model the data (e.g. Hennig et al. 2001; Sar & D'Souza 2001). While this approach enables accurate modeling of the observed adsorption behavior in the experimental systems, it cannot be used to quantitatively model the extent of adsorption that occurs in systems that differ significantly from those investigated in the laboratory.

Fowle et al. (2000) and Haas et al. (2001) each invoke a surface complexation approach to determine the dominant U adsorption mechanisms onto bacterial surfaces and to quantify the thermodynamic stabilities of the important bacterial surface species. Fowle et al. (2000) studied U adsorption onto the Gram-positive bacterial species *Bacillus subtilis*, exclusively under low pH conditions where the un-complexed uranyl ion dominates the aqueous speciation. Based on their modeling results, Fowle et al. (2000) concluded that un-complexed uranyl adsorbs onto protonated phosphoryl sites and onto deprotonated carboxyl sites on the cell wall under these pH conditions, a result confirmed by x-ray absorption spectroscopy experiments conducted under similar pH conditions (Kelly et al. 2002). Haas et al. (2001) measured uranyl adsorption onto the Gram-negative bacterium *Shewanella putrefaciens*. Adsorption experiments encompassed the pH range 1–10 and the best-fitting thermodynamic models of the data invoked adsorption of the uncomplexed uranyl ion and a uranyl hydroxide species onto the bacterial surface.

The stability constants from the Fowle et al. (2000) and Haas et al. (2001) studies could be applied to more complex systems, provided that the proposed reaction stoichiometries are correct. There are significant differences between the two proposed surface complexation models, both in reaction stoichiometry and in calculated stability constant values. This variation could be due to differences in the Gram-negative and Gram-positive cell wall structures. However, the two studies also employed different pretreatment procedures for the experimental bacterial suspension. Acid wash pretreatments of the bacterial cells can increase the extent of metal adsorption (Borrok et al., 2003, submitted).

We measured aqueous U adsorption onto *B. subtilis* in the pH range 4–9 to investigate the role of acid exposure to U sorption and determine the effect of aqueous hydroxide and carbonate complexation on uranyl adsorption to a Gram positive bacterium. The purpose of this study is to provide constraints on the potentially complicated mechanisms for U adsorption onto a common Gram-positive soil bacterium under circumneutral pH conditions, and to quantify the thermodynamic stability of the important U-bacterial surface complexes.

2 THERMODYNAMIC MODELING

We apply a surface complexation approach to model the adsorption of U to the *B. subtilis* cell wall. Using the discrete site, non-electrostatic model of Fein et al. (2003, submitted) to describe the proton-active sites on the bacterial surface, proton binding is ascribed to the reactions of the following stoichiometry (1):

$$R - AH^0 \longleftrightarrow R - A^- + H^+ \tag{1}$$

Where R is a bacterium to which a proton-active functional group type, A, is attached. Fein et al. 2003, submitted) invoke a 4-site model (4 different types of functional groups, each with a distinct acidity constant) to account for proton binding, and use titration data to constrain the acidity constant values and the site concentrations for each site type. In this approach, uranyl binding can be expressed as shown below (2) (written for a generic surface functional group, A).

$$UO_2(OH)_x(CO_3)_y^{(2-x-2y)} + R - AH_z^{(z-1)} \longleftrightarrow R - AH - UO_2(OH)_x(CO_3)_y^{1-x-2y+z} \tag{2}$$

We test the ability of a range of stoichiometries to account for the observed adsorption behavior, including when x = 1,2,3,5,7; y = 1,2,3; and z = 0,1. Although we have no spectroscopic data to help constrain the important binding mechanisms, adsorption measurements conducted as a function of pH and solute:sorbent ratio can provide rigorous constraints on the stoichiometry and thermodynamic stability of the important bacterial surface species. Each reaction stoichiometry that we test yields a different expected adsorption behavior as functions of pH and solute:sorbent ratio, so the experimental data provide constraints on the reaction stoichiometries. Hydroxide and carbonate complexation reactions (e.g., reactions shown below for just one each of the possible uranyl hydroxide (3) and carbonate complexes (4) that form in solution) compete with the bacterial surface for the uranyl ion:

$$UO_2^{2+} + H_2O \longleftrightarrow UO_2OH^+ + H^+ \tag{3}$$

$$UO_2^{2+} + CO_3^{2-} \longleftrightarrow UO_2CO_3^{\circ} \tag{4}$$

We account for the aqueous complexation reactions in our thermodynamic models to determine if the calculated stability constants that we obtain for the bacterial surface complexes can quantitatively account for the observed competitive effect. We use FITEQL (Westall 1982) to test the goodness of fit of each possible adsorption model, taking the model with the lowest variance parameter, V(Y), as the best-fitting model.

3 EXPERIMENTAL PROCEDURES

B. subtilis cells were cultured and prepared following the procedures outlined in Fein et al. (1997) and Fowle & Fein (2000). Cells were maintained on agar plates consisting of trypticase soy agar with 0.5% yeast extract added. Cells for the adsorption experiments were grown by first inoculating a test-tube containing 3 ml of trypticase soy broth and 0.5% yeast extract, and incubating it for 24 h at 32°C. The 3 ml bacterial suspension was then transferred to a 1 L volume of trypticase soy broth, also with 0.5% yeast extract, for another 24 h on an incubator shaker table at 32°C. Cells were pelleted by centrifugation at 4500 rpm for 15 min, agitated for 30 min with 0.03 M HNO_3, and rinsed seven times with 0.1 M $NaClO_4$. Additional experiments were performed with the same harvesting procedure excluding the acid wash step in the cell preparation. In both types of experiments, the bacteria were then pelleted by centrifugation at 7500 rpm for 60 min to remove excess water to obtain a wet weight so that suspensions of known bacterial concentration could be created. The ratio of this wet weight to dry weight is 10:1 (Fein & Delea 1999).

The adsorption experiments were conducted as functions of pH and solid:solute ratio in batch reaction vessels at 25 ± 1°C, using a 0.1 M $NaClO_4$ electrolyte solution to buffer ionic strength. Bacteria were suspended in a $10^{-5.37}$ M U-bearing electrolyte solution diluted from a 1000 ppm atomic absorption spectroscopy standard. This U concentration was chosen, based on control experiments, to avoid precipitation of any U mineral phases. These control experiments were performed in the same manner as the bacterial adsorption experiments described below, except without bacteria in order to test for loss of U due to precipitation or adsorption to the reaction vessel. Control experiments using polypropylene reaction vessels revealed significant U loss from solution with starting concentrations above 1 ppm. The loss of U was most likely due to adsorption onto the reaction vessel and/or precipitation of schoepite. Experiments with 1 ppm U run in polypropylene reaction vessels also exhibited some loss of U, however we did not observe any significant loss of U in the same 1 ppm control experiment run in Teflon reaction vessels.

For the bacteria-bearing experiments, we divided the bacterial suspension into separate 10 ml Teflon centrifuge tubes and adjusted the pH of the solution in each to be between pH 4 and 9 using small volumes (less than 1% of total volume) of concentrated NaOH and HNO_3. Reaction vessels were slowly agitated end over end for 2 h and the equilibrium pH was measured. Control experiments demonstrated that filtration caused a significant loss of aqueous U to the filtration apparatus, so bacterial separation from the experimental solution was conducted by high speed centrifugation. Samples were centrifuged for 15 min at 13,500 rpm to pellet the bacteria, approximately 6 ml of solution was pipetted off the top of the supernatant so as not to disturb the bacterial pellet, and each sample was acidified with 15 μL of 15 M HNO_3 to maintain the U in solution prior to analysis. Analysis for the total aqueous U concentration in the supernatant of each sample was performed by inductively coupled plasma optical emission spectroscopy with matrix-matched standards.

4 RESULTS

The experimental results indicate that U has a high affinity to adsorb onto *B. subtilis*from pH 4 to 6.5. The extent of adsorption plateaus between pH 5 and 6, and above pH 6.5 adsorption decreases markedly with increasing pH. The concentration of U bound to the bacteria is strongly dependent on the solid:solute ratio (Fig. 1a). With 0.5 g/L bacteria, we observed nearly 95% removal of U from solution. With a lower bacterial concentration of 0.125 g/L, the maximum extent of U adsorption observed was 70%. In addition, the acid-washed bacteria exhibited increased adsorption from pH 4 to 6.5 relative to the non-acid washed bacteria. The 0.125 g/L acid washed bacteria exhibited nearly 20% more adsorption than the non-acid washed bacteria (Fig. 1b). Above pH 6.5, adsorption behaviors were similar regardless of whether an acid wash pretreatment was performed or not.

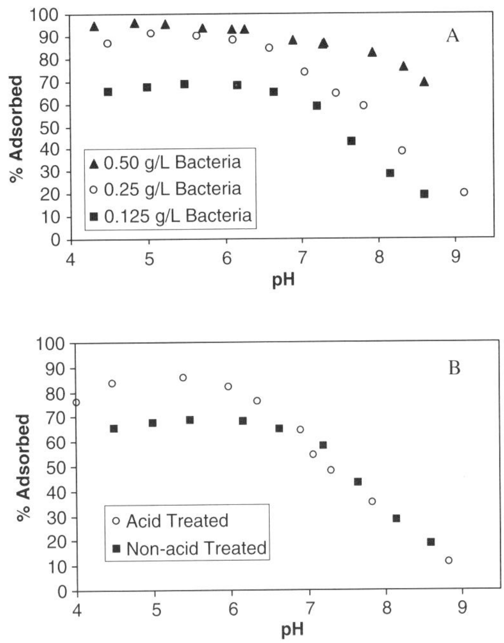

Figure 1. a: U adsorbed onto 0.125, 0.25, and 0.5 g/L of non-acid treated bacteria, b: U adsorbed onto 0.125 g/L acid treated and non-acid treated bacteria.

5 DISCUSSION

The high amount of U adsorption that we observed in the mid-pH range most likely indicates that more than just the uranyl ion adsorbs onto the cell wall. Uranyl-hydroxide species dominate in this pH range and they likely are involved in the adsorption reactions that occur between pH 5 and 6. Above pH 6, the negatively charged carbonate complexes begin to dominate the speciation and we see a marked decrease in adsorption, likely due to competition for the available uranyl ions between the negatively charged bacterial surfaces and the aqueous carbonate species. The general decrease in adsorption under high pH conditions suggests that the uranyl carbonate species do not adsorb as effectively as does the uranyl ion itself or its hydroxyl complexes.

The difference in the adsorptive capacity of the acid washed and non-acid washed bacteria is likely due to the removal of residual metals bound to the surface from the growth medium (e.g., Borrok et al. 2003, submitted). Acid washing the cells removes cell-wall bound metals and thereby effectively increases the sites available for adsorption. Bacteria in natural environments are likely better represented as non-acid washed cells. Therefore, the extent of bacterial adsorption that occurs in nature is likely closer to that observed in the non-acid washed bacterial experiments.

Our thermodynamic modeling results indicate that uranyl ion adsorption alone is not likely to be the only important aqueous U species involved in U adsorption onto the *B. subtilis* surface. The abundance of UO_2^{2+} decreases markedly above pH 5, yet we observed high, and constant, extents of adsorption to at least pH 6.5. Our models suggest that uranyl-hydroxide complexes become responsible for the overall U adsorption that we observed under these higher pH conditions. A similar shift from bare cation adsorption to cation-hydroxide adsorption was also observed by Fowle & Fein (2001) for Cu adsorption onto *B. subtilis* under similar pH conditions. Our modeling results also indicate that the competition between the bacterial surface and aqueous carbonate for the available uranyl can be quantitatively modeled using the thermodynamic stability constants for the U-bacterial surface complexes derived in this study.

6 CONCLUSIONS

U speciation under circumneutral pH conditions greatly affects U adsorption onto bacterial surfaces. We observed significant amounts of U adsorption up to near neutral pH conditions. Above pH 6.5 negatively charged uranyl-carbonate complexes begin to dominate aqueous speciation and adsorption onto the bacteria decreases markedly. Our thermodynamic modeling of the U adsorption data not only constrains the stoichiometry of the important uranyl adsorption reactions, but it also yields thermodynamic stability constants for the corresponding U-bacterial surface complexes. These constants can be used to quantitatively model the competition between aqueous carbonate complexes and the bacterial surface sites for available U under conditions more realistic and complex than those studied here.

REFERENCES

Borrok, D., Fein, J.B., Tischler, M., O Loughlin, E., Meyer, H., Liss, M. & Kemner, K.M. 2004 (submitted). The effect of acidic solutions and growth conditions on the adsorptive properties of bacterial surfaces. *Chem. Geol.*

Fein, J.B. 2000. Quantifying the effects of bacteria on adsorption reactions in water-rock systems. *Chem. Geol.* 169: 265–280.

Fein, J.B., Boily, J.F. & Yee, N. 2004 (submitted). Potentiometric titrations of *Bacillus subtilis* cells and a comparison of modelling approaches. *Geochim. Cosmochim. Acta.*

Fein, J.B., Daughney, C.J., Yee, N. & Davis, T.A. 1997. A chemical equilibrium model for metal adsorption onto bacterial surfaces. *Geochim. Cosmochim. Acta* 61(16): 3319–3328.

Fein, J.B. & Delea, D. 1999. Experimental study of the effect of EDTA on Cd adsorption by *Bacillus subtilis*: a test of the chemical equilibrium approach. *Chem. Geol.* 161: 375–383.

Fowle, D.A. & Fein, J.B. 2000. Experimental measurements of the reversibility of metal-bacteria adsorption reactions. *Chem. Geol.* 168: 27–36.

Fowle, D.A. & Fein, J.B. 2001. Quantifying the effects of *Bacillus subtilis* cell walls on the precipitation of copper hydroxide from aqueous solution. *Geomicrobiol. J.* 18(1): 77–91.

Fowle, D.A., Fein, J.B. & Martin, A.M. 2000. Experimental study of uranyl adsorption onto *Bacillus subtilis*. *Environ. Sci. Technol.* 34: 3737–3741.

Haas, J.R., Dichristina, T.J. & Wade, R., Jr. 2001. Thermodynamics of U(VI) sorption onto *Shewanella putrefaciens*. *Chem. Geol.* 180: 33–54.

Hennig, C., Panak, P.J., Reich, T., Roberg, A., Raff, J., Selenska-Pobell, S., Matz, W., Bucher, J.J., Bernhard, G. & Nitsche, H. 2001. EXAFS investigation of uranium(VI) complexes formed at *Bacillus cereus* and *Bacillus sphaericus* surfaces. *Radiochim. Acta* 89: 625–631.

Kelly, S.D., Kemner, K.M., Fein, J.B., Boyanov, M.I., Bunker, B.A. & Yee, N. 2002. X-Ray adsorption fine structure determination of pH-dependent U-bacterial cell wall interactions. *Geochim. Cosmochim. Acta* 66(22): 3855–3871.

Sar, P. & D'Souza, S.F. 2001. Biosorptive uranium uptake by a *Pseudomonas* stain: characterization and equilibrium. *J. Chem. Technol. Biotechnol.* 76: 1286–1294.

Westall, J.C. 1982. FITEQL, A computer program for determination of chemical equilibrium constants from experimental data. Dept. of Chem. Oregon St. Univ.

Water–Rock Interaction, Wanty & Seal II (eds)

Evaluation of Sn^{IV} sorption on Callovo-Oxfordian from equilibrated sedimentary waters

C. Latrille & J. Ly
Commissariat à l'Energie Atomique, DEN Saclay/SECR/L3MR, Gif sur Yvette, France

ABSTRACT: In order to test the ability of the Callovo-Oxfordian argillite to retard the movement of soluble Sn at ultra-trace levels, distribution coefficient (Kd) measurements of Sn^{IV} on two clay-rich samples were acquired in typical sedimentary water. The results show large Kd values (1.4×10^4 to 2.4×10^4 mL/g), which are unaffected by moderate variation in the water composition and the proportion of reactive mineral phases in the suspensions.

1 INTRODUCTION

The French concept of deep geological disposal is based on optimizing the isolation of nuclear waste from the human environment. This concept utilizes multiple barriers composed of a container, a clay-rich barrier, and the host rock for the disposal vault. The rock acts as a natural barrier to the transfer of dissolved radioactive contaminants. Interactions between the dissolved contaminants and the mineral surfaces of the host rock retard the movement of contaminants and allow their concentrations in aqueous media to remain within acceptable limits. Many research programs have addressed the behavior of dissolved actinides or fission products in such environments. However, sorption experiments with Sn are poorly represented, being limited by the relatively low solubility of Sn in aqueous solution in the pH range of interest and by the availability of free Sn radioisotopes (Ticknor et al. 1996; Legoux et al. 1992).

This study examines the retention of Sn^{IV} by argillites simulating the host rock. The results are presented as distribution coefficient (Kd) measurements from waters representative of the equilibrium pore water. Batch experiments were conducted to examine the effect of a variation of water composition on Sn^{IV} sorption. Experimental conditions are focused on maintaining the oxidation–reduction conditions of the natural environment and the solubility of Sn.

2 MATERIALS AND METHODS

2.1 *Materials*

2.1.1 *Solids*

The two argillites used in these experiments are from the Callovo-Oxfordian formation (Meuse/Haute-marne, Eastern France). They were chosen for their distinct proportions of minerals. The first argillite (A1) is dominantly composed of carbonate minerals (10–23% clay minerals and 48–64% carbonates). The second argillite (A2) contains dominantly clay minerals (37–55% clay minerals and 27% carbonates) (Cassagnabère et al. 2001). These rocks were grounded in an agate mortal and sieved at 63 µm. The fraction less than 63 µm was retained for the experiments.

2.1.2 *Solutions*

The influence of water composition on the retention of Sn^{IV} was examined using eight synthetic waters for determination of adsorption isotherms on each solid. The first water (W1) was based on the composition of a representative natural pore water at the site. The seven other synthetic waters (MD1–MD7) compositions had a range of chloride concentrations (20×10^{-3} to 71.7×10^{-3} mol/L), pH (7.2–7.7) and $p(CO_2)$ (3.09×10^{-3} to 7.63×10^{-4} atm) consistent with *in situ* equilibrated sedimentary waters, using a Doehlert experimental design matrix (Doehlert 1970) and the geochemical equilibrium model of

Table 1. Concentration in Sn introduced in suspensions of A1 and A2 samples.

Concentration (mol/L)	Suspensions				
[Sn] total	1.67×10^{-8}	5.58×10^{-9}	1.67×10^{-9}	5.58×10^{-10}	1.67×10^{-10}
$[^{113}Sn]$	9.83×10^{-12}	3.28×10^{-12}	9.83×10^{-13}	3.28×10^{-13}	9.83×10^{-14}

Beaucaire et al. (2000). The $p(CO_2)$ was controlled by bubbling appropriate CO_2-argon gas mixtures through solutions.

2.1.3 *Radiotracer ^{113}Sn mother-solution*

The ^{113}Sn used was 10 μg/g $SnCl_4$ dissolved in 6 mol/L HCl and obtained from Cerca, Pierrelatte, (France). The ^{113}Sn was combined with stable Sn in a molar ratio of 5.87×10^{-4}.

2.2 *Methods*

Argillite powder (0.1 g) was introduced to 5 mL of synthetic water in a polycarbonate centrifuge tube used as the reaction vessel. The sample (A1 or A2) was equilibrated with each of the eight solutions in four successive equilibrations with the synthetic water. Equilibrium pH values of the suspensions were between 7.2 and 7.8. Measured Eh values in A1 and A2 suspensions were both +220 mV at 22°C, in agreement with stability of Sn^{IV} (Séby et al. 2001).

The radionuclide was introduced in known amounts, as small aliquots of solutions previously prepared by dilution of a stock solution in deionized water and appropriately neutralized. The total Sn introduced in each tube was calculated in order to be under the solubility limits of $Sn(OH)_4$ (am) and SnO_2 (cr) (Table 1).

The tubes were then mechanically agitated for 7 days and centrifuged at 40,000 rpm for 20 min. An aliquot of solution (1 mL) was taken for γ radiation measurement using a gamma-counter.

Values of the distribution coefficient Kd (mL/g) are calculated by the equation:

$$Kd = \left(\frac{Ao}{Aeq} - 1\right) \times \left(\frac{V}{m}\right)$$

where Ao = initial activity; Aeq = activity at equilibrium; V = volume of suspension in mL; and m = solid mass in g.

The values of Ao and Aeq are proportional to the initial concentration and the equilibrium concentration of the radionuclide. The concentration of total Sn is calculated from the ^{113}Sn/Sn total ratio.

3 EXPERIMENTAL RESULTS

Kd measurements of Sn^{IV} on A1 solid with the eight synthetic waters are presented in Figure 1. Among the five initial Sn concentrations investigated, only two or three determinations are above analytical detection limit. The other results are not significant and thus not shown in Figure 1. All Kd values lie between 5.3×10^3 and 2.4×10^4 mL/g. The average value is 1.4×10^4 mL/g. The relative standard deviation of Kd from the average value is 43% (0.4 in log(10) units).

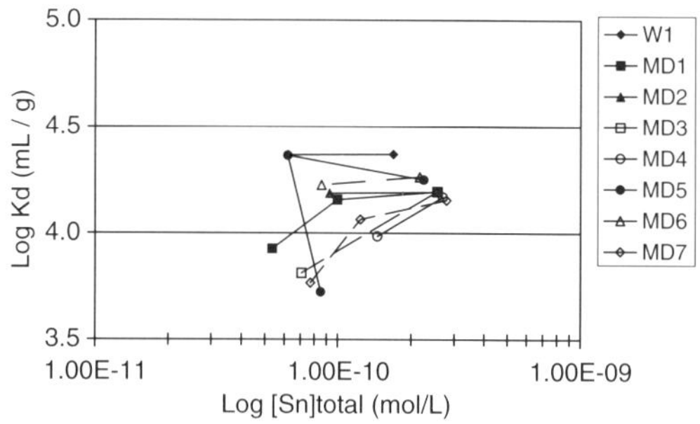

Figure 1. Kd vs. total Sn^{IV} on A1 argillite in synthetic waters W1 and MD1 to MD7.

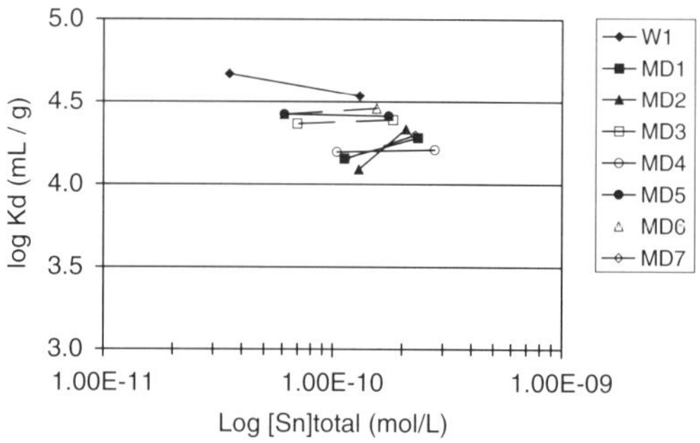

Figure 2. Kd vs. total Sn^{IV} on A2 argillite in synthetic waters W1 and MD1 to MD7.

Kd vs total Sn^{IV} on A2 solid with synthetic waters W1 and MD1 to MD7 are presented in Figure 2. The Kd distributions in each water show only two significant points for the same reason as mentioned before. Kd values lie between 1.2×10^4 and 4.7×10^4 mL/g, with the average Kd value 2.1×10^4 mL/g. Kd distribution acquired with the W1 water seems to be above the others. However, the Kd values vary up to one half log unit, which is greater than the observed deviations. So, no significant distinction could be made between all Kd values.

4 CONCLUSIONS

Kd values for Sn^{IV} at ultratrace levels on the two Callovo-Oxfordian argillites in typical sedimentary waters are very large. These results indicate that these argillites should effectively retard the movement of Sn under similar natural conditions. Similar results were acquired on montmorillonite (Ticknor et al 1996) and other natural materials mixed with phyllosilicates (Legoux et al. 1992), although experimental conditions were slightly different.

For the range of compositions examined, which correspond to natural waters equilibrated with the rock, no significant variation of the distribution coefficient was observed.

Comparison between samples A1 and A2 showed no sensitivity of the Sn^{IV} retention at trace levels for different mineral proportions.

The present work shows that retention of Sn^{IV} at ultra-trace levels on Callovo-Oxfordian argillite is strong and largely insensitive to chemical and mineralogical variations observed in the formation.

ACKNOWLEDGEMENTS

The authors wish to thank the Agence nationale pour la gestion des déchets radioactifs (ANDRA) for funding this study.

REFERENCES

Beaucaire, C., Pitsch, H., Toulhoat, P., Motellier, S. & Louvat, D. 2000. Regional fluid characterisation and modelling of water–rock equilibrium in the boom clay Formation and in the Rupelian aquifer at Mol, Belgium. *Appl. Geochem. 15*: 667–686.

Cassagnabère, A., Bouchet, A., Rassineux, F., Sanmartino, S., Parneix, J-C. Marchand, D. & Caillaud, J. 2001. Laboratoire de recherche souterrain de Meuse/Haute-marne. Analyses minéralogiques et géochimiques du forage est 205. Rapport de mesures final. Contrat de l'ANDRA ARZ LC n°015794. Analytical report of the Company Etudes Recherches Matériaux. Etudes Recherches Matériaux, Poitiers, France, p. 565.

Doehlert, D.H. 1970. Uniform shell designs. *Appl. Statistics, 19*: 231–239.

Legoux, Y., Blain, G., Guillaumont, R., Ouzounian, G., Brillard, L. & Hussonois, M. 1992. Kd measurements of activation, fission and heavy elements in water/solid phase systems. *Radiochim. Acta 58/59*: 211–218.

Séby, F., Potin-Gautier, M., Giffaut, E. & Donard, O.F.X. 2001. A critical review of thermodynamic data for inorganic tin species. *Geochim. Cosmochim. Acta* 65(18): 3041–3053.

Ticknor, K.V., Vandergraaf, T.T., McMurry, J., Boisvenue, L. & Wilkin D.L. 1996. Parametric studies of factors affecting Se and Sn sorption, AECL, Whiteshell Laboratories, Pinawa, Manitoba, Canada: p.59.

Water–Rock Interaction, Wanty & Seal II (eds)

Influence of humic acids on the migration of $U^{(IV/VI)}$ in quartz sand

J. Mibus, S. Sachs, C. Nebelung & G. Bernhard
Forschungszentrum Rossendorf e.V., Institute of Radiochemistry, Dresden, Germany

ABSTRACT: This study investigated the migration behavior of U^{IV} and U^{VI} in the system humic acid/U/quartz sand by means of column experiments. Humic acid exhibits a significant influence on the transport of U^{IV} and U^{VI}. In case of U^{VI} a clear mobilizing effect was observed. There are strong indications for a similar influence of humic acid on U^{IV} transport. The two redox species of U differ slightly in their migration behavior in the presence of humic acid. This is mainly reflected in a difference of their total recoveries of approximately 25%.

1 INTRODUCTION

The reliable long-term risk assessment for potential nuclear waste repositories and for the remediation of contaminated areas and facilities from the former U mining and milling in Saxony and Thuringia, Germany, requires detailed knowledge of the migration behavior of actinides in natural aquifer systems. It is necessary to know processes and substances that can influence the migration of actinides and to understand the interaction between actinides and relevant rock materials.

Humic acids (HA), organic macromolecules ubiquitous in natural systems, play an important role in the interaction processes of actinide ions. HA are soluble in the pH range of natural waters and possess a high complexing capacity toward metal ions. They are characterized by pronounced redox properties, which can influence the oxidation state of metal ions. Due to these properties HA can affect the speciation of metal ions, e.g. actinide ions, and therefore the migration of these pollutants in the environment (Choppin 1999).

The influence of HA on the migration of actinides in the environment can be studied in laboratory column experiments. Column experiments with natural sand and groundwaters rich in humic substances have demonstrated that a certain fraction of the actinides present (e.g., U^{VI}, Artinger et al. 2002; Am^{III}, Artinger et al. 1998; and $Np^{IV/V}$, Artinger et al. 2000) migrate as humic colloid-borne components as fast as the groundwater flow. This colloid-borne migration is controlled by kinetic processes.

The present study focuses on the influence of HA on the migration of U^{VI} and U^{IV} in a quartz sand system at a laboratory scale. The migration of U^{VI} was studied in the presence and absence of HA. In addition, the HA migration under the applied experimental conditions was examined. To characterize the migration behavior of U under reducing conditions we studied the migration of the redox couple $U^{IV/VI}$ in the presence of HA.

2 EXPERIMENTAL

2.1 *Experimental set-up*

Migration experiments were performed in a glove box under a nitrogen atmosphere. The experimental setup is shown in Figure 1.

Columns 250 mm in length and 50 mm in diameter and tightly packed with quartz sand were used. A marine fine sand, Gaussian-distributed in particle size, with a mean grain diameter of 153 μm from

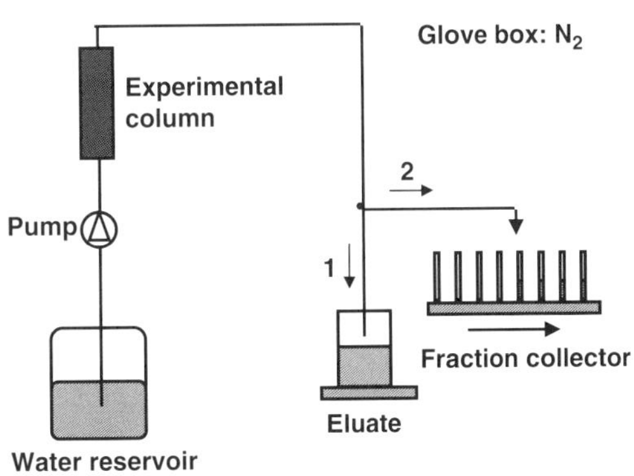

Figure 1. Set-up of the column experiments.

Table 1. Experimental conditions and transport parameters of the column experiments.

Experiment	Impulse (mL)	HA (mg/L)	U (mol/L)	Darcy velocity (m/s)	Effective pore volume (mL)	Dispersion coefficient (m^2/s)
1	660	49.9	–	$3.3 \cdot 10^{-6}$	181	$5.5 \cdot 10^{-5}$
2	716	50.3	–	$4.4 \cdot 10^{-6}$	181	$5.5 \cdot 10^{-5}$
3	690	–	$1.1 \cdot 10^{-5}$	$3.4 \cdot 10^{-6}$	181	$6.0 \cdot 10^{-5}$
4	457	49.9	$1.0 \cdot 10^{-5}$	$3.3 \cdot 10^{-6}$	178	$6.1 \cdot 10^{-5}$
5	462	50.0	$1.0 \cdot 10^{-6}$	$3.6 \cdot 10^{-6}$	173	$5.0 \cdot 10^{-5}$

Heerlen, Netherlands, was chosen for its chemical purity ($Fe_2O_3 \leq 0.01\%$, $Al_2O_3 \leq 0.05\%$, $TiO_2 \leq 0.02\%$). Prior to use, the quartz sand was washed with Milli-Q water and annealed for 6 h at 700°C. After packing, the sand columns were equilibrated with degassed $NaClO_4$ solution (0.1 M, pH 7.5) over a period of 3 weeks.

All experiments were carried out at room temperature in the absence of light, applying degassed solutions. Continuous pulse injection was applied. The tracer impulse was followed by a multiple elution of the column with 0.1 M $NaClO_4$ (pH 7.5). For the determination of the hydraulic and transport parameters of the columns, tritiated water (HTO) was used as conservative tracer. Each experimental condition was studied in a separate column in order to avoid mutual interferences. The transport parameters and the experimental conditions for all experiments are listed in Table 1 and demonstrate good reproducibility of the column packing.

2.2 *HA migration*

The HA migration in the quartz sand system in the presence and absence of U was studied applying synthetic HA type M42 (Pompe et al. 1998). This shows functional and structural properties that are comparable to natural HA. In addition, synthetic HA type M42 exhibits a U^{VI} complexation behavior that is comparable to natural HA (Pompe et al. 1998). In order to perform a very precise HA measurements in environmentally relevant concentrations we spiked the HA solutions with ^{14}C-labeled synthetic HA type M42 (2.38 MBq/g, Pompe et al. 2001). This ^{14}C-labeled HA was synthesized according to the unlabeled product, however, applying uniformly ^{14}C-labeled glutamic acid as a precursor substance. Both, the unlabeled and the ^{14}C-labeled synthetic HA show comparable properties (Pompe et al. 2001).

For the study of the HA migration in the absence of U (experiments 1 and 2), HA solutions with a concentration of 50 mg/L (specific activity: 70 kBq/L) were prepared by dissolving unlabeled and ^{14}C-labeled synthetic HA type M42 in 0.1 M $NaClO_4$ solution. The pH values of these solutions were adjusted to pH 7.5 using dilute $HClO_4$ and NaOH solutions.

2.3 *U^{VI} migration in the absence and presence of HA*

The U^{VI} migration in the absence of HA (experiment 3) was studied using a 1.1×10^{-5} M $UO_2(ClO_4)_2$ (0.1 M $NaClO_4$, pH 7.5) solution.

An acidic $UO_2(ClO_4)_2$ solution was added to a ^{14}C-labeled HA solution in order to determine the U^{VI} migration in the presence of HA (experiment 4). The final uranyl humate solution with a HA concentration of 50 mg/L (^{14}C: 70 kBq/L) and a U concentration of 1.0×10^{-5} M (0.1 M $NaClO_4$) was spiked with 40 kBq/L ^{234}U containing traces of ^{232}U. Before the injection into the column this solution was equilibrated for 98 h.

2.4 *U^{IV} migration in presence of HA*

For the study of the U^{IV} migration in the presence of HA (experiment 5), U^{IV} was prepared by electrochemical reduction of an acidic uranyl (^{234}U) nitrate solution. The absence of U^{VI} in the U^{IV} stock solution was confirmed by time-resolved laser-induced fluorescence spectroscopy. For experiment 5, the acidic $^{234}U^{IV}$ solution was added to a ^{14}C-labeled HA solution. The final concentrations of the U humate solution (0.1 M $NaClO_4$, pH 7.5) amounted to 50 mg/L HA (^{14}C: 70 kBq/L) and 1.0×10^{-6} M ^{234}U (60 kBq/L). This solution was equilibrated for 2 h prior to the start of the experiment.

In order to differentiate between U^{IV} and U^{VI}, redox speciation was determined by liquid–liquid extraction using TTA (thenoyltrifluoracetone; Bertrand & Choppin 1982). The extraction yield was found to be 90%. The standard deviation amounts to 5%.

2.5 *Measurement of breakthrough curves*

Breakthrough curves of HTO, HA and U were measured by fraction analysis using liquid scintillation counting (LSC) for HTO, ^{14}C and ^{234}U, and ICP-MS for ^{238}U. The LSC spectra were deconvoluted to determine the activity contribution of ^{232}U daughter nuclides.

3 RESULTS AND DISCUSSION

3.1 *Speciation of U*

The interpretation of the experimental data requires detailed knowledge on the species distribution. In a 10^{-5} M carbonate-free U^{VI} solution at pH 7.5 the dominating U^{VI} species is $UO_2(OH)_2(aq)$. In the presence of HA about 90% of U^{VI} is complexed as a ternary U^{VI} humate complex ($UO_2(OH)HA(I)$). At a redox potential of 50 mV the U speciation in carbonate-free solution is dominated by 84% $UO_2(OH)_2(aq)$, with 11% $U(OH)_4(aq)$ and 4% $UO_2(OH)_3^-$. No species calculations in the presence of HA are possible due to the lack of thermodynamic data for the U^{IV} HA complexation.

In the studied system HA controls the redox potential. A redox potential of about 50 mV can be derived for the studied system from literature data of Baraniak et al. (1998).

3.2 *HA migration in the absence of U (experiments 1 and 2)*

Evaluating the HA breakthrough curves, the retardation factor R_f, which represents the quotient of the effluent volume and the effective pore volume, as well as the total recovery of HA were determined (Table 2). The R_f values indicate only slightly retarded transport of the HA compared to HTO (R_f: 1). Thus, there are no significant interactions between HA and the quartz sand surface, due to the negative charge of both species at pH 7.5. Batch experiments confirm this result. The detected R_f implies that there are no grain size exclusion effects, which would be reflected in R_f values less than 1. The HA recovery of about 0.9 points to irreversible immobilization processes, e.g. filtration effects.

3.3 *U^{VI} migration in the absence of HA (experiment 3)*

The U^{VI} migration in the absence of HA is characterized by a strong retardation, which is reflected in the R_f value greater than 1 (cf. Table 2 and Fig. 2). The shape of the breakthrough curve indicates kinetically controlled processes (Toride et al. 1995). The observed tailing indicates a slow release of U^{VI} from the quartz surface even after five eluted pore volumes. The low U^{VI} recovery can be caused by precipitation or sorption processes. At a U^{VI} concentration of 10^{-5} M, the solution is oversaturated with respect to the solid phase schoepite. However, it

Table 2. Retardation factors and total recoveries of the different injected species.

Experiment	Species	R_f	R^*_{tot}
1	HA	1.02	0.91
2	HA	1.02	0.92
3	U^{VI}	2.09	0.43
4	HA	1.05	0.85
4	U^{VI}+HA	1.19	0.80
5	HA	1.04	0.90
5	U^{VI}+HA	1.11	0.90
5	U^{IV}+HA	1.07	0.66

* R_{tot}: total recovery of the injected species.

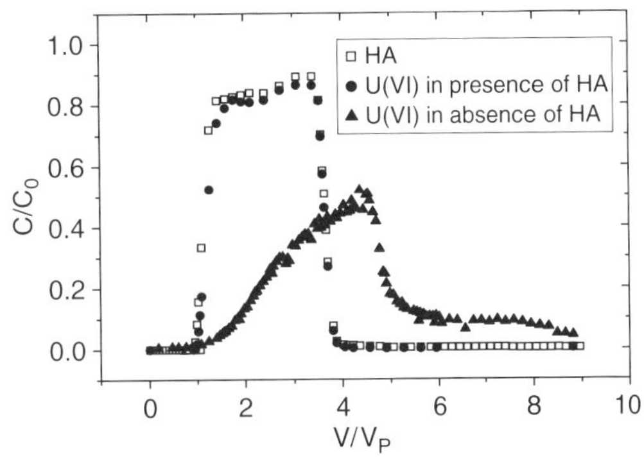

Figure 2. Breakthrough curves of HA and U^{VI} in experiments 3 and 4. C: measured and C_0: initial concentration; V: effluent volume; V_P: effective pore volume.

is not clear if this solid is formed over the residence time of U in the system.

3.4 *U^{VI} migration in the presence of HA (experiment 4)*

HA-bound U^{VI} exhibits a completely different breakthrough behavior, which is depicted in Figure 2. In the presence of HA the U^{VI} transport is accelerated, and the U^{VI} recovery is increased. This behavior is attributed to the formation of soluble U^{VI} humate complexes. However, for both components, HA and U^{VI}, slightly different R_f values and recoveries were detected (cf. Table 2). These effects are caused by association/dissociation processes in the system HA-U^{VI}-quartz. A certain part of U^{VI} dissociates from the soluble humate complex and adsorbs onto the quartz surface. Comparing the U^{VI} and HA recoveries one can conclude that about 5% of U^{VI} is irreversibly bound on the quartz surface. Considering the results

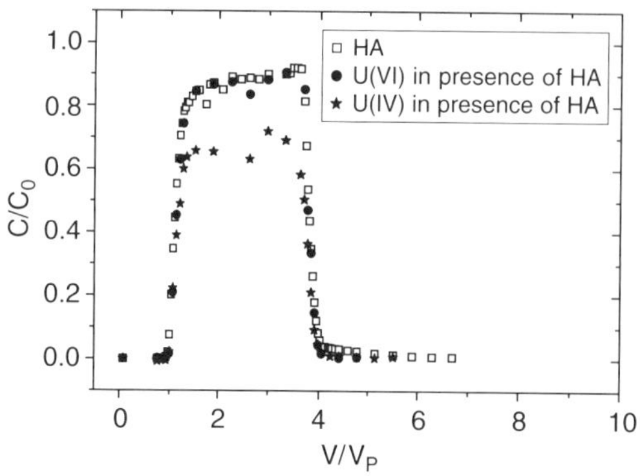

Figure 3. Breakthrough curves of HA, U^{IV}, and U^{VI} in experiment 5. C: measured and C_0: initial concentration; V: effluent volume; V_P: effective pore volume.

reported by Artinger et al. (2002), it can be assumed that these processes are kinetically controlled.

3.5 *Migration of the U^{IV}/U^{VI} couple in the presence of HA (experiment 5)*

The distribution of U^{IV} and U^{VI} in the starting solution amounts to 8% and 92%, respectively. This is comparable to the calculated redox species distribution at 50 mV mentioned above. This redox speciation was stable over the duration of the experiment. Moreover, this initial ratio of U^{IV} and U^{VI} was also found in the effluent solutions. This indicates that no oxidation processes occurred in the column.

Figure 3 shows the breakthrough curves for HA, U^{IV} and U^{VI}. Based on these results, R_f values and total recoveries (listed in Table 2) are derived. As with the case of U^{IV}, the association/dissociation processes mentioned above control the transport behavior in the presence of HA ($R_f > 1$). Although U^{IV} and U^{VI} are characterized by different thermodynamic properties (Choppin 1983) no significant differences in their R_f values were observed under the applied experimental conditions.

A comparison of the U balances in experiments 4 and 5 shows that a decreased total U concentration results in a slightly higher recovery. In experiment 5 the total recoveries of both U^{VI} and HA agree very well. This fact supports a HA-bound U^{VI} transport.

In spite of the increased uncertainty of the extraction data, the U^{IV} recovery is lower than that of U^{VI}. This result indicates a stronger interaction of U^{IV} with the quartz sand surface and thus a more distinct immobilization. Nevertheless, there are strong indications that HA has a mobilizing effect on the U^{IV} transport in the investigated system due to its complexing ability.

4 CONCLUSIONS

The HA affects the migration behavior of both U^{IV} and U^{VI}. In presence of HA, U^{VI} is significantly mobilized. A similar effect is likely for U^{IV}. Both U redox species exhibit a different migration behavior, in particular reflected in their effluent balances.

The system under investigation represents a strong simplification of natural aquifers in view of geology and solution composition. More complex compositions are typical in natural aquifer systems. Strong reducing conditions can occur in deep groundwaters. There, the tetravalent actinides become more important. However, the performed experiments are suitable in order to identify sensitive relationships between actinides, humic acids and rock matrix and to establish more sophisticated concepts for understanding and modeling of complex migration processes.

ACKNOWLEDGEMENTS

This work was financially supported by the Bundesministerium für Wirtschaft und Arbeit under contract No. 02 E 9299. We thank V. Brendler for the thermodynamic calculations, J. Tutschku for the electrochemical reduction of U^{VI}, and U. Schaefer for ICP-MS measurements. The authors gratefully acknowledge S. Brockmann, R. Ruske, and M. Meyer for their technical assistance.

REFERENCES

Artinger, R., Kienzler, B., Schuessler, W. & Kim, J.I. 1998. Effects of humic substances on the ^{241}Am migration in a sandy aquifer: batch and column experiments with Gorleben groundwater/sediment systems. *J. Contam. Hydr.* 35: 261–275.

Artinger, R., Seibert, A., Marquardt, C.M., Trautmann, N., Kratz, J.V. & Kim, J.I. 2000. Humic colloid-borne Np migration: influence of the oxidation state. *Radiochim. Acta* 88: 609–612.

Artinger, R., Rabung, T., Kim, J.I., Sachs, S., Schmeide, K., Heise, K.H., Bernhard, G. & Nitsche, H. 2002. Humic colloid-borne migration of uranium in sand columns. *J. Contam. Hydr.* 58: 1–12.

Baraniak, L., Mack, B., Abraham, A. & Neubert, H. 1998. Untersuchung des Einflusses der in Grubenwässern gelösten organischen Verbindungen auf den Valenzzustand von Radionukliden und Schermetallen. *Final Report, Project No. 4–7541.83-FZR/512.* Forschungszentrum Rossendorf, Rossendorf, Germany: Institute of Radiochemistry.

Bertrand, P.A. & Choppin, G.R. 1982. Separation of actinides in different oxidation states by solvent extraction. *Radiochim. Acta* 31: 135–137.

Choppin, G.R. 1983. Solution chemistry of the actinides. *Radiochim. Acta* 32: 43–53.

Choppin, G.R. 1999. Role of humics in actinide behavior in ecosystems. In G.R. Choppin & M.Kh. Khankhasayev (eds.), *Chemical Separation Technologies and Related Methods of Nuclear Waste Management:* 247–260. Kluwer Academic Publishers.

Pompe, S., Brachmann, A., Bubner, M., Geipel, G., Heise K.H., Bernhard, G. & Nitsche, H. 1998. Determination and comparison of uranyl complexation constants with natural and synthetic humic acids. *Radiochim. Acta* 82: 89–95.

Pompe, S., Bubner, M., Meyer, M., Nicolai, R., Heise, K.H. & Bernhard, G. 2001. Synthesis of ^{14}C-labeled and unlabeled synthetic humic acid type M42. In Th.Fanghänel (ed.), *FZR-318, Annual Report 2000:* 31.Forschungszentrum Rossendorf, Rossendorf, Germany: Institute of Radio-chemistry.

Toride, N., Leji, F.J. & van Genuchten, M.Th. 1995. The CXTFIT code for estimating transport parameters from laboratory or field tracer experiments. version 2.1. *Research Report No. 137.* U.S. Salinity Laboratory, Agricultural Research Service, Riverside, California: U.S. Department of Agriculture.

Water–Rock Interaction, Wanty & Seal II (eds)
© 2004 Taylor & Francis Group, London, ISBN 90 5809 641 6

Interactions of the U mining waste pile isolate *Bacillus sphaericus* JG-A12 with U

J. Raff, M. Merroun & S. Selenska-Pobell
Institute of Radiochemistry, Research Center Rossendorf, Dresden, Germany

A. Rossberg
ESRF, Grenoble, France

U. Soltmann & H. Böttcher
GMBU, Department of Functional Layers, Dresden, Germany

ABSTRACT: Apart from chemical and physical factors influencing mobility of radionuclides, microbial metabolism also plays a decisive role in enhancing or retarding migration effects. Bacteria are able to reduce or oxidize metals and also to accumulate and bind them. In this work biosorption of U by cells, spores and S-layers of a U mining waste pile isolate *Bacillus sphaericus* JG-A12 was investigated. Furthermore, these three bacterial structures were used for construction of porous biocomponent-silicate hybrids referred to as biological ceramics (biocers) for bioremediation. The results show the highest binding capacity for the spores followed by the cells and the S-layer sheets. The silicate matrix itself binds only small amounts of U. Sol-gel encapsulation does not affect the U binding properties of the cells or the S-layer sheets but strongly reduces the binding capacity of the spores. Spectroscopic analyses of the U complexes demonstrate that U is bound to phosphate and carboxyl groups of the biocomponents and to silanol groups of the silicate matrix.

1 INTRODUCTION

After a long-lasting period of U mining and production in Saxony and Thuringia, mines have fallen into disuse. Soil, subsoil and waste waters of the former mines and production sites still contain high concentrations of heavy metals and radionuclides. All these sites and extant waste waters have to be cleaned before a renewed use will be possible. Current methods for cleaning contaminated effluents (i.e. ion exchange, precipitation, adsorption or electrochemical procedures) are only useful for high concentrations of metals but unsuitable for low concentrations. Besides chemical and physical procedures, the biota offer possibilities for removing toxic metals. For instance, bacteria in highly polluted environments often possess protective mechanisms against toxic metals, based on biotransformations, accumulation, sorption and mineralization. Comprehensive studies of the interactions of bacteria with metals are necessary prior to development of effective bioremediation strategies. As already demonstrated, the natural isolate *B.sphaericus* JG-A12, recovered from the U waste pile "Haberland" situated near the town of Johanngeorgenstadt, Saxony, Germany, selectively and reversibly binds U, Cu, Pb, Al, and Cd (Selenska-Pobell et al. 1999). Additionally, *B. sphaericus* JG-A12 possesses a surface layer (S-layer) protein with an unusual structure (Pollmann, in press). This outermost cell membrane, with a highly ordered lattice structure plays a decisive role in the interaction of bacteria with compounds in the environment. As native microbial biomass or purified proteins are not suitable as a binding matrix in large bioreactors, immobilization of the biocomponent is of fundamental importance. It was demonstrated that by using a sol-gel technique various biomolecules or microbial cells can be embedded in SiO_2 ceramics without loosing their activity (Böttcher 2000). The resulting biological ceramics (biocers) combine high mechanical stability, biocompatibility and controlled matrix porosity of the inorganic silicate with the capacity of bacteria to detoxify heavy metals and radionuclides by biosorption and therefore offer the opportunity to construct selective metal binding filters (Raff et al. 2003; Soltmann et al. 2003).

This paper reports U sorption of biocers containing cells, spores and purified S-layer of *B. sphaericus* JG-A12 and of the free biocomponents as a function of time and preparation conditions. The atomic structure of the U complexes formed was investigated by extended X-ray absorption fine structure (EXAFS) spectroscopy.

2 EXPERIMENTAL PROCEDURES

2.1 *Bacterial growth*

B. sphaericus was grown on nutrient broth ($8\,g\,l^{-1}$, Difco) at pH 7.0 and 30°C. Cells were harvested at the late exponential growth phase.

2.2 *Preparation of spores*

For sporulation the nutrient broth medium was supplemented with $10\,mg\,l^{-1}$ $MnSO_4 \times H_2O$. Spores were harvested by centrifugation at 10,000 G for 20 min, resuspended in ultra pure water at 4°C and treated alternately with $0.2\,mg\,l^{-1}$ lysozyme and $0.1\,mg\,l^{-1}$ trypsin (Boehringer, Ingelheim) until complete lysis of the cells and the cell wall fragments.

2.3 *Isolation of surface layer protein*

Harvested cells were suspended in 50 mM TRIS-HCl, 1 mM $MgCl_2 \times 6\,H_2O$, 3 mM NaN_3, pH 7.5 (standard buffer) and separated from flagella by homogenization in a rotating-blade blender IKA T8 (IKA Labortechnik, Stauffen) for 10 min on ice at maximum speed. Flagella free cells were resuspended 1:1 in standard buffer and mixed with a few crystals of DNAse II and RNAse. Cells were disintegrated in a Microfluidizer 110 L (Microfluidics Corporation, Newton) at 4°C, a pressure of 960 bar and three passes. Unbroken cells were removed by centrifugation at 6,000 G for 10 min at 4°C. Using 1% Triton X-100 in standard buffer plasma membranes were solubilized for 10 min at room temperature (RT). Peptidoglycan was lysed incubating samples in buffer containing $0.2\,mg\,ml^{-1}$ lysozyme for 6 h at 30°C. For further purification of the S-layer, protein suspensions with concentration of $20\,mg\,ml^{-1}$ were mixed 1:5 with 6 M guanidine hydrochloride in 50 mM TRIS (pH 7.2) and stirred at RT for 6 h. After centrifugation at 12,400 G for 60 min at 4°C the supernatant was dialyzed two times against 2 liters 10 mM $CaCl_2$ for a total of 24 h at 4°C. Dialysis tubing with a molecular weight cut-off of 50 kDa was used. Reassembled S-layer sheets were harvested by centrifugation at 12,400 G for 60 min at 4°C, suspended in 10 mM $CaCl_2$ and stored at 4°C.

2.4 *Crosslinking of S-layer*

To stabilize the square lattice structure of native and recrystallized S-layer sheets at acidic pH, the protein monomers were crosslinked using 1-Ethyl-3-(N,N′-dimethyl-aminopropyl)-carbodiimid (EDC, Sigma-Aldrich Chemie, Taufkirchen; Hoare 1967). S-layers were incubated with $30\,mg\,ml^{-1}$ EDC in buffer for 48 h at 20°C. Crosslinked S-layers were harvested by centrifugation at 12,400 G for 30 min at 4°C, washed with distilled water and stored at 4°C.

2.5 *Preparation of aqueous nanosols and biocers*

Silica sols were prepared as previously described by Soltmann et al. (2003). For biocers 40 ml aqueous silica sol was adjusted to pH 7 with NaOH and mixed with biocomponent suspensions in the following amounts: $9.35\,mg\,ml^{-1}(= 6.5 \times 10^9)$ *B. sphaericus* cells, $3.96\,mg\,ml^{-1}$ $(= 6.5 \times 10^9)$ spores or 9.57 mg ml^{-1} S-layer. Sorbitol (20% w/w SiO_2) was added to some silica sols to achieve a higher porosity. The biosol was poured into trays to a layer thickness of circa 7 mm. Gelling occurs shortly after neutralization and addition of the biocomponents. All gels were aged for 3 d at 4°C, cut into small pieces and air dried at RT or by freeze drying. The dry gels were sieved to particles with a size of 355–500 μm.

2.6 *Sorption and desorption experiments*

Sorption experiments were carried out with 200 mg dry wt of xerogel or biocer particles containing 36.4 mg *B. sphaericus* cells ($= 2.6 \times 10^{10}$ particles), 17.23 mg spores ($= 2.8 \times 10^{10}$) or 36.4 mg S-layer protein and with free cells, spores and S-layer as well. All samples were incubated with 35 ml 0.9% $NaClO_4$ at pH 4.5 with 9×10^{-4} M $UO_2(NO_3)_2 \times 6H_2O$ (Fluka, Deisenhofen) at 30°C for 48 h. Desorption experiments were carried out by washing two times with 0.9% $NaClO_4$ (pH 4.5) in 35 ml and 0.5 M citric acid, trisodium salt (pH 4.5) at 30°C for 24 h.

2.7 *Extended X-ray absorption fine structure (EXAFS) spectroscopy*

U L_{III}-edge X-ray absorption spectra were collected in fluorescence modeat the Rossendorf Beamline at the European Synchrotron Radiation Facility (Grenoble) using the Si(111) double-crystal monochromator at RT. The energy was calibrated by measuring the Y K-edge transmission spectrum of a yttrium foil and defining the first inflection point as 17,038 eV. Eight spectra for each sample were recorded. All samples were measured as dry samples.

3 RESULTS AND DISCUSSION

3.1 *The S-layer of Bacillus sphaericus JG-A12*

Bacillus sphaericus JG-A12 cells possess an S-layer protein, which forms a tetragonally arranged structure on the surface of the cell wall with a lattice constant of 12.5 nm (Raff 2002). The molecular weight of the monomers was estimated to be 126 kDa. Its primary structure differs significantly from all S-layer proteins previously known, but shows significant similarity to that of the S-layer protein of *B. sphaericus* NCTC 9602 also analyzed in this work (Pollmann, in press). Interestingly, both S-layers are phosphorylated but those of JG-A12 possess five to six times more phosphorous (Raff 2002). The functions of the S-layers are not uniform and well understood. In the case of environmental isolates, the S-layer may function as a molecular sieve, ion trap or protective shell (Beveridge 1979; Sàra & Sleytr 1987). It was demonstrated that they can bind heavy metals (Shenton et al. 1997; Brown et al. 1998), which makes them very suitable for the construction of metal binding filters.

3.2 *Embedding of biocomponents*

Reliable immobilization of biocomponents without influencing binding capacities is fundamental for multiple application as metal binding matrix. Using sol-gel techniques complete embedding and uniform distribution of cells, spores and S-layer in a porous silicate matrix was achieved, confirmed by scanning electron microscopy (Raff et al. 2003). In addition to immobilization, accessibility of the biological binding sites must be ensured. During air drying the silicate matrix shrinks, which results in a reduction of the pore size and therefore a reduced accessibility. To prevent the latter, water soluble compounds such as sorbitol were added to the silica sol or freeze drying of biocers was performed. Leaching of the sorbitol from the particles after drying or sublimation of ice during the freeze-drying process leads to a more porous structure and less shrinkage.

3.3 *U binding*

U sorption of used biocomponents, the corresponding biocers and of the xerogel was compared in order to investigate the influence of the embedding procedure on the U binding capacity of the biocomponents and on the binding kinetics. Based on dry weights, spores possess the highest binding capacity with 21.9 mg U per 200 mg dry weight followed by cells which bind 12.8 mg, S-layer with 3.9 mg and xerogel with 2.7 mg (Fig. 1). 200 mg spores are equal to 3.3×10^{11} particles and 200 mg of cells to 1.4×10^{11}. 200 mg air

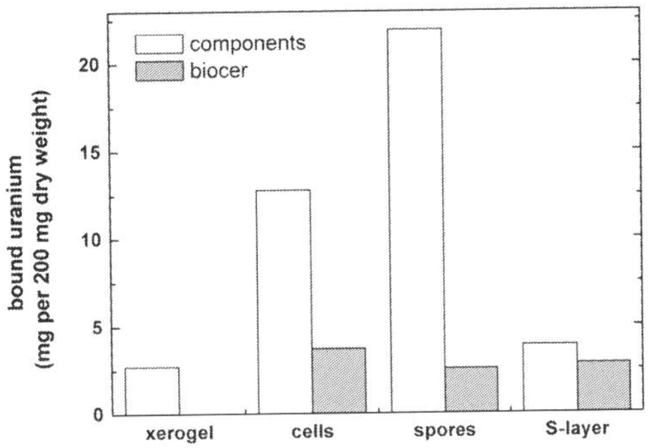

Figure 1. U sorption of 200 mg dry weight of the single components and of the corresponding biocers.

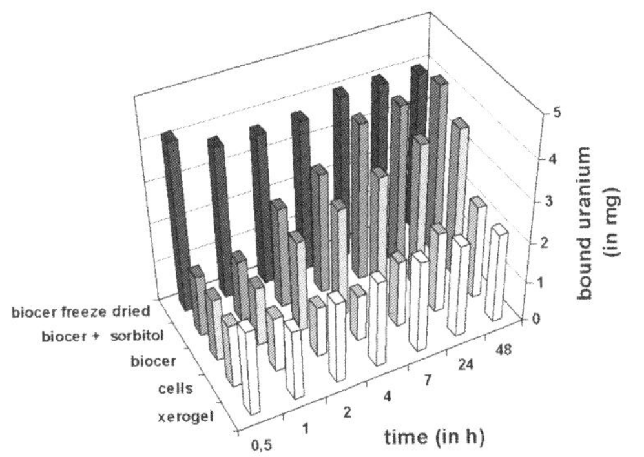

Figure 2. U sorption by 200 mg dw of the biocer types, 36 mg dw cells, and 164 mg dw xerogel (dw = dry weight).

dried biocers were obtained by embedding 36.4 mg cells, 17.23 mg spores or 36.4 mg S-layer. Immobilization of the cells has no negative effects on their binding capacity. 200 mg cell biocer binds 3.7 mg of U. The same applies for the S-layer biocer which binds 2.8 mg.

Only spore biocer possesses a lower binding capacity in comparison to the free spores. An explanation for this effect may be the low water content of spores. While cells, S-layers and the sol have high water contents and shrink during the drying process, the spores keep their original structure. The latter leads, in the case of spore-biocers, to a densification of the silicate matrix and to a decrease of the access to the binding sites.

3.4 *Influence of porosity on binding kinetics*

The higher porosity of freeze-dried biocers or biocers derived from nanosols containing sorbitol leads to faster U binding kinetics (Fig. 2).

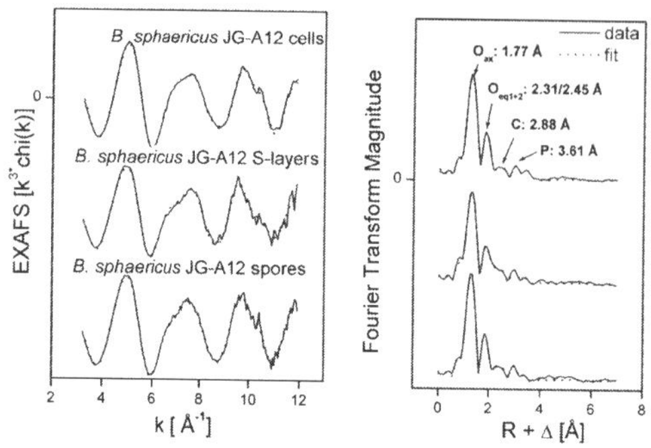

Figure 3. U EXAFS spectra (left) and their corresponding Fourier transforms (FT) of U bound by *B. sphaericus* JG-A12 cells, spores and its native S-layer.

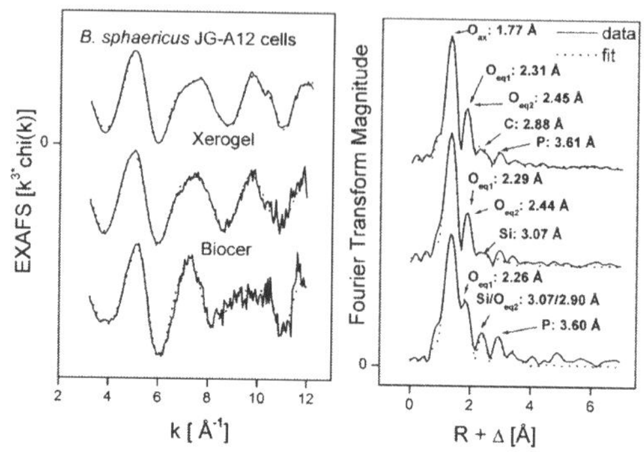

Figure 4. EXAFS spectra (left) and their corresponding Fourier transforms of U complexes formed by *B. sphaericus* JG-A12 cells, xerogel and the cell biocer.

As evident from the results presented in Figure 2, the fine structure of the biocers affects the sorption process significantly.

Using freeze-dried biocers the maximal amount of U is already bound after 30 min. In the case of sorbitol-supplemented biocers, the effect on the binding kinetics was only marginal. In addition, biocers with higher porosity show higher binding capacity for U. Besides that, bound U can be completely removed from all types of biocers by washing with 0.5 M citric acid for 24 h, which allows them to be reused.

3.5 *EXAFS spectroscopic analyses*

Extended X-ray absorption fine structure spectroscopy was used to identify the functional groups involved in the U binding based on chemical models. Preliminary results show that there are no structural differences between the U complexes formed by *B. sphaericus* cells, spores and S-layer (Fig. 3). The U is coordinated by two axial oxygen atoms (O_{ax}) at a distance of 1.77 Å. The second and third shells are from two equatorial oxygens (O_{eq1} and O_{eq2}) with distances of 2.31–2.34 and 2.45–2.50 Å. All Fourier transforms contain a fourth shell that originates from C neighbors at 2.88 Å. This distance is typical for carbonate or carboxylic groups coordinated to U(VI) in a bidentate fashion. The last shell arises from P neighbors at a distance of 3.61 Å. In summary, in all cases U is bound by carboxyl and phosphate groups.

X-ray absorption spectra of U bound by cells and xerogel confirm the results of sorption experiments. U was bound by free cells and by the silicate matrix itself (Fig. 4). The first peak of the Fourier transformed EXAFS spectra corresponds to the scattering contribution of an axial oxygen with a bonding length of 1.77 Å. The equatorial shell of the U-xerogel complex shows a splitting of bond distances into 2.29 Å and 2.44 Å. The second oxygen atom arises from water molecules. The last peak can be attributed to a U-Si interaction at a distance of 3.07 Å. These results indicate U(VI) binding by the xerogel via silanol groups. In the case of cell biocer, silicate groups may be involved in U(VI) coordination whereas the third peak could be fitted by a U-Si interaction at a distance of 3.07 Å or a U-O_{eq2} contribution at 2.90 Å. But mainly U was bound by phosphate groups of the cells, reflected by the fourth peak which originates from P neighbors at 3.60 Å.

4 CONCLUSIONS

The U mining waste pile isolate *B. sphaericus* JG-A12 is a spore forming bacterium, which is covered by a surface layer protein. The S-layer protein is phosphorylated and differs significantly in its primary structure from the S-layer proteins known to date. Not only the S-layer but also cells and spores of *B. sphaericus* bind large amounts of U selectively and reversibly via carboxyl and phosphate groups. Using sol-gel techniques, all biocomponents were completely embedded in a porous silicate matrix without influencing the binding capacity of the cells and the S-layer for U. Reduction of U binding capacity of immobilized spores may be due to a higher density of the biocers and therefore lower accessibility of binding sites. Freeze drying or adding water soluble agents such as sorbitol during the preparation accelerates U binding and increases the amount of bound metal.

Due to the mechanical stability, the high U binding capacity as well as the reusability, sol-gel embedded *B. sphaericus* JG-A12 cells and S-layer sheets are suitable for construction of intelligent filters for bioremediation.

ACKNOWLEDGEMENTS

Financial support of the German Research Community (DFG) and the German Ministry for Education and Research (BMBF) is gratefully acknowledged.

REFERENCES

Beveridge, T.J. 1979. Surface Arrays on the wall of *Sporosarcina ureae*. *J. Bacteriol.* 139 (3): 1039–1048.

Böttcher, H. 2000. Bioactive Sol-Gel Coatings. *J. Prakt. Chem.* 342: 427–436.

Brown, D.A., Beveridge, T.J., Keevil, C.W. & Sherriff, B.L. 1998. Evaluation of microscopic techniques to observe iron precipitation in a natural microbial biofilm. *FEMS Microbial. Ecol.* 26: 297–310.

Pollmann, K., Schnorpfeil, M., Radeva, G. & Selenska-Pobell, S. 2004. Molecular analyses of the S-layer genes and proteins of *Bacillus sphaericus* JG-A12 and NCTC 9602. *FZR Report* (in press).

Raff, J. 2002. *Interaction of surface layer proteins of bacteria from uranium mining waste piles with heavy metals*. PhD thesis, University of Leipzig.

Raff, J., Soltmann, U., Matys, S., Selenska-Pobell, S., Böttcher, H. & Pompe, W. 2003. Biosorption of Uranium and Copper by Biocers. *Chem. Mater.* 15: 240–244.

Sára, M. & Sleytr, U.B. 1987. Charge distribution on the S-layer of *Bacillus stearothermophilus* NRS 1536/3c and importance of charged groups for morphogenesis and function. *J. Bacteriol.* 169(6): 2804 2809.

Selenska-Pobell, S., Panak, P., Miteva, V., Boudakov, I., Bernhard, G. & Nitsche, H. 1999. Selective accumulation of heavy metals by three indigenous *Bacillus* strains, *B. cereus*, *B. megaterium* and *B. sphaericus*, from drain waters of a uranium waste pile. *FEMS Microbiol. Ecol.* 29: 59–67.

Shenton, W., Pum, D., Sleytr, U.B. & Mann, S. 1997. Synthesis of cadmium sulphide superlattices using self-assembled bacterial S-layers. *Nature* 389: 585–587.

Soltmann, U., Raff, J., Selenska-Pobell, S., Matys, S., Pompe, W. & Böttcher, H. 2003. Biosorption of heavy metals by sol-gel immobilized *Bacillus sphaericus* cells, spores and S-layers. *J. Sol-Gel Sci. Technol.* 26: 1209–1212.

Water–Rock Interaction, Wanty & Seal II (eds)

U, Th, Eu and colloid mobility in a granite fracture under near-natural flow conditions

T. Schäfer, H. Geckeis, M. Bouby & Th. Fanghänel
Forschungszentrum Karlsruhe, Institut für Nukleare Entsorgung (INE), Karlsruhe, Germany

ABSTRACT: Laboratory core migration experiments were performed by injecting 10^{-6} mol · L^{-1} U^{VI}, 10^{-8} mol · L^{-1} Th^{IV} and 10^{-8}mol · L^{-1} Eu^{III} solution without (cocktail I) and with addition of 2 mg · L^{-1} bentonite colloids (cocktail II) into a natural fractured granite of the Grimsel Test Site (GTS, central Swiss Alps). Flow velocity and therefore tracer contact time with the fracture surface were varied. A partly unretarded U^{VI} migration observed at short contact times (1–2 h) in both field (Geckeis et al., accepted) and laboratory experiments (this study) diminished under near-natural groundwater velocities (~46 m · yr^{-1}). A weak retardation (R_f = 19.5) is found under those conditions. AsymFFFF and ultrafiltration measurements revealed a colloid association of Th^{IV} and Eu^{III} in both spiked natural ground waters with and without bentonite addition. Nevertheless, Th^{IV} and Eu^{III} (cocktail I) were completely retarded in the core experiments. Only at quite high groundwater velocities (94 m · yr^{-1}) in the presence of bentonite colloids (cocktail II) a partially bentonite colloid facilitated transport was observed, with 27% and 37% mass recoveries, respectively.

1 INTRODUCTION

Transport of dissolved and colloidal species in saturated porous media or fractured granite host rock has been an important field of research over the last decade from both the experimental and the modeling viewpoint (Grolimund et al. 1998; Oswald & Ibaraki 2001; Schäfer et al. 2003). Large granite block scale experiments (Vilks & Baik 2001) as well as field migration studies (Hauser et al. 2002; Möri et al., 2003a) have demonstrated the mobility of inorganic colloids. However, the sensitivity of colloidal transport on average flow velocity, colloid size and colloid concentration (Vilks & Baik 2001) illustrate the general problem of field-scale experiments. Groundwater flow velocities or colloid concentrations are frequently much higher than in natural hydraulic conditions (e.g. flow velocity <2 m · yr^{-1} in the Whiteshell research area).

The in situ colloid migration experiments performed at the Grimsel Test site (GTS, central Swiss Alps) within the CRR (**C**olloid **R**adionuclide **R**etention) international joint project have demonstrated the mobility of injected Febex (**f**ull-scale **e**ngineered **b**arrier **ex**periment; Villar et al. 1998) bentonite colloids in a 2.23 m and 5 m length test shear zone under the given hydrodynamic conditions (Hauser et al. 2002; Möri et al. 2003a). Furthermore these detailed investigations have shown even without colloid addition a partly colloid-mediated transport of tri- and tetravalent radionuclides via interaction with groundwater colloids (Geckeis et al., accepted). Furthermore, dissolved Np^{IV} and U^{VI} revealed a partly unretarded transport under the given experimental conditions. A close inspection of the results obtained in those field studies suggests an important influence of sorption kinetics on radionuclide migration.

The aim of the ongoing laboratory studies presented in this paper is to elucidate the effect of groundwater velocity or fracture residence time on the mobility of dissolved species, colloids and colloid-associated lanthanides/radionuclides in order to investigate the effect of near natural flow conditions.

2 MATERIALS AND METHODS

Laboratory migration studies were performed under an argon atmosphere in a glove box. The core used for the laboratory migration studies was provided by Nagra (core CRR-FZK-2). A core of 0.315 m length and a slab of 0.02 m length, later referred to as core #2 and core #3 were used. The groundwater used throughout the experiments was sampled under inert gas conditions in 50 L Teflon-coated stainless steel barrels from the borehole (BOMI 87.010) in the Grimsel underground laboratory. The composition of the sampled groundwater used for the experiments is listed in Table 1. Laser-induced breakdown detection

Table 1. Grimsel groundwater composition (borehole BOMI 87.010).

Element	Concentration (μmol/L)	Element (μmol/L)	Concentration
Na	530	Cl	173
K	4.1	SO_4^{2-}	50.8
Li	11.3	NO_3^-	2.1
Ca	203	PO_4^{3-}	n.d.
Mg	0.32	F^-	340
Sr	2.25	Fe	n.d.
Al	0.42	Mn	29.0
DOC	$1.2 \pm 0.2\,mg \cdot L^{-1}$	DIC:	$3.4\,mg \cdot L^{-1}$
pH	9.6		

(LIBD) colloid measurements of the barreled groundwater revealed a number weighted average diameter of 91 nm and a very low colloid concentrations of $6.8\,\mu g\,L^{-1}$. Injection was performed by 100 μL Dirac-type pulse or permanent injection.

Febex bentonite colloid suspension was prepared according to (Möri et al. 2003a) in natural Grimsel groundwater. The average number weighted size of bentonite colloids determined by LIBD is 105 nm and the intensity weighted average hydrodynamic diameter d_H measured by photon correlation spectroscopy (PCS) is 202 ± 43 nm. These results demonstrate the rather broad bentonite colloid size distribution (Plaschke et al. 2001). Prior to column injection the lanthanide cocktails were prepared by adding aliquots of U, Th and Eu 1000 ppm stock solutions to Grimsel groundwater and readjusting the pH with NaOH to pH 9.6. Two injection cocktails were prepared:

Cocktail I: $1 \cdot 10^{-6}$ mol/L U; $1 \cdot 10^{-8}$ mol/L Th; $1 \cdot 10^{-8}$ mol/L Eu.

Cocktail II: $1 \cdot 10^{-6}$ mol/L U; $1 \cdot 10^{-8}$ mol/L Th; $1 \cdot 10^{-8}$ mol/L Eu $+ 2\,mg \cdot L^{-1}$ bentonite colloids.

Cocktail characterization included ultrafiltration using polyethersulfone membranes of a nominal molecular weight cutoff ranging from 10^3 to 10^6 Dalton (Filtron Co., Microsep Microconcentrators). Aliquots of the non-filtered cocktails were characterized by Asymmetrical Flow Field Flow Fractionation AsymFFFF (Postnova Analytics, HRFFF 10.000 AF4) coupled to ICP-MS (Bouby et al., subm.). Column outlet fractions of the migration experiments as well as ultrafiltration aliquots were analysed by ICP-MS (Perkin Elmer ELAN 6000).

3 RESULTS AND DISCUSSION

3.1 *Colloid and radionuclide/lanthanide characterization*

The ultrafiltration of the initial cocktail with and without bentonite colloid addition shows clearly that

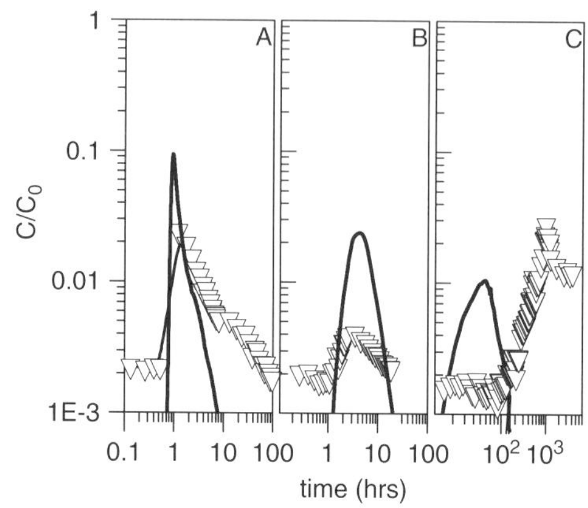

Figure 1. Pulse and permanent injection breakthrough curves of tritiated water (solid line) and ^{238}U (open triangles). A: 100 μL pulse in 2 cm long fracture (core #3) $v_a = 0.52\,m \cdot d^{-1}$, B: 100 μL pulse in 2 cm long fracture (core #3) $v_a = 0.26\,m \cdot d^{-1}$, C: 144 mL permanent injection in a 31.5 cm long fracture (core #2) $v_a = 0.128\,m \cdot d^{-1}$.

U^{VI} can be found in both cocktails as dissolved species with >70% passing the 1 kDa filter.

In the case of Th and Eu only 50% and 30%, respectively, passed the 1000 kDa filter without bentonite addition, indicating that a major fraction of these elements is associated with larger colloidal phases in the groundwater. The spiked concentration chosen for Th and Eu are close to the solubility limit calculated under Grimsel groundwater conditions, which are $\sim 10^{-8}$ and 10^{-7}, respectively (Duro 1999). Together with the estimated available sorption sites of $1 \cdot 10^{-10}\,mol \cdot L^{-1}$ of natural groundwater colloids (2 sites $\cdot\,nm^{-2}$ site density) a formation of Th/Eu eigencolloids cannot be excluded a priori. Ultrafiltration results of cocktail II with $2\,mg \cdot L^{-1}$ added Febex bentonite colloids reveal a clear shift to larger colloid sizes with only ~20% of Th and Eu passing the 1000 kDa filter ($d_H \sim 20$–40 nm). These results are corroborated by AsymFFFF-ICP-MS analysis, were Eu and Th elute with Al representing the added bentonite colloids of 110 nm mean colloid size. In the case of U only a minor fraction elutes together with bentonite colloids showing a negligible association with colloids, whereas the major part elutes in the void peak as dissolved species.

3.2 *Breakthrough behavior of U, Th and Eu (cocktail I)*

In the pulse injection experiments made in the absence of bentonite colloids under variation of the flow velocity only a breakthrough of U could be detected (Fig. 1), whereas Th and Eu did not show an unretarded or retarded breakthrough in this experimental set-up.

At a residence time of 1 h, similar to that in field experiment (Geckeis et al., accepted), the U recovery of >90% (Fig. 1A) was comparable to that observed in the field and showed an almost unretarded peak (retardation factor R_f of ~1.3). In addition, the more pronounced tailing of the U^{VI} breakthrough compared to that of the conservative tracer and the additional hump indicate a retarded fraction with an R_f of ~15. Changing the flow velocity and therefore the fracture tracer residence time to 5 h decreased the unretarded fraction to 14% (Fig. 1B). A retarded fraction could not be resolved from the background signal in this experiment. To study the U^{VI} breakthrough behavior under near-natural flow conditions (56 h residence time or ~45 $m \cdot yr^{-1}$) a long pulse injection experiment in core #2 was performed. In this experiment (Fig. 1C) no unretarded U^{VI} fraction could be observed and only a retarded fraction with an R_f of ~20 was detectable. The U recovery after 80 d is >7.2%. The calculated distribution coefficient K_D based on the experimentally determined retardation factor R_f with the following formula:

$$R_f = 1 + \frac{\rho_b \cdot K_D}{\Theta} \qquad (1)$$

where ρ_b is the bulk density (2.6 $g\,cm^{-3}$) and Θ the matrix porosity (0.2; increased matrix porosity; acc. to Möri et al. 2003b) revealed K_D values of 0.7 to 1.1 for the retarded U^{VI} fraction. These values are in fair agreement with batch experiment derived K_D values on fracture infill material of approximately 5 (Geckeis et al., accepted). The existence of an unretarded U^{VI} fraction at increased groundwater flow is thus due to the low collision probability of dissolved species with the fracture surfaces. The weak U^{VI} sorption to surfaces can be explained by the presence of anionic carbonato complexes (Duro 1999). Similar behavior was observed for Cs^+ and Sr^{2+} in the Migration Experiment at the GTS (Hoehn et al. 1998). The absence of a unretarded or retarded Th and Eu breakthrough in the pulse experiments is in accordance with the calculated retardation factors R_f of ~100 for Th^{IV} and ~600 for Eu^{III} taking into account the K_D values published in Geckeis et al. (accepted).

Therefore, a permanent injection experiment was performed, and Eu and Th breakthrough showed insignificant differences in the leading edge of the breakthrough curve compared to the conservative tracer (Fig. 2), but exhibited a steeper concentration decline to background values after a switch back to unspiked Grimsel groundwater. Furthermore, the breakthroughs of Eu and Th exhibit a plateau of C/C_0 values in the range of 0.35 to 0.39 (Fig. 2). This indicates the establishment of a steady-state condition. Either because of the kinetically controlled dissociation from colloids or the time dependent attachment of colloids to the fault gauge material, only 60–65% of the Th and Eu are retained in the fracture under steady-state conditions. The rapid decline of the Th and Eu breakthrough curves, steeper than that of tritiated water, demonstrate their colloidal nature (Oswald & Ibaraki 2001). The overall recovered unretarded fractions of Eu and Th transport are 23.9% and 27.9%, respectively. Th and Eu recovery is in quite good agreement to published values of Möri et al. (2003a) for Am^{III} and Pu^{IV} of 34 ± 4% and 21 ± 3% in the field experiment under comparable tracer residence times without bentonite colloid addition.

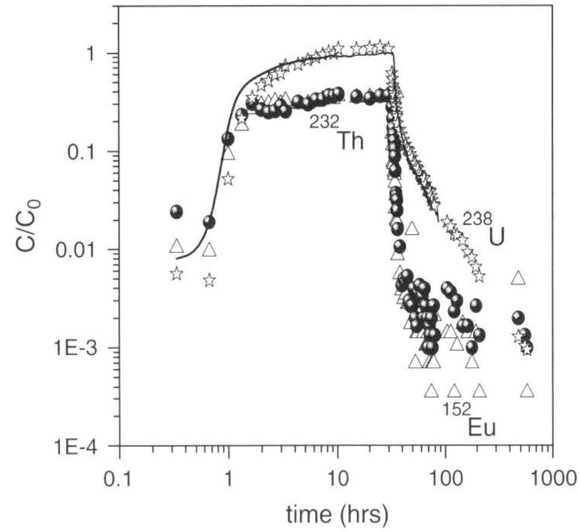

Figure 2. Permanent injection breakthrough curves of the conservative tracer tritium (solid line) and U (open stars), Th (filled circles) and Eu (open triangles).

3.3 *Colloid breakthrough behavior (cocktail II)*

The CRR field experiments (Geckeis et al., accepted; Möri et al. 2003a) have demonstrated that an addition of 20 $mg \cdot L^{-1}$ bentonite colloids significantly increased the mobility of tri- and tetravalent radionuclides under the given hydrodynamic conditions in a 2.23 m long fracture. In the laboratory experiments we decreased the flow velocity to approximately 95 $m \cdot yr^{-1}$ (Fig. 3). The results of these column experiments show that the U^{VI} breakthrough was unaffected by the addition of 2 $mg \cdot L^{-1}$ bentonite colloids, as expected, and the 11.2% recovery of the unretarded fraction was comparable to the results of previous laboratory experiment (14.3% recovery) under the same hydraulic conditions without bentonite.

The major difference in this experiment was the detection of an unretarded Th and Eu breakthrough with the peak maximum slightly ahead of the conservative tracer. The recoveries were determined to be 27% (Th) and 37% (Eu), which are slightly higher

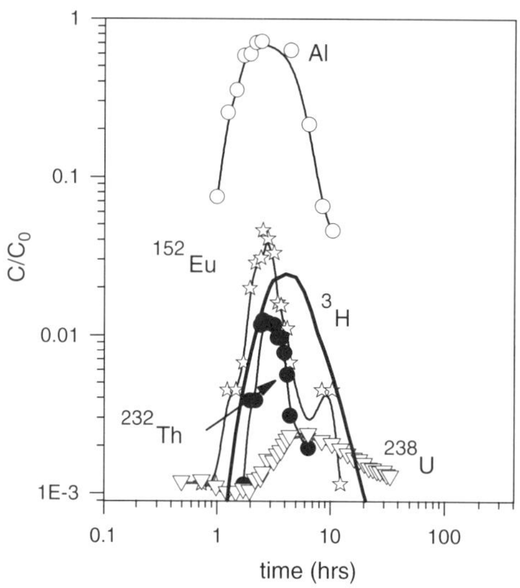

Figure 3. Breakthrough curves obtained after 100 μL pulse injection of cocktail II. Tritiated water is indicated by a solid line, U as open triangles, Eu as open stars, Th as filled circles and Al as open circles. (2 cm long fracture (core #3) $v_a = 0.26\,m \cdot d^{-1}$).

than the recoveries found for these elements under higher flow velocities in the permanent injection experiment (Fig. 2), but significantly lower than the recoveries of 70 ± 15% (Am^{III}) and 86 ± 9% (Pu^{IV}) observed in the CRR experiment with $20\,mg \cdot L^{-1}$ bentonite colloid addition (Möri et al. 2003a). A quantification of the mobile bentonite colloid fraction was not possible due to a significant mobilization of fracture derived natural colloids increasing the Al concentration background scatter to approximately 100 ppb in the column outlet before the experimental start compared to ~11 ppb found in the natural Grimsel groundwater (Fig. 3).

4 CONCLUSIONS

In this paper, laboratory core migration experiments are presented under variation of linear flow velocity to investigate kinetic aspects of colloidal and dissolved radionuclide migration. Compared to the CRR (**C**olloid **R**adionuclide **R**etention) field migration studies (Möri et al. 2003a) the linear flow velocity could be reduced to near-natural values as low as $46\,m \cdot yr^{-1}$. The following conclusions can be drawn:

- The partly unretarded breakthrough of U^{VI} observed in the field experiments diminishes with increasing tracer residence time and only a retarded breakthrough was detected.
- Th^{IV} and Eu^{III}, serving as analogues for Pu^{IV} and Am^{III}, respectively, revealed a partly unretarded bentonite colloid associated transport under a flow velocity of $94\,m \cdot yr^{-1}$, with a clearly lower recovery as compared to the field experiments.

In general, the results presented call for further studies focused on: (a) the mechanism of colloid transport and retention under near-natural groundwater flow conditions in an undisturbed granite fracture and (b) the kinetics of colloid-radionuclide sorption/desorption under repository relevant time scales.

REFERENCES

Bouby, M., Geckeis, H., Ngo Manh, T., Yun, J.I., Dardenne, K., Schäfer, T., Walther, C., Kim, J.I. submitted. Laser-Induced Breakdown Detection (LIBD) Combined with Flow Field-Flow Fractionation: Application to the Characterization of Polystyrene and Iron Oxi/hydroxide Colloids. *J. Chrom. A.*

Duro, L. 1999. Prediction of the solubility and speciation of radionuclides in Febex and Grimsel waters. *Nagra Aktennotiz AN 99–218*, Nagra, Wettingen, Switzerland.

Geckeis, H., Schäfer, T., Hauser, W., Rabung, T., Missana, T., Degueldre, C., Möri, A., Eikenberg, J., Fierz, T., Alexander, W.R. accepted. Results of the Colloid and Radionuclide Retention experiment (CRR) at the Grimsel Test Site (GTS), Switzerland- Impact of reaction kinetics and speciation on radionuclide migration. *Radiochim. Acta.*

Grolimund, D., Elimelich, M., Borkovec, M., Barmettler, K., Kretzschmar, R., Sticher, H. 1998. Transport of in situ mobilized colloidal particles in packed soil columns. *Environ. Sci. Technol.* 32: 3562–3569.

Hauser, W., Geckeis, H., Kim, J.I., Fierz, T. 2002. A mobile laser-induced breakdown detection system and its application for the in situ monitoring of colloid migration. *Colloids Surf. A* 203(1–3): 37–45.

Hoehn, E., Eikenberg, J., Fierz, T., Drost, W., Reichlmayr, E. 1998. The Grimsel Migration Experiment: field injection-withdrawal experiments in fractured rock with sorbing tracers. *J. Contam. Hydrol.* 34(1–2): 85–106.

Möri, A., Alexander, W.R., Geckeis, H., Hauser, W., Schäfer, T., Eikenberg, J., Fierz, T., Degueldre, C., Missana, T. 2003a. The colloid and radionuclide retardation experiment at the Grimsel Test Site: influence of bentonite colloids on radionuclide migration in a fractured rock. *Colloids Surf. A* 217(1–3): 33–47.

Möri, A., Mazurek, M., Adler, M., Schild, M., Siegesmund, S., Vollbrecht, A., Ota, K., Ando, T., Alexander, W.R., Smith, P.A., Haag, P., Bühler, C. 2003b. Grimsel Test Site Investigation Phase IV (1994–1996). The Nagra-JNC in situ study of safety relevant radionuclide retardation in fractured crystalline rock IV: The in situ study of matrix porosity in the vicinity of a water conducting fracture. *Nagra Technical Report NT 00–08*, pp. 91, Nagra, Wettingen, Switzerland.

Oswald, J.G., Ibaraki, M. 2001. Migration of colloids in discretely fractured porous media: effect of colloidal matrix diffusion. *J. Contam. Hydrol.* 52(1–4): 213–244.

Plaschke, M., Schäfer, T., Bundschuh, T., Manh, T.N., Knopp, R., Geckeis, H., Kim, J.I. 2001. Size characterization of bentonite colloids by different methods. *Anal. Chem.* 73(17): 4338–4347.

Schäfer, T., Artinger, R., Dardenne, K., Bauer, A., Schuessler, W., Kim, J.I. 2003. Colloid-borne Americium migration in Gorleben groundwater: Significance of iron secondary phase transformation. *Environ. Sci. Technol.* 37(8): 1528–1534.

Vilks, P., Baik, M.H. 2001. Laboratory migration experiments with radionuclides and natural colloids in a granite fracture. *J. Contam. Hydrol.* 47(2–4): 197–210.

Villar, M.V., Martin, P.L., Pelayo, M., Ruiz, B., Rivas, P., Alonso, E., Lloret, A., Pintado, X., Gens, A., Linares, J., Huertas, F., Caballero, E., Jimenz de Cisneros, C., Obis, J., Perez, A., Velasco, J. 1998. FEBEX bentonite: origin, properties and fabrication of blocks (full-scale engineered barriers experiment in crystalline host rock). *Enresa Technical Report 05/98*, pp.230, Enresa, Barcelona, Spain.

Water–Rock Interaction, Wanty & Seal II (eds)
© 2004 Taylor & Francis Group, London, ISBN 90 5809 641 6

Pu^{V} removal from solution by oxidation and adsorption onto manganese dioxide

S.A. Stout, S.D. Reilly, D.M. Smith & M.P. Neu
Los Alamos National Laboratory, Los Alamos, NM, USA

ABSTRACT: To better understand the influence of manganese minerals on contaminant plutonium speciation, solid/solution phase distribution, and overall fate and transport, interactions between Pu and redox-active synthetic birnessite, δ-MnO_2, are being studied. Optical absorbance spectra and radioanalytical counting indicate that in the presence of δ-MnO_2 Pu^{V}(aq) is rapidly oxidized to Pu^{VI}(aq) and removed from solution under all conditions studied. X-ray adsorption spectra show that the Pu is adsorbed onto the δ-MnO_2 as Pu^{VI}. The model that best matches the EXAFS spectra is based on surface-bound mononuclear dioxo Pu^{VI}. Adsorption isotherms show classical cation-mineral behavior, with maximum Pu^{VI} adsorption between pH 4 and 7, and the adsorption edge shifting to higher pH with increasing $[Pu^{VI}]_{init}$. Nearly all measured and predicted Pu^{V}-mineral interactions invoke Pu^{IV} adsorption or precipitation. Our results suggest that common Mn^{IV}-containing minerals have the potential to stabilize oxidized forms of plutonium and that transport models should be expanded to include Pu^{VI}-mineral surface complexation reactions.

1 INTRODUCTION

Actinide contaminants at nuclear material production and reprocessing sites are likely to be released to the far-field slowly and at low concentrations. We increasingly rely on reactive transport models to estimate the overall rates of actinide migration. Predicting the speciation and mobility of Pu is particularly complicated because of its rich solution chemistry and the lack of detailed mineral and sorption studies. In aqueous solution, Pu can co-exist in the III–VI oxidation states. Under groundwater conditions Pu^{IV} and Pu^{V} are the most stable states, with the minor Pu^{V} fraction generally remaining in solution and Pu^{IV} primarily present in the solid phase as sparingly-soluble intrinsic oxyhydroxides and surface-adsorbed species. However, under some conditions colloidal Pu^{IV} species can predominate, and under others Pu^{III} and Pu^{VI} species constitute significant fractions of the total plutonium. Most mineral sorption studies, primarily focused on iron oxyhydroxides, silica, and aluminosilicates, show that surface-adsorbed or precipitated Pu^{IV} forms, independent of initial oxidation state. In contrast, recent studies indicate that some redox-active mineral phases, such as birnessite, have the potential to oxidize and preferentially adsorb $Pu^{V/VI}$.

High surface area manganese oxides are common vadose zone soil and sediment components. Birnessite is composed of layers of octahedrally coordinated Mn^{IV}, with vacancies occurring every 6th Mn within the layers. The presence of pH-dependent amphoteric surface sites creates a layer charge that is balanced by surficial and interlayer cations.

Previous studies of plutonium solution species interacting with δ-MnO_2 have suggested that multiple oxidations states/species persist in the resultant solid phase. For example, Keeney-Kennicutt & Morse (1985) reported that at low MnO_2 to water ratios ($5\,m^2$ of surface area per 1 L of H_2O) Pu added to mineral suspensions as Pu^{V} was adsorbed primarily as Pu^{V} (80%) with Pu^{VI} also present (20%). At higher MnO_2 to water ratio ($100\,m^2$/1L), results indicated a mixture of 50% Pu^{IV}, 30% Pu^{V}, and 20% Pu^{VI}. Although the oxidation state determination was indirect, Duff et al. (1999) reported that in a suspension of zeolitic tuff Pu^{V} is removed from solution via preferentially adsorbing onto MnO_2in an oxidized state (V or VI). For comparison, when Pu is combined with predominantly $Mn^{II/III}$ oxide/oxyhydroxide minerals Pu^{VI} is reduced to Pu^{IV} and adsorbed onto the minerals in this same oxidation state (Shaughnessey et al. 2003).

2 PU^V IN CONTACT WITH δ-MNO_2

Delta-MnO_2 was prepared according to the procedure of McKenzie (1971). Concentrated HCl (100 mL) was added to a boiling 0.4 M solution of $KMnO_4$ prepared in 1.5 L of deionized H_2O. The resulting finely-divided brown δ-$MnO_2$was separated from solution by vacuum filtration. The product was washed repeatedly under vacuum with deionized H_2O to remove KCl and unreacted $KMnO_4$. The δ-$MnO_2$was freeze-dried in a lyophilizer and stored at ambient conditions prior to use in adsorption experiments. Powder X-ray diffraction indicated that the solid was a pure phase δ-$MnO_2$with *d*-spacings of 7.24, 3.65, 2.43, and 1.41 Å. The surface area of the δ-$MnO_2$was determined to be $214 \pm 1\ m^2/g$ using the EGME method, and $41.1 \pm 0.2\ m^2/g$ using the N_2 BET method. Characterization data compare well with those reported previously for δ-MnO_2 synthesized in a similar manner.

Pu^V solutions (3.1×10^{-4} M) were equilibrated at pH = 4.4, 6.2, and 7.7 with δ-MnO_2 (1 mg/ml) for 24 h. Pu^V is rapidly oxidized in solution to Pu^{VI} as evidenced by absorbance bands in the UV/Vis spectra characteristic of Pu^{VI} present at 520, 615, and 830 nm and the absence of a Pu^V characteristic absorbance at 569 nm (Fig. 1). Similar results were observed at higher pH, although as expected absorbance bands characteristic of PuO_2OH^+ and $PuO_2(OH)_2$ were also present due to hydrolysis of Pu^{VI} in solution. As the Pu^{VI}/δ-MnO_2 contact time increased, the intensity of the 830 nm absorbance band decreased and no additional absorbance bands were observed, indicating other solution species were not formed. The concentration of Pu^{VI} in solution decreased rapidly.

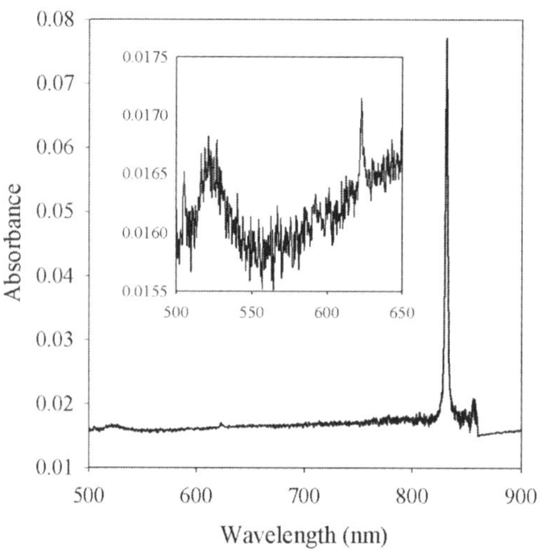

Figure 1. UV/V is spectra for a solution, initially $[Pu^V] = 3.1 \times 10^{-4}$ M, in contact with δ-MnO_2 at pH 4.4, showing exclusively absorbance bands characteristic of Pu^{VI}.

3 X-RAY ABSORPTION STUDIES

Previously, it was assumed that Pu (initially Pu^V) associates with MnO_2 phases in a reduced form (III or IV); however, here we report that Pu is adsorbed to the mineral surface in an oxidized form (VI). Figure 2 shows the XANES spectra for Pu^V reacted with δ-MnO_2 along with standard spectra for $Pu^{IV}O_2$ and $K_3Pu^VO_2(CO_3)_2$. The edge energies for tetravalent PuO_2 and pentavalent $K_3PuO_2(CO_3)_2$ standards increase with increasing formal oxidation state. Pu^V and Pu^{VI} have a distinct shoulder in the near-edge region and a less intense white line compared with Pu^{VI} (Fig. 2, inset). The edge energies for Pu adsorbed onto MnO_2 are higher than that observed for the Pu^{IV} standard and possess a shoulder in the near-edge indicative of Pu^{VI}.

Features in the EXAFS spectra for Pu^V reacted with δ-MnO_2 are consistent with the near-edge spectra, indicating that the Pu is not bound as Pu^V, but as Pu^{VI}. The Pu coordination includes the dioxo moiety, O = Pu = O, with a Pu-O distance of 1.76 Å (compared with 1.85 Å expected for Pu^V dioxo species). Additional atomic shells include approximately four Pu-O bonds with distances of 2.40 Å, consistent with two types of oxygens – OH_2 and O-Mn, and one to two Pu-Mn with a distance of 3.40–3.44 Å. The Pu-Mn distance reported here is consistent with site-specific, inner-sphere adsorption at low concentration.

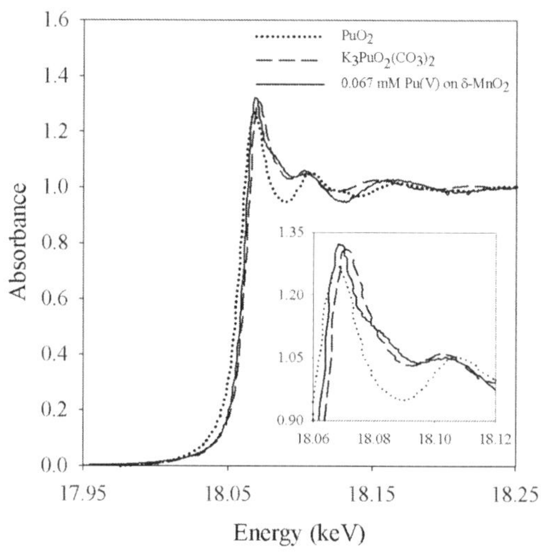

Figure 2. The L_3-edge XANES spectra for Pu adsorbed δ-MnO_2.

At higher $[Pu^{VI}]_{init}$ the surficial and interlayer binding sites may become saturated forcing the Pu^{VI} to bind to less favorable edge sites.

These Pu-O bond distances are longer than those reported by Shaughnessy et al. (2003) for Pu^{VI} reacted with MnOOH and Mn_3O_4 at pH 5, 2.30 Å and 2.32 Å, respectively. The Pu-Mn interatomic distances calculated from our EXAFS data were also longer than those reported by Shaughnessy et al. (2003), 3.37 Å for both minerals at pH 5. These differences can be explained by binding site differences in the phases, pH differences (resulting in a greater contribution from Pu-O-Mn than Pu-OH_2) and bonding differences between $Pu^{VI}$and Pu^{IV}.

4 ADSORPTION ISOTHERMS

Batch adsorption experiments were conducted from pH 2 to 8 at initial Pu^{VI} concentrations of 2.5×10^{-5}, 1.0×10^{-5}, and 1.0×10^{-6} M and $[MnO_2] = 1$ mg/mL. Ammonium nitrate was added to the samples to adjust the ionic strength to 0.10 M and the samples were sparged with Ar to minimize absorption of atmospheric CO_2. The concentration of Pu remaining in solution following equilibration was determined by liquid scintillation counting.

Figure 3 shows the adsorption of Pu^{VI} onto δ-MnO_2 as a function of pH and initial Pu concentration after 24 h equilibration. Previous kinetics experiments indicate that Pu is rapidly removed from solution by δ-MnO_2. Although the point of zero charge (PZC) for our δ-MnO_2 was not determined experimentally, literature values for δ-MnO_2 prepared in a similar manner are between 2.0 and 2.8 (Fendorf et al. 1992; Kim et al. 2002). Very little Pu is adsorbed onto the surface of the δ-MnO_2 near the PZC regardless of the initial $[Pu^{VI}]$. At pH near the PZC, few negative surface sites are present and presumably PuO_2^{2+} must compete with protons for the available binding sites. The percentage of Pu removed from the solution increases as the surface of the oxide become more negatively charged until a maximum is reach between pH 4 and 6, where almost 100% of the Pu is removed from solution. The percentage of Pu removed decreases at pH > 6, approximately the same pH where the neutral species, $PuO_2(OH)_2$, becomes predominant in solution. The adsorption of Pu^{VI} onto δ-MnO_2 correlates well with the hydrolysis behavior of Pu^{VI} (Fig. 4) and displays classic cation sorption behavior; the adsorption edge shifts toward higher pH as the initial concentration of Pu^{VI} in solution increases. A surface complexation model for the pH-dependent $Pu^{V,VI}$ adsorption onto δ-MnO_2 is being developed.

5 CONCLUSIONS

Manganese dioxide oxidizes Pu^V in solution to Pu^{VI}, which is then rapidly adsorbed onto the mineral, forming an inner-sphere surface complex. The observed adsorption behavior correlates well with $Pu^{VI}O_2^{2+}(aq)$ hydrolysis. Given the prevalence of potentially oxidizing Mn-containing minerals and the dearth of data for Pu^V and Pu^{VI}-mineral species in radionuclide migration models, current estimates of Pu mobility under oxic conditions may be inaccurate.

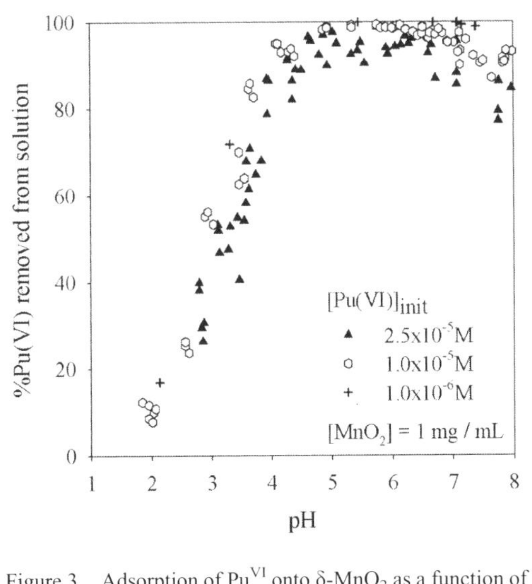

Figure 3. Adsorption of Pu^{VI} onto δ-MnO_2 as a function of pH and initial Pu^{VI} concentration after 24 hrs contact time.

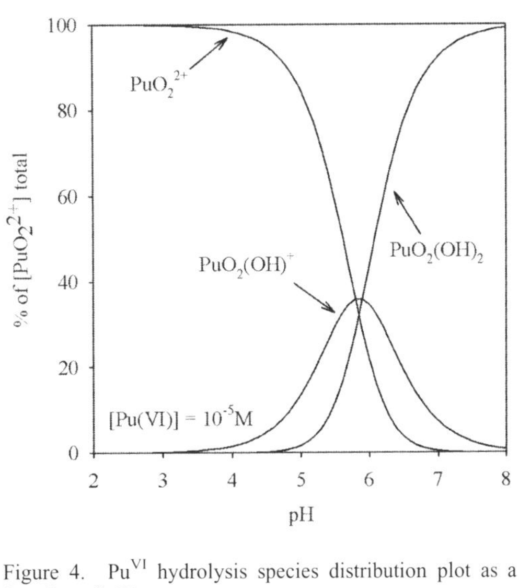

Figure 4. Pu^{VI} hydrolysis species distribution plot as a function of pH.

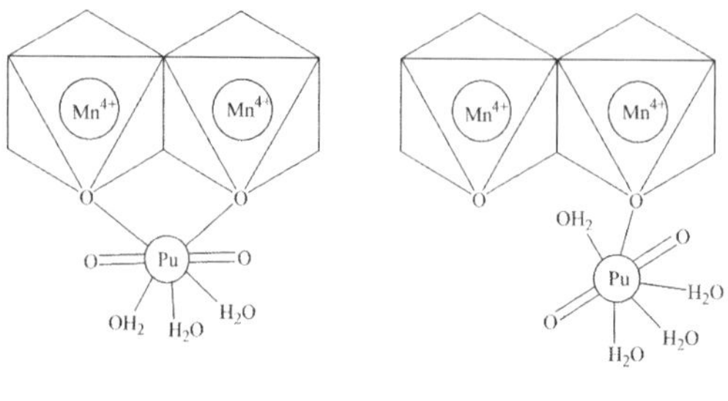

Figure 5. Potential coordination of Pu^{VI} to the δ-MnO_2.

The X-ray absorption results presented here suggest that Pu^{VI} forms either a monodentate or bidentate inner-sphere complex with the surface of the δ-MnO_2 (Fig. 5). We propose the following reactions for the oxidation of Pu^V to Pu^{VI} at the surface of δ-MnO_2:

$$PuO_2^+ + \delta\text{-}Mn^{IV}O_2 \rightarrow PuO_2^{2+} + Mn^{III}OOH \quad (1)$$

$$2PuO_2^+ + \delta\text{-}Mn^{IV}O_2 \rightarrow 2PuO_2^{2+} + Mn^{2+} \quad (2)$$

Reaction 1 represents the transfer of one electron with the formation of a Mn^{III} intermediate, which may result in oxidation of a second Pu^V. Reaction 2 represents the oxidation of two $Pu^VO_2^+$ cations and a release of Mn^{II} into solution.

Previous studies involving transition metals have proposed that the cations bind to the surface of δ-MnO_2 prior to oxidation. In contrast, our results show oxidation of Pu^V to Pu^{VI} in solution. Another possible mechanism is that Pu^V is oxidized by Mn^{III} cations bound above/below the layer vacancies on the surface of the MnO_2. This reaction would result in the release of Mn^{II} into solution along with Pu^{VI}. Chalcophanite, a structural analog of birnessite, consists of sheets of $Mn^{IV}O_6$ octahedra alternating with layers of water and Zn cations, which share three corners with the Mn octahedra and are bound above and below the layer vacancies (Post & Appleman 1988). Manceau et al. (2000) reported that Cu adsorbs onto birnessite in a similar manner.

Redox-active transition metals, may be oxidized at the surface, then incorporated into the mineral structure. For example, Manceau et al. (1997, 2000) reported that Co^{II} was oxidized at the surface of MnO_2 followed by incorporation of Co^{III} into the layer vacancies. This process involves the transfer of one electron, and since Mn^{IV} is not converted to Mn^{II}, would not require reductive dissolution of the δ-MnO_2.

Manceau & Charlet (1992) reported a three step mechanism for Cr^{III} oxidation by birnessite: 1. migration of Cr^{III} into octahedral vacancies in the MnO_2 structure; 2. electron transfer and intermediate species such as Mn^{III}, Cr^{IV}, and Cr^V; 3. desorption of reaction products, i.e. Mn^{II} and Cr^{VI}. Banerjee & Nesbitt (1999) reported that the electron transfer involves the reduction of Mn^{IV} to Mn^{III} resulting in the formation of a surface oxide, probably MnOOH, followed by reduction of the Mn^{III} to Mn^{II}, which is lost to solution. Cr^{III} undergoes oxidation by the loss of one electron to Mn ions resulting in the formation of Cr^{IV} and Cr^V intermediates, and finally Cr^{VI}. Fendorf et al. (1992) reported a similar mechanism described by the following reaction:

$$Cr^{3+} + 1.5\delta\text{-}MnO_2 \rightarrow HCrO_4^- + 1.5Mn^{2+} \quad (3)$$

Mechanisms for the oxidation and potential incorporation of actinyl species into the MnO_2 structure will be constrained by the geometry and coordination properties of these large and linear dioxo cations. XPS studies of MnO_2 following reaction with Pu^V and Pu^{VI}, aimed at detecting surface or near surface Mn^{III} and/or Mn^{II}, are in progress. Comparative batch adsorption studies of δ-MnO_2 with UO_2^{2+}, NpO_2^+ and Pu^{IV} species are expected to reveal the limit of the minerals oxidizing potential *in situ* as well as provide additional thermodynamic and mechanistic information.

ACKNOWLEDGEMENTS

Research was supported by the Environmental Management Science Program, OBER-ERSD, Office of Science, U.S. DOE. X-ray absorption measurements were made at the Advanced Photon Source in collaboration with L. Soderholm and S. Skanthakumar.

REFERENCES

Banerjee, D. & Nesbitt, H.W. 1999. Oxidation of aqueous Cr(III) at birnessite surfaces: Constraints on reaction mechanism. *Geochim. Cosmochim. Acta* 63(11/12): 1671–1687.

Duff, M.C., Hunter, D.B., Triay, I.R., Bertsch, P.M., Reed, D.T., Sutton, S.R., Shea-McCarthy, G., Kitten, J., Eng, P., Chipera, S.J. & Vaniman, D.T. 1999. Mineral associations and average oxidation states of sorbed Pu on tuff. *Environ. Sci. Technol.* 33(13): 2163–2169.

Fendorf, S.E., Fendorf, M., Sparks, D.L. & Gronsky, R. 1992. Inhibitory mechanism of Cr(III) oxidation by δ-MnO_2. *J. Colloid Interf. Sci.* 153: 37–54.

Keeney-Kennicutt, W.L. & Morse, J.W. 1985. The redox chemistry of $Pu(V)O_2^+$ interaction with common mineral surfaces in dilute-solutions and seawater. *Geochim. Cosmochim. Acta* 49(12): 2577–2588.

Kim, J.G., Dixon, J.B., Chusuei, C.C. & Deng, Y. 2002. Oxidation of chromium (III) to (VI) by manganese oxides. *Soil Sci. Soc. Am. J.* 66: 306–315.

Manceau, A. & Charlet, L. 1992. X-ray absorption spectroscopic study of the sorption of Cr(III) at the oxide-water interface: I. Molecular mechanism of Cr(III) oxidation on Mn oxides. *J. Colloid Interf. Sci.* 48: 425–442.

Manceau, A., Drits, V.A., Silvester, E., Bartoli, C. & Lanson, B. 1997. Structural mechanism of Co^{2+} oxidation by the phyllomanganate buserite. *Am. Mineral.* 82 (11/12): 1150–1175.

Manceau, A., Schegel, M.L., Musso, M., Sole, V.A., Gauthier, C., Petit, P.E. & Trolard, F. 2000. Crystal chemistry of trace elements in natural and synthetic goethite. *Geochim. Cosmochim. Acta.* 64: 3643–3661.

Manning, B.A., Fendorf, S.E., Bostick, B. & Suarez, D.L. 2002. Arsenic(III) oxidation and arsenic(V) adsorption reactions on synthetic birnessite. *Environ. Sci. Technol.* 36(5): 976–981.

McKenzie, R.M. 1971. Synthesis of Birnessite, Cryptomelane, and some other oxides and hydroxides of manganese. *Min. Mag.* 38: 493–502.

Post, J.E. & Appleman, D.E. 1988. Chalcophanite, $ZnMn_3O_7 \cdot 3H_2O$: New crystal-structure determinations. *Am. Mineral.* 73: 1401–1404.

Shaughnessy, D.A., Nitsche, H., Booth, C.H., Shuh, D.K., Waychunas, G.A., Wilson, R.E., Gill, H., Cantrell, K.J. & Serne, R.J. 2003. Molecular interfacial reactions between Pu(VI) and manganese oxide minerals manganite and hausmannite. *Environ. Sci. Technol.* 37(15): 3367–3374.

Stouff, P. & Boulègue, J. 1988. Synthetic 10-Å and 7-Å phyllomanganates: Their structures as determined by EXAFS. *Am. Mineral.* 73: 1162–1169.

Water – Rock Interaction, Wanty & Seal II (eds)
© 2004 Taylor & Francis Group, London, ISBN 90 5809 641 6

The role of secondary mineral precipitates on radionuclide sequestration at the Hanford Site

W. Um, R.J. Serne, S.B. Yabusaki & V.L. Freedman
Pacific Northwest National Laboratory, Richland, WA, USA

S.D. Samson & K.L. Nagy
University of Illinois at Chicago, Chicago, IL, USA

ABSTRACT: The effects of secondary mineral precipitates on radionuclide sequestration at the Hanford Site were investigated by reacting quartz and Hanford sediment (Warden soil) with caustic solution of high ionic strength (2 M $NaNO_3$), high pH (~13), high temperature (~90°C), and dissolved Al (0.01 M $Al(NO_3)_3$). Continuous Si dissolution and concomitant secondary mineral precipitation were the principal reactions observed. Nitrate-cancrinite was the dominant secondary precipitate on mineral surfaces after 3–10 d reaction time. The presence of dissolved Al in the simulated tank fluid was found to depress the net Si dissolved concentration. Based on batch equilibrium sorption results, secondary precipitates (cancrinite) at the primary mineral surfaces enhanced the sorption capacity of typical Hanford sediment for radionuclides of major concern at the Hanford Site such as ^{129}I, ^{79}Se, ^{99}Tc, and ^{90}Sr.

1 INTRODUCTION

Large volumes of high-level nuclear waste (HLW) were generated from plutonium production and separation processes at the Hanford Site, Washington, during the Cold War. Because of certain chemical processes (e.g., PUREX and REDOX) used in Pu extraction from irradiated spent nuclear fuel rods and the decay of short half-life radionuclides, various forms of waste, including sludge, salt cake, and liquid, exist in the tanks under conditions of high pH (>10–14), high ionic strength (2 to 16 M $NaNO_3$), and high temperature (>60–110°C) with high concentrations of dissolved aluminum and phosphate.

Leaks, however, of HLW from perhaps as many as 67 single-shell tanks at the Hanford Site have been detected since the late 1950s. Additionally, low-level wastes were dumped directly to ground without any processing in the early years, therefore soils in the vadose zone and groundwater are expected to be severely contaminated by radionuclides including ^{129}I, ^{99}Tc, ^{79}Se, ^{237}Np, ^{90}Sr, ^{137}Cs, and $^{233,235,238}U$. Because these radionuclides may travel through vadose zone soils and reach groundwater leading to potential health risks to human beings, an understanding of the effects of caustic leachates on the soils is essential to predicting the migration and the sequestration of radionuclides at the Hanford Site.

Previous laboratory experiments showed that simulated caustic tank waste leachates dissolved primary minerals resulting in the release of Si to solution and the secondary precipitation of cancrinite, a zeolite-like mineral (Chorover et al. 2003). Cancrinite is thought to form on quartz surfaces, the primary mineral of the Hanford soil, as a result of contact with caustic tank waste leachates, and uptake of Cs and Sr within the newly precipitated zeolite structure has been studied using simulated tank waste leachate (Chorover et al. 2003). There are, however, few studies on the specific effects of secondary mineral precipitates such as cancrinite on the sequestration of radionuclides. Because leaks from the underground storage tanks have occurred, zones of altered primary minerals and neo-formed secondary precipitates are likely present. The migration and degree of retardation of key radionuclides, driven by natural recharge, past water pipeline leaks and future tank sluicing activities, from and through these altered zones of secondary precipitates, which are cementing grains and providing new sorption sites, require more study. In this study, batch dissolution and sorption experiments were conducted to increase our understanding of the effects of these secondary precipitates on the sequestration of radionuclides at the Hanford Site.

2 MATERIALS AND METHODS

2.1 *Materials*

Quartz sand (Aldrich) was pretreated with acid washing (100 g of sand in 100 mL of concentrated H_2SO_4 for 2 days), rinsed several times with NANOpure^{TM} water, and oven-dried at 85°C for 48 h. Powder X-ray diffraction (XRD) analysis revealed no mineral phases other than SiO_2. Warden silt loam (Warden soil) without pretreatment was used to represent the untreated Hanford sediment. Warden silt loam is one of the most widespread soils at the Hanford Site and is dominated by quartz and feldspar with minor amounts of calcite and mica. All solid materials used in the experiments were sieved to less than 250 μm. Simulated tank leachates were prepared using reagent grade chemicals and consisted of 2 M $NaNO_3$, 1.04 M NaOH, and 0.01 M $Al(NO_3)_3$.

2.2 *Dissolution experiments*

The batch dissolution experiments were prepared in 60-mL polypropylene bottles. In each bottle, 3 g of solid were mixed with 60 mL of synthesized tank waste. A series of bottles was prepared for each study to provide separate samples after 12 h, 1, 3, 7, 14, 21, and 30 d. The bottles were placed in an oven with a constant temperature of 90°C and shaken gently twice a day to ensure complete mixing. At the proposed collection time, the bottles were removed and placed in an ~80°C water bath. The pH was measured at high temperature using an Accumet pH meter with a solid-state electrode and the measured pH values were corrected to 90°C using the temperature dependence of the dissociation constant of water. After measuring the pH, the supernatant was filtered through a 0.45-μm PTFE syringe filter, diluted with NANOpure^{TM} water, and acidified with concentrated HNO_3 to prevent precipitation. Analyses of Si and Al concentrations were conducted using inductively coupled plasma optical emission spectrometry (ICP-OES).

After finishing experiments, the solid particles were separated from the remaining solution by filtration, washed several times with ethanol (5 g of solid in 500 mL of ethanol) to remove the sodium salts, and then dried at room temperature. The mineralogy of the secondary precipitates was identified by X-ray diffraction (XRD) on a Philips PW3040/00 X'pert MPD system with Cu Kα radiation. Information about the morphology and chemical composition of the precipitates was obtained by field emission scanning electron microscopy (FESEM) and energy dispersive X-ray spectroscopy (EDS) using LEO982 FESEM system with an Oxford ISIS energy dispersive X-ray detector.

2.3 *Sorption experiments*

Batch sorption experiments with reacted adsorbents and radionuclides of major concern at the Hanford Site, such as ^{99}Tc, ^{129}I, ^{79}Se, and ^{90}Sr, were conducted to investigate the possible sequestration of radionuclides by the newly formed secondary precipitates. Quartz and Warden soil reacted with caustic solutions at 90°C for 14 d were used as adsorbents. Radionuclides, including ^{125}I and ^{75}Se as analogues for ^{129}I and ^{79}Se, respectively, were used for sorption experiments on both reacted and unreacted quartz and Warden soil to compare the K_d values, a partition distribution coefficient between solution and solid phases. Single crystals of nitrate-cancrinite synthesized at high temperature were also used as an adsorbent for radionuclide sorption (Buhl et al. 2000).

Batch sorption experiments were conducted in 15-mL polypropylene centrifuge test tubes by mixing adsorbents and uncontaminated Hanford groundwater ($I = 0.0094$ M and $pH = 7.5$) spiked with known activities of radionuclides. Solids concentrations of 100 g/L and 5 g/L were used for anion (^{99}Tc, ^{125}I, ^{75}Se) and cation (^{90}Sr) radionuclides, respectively. The spiked groundwater for each radionuclide was added to the adsorbent and the test tubes were placed on a slow-moving platform shaker for 7 d. A blank sample containing no solid was also prepared in an identical manner to determine the initial activity of each radionuclide and to assure that there was no radionuclide sorption to the test tubes. After seven d of contact, the suspensions were centrifuged and the supernatants were filtered through 0.45-μm filters. Sorption of radionuclides on the test tubes and filters was negligible. The final activities of ^{125}I, ^{99}Tc, ^{75}Se, and ^{90}Sr were determined using either Wallac® 1415 liquid scintillation counting (LSC) or a Wallac® 1480 Wizard^{TM} 3-inch NaI automatic gamma detector.

3 RESULTS AND DISCUSSION

3.1 *Batch dissolution and precipitates*

When the caustic solution contacted the different solid samples at high temperature, the subsequent silica concentration in solution increased from the dissolution of quartz and perhaps other aluminosilicate minerals. Dissolved silica concentrations increased until the solutions reached supersaturation with respect to an aluminosilicate phase and precipitation occurred. Dissolved silica (Si) and alumina (Al) concentrations in Quartz sample are shown in Figure 1. The dissolved Al concentration was constant throughout day 1 suggesting there was no precipitation at reaction times less than 1 d. After 1 d of reaction time, however, the Al concentration decreased, reaching zero after 14 d, consistent with

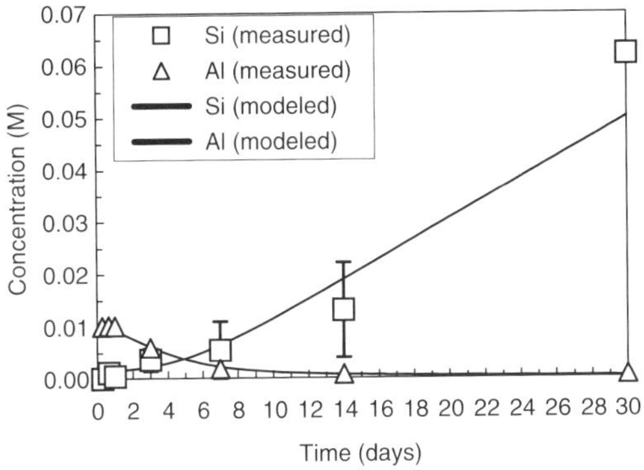

Figure 1. Dissolved Si and Al concentrations with time for quartz reacted with caustic solution. Theoretical Si and Al dissolved concentrations were predicted using Pitzer's model.

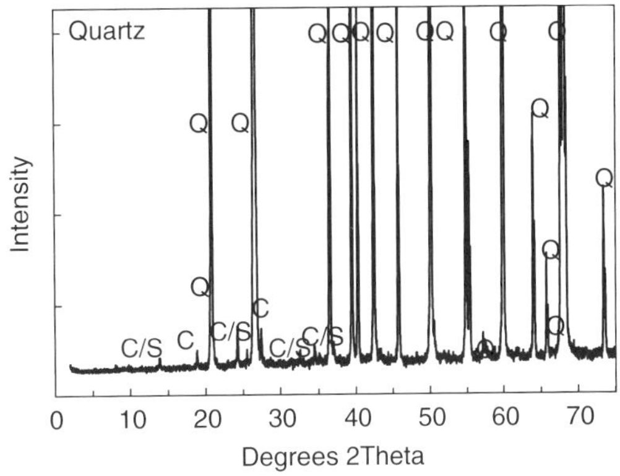

Figure 2. XRD patterns of reacted quartz sample after 14 d reaction (Q: quartz, C: cancrinite, S: sodalite).

near complete consumption of dissolved Al to form secondary precipitates. In contrast, the dissolved Si concentration increased with reaction time. The Si release rate (the slope of the dissolved Si concentration as a function of reaction time) for the quartz sample with added Al increased during the initial 0–3 d, decreased from 3–10 d, and then increased again after 12 d. The period of reduced Si dissolution rate during 3–10 d is considered to be the main phase of secondary precipitation. Initially, dissolved Si forms secondary precipitates with dissolved Al. After precipitation completely consumes the dissolved Al, dissolution of Si, rather than precipitation, again becomes the dominant process.

Si dissolved from quartz in the tank simulant without added Al was present at much higher concentrations than Si dissolved from quartz using tank simulant with added Al, indicating that the presence of dissolved Al reduced Si dissolution from quartz (data not shown). Although the presence of alkaline earth cations such as Na has been shown to increase the dissolution rate of quartz (Dove 1999), the inhibition of Si dissolution by Al species has also been widely reported in quartz and other aluminosilicate mineral dissolution experiments (Oelkers & Gislason 2001). The dissolution rate of Si is proportional to the concentration of deprotonated silanol groups on the surface (House 1994), and adsorption of Al species on quartz surfaces results in a slow hydrolysis step in the dissolution process (Ganor & Lasaga 1998) thus inhibiting dissolution. Surface characterization of quartz as a function of reaction time indicated qualitative evidence of different dissolution mechanisms for tank simulant with and without added Al. Concentrations of Si dissolved from quartz leached with tank simulant without added Al reflected an almost linear dissolution rate except in the early stages. The parabolic rate found in the early stages is attributed to the rate-controlling dissolution step of diffusion through a surface layer.

3.2 *Characteristics of secondary precipitates*

X-ray diffraction patterns of quartz sample after 14 d reaction time are shown in Figure 2. The patterns reveal the presence of a secondary precipitate identified as cancrinite. Cancrinite is a feldspathoid and has a hexagonal crystal system with space group $P6_3$. Another potential sodium aluminosilicate phase is sodalite, which has a cubic crystal structure with space group P43n (*9*, *27*). Because cancrinite and sodalite consist of AB and ABC sequences of 6 and 12-membered ring planes, respectively, the structural frameworks are similar. Both cancrinite and sodalite have a strong reflection at d = 3.66 Å (24.3 degrees 2Θ) and minor peaks at d = 6.33 Å (13.9 degrees 2Θ), d = 2.74 Å (32.6 degrees 2Θ), and d = 2.63 Å (34.1 degrees 2Θ), but cancrinite can be distinguished from sodalite by a strong 101-reflection at d = 4.69 Å (18.9 degrees 2Θ) and a 211-reflection at d = 3.24 Å (27.5 degrees 2Θ) (*9*, *11*, *27*). Because quartz sample reacted with synthetic tank leachates for 14 d showed typical cancrinite XRD peaks at d = 4.69 Å and 3.24 Å, cancrinite is considered to be the dominant species in the secondary precipitates, even if minor concentrations of sodalite coexist with the cancrinite.

FESEM analysis revealed more evidence for cancrinite precipitates on the mineral surfaces (not shown in here). Platy crystals were initially formed on pits or etches of quartz surfaces after 1 d reaction and ball-shaped aggregates with radiating needle-shaped hexagonal prisms were grown on the surfaces or from the Al deposits adsorbed on quartz surfaces during 3–7 d reaction.

Table 1. Distribution coefficients of key radionuclides.

Radionuclides	Adsorbents	K_d(ml/g)#	pH
^{125}I	Reacted Warden	1.85	10.4
^{125}I	Unreacted Warden	0.0	7.5
^{125}I	Reacted Quartz	0.0	9.8
^{125}I	Unreacted Quartz	0.0	7.5
^{125}I	Nitrate-cancrinite*	2.47	8.3
^{75}Se	Reacted Warden	31.0	10.4
^{75}Se	Unreacted Warden	4.11	7.6
^{75}Se	Reacted Quartz	0.10	9.8
^{75}Se	Unreacted Quartz	0.06	7.5
^{75}Se	Nitrate-cancrinite*	917.5	8.2
^{99}Tc	Reacted Warden	0.23	10.4
^{99}Tc	Unreacted Warden	0.0	7.5
^{99}Tc	Reacted Quartz	0.0	9.5
^{99}Tc	Unreacted Quartz	0.0	7.6
^{99}Tc	Nitrate-cancrinite*	1.81	7.8
^{90}Sr	Reacted Warden	90104	10.5
^{90}Sr	Reacted Warden	85086	8.5
^{90}Sr	Unreacted Warden	12.8	7.5
^{90}Sr	Reacted Quartz	68.3	9.7
^{90}Sr	Unreacted Quartz	10.2	7.4
^{90}Sr	Nitrate-cancrinite*	139321	8.0

Average values of two replicates.
* Synthesized nitrate-cancrinite (Buhl et al. 2000).

3.3 *Sorption experiments*

The presence of secondary minerals precipitated on mineral grains after contact with caustic waste fluids might have a significant role in the immobilization and ultimate fate of key radionuclides in the vadose zone sediments at the Hanford Site. Because leaking solution was subsequently introduced, contaminant radionuclides might have a contact with newly formed secondary minerals.

Distribution coefficients (K_ds) for key radionuclides, including ^{125}I, ^{75}Se, ^{99}Tc, and ^{90}Sr, for various adsorbents are shown in Table 1. Because reacted Warden soil and quartz showed higher sorption capacities for radionuclides than unreacted Warden soil and quartz, secondary mineral precipitates are expected to decrease the mobility of radionuclides.

Anionic radionuclides (^{125}I, ^{75}Se, ^{99}Tc) normally show low sorption affinities on unreacted quartz and Warden soil resulting in high mobility in the subsurface environment at the Hanford Site. However, caustic reacted sediment samples showed increased sorption capabilities, especially for reacted Warden soil, even at high pH. Radioactive Se showed one order of magnitude higher sorption onto reacted Warden soil as opposed to untreated Warden soil. A cationic radionuclide, ^{90}Sr, showed a tremendous increase in sorption affinity on reacted Warden soil. The increased sorption was attributed to the high cation exchange capacity of cancrinite. Synthesized nitrate-cancrinites showed much higher sorption capacities for all key radionuclides, suggesting that potential secondary precipitates such as cancrinite on the mineral surfaces could sequester radionuclides and thus result in decreased mobility in the Hanford vadose zone.

4 CONCLUSIONS

The precipitation of cancrinite onto the primary Hanford mineral surfaces was found after contact with caustic tank fluids. The results obtained from this research identify caustic fluid induced dissolution/precipitation reactions that lead to formation of secondary minerals with large sorption affinities for several high-risk radionuclides. Should the zeolitic secondary minerals remain stable in the Hanford vadose zone, they provide a potentially favorable contaminant sequestration mechanism that has not been previously considered in risk analysis at the Hanford Site.

REFERENCES

Buhl, J.-C., Stief, F., Fechtelkord, M., Gesing, T.M., Taphorn, U. & Taake, C. 2000. Synthesis, x-ray diffraction and MAS NMR characteristics of nitrate cancrinite $Na_{7.6}[AlSiO_4]_6(NO_3)_{1.6}(H_2O)_2$. *J. Alloys Compounds* 305: 93–102.

Chorover, J., Choi, S., Amistadi, M.K., Karthikeyan, K.G., Crosson, G. & Mueller, K.T. 2003. Linking cesium and strontium uptake to kaolinite weathering in simulated tank waste leachate. *Environ. Sci. Technol.* 37: 2200–2208.

Dove, P.M. 1999. The dissolution kinetics of quartz in aqueous mixed cation solutions. *Geochim. Cosmochim. Acta* 63: 3715–3727.

Ganor, J. & Lasaga, A.C. 1998. Simple mechanistic models for inhibition of a dissolution reaction. *Geochim. Cosmochim. Acta* 62: 1295–1306.

House, W.A. 1994. The role of surface complexation in the dissolution kinetics of silica: Effects of monovalent and divalent ions at 25°C. *J. Colloid Interf. Sci.* 163: 379–390.

Oelkers, E.H. & Gislason, S.R. 2001. The mechanism, rates and consequences of basaltic glass dissolution: I. An experimental study of the dissolution rates of basaltic glass as a function of aqueous Al, Si and oxalic acid concentration at 25°C and pH = 3 and 11. *Geochim. Cosmochim. Acta* 65: 3671–3681.

Complexity of mineral surfaces: experimental and theoretical studies

Water–Rock Interaction, Wanty & Seal II (eds)
 ISBN 90 5809 641 6

Etch pit morphology and Mg inhibition of calcite dissolution

R.S. Arvidson, K.J. Davis, M. Collier & A. Luttge
Department of Earth Science MS-126, Rice University, Houston, TX, USA

J.E. Amonette
Environmental Molecular Sciences Laboratory, Pacific Northwest National Laboratory, Richland, WA, USA

ABSTRACT: Mg inhibition of calcite dissolution is observed in simple carbonated solutions in single crystal experiments. Although inhibition of the overall rate occurs at relatively low Mg concentrations, increasing concentrations give rise to progressive changes in the morphology of etch pits. These observations can be understood in the context of changes in the relative rates of double kink nucleation and single kink propagation rates on step edges, tentatively attributed to site-specific surface adsorption of $MgHCO_3^+$ or $MgCO_3^\circ$.

1 INTRODUCTION

1.1 *Recent progress in carbonate mineral dissolution kinetics*

Significant progress has been made in understanding the mechanistic details of dissolution dynamics on the (104) cleavage surface of carbonate minerals. Much of this recent work has focused on AFM observations of the rate of step and kink movement (e.g. Liang et al. 1996a, b, Jordan & Rammensee 1998, Lea et al. 2001, Higgins et al. 2002). Observations of dissolution rates over larger lateral scales have also provided basic insight into the relationships of etch pit development and defect distribution (MacInnis & Brantley 1992, 1993, Luttge et al. 2003), as well as comparison with the overall "bulk" rate measured in mineral powders (Arvidson et al. 2003). In addition, considerable insight has been generated by integrating surface complexation models and observations (Van Cappellen et al. 1993, Stipp et al. 1994, Pokrovsky et al. 1999, Fenter et al. 2000, Pokrovsky & Schott 2002) with an atomistic description of surface topography and measurement of step velocities (Duckworth & Martin 2003). This general approach also has the potential to yield substantial detail regarding inhibition mechanisms in dissolution reactions as well. In this brief paper, we describe preliminary experimental observations of magnesium inhibition of calcite dissolution, and suggest how this inhibition may be related mechanistically to site-specific surface properties.

1.2 *Previous work on Mg inhibition of calcite dissolution*

Previous data on magnesium inhibition of calcite dissolution kinetics have been derived almost entirely through conventional experiments using mineral powders. These early data in general confirm inhibition, but differ in terms of the concentration required to impose a given reduction in rate. These variations most likely reflect the inherent diversity of experimental conditions: pH, ΔG_r, ionic strength, composition of supporting electrolytes, solid phase characteristics, and other factors (e.g. Berner 1967). Subsequent to work on trace metal inhibition by Terjesen et al. (1961) and Nestaas & Terjesen (1969), Sabbides & Koutsoukos (1995) observed dissolution to be inhibited at Mg^{2+} concentrations between 40 and 460 µM (rates reduced by 0.3 and 0.5 $\log_{10}$ units, respectively) in experiments performed close to calcite saturation. In contrast, although Gutjahr et al. (1996) found a reduction in calcite *growth* rates, they found Mg to have no effect on dissolution at concentrations of 300 µM $Mg(NO_3)_2$. In dilute, mixed Mg^{2+}–Ca^{2+} solutions, Sjöberg (1978) concluded inhibition of calcite dissolution to be sufficiently complex as to preclude incorporation of both these components in a general rate equation. Compton & Brown (1994) found a significant reduction in rate in the presence of 10–80 mM Mg^{2+}, (~50% at 40 mM, pH 8–9), which they found to be consistent with a Langmuir-Volmer adsorption isotherm. Alkattan et al. (2002) studied dissolution of calcite *single crystals* in

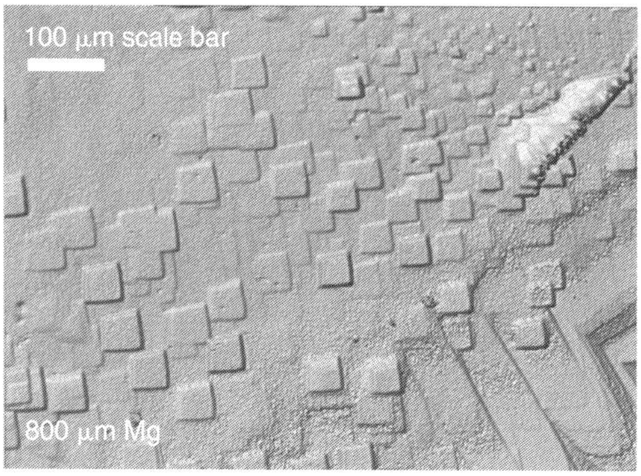

Figure 1. VSI data (10× magnification, white light) from coalescing etch pits on calcite (104) surface, reacted in 4.4 mM $NaHCO_3$, 800 μM $MgCl_2$, maintained at pH ~ 9, $PCO_2 \sim 10^{-3.5}$, 25°C. Positive step directions $[-441]_+$ and $[48-1]_+$ are oriented to the right and down, respectively.

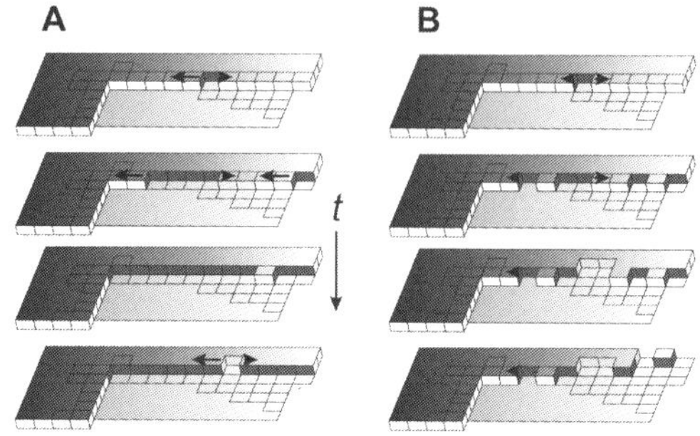

Figure 2. Evolution of step edges showing relationships between double kink nucleation rate (R_{KK}), kink migration rate (R_K), kink density, and step movement (adapted from Liang et al. 1996a). Time t increases from top to bottom. (A) In order for steps to remain straight, kink density must remain low: either no new double kinks can nucleate prior to the exit of single kinks at the step terminus (self annihilation), or single kinks formed from adjacent double kinks must annihilate each other (kink–kink annihilation). In both cases, R_K must remain high relative to R_{KK}. (B) Rough, irregular step profiles should thus result from inhibition of R_K relative to R_{KK}. "Curved" step profiles can be produced by a coupled reduction in R_K relative to R_{KK} and a pronounced anisotropy in single kink migration rates, i.e. $R_{K+} \neq R_{K-}$.

rotating disk experiments and found no inhibition at $MgCl_2$ concentrations up to 0.1 molal (pH 1–3).

2 METHODS

We measured Mg inhibition of calcite dissolution on single crystal (104) surfaces by two approaches. Absolute surface-normal retreat rates were derived from time-lapse changes in surface topography by *ex situ* vertical scanning interferometry (VSI), after flowing reactant solution over this surface in a fluid microcell (for details, see Luttge et al. 1999, 2003, Arvidson et al. 2003). This method allowed quantitative mapping of relatively large areas of the mineral surface at subnanometer vertical resolution (Fig. 1). The second approach involved monitoring similarly prepared calcite surfaces with reactant solutions in an AFM flow cell (Digital Instruments Nanoscope III Multimode SPM, operated in contact mode). All experiments used cleavage rhombs derived from high purity synthetic calcite. Carbonated reactant solutions were prepared from Na_2CO_3 or $NaHCO_3$ (4.4 meq/L) continuously equilibrated with atmospheric CO_2 (pH 8.8–9.0, 25°C; *cf.* Liang et al. 1996b, Lea et al. 2001). Mg was added (as $MgCl_2$) in concentrations of 0, 10, 50, 300, and 800 μM. Additional experiments were also performed as a function of pH in N_2-sparged (CO_2-free) solutions at the same ionic strength and $MgCl_2$ concentrations.

3 RESULTS AND DISCUSSION

In our VSI runs, we observed that inhibition occurs in carbonated solutions (pH ~ 9) at relatively low concentrations: relative to the rate in Mg-free $NaHCO_3$ solution, the addition of 10 μM $MgCl_2$ reduces the dissolution rate by ~80%, and 300 μM reduces the rate by ~90%. At higher concentrations, under conditions in which the solution is supersaturated with respect to pure magnesite, we observed that etch pit morphology departed significantly from the symmetrical rhombic shape observed in Mg- and CO_2-free solutions. Etch pit morphologies in carbonated Mg-free solutions (pH ~ 9) showed rounding at the intersection of the obtuse (+) step edges, such that the radius of the (+) steps decrease toward their common terminus. Lea et al. (2001) showed this morphology to be associated with a reduction in (+) step velocity, and attributed this inhibition to probable adsorption of carbonate ion at (+,+) kink sites. In our experiments, carbonated solutions having the highest (~800 μM) Mg concentrations produced (+,+) step intersections that remained relatively sharp but instead curved toward the (+,−) step intersections (Fig. 1). These changes implied changes in kink density along step edges, and variations in the kink nucleation rate.

Step curvature and roughness can be understood in the context of the rate of two related processes: double kink nucleation (R_{KK}) and single kink propagation of either positive (R_{K+}) or negative (R_{K-}) kink sites. As shown in Figure 2, if R_{KK} is slow relative to R_K, then kinks will tend to annihilate one another before new kinks can form. Conversely, if R_K

is reduced to the point where R_{KK} is competitive, then existing kinks may be preserved prior to the next cycle of kink nucleation. Liang et al. (1996a) elegantly demonstrated that simple kink annihilation at the step terminus was insufficient to explain both straight step morphology and constant step velocity, and postulated that the persistence of straight steps must reflect a balance between R_{KK} and the annihilation of adjacent kink pairs.

Morphologic changes occur at Mg concentrations substantially greater than those required to induce significant overall inhibition. Because movement of the "fast" (+) steps is responsible for much of the bulk dissolution (Liang et al. 1996b), inhibition at these sites may occur with relatively little inhibitor addition, similar to the effect observed for dissolved carbon and Mn^{2+} (Lea et al. 2001, Arvidson et al. 2003). It is unlikely that these morphologies reflect simple adsorption of dehydrated Mg^{2+} ions at specific kink sites, as they were not observed in CO_2-free solutions (pH 8–11+). Formation of ion pairs $MgCO_3^{\circ}$ or $MgHCO_3^{+}$ scavenge free carbonate ion from solution, and thus may reduce the carbonate ion concentration at kink sites. These components (as well as Mg^{2+}) may also form significant surface complexes as well, and their interactions result in overall inhibition (Arvidson et al., in prep.).

The basic anisotropy of obtuse versus acute step velocities on the calcite surface may reflect a complex interaction between the energetics of removal of cationic and anionic components broadly mediated by dominant surface complex populations ($>MOH_2^+$ versus $>CO_3H$; Duckworth & Martin, 2003). Because of the oblique orientation of carbonate groups with respect to the (104) surface, these sites differ in terms of coordination. For example, although removal of surface calcium demands cleavage of Ca–O bonds, the O–O distances within the surface-truncated Ca-octahedra vary according to their crystallographic orientation at (+,+), (+,−) and (−,−) metal sites (Fig. 3).

The relationship of step movements and inhibition can also be clarified by study of time-lapse AFM data. Figure 4 shows a sequence in which a previously developed etch pit, grown in carbonated Mg-free solution, is the site of new etch pit formation upon introduction of 800 μM carbonated $MgCl_2$ into the flow cell. This sequence definitively shows that (1) the (+) steps are severely inhibited after Mg introduction, with little measurable advance, and (2) growth of new etch pits occurs by advance of (−) steps. Although dissolved carbon ("CO_3", Fig. 4) reduces (+) step velocities (Lea et al. 2001), the fact that step curvature towards the (+,+) corner, previously developed under Mg-free conditions, is retained, while newly formed pits at 800 μM $MgCl_2$ show relatively sharp (+,+) corners, suggests that

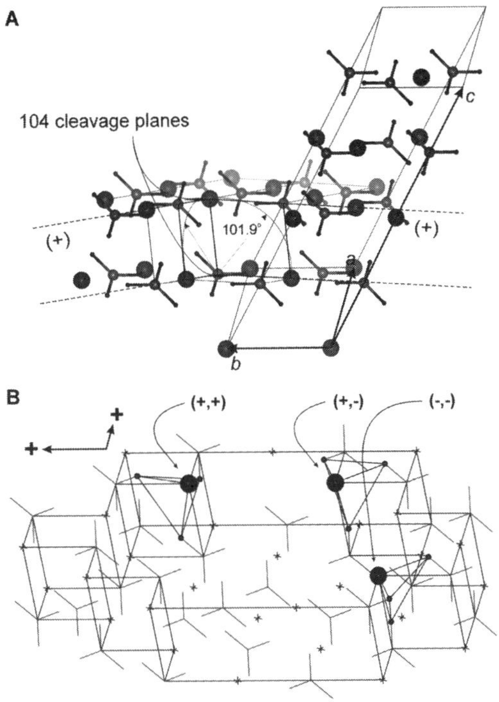

Figure 3. Calcite 104 cleavage surface. (A) Orientation of cleavage planes, hexagonal unit cell, obtuse (+) step faces, Ca atoms (large gray), and trigonal CO_3 groups. Out-of-plane, in-plane, and below-plane oxygens coordinate each carbon atom. (B) Ca–O coordination at kink sites, showing monatomic etch pit (black outline), trigonal CO_3 groups (gray), and Ca atoms (shown either as large black atoms, stars, or omitted for clarity). The length of Ca–O bonds between surface calciums (large black) and coordinating oxygens (small gray) at different kink sites have the same length. However, oxygen–oxygen distances within these truncated coordination octahedra are not equivalent: oxygens coordinating Ca at the (+,+) surface site all have O–O distances of 3.26 Å, whereas O–O distances associated with (+,−) and (−,−) sites are mixed (3.26 and 3.41 Å).

the (+) steps are fully passivated, and that the rates of both double kink nucleation and single kink migration along either $[48\bar{-}1]_+$ or $[-441]_+$ are low. Curvature of the (+) steps may instead reflect ineffective kink removal at the (−,+) corner, implying that the rate of single kink migration (R_{K+}) along the (−) steps is inhibited in the (+) direction as well. However, for the (−) steps to remain straight (Fig. 4) also implies that $R_K >> R_{KK}$.

4 CONCLUSIONS

We have documented magnesium inhibition of calcite dissolution in alkaline solutions far from

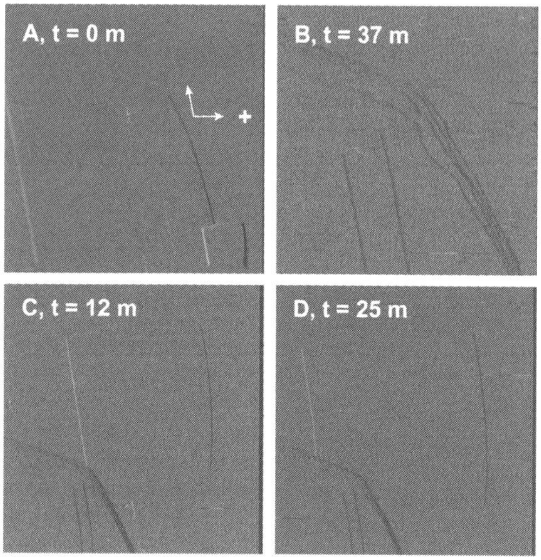

Figure 4. AFM reaction sequence. (A) 10 × 10 μm image of large etch pit previously formed in Mg-free solution, showing "CO_3" inhibition of obtuse (+) step directions by rounding towards the (+,+) intersection. Etch pit at lower right formed immediately after introduction of 800 μM $MgCl_2$. (B) 1 × 1 μm image detail of (+,+) corner (left of arrows in A), contrasting rough step edges of "CO_3"-inhibited etch pit with those of newly formed pit (appearing after $MgCl_2$ introduction) having sharp (+,+) corners. (C, D) Growth of etch pit via advance of acute (−) steps in 800 μM $MgCl_2$ solution, showing coalescence with static (+,+) intersection of older etch pit. Scan field is 3 × 3 μm.

equilibrium. As is the case for dissolved carbon addition, inhibition from increasing Mg concentration is associated with unique etch pit morphologies, reflecting possible changes in the relative rates of double kink nucleation and single kink migration. These changes are also linked to dissolved carbon concentration, and may reflect interaction of Mg ion pairs with the surface. These relationships will be explored in future modeling work.

ACKNOWLEDGEMENTS

This work was supported by the Office of Basic Energy Science, Geosciences Research Program, US Department of Energy. PNNL is a multiprogram national laboratory operated by Battelle Memorial Institute for the US Department of Energy under Contract DE-AC06-76RL0 1830.

REFERENCES

Alkattan, M., Oelkers, E.H., Dandurand, J.-L. & Schott, J. 2002. An experimental study of calcite dissolution rates at acidic conditions and 25°C in the presence of $NaPO_3$ and $MgCl_2$. *Chem. Geol.* 190: 291–302.

Arvidson, R.S., Ertan, I.E., Amonette, J.E. & Luttge, A. 2003. Variation in calcite dissolution rates: A fundamental problem? *Geochim. Cosmochim. Acta* 67: 1623–1634.

Berner, R.A. 1967. Comparative dissolution characteristics of carbonate minerals in the presence and absence of aqueous magnesium ion. *Am. J. Sci.* 265: 45–70.

Compton, R.G. & Brown, C.A. 1994. The inhibition of calcite dissolution/precipitation: Mg^{2+} cations. *J. Colloid. Interface Sci.* 165: 445–449.

Duckworth, O.W. & Martin, S.T. 2003. Connections between surface complexation and geometric models of mineral dissolution investigated for rhodochrosite. *Geochim. Cosmochim. Acta* 67: 1787–1801.

Fenter, P., Geissbuhler, P., DiMasi, E., Srajer, G., Sorensen, L.B. & Sturchio, N.C. 2000. Surface speciation of calcite observed *in situ* by high-resolution X-ray reflectivity. *Geochim. Cosmochim. Acta* 64: 1221–1228.

Gutjahr, A., Dabringhaus, H. & Lacmann, R. 1996. Studies of the growth and dissolution kinetics of the $CaCO_3$ polymorphs calcite and aragonite: II. The influence of divalent cation additives on the growth and dissolution rates. *J. Cryst. Growth* 158: 310–315.

Higgins, S.R., Jordan, G. & Eggleston, C.M. 2002. Dissolution kinetics of magnesite in acidic aqueous solution: A hydrothermal atomic force microscopy study assessing step kinetics and dissolution flux. *Geochim. Cosmochim. Acta* 66: 3201–3210.

Jordan, G. & Rammensee, W. 1998. Dissolution rates of calcite (10–14) obtained by scanning force microscopy: Microtopography-based dissolution kinetics on surfaces with anisotropic step velocities. *Geochim. Cosmochim. Acta* 62: 941–947.

Lea, A.S., Amonette, J.E., Baer, D.R., Liang, Y. & Colton, N.G. 2001. Microscopic effects of carbonate, manganese, and strontium ions on calcite dissolution. *Geochim. Cosmochim. Acta* 65:369–379.

Liang, Y., Baer, D.R., McCoy, J.M. & LaFemina, J.P. 1996a. Interplay between step velocity and morphology during the dissolution of $CaCO_3$ surface. *J. Vac. Sci. Technol. A*14: 1363–1375.

Liang, Y., Baer, D.R., McCoy, J.M., Amonette, J.E. & LaFemina, J.P. 1996b. Dissolution kinetics at the calcite-water interface. *Geochim. Cosmochim. Acta* 60: 4883–4887.

Luttge, A., Bolton, E.W. & Lasaga, A.C. 1999. An interferometric study of the dissolution kinetics of anorthite: the role of reactive surface area. *Am. J. Sci.* 299: 652–678.

Luttge, A., Winkler, U. & Lasaga, A.C. 2003. An interferometric study of the dolomite dissolution: A new conceptual model for mineral dissolution. *Geochim. Cosmochim. Acta* 67: 1099–1116.

MacInnis, I.N. & Brantley, S.L. 1992. The role of dislocations and surface morphology in calcite dissolution. *Geochim. Cosmochim. Acta* 56: 1113–1126.

MacInnis, I.N. & Brantley, S.L. 1993. Development of etch pit size distributions on dissolving minerals. *Chem. Geol.* 105: 1–49.

Nestaas, I. & Terjesen, S.G. 1969. The inhibiting effect of scandium ions upon the dissolution of calcium carbonate. *Acta Chem. Scand.* 23: 2519–2531.

Pokrovsky, O.S., Schott, J. & Thomas, F. 1999. Processes at the magnesium bearing carbonates/solution interface. I. A surface speciation model for magnesite. *Geochim. Cosmochim. Acta* 63:863–880.

Pokrovsky, O.S. & Schott, J. 2002. Surface chemistry and dissolution kinetics of divalent metal carbonates. *Environ. Sci. Technol.* 36: 426–432.

Sabbides, G. & Koutsoukos, P.G. 1995. Dissolution of calcite carbonate in the presence of magnesium and orthophosphate. In Zahid Amjad (ed.), Mineral Scale Formation and Inhibition: 73–86. New York: Plenum Press.

Sjöberg, E.L. 1978. Kinetics and mechanism of calcite dissolution in aqueous solutions at low temperatures. *Acta Univ. Stockh. Contrib. Geol.* 332: 1–92.

Stipp, S.L.S., Eggleston, C.M. & Nielsen, B.S. 1994. Calcite surface structure observed at microtopographic and molecular scales with atomic force microscopy (AFM). *Geochim. Cosmochim. Acta* 58: 3023–3033.

Terjesen, S.G., Erga, O., Thorsen, G. & Ve, A. 1961. II. Phase boundary processes as rate determining steps in reactions between solids and liquids: The inhibitory action of metal ions on the formation of calcium bicarbonate by the reaction of calcite with aqueous carbon dioxide. *Chem. Eng. Sci.* 74: 277–288.

Van Cappellen, P., Charlet, L., Stumm, W. & Wersin, P. 1993. A surface complexation model of the carbonate mineral-aqueous solution interface. *Geochim. Cosmochim. Acta* 57: 3505–3518.

Water–Rock Interaction, Wanty & Seal II (eds)
© 2004 Taylor & Francis Group, London, ISBN 90 5809 641 6

Using time-lapse ATR-FTIR spectroscopy to investigate reactions at the water–rock interface

M.J. Borda & D.L. Sparks
Department of Plant and Soil Sciences, University of Delaware, Newark, DE, USA

M.A.A. Schoonen
Department of Geosciences and Center for Environmental Science, Stony Brook University, NY, USA

D.R. Strongin
Department of Chemistry, Temple University, Philadelphia, PA, USA

ABSTRACT: The following paper reviews the use time-lapse attenuated total reflectance (ATR) Fourier transform infrared (FTIR) spectroscopy in several water-rock interaction studies.

1 INTRODUCTION

In the environment, a number of processes control the fate and transport of species. These processes include sorption and desorption reactions, precipitation, complexation, dissolution, and redox reactions. For decades, macroscopic studies have been performed to study the kinetics of these processes. Because a number of these important reactions occur rapidly and may never attain equilibrium, it becomes increasingly more important to study the kinetics of the reactions to ascertain the mechanism(s) that control them. The following abstract will focus on three distinct systems where time-lapse attenuated total reflectance (ATR) Fourier transform infrared (FTIR) spectroscopic kinetic studies have been employed to obtain mechanistic information. The ability to not only collect rate information but also chemical speciation and coordination information establishes ATR-FTIR spectroscopy as a powerful tool for mineral–water interface studies. This technique is particularly informative due to the ability to specifically analyze adsorbed species even in the presence of aqueous species. For example, the oxalate study reviewed in this abstract was possible because oxalate adsorbed at the goethite surface was analyzed without interference from aqueous oxalate.

The systems to be reviewed are the oxidative dissolution of pyrite (FeS_2) by ferric iron (Fe^{3+}; Borda et al. 2003) and molecular oxygen (O_2; Borda et al. 2004) and the heterogeneous photolysis of oxalate ($C_2O_4^{2-}$) at the goethite (α-FeOOH) surface. Current research on manganese dioxide (MnO_2) and its influence on the redox transformations of arsenic species in the environment will also be discussed.

2 BACKGROUND

2.1 *FeS_2 oxidation*

Over the past several decades, numerous laboratories have studied the oxidation of pyrite (FeS_2) (Oscarson et al. 1983; McKibben & Barnes 1986; Moses & Herman 1991; Williamson & Rimstidt 1994; Scott & Morgan 1995; Kamei & Ohmoto 2000; Schoonen et al. 2000). Pyrite oxidation is the primary source of acid mine drainage (AMD) and a poses a significant threat to environments that are impacted. Our interests lie in establishing the effects that variables such as oxidant, pH and light have on the pyrite oxidation reaction. These studies provide a molecular scale insight into this process and form the basis for new abatement strategies (Elsetinow et al. 2002).

The oxidation of the disulfide group (S_2^{2-}), to form sulfate (SO_4^{2-}), during pyrite oxidation is thought to proceed via a step-wise transfer of electrons. It is anticipated that the overall reaction must require the production of sulfur species with oxidation states between S^{-I} and S^{VI}.

A number of studies have used spectroscopy to investigate the oxidation of pyrite; however, no direct, *in situ*, real-time spectroscopic investigation has been performed. To accomplish the research goals described above, time-lapse ATR-FTIR spectroscopy was used to investigating the oxidative dissolution of pyrite.

2.2 *Heterogeneous oxalate photolysis*

Ferric oxalate solutions are a commonly used actinometer in chemistry to measure the photon flux of light sources. The basis of the technique is deriving the rate of photolysis of oxalate. From this rate, a photon flux can be obtained. Our interest was in analyzing the kinetics of the homogenous photolysis of oxalate and comparing it to the heterogeneous photolysis of oxalate on goethite as a proof of concept for a novel ATR photochemical cell design.

2.3 *As oxidation on MnO_2*

Manganese oxides are arguably one of the most reactive and important natural materials that impact metal(loid) speciation and ultimately toxicity and bioavailability. For example, well-studied oxyanions containing arsenic show a substantial interaction with Mn-oxide minerals in soils. In particular, it has been shown that Mn oxides in natural settings are capable of oxidizing arsenite (As^{3+}) to arsenate (As^{5+}), thereby generating a far less mobile and toxic species (Oscarson et al. 1983; Scott & Morgan 1995). Our interest is in establishing the electron transfer mechanism for the oxidation of As^{3+} to As^{5+} on MnO_2 through analysis of the initial As^{3+} complex at the MnO_2 surface and the use of rapid scan ATR-FTIR spectroscopy to investigate the kinetics of the oxidation reaction with a time resolution previously unattainable. Based on previous batch studies, it is anticipated that a majority of the As-oxidation reaction will occur in the first several minutes.

3 METHODOLOGY

3.1 *Pyrite oxidation studies*

The step-wise oxidation of the disulfide group during pyrite oxidation by ferric iron was investigated using *in situ* ATR-FTIR spectroscopy. Pyrite oxidation by ferric iron was investigated at pH 2.0, in the dark, and with both ultraviolet and visible wavelength illumination. Pyrite oxidation by molecular oxygen was investigated at pH 2.5 and pH 5.6, in the dark, and with ultraviolet light.

3.2 *Oxalate studies*

A solution experiment was run using a 0.12 M solution of potassium oxalate ($K_2C_2O_4$) and 0.02 M ferric sulfate ($Fe_2(SO_4)_3$). This solution was investigated in the dark and during illumination with UV light. Second, a 50 μM solution of oxalate was adsorbed to a goethite substrate and illuminated with UV light.

3.3 *In situ FTIR spectroscopy*

Experiments were run using a Nicolet Nexus 670 FTIR spectrometer fitted with a Spectra-Tech™ Continuously Variable Angle Model 300 ATR Accessory and a liquid N_2 cooled MCT-A detector. A custom-built Teflon™ flow-through reaction cell was designed for our laboratory by Spectra-Tech™ to be used with the ATR stage.

A germanium (Ge) ATR crystal (45° parallelogram; 25 × 10 × 3 mm) was used and pressure fitted against the Teflon™ reaction cell using the cell edge as a seal. In all of the experiments, the ATR crystal was coated with mineral slurries in accordance with the method of Awatani et al. (1998). All spectra were collected utilizing 128 co-added scans at 1 cm^{-1} resolution.

Light was directed on the sample using a 35 W deuterium/tungsten lamp (Analytical Light Instruments Inc. DT1000). The lamp was equipped with optional 200–400 nm ultraviolet (UV) wavelength output and 400–800 nm visible wavelength output.

3.4 *Advances being utilized in MnO_2/As studies*

The ATR technique will be performed on a Pike Technology™ flow-through, liquid jacked, temperature controlled, horizontal ATR accessory. Reaction temperatures will be controlled using a Fischer Scientific Isotemp re-circulating heater/chiller. Experimental solutions will either be injected into the ATR cell using airtight syringes or flowed-through using an HPLC pump.

Utilizing Nicolet SERIES rapid scan software, ATR-FTIR scans will be performed at a rate of 75 scans per second so that *in situ* reactions can be performed and kinetic data can be obtained through the collection of time-lapse spectra. To collect these spectra more rapidly, the resolution and number of co-added scans will be lowered. Because a detailed investigation into the adsorbed complex will be performed separately, the lower resolution scans will be more than adequate to establish the rate of appearance of the reaction product, the As^{5+} species. Oxidation experiments will be performed over a range of temperatures utilizing the temperature control ability of the ATR cell, in order to establish reaction rates at different temperatures. Using Arrhenius-type plots, the activation energy for the process will be obtained.

4 RESULTS AND DISCUSSION

4.1 *Oxidation with ferric iron*

The relative rate differences among three oxidation experiments are shown in Figure 1, based on

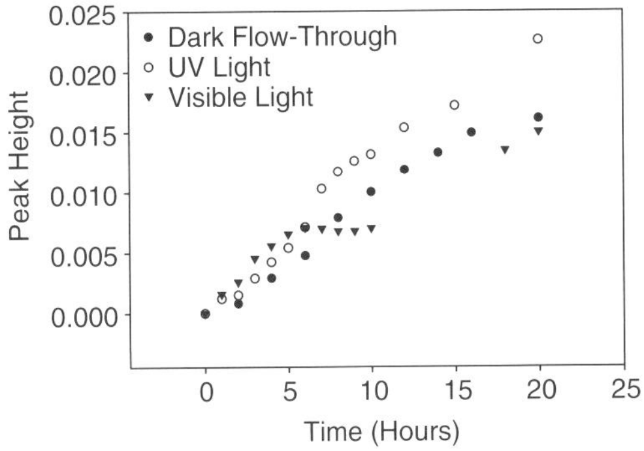

Figure 1. Plot showing results for the oxidation rate of pyrite using Fe^{3+} as the oxidant. Three experimental conditions, dark, UV light and visible light are shown. The peak height of thiosulfate associated with pyrite oxidation is plotted versus reaction time.

the relative peak height for thiosulfate. These observations are consistent with previous macroscopic batch aqueous experiments that showed the rate of oxidation of pyrite by Fe^{3+} is increased with light (Schoonen et al. 2000). Due to uncertainties with peak height measurements, absolute rates cannot be calculated from the IR spectra. Discrepancy between the rates for UV light and visible light may stem from differences in photon flux from the light source. The initial rates for both UV and visible light are greater than those determined in the dark experiment. The break in the oxidation experiment with visible light could suggest a change in mechanism or a fluctuation in the source.

4.2 *Oxidation with molecular oxygen*

Figure 2 is a plot of peak height versus time based on the relative peak height of the sulfate peak. There is a dramatic difference in the rate between the pH 5.6 and pH 2.5 experiments that may result from the strong pyrite oxidation rate dependence on pH (Schoonen et al. 2000). The rate shown for both the dark and ultra-violet light experiments at pH 5.6 show a correlation, however, this observation is tenuous. Although the relative rates of sulfate production estimated from the ultraviolet and dark experiments at pH 5.6 are in good agreement, the sulfate peak may be measuring more than simply the relative production rate of sulfate through the oxidation reaction. Ferric oxide patches at pH levels above 3.5 may form at the pyrite surface. Due to readsorption of sulfate to these patches, the sulfate peak may more accurately measure the capacity for the oxide patches to adsorb sulfate than the rate of production of sulfate.

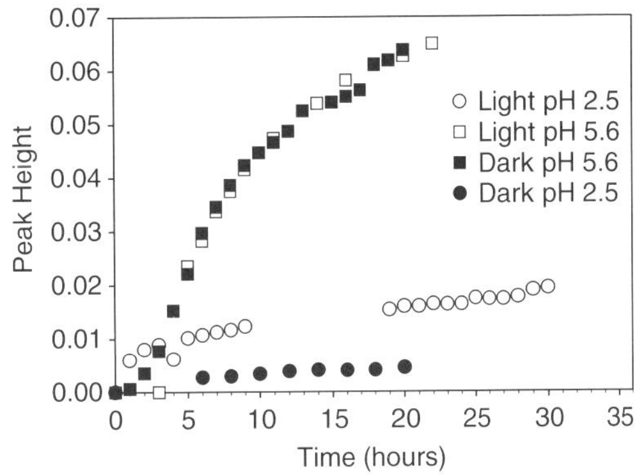

Figure 2. Plot showing results for the oxidation rate of pyrite using O_2 as the oxidant. Four experimental conditions, dark (pH 5.6), UV light (pH 5.6), dark (pH 2.5) and UV light (pH 2.5) are shown. The peak height of thiosulfate associated with pyrite oxidation is plotted versus reaction time.

4.3 *Oxalate photolysis*

From a plot of the natural log of the peak height for oxalate versus time, it can be shown that the heterogeneous photolysis of oxalate on FeOOH exhibits first-order rate kinetics (Fig. 3, bottom panel). The kinetics in this experiment appear to be more complicated than in the solution experiment. This is expected due to several pathways being available with the presence of the oxide surface. These results illustrate the importance of directly measuring surface bound oxalate rather than the combination of surface and aqueous oxalate measured in batch studies. This allows one to separate the homogeneous photolysis effects in a heterogeneous experiment.

5 SUMMARY

ATR-FTIR is well suited to study reactions that involve the oxyanions of a number of elements. For example, a number of environmentally important species are detectable in the mid-IR including the oxyanions of, phosphorous, sulfur, arsenic, chromium, boron, and selenium. A large number of environmentally important organic species are also detectable in the mid-IR. Using the growth, or reduction, of specific peaks as a function of time it is possible to follow the kinetics of reactions that involve the aforementioned oxyanions. This review illustrates some of the studies that have been performed using this powerful technique and the current studies on the MnO_2/As system point to frontiers in the study of rapid mineral-water interface reactions using ATR-FTIR spectroscopy.

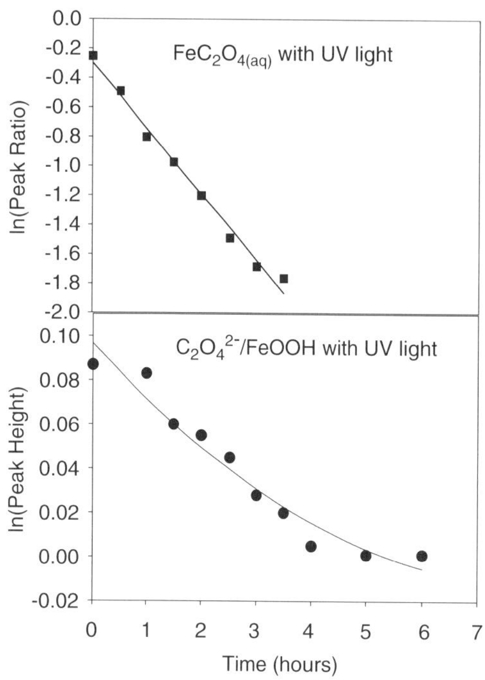

Figure 3. Plots showing the results of the homogenous photolysis of a ferrioxalate solution (top panel) and the heterogeneous photolysis of oxalate adsorbed to goethite (bottom panel). The natural log of the ratio of the peak height of an oxalate peak to a static sulfate peak was plotted versus time in the homogeneous experiment. The natural log of the peak height of an oxalate peak was used in the heterogeneous experiment.

REFERENCES

Awatani, T. & McQuillan, A. 1998. Adsorbed thiosulfate intermediate of cadmium sulfide aqueous photocorrosion detected and characterized by in situ infrared spectroscopy. *J. Phys. Chem.* B 102: 4110–4113.

Borda, M.J., Strongin, D.R. & Schoonen, M.A. 2003. A vibrational spectroscopic study of the oxidation of pyrite by ferric iron. *Am. Mineral.* 88: 1318–1323.

Borda, M.J., Strongin, D.R. & Schoonen, M.A. 2004. A vibrational spectroscopic study of the oxidation of pyrite by molecular oxygen. *Geochim. Cosmochim. Acta* in press.

Elsetinow, A.R., Strongin, D.R., Borda, M.J. & Schoonen, M.A.A. 2002. Suppression of pyrite oxidation in acidic aqueous environments using lipids having two hydrophobic tails. *Adv. Environ. Res.* in press.

Kamei, G. & Ohmoto, H. 2000. The kinetics of reactions between pyrite and O_2-bearing water revealed in *in situ* monitoring of DO, Eh, and pH in a closed system. *Geochim. Cosmochim. Acta* 64(15): 2585–2601.

McKibben, M. & Barnes, H. 1986. Oxidation of pyrite in low temperature acidic solutions: Rate laws and surface textures. *Geochim. Cosmochim. Acta* 50: 1509–1520.

Moses, C.O. & Herman, J.S. 1991. Pyrite oxidation at circumneutral pH. *Geochim. Cosmochim. Acta* 55: 471–482.

Oscarson, D.W., Huang, P.M., Liaw, W.K. & Hammer, U.T. 1983. Kinetics of Oxidation of Arsenite by Various Manganese Dioxides. *Soil Sci. Soc. Am. J.* 47(4): 644–648.

Schoonen, M.A.A., Elsetinow, A., Borda, M. & Strongin, D.R. 2000. Effect of temperature and illumination on pyrite oxidation between pH 2 and 6. *Geochem. Trans.* 4.

Scott, M.J. & Morgan, J.J. 1995. Reactions at oxide surfaces. 1. Oxidation of As(III) by synthetic birnessite. *Environ. Sci. Technol.* 29(8): 1898–1905.

Williamson, M.A. & Rimstidt, J.D. 1994. The kinetics and electrochemical rate-determining step of aqueous pyrite oxidation. *Geochim. Cosmochim. Acta* 58(24): 5443–5454.

Water–Rock Interaction, Wanty & Seal II (eds)
© 2004 Taylor & Francis Group, London, ISBN 90 5809 641 6

Galena dissolution in acidic pH at 25°C

J. Cama, P. Acero & C. Ayora
Institute of Earth Sciences "Jaume Almera", CSIC, Barcelona, Catalonia, EU

ABSTRACT: Using non-stirred flow-through experiments, the steady-state dissolution rate of galena (PbS) at pH 3 under oxygen saturated atmosphere and 25°C based on Pb release (R_{Pb}) is $-1.25 \pm 0.18 \times 10^{-10}$ mol $m^{-2} s^{-1}$ when normalized to initial surface area. The dissolution rate is compared to that obtained by dissolving galena of a pyritic sludge from Aznalcóllar mine waste (SW Spain). An in situ AFM experiment at pH 3 and 1 under oxygen saturated atmosphere and 23 ± 2°C was carried out to examine the surface evolution during dissolution, and reveals that the galena dissolution rate increases by decreasing pH. The dissolution rates based on Pb release normalized to geometric surface area are $-5.13 \pm 0.76 \times 10^{-7}$ and $-3.40 \pm 0.51 \times 10^{-6}$ mol $m^{-2} s^{-1}$, respectively.

1 INTRODUCTION

Galena (PbS) is found in the Iberian Pyrite Belt, a massive sulfide district in SW Spain that mainly consists of pyrite and chalcopyrite. The river waters of the area are very acidic and highly metal contaminated. Remarkably, the amounts of Pb in river waters of the area are rather low, at least one order of magnitude less than those of Cu and Zn, taking into account that the amounts of Cu-Zn-Pb bearing phases in the sulfide ores are on average similar. Dissolution and precipitation reactions play an important role in the sulfide weathering, which is the overall process responsible for the metal release. Distinct dissolution rates of the sulfides making up the deposit (pyrite, chalcopyrite, sphalerite, galena) must generate different metal releases to river waters. In particular, the galena dissolution reaction must be directly involved in the Pb release, resulting in low lead concentration into the river waters. Considering the low Pb concentration in the river waters, the galena dissolution rate might be slower than the dissolution rates of chalcopyrite and sphalerite under the same environmental conditions. In the present paper, we present preliminary results of galena dissolution obtained by dissolving galena either as pure phase or as a minor phase (0.8 wt. %) of a pyritic sludge (Domènech et al. 2002) by means of non-stirred flow-through experiments. Sphalerite and chalcopyrite dissolution rates are also estimated from the pyritic sludge dissolution to compare to galena dissolution rates at pH 3 and 25°C. An in-situ AFM experiment has been carried out to examine the evolution of a cleaved galena surface at pH 3 and 1. The galena steady-state dissolution rates obtained via in situ AFM experiments are based on the output solution chemistry.

2 MATERIALS AND METHODS

2.1 *Sample characterization*

The galena sample used in the present study is from Mine Mineralògica, Bellmunt del Priorat (Catalànids range). Powder XRD patterns showed the sample consists mainly of galena (PbS) and minor amounts of cerussite ($PbCO_3$) (ca. 5%). For the in situ AFM experiment, a galena fragment with preferential crystallographic orientation {100} was used. The geometric surface area of the fragment is estimated. For the flow-through experiment, some mineral fragments were ground in an agate mortar and pestle to obtain a powder fraction between 10 and 100 μm. The BET-determined initial surface area of the galena powder was $0.3 \pm 10\%$ $m^2 g^{-1}$ using 5-point N_2 adsorption isotherms. The temperature of sample degassing is 125°C for 20 min. The pyritic sludge sample used in the present study is well characterized by Domènech et al. (2002).

2.2 *Experimental set up and analysis*

Non-stirred flow-through experiments immersed in a water-bath have been used to calculate galena steady-state dissolution rate. A detailed description of the experimental setting can be found in Cama et al. (2002).

In the flow-through experiments, as the steady state was attained, the dissolution rate, R, (mol m^{-2} s^{-1}) was based on the release of Pb and S according to the expression

$$\nu_j R = -\frac{q}{A}(C_{j,out} - C_{j,inp}) \quad (1)$$

where $C_{j,inp}$ and $C_{j,out}$ are the concentrations of component j (Pb or S) in the input and the output solutions, respectively (mol m^{-3}), ν_j is the stoichiometry coefficient of j in the dissolution reaction, t is time (s), A is the reactive surface area (m^2) and q is the fluid volume flux through the system ($m^3 s^{-1}$). Note that in this formalism, the rate is defined to be negative for dissolution and positive for precipitation. The non-stirred flow-through experiments have been performed at $25 \pm 0.01°C$, input pH 3 (HCl) and O_2 saturated atmosphere. The in situ AFM experiments were carried out using a Molecular Imaging microscope equipped with Digital Instrument Software at room temperature ($23 \pm 2°C$). A transparent polycarbonate cell was designed to hold the fragment. The surface area explored was less than 5 μm squared. In the in situ AFM experiment, the flow rate was maintained at 0.40 ± 0.05 mL min^{-1}, yielding a residence time of 1.5 min in the cell.

Total Pb and S and trace metal elements from the collected output solution were analyzed with ICP-AES and ICP-MS, respectively. Input and output solution pH was measured with a Crison combined glass electrode at room temperature. Eh of the output solutions while passing through a by-pass cell connected to the flow-through reactor outlet was measured with a Crison platinum electrode ORP. Solutions were calibrated with Crison standard solutions of 220 and 468 mV.

3 RESULTS AND DISCUSSION

3.1 *Flow-through experiments*

In the flow-through experiment with galena as starting material, release of Pb and S is higher at the onset of the experiment. Afterwards Pb and S concentration decrease until the steady state is reached (Fig. 1). The steady state condition lasted at least 100 h. The output pH varied slightly during the experiment to be 3.00 ± 0.06. The Eh ranged between 0.3 and 0.32 V through the entire experiment. The stoichiometric ratio between Pb and S ($SR_{Pb/S}$) is defined as the ratio between the release of Pb and the release of S at steady state. For the entire experimental run the output S concentration is lower than the output Pb concentration, yielding a $SR_{Pb/S}$ stoichiometric ratio of approx. 4 (Fig. 2). Through the experiment the galena dissolution is taking place in conditions of undersaturation with respect to galena (PbS), anglesite ($PbSO_4$), native sulphur (S) and cerussite ($PbCO_3$) at Eh values ranging from 0.3 to 0.32 V. The high release of Pb and S at the onset of the experiment is caused by the fast dissolution of galena microparticles attached onto the grains, and part of the Pb release is also due to the fast dissolution of cerussite ($PbCO_3$) at this pH. Afterwards, the Pb and S concentrations gradually decrease to steady state. The depletion of sulfur through the entire experiment cannot be due to precipitation of native sulfur onto the galena surface because the dissolution occurs at undersaturated conditions with respect to S. Also, the sulfur depletion cannot be associated to $PbSO_4$ precipitation because only the S concentration is

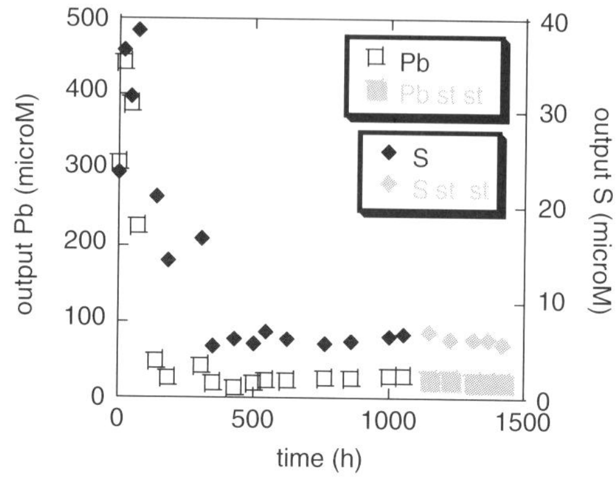

Figure 1. Variation in Pb and S concentrations as a function of time in the flow-through experiment at pH 3 with galena as starting material. St st denotes output concentrations in steady state.

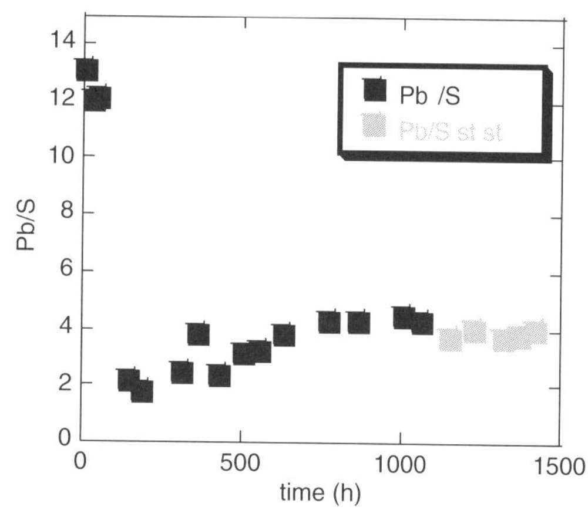

Figure 2. Variation in Pb/S ratio as a function of time in the flow-through experiment at pH 3 with galena as starting material. At steady state (st st) the average Pb/S ratio is approx. 4.

depleted and the solution is undersaturated with respect to anglesite, calculated using the EQ3NR code (Wolery 1992). A plausible explanation for sulphur depletion may be that assuming that the overall reaction accounting for galena dissolution is

$$PbS(s) + 2H^+ \leftrightarrow Pb^{2+} + H_2S(aq) \qquad (2)$$

an amount of aqueous sulphur is partially converted to $H_2S(g)$. The galena dissolution rate was obtained based on the release of Pb ($Rate_{Pb}$) at steady state by normalizing to the initial surface area. The average dissolution rate was $-1.25 \pm 0.31 \times 10^{-10}$ mol m^{-2} s^{-1}.

The flow-through experiment with a pyritic sludge sample was run for 3000 h assuming that 0.8 wt. % of the sample is galena (Domènech et al. 2002). Based on Pb output concentration after 1500 h, steady state was attained and the calculated dissolution rate is $-1.22 \pm 0.35 \times 10^{-10}$ mol m^{-2} s^{-1} as the specific surface area of galena ($0.7 \pm 0.07\, m^2 g^{-1}$) was used for normalization. Therefore, based on Pb release, the dissolution rate obtained by dissolving galena in the pyritic sludge is the same within error as that obtained by dissolving galena as starting material. Likewise, based on the Zn and Cu output concentrations and the respective amounts of sphalerite (1.4 wt. %) and chalcopyrite (0.6 wt. %) in the pyritic sludge, the steady-state dissolution rates for sphalerite and chalcopyrite are $-2.92 \pm 0.44 \times 10^{-11}$ and $-4.48 \pm 0.67 \times 10^{-11}$ mol m^{-2} s^{-1}, respectively, as normalized to their specific surface area (1.4 ± 0.2 and $1.2 \pm 0.2\, m^2 g^{-1}$, respectively).

The galena dissolution rate is faster than those of sphalerite ($Rate_{Zn}$) and chalcopyrite ($Rate_{Cu}$) at pH 3 as was obtained in Rimstidt et al. (1994) with the presence of ferric iron in solution at pH approx. 2. Therefore, the low Pb content in the river waters of the Iberian Pyrite Belt may not be due to merely lower dissolution rate of galena. Other phenomenon, like a possible inhibitory effect of the iron content in the river waters on the galena dissolution, may play an important role.

3.2 *In situ AFM experiment*

The in situ AFM experiment consisted of two stages (pH 3 and 1). Output Pb and S concentrations obtained were used to evaluate the evolution of the experiment and to determine possible attainment of steady state. The resulting average dissolution rates are only based on lead release because a non-stoichiometric behavior is also observed ($SR_{Pb/S} = 5.1$ and 4 at pH 3 and 1, respectively). In the first stage at pH 3, the output concentration approached steady state after 30 min. Then the second stage was initiated at pH 1 causing an increase of the output concentrations until steady state was attained. According to

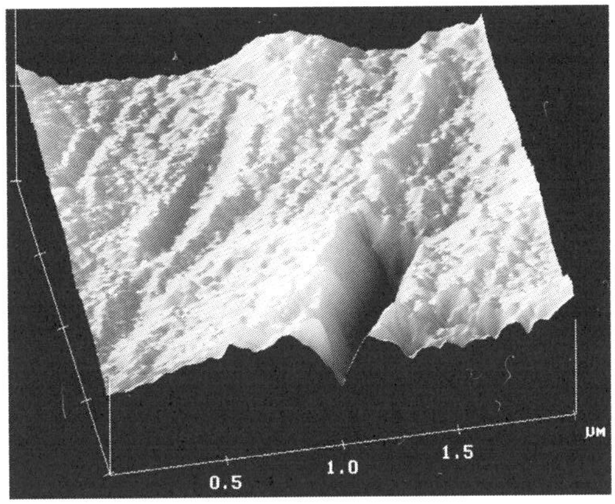

Figure 3. AFM image in 3-D mode of galena surface of the in-situ AFM experiment at pH 1. Surface roughness is caused by formation of protrusions during dissolution.

eq. (1), the corresponding dissolution rate values based on Pb release ($Rate_{Pb}$) were $-5.13 \pm 0.76 \times 10^{-7}$ and $-3.40 \pm 0.51 \times 10^{-6}$ mol m^{-2} s^{-1}, respectively, normalized to the geometric surface area (A) of the fragment, which is the sum of the face areas that interact with the solution (all faces except that affixed to the bottom cell).

At the pH range of the in situ AFM experiment conducted in the present study, the primary microtopographic feature observed during {100} surface dissolution has been the formation of protrusions over the cleaved surface (Fig. 3) as De Giudici & Zuddas (2001) observed in their in situ AFM experiments with galena cleavage surfaces.

Based only on the dissolution rates obtained at pH 3 and 1, assuming that the dissolution rate is proportional to a fractional power of the hydrogen-ion activity because the proton promoted mechanism may dominate the galena dissolution rate under acidic condition, the galena dissolution-rate dependence on pH is estimated according to the empirical equation

$$Rate = k_{H^+} \cdot a_{H^+}^{n_{H^+}}$$

where a_{H^+} is the activity of protons in the solution, n_{H^+} is the order of the reaction with respect to H^+ and k_{H^+} is a rate coefficient. The linear regression yields $n_{H^+} = 0.56$. Hsieh & Huang (1989) also reported an increase of dissolution rate by decreasing the pH. The difference in the dissolution rate value at pH 3, obtained from the non-stirred flow-through and the in situ AFM experiments, highlights the importance of the value used to normalize the dissolution rates, e.g. the use of the BET surface area or the geometric surface area. The former would represent the bulk dissolution, taking place in the overall surface of galena grains, whereas the latter would result from

Figure 4. SEM photograph of the surface of a galena fragment. Typical features of a cleaved face as terraces, borders and kinks are shown.

considering flat galena surfaces. A SEM photography of a fragment surface of galena (Fig. 4), similar to that utilized in the in situ AFM experiment, shows distinct types of surface features, such as borders, terraces or kinks with presumably different reactivity that likely would yield different dissolution rates. More in situ AFM experiments may resolve the evolution of the different surface features and the related dissolution mechanisms.

4 CONCLUSIONS

By means of non-stirred flow-through experiments with galena as starting material, the galena steady-state dissolution rate, based on Pb release (R_{Pb}), is $-1.2 \pm 0.18 \times 10^{-10}$ mol m^{-2} s^{-1} at pH 3 and 25°C and oxygen saturated atmosphere. Likewise, the dissolution of the Aznalcóllar pyritic sludge with a galena content of 0.8 wt. % at the same conditions results in the same dissolution rate within error. Furthermore, the dissolution rates of sphalerite and chalcopyrite (minor components of the pyritic sludge), based on Zn and Cu release, are $-2.92 \pm 0.44 \times 10^{-11}$ and $-4.48 \pm 0.67 \times 10^{-11}$ mol m^{-2} s^{-1}, respectively, resulting in lower dissolution rates of sphalerite and chalcopyrite than that of galena. Therefore, the low content of Pb in river waters of the Iberian Pyrite Belt, compared to that of Zn and Cu, cannot be explained by assuming a lower dissolution rate of galena, at least at pH 3. The galena dissolution reaction at pH 3 and 1 is not stoichiometric as sulfur depletion is observed during the flow-through and in situ AFM experiments. Precipitation of native sulfur or anglesite onto the galena surface is ruled out because the dissolution is occurring at conditions of undersaturation with respect to both phases. Moreover, the inspection of the galena surface via in situ AFM experiment confirms the absence of S-bearing stable phases onto the galena substrate. Based on the Pb released at steady state during the in situ AFM experiment, and normalizing to geometric surface area, the galena dissolution rate decreases with increasing pH. The order of the reaction with respect to H^+ (eq. 2), obtained by linear regression of log R_{Pb} versus pH (only pH 1 and 3), is $n_{H^+} = 0.56$.

REFERENCES

Cama, J., Metz, V. & Ganor, J. 2002. The effect of pH and temperature on kaolinite dissolution rate under acidic conditions. *Geochim. Cosmochim. Acta* 66: 3913–3926.

De Giudici, G. & Zuddas, P. 2001. *In situ* investigation of galena dissolution in oxygen saturated solution: Evolution of surface features and kinetic rate. *Geochim. Cosmochim. Acta* 65: 1381–1389.

Domènech, C., De Pablo, J. & Ayora, C. 2002. Oxidative dissolution of pyritic sludge from the Aznalcóllar mine (SW Spain). *Chem. Geol.* 190: 339–353.

Hsieh, Y.H. & Huang, C.P. 1989. The dissolution of PbS(s) in dilute aqueous solutions. *J. Colloid Interf. Sci.* 131: 537–549.

Rimstidt, J.D., Chermak, J.A. & Gagen, P.M. 1994. Rates of reaction of galena, sphalerite, chalcopyrite, and arsenopyrite with Fe(III) in acidic solutions. In C.N. Alpers & D.W. Blowes(eds.): *Environmental Geochemistry of Sulfide Oxidation*, ACS Symp. Ser. 550, 1–13. American Chemical Society.

Wolery, T.J. 1992. EQ3NR, a computer program for geochemical aqueous speciation-solubility calculations: theoretical manual, User's guide, and related documentation (version 7.0), Lawrence Livermore National Lab, Berkeley, CA.

Water–Rock Interaction, Wanty & Seal II (eds)
© 2004 Taylor & Francis Group, London, ISBN 90 5809 641 6

Influence of nanoscale porosity on fluid behavior

D.R. Cole, M.S. Gruszkiewicz, J.M. Simonson, A.A. Chialvo, Y.B. Melnichenko, G.D. Wignall, G.W. Lynn, J.S. Lin, A. Habenschuss, B. Gu, K.L. More & T.D. Burchell
Oak Ridge National Laboratory, Oak Ridge, USA

A. Striolo, Y. Leng & P.T. Cummings
Vanderbilt University, Nashville, USA

W.T. Cooper & M. Schilling
Florida State University, Tallahassee, USA

K.E. Gubbins
North Carolina State University, Raleigh, USA

H. Frielinghaus
Institut für Festkorperförschung, Jülich, Germany

ABSTRACT: Fluids including hydrocarbons, aqueous solutions, and gaseous species (e.g. CO_2, CH_4) can occupy the pores or fractures within solid matrices. The size, distribution and connectivity of these confined geometries, the chemistry of the solid, the chemistry of the fluids and their physical properties collectively dictate how fluids migrate into and through these micro- and nano-environments, wet and ultimately react with the solid surfaces. The purpose of this project is to demonstrate the application of state-of-the-art experimental, analytical and computational tools to assess key features of the fluid-matrix interaction. Our multidisciplinary approach combines neutron and X-ray scattering experiments, FTIR and NMR spectroscopy, molecular dynamics simulations, and thermodynamic measurements to quantitatively assess molecular properties of fluids confined to well-characterized porous media, subjected to a wide range of experimental conditions.These studies are providing a molecular-level understanding of how intrinsically different fluids behave in confined geometries compared to bulk systems.

1 INTRODUCTION

Fluids containing inorganic and organic solutes (including hydrocarbons) and gaseous species (e.g. CO_2, CH_4) can occupy the pores or fractures of numerous types of complex heterogeneous solids. The porosity (void) volumes within the solids span wide length scales including micro-, meso- and macroporous regimes ($d/Å < 20$, $20 < d/Å < 500$, and $d/Å > 500$, respectively, as defined by IUPAC). These solid materials include such practical systems as supported catalysts, ceramics and composites, glasses, rock, soil, and bone. A number of factors dictate how fluids, and with them reactants and products of intra-pore transformations, migrate into and through these nano-environments, wet and ultimately adsorb and react with the solid surfaces. These include the size, shape, distribution and interconnectivity of these confined geometries, the chemistry of the solid, the chemistry of the fluids and their physical properties (Cole & Chakraborty 2001). The dynamical behavior of fluids and gases contained within solids is controlled by processes occurring at the interface between the various phases (e.g. water-water, water-solute, water-solid, solute-solid etc.), as well as the rates of supply and removal of mobile constituents. The richness and complexity of fluid behavior (e.g. phase transitions, molecular orientation and relaxation, diffusion, adsorption, wetting, capillary condensation, etc.) in confined geometries only underscores the need to adopt a multidisciplinary approach when trying to quantify this behavior regardless of the fluid type or nature of the porous medium. If properly calibrated and scaled, an atomistic or molecular understanding of fluid-solid interaction may provide quantitative insight into the behavior of systems at the macroscopic level.

2 APPROACH

There is general agreement that the collective structure and properties of bulk fluids are altered by confinement between two surfaces or in narrow pores due to the interplay of the intrinsic length scales of the fluid and the length scale due to confinement (Idziak & Li 1998). Also crucial to the molecular behavior of fluids is the geometry of the pore which can include simple planar walls as encountered in micas; slits, cylinders, and spheres; and spheres linked with cylinders as observed in zeolites (Alejandre et al. 1996). Other factors that contribute to the modification of fluid properties include the randomness of the matrix and the connectivity of the pore network (Pitard et al. 1996).

In order to assess key features of the fluid-matrix interaction at the nanoscale, a multidisciplinary approach was taken that addressed the following areas:

(a) Characterization of the physical nature (e.g. size, shape, fractal geometry, etc.) of the pores using various types of microscopy (SEM; TEM), scattering (small-angle X-ray and neutron), and gas sorption methods, (b) Determine the thermodynamic properties of water in confined geometries using the isopiestic method over a wide ranges of temperature and pressure, (c) Determine the local short-duration dynamics and longer-scale hindered motion of water, CO_2 and hydrocarbons through use of complementary techniques including neutron scattering, NMR, and FTIR spectroscopy, as a function of pressure and temperature, (d) Apply, refine and, where necessary, develop molecular-based simulations methods that describe the properties and structure of confined fluids. Selected aspects of these activities are highlighted below.

3 SOLIDS CHARACTERIZATION

The physical characterization of solids is a crucial step that sets the stage for how variations in pore size, roughness, distribution, connectivity, composition (hydrophobic versus hydrophilic), surface charge, and structure (i.e. crystalline versus amorphous) influence molecular behavior of confined fluids. Table 1 summarizes the different porous solids selected for experimentation and/or modeling, their pore shapes and widths, and surface chemistry.

The solids we selected for study exhibit many of the following characteristics: (a) sparingly soluble at the pressure-temperature-fluid composition conditions of the experiments, (b) suitable for detailed characterization of pore features such as size, shape, distribution, roughness, etc., (c) susceptible to chemical manipulation from hydrophobic to hydrophilic by chemical pretreatments, and (d) limited crystallinity to minimize the impact on small-angle scattering. In addition to the more conventional methods typically used to characterize porous matrices (SEM, TEM), we used a number of scattering and spectroscopic techniques. While not yielding direct imaging information, scattering and diffraction experiments can provide global statistical information about interfaces (e.g. total porosity, pore surface area, pore length scale, skeletal density, fractality) over an enormous range of length scales (Å to μm), represented by reciprocal or Q-space (where $Q = 4\lambda^{-1}\sin\theta$ is the momentum transfer, 2θ is the angle of scattering, and λ is the wavelength). The scattering intensity, I(Q), is measured as a function of momentum transfer, and exploits the difference in scattering power (contrast) between the solid matrix, and air or fluid occupying the pore space.

An example of results from small-angle neutron scattering (SANS) is shown in Figure 1 for three different porous silica sol–gel glasses. This plot reveals that the correlation peaks (vertical arrows) shift to successively longer distances, which is indicative of variations in the dehydration process designed to produce the different pore sizes. The glass with 75 Å pores yields the minimal excess scattering in the small-angle region, which makes it better suited for studying the influence of confinement on liquid-gas critical behavior. For the most part, the disordered porous silica glasses have rather smooth pores that are uniformly distributed throughout the solid. The porous carbons have more wide-ranging pore sizes, and exhibit varying degrees of crystallinity depending on the temperature of burn-off and its duration used to activate the solid.

Table 1. Porous solids used in experiments and modeling.

Solid	Surface	Pore shapes	Pore widths (nm)
Porous glass	O, Si	Cylinder	2–8
Silica sol-gels	O, Si	Cylinder	2.5–20
Silica aerogel	O, Si	Irreg. cyl.	6–7
Activated C fibers	C (O, H)	Varied	0.4–1.0
C nanotubes	C (O, H)	Cylinder	1.0–1.6
Na-K Zeolites	K, Na, Si, Al, O	Cylinder, cage	0.3–0.8
Montmorillonite	Na, Si, Al, O	Slit, pillars	0.5–0.8
Muscovite (2M1)	K, Si, Al, O	Slit	0.6–1.0

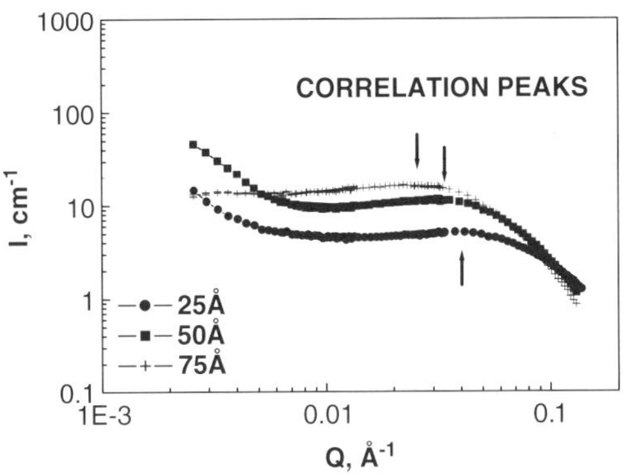

Figure 1. SANS from various silica sol–gel glasses.

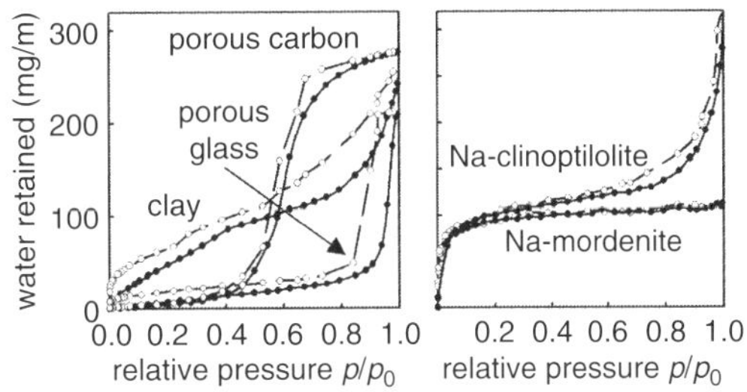

Figure 2. Adsorption (solid) and desorption (dashed) isotherms of water measured for a variety of porous solids at 105°C.

4 WATER SORPTION

The total adsorption capacity of a porous or fractured solid matrix depends on the total exposed solid surface area, the shape and size distribution of voids in the matrix, and the nature of the interaction between fluid molecules and the solid surface, and between the molecules themselves. The competition between liquid–solid and liquid–liquid interactions leads to monolayer and cluster formation at low pressure, and capillary condensation at high pressures. In order to understand these effects in a fluid with strong intermolecular interactions, adsorption/desorption isotherms of water were measured for 15 solid samples (e.g. porous silica, carbon monoliths, zeolites, clay) at 105, 150, and 200°C from vacuum to saturated vapor pressure in an isopiestic apparatus. This apparatus is unique in its capability for precise weighing of samples at elevated temperature and pressure between vacuum and 4 MPa. This gravimetric method, in contrast to the more common volumetric techniques, allows for continuous monitoring of the changes of the adsorbent itself due to degassing, oxidation, or decomposition (Gruszkiewicz et al. 2001).

Figure 2 shows representative water sorption results at 105°C from vacuum to saturated vapor pressure (0.121 MPa) for several of the porous solids. At low pressure the isotherms reflect competition between liquid–solid and liquid–liquid interactions (forming the monolayer and subsequent clusters or layers, pore filling). The behavior at higher pressures reflects the impact pore size and structure have on capillary condensation and possible hysteresis.

Both zeolites exhibit very steep isotherms at low pressure followed by relatively flat trends at moderate to high pressures. This behavior is indicative of strong water-solid intermolecular interaction dominated by mono-layering. The porous carbon and silica glass have rather flat slopes at low pressure followed by steep trajectories indicating weak water-solid interaction and capillary pore filling. The montmorillonite exhibits a steady increase in water uptake suggestive of multi-layering characteristic of dipolar electrostatic interaction. Results such as these can form the basis of testing various MD simulation schemes.

5 FLUID STRUCTURE AND DYNAMICS

Confinement in porous media leads to perturbation of structures and motion of molecules such as water. Properties affected include vibrational, rotational and translational motion, adsorption/condensation, phase equilibria, critical phenomena, and hydrogen bonding. Because molecular motion and structure occur over widely different time and length scales, respectively, we used a number of complementary techniques to address molecular behavior in a variety of systems: (a) water and electrolytes (LiCl, $CaCl_2$, $NdCl_3$) in porous silica glass using QENS, (b) water, chloroform and soil organic matter in porous silica glass with NMR, (c) CO_2 phase behavior and critical phenomena in silica aerogel using SANS, (d) and water and aniline in montmorillonite with FTIR. One example of this effort is discussed below.

Phase transitions can be significantly modified by confinement and disorder. The liquid–vapor behavior in porous systems with quench disorder, such as aerogels, is of particular interest because it may provide an experimental test of which theoretical model best describes critical phenomena in confined fluids. Currently, there are two competing theoretical models – the Random Field Ising model and the "single pore" model (Melnichenko et al., in press). The experimental efforts to verify the validity of the models for confined binary liquid solutions have been obscured by experimental complications caused by the local composition gradients in pores as well as sluggish kinetics in the liquid–liquid region. SANS

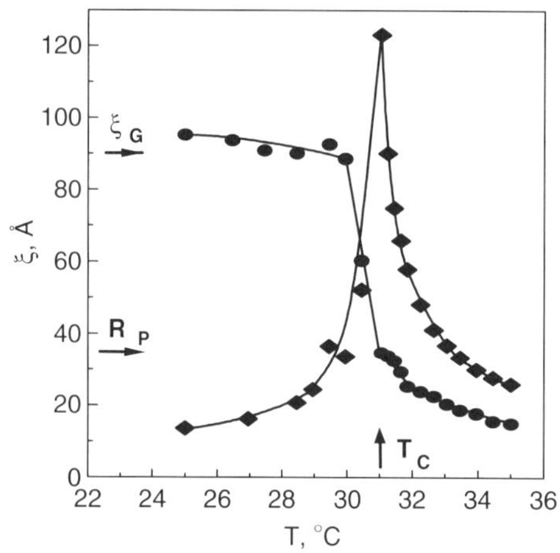

Figure 3. Correlation length of density fluctuations of CO_2 for bulk (◆) and confinement (•). R_p = pore radius, ξ_G = aerogel correlation length, T_C = critical temperature.

is particularly well suited to study the critical phenomena in small pores due to a short wavelength of neutron radiation (~ several Å), which allows for extracting information on the correlation length of the order of a nanometer and higher. In our experiments silica aerogel (~96% porosity) was filled with CO_2 at different temperatures and pressures appropriate for bracketing the critical region. SANS measurements (Fig. 3) of the correlation length of the density fluctuations (ξ) indicate that quench disorder works to suppress density fluctuations in the critical region. The correlation length does not exceed the characteristic pore size which provides strong evidence in favor of the single pore model (Melnichenko et al., in press).

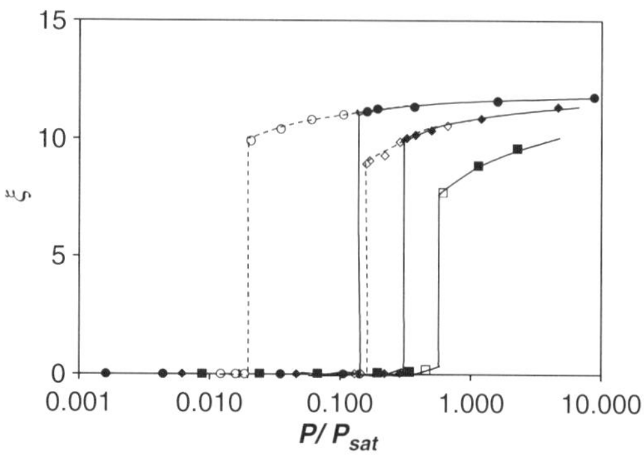

Figure 4. GCMC results for SPC/E water in 1.0 nm wide slit-carbon pores. The number of adsorbed water molecules per nm^2 of porous surface, ξ, is plotted as a function of reduced water, P/P_{sat}. Circles: 298 K, diamonds: 373 K, squares: 473 K, solid: adsorption, open: desorption. (Striolo et al., in press).

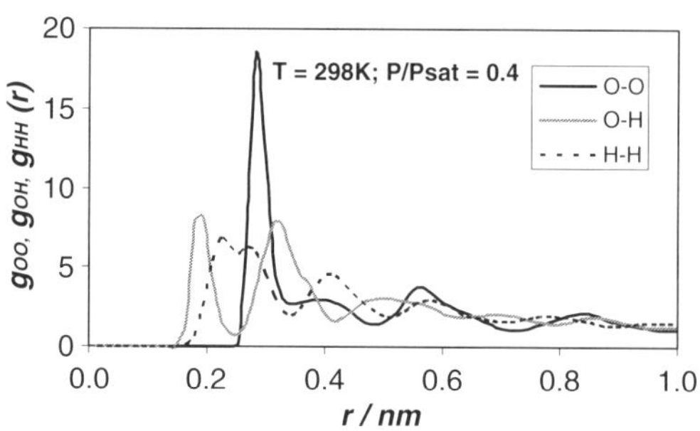

Figure 5. In-plane O-O, O-H, and H-H radial distribution functions (RDF) for SPC/E water in 0.8 nm carbon slit-pores.

6 SIMULATION STUDIES

Experimental information determined from characterization of the porous media, adsorption-desorption isotherms, and scattering and spectroscopic assessments of fluids at different length and time scales provide crucial synergistic input to molecular-based simulations. Our efforts have focused on the behavior of water in two very different solids, porous carbons and muscovite. In the former we used the SPC/E model for water, whereas in the latter we used TIP4P. The grand canonical Monte Carlo method (GCMC) was used to simulate water in carbon-slit pores with widths of 0.6, 1.0 and 1.6 nm and in single-walled carbon nanotubes with diameters of 1.36 and 1.63 nm in the temperature range 298–530 K. A comparison of experimental sorption results (Fig. 2) and the simulations (Fig. 4) indicates that because of a competition between strong water-water and weak water-carbon interactions, adsorption isotherms are characterized by negligible uptake at low pressures, by pore filling following a capillary-like condensation mechanism, and by adsorption/desorption hysteresis loops. As temperature increases, the relative pressure at which pore filling occurs increases and the size of the hysteresis loops decreases.

Differences between simulated and experimental results are primarily due to heterogeneous chemical composition (O, H in C fibers), pore connectivity, and non-uniform pore-size distribution, which are not accounted for in the simulation models. Water confined to 0.8–1 nm slit pores undergoes a disorder-to-order transition at P lower than saturation at 298 K. The in-plane RDFs (Fig. 5) suggest the formation of a peculiar structure characterized by long-range spatial correlations.

7 SUMMARY

A multidisciplinary, multi-institutional effort has addressed a number of key issues relevant to water/solid interaction at the nanoscale. Contributions have been made in the characterization of numerous complex nanoporous solids with a variety of techniques including neutron scattering. Advanced sorption, scattering and spectroscopic methods have documented the behavior of water and CO_2 in confined geometries. MD simulations have provided the theoretical underpinning of the experimental results.

ACKNOWLEDGEMENTS

Sponsored by Laboratory Directed Research and Development Program (ORNL) and the Division of Chemical Sciences, Geosciences and Biosciences, Office of Basic Energy Sciences, U. S. DOE under contract DE-AC05-00OR22725, Oak Ridge National Laboratory, managed and operated by UT-Battelle, LLC.

REFERENCES

Alejandre, J., Lozada-Cassou, M. & Degreve, L. 1996. Effect of pore geometry on a confined hard sphere fluid. *Molec. Phys.* 88: 1317–1336.

Cole, D. R. & Chakraborty, S. 2001. Rates and mechanisms of isotopic exchange. In J. W. Valley, J. W. & D. R. Cole (eds.), *Stable Isotope Geochemistry, Rev. Min. Geochem.* 44: 84–223.

Gruszkiewicz, M. S., Horita, J., Simonson, J. M., Mesmer, R. E. & Hulen, J. B. 2001. Water adsorption at high temperature on core samples from the Geysers geothermal field, California, USA. *Geothermics* 30: 269–302.

Idziak, S. H. J & Li, Y. 1998. Scattering studies of complex fluids in confinement. *Curr. Opin. Colloid Interface Sci.* 293–298.

Melnichenko, Y. B., Wignall, G. D., Cole, D. R., & Frielinghaus, H. in press. Density fluctuations near the liquid-gas critical point of a confined fluid. *Phys. Rev. E.*

Pitard, E., Rosinberg, M. L. & Tarjus, G. 1996. Thermodynamics of fluids in disordered porous materials. *Molec. Sim.* 17: 399–419.

Striolo, A., Gubbins, K. E., Gruszkiewicz, Cole. D. R., Simonson, J. M., Chialvo, A. A., Cummings, P. T., Burchell, T. D. & More, K. L. in press. Effect of temperature on water adsorption in porous carbons. *Fluid Phase Equil.*

Water–Rock Interaction, Wanty & Seal II (eds)
© 2004 Taylor & Francis Group, London, ISBN 90 5809 641 6

Coupled dissolution/precipitation rates in the system $CaCO_3$-$CdCO_3$

P. Cubillas & M. Prieto
Departamento de Geología, Universidad de Oviedo, Oviedo, Asturias, Spain

S. Köhler & E.H. Oelkers
Géochimie, Toulouse, France

ABSTRACT: Dissolution rates of calcite, aragonite, and ground bivalve shells, in the presence and absence of aqueous Cd were measured in mixed-flow reactors. Experiments in Cd-free solutions indicate that geometric surface area normalized dissolution rates of (a) abiogenic and biogenic aragonite and (b) abiogenic and biogenic calcite can each be described as a single function of chemical affinity. Experiments in Cd-bearing solutions allowed characterization of (a) Cd inhibition of $CaCO_3$dissolution rates, (b) otavite ($CdCO_3$) precipitation rates as a function of chemical affinity, and (c) the effect of surface precipitates on the dissolution rate. Cd inhibits aragonite dissolution rates by less than 40%, whereas it inhibits calcite dissolution by an order of magnitude or more. Otavite precipitation rates are $\sim 7 \times 10^{-11}$ mol/cm^2/s at far-from-equilibrium conditions decreasing to 0 at equilibrium. Surface precipitates are found not to affect dissolution rates until they cover more than 90% of the original surface.

1 INTRODUCTION

As calcium carbonate is one of the most widespread minerals on the Earth's surface, its interactions with metal-bearing aqueous solutions have a major role in the control in the composition of natural waters. As such, these interactions have received considerable attention in the literature (Lorens 1981, Davis et al. 1987, Zachara et al. 1991, Stipp et al. 1992, Reeder 1996, Tesoriero & Pankow 1996). Moreover it has been demonstrated that the interaction of Cd bearing solutions with biogenic carbonates surfaces can efficiently remove this toxic metal from natural solutions (Prieto et al. 2003). This study is motivated by both the desire to better quantify these interactions and to determine experimentally the effect of the presence of mineral precipitates on mineral dissolution rates.

2 EXPERIMENTAL

Dissolution and precipitation experiments were performed on abiogenic aragonite and calcite, and three biogenic carbonates belonging to the Phyllum *Mollusca* and Class *Bivalva* and to which are referred to by their common names of clam, cockle, and mussel. Samples were cleaned using $NaOH_{(aq)}$ solutions to remove organics, grounded in a metallic mortar, sieved to obtain the 1–1.5 mm fraction, and ultrasonically cleaned. Mineralogical compositions, obtained by XRD, show that the clam and cockle shells are composed of 100% aragonite and the mussel of a mixture of 90% calcite and 10% aragonite. The BET surface areas of biogenic carbonates are more than 1 order of magnitude bigger than those of the abiogenic carbonates. Geometric surface areas were calculated for all solids assuming they are smooth spheres.

Two types of experiments were performed: (1) dissolution in pH 4 HCl inlet solutions and (2) coupled dissolution-precipitation in pH 5–6 $CdNO_3$ inlet solutions. Both types of experiments were performed in mixed-flow reactors consisting of Azlon® beakers with a Teflon lid stored in a controlled temperature room (25 ± 2°C). Continuous stirring was achieved using floating Teflon-covered magnetic stirrers. Reactive solutions were injected into this reactor using a Gilson peristaltic pump. The output solutions left the reactor through a 0.45 μm nitrocellulose filter. Experiments were initiated by placing from 0.5 to 2 g of powder into the reactor. Outlet solutions were analyzed for pH, alkalinity, Ca, and Cd in the case of the dissolution-precipitation experiments. Coupled dissolution-precipitation experiments were carried out with inlet Cd concentrations varying from 1 to

50 ppm. Flow rate and initial mass of the solid was varied to obtain different saturation states within the reactor in both types of experiments.

3 RESULTS AND DISCUSSION

3.1 *Dissolution experiments*

Dissolution rates were calculated using steady-state Ca outlet concentrations; rates normalized to both BET surface areas (r_{BET}) and geometric surface areas (r_{gsa}) were computed. r_{BET} were found to be strongly dependent on the nature of the reacting solid; r_{BET} of the abiogenic samples was one order magnitude faster than their biogenic counterparts. In contrast, r_{gsa} of cockle, clam, and abiogenic aragonite were found to be broadly consistent with each other, as can be seen in Figure 1. The observation that r_{gsa} rather than r_{BET} provides a more consistent description of dissolution rates has also been argued by Keir (1980) and Walter & Morse (1984). Walter & Morse (1984) concluded that only a small percentage of the BET measured surface area was reactive. Measured r_{gsa} of calcite are approximately 30% slower than those of aragonite, which is consistent with the results of Busenberg & Plummer (1986).

Consistent with previous work, (Ohara & Reid 1973, Morse 1978, 1983, Keir 1980, Walter & Morse 1984, 1985, Walter 1985) r_{gsa}were fit as a function of Ω, the saturation index according to:

$$r_{gsa} = k(1 - \Omega)^n$$

where k stands for a rate constant. r_{gsa} for aragonite, cockle, and clam can be described using $k = 2.69 \pm 0.5 \times 10^{-10}$ mol/cm²/s and $n = 0.86 \pm 0.11$; r_{gsa} for calcite can be described using $k = 1.82 \pm 0.2 \times 10^{-10}$ mol/cm²/s and $n = 1.25 \pm 0.16$. Mussel shell dissolution is inconsistent with both equations, due to the presence of both calcite and aragonite in its structure.

3.2 *Influence of $[Cd]_{aq}$ on the dissolution of abiogenic and biogenic carbonates*

The r_{gsa} values for calcium carbonates were measured as a function of inlet Cd concentration. Constant affinity aragonite, clam, and cockle rates decreased by ~40% with increasing aqueous Cd concentration from 0 to 1.6×10^{-4} mol/kg. In contrast, calcite is more strongly affected by the presence of Cd; calcite dissolution rates decrease by more than 95% with increasing aqueous Cd concentration from 0 to 1×10^{-4} mol/kg. These results suggest that different mechanisms for Cd inhibition of aragonite versus calcite dissolution. For the case of aragonite, clam, and cockle, inhibition is probably due to the

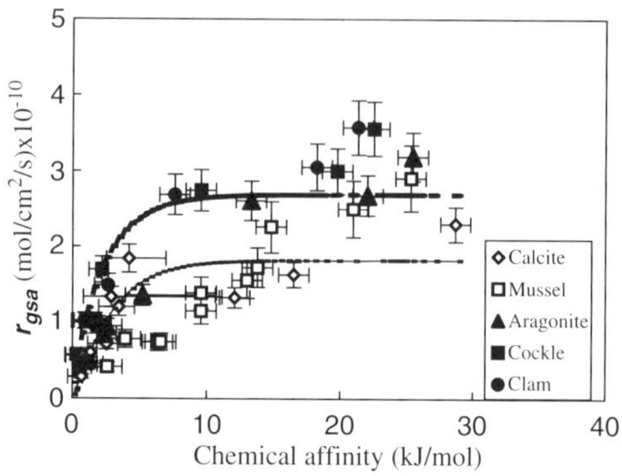

Figure 1. r_{gsa} as a function of chemical affinity for the 25°C dissolution of various calcium carbonates at neutral to basic pH measured in the present study.

blocking of reactive sites by the attaching of Cd ions, whereas in the case of calcite it may be due to the epitaxial growth of few unit cells of otavite-calcite solid solution upon the surface, which effectively stops calcite dissolution (Prieto et al. 2003). Interestingly mussel dissolution is inhibited in similar way as the aragonite, clam, and cockle. This could be due to the fact that the crystal faces exposed on the mussel microstructure do not favor epitaxial growth.

3.3 *Influence of the surface precipitate on the dissolution rates*

Long duration coupled dissolution/precipitation experiments were performed to quantify the effect of otavite precipitation on aragonite/bivalves dissolution rates. Otavite precipitation diminishes aragonite/bivalves dissolution rates until dissolution essentially stops. Otavite precipitation is also arrested as stopping aragonite dissolution removes the source of aqueous carbonate. SEM observations indicate that otavite precipitates uniformly over all $CaCO_3$ surfaces. Comparing calcium carbonate dissolution rates in the presence of otavite precipitates with corresponding rates in otavite-free systems can be used to estimate the fraction of reactive surface area still participating in dissolution. A representative result is shown in Figure 2, which suggests that an average otavite depth of 5–6 μm is required to arrest aragonite dissolution. This result is also in agreement with SEM observations carried out of the post-experiment samples, as can be seen in Figure 3. Intermediate duration experiments, where only 90% of the aragonite surface was covered by otavite showed negligible aragonite dissolution inhibition (Fig. 4). A possible explanation of the inefficiency of otavite surface precipitates in arresting aragonite dissolution is a misfit between the dissolving aragonite and the

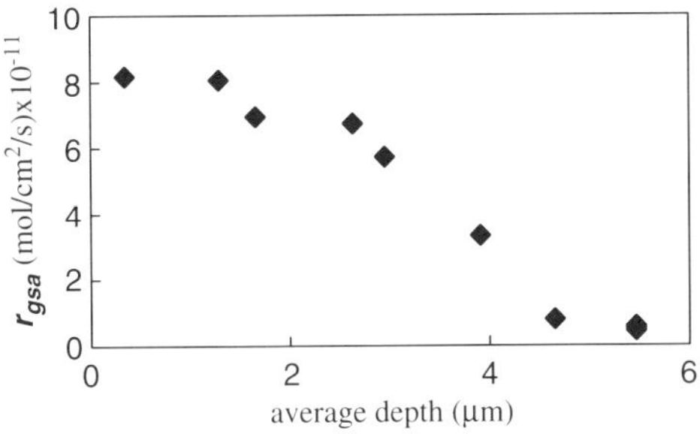

Figure 2. Representative example of clam dissolution rates as a function of the average thickness of a precipitated otavite layer.

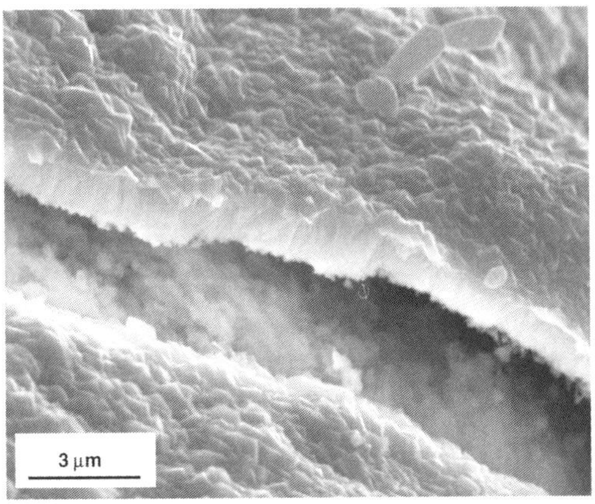

Figure 3. SEM microphotograph showing the layer of otavite precipitate formed on a surface of a dissolving clam shell.

Figure 4. Backscattered electron SEM photomicrograph of a clam grain nearly covered by an otavite layer during an intermediate term coupled dissolution/precipitation experiment. The original surface of the shell appears as gray.

otavite precipitate, dissolution occurs beneath the precipitated otavite at the same rate as in the absence of precipitates.

4 CONCLUSIONS

The major conclusions of this study are:

1. Geometric surface area provides a better proxy than BET surface area to describe calcium carbonate reactive surface area.
2. In the absence of precipitation, Cd is a far more effective inhibitor of calcite than aragonite dissolution. Calcite dissolution rates decrease by more than 95% in the presence of 10^{-4} mol/kg aqueous Cd.
3. Otavite precipitates only affect aragonite and ground shell dissolution rates when they cover more than 90% of the original surface; it arrests precipitation only after a 5 micrometer encompassing layer has formed.

REFERENCES

Busenberg, E. & Plummer, L.N. 1986. A comparative study of the dissolution and crystal growth kinetics of calcite and aragonite. In F.A. Mumptom (ed.), Studies Diagenesis. *U.S. Geol. Surv. Bull.* 1578: 139–168.

Davis, J.A., Fuller, C.C. & Cook, A.D. 1987. Mechanisms of trace metal sorption by calcite: Adsorption of Cd^{2+} and subsequent solid solution formation. *Geochim. Cosmochim. Acta* 51: 1477–1490.

Keir, R.S. 1980. The dissolution kinetics of biogenic calcium carbonates in seawater. *Geochim. Cosmochim. Acta* 44: 241–252.

Morse, J.W. 1978. Dissolution kinetics of calcium carbonate in seawater. VI. The near-equilibrium dissolution kinetics of calcium carbonate-rich deep sea sediments. *Amer. J. Sci.* 278: 344–353.

Morse, J.W. 1983. The kinetics of calcium carbonate dissolution and precipitation. *Rev. Min.* 11: 227–264.

Lorens, R.B. 1981. Sr, Cd, Mn, and Co distribution coefficients in calcite as a function of calcite precipitation rate. *Geochim. Cosmochim. Acta* 45: 553–561.

O'Hara, M. & Reid, R.C. 1973. *Modeling Crystal Growth Rates from Solution*. New Jersey: Prentice-Hall.

Prieto, M., Cubillas, P. & Fernández-González, A. 2003. Uptake of dissolved Cd by biogenic and abiogenic aragonite: a comparison with sorption onto calcite. *Geochim. Cosmochim. Acta* 67: 3859–3869.

Reeder, R.J. 1996. Interaction of divalent cobalt, zinc, cadmium, and barium with the calcite surface during layer growth. *Geochim. Cosmochim. Acta* 60: 1543–1552.

Stipp, S.L., Hochella, M.F., Parks, G.A. & Leckie, J.O. 1992. Cd^{2+} uptake by calcite, solid-state diffusion, and the formation of solid-solution: Interface processes observed with near-surface sensitive techniques (XPS, LEED, and AES). *Geochim. Cosmochim. Acta* 56: 1941–1954.

Tesoreiro, A.J. & Pankow, J.F. 1996. Solid solution partitioning of Sr^{2+}, Ba^{2+}, and Cd^{2+} to calcite. *Geochim. Cosmochim. Acta* 60: 1053–1063.

Walter, L.M. & Morse, J.W. 1984. Reactive surface are of skeletal carbonates during dissolution: Effect of grain size. *J. Sediment. Petrol.* 54: 1081–1090.

Walter, L.M. & Morse, J.W. 1985. The dissolution kinetics of shallow marine carbonates in seawater: a laboratory study. *Geochim. Cosmochim. Acta* 49: 1503–1513.

Walter, L.M. 1985. Relative reactivity of skeletal carbonates during dissolution: Implications for diagenesis. In N. Schneidermann & P.M. Harris (eds.), *Carbonate Cements. Society of Economic Paleontologist and Mineralogists, Special Publication* 36: 3–16.

Zachara, J.M., Cowan, C.E. & Resch, C.T. 1991. Sorption of divalent metals on calcite. *Geochim. Cosmochim. Acta* 55: 1549–1562.

Water–Rock Interaction, Wanty & Seal II (eds)

Interaction between galena surface and dissolved organic ligands: a chemical kinetic study

G. De Giudici, L. Fanfani & P. Lattanzi
Dipartimento di Scienze della Terra, Università di Cagliari, Cagliari

ABSTRACT: A flow-through reactor was used to investigate the kinetics of a galena grain dissolution in conditions that partially mimic mineral/water interaction in soils, namely acidic (HCl) and oxygen saturated solutions with and without addition of acetate-oxalate anions. By comparing inorganic and organic dissolution kinetics, evidence was found for acetate catalysis of the overall reaction. Calculated activation energies show a clear dependence on solution pH and organic anion addition, suggesting a change in the surface speciation.

1 INTRODUCTION

Understanding mechanisms that rule the cycling of metals between geosphere and biosphere is central to develop both thermodynamic prediction models on mineral stability, and techniques of bio-remediation for polluted sites such as abandoned mine areas. Mass exchange between soils and vegetation takes place at the interface between mineral surfaces and root exudates. Organic acids are the main biochemical component of root exudates (Drever & Vance 1994; Lopez-Bucio et al. 2000), and can be found in significantly high concentrations in both hydrothermal and sedimentary environments. In the past decades, the geochemical role of organic anions in ground waters has attracted considerable attention. Accurate measurements for a wide range of temperatures and ionic strength conditions indicate significantly high stability constants of metal-organic complexes (cf. Hennet et al. 1988; Giordano 1989; Harrison & Thyne 1992). More recently, the effect of organic ligands on mineral surfaces was investigated to gain insight into the complexity of organic ligand activity at the mineral-water interface. A fairly large body of literature is available for silicate and oxide minerals (cf. Ganor et al. 2001; Cheah et al. 2003). By contrast, little is known about the interaction between metal sulfide surfaces and organic ligands at Earth surface conditions.

The dissolution reaction of galena in acidic and oxygen-saturated solutions has been described classically by a first step of hydrogen ion adsorption followed by oxygen adsorption. This results in the release of Pb and S species from the surface to the bulk solution (Fornasiero et al. 1994). In the presence of an organic ligand, the overall process of galena dissolution can be represented by the following mass balance:

$$PbS + nO_2 + mL \rightarrow Pb^{2+} + \Sigma Pb_L + \Sigma S_{tot} \qquad (1)$$

where ΣPb_L accounts for the sum of the different Pb-organic complexes in solution, and ΣS_{tot} stands for the sum of different sulphur species. The coefficients n and m depends on the particular species involved in the reaction and electroneutrality is obtained by balancing charge of Pb ions with S species and stoichiometry of complexes. The activity of these complexes is expected to result in increased amounts of dissolved lead sulfide.

Apparent activation energy represents the energy barrier that limits the probability that a reaction takes place. For heterogeneous reactions taking place in presence of an adsorbent, the apparent activation energy is the sum of the intrinsic value of activation energy plus the enthalpy change pertaining to the adsorption reaction (cf. Casey & Sposito 1992). Hence, exothermic adsorption reactions lower the energy barrier for the overall reactions, and catalyse the reaction.

In this work the role that hydrochloric and organic anions play in the reaction of galena dissolution is investigated by chemical kinetics. It will be shown that (1) the observed catalytic effect of HCl is higher than that observed in the presence of acetate in solution, and that (2) activation energies clearly depend on the pH.

2 EXPERIMENTAL METHODS

A grain of natural galena was cleaved in the laboratory and then mounted in a flow-through reactor made of Teflon especially built for these experiments. The reaction solution was renewed at a controlled rate (70 μL per min) by means of a peristaltic pump and lasted about 1 min. The temperature was held constant to ±0.1 K by a thermostatic bath. The experimental conditions were T = 298, 323 and 348 K, solution pH ≈ 1.2, 4.6, 5.8 (HCl), and saturation with respect to atmospheric oxygen. In some of the experiments, the effect of organic anions was investigated by adding given amounts of Na-acetate or Na-oxalate to the initial solution, and then adjusting the pH by addition of HCl. The outflowing solutions were sampled and the lead concentration analysed by differential pulse anodic stripping voltammetry (MOhm 747/746). Dissolution rates (mol m^{-2} sec^{-1}) were calculated by the formula:

$$Rate = C_{Pb}\frac{Q}{A} \qquad (2)$$

where C_{Pb} is the measured total lead concentration, Q the solution flow, and A the total surface area as a geometric area obtained by macroscopic and microscopic (Atomic Force Microscopy) measurements. The lead concentrations used to calculate dissolution rates at stationary states were constant for at least two days. The uncertainty in rate values was calculated to be about 15%, mainly due to the error in the measurement of A. It should be pointed out that all the values of dissolution rates presented here were obtained on a set of experiments made with the same grain, where only the initial solutions and temperature conditions were changed. The initial solutions were prepared by adding the desired amounts of HCl, Na-acetate or Na-oxalate to MilliQ water (Millipore). The solution pH was measured before and after the interaction, whereas the electrode was calibrated against the NIST buffer standard. The concentrations of the organic free anions were calculated on the basis of the dissociation constants of the acetic acid ($\log K_1 = -4.75$) and the oxalic acid ($\log K_1 = -0.673$; $\log K_2 = -4.312$. The experiments hereafter discussed were repeated two times, with replicate values in agreement.

3 RESULTS AND DISCUSSION

3.1 *Dissolution rate vs. reaction time*

As documented by the measured lead concentration in solution flowing out from our flow-through reactor, the reactivity of a freshly cleaved surface interacting with solutions typically shows a decrease with interaction time. Figure 1 illustrates the initial dissolution of galena in presence of 0.028 mol/L of free acetate anion at pH 3.6. It can be observed that the lead concentration decreases by a factor of about 7 in one day and reached a constant value for the remaining two days of the experimental time. The same decrease was observed in all the newly started experiments (ten in all) of galena dissolution. Given the fast solution renewal of our flow-through reactor, the interacting solutions are very far from saturation with respect to both lead sulfide and lead sulfate, and the solution saturation state decreases with interaction time (Table 1). Hence, the resulting decrease in dissolution rates is only due to a change at the dissolving surface. An in-situ electrochemical Scanning Tunnelling Microscopy (STM) investigation by Higgins & Hamers (1996) showed evidence of the formation of S islands on a (001) galena surface dissolving in oxidative conditions. A portion of the same sample studied here was investigated by liquid-cell Atomic Force Microscopy (AFM). It was found that the galena dissolution in acidic and oxidative solution takes place via an evolution of both surface features and dissolution rates, whereas the dissolution

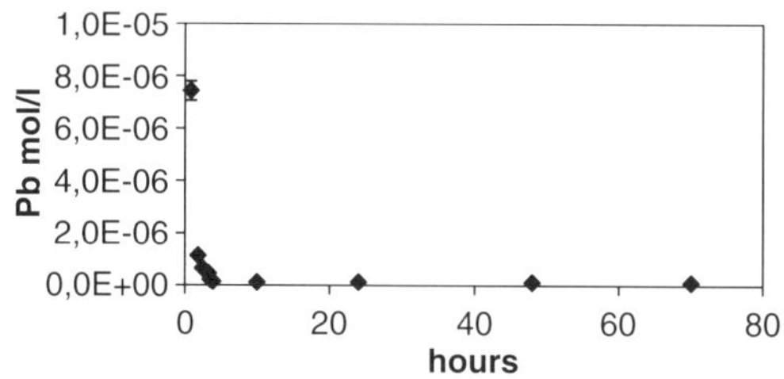

Figure 1. Lead concentration as a function of interaction time in 0.028 mol/L acetate solution (pH = 3.6).

Table 1. Example of the initial condition of a dissolution run. Total S was assumed to be equal to measured Pb, and the concentrations of S species were then calculated for equilibrium conditions. The values of the saturation state with respect to galena are $<10^{-8}$ and not reported in the table.

Time (hours)	Pb^{2+} mol L^{-1}	Saturation $PbSO_4$
0.83	7.43×10^{-6}	3.4×10^{-3}
1.83	1.13×10^{-6}	7.9×10^{-5}
2.45	6.47×10^{-7}	2.6×10^{-5}
3.28	4.54×10^{-7}	1.3×10^{-5}
3.51	2.22×10^{-7}	3.0×10^{-6}
3.89	1.45×10^{-7}	1.3×10^{-6}
10	1.20×10^{-7}	8.9×10^{-7}
24	1.25×10^{-7}	9.6×10^{-7}
48	1.15×10^{-7}	8.2×10^{-7}
70	1.10×10^{-7}	7.5×10^{-7}

Table 2. Measured dissolution rates(mol m^{-2} sec^{-1}) and calculated activation energies. The error in the dissolution rates is of about 15%.

HCl pH = 1.2	HCl pH = 4.6	Run	Water pH = 5.8	Acetate 0.02 mol l^{-1} pH = 4.3	Acetate 0.28 mol l^{-1} pH = 4.6	Oxalate 0.05 mol l^{-1} pH = 4.7
3.7×10^{-7}	3.1×10^{-9}	Rate 298 K	1.4×10^{-9}	5.1×10^{-9}	1.0×10^{-8}	2.2×10^{-9}
5.0×10^{-7}	7.8×10^{-9}	Rate 323 K	9.0×10^{-9}	1.7×10^{-8}	3.9×10^{-8}	1.3×10^{-8}
7.7×10^{-7}	4.4×10^{-8}	Rate 348 K	2.6×10^{-8}	3.1×10^{-8}	7.4×10^{-8}	2.8×10^{-8}
12.4 ± 2.1	44.4 ± 4	Ea kJ mol^{-1}	51.4 ± 4.4	31.4 ± 3.2	34 ± 3.4	44.5 ± 4.1

of a surface coating probably made of S nanophases, represents the rate-limiting step (De Giudici & Zuddas 2001).

3.2 *Catalysis and effect of temperature on dissolution rates*

All the dissolution rates presented here refer to the interaction of the same grain with different solutions. A freshly cleaved grain was held in interaction with acidic (HCl, pH = 1.2) and oxygen saturated solution for several weeks. The experimental conditions (e.g. pH, T and addition of organic anions) were then changed, and Pb concentrations periodically monitored. For each experimental condition, lead concentration showed generally a parabolic decrease by a factor 2, and were assumed to be stationary if constant for at least three days. Measured stationary Pb concentrations were used to calculate steady state dissolution rates by Equation 2.

The effects of HCl alone and of the addition of organic anions on the dissolution process were investigated in separate experiments. At pH = 4.6, the dissolution rates in solution containing 0.28 and 0.02 mol L^{-1} of free acetate were higher than those in a HCl solution by a factor in the range 1.6–3.3 (Table 2). However, at relatively high temperatures, acetate ion accelerates the reaction only for relatively high concentrations (compare dissolution rates for acetate pH 4.3 and 4.6, and HCl pH 4.6). Free oxalate has no effect on the velocity of the galena dissolution process. Dissolution rates at pH = 1.2 (HCl) are higher by more than one order of magnitude than those measured in the presence of a similar amount (0.28 mol L^{-1}) of free acetate anion at pH 4.6. This clearly indicates that hydrogen ion has a stronger effect than acetate anion.

The knowledge of the effect of temperature on dissolution kinetics provides unique information on both the reaction mechanism (hydrogen ion adsorption enthalpy) and the kinetic regime. Activation energy is described by the following equation:

$$Rate = Ae^{-E_a/RT} \qquad (3)$$

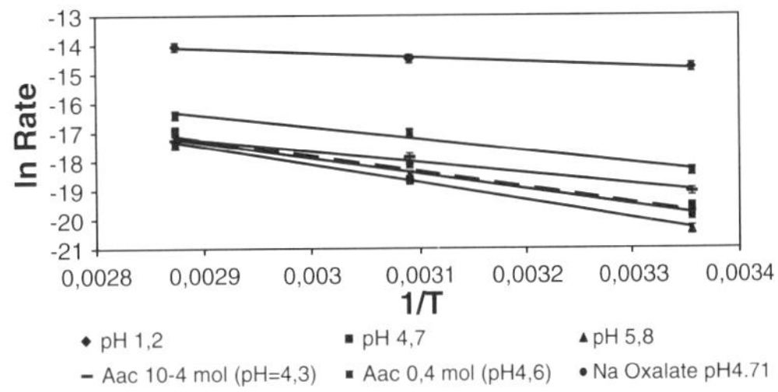

Figure 2. Arrhenius plot of measured dissolution rates.

where A is the pre-exponential factor, E_a the activation energy, R the gas constant, and T the temperature. The Arrhenius plot indicates that dissolution rates in solution of both the hydrochloric acid and organic anions follow an Arrhenius behavior (Fig. 2). Activation energies were then calculated by fitting data in the Arrhenius plot (Table 2).

For dissolution in HCl solution, an increase in pH from 1.2 to 5.8 results in an increase in activation energy by a factor of about 4. At solution pH = 4.6, the activity of acetate results in a decrease in activation energy by about 10 kJ mol^{-1} compared to the value calculated for the HCl and Na-oxalate solution at the same pH. The measured activation energy in presence of a lower concentration of acetate anion, i.e. 0.018 mol L^{-1}, does not change significantly compared to the higher acetate concentration. This suggests that the the reaction mechanism should not be affected by the organic anion concentration.

3.3 *Reaction kinetics as described by solution chemistry*

In the oxidative condition, the galena dissolution process needs several separate steps to allow the exchange of eight S-electrons. Hence, several metastable S species can be involved in the dissolution process and could be formed already at the mineral-water interface. X-ray photoelectron spectroscopy gives a coherent picture of the chemical composition of a galena surface. Other than lead oxide and lead hydroxide, sulfate, sulfur and thiosulfate can be found at a galena surface after reaction with acidic and oxidative solutions (Fornasiero et al. 1994). Kartio et al. (1996) investigated the electrochemical behaviour of galena surfaces in an acetate buffer at pH 4.6 by synchrotron radiation XPS. From the high intensity signal S_{2p}, they found evidence for the structural reconstruction of reacted galena, which corresponds to the appearance of sulfur or polysulfides, depending on the applied electrochemical potential. Unlikely, a galena surface interacting with acidic and oxidative solutions evolves during the reaction time and tends to become rough and coated by nanophases (Kim et al. 1995, De Giudici & Zuddas 2001) that can make it very difficult to measure any additional differences in surface features. Hence, the present molar scale measurements provide information that cannot be easily explored by measurement at the molecular scale by AFM or different surface techniques requiring flat surfaces.

Activation energies allow the determination of the kinetic regime under which a chemical reaction takes place. Lasaga (1984, 1998) proposed that (1) a kinetic regime corresponds to the diffusive processes below the activation energy value of 20 kJ mol^{-1}, (2) above values of 70–80 kJ mol^{-1}, the regime is controlled by surface reaction, and (3) between 20 and 70–80 kJ mol^{-1}, the regime is controlled by a mixture of diffusion and surface reaction. As defined at the molar scale, galena dissolution at pH = 1.2 takes place under the control of diffusive processes, whereas at higher pH values, the kinetic regime is a mixture.

The apparent activation energy depends not only on the intrinsic variation of the kinetic constant, but also on the adsorption enthalpy of the hydrogen ion, i.e. the density of reactive protonated sites (Casey & Sposito 1992), or other adsorbate species. The apparent activation energy should hence provide direct information on the molecular mechanism of galena dissolution. The decrease in apparent activation energy observed in presence of acetate indicates an exothermic reaction of acetate adsorption onto dissolving galena surface. However, the activation energy is not decreased as a result of oxalate addition. We propose that the observed variation of activation energies corresponds to a true change in the reaction mechanism. This change could result in different surface complexes that depend on the solution pH, redox processes, and the selective adsorption of organic ligands. Clearly, future molecular scale measurements are needed to elucidate the dependence of the pH on chemical speciation and structural change at the galena surface.

ACKNOWLEDGEMENT

This work is part of a MIUR funded national project (PRIN 2002; national coordinator G. Artioli).

REFERENCES

Casey, W.H. & Sposito, G. 1992. On the temperature dependence of the mineral dissolution rate. *Geochim. Cosmochim. Acta* 56: 3825–3830.

Cheah, H.S., Kraemer, S.M., Cervini-Silva, J. & Sposito, G. 2003. Steady state dissolution kinetics of goethite in the presence of desferrioxamine B and oxalate ligands: implications for the microbial acquisition of iron. *Chem. Geol.* 198: 63–75

De Giudici, G. & Zuddas, P. 2001. In situ investigation of galena dissolution in oxygen-saturated solutions: evolution of surface features and kinetic rate. *Geochim. Cosmochim. Acta* 65: 1381–1389.

Drever, J.L. & Vance, G.F. 1994. Role of soil organic acids in mineral weathering processes. In E.D. Pittman & M.D. Lewan (eds.), *Organic Acids in Geological Processes* 138–161. Berlin: Springer.

Fornasiero, D., Fengsheng, L., Ralston, J. & Smart, R. 1994. Oxidation of galena surfaces. I. X-Ray photoelectron spectroscopic and dissolution kinetic studies. *J. Colloid Interf. Sci.* 164: 333–344.

Ganor, J., Sharon, N. & Cama, J. 2001. The effect of kaolinite (bio)degradation at 25°C, and possible implications for adsorption isotherm measurements. *Chem. Geol.* 177: 431–432.

Giordano, T.H. 1989. Anglesite ($PbSO_4$) solubility in acetate solutions: the determinations of stability constants for lead acetate complexes to 85°C. *Geochim. Cosmochim. Acta* 53: 359–366.

Harrison, W.J. & Thyne, G.D. 1992. Predictions of diagenetic reactions in the presence of organic acids. *Geochim. Cosmochim. Acta* 56: 565–586.

Hennet, R.J., Crerar, D.A. & Schawrtz, J. 1988. Organic complexes in hydrothermal systems. *Econ. Geol.* 83: 742–764.

Higgins, R.S. & Hamers, R.J. 1996. Chemical dissolution of the galena (001) surface observed using electrochemical scanning tunneling microscopy. *Geochim. Cosmochim. Acta* 60: 3067–3073.

Kartio, I., Laajalehto, K., Kaurila, T. & Suoninen, E. 1996. A study of galena (PbS) surfaces under controlled potential in pH 4.6 solutions by synchrotron radiation excited photoelectron spectroscopy. *Appl. Surf. Science* 96: 167–177.

Kim, B., Hayes, R.A., Prestidge, C.A., Ralston, J. & Smart, R. St.C. 1995. Scanning tunneling microscopy studies of galena: the mechanism of oxidation in aqueous solution. *Langmuir*11: 2554–2562.

Lasaga, A. 1984. Chemical kinetics of water-rock interactions. *J. Geophys. Research* 89: 4009–4025.

Lasaga, A. 1998. *Kinetic Theory in the Earth Sciences*. Princeton: Princeton Univ. Press.

Lopez-Bucio, J., Nieto-Jacobo, M.F, Ramirez-Rodriguez, V. & Herrera-Hestrella, L. 2000. Organic acid metabolism in plant from adaptive physiology to transgenic varieties for cultivation in extreme soils. *Plant Sci.* 160: 1–13.

Water–Rock Interaction, Wanty & Seal II (eds)
© 2004 Taylor & Francis Group, London, ISBN 90 5809 641 6

New insights into the orthoclase dissolution mechanism with X-ray reflectivity

P. Fenter, C. Park & L. Cheng
Argonne National Laboratory, Argonne, Illinois, USA

Z. Zhang
Northwestern University, Evanston, Illinois, USA

N.C. Sturchio
University of Illinois at Chicago, Chicago, Illinois, USA

ABSTRACT: The dissolution of feldspar minerals has been extensively studied over the past few decades. Here we describe the use of X-ray reflectivity to probe dissolution of single cleavage surfaces to obtain new and precise information concerning the dissolution process and dissolution kinetics of individual crystal faces. This approach is beginning to reveal new insights into the dissolution mechanism of orthoclase.

1 INTRODUCTION

1.1 *Chemical weathering of feldspar*

Chemical weathering of feldspars has been studied for many years using a variety of macroscopic and microscopic approaches to gain insight into the associated microscopic processes, kinetics and energetics. Dissolution is generally thought to proceed by a surface-controlled process involving disruption of the aluminosilicate network through elementary hydrolysis reactions (see Blum & Stillings 1995, and references therein). Theoretical studies relate these macroscopic observations to microscopic processes. Most of this work has used the analogy with simpler mineral systems in which the dissolution rate is related to protonation/deprotonation reactions that occur at the surface (Walther 1997, Brantley & Stillings 1997). Quantum mechanical calculations of idealized aluminosilicate clusters have been performed to test the relative proton affinities of different bonds (Kubicki 1996).

1.2 *X-ray reflectivity as a probe of dissolution*

Surface dissolution processes can be characterized by time-resolved measurements of X-ray reflectivity at the "anti-Bragg" condition (Teng et al. 2001, Fenter et al. 2003a). At this condition, the angle of incidence, α, with respect to the surface plane is specified by Bragg's law, $n\lambda = 2d_{001}\sin(\alpha)$, where $n = 1/2$ for the first anti-Bragg condition, λ is the X-ray wavelength, and d_{001} is the Bragg plane spacing of the orthoclase (001) plane. At this scattering condition, there is high sensitivity to changes in the mineral *surface* topography due to destructive interference between X-rays that are reflected from neighboring terraces that differ in height by a single layer spacing (e.g. $d_{001} = 6.459$ Å for the (001) surface), as shown in Figure 1. The destructive interference at this anti-Bragg condition results in a strong variation of the reflectivity as a function of the occupation of the outermost layer. The reflectivity is maximized when the outermost orthoclase layer is fully occupied (i.e. smooth surfaces). Instead the reflectivity is exactly zero if the outermost layer is

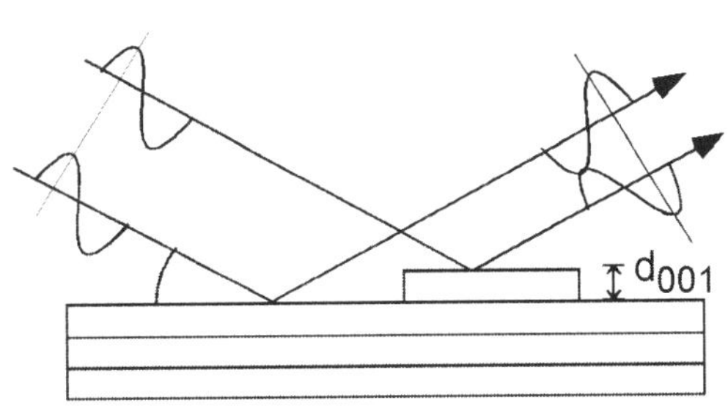

Figure 1. Schematic of the anti-Bragg condition, whereby X-rays reflected by neighboring terraces destructively interfere.

half-occupied, acting as an "anti-reflective coating". Therefore, it is reasonable to think that the reflectivity may oscillate as each layer is removed, providing a simple and direct way to measure dissolution rates in situ and in real-time. This approach has been used extensively to understand mechanisms of metal and semiconductor growth (van der Vegt et al. 1992). Because this approach is sensitive to the amount of material removed, it measures, in essence, the same quantity that is determined by traditional powder dissolution measurements.

2 RESULTS

2.1 *Dissolution kinetics*

A series of real-time reflectivity measurements during dissolution of the orthoclase (001) surface for different temperatures at pH = 12.9 are shown in Figure 2 (Fenter et al. 2003a). These data have similar slightly-damped oscillatory reflectivity patterns over the temperature range studied (50 to 83.3°C), although the time between subsequent reflectivity maxima varies strongly with temperature. These data show the expected behavior as each layer is removed from the surface by dissolution: the reflectivity minima correspond to surfaces that are terminated by half occupied layers, and the reflectivity maxima correspond to fully occupied layers (first maximum in each data set indicated by an arrow in Figure 2). The dissolution rate is most simply estimated by noting the time at the first reflectivity maxima, t_{ML}. From this, the dissolution rate is (1 layer of orthoclase)/t_{ML} (expressed in units of monolayers per hour), or $1.66 \times 10^{-9}/t_{ML}$ (expressed in $mol/m^2/sec$) with t_{ML} expressed in hours in the last expression. These data directly show a substantial change in the dissolution rate as a function of temperature at pH = 12.9.

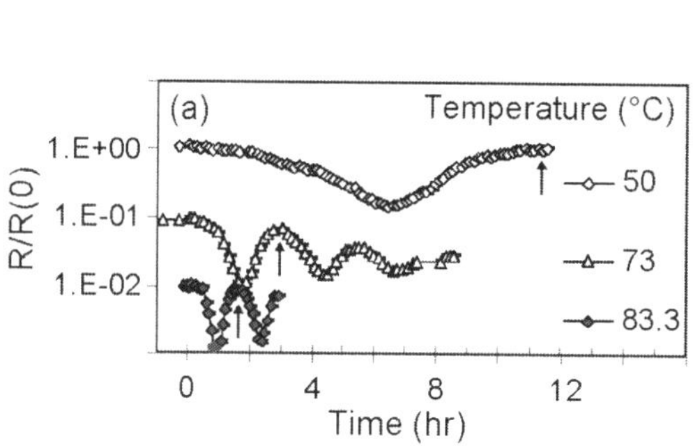

Figure 2. X-ray reflectivity measurements of the orthoclase(001) surface as a function of time during dissolution at pH 12.9 at selected temperatures. Removal of a single orthoclase layer is completed at times indicated by vertical arrows. (data sets offset by factors of 10 for clarity). (Adapted from Fenter et al. 2003a).

The systematic variation of the dissolution rate, $1/\tau$, as a function of inverse temperature is shown in Figure 3a for the (001) surface at pH 12.9, and in Figure 3b for the (001) and (010) surfaces at pH 1.1 (Fenter et al. 2003a). These data are all consistent with Arrhenius behavior [i.e. rate $\sim \exp(-\Delta E_{app}/RT)$, where ΔE_{app} is the apparent activation energy] albeit with significantly different slopes that are proportional to the apparent activation energies for dissolution at each pH and for each crystal orientation. At pH 12.9, we find a best-fit apparent activation energy of $\Delta E_{app} = 65 \pm 7$ kJ/mol for the (001) surface. At pH = 1.1, we find apparent activation energies of $\Delta E_{app} = 87 \pm 7$ kJ/mol and 41 ± 7 kJ/mol for the (001) and (010) surfaces, respectively. The apparent activation energies for dissolution of these two surfaces differ by a factor of ~2, in spite of their close structural similarity. Recent high resolution X-ray reflectivity studies have directly shown that these two surfaces have very similar terminations (Fenter et al. 2003b). This unexpected sensitivity of the apparent activation energy to the orthoclase surface orientation therefore suggests that the dissolution mechanism of orthoclase surfaces may be more complex than previously imagined.

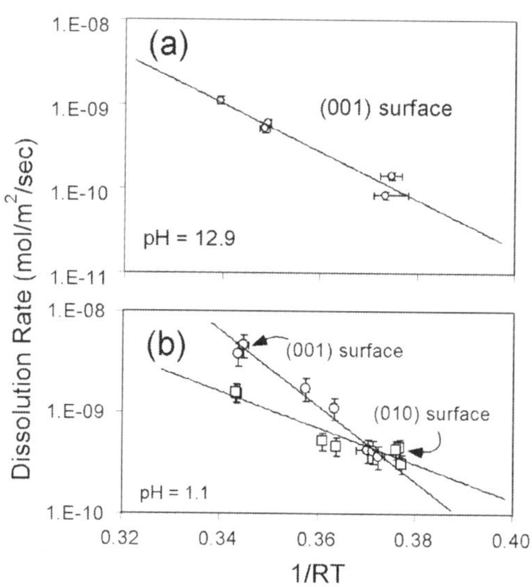

Figure 3. Dissolution kinetics measured by X-ray reflectivity at the orthoclase (001) and (010) surfaces at (a) pH = 12.9, and (b) pH = 1.1. The linear relationship between $\log(R_d)$ and 1/RT indicates Arrhenius-like behavior, but with apparent activation energies at pH 1.1 that differ in magnitude by a factor of ~2. (Adapted from Fenter et al. 2003a).

2.2 *New results*

To begin to understand the dissolution mechanism at acidic pH, we have made systematic measurements of the dissolution rate as a function of temperature ($31.5 \leq T \leq 76°C$) and pH ($0 \leq pH \leq 3.5$). If we assume the generic variation of dissolution rates as a function of pH and T, $R_{dis} \sim exp(-\Delta E_{app}/RT)\ 10^{-\alpha pH}$, we find values of ΔE_{app} and α that are consistent with powder dissolution measurements of albite (Chen & Brantley 1997). This confirms that we are measuring same process that is probed by traditional powder dissolution measurements. However, there are substantial discrepancies between the measured rates and this simple functional form that are outside of the experimental error of the X-ray reflectivity measurement. These observations suggest that a more detailed mechanistic description of the variation of dissolution rate is needed. Comparison of these data with predictions of a surface complexation model is underway. This comparison will be used to test the applicability of various models of mineral dissolution and to obtain new insight into the essential characteristics of the mechanism by which orthoclase dissolves at acidic pH.

REFERENCES

Blum, A.E. & Stillings, L.L. 1995. Feldspar dissolution kinetics. In A.F. White & S.L. Brantley (eds.), *Chemical Weathering Rates of Silicate Minerals, Rev. Mineral.* 31: 291–351. Washington, DC: Mineralogical Society of America.

Brantley, S.L. & Stillings, L.L. 1997. Reply to comment: Feldspar dissolution at 25°C and low pH. *Am. J. Sci.* 297: 1021–1032.

Chen, Y. & Brantley, S.L. 1997. Temperature- and pH-dependence of albite dissolution rate at acid pH. *Chem. Geol.* 135: 275–290.

Fenter, P., Teng, H., Geissbühler, P., Hanchar, J.M., Nagy, K. & Sturchio, N.C. 2000. Atomic-scale structure of the orthoclase (001)-water interface measured with high-resolution X-ray reflectivity. *Geochim. Cosmochim. Acta* 64: 3663–3673.

Fenter, P., Park, C., Cheng L., Zhang, Z., Krekeler, M.P.S. & Sturchio, N.C. 2003a. Orthoclase dissolution kinetics probed by in situ X-ray reflectivity: effects of temperature, pH and crystal orientation. *Geochim.t Cosmochim. Acta* 67: 197–211.

Fenter, P., Cheng L., Park, C., Zhang, Z. & Sturchio, N.C. 2003b. Structure of the orthoclase (001) and (010)-water interfaces using high-resolution X-ray reflectivity. *Geochim. Cosmochim. Acta*, 67: 4267–4275.

Kubicki, J.D., Blake G.A. & Apitz, S.E., 1996. *Ab initio* calculations on aluminosilicate Q^3 species: Implications for atomic structures of mineral surfaces and dissolution mechanisms of feldspars. *Am. Mineral.* 81: 789–799.

Teng, H.H., Fenter, P., Cheng, L. & Sturchio, N.C. 2001. Resolving orthoclase dissolution mechanisms with atomic force microscopy and X-ray reflectivity. *Geochim. Cosmochim. Acta* 65: 3459–3474.

Walther, J.V. 1997. Comment: Feldspar dissolution at 25 degrees C. and low pH. *Am. J. Sci.* 297: 1012–1021.

van der Vegt, H.A., van Pinxteren, H.M., Lohmeier, M., Vlieg, E. & Thornton, J.M.C. 1992. Surfactant-induced layer-by-layer growth of Ag on Ag(111). *Phys. Rev. Lett.* 68: 3335–3338.

Water–Rock Interaction, Wanty & Seal II (eds)

Chemical alteration of feldspar: a comparative study using SIMS and HRTEM/EFTEM

R. Hellmann
LGIT, Univ. Grenoble I, CNRS, OSUG, Grenoble, France

J.-M. Penisson
DRFMC, CEA-Grenoble, Grenoble, France

R.L. Hervig
Center for Solid State Science, Arizona State University, Tempe, AZ USA

J.-H. Thomassin & M.-F. Abrioux
UMR CNRS 6532, LMGE – ESIP – Université de Poitiers, Poitiers, France

ABSTRACT: Here we report on the structural and chemical changes that occur at the surface and near surface of labradorite after dissolution at acid pH. These changes were measured with two different analytical techniques: secondary ion mass spectrometry (SIMS) and high resolution and energy filtered transmission electron microscopy (HRTEM and EFTEM, respectively). The main purpose of this comparative study is to show the similarities and differences in the data generated by these two analytical techniques. The SIMS results are in general compatible with the leached layer theory of dissolution, which is based on the preferential leaching of certain elements from the non-altered mineral. On the other hand, the higher resolution HRTEM/EFTEM data point to another mechanism of dissolution, one based on interfacial dissolution-reprecipitation.

1 INTRODUCTION

During water–rock interactions the fluid-solid interface plays the primordial role in the exchange of energy and matter between the aqueous fluid and the component minerals undergoing hydrolysis. Understanding how fluids interact with mineral surfaces requires analytical tools for measuring changes in the structure and chemistry of the surface and near-surface of the mineral; commonly used techniques include electron microscopy, X-ray diffraction and reflectivity, spectroscopy, and ion beam methods. Inherent to each technique are specific advantages and limitations with respect to the measurement of the structure or chemistry of hydrolyzed mineral surfaces. Among the key analytical parameters of any given technique are its elemental sensitivity and specificity, the spatial resolution (i.e. lateral, as well as depth), and the analysis area and volume. In this communication we compare the application of two different analytical techniques that were used to study the chemical alteration of feldspar.

We measured the structural and chemical changes that occur at the surface and near surface of labradorite after hydrolysis at 22°C and pH 1 (the data are based in part on results given in Hellmann et al. 2003). The hydrolyzed feldspar was analyzed with two very different techniques: secondary ion mass spectrometry (SIMS) and high resolution and energy filtered transmission electron microscopy (HRTEM and EFTEM, respectively). SIMS was employed to obtain chemical depth profiles perpendicular to the altered surface. HRTEM and EFTEM were used to determine structural and chemical changes of the altered region (extending from the surface through the altered region to the non-altered mineral); however, in this case, the samples were prepared in cross section by an ultramicrotome.

Analyses of the same altered labradorite sample by SIMS and HRTEM/EFTEM yielded similar and complementary data. Nonetheless, there were important differences in the two data sets, which are primarily a function of the resolution of the two techniques. By comparing these two techniques (including sample preparation) and their respective results, we will show the importance that the analytical method has with respect to the interpretation of results and how this affects the proposed mechanism of dissolution.

2 SAMPLE PREPARATION AND INSTRUMENTAL ANALYSES

Samples of labradorite ($Na_{0.39}Ca_{0.59}Al_{1.59}\ Si_{2.41}O_8$) were dissolved in HCl/H_2O solutions at pH 1 (500 h) at 22 ± 2°C in flow-through reactors (for details, see Hellmann et al. 2003).

SIMS presents many advantages as an analytical tool for studying mineral surface alteration; perhaps its greatest asset is its versatility, because nearly all elements can be measured at trace concentrations (including their most important isotopes) to depths on the order of several hundred nm. Perhaps one of the most important elements that can be measured by SIMS is H, given that H^+, H_2O, and OH^- are the key aqueous species involved in bond hydrolysis (note that H cannot be measured by EFTEM). The depth resolution of SIMS generated profiles decreases with increasing depth, due mostly to the deleterious effects of ion sputtering and crater formation; this is a major limitation when profiles intercept chemical or structural interfaces at significant depths. An analytical artifact associated with SIMS and other ion beam methods is the tendency of ion beams to broaden the apparent width of interfaces, especially at depth (Hues & Williams 1986, Barbour & Doyle 1995). The exact degree of interface broadening is dependent on the method, analytical conditions, and the nature of the sample.

SIMS analyses were carried out with a Cameca IMS 3f instrument using a $^{16}O^-$ primary ion beam with the following instrument settings: 25 nA beam current, 12.6 kV accelerating voltage, 125 × 125 μm raster area. Profile depths were determined from sputtering rates that were calibrated by the measurement of crater depths (Alpha step 200 Tencor Instruments profilometer). The reproducibility of crater depth measurements is approximately ±20–30 nm. Using a $^{16}O^-$ primary beam, the sputtering rate was determined to be ≈2.3x faster in the altered zone than in non-altered labradorite.

High resolution images of the altered/non-altered interface were obtained with a JEOL 3010 TEM operated at 300 kV; its point resolution is 0.17 nm. EFTEM provides a convenient method for obtaining 2-D spatial distributions (i.e. concentration maps) of any chosen element at nm-resolution (with the exception of H). The ability to obtain nanometer-resolution chemical information is by far the greatest asset of this technique. In our study we determined jump ratio images based on electron energy loss spectroscopy (EELS) spectra that were obtained with a Gatan imaging filter attached to a JEOL 3010 TEM operated at 300 keV, with a dispersion = 0.2 eV/channel, and a collection angle β = 6.5 mrad. A jump ratio image is calculated by dividing the post-edge image by the pre-edge image. Each image is obtained by selecting a (20–30 eV width) part of the chosen core-loss ionization edge of the particular element (Ca $L_{2,3}$, O K, Al K, Si K edges used in this study).

3 RESULTS

3.1 *Elemental depth profiles measured by SIMS*

Chemical depth profiles of a sample reacted at pH 1 (Fig. 1) measured by SIMS, presented in terms of raw ion counts, reveal that the altered zone is depleted in Ca, K, Na, and Al, and enriched in H, Si, and B, as compared to the non-altered labradorite (boron was present as an impurity in the aqueous solutions). The profiles for depleted and enriched elements are sigmoidal and anti-correlated. The SIMS results we report here are in good accord with previous SIMS and other ion beam and spectroscopy measurements of acid pH-altered near-surface zones on feldspars (e.g. Casey et al. 1989a, Hellmann et al. 1990, Petit et al. 1990, Inskeep et al. 1991, Muir & Nesbitt 1997, Schweda et al. 1997, Nesbitt & Skinner 2001).

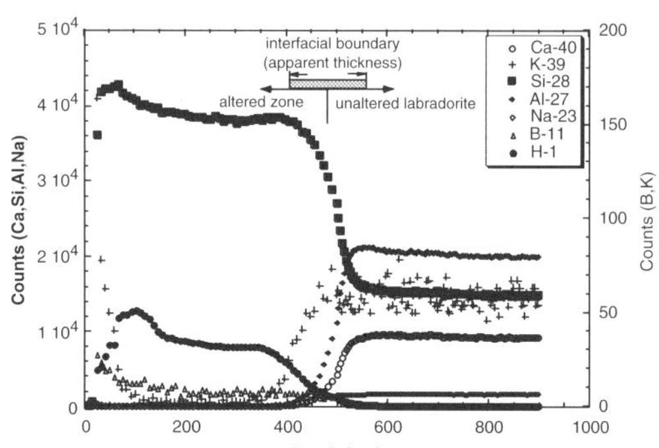

Figure 1. SIMS profile showing raw ion counts vs. depth for a labradorite sample hydrolyzed at pH 1 and 22°C. The apparent interface between the altered zone and the non-altered mineral occurs at 480–500 nm. The thickness of the interfacial boundary ranges from 100–150 nm, depending on the element. Note that the altered zone is depleted in Ca, K, Na, Al, and enriched in Si, H, and B (the latter is present as an impurity in the hydrolyzing solution). The anti-correlated nature of the profiles for depleted and enriched elements is typically used as evidence for an interdiffusion process, which is an integral part of the leached layer hypothesis.

The average inflection point for all elements is located at approximately 480–500 nm, which corresponds to the depth measured from the surface to the interface between the altered zone and the non-altered labradorite. Note that the inflection points for the depleted elements are not the same, varying from 430 nm (K) to 500 nm (Na, Ca); the reason for this variability is not known. The apparent thickness (i.e. chemical gradient) of the interfacial boundary, as defined by the sigmoidal part of the profiles, is 100–150 nm (depending on the element). A significant proportion of the measured interfacial thickness is an artifact of the previously mentioned phenomenon of ion beam-induced interface broadening.

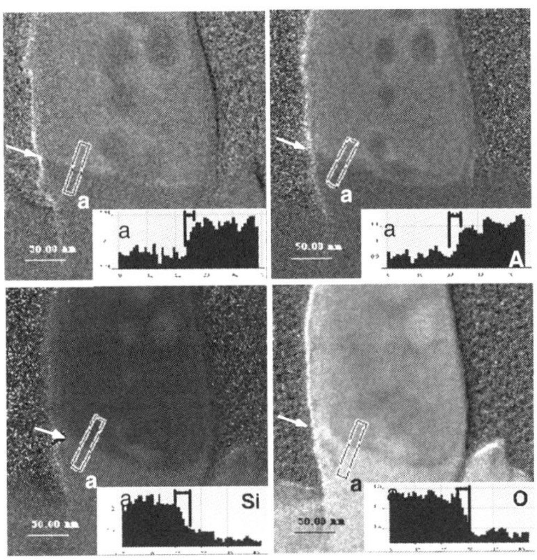

Figure 2. Single lamella of labradorite altered at pH 1 prepared in cross section with an ultramicrotome (using the same sample as in Figure 1). The four jump-ratio EFTEM images represent qualitative chemical concentration maps of Ca, Al, Si, and O (left to right, top to bottom, respectively; lighter tones represent higher concentrations; scale bar in lower left corner is 50 nm). In each case, the interface between the non-altered mineral (top) and altered zone (bottom) is very sharp and distinct (see white arrows). The insets show chemical profiles based on the diagonal rectangular areas; in each inset the profile extends (left to right) from the altered zone (bottom half of lamella) to the non-altered mineral. In contrast to the SIMS depth profile shown in Figure 1, the interfacial thicknesses are very sharp, and range between 3.8 and 6.8 nm (delimited by small bracketing lines in each inset), depending on the element. The sharpness of the spatially coincident chemical interfaces (which are also coincident with an even sharper structural interface- not shown) is consistent with an interfacial dissolution-reprecipitation mechanism, and not leached layer formation (see text for details). Figure 2 is modified from Hellmann et al. (2003).

3.2 *The altered/non-altered interface measured by HRTEM and EFTEM*

As already mentioned, the electron transparent ultrathin sections of the altered zone examined by HRTEM/EFTEM were cut in cross section; this presents an important advantage because this provides a method for examining the interface directly, thus avoiding the deleterious effects of ion bombardment and cratering. In addition, surface rugosity (which is very often present after hydrolysis) does not affect the results. Thus, the ultramicrotome sample preparation technique used in this case presents several distinct advantages over ion beam methods.

In general, our SIMS and EFTEM data are complementary and show the same chemical trends and depths of alteration. Nonetheless, there is a crucial difference between the data sets that is due to the inherently higher resolution of the EFTEM technique. The true chemical gradients of the interfacial boundary are significantly less thick than those estimated from the SIMS results, based on chemical profiles of the ultra-thin sections that we measured with EFTEM. EFTEM jump-ratio images of a typical interfacial region (reacted at pH 1) on a single lamella portray qualitatively the concentrations of Ca, Al, Si, and O (Fig. 2). Sharp, step-like changes in the concentrations of these four elements occur exactly at the same interfacial boundary that marks the structural transition between the amorphous, altered zone and the non-altered, crystalline labradorite, based on HRTEM (not shown). Chemical profiles across and perpendicular to the interfacial region (Fig. 2) give the following interfacial boundary thicknesses (in nm): Ca $\approx$ 3.8; Al $\approx$ 5.1; Si $\approx$ 6.8; O $\approx$ 4.9.

4 DISCUSSION AND CONCLUSIONS

If we examine the SIMS results in Figure 1, these are at first glance compatible with the leached layer hypothesis that is currently invoked to explain aluminosilicate mineral dissolution at acid pH. Very briefly, this theory is based on the formation of an amorphous, altered near-surface zone that represents a relict 'structure', composed primarily of Si and O, that is created after certain elements (e.g. Na, K, Ca, and Al) are preferentially released from the non-altered mineral. Silanol groups in the leached layer may undergo condensation reactions (Casey et al. 1989a,b, Hellmann et al. 1997). The preferential release of elements occurs by interdiffusion with H^+ from the aqueous solution, thus keeping the leached layer electrically neutral. The anti-correlated, sigmoidal depth profiles shown here and in other studies appear to fit (but only qualitatively) with interdiffusion models (e.g. see details in Hellmann 1997, Hellmann et al. 2003).

On the other hand, the HRTEM/EFTEM results show that the structural and chemical gradients (Fig. 2) are spatially coincident and extremely sharp (structural gradient <2 nm, chemical gradients: 4–7 nm). These interfacial gradients are too sharp to have been produced by a volume interdiffusion process associated with leached layer formation; this is supported by interdiffusion modeling presented in Hellmann et al. (2003). In addition, the true interfacial gradients are in reality probably even less than those presented here (i.e. sharper gradients, see discussion in Hellmann et al. 2003).

Even though the sharpness of the interfacial gradients is the primary reason for the non-applicability of the leached layer theory, we also cite the following other reasons: 1. the enrichment of Si and O in the altered zone, which is difficult to explain based on preferential leaching; 2. equivalent depths of depletion for fast (Na, K, Ca) and slow (Al) diffusing species, based on leached layer formation and a volume interdiffusion process; and 3. the presence of B in the SIMS profiles after treatment with a strong B-complexing agent (i.e. identical profiles before and after treatment).

The mostly likely explanation for the results we show here is that the altered zones form via an interfacial dissolution-reprecipitation process. The interfacial nature of this process needs to be stressed because we hypothesize that bond breaking (dissolution) and bond re-making (reprecipitation) occur in a thin solution film directly in contact with the interface. Thus, supersaturation occurs in an interfacial thin film, and not in the bulk solution. During hydrolysis of the labradorite the bulk solution was undersaturated with respect to all possible secondary phases (e.g. silica gel). The precipitate corresponds to a hydrated silica gel, based on the SIMS and EFTEM results. This altered zone most probably has a very high porosity (e.g. Casey et al. 1989b, Putnis 2002), thus allowing non-diffusion limited transport of dissolution products and aqueous species between the interface and the external fluid.

The absence of leached layers is crucial to the mechanistic pathway of the intrinsic dissolution process. The interfacial solution-reprecipitation mechanism implies that the intrinsic dissolution process is stoichiometric at acid pH conditions. It implies a single, pH-invariant mechanism that controls the release of all elements comprising the feldspar structure. Any measured deviations from stoichiometric dissolution behavior (e.g. as measured by solution analyses) can be attributed to precipitation that occurs after the initial, stoichiometric detachment of elements from the non-altered matrix.

REFERENCES

Barbour, J.C. & Doyle, B.L. 1995. Elastic recoil detection: ERD (or Forward Recoil Spectrometry: FRES). In J.R. Tesmer, M. Nastasi, J.C. Barbour, C.J. Maggiore & J.W. Mayer (eds), *Handbook of Modern Ion Beam Analysis*: 83–138, Pittsburgh (USA): Materials Research Society.

Casey, W.H., Westrich, H.R., Arnold, G.W. & Banfield, J.F. 1989a. The surface chemistry of dissolving labradorite feldspar. *Geochim. Cosmochim. Acta* 53: 821–832.

Casey, W.H., Westrich, H.R., Massis, T., Banfield, J.F. & Arnold, G.W. 1989b. The surface of labradorite feldspar after acid hydrolysis. *Chem. Geol.* 78: 205–218.

Hellmann, R. 1997. The albite-water system Part IV. Diffusion modeling of leached and hydrogen-enriched layers. *Geochim. Cosmochim. Acta* 61: 1595–1611.

Hellmann, R., Dran, J.-C. & Della Mea, G. 1997. The albite-water system Part III. Characterization of leached and hydrogen-enriched layers formed at 300°C using MeV ion beam techniques. *Geochim. Cosmochim. Acta* 61: 1575–1594.

Hellmann, R., Eggleston, C.M., Hochella, M.F. Jr. & Crerar, D.A. 1990. The formation of leached layers on albite surfaces during dissolution under hydrothermal conditions. *Geochim. Cosmochim. Acta* 54: 1267–1281.

Hellmann, R., Penisson, J.-M., Hervig, R.L., Thomassin, J.-H. & Abrioux, M.F. 2003. An EFTEM/HRTEM high resolution study of the near surface of labradorite feldspar altered at acid pH: evidence for interfacial dissolution-reprecipitation. *Physics Chemistry Minerals* 30: 192–197.

Hues, S.M. & Williams, P. 1986. Oxygen-induced segregation effects in sputter depth-profiling. *Nuclear Instruments and Methods in Physics Research* B15: 206–209.

Inskeep, W.P., Nater, E.A., Bloom, P.R., Vandervoort, D.S. & Erich, M.S. 1991. Characterization of laboratory weatheredlabradorite surfaces using X-ray photoelectron spectroscopy and transmission electron microscopy. *Geochim. Cosmochim. Acta* 55: 787–800.

Muir, I.J., Bancroft, G.M., Shotyk, W. & Nesbitt, H.W. 1990. A SIMS and XPS study of dissolving plagioclase. *Geochim. Cosmochim. Acta* 54: 2247–2256.

Nesbitt, H.W. & Skinner, W.M. 2001. Early development of Al, Ca, and Na compositional gradients in labradorite leached in pH 2 HCl solutions. *Geochim. Cosmochim. Acta* 65: 715–727.

Petit, J.-C., Dran, J.-C. & Della Mea, G. 1990. Energetic ion beam analysis in the Earth sciences. *Nature* 344: 621–626.

Putnis, A. 2002. Mineral replacement reactions: from macroscopic observations to microscopic mechanisms. *Mineralogical Magazine* 66: 689–708.

Water–Rock Interaction, Wanty & Seal II (eds)
 ISBN 90 5809 641 6

Kinetics of elementary steps on cleaved dolomite surfaces in undersaturated alkaline aqueous solutions

S.R. Higgins & X. Hu
Department of Chemistry, Wright State University, Dayton, OH, USA

K.G. Knauss
Earth Sciences Division, Lawrence Livermore National Laboratory, Livermore, CA, USA

ABSTRACT: The kinetics of dissolution from step edges on dolomite ($CaMg(CO_3)_2$) cleavage surfaces were investigated with in-situ atomic force microscopy in alkaline (pH 9) aqueous solutions as a function of saturation state. The step kinetics were clearly non-linear versus saturation state and were found to be qualitatively consistent with a step kinetic model derived for AB-type crystals. The kinetic ionic ratio, r_i, from this model was found to be on the order of 10^{-5}. This result provides microscopic evidence for large differences between cation and anion dynamics at step edges, pointing to the difficulty in low-temperature growth of dolomite. Further results from experiments in which the activity ratio in solution $((Ca^{2+})+(Mg^{2+}))/(CO_3^{2-})$ was varied suggested that the AB-type model is a useful conceptual model even when applied to the more complicated ABX_2-type problem of dolomite.

1 INTRODUCTION

Dolomite ($R\bar{3}$, $CaMg(CO_3)_2$) is commonly found in sedimentary rocks and there is evidence to suggest that dolomite precipitated from seawater in the past, but is rarely found to precipitate from natural aqueous systems at present even though there exists a favorable thermodynamic driving force. This observation forms the basis of the so-called "dolomite problem". Several explanations have been suggested including the idea of an ancient natural catalyst that is no longer present in sufficient quantities to affect dolomite precipitation. For example, Brady et al. (1996) suggested that sulfate may catalyze dolomite growth possibly by accelerating the rate of dehydration of surface adsorbed Mg^{2+} ions. The fact that dolomite does not form readily in clean laboratory experiments at low temperature was used by these authors to rule out the possibility that a modern inhibitor present in today's natural waters could explain the dolomite problem.

The problem of dolomite crystallization may be alternately addressed by characterization of microscopic dolomite surface properties such as the kinetics of step motion and surface microtopography because these properties may be compared to theoretical model predictions based on detachment and attachment of fundamental "growth units" in addition to being compared to *ab initio* theoretical predictions of hydration and defect formation energies. (de Leeuw 2002, de Leeuw & Parker 2001, Titiloye et al. 1998, Wright et al. 2002, Zhang & Nancollas 1990a, Zhang & Nancollas 1998).

The dissolution kinetics of dolomite under acidic conditions have been characterized by several researchers (Busenberg & Plummer 1982, Chou et al. 1989, Gautelier et al. 1999, Herman & White 1985, Lüttge et al. 2003, Orton & Unwin 1993, Pokrovsky et al. 1999), yet the number of these investigations is still small compared to studies of calcite dissolution. Most of the dolomite studies were performed under conditions of relatively high undersaturation and the primary emphasis of these works was to characterize acidic dissolution properties. Because dolomite and other carbonates are not easily formed under acidic conditions, this study focused on dolomite surface kinetics in alkaline solutions so that near-equilibrium undersaturated conditions were achievable.

2 MATERIALS AND METHODS

Solutions were prepared using de-ionized water (18 MΩ-cm resistivity), high-purity $MgCl_2$, $CaCl_2$, $NaHCO_3$, NaCl, NaOH and HCl to reduce divalent cation contamination of the aqueous solutions, a problem purported in the of use reagent-grade chemicals for calcite growth. (Teng et al. 1999) Solutions were prepared immediately prior to each experiment by adding appropriate amounts of the salts from stock solutions to air-equilibrated DI H_2O followed by titration of the solution to pH 9.0 ± 0.1 using a dilute NaOH solution. The ionic strength was approximately fixed at 0.10 M. We verified that the kinetics of re-equilibration of the solution with the atmospheric CO_2 was slow (>2 hrs without stirring) so that the titrated solution could be drawn into a syringe, free of air, permitting generation of solutions of varying undersaturation at a fixed pH. Activities of Ca^{2+}, Mg^{2+}, and CO_3^{2-} were estimated using MINTEQ software by treating the system as closed to the atmosphere after establishing the total inorganic carbon in the initial solution. The measured pH of this initial solution agreed well with the MINTEQ predicted pH and the amount of NaOH necessary to bring the pH to 9.0 was in agreement with our treatment of the system as closed during the titration procedure. The saturation state, S, of the aqueous solutions were defined by the following:

$$S = \left(\frac{(Ca^{2+})(Mg^{2+})(CO_3^{2-})^2}{K_{sp}(dolomite)} \right)^{\frac{1}{4}} \quad (1)$$

The quantities in parentheses are activities of the species and the exponent of 1/4 was used in accordance with the number of ions in the formula unit. (Zhang & Nancollas 1990b). All experiments were conducted at $27 \pm 1°C$. Dolomite crystals were optical-grade, originating in Bahia, Brazil (Ward's Earth Science). Each experiment began with a freshly cleaved **10$\bar{1}$4** surface.

The AFM was a Molecular Imaging PicoSPM II and was operated in contact-mode with 0.2 N/m nominal spring constant uncoated Si cantilevers. The cleaved dolomite samples (typically 5 mm × 5 mm × 1 mm) were mounted on a glass coverslip using a small amount of cyanoacrylate adhesive on the underside of the crystal. The samples were then placed on the AFM sample plate and held in place with a customized Kel-F fluid cell. The open fluid cell was designed with fluid inlet and outlet ports for continuous flow experiments using dual syringe pumps. The nominal volume of the fluid cell was 1.2 mL and flow rates of 10 μL/s were found to be sufficient to observe step velocities not limited by the rate of fluid flow.

3 RESULTS AND DISCUSSION

In Figure 1 are representative images showing the general shapes of etch pits observed on the dolomite cleavage surfaces at pH 9. These pits, in contrast to the familiar rhombic shaped pits observed on calcite, often displayed a parallelogram shape, having two longer, parallel edges. This general observation may be explained by the lack of a c-glide plane in the dolomite structure in contrast to the higher symmetry in the calcite structure. One consequence of the lower space group symmetry in dolomite includes p1 plane group symmetry on the **10$\bar{1}$4** surface. This results in all four "rhombic" steps being structurally nonequivalent and thus four unique step velocities are possible. In contrast to these results at alkaline pH, a recent study by Lüttge et al. (2003) reported roughly circular or elongated dolomite etch pits whereas Orton & Unwin (1993) showed more polygonized etch structures in strongly acidic solution (pH 3–4). AFM studies of natural dolomites by Kessels et al. (2000) suggested that the composition and defects in various materials has a dramatic influence on the surface microtopography.

Figure 1. In-situ AFM image (2 μm × 2 μm) showing the dolomite surface morphology at $S = 0.45$, $pH = 9.0$, $r_{C/A} = 1$. Most step edges are a single elementary layer in height (~2.9 Å). The c-axis runs SE to NW in the image and comes out through the plane of the page along this path. Two acute steps are labeled for clarity of step designation.

Figure 2 shows the results of step velocity measurements (made as the sum of the two opposing step velocities ($[48\bar{1}]_+ + [48\bar{1}]_-$ and $[\bar{4}41]_+ + [\bar{4}41]_-$)) versus $f(S) = (S-1)(2-S)^{-1/2}$. Comparison of $48\bar{1}$ steps with $\bar{4}41$ steps reveals $48\bar{1}$ steps dissolve, on average, at about twice the velocity of the $\bar{4}41$ steps, resulting in elongated etch pits. The large squares represent data from stoichiometric solutions (i.e. equal cation, $(Mg^{2+} + Ca^{2+})$ and anion (CO_3^{2-})

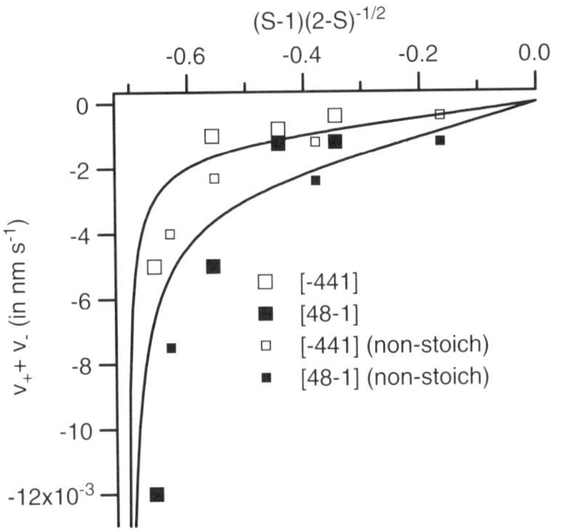

Figure 2. Summed step velocity as a function of undersaturation. The undersaturation function is described in further detail in Zhang & Nancollas (1998).

activities, $r_{C/A} = 1$, see Eqn. (2)). For this discussion, we will define the cation to anion ratio $r_{C/A}$ as:

$$r_{C/A} = \frac{(Ca^{2+}) + (Mg^{2+})}{(CO_3^{2-})} \tag{2}$$

The step kinetic model of Zhang & Nancollas (1990a) predicts a linear relationship between step velocity and f(S) when the kinetic ionic ratio, r_i is equal to unity where r_i is defined as:

$$r_i = \frac{k_A(A)}{k_B(B)} \tag{3}$$

In Equation 3, k_i are the first order rate constants for integration of ions into the crystal. The observed relationship in the experimental data is clearly not consistent with this linear model if we assume $(A) = (Ca^{2+}) + (Mg^{2+})$. The solid lines are approximate best fits to the large squares using the Zhang & Nancollas (1998) model, incorporating a correction function that depends on r_i and S. For both solid lines, $r_i = 5 \times 10^{-6}$ was used, suggesting that $k_A/k_B \sim 10^{-5}$ which is an order of magnitude estimate and is very small due to the extremely sluggish dehydration and ligand exchange kinetics of Mg^{2+} ions.(e.g. Pokrovsky & Schott (2002)). However, as pointed out by Zhang & Nancollas (1998), even in solutions with $r_{C/A} = 1$, r_i is likely to deviate significantly from unity. Therefore, non-linear step kinetics are expected and there is not a need to invoke a step roughening argument.

The smaller squares in Figure 2 are step velocity measurements made from non-stoichiometric solutions (i.e. with $r_{C/A} \sim 3$, an excess of cations over anions). These non-stoichiometric solutions resulted in an increase in step velocity compared to the step velocities in the stoichiometric solutions. If we assume that the attachment kinetics k_A of Mg^{2+} ions are responsible for the very small r_i value, an increase in $r_{C/A}$ from unity should increase r_i and the step velocities should increase. In other words, the knee in the model fit in Figure 2 will become less sharp as r_i increases toward unity and this is consistent with what is observed experimentally in Figure 2.

However, considering the assumption that Mg^{2+} attachment and detachment at kink sites, which involves dehydration and hydration, respectively, is limiting the rate of step motion, kinks will most often (on a statistical basis) be terminated by Mg^{2+} ions and an increase of (Mg^{2+}) might be expected to decrease the tendency for these ions to detach from kinks in undersaturated solution and therefore, decrease the rate of step motion. The error in this reasoning stems from the assumption that Mg^{2+} ion dynamics are very slow and essentially irreversible. Once an Mg^{2+} ion detaches from a kink site, the next site exposed, whether it be a CO_3^{2-} or Ca^{2+} site will undergo multiple exchanges and only when the next Mg^{2+} ion detaches will the kink be considered to have retreated by a full unit cell. We have applied a model developed for an AB-type crystal to a problem involving formally an ABX_2-type crystal. In addition to the affects of $r_{C/A}$ on the step kinetics, a more rigorous approach including the affects of $r_{A/B}$ (e.g. Ca^{2+}:Mg^{2+}) should be considered.

The effects of dissolved crystal cations on the dissolution rate of dolomite were investigated by Busenberg & Plummer (1982) and Gautelier et al. (1999). Busenberg & Plummer (1982) showed a decrease in rate with increasing cation concentration. Gautelier et al. (1999) reported negligible effect of Ca^{2+} and Mg^{2+} acitivites on dissolution rates. Whereas these results do not agree with our observations, these earlier experiments were performed under conditions of low pH in general and are not directly comparable to our observations.

In our application of the Zhang & Nancollas (1998) model, we held K_{sp} fixed. This restriction limits the flexibility of the model fit. Using a modified K_{sp} based on the step velocity data would imply a different "K_{sp}" for different step edges. Because interfacial step edge energies are likely to be a function of step orientation, it seems plausible that each step would then have its own "K_{sp}". If we treat the system in such a manner, then when thermodynamic equilibrium is achieved, hypothetically, perhaps none of the observed rhombic step directions would exist. An equilibrium condition may be considered to exist along a step when the net rate of kink propagation (also known as the kink velocity) is zero. In other words, any kinks along the step simply fluctuate

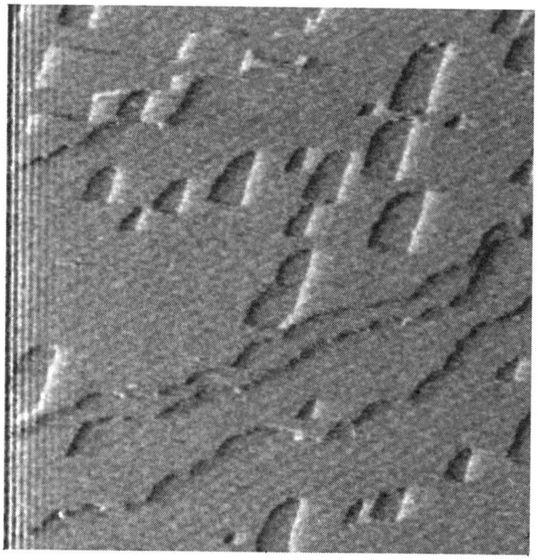

Figure 3. In-situ AFM image (2 μm × 2 μm) showing the pit morphology that develops when flow is stopped in the fluid cell. The sample orientation is identical to that of Figure 1.

about a "mean" position. Kinks may still be created and annihilated if the fluctuation amplitude is large enough to permit communication between kinks. Due to the step structure on the rhombohedral carbonates, oppositely facing kinks on a rhombic step are likely to have different kink formation energies suggesting that the equilibrium step direction will favor the presence of low energy kinks over that of high energy kinks. Steps observed under conditions of net dissolution may not correspond to the steps of minimum interfacial edge energy. To address this possibility, we performed additional experiments where a flow through dissolution process was interrupted by stopping flow through the fluid cell. In this experiment, the system drifted toward equilibrium and the etch pits developed curved acute step edges (Fig. 3), suggesting that acute, rhombic steps are not stable very near equilibrium.

4 SUMMARY

The experimental results from in-situ AFM of dolomite surfaces under alkaline aqueous solutions showed that the rate of step retreat on dolomite varied in a highly non-linear fashion when compared to the behavior expected from an AB-type crystal with similar cation and anion attachment kinetics. For stoichiometric solutions, the kinetic ionic ratio was estimated to be 10^{-5}, suggesting the non-linearity arises from differences in Mg^{2+} and Ca^{2+} hydration and/or ligand exchange kinetics. The implications of this result are that dolomite dissolution and precipitation kinetics are expected to behave linearly only with high cation to anion activity ratios in solution.

ACKNOWLEDGMENT

SRH and KGK thank the US Dept. of Energy, Basic Energy Sciences for financial support.

REFERENCES

Brady, P.V., Krumhansl, J.L. & Papenguth, H.W. 1996. Surface complexation clues to dolomite growth. *Geochim. Cosmochim. Acta* 60: 727–31.

Busenberg, E. & Plummer, L.N. 1982. The kinetics of dissolution of dolomite in CO_2-H_2O systems at 1.5 to 65°C and 0 to 1 atm PCO_2. *Am. J. Sci.* 282: 45–78.

Chou L., Garrels, R.M. & Wollast, R. 1989. Comparative study of the kinetics and mechanisms of dissolution of carbonate minerals. *Chem. Geol.* 78: 269–282.

De Leeuw, N.H. 2002. Surface structures, stabilities, and growth of magnesian calcites: A computational investigation from the perspective of dolomite formation. *Am. Mineral.* 87: 679–689.

De Leeuw, N.H. & Parker, S.C. 2001. Surface-water interactions in the dolomite problem. *Phys. Chem. Chem. Phys.* 3: 3217–3221.

Gautelier, M., Oelkers, E.H. & Schott, J. 1999. An experimental study of dolomite dissolution rates as a function of pH from −0.5 to 5 and temperature from 25 to 80°C. *Chem. Geol.* 157: 13–26.

Herman, J.S. & White, W.B. 1985. Dissolution kinetics of dolomite: Effects of lithology and fluid flow velocity. *Geochim. Cosmochim. Acta* 49: 2017–2026.

Kessels, L.A., Sibley, D.F. & Nordeng, S.H. 2000. Nanotopography of synthetic and natural dolomite crystals. *Sedimentology* 47: 173–186.

Lüttge, A., Winkler, U. & Lasaga, A.C. 2003. Interferometric study of the dolomite dissolution: A new conceptual model for mineral dissolution. *Geochim. Cosmochim. Acta* 67: 1099–1116.

Orton, R. & Unwin, P.R. 1993. Dolomite dissolution kinetics at low pH: A channel-flow study. *J. Chem. Soc. Faraday Trans.* 89: 3947–3954.

Pokrovsky, O.S. & Schott, J. 2002. Surface chemistry and dissolution kinetics of divalent metal carbonates. *Environ. Sci. Technol.* 36: 426–432.

Pokrovsky, O.S., Schott, J. & Thomas, F. 1999. Dolomite surface speciation and reactivity in aquatic systems. *Geochim. Cosmochim. Acta* 63: 3133–3143.

Teng, H.H., Dove, P.M., & DeYoreo, J.J. 1999. Reversed calcite morphologies induced by microscopic growth kinetics: Insight into biomineralization. *Geochim. Cosmochim. Acta* 63: 2507–2512.

Titiloye, J.O., De Leeuw, N.H. & Parker, S.C. 1998. Atomistic simulation of the differences between calcite and dolomite surfaces. *Geochim. Cosmochim. Acta* 62: 2637–2641.

Wright K., Cygan, R.T., & Slater, B. 2002. Impurities and nonstoichiometry in the bulk and on the (1014) surface of dolomite. *Geochim. Cosmochim. Acta* 66: 2541–2546.

Zhang, J. & Nancollas, G.H. 1990a. Kink densities along a crystal surface step at low temperatures and under nonequilibrium conditions. *J. Cryst. Growth* 106: 181–190.

Zhang, J. & Nancollas, G.H. 1998. Kink density and rate of step movement during growth and dissolution of an AB crystal in a nonstoichiometric solution. *J. Colloid Interf. Sci.* 200: 131–145.

Zhang, J.-W. & Nancollas, G.H. 1990b. Mechanisms of growth and dissolution of sparingly soluble salts. In M. F. Hochella & A. F. White (eds.), *Mineral-Water Interface Geochemistry, Rev. Mineral.* 23. Washington, DC: Mineralogical Society of America.

Water–Rock Interaction, Wanty & Seal II (eds)
© 2004 Taylor & Francis Group, London, ISBN 90 5809 641 6

Role of molecular oxygen in the dissolution of siderite and rhodochrosite

S.T. Martin* & O.W. Duckworth†
Division of Engineering and Applied Sciences, Harvard University, Cambridge, MA, USA
†Current address: Department of Civil and Environmental Engineering, University of California, Berkeley, CA, USA

ABSTRACT: The dissolution of siderite ($FeCO_3$) and rhodochrosite ($MnCO_3$) under oxic and anoxic conditions is investigated at 298 K. The anoxic dissolution rate of siderite is $10^{-8.65}$ mol $m^{-2}\,s^{-1}$ for $5.5 < pH < 12$ and increases as $[H^+]^{0.75}$ for $pH < 5.5$. Under oxic conditions the dissolution rate decreases to below the detection limit of 10^{-10} mol $m^{-2}\,s^{-1}$ for $6.0 < pH < 10.3$, and hillock precipitation preferential to steps is observed in concurrent atomic force micrographs. For $5.8 < pH < 7.7$ under oxic conditions, the AFM micrographs show a tabular precipitate growing by preferential expansion along the *a*-axis. For $pH > 7.7$ under oxic conditions, the dissolution rate decreases from $10^{-8.45}$ to $10^{-9.0}$ mol $m^{-2}\,s^{-1}$. Flattened hillock precipitates grow across the entire surface without apparent morphological influence by the underlying rhodochrosite surface.

1 INTRODUCTION

Rhodochrosite and siderite are important minerals of reduced metal ions that occur in terrestrial and marine deposits and sedimentary rocks. Their solubility limits aqueous Fe and Mn concentrations under anoxic conditions in many diverse environments. Under oxic and mildly alkaline conditions, $FeCO_3$ and $MnCO_3$ solubility is complicated by oxidation of the released $Fe^{2+}(aq)$ and $Mn^{2+}(aq)$ and subsequent precipitation of Fe^{III} and $Mn^{III/IV}$ (hydr)oxide minerals, which are orders of magnitude less soluble than the parent carbonate material. The interconversions among minerals are episodic and seasonal, varying with temporal changes in local pE and pH. Solubility may be further complicated by morphology: hysteresis in dissolution behavior occurs when (hydr)oxide minerals precipitate as insoluble coatings on the surfaces of carbonate grains and thus kinetically inhibit dissolution even in cases where carbonate dissolution is highly favored thermodynamically. Dissolution and precipitation are significant in the biogeochemical cycling not only of Fe and Mn but also of other metals (e.g. U, Co, Pb, Ni, Cu, and Zn) that coprecipitate with Fe and Mn (hydr)oxides.

In the natural environment, cycling of Fe and Mn occurs at oxic/anoxic interfaces where precipitation and dissolution occur simultaneously. Accurate understanding of biogeochemical cycling thus requires a quantitative description of the effects of O_2. Molecular oxygen (O_2) plays a critical role in determining the bioavailability of Fe and Mn due to the great solubility difference between $Fe^{III}/Mn^{III/IV}$ (hydr)oxides and Fe^{II}/Mn^{II} carbonates. If equilibrium conditions prevailed, understanding the effects of O_2 would be straightforward: a thermodynamic analysis would provide both the mix of solids present and the aqueous (i.e. bioavailable) concentrations of Fe and Mn species. However, the equilibrium treatment fails for two important reasons. First, surface-controlled dissolution and precipitation rates may be slow. Second, morphological features such as precipitation of an oxide supralayer may armor the carbonate surface and significantly reduce dissolution rates. Understanding the effects of O_2 on the dynamics of dissolution requires a kinetic chemical transport description.

In this study, we quantify the dissolution rates of the (1014) face of $FeCO_3$ and $MnCO_3$ under oxic and anoxic conditions for $1.5 < pH < 12$ at 298 K. Concurrently, atomic force microscopy is utilized to observe changes in the morphology of pits and the appearance of secondary mineral precipitates. X-ray photoelectron spectroscopy is employed to identify the chemical composition of these (hydr)oxide precipitates. We find that thermodynamic analyses rationalize and bring cohesion to a variety of complex and interrelated data sets (e.g. the effects of $P\text{-}O_2$ and pH on a time series of AFM images, dissolution rates, and XPS spectra).

2 METHODS

2.1 *Materials*

Crystals of rhodochrosite (sample #96030, Colorado) and siderite (sample #134827, Ivigtut, Greenland) are obtained from the Harvard University Mineralogical Museum. Mineral identity and the absence of other phases (nominal detection limit = 2%) are confirmed with X-ray diffraction (Phillips X-ray diffractometer). Proton-induced X-ray emission spectroscopy (PIXE, Cambridge accelerator for materials science (CAMS), Harvard University) is utilized to determine the concentrations of cation impurities in the surface regions of the samples. PIXE analysis of siderite shows an Fe purity of 95.35%, with 4.47% Mn and trace amounts (<0.04%) of Zn, Ca, and Sr. Other cations, including Cd, Ni, and Pb, are below the minimum detection limit, which ranges from 0.001–0.04% cation mole percent.

The $MnCO_3$ and $FeCO_3$ crystals are cleaved with a razor blade along the $\{10\bar{1}4\}$ plane (perfect cleavage). We employ the Miller indices conventions of O'Keeffe & Hyde (1996). The cleaved samples are pressed into warmed dental modeling wax (Cavex, Holland) for attachment to a steel puck. The wax covers the edges of the sample so that only the imaged surface dissolves. Control experiments conducted with calcite yield step velocities, pit morphologies, and dissolution rates that agree well with prior studies. A new surface is prepared for each experiment. The geometric surface area (A) is calculated by employing *Photoshop* (Adobe Systems, San Jose, CA) to analyze digital photographs.

2.2 *Simultaneous AFM/flow-through reactor experiments*

A crystal affixed to a steel puck is placed on the end of the AFM piezoelectric scanner, and a drop of deoxygenated solution is placed on the sample. This step is necessary to avoid bubbles that prevent imaging in the fluid cell. The fluid cell is immediately placed over the sample to minimize exposure to atmospheric oxygen, which causes detectable surface oxidation within 30 min at neutral pH. A dissolution liquor is pumped through the fluid cell at 0.01–1.5 mL min^{-1}. At a chosen time, the syringe is switched so as to change the pH and pE conditions in the cell.

An atomic force microscope is employed to image the surface in contact mode (Nanoscope IIIa Multimode SPM, Digital Instruments) at 1 to 4 Hz with oxide-sharpened silicon nitride probes (nominal radius of curvature of 5–40 nm and force constant of 0.12 $N\,m^{-1}$) while solution flows through the AFM cell. Images collected in the same scan direction (i.e., exclusively upscans or downscans) are employed for analysis. Possible artifacts induced by imaging, such as scan fields, are examined by periodically scanning normal to the original direction and by comparing images where the tip has rastered for several hours (e.g. during an experiment to collect a time series of images) to an image where the tip has scanned only once.

2.3 *Dissolution rates*

Dissolution rates are calculated from the metal ion flux through the fluid cell, as follows:

$$R = \frac{\Delta M_T\, q}{A} \qquad (1)$$

where R ($mol\,m^{-2}\,s^{-1}$) is the dissolution rate, ΔM_T (M) is the difference in Fe_T or Mn_T between the influent and the effluent, q is the flow rate ($L\,s^{-1}$), and A is the geometric surface area (m^2) of the sample. In our experiments, the inlet concentration is below the limit of detection (i.e. $Mn_T < 2$ nM and $Fe_T < 4$ nM) while the effluent concentration varies from 2 to 4000 nM. In control experiments, we verify that the dissolution rate is independent of q.

3 RESULTS AND DISCUSSION

Dissolution rates R_{an} and R_{ox} are summarized in Figure 1 for $FeCO_3$ and $MnCO_3$. For $FeCO_3$ (Figure 1A), $R_{an} = 10^{-8.65}$ $mol\,m^{-2}\,s^{-1}$ for $5.5 < pH < 12$ and increases steadily for $pH < 5.5$. For $pH < 5$, the presence of oxygen does not affect the dissolution rate (i.e., $R_{an} = R_{ox}$). However, for $6 < pH < 10.3$, R_{ox} is below the limit of detection. At $pH > 10.3$, R_{ox} is as high as $10^{-7.5}$ $mol\,m^{-2}\,s^{-1}$, which is greater than under the corresponding anoxic conditions. For $MnCO_3$ (Figure 1B), $R_{an} = 10^{-8.45}$ $mol\,m^{-2}\,s^{-1}$ for $7 < pH < 11$ and increases steadily for $pH < 7$. For $pH < 7.7$, the presence of oxygen does not affect the dissolution rate (i.e., $R_{an} = R_{ox}$). For $pH > 7.7$ R_{ox} is reduced 80% with respect to R_{an}.

Figure 2 shows how exposure to $O_{2(aq)}$ in circumneutral to mildly alkaline pH affects the $(10\bar{1}4)$ face of $MnCO_3$ at the mesoscale (40 μm × 40 μm). A fresh $MnCO_3$ surface imaged in anoxic solution at pH = 3.7 is shown in Figure 2A, and a step and terrace morphology typical of the $(10\bar{1}4)$ face of $MnCO_3$ is evident. The surface is subsequently exposed to oxygen-saturated solution at pH = 6.0 (Figure 2B-D) and pH = 9.9 (Fig. 2E). In 2B after 165 min, a 2 nm discontinuous flat layer initiating from several islands grows over the terrace region. Earlier micrographs at 30 and 135 min (2C & 2D) show in more detail the time-dependent growth of the layer. The discontinuous layer does not grow over pits. Growth of this tabular precipitate also occurs preferentially

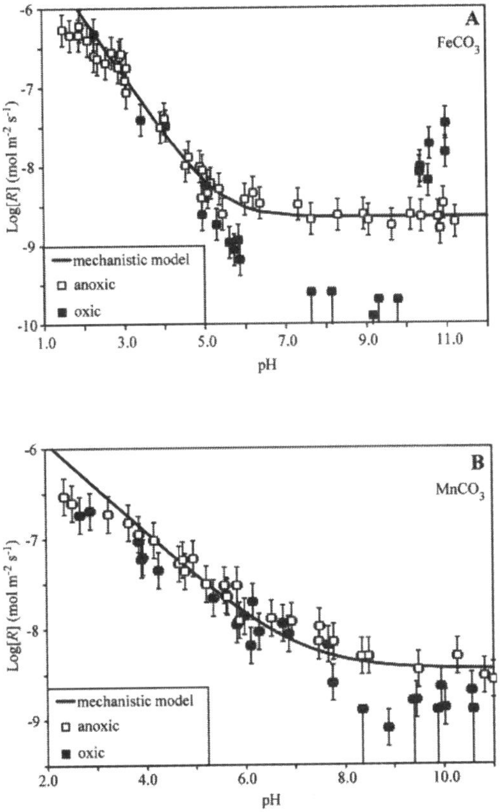

Figure 1. Dissolution rate of the (1014) faces of (A) $FeCO_3$ and (B) $MnCO_3$ as a function of pH under oxic and anoxic conditions. A solid line shows a fit by a mechanistic model to the anoxic data (see text). Uncertainty bars that extend to the x-axis indicate upper limits for dissolution rates (i.e. Fe_T or Mn_T is below the limit of detection of our analytical method).

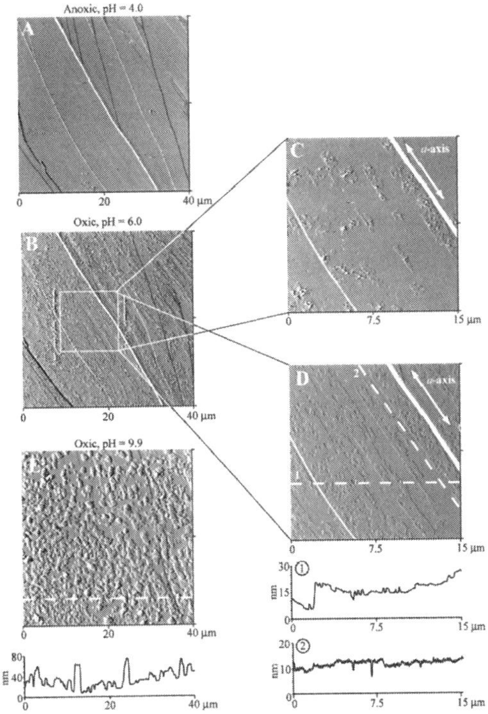

Figure 2. AFM deflection mode micrographs and height mode cross-sections showing the effect of O_2at $pH = 6.0$ and 10.0 on the (1014) face of $MnCO_3$. (A) Anoxic solution at $pH = 4.6$. (B) A 2 nm discontinuous layer grows over the terrace region following exposure to oxygen-saturated solution at $pH = 6.0$ for 165 min. Earlier micrographs at 30 and 135 min (C and D) show in more detail the time-dependent growth of the layer. (E) Following exposure to O_2 at $pH = 9.9$ for 120 min, the surface is covered with a patchy layer of hillocks 20–25 nm high by 1000–2000 nm diameter.

along the crystallographic a-axis. Images 2C and 2D occur within a scan field (cf. debris in Figure 2B), but an inspection of the image beyond the field shows the same basic morphological features. This tabular precipitate occurs for $5.8 < pH < 7.7$.

In 2E, which is captured after 2 hr of exposure to $O_2(aq)$ at $pH = 9.9$, the surface is covered with a patchy layer of flattened hillocks, which are 10–25 nm high by 500–1000 nm across. They have an apparent contact angle of $4 \pm 2°$. These contact angles are possibly a convolution of the AFM tip with the hillock. Nevertheless, they remain useful for comparison of the hillock morphologies occurring on $FeCO_3$ and $MnCO_3$. Images with hillocks are characteristic of $pH > 7.7$ under oxic conditions. The initial surface for 2E (not shown) is similar to 2A. Figure 1B shows a large decrease in the dissolution rate concurrent with the AFM imaging ($pH > 7.7$).

4 CONCLUSIONS

Under anoxic conditions, $FeCO_3$ and $MnCO_3$dissolve chemically by parallel proton- and water-promoted pathways and physically by pit formation and step retreat. $FeCO_3$ and $MnCO_3$solution concentrations are always far from equilibrium for the entire pH and P-O_2range of the conditions of our study. The dissolution and precipitation reaction mechanisms under oxic conditions are summarized in Figure 3. Under acidic oxic conditions, dissolution occurs by the same mechanism as the corresponding anoxic case. Under oxic conditions and neutral to mildly alkaline pH, surface precipitates and $R_{ox} < R_{an}$ rate are observed for both minerals. Thermodynamic calculations rationalize the reduction in the dissolution rate: O_2 reduces R at pH values where the presence of O_2

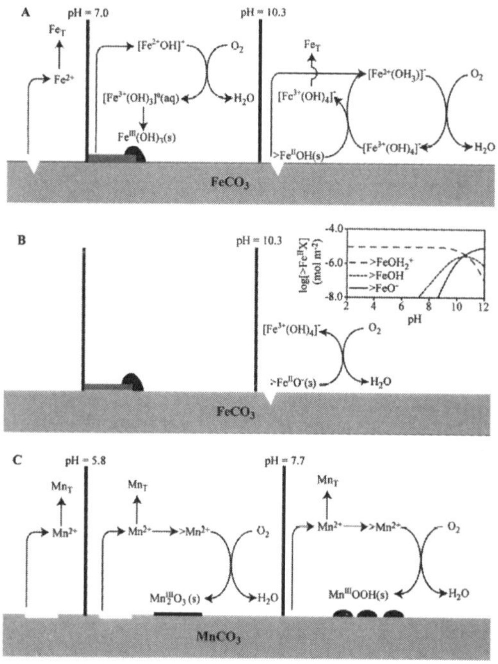

Figure 3. Depiction of the proposed dissolution and precipitation reaction mechanisms active under oxic conditions over a range of pH regimes for (A, B) $FeCO_3$ and (C) $MnCO_3$. Under anoxic conditions, dissolution reactions at all pH values are identical to those shown for low pH oxic conditions. Insert *B*: Fe^{II} hydroxyl speciation for experimental conditions.

yields saturation of (hydr)oxide solids. XPS results and thermodynamic calculations implicate $Fe(OH)_3$ as the hillock precipitate on $FeCO_3$ for $7 < pH < 10.3$, Mn_2O_3 as the tabular precipitate on $MnCO_3$ for $5.8 < pH < 7.7$, and MnOOH as the hillocks on $MnCO_3$ for $pH > 7.7$. Due to a low affinity of iron ions or hydroxide monomers for the terrace region of the surface, iron hydroxide hillock precipitates form at steps on $FeCO_3$. In contrast, the lack of site specificity of the manganese oxide hillock precipitate indicates a strong affinity of manganese ions or oxide monomers for the $MnCO_3$ surface. For $FeCO_3$ at $pH > 10.3$, $R_{ox} > R_{an}$ due to by oxidative attack by aqueous O_2 or $[Fe^{3+}(OH_4)]^-$.

Our results show that molecular oxygen is a critical factor for quantitative treatments of the dissolution and precipitation of transition metal carbonates and (hydr)oxides for $pH > 6$. Because carbonate minerals occur in environments undergoing both gradual (e.g. seasonal) and abrupt (e.g. episodic) changes in pH and pE, (hydr)oxide patinas occur widely and, when they occur, regulate the release rate of transition metal micronutrients to the aqueous phase. The apparent macroscopic dissolution rate is slowed. Future work is needed to investigate the properties and persistence of the (hydr)oxide patinas in variable chemical environments.

ACKNOWLEDGEMENTS

OWD was funded by a Sandia National Laboratories Campus Executive Fellowship. We are grateful for support received from the New York Community Trust Merck Fund and from the Chemical Sciences, Geosciences, and Biosciences Division, Office of Basic Energy Sciences, Office of Science, U.S. Department of Energy.

Water–Rock Interaction, Wanty & Seal II (eds)
© 2004 Taylor & Francis Group, London, ISBN 90 5809 641 6

Cation sorption at the muscovite–water interface using X-ray reflectivity

K.L. Nagy & N.C. Sturchio
Department of Earth and Environmental Sciences, University of Illinois at Chicago, Chicago, IL, USA

M.L. Schlegel
Commissariat à l'Energie Atomique de Saclay, Gif-sur-Yvette, France

P. Fenter
Environmental Research Division, Argonne National Laboratory, Argonne, IL, USA

ABSTRACT: High-resolution in-situ X-ray reflectivity measurements were used to characterize the sorption of K, Cs, Ca, and Ba from chloride solutions and Zn from chloride, nitrate, and trifluorosulfonate (triflate) solutions onto the basal surface of muscovite. X-ray reflectivity data were collected through the specular reflection condition and converted to electron-density profiles perpendicular to the interface by applying atomistic structural models that included surface relaxation, sorbate concentration and position, water structure, and surface roughness. Cations were attached to the basal surface dominantly in the first adsorbed layer, which has the highest electron density of the solution phase. All derived heights imply that no separate water layer exists between the sorbates and mineral surface. Water layering above the sorbed cation layer and minimal relaxation of the bulk muscovite were observed.

1 INTRODUCTION

Phyllosilicate basal surfaces sorb contaminants, natural organic matter, and life-sustaining nutrients in natural systems. Despite our understanding of sorption from a macroscopic perspective on micas and clays, little is known about atomic-scale sorption mechanisms. We have used X-ray reflectivity measurements of the muscovite(001)-water interface in KCl, CsCl, $CaCl_2$, and $BaCl_2$ solutions (0.01 and 0.5 M) as well as in a series of Zn-anion solutions to determine the distribution of ions and water perpendicular to the interface, the extent of organization of water molecules away from the mineral surface, and the extent of relaxation of crystal atoms at the surface.

2 MATERIALS AND METHODS

Freshly-cleaved muscovite ($KAl_3Si_3O_{10}(OH)_2$) sheets (Asheville Schoonmaker Mica Company, ASTM-V1 "scratch free" grade) were immersed in solutions for at least 1 hour prior to mounting in a thin film cell. Solutions of 0.01 m (K^+, Cs^+, Ca^{2+}, Ba^{2+}, and Zn^{2+}), and 0.5 m (K^+, Cs^+, and Ca^{2+}) were examined. The zinc solutions were prepared from a series of anionic salts including chloride, nitrate, and trifluorosulfonate (triflate). X-ray reflectivity data were collected on wet surfaces at the 12-BM, 12-ID, and 11-ID stations (BESSRC-CAT, Advanced Photon Source, Argonne National Laboratory). Reflected X-ray intensity as a function of momentum transfer Q ($=(4/\lambda)\sin(\theta)$) was measured using rocking-curve scans through the specular reflection condition followed by background subtraction and area integration of the rocking-curve peaks. Measured reflected X-ray intensities were reproduced on different muscovite samples and in one case, by reversibly changing the chloride solutions in contact with a single muscovite surface. Reflectivity data were obtained at high resolution (~1.3 Å) and converted to electron-density profiles perpendicular to the interface by applying atomistic structural models that included surface relaxation, sorbate concentration and position, water structure, and surface roughness. The fitted models for the electron density profiles were robust. Derived electron-density profiles in the solution were independent of the exact atomistic composition of individual layers of atoms. Interpretation of the composition of the individual layers in the electron density profiles is based on progressive examination and comparison of simple (water alone;

Cheng et al. 2001) through more complex systems (monovalent and then divalent cations).

3 RESULTS AND DISCUSSION

Reflectivity data for K^+ and Cs^+ sorption from 0.01 m solutions (Fig. 1) show distinct differences, for example from $Q = 0.75$ Å^{-1} to $Q = 1$ Å^{-1}. Conversion of the reflectivity data to electron density profiles (Figs 2a, b) shows that the electron density is highest in the adsorbed layer closest to the surface, indicating that the cations are concentrated in this layer. In addition, the layer nearest the surface has a higher electron density for solutions containing the cation with the higher Z value (Cs^+). Similar observations were made for the experiments using $CaCl_2$ and $BaCl_2$ (Figs 2c, d).

In all solutions, cations were attached to the (001) surface dominantly in the first adsorbed layer. The derived heights of K and Cs ions above the plane of surface tetrahedral oxygens, 1.67(6) and 2.15(9) Å, respectively, differ in magnitude by their crystallographic radii and correspond closely to values found in bulk K- and Cs-mica structures. The divalent ion heights 2.5(1) and 2.04(5) Å for Ca and Ba, respectively, are counter to the trend expected based on crystallographic radii but instead are consistent with one based on the difference in hydrated ion radii. All derived ion heights imply that there is no water layer between the adsorbed ion and the muscovite surface. Interpretations of the derived electron density profiles indicate that monovalent K^+ and Cs^+ cosorbed with some water molecules and that divalent Ba^{2+} and Ca^{2+} sorbed either with water

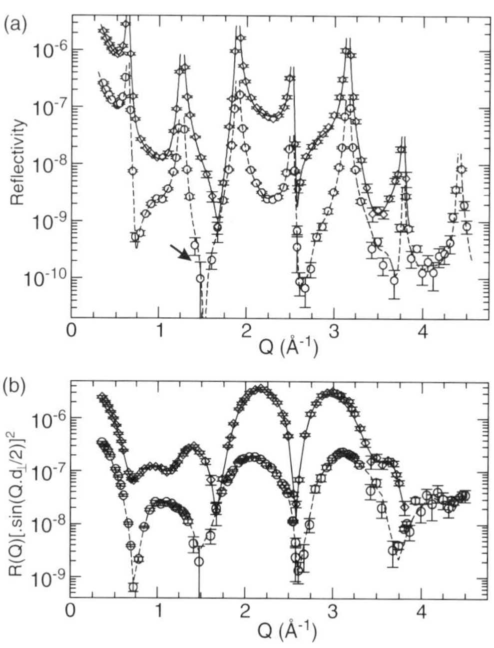

Figure 1. (a) Reflectivity data for muscovite in 0.01 m KCl (○) and 0.01 m CsCl (◇). Note that the amplitude for CsCl has been multiplied by 10. Also shown are the best-fits for KCl (dashed line) and CsCl (solid line). Only the tails of the Bragg peaks are shown. (b) Normalized reflectivity data $R(Q).[Q\sin(Qd_\perp/2)]^2$ (where d is the vertical distance between two muscovite layers) and best fit models. Same symbols as in (a).

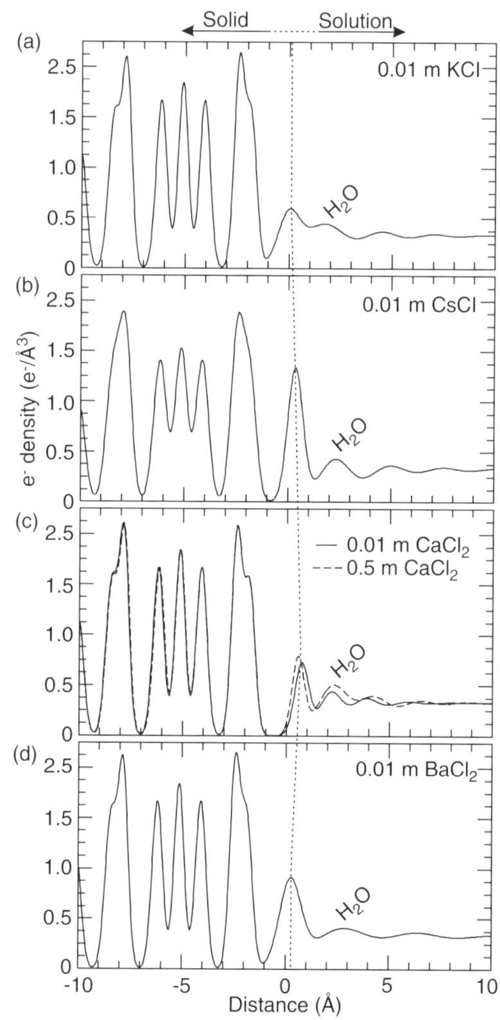

Figure 2. Plots of the laterally averaged electron density profiles in the direction normal to the surface, as obtained from the best fits results, for (a) 0.01 m KCl, (b) 0.01 m CsCl (), (c) 0.01 m $CaCl_2$ (solid line) and 0.5 m $CaCl_2$ (dashed line) and (e) 0.01 m $BaCl_2$.

or possibly with Cl^-. In the latter case, the sorbed complex may be a monovalent cation-chloride complex. Recent work on sorption at the mica basal surface indicates that anions such as Cl^- may cosorb with some cations (Xu & Salmeron 1998, Gier & Johns 2000). Evidence from cation exchange measurements suggests that the same cosorption mechanism occurs on smectite clay (Sposito 1991).

Damped oscillations of water density away from the first sorbed layer were observed in the models in agreement with previous results on the muscovite (001) surface in pure water (Cheng et al. 2001). Minimal relaxation was observed in the bulk muscovite near the interface, indicating that sorption is largely in response to the permanent negative charge of the tetrahedral sheets, as expected.

Experiments using the Zn-anion solutions were designed to further assess the possible presence of monovalent cation-chloride complexes at the mica basal surface. The selected anions exhibit a range of complexing behavior with free Zn in solution as well as a range in size. Both characteristics place constraints on possible interpretations of the electron density profiles. Results suggest that a monovalent species, $ZnCl^+$, is likely not the dominant species in the first adsorbed layer in the 0.01 m Zn-chloride solution. However, the chemistry of Zn is distinct from that of the alkali and alkaline earth cations and may result in sorption behavior that cannot be generalized to all divalent cations. Further investigation of sorption of cation-anion complexes at the mica surface is needed to resolve this question.

4 CONCLUSIONS

X-ray reflectivity is unique in its ability to determine the sorptive behavior of both ions and water molecules at mineral surfaces. Thus, the extent of the inner-sphere or outer-sphere nature of sorbates can be assessed directly. The mica basal plane surface poses an extreme end member of surface types in that the charge on this surface is permanently negative and not affected by solution pH (or the protonation state of surface hydroxyls which are nonexistent unlike for framework silicates such as quartz or feldspar). Sorption on this surface indicates that monovalent cations have an inner-sphere sorption mechanism as expected, but that divalent cations, with more negative hydration enthalpies, are sorbed with a more inner-sphere nature than anticipated. In addition, interpretation of electron density profiles derived from the measured data indicate that divalent cations may sorb, at least in part, as monovalent cation-anion complexes, and this sorption mechanism deserves further investigation.

ACKNOWLEDGEMENTS

This work was funded by the Geosciences Research Program, Office of Basic Energy Sciences, US Department of Energy.

REFERENCES

Cheng, L., Fenter, P., Nagy, K.L., Schlegel, M.L. & Sturchio, N.C. 2001. Molecular-scale density oscillations in water adjacent to a mica surface. *Phys. Rev. Lett.* 87: 156103.

Gier, S. & Johns, W.D. 2000. Heavy metal-adsorption on mica and clay minerals studied by X-ray photoelectron spectroscopy. *Appl. Clay Sci.*16: 289–299.

Sposito, G. 1991. Effect of chloride ions on sodium-calcium and sodium-magnesium exchange on montmorillonite. *Soil Sci. Soc. Am. J.* 55: 965–967.

Xu, L. & Salmeron, M. 1998. An XPS and scanning polarization force microscopy study of the exchange and fmobility of surface ions on mica. *Langmuir* 14: 5841–5844.

Water–Rock Interaction, Wanty & Seal II (eds)
© 2004 Taylor & Francis Group, London, ISBN 90 5809 641 6

Competitive arsenate and silicate sorption by thermal water scales

C.E. Tommaseo & M. Kersten
Geoscience Institute, Gutenberg-University Mainz, Germany

ABSTRACT: Thermal waters of Wiesbaden spa contain upto 20 μM arsenic and 1 mM silicic acid. Arsenic accumulates at concentrations upto 2 wt. % through scale deposits on wellhead stones. Microprobe analysis revealed the arsenic to be associated together with silica to a X-ray amorphous hydrous ferric oxide (HFO) phase. As and Si K-edge EXAFS analyses revealed that both elements were bound in form of arsenate and silicate tetrahedral on the same inner sphere bidentate binuclear surface ^2C complexation sites. However, arsenate additionally forms bidentate mononuclear ^1E surface complexes whereas silicate does not. Surface complexation modeling could overestimate competitive adsorption effects if this site heterogeneity is not considered appropriately.

1 INTRODUCTION

Arsenic is an important trace metal affecting environmental and human health when present in hydrogeological systems, not necessarily related to pollution such as at sites of hydrothermal water outcrops. Batch equilibrium experiments have shown that both arsenate and arsenite adsorb to surfaces of a variety of aquifer materials, including Fe and Al oxyhydroxides and clay minerals. However, arsenic surface complexes created with pure adsorbate-adsorbent systems in laboratories cannot be compared readily to natural ones, because of competing effects of, e.g. silicate and other oxyanions ubiquitous in natural waters. Water chemistry experiments suggest that there is strong inhibition of arsenate adsorption on HFO surface sites at dissolved Si concentrations above 0.1 mM, which may hinder groundwater remediation (Su & Puls 2003). However, despite that several studies have been devoted to adsorption of aqueous silica onto HFO, little is known about the molecular structure of the silicate surface complexes as compared to analogue As studies. Adsorption equilibrium experiments, FTIR, and ^{29}Si NMR studies have indicated that silicate likely forms inner sphere adsorbate complexes on goethite surfaces (Glasauer et al. 1999), though their precise structure is still unknown. We will present in this paper new molecular structural data derived from As, Fe and Si K-edge X-ray absorption spectroscopy (XAS) analyses for both arsenate and silicate surface complexes formed on the same natural HFO material.

2 MATERIALS AND METHODS

2.1 *Study site*

An active thermal and mineral water spring system emerges along the southern margin of the Rhenish Massif in Central Europe which controls the general alignment of the springs. The city of Wiesbaden covers part of this relatively large fault system, with 27 thermal water outcrops captured by wells or spring chambers discharging throughout the spa town. Total discharge per year is estimated to be over 1 million m^3, which makes Wiesbaden one of the most famous European health spas known since Roman times. The most prominent thermal water spring is the "Kochbrunnen" well, capturing a spring at 43 m depth with a wellhead temperature of 66°C. Geochemically, the thermal waters are of NaCl-type, with 80 mg L^{-1} H_2SiO_3, $pH = 6.0 \pm 0.1$ (at the in-situ temperature), and total dissolved arsenite concentrations of 0.2 mg L^{-1}. Kochbrunnen wellhead scale sampling was performed by scraping subsamples from the brownish deposit, which were found by XRF to consist on average of 74 wt.% $CaCO_3$, 22 wt.% $Fe(OH)_3$, 3.4 wt.% SiO_2, 0.1 wt.% MnO_2, and 0.5 wt.% As (Schwenzer et al. 2001). Less than 0.01 wt.% organic carbon was found in the samples, which indicates that the scales were not precipitated within microbial mats as reported for other hydrothermal systems.

2.2 *XAS analyses*

XAS spectra at the Fe and As K-edges were collected at HASYLAB (Hamburg), whereas those of the Si

K-edge at ELSA (Bonn). EXAFS spectra were analyzed according to standard procedures and software (WinXas). In brief, upon normalization and background correction of the spectra, radial structure functions (RSFs) were obtained by Fourier-transforming EXAFS spectra with a Kaiser window. Interatomic distances, *R*, and numbers of atoms in the two nearest atomic shells, *N*, around sorbed metals were determined by fitting Fourier back-transformed (FT^{-1}) RSF peaks using theoretical phase shifts and amplitude functions. Functions for the As-O and As-Fe pairs were obtained from scorodite, and those for Fe-O and Fe-Fe from goethite reference minerals. The function for the multiple scattering paths (such as Fe-Si-O-Fe) are obtained from epidote, in which the silicate tetrahedron is double corner sharing two iron octahedra. Nontronite was also used as a reference mineral to evaluate ΔE_0 fit values for the natural sample.

3 RESULTS AND DISCUSSION

3.1 *Chemical characterization*

Microprobe element maps of the natural scale sample showed clear separation of the Ca- and Fe-rich phases, Si and As being detectable only in the latter. This suggests that aragonite plays an at best minor role in As trapping. The fine-grained Fe oxide phase was found mainly in the form of dendritic textures at the abundant intergranular spaces between the aragonite crystals. This X-ray amorphous Fe oxide phase appears to be the only substrate for both As and Si sorption or co-precipitation. Microprobe analyses of this phase alone gave up to 2% As, 0.1% Mn, and 8% Si concentrations.

3.2 *XAS results for Fe and Si*

The Fe K-EXAFS spectrum and 2nd shell RSF for the natural HFO sample is similar to that of 2-line ferrihydrite, the unknown and reference spectra differing only by their amplitude. The FT's were executed on k^1- and k^3-weighted χ-spectra. These different k-weightings present the opportunity to reinforce alternatively the relative contribution of either the low *k* or the high *k* region in the FT. The two separate treatments may reveal weak atomic contributions that otherwise would be overlooked, because the low *k* region is dominated by the contribution of "light" element neighbors (i.e. O and Si), and by multiple scattering (MS) paths of the photoelectron as well, whereas the high *k* region is dominated by "heavy" elements (i.e. Fe). Better agreement between experiment and theory was obtained by assuming 1.1 Fe neighbors at 3.06 Å, 1.3 Fe neighbors at 3.44 Å, and by introducing in the fit the double scattering path of Fe-Si-O-Fe at 3.12 Å (Fig. 1). This contribution of multiple scatterings of Fe-Si-O-Fe may be due to an arrangement of these atoms in an approximately collinear way. In such cases, the outgoing photoelectron is strongly forward-scattered by the intervening atom, resulting in a significant amplitude enhancement. In fact, both the amplitude and the phase are modified by the intervening atom for bond angles ranging from 180° to about 75°. The effect, however, drops off very rapidly for bond angles smaller than ca. 150°. For these systems (like in our case), it is necessary to take into account multiple scatterings involving the intervening atoms.

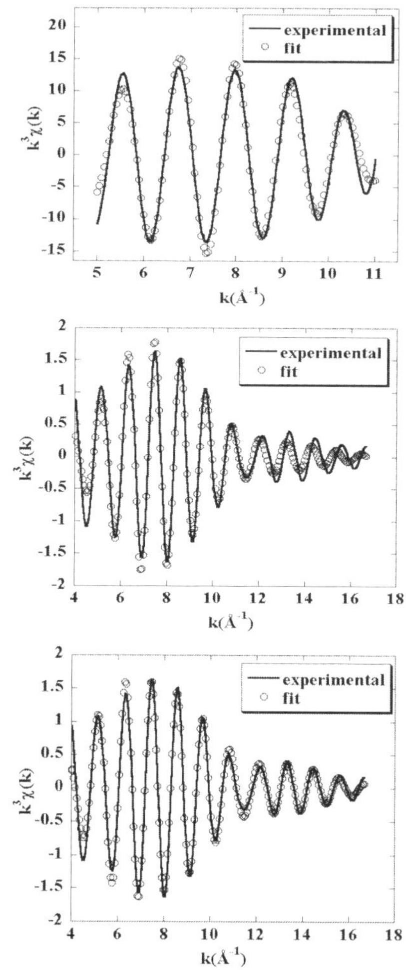

Figure 1. Fourier filtered 2nd shell Fe K-edge EXAFS of the scale sample due to both edge- and corner-sharing Fe-Fe contributions, and its fit without (above) and with (middle) additional contribution of Si neighbors, and the 2nd shell Si K-edge EXAFS fitted by the Si-Fe contribution (below).

Evidence of the additional Si-Fe bond was obtained by measuring Si K-edge EXAFS spectra, where the light absorber element is surrounded favorably by the much heavier second-shell elements. The second peak in the Si K-edge RSF is in fact due to Si-Fe correlations. Curve-fitting analysis in the *k*-range of 5–11 Å^{-1} was performed by back-transforming this second RSF peak into frequency *k*-space (Fig. 1c). The good fit resulted in a Si-Fe distance of approx. 3.13 Å, which is close to those found for dissolved Fe silicate complexes in Si-bearing Fe nitrate and chloride solutions of varying pH values (3.18 Å, Pokrovski et al. 2003), and those for Fe-P found in Fe^{III} phosphate minerals and aqueous solution (~3.20 Å, Rose et al. 1997). Note that in particular the latter could only be confirmed using similarly both Fe and P K-edge EXAFS analysis. The thus derived first and higher shell structural parameters (Table 1) can now be used to constrain by a topological approach the types of linkages between the SiO_4 tetrahedra and the ferrihydrite surface present in the scale sample. This is possible because there are a finite number of ways in which SiO_4 tetrahedra can be linked to the $Fe(O,OH)_6$ octahedron, and each type of linkage results in a characteristic Si-Fe distance. Both the interatomic distances and the coordination number $N \sim 2$ obtained for the Si-Fe shell are consistent with the formation of a bidentate binuclear ^2C surface complex, whereby one silicate tetrahedron is bridging by its corner the apices of the two corner-sharing iron octahedra distant 3.45 Å from each other.

3.3 *XAS results for As*

The position of the half height of its K-edge jump (11869.6 ± 0.3 eV) revealed that As is present in the tetrahedral coordination of arsenate rather than arsenite with oxygen. A one shell least square fit of the peak A (1st shell) of the As K-edge EXAFS for the Kochbrunnen sample (Fig. 2a) yields four next oxygen neighbors at a distance of 1.68 Å, consistent with the fourfold oxygen-coordinated arsenate tetrahedron structure. The presence of an organized second shell is indicated by existence of a relatively intense second peak B in the RSF, and therefore the existence of a local order extending beyond the first oxygen shell. However, the imaginary part of this peak is shifted to a lower distance relative to those of the reference mineral scorodite. This is critical in that As-Fe pairing distances in the natural sample are smaller than As-Fe distances in scorodite. By using theoretical phase and amplitude functions extracted by the FEFF8 code from the local structure around As in the reference mineral scorodite, a good-quality fit is obtained for the scale sample using 1.0 Fe atom at 2.82 Å and 1.0 Fe atom at 3.25 Å distance, resulting in a $\sigma^2 = 0.003$ (Table 1, Fig. 2b).

The shorter As-Fe distance of 2.82 Å is associated with the edge-sharing bidentate mononuclear inner sphere surface complex ^1E, whereas a longer distance of 3.25 Å corresponds to a bidentate binuclear ^2C complex similar to that found already for silicate (Fig. 3). Whereas it is known from literature that a binuclear arsenate complex is favored in cases where this oxyanion is co-precipitated with ferric ion, for our natural Si-bearing scale sample a better fit was obtained using both As-Fe distances.

Table 1. EXAFS results for some of the As-Si-Fe-O shells.

Shell	R (Å)	N	σ^2	ΔE_0
As-O	1.68	3.4	0.004	8.6
As-Fe	2.82	0.9	0.003*	7.0*
As-Fe	3.25	1.1	0.003*	7.0*
Si-O	1.63	3.8	0.004	2.3
Si-Fe	3.13	1.9	0.003	4.0

* Correlated during fit.

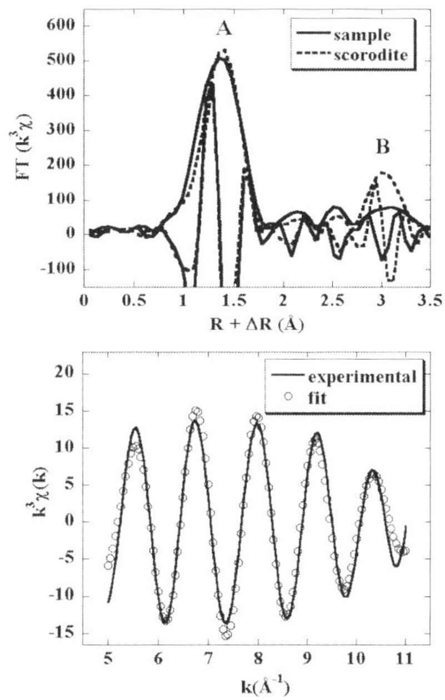

Figure 2. (a) As K-edge EXAFS RSF's of the natural scale sample compared with that of scorodite weighted with k^3 (above, not corrected for phase shift). The envelope curve is the magnitude of the Fourier transforms, and the dotted curves are the imaginary parts of the transforms. Peaks A and B correspond, respectively, to the contribution of O and Fe shells. (b) Best two-shell spectral fit of the 2nd shell due to the two As-Fe contributions.

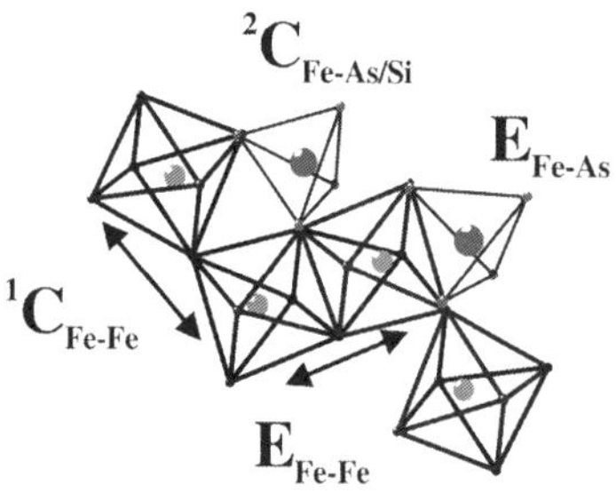

Figure 3. Structural model of the silicate and arsenate tetrahedra linkages to $Fe(O/OH)_6$ octahedra at the ferrihydrite surface. Both tetrahedra are forming a bidentate binuclear ^2C complex corner-bridging two iron octahedra at 3.46 Å distance from each other. Arsenate tetrahedra only are additionally forming a bidentate mononuclear ^1E edge-sharing complex.

4 CONCLUSIONS

The EXAFS data evidence the ability of silicate to compete with arsenate for the same ^2C surface sites. However, the predominance of silicate in the thermal water system leads to a significant shift of the arsenate binding from the common bidentate binuclear ^2C to the bidentate mononuclear ^1E surface sites. The ratio of available ^1E to ^2C surface sites rises to a maximum for colloidal HFO, because free edges are located only at the termination positions of the chains of edge-sharing (^1E) $Fe(O,OH)_6$ octahedra cross-linked by double corners (Fig. 3). The extreme shortness of these dioctahedral chains in HFO colloids results in a tremendous increase in the available ^1E chain termination sites. The site heterogeneity effect might therefore be more relevant for amorphous HFO exposed to both dissolved Si and As during oxide precipitation, than for crystalline FeOOH phase formed before addition of Si and As, because of the relative scarcity of free edge sites for ^1E complex formation on well-developed needle-shaped goethite crystals. In the latter, ^1E sites are limited to the small chain termination (001) planes, and to crystal defects or cavities on the dioctahedral (hk0) planes. Both the existence and the relative prevalence of edge over corner surface linkages are crucial for understanding adsorption behavior of As in the presence of competing oxyanions.

The site heterogeneity effect creates the need to involve more than one binding site in arsenate adsorption modeling on colloidal oxide surfaces in natural aquatic environments. Oxyanions competing for the ^2C surface sites such as silicate therefore stabilize the alternative bidentate mononuclear complex of arsenate, a mechanism by which silicate can be outcompeted. It is not yet clear which of the binding sites is stronger, but such a shift between the binding sites clearly cannot be considered by a competitive adsorption model based on single oxyanion system model data. A poor agreement between calculated and observed arsenate removal in the presence of silicate is then to be expected, because the adverse effect of silicate would be overpredicted (Swedlund & Webster 1999, Meng et al. 2000). A better agreement can be expected with surface complexation models constructed to account for structurally defined surface functional groups.

ACKNOWLEDGEMENTS

This work was supported by German Science Foundation grant Ke 508/6, together with access time at XAS beamlines of both ELSA and HASYLAB. We gratefully acknowledge K. Attenkofer, R. Chauvistré, B. Frenzel, H. Fischer, H. Modrow, L. Tröger, and E. Welter, who helped us with the beamlines.

REFERENCES

Glasauer, S.M., Friedl, J. & Schwertmann, U. 1999. Properties of goethites prepared under acidic and basic conditions in the presence of silicate. *J. Colloid Interf. Sci.*216: 106–115.

Meng, X., Bang, S. & Korfiatis, G.P. 2000. Effects of silicate, sulfate, and carbonate on arsenic removal by ferric chloride. *Wat. Res.*34: 1255–1261.

Pokrovski, G.S., Schott, J., Farges, F. & Hazemann, J.-L. 2003. Iron (III)-silica interactions in aqueous solution: Insights from X-ray absorption fine structure spectroscopy. *Geochim. Cosmochim. Acta* 67: 3559–3573.

Rose, J., Flank, A.M., Masion, A., Bottero, J.Y. & Elmerich, P. 1997. Nucleation and growth mechanisms of Fe oxyhydroxide in the presence of PO_4 ions. 2. P K-edge EXAFS study. *Langmuir* 13: 1827–1834.

Schwenzer, S. P., Tommaseo, C.E., Kersten, M. & Kirnbauer, T. 2001. Speciation and oxidation kinetics of As in thermal springs of Wiesbaden spa, Germany. *Fresenius J. Anal. Chem.*371: 927–933.

Su, C. & Puls, R.W. 2003. In situ remediation of arsenic in simulated groundwater using zerovalent iron: laboratory column tests on combined effects of phosphate and silicate. *Environ. Sci. Technol.*37: 2582–2587.

Swedlund, P.J. & Webster, J.G. 1999. Adsorption and polymerization of silicic acid on ferrihydrite and its effect on arsenic adsorption. *Wat. Res.*16: 3413–3422.

Water–Rock Interaction, Wanty & Seal II (eds)
© 2004 Taylor & Francis Group, London, ISBN 90 5809 641 6

Temperature-effects and structure at the rutile-water interface

D.J. Wesolowski, L.M. Anovitz, P. Bénézeth, A.A. Chialvo & D.A. Palmer
Oak Ridge National Laboratory, Oak Ridge, TN, USA

P. Fenter, L. Cheng & N.C. Sturchio
Argonne National Laboratory/UIC, Argonne, IL, USA

Z. Zhang & M.J. Bedzyk
Northwestern University, Evanston, IL, USA

J.D. Kubicki, M.V. Fedkin, S.N. Lvov & D. Sykes
Penn State University, State College, PA, USA

P.T. Cummings
Vanderbilt University, Nashville, TN, USA

M.K. Ridley
Texas Tech University, Lubbock, TX, USA

M.L. Machesky
Illinois State Water Survey, Champaign, IL, USA

M. Předota
University of South Bohemia, České Budějovice, Czech Republic

A.V. Bandura
St. Petersburg State University, St. Petersburg, Russia

ABSTRACT: A multidisciplinary team is investigating the electrical double layer at the rutile-water interface, employing: pH titrations and microelectrophoresis measurements on powder suspensions from room temperature to 200–300°C; X-ray standing wave and crystal truncation rod studies of submerged (110) crystal surfaces; and MD simulations of *ab initio*-optimized (110) surfaces in contact with bulk water and ions at liquid density. Cation sorption affinities on rutile powders with (110) dominant, increase much more rapidly with increasing temperature than the corresponding decrease in the pH of zero surface charge. We have found very weak interaction of monovalent anions (Br^-, Cl^-, Tr^-) with the (110) surface, but site-specific, inner-sphere binding of all cations studied (Na^+, Rb^+, Ca^{2+}, Sr^{2+}, Co^{2+}, Ni^{2+}, Zn^{2+}, Y^{3+}, Nd^{3+}).

1 INTRODUCTION

The electrical charge developed on oxide and silicate particles in aqueous solutions, and the consequent stabilities and transport properties of colloids, have been extensively studied by pH titrations and electrokinetic measurements of particle suspensions in aqueous electrolyte solutions. However, very few studies have been conducted at elevated temperatures, and almost none above 95°C (Kosmulski 2001).

In recent years, our research team has reported the pH-dependent surface charge and pH_{pzc} (point of zero charge) of rutile (α-TiO_2) and magnetite (Fe_3O_4) in "indifferent" 1:1 electrolytes by potentiometric titration (Machesky et al. 1998, Ridley et al. 2002, Wesolowski et al. 2000) and the zeta potential and pH_{iep} (isoelectric pH) of rutile and baddaleyite (ZrO_2) by microelectrophoresis (Fedkin et al. 2003, Zhou et al. 2003) from room temperature to 200–300°C. We found similar pH_{pzc} and pH_{iep} values for the common mineral rutile, and our combined results indicate that for these oxides, with pH_{pzc}'s of 5.3–7.0 at 25°C, the pH_{pzc} and pH_{iep} decrease with increasing temperature at the same rate as that of the neutral pH ($1/2pK_w$). Kulik (2000) has argued that this relationship will not hold for minerals with much higher or lower pH_{pzc}'s, but this remains to be experimentally proven. Machesky et al. (2001) presented a method for extrapolating to 300°C the structure-specific protonation constants of surface oxygens derived at room temperature from the revised MUSIC Model (Hiemstra et al. 1996), and demonstrated that this approach reproduces the pH_{pzc} values for rutile and magnetite to 250 and 290°C, within experimental error.

It has long been recognized that Gouy-Chapman-Stern type site complexation models (SCM's) are readily able to rationalize titration and electrokinetic data on surface charging and ion adsorption, but that the best-fit model parameters (proton binding constants and stoichiometries, Stern layer capacitances, specific ion adsorption constants, Stern layer dielectric constants, site densities, etc.) are highly covariant, and thus ambiguously-defined. In an attempt to develop a quantitative molecular-level understanding of the structure and reactivity of the electrical double layer (EDL) at the mineral-water interface, we have initiated a series of coordinated studies, in which pH-titration and electrokinetic data from powder suspensions are coupled with synchrotron X-ray standing wave (XSW) and crystal truncation rod (CTR) studies (Fenter 2002, Bedzyk & Cheng 2002) of single crystal surfaces in contact with electrolyte solutions, as well as *ab initio* quantum calculations and molecular dynamics (MD) simulations of the mineral-solution interface, since none of these techniques alone is currently sufficient to treat the inherent complexity of the EDL.

2 RESULTS

2.1 *pH titrations of ion adsorption*

The sorption of ions on mineral surfaces is of great interest in modeling water–rock interactions, but there have been virtually no quantitative experimental studies of the pH dependence of ion adsorption at temperatures above about 75°C, with the exception of the work of Pivovarov (2001) on Cd^{2+} adsorption on hematite to 100°C, and our research group's studies of Ca^{2+} adsorption on rutile (Ridley et al. 1999, 2004) to 250°C. In Figure 1, we present these and additional unpublished data on the sorption of multivalent cations on rutile powders in 0.03–0.3 molal solutions of NaCl, RbCl and sodium trifluoromethanesulfonate (NaTr), the latter salt chosen

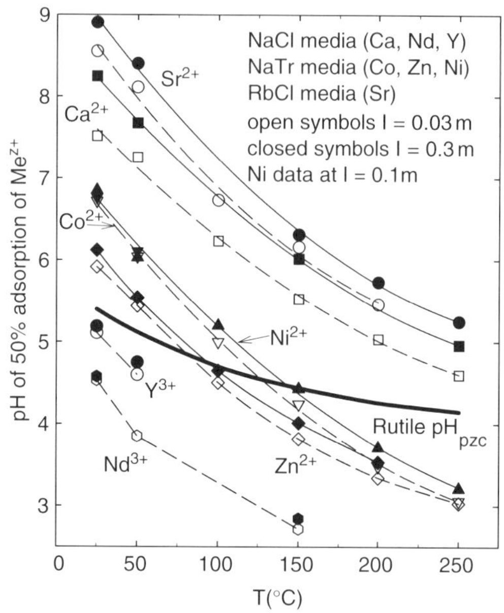

Figure 1. pH of 50% adsorption of Me^{z+} on rutile powder surfaces.

to avoid extensive chloride ion-pair formation with the transition metals.

Our new results are presented in the form of the pH at which 50 mole % of a trace level of the multivalent ion in the starting solution is adsorbed on the surface. All of the experiments were conducted with rutile powders similar to those described by Machesky et al. (1998) and Ridley et al. (2002), typically with approximately 1 gram of the powder (15–17 m^2/g surface area) in contact with about 50 grams of the solution containing 0.001 molal Me^{z+} at the specified stoichiometric molal ionic strength (calculated neglecting ion pair formation involving background electrolyte ions). Separate experiments were conducted to determine (a) the proton release associated with cation adsorption, and (b) the concentration of dissolved cation as a function of pH. The $pH_{50\%}$ values approximate equilibrium partitioning constants, though they ignore electrostatic effects, and they vary somewhat with solid/solution ratio and initial Me^{z+} concentration. However, this observation provides a model-independent measure of the relative Me^{z+} sorption affinities.

The $pH_{50\%}$ values exhibit similar slopes and only modest ionic strength dependencies, which are larger for the more weakly-bound cations (Sr^{2+}, Ca^{2+}). Furthermore, the $pH_{50\%}$ values decrease much more rapidly with increasing temperature than the pH_{pzc} of the solid in the absence of these strong absorbers, shown as the heavy curve in the figure. This means that the surface affinities of the ions increase strongly with increasing temperature for a given pH relative to the pH_{pzc}, and we expect this to be a common feature of cation interaction with oxide surfaces, similar to the temperature trend of aqueous charge-neutralization reactions. In fact, the $pH_{50\%}$ curves in Figure 1 closely parallel the first hydrolysis constants of Me^{z+}, though the $Me(OH)^{z-1}$ species in bulk solution become significant only at pH's several log units more basic than the $pH_{50\%}$ values.

2.2 *Synchrotron X-ray studies*

We have obtained XSW and CTR data at 25°C for NaCl, NaBr, RbCl, $SrCl_2$, $ZnTr_2$ and YCl_3 solutions in contact with rutile (110) single crystal surfaces. These methods rely on strong X-ray scattering and/or fluorescence of the heavier species (Rb^+, Sr^{2+}, Zn^{2+}, Y^{3+}, Br^-) to determine whether or not these ions, as well as water molecules in the interfacial region, occupy ordered positions relative to one another and/ or the underlying Ti-O atomic array. Analysis of our current data (Zhang et al. 2004) indicates that Br^- does not significantly interact, even with positively-charged rutile surfaces, but that all cations sorb on neutral or negatively-charged surfaces as inner sphere species at specific sites relative to the atomic structure of rutile (110), as shown in Figure 2. The larger Rb^+, Sr^{2+} and Y^{3+} ions all sorb at a "tetradentate" site involving two "bridging" surface oxygens, each bound to two underlying surface Ti atoms, and two "terminal" oxygens, each bound to a single underlying surface Ti. However, Zn^{2+}, with a much smaller ionic radius, preferentially forms "monodentate" bonds approximately above bridging oxygens, with a subordinate "bidentate" species lying above and between two terminal oxygens.

Sorbed Zn^{2+} approximately occupies Ti lattice-equivalent sites, as was also suggested by Towle et al. (1999) for Co^{2+} binding on rutile (110) from grazing-incidence XAFS studies. CTR measurements in pure water and neutral to high pH RbCl solutions also indicate that there is little or no ordering of water molecules beyond the first layer above the surface oxygens, and that sorbing cations displace sorbed water molecules. Water molecules in the first layer also occupy specific sites relative to the rutile lattice. XSW studies of Sr^{2+} and Zn^{2+} in solutions with ionic strengths ranging over 4 orders of magnitude revealed little or no change in the sorption positions or layer coverages of these ions.

2.3 *Ab Initio, MD and MUSIC/SCM modeling*

We have developed an *ab initio*-optimized MD simulation model (Bandura & Kubicki 2003, Bandura et al. 2004, Predota et al. 2004a, b) of the atomically-detailed rutile (110) surface in contact with SPC/E water, a rigid, non-polarizable, non-dissociating water model which accurately reproduces the phase envelope, H-bonding structure and dielectric properties of real water and electrolyte solutions over wide ranges of temperature, pressure and composition. The details of this model are presented elsewhere in this volume by Kubicki et al.

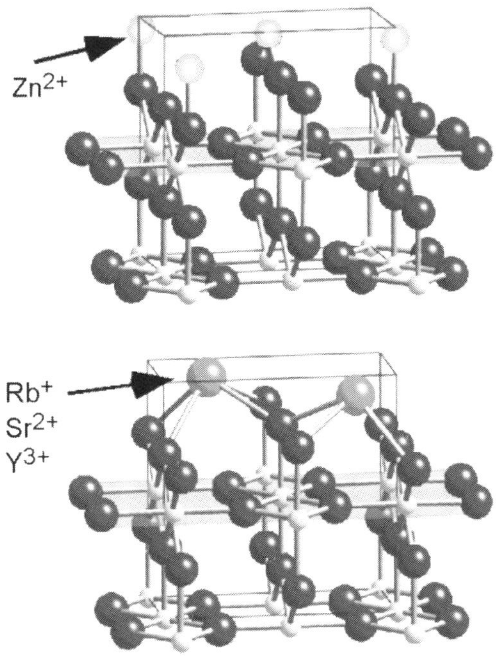

Figure 2. Summary of cation inner sphere sorption sites identified from XSW and CTR studies of rutile (110) in water at room temperature. Light grey balls are Ti atoms, black balls oxygen, and grey balls sorbed cations. The Ti-O surface plane is shaded, with bridging and terminal surface oxygens protruding above this plane.

The distribution of Rb^+ and Sr^{2+} ions at the interface between SPC/E water and the model surface proved to be virtually identical to that determined by XSW and CTR (Fig. 2). For Zn^{2+}, the MD simulations give approximately the same sorption positions, while the heights above the Ti-O plane are somewhat different than indicated from XSW. This minor discrepancy is currently being investigated. Like the X-ray results, the MD simulations also indicate that the solution rapidly approaches bulk water structure and mobility beyond the first layer of sorbed ions and water molecules. Na^+, Ca^{2+} and Cl^- do not scatter X-rays sufficiently to allow easy imaging by XSW and CTR techniques. However, MD simulations indicate that $Na^+$and Ca^{2+} also occupy the tetradentate sorption site, while no significant concentration or ordering of Cl^- was identified, indicating no "coadsorption" of Cl^- accompanying multivalent cation adsorption.

The tetradentate site geometry was incorporated into a modified Basic-Stern SCM applied to our Ca^{2+} (Ridley et al. 2004) and Sr^{2+} sorption studies on rutile powders, and it is demonstrated that a number of other sorption stoichiometries (including mono- and bidentate binding) provide equally good fits of the titration data. Only the X-ray and *ab initio*/MD input enable unambiguous determination of the actual atomic structure and the reaction stoichiometry. MD simulations at elevated temperatures indicate no significant changes in ion sorption height or latter position, while the SCM indicates a regular increase in the Stern layer capacitance with increasing temperature. We are now attempting to rationalize this in terms of the Stern layer dielectric constant, using water structure input from the simulations. Interestingly, incorporation of *ab initio* bond lengths of surface oxygens into the revised MUSIC model results in a calculated pH_{pzc} for the (110) surface of 4.9 at 25°C, which is in perfect agreement with new measurements of second harmonic generation (SHG) at (110) single crystal surfaces in $NaNO_3$ solutions as a function of pH (Fitts et al. 2004). However, the simulations do not indicate a "shear plane" far from the crystal surface, as is suggested from our microelectrophoresis results (Fedkin et al. 2003).

3 CONCLUSIONS

Monovalent anions have a weak affinity for rutile surfaces, but all cations studied, including the alkali metals purported to constitute "indifferent" 1:1 electrolyte species, form site-specific inner-sphere complexes with rutile (110), dictated by the structure of the surface Ti and O atoms and the cation radius. This supports the recent hypothesis of Sverjensky (2001) that oxides like rutile with high bulk dielectric constants should interact with aqueous cations according to their bare ionic radii, rather than their hydrated radii, as would be expected for low-dielectric solids like quartz. Increasing temperature greatly enhances the sorption affinities of cations for rutile surfaces, making interfacial reactions even more significant under hydrothermal conditions. It is shown that Gouy-Chapman-Stern SCM's are compatible with X-ray and MD results, but not sufficient to identify the atomic structure of the EDL without such input. Models of surface protonation are also improved by atomistic computational and experimental input, and it is clear that generic surface site models will yield to structurally detailed models as more such information becomes available.

ACKNOWLEDGEMENTS

This research was sponsored by the U.S. Department of Energy, Office of Basic Energy Sciences, in the project "Nanoscale Complexity at the Oxide-Water Interface" (ERKCC41).

REFERENCES

Bandura, A.V. & Kubicki, J.D. 2003. Derivation of force field parameters for TiO_2-H_2O systems from ab initio calculations. *J. Phys. Chem. B* 107: 11072–11081.

Bandura, A.V., Sykes, D. & Kubicki, J.D. 2004. Adsorption of water on the TiO_2 (rutile) [110] surface: A comparison of DFT and embedded cluster calculations. *J. Phys. Chem. B* (in press).

Bedzyk, M.J. & Cheng, L. 2002. X-ray standing wave studies of mineral surfaces: Principles and Applications. In P.A. Fenter, M.L. Rivers, N.C. Sturchio and S.R. Sutton (eds), *Applications of Synchrotron Radiation in Low-Temperature Geochemistry and Environmental Science*: *Rev. Mineral. & Geochem.* 49: 221–266.

Fedkin, M.V., Zhou, X.Y., Kubicki, J.D., Bandura, A.V., Lvov, S.N., Machesky, M.L. & Wesolowski, D.J. 2003. High temperature microelectrophoresis studies of the rutile/aqueous solution interface. *Langmuir* 19: 3797–3804.

Fenter, P.A. 2002. X-ray reflectivity as a probe of mineral-fluid interfaces. In P.A. Fenter, M.L. Rivers, N.C. Sturchio and S.R. Sutton (eds), *Applications of Synchrotron Radiation in Low-Temperature Geochemistry and Environmental Science. Rev. Mineral. Geochem.* 49: 149–220.

Fitts, J.P., Eisenthal, K., Heinz, T., Kubicki, J.D., Machesky, M.L. & Wesolowski, D.J. 2004. Second harmonic generation at the water-rutile (110) interface: (manuscript in preparation).

Hiemstra, T., Venema, P. & Van Riemsdijk, W.H. 1996. Intrinsic proton affinity of reactive surface groups of metal (hydr)oxides: the bond valence principle. *J. Colloid Interf. Sci.* 184: 680–692.

Kosmulski, M. 2001. *Chemical Properties of Material Surfaces*. New York, Marcel Dekker.

Kulik, D.A. 2000. Thermodynamic properties of surface species at the mineral-water interface under hydrothermal conditions: A Gibbs energy minimization single-site $2pK_A$ triple-layer model of rutile in NaCl electrolyte to 250°C, *Geochim. Cosmochim. Acta* 64: 3161–3179.

Machesky, M.L., Wesolowski, D.J., Palmer, D.A. & Hayashi, K.-I. 1998. Potentiometric titrations of rutile suspensions to 250°C. *J. Colloid Interf. Sci.* 200: 298–309.

Machesky, M.L., Wesolowski, D.J., Palmer, D.A. & Ridley, M.K. 2001. On the temperature dependence of intrinsic surface protonation equilibrium constants: an extension of the refined MUSIC Model. *J. Colloid Interf. Sci.* 239: 314–327.

Pivovarov, S. 2001. Adsorption of cadmium onto hematite: Temperature dependence. *J. Colloid Interf. Sci.* 234: 1–8.

Ridley, M.K., Machesky, M.L., Wesolowski, D.J. & Palmer, D.A. 1999. Ca(II) adsorption at the rutile-water

interface: A potentiometric study in NaCl media to 250°C. *Geochim. Cosmochim. Acta* 63: 3087–3096.
Ridley, M.K., Machesky, M.L., Palmer, D.A. & Wesolowski, D.J. 2002. Potentiometric studies of the rutile-water interface: hydrogen-electrode concentration cell versus glass-electrode titrations. *Colloids and Surfaces A* 204: 295–308.
Ridley, M.K., Machesky, M.L., Wesolowski, D.J. & Palmer, D.A. 2004. Modeling the surface complexation of calcium at the rutile-water interface to 250°C. *Geochim. Cosmochim. Acta* (in press).
Predota, M., Bandura, A.V., Cummings, P.T., Kubicki, J.D., Wesolowski, D.J., Chialvo, A.A. & Machesky, M.L. 2004a. Electrical double layer at rutile surfaces. I. Structure of surfaces and interfacial water from molecular dynamics using ab initio potentials. *J. Phys. Chem. B* (in press).
Predota, M., Cummings, P.T., Zhang, Z., Fenter, P. & Wesolowski, D.J. 2004b. Electrical double layer at rutile surfaces. II. Adsorption of ions from molecular dynamics simulations. *J. Phys. Chem. B* (in press).
Sverjensky, D.A. 2001. Interpretation and prediction of triple-layer model capacitances and the structure of the oxide-electrolyte-water interface. *Geochim. Cosmochim. Acta* 65: 3643–3655.
Towle, S.N., Brown, G.E., Jr. & Parks, G.A. 1999. Sorption of Co(II) on metal oxide surfaces I. Identification of specific binding sites of Co(II) on (110) and (001) surfaces of TiO_2 (rutile) by grazing-incidence XAFS spectroscopy. *J. Colloid Interf. Sci.* 217: 299–311.
Wesolowski, D.J., Machesky, M.L., Palmer, D.A. & Anovitz, L.M. 2000. Magnetite surface charge studies to 290°C from in situ pH titrations. *Chem. Geol.* 167: 193–229.
Zhang, Z., Fenter, P., Cheng, L., Sturchio, N.C., Bedzyk, M.J., Predota, M., Bandura, A.V., Kubicki, J.D., Lvov, S.N., Cummings, P.T., Chialvo, A.A., Ridley, M.K., Bénézeth, P., Anovitz, L.M., Palmer, D.A., Machesky, M.L. & Wesolowski, D.J. 2004. Ion adsorption at the rutile-water interface: Linking molecular and macroscopic properties. *Langmuir* (in press).
Zhou, X.Y., Wei, X.J., Fedkin, M.V., Strass, K.H. & Lvov, S.N. 2003. A zetameter for microelectrophoresis studies of the oxide/water interface at temperatures to 200°C. *Rev. Scientific Instr.* 74: 2501–2506.

Water–Rock Interaction, Wanty & Seal II (eds)
© 2004 Taylor & Francis Group, London, ISBN 90 5809 641 6

The surface chemistry of carbonates, a new approach

M. Wolthers
Department of Earth Sciences – Geochemistry, Faculty of Geosciences, Utrecht University, Utrecht, the Netherlands; Environmental Geochemistry Group, LGIT-OSUG, University of Grenoble, Grenoble, France

L. Charlet
Environmental Geochemistry Group, LGIT-OSUG, University of Grenoble, Grenoble, France

P. Van Cappellen
Department of Earth Sciences – Geochemistry, Faculty of Geosciences, Utrecht University, Utrecht, the Netherlands

ABSTRACT: The Charge Distribution-MUltiSite Ion Complexation (CD-MUSIC) model was applied to describe the surface chemistry of calcite. Protonation reactions and proton affinity constants were predicted and the resulting surface model was compared to the existing Surface Complexation (SC) model. The CD-MUSIC model approach results in a surface charge pH-dependence that fits the point of zero proton charge (pH_{ZPC}) values as reported from electrophoretic measurements.

1 INTRODUCTION

Carbonates play an important role in regulating the chemistry of aquatic environments from lakes to oceans, aquifers to hydrothermal systems, and soils to sediments under oxic to anoxic conditions. Through mineral surface processes such as dissolution, precipitation and adsorption, carbonates affect the biogeochemical cycles of Ca, Mg, Fe and C, major components of carbonate minerals, as well as H, P and trace elements such as radionuclides.

In order to understand and describe the reactivity of the carbonate–water interface, Van Cappellen et al. (1993) proposed a Surface Complexation model for divalent metal carbonates. This model is based on the modeling of experimentally determined surface charge curves from acid/base titrations of rhodochrosite ($MnCO_3$) and siderite ($FeCO_3$) suspensions and on data from electrokinetic and flotation studies of calcite ($CaCO_3$) suspensions. The model was validated by comparison to spectroscopic measurements of the surface group concentrations over a large pH range (3 to 12; Chiarello et al. 1993; Pokrovsky et al. 2000). Table 1 summarizes the parameters from the SC model for carbonate mineral surfaces.

However, as Van Cappellen et al. pointed out, the model represents a first-order description of the chemistry of the carbonate–aqueous solution interface, and does not take into account surface site heterogeneities. In the present study, an alternative model has been developed, describing the surface reactivity based on crystallographic data using the CD-MUSIC model approach (e.g. Hiemstra & Van Riemsdijk 1996). In the CD-MUSIC model, the bond valence theory is used to describe the basic charging of mineral surfaces (Hiemstra et al. 1989a,b). Initially developed and refined for simple metal (hydr)oxides (Hiemstra et al. 1989a,b, Hiemstra & Van Riemsdijk, 1996, Hiemstra et al. 1996), it has recently been applied successfully to the more complex phyllosilicates (Tournassat et al., in press). Here, it will be applied to theoretically describe the surface charge pH dependence of carbonate minerals. The CD-MUSIC

Table 1. Surface complexation model for calcite. Log K values are at 25°C, 1 atm, $I = 0$ (Van Cappellen et al. 1993).

Surface reactions	Log K
$\equiv CO_3^- + H^+ \leftrightarrow \equiv CO_3H^0$	+4.9
$\equiv CO_3Ca^+ + H^+ \leftrightarrow \equiv CO_3H^0 + Ca^{2+}$	+2.8
$\equiv CaO^- + H^+ \leftrightarrow \equiv CaOH^0$	+17
$\equiv CaOH^0 + H^+ \leftrightarrow \equiv CaOH_2^+$	+12.2
$\equiv CaOH^0 + CO_2 \leftrightarrow \equiv CaHCO30$	6.0
$\equiv CaOH^0 + CO_2 \leftrightarrow \equiv CaCO_3^- + H^+$	−2.6

model is then compared to the SC model, which shall be used to describe radionuclide sorption processes at the calcite surface.

2 METHOD

The surface chemistry model is based on ideal, unrelaxed surfaces and therefore should be considered as a preliminary description of the actual surface, as the composition of the solution will affect the surface characteristics. Nevertheless, XPS and LEED studies (Stipp & Hochella 1991) have shown the presence of an ordered calcite surface at least 1 nm thick that is very similar to the bulk lattice. Hence, approximating the calcite surfaces based on the distances and bond angles present within the bulk structure is a valid first-order approach. Koretsky et al. (1998) tested five different methods to calculate the number of surface sites on each surface of various minerals and compared the results for each method to available experimental estimates. Estimates based on the number of broken bonds gave the best agreement with site densities determined using the tritium exchange method. In this method, each broken bond on a near-surface atom was considered as a surface re active group. This method will be followed here to calculate the site densities for carbonate surfaces. The calculated site densities and stoichiometries will be compared to experimentally determined values reported in the literature.

To determine the surface structures, the calcite mineral structure was taken from Diamond©. The orientation of cleavage and growth planes are well established experimentally (e.g. Reeder & Grams 1987; Stipp & Hochella 1991; Chiarello et al. 1993; Paquette & Reeder 1995; Pokrovsky et al. 2000; Elzinga & Reeder 2002). Additionally, atomistic simulation methods have been used to calculate and compare the surface structure and energies of various surfaces of calcite (Titiloye et al. 1998). The calcium and carbonate-terminated {1014} surface of calcite is the most stable crystal plane and will dominate the predicted equilibrium morphology. Therefore, the surface site types for calcite will be derived from an ideal cleavage rhombohedron with {1014} faces and carbonate groups at the corners (e.g., Fig. 1). Surface sites present on other important faces, as shown by sector zoning, shall be explored. In the CD-MUSIC model, the site types are constrained by the crystal structure, so they are not free fitting parameters within the model. The formal charges and proton affinities of the sites have been calculated using the CD-MUSIC model (Hiemstra & Van Riemsdijk 1996, Hiemstra et al. 1996). For the surface charge pH-dependence calculations (Fig. 2), a size of 0.04 μm for the rhombohedra was assumed.

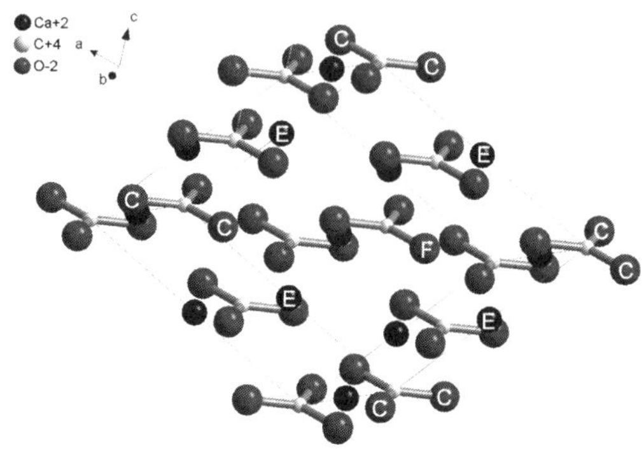

Figure 1. Sketch of the cleavage rhombohedron of calcite, consisting of {1014} faces. Rib lengths are 6.399 Å. For one face, the corner (C), egde (E) and face (F) sites are indicated for the oxygen in CO_3 and the calcium in ≡CaO groups.

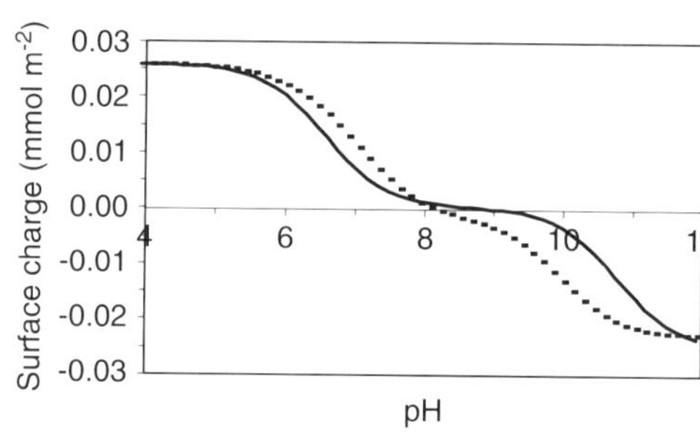

Figure 2. Preliminary surface charge pH-dependence calculated by the CD-MUSIC model (black line), compared to the surface charge pH-dependence calculated by the SC model (dashed line). The calcite surface charge was calculated using either the SC parameters or the CD-MUSIC parameters in the constant capacitance model in MINEQL+. Modeling was performed with the following conditions: specific surface area = $55\,m^2\,g^{-1}$ morphology = ideal cleavage rhombohedra; capacitance $k = 30\,F\,m^{-2}$; $I = 0.032\,M$; $[Ca^{2+}] = [CO_3^{2-}] = 10^{-3}\,M$. For the CD-MUSIC curve, corner sites were used as representing all sites with respect to actual charges, and for calcium and carbonate sorption, the SC reactions were used.

3 RESULTS AND DISCUSSION

The protonation reactions and the proton affinity constants predicted using the CD-MUSIC model are listed in Table 2. The formal charges for $\equiv CO_3$ and ≡CaO vary depending on their position at the surface. This is due to a difference, with position, in coordination of the surface carbon or calcium to the bulk lattice. As can be seen in Table 2, the protonation affinities of the $\equiv CO_3$ and ≡CaO groups do not vary with position, in contrast to what has been observed,

Table 2. Protonation reactions and proton affinities predicted from applying the CD-MUSIC model to calcite.

Type	Surface reactions	Log *K*
Face	$\equiv CO_3^{-1/3} + H^+ \leftrightarrow \equiv CO_3H^{+2/3}$	+4.4
Edge	$\equiv CO_3^{-2/3} + H^+ \leftrightarrow \equiv CO_3H^{+1/3}$	+4.4
Corner	$\equiv CO_3^- + H^+ \leftrightarrow \equiv CO_3H^0$	+4.4
	$\equiv CO_3H^0 + H^+ \leftrightarrow \equiv CO_3H_2^+$	−7.5
Face	$\equiv CaO^{-5/3} + H^+ \leftrightarrow \equiv CaOH^{-2/3}$	+25.0
	$\equiv CaOH^{-2/3} + H^+ \leftrightarrow \equiv CaOH_2^{+1/3}$	+13.1
Edge	$\equiv CaO^{-4/3} + H^+ \leftrightarrow \equiv CaOH^{-1/3}$	+25.0
	$\equiv CaOH^{-1/3} + H^+ \leftrightarrow \equiv CaOH_2^{+2/3}$	+13.1
Corner	$\equiv CaO^- + H^+ \leftrightarrow \equiv CaOH^0$	+25.0
	$\equiv CaOH^0 + H^+ \leftrightarrow \equiv CaOH_2^+$	+13.1

for example, for goethite (Hiemstra & Van Riemsdijk 1996, Hiemstra et al. 1996). This is due to the constant mono-coordination of the oxygen within the $\equiv CO_3$ and $\equiv CaO$ groups to either calcium or carbon, independent of surface site position. Since the coordination of the oxygen does not change with site position, its proton affinity does not change.

Comparison of the proton affinities from the SC model (Table 1) and the predicted proton affinities from the CD-MUSIC model (Table 2) shows that the affinities are close for the first protonation step of the $\equiv CO_3$ group and the second of the $\equiv CaO$ group. For the first protonation step of the $\equiv CaO$ group, the affinities do not agree. However, the large proton affinity of the $\equiv CaO$ group in both models indicates that, within the pH range 2–12, the deprotonated $\equiv CaO$ group is insignificant.

To facilitate comparison of the two models, the surface charge pH dependence of calcite was calculated using both models. The resulting curves are depicted in Figure 2. For calcite, no acid/base titration data are known at present. Electrophoretic measurements indicate a pH_{ZPC} varying from 7 to 10.8 (e.g. Somasundaran & Agar 1967; Mishra 1978; Amankonah & Somasundaran 1985; Thompson & Pownall 1989; Cicerone et al. 1992). The SC model was fit to a pH_{PZC} of 8.2 (Mishra 1978), using the surface site density as calculated from crystallography (Van Cappellen et al. 1993). In contrast, the CD-MUSIC model gives a prediction of the surface charge, based on crystallographic information and the proton affinity calculations, that is in good agreement with the previously reported pH_{ZPC} values. In fact, the surface charge curve from the CD-MUSIC model is flat around pH_{ZPC}, which may be an explanation of the large range of pH_{ZPC} observed in electrophoretic measurements.

It must be noted that the CD-MUSIC model presented here is a preliminary model. Ongoing work on the model encompasses including the sorption reactions of carbonate and calcium, which are also potential-determining ions for the pristine calcite surface. Also, the model will be applied to other carbonate minerals, such as rhodochrosite, for which surface acid/base titration data are available, to test the model approach. Furthermore, data on radionuclide adsorption onto calcite are being reviewed and shall be described using the model. The importance of specific adsorption on various faces shall be discussed.

4 CONCLUSIONS

The predicted proton affinities of the calcite surface by the preliminary CD-MUSIC model approach results in a surface charge pH-dependence that fits the pH_{ZPC} values as reported from electrophoretic measurements. The flat surface charge curve around observed pH_{ZPC} values may be an explanation of the large range of pH_{ZPC} values reported in literature.

The proton affinities of the surface sites do not vary with position at the surface, due to the constant mono-coordination of the surface oxygen atoms to either calcium or carbon. The formal charges of the surface $\equiv CO_3$ and $\equiv CaO$ groups do vary with position due to the change in coordination of these groups to the bulk lattice.

ACKNOWLEDGEMENT

This research was supported by the French national radioactive waste management agency ANDRA.

REFERENCES

Amankonah, J.O. & Somasundaran, P. 1985. Effects of dissolved mineral species on the electrokinetic behavior of calcite and apatite. *Colloids Surfaces* 15: 335–353.

Chiarello, R.P., Wogelius, R.A. & Sturchio, N.C. 1993. *In situ* synchrotron X-ray reflectivity measurements at the calcite-water interface. *Geochim. Cosmochim. Acta* 57(16): 4103–4110.

Cicerone, D.S., Regazzoni, A.E. & Blesa, M.A. 1992. Electrokinetic properties of the calcite/water interface in the presence of magnesium and organic matter. *J. Colloid Interf. Sci.* 154(2): 423–433.

Davis, J.A., Fuller, C.C. & Cook, A.D. 1987. A model for trace metal sorption processes at the calcite surface: adsorption of Cd^{2+} and subsequent solid solution formation. *Geochim. Cosmochim. Acta* 51: 1477–1490.

Elzinga, E.J. & Reeder, R.J. 2002. X-ray absorption spectroscopy study of Cu^{2+} and Zn^{2+} adsorption complexes at the calcite surface: Implications for site-specific metal incorporation preferences during calcite crystal growth. *Geochim. Cosmochim. Acta* 66(22): 3943–3954.

Hiemstra, T. & Van Riemsdijk, W.H. 1996. A surface structural approach to ion adsorption: the charge

distribution (CD) model. *J. Colloid Interf. Sci.* 179: 488–508.
Hiemstra, T., De Wit, J.C.M. & Van Riemsdijk, W.H. 1989a. Multisite proton adsorption modeling at the solid/solution interface of (Hydr)oxides: a new approach. II Application to various important (Hydr)oxides. *J. Colloid Interf. Sci.* 133(1): 105–117.
Hiemstra, T., Van Riemsdijk, W.H. & Bolt, G.H. 1989b. Multisite proton adsorption modeling at the solid/solution interface of (Hydr)oxides: a new approach. I. Model description and evaluation of intrinsic reaction constants. *J. Colloid Interf. Sci.* 133(1): 91–104.
Hiemstra, T., Venema, P. & Van Riemsdijk, W.H. 1996. Intrinsic proton affinity of reactive surface groups of metal (hydr)oxides: the bond valence principle. *J. Colloid Interf. Sci.* 184: 680–692.
Koretsky, C.M., Sverjensky, D.A. & Sahai, N. 1998. A model of surface site types on oxide and silicate minerals based on crystal chemistry: implications for site types and densities, multi-site adsorption, surface infrared spectroscopy, and dissolution kinetics. *Am. J. Sci.* 298: 349–438.
Mishra, S.K. 1978. The electrokinetics of apatite and calcite in inorganic electrolyte environment. *Int. J. Min. Proc.* 5: 69–83.
Paquette, J. & Reeder, R.J. 1995. Relationship between surface structure, growth mechanism, and trace element incorporation in calcite. *Geochim. Cosmochim. Acta* 59: 735–749.
Pokrovsky, O.S., Mielczarski, J.A., Barres, O. & Schott, J. 2000. Surface speciation models of calcite and dolomite/aqueous solution interfaces and their spectroscopic evaluation. *Langmuir* 16(6): 2677–2688.
Reeder, R.J. & Grams, J.C. 1987. Sector zoning in calcite cement crystals: Implications for trace element distributions in carbonates. *Geochim. Cosmochim. Acta* 51(2): 187–194.
Somasundaran, P. & Agar, G.E. 1967. The zero point of charge of calcite. *J. Colloid Interf. Sci.* 24: 433–440.
Stipp, S.L. & Hochella, M.F.J. 1991. Structure and bonding environments at the calcite surface as observed with X-ray photoelectron spectroscopy (XPS) and low energy electron diffraction (LEED). *Geochim. Cosmochim. Acta* 55: 1723–1736.
Thompson, D.W. & Pownall, P.G. 1989. Surface electrical properties of calcite. *J. Colloid Interf. Sci.* 131(1): 74–82.
Titiloye, J.O., de Leeuw, N.H. & Parker, S.C. 1998. Atomistic simulation of the differences between calcite and dolomite surfaces. *Geochim. Cosmochim. Acta* 62(15): 2637–2641.
Tournassat, C., Ferrage, E., Poinsignon, C. & Charlet, L. in press. The titration of clay minerals. Part II. Structural based model and implications on clay reactivity. *J. Colloid Interf. Sci.*
Van Cappellen, P., Charlet, L., Stumm, W. & Wersin, P. 1993. A surface complexation model of the carbonate mineral-aqueous solution interface. *Geochim. Cosmochim. Acta* 57: 3505–3518.

Water–Rock Interaction, Wanty & Seal II (eds)

Modifications of mineral surfaces in hydrothermal solutions and their reaction kinetics

X. Zhang
School of Material Science and Technology, China University of Geosciences; and Chinese Academy of Geological Sciences, Beijing, China

R. Zhang
Laboratory of Geochemical Kinetics, Institute of Mineral Resources, Chinese Academy of Geological Sciences, Beijing, China

ABSTRACT: Detail study of mineral surfaces in hydrothermal reactions provides information on the kinetic process of mineral dissolution. We have already measured reaction rates for a variety of minerals: zeolite, carbonate (rhodochrosite, dolomite), albite, actinolite, and fluorite in various solutions at high temperatures, and conducted associated studies of mineral surfaces by using SEM, XPS, SIMS, etc. Experiments investigated surface reactions of rhodochrosite and zeolite with water and with low pH solutions at room temperature, and albite, dolomite, actinolite and fluorite dissolution at elevated temperatures. The results show that incongruent dissolution of minerals was related to surface chemical modifications.

1 INTRODUCTION

We conducted experiments on zeolite, albite, dolomite, $MnCO_3$ and actinolite dissolution in water and in acidic solutions. We have found that these minerals exhibit incongruent dissolution in most cases. The Na, Al, Si release rates of zeolite and albite, and the Ca, Fe, Mg, and Mn release rates of carbonate in water and various solutions were measured as function of temperature, flow velocity, pH and solution composition in the reaction system. Also, we have performed complementary studies between measurement of dissolution rates and characterization of surface modifications.

2 EXPERIMENTAL APPROACH

The equipment used during the laboratory experiments duplicated that used in previous studies and included a CSTR reactor, liquid pump, back pressure regulator, temperature controller, heat source (furnace), pressure gauge, electrical conductivity detector and computer.

Natural zeolite crystals were obtained from a pegmatite in Xinjiang district. Other minerals were obtained from the Geological Museum, Beijing. Mineral particles were crushed in a shatter box and sieved to 20 to 40 mesh, then the pure mineral crystals were hand picked under a microscope. Mineral particles were crushed again in a shatter box and sieved to 20 to 120 mesh. These particles were cleaned ultrasonically with acetone to remove fine particles, rinsed with distilled water and dried at 70 to 80°C. Surface areas of representative samples were then determined by the single point (Kr-He) BET method.

Mineral (zeolite or rhodochrosite) particles were introduced to a mesh bag, which was put in the pressure vessel. De-ionized and de-gassed water was introduced into the vessel from the bottom to the top at different flow rates (0.5–3 mL/min). Output solutions were sampled and the composition of these solutions was analyzed. The continuous flow reactor incorporates a continuous input of aqueous solutions through the sample bag, and a resulting continuous solution output.

Within the CSTR reactor, a transient material balance gives:

$$dC/dt = \text{Rate } viA/V - q/V(C_i - Ci^0)$$

where C_i refers to the output concentration; Ci^0 is input concentration of input solutions; A is total surface area, V is volume of pressure vessel, q is the flow flux (flow velocity in some cases); r is the reaction rate (bulk reaction rate (mol./min,) or surface reaction rate $(mol/\min/m^2)$. The vi is the stoichiometric coefficient of ith reaction. The output solutions pass through an electrical conductivity detector, and a computer monitor displays the continuous variation of electrical conductivity of the fluids in the reaction system, which indicates if the system is at a steady state or not.

If $dC/dt = 0$, the system is at a steady state. At this time, the product solution of the system is sampled (Zhang et al. 1990, 1992). The experiments for zeolite (or rhodochrosite) were carried out under uniform conditions at the temperature range of 15° to 18°C, a constant pressure 1 atm, and a flow velocity varying from 0.5 to 3.0 mL/min.

Through examination of the residence time, mineral surface area, and the concentration of dissolved material in the input and output solutions, the authors were able to measure the reaction rates. Dissolution experiments for dolomite, albite and actinolite were performed using packed bed reactor of an open system, in a temperature range from 25 to 400°C.

3 EXPERIMENTAL RESULTS

The experimental results are presented in term of the release rate of Na, Al, Si for zeolite dissolution. The release rates were obtained at a constant temperature and pressure; and under a variety of hydrodynamic conditions, i.e. varied flow velocities.

3.1 *Release rate of Na, Al and Si during zeolite dissolution*

The experiments were carried out using a CSTR reactor (continuous stirred tank reactor) in an open system. Zeolite dissolution rates in the flow system of the different aqueous solutions, regardless of different flow rates, were measured repeatedly. Zeolite dissolution rates in water and in acidic solutions (pH = 2.45 and pH = 3.26) have been measured in a CSTR open-flow reactor. Zeolite dissolutions in water and in acidic solutions were incongruent in most cases. Results indicate zeolite dissolution rates vary with pH of the input solutions, and the flow velocities (Fig. 1).

Na, Al and Si release rates in aqueous solutions were measured as functions of flow rate in reaction systems. The dissolution rates increased as the flow velocities were increased from lower to higher. However, under constant hydrodynamic conditions (i.e. constant flow velocity) dissolution rates rose after the pH decreased. We have found that the release rate of Si was not similar to that of Al and Na as pH varied.

Dissolution rates increased as the pH of input solutions decreased from 7 to 2.45 and rose when the

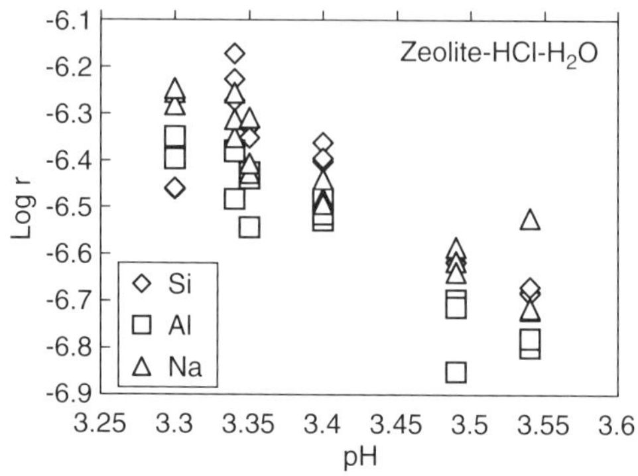

Figure 1. Logarithms of dissolution rates of zeolite vs pH of the output solutions during mineral reacted with low pH solutions.

flow velocity increased. The strong dependence on pH in zeolite hydrolysis could be expressed by

$$-r = kS(a_H^+)_m/(a_i)^n \qquad (1)$$

where $-r$ is dissolution rate (mol/min); k is the rate constant; S is the surface area of the minerals; m and n are the orders of the rates; a_H^+ is hydrogen ion activity in solution; a_i is the activity of aqueous species, such as Si, Al and Na in solution.

At far-from-equilibrium conditions, dissolution rates were found to vary linearly with a_i^n, where a_i values were designated as the activities of aqueous silica, aluminum and sodium, and n is the rate order.

Because of incongruent dissolution, the release rates for Si are different from those for Al or Na in most cases. The n of the release rate of Si is different from those of Al or Na.

Based on the relation between dissolution rate (release rate for Na, Al and Si) and the concentration of the output solution, and the relations between logarithm values of dissolution rates and pH values, the zeolite dissolution rate law can be derived. The simplified rate can be expressed as follows (Aagaard & Helgeson 1982):

$$-r = k\Pi a_I^n[1 - \exp(-A/\sigma RT)] \text{ (far from equilibrium)}$$
$$-r = k\Pi a_I^n \text{ (near to equilibrium)}$$

where, $-r$ is the reaction rate (mol/min, or mol/min/m^2), α_i is activity of aqueous species, $-A$ is the chemical affinity, $-n_{ij}$ is the stoichiometric coefficients of the jth reaction, Πa_I^n expresses the activity product from n aqueous species; σ is the stoichiometric number. In fact, the value of $\exp(-A/\sigma RT)$ could be negative in the far from equilibrium state case.

Comparison of albite and actinolite dissolution, albite dissolution is always incongruent in most cases except for 300°C. The release Al/Si (or Na/Si) ratio

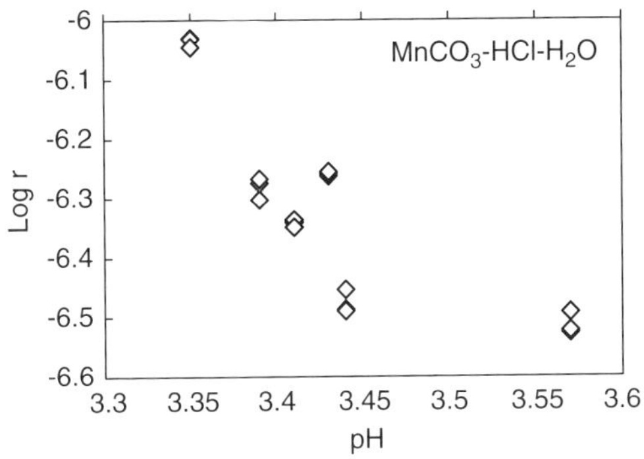

Figure 2. Logarithms of dissolution rates of rhodochrosite vs pH of the output solutions during mineral reaction with low pH solutions.

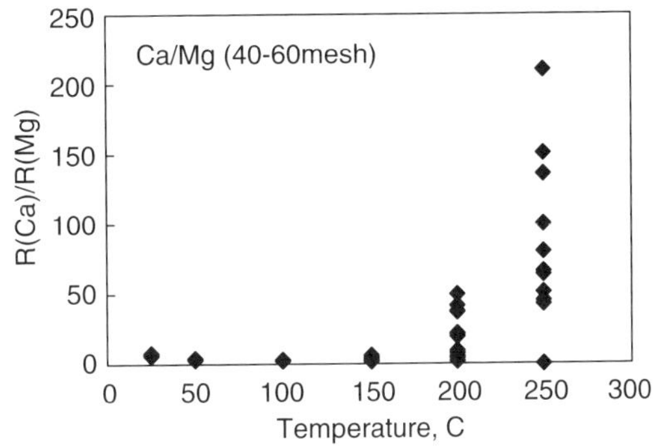

Figure 3. Molal concentration ratio of Ca/Mg in the dolomite – water system vs temperatures.

for albite in water changed from positive to negative as temperature increased at from 25 to 400°C.

For incongruent dissolution of actinolite, with increasing of temperature from 25 to 300°C, the release rate of Si is lower than those of Al, Ca, Mg and Na. As temperature continued increasing (>300°C), the Si release rate became higher than the others.

3.2 *Carbonate dissolution*

For dissolution of carbonates composed of Mn, Ca, Mg, the release rates of Mn, Ca, Mg are also different in most cases. The release rates of Mn from rhodochrosite in aqueous solutions were measured as functions of flow rates in reaction systems, as well as functions of input solution pH. Experimental results indicate that rhodochrosite dissolution rates vary with pH of the input solutions and the flow velocities. Dissolution rates increased as the pH of input solutions decreased from 7 to 2.45 and increased when the flow velocities increased. The strong dependence on pH in rhodochrosite hydrolysis could be also expressed using equation 1.

The rhodochrosite dissolution is described by plotting the output concentration of Mn against average resident time or by plotting logarithm of the dissolution rate against pH at each flow velocity (Fig. 2).

For dolomite dissolution, release rates of Mg are less than those of Ca at high temperatures. Because temperature increased from 25 to 300°C, the release Ca/Mg ratio also increased. Dissolution of dolomite is also incongruent in most cases and varied with temperature, pH and nature of aqueous solutions (Fig. 3).

4 SURFACE MODIFICATION

Incongruent dissolution is usually caused by the chemical modifications of mineral surfaces and the changes in leached surface layer during aqueous solution-mineral interactions. Observations comparing fresh mineral surfaces with reacted surfaces were made using SEM (scanning electronic microscopy), XPS (X-ray photo spectrometry) and SIMS (second ion mass spectroscopy).

The SIMS is a new technique to characterize mineral surfaces after determining reaction rates. The SIMS study indicates that concentrations of Al, Na, Si, Ca, Mn, H, Cl and F in the leached layer of zeolite, albite, rhodochrosite and fluorite surfaces, in 1000 Å thicknesses, changed with distance from surface to inside of the crystal. Distributions of Mn, C, Cl, H in leached layer of rhodochrosite and Ca, F, H, Cl in leached surface layer of fluorite both have a non-linear behavior.

The SIMS analyses of rhodochrosite reacted with HCl solution (pH of 2.47) were collected with Cs ion-sputtering intervals of different times. In fact, these data represent a compositional depth profile.

It has been found that Cl^- and proton (H^+) penetrate into the surface of the mineral after reaction, and Mn is removed from the surface layer. The Mn, H^+, and Cl^- distribution curves in the depth profile are characterized by non-linear features and the curves are waved in the depth profile.

The SEM study of zeolite, rhodochrosite, fluorite and albite showed that a smooth mineral surface appeared as mineral reacted with acidic solutions (pH = 2–4), but a rough surface resulted from reaction with water (Figs. 4 & 5).

On the basis of transition state theory, the authors have calculated values for surface activation energy E_a, activation entropy, $\Delta S^{\pm}$, and activation enthalpy, $\Delta H^{\pm}$. When water reacts with fluorite, the activated complex formed on surface is loosely bound, and the entropy decrease is small. In contrast, when HCl–H_2O reacted with fluorite, the activated complex on the surface is tightly bound and $\Delta S^{\pm}$ is a large negative value.

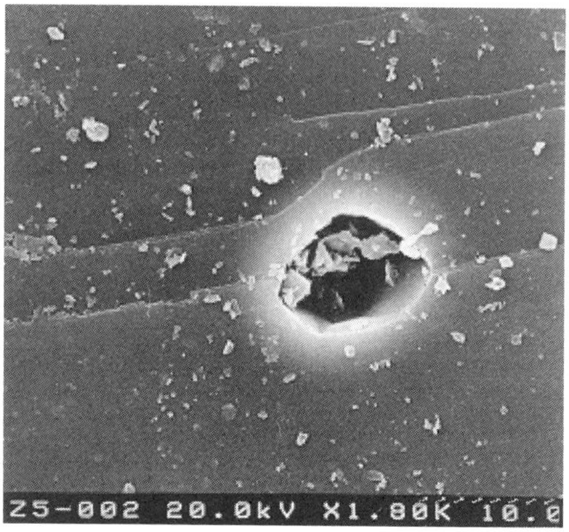

Figure 4. Zeolite surface after reacted with water. A big hole occurred in a roughly surface.

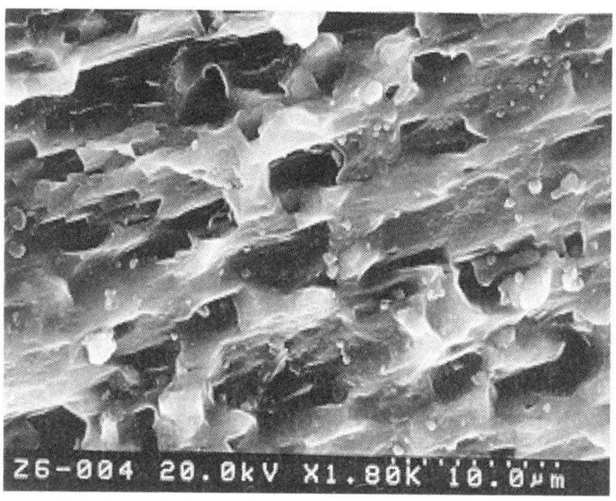

Figure 5. Zeolite surface after reacted with low pH solution. Smooth surface shows a crystalline structure.

Experimental studies suggest the changes in the stoichiometry of mineral dissolution can be determined as a function of time when the mineral is dissolved under constant physical and chemical conditions. For fluorite – water reactions, as the reaction time increased to 465 hours, the surface modification became stronger and stronger (etch pits and line erosion). Simultaneously, as the dissolution rates were measured, we also found that the dissolution rate of fluorite varied and decreased gradually.

5 CONCLUSIONS

Experiments tested the surface of rhodochrosite and zeolite reacted with water and with low pH solutions at room temperature, and albite, dolomite, actinolite and fluorite dissolution was studied at elevated temperatures. The results show that the release rates of Si, Al, Ca, Mg and Na from silicate minerals (zeolite, albite and actinolite) are different in most cases, i.e. dissolutions of albite, actinolite and zeolite are often incongruent. The incongruent dissolution of minerals was related to surface chemical modifications.

ACKNOWLEDGEMENTS

Part of this paper is based on Ms Zhang Xuetong's thesis of Beijing Union University. We would like to thank the Ministry of Science and Technology and the Ministry of Land and Resources for supporting our project: 9502215, and Chinese NSF, Grant no. 29673008.

REFERENCES

Aagaard, P. & Helgeson, H.C. 1982. Thermodynamic and kinetic constraints on reaction rates among minerals and aqueous solutions. I. Theoretical considerations. *Am. J. Sci.* 282: 237–285.

Zhang, R., Posey-Dowty, J., Hellmann, R., Borcsik, M., Crerar, D. & Hu, S. 1990. Kinetics of mineral – water reactions in hydrothermal flow systems at elevated temperatures and pressures. *Science in China (series B)* 33(9): 1136–1152 (English edition).

Zhang, R. & Hu, S. 1992. *Chemical Kinetics of Minerals in Hydrothermal Systems and Mass Transfer*. Beijing: Science Publishing House.

Zhang, R., Hu, S., Tong, J. & Jiang, L. 1998. *Mineral -Fluid Reaction Kinetics in Open Systems*. Beijing: Science Publishing House.

Zhang, R. & Hu, S. 1996. Reaction kinetics of fluorite in flow systems and surface chemistry. *Science in China (Series D)* 39(6): 561–575.

Zhang, X., Zhang, R., Shumin, H.U. & Wenbing, Y.U. 2000. Mineral surface after reacted with aqueous solution at high temperatures and pressures. *Acta Geol. Sinica* 74(2): 406–411.

Weathering studies at all space and time scales

Water–Rock Interaction, Wanty & Seal II (eds)
© 2004 Taylor & Francis Group, London, ISBN 90 5809 641 6

Cosmogenic nuclide methods for measuring long-term rates of physical erosion and chemical weathering

James W. Kirchner & Clifford S. Riebe
Department of Earth and Planetary Science, University of California, Berkeley, CA, USA

Robert C. Finkel
Lawrence Livermore National Laboratory, Livermore, CA, USA

ABSTRACT: Understanding the evolution of geochemical and geomorphic systems requires measurements of long-term rates of physical erosion and chemical weathering. Erosion and weathering rates have traditionally been estimated from measurements of sediment and solute fluxes in streams. However, modern sediment and solute fluxes are may be decoupled from long-term rates of erosion and weathering, due to storage or remobilization of sediment and solutes upstream from the sampling point. Recently, cosmogenic nuclides such as ^{10}Be and ^{26}Al have become important new tools for measuring long-term denudation rates. Cosmogenic nuclides are produced *in situ* within mineral grains by secondary cosmic radiation. Because this cosmic ray flux is attenuated by the mass overlying the sample, the accumulated cosmogenic nuclide concentration in a sample records how rapidly the overlying mass (and thus the cosmic ray shielding) has been removed, either by physical erosion or chemical weathering. Thus cosmogenic nuclides can be used to measure the total denudation flux (the sum of the rates of physical erosion and chemical weathering). We have recently shown how this total denudation flux can be partitioned into its physical and chemical components, using the enrichment of insoluble tracers (such as Zr) in regolith relative to its parent rock. Thus by combining cosmogenic nuclide measurements with the bulk elemental abundance of rock and soil, we can measure rates of chemical weathering averaged over 1000- to 10,000-year time scales.

We have measured long-term chemical weathering rates for 43 sites in granitic terrain, encompassing widely varying climates and denudation rates: mean annual temperatures vary from 2 to 32°C, average annual precipitation spans a 20-fold range (from 22 to 420 cm/yr), and denudation rates vary by 32-fold across our sites. Our data show that chemical weathering rates are tightly coupled with physical erosion rates, such that the relationship between climate and chemical weathering rates may be obscured by site-to-site differences in the rate that minerals are supplied to soil by physical erosion of rock. The relative importance of chemical weathering can be quantified using the "Weathering Intensity Factor" (WIF), the ratio of the chemical weathering rate to the physical erosion rate. WIF's increase systematically with average annual precipitation and mean annual temperature, both for soils as a whole, and for individual elements including Si, Na, and Ca. Over 60% of the variance in WIF's can be explained by a simple Arrhenius-like relationship based on mean annual temperature and average annual precipitation. Coupling this Arrhenius relationship with our measurements of long-term erosion rates, we obtain a simple prediction equation that explains over 80% of the variance in long-term chemical weathering rates. The temperature-dependence of WIF is roughly half of what one would expect from laboratory measurements of activation energies for feldspar weathering and previous inter-comparisons of short-term average weathering rates from the field. We suggest that this result may reflect biologically mediated weathering, in which vegetation and microbes effectively catalyze weathering reactions and thus lower their apparent activation energies. Our results imply that the strength of climate change feedbacks between temperature and silicate weathering rates may be weaker than previously thought, at least in actively eroding, unglaciated terrain similar to our study sites. Our results further indicate that chemical weathering rates may often be limited by the rates that fresh minerals are supplied to soils by erosion, implying that tectonic uplift may be an important regulator of long-term chemical weathering rates in mountainous, granitic landscapes.

Water–Rock Interaction, Wanty & Seal II (eds)
© 2004 Taylor & Francis Group, London, ISBN 90 5809 641 6

Laboratory study of chemical weathering of Middle Ordovician black shales

P.A. Abanda & R.E. Hannigan
Program for Environmental Science, Arkansas State University, State University AR, USA

ABSTRACT: To understand the controls of rare earth element (REE) release during weathering, and to see if there are any differences in the weathering products of black shales of different thermal grades, we carried out a laboratory study using buffered weathering solutions at pH 5 and studying the release of REE over time. Our results show that concentrations of the REE are relatively low in all samples when compared to average shale composition. However we observed a pattern of enrichment of the HREE (Dy-Lu) in all samples, which is consistent with breakdown of carbonate minerals. Samples of higher grade of thermal maturity (post mature), which have lower CIA and CIW values (high CaO), tend to show higher elemental concentrations compared to those of lower grade over same reaction period.

1 INTRODUCTION

1.1 *Rare earth elements*

Several studies have shown that the rare earth elements (REE) undergo significant fractionation during weathering of the parent rock (Nesbitt 1979; Duddy 1980; Harris 1985). Since most studies of provenance of sedimentary rocks rely on the assumption that the composition of these rocks reflect the average composition of their source material, (Taylor & McLennan 1985; McLennan 1989) fractionation during weathering may affect the chemistry of sedimentary material and may lead to erroneous provenance interpretations. There has been some controversy regarding the chemical changes that take place in black shales during early diagenesis. For example, Chaudhuri & Cullers (1979), working on sediment cores from the Gulf coast of the US, suggested that no significant changes in the relative abundance of REEs occurred that could be ascribed to early diagenetic reactions. Hannigan & Basu (1998) suggested that significant fractionation/redistribution of the REE, especially the light REE (LREE; La – Nd) does take place during dissolution and precipitation of diagenetic minerals. Processes such as incorporation of REE into diagenetic minerals or migration within pore fluids have been proposed as processes that fractionate the REE in black shales (Nesbitt 1979; Lev et al. 1999).

1.2 *Geologic setting*

The research presented here is part of a larger study assessing the partitioning of trace elements, including the REE, between different shale components (carbonate, organic, sulfide, phosphate, and silicate fractions). We focus our attention on the trace element geochemistry of the Middle Ordovician Utica Shale magnafacies of Quebec, Ontario and New York (Hannigan & Basu, 1998 and references therein). These units were deposited in the Taconic Foreland Basin and have been constrained in time by graptolite biostratigraphy and bentonite chrono-stratigraphy. The magnafacies thermal grade varies geographically: Quebec (immature: T_{max} 20 to 50°C), Ontario (mature: T_{max} 50 to 140°C), and New York (post mature: $T_{max} > 200$°C; Hannigan & Basu 1998).

The objectives of this study are:

- Do rocks of different thermal grades weather differently?
- Are there differences in the chemistry of the weathering products within and between localities (Quebec, Ontario and New York)?
- How do differences in thermal grade, mineralogy, and major element composition affect the release of trace elements during water rock interaction at constant pH?

2 METHODS

2.1 *Batch weathering experiments*

We prepared buffer solutions pH 5 according to method outlined in the CRC handbook (we used ultrapure HCl in preparing buffer). For each sample, 2 g of <1 mm sized rock chips was weighed and transferred into 100 mL PTFE bottles. 25 mL of buffered weathering solution was added to each sample and each bottle was placed on a Lab-Line® orbital shaker at moderate speed for the length of the experiment (80 hours). 2 mL of weathering fluid was periodically removed from each reaction vessel and stored at 4°C for analysis.

2.2 *Geochemistry*

Metal concentrations were measured by dynamic reaction cell inductively coupled plasma mass spectrometry (ICP-MS; PerkinElmer DRCII) following EPA 200.8 methodology. Standards were prepared in a linear range from 1 ug/L to 100 mg/L. Each sample and standard contained 40 ug/L ^{6}Li, ^{45}Sc, ^{75}Ge, ^{115}In, ^{169}Tm, and ^{209}Bi. 40 ug/L ^{196}Au was added to solutions to stabilize Hg. NIST 1640 was measured as an unknown to monitor external precision. Trace element concentrations were within 3% of the reported values of these standards. Relative standard deviations for sample concentrations are better than 5% for all analytes.

Clay analyses were conducted on the clay size fraction using X-ray diffraction according to the method of Moore and Reynolds (1989). Major element analyses were conducted by X-ray fluorescence spectroscopy method XRF100 with lowest detection limits of 0.1% for all majors and loss on ignition at the SGS analytical laboratories in Canada.

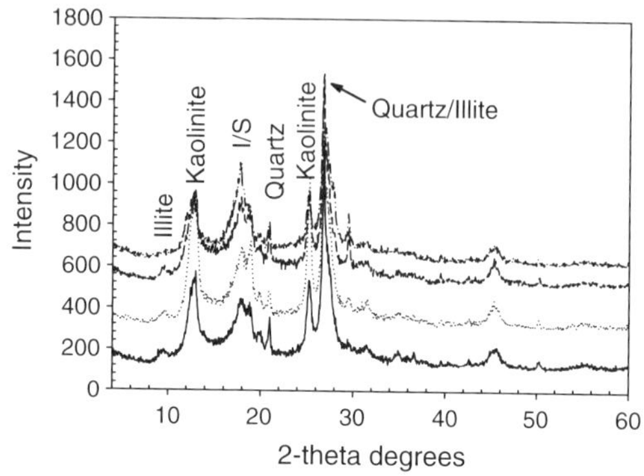

Figure 1. Clay mineralogy of shale samples weathered in this study. Moving from bottom to top samples are Quebec (immature), Ontario (mature) and the top two samples are New York (post-mature). I/S is mixed layer illite and smectite.

Table 1. Average chemical index of alteration (CIA) and chemical index of weathering (CIW) calculated from major element data. Calcium oxide (CaO). All data in percent (%).

Sample	CIA	CIW	CaO
Quebec (immature, n = 10)	55.6	64.4	7.7
Ontario (mature, n = 13)	43.4	52.1	14.4
New York (post mature n = 6)	19.8	22.1	31.8
NASC	57.9	64.3	5.0

3 RESULTS AND DISCUSSION

3.1 *Clay mineralogy*

The bulk of the adsorbed metals will be associated with the clay and organic fractions. In order to interpret the batch weathering data in terms of host fractions we measured the clay mineralogy of shale samples used in our weathering experiments. Clay mineralogical analyses (Fig. 1) show that the major clay minerals in all three groups of samples include illite, illite/smectite mixed layer clay, quartz and kaolinite. The abundance of the different clay minerals varies between thermal maturation levels and geographic position.

3.2 *Chemical index of alteration*

We calculated the average chemical index of alteration (CIA) and chemical index of weathering (CIW) for each shale sample used in our weathering experiments from whole rock major element data (Table 1). For comparison we provide major element data for North American Shale Composite (NASC; Gromet et al. 1984). Chemical index of alteration was calculated as:

$$(Al_2O_3/(Al_2O_3 + Na_2O + K_2O + CaO)) * 100\%$$

while CIW was calculated as:

$$(Al_2O_3/(Al_2O_3 + Na_2O + CaO)) * 100\%$$

with the exclusion of the alkali oxide K_2O. Figure 2 shows the relative variations in CIA and CIW between regions of different thermal grade as compared to NASC.

The CIA and CIW values in the mature and post mature samples are low when compared to NASC values. The results suggest a highly unaltered source material. These results are in contradiction to the clay mineral data. The observed differences in CIA within the three groups of samples may be due to difference in degree of alteration of the source material or may be related to degree of diagenesis. A plausible explanation for the low CIA and CIW values of the mature and post-mature samples may be due to the high alkaline (CaO) content of the samples. The Utica Shale magnafacies of New York and Ontario is comprised of black shale interbedded with limestone (Hannigan & Basu 1998). The differences in facies

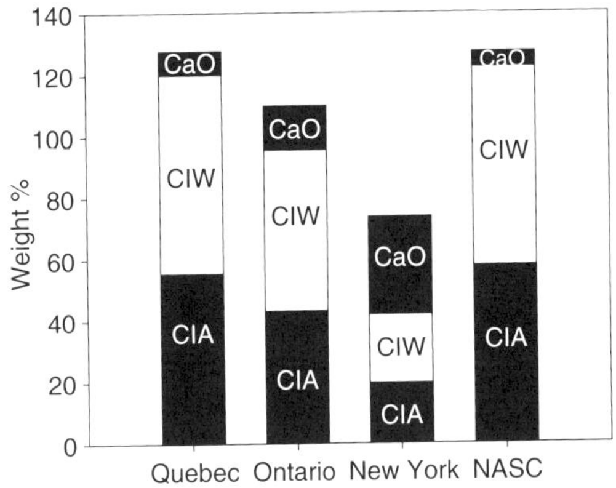

Figure 2. Comparison of CIA, CIW and CaO of three groups of samples with NASC.

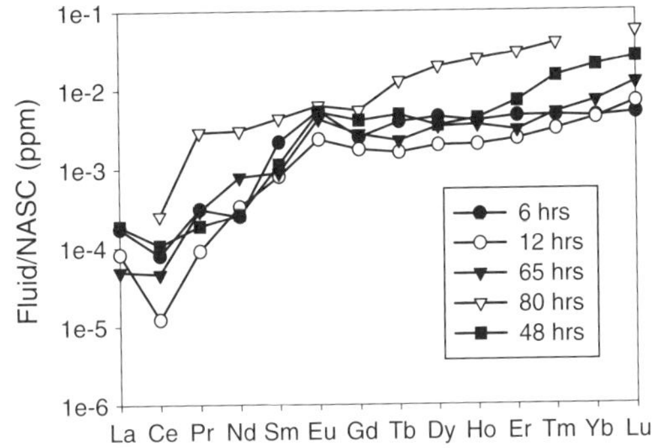

Figure 3. REE abundance patterns of weathering fluids normalized to NASC. The HREE enrichments indicate carbonate weathering and clearly demonstrate incongruent weathering of the shales. These data are from the Quebec experiments but all samples show similar trends.

between the regions likely contributed to the high CaO values. The Quebec facies, on the other hand, is relatively poor in carbonate when compared to New York and Ontario samples.

3.3 *Chemical weathering and REE*

Previous research on the REE chemistry of the Utica Shale magnafacies (Hannigan & Basu 1998; Peucker-Ehrenbrink & Hannigan 2000; Hannigan & Sholkovitz 2001; Abanda & Hannigan 2002) show that the REE are bound in the mineral fraction and are not easily released into solution during batch weathering unless fluids are acidified and mechanical and chemical weathering is simulated. In our experiments the reaction period was less than a week, which may not be enough time for complete breakdown of the host sample. However because our experiments were carried out under near-neutral conditions we are able to assess the mobility of the REE under conditions closer to those experienced in nature. A trend of enrichment of the heavy REE (HREE; Dy-Lu) in fluids derived from weathering of immature, mature and post-mature samples and can be attributed to the weathering of the carbonate fraction and release of the HREE into solution (Fig. 3). At pH 5 of the reaction (pH 5) carbonate is easily weathered. Physical and chemical isolation and analyses of mineral fractions of black shales from the Utica shale magnafacies of Quebec, Ontario and New York, (Abanda & Hannigan 2002) showed that the carbonate fraction of black shales is enriched in HREE relative to NASC, a pattern that is consistent with that of REE chemistry in sea water (Elderfield & Greaves 1982; Pipegras & Jacobsen 1992) The concentrations of the REE in all weathering extracts were exceptionally low and ranged from 54.3–105.5 ppb with the highest concentrations occurring in the fluids from the post mature experiments.

4 CONCLUSIONS

Laboratory studies of chemical weathering at near-neutral pH, can be very useful in understanding natural weathering processes and metal release in black shales. Weathering of black shales of different thermal grade results in REE composition which when normalized to NASC, shows a pattern of enrichment of the HREE relative to the LREE. Post mature black shales of the Utica shales magnafacies of New York tend have a high CaO content and low CIA and CIW values relative to Quebec, Ontario and NASC samples.

REFERENCES

Abanda, P.A. & Hannigan, R.E. 2002. Rare earth elements in black shales: isolation and analysis of mineral fractions. *Abstracts with Program; Geol. Soc. Am. Ann. Meeting, Denver, CO, 27–30 October 2002.*

Chaudhuri, S. & Cullers, R.L. 1979. The distribution of rare-earth elements in deeply buried Gulf Coast sediments. *Chem. Geol.* 24: 327–338.

Duddy, I.R. 1980. Redistribution and fractionation of the rare earth and other elements in weathering profiles. *Chem. Geol.* 30: 363–381.

Elderfield, H. & Greaves, M.J. 1982. The rare earth elements in seawater. *Nature* 296: 214–219.

Gromet, L.P., Dymek, R.F., Haskin, L.A. & Korotev, R.L. 1984. The "North American Shale Composite": its compilation, major and trace element characteristics. *Geochim. Cosmochim. Acta* 37: 419–438.

Hannigan, R.E. & Basu, A. 1998. Late Diagenetic Trace Element Remobilization in Organic-Rich Black

Shales of the Taconic Foreland Basin of Quebec, Ontario and New York. In J. Schieber, W. Zimmerle & P. Sethi (eds.), *Mudstones and Shales: Recent Progress in Shale Research* Berlin: Schweizerbart sche Verlagsbuchhandlung.

Hannigan, R.E. & Sholkovitz, E.R. 2001. The development of middle rare earth element enrichments in freshwaters: weathering of phosphate minerals. *Chem. Geol.* 175: 495–508.

Harris, C. 1985. Guano-derived rare earth-rich phosphatic amygdales in gabbroic inclusions from Ascension Island. *Earth Plan. Sci. Lett.* 72: 141–148.

Lev, S.M., McLennan, S.M. & Hanson, G.N. 1999. Mineralogic controls on REE mobility during black shale diagenesis. *J. Sed. Res. Part A69*: 1071–1082.

McLennan, S.M. 1989. Rare earth elements in sedimentary rocks: Influence of provenance and sedimentary processes. In *Reviews in Mineralogy Volume 21: Mineralogical Society of America*. pp. 169–200.

Moore, D.M. & Reynolds, R.C. Jr. 1989. *X-ray Diffraction and the Identification and Analysis of Clay Minerals*. Oxford: Oxford University Press.

Nesbitt, H.W. 1979. Mobility and fractionation of rare earth elements during weathering of a granodiorite: *Nature* 279: 206–210.

Peucker-Ehrenbrink, B. & Hannigan, R.E. 2000. Effects of black shale weathering on the mobility of rhenium and platinum group elements. *Geol.* 28:475–478

Pipegras, D.J. & Jacobsen, S.B. 1992. The behaviour of rare earth elements in seawater: precise determination of variations in the North Pacific water column. *Geochim. Cosmochim. Acta* 56: 1851–1862.

Taylor, S.R. & McLennan, S.M. 1985. *The Continental Crust: Its Composition and Evolution*. Oxford: Blackwell Scientific.

Water–Rock Interaction, Wanty & Seal II (eds)
 ISBN 90 5809 641 6

Chemical weathering along a deposequence of glacial loess-derived soils in Alaska

S.P. Anderson
Institute of Arctic and Alpine Research, University of Colorado, Boulder, CO, USA

D.H. Mann
Institute of Arctic Biology, University of Alaska, Fairbanks, AK, USA

A.E. Blum
U.S. Geological Survey, Boulder, CO, USA

ABSTRACT: The geochemical influence of glaciers extends beyond ice margins into areas where glacial sediments are deposited. Glacial erosion produces large amounts of silt-sized sediments. Silt and finer material are winnowed from glacial sediments by the wind and deposited as loess across vast areas. On land, loess deposition has the potential to increase chemical weathering fluxes from existing soils. We sampled six soil profiles developed on Donnelly age glacial outwash (ca. 14 ^{14}C ka) along the Delta River in central Alaska. The loess cap varied from ~0.3 m to >2 m thick in these soils. We will present a chemical inventory that incorporates inputs from loess deposition and outputs due to chemical weathering. Our goal is to assess the CO_2 consumption associated with silicate chemical weathering in similarly aged soils affected by varying amounts of loess input.

1 WEATHERING AND GLACIERS

1.1 *The unfulfilled expectation*

High physical erosion rates are thought to enhance chemical weathering fluxes (Gaillardet et al. 1999; Riebe et al. 2001). Based on a survey of sediment evacuation rates, Hallet et al. (1996) argued that erosion under temperate glaciers can exceed by one or more orders of magnitude even the steep mountainous watersheds (unglaciated) identified by Milliman & Syvitski (1992) as the most important contributors of sediments to the oceans. Yet total solute fluxes from glaciers are not particularly high, and products of silicate mineral weathering are markedly low in glacial streams (Anderson et al. 1997; Tranter et al. 2002). In other words, physical weathering is rapid, but chemical weathering is subdued underneath active glaciers.

Glaciers produce sediment with grain sizes ranging from clay to boulders. A prominent grain size mode in basal till (sediment deposited directly beneath a glacier) is found in silt size classes; sediment in the silt and clay size ranges commonly comprises at least 20–30% of the sediment. Because this fine sediment is produced by crushing and abrasion (Haldorsen 1981), its mineralogy closely mirrors that of the bedrock (Nesbitt & Young 1996). Thus, glacier erosion produces more reactive mineral surface area per unit of erosion than other bedrock erosion processes (Anderson, in review).

1.2 *Loess and weathering*

A portion of the silt produced by glacier erosion is entrained from outwash plains by wind and distributed in widespread loess deposits. Most loess is deposited where vegetation provides roughness elements that reduce wind speeds (Pye 1987). We hypothesize that chemical weathering fluxes, particularly silicate chemical weathering fluxes, are enhanced in soils on which loess is deposited. Since vegetation (Moulton & Berner 1998) and high mineral surface area both promote chemical weathering, addition of fresh rock flour – loess – to existing forest soils should enhance weathering fluxes. A 5 cm addition of loess can double the mineral surface area of a 1 m thick sandy soil. Therefore the potential impact of loess additions on weathering fluxes is tremendous.

Recognizable loess deposits cover 10–15% of the global land surface (Pye 1987), and trace amounts of loess are deposited worldwide. Loess is the "most ubiquitous" Quaternary deposit in Alaska (Péwé 1975). We speculate that during periods of Earth history when loess deposition was widespread, the enhanced silicate weathering caused by fresh loess additions to global soils was a significant sink for atmospheric CO_2.

2 FIELD SITE

To test our hypothesis, we sampled six soil profiles along a loess "deposequence", a series of soils whose other state factors (e.g. age, topography, vegetation, parent material) are held constant but have experienced differing amounts of loess input. The study soils are subjected to complete elemental mass balance analysis (Brimhall & Dietrich 1987). The soils are all developed on a contemporaneous geomorphic surface, outwash of Donnelly age (~12–14 ^{14}C ka; Péwé & Reger 1983) along the Delta River in Alaska. Because loess deposition rate varies spatially, the capping loess varies in thickness from profile to profile. Péwé & Reger (1983) report loess deposition rates in this area ranging from 0.2 to 2 mm/yr over the last 7000 years, although our measurements of loess thicknesses suggest that deposition rates are commonly less than 0.2 mm/yr. In addition to soils, we sampled modern silt in transport in the Delta River and Tanana River by collecting recent high water deposits.

The Delta River flows north from the Alaska Range. Its headwaters include a number of large glaciers, including the Black Rapids Glacier, and is underlain by schist and granodiorite (Nokleberg et al. 1992). Donnelly age outwash covers an area extending northward ~25 km from the Donnelly moraine to the Tanana River. Mean annual temperature is −2.2°C, and mean annual precipitation is 300 mm (at Big Delta FAA station, at the northern portion of the field area; data from www.cdc.noaa.gov).

3 SOILS

Donnelly outwash is sandy cobble gravel, with a mean grain size of 3–10 mm, while the loess cap has a mean grain size of about 30 mm (Péwé et al. 1969). The thickest post-Donnelly silt accumulations occur along the Delta River bluffs, particularly upstream of the Donnelly end moraine. We sampled one profile in a relatively high loess accumulation area. The layered silt was 2.3 m thick above outwash gravels. This profile contained abundant charcoal, which we sampled for ^{14}C dating. The Jarvis Creek ash, a locally extensive ash deposited 3.6 ka B.P., was found 0.9 m below the surface. Weathering within this profile appeared minimal, although several buried soils could be identified on the basis of buried organic horizons.

The soil with the thinnest loess cap in our transect was found near the terminus of the Donnelly moraine. The loess cap on this profile was 0.25–0.35 m thick. Brightly oxidized bands were found at the base of the loess cap, above what is probably a groundwater E (leached) horizon.

In each soil, our profile extends at least 0.5 m into the underlying outwash material so that we can compare weathering fluxes from uniform columns of parent material. We saw little difference in the visual appearance of weathering of the underlying outwash in these profiles.

4 FUTURE WORK

The samples we collected will be analyzed for elemental mass balance analysis, using the formalism of Brimhall & Dietrich (1987). We will modify the mass balance to account for loess additions to the profiles, which will be quantified with ^{14}C dates on charcoal collected from the profiles. Our goal is to quantify weathering mass losses from these accumulating soil profiles, specifically looking at the role of varying loess deposition rates on the weathering fluxes. Visually, the profiles with thinner loess caps appeared more weathered than the profile with the thickest loess cover. The mass balance analysis will allow quantification of the integrated weathering mass losses from the range of loess deposition rates, and assessment of the impact of this weathering on atmospheric CO_2 consumption.

ACKNOWLEDGEMENTS

This work is funded by NSF-0345136 and NSF-0240919. We thank Dr. P. Groves for logistical field support.

REFERENCES

Anderson, S.P. in review. Glaciers show direct linkage between erosion and chemical weathering. *Geomorphology*.

Anderson, S.P., Drever, J.I. & Humphrey, N.F. 1997. Chemical weathering in glacial environments. *Geol.* 25(5): 399–402.

Brimhall, G.H. & Dietrich, W.E. 1987. Constitutive mass balance relations between chemical composition, volume, density, porosity, and strain in metasomatic hydrochemical systems: Results on weathering and pedogenesis. *Geochim. Cosmochim. Acta* 51(3): 567–587.

Gaillardet, J., Dupré, B., Louvat, P. & Allègre, C. 1999. Global silicate weathering and CO_2 consumption rates deduced from the chemistry of large rivers. *Chem. Geol.* 159: 3–30.

Haldorsen, S. 1981. Grain-size distribution of subglacial till and its relation to glacial crushing and abrasion. *Boreas* 10: 91–105.

Hallet, B., Hunter, L. & Bogen, J. 1996. Rates of erosion and sediment evacuation by glaciers: A review of field data and their implications. *Global and Planetary Change* 12: 213–235.

Milliman, J.D. & Syvitski, J.P.M. 1992. Geomorphic/tectonic control of sediment discharge to the ocean: The importance of small mountainous rivers. *J. Geol.* 100: 525–544.

Moulton, K.L. & Berner, R.A. 1998. Quantification of the effect of plants on weathering: Studies in Iceland. *Geol.* 26(10): 895–898.

Nesbitt, H.W. & Young, G.M. 1996. Petrogenesis of sediments in the absence of chemical weathering: Effects of abrasion and sorting on bulk composition and mineralogy. *Sedimentology* 43: 341–358.

Nokleberg, W.J., Aleinikoff, J.N., Lange, I.M., Silva, S.R., Miyaoka, R.T., Schwab, C.E. & Zehner, R.E. 1992. Preliminary geologic map of the Mount Hayes Quadrangle, Eastern Alaska Range, Alaska. *U.S. Geological Survey Open File Report* 92–594.

Péwé, T. 1975. Quaternary Geology of Alaska. *U.S. Geological Survey Professional Paper 835.*

Péwé, T.L., Church, R.E. & Andresen, M.J. 1969. *Origin and paleoclimatic significance of the large-scale patterned ground in the Donnelly Dome Area, Alaska.* Special Paper No. 103. Boulder: Geological Society of America.

Péwé, T.L. & Reger, R.D. 1983. Delta River Area, Alaska Range. In T.L. Péwé & R.D. Reger (eds), *Guidebook to permafrost and Quaternary geology along the Richardson and Glenn Highways between Fairbanks and Anchorage, Alaska:*47–135. Fairbanks, AK: Department of Natural Resources, State of Alaska.

Pye, K. 1987. *Aeolian Dust and Dust Deposits.* London: Academic Press.

Riebe, C., Kirchner, J., Granger, D. & Finkel, R. 2001. Strong tectonic and weak climatic control of long-term chemical weathering rates. *Geol.* 29(6): 511–514.

Tranter, M., Huybrechts, P., Munhoven, G., Sharp, M.J., Brown, G.H., Jones, I.W., Hodson, A.J., Hodgkins, R. & Wadham, J.L. 2002. Direct effect of ice sheets on terrestrial bicarbonate, sulphate and base cation fluxes during the last glacial cycle: minimal impact on atmospheric CO_2 concentrations. *Chem. Geol.* 190: 33–44.

Water–Rock Interaction, Wanty & Seal II (eds)
© 2004 Taylor & Francis Group, London, ISBN 90 5809 641 6

Geochemical and isotopic constraints on the sources of arsenic in groundwater from Coastal Maine, New England

R.A. Ayuso & N.K. Foley
U.S. Geological Survey, Reston, Virginia

ABSTRACT: Anomalously high contents of arsenic are found in groundwater of Coastal Maine. Major and trace element compositional variations of sulfidic bedrock, sediments, soils, and arsenic-bearing minerals from mines and disseminated in rocks were studied to identify the source of arsenic in groundwater. Our inventory identifies possible arsenic sources and establishes pathways by which arsenic is transported from arsenic-bearing minerals (arsenian pyrite, arsenopyrite, orpiment, arsenolite, etc.), via sulfide oxidation or carbonation reactions, into various generations of secondary minerals (goethite, jarosite, and natrojarosite, etc.). Pb isotopic compositions of sulfides and secondary minerals produced by weathering are similar. Isotopic variations help to establish the compositions of fluids in which the secondary minerals precipitated and point to predominant control from the natural environment. A local contribution as a result of anthropogenic activities is also possible.

1 INTRODUCTION

Portions of coastal New Hampshire and Maine are increasingly recognized as areas of concern to human health because of high contents of arsenic in groundwater (Welch et al. 2000). Arsenic levels there may exceed the maximum contaminant level recently established by the U.S. Environmental Protection Agency (10 µg/L). Despite intense scrutiny, however, mechanisms controlling the input of arsenic into groundwater and the ultimate source of arsenic have remained enigmatic and controversial. Both natural and anthropogenic sources have been suggested.

Recent work in eastern New England has shown that drinking water with high arsenic contents is associated with bedrock aquifers, in contrast to drinking water from unconsolidated bedrock aquifers, which have lower arsenic contents (Ayotte et al. 1999; Ayotte et al. 2003). In particular, bedrock wells in variably calcareous (5–50% calcite) metasedimentary rocks in coastal New Hampshire and Maine have indicated a strong link to groundwater with high arsenic contents (>10 µg/L; Ayotte et al. 2003). Geological characterization of calcareous metasedimentary rocks obtained from drill cores shows that fractures associated with aquifers have sparse arsenian pyrite, pyrrhotite, and chalcopyrite (plus carbonate). Notably, neither the bulk rocks (as up to ~40 ppm) nor the minerals (e.g., pyrite, as up to 3000 ppm) contain exceptionally high arsenic (Ayuso et al. 2003a, b).

Recent studies in New Hampshire have linked the highest groundwater arsenic concentrations to areas containing abundant pegmatites and associated granitic rocks (Peters et al. 1999), in contrast to earlier studies that related the high contents of arsenic to human activities (Boudette et al. 1985).

As in New Hampshire, in eastern Maine a complex interplay of possible arsenic sources has been recognized (Fig. 1). In some cases, arsenic sources appear to be dominated by anthropogenic contributions (Marvinney et al. 1994), bedrock (Loiselle et al. 2001; Ayuso & Foley 2002; Ayuso et al. 2002, 2003a; Foley et al. 2002; Reeve et al. 2001, 2002; Robinson et al. 2002;) or both (e.g., Loiselle et al. 2002; Robinson & Ayotte 2002; Robinson & Ayuso, in press).

Geochemical and isotopic investigations are essential to evaluate pathways by which arsenic becomes concentrated in bedrock weathering products such as mineral coatings on rock surfaces, and particularly along fractures in aquifers. Our detailed study in Coastal Maine (Fig. 1) is centered on a densely populated area known to enclose numerous groundwater wells having anomalously high arsenic contents (Reeve et al. 2001, 2002; G. Lipfert, personal

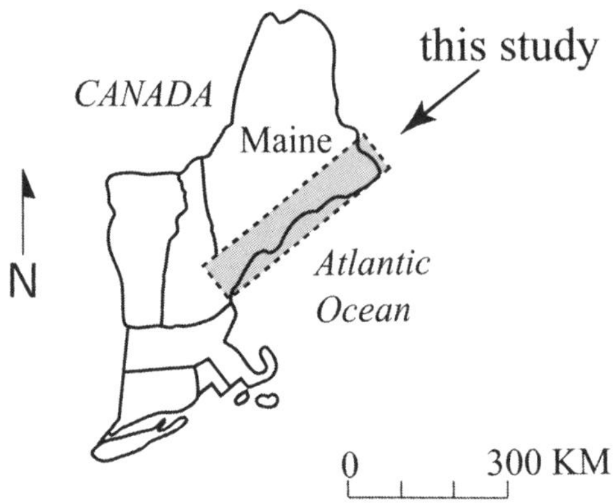

Figure 1. Map showing location of study.

comm. 2003). Bedrock in our focus area in Coastal Maine is variably metamorphosed Ordovician Fe-rich sulfidic schist (up to sillimanite-cordierite-biotite gneiss) that includes pelitic beds, sandy siltstones, calcareous sandstone, and rare limestone, and which was intruded by diorite and muscovite granite (Stewart 1998).

This study seeks to provide an inventory of naturally occurring arsenic in rock samples from surface outcrop and drill core, in addition to soils and sediments. The objective is to examine which of the possible natural and anthropogenic sources may have contributed to the anomalous arsenic contents in groundwater. Among the geological candidates are widely distributed point sources including sulfide mines and areas of sulfide-rich schists, gneisses, and granitic pegmatites. Moreover, the potential impact of arsenic resulting from the effects of land use, particularly associated with heavy historical application of pesticides and herbicides in New England, for example, can be considerable (D'Angelo et al. 1996).

2 ANALYTICAL PROCEDURES

Bulk chemical analyses of rocks ($n = 350$) and separates of minerals ($n = 56$) were done by WD-XRF (major elements), ICP-MS (trace elements), and INAA (rare earth elements). Pb isotopic compositions ($n = 86$) were determined at the U.S. Geological Survey, Reston, by thermal ionization mass spectrometry (TIMS). Samples included bulk and acid-leached aliquots of bedrock, soil, and stream sediments, as well as various sulfides, oxides and oxy-hydroxides that formed along rock surfaces, joints, crevices or fractures.

3 RESULTS AND DISCUSSION

Bedrock occurrences represented by bulk rock samples have a wide range in As contents. For example, in particular areas of Coastal Maine where As contents in groundwater is elevated (up to 1940 μg/L; Reeve et al. 2002; G. Lipfert, personal comm. 2003), sulfidic (arsenian pyrite dominated) bedrock and sulfidic drill core samples have As contents as high as 190 ppm (average As ~32 ppm; CaO ~1.31 wt. %, up to 5.58 wt. %). Bulk compositions of calcite-bearing soils from various horizons (and their leach compositions) closely parallel the arsenic contents of bedrock protoliths (calcite < 10%) and range in As up to 657 ppm (most bulk soil samples have As <100 ppm; leach samples up to 1860 ppb).

Arsenic-bearing minerals, including arsenides, sulfides, sulfosalts, and secondary Fe- and Mn-oxyhydroxides, are widely distributed in bedrock, mineralized zones, and soils of the study area. Primary As-bearing minerals include pyrite (up to 4 wt.%), pyrrhotite (up to 0.5 wt.% As in $Fe_{1-x}S$), löllingite, realgar, cobaltite, arsenopyrite, cobaltian arsenopyrite (to 8.4 wt.% Co), and tennantite. As and Pb contents of the secondary minerals are highly variable (Foley et al. 2002). For example, in the sulfidic schists, pyrites range from ~20 ppm to ~2000 ppm As and ~10 ppm to >400 ppm Pb, but pyrites from the mining areas are substantially higher in As and Pb and range to thousands of ppm (Foley et al. 2002). Supergene minerals in weathered rocks and soils act as intermediate As sources that can influence the As contents of groundwater. These minerals include orpiment, arsenolite-minerals, scorodite ($FeAsO_4 2H_2O$) and secondary arsenopyrite, pyrite, marcasite, Fe-Mn-oxyhydroxides, and possibly Co-Ni-arsenates and calcium-arsenates (rauenthalite, phaunouxite?).

Weathering of As-poor pyrrhotite produces complex mixtures of secondary pyrite + marcasite, Fe-oxyhydroxides, and secondary salts (e.g., ferrihydrite, rozenite, and melanterite). Weathering of pyrite, löllingite, realgar, arsenopyrite, cobaltite, and other minerals generates acid via oxidation and releases As, Co, Ni, Pb, etc., with subsequent pH-dependent sorption on co-precipitated Fe-oxyhydroxides. Leaching of arsenolite, orpiment and As-sulfosalt minerals by bicarbonate-bearing fluids, which causes the release of As, but not Fe, into the groundwater system, has also been suggested as a mechanism to acquire As in groundwater in coastal Maine (Foley et al. 2002). Reaction of carbonate ion in groundwater with arsenic-bearing sulfide minerals in aquifer rocks (a process suggested by Kim et al. 2000) could contribute, in part, to the strong correlation of As and pH, and the decoupling of As and Fe. That carbonation of sulfides is potentially a relevant

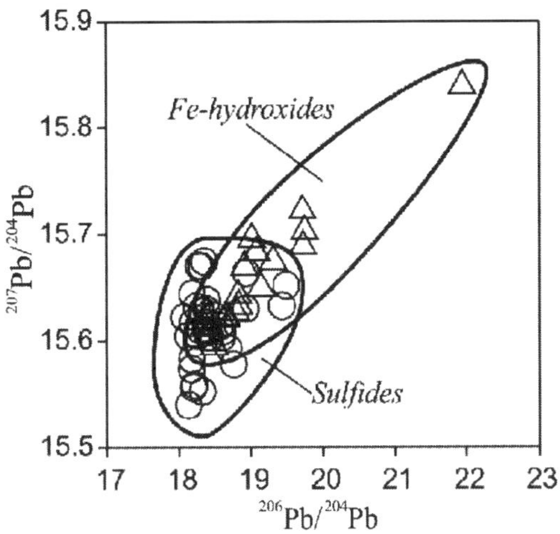

Figure 2. Pb isotope plot showing sulfides and Fe-hydroxides from Coastal Maine.

mechanism to release arsenic from aquifer rocks has also been recognized in eastern New England (Ayotte et al. 2003), in addition to arsenic desorption from iron oxides and from various clay minerals (Ayotte et al. 2003; Welch & Ayotte 2002).

The overall similarity of isotopic compositions for the sulfides ($^{206}Pb/^{204}Pb = 18.073–19.489$; $^{207}Pb/^{204}Pb = 15.539–15.675$; and $^{208}Pb/^{204}Pb = 37.947–39.102$) and secondary minerals (goethite, jarosite, and natrojarosite, rozenite, and melanterite): $^{206}Pb/^{204}Pb = 18.356–21.945$; $^{207}Pb/^{204}Pb = 15.595–15.839$; $^{208}Pb/^{204}Pb = 38.169–39.162$) can be petrographically and chemically attributed to sulfide weathering reactions (Fig. 2). Isotopic compositions of the secondary minerals plot on uranogenic and thorogenic Pb diagrams as broad fields that extensively overlap but extend to more radiogenic compositions. Similarity of isotopic compositions provides evidence that fluid-mineral reactions leading to the decomposition of the sulfides released metals and As, and imprinted the Pb isotopic signatures of the sulfides on the secondary minerals, reflecting the compositions of coexisting hydrous fluids.

Although the contribution from bedrock appears dominant, the secondary minerals also may indicate a contribution from a more radiogenic Pb source. For example, geochemical and lead isotope studies of stream sediments in New England indicate that anthropogenic activities (historical use of arsenical pesticides) can be locally important contributors to the As budget (Robinson & Ayuso, in press). Factor analysis on metal concentrations and lead isotopic compositions in stream sediment samples categorized according to regional land use settings link agricultural areas that used arsenical pesticides and stream sediments containing high arsenic and lead concentrations. Lead isotope compositions for stream sediments generally overlap the compositions of bedrock sulfides and their weathering products, but sediments from areas associated with heavy agricultural use point to a significant local contribution from industrial lead, and possibly from the pesticides.

4 CONCLUSIONS

Isotopic compositions of the sulfides in bedrock can account for most of the variations in the secondary minerals as a result of weathering, but a more radiogenic Pb contribution may also be present. Pb isotopic variations of the Fe-oxyhydroxides also constrain the isotopic compositions of the fluids in which the Fe-oxyhydroxides crystallized. Possible Pb contributions from anthropogenic sources cannot be disregarded, including remnants of older industrial Pb and Pb-rich pesticides in soils and sediments.

ACKNOWLEDGEMENTS

We thank J.D. Ayotte, G. Lipfert, R. Marvinney, A. Reeve, and G. Robinson, Jr., for discussions. We are grateful to Greg Wandless, John Jackson, Eric Binnie, Jeremy Dillingham, and Nicole West for laboratory and field support.

REFERENCES

Ayotte, J.D., Montgomery, D.L., Flanagan, S.M. & Robinson, K.W. 2003. Arsenic in groundwater in eastern New England: Occurrence, controls, and human health implications. *Environ. Sci. Technol.* 37: 2075–2083.

Ayotte, J.D., Nielsen, M.G., Robinson, G.R., Jr. & Moore, R.B. 1999. Relation of arsenic, iron, manganese in ground water to aquifer type, bedrock lithogeochemistry, and land use in the New England coastal basins. U.S. Geological Survey *Water Resources Investigations Report 99–4162.*

Ayuso, R.A. & Foley, N.K. 2002. Arsenic in New England: Mineralogical and geochemical studies of sources and enrichment pathways. U.S. Geological Survey *Open-File Report 02–454.* (http://pubs.usgs.gov/of/2002/of 02–454/).

Ayuso, R.A., Foley, N.K., Ayotte, J.D., Lyon, A., Burns, J., Marvinney, R.G., Reeve, A.S. & Robinson, G.R., Jr. 2002. Lead isotopes, arsenic sources and enrichment pathways linking sulfides from mines and unmineralized rocks to secondary iron oxides, Coastal New England. *A multidisciplinary scientific conference, New Hampshire Consortium on Arsenic*, May 29–31, 2002, p. 29.

Ayuso, R.A., Foley, N.K., Dillingham, J.D., West, N., Robinson, G.R, Jr. & Ayotte, J.D. 2003a. Geological characterization of Great Bay Drill Holes, New Hampshire: Preliminary results. U. S. Geological Survey *Open File Report*, in press.

Ayuso, R.A., Foley, N.K., Robinson, G.R., Jr., Dillingham, J.D., Wandless, G., West, N., Lipfert, G. & Reeve, A.S. 2003b. Geologic sources of arsenic in Coastal Maine, New England: from bedrock sulfides to secondary minerals (abs.). *Geol. Soc. Am.*, in press.

Boudette, E.L., Canney, F.C., Cotton, J.E., Davis, R.I., Ficklin, W.H. & Motooka, J.M. 1985. High levels of arsenic in ground waters of southeastern New Hampshire – a geochemical reconnaissance. U.S. Geological Survey *Open File Report 85–202*, 20 p.

D'Angelo, D., Norton, S.A. & Loiselle, M.C. 1996. Historical Uses and Fate of Arsenic in Maine. *Water Research Institute Completion Report 1986*. University of Maine. Orono, Maine. 24p.

Foley, N.K., Ayuso, R.A., Ayotte, J.D., Marvinney, R.G., Reeve, A.S. & Robinson, G.R., Jr. 2002. Mineralogical pathways for arsenic weathering meta-shales: an analysis of regional and site studies in the northern Appalachians. *A multidisciplinary scientific conference, New Hampshire Consortium on Arsenic*, May 29–31, 2002, p. 35.

Kim, M., Nriagu, J. & Haack, S. 2000. Carbonate ions and arsenic dissolution by groundwater. *Environ. Sci. Technol.* 34:3094–3100.

Loiselle, M.C., Marvinney, R.G. & Smith, A.E. 2001. Spatial distribution of arsenic in groundwater in Maine (abs.). *Geol. Soc. Am.*, 33: 54.

Loiselle, M.C., Marvinney, R.G. & Smith, A.E. 2002. Arsenic in ground water wells in Maine. *A multidisciplinary scientific conference, New Hampshire Consortium on Arsenic*, May 29–31, 2002, p. 16.

Marvinney, R.G., Loiselle, M.C., Hopeck, J.T., Braley, D. & Krueger, J.A. 1994. Arsenic in Maine ground water: an example from Buxton, Maine. In*: Proceedings of the 1994 Focus Conference on Eastern Regional Ground Water Issues*. National Ground Water Association 701–715.

Peters, S.C., Blum, J.D., Klaue, B. & Karagas, M.R. 1999. Arsenic occurrence in New Hampshire drinking water. *Environ. Sci. Technol.* 33: 1328–1333.

Reeve, A.S., Horesh, M., Warner, B. & Yates, M. 2001. Geochemical evaluation of sources for arsenic in ground water, Northport, Maine (abs.). *Geol. Soc. Am.* 33:61.

Reeve, A.S., Horesh, M., Marvinney, R. & Ayuso, R. 2002. Arsenic geochemistry in a bedrock aquifer, Northport, Maine. *A multidisciplinary scientific conference, New Hampshire Consortium on Arsenic*, May 29–31, 2002, p. 20.

Robinson, G.R., Jr. & Ayotte, J.D. 2002. Spatial associations between arsenic in groundwater, sediments, bedrock and agricultural landuse in New England. *A multidisciplinary scientific conference, New Hampshire Consortium on Arsenic,* May 29–31, 2002, p. 21.

Robinson, G.R., Jr., Ayotte, J.D., Montgomery, D.L. & DeSimone, L.A. 2002. Lithogeochemical character of near-surface bedrock in the New England Coastal Basins. U.S. Geological Survey *Open-File Report 02–7.*

Robinson, G.R. & Ayuso, R.A., in press, Use of spatial statistics and isotopic tracers to measure the influence of arsenical pesticide use on stream sediment chemistry in New England. *Appl. Geochem.*

Stewart, D.B. 1998. Geology of the northern Penobscot Bay, Maine. U.S. Geological Survey, *Miscellaneous Investigations Series Map I-2551.*

Welch, A.H. & Ayotte, J.D. 2002. National to local scale cycling of arsenic in ground water: *A multidisciplinary scientific conference, New Hampshire Consortium on Arsenic*, May 29–31, 2002, p. 25.

Welch, A.H., Westjohn, D.B., Helsel, D.R. & Wanty, R.B. 2000. Arsenic in ground water of the United States: occurrence and geochemistry. *Groundwater* 38: 589–604.

Water–Rock Interaction, Wanty & Seal II (eds)
 ISBN 90 5809 641 6

Organic matter weathering and atmospheric oxygen: a field and modeling study of black shale oxidation

R.A. Berner, E.W. Bolton & R.A. Wildman
Yale University, Dept. of Geology and Geophysics, New Haven, CT, USA

S.T. Petsch
University of Massachusetts, Department of Geosciences, Amherst, MA, USA

ABSTRACT: Oxidation during weathering of ancient organic matter is one of the major controls on atmospheric oxygen levels over geologic time. In an attempt to quantify this process, we have studied organic matter weathering in a near surface environment. The field study is based on core samples of a Devonian black shale from central-eastern Kentucky. Various depth intervals were analyzed for organic C, N and S, the spatial distribution of elements by SEM mapping, BET surface area, porosity, and permeability. These results were used in the creation of a numerical model for kinetically controlled oxidation of the organic matter within the shale (based on kinetics from Chang & Berner 1999). The one-dimensional model includes erosion, oxygen diffusion in the vadose zone, and the oxidation reaction of organic matter. The model provides a convenient means of exploring the influence of variable rates of erosion, oxygen level, and physical and chemical characteristics of the shale on organic matter oxidation. We conclude that oxygen consumption at most field sites should be erosion rate limited.

1 INTRODUCTION

1.1 *Motivation and general context*

The two most important processes affecting the long-term evolution of atmospheric oxygen are the burial of organic matter (OM), and its oxidation during erosion (Ebelman 1845; Holland 1984). Some basic questions yet unanswered in the latter process are: What is the effect of erosion rates on oxygen levels? Is all of the organic matter consumed before being unearthed, or are there finite amounts of organic matter exposed by erosion that are subsequently transported and reburied before being oxidized? If all the OM is oxidized before reaching the surface, oxygen would be consumed at a rate controlled by both erosion rates and the quantity of OM exhumed. On the other hand, if some of the OM is eroded and deposited elsewhere, the rate of consumption of oxygen would depend on the concentration of oxygen in the atmosphere. Both rate controlling mechanisms could be represented in distinct field settings. Quantification of field studies and comparison with numerical modeling can elucidate these mechanisms.

Weathering of organic matter consumes oxygen: cf. Berner & Canfield (1989)

$$CH_2O + O_2 \rightarrow CO_2 + H_2O \tag{1}$$

(The reverse reaction represents burial). Pyrite oxidation during weathering also consumes oxygen and plays an important secondary role in controlling atmospheric O_2 as well as the pH of weathering solutions.

The goal of this research was to gain a deeper understanding of organic matter weathering by means of a combined field and modeling study. By this approach we may assess the magnitude of feedback between organic matter weathering and atmospheric oxygen level. This study extends the field and modeling studies of Petsch (2000) and the weathering component of Lasaga & Ohmoto (2002) to include better estimates of: the texture of weathering organic matter, the rate of transport of O_2 into rocks undergoing weathering, and the spatial and temporal resolution of the near-surface boundary layer.

2 METHODS

2.1 *Field and laboratory methods*

A 13 m vertical core was taken from the New Albany formation (Upper Devonian, 365 Ma, Clay City, KY), where the strata are essentially horizontal. The core location is very near a horizontal layer taken from the same site and studied previously by Petsch et al. (2000; 2001a, b). The top 254 cm of the core were collected with a split-spoon auger powered by a percussive hammer. Below that depth, the core was collected with a rotary diamond-bit drilling corer.

Total carbon, total nitrogen, and total sulfur data were determined on powdered rock samples in triplicate with an elemental analyzer housed at the University of Massachusetts. Organic C was determined by the removal of inorganic C by acidification (Verardo et al. 1990). Organic N and S were determined with the elemental analyzer on kerogen isolates via the method of Petsch et al. (2000), and inorganic S, present as pyrite, was calculated by difference.

The spatial distribution of organic matter and pyrite in the shale was examined via backscattered-electron (BSE) imaging, elemental x-ray mapping, and energy-dispersive spectral (EDS) data collection using a JEOL JXA-8600 electron microprobe housed in the Yale University Department of Geology and Geophysics.

Porosity measurements were performed by comparing dry and wet weights. Permeability of the shale with respect to water was evaluated at various depths of the core in the Laboratory for Rock Mechanics at MIT in a custom built permeameter (Bernabe 1987; Mok et al. 2002). Surface area analysis was conducted at Yale using a 5-point BET analysis with N_2 gas.

2.2 *Modeling formulation*

For the present model we assumed: a porous medium with an interconnected gas phase, adsorbed water films on solid phase particles, Henry's law fractionation of oxygen between gaseous and aqueous phases, and plate-like organic matter particles with aspect ratio $\alpha = 5$ (width and length)/(thickness). Diffusive oxygen transport occurs in the porous medium with an effective diffusion coefficient D_e that includes effects of tortuosity T^* (<1), the porosity ϕ values in the low permeability matrix, and the diffusivity of oxygen in air ($D_e = \phi_{air} T^* D_{air}$). The rate of oxygen consumption and organic matter oxidation includes Michaelis-Menten kinetics based on Chang (1997). For the evolution of oxygen concentration in the pores, a, and the organic matter concentration in the sediment, g, we write:

$$\frac{\partial a}{\partial t} = \frac{\partial}{\partial x}\left(D_e \frac{\partial a}{\partial x}\right) - \frac{R^*_{max} a g}{K_m + a}\left(\frac{1}{\phi}\right) \tag{2}$$

$$\frac{\partial g}{\partial t} = -\omega \frac{\partial g}{\partial x} - \frac{R^*_{max} a g}{K_m + a} \tag{3}$$

where the effective rate constant R^*_{max} also includes geometric factors for the organic matter platelets and Henry law partitioning between the gas phase and aqueous films. Here t is time, ω is the erosion rate, x is the distance in the core from the surface, and K_m is another kinetic rate coefficient.

3 RESULTS

3.1 *Physical properties*

We estimated the permeability at field conditions using linear equations fitted to the data and extrapolated to an effective pressure of 1.4 bars (the effective pressure due to a natural confining pressure of 10 m of shale and minimal fluid pressure). The water permeability ranged between $6.5 \times 10^{-18}\,m^2$ and $7.7 \times 10^{-17}\,m^2$. An air permeability measurement of one sample was about 10 times larger than that of water, consistent with the fine grained nature of the sediments. BET surface areas ranged from 25 to $40\,m^2/g$.

3.2 *Chemical and mineralogical properties*

Figure 1 shows the concentration of organic carbon and pyrite sulfur measured from the core samples. The organic C is slightly elevated at the surface due to modern C, while ancient C remains at depth. The most noteworthy aspect of the data of Figure 1 is the sudden drop in organic carbon over a narrow distance in the rock. This is found in the core at around 2.5 m depth. This suggests that most organic matter oxidation occurs rapidly at a "front" where incoming O_2

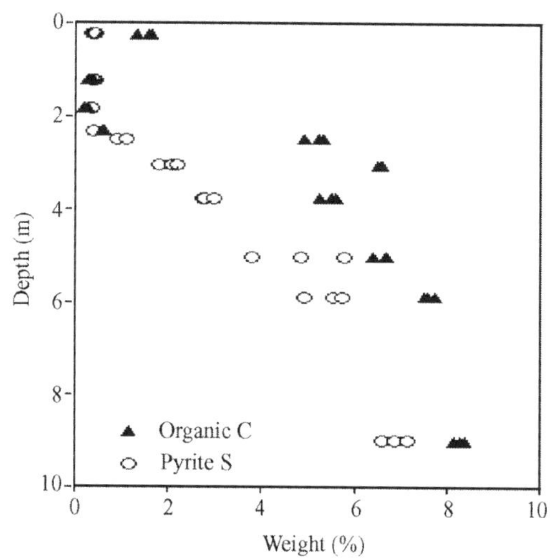

Figure 1. The concentration of organic carbon and pyrite sulfur measured from the core samples.

first contacts the organic matter. Beyond the front, oxidation is less rapid as indicated by the concentration of organic C rising slightly with distance and approaching a constant value, characteristic of the unaltered shale.

Pyrite shows a continuous concentration gradient, decreasing towards the carbon "front" where the pyrite S concentration falls to very low levels. Loss of pyrite in the presence of abundant organic matter indicates that pyrite oxidizes faster than the organic matter. A reasonable scenario for a well developed weathering horizon is that inward migrating gaseous or dissolved O_2 reacts first with modern soil organic matter and then with any remaining shallow pyrite. Once the pyrite is gone the O_2 attacks organic matter at the same level resulting in a carbon reaction front. A small amount of residual O_2 migrates beyond the front to react with pyrite resulting in loss of some pyrite but little loss of organic matter.

Minerals identified by EDS include very fine-grained potassic aluminosilicates (clays), quartz, and pyrite, in addition to organic matter. Element mapping indicates that shale organic matter is present mainly as 0.5–50 µm discoidal clots and discontinuous laminae ("stringers") and not mainly as monomolecular coatings on clay grains (cf. Kennedy et al. 2002).

3.3 *Numerical modeling results*

We conducted a large number of numerical simulations using the model described above. Figure 2 shows the resulting steady-state vertical profile for OM (solid curve, note $2\times$ scale factor) and oxygen concentration (dashed) for an erosion rate of 5×10^{-5} m/yr. This model assumed the current atmospheric oxygen level at the surface, 10% organic C at depth, porosity fraction 0.05, tortuosity $T^*=0.2$, OM platelet thickness $d=5$ µm. The solution achieved 95% approach to steady state after 188,000 years. The characteristic front depth is 2.5 m, very similar to that measured in the core (Fig. 1). Note the dramatic decrease in oxygen concentration with increasing depth. Most of the oxidation occurs at low oxygen levels. The constant gradient of oxygen near the surface maintains a constant diffusive flux toward the region of consumption. This simulation did not contain pyrite; however we expect closer agreement with the observed profiles for simulations that include pyrite oxidation.

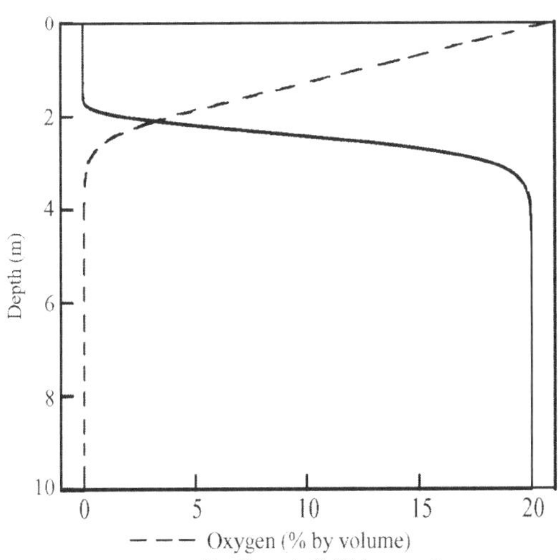

Figure 2. The resulting steady-state vertical profile for OM (solid curve, note $2\times$ scale factor) and oxygen concentration (dashed) for an erosion rate of 5×10^{-5} m/yr. See text.

The time to steady state for various simulations performed so far ranged from 400 yrs to 380,000 yrs. Time to steady state increases with increasing atmospheric oxygen levels and other factors that favor deeper buried fronts. The shortest time to steady state was observed for cases where finite amounts of OM remain at the eroding surface, as favored by rapid erosion rates. Erosion rates larger than 1×10^{-4} m/yr typically yielded finite amounts of OM remaining at the surface. However, microbial acceleration of oxidation (cf. Petsch et al. 2001a) would favor complete OM oxidation at considerably higher erosion rates.

4 CONCLUSIONS

These results suggest that organic matter weathering in shales can be treated in terms of the reaction of organic matter with gaseous and dissolved O_2. Once a weathering profile is developed the inward or downward migrating O_2 reacts first with modern soil organic matter and subsequently with low concentrations of remaining pyrite and ancient organic matter before reaching the "front". Pyrite apparently reacts faster with O_2 than does organic matter as evidenced by reduced pyrite concentrations accompanying high organic matter concentrations at the front. Upon further inward migration lower levels of O_2 react with higher concentrations of pyrite and organic matter. These data provide an important basis for the theoretical modeling of organic matter oxidation during weathering.

The numerical modeling exhibits an organic matter "front" at a similar depth as the field site when using the regional average erosion rate of 5×10^{-5} m/yr. These results, along with simulations at various erosion rates, suggest that in most field settings oxygen would be consumed at a rate independent of the level of atmospheric O_2 and directly proportional to "erosion rate and the organic matter content of the buried sediments. Only in very rapidly eroding environments should significant quantities of ancient organic matter be eroded and reburied before being oxidized, a process that would consume oxygen at rates dependent on atmospheric concentration.

ACKNOWLEDGEMENTS

We thank U. Mok and B. Evans of the MIT Rock Mechanics Lab for assisting with the measurements of porosity and permeability. We thank J.O. Eckert for making images and microprobe measurements at Yale. We wish to acknowledge helpful discussions with Jay Ague. We also thank Tom Dobkowski and Mike Breeding for help with the thin sections and epoxy plugs, respectively. We received research support from NSF-EAR-0104797 and DOE-BES grant DE-FG02-01ER15216. Thanks to Lillie Jaffe and Bernhard Peucker-Ehrenbrink for information regarding the field area.

REFERENCES

Bernabe, Y. 1987. A wide range permeameter for use in rock physics: *Intl. J. Rock Mech. and Mining Sci. & Geomechanics Abstracts* 24: 309–315.

Berner, R.A. & Canfield, D.E. 1989. A model for atmospheric oxygen over Phanerozoic time. *Am. J. Sci.* 289: 333–361.

Chang, S. & Berner, R.A. 1998 Humic substance formation via the oxidative weathering of coal. *Environ. Sci. Tech.* 32: 2883–2886.

Chang, S. & Berner, R.A. 1999. Coal weathering and the geochemical carbon cycle *Geochim. Cosmochim. Acta* 63: 3301–3310.

Chang, S. 1997. *Coal weathering and the geochemical carbon cycle,* Ph.D. Dissertation, Yale University.

Ebelmen, J.J. 1845. Sur les produits de la décomposition des especes minérales de la famille des silicates, *Ann. des Mines* 7: 3–66.

Holland, H.D. 1984. *The Chemistry of the Atmosphere and Oceans*. New York: Wiley.

Kennedy, M.J., Pevear, D.R. & Hill, R.J. 2002. Mineral surface control of organic carbon in black shale. *Science* 295: 657–660.

Lasaga, A.C. & Ohmoto, H. 2002.The oxygen geochemical cycle: dynamics and stability. *Geochim. Cosmochim. Acta* 66(3): 361–381.

Mok, U., Bernabé, Y. & Evans, B. 2002. Permeability, porosity and pore geometry of chemically altered porous silica glass, *J. Geophys. Res.,* 107 (0), 10, 1029/200 1JB000247.

Petsch, S.T. 2000. *A study on weathering of organic matter in black shales and implications for the geochemical cycles of carbon and oxygen,* Ph.D. Dissertation, Yale University.

Petsch, S.T., Berner, R.A. & Eglinton, T.I. 2000. A field study of the chemical weathering of ancient sedimentary organic matter: *Org. Geochem.* 31: 475–487.

Petsch, S.T., Eglinton, T.I. & Edwards, K.J. 2001a. ^{14}C-dead living biomass: Evidence for microbial assimilation of ancient organic matter during shale weathering: *Science* 292: 1127–1131.

Petsch, S.T., Smernik, R.J., Eglinton, T.I. & Oades, J.M. 2001b. A solid state ^{13}C-NMR study of kerogen degradation during black shale weathering: *Geochim. Cosmochim. Acta* 65: 1867–1882.

Verardo, D.J., Froelich, P.N. & McIntyre, A. 1990. Determination of organic carbon and nitrogen in marine sediments using the Carlo Erba NA-1500 Analyzer: *Deep-Sea Res. Part A – Oceanog. Res. Papers* 37: 157–165.

Water–Rock Interaction, Wanty & Seal II (eds)

Mineral weathering sources of calcium in northeastern USA forests

J.D. Blum & C.A. Nezat
The University of Michigan, Ann Arbor, MI, USA

R.D. Yanai
State University of New York, Syracuse, NY, USA

S.P. Hamburg
Brown University, Providence, RI, USA

M.A. Arthur
University of Kentucky, Lexington, KY, USA

ABSTRACT: Acid rain and repeated forest harvest are thought to threaten the supply of calcium available to meet the nutrient demands of some forested ecosystems. Based on sequential extractions of soils and measurement of their major and trace element composition, we have identified apatite (Ca phosphate) as an important source of calcium in the soils of the Hubbard Brook Experimental Forest in New Hampshire, USA (Blum et al. 2002; Hamburg et al. 2003). Approximately 20% of the calcium that has been released by long-term weathering of soils at Hubbard Brook came from dissolution of apatite (Nezat et al. 2004) and the present reservoir of Ca in soil apatite is 65 g Ca m^{-2} in E and B horizons and 6 g Ca m^{-2} per cm of depth in the C-horizon. Based on our studies of Ca budgets in forests of variable ages near Hubbard Brook, we have further proposed that apatite weathering is accelerated in young forest stands, perhaps because there is a high demand for P during the first few decades after forest harvest (Hamburg et al. 2003). To assess whether our observations on the importance of apatite in the vicinity of Hubbard Brook are applicable regionally, we characterized Ca pools in soil parent materials from ten sites in New York, Pennsylvania, and Maine. We demonstrate using Ca and P analyses of sequential soil extractions that apatite is an important pool of Ca in all soils that we studied that are developed on crystalline rocks of various types over a broad area. However, apatite was not found to be a significant source of Ca on soils developed from sedimentary parent materials of the Allegheny Plateau.

1 INTRODUCTION

We commonly think of N and P as the nutrient elements most likely to limit the growth of forests, with N limitation most common in temperate and boreal ecosystems, and P limitation most common in the tropics where soils are very highly weathered. However, N is not as scarce as it was before the industrial revolution and the advent of air pollution. Nitrogen deposition from the atmosphere fertilizes forests, and in excess can "saturate" plant demand (Aber et al. 2003). One consequence of acid rain and elevated N deposition is that nitrate not used by plants is leached from soils, along with base cations such as Ca, Mg, and K. Ultimately, exchangeable pools of base cations in soils can be depleted (Shortle & Bondietti 1992), which is also reflected in declining concentrations of these elements in stream water (Kirchner & Lyderson 1995), even when N deposition rates are reduced (Likens et al. 1998).

Forest decline has been linked to base cation deficiencies in a variety of systems. In the northeastern United States, red spruce decline has been linked to Ca loss from foliage (DeHayes et al. 1999, Kobe et al. 2002). In Pennsylvania, sugar maple mortality has been related to Ca, Mg, Mn and K deficiencies and other predisposing factors (Long et al. 1997; Horsley et al. 2000). There is growing concern that base cation depletion of soils as a result of leaching losses or Ca removals through harvesting may threaten forest productivity (Huntington et al. 2000). The amount of nutrients removed in harvest can be large relative to plant-available "exchangeable" pools. According to an analysis of sites where ecosystem budgets were available, Ca is the nutrient predicted to be most rapidly depleted by repeated harvest and associated leaching losses (Federer et al. 1989). At the Hubbard Brook Experimental Forest, New Hampshire, USA, where nutrient cycling in forests has been studied for 35 years, the "exchangeable" pool of Ca, as defined by a neutral-salt extraction, is so small that it is insufficient to replace the amount of Ca removed in harvested material in one forest rotation (Likens et al. 1998).

In response to concerns about acid rain, cation depletion, and repeated harvest threatening Ca supplies, we undertook a series of studies directed at improving our understanding of Ca cycling and Ca supply in northern forests. We studied patterns of Ca cycling in different aged forests in the White Mountain National Forest, New Hampshire. We analyzed elemental ratios and Sr isotopes in sequential extracts of soils from the Hubbard Brook Experimental Forest, also in New Hampshire, which showed that apatite is an important Ca pool, especially at depth (Blum et al. 2002; Hamburg et al. 2003; Nezat et al. 2004). Finally, in the study reported here we analyzed soil parent materials from ten additional sites in Maine, New York, and Pennsylvania, to determine the potential importance of apatite to Ca supply across the region.

2 IMPLICATIONS OF APATITE WEATHERING

The weathering of apatite has mostly been studied because of its importance to P cycling: apatite is the most common P-bearing mineral in virtually all rock types and is the ultimate source of almost all P in the forest soils of the northeast. Apatite has not previously been considered in calculations of Ca input from weathering in forest ecosystems because the vast majority of Ca in rocks is generally contained in other minerals, such as feldspars, pyroxenes, and amphiboles in igneous rocks and calcite or dolomite in carbonate rocks. But the fact that apatite is much more readily weathered than the silicate minerals with which it is commonly found means that it might be important as a source of available Ca capable of meeting forest demand, at least in young soils such as those affected by recent glaciation.

The addition of apatite to our thinking about Ca cycling has some important implications for ecosystem nutrient cycling and for forest management. Estimates of Ca weathering rates at Hubbard Brook have assumed that plagioclase dominates the weathering flux of both Ca and Na in crystalline rocks. With this assumption, Na fluxes can be used to estimate the proportion of Ca in a stream that is derived from weathering versus depletion of the cation exchange pool (see Bailey et al. 2003 for discussion). Approximately 20% of the Ca in soil parent materials at Hubbard Brook is held in apatite, and at least 20% of the Ca lost by weathering over the age of the soil came from dissolution of apatite (Nezat et al. 2004). The loss of apatite from the upper part of the soil profile means that significant amounts of Ca have been weathered without producing Na, and the supply of Ca from weathering has, therefore, been significantly underestimated.

3 REGIONAL STUDY

In this study we ask the question whether apatite is an important source of Ca across the region, rather than a peculiarity of the Hubbard Brook soils we have studied. We collected soil samples from eight sites in glaciated granitoid terrain, in New York and Maine, and two sites in the Allegheny Plateau of Pennsylvania, which have soils derived from sedimentary rocks (Fig. 1). At each site, C-horizon samples were collected from a single soil pit, air-dried, sieved, and subjected to the series of extractions described in the Figure 2 caption.

As at Hubbard Brook, the exchangeable Ca pool was the smallest Ca pool in all ten samples, and the largest pool of Ca was the "residual" pool, extracted by total digestion after removal of the more available Ca (Fig. 2a). The sites in New York and Maine had more Ca in every fraction than the two sites in

Figure 1. Map of Northeastern USA states with locations of forest soil sample sites

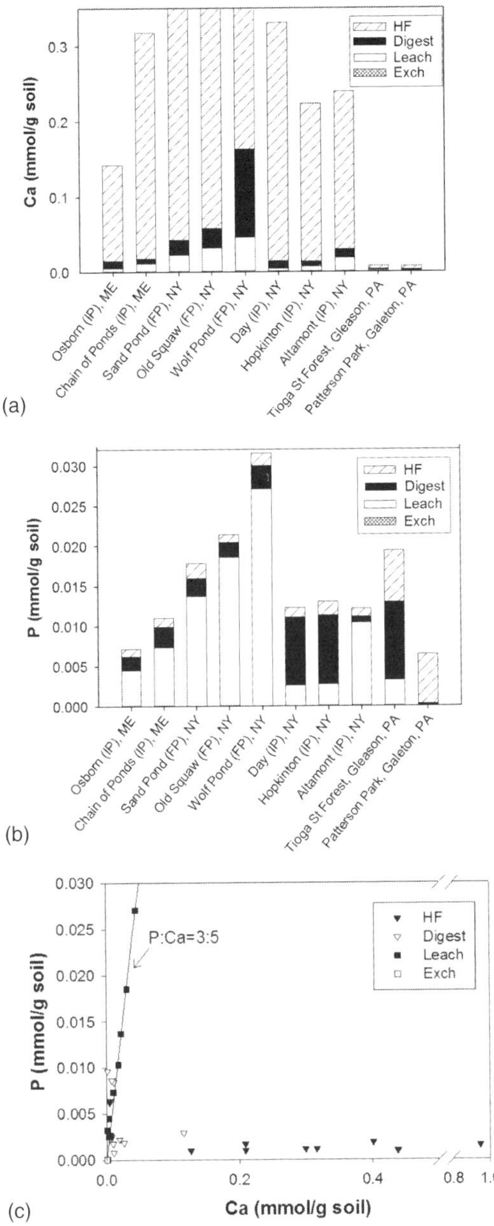

Figure 2. Concentrations of (a) Ca, (b) P and (c) Ca versus P in four sequential, operationally defined pools in ten C-horizon soil samples collected in New York, Maine, and Pennsylvania. "Exch" is the 1 M ammonium chloride exchange fraction, "Leach" is a 1 M cold nitric acid leach, "Digest" is a 1 M hot nitric acid digest, and "HF" is a hot concentrated hydrofluoric and nitric total digest of the remaining silicate material.

Pennsylvania. The "leach" pool extracted by 1 M nitric acid, which is dominated by apatite, was significant in all of the sites in New York and Maine, but nearly absent from the sites in Pennsylvania. The same pattern appears in the P concentrations of the four extractions (Fig. 2b). The ratio Ca:P in the "apatite" pool confirms its identity, as it is indistinguishable from the 3:5 molar ratio in apatite (Fig. 2c).

The two sites in Pennsylvania were selected to bracket the southern extent of the Wisconsin glaciation. Sugar maple decline has been observed to occur on unglaciated soils of the Allegheny Plateau (Horsley et al. 2000), presumably because Ca depletion is more severe on older soils, while glaciers tend to recreate the soils providing fresh minerals for weathering. Therefore, we expected to see greater amounts of weatherable (leachable) apatite at the unglaciated site, in Galeton PA. The fact that both Galeton and Gleason had very little apatite, with Galeton having even less than Gleason, suggests that the parent material is very poor in apatite and thus glaciation has little effect on its availability. With further study the importance of apatite across the landscape could probably be predicted using bedrock geology and patterns of glacial deposition.

4 CONCLUSIONS

Young forest stands appear to be more effective than old stands at obtaining Ca from the mineral soil. A key question is whether this additional Ca comes from a sustainable source, such as mineral weathering, or whether it signals the accelerated depletion of a limited available pool. The answer may depend on the parent material, such that mapping the distribution of soil properties is essential to predicting the risk of Ca depletion by acid rain and repeated forest harvest. Apatite is a previously under-appreciated, biologically significant and readily weathered source of Ca, which is likely to be important in meeting the Ca needs of forests growing on young soils formed in granitoid parent materials. Knowing what controls rates of mineral weathering, including the possible response of apatite weathering to P limitation, and the role of forest disturbance both anthropogenic and natural, will be key to the development of sustainable forest management. Geological differences across the northeastern USA are likely to impact Ca availability and best management practices.

ACKNOWLEDGEMENTS

For site identification and access, we thank R Dziengaleski and D Osterberg, Finch Pruyn &

Company; T Hall, Pennsylvania Department of Conservation and Natural Resources; and P Malerba and Cathy Irwin, International Paper. A Graham, R Maher, B Park, A Smith, and J Borza collected the samples. Financial support was provided by the Agenda 2020 collaboration between Forest Industry and the US Forest Service and by the National Science Foundation (DEB 0235650).

REFERENCES

Aber, J.D., Goodale, C.L., Ollinger, S.V., Smith, M.L., Magill, A.H., Martin, M.E., Hallett, R.A. & Stoddard, 2003. Is nitrogen deposition altering the nitrogen status of northeastern forests? *Bioscience* 53: 375–389.

Bailey, S.W., Buso, D.C. & Likens, G.E. 2003. Implications of sodium mass balance for interpreting the calcium cycle of a forested ecosystem. *Ecology* 84: 471–484.

Blum, J.D., Klaue, A., Nezat, C.A., Driscoll, C.T., Johnson, C.E., Siccama, T.G., Eagar, C., Fahey, T.J. & Likens, G.E. 2002. Mycorrhizal weathering of apatite as an important calcium source in base-poor forest ecosystems. *Nature* 417: 729–731.

DeHayes, D.H., Schaberg, P.G., Hawley, G.J. & Strimbeck, G.R. 1999. Acid rain impacts on calcium nutrition and forest health. *Bioscience* 49: 789–800.

Federer, C.A., Hornbeck, J.W., Tritton, L.M., Martin, C.W., Pierce, R.S. & Smith, T.S. 1989. Long term depletion of calcium and other nutrients in eastern US forests. *Environmental Management* 13: 593–601.

Hamburg, S.P., Yanai, R.D., Arthur, M.A., Blum, J.D. & Siccama, T.G. 2003. Biotic control of calcium cycling in northern hardwood forests: acid rain and aging forests. *Ecosystems* 6: 399–406.

Horsley, S.B., Long, R.P., Bailey, S.W., Hallett, R.A. & Hall, T.J. 2000. Factors associated with the decline disease of sugar maple on the Allegheny Plateau. *Can. J. For. Res.* 30: 1365–1378.

Huntington, T.G., Hooper, R.P., Johnson, C.E., Aulenbach, B.T., Cappellato, R. & Blum. A.E. 2000. Calcium depletion in a southeastern United States forest ecosystem. *Soil Sci. Soc. Am. J.* 64: 1845–1858.

Kirchner, J. W. & Lyderson, E. 1995. Base cation depletion and potential long-term acidification of Norwegian catchments. *Environ. Sci. Tech.* 29: 1953–1960.

Kobe, R.K., Likens, G.E. & Eager, C. 2002. Tree seedling growth and mortality responses to manipulations of calcium and aluminum in a northern hardwood forest. *Can. J. For. Res.* 32: 954–966.

Likens, G.E., Driscoll, C.T., Buso, D.C., Siccama, T.G., Johnson, C.E., Lovett, G.M., Fahey, T.J., Reiners, W.A., Ryan, D.F., Martin, C.W. & Bailey, S.W. 1998. The biogeochemistry of calcium at Hubbard Brook. *Biogeochemistry* 41: 89–173.

Long, R.P., Horsley, S.B. & Lilja, P.R. 1997. Impact of forest liming on growth and crown vigor of sugar maple and associated hardwoods. *Can. J. For. Res.* 27: 1570–1573.

Nezat, C.A., Blum, J.D., Klaue, A., Johnson, C.E. & Siccama, T.G. 2004. Influence of landscape position and vegetation on long-term weathering rates at Hubbard Brook, New Hampshire, USA. *Geochim. Cosmochim. Acta*, in press.

Shortle, W.C. & Bondietti, E.A. 1992. Timing, magnitude, and impact of acidic deposition on sensitive forest sites, *Water Air Soil Pollut.* 61: 253–267.

Water–Rock Interaction, Wanty & Seal II (eds)
© 2004 Taylor & Francis Group, London, ISBN 90 5809 641 6

Calcium stable isotope evidence for three soil calcium pools at a granitoid chronosequence

T.D. Bullen, J.A. Fitzpatrick, A.F. White, M.S. Schulz & D.V. Vivit
Water Resources Division, U.S. Geological Survey, Menlo Park, CA, USA

ABSTRACT: Calcium isotope compositions of porewaters collected from soil-weathered sediment profiles at a marine terrace chronosequence along the central California coast reveal the existence of three distinct soil calcium pools in this granitoid mineral weathering system. A "mineral Ca" pool is required to explain the δ^{44}Ca of porewater collected from the deep, primary mineral-rich zone. An "organically-complexed Ca" pool, derived from biologic cycling within the dominant grass ecosystem, is required to explain both exceptionally light Ca in shallow porewater collected at the youngest terraces and strongly elevated Ca/Sr ratios in the uppermost soils at all terraces. An "exchangeable Ca" pool is required to explain the trend of increasing δ^{44}Ca of shallow porewater toward seawater values with increasing terrace age. Increasing Ca isotope disequilibrium between the porewaters and labile soil Ca pools with increasing terrace age demonstrates the inefficiency of cation exchange processes in this weathering environment.

1 INTRODUCTION

Calcium (Ca) is an important product of mineral weathering and an essential macronutrient for plants. Previous assessments of Ca dynamics in watersheds have largely been accomplished by simple mass balance approaches (i.e., input-output budgets) or by using surrogate tracers such as strontium (Sr) isotopes that are assumed to behave similarly to Ca in biogeochemical cycles. However, classic input-output budgets only provide a net evaluation of watershed-scale mass balance, and cannot provide information on details such as preferential retention or loss of Ca from discrete components within the soil-ecosystem framework. Realizing this limitation, Kennedy et al. (2002) used an enriched Sr isotope tracer approach to understand Ca dynamics at a pristine watershed in southern Chile, and concluded that organically-complexed Ca was the main source of Ca nutrition to the forest.

A number of recent studies have demonstrated that Ca and Sr may not behave similarly particularly in biogeochemical cycles, and thus our ability to understand Ca dynamics may be confounded by assuming that Sr isotopes provide a robust surrogate tracer. The development of tracers that provide a more direct assessment of Ca behavior is required if we are to understand the details of Ca dynamics in hydrologic and biogeochemical systems. Ca has six stable isotopes spanning a large mass range (^{40}Ca, ^{42}Ca, ^{43}Ca, ^{44}Ca, ^{46}Ca and ^{48}Ca), and thus there is the potential for variations in Ca isotope composition to record information about sources of, and processes involving Ca. Based largely on the observation of Skulan et al. (1997) that biological processes are able to fractionate Ca isotopes, interest in the field of Ca isotope geochemistry has recently exploded, particularly with respect to oceanic paleo-thermometry (Gussone et al. 2003), food-web dynamics (Skulan et al. 1997) and bone mineral balance (Skulan & DePaolo 1999). However, only exploratory work has been accomplished to date on the behavior of Ca isotopes during mineral weathering, soil development and ecosystem-level processes in forested and manipulated watersheds. Regardless, it is reasonable to assume that biologic processing of Ca will result in variations of Ca isotope compositions that may be useful for identifying Ca sources, sinks and transport pathways in watersheds and associated ecosystems.

Here we present an assessment of Ca isotope systematics in soils, soil porewaters and grasses at a soil chronosequence developed on granitic materials exposed on a series of marine terraces. This setting is particularly suitable for testing the ability of Ca isotopes to provide useful information about Ca dynamics on a watershed scale due to the influence of multiple potential sources of Ca (i.e., Ca-mineral weathering vs. seawater-dominated precipitation), similar grass-dominated ecosystem development and consistent anthropogenic impact (i.e., tree clearing for pasturing) across the chronosequence.

2 SITE CHARACTERISTICS AND ANALYTICAL TECHNIQUES

The marine terrace chronosequence is situated just west of Santa Cruz, CA, USA along the Pacific Ocean coast, an area that is characterized by a fairly wet Mediterranean climate (Bradley 1957). The five terraces that comprise the chronosequence are made up of reworked granitoid minerals (plagioclase, K-feldspar, hornblende, biotite, apatite, quartz) derived from the underlying Miocene Santa Margarita sandstone and Santa Cruz mudstone and originally from Cretaceous granitoids such as those in the Santa Cruz Mountains. The terrace surfaces have been dated using several techniques (see Perg et al. 2001 for pertinent references) and range in age from 65–226 ka. The data discussed here represent samples collected from the 2nd through 5th terraces that range in age from 92–226 ka; samples from the 1st terrace (~65 ka) have yet to be analyzed. At each terrace, a profile of soil and weathered sediment samples was collected by hand auger. Porewater samples were collected using tension water samplers, and were filtered using a 0.45 μm syringe filter and acidified to pH = 2 with Teflon-distilled HNO_3. A grass sample was collected at each terrace.

Soil samples were wet-sieved in distilled water to provide the <2 mm fractions, air-dried and then subjected to a three-step leaching/digestion procedure. First, the "exchangeable" cation fraction was removed using 1*N* ammonium acetate. Second, following rinsing with distilled water, the "acid-leachable" fraction was removed using 1*N* HNO_3. Third, following rinsing with distilled water, residual samples were air-dried, ashed at 650°C in a muffle furnace and totally digested in a mixture of HCl, HNO_3 and HF in a sealed Teflon beaker on a warm plate (30°C) to produce the "digestible" fraction. Grass samples were totally digested in hot (150°C), concentrated HNO_3. All samples were reserved as a solution for analysis of chemical and Ca isotope composition.

Ca, Si and Sr concentrations of the reserved solutions were measured with a Perkin Elmer Elan 6000 inductively coupled plasma mass spectrometer, using synthetic multi-element solutions as standards and Ge as the internal standard. Internal precision for these elements is better than 2%. For determination of Ca isotope composition, sufficient liquid to provide 750 ng of Ca from each sample solution was mixed with a set amount of ^{42}Ca–^{48}Ca "double-spike" solution and then processed using cation exchange chromatography to obtain a purified Ca separate. Ca isotope compositions were measured using a Finnigan MAT 261 thermal ionization mass spectrometer, in peak hopping mode. Addition of the double spike allowed for correction of both isotopic fractionation during resin processing and machine discrimination during mass spectrometry. Ca isotope compositions are given in terms of $\delta^{44}Ca$, the per mil deviation of the $^{44}Ca/^{40}Ca$ ratio of the sample compared to that of seawater, the arbitrary delta basis. Using this notation, $\delta^{44}Ca$ of seawater has a value of zero. Precision of $\delta^{44}Ca$ at the 2σ-level is ±0.15‰ or better for all samples discussed here.

3 RESULTS AND DISCUSSION

Ca is efficiently removed from the soil-weathered sediment profile as weathering proceeds, due to dissolution of Ca-bearing minerals such as plagioclase, hornblende and apatite. Ca/Si ratios of bulk samples progressively decrease up-profile from un-weathered sediment values to a minimum at about 150 cm depth, as shown in Fig. 1a that displays the profile collected at the 3rd terrace as an example. However, above this depth the trend reverses and Ca/Si ratios progressively increase up-profile to the surface. Other alkali earth elements such as Sr (and Mg) do not display this shallow-level enrichment, as shown in Fig. 1b for the profile at the 3rd terrace. In fact, Ca/Sr ratios of bulk samples increase from a minimum at about 200 cm depth toward values measured in the grass sample collected from the terrace surface, suggesting enrichment of the shallow soils by biogenic Ca.

We envision three potential sources of Ca in this system: primary minerals in sediment, seawater-dominated precipitation and plants (i.e., grasses). Of these source pools, seawater has the heaviest Ca ($\delta^{44}Ca = 0$‰), grasses have the lightest Ca ($\delta^{44}Ca \sim -2.1$‰ across the chronosequence), and the least weathered sediment samples (i.e., the deepest samples collected from each profile) have Ca that is intermediate in isotopic composition ($\delta^{44}Ca \sim -0.85$‰ across the chronosequence). We point out that Sr isotopes used as a proxy for Ca are capable of revealing only the sediment mineral and seawater sources, and are unable to uniquely distinguish biogenic sources of Ca (Fitzpatrick et al. 2002).

As shown in Fig. 2, porewater samples collected from three depths at each terrace (i.e., "shallow" (~15 cm depth); "intermediate" (at the depth of the Ca/Si minimum, ~200 cm depth); "deep" (at the depth of auger refusal in sediment ~700 cm depth) have $\delta^{44}Ca$ values within this range. $\delta^{44}Ca$ of the deep porewaters is essentially constant across the chronosequence and similar to that of the bulk unweathered sediment, indicating the importance of Ca-mineral

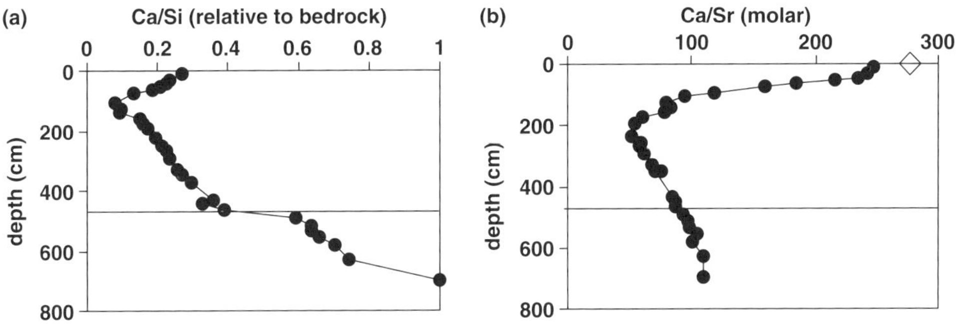

Figure 1. Chemical variations in bulk samples from soil and weathered sediment profile collected at the 3rd terrace. Solid line represents soil-weathered sediment interface. a) Ca/Si ratios compared to unweathered sediment values; b) molar Ca/Sr ratios. Diamond symbol represents grass sample collected from terrace surface.

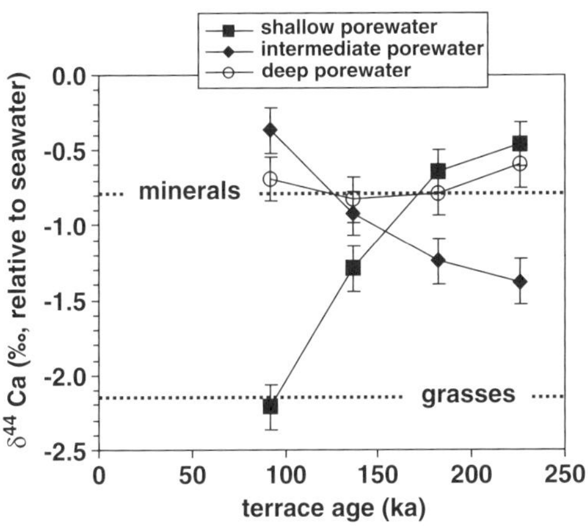

Figure 2. δ^{44}Ca of shallow, intermediate and deep porewater samples collected from the 2nd, 3rd, 4th and 5th terraces. Error bars on data points show maximum uncertainty at 2σ level; uncertainty for any single point lies within this range.

dissolution on porewater compositions in the deep system. δ^{44}Ca of the shallow and intermediate-depth porewaters is variable across the chronosequence, and the temporal evolution of δ^{44}Ca of the shallow porewaters is strikingly opposite to that of the intermediate-depth porewaters. Shallow porewater at the 2nd terrace has δ^{44}Ca that is identical to that of the grasses at the site; with increasing terrace age, δ^{44}Ca of shallow porewater increases toward the composition of seawater. In contrast, intermediate-depth porewater at the 2nd terrace has δ^{44}Ca that is close to that of seawater; with increasing terrace age, δ^{44}Ca of intermediate-depth porewater decreases toward the composition of the grasses.

The porewater data can be interpreted by envisioning three distinct soil Ca pools, referred to here as the "mineral", "organically-complexed" and "exchangeable" pools. The mineral pool of Ca is supplied through dissolution of primary sediment minerals, and is the primary Ca source in the deep system. The organically-complexed pool of Ca is required to explain the light Ca isotope compositions of shallow porewaters at the youngest terraces and intermediate-depth porewaters at the older terraces relative to that of primary minerals, as well as the consistently light Ca isotope composition of grasses across the chronosequence. We suspect that the organically-complexed Ca is derived from biologic cycling within the dominant grass ecosystem, due to the similarity of δ^{44}Ca of the grasses to that of the shallow porewaters. Concentration of organically-complexed Ca in the shallow soils provides a convenient mechanism to explain strongly increasing Ca/Sr_{molar} ratios (and Ca/Mg_{molar} ratios) of bulk soil samples from about 200 cm depth to the surface at all terraces (Fig. 1b), assuming that the grasses organically complex Ca but not Sr (or Mg). Preferential complexation of Ca relative to other alkali earth elements, for example as calcium oxalate, has been demonstrated for other ecosystems (e.g., Dutton & Evans 1996) and may pertain to the grass ecosystem at this chronosequence.

The exchangeable pool of Ca is more difficult to characterize, but is required to explain increasing δ^{44}Ca of the shallow porewaters with increasing terrace age. Specifically, Fig. 1a shows that primary mineral Ca has been effectively weathered out of the shallow soils (i.e., bulk-soil Ca/Si ratios trend toward zero with decreasing depth in the intermediate soil-depth weathering zone), and it is reasonable to expect increasing cation exchange capacity in the shallow soils as they evolve. The primary input of Ca to the

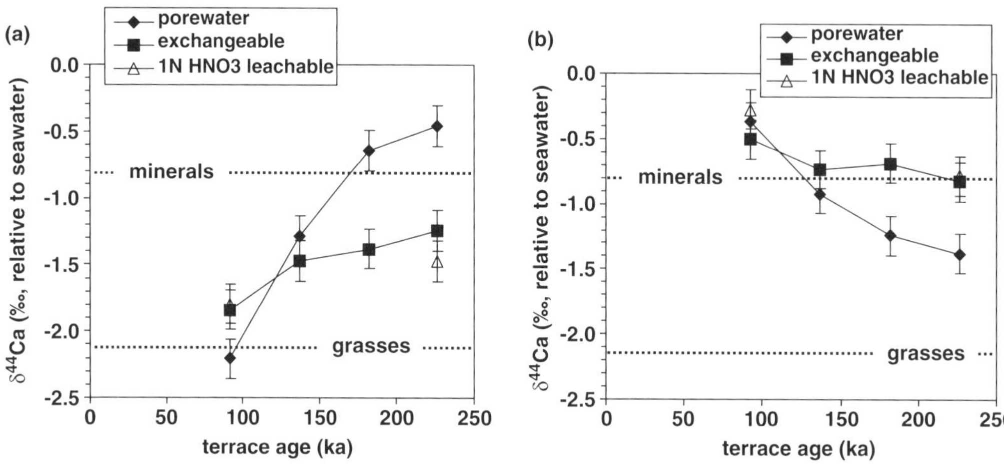

Figure 3. $\delta^{44}Ca$ of porewater samples compared to exchangeable and acid-leachable fractions of soils from (a) "shallow" and (b) "intermediate" depths. Error bars on data points represent 2σ uncertainties as in Fig. 2.

exchangeable reservoir of the shallow soil system is through precipitation and dry deposition, which is dominated by seawater and its relatively heavy Ca at this site. Thus exchange sites in the shallow soils should be progressively enriched with seawater Ca as the terrace ages. On the other hand, $\delta^{44}Ca$ of exchangeable pools at depth will vary depending mainly on both contributions of Ca from mineral weathering and the extent of interaction of infiltrating soil waters with the exchange sites (Bullen et al. 1997).

In Fig. 3, $\delta^{44}Ca$ of porewaters from shallow and intermediate depths are compared to $\delta^{44}Ca$ of exchangeable and leachable soil fractions at those depths. There is isotopic dis-equilibrium between Ca in the porewaters and Ca on the labile soil sites, and the extent of dis-equilibrium increases with increasing terrace age and, presumably, cation exchange capacity. Apparently, cation exchange is an inefficient, non-instantaneous process in these soils, a feature in common with soils comprising an analgous chronosequence developed on granitoid alluvium near Merced, CA that is set in a somewhat more arid climatic regime (White et al. 1996; Bullen et al. 1997). Modeling the development of the cation exchange pool should provide constraint on sources, sinks and fluxes of Ca during the evolution of this marine terrace chronosequence.

4 SUMMARY

Determination of Ca isotope compositions of soils and porewaters at this chronosequence has provided new insights into the complexity of Ca dynamics during weathering processes, soil development and ecosystem nutrient cycling. We recognize an important role for organically-complexed Ca, which is probably derived from biologic cycling within the dominant grass ecosystem. Consideration of this additional component at other sites should lead to a better understanding of sources of, and stresses on Ca nutrition for plants and the ecosystem as a whole.

REFERENCES

Bradley, W.C. 1957. Origin of marine terrace deposits in the Santa Cruz area, CA. *Geol. Soc. Am. Bull.* 68: 421–444.

Bullen, T.D., White, A.F., Blum, A.E., Harden, J. & Schulz, M.S. 1997. Chemical weathering of a soil chronosequence on granitoid alluvium: II. Mineralogic and isotopic constraints on the behavior of strontium. *Geochim. Cosmochim. Acta* 61(2): 291–306.

Dutton, M.V. & Evans, C.S. 1996. Oxalate production by fungi: its role in pathogenicity and ecology in the soil environment. *Can. J. Microbiol.* 42: 881–895.

Fitzpatrick, J.A., Bullen, T.D., Vivit, D.V. & White, A.F. 2002. A Strontium Isotope Reconnaissance of a Marine Terrace Chronosequence in Central California, USA. Eos Trans. AGU, 83(47), Fall Meet. Suppl., Abstract H61C-0807.

Gussone, N., Eisenhauer, A., Heuser, A., Dietzel, M., Bock, B., Bohm, F., Spero, H., Lea, D., Bijma, J. & Nagler, Th. 2003. Model for kinetic effects on calcium isotope fractionation ($\delta^{44}Ca$) in inorganic aragonite and cultured planktonic foraminifera. *Geochim. Cosmochim. Acta* 67(7): 1375–1382.

Kennedy, M.J., Hedin, L.O. & Derry, L.A. 2002. Decoupling of unpolluted temperate forests from rock nutrient sources revealed by natural $^{87}Sr/^{86}Sr$ and ^{84}Sr tracer addition. *PNAS Geology* 99(15): 9639–9644.

Perg, L.A., Anderson, R.S. & Finkel, R.C. 2001. Use of a new ^{10}Be and ^{26}Al inventory method to date marine terraces, Santa Cruz, California, USA. *Geology* 29(10): 879–882.

Skulan, J., DePaolo, D.J. & Owens, T.L. 1997. Biological control of calcium isotopic abundances in the global calcium cycle. *Geochim. Cosmochim. Acta* 61: 2505–2510.

Skulan, J. & DePaolo, D.J. 1999. Calcium isotope fractionation between soft and mineralized tissues as a monitor of calcium use in vertebrates. *PNAS Biochemistry* 96(24): 13709–13713.

White, A.F., Blum, A.E., Schulz, M.S., Bullen, T.D., Harden, J.W. & Peterson, M.L. 1996. Chemical weathering of a soil chronosequence on granitoid alluvium: I. Quantification of mineralogical and surface area changes and calculation of primary silicate reaction rates. *Geochim. Cosmochim. Acta* 60: 2533–2550.

Water–Rock Interaction, Wanty & Seal II (eds)
© 2004 Taylor & Francis Group, London, ISBN 90 5809 641 6

Mineral dissolution at the granite–saprolite interface

H.L. Buss & S.L. Brantley
Department of Geosciences, The Pennsylvania State University, University Park, PA, USA

P.B. Sak
Department of Geology, Bucknell University, Lewisburg, PA, USA

A.F. White
US Geological Survey, Menlo Park, CA, USA

ABSTRACT: The Rio Icacos watershed in Puerto Rico's Luquillo Experimental Forest is located on the Tertiary Rio Blanco quartz diorite bedrock mantled by 200–800 cm of saprolite. Previous workers have documented large differences in mineralogy and chemistry between the bedrock and the overlying saprolite. Those studies have predominately focused on wholesale changes between the bedrock and the saprolite. Herein we present a detailed examination of the weathering across the bedrock-saprolite interface. At the study site, the bedrock weathers spheroidally, forming corestones surrounded by 20–60 cm of concentric layers of weathering rock. We are exploring the chemistry and mineralogy of this complex interface between the corestones and saprolite at the microscale using inductively coupled plasma-mass spectrometry (ICP-MS), optical microscopy, scanning electron microscopy (SEM), energy dispersive X-ray spectroscopy (EDS), and electron probe microanalysis (EPMA) to characterize the chemical and physical processes that propagate the weathering front. Preliminary results from EPMA and SEM indicate that weathering is localized in cracks and along the edges of the weathering rock layers. At the outer edge of the corestone and along cracks within the first spheroidal layer, both plagioclase and hornblende release Al in the presence of water and a hydrous aluminosilicate phase precipitates in the void space. ICP-MS results are used to calculate the open-system mass transport (loss or gain) of the elements with respect to the corestone, accounting for density differences across the sampled section. These calculations show that most of the spheroidal layers have a bulk chemistry similar to the corestone but slightly depleted in most elements, with the largest chemical differences in the layer adjacent to the saprolite. Across the interface zone, the average order of mobility of the cations is $Ca > P > Na > Si > Al > Fe > Mg \approx Mn > K > Ti$.

1 INTRODUCTION

Little is known about how cohesive bedrock dissolves and disaggregates into saprolite. The rate of advance of the bedrock-saprolite interface into the subsurface depends upon water permeation and mineral dissolution. To study these phenomena, we investigate granite weathering in the Luquillo Experimental Forest (LEF) of Puerto Rico. The Rio Icacos watershed in the LEF has the highest documented chemical weathering rate of granitic rocks in the world (White et al. 1998). The Guaba Ridge in the Rio Icacos watershed is composed of Tertiary age Rio Blanco quartz diorite bedrock capped with 200–800 cm of saprolite and topped by 50–100 cm of soil. Weathering of the bedrock to saprolite in the Rio Icacos watershed has been studied and the processes by which the plagioclase, biotite, and quartz weather have been established (Murphy et al. 1998; White et al. 1998; Schulz & White 1999; White et al. 2001; White 2002; Turner et al. 2003; Riebe et al. 2003). However, the processes controlling hornblende dissolution from the bedrock has yet to be systematically investigated. Hornblende is ubiquitous in the bedrock but absent in the saprolite. Previous mineral weathering studies at this site have either not reported (Murphy et al. 1998; White et al. 1998; Schulz & White 1999) or characterized (Turner et al. 2003) partially weathered hornblende. The absence of hornblende within the saprolite suggests that hornblende must weather completely at the bedrock-saprolite interface. However, the interface is not simple, but is comprised of about 20–60 cm of spheroidal weathering layers surrounding corestones. Each of these closely spaced layers of intact, but weathered, bedrock is about 2–6 cm thick.

Corestones are residual bedrock that is weathered in place into boulders (Linton 1955; Ollier 1971). *Exfoliation* describes the formation of sheets of rock by any weathering process although the term is more commonly used for large scale sheeting formed by tectonic forces, unloading, or confining stress. *Spheroidal weathering* is a specific type of exfoliation in which corestones are surrounded by concentric layers of weathering rock (Ollier 1971). Spheroidal weathering is synonymous with *concentric* or *onion-skin weathering* although Twidale (1968) differentiated thin sheeting as spheroidal and thick sheeting as onion-skin weathering. The sheets formed by spheroidal processes are variously called sheets, layers, shells, plates, scales, or rindlets (Bisdom 1967; Ollier 1971; Fritz & Ragland 1980). The term *rindlets* was used by Turner et al. (2003) to describe layers of weathering rock within the Rio Icacos saprolite. Although those rindlets are related to the present study, they are not part of the corestone-saprolite interface, but are isolated within the saprolite. Therefore, we use different terminology here to identify the spheroidal or onion-skin layers surrounding the corestones as distinct from the rindlets of Turner et al. (2003).

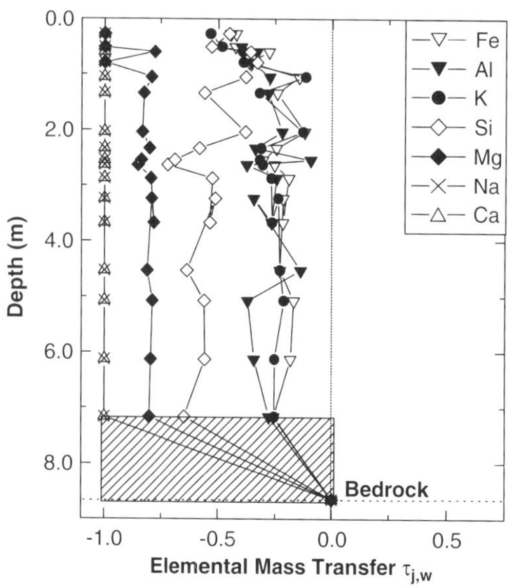

Figure 1. Elemental mass transfer with depth in the Rio Icacos saprolite adapted from White et al. (1998). The shaded region indicates the approximate zone sampled for the current study.

The abrupt changes in mineralogy and chemistry between the bedrock and saprolite reported in previous work at the Rio Icacos site span the onion-skin interface zone (Fig. 1). Herein, we present detailed chemical and mineralogical microanalyses of the onion-skin weathering zone using ICP-MS, optical microscopy, SEM, EDS, and EPMA. Our approach is to analyze chemical weathering at the site in a manner similar to previous workers (White et al. 1998; White et al. 2001; White 2002; Riebe et al. 2003; Turner et al. 2003) but at a finer scale (Fig. 1). In this way we plan to examine processes by which hornblende weathers and by which the complex onion-skin weathering front forms and propagates.

2 METHODS

A suite of 10 samples was collected at a road cut in the LEF by carefully removing a complete 49 cm continuous section of onion-skin layers and part of the adjacent corestone (Fig. 2). The 200 cm of overlying saprolite was sampled every 10 cm from the top onion-skin layer up to the ground surface.

Thin-sections were prepared from each sample set, straddling the edges of the weathering layers. Minerals within the thin sections were identified using optical microscopy and EPMA (Cameca SX-50). Backscattered electron maps of the thin-sections were made on an FEI Quanta 400 SEM.

Bulk elemental analysis of both major and minor elements was performed by ICP-MS of pulverized and sieved samples from each of the weathering layers. Mass transfer (loss or gain from the corestone)

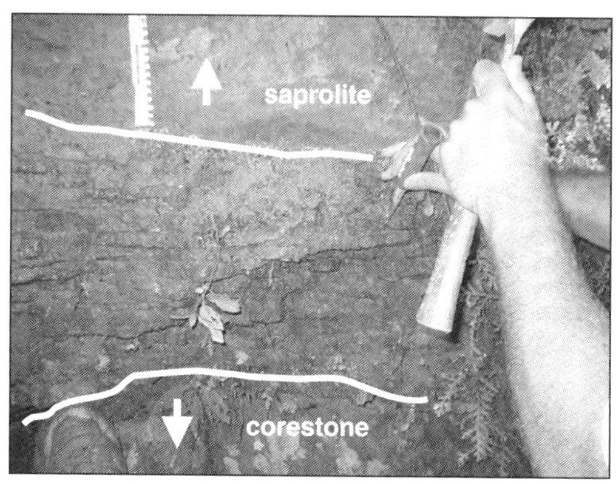

Figure 2. A continuous 49 cm section of the onion-skin layers (between the white lines) at the saprolite-corestone interface was carefully sampled. The upper portion of the corestone and every 10 cm of the entire 2 m of overlying saprolite were also sampled. Arm of A.F. White shown for scale. A schematic diagram of this sampled section is shown in Figure 4.

of individual elements relative to Ti was calculated for each section of weathered rock layers (Brimhall & Dietrich 1987). Mass transport in an open-system can be quantified when the parent material is homogenous (Rio Blanco quartz diorite), of uniform age (Tertiary), and contains an inert component (Ti) present in both the parent and product material (Chadwick et al. 1990). The immobility of Ti in the Rio Icacos watershed was determined by White et al. (1998) and confirmed by our calculations of volumetric strain with respect to Ti. Mass transfer (gain or loss) of an element j from a sample i, relative to the parent rock is calculated as:

$$\tau_{i,j} = \left(\frac{\rho_w C_{j,w}}{\rho_p C_{j,p}}\right)(\varepsilon_{Ti,w} + 1) - 1 \tag{1}$$

where, $\varepsilon_{Ti,w}$ is the volumetric strain for Ti, ρ_w is dry bulk density of the weathered layer, ρ_p is dry bulk density of the parent rock (corestone), $c_{j,w}$ is mass concentration of element i in the weathered layer, and $c_{j,p}$ is mass concentration of element i in the corestone. The more negative the value of $\tau_{i,j}$ the more mobile the element, such that if $\tau_{i,j} = -1$, element j is entirely lost. If $\tau_{i,j} > 0$, a net gain of element j relative to the corestone is indicated.

3 RESULTS AND DISCUSSION

In thin-section, no porosity is evident in the corestone (Sample 10), whereas, the first onion-skin layer (Sample 8) above the corestone contains several channels (Figs. 3 & 4). The channels cut through all minerals without preference; therefore, control of channel formation does not appear to be chemical or mineralogical but rather physical. Minerals along channels are broken up by multiple cracks and have more void space in and around them (Fig. 5), suggesting that once channels exist, they provide wetted surface area that allows water to penetrate and weather minerals further into the rock. In this manner, a positive feedback loop is established in which void space permits water infiltration, which weathers minerals to create more void space.

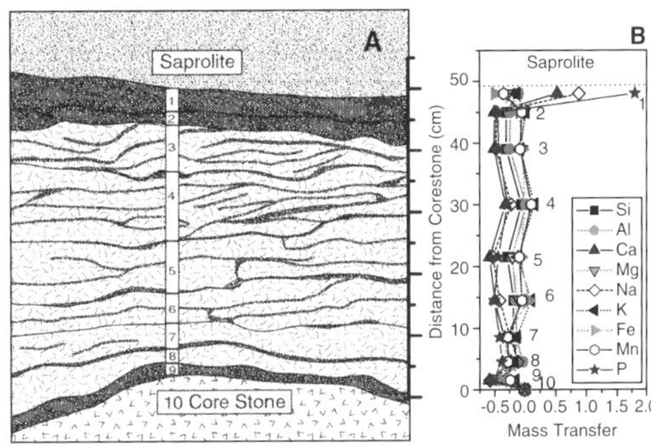

Figure 4. (A) Schematic drawing of sampled section (also shown in Fig. 2). Dark patterned areas are more friable than the light patterned onion-skin layers, but are not entirely weathered to saprolite. The onion-skin layers in the mid-section of the diagram are weathered but cohesive rock. Numbers correspond to locations of samples analyzed by ICP-MS from which loss or gain of elements from the corestone (mass transfer) was calculated shown as in (B) at the same scale as (A).

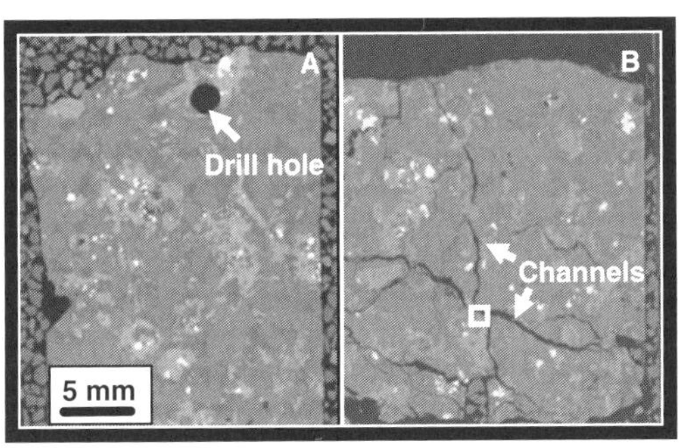

Figure 3. Backscattered electron maps of thin-sections. The top edge of the thin-section in (A) is the edge of the corestone nearest the ground surface (Sample 10, see Fig. 4). No secondary porosity is evident in the corestone. The thin-section in (B) entirely spans the first onion-skin layer above the corestone (Sample 8, see Fig. 4) and is oriented such that the top is nearest the ground surface and the bottom is towards the corestone. Several channels cross-cut this onion-skin layer. The box indicates the area shown in Figure 5.

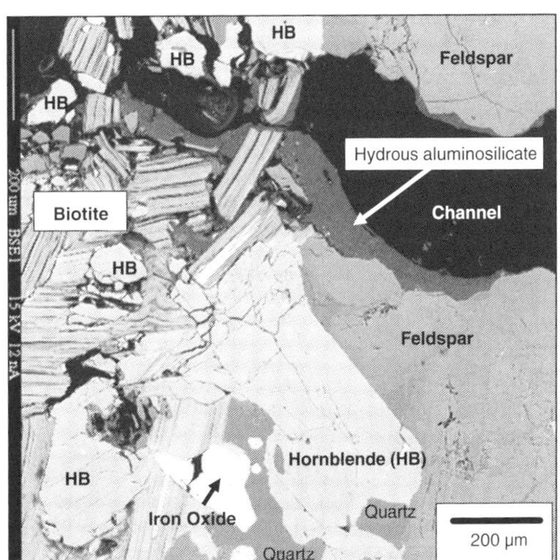

Figure 5. Backscattered electron image of a region surrounding one of the channels seen in Figure 3B.

Lining the channels of sample 8, EPMA revealed a hydrous aluminosilicate material (kaolinite?) not present elsewhere in the samples (Fig. 5). Clay minerals are found only near void spaces such as channels, cracks, or edges of the rock layers. In Sample 8, plagioclase crystals adjacent to clays or void space have less Ca and Al relative to Si than plagioclase not bordering clays or void space, as analyzed by EPMA. Similarly, hornblende crystals adjacent to void space or clays have slightly less Na and Al relative to Si than hornblende not in contact with void space or clays. Evidently the channels allow fluids to infiltrate the rock and are the focus of incongruent dissolution and precipitation of secondary phases. It remains unclear which phase weathers first.

The Al-rich material lining the channels may be precipitated from the Al released from weathered plagioclase and hornblende. Average atomic ratios of Al/Si in unweathered minerals (0.58, 0.16) are larger than the Al/Si in weathered minerals (0.49, 0.12) for plagioclase and hornblende, respectively. The higher Al content and abundance ($\sim$56 vs. $\sim$6 wt.% of the bedrock mineralogy; White et al. 1998) of plagioclase compared to hornblende indicate that plagioclase is therefore likely the primary contributor to the Al in the precipitate within the channels.

"Micro-cracks" have been related to spheroidal weathering in granites by Bisdom (1967) who suggested that chemical and physical processes widen pre-existing fractures and cleavages and form new cracks along zones of weakness. Unlike that study, our samples contain no pre-existing fractures or cleavages and weathering-induced cracks appear to cut through crystals indiscriminately (Figs. 3 & 5).

Bisdom (1967) suggests that swelling and contraction creates and widens micro-cracks. Also, expansion of biotite to hydrobiotite or clays can fracture granite to form grus (Eggler et al. 1969; Isherwood & Street 1976) and has been implicated in spheroidal weathering (Chapman & Greenfield 1949). However, this mechanism has been disputed because spheroidal weathering usually occurs isovolumetrically and often at depths where expansion is improbable due to overlying rock (Ollier 1971).

Ollier (1971) also states that crack formation is likely the first step in the spheroidal weathering process and mentions the role of infiltrating water in transporting material to maintain constant volume during weathering. This is consistent with the isovolumetric weathering at our site indicated by White et al. (1998) and our mass transfer calculations.

Mass transfer calculations show that the average order of mobility of cations in the interface zone relative to the corestone is Ca > P > Na > Si > Al > Fe > Mg $\approx$ Mn > K > Ti $\approx$ 0. This is different from what is seen in the saprolite Na $\approx$ Ca > Mg > Si > K > Al > Fe > Ti $\approx$ 0 (White et al. 1998). Mass transfer calculations also show that most of the onion-skin layers are only slightly depleted in most elements relative to the corestone (Fig. 4). The largest bulk chemical differences relative to the corestone are found in the 3 cm thick layer (Sample 1) adjacent to the saprolite. This friable, highly weathered layer is very different in texture from the relatively intact, rocky onion-skin layers below. This boundary layer has a net gain of P, Ca, and Na, the elements that are most depleted in the rocky layers below (Fig. 4). The large differences observed in this 3 cm layer indicate that the boundary between the saprolite and bedrock chemistry may be significantly more abrupt than the 150–200 cm transition zone recognized by others (Fig. 1; White et al. 1998; Riebe et al. 2003).

4 CONCLUSIONS

The spheroidal weathering zone between the saprolite and corestones in the Rio Icacos watershed is physically and chemically complex. Preliminary results suggest initial physical control of weathering exhibited by crack formation in the layers nearest the corestone and large chemical changes in the layers adjacent to the saprolite. These results imply that weathering starts locally near the corestone and intensifies at the outer edges of the spheroidal weathering zone. Further analyses of the weathering layers, corestones, and saprolite using EPMA, ICP-MS, XPS, SEM, and EDS are underway and will be presented. Questions remain about the relative order and rates of hornblende and plagioclase weathering, weathering mechanisms, and controls on crack formation. We seek to answer these questions and to explain the processes that govern the character and propagation of the weathering front via detailed investigations of the interface between the highly weathered saprolite and the unweathered bedrock.

ACKNOWLEDGEMENTS

We thank J. Troester and D. Mericle for field support, M. Angelone for EPMA assistance. Funding provided by NSF-IGERT grant DGE-9972759.

REFERENCES

Bisdom, E.B.A. 1967. The role of micro-crack systems in the spheroidal weathering of an intrusive granite in Galicia (NW Spain). *Geologie en Mijnbouw* 46: 333–340.

Brimhall, G.H. & Dietrich, W. E. 1987. Constitutive mass balance relations between chemical composition, volume, density, porosity, and strain in metasomatic

hydrochemical systems. *Geochim. Cosmochim. Acta* 51: 567–587.
Chadwick, O.A., Brimhall, G.H. & Hendricks, D.M. 1990. From black box to a grey box: a mass balance interpretation of pedogenesis. *Geomorphology* 3: 369–390.
Chapman, R.W. & Greenfield, M.A. 1949. Spheroidal weathering of igneous rocks. *Am. J. Sci.* 247: 407–429.
Eggler, D.H., Larson, E.E. & Bradley, W.C. 1969. Granites, grusses, and the Sherman erosion surface, southern Laramie Range, Colorado-Wyoming. *Am. J. Sci.* 267: 510–522.
Fritz, S.J. & Ragland, P.C. 1980. Weathering rinds developed on plutonic igneous rocks in the North Carolina piedmont. *Am. J. Sci.* 280: 546–559.
Isherwood, D. & Street, A. 1976. Biotite-induced grussification of the Boulder Creek Granodiorite, Boulder County, Colorado. *Geol. Soc. Am. Bull.* 87: 366–370.
Linton, D.L. 1955. The problem of tors. *Geog. J.*121 470–487.
Murphy, S.F., Brantley, S.L., Blum, A.E., White, A.F. & Dong, H. 1998. Chemical weathering in a tropical watershed, Luquillo Mountains, Puerto Rico: II. Rate and mechanism of biotite weathering. *Geochim. Cosmochim. Acta* 62: 227–243.
Ollier, C.D. 1971. Causes of spheroidal weathering. *Earth-Sci. Rev.*7: 127–141.
Riebe, C.S., Kirchner, J.W. & Finkel, R.C. 2003. Long-term rates of chemical weathering and physical erosion from cosmogenic nuclides and geochemical mass balance. *Geochim. Cosmochim. Acta* 67(22): 4411–4427.
Schulz, M.S. & White, A.F. 1999. Chemical weathering in a tropical watershed, Luquillo Mountains, Puerto Rico: III. Quartz dissolution rates. *Geochim. Cosmochim. Acta* 63: 337–350.
Turner, B.F., Stallard, R.F. & Brantley, S.L. 2003. Investigation of in situ weathering of quartz diorite bedrock in the Rio Icacos basin, Luquillo Experimental Forest, Puerto Rico. *Chem. Geol.* 202: 313–341.
Twidale, C.R. 1968. *Geomorphology, with special reference to Australia*. Melbourne: Nelson.
White, A.F., Blum, A.E., Schulz, M.S., Vivit, D.V., Stonestrom, D. A., Larsen, M., Murphy, S. F. & Eberl, D. 1998. Chemical weathering in a tropical watershed, Luquillo Mountains, Puerto Rico: I. Long-term versus short-term weathering fluxes. *Geochim. Cosmochim. Acta* 62: 209–226.
White, A.F., Bullen, T.D., Schulz, M.S., Blum, A.E., Huntington, T.G. & Peters, N.E. 2001. Differential rates of feldspar weathering in granitic regoliths. *Geochim. Cosmochim. Acta* 65: 847–869.
White, A.F. 2002. Determining mineral weathering rates based on solid and solute weathering gradients and velocities: application to biotite weathering in saprolites. *Chem. Geol.*190: 69–89.

Water–Rock Interaction, Wanty & Seal II (eds)
© 2004 Taylor & Francis Group, London, ISBN 90 5809 641 6

Chemical denudation in the western Indies: preliminary data indicate relatively low denudation rates

J. Gaillardet, S. Rad, P. Louvat & C. Allègre
Laboratoire de géochimie-cosmochimie, Groupe de Recherche sur l'Erosion et les Eaux Continentales, Institut de Physique du Globe de Paris, Paris, France

ABSTRACT: We report here preliminary data on river chemistry on Martinique and Guadeloupe Islands. These Islands belong to the Western Indies volcanic arc and are characterized by explosive active volcanism (Soufrière de Guadeloupe and Montagne Pelée). River chemistry shows a limited influence of hydrothermal springs, especially for the oldest part of the Islands. Here, we give a first order estimate of chemical denudation rates based on the analysis of river chemistry.

1 INTRODUCTION

Previous studies (e.g. Louvat & Allègre 1997) have pointed out the major influence of basalt weathering for the global C cycle. A rough estimate made recently by Dessert et al. (2003) lead to the conclusion that although basaltic rocks outcrop on about 5% of the Earth's surface, the CO_2 consumption flux that results from the chemical interactions between basalts and surface waters may be on the order of 30% of the total CO_2 consumed by silicate weathering on Earth. However, most of the data available on the chemistry of rivers draining volcanic rocks are for rivers draining basalts (Deccan Traps, Reunion, Azores, Iceland). The volcanic terranes of subduction zones have garnered little attention for river chemistry purposes. Especially, the chemical denudation rates of andesites and other arc-related lithologies are poorly documented. There is a clear need for more data for such geological settings, which are a preferential locus of mantle-sediments and crust coupling and where new continental crust is formed.

This paper is the first report of a broad initiative to study erosion (on both mechanical and chemical processes) in the western Indies. The volcanic arc of the western Indies presents several interesting features that make it a good natural laboratory for understanding the "Earth's weathering engine". In those islands where demographic pressure is becoming more and more important, there is a clear need to better understand water quality issues and predict the changes that are occurring on both the quality and quantity of water. In the perspective of developing a worldwide network of hydrosystems and an interdisciplinary and concerted program on chemical weathering at all scales and with different approaches (for the U.S. see the Weathering System Science Consortium, http://www.wssc.psu.edu), the western Indies could represent a normative site for humid tropical weathering systems in a subduction context.

2 MAJOR CLIMATIC, GEOMORPHIC AND SOIL FEATURES OF THE WESTERN INDIES

The volcanic style in the western Indies is explosive volcanism. The most recent eruptions occurred on Montserrat Island in 1991 and in Martinique (Montagne Pelée) in 1902 (Fig. 1). A gradient of age can be found among the different islands of the western Indies going back to the Mid-Tertiary terranes of Southern Martinique Island. In Guadeloupe Island, there is a regular increase in age from the active volcano of La Soufrière (in the South) to the pre-Quaternary terranes to the north.

Rock types are mainly quartz-poor andesites and pyroclastic formations. Preferential erosion of the latter leads to a dissected relief with steep slopes for the andesitic remnants. On the youngest parts of the islands, permeable pyroclastic formations dominate.

Precipitation is important; local gradients in precipitation occur with more rain on the eastern coasts (exposed to the trade winds) and lower precipitation on the Western coasts. At the top of the volcanoes, exceptional values of precipitation rates are observed.

Figure 1. The volcanic arc of the western Indies.

For example, the top of the volcano Soufrière of Guadeloupe has 11 meters per year of precipitation. The dryer parts of western Indies receive 500 mm of rainwater per year.

A major feature of the western Indies is soil development. Deep saprolites are found on Martinique and Guadeloupe Islands even for terranes formed a couple of thousand years ago. This observation indicates that propagation of a saprolite front has been very rapid. A significant part of the upper soils results from weathering of unconsolidated pyroclasts deposited during major cataclysmic eruptions, as indicated by the presence of bipyramidal quartz. A major mode of physical erosion seems to be landsliding, but that phenomenon remains essentially unquantified. Most landslides and transport of solid material by rivers occur sporadically and preferentially during the rainy season (July to November), induced by tropical storms.

3 SAMPLING AND ANALYSIS

We report here analyses for the dissolved load of the major rivers of Guadeloupe and Martinique rivers. Samples were filtered through 0.2 μm filters on site and analyzed for major elements and alkalinity by ionic chromatography. Discharge data were not measured on site and only the annual average data are used here. The influence of variations in discharge (and associated variations in dissolved concentrations) will be a major concern of our future investigations. The numbers for chemical denudation rates are thus to be considered as first order estimates.

4 RESULTS

Most of the rivers from the western Indies show a total dissolved load of about 500 to 600 mg/l. A few rivers, principally located on the Soufrière volcano carry a greater dissolved load (up to 1800 mg/l). In most rivers, the dominant cation is Na and the mean Ca/Na molar ratios (not corrected for sea salts) are between 0.2 to 1.5. Chloride concentrations are significantly higher than those found in other volcanic islands, and average 300 μmol/l. Rivers of high runoff exhibit the lowest Cl concentrations (150 μmol/l), while rivers characterized by lower runoff exhibit higher concentrations (350 μmol/l).This contrast corresponds to a gradient in precipitation and evapotranspiration. The anionic charge of the rivers is dominated by bicarbonate and sulfate ions. Chloride ion is never dominant. The ratio of SO_4^{2-} concentrations to $HCO_3^- + SO_4^{2-}$ concentrations ranges between a few percent to 100%. The mean value is close to 20–30%.

Bicarbonate concentrations are lower in Guadeloupe rivers, ranging from 300 to 1000 μmol/l (mean value of 450 μmol/l), than for Martinique Rivers where HCO_3^- concentrations vary from 100 to 4000 μmol/l (mean value 800 μmol/l). The highest concentrations are found in the Rivière Blanche, which partly drains the "coulée blanche" lava flow of 1902.

We calculated cationic denudation rates after correcting the concentrations measured in the rivers from atmospheric input (Cl-based method). Atmospheric CO_2 consumption rates were calculated based on the concentration of HCO_3^- in the rivers. Mean annual runoff was used. A correlation was observed between cationic denudation rates and CO_2 consumption rates except for the rivers having high sulfate concentrations.

5 DISCUSSION AND CONCLUSIONS

5.1 *Sulfide oxidation vs. CO_2 consumption*

The early stages of chemical denudation preferentially use sulfuric acid as a H^+-promoting agent. Sulfuric acid is derived from the oxidative weathering of volcanic and hydrothermal sulfides. This is especially evident on La Soufrière volcano. The Montagne Pelée volcano shows much less influence of sulfide oxidation. On a long-term perspective, chemical denudation of the Soufrière volcano is not a sink of CO_2 but rather a source. The older parts of Martinique and Guadeloupe Islands are dominated by CO_2-driven chemical weathering reactions. A couple of hydrothermal deposits or thermal springs may contribute locally, but most of the cations are equilibrated by atmospheric-derived HCO_3^- for bedrock ages greater than a few hundred thousand years.

5.2 *Effect of runoff*

While runoff values for the different rivers studied show a large range of variations, no simple trend of either concentrations or fluxes (of cations or CO_2 consumption) can be seen in the data. This result shows that runoff is not the dominant controlling factor in Martinique and Guadeloupe Islands. River chemistry is highly sensitive to other parameters that overprint the role of runoff. Even a low contribution of hydrothermal precipitates is able to significantly impact the river chemistries.

5.3 *Effect of time*

Time is one of the most important factors influencing the chemistry of rivers in the western Indies. There is a sharp decrease in chemical denudation rates with the age of the bedrock. The highest chemical denudation rates in the western Indies rivers are for the rivers draining the pyroclastic terranes of the Montagne Pelée volcano. Data from the groundwaters flowing on the slopes of the volcano show high cation concentrations and alkalinity values probably caused by enhanced residence times of water and higher temperatures compared to surface waters. Even a small contribution of flank aquifers to rivers would considerably increase the chemical load of rivers. This is probably the case for the rivers draining the Montagne Pelée.

5.4 *Weathering rates and CO_2 consumption rates are low*

Decoupling the influence of the above parameters is crucial and needs the help of specific chemical and isotopic tracers as well as temporal surveys. However, the major result of the present study is that despite optimum climatic conditions, the cationic denudation rates of Guadeloupe Island are low compared to those of other volcanic areas covering a range of climates (data from Louvat & Allègre 1997, 1998; Dessert et al. 2003). Cationic denudation rates of Guadeloupe rivers (mean temperature of 25°C) are

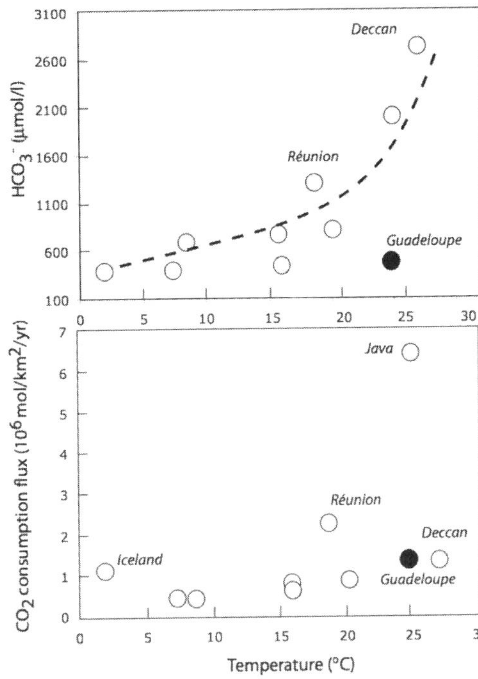

Figure 2. Alkalinity concentration and CO_2 consumption fluxes as a function of temperature deduced from river chemistry of a panel of rivers draining volcanic rocks and submitted to different climates.

on the order of magnitude of those of Réunion Island (mean temperature of 20°C), and 3 times lower than those of Java Island (27°C). Due to the contribution of sulfate ions to the anion budget, alkalinity and CO_2 consumption rates found in the western Indies are low. Figure 2 shows HCO_3^- concentrations and fluxes as a function of temperature. Compared to Reunion Island, CO_2 consumption rates of Guadeloupe Island are 3 times lower and comparable with the chemical denudation rates of Icelandic rivers. The present data from Guadeloupe Island add some scatter in the global correlation reported by Dessert et al. (2003) between chemical denudation rates of volcanic areas (or alkalinity concentrations) and temperature. If these data are representative of the actual situation on Guadeloupe Island, then CO_2 consumption by volcanic rock chemical denudation does not depend solely on temperature.

There are several possible explanations for the low denudation and CO_2 consumption rates in the western Indies. Apart from the geological structure, the importance of sulfide oxidation, the volcanic rock types, the main difference between La Réunion and Guadeloupe Islands is the existence of thick soils and deep saprolites on Guadeloupe. Waters from La Montagne Pelée, where soils have not had time to develop, show higher CO_2 consumption rates.

Our conclusion is that more work is needed to characterize the role of different factors that influence river chemistry of subduction-zone islands. This effort requires better hydrological control, the search for proxies of hydrothermal processes and a better understanding of the feedbacks between soil and river water chemistry. The dynamics of soil formation are a major concern. The question of knowing why saprolite and soil development are so different from one place to another is probably one of the most important outstanding questions that requires a multidisciplinary approach, including geomorphologists, soil scientists, hydrologists and biogeochemists. The western Indies are a good place to focus our efforts.

REFERENCES

Louvat, P. & Allègre, C.J. 1997. Present denudation rates on the Island of Réunion determined by river chemistry. *Geochim. Cosmochim. Acta* 61: 3645–3669.

Louvat, P. & Allègre, C.J. 1998. Riverine erosion rates on Sao Miguel volcanic island, Azores Archipelago. *Chem. Geol.* 148: 177–200.

Dessert, C., Dupré, B., Gaillardet, J., François, L. & Allègre, C.J. 2003. Basalt weathering laws and the impact of basalt weathering on the global carbon cycle. *Chem. Geol.* 202: 257–273.

Water–Rock Interaction, Wanty & Seal II (eds)
ISBN 90 5809 641 6

Contributions of chemical weathering and physical erosion of soil to landscape lowering in a granitic terrain

E.G. Green, W.E. Dietrich & J.F. Banfield
University of California, Berkeley, California, USA

S.A. Welch
Australian National University, Canberra, Australian Capitol Territory, Australia

ABSTRACT: We measured the contribution of weathering to landscape lowering across a retreating escarpment in southern NSW, Australia. On convex hilltops, where transport of particles is dominated by soil creep, we determined that significant solute loss occurs in the soil. The contribution of erosion to landscape lowering varies significantly with position relative to the escarpment and total denudation rate, but net chemical weathering rates are indistinguishable.

1 INTRODUCTION

Two main processes, chemical weathering and physical erosion, control the removal of mass from hillslopes. Chemical weathering of minerals at the Earth's surface controls the amount of solutes in watersheds, rivers, and the oceans, as well as the formation of clays. Chemical weathering is also the primary control on macronutrients (other than C and N) that are supplied to a soil environment. The concentrations of Ca, Mg, K, and PO_4, as well as micronutrients including Fe are also determined by chemical weathering rates. The rate at which soil minerals dissolve directly influences the soil chemistry in a given ecosystem. An understanding of how minerals dissolve in a soil environment can yield important information about how these environments will respond to changes in temperature, precipitation, and overall climate.

Physical erosion is the mechanism by which soil particles are ultimately transported off of a hillslope. Bioturbation is one mechanism that converts saprolite to soil and is the driving force for movement of soils downslope. Cosmogenic nuclide accumulation can be used to measure the flux of soil mass on a given hillslope, some unknown portion of which is due to chemical weathering.

By combining soil erosion rates with a mass-balance approach to chemical weathering in soils, it is possible to measure both chemical weathering rates for various elements present in the soil, as well as estimate the rate of physical erosion occurring on the hillslope.

At a field site in NSW, Australia, Heimsath et al. (2000) found that the soil production rates decrease exponentially with increasing soil depth. In this study, we use the mass removal described by the soil production function to measure the chemical weathering rates in both the soil and saprolite. We accomplish this by combining a mass balance approach to soil weathering, as outlined by (Brimhall & Dietrich 1987) and the soil production function as a measurement of mass flux.

2 METHODS

2.1 *Field area*

The Great Escarpment in southeastern Australia runs approximately parallel to the eastern coastline, at distances between 50 and 200 km from the shore. The ~900 m escarpment separates low-relief highlands to the west from the coastal belt. The escarpment is thought to have formed as a result of initiation of the rift that formed the Tasman Sea approximately 80 Mya. Since that time, the escarpment has been slowly propagating inward (Ollier 1982). Several hundred kilometers north of the field site for this study, cosmogenic nuclide analyses of erosion rates on either side of the escarpment indicate the highlands to the west are eroding up to 20 times slower than the lowlands east of the escarpment (Weissel & Seidl 1998).

Nunnock River (NR), the field site for this study, is located at the base of the Great Escarpment.

Previous workers studied the Bemboka site, located at the top of the escarpment (Banfield & Eggleton 1989; Taunton et al. 2000a & b). Both field sites are located within the Bemboka granodiorite. Detailed mineralogical and geochemical studies at the Bemboka site have shown that no Ce loss occurs in the saprolite, but it is accumulated in the soil (i.e. Ce is a conservative tracer, Taunton et al. 2000b).

The mean annual precipitation at both the Nunnock river and Bemboka sites is ~80 cm/yr, and the average annual temperature is ~12°C (Bureau of Meteorology, Australia, 2003). The hillslope gradient on the eastern face of the Nunnock River, where the sampling for this study occurred, is between 0.23 and 0.25. A tributary of the Nunnock River bounds the steeper western side of the hillslope. The gradient on the west side is 0.35. We attribute this difference in slope (Fig. 1) to rapid incision along the fault that is exposed in the creek bed.

2.2 *Sample collection*

Samples were collected at the Nunnock River site from the soil-mantled hillslope at a variety of soil transport distances (calculated as distance along greatest gradient from the ridge) and depths. The locations of sampled soil pits, up to 1 m deep, are indicated by the diamonds on Figure 1. Four transects were sampled. The first runs along the ridge for 117 m. The second and third transects parallel the direction of soil transport and the greatest gravitational gradient. A fourth transect was excavated by a backhoe, and extended perpendicular to the ridge for 18 m. The area exposed by the trench included places up to 50 cm below the soil/saprolite boundary at the crest of the ridge, and soils greater than 80 cm deep at the boundary between convex and concave (swale) regions of the hillslope. This sampling scheme allowed us to collect samples from soils which have undergone little or no transport (along the ridge crest) as well as soils which have been transported a variety of distances downslope. Fresh rock samples were taken from emergent tors on a nearby slope. The bulk geochemistry of each sample was measured using X-ray fluorescence (XRF).

2.3 *Measurement of bulk chemistry*

Soil samples were oven-dried at 105°C for 24 hours and then prepared for analysis by XRF. Samples were powdered in a tungsten carbide mill for 5 minutes, and then ashed in ceramic crucibles at 500°C to remove organic material. Trace element concentrations were measured from pressed powder pellets bound with boric acid, while major element compositions (Si, Al, Ca, Na, K, Ti, Fe, Mn, and Mg) were measured from glass discs fused with lithium tetraborate.

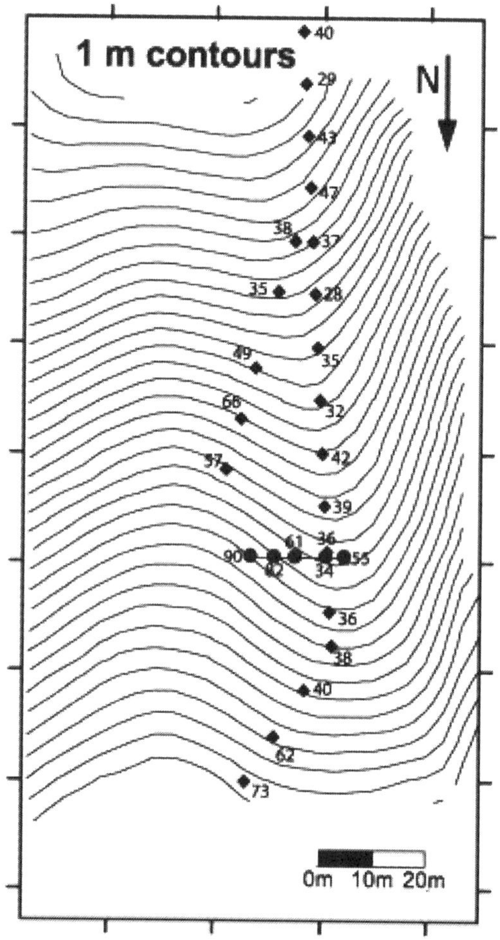

Figure 1. Contour map of the hillslope showing location of soil pit transects and trench. Numbers next to pits are soil depths (cm). Data for contour map from Heimsath et al. (2000).

2.4 *Laser ablation ICP-MS*

Samples were analysed by laser ablation ICP-MS using an ArF Excimer laser (193 nm), with a sample cell connected to an Agilent HP 7500s ICP-MS. Sample ablation was conducted under He and the ablated material was carried to the ICP-MS in a mixed stream of Ar and He. Data was collected from several traverses across polished rock and mineral mounts using a 50 μm spot size.

2.5 *Theory and calculations*

We use two factors to describe the extent of weathering that a sample has undergone. The degree of accumulation of an immobile element relative to its fresh rock concentration is an indicator of how weathered the soil sample is, and an expression for this, the Mass Loss Factor (MLF) is defined as:

$$MLF = \frac{[Zr_{rock}]}{[Zr_{soil}]} - 1 \qquad (1)$$

.

The more negative an MLF value is, the greater the percentage of overall removal of mass by chemical weathering. Using the accumulation of Zr in a weathered sample, it is also possible to write an expression that relates how much a specific element is depleted relative to the fresh rock, as shown in Equation 2:

$$\tau_X = \frac{[Zr_{rock}]}{[Zr_{soil}]}\frac{[X_{soil}]}{[X_{rock}]} - 1 \qquad (2)$$

Positive values of τ indicate that a sample is enriched in that element relative to the fresh rock, while negative values indicate a net loss of mass due to weathering. We can use τ to calculate the bulk weathering rates from the hillslope as well:

$$\text{Weathering Rate}_X = \frac{\rho_{rock}[X_{rock}]\tau_X h}{\text{soil residence time}} \qquad (3)$$

where h represents average soil thickness, and ρ_{rock} is the density of the parent material. The soil residence time equals the soil thickness divided by the landscape lowering rate (soil residence time $= h/\varepsilon$) where ε is the expression for the dependence of the soil production function on the depth of soil previously defined for Nunnock River (Heimsath et al. 2000).

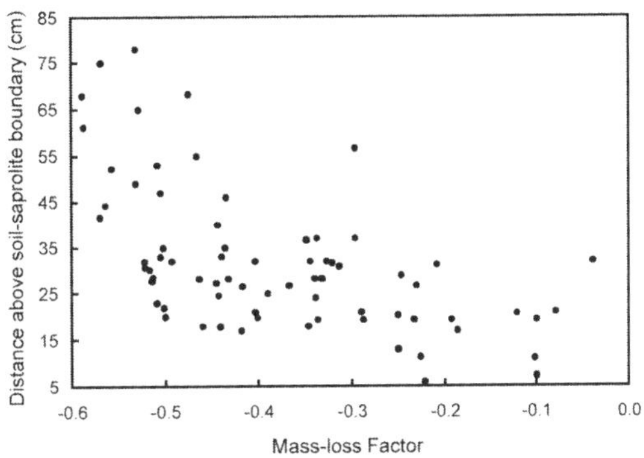

Figure 2. Mass Loss Factor vs. distance from soil/saprolite boundary.

Table 1. Mass Loss Factor and τ of Si for various locations at Nunnock river.

Sample location	MLF	Average τ_{Si}
Saprolite	−0.19	−0.17
Boundary soil	−0.20	−0.22
Swale soil	−0.53	−0.48

3 RESULTS

3.1 *MLF and τ*

Mass Loss Factor values were calculated for each sample from the Nunnock River site, and plotted with respect to distance from the soil/saprolite boundary (Fig. 2). This plot shows that MLF values generally decrease as distance from the soil/saprolite boundary increases, indicating that the mass removal by chemical weathering is also increasing (details will be reported in Green et al., in prep). Values for the MLF in the saprolite near the surface are around 0.2, indicating that approximately 20% of the mass of the original material has been lost before conversion of saprolite to soil. In addition, we find that MLF values reach as great as 0.6 in soils that have been transported approximately 60 m downslope.

The distribution patterns of τ support those for the MLF. As distance from the soil-saprolite boundary increases, values of τ for all major elements with the exception of manganese become increasingly negative. Values of τ for Si range from −0.2 in the saprolite to −0.6 in soils in the swale (see Table 1). This indicates that well over half of the silica loss occurs in the soil. The values of τ for other major elements also become increasingly negative as distance from the boundary increases, indicating that the majority of weathering takes place in the soil region.

3.2 *Chemical weathering rates, and contributions to landscape lowering*

At Nunnock River, the dependence of soil production rates on soil thickness implies that the overall erosion rates vary across the hillslope (Heimsath et al. 2000). At the top of the ridge at NR, where soils are only 30 cm thick, erosion rates are highest, $30\,\text{m} \cdot \text{Ma}^{-1}$. In contrast, soils nearer the swale are up to 80 cm thick and are eroded much more slowly, $\sim 11\,\text{m} \cdot \text{Ma}^{-1}$. Using these erosion rates, we calculate the soil residence times to vary between 10 and 80 kyr. We calculated the weathering rate of Si from soils just prior to entry into the swale (where a new transport mechanism dominates) to be $13\ \text{tons} \cdot \text{km}^{-2} \cdot \text{yr}^{-1}$.

The degrees to which solute loss contributes to mass removal from the hillslopes were calculated for both the Nunnock River (at several locations across the hillslope) and the Bemboka field sites, and are shown in Table 2. At the Bemboka field site, where catchment-averaged denudation rates are lower (Heimsath et al. 2001), the contribution of chemical

Table 2. Comparison of absolute lowering rates and contributions of physical and chemical processes towards lowering from Nunnock river and Bemboka sites.

Site	Lowering rate $m \cdot Myr^{-1}$	Physical $tons \cdot km^{-2} \cdot yr^{-1}$ %	Chemical %	Si loss rate
Bem	9	23	77	14
NR (ridge)	30	64	36	7
NR (swale)	11	44	56	13

weathering to mass removal is much greater (Table 2) than at any point at Nunnock River.

The percentage of landscape lowering due to erosional losses is higher where erosion rates are higher (Table 2). It is interesting to note that the annual rates of silica loss from both sites are indistinguishable, despite the difference in erosion rate and total landscape lowering rate.

The distance to the saprolite/granite boundary was measured to be ~18 m in a drill core. The boundary was observed to have low topography and similar depth in several nearby road cuttings. Using an average saprolite thickness of 18 m, an estimated average surface area for the saprolite and soil of $0.1\ m^2 \cdot cm^{-3}$, and assuming the system is in near steady state, we calculated a rock dissolution rate based on silica loss of $2 \cdot 10^{-19} mol \cdot cm^{-2} \cdot sec^{-1}$.

3.3 *Laser-Ablation ICP-MS*

Grains of feldspar, quartz, and biotite were analyzed with Laser Ablation Inductively Coupled Plasma Mass Spectrometry (ICP-MS). Results show that Ga is a relatively immobile element and is concentrated in secondary Fe-Al phases. However, it is depleted relative to Al, indicating that it is more mobile than Al. This would not be expected if complexation by organic molecules were significant. Although Ce is locally enriched relative to the other lanthanides in a small number of cases, typically Ce and other lanthanide concentrations are correlated. This distinguishes weathering at Nunnock River from that occurring in the Bemboka profiles studied previously (Taunton et al. 2000a; 2000b, Banfield & Eggleton 1989).

4 CONCLUSIONS

By integrating results from geomorphological and geochemical studies, we have determined relative contributions of solute loss and physical erosion to landscape lowering for two landscapes separated by an escarpment. We showed that solute loss is significant in both cases, and physical erosion is more important at the Nunnock River field site than at the Bemboka site. The higher physical erosion rate correlates with higher overall denudation rates, whereas chemical weathering rates are similar. The mass loss due to chemical weathering accounts for up to 77% of total denudation rate. At both field sites the majority of solute loss occurs in the soil. The total silica loss rates are comparable to those reported from other granitic terrains worldwide (Stonestrom et al. 1998) and yield an estimated granite weathering rate based on silica release on the order of $2 \cdot 10^{-19}$ $mol \cdot cm^{-2} \cdot sec^{-1}$.

REFERENCES

Banfield, J.F. & Eggleton, R.A. 1989. Apatite Replacement and Rare-Earth Mobilization, Fractionation, and Fixation During Weathering. *Clays Clay Mins.* 37: 113–127.

Brimhall, G.H. & Dietrich, W.E. 1987. Constitutive Mass Balance Relations between Chemical-Composition, Volume, Density, Porosity, and Strain in Metasomatic Hydrochemical Systems – Results on Weathering and Pedogenesis. *Geochim. Cosmochim. Acta* 51: 567–587.

Heimsath, A.M., Chappell, J., Dietrich, W.E., Nishiizumi, K., & Finkel, R.C. 2000. Soil production on a retreating escarpment in southeastern Australia. *Geol.* 28: 787–790.

Heimsath, A.M., Chappell, J., Dietrich, W.E., Nishiizumi, K., & R.C. Finkel 2001. Late Quaternary erosion in southeastern Australia: a field example using cosmogenic nuclides. *Quaternary International* 83–5: 169–185.

Ollier, C.D. 1982. The Great Escarpment of Eastern Australia: tectonic and geomorphic significance. *J. Geol. Soc. Australia* 29: 13–23.

Stonestrom, D.A., White, A.F., & Akstin, K.C. 1998. Determining rates of chemical weathering in soils – solute transport versus profile evolution. *J. Hydrol.* 209: 331–345.

Taunton, A.E., Welch, S.A., & Banfield, J.F. 2000a. Geomicrobiological controls on light rare earth element, Y and Ba distributions during granite weathering and soil formation. *J. Alloys Compounds* 303: 30–36.

Taunton, A.E., Welch, S.A., & Banfield, J.F. 2000b. Microbial controls on phosphate and lanthanide distributions during granite weathering and soil formation. *Chem. Geol.* 169: 371–382.

Weissel, J.K. & Seidl, M.A. 1998. Inland propagation of erosional escarpments and river profile evolution across the southeast Australian passive continental margin. In K.J. Tinkler and E.E. Wohl (eds.) *Rivers over rock: Fluvial processes in bedrock channels*: 189–206. Washington: American Geophysical Union Monograph.

Water–Rock Interaction, Wanty & Seal II (eds)
© 2004 Taylor & Francis Group, London, ISBN 90 5809 641 6

Tracing silicate weathering and terrestrial silica cycling with Ge/Si ratios

A.C. Kurtz
Boston University, Department of Earth Sciences, Boston, MA

L.A. Derry
Cornell University, Department of Earth and Atmospheric Sciences, Ithaca, NY

ABSTRACT: Germanium follows silicon through its biogeochemical cycle, but Ge/Si ratios are fractionated by several processes in the weathering environment, potentially providing insight on silicate weathering processes and silica cycling. We present results from ongoing studies of Ge/Si behavior during weathering of basaltic (Hawaii) and granitoid (Puerto Rico) parent materials. Ge/Si ratios in streamwaters trace delivery of silica sourced from primary weathering, dissolution of secondary soil minerals, and plant-cycled silica to streams. Our initial results indicate that streamwater Ge/Si ratios at Luquillo, Puerto Rico reflect a mixture of silica derived from various weathering reactions, as predicted by earlier models of Ge/Si behavior. In contrast, streamwater Ge/Si ratios in Hawaii are strongly overprinted by export of biologically cycled silica.

1 INTRODUCTION

Germanium has been termed a "pseudo-isotope" of silicon, based on their similar chemical properties. Ge and Si, both group IV elements, have identical outer electron structures and have remarkably similar chemical characteristics. The similarity between tetrahedral Ge-O (1.75 Å) and Si-O (1.64 Å) bond lengths (Martin et al. 1996) allows Ge to substitute readily for Si in the tetrahedral site of silicate minerals, an example of Goldschmidt's principle of "camouflage". Germanium normally occurs as a trace element, averaging ~1 ppm in crustal rocks. Fresh igneous rocks do not vary widely in their Ge/Si ratios; the molar Ge/Si ratio in most rocks is 1 to 3 μmol/mol (DeArgollo & Schilling 1978; Bernstein 1985). Soils generally have higher Ge/Si ratios than the parent material on which they develop (Murnane & Stallard 1990; Kurtz et al. 2002). This primarily reflects partitioning of Ge into secondary soil clays, although pedogenic Fe-oxyhydroxides may play some role in sequestering Ge in soils (Kurtz et al. 2002). Ge/Si ratios in streamwaters are generally lower than bedrock ratios (Mortlock & Froelich 1987; Murnane & Stallard 1990; Chillrud et al. 1994; Anders et al. 2003.) Variability in streamwater Ge/Si ratios may reflect "intensity" of weathering in a watershed (Murnane & Stallard 1990; Froelich et al. 1992).

2 Ge/Si RATIOS IN SOILS AND PLANTS

2.1 *Basalt weathering: Hawaii*

We measured Ge/Si ratios in soils developed on a chronosequence in the Hawaiian Islands (Kurtz et al. 2002). Basaltic parent material at these sites ranges in age from 0.3 ka to 4.1 Ma. All sites are located near 1200 meters elevation, receive approximately 250 cm/yr rainfall, and are dominated by the same tree species. Degree of soil development differs substantially across the age gradient. Soils developed on the youngest sites (0.3 and 2.1 ka) have lost very little Si, while soils greater than 20 ka have lost as much as 95% of their original Si.

Unweathered Hawaiian basalts have fairly uniform Ge/Si ratios averaging 2.5 μmol/mol (DeArgollo & Schilling 1978). Soils developed on young parent material (0.3 ky and 2.1 ky) have Ge/Si only marginally enriched (2.9×10^{-6}) over unweathered Hawaiian basalts. Soils developed on older (150 ky and 4100 ky), more deeply weathered parent material, have highly fractionated Ge/Si ranging from 10×10^{-6} to 21×10^{-6}. Soil Ge concentrations increase with Si, and decrease with Fe, suggesting that Ge sequestration is related to accumulation of secondary soil silicates, rather than retention in soil Fe oxyhydroxides. Sequential extractions of these soils suggest that Ge/Si fractionation takes place by Ge sequestration during the initial precipitation of secondary soil aluminosilicates (principally allophane).

2.2 *Granite weathering: Puerto Rico*

As a complement to our field studies in Hawaii, we are investigating Ge/Si fractionation during weathering of granitic rocks under similar tropical climate conditions at a site in the Luquillo Experimental Forest in Puerto Rico (White et al. 1998). We find that Ge/Si ratios are fractionated during weathering of granitic rocks such that bulk soils and saprolites have higher Ge/Si ratios than the parent rock. The Luquillo quartz diorite has a Ge/Si ratio of 2.4 μmol/mol, while soil and saprolite Ge/Si ratios range from 2.6 to 3.6 μmol/mol (Kurtz et al. 2002).

Minerals comprising granitic rocks have distinct Ge/Si ratios and weather at different rates. Relative mineral weathering rates exert a first-order control on Ge/Si in granitic soils and saprolites. Luquillo parent material is composed dominantly of plagioclase (Ge/Si = 3.1 μmol/mol) and quartz (0.6 μmol/mol), with biotite (6.1 μmol/mol) and hornblende (6.6 μmol/mol) comprising less than 10% of the parent rock (White et al. 1998). Like the primary igneous minerals, Luquillo saprolite minerals have distinct Ge/Si ratios. White et al. (1998) argued that the dominant reaction in the bedrock to saprolite transition is the incongruent weathering of plagioclase to form kaolinite. Our data suggest that Ge is strongly partitioned into the solid during precipitation from kaolinite-saturated porewaters. Bulk soil and saprolite Ge/Si ratios reflect a mixture of quartz (0.6 μmol/mol) and kaolinite (5.9 μmol/mol).

2.3 *Phytolith Ge/Si*

Plants take up significant quantities of silica from soil solution and form phytoliths, opaline precipitates normally found in leaf tissue. Litter fall recycles this silica to the soil. Recent work has suggested that in some environments, the "internal" soil-plant silica cycle is rapid compared to the weathering input and hydrologic output of silica (Alexandre et al. 1997; Meunier et al. 1999). Hydroponic experiments with wheat demonstrate that some plants discriminate against Ge during uptake and precipitation of silica. This biological fractionation is illustrated by opal phytolith Ge/Si ratios of 0.05 μmol/mol, significantly lower than the ratio in the water in which they grow (0.4 μmol/mol). We are investigating the role of biological fractionation in influencing Ge/Si ratios in our Hawaii and Puerto Rico study sites by isolating phytolith opal from living plants. Ferns at our youngest Hawaii chronosequence site have the highest silica concentrations (up to 5 wt%), and very low Ge/Si ratios (0.05 μmol/mol), again suggesting significant discrimination against Ge during plant silica uptake. Ge/Si ratios in ferns growing in silica-depleted soils at the 150 ka chronosequence site contain between 0.5 and 1.5 wt% opal, with Ge/Si ratios between 0.1 and 0.4 μmol/mol. Ten plant species from the Luquillo site have between 0 and 4.4 wt% opal, with Ge/Si ratios ranging from 0.05 to 0.6 μmol/mol. Phytolith opal has Ge/Si ratios consistently lower than any soil mineral component at either our Hawaii or Luquillo sites.

3 Ge/Si RATIOS IN WATERS

3.1 *Soilwaters*

Froelich et al. (1992) expanded on work by Murnane & Stallard (1990) and proposed that Ge/Si ratios in streamwaters could be explained by two-component mixing. In this model, a high Si, low Ge/Si component reflects incongruent weathering of primary silicates. The second component, with low Si, high Ge/Si is derived from dissolution of Ge-enriched secondary minerals under more intense chemical weathering. The Ge/Si ratio appearing in streams is a mixture of these two sources and varies as a function of weathering intensity. If this is true, the Ge/Si ratio offers a powerful way to trace watershed-integrated weathering reactions by measuring streamwater ratios. We evaluate this model by measuring Ge/Si ratios in soil water, and comparing these to ratios measured in soil solids and streamwaters.

Soil and saprolite porewater compositions at the Luquillo site were measured on waters sampled from a suite of suction lysimeters installed in three profiles at depths of 15 cm to near the saprolite bedrock interface (up to 800 cm; White et al. 1998). Solution silica concentrations and Ge/Si ratios vary with depth (Fig. 1) and are explainable in terms of weathering reactions occurring in different parts of the saprolite profile. Water from near the saprolite-bedrock interface at the LG-1 site has the highest Si concentration of any porewater sample in this site (195 μmol/mol) and a low Ge/Si ratio (1 μmol/mol). This composition is expected as a result of weathering of plagioclase to form Ge-enriched kaolinite, while releasing some silica to solution. Higher in the saprolite profile, solution Ge/Si ratios vary between 2.3 and 3.3, consistent with dissolution of quartz and biotite (White et al. 1998). In the top 15 cm, Ge/Si ratios rise to 4.6 μmol/mol, consistent with dissolution of kaolinite (White et al. 1998.) Although plants at the Luquillo site contain abundant opal with very low Ge/Si ratio, there is no evidence in porewater compositions for very high Si, very low Ge/Si ratios expected from dissolution of plant opal.

Soil solutions from the Hawaii chronosequence sites were measured by the "saturation paste" method

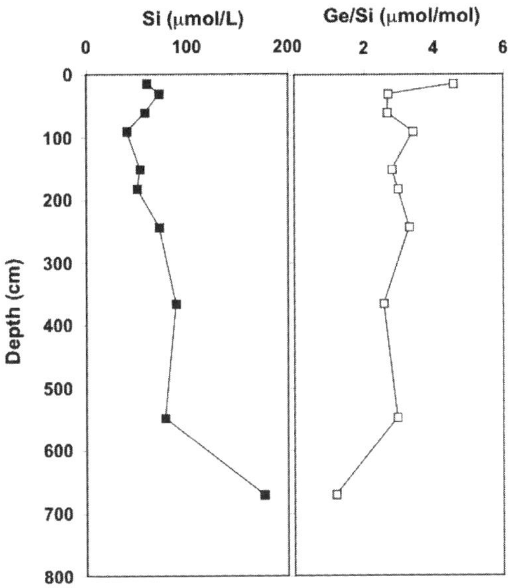

Figure 1. Depth profiles of [Si] and Ge/Si from lysimeters installed at the LG-1 (ridgetop) site at the Luquillo, Puerto Rico WEBB research site. High [Si] and low Ge/Si at depth reflects weathering of plagioclase to kaolinite. High Ge/Si at 15 cm reflects weathering of kaolinite. Ge/Si ratios at intermediate depths reflect dissolution of biotite and quartz.

(Lathja et al. 1999). At all sites, soilwater silica concentrations are highest in the top 20 cm, with values >500 μmol in some cases. Soilwater Ge/Si ratios from shallow horizons are consistently less than 1 μmol/mol, and average 0.3 μmol/mol. The combination of high Si, low Ge/Si waters is best explained by recycling of plant opal phytoliths in the top 20 cm of the soil profiles. We have not found evidence for direct water-basalt interactions that produce similar solutions, even in young, glassy soils. Ge/Si in soil waters from below 20 cm depth have low [Si] and high Ge/Si (1.5 to 5 μmol/mol) Many soil water samples from deeper than 20 cm are near gibbsite-kaolinite equilibrium (~40 μM Si).

3.2 *Silica sources in streamwaters*

The Luquillo site is instrumented with automatic streamwater samplers, allowing us to examine variations in streamwater Ge/Si ratios with individual discharge events. Ge/Si ratios measured at base flow in two gauged streams (Quebrada Guaba and Rio Icacos) at the Luquillo site are typically 0.5 μmol/mol, with dissolved Si concentrations of ~320 μmol/L (Fig. 2). This composition is consistent with incongruent weathering of plagioclase to form kaolinite, and is similar to the composition we find in lysimeter samples near the saprolite-bedrock interface. We have thus far measured streamwater Ge/Si ratios from one storm event on Rio Icacos. Si decreases and Ge/Si rises sharply as discharge increases, to a value of 2.3 μmol/mol, similar to the composition typical of most saprolite porewaters. We interpret this pattern as changing sources of silica during a storm event, with plagioclase weathering at the saprolite-bedrock interface the dominant source, switching to biotite-quartz sourced Si being flushed out of saprolite porewaters during storms.

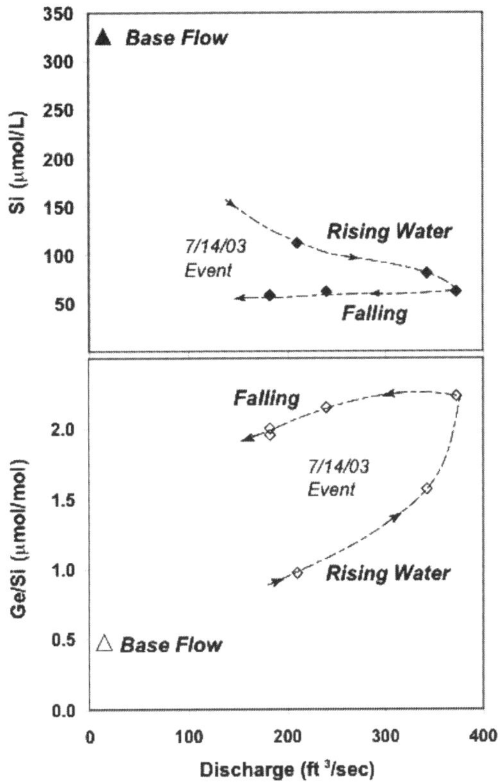

Figure 2. Streamwater [Si] and Ge/Si from Rio Icacos, a stream draining the Luquillo research site. At base flow, [Si] and Ge/Si ratios are consistent with a primary weathering source. With increasing discharge, [Si] decreases and Ge/Si increases to values similar to porewater compositions.

In Hawaii we see a similar pattern in terms of high Si, low Ge/Si ratios at low discharge, but the Si concentrations are much higher (up to 500 μmol/L), and the Ge/Si ratios as low as 0.3 μmol/mol. In contrast to Luquillo, our analysis of soilwater compositions indicates that this component must come from dissolution of plant opal in the top 20 cm of the soil profile. We find no evidence of a low Ge/Si water component resulting from incongruent dissolution of primary silicates, even in young soils that still contain volcanic glass. As at Luquillo, streamwater Ge/Si ratios rise with increasing discharge, toward a ratio consistent with typical soilwater compositions. However, Hawaiian streamwater Ge/Si ratios appear to trace mixing between phytolith-derived Si sourced in surface soils, and secondary mineral-derived Si sourced in deep soils. A compilation of published Ge/Si data from USGS-gauged Hawaiian streams (Mortlock & Froelich 1987) indicates that all are dominated by this low Ge/Si, apparently phytolith-derived source of Si. Only watersheds draining well-developed soils ever show high Ge/Si ratios, and only during periods of high discharge.

4 CONCLUSIONS

Ge/Si ratios are useful in tracing weathering reactions and pathways in the terrestrial silica cycle. Soils have high Ge/Si ratios relative to bedrock due largely to retention of Ge by secondary aluminosilicates. Understanding the reactions that give rise to soilwater Ge/Si ratios allows us to trace silica sources in streamwaters. At the Luquillo site, streamwater Ge/Si ratios are consistent with the two-component mixing model proposed by Froelich et al. 1992. Information from the Ge/Si system will ultimately be a useful addition to geochemical hydrograph separation models based on Si and oxygen isotopes. In Hawaii, streamwater Ge/Si ratios suggest a surprisingly large contribution from plant-cycled silica mixing with silica derived from dissolution of secondary clays. If our interpretations are correct, 80% of the Si delivered to the Hawaiian streams we studied was delivered via the biological pool. Although we find similarly high levels of Si in plants at the Luquillo site, we don't see evidence of plant-cycled Si either in Luquillo soil waters or in stream waters perhaps suggesting a tighter biological silica cycle.

REFERENCES

Alexandre, A., Meunier, J.-D., Colin, F. & Koud, J.-M. 1997. Plant impact on the biogeochemical cycle of silicon and related weathering processes. *Geochim. Cosmochim. Acta* 61: 677–682.

Anders, A.M., Sletten, R.S., Derry, L.A. & Hallet, B. 2003. Germanium/silicon ratios in the Copper River Basin, Alaska: Weathering and partitioning in periglacial versus glacial environments. *J. Geophys. Res.* 108, doi 10.1029/2003JF000026.

Bernstein, L.R. 1985. Germanium geochemistry and mineralogy. *Geochim. Cosmochim. Acta* 49: 2409–2422.

Chillrud, S.N., Pedrozo, F.L., Temporetti, P.F., Planas, H.F. & Froelich, P.N. 1994. Chemical weathering of phosphate and germanium in glacial meltwater streams: Effects of subglacial pyrite oxidatation. *Limnol. Oceanogr.* 39: 1130–1140.

DeArgollo, R. & Schilling, J.-G. 1978. Ge-Si and Ga-Al fractionation in Hawaiian volcanic rocks. *Geochim. Cosmochim. Acta* 42: 623–630.

Froelich, P.N., Blanc, V., Mortlock, R.A., Chillrud, S.N., Dunstan, W., Udomkit, A. & Peng, T.-H. 1992. River fluxes of dissolved silica to the ocean were higher during the glacials: Ge/Si in diatoms, rivers, and oceans. *Paleoceanography* 7: 739–768.

Kurtz, A.C., Derry, L.A. & Chadwick, O.A. 2002. Germanium-silicon fractionation in the weathering environment. *Geochim. Cosmochim. Acta* 66: 1525–1537.

Lajtha, K., W M Jarrell, W.M., Johnson, D.W., and Sollins, P., 1999, Collection of soil solution, in Robertson, G.P., Colemna, D.C., Bledsoe, C.S., and Sollins, S., eds., Standard Soil Methods for Long-Term Ecological Research, Oxford Univ Press, p. 166–182.

Martin, F., Ildefonse, P., Hazemann, J.-L., Petit, S., Grauby, O. & Decarreau, A. 1996. Random distribution of Ge and Si in synthetic talc: an EXAFS and FTIR study. *European J. Mineral.* 8: 289–299.

Meunier, J.D., Colin, F. & Alarcon, C. 1999. Biogenic silica storage in soils. *Geology* 27: 835–838.

Murnane, R.J. & Stallard, R.F. 1990. Germanium and silicon in rivers of the Orinoco drainage basin. *Nature* 344: 749–752.

Mortlock, R.A. & Froelich, P.N. 1987. Continental weathering of germanium: Ge/Si in the global river discharge. *Geochim. Cosmochim. Acta* 51: 2075–2082.

White, A.F., Blum, A.E., Schulz, M.S., Vivit, D.V., Stonestrom, D.A., Larsen, M. & Murphy, S.F. 1998. Chemical weathering in a tropical watershed, Luquillo Mountains, Puerto Rico: I. Regolith weathering rates. *Geochim. Cosmochim. Acta* 62: 209–226.

Water–Rock Interaction, Wanty & Seal II (eds)
© 2004 Taylor & Francis Group, London, ISBN 90 5809 641 6

Dissolution of Cr-containing muscovite under oxidizing and non-oxidizing conditions

G. Lee & J.G. Hering
Environmental Science & Engineering, California Institute of Technology, CA, USA

ABSTRACT: Flow-through experiments were conducted to study the dissolution of fuchsite, a Cr-rich variety of muscovite under oxidizing and non-oxidizing conditions at pH 2 and 9. Under non-oxidizing conditions, the release rates of Cr were lower than those of Al and Si even when all release rates were normalized to the elemental composition of the solid. Under oxidizing conditions, the normalized Cr release rate increased significantly at pH 9 and was comparable to that of Al. At pH 2, however, the presence of the strong oxidant ($HOCl/OCl^-$) did not enhance Cr release possibly due to adsorption of the product Cr^{VI}.

1 INTRODUCTION

Chromium in groundwater occurs primarily in the +III and +VI oxidation states. Chromium(VI) is the thermodynamically-stable species in the presence of atmospheric oxygen except at very low pH. The chemical properties and reactivities of Cr^{III} and Cr^{VI} are quite distinct and the mobility of Cr depends strongly on its redox speciation. In the United States, Cr is the third most common pollutant at hazardous waste sites (National Research Council 1991). In addition to anthropogenic contamination, Cr of natural origin has been found in groundwater in arid regions at concentrations in the tens of ppb as Cr^{VI}. Previous studies have indicated that high concentrations of Cr in groundwater are often associated with ultramafic rocks. Little is known of the mechanisms and rates of Cr release despite extensive studies of aluminosilicate mineral dissolution because these studies focused primarily on the release of major constituents of the solid. In this study, rates of Cr release were examined using fuchsite, a Cr-rich muscovite, as a model mineral under a variety of conditions.

2 MATERIALS AND METHODS

2.1 *Solid phases*

Brazilian fuchsite was obtained from the Smithsonian Institution, Washington, D.C. Samples were ground and sieved and the 125–88 μm size fraction was washed and cleaned of ultrafine particles. The elemental composition of the fuchsite samples is reported in Table 1. Surface area of the fuchsite samples was determined by BET N_2-adsorption.

Table 1. Chemical composition of fuchsite samples.

	S1	S2	S3
SiO_2	46.26	46.64	45.60
Al_2O_3	32.39	31.60	30.92
CaO	0.02	0.03	0.16
MgO	0.04	<0.0	0.18
Na_2O	0.46	0.93	1.23
K_2O	9.90	8.95	8.82
Fe_2O_3*	1.40	1.06	2.51
MnO	0.01	<0.01	0.05
TiO_2	5.03	6.15	5.15
P_2O_5	0.09	0.06	0.12
Cr_2O_3	0.71	0.49	0.96
LOI**	4.25	4.25	4.05
Sum	100.6	100.2	99.76

* as total Fe.
** Loss on ignition.

2.2 *Flow-through dissolution experiments*

Experiments were conducted using two polymethylmethacrylate stirred flow-through reactors (~50 mL) as described previously (Kraemer & Hering 1997; Giammar & Hering 2002). Experiments were conducted at the ambient temperature of the laboratory (23.0 ± 0.4°C). Two reactors were run parallel at the same pH (2 or 9) for oxidizing and non-oxidizing

Table 2. Experimental conditions for fuchsite dissolution (flow-through) and Cr^{VI} sorption onto fuchsite (batch).

Dissolution	Influent solution	pH	Sample	A (m^2/g)	Solid conc. (g/L)	Flow rate (mL/hr)	Reaction time (hr)
Non-oxidizing	0.01 M $NaNO_3$	2.00 ± 0.02	S1	0.890	18.28	1.47 ± 0.08	473.0
Oxidizing	+1 mM NaOCl	2.01 ± 0.02	S1	0.890	18.45	1.68 ± 0.10	473.0
Non-oxidizing	+0.01 M $Na_2B_4O_7$ + 200 μM EDTA	9.09 ± 0.03	S3	0.906	19.17	1.48 ± 0.31	359.5
Oxidizing	+0.01 M $Na_2B_4O_7$ + 1 mM NaOCl	9.10 ± 0.03	S3	0.906	19.35	1.77 ± 0.11	359.5

Sorption	Sample	$[Cr^{VI}]_T$ (μM)	Solid conc. (g/L)	Final pH	A (m^2/g)	Equilibration time (min)
pH edge	S2	1.93	4.01	6.88 – 1.82	1.199	5–10
Isotherm	S1	0–1.48	20.04 ± 0.03	1.98 ± 0.06	0.890	30

conditions (see Table 2 for details). Flow was controlled by a peristaltic pump and stirring was provided by a Teflon stir bar and a magnetic stirring plate. The effluent was passed through double 0.2 μm membrane filters (Isopore™, Millipore) attached to a 47 mm polypropylene filter holder capping the reactor. Total dissolved Cr, Al, Si and Fe concentrations in the effluent were determined by inductively coupled plasma mass spectrometry (ICP-MS).

The steady-state release rate, R ($mol \cdot m^{-2} s^{-1}$) of each element from the solid was calculated using the following equation:

$$R = \frac{1}{S \cdot A} \cdot \frac{d[M]_{diss}}{dt} = \frac{1}{S \cdot A} \cdot [M]_{diss} \cdot \frac{q}{3600 \cdot V} \quad (1)$$

where $[M]_{diss}$ is total dissolved metal concentrations [$mol \cdot L^{-1}$] in the effluent, S is the solid concentration [$g \cdot L^{-1}$], A is the specific surface area of the solid [$m^2 \cdot g^{-1}$], q is the flow rate [$mL \cdot h^{-1}$] and V is the reactor volume [mL]. Steady-state release rates were determined after metal concentrations in the effluent reached a constant value.

2.3 *Chromate sorption experiments*

Sorption of Cr^{VI} on fuchsite (125–88 μm fraction) was determined in batch systems as a function of both pH and variable sorbate concentration (see Table 2 for experimental conditions). For pH experiments, the solid was added to a solution of Cr^{VI} in 0.01 M $NaNO_3$ at pH ~ 7. Subsequently, the suspension was magnetically stirred and the pH was lowered stepwise by addition of 0.1 or 1.0 M HNO_3 to a final pH of 2. At each step, a sample was taken from the suspension 5 to 15 min after the pH stabilized.

The sorption isotherm for Cr^{VI} was obtained at pH ~ 2. Aliquots of Cr^{VI} stock solutions were added to fuchsite suspensions to yield the desired final concentrations of $[Cr^{VI}]_T$. The suspensions were then equilibrated for 30 min. The short reaction time was chosen in order to minimize fuchsite dissolution. All samples were filtered through 0.2 μm filters and were acidified to 1% HNO_3 for measurement of total dissolved metal concentrations using ICP-MS.

3 RESULTS AND DISCUSSION

3.1 *Dissolution of fuchsite*

Release rates of metals from fuchsite at both pH 2 and 9 reached steady state in about 12 d. The steady-state release rates of Cr, Al, Si and Fe normalized to their contents in the original solid are shown in Fig. 1. Under non-oxidizing conditions, the release rates of Cr were lower than those of major constituents such as Al and Si at both pH 2 and 9. Previous crystallographic and spectroscopic studies of Cr-containing muscovite have shown that the structural Cr is trivalent and that >99.5% of the Cr occupies octahedral sites by substituting for Al^{III}(Brigatti et al. 2001; Cardelli et al. 2003). However, the release rates were higher for Al than for Cr by a factor of 2.3 at pH 2 and a factor of 10 at pH 9. Release rates for Al and Si were comparable at pH 9 but, at pH 2, the release rate of Al was higher than that of Si by a factor of 1.7. The release rates of Si and/or Al from muscovite determined in previous studies (Nickel 1973; Lin & Clemency 1981; Knauss & Wolery 1989; Kalinowski & Schweda 1996; Maurice et al. 2002) were comparable to those obtained in this study.

Under oxidizing conditions, nearly stoichiometric release of Al, Si, and Fe was observed at both pH values but the release rates of these constituents were lower than those observed under non-oxidizing conditions. The effect of oxidation of structural Fe^{2+} on the dissolution rates of silicate minerals has been

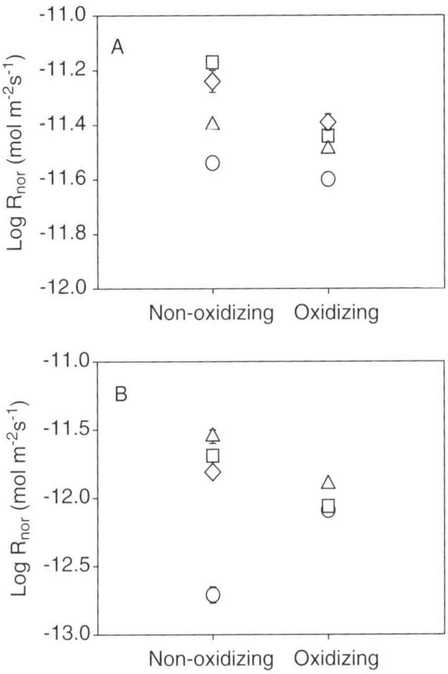

Figure 1. Steady-state release rates of metals under oxidizing or non-oxidizing conditions. Release rate of each metal was normalized by the content of the metal in the original solid. Release rate at (A) pH 2, (B) pH 9. Symbols: (circle) Cr, (square) Al, (triangle) Si, (diamond) Fe. Experimental conditions are given in Table 2.

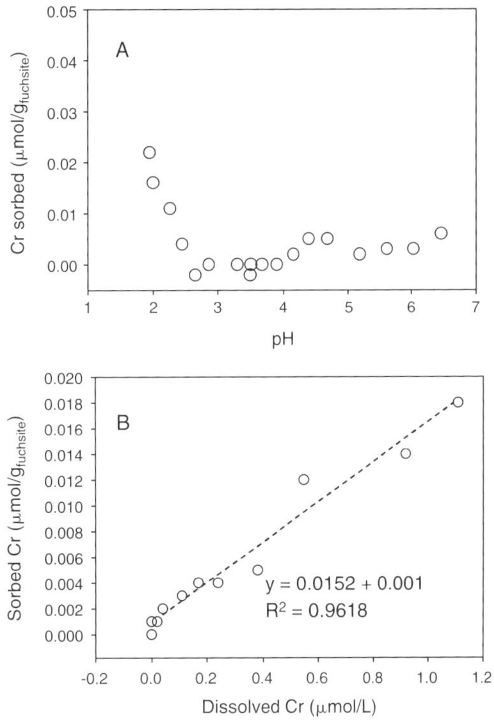

Figure 2. Sorption of Cr^{VI} on fuchsite. (A) pH-edge with $[Cr^{VI}]_T = 1.93\ \mu M$, solid (S2) = 4.01 g/L. (B) Sorption isotherm with $[Cr^{VI}]_T = 0 - 1.48\ \mu M$, solid (S1) = 20.04 g/L, pH = 2. Experimental conditions are given in Table 2.

examined previously (Schott & Berner 1983; White & Yee 1985; Hoch et al. 1996). Schott & Berner (1983) reported that the dissolution rate of bronzite was slower under oxic than anoxic conditions. Casey and co-workers have argued that dissolution rates correspond to the kinetics of water loss from the hydration sphere of the corresponding dissolved cation and thus that the Fe^{3+}-O bond is inherently slower to react than the Fe^{2+}-O bond (Casey & Westrich 1992; Casey et al. 1993). Hoch et al. (1996), on the contrary, found that release of Si from pyroxene was enhanced at higher dissolved O_2 concentrations in the dissolution medium. White & Yee (1985) suggested a mechanism for the oxidative dissolution as coupled electron-cation transfer in which release of a cation would be enhanced upon oxidation of structural Fe^{2+} to achieve charge balance in the solid. In their study, however, the release rates of Al, Si and Fe were not determined. At this point, the effect of the oxidation of a trace constituent of a solid on the release rates of the major constituents is not clearly understood.

Under oxidizing conditions, the release rate of Cr increased substantially at pH 9 but decreased slightly at pH 2 (Fig. 1). The oxidant used in this study, $HOCl/OCl^-$, provides a strong thermodynamic driving force for the oxidation of Cr^{III} to Cr^{VI} at both pH 2 and 9. The reaction does become more favorable with increasing pH and it is possible that the kinetics of the reaction are correspondingly pH dependent. It is also possible that the sorption of the product Cr^{VI} on the reacting fuchsite (see section 3.2) inhibits further dissolution of the solid. Such inhibition appears to be primarily responsible for the very strong pH dependence observed in the oxidative dissolution of $Cr(OH)_{3(s)}$ (Lee & Hering, unpublished). In those batch experiments, however, conditions allowed near saturation of the $Cr(OH)_{3(s)}$ surface by Cr^{VI} at pH 2.

3.2 *Sorption of Cr^{VI} on fuchsite*

Under oxidizing conditions, the product of the dissolution reaction, Cr^{VI}, can adsorb onto the residual solid, particularly in the acidic range. Sorption of Cr^{VI} on fuchsite was observed only for pH ≤ 2.5 (Fig. 2A). At pH 2, the amount of Cr^{VI} sorbed increased with increasing dissolved Cr^{VI} concentrations (Fig. 2B). Under flow-through conditions, it would be expected that a pseudo-equilibrium between dissolved and adsorbed Cr^{VI} would be established even as the dissolution reaction progressed. The flux of dissolved Cr from the flow-through reactor should thus correspond to the rate of fuchsite dissolution (as in the case without adsorption of the reaction product). At pH 2 under oxidizing conditions, the steady-state dissolved Cr concentration was 0.20 μM. The corresponding pseudo-equilibrium concentration of adsorbed Cr^{VI} would be 0.004 μmol/g (cf. Fig. 2B) or 0.07 μM at the solid concentration in the flow-through experiment. Although this sorbed concentration is clearly well below surface saturation, it may inhibit the reaction of the oxidant at the fuchsite surface to some extent.

4 CONCLUSIONS

When fuchsite dissolved under non-oxidizing conditions, Cr was released more slowly than the major constituents Al and Si at both pH 2 and 9. The strong oxidant $HOCl/OCl^-$ increased the rate of Cr release at pH 9 but not at pH 2; rates of Al and Si release decreased in the presence of the oxidant at both pH values. Congruent release of Cr with Al and Si was observed only under oxidizing conditions at pH 9. At pH 2, adsorption of Cr^{VI} on fuchsite was observed and may have inhibited the dissolution rate. These results illustrate the potential significance of oxidative dissolution processes not only for the release of Cr^{VI} from Cr^{III} -containing minerals but also for the rate of bulk mineral dissolution. In addition, sorption of Cr^{VI} released by oxidative dissolution may influence both the accumulation of Cr^{VI} in the dissolved phase and the rate of mineral dissolution.

REFERENCES

Brigatti, M.F., Galli, E., Medici, L., Poppi, L., Cibin, G., Marcelli, A., & Mottana, A. 2001. Chromium-containing muscovite: crystal chemistry and XANES spectroscopy. *Eur. J. Mineral* 13: 377–389.

Cardelli, A., Cibin, G., Benfatto, S., Della Longa, S., Brigatti, M.F. & Marcelli, A. 2003. A crystal-chemical investigation of Cr substitution in muscovite by XANES spectroscopy. *Phys. Chem. Minerals* 30: 54–58.

Casey, W.H., Banfield, J.F., Westrich, H.R. & McLaughlin, L. 1993. What do dissolution experiments tell us about natural weathering. *Chem. Geol.* 105: 1–15.

Casey, W.H. & Westrich, H.R. 1992. Control of dissolution rates of orthosilicate minerals by divalent metal-oxygen bonds. *Nature*355: 157–159.

Giammar, D.E. & Hering, J.G. 2002. Equilibrium and kinetic aspects of soddyite dissolution and secondary phase precipitation in aqueous suspension. *Geochim. Cosmochim. Acta* 66(18): 3235–3245.

Hoch, A.R., Reddy, M.M. & Drever, J.I. 1996. The effect of iron content and dissolved O_2 on dissolution rates of clinopyroxene at pH 5.8 and 25°C: preliminary results. *Chem. Geol.* 132: 151–156.

Kalinowski, B.E. & Schweda, P. 1996. Kinetics of muscovite, phlogopite, and biotite dissolution and alteration at pH 1–4, room temperature. *Geochim. Cosmochim. Acta* 60: 367–385.

Knauss, K.G. & Wolery, T.J. 1989. Muscovite dissolution kinetics as a function of pH and time at 70°C. *Geochim. Cosmochim. Acta* 53: 1493–1501.

Kraemer, S.M. & Hering, J.G. 1997. Influence of solution saturation state on the kinetics of ligand-controlled dissolution of oxide phases. *Geochim. Cosmochim. Acta* 61: 2855–2866.

Lin, F.-C. & Clemency, C.V. 1981. The kinetics of dissolution of muscovites at 25°C and 1 atm CO_2 partial pressure. *Geochim. Cosmochim. Acta* 45: 571–576.

Maurice, P.A., McKnight, D.M., Leff, L., Fulghum, J.E. & Gooseff, M. 2002. Direct observations of aluminosilicate weathering in the hyporheic zone of an Antarctic Dry Valley stream. *Geochim. Cosmochim. Acta* 66: 1335–1347.

National Research Council. 1991. *Environmental Epidemiology Vol. 1, Public Health and Hazardous Wastes.* Washington DC: National Academy Press.

Nickel, E. 1973. Experimental dissolution of light and heavy minerals in comparison with weathering and intrastratal solution. *Contrib. Sediment.* 1: 1–68.

Schott, J. & Berner, R.A. 1983. X-ray photoelectron studies of the mechanisms of iron silicate dissolution during weathering. *Geochim. Cosmochim Acta* 47: 2233–2240.

White, A.F. & Yee, A. 1985. Aqueous oxidation-reduction kinetics associated with coupled electron-cation transfer from iron-containing silicates at 25°C. *Geochim. Cosmochim. Acta* 49: 1263–1275.

Water–Rock Interaction, Wanty & Seal II (eds)

Connecting the molecular – with the macro-scale: an integrated approach towards a quantitative understanding of crystal dissolution

A. Lüttge
Department of Earth Science MS-126, Rice University, Houston, USA

ABSTRACT: We use an integrated approach consisting of Monte Carlo simulations, ab initio and density functional theory (DFT) calculations, bulk dissolution measurements, and direct surface observations conducted with atomic force microscopy (AFM) and vertical scanning interferometry (VSI) techniques to explore the dissolution kinetics of crystalline materials over a wide range of length and time scales ranging from 10^{-10} to 10^{-2} meters and minutes to years. Our approach is different from established models based on either molecular precursor complexes or surface adsorption and is based on the assumption that crystal dissolution kinetics can be treated as a stochastic many-body problem. An increasing number of experimental and theoretical results demonstrate that our approach is a successful strategy to address in detail topics like the understanding and quantification of reactive surface area, as well as the dependence of dissolution rate on the distance from equilibrium or reaction affinity. In addition, we focus on the importance of a kinetically defined dissolution product. A quantitative understanding of these topics is likely to be a prerequisite for our attempt towards a detailed understanding of fluid–solid interactions in the presence of microorganisms and biofilms.

1 INTRODUCTION

The rapidly increasing problems of pollution, waste management, rising temperatures, and loss of soil that most modern societies and the global community are facing has led to an intensive research effort on mineral dissolution kinetics over the last two decades. As a result, significant progress has been made in measuring mineral dissolution rates that are important in quantifying the weathering of rocks (e.g. White & Brantley 1995; Oelkers 2002; Lüttge et al. 2003 and references therein). Various rate laws have been developed that allow us today to predict dissolution rates at conditions beyond the measured data (e.g. Lasaga 1998). In addition, new powerful experimental and analytical techniques like atomic force microscopy (AFM) and X-ray photoelectron spectroscopy (XPS) have been developed and applied in research studies to tackle the problems in more detail (e.g. Liang et al. 1996a,b; Jordan & Rammensee 1998; Jordan et al. 2001; Higgins et al. 2002). However, some problems remain still unsolved, e.g., the definition and quantification of reactive surface area, the dependence of dissolution rate on the saturation state at conditions close to equilibrium, and the related challenge of applying laboratory results to natural systems (e.g. Hochella & Banfield 1995).

A more recent development is our recognition that microorganisms like bacteria, algae, and fungi participate often heavily in fluid–solid interactions. For example, Lüttge & Conrad (2004) demonstrate that *Shewanella oneidensis* MR-1 controls the dissolution kinetics of calcite and other carbonate minerals by suppressing the formation of etch pits under certain laboratory conditions (compare Fig. 1, Davis & Lüttge, unpub.). From the rapidly growing literature on these issues it is not difficult to predict that our understanding of the attachment of microorganisms to mineral surfaces and their role in the water-solid interactions will continue to be of central importance. Our attempt to include microorganisms into the great picture of water–rock interactions will add another dimension of complexity to the task. It is most important to solve the open questions related to the abiotic dissolution of crystalline matter, because the abiotic dissolution system will have to serve as a base or a reference for the biotically controlled dissolution processes. Here I suggest an integrated approach of computer simulations and experimental/analytical techniques that allow the investigation of crystal dissolution from the molecular to the millimeter scale. An important question that I then discuss below is if crystal dissolution is controlled by a single (elementary) reaction or if dissolution can be treated

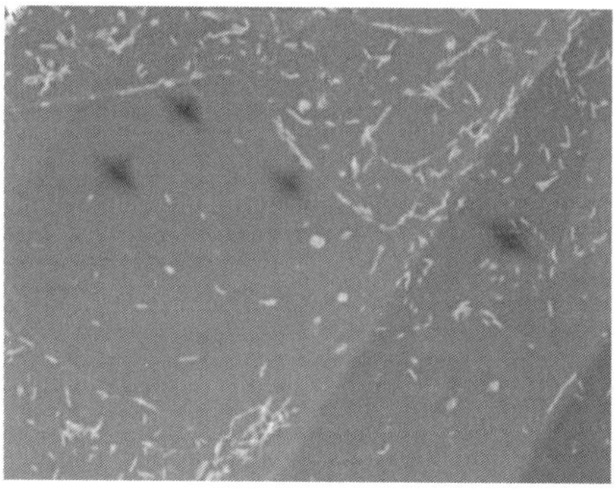

Figure 1. VSI height map of *S. oneidensis* MR-1 on a calcite cleavage surface. Etch pits (dark spots) are developing exclusively at parts of the surface that are not covered by MR-1.

successfully as a stochastic many-body problem. The answer to this question is central to the development of a fundamental theory on crystal dissolution and our approach to quantify surface controlled dissolution rates.

2 METHODS

I am using a combination of computer simulations and experimental/analytical techniques to study the dissolution of aluminosilicates, carbonates, and barite (e.g. Lüttge et al. 1999, 2003; Lasaga & Lüttge 2001, 2003, 2004a,b; Arvidson et al. 2003, 2004a,b; Lüttge 2004). Currently, there is no experimental technique available that allows us to directly observe the dynamics of molecules at the fluid solid interface. *Ab initio* techniques in combination with density functional theory (DFT) have proven useful in studying reaction pathways and calculating important parameters like adsorption energies and activation energies that are necessary to break and make certain bonds, and to optimize molecular configurations. However, large and/or complex systems like crystal surfaces that incorporate steps and dislocations can not currently be studied with high level theoretical models due to limited computer speed. Therefore, I am using Monte Carlo (MC) techniques that can simulate both elementary reactions, i.e., bond-breaking and bond-forming, and basic reactions, i.e., dissolution of surface units, adsorption and incorporation of units from the solution, as well as the mobility of molecules at the crystal surface. Current models are advanced enough to incorporate the full crystal structure to calculate the interactions of neighboring atoms as well as possible defects of the crystal lattice. This approach is different from models based on either molecular precursor complexes or adsorption (e.g. Oelkers 2002). I use results from *ab initio* and DFT

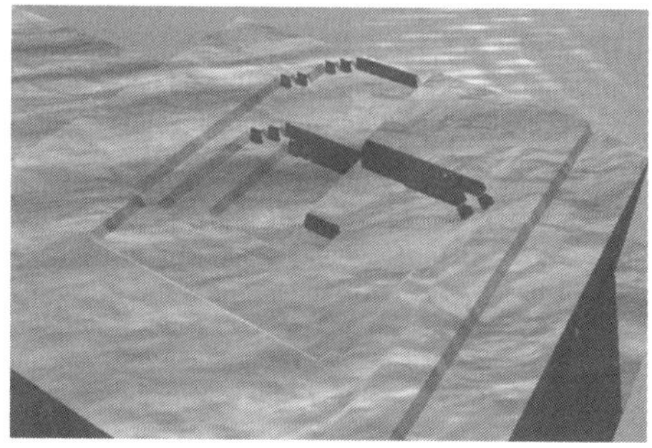

Figure 2. Surface map drawn from MC simulation results: An etch pit is developing along a screw-dislocation.

studies to improve the MC simulations (e.g. Cygan & Kubicki 2001, and citations therein; Xiao & Lüttge 2002). Based on MC calculations I am able to make predictions of how a dissolution process will occur (e.g. Fig. 2). The prediction can then be tested experimentally. Tests are conducted by using AFM (Digital Instruments Nanoscope III Multimode SPM, operated in contact mode) and VSI techniques (MicroXAM MP8, ADE-Phase Shift) that allow the direct observation and quantification of sample surface topographies and their changes as a function of time. AFM and VSI are complementary techniques that can be linked successfully. This link can be further improved by the development and application of super-resolution techniques for VSI (Sawyer et al. 2002). The use of VSI provides the field of view that allows us to observe relatively large parts of the crystal surface at the same time and in a next step to link these observations to bulk dissolution rates measured in the field and the laboratory. Thus, our approach provides a link between (sub)microscopic and macroscopic scale observations of the processes that govern crystal dissolution kinetics.

3 RESULTS AND DISCUSSION

3.1 *Results of Monte Carlo simulations*

Our computer simulations of crystal dissolution kinetics have already led to some interesting results – most important to the prediction of the dissolution stepwave mechanism (Lasaga & Lüttge 2001). This model recognizes that the surface-normal retreat of the crystal surface is not always directly governed by the formation and coalescence of etch pits. In contrast, I suggested that trains of steps are generated at the outskirts of the etch pits. These steps travel laterally across the crystal surface and each step lowers the crystal surface by an amount equal to its step height causing the surface-normal retreat.

An important advantage of model simulations is the fact that processes can be explored easily at conditions that are difficult or even impossible to generate in the laboratory. A good example is the situation in which dissolution occurs close to equilibrium. These conditions usually result in very slow reaction rates that are very difficult to measure experimentally. For example, feldspar dissolution rates are often too slow to be measured even with a high-temperature AFM. Therefore, I am exploring plagioclase dissolution kinetics as a function of ΔG, i.e., the distance from equilibrium with our MC simulations, and I have (re)derived the solubility product – which is well known in thermodynamic concepts – from a purely kinetic approach.

The generation of surface maps (Fig. 2) calculated by MC simulations is useful because such images can be compared with direct observational techniques such as AFM and VSI. This possibility is particularly helpful if I intend to study the inhibitory or catalytic effect of certain molecules and complexes present in the solution. The inhibitory effects are often not only reflected by a decrease in reaction rate, they are also reflected by changes in the surface topography, e.g., the shape of etch pits.

3.2 *Results of AFM and VSI studies*

AFM and VSI techniques allow the direct observation and quantification of the crystal surface that is dissolving (e.g. MacInnis & Brantley 1992, 1993; Lüttge et al. 1999, 2003; Teng 2004). Processes at the molecular scale, e.g., changes in the rate of kink-site formation, will result in changes in step velocities. These velocities and also step densities on the crystal surface can be quantified by AFM. Any relative change in velocities occurring in different directions will result in the change of the shape of etch pits. Therefore, our experimental approach is capable of determining changes in surface features caused by inhibitors or catalysts even if the absolute rates do not change significantly. This is particularly important at conditions close to equilibrium, where growth or dissolution rates are acutely sensitive to the presence of inhibitors. Due to its larger field of view VSI allows the observation of many surface features and there interactions at the same time (Fig. 3A, B). Consequently, changes at the step velocity level can be linked with bulk dissolution rates.

As an example, Fewless & Lüttge (2003) were recently able to link step velocities and densities on a barite cleavage surface measured by AFM with surface-normal retreat rates measured by VSI. The results from direct observations of the response of surface topography to the dissolution process can be used to test for example the dissolution stepwave mechanism predicted by Lasaga & Lüttge (2001, 2003).

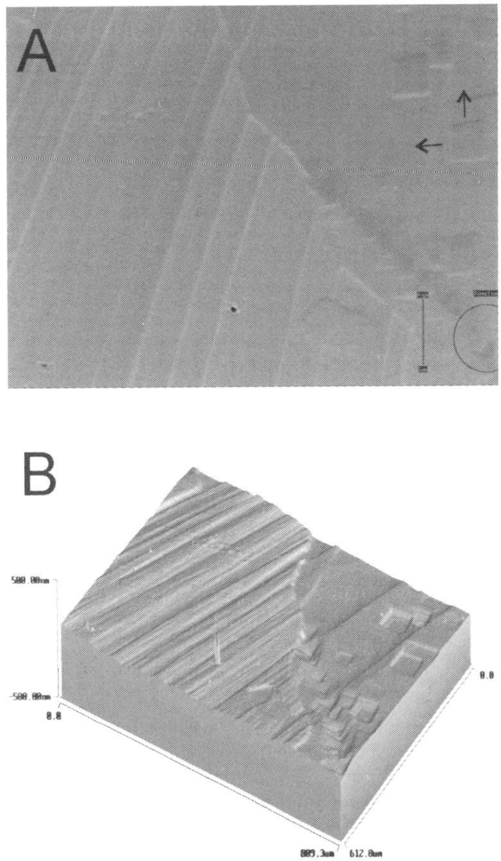

Figure 3. Modification of etch pit morphologies by magnesium inhibition (VSI data), shown in 2 – (A) and 3 – dimensional rendering (B). Attachment of Mg at obtuse (+) step edges inhibits overall calcite dissolution in carbonated solutions, producing deformed etch pits in which growth of pits is accomplished by acute (arrows) step advance only. Images are 800 μm in *x*-direction (Arvidson et al. 2003).

3.3 *Crystal dissolution treated as a many-body problem*

A prerequisite for our approach is the assumption that crystal dissolution can be treated as a many-body problem and therefore as a stochastic problem. This assumption will be discussed in detail in a subsequent paper. It is usually not disputed that surface controlled dissolution rates depend on the *reactive* surface area that is different from the total (BET) surface area. The reactivity of surface area depends on the number, type, and distribution of sites at a surface and is defined by the bond-energies. If that would be not true then all crystal surfaces of the same type should produce the same dissolution rate, i.e., an atomically flat surface should dissolve with the same rate as a very rough surface of the same total size and type. I know that this is not the case. Thus, I must treat and monitor the surface molecules in their specific situation. Monte Carlo methods can do that. The sum of all configurations, i.e., the sites, will determine the crystal surface that is exposed to the solution and its reactivity.

4 CONCLUSIONS

I have established an integrated approach of Monte Carlo based computer simulations and experimental/analytical techniques that is capable of studying crystal dissolution mechanisms and kinetics quantitatively. The ability to link the results obtained by the different "tools" at different time and length-scales leads to new insight into key questions of crystal dissolution kinetics.

REFERENCES

Arvidson, R.S., Ertan, I.E., Amonette, J.E. & Lüttge, A. 2003. Variation in calcite dissolution rates: A fundamental problem? *Geochim. Cosmochim. Acta* 67: 1623–1634.

Arvidson, R.S., Collier, M., Davis, K.J., Vinson, M.D., Amonette, J.E. & Lüttge, A. 2004a, submitted. The role of magnesium in calcite dissolution.

Arvidson, R.S., Beig, M.S. & Lüttge, A. 2004b, in press. Variation in calcite dissolution rates: A fundamental problem? *Geochim. Cosmochim. Acta* 67: 1623–1634.

Cygan, R.T. & Kubicki, J.D. (eds.) 2001. Molecular modeling theory: Applications in the geosciences. *Revs. in Mineral. V. 42*. Washington, D.C.: Mineral. Soc. of Amer. and Geochem. Soc.

Fewless, T.A. & Lüttge, A. 2003. Connecting the molecular- with the macro-scale: Evidence for the stepwave dissolution model. AGU 2003 Fall meeting, San Francisco, CA (December 8–12, 2003); EOS Transactions Suppl.

Higgins, S.R., Jordan, G. & Eggleston, C.M. 2002. Dissolution kinetics of magnesite in acidic aqueous solution: A hydrothermal atomic force microscopy study assessing step kinetics and dissolution flux. *Geochim. Cosmochim. Acta* 66(18): 3201–3210.

Hochella Jr., M.F. & Banfield, J.F. 1995. Chemical weathering of silicates in nature: A microscopic perspective with theoretical considerations. In: A.F. White & S.L. Brantley (eds.). *Chemical weathering rates of silicate minerals. Revs. in Mineral. 31*: 353–406. Washington D.C.: Mineralogical Society of America.

Jordan, G. & Rammensee, W. 1998. Dissolution rates of calcite (10–14) obtained by scanning force microscopy: Microtopography-based dissolution kinetics on surfaces with anisotropic step velocities. *Geochim. Cosmochim. Acta* 62: 941–947.

Jordan, G.H., Higgins, S.R., Eggleston, C.M., Knauss, K.G. & Schmahl, W.W. 2001. Dissolution kinetics of magnesite in acidic aqueous solution, a hydrothermal atomic force microscopy (HAFM) study: step orientation and kink dynamics. *Geochim. Cosmochim. Acta* 65: 4257–4266.

Lasaga, A.C. 1998. *Kinetic theory in the earth sciences*. Princeton, 811 pp.

Lasaga, A.C. & Lüttge, A. 2001. Variation of crystal dissolution rate based on a dissolution stepwave model. *Science*, 291: 2400–2404.

Lasaga, A.C. & Lüttge, A. 2003. A model for crystal dissolution. *Eur. J. Mineral.* 15: 603–615.

Lasaga, A.C. & Lüttge, A. 2004a, in press. A fundamental approach to mineral dissolution kinetics. *Am. Mineral.*

Lasaga, A.C. & Lüttge, A. 2004b, accepted. A fundamental approach to mineral dissolution kinetics II. *Eur. J. Mineral.*

Liang, Y., Baer, D.R., McCoy, J.M., & LaFemina, J.P. 1996a. Interplay between step velocity and morphology during the dissolution of $CaCO_3$ surface. *J. Vac. Sci. Technol. A* 14(3): 1363–1375.

Liang, Y., Baer, D.R., McCoy, J.M., Amonette, J.E. & LaFemina, J.P. 1996b. Dissolution kinetics at the calcite-water interface. *Geochim. Cosmochim. Acta* 60: 4883–4887.

Lüttge, A. 2004, in press. Crystal Dissolution Kinetics studied by Vertical Scanning Interferometry (VSI): A Brief Review and Outlook. *In*: Liu, X.-Y. (ed.) *From Solid-Fluid Interfacial Structure to Nanostructural Engineering. Series on Nanoscience and Technologies*. Kluwer.

Lüttge, A., Bolton, E.W. & Lasaga, A.C. 1999. An interferometric study of the dissolution kinetics of anorthite: the role of reactive surface area. *Am. J. Sci.* 299: 652–678.

Lüttge, A. & Conrad, P.G. 2004, in press. Bacterial control of calcite dissolution kinetics. *Applied and Environmental Microbiology*.

Lüttge, A. & Lasaga, A.C. 2004, subm. Fundamental crystal dissolution kinetics: Feldspar dissolution. *Chem. Geol.*

Lüttge, A., Winkler, U. & Lasaga, A.C. 2003. An interferometric study of dolomite dissolution: A new conceptual model for mineral dissolution. *Geochim. Cosmochim. Acta* 67: 1099–1116.

MacInnis, I.N. & Brantley, S.L. 1992. The role of dislocations and surface morphology in calcite dissolution. *Geochim. Cosmochim. Acta* 56: 1113–1126.

MacInnis, I.N. & Brantley, S.L. 1993. Development of etch pit size distributions on dissolving minerals. *Chem. Geol.* 105: 31–49.

Oelkers, E.H. 2002. General kinetic description of multioxide silicate mineral and glass dissolution. *Geochim. Cosmochim. Acta* 65: 3703–3720.

Sawyer, D.S., Lüttge, A. & Conrad, P.G. 2002. Towards a High Resolution Vertical Scanning Interferometer: Applications for Fluid-Solid-Microbe Interactions. GSA 2002 Fall meeting, Denver, CO (October 27–30, 2002).

Teng, H.H. 2004. Controls by saturation state on etch pit formation during calcite dissolution, *Geochim. Cosmochim. Acta* 68: 253–262.

White, A.F. & Brantley, S.L. (eds) 1995. Chemical weathering rates of silicate minerals. *Revs. in Mineral. V. 31*. Washington, D.C.: Mineralogical Society of America.

Xiao, Y. & Lüttge, A. 2002. Solvated *ab initio* and density functional theory (DFT) modeling of mineral-fluid surface reactions: Towards a better understanding of aluminosilicate dissolution mechanisms. GSA 2002 Fall meeting, Denver, CO (October 27–30, 2002).

Water–Rock Interaction, Wanty & Seal II (eds)
 ISBN 90 5809 641 6

Field vs. lab pH in remote wells, Konza Prairie LTER site, and implications for limestone weathering

G.L. Macpherson
University of Kansas, Lawrence, Kansas, USA

ABSTRACT: Ground waters from shallow, horizontal, thin limestones in the central USA are Ca-HCO_3 type waters with moderately high alkalinity (>5 mmol) and low SO_4 (<0.5 mmol). Because of their remote location, wells are bailed to collect samples, making accurate pH measurements difficult. This 13-month study compares field-measured pH with lab-measured pH that was corrected for controlled degassing of CO_2. Lab pH was not always higher than field pH, and the difference between the two was greatest in the winter to spring months, when calculated P-CO_2 values were lowest, dissolved ions concentrations were lowest, and meteoric precipitation was lowest. Saturation index calculations show that the ground water from the well used for this study was oversaturated with respect to calcite during late winter to spring months, despite the lower dissolved ions concentrations, and close to saturation or undersaturated during the summer months. These findings suggest that equilibrium with calcite is accelerated during periods of high P-CO_2 and impeded during periods of low P-CO_2, and that most of the weathering of the limestone aquifers takes place during the summer months.

1 INTRODUCTION

Measuring accurate pH of ground water is a challenge under any circumstances, but wells in areas not accessible to a source of power add additional difficulties. Sampling of such wells is typically done using a bailer, which results in greater exposure of ground water to the atmosphere and greater chance of temperature change during sampling than would be true of a mechanically pumped well with water directed into a flow-through cell. Correct pH determination is essential to correct calculation of the partial pressure of carbon dioxide and carbonate-mineral saturation indices, and these in turn are needed to develop an understanding of seasonally-dependent weathering processes and rates.

Chemical weathering rates are known to vary on many time scales, and denudation rates of carbonate-dominated watersheds depend upon understanding the temporal patterns of calcite saturation in ground and surface water. Oversaturation of ground water with respect to calcite has been attributed to mixing of chemically different end members (Back & Hanshaw 1970; Plummer 1975; Wicks & Herman 1996) or CO_2-degassing of water (Hanor 1978). Undersaturation of ground water with respect to calcite has been attributed to water residence time shorter than chemical reaction time (Langmuir 1971) or mixing (Thraillkill 1968). This study addresses patterns of calcite saturation state and partial pressure of CO_2 (P-CO_2) in thin limestone aquifers.

The Konza Prairie NSF LTER Site (Konza), field site for this study (Fig. 1), is a dissected upland underlain by nearly flat-lying, alternating Permian limestone and shale layers. The limestone layers

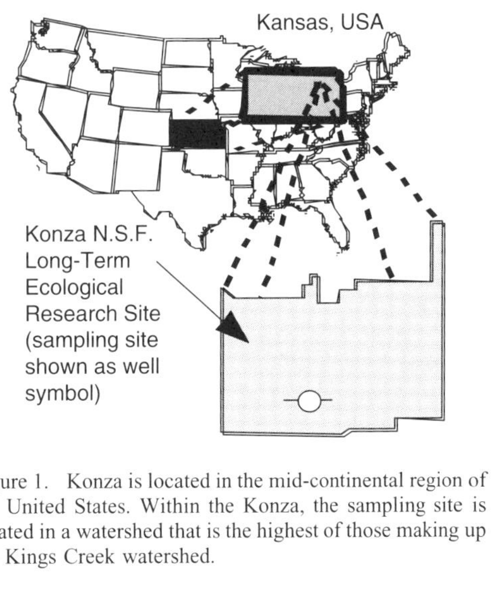

Figure 1. Konza is located in the mid-continental region of the United States. Within the Konza, the sampling site is located in a watershed that is the highest of those making up the Kings Creek watershed.

behave as aquifers, and hydraulic conductivity is related to fracture development and amount of fine-grained silicilastic or carbonate material. Konza is an undisturbed tallgrass prairie located in the north-eastern part of the state of Kansas, in the central part of the USA. Climate is temperate mid-continental, with average annual air temperature of 13°C, meteoric precipitation of 835 mm, and annual potential evapotranspiration (AET) about 1250 mm. Intense rainfall thunderstorms (>56 mm/hr) are common (250–300 hours/yr). Meteoric precipitation is strongly seasonal, with 75% of precipitation falling during the growing season (Hayden 1998).

Ground water chemistry at Konza is typical of water that reacts primarily with limestone, having relatively high dissolved Ca^{2+} and HCO_3^- and having much lower concentrations of other solutes (Macpherson 1996). Ground water temperature is nearly constant through the year, varying only ±1 to 2°C from 13°C.

One watershed is currently instrumented with 35 PVC wells completed with 30-cm long screens. Five of these wells are sampled regularly for water chemistry; four of these were the target of an intensive project to determine *in situ* pH during 1998–2000. This paper will focus on results from one of these wells, which is 12.5 m deep. The study period reported here includes a total of 13 months.

2 METHODS

Wells were sampled approximately monthly by bailing to remove at least two well-plus-gravel pack volumes of water before sampling. Samples were removed from the well using a clean, one-liter Teflon® bailer suspended using Teflon®-coated steel wire. A full bail was directed into a N_2-filled glass vessel using a Teflon bottom-emptying device with control valve and 10-cm exit tube so that it overfilled the vessel. Some of the remaining portion of the bail was emptied into a N_2-filled glass bottle with Teflon®-lined lid, again using the bottom-emptying device which permitted filling from the bottom of the bottle. This bottle was filled approximately half-full, then capped tightly and stored in an ice chest. pH was measured immediately in the field using a Fisher Accumet 1003 meter with an Accumet gel-filled combination pH (Ag-AgCl)-reference-temperature electrode. The meter was calibrated before measurement and calibration solutions were remeasured before and after ground water measurement as a check on electrode response.

Alkalinity was titrated in the lab within three days of sample collection using 0.02 N H_2SO_4 and using the same pH meter and electrode used in the field. Eight years of sampling from this site showed that samples can be held at room temperature at least one week before alkalinity is altered by precipitation of carbonates. Anions were determined on field-filtered samples (0.45 μm) by ion chromatography (Dionex 4000i); cations were determined by ICPAES (traceable to NIST 1643d) on field-filtered (0.45 μm) and lab-acidified samples (final acidification level was 2% by volume, HNO_3). Charge balances on analyses (cations/anion meq) were within the range of 1 ± 0.03.

In the lab, bottled samples were stored in a refrigerator (not frost-free; temperature around 8°C) for a period of about one week. Bottles were insulated and then removed from the refrigerator. The bottle seal was broken, pH probe (same unit used in the field) inserted and then the bottle was resealed around the probe using Parafilm®. pH of the water in the bottles was then measured every 30 seconds until it stabilized, usually in about 8 to 11 minutes (Fig. 2). Stabilization time reflected the electrode response time to the cold temperature of the water (approximately 8°C), and not CO_2-equilibration.

pH measurements of the lab samples were adjusted according to the following. Because the atmosphere in the collection bottle was initially pure N_2, degassing of CO_2 occurred during the refrigeration period. Because the volume of the bottle was determined prior to sampling and the volume of water collected in the bottle determined, the headspace volume was known. The measured pH of the sample in the bottle and the alkalinity of the sample reflected the dissolved CO_2 content of the water and the P-CO_2 of the headspace. The amount of CO_2 in the headspace was

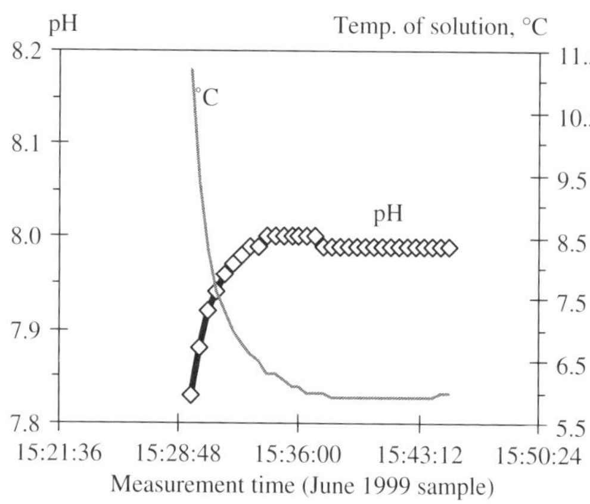

Figure 2. Example of pH equilibration in the lab in a bottle filled with N_2 and then partially filled with ground water, stored in the refrigerator, and then measured after CO_2 degassing was presumed complete. Measurements continued until temperature began to increase rather than decrease and pH and temperature changed less than 0.02 pH units and 0.1°C, respectively, over a period of two minutes.

calculated and added to the total dissolved carbon. PHREEQC 2.7 with the PHREEQC thermodynamic data base was used to correct pH in the samples to the *in situ* ground-water temperature (13°C); the adjusted total carbon and the titrated alkalinity along with all major and some minor dissolved solutes were among the input parameters. The *in situ* pH and partial pressure of CO_2 (P-CO_2) were calculated, and the solutions were tested for saturation state of carbonates and other minerals (not discussed here).

3 RESULTS

The difference between the lab-measured and field-measured pH was greatest when P-CO_2 was lowest (Fig. 3) and P-CO_2 was lowest during the winter to spring (December through April or May; Fig. 4). This same pattern held for the three other wells (not shown) that were sampled and analyzed in the same way during the same time period.

The saturation index for calcite (Fig. 4) showed that the ground waters were oversaturated with respect to calcite during the winter to spring (when P-CO_2 was lowest) and closer to saturation or undersaturated in the summer (June through August; when P-CO_2 was highest).

Dissolved calcium was lowest in late winter to spring, when ground water was most oversaturated with respect to calcite, and highest when calcite was near saturation (Fig. 5). These two parameters are not precisely in phase, however, as early summer (June) may coincide with high dissolved calcium and relatively high calcite saturation index. The general seasonal pattern of higher solute concentrations during the summer to fall and lower concentrations during the winter holds for all dissolved ions, although peak concentrations in Mg, SO_4 and Cl may be earlier (July) than other solutes (September). The pattern of seasonal variation in solute concentration was consistent over the overlapping portions of the two years sampled, with solutes having peak concentrations at nearly the same time during the same month in both years.

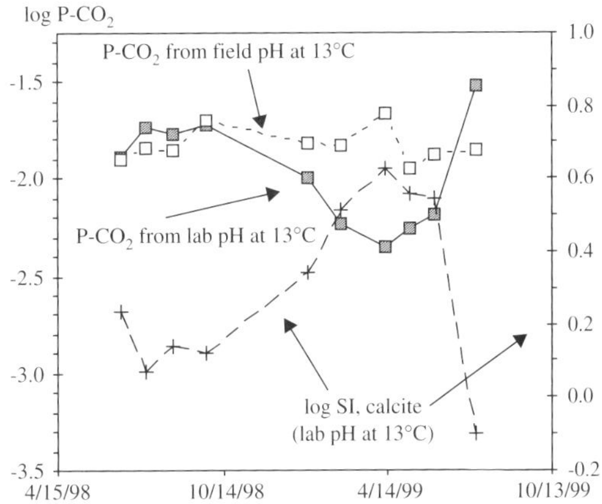

Figure 4. Ground water samples were closest to calcite saturation during the summer when P-CO_2 was highest, and farthest from saturation (oversaturated) when P-CO_2 was lowest. This suggests a kinetic control on rock weathering, driven by CO_2.

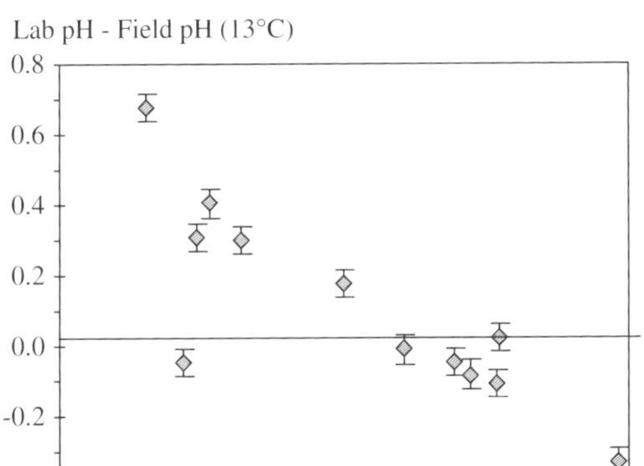

Figure 3. The difference between laboratory- (see text) and field-measured pH, both corrected to ground water *in situ* temperature of 13°C, is not consistently positive, but varies inversely with P-CO_2.

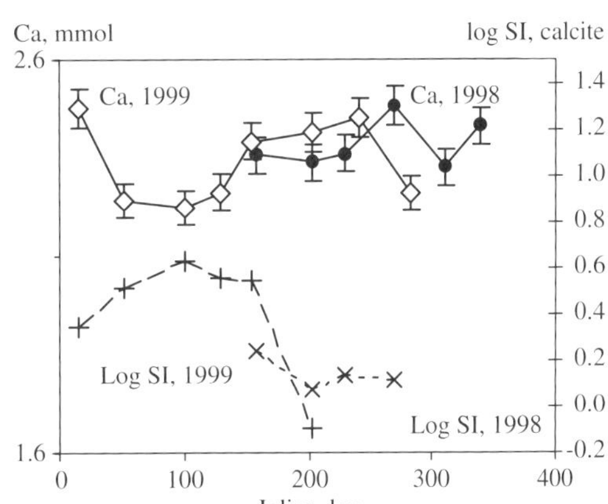

Figure 5. Solutes concentrations varied through the year on the same approximate cycle for the two years studied. Ground water samples were farthest from saturation with respect to calcite when solute concentrations were lowest.

4 INTERPRETATIONS

Correction of the carefully determined lab pH to *in situ* ground water temperature resulted in a smoother temporal trend of calculated P-CO_2 than P-CO_2 calculated from the field-measured pH at 13°C (Fig. 4), and one that reflected the expected seasonal variation in below-ground CO_2. However, log P-CO_2 was usually lower than expected (−1.62) for the AET of the area, the expected value calculated using the following equation (White 1988):

$$\log P\text{-}CO_2 = -3.47 + 2.09(1 - e^{(-0.00172AET)})$$

P-CO_2 changed in a predictable way during the year, increasing in the summer and decreasing at other times. Similar seasonal variations in soil CO_2 have been demonstrated in a sandy unsaturated zone (Hendry et al. 1999).

During times of high P-CO_2 values at this site, kinetic barriers to achieving calcite equilibrium were apparently overcome, as evidenced by the coincidence of high P-CO_2 with closest approach to calcite saturation (Fig. 4). These times coincided with lower concentrations of all solutes.

A suggested working hypothesis to explain these patterns is as follows.

- Spring recharge dilutes ground water, lowering calcite saturation indices. P-CO_2 is relatively low because soil microbial activity is still suppressed due to cool air temperatures.
- Summer recharge (less than spring recharge), charged with CO_2 from microbial activity, achieves near equilibrium with the limestone aquifers. P-CO_2 values are highest and solute concentrations are nearly highest because chemical weathering of limestone aquifers is highest; still near-saturation with respect to calcite is maintained.
- Fall recharge (less than summer recharge) may dilute solute concentrations in these limestone aquifers somewhat, but soil CO_2 production is waning and calcite saturation indices are increasing. Dissolved silica is highest at this time.
- Winter is a time of very little ground water recharge, and P-CO_2 values decrease and solute concentrations change little, resulting in increasing oversaturation with respect to calcite.

5 CONCLUSIONS

pH of bailed ground water can be successfully measured using a headspace technique. Using this technique, P-CO_2 values and calcite saturation states were determined for ground water from a shallow, thin limestone aquifer. Results show that there is a relatively short period during a typical year when limestone weathering is active (late spring to early summer), and that when ground water has low P-CO_2 (winter), equilibration with calcite is inhibited. The proposed yearly pattern will likely change if the region's climate changes from its present pattern of spring to summer recharge from episodic intense storms, and minor recharge during the fall.

ACKNOWLEDGEMENTS

Special thanks to Ryan Coiner for field work that made this project possible and J.R. Rogers for discussion. This project was funded by the NSF LTER program (BSR-9011662, IBN-9632851), the KU Geology Endowment Fund, and the KU Dept. of Geology.

REFERENCES

Back, W. & Hanshaw, B.B. 1970. Comparison of chemical hydrogeology of the carbonate peninsulas of Florida and Yucatan. *J. Hydrol.* 10: 330–368.

Hanor, J.S. 1978. Precipitation of beachrock cements – Mixing of marine and meteoric waters *vs.* CO_2-degassing. *J. Sed. Pet.* 48: 489–501.

Hayden, B.P. 1998. Regional Climate and the distribution of tallgrass prairie. In A.K. Knapp, J.M. Briggs, D.C. Hartnett, & S.L. Collins (eds.), *Grassland Dynamics – Long-Term Ecological Research in Tallgrass Prairie*: 19–34. New York: Oxford.

Hendry, M.J., Mendoze, C.A., Kirkland, R.A. & Lawrence, J.R. 1999. Quantification of transient CO_2 production in a sandy unsaturated zone. *Wat. Res. Research* 35(7): 2189–2198.

Langmuir, D. 1971. The geochemistry of some carbonate ground waters in central Pennsylvania. *Geochim. Cosmochim. Acta* 35: 1023–1045.

Macpherson, G.L. 1996. Hydrogeology of thin limestones – the Konza Prairie Long-Term Ecological Research Site, northeastern Kansas. *J. Hydrol.* 186: 191–228.

Plummer, L.N. 1975. Mixing of seawater with calcium carbonate water. *Geol. Soc. Am. Mem. 142.*

Thraillkill, J. 1968. Chemical and hydrologic factors in the excavation of limestone caves. *Geol. Soc. Am. Bull.* 79: 19–45.

White, W.B. 1988. *Geomorphology and hydrology of karst terrains*. New York: Oxford.

Wicks, C.M. & Herman, J.S. 1996. Regional hydrogeochemistry of a modern coastal mixing zone. *Wat. Res. Research* 32(2): 401–407.

Sphalerite weathering kinetics: effect of pH and particle size

M.E. Malmström & C. Collin
Industrial Ecology, Chemical Engineering and Technology, Royal Institute of Technology, Stockholm, Sweden

ABSTRACT: We present preliminary results of abiotic rates of element release from sphalerite ($Zn_{1-x}Fe_xS$) over ~50 days of experiments at $22 \pm 2°C$ in air-purged batch reactors at specified pH ($1 \leq pH \leq 7$) in perchlorate medium. The release of elements was found to be strongly non-stoichiometric with preferential release of metal ions (Zn and Fe) as compared to sulfur, particularly at low pH. For the metals, the release rates declined with time, such that long-term experiments are required for obtaining rates that are relevant for assessing e.g. the acid mine drainage problem. For all elements, the mass-normalized release rates at the end of the experiments were found to be only weakly dependent on pH ($1 \leq pH \leq 4$) and on particle size (35–500 μm).

1 INTRODUCTION

The main environmental concern associated with mining waste is the production and release of acid mine drainage (AMD), which is characterized by high concentrations of acidity and dissolved heavy metals and metalloids. While dissolution of pyrite, FeS_2, is the main source of acidity, ferrous iron, and sulfate, heavy metals are also released during weathering of other sulfide minerals, such as sphalerite ($Zn_{1-x}Fe_xS$), occurring in minor amounts in the waste. Conceptual understanding and quantification of sulfide mineral weathering processes is thus a prerequisite for ability to understand and predict the composition of AMD.

The rate of sulfide mineral weathering in nature and in mining waste is sufficiently slow to, in many cases, be limited by the surface reaction. Sphalerite undergoes weathering in parallel mechanisms, using dissolved molecular oxygen or ferric iron as the oxidant (Eqs 1 and 2, respectively; Nordstrom & Alpers 1999). At acidic conditions, it may also undergo non-oxidative dissolution (Eq. 3).

$$Zn_{1-x}Fe_xS_{(s)} + O_{2(aq)} \rightarrow (1-x)Zn^{2+} + xFe^{2+} + SO_4^{2-} \quad (1)$$

$$Zn_{1-x}Fe_xS_{(s)} + 8Fe^{3+} + 4H_2O \rightarrow (1-x)Zn^{2+} + (8+x)Fe^{2+} + SO_4^{2-} + 8H^+ \quad (2)$$

$$Zn_{1-x}Fe_xS_{(s)} + 2H^+ \rightarrow (1-x)Zn^{2+} + xFe^{2+} + H_2S_{(g)} \quad (3)$$

While the rate of abiotic pyrite oxidation has been reported by many authors (see e.g. review by Williamson & Rimstidt 1994), fewer studies have focused on the weathering kinetics of minor sulfide minerals, such as sphalerite (e.g. Scharer et al. 1994; Rimstidt et al. 1994; Weisener et al. 2003).

In this paper, we present preliminary results from a study of sphalerite weathering kinetics aiming at:

- Quantifying the stoichiometry and rate of element release from sphalerite at pH values that are relevant for mine waste deposits;
- Determining the reaction order with respect to protons;
- Determining the effect of particle size on the mass-normalized element release rates.

2 METHODS

2.1 *Mineral specimen and powder preparation*

A low Fe sphalerite from Elmwood Mine in Smith Co., Tennessee, USA, was used in this experiment. A cm × cm × cm piece was crushed in a ceramic mortar and dry-sieved to obtain the 35–71, 71–125, 125–250, and 250–500 μm size fractions. Ultrafine particles were removed by repeated ultrasonication in ethanol and samples were gently dried and stored at

5°C before use. The powder was found to be pure sphalerite within the detection limits of X-ray powder diffraction (1–5%).

2.2 *Experimental procedures and analyses*

A known amount (~0.5 g) of sphalerite from a certain particle size fraction was reacted with 200 mL of solution of specified composition and pH (1, 2, 3, 4, and 7) in polypropylene batch reactors. Except for the pH 1 solution, which was 0.1 M $HClO_4$, the solutions had the composition $[H^+]+[Na^+]=[ClO_4^-]=$ 0.010 M. All reagents used were of p.a. grade or better. The reactors were continuously purged with water saturated air and agitated at ~275 min^{-1}.

Aqueous samples were removed on 10 occasions over the 53 days of experiments, filtered, and acidified to pH < 1 using suprapur HNO_3 and stored at 5°C until analysis, using ICP-AES (Varian Vista AX, CCD). The pH and the redox potential were determined in separate sub-samples using an Orion Ross™ combined pH electrode and an Orion Epoxy Sure-Flow® combination redox/ORP electrode, respectively. The pH values were stable at 1.05, 2.02, 3.05, 4.10 and 7.21 and redox potentials were 400–600 mV/SHE in the various reactors over the course of the experiments.

3 RESULTS AND DISCUSSION

3.1 *General release behavior*

Figure 1 shows the cumulative, mass-normalized release of elements to solution as function of time for an experiment at pH 3 using the size fraction 125–250 µm. The release is highly non-stoichiometric, with preferential release of metals (Zn and Fe) as compared to sulfur. Such apparent non-stoichiometric release of elements has previously been reported for sphalerite, and may reflect accumulation of a layer of elemental sulfur or polysulfides on the mineral particle surface (e.g. Weisener et al. 2003, and others) and/or degassing of $H_2S_{(g)}$, subsequent to non-oxidative dissolution of sphalerite at the acidic pH in this experiment (cf. Eq. 3). The latter hypothesis was supported by an observed smell of $H_2S_{(g)}$ at the start of the experiments.

The slopes of the release curves (Fig. 1) indicate the mass-normalized element release rates; for Zn and Fe it is evident that the release was faster in the beginning of the experiments and leveled off with time. This is often observed in weathering experiments and has been attributed to dissolution and depletion of ultrafine particles or more reactive surface sites. For sulfide minerals that dissolve incongruently, it may also be due to increasing diffusion

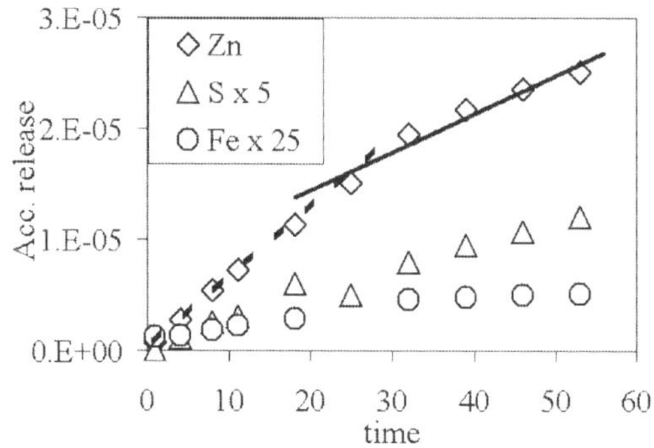

Figure 1. Cumulative, mass-normalized release [mol g^{-1}] of Zn, S, and Fe from sphalerite of the 125–250 µm particle size fraction at pH 3 versus time [day^{-1}]. The dotted and full lines indicate linear regression on the Zn data to obtain initial and final release rates, respectively. NB: For clarity the S and Fe data has been magnified with the factor 5 and 25, respectively.

resistance of the accumulating layer of e.g. polysulfide at the particle surface (Weisener et al. 2003).

In the experiment in Figure 1, S release was fairly linear with time; in some of the other experiments, however, a phase of slow release was followed by a phase of faster release. Initial and final release rates were thus obtained by linear regression of subsets of data from the beginning and the end of the experiments, respectively. We note that the final rates are more relevant for e.g. the acid mine drainage problem, which has a characteristic time scale of decades to centuries.

3.2 *Particle size dependence*

Figure 2 shows the final mass-normalized release rates of elements at pH 3 as a function of the logarithm of the median particle radius. As seen from the figure, the release rate showed a low dependence on particle size. This is in contrast to what is expected for a surface controlled reaction. If the release rate scales with the geometric surface area, an inverse dependence on the particle radius is expected (symbolized by the dotted line in Fig. 2). In some cases, such increase in rate with decreasing particle size is pronounced by a higher reactivity of fine particles, due to surface free energy effects (e.g. Hodson 2002).

The results in Figure 3 also indicate that the release stoichiometry at the end of the experiments is fairly constant over the different particle size fractions. The seemingly constant mass-normalized rate of release may be related to, for example, different surface roughness or contribution from internal surfaces between the different particle sizes and is currently under further investigation.

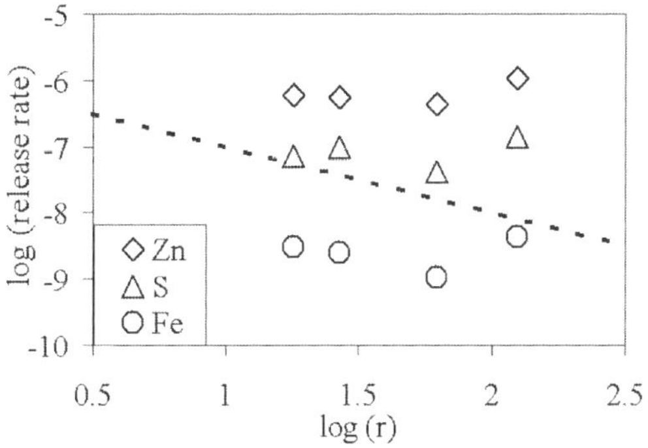

Figure 2. Logarithm of the mass-normalized final release rate [mol g^{-1} day^{-1}] of elements from sphalerite at pH 3 as function of the logarithm of the median particle radius [μm]. The dotted line indicates the expected variation of release rate with particle size, based on the geometric surface area of spherical or cubic particles, for a surface reaction controlled process.

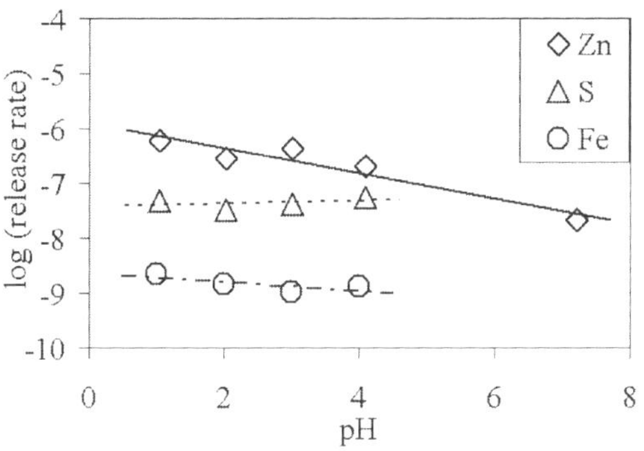

Figure 3. Logarithm of the mass-normalized, final release rates [mol g^{-1} day^{-1}] of elements from the 125–250 μm particle size fraction of sphalerite versus pH. The lines represent linear regressions on the data to obtain reaction orders (~0.2 for Zn, ~0.1 for Fe, and ~0 for S) with respect to protons.

3.3 *pH dependence*

Figure 3 shows the mass-normalized release rates of elements from the 125–250 μm size fraction of the sphalerite as a function of pH. For S and Fe, the concentrations were below detection limit in the reactor at pH ~7. For the other conditions, the release rates were found to be only weakly dependent on pH, with apparent reaction orders with respect to protons of ~0.2 for Zn, ~0.1 for Fe, and ~0 for S. While the pH dependence of sphalerite dissolution has not previously been reported in the literature, these results are consistent with the fractional and low reaction order with respect to protons for pyrite (e.g. Williamson & Rimstidt 1994).

The data in Figure 3 also indicate that the sphalerite dissolves incongruently at least in the pH region $1 \leq pH \leq 4$, and that the reaction is more strongly non-stoichiometric at low pH.

3.4 *Comparison to literature data*

The data at pH 1 in Figure 3 corresponds to final surface area normalized release rates (based on a geometric specific surface area; $A_{geo} = 0.01\ m^2/g$ for the 125–250 μm size fraction) for Zn and S of 6×10^{-10} mol m^{-2} s^{-1} and 5×10^{-11} mol m^{-2} s^{-1}, respectively. The corresponding initial release rates were 4×10^{-9} mol m^{-2} s^{-1} and 1×10^{-10} mol m^{-2} s^{-1}. For Zn, our initial release rate is intermediate the initial (5.75×10^{-9} mol m^{-2} s^{-1}) and final (1.14×10^{-9} mol m^{-2} s^{-1}) rates reported by Weisener et al. (2003; their Table 2) for sphalerite dissolution at similar conditions (25°C, pH 1, $HClO_4$ medium, 38–75 μm particle size fraction). In their study, the initial rate was observed during the first ~3 h of reaction, which was not addressed in our study. Their final rates were obtained during the 144 h of experiment, well within the fast (initial) dissolution phase of our experiments. While these two experiments have been designed to observe the kinetics in different time regimes, and thus the results are not completely comparable, it seems that, within a factor of ~4, our results agree to those previously reported by Weisener et al. (2003). This difference in rate is considered small, given that different size fractions and mineral specimen were used in the two different studies.

4 CONCLUSIONS

We present preliminary results from studies of abiotic sphalerite ($Zn_{1-x}Fe_xS$) weathering kinetics over different particle size fractions (35–71, 71–125, 125–250, and 250–500 μm), using molecular oxygen as the primary oxidant. The experiments were run over ~50 days at 22 ± 2°C in air-purged batch reactors at specified pH (1, 2, 3, 4, and 7). For most of the elements and conditions, two periods with different release rates were observed. We conclude that long-term experiments are necessary for quantifying release kinetics that is relevant for e.g. addressing the acid mine drainage problem. For all elements, the mass-normalized release rates after ~50 days of experiment were only weakly dependent on pH ($1 \leq pH \leq 4$), which is consistent with what has previously been reported for pyrite (Williamson & Rimstidt 1994).

The release of elements was found to be strongly non-stoichiometric with preferential release of Zn and Fe over S over the studied pH range and for all size fractions; this may indicate accumulation of sulfur at the particle surface or degassing of H_2S subsequent to acid (non-oxidative) dissolution of sphalerite. Final release rates were only weakly dependent on particle size (35–500 μm), an effect that, for example, may be due to different surface roughness or contributions from internal particle surfaces between the different size fractions. Further investigations addressing the stoichiometry of the weathering process and the particle size effect on the release rates are required for a conceptual understanding of the weathering kinetics, such as necessary for prediction of element release rates from sphalerite in environmental applications.

ACKNOWLEDGEMENTS

This project was financed by FORMAS, the Swedish Research Council for Environmental, Agricultural and Spatial Planning.

REFERENCES

Hodson, M.E. 2002. Variation in element release rates from different mineral size fractions from the B horizon of a granitic podzol. *Chem. Geol.* 190: 91–112.

Nordstrom, D.K. & Alpers, C.N. 1999. Geochemistry of acid mine waters. In G.S. Plumlee & M.J. Logsdon (eds.), *The environmental geochemistry of mineral deposits, Reviews in Economic Geology*: vol. 6A: 133–160. Chelsea (USA): Soc. Econ. Geologists.

Rimstidt, J.D., Chermak, J.A. & Gagen, P.M. 1994. Rates of reaction of galena, sphalerite, chalcopyrite and arsenopyrite with Fe(III) in acidic solutions. In C.N. Alpers & D.W. Blowes (eds.), *Environmental geochemistry of sulfide oxidation, ACS symposium series*: vol. 550: 2–13, Washington: Am. Chem. Soc.

Scharer, J.M., Nicholson, R.V., Halbert, B. & Snodgrass, W.J. 1994. A computer program to assess acid generation in pyritic tailings. In C.N. Alpers & D.W. Blowes (eds.), *Environmental geochemistry of sulfide oxidation, ACS symposium series*: vol. 550: 132–152, Washington: Am. Chem. Soc.

Weisener, C.G., Smart, R.St.C. & Gerson A.R. 2003. Kinetics and mechanisms of the leaching of low Fe sphalerite. *Geochim. Cosmochim. Acta* 67: 823–830.

Williamson, M.A. & Rimstidt, J.D. 1994. The kinetics and electrochemical rate-determining step of aqueous pyrite oxidation. *Geochim. Cosmochim. Acta* 58: 5443–545.

Water–Rock Interaction, Wanty & Seal II (eds)
© 2004 Taylor & Francis Group, London, ISBN 90 5809 641 6

Processes controlling weathering rind advancement on Costa Rican basalt clasts

A.K. Navarre & S.L. Brantley
Department of Geosciences, The Pennsylvania State University, University Park, PA, USA

P.B. Sak
Geology Department, Bucknell University, Lewisburg, PA, USA

ABSTRACT: Soils developing on the 120 ka terrace of the Rio Parrita near the Pacific coast of Costa Rica contain basalt clasts with well-developed weathering rinds. These rinds are the result of *in situ*, isovolumetric chemical weathering of the parent clast. This weathering rind system can be viewed as a miniature saprolite-forming environment where the processes occurring at the rind/core interface are analogous to processes occurring at larger landscape scales at the saprolite/bedrock interface. Physical and chemical characteristics of the rind/core interface have been examined in two clasts with different textures using scanning electron microscopy (SEM), SEM energy dispersive X-ray (SEM-EDX), and computed tomography X-ray (CT) imaging. Dissolution of plagioclase at this boundary is clearly evident. The migration of the core/rind boundary appears to depend in part upon the dissolution of plagioclase and resulting formation of secondary porosity at the interface.

1 INTRODUCTION

Accurate long term silicate dissolution rates are required for accurate quantification of sequestration of atmospheric CO_2, soil nutrient cycling, and large-scale physical erosion. Historically, long term dissolution rates have been constrained by soil/saprolite formation studies (e.g. White et al. 1996). White (2002) has proposed a systematic method for calculating long-term weathering rates in soil profiles where the rate of silicate mineral weathering is related to the thickness of a depletion zone just above the saprolite/bedrock interface. This depletion zone is defined by the gradual loss of an index mineral, usually plagioclase, from zero to original parent material concentrations. When depth (y-axis) is plotted against concentration of index mineral (x-axis) the resultant slope of the weathering gradient can be used to calculate the rate of mineral dissolution. If the age of the soil surface is known, a rate of saprolite formation can be calculated and a mineral dissolution rate determined. This model depends on the rate of saprolite/bedrock interface migration, but the processes controlling the migration are not well understood. Herein we use weathered basaltic clasts from a 120 ka alluvial terrace exposed along the central Pacific coast of Costa Rica to quantify rates of silicate dissolution. In these samples, the plagioclase depletion zone is thin (mm-scale) as compared to the thick (~5 m) depletion zone reported for the Panola granite (White 2002).

2 WEATHERING RIND SYSTEM

Weathering rinds are permeable crusts enriched in immobile oxides (i.e. Fe_2O_3, TiO_2, and Al_2O_3) that envelop unweathered, non-porous basaltic cores. The *in situ* conversion of unweathered parent material to porous rind material is analogous to the saprolite/rock transition with two important differences: (1) outer surface of the weathering rind shows no evidence of erosion (Colman 1982; Sak et al. in press) and (2) four of the five soil-forming factors (topography, climate, biologic influences, and time) are identical for different clasts weathering in the same soil. In the study area, mean annual precipitation and mean annual temperature for the period extending from 1941–1982 are 3085 mm and 27.3°C, respectively (Instituto Meteorológico Nacional de Costa Rica 1992, unpublished data) and paleoclimatic investigations suggest climatic stability throughout the late Pleistocene (Gomez 1986). Clasts used in this study were collected from a 120 ka alluvial fill terrace

(Sak et al., in press); therefore, we assume that the climate, topography, and time are constant between samples. Biologic activity that may alter the rates of mineral dissolution is unknown, but is assumed constant throughout each clast and within the terrace profile. Thus weathered clasts provide a suitable system to examine the processes of rind (saprolite) advance in the absence of physical weathering and large-scale heterogeneities that complicate catchment-scale field investigations (e.g. White 1995).

Volumetric changes across the core/rind interface are estimated from the strain, $\varepsilon_{i,w}$, (Brimhall & Dietrich 1987; Chadwick et al. 1990; Brimhall et al. 1991):

$$\varepsilon_{i,w} = \frac{\rho_p C_{i,p}}{\rho_w C_{i,w}} - 1 \quad (1)$$

where ρ_p is dry bulk density (g cm^{-3}) for the core, ρ_w is dry bulk density (g cm^{-3}) of rind, $C_{i,x}$ is mass concentration (percent) of an inert element i, in this case Ti in the core ($C_{i,p}$) and the rind ($C_{i,w}$).

The loss or gain of an element j from the rind, $\tau_{j,w}$ is calculated as:

$$\tau_{j,w} = \frac{\rho_p C_{i,p}}{\rho_w C_{i,w}} \left(\varepsilon_{i,w} + 1\right) - 1 \quad (2)$$

where, $\varepsilon_{i,w}$ is the volumetric strain for Ti as determined in (1) (Figure 1). When $\tau_{j,w} = 0$, the element is immobile. If $\tau_{j,w} = -1$ then element j is completely lost during weathering. Thus the more negative the value of $\tau_{j,w}$, the more mobile the element (Fig. 1). Furthermore, the long-term weathering flux per unit time, $Q_{j,r}$ (mol m^{-2} s^{-1}), is defined as:

$$Q_{j,r} = \left(L_j t^{-1}\right) 10^4 \quad (3)$$

where t is time (s^{-1}). The elemental and mineralogic loss across the rind, L_j (mol cm^{-2} geometric area of rock surface), is calculated for samples by integrating $\tau_{j,w}$ over the distance from the core-rind boundary, z (Merritts et al. 1992)

$$L_j = \left(\rho_p \frac{c_{j,p}}{M_j}\right) \int_{z=0}^{z=d} \tau_{j,w}\, dz \quad (4)$$

where d is the total distance from the core-rind boundary (0.9 cm for the fine grained sample and 1.6 cm for the coarse grained sample), M_j is the atomic weight of element j, and ρ_p is the core density (2.8 g cm^{-3}). For 1 cm^2 of rind area, the resulting values of L_j, integrating across the rind of range from 1.76×10^{-3} to 5.56×10^{-6} mol cm^{-2} for Si and Mn,

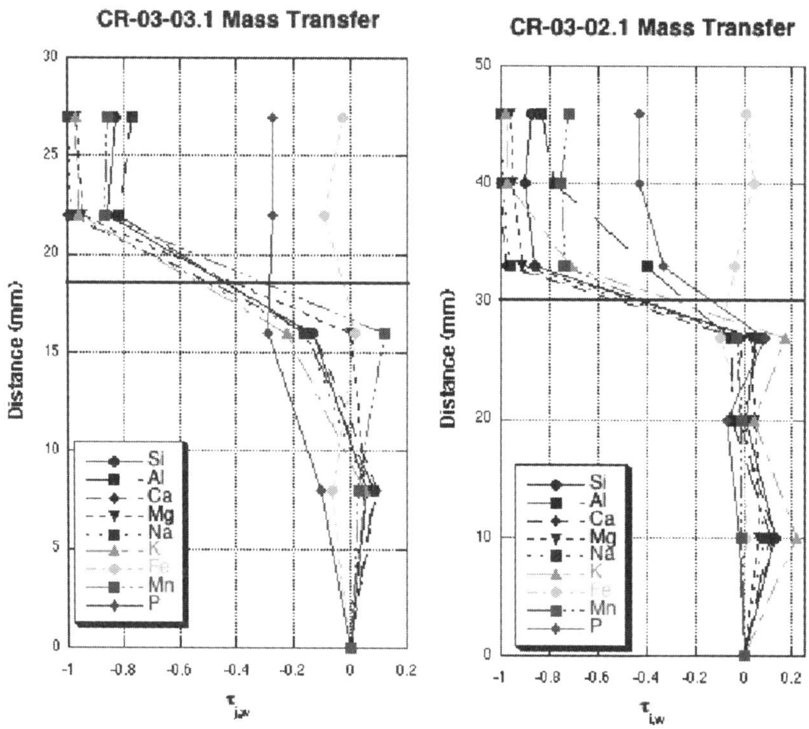

Figure 1. Elemental mass transfer for two basalt clasts. Negative values of τ indicate loss of that element during weathering.

respectively in the fine grained sample and 2.04×10^{-2} to 8.37×10^{-6} mol cm^{-2} for Si and P, respectively in the coarse grained sample. The long term Si flux rates are similar to rates reported for the Icacos granite in Puerto Rico where the weathering surface is 200 ma (White et al. 1998). Like the Costa Rica site the Puerto Rico site is located in an area of tropical climate. Average annual temperatures at both sites are above 22°C, but the Puerto Rico site receives 1150 mm more precipitation per year than the Costa Rica site (4200 mm/yr and 3055 mm/yr, respectively).

An advantage to studying weathering rind formation is that the transition zone between parent and weathered rind is confined to a narrow (<500 μm wide) zone (Fig. 2). The weathering rind system is well-constrained and provides a setting where the rate of advance of the rind/core interface is isolated and a physicochemical model can be developed to predict mechanisms that control the rate of advance of the core/rind (regolith/bedrock) interface.

Parent material composition and mineral surface texture of the basalt clasts has been determined by ICP-MS, petrographic, and scanning electron microscopy analyses (SEM; Sak et al., in press).

Porosity of the basaltic core is generally <1% compared to 50% porosity in the rind (Sak et al., in press). Porosity is required for fluid migration into the non-porous core where it can interact with minerals. Core porosity may be either primary or secondary, where secondary porosity is formed during *in situ*, isovolumetric weathering of plagioclase phenocrysts within the core. A similar mechanism is observed in the Panola Granite (GA) where rapid dissolution of disseminated calcite results in the formation of secondary porosity enabling reactants to migrate further into the bedrock in turn accelerating plagioclase

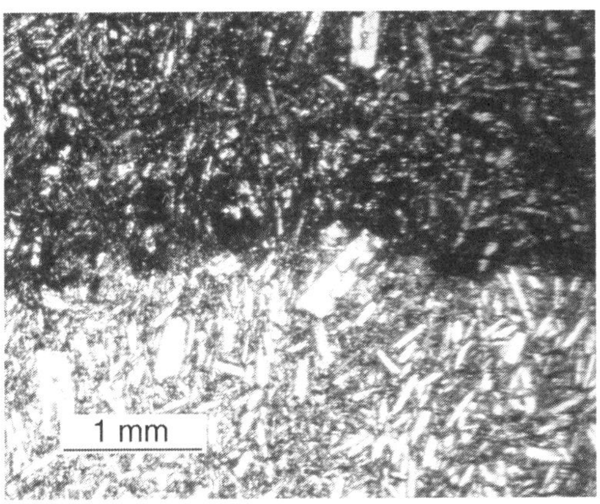

Figure 2. Photomicrograph of fine grained basalt clast. The interface between the core (light) and rind (dark) is abrupt.

dissolution (White et al. 2001). In both the Costa Rican basalts and the Panola granite, the amount of secondary porosity may be a function of grain size and modal abundance of the most rapidly dissolving phase in the parent material. Increasing pore space characteristic of weathering rind development results in the development of a density profile across the depletion zone.

3 METHODS

Two basalt clasts with differing textures were analyzed using petrographic, SEM, and SEM energy dispersive X-ray (SEM-EDX) to assess the role of textural variability on the development of secondary porosity in the rind. Samples were collected from terraces of the same age and approximately the same elevation. Distance between the sampled terraces is less than 1 km. Computed tomography X-ray technology has been used to map density throughout a clast. This instrument creates a 3-D map of density from individual X-ray pictures. Actual density and changes in density can be determined on less than a millimeter scale (Fig. 3). These data will allow for accurate, consistent quantification of the thickness of both the weathering rind and depletion zone.

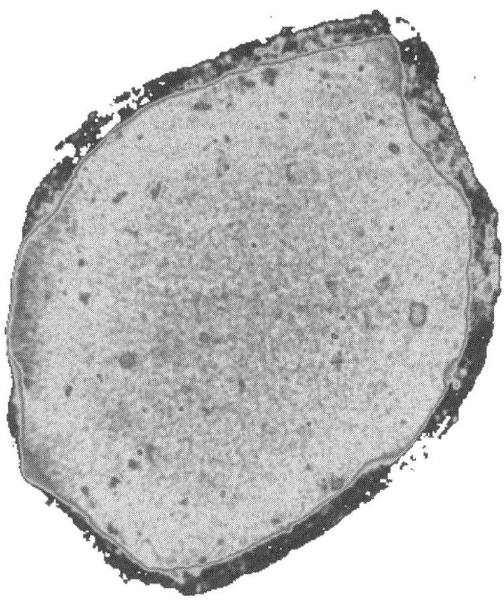

Figure 3. CT image of basalt clast approximately 7 cm in diameter. Changes in color represent changes in density. Outer dark later is the low density weathering rind. Inhomogeneities in the core are the result of density differences between mafic phenocrysts and fine grained matrix.

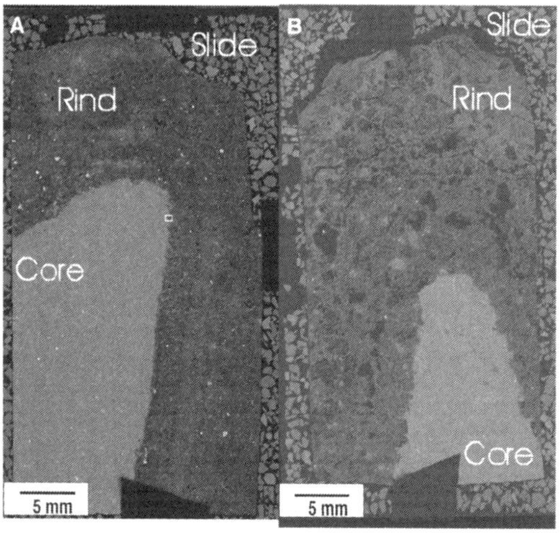

Figure 4. SEM backscatter images of two basalt clasts. (A) is much finer grained then (B) and also has a more abrupt transition from core to rind.

4 RESULTS AND DISCUSSION

SEM backscatter images of two clasts suggest that characteristics of the rind/core interface may vary with grain size (Fig. 4). Dissolution of plagioclase and other minerals was seen at the rind/core interface in thin sections of both samples, but plagioclase dissolution was also observed coreward of the interface. The dissolution of these crystals increases surface area available for dissolution reactions and the pore space created provides a pathway for water infiltration to these surfaces (Fig. 5). In addition, SEM-EDX element maps show Ca and Na depletion in the rind, consistent with open-system mass balance and elemental flux calculations. The interface observed on the clast with large (<2 mm long) euhedral plagioclase phenocrysts is very diffuse when compared to the abrupt interface (<100 μm wide) in the clast with small (<750 μm long) plagioclase lathes (Fig. 4). Thickness of the weathering rind also appears to increase as grain size increases. Measured rind thicknesses were 2–2.1 cm and 0.9–1.1 cm for the coarse and fine grained basalt clasts, respectively. The error on these measurements is unknown because the rind material is unconsolidated and difficult to sample without leaving material behind. Preliminary examination of clasts indicates that porosity within the depletion zone controls the width of the transition zone between the unweathered core and the rind. Roughness of the interface surface is markedly higher in the coarser grained clast. An increase in roughness of this internal surface increases the surface area exposed to infiltrating water, possibly leading to higher rates of interface migration.

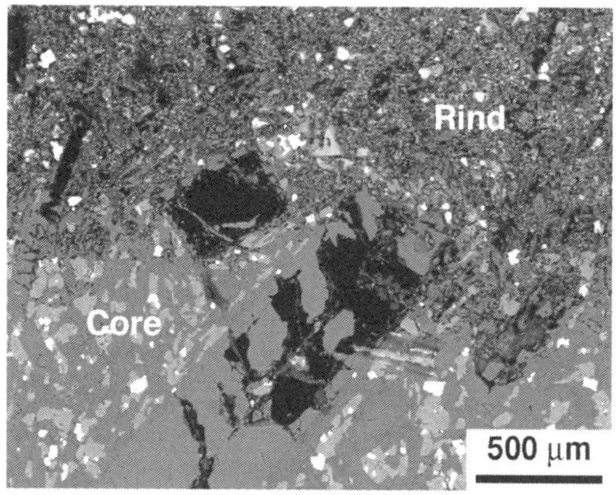

Figure 5. Dissolving plagioclase grain at rind/core interface shown in SEM backscatter image. As these grains dissolve more surface area is exposed and available for dissolution to take place.

The interface can be imaged by mapping changes in density with X-ray CT scanning techniques to provide insight into the characteristics of this surface.

5 CONCLUSIONS

Development of weathering rinds on basaltic clasts provides a suitable system to investigate the intrinsic rates of saprolite advance in a field setting. The primary control appears to be reactive surface area, which increases as a function of secondary porosity development at the saprolite/bedrock interface during plagioclase dissolution. If the weathering rind is dominated by secondary porosity within the depletion zone, controls on the rate of formation of this porosity become primary controls on the migration of the rind/core boundary. Since weathered clasts are small scale proxies for saprolite formation, results may shed insight into processes occurring at the saprolite/bedrock interface.

ACKNOWLEDGEMENTS

We thank A. Grader and P. Halleck at the Center for Quantitative Imaging, Pennsylvania State University for CT images of basalt clast and C. Daniel (Bucknell University) for assistance with SEM-EDX imaging.

REFERENCES

Brimhall, G.H., Chadwick, O.A., Lewis, C.J., Compston, W., Williams, I.S., Danti, K.J., Dietrich, W.E., Power, M.E., Hendricks, D. & Bratt, J. 1991. Deformational mass

transport and invasive processes in soil evolution. *Science* 255: 695–702.
Brimhall, G.H. & Dietrich, W.E. 1987. Constitutive mass balance relations between chemical composition, volume, density, porosity, and strain in metasomatic hydrochemical systems. *Geochim. Cosmochim. Acta* 51: 567–587.
Chadwick, O.A., Brimhall, G.H. & Hendricks, D.M. 1990. From black box to a grey box: a mass balance interpretation of pedogenesis. *Geomorph.* 3: 369–390.
Colman, S.M. 1982. Clay mineralogy of weathering rinds and possible implications concerning the sources of clay minerals in soil. *Geology* 10: 370–375.
Gómez, L.D. 1986. Vegetación de Costa Rica: Apuntes para una biogeografía costarricense, San José, Costa Rica, Editorial Universidad Estatal a Distancia.
Merritts, D.J., Chadwick, O.A., Hendricks, D.M., Brimhall, G.H. & Lewis, C.J. 1992. The mass balance of soil evolution on late Quaternary marine terraces, Northern California. *Geol. Soc. Am. Bull.* 104: 1456–1470.
Sak, P.B., Fisher, D.M., Gardner, T.W., Murphy, K. & Brantley, S.L. (in press) Rates of weathering rind formation on Costa Rican Basalt. *Geochim. Cosmochim. Acta*.
White, A.F. 1995. Chemical weathering rates of silicate minerals in soils. In A.F. White & S.L. Brantley (eds.), *Chemical Weathering Rates of Silicate Minerals*. Washington D.C.: Mineralogical Society of America.
White, A.F., Blum A.E., Schulz, M.S., Bullen, T.D., Harden, J.W. & Peterson M.L. 1996. Chemical weathering rates of a soil chronosequence on granitic alluvium: I. Quantification of mineralogical and surface area changes and calculation of primary silicate reaction rates. *Geochim. Cosmochim. Acta* 60: 2533–2550.
White, A.F., Blum, A.E., Schulz, M.S., Vivit, D.V., Stonestrom, D.A., Larsen, M., Murphy, S.F. & Eberl, D. 1998. Chemical weathering in a tropical watershed, Luquillo Mountains, Puerto Rico: I. Long-term versus short-term weathering fluxes. *Geochim. Cosmochim. Acta* 62: 209–226.
White, A.F., Bullen, T.D., Schultz, M.S., Blum, A.E., Huntington, T.G. & Peters, N.E. 2001. Differential rates of feldspar weathering in granitic regoliths. *Geochim. Cosmochim. Acta* 65: 847–869.
White, A.F. 2002. Determining mineral weathering rates based on solid and solute weathering gradients and velocities: application to biotite weathering in saprolites. *Chem. Geol.* 190: 69–89.

Water–Rock Interaction, Wanty & Seal II (eds)
© 2004 Taylor & Francis Group, London, ISBN 90 5809 641 6

Origin and mixing of groundwaters from Lupin (Canada): Nd isotope evidence

P. Négrel & J. Casanova
BRGM, Water Division, Orléans, France

R. Blomqvist
GTK, Nuclear Waste Disposal Research, Espoo, Finland

ABSTRACT: Nd contents and Nd isotopes were determined on surface waters, co-existing ice and groundwaters in a Canadian frozen mine (Lupin, Northwest Territories). In this context, Nd isotope ratios are used to establish the degree of water-rock interaction and to clarify mixing processes within the groundwaters. Variations in Nd isotopes provide new information on the mixing history in deep groundwaters in the Canadian Shield.

1 INTRODUCTION

The occurrence of highly saline waters in deep geological formations is a world-wide phenomenon and such saline waters, with total dissolved solids up to $350\,g\,l^{-1}$ are encountered either in shields or sedimentary provinces and have been investigated during the past decades using a large variety of geochemical tools including major, trace elements and isotopes (Frape et al. 1984; Edmunds et al. 1984; Nurmi et al. 1988; Douglas et al. 2000; Grobe et al. 2000; Kloppmann et al. 2001; Cai et al. 2001; Négrel et al. 2003).

Isotopic studies provide valuable information for estimating the origin and age of dissolved solids of deep groundwaters, including highly saline waters, and to draw a comprehensive picture of the hydrologic system, as water in different environments develops isotopic fingerprints with which it can be identified and its origins traced (Bottomley et al. 1999; Kloppmann et al. 2001; 2002). Furthermore, the variations of radiogenic isotope (Sr, Nd, etc.) within a hydrologic system can provide information about the sources of these elements and the different mixing processes involved (McNutt et al. 1990; Peterman & Wallin 1999; Négrel et al. 2001; 2003).

Although such saline waters have been intensively investigated by many researchers from different countries, the origin of their salinity and therefore the nature of the processes involved in their formation, is still a subject of debate. Most autochthonous hypotheses for the origin of brines in the Shields are grounded on $^{87}Sr/^{86}Sr$ and suggest water/rock interaction on the long term such as silicate mineral hydrolysis (McNutt et al. 1990; Peterman & Wallin 1999; Négrel et al. 2003). On the other hand, the allochthonous sources of salinity refer to seawater, either though evaporation or freezing (Bottomley et al. 1999; Starinsky & Katz 2003).

In this paper, we report both the contents and the isotopic composition of Nd in surface and groundwater collected in the Lupin hydrologic system in the Northwest Territories, Canada. The Nd isotope ratios are used to complement the degree of water–rock interaction and clarify mixing processes within the groundwaters.

2 SAMPLES AND ANALYSIS

The Lupin site is located 80 km south of the Arctic Circle and about 400 km north-northeast of Yellowknife, Northwest territories (Fig. 1, Robert & Poulsen 1997). The Lupin deposit is included within the Yellowknife Supergroup of supracrustal rocks including a sequence of amphibolite-grade metaturbidites and iron formation (Robert & Poulsen 1997). Three rock samples were analysed, including quartzite and phyllite. One main feature of Lupin is its location in a region of continuous permafrost. The permafrost conditions in the topmost 500 m have preserved groundwaters in the rocks as ice that can actually be sampled in the galleries at the outlet of the main fractures. Below this level, free groundwaters are present down to the 1130 m level.

Samples collected in Lupin consist of snow, lake dilute water, ice in the topmost 400 m and saline groundwaters as deep as 1130 m in the mine. Precleaned polyane bags were used for snow and ice collection and all water samples used for Nd content and isotope measurements and cation determination were filtered on site through 0.2 μm cellulose acetate filters and stored in precleaned polypropylene bottles, and acidified to pH 2 with ultrapure HNO_3. One bottle for each sample was not acidified and was used for anion determination. Chemical analysis of water samples was carried out by HPLC for anions and inductively coupled plasma mass spectrometry (ICP-MS) for Nd contents.

The Nd isotopic compositions were determined following the method of Négrel et al. (2000). Briefly, Nd was extracted using ferric hydroxide co-precipitation and separated using classical cation exchange and HDEHP reverse chromatography with a total procedural blank less than 100 pg. The Nd isotopic ratios were measured using a Finnigan MAT 262 multiple collector mass spectrometer with double Re filaments. The $^{143}Nd/^{144}Nd$ ratios were normalised to a $^{146}Nd/^{144}Nd$ of 0.7219, adjusted to the La Jolla standard value of 0.511860 and expressed as $\varepsilon_{Nd}(0)$, which represents the deviation in parts per 10^4 (ε units) from $^{143}Nd/^{144}Nd$ in a chondritic reservoir (present day value of 0.512636. Repeated measurements of the La Jolla international standard during the period of analysis yielded a mean $^{143}Nd/^{144}Nd$ of 0.511870 ± 0.000020 (2σ, n = 10).

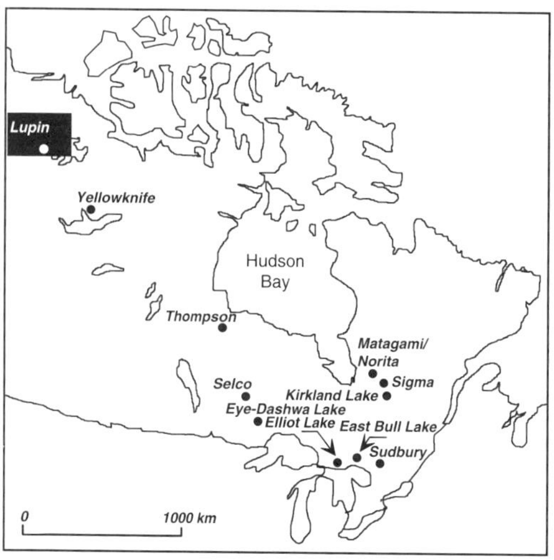

Figure 1. General location map of the Lupin site in the Northwest territories (Canada).

Nd isotopic compositions were measured on three samples of bedrock following standard acid-dissolution procedures of powdered bedrock samples and chemical separation described for waters.

3 RESULTS AND DISCUSSION

3.1 *Cl variations*

Chloride is one of the most conservative species and is frequently used either as a substitute for salinity (e.g. TDS) or referred to for element ratios, e.g. Br/Cl, Na/Cl, etc. (Frape et al. 1984; Edmunds et al. 1984; Nurmi et al. 1988). The chloride concentration in groundwaters increases with depth, as well as other constituents, with increasing TDS. This is illustrated in Figure 2 where the Cl contents are plotted versus depth in the Yellowknife area (Frape et al. 1984; McNutt et al. 1990; Bottomley et al. 1999; Douglas et al. 2000). The data from Lupin, including snow, ice and saline groundwaters' are also plotted on this figure. Yellowknife and Lupin are separated by 400 km and fall in the Slave province (Robert & Poulsen 1997). A general trend showing an increase of Cl contents vs. depth is evidenced and the scattering of some Cl values occurring mainly at 700 m and 1300–1400 m depth roughly reflects intrusion of dilute waters through mining activities (Frape et al. 1984; McNutt el al. 1990; Bottomley et al. 1999;

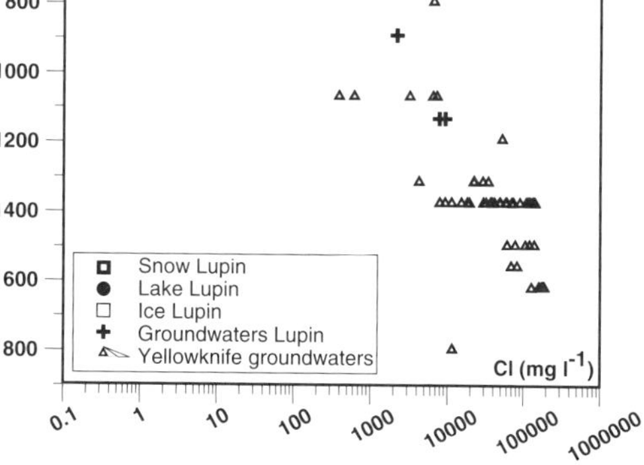

Figure 2. Plot of the Cl (mg L^{-1}) contents in Yellowknife groundwaters, stream waters and Lupin snow, lake, ice and groundwaters vs. depth (m).

Douglas et al. 2000). The Lupin data set also describes a Cl enrichment from the most diluted samples (snow and ice) in the topmost 400 m down to deep groundwaters. The snow and lake samples have similar Cl contents, lower than that of stream waters given by Millot et al. (2002). The Cl contents of ice bounded groundwater and stream waters are very similar. The Cl contents of groundwater from Yellowknife and Lupin are in the same range.

3.2 *Bedrocks*

Both quartzite and phyllite have negative $\varepsilon_{Nd}(0)$, in the range −28.29 to −32.19, respectively, associated with Nd contents ranging from 21 ppm to 34 ppm.

3.3 *Neodymium in surface- and groundwaters*

The Nd contents of Lupin snow and lake, ice are respectively 0.01 and 0.08 $\mu g\ L^{-1}$. In the ice-bound groundwater, a large range in the Nd contents is observed (0.01–5.73 $\mu g\ L^{-1}$). In the groundwaters the Nd contents range from 0.07 to 12.7 $\mu g\ L^{-1}$. The contents of Nd in the investigated waters vary over several orders of magnitude and fall within the ranges observed by several authors studying rare earth elements in groundwaters from silicate bedrocks (Smedley 1991; Banks et al. 1999; Négrel et al. 2000; 2001; Négrel & Petelet-Giraud 2002).

Figure 3 illustrates the relationships between Nd and Cl contents in the snow, lake, ice and groundwaters from Lupin. Notwithstanding the large increase of Cl contents, the lack of a clear relationship between Nd and Cl contents in the whole sample set and within the groundwaters seems to indicate that the dissolved Nd concentrations are not initially related to the salinity of the water as demonstrated by Négrel et al. (2000; 2001).

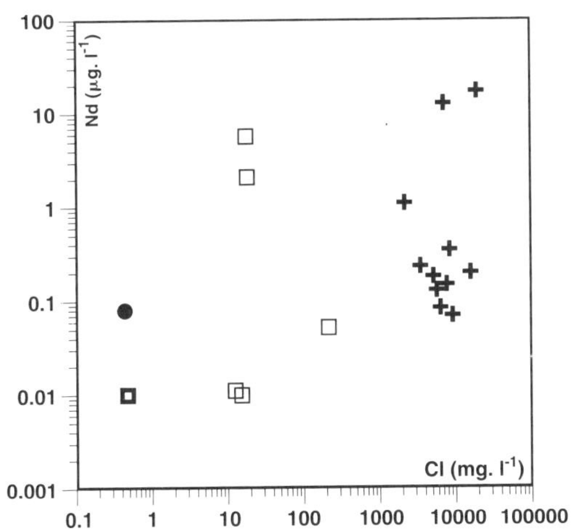

Figure 3. Plot of Cl contents (in mg L^{-1}) vs. Nd contents (in $\mu g\ L^{-1}$) in the samples from Lupin. Symbols as in Figure 2.

3.4 *Neodymium isotopes*

Up to now, most of the studies with Nd isotopes showed small differences between $\varepsilon_{Nd}(0)$ values for both the dissolved and suspended phases found within the same rivers, suggesting that the Nd isotopic composition can be used as an indicator of the weathered parent rock. Négrel et al. (2000) studying mineralised waters in the Massif Central (France) and Négrel & Petelet-Giraud (2002) studying ground- and surface waters in French Guiana evidenced that both the waters and silicate parent rocks have similar $\varepsilon_{Nd}(0)$ signatures.

The isotopic composition of Nd in the groundwaters from Lupin ranges from $\varepsilon_{Nd}(0) = -23.31$ to −30.86 with most of the values in the range of the parent rocks (−28.29 to −32.19).

Figure 4 illustrates the variation of Nd contents and $\varepsilon_{Nd}(0)$ versus depth. Two samples, from around

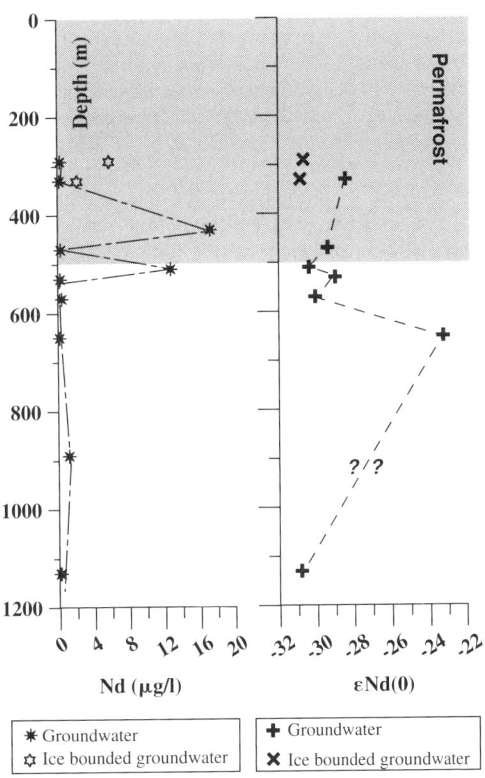

Figure 4. Plot of Nd (µg/L) contents and $\varepsilon_{Nd}(0)$ measured in the Lupin ice and groundwaters vs. depth (m).

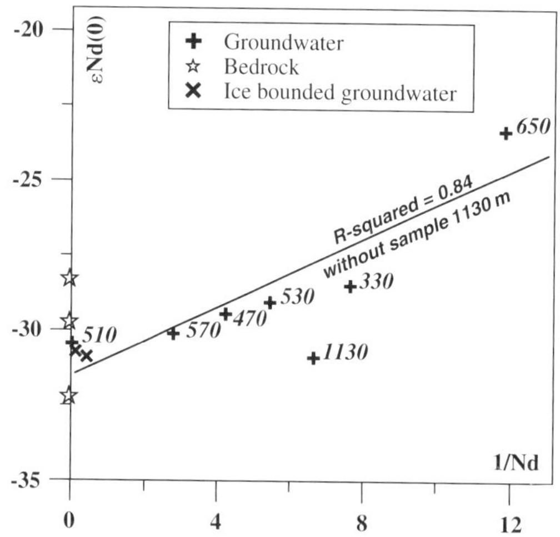

Figure 5. Plot of the $\varepsilon_{Nd}(0)$ vs. the 1/Nd ratio in the samples from Lupin. Symbols as in Figure 2, numbers refer to depth (m).

430 m and 510 m, within and at the permafrost limit, show a large increase in Nd content while thereafter the Nd contents remain similar. Conversely, the $\varepsilon_{Nd}(0)$ gradually decreases from −28.46 (at 330 m depth) to −30.08 at 570 m, shows a large increase up to −23.31 at 650 m depth and then becomes more negative at greater depths (−30.86 at 1130 m).

The mixture of two or more components and the end-member water compositions can be evaluated by using a classical diagram involving the isotopic composition and the inverse of the content (Négrel et al. 2001; 2003). In terms of mixing lines (Fig. 5), one trend (R^2 close to 0.84) extends from the bedrock to the 650 m water sample and includes all groundwater samples except the 1130 m depth. This trend is consistent with two-component mixing between a high radiogenic – low 1/Nd end-member, with an $\varepsilon_{Nd}(0)$ similar to the bedrock, and a lower radiogenic – high 1/Nd end-member which could be consistent with the surface water (1/Nd in the lake close to 12.5). This second end-member, with a 1/Nd higher than 12 induces a dilution of deeper groundwaters that interacted with the bedrock. Groundwater at 1130 m depth, with a $\varepsilon_{Nd}(0)$ similar to the bedrock should have experienced another history, as Casanova et al. and Négrel et al. (both submitted) suggested with B and Sr isotopes, respectively.

4 CONCLUSIONS

Nd isotope analyses of groundwaters characteristic of the Lupin site suggest a variable origin in which mixing processes and water-rock interaction have both contributed to the observed values.

Mixing processes are clearly demonstrated by plotting the εNd(0) and 1/Nd ratios. The mixing line defines one end-member that corresponds to the interaction with the bedrock, the other one should be related to dilution by surface waters. The Nd isotopes confirm some results obtained using B and Sr isotopes, and detail some processes leading to better characterisation of water-rock interactions.

REFERENCES

Banks, D., Hall, G., Reimann, C., & Siewers, U. 1999. Distribution of rare earth elements in crystalline bedrock groundwaters: Oslo and Bergen regions, Norway. *Appl. Geochem.* 14: 27–39.

Bottomley, D.J., Katz, A., Chan, L.H., Starinsky, A., Douglas, M., Clark, I.D., & Raven, K.G. 1999. The origin and evolution of Canadian Shield brines: evaporation or freezing of seawater? New lithium isotope and geochemical evidence from the Slave craton. *Chem. Geol.* 155: 295–320.

Cai, C., Franks, S.G., & Aagaard, P. 2001. Origin and migration of brines from Paleozoic strata in Central Tarim, China: constraints from 87Sr/86Sr, D, 18O and water chemistry. *Appl. Geochem.* 16: 1269–1284.

Casanova, J., Négrel, P., & Blomqvist, R. Submitted. Boron isotope fractionation in groundwaters as an indicator of permafrost past conditions in the fractured crystalline bedrock of the Fennoscandian Shield. *Water Research.*

Douglas, M., Clark, I.D., Raven, K. & Bottomley, D. 2000. Groundwater mixing dynamics at a Canadian Shield mine. *J. Hydrol.* 235: 88–103.

Edmunds, W.M., Andrews, J.N., Burgen, W.G., Kay, R.L.F. & Lee, D. 1984. The evolution of saline and thermal groundwaters in the Carnmenellis granite. *Min. Mag.* 48: 407–424.

Frape, S.K., Fritz, P. & McNutt, R.H. 1984. Water-rock interaction and chemistry of groundwaters from the Canadian Shield. *Geochim. Cosmochim. Acta* 48: 1617–1627.

Grobe, M., Machel, H. G. & Heuser, H. 2000. Origin and evolution of saline groundwater in the Münsterland Cretaceous Basin, Germany: oxygen, hydrogen, and strontium isotope evidence. *J. Geochem. Explor.* 69–70: 5–9.

Kloppmann, W., Négrel, P., Casanova, J., Klinge, H., Schelkes, K. & Guerrot, C. 2001 Halite dissolution derived brines in the vicinity of a Permian salt dome (N German Basin). Evidence from boron, strontium, oxygen and hydrogen isotopes. *Geochim. Cosmochim. Acta* 65: 4087–4101.

Kloppmann, W., Girard, J.P. & Négrel, P. 2002. Exotic stable-isotope compositions of saline waters and brines from the crystalline basement. *Chem. Geol.* 184: 49–70.

McNutt, R.H., Frape, S.K., Fritz, P., Jones, M.G. & MacDonald, I.M. 1990. The $^{87}Sr/^{86}Sr$ values of Canadian Shield brines and fracture minerals with applications to groundwater mixing, fracture history, and geochronology. *Geochim. Cosmochim. Acta* 54: 205–215.

Millot, R., Gaillardet, J., Dupré, B. & Allègre, C.J. 2002. The global control of silicate weathering rates and the coupling with physical erosion: new insights from rivers of the Canadian Shield. *Earth Plan. Sci. Let.* 196: 83–98.

Négrel, P., Casanova, J. & Blomqvist, R. 2001. Nd isotope variation in groundwater and mixing phenomena from Palmottu (Finland). *Water Research* 35: 1617–1623.

Négrel, P., Guerrot, C., Cocherie, A., Azaroual, M., Brach, M. & Fouillac, C. 2000. Rare earth elements, neodymium and strontium isotopic systematics in mineral waters: evidence from the Massif Central, France. *Appl. Geochem.* 15: 1345–1367.

Négrel, P. & Petelet-Giraud, E. 2002. Rare earth elements and neodymium isotopic systematics in the groundwaters of French Guiana. 12th V.M. Goldschmidt Conference, Davos (Suisse). *Geochim. Cosmochim. Acta* 66: 548.

Négrel, P., Casanova, J., Blomqvist, R., Kaija, J. & Frape, S.K. 2003. Strontium isotopic characterisation of the Palmottu hydrosystem (Finland): Water-rock interaction and geochemistry of groundwaters. *Geofluids* 3: 161–175.

Négrel, P. & Casanova, J. Submitted. Comparing the Sr isotopic signatures in brines between the Canadian and Fennoscandian Shields. *Appl. Geochem.*

Nurmi, P.A., Ilmo, T.K. & Lahermo, W. 1988. Geochemistry and origin of saline groundwaters in the Fennoscandian shield. *Appl. Geochem.* 3: 185–203.

Peterman, Z.E. & Wallin, B. 1999. Synopsis of strontium isotope variations in groundwater at Äspö, southern Sweden. *Appl. Geochem.* 14: 939–951.

Robert, F. & Poulsen, K.H. 1997. World-class Archaean gold deposits in Canada: an overview. *Austral. J. Earth Sci.* 44: 329–351.

Smedley, P.L. 1991. The geochemistry of the rare earth elements in groundwater from the Carnmenellis area, southwest England. *Geochim. Cosmochim. Acta* 55: 2767–2779.

Starinsky, A. & Katz, A. 2003. The formation of natural cryogenic brines. *Geochim. Cosmochim. Acta* 67: 1475–1484.

Water–Rock Interaction, Wanty & Seal II (eds)
© 2004 Taylor & Francis Group, London, ISBN 90 5809 641 6

The significance of suspended material in the chemical transport in rivers of NE Iceland

E.H. Oelkers
Géochimie, CNRS URM, Toulouse, France

S.R. Gislason & E.S. Eiriksdottir
Science Institute, Reykjavik, Iceland

S.O. Elefsen & J. Hardardottir
Hydrological Service of National Energy Authority, Reykjavik, Iceland

ABSTRACT: Co-existing suspended material and water samples have been collected regularly from 8 catchments in NE Iceland over a four year period. Detailed chemical characterization of these samples has been performed to assess the role of suspended material in the chemical transport of the elements. The suspended material in these catchments is similar to that of the host rock, which itself is over 90% basaltic glass in composition. Comparison of the bulk chemistry of this collected suspended material with the chemistry of the dissolved load demonstrates that suspended material dominates the transport of all elements; and suspended material flux ranges from 4 to 10,000 times the flux of the dissolved load, depending on the element. Preliminary calculations based on laboratory measured dissolution rates suggest that the ocean acquires the majority of its chemical fluxes of most elements from the dissolution of this suspended material following its arrival in the ocean, rather than from dissolved chemical transport. Suspended material desorption experiments suggest that transport of elements adsorbed onto the surface of suspended material may be significant for the transport of some elements including B, As, Cd, and Mo.

1 INTRODUCTION

The present study focused on suspended material transport in 8 catchments in NE Iceland. The rivers in NE Iceland were chosen for this study because they 1) drain almost exclusively basalt/basaltic glass catchments, 2) experience limited but variable biological activity, 3) drain catchments of variable glacier cover, 4) are unpolluted, and 5) have regular predictable water flow variations, allowing ready sampling during both flood events and at low flow times. Consequently these rivers have a relatively simple chemistry and may be good proxies for the chemical functioning of other rivers draining either glacial terrains or basaltic catchments.

The Hydrological Service of the National Energy Authority has monitored continuously the discharge and suspended load mass of most of these catchments for more than 30 years (Pálsson & Vigfússon 1996; Aðalsteinsson 2000; Snorrason 2002). The total organic and inorganic dissolved and suspended load mass, and the flux of elements due to sorption onto suspended load were monitored over a four-year period (Gislason et al. 2000, 2001, 2002 a, b, c & d). GIS modeling was used to assess the relative importance of climate, runoff, rock age, relief, glacier, soil organic carbon, and vegetative cover on these fluxes (Kardilijov et al. 2002a & b). The purpose of this communication is to present some of the highlights of this study, and in particular to illuminate the role of suspended material in the chemical flux from the continents to the oceans.

2 CHARACTERIZATION OF SUSPENDED MATERIAL

The suspended material in these rivers is over 90 percent basaltic glass. The chemistry of this material is therefore impressively uniform. The chemical composition, as measured by ICP/MS after fusion, shows it to have typical basaltic glass composition;

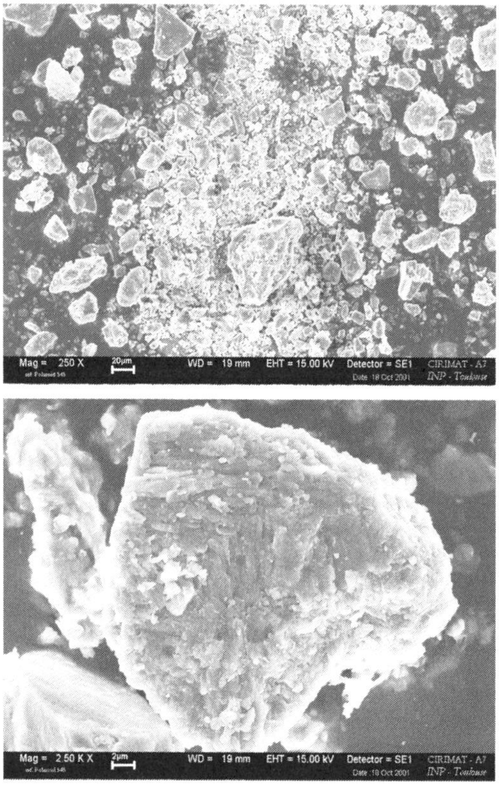

Figure 1. SEM photomicrographs of suspended matter from Jökulsá á Dal at Hjarðarhagi.

48 ± 2 weight percent SiO_2 and 14 ± 1 weight percent Al_2O_3.

The specific surface area of the selected samples of suspended matter taken from the top 50 cm of 4 glacier-fed rivers was measured using the B.E.T. method. These measurements were performed at the Pennsylvania State University, University Park, Pennsylvania, U.S.A. and the Paul Sabatier University, Toulouse, France. Resulting specific surface area of the suspended matter ranges from 10 to 80 m^2/g, with an average value of ~30 m^2/g.

Selected suspended material samples were studied with the aid of Scanning Electron Microscope (SEM). Several examples are illustrated in Figure 1. Suspended material samples appear to be predominantly agglomerates of basaltic glass having a typical size of ~10 µm. Considerable fine grains, less than 2 µm in diameter, are found adhering to the larger grains. The enlarged image of a single grain illuminates the origin of the high surface area of this suspended material. This grain exhibits an extreme surface roughness likely due to the consequences of mechanical mixing within the rivers. X-ray photoelectron spectroscopy (XPS) of the surfaces suggests minor Fe enrichment; fresh glass has 2.6 weight percent Fe but the suspended load has 3.4 weight percent Fe. This observation suggests some minor Fe (oxy)hydroxide phases may be present, perhaps cementing together the basaltic glass aggregates.

Selected suspended material samples were analyzed via IR-DRIFT spectra. DRIFT spectra were recorded with 200 scans at 2 cm^{-1} resolution using a Bruker IFS 55 spectrophotometer equipped with a diffuse reflectance attachment. All spectra exhibit similar features: a broad band around 3,400 cm^{-1} corresponding to OH bending vibrations; OH stretching vibrations at 1635 cm^{-1}, and a peak at around 1010–1020 cm^{-1}, similar to that found on freshly ground basaltic glass from Iceland. The last peak may also be characteristic for iron oxide or goethite surface coatings. No kaolinite, smectite or montmorillonite peaks were recognized. Overall, the DRIFT spectra of suspended material show it to be very similar to ground basaltic glass from the same region, with, probably, some excess Fe oxy-hydroxides on the surfaces. No differences in spectra of samples collected at various time of the year or at various places on the same river have been observed. Note that no adsorbed organic material on the surface of river particles was detected. This finding is in contrast to observations in temperate rivers (Lartiges et al. 2001). It is consistent, however, with quite low concentration of dissolved organic carbon (<0.2 mg C/l) and low primary productivity in these boreal suspended loaded rivers (Gislason et al. 2002d).

3 SUSPENDED VERSUS DISSOLVED TRANSPORT OF ELEMENTS

Bulk chemical analyses of over 50 elements of both suspended material and co-existing waters indicate that elemental transport is dominated by suspended material flux, though the degree depends on the element in question. A representative example of these relative fluxes of some elements at one sampling site over a two year period is shown in Figure 2. Al and Fe are highly insoluble in dilute solutions at neutral pH, the suspended material flux of these elements tends to be over 1000 times that of the dissolved flux. In contrast, soluble elements such as Na and Ca are less dominated by suspended material flux. Nevertheless far more Na and Ca are transported by suspended material flux than by dissolved flux.

4 ADSORBED TRANSPORT

The high BET surface area of the suspended material suggests that chemical transport of chemical species

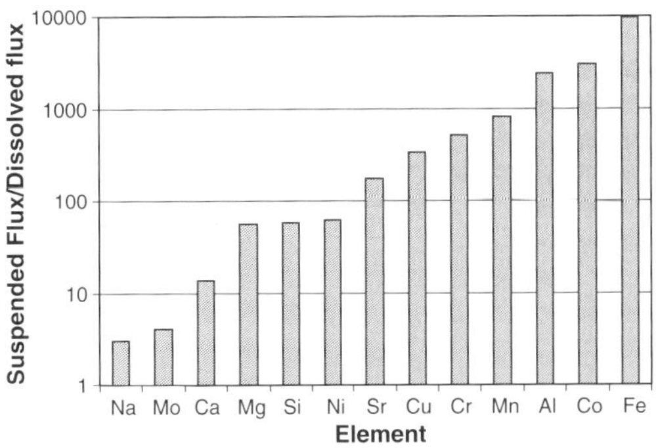

Figure 2. Total molar flux of selected elements in the suspended material divided by that of the dissolved load in the Jökulsá á Dal River at Bru from 21 November 2000 to 22 October 2002.

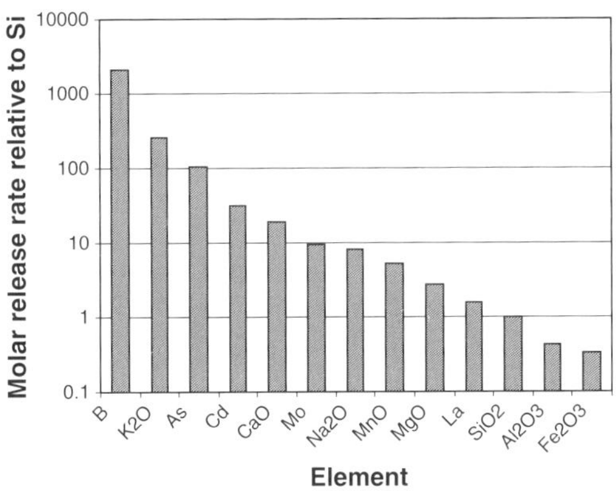

Figure 3. Molar release rate of elements relative to the bulk composition of the suspended material normalized to Si during a desorption experiment of sample 02-A025 at pH 4. This suspended material was sampled at Jökulsá á Dal River water at Bru, 13 August 2002.

adsorbed to the surface of this suspended material may also be significant. For example, the average mass of suspended material in the Jökulsá á Dal River at Bru during 2000–2002 was 0.57 g of suspended material per kg of water. Multiplying this by an average specific surface area of $30\,m^2/g$ leads to an average of $17\,m^2$ of suspended material surface area per kg of river water. Assuming a surface site density equal to that of quartz of 8 sites per square nanometer, an average of 0.226 mmol of surface sites per kg are estimated to be present in Jökulsá á Dal River water at Bru. This value is substantially greater than the average dissolved concentrations of the major elements in the dissolved phase. For example, the average dissolved Na, Si, and Ca concentration over this period was 0.14, 0.08, and 0.08 mmol/kg, respectively. To assess the potential role of adsorbed chemical transport, selected suspended material samples were desorbed in both pH 4 solutions and 0.5 m NaCl solutions. A representative example of the results of these titrations can be seen in Figure 3. Results in Figure 3 are normalized to the release rate of Si to distinguish the effects of suspended material dissolution from those of desorption. These results, which are typical for all the suspended material analyzed, show strong desorption of B, K, Na, Ca, and As relative to Si suggesting their adsorption to the suspended material. The apparent retention of other elements like Al and Fe may be due to their supersaturation in these desorption experiments. These results suggest that adsorbed transport may be very significant for several trace elements like B, As, Cd, and Mo, and less so for the major elements Na, K, and Ca.

5 SUSPENDED MATERIAL DISSOLUTION IN THE OCEAN

Suspended material will dissolve following its arrival into the deltas. Taking account of basaltic glass dissolutions rates reported by Oelkers & Gislason (2001) and Gislason & Oelkers (2003), basaltic glass dissolution rates in seawater are estimated to be 2.34×10^{-16} mol/cm(BET)^2/s. Taking account of an average suspended material surface of $30\,m^2/g$, the dissolution rate of the suspended material in sea water can be computed. The results of this preliminary calculation indicate that 0.05 percent of this suspended material is expected to dissolve into seawater each day, and 1.6 percent can be expected to dissolve over the course of a month. If the suspended material is in contact with seawater for the month following its arrival in the ocean, and stoichiometric dissolution occurs, 1.6 percent of the mass of the each element in the suspended material would be added to the ocean. It can be deduced from Figure 2 that under these circumstances the input to the oceans stemming from suspended material flux will exceed that stemming from dissolved transport for many major and minor elements.

6 CONCLUSIONS

The results of a four-year study of suspended material transport in the rivers of northwest Iceland show:

a) Suspended material flux dominates the flux of all elements from the continents to the oceans.
b) Substantial dissolution of this material may be expected to dissolve after it arrives in the ocean. The extent of this dissolution may be sufficiently

rapid that chemical input to the oceans may be dominated by the dissolution of suspended material rather than dissolved chemical fluxes from rivers.

c) The number of adsorption sites on suspended material may be sufficiently large such that adsorbed transport may be important in the case of several elements such as B, As, and Mo.

As transport of these rivers may be representative of that in other basaltic-rich high topography systems, such as are found in Southeast Asia/Oceania, which have been recently been shown to contribute notably to the global elemental erosion budget (Dessert et al. 2003), suspended material transport may also prove to be significant to these global fluxes.

ACKNOWLEDGEMENT

This research was supported by the Icelandic Science foundation, contract 99668-0000, The National Power Company, The Environment and Food Agency, The National Energy Authority and the Centre Nationale de la Recherche Scientifique (CNRS).

REFERENCES

Aðalsteinsson H. 2000. Aurframburður á Eyjabökkum. OS-2000/071, Orkustofnun Reykjavík.

Dessert, C., Dupre, B., Gaillard, J., Francois, L. & Allegre, C.J. 2003. Basal weathering rate laws and global geochemical cycles *Chem. Geol.* 202: 257–273.

Gíslason S.R., Snorrason, Á., Eiríksdóttir, E.S., Elefsen, S.O., Gunnarsson, Á., Torsander, P. & Óskarsson, N.O. 2000. Efnasamsetning, rennsli og aurburður straumvatna á Austurlandi, I. Gagnagrunnur Raunvísindastofnunar og Orkustofnnunar. Raunvísindastofnun, RH-12-2000, 50 pp.

Gíslason S.R., Snorrason, Á., Eiríksdóttir, E.S., Elefsen, S.O., Gunnarsson, A., Hreinsson, E.O., Torsander P. & Óskarsson, N.O. 2001. Efnasamsetning, rennsli og aurburður straumvatna á Austurlandi, II. Gagnagrunnur Raunvísindastofnunar og Orkustofnnunar. Raunvísindastofnun, RH-5-2001, 75 pp.

Gíslason S.R., Oelkers E.H. & Wolff-Boenisch D. 2002a. The consequences of basaltic glass dissolution. *Abstract Volume, 6th International Symposium on the Geochemistry Of The Earth's Surface (GES-6)* 20–24 May, 2002, Honolulu, Hawaii, pp. 27–30.

Gíslason S.R., Kardjilov, M.I., Gísladóttir, G., Eiríksdóttir, E.S., Sigfússon, B., Elefsen, S.Ó., Snorrason, Á, Wolff-Boenisch, D., Oelkers, E.H., & Torsander P. 2002b. A quantitative field based study of basalt/basaltic glass weathering and its role in carbon fixation. *Geochim. Cosmochim. Acta Special Supplement. Goldschmidt Conference Abstracts 2002*: A275.

Gíslason S.R., Snorrason, A., Eiríksdóttir, E.S., Sigfússon, B., Elefsen, S.O., Harðardóttir, J., Gunnarsson, A., Hreinsson, E.O., Torsander, P., Kardjilov, M.I. & Óskarsson, N.O. 2002c. Efnasamsetning, rennsli og aurburður straumvatna á Austurlanndi. Vorráðstefna 2002. Ágrip erinda og veggspjalda. Jarðfræðafélag Íslands, pp. 45–48.

Gíslason, S.R., Snorrason, Á., Eiríksdóttir, E.S., Sigfússon, B., Elefsen, S.O., Harðardóttir, J., Gunnarsson, Á., Hreinsson, E.O., Torsander, P., Kardjilov, M.I., & Óskarsson, N.O. 2002d. Efnasamsetning, rennsli og aurburður straumvatna á Austurlandi, III. Gagnagrunnur Raunvísindastofnunar og Orkustofnnunar. Raunvísindastofnun, RH-11-2002, 96 pp.

Gíslason, S.R. & Oelkers, E.H. 2003 Mechanism, rates, and consequences of basaltic glass dissolution: II. An experimental study of the dissolution rates of basaltic glass as a function of pH and temperature. *Geochim. Cosmochim. Acta* 67: 3817–3832.

Kardjilov, M.I., Gísladóttir G. & Gíslason S.R. 2002a. GIS modelling of organic and inorganic matter transport in Northeast Iceland. Ágrip erinda og veggspjalda. Jarðfræðafélag Íslands, p. 29.

Kardjilov, M.I., Gísladóttir, G. & Gíslason, S.R. 2002b. GIS modeling of organic matter and rock weathering and present fixation of carbon by vegetation in Northeast Iceland. Ágrip erinda og veggspjalda. Landafræðifélag Íslands, p. 17.

Lartiges, B.S., Deneux-Mustin, S., Villemin, G., Mustin, C., Barres, O., & Babut, M. 2001. Composition, structure, and size distribution of suspended particulates from the Rhine River. *Water Research:* 35:808–816

Oelkers, E.H. & Gíslason, S.R. 2001. The mechanism, rates and consequences of basaltic glass dissolution: I. An experimental study of the dissolution rates of basaltic glass as a function of aqueous Al, Si, and oxalic acid concentration at 25°C and pH 3 and 11. *Geochim. Cosmochim. Acta* 65: 3671–3681.

Pálsson, S. & Vigfússon, G.H. (1996). Gagnasafn aurburðarmælinga 1963–1995, Orkustofnun OS-96032/VOD-05 B, Orkustofnun, Reykjavík, 270 bls.

Snorrason, Á. 2002. The Database of the Hydrological Survey, National Energy Authority, Reykjavík, Iceland.

Water–Rock Interaction, Wanty & Seal II (eds)
© 2004 Taylor & Francis Group, London, ISBN 90 5809 641 6

How potent is soil CO_2 as an acidification and weathering agent?

Neung-Hwan Oh
Yale University, New Haven, CT, USA

Daniel D. Richter
Duke University, Durham, NC, USA

ABSTRACT: Soil acidification is a fundamental process of the earth's environment. A major source of soil acidity is soil CO_2 and carbonic acid, derived from root and microbial respiration. Here we demonstrate with column-leaching experiments and contrasting soils, that soil CO_2 may be a much more potent agent of soil acidification and weathering than is generally appreciated, capable of displacing nearly all exchangeable base cations from highly weathered soils and even elevating Al concentrations in H_2CO_3-acidified soil water. The potency of H_2CO_3 is attributed to the low $pK_{a,1}$ of H_2CO_3 (3.76 at 25°C), which contrasts greatly with the oft-cited $pK_{a,1}$ of 6.36 at 25°C for H_2CO_3* (the combination of $CO_2(aq)$ and H_2CO_3). This distinction is of fundamental significance to soil systems affected by high partial pressures of CO_2. Results have implications for general concepts of weathering and pedogenic process and for global changes associated with rising concentrations of atmospheric CO_2. Modeling results suggest that a doubling of atmospheric CO_2 may increase acid inputs from carbonic acid leaching by up to 50% and that increases in atmospheric CO_2 since the interglacial have gradually acidified soils throughout the world.

Water–Rock Interaction, Wanty & Seal II (eds)
© 2004 Taylor & Francis Group, London, ISBN 90 5809 641 6

Bound water as an accumulator of solar energy by weathering products

S.L. Shvartsev
Tomsk Affiliated Branch of the Institute of Oil and Gas Geology, Siberian Branch of RAS, 3 prosp. Akademicheskii, Tomsk, Russia

ABSTRACT: The disagreements on the energetics of exogenous processes have been analyzed in the context of modern views on water-rock interaction. Results of thermodynamic analysis of weathering reactions are reported. The formation of supergene clays involves two geological processes: hydrolysis of aluminosilicates and binding of large amounts of water. Relative to free water, the bound water has a higher free energy of formation. This difference considerably changes the energetics of hydrolysis reactions, which actually are all endothermic. Physically bound water and, therefore, clays, bear large stocks of accumulated solar energy. This fact proves that oriented inner evolution and self-organization took place in inorganic matter like those known in biogenic systems, with a genetic relationship existing between them.

1 INTRODUCTION

Problems of energetics of sedimentary processes are quite complicated and poorly studied. The most complete review of them has recently been published by Timofeev et al. (1989). These authors concluded that the "sedimentary sequence of rocks is to be considered a carrier of large reserves of accumulated solar energy received by the Earth for its long geological history" (p. 193). They believed that solar energy is accumulated by plant and animal remains, by surficial dispersion of a huge mass of monolithic rocks into fine-grained rocks, and by entrapment of solar energy by crystal lattices of some minerals.

Accumulation of solar energy by weathering products (clays) has been studied by many famous scientists: P.N. Panyukov, N.V. Belov, V.I. Lebedev, L.V. Pustovalov, A.I. Perei man, P.F. Shvetsov, V.I. Molchanov, et al. They proposed some mechanisms of this accumulation. Thus, Panyukov believed that the "energization" of rocks occurred in the process of mechanical disintegration (crushing). The same idea was later developed by Molchanov (1981). The hypothesis of geochemical accumulators according to which the energy is accumulated at the expense of transition of Al from the surrounding tetrahedra by oxygen atoms to octahedra on weathering was proposed by Belov (1952) and Lebedev (1954). Unfortunately, no evidence of energy accumulation by this mechanism has been established until now.

At the same time, analysis of reaction energy of rock weathering (hydrolysis of aluminosilicates) shows quite an intricate pattern. Zverev (1983) reports calculated free energies of formation of 60 hydrolysis reactions of aluminosilicates, of which 40% are endothermic, and 60% exothermic. Timofeev et al. (1989) stated that the "majority of solid solution reactions proceeding in the hypergenesis zone are of an exothermic nature" (p. 97). It is generally agreed that natural processes in abiogenic systems tend towards free energy minimization (Dirichlet principle), which controls the external conditions of physicochemical systems and implies the maximization of total energy of interatomic bonds. The principle of free energy minimization is fundamental to physicochemical modeling of geological processes, such as weathering, sedimentation, lithogenesis, hydrothermal and metasomatic processes, ore formation, etc. (Karpov 1981; Anderson 1996).

To solve this problem is now especially important in connection with the development of synergetics (Prigogine & Stengers 1984), recognition of mechanisms of self-organization of geologic systems (Letnikov 1992), in particular in water-rock systems (Shvartsev 2001). This paper is aimed at justifying a special role of physically bound water in the energy budget of reactions of chemical weathering.

2 THERMODYNAMIC ANALYSIS OF ALUMINOSILICATE WEATHERING REACTIONS

First, we will outline systematics of exo- and endothermic reactions of aluminosilicate hydrolysis, taking into account the stage of weathering. For this purpose we will calculate free energies of formation and thermodynamic constants of some reactions, chiefly weathering.

Using the mutually concordant values of free energy of formation of mineral compounds and dissolved ions (Robie & Hemingway 1995), we revised type reactions of weathering and recalculated the free energy of their formation. A total of 100 reactions have been studied, and most of them appeared to be endothermic rather than exothermic as they were earlier thought to be. A significant dependence between the energetics of reactions and stages of weathering was found. Thus, all reactions of formation of Al and Fe hydroxides are endothermic. At the same time, reactions of kaolinization are energetically more heterogeneous and about 50% of them are endothermic and 50% exothermic.

The next stage of weathering leads to the formation of hydromicas and various montmorillonites, and is dominated by endothermic reactions which, according to our data, make up 95%, with exothermic reactions no more than 5%. A different ratio is observed at the stage of carbonatization, when, along with montmorillonites, Ca carbonates form. In this case, exothermic reactions constitute no less than 90% of reactions.

By analyzing the recognized dependences of reaction energetics on stages of weathering, we have established that energetics of weathering is controlled by the amount of water bound by secondary products. If the products bind much water, their formation reactions are mostly endothermic. Montmorillonite and hydromicas, products of endothermic reactions, contain the greatest amounts of bound water. The same is true for hydroxides whose dehydration and crystallization lead to a decrease in free energy of formation. What is the nature of the relationship between energy of reaction and water amount in its products? We will consider this question in more detail.

3 ENERGETICS OF PHYSICALLY BOUND WATER

It is known that when formed, clay minerals physically and chemically bind a great amount of water. The smaller the size of particles, the more water they contain. The greatest amount of water is contained in colloid formations, with the size of particles less than 1 μm. Clay minerals, whose particles are commensurate with colloids, are also rich in this water (Table 1).

The reported data show that all clay minerals contain a great amount of water, though the ratio of forms is different. Thus, kaolinite is richest in constitutional water, and montmorillonite, in crystallization, tightly and loosely bound. Hydromica occupies an intermediate position, except of moisture capacity, which is minimal in this case. It is noteworthy, however, that an extremely dense sample of old clay was chosen for experiments. It has a high maximal molecular moisture capacity, and the share of hygroscopic water in it makes up 36–38% for kaolinite, 52–62% for hydromica, and 77–100% for montmorillonite. Thus, all mineral kinds of clays contain sufficient amount of physically bound water. Its content in kaolinites and hydromicas is 20–25 wt. %, and in montmorillonites 30–50%.

Physically bound water considerably differs in properties from free water. Its density is 1.2–1.4 g/cm^3; viscosity is considerably higher than in free water; modulus of rigidity in a Film 0.09 μm thick is 0.2 MPa, i.e., 300 times as low as that parameter of lead; dielectric permeability in a 0.07 μm-thick film decreases to 4.5 and even down to 2.0–2.2; temperature of freezing of this water in clay grounds varies from −30 to −50°C; heat conductivity in a 0.05 μm-thick film is two orders of magnitude as great as heat conductivity in volume, etc. (Sergeev et al. 1983; Shvartsev 1996). These data suggest that physically bound water differs in its properties from water in volume, including free energy of formation.

The last inference has received direct experimental support. Thus, a group of Belgian researchers (Fripiat et al. 1965), using methods of magnetic resonance, conductometry, and dielectrometry, has established that the water adsorbed by Na-Ca montmorillonite is characterized by an extremely high constant of dissociation ($10^{-7.15}$), i.e., exceeding the constant of free water by nearly seven orders of magnitude.

Using these data and the van't Hoff equation giving the temperature dependence of the equilibrium constant of a reaction, we have calculated values of

Table 1. Amount of bound water in clay rocks of various mineral composition, %.

	Water type		Water capacity			Typical water amount	
Clay	Constitutional	Crystallization	Maximum hygroscopic	Maximum molecular	Bulk	H_2O	H_2O^+
Kaolinite	27.5	1.48	7.8–9.0	21.4–23.7	43	0.2–2.6	13.5–14.6
Hydromicaceous	9.0–9.3	1.23	5.9–8.3	11.5–13.2	29	0.6–2.9	3.6–11.8
Montmorillonite	9.2–9.4	13.8	38–40	38–52	71	8.2–17.4	3.9–14.5

the parameters characterizing the absolute neutrality of bound water in montmorillonite (Table 2).

Thus, physically bound water drastically differs from free water in many parameters (pH, Eh, PO_2, PH_2, density, etc.). It is especially important that physically bound water is characterized by considerably higher values of free energy of formation (ΔG^0_{form}). Of course, the obtained values of ΔG^0_{form} are not to be absolutized, since they are different in different types of clay-bound water, but it is crucial that in any case they are higher than the relevant values in the volume water. It is also obvious that the closer the water layer to the clay particle, the stronger it is retained by surficial forces and the higher energy of formation it has.

What is the source of energy of this water? As we deal with an open system, whose development is determined chiefly by the supply of solar energy to the zone of weathering, it is natural that the transition to a higher energetic level on binding water by clays is also provided by solar energy. Of course, in each particular case the energy sources may be different (thermal, surface moistening, oxidation of organics and sulfides, radioactivity, etc.), but all these sources except radioactive are initially solar.

Hence, when binding a huge amount of water in the process of their formation, clays simultaneously transfer it to a higher energetic level, thus becoming carriers of huge reserves of accumulated solar energy.

4 ENERGETICS OF CLAY FORMATION REACTIONS

Thus, clay formation on weathering includes two different geological phenomena: hydrolysis of aluminosilicates with the formation of a clay framework and binding of free water by the fresh clays, which brings it to a higher energy level. The first process is presently described by thermodynamic methods quite reliably, whereas the second process is described only partly. The case is that to estimate the free energy of formation of any mineral, its ultimately dewatered varieties are used. Even crystallization water, which is a formula component, is ignored in calculation of free energy of its formation. At the same time, not only crystallization but also hygroscopic water which cannot be removed without a complete destruction of crystal lattice of mineral is always present in a certain amount, affecting the free energy of formation of corresponding compound. The higher content of such water in a mineral, the higher free energy of its formation. In our opinion, this explains the discrepancy in energetics of reactions of hydrolysis of aluminosilicates at different stages of their weathering. Formation of hydroxides, montmorillonites, and illites, which contain much physically bound water, is accompanied by energy accumulation, whereas in formation of kaolinites and calcites, which contain little crystallization water, this phenomenon is expressed weakly.

Table 2. Some physical and thermodynamic properties of free and bound water in a state of absolute neutrality (Shvartsev et al. 1982).

Property	Water: free	Water: physically bound
pH	7.0	3.6
Eh, V	0.4	−0.15
Partial pressure, atm		
O_2	$10^{-27.86}$	$10^{-5.86}$
H_2	$10^{-27.56}$	$10^{-5.56}$
Constant of dissociation	$10^{-14.0}$	$10^{-7.15}$
decomposition	$10^{-83.1}$	$10^{-17.0}$
ΔG^0_{form}, kcal/mol		
H_2O	−56.69	−37.7
OH^-	−37.595	−6.1
Density, g/cm	1.0	1.4
Electrical constant	78.3	10–20

The effect of energy growth of physically bound water can also be explained from the classical point of view. The mechanical crushing of any mineral is known to be accompanied by an increase in free surface and growth of energy which is absorbed from the environment (Molchanov 1981). The moistening of the newly formed surface is accompanied by heat liberation (Timofeev et al. 1989). It is obvious that, at least, partly the energy liberated on moistening is accumulated by water, which thus is brought to a higher energy level. Hence, energy is accumulated not by crystal lattice but by water, covering any free surface of a solid. Once more, we see that water is a substance which is subject to rearrangement and can bind energy partly or completely, as it occurs on vapor formation. Thus, physically bound water is a specific phase state, which differs in structure and free energy of formation. Owing to this water, the system acquires a higher energy level, which is better adapted to a medium in which a new system, water-clay, appears instead of the system water-endogenous aluminosilicate (Shvartsev 1998).

Therefore, the formation of physically bound water in clays should be taken into account as an independent process in calculation of energy of hydrolysis reactions. We suggest introducing into hydrolysis reactions an additional amount of water which is bound by a particular mineral. For example, kaolinite contains about 10–12% firmly bound water, energetically different from free water, which is to be reflected in hydrolysis reaction. Usually this reaction has been written as:

$$CaAl_2Si_2O_8 + 3H_2O + 2CO_2 = Al_2Si_2O_5(OH)_4 + Ca^{2+} + 2_3^- \quad (1)$$

We to take into account the processes of formation of the bound water.

$$CaAl_2Si_2O_8 + 4.5H_2O + 2CO_2 = Al_2Si_2O_5(OH)_4 + 1.5H_2O^* + Ca^{2+} + 2_3^- \quad (2)$$

where H_2O^* is the kaolinite-bound water, which does not enter the crystal lattice but differs from free water in energy.

Given the energy of formation of this water is −37.7 kcal/mol, ΔG^0 of the above reaction will be 23.65 kcal/mol, i.e., the reaction becomes endothermic rather than exothermic.

Montmorillonite firmly binds no less than 30–35% water. Therefore, (13) from Table 2 is to be rewritten as follows:

$$1.83CaAl_2Si_2O_8 + 6H_2O + 1.66CO_2 = Ca_{0.17}Al_{2.33}Si_{3.67}O_{10}(OH)_2 + 3H_2O^* + 1.66CaCO_3 + 1.34Al(OH)_3 \quad (3)$$

.

In this case, ΔG^0 of reaction is 38.29 kcal/mol, not −18.68 kcal/mol, and for the same reasons the reaction itself is endothermic. There are many other examples of this kind.

Therefore, taking into account the effect of binding water by clay minerals, all reactions of hydrolysis at all stages of weathering appear to be physically endothermic rather than exothermic as thought earlier. Thus, the rock weathering processes obey the laws of generation of new dissipative structures under conditions far from equilibrium (Prigogine & Stengers 1984), which accumulate energy in the process of formation and, hence, decrease entropy.

5 CONCLUSIONS

As clays form they bind a great amount of water, which in its physical and thermodynamic properties differs considerably from free water, thus accumulating solar energy. As a result, according to Le Chatelier's rule, products form on weathering which, absorbing energy, makes the action of external factors weaker.

In the context of hydrodynamics, the formation of supergene clays is to be considered a result of two processes: hydrolysis of endogenous aluminosilicates and binding considerable amounts of water, which is brought to a higher energy level. With these two processes taken into account, all reactions of hydrolysis of aluminosilicates are endothermic rather than exothermic.

The system water-endogenous aluminosilicates develops to form hydrolysis products with a higher energy as compared with the initial substances, i.e., the energy of the system on weathering increases, whereas the entropy correspondingly decreases, which leads, in the context of modern synergetics, to the formation of principally new dissipative structures under the conditions far from equilibrium, which are higher-organized and more complicate in the structural, compositional, and information aspects.

ACKNOWLEDGEMENTS

This work was supported by grant 00-05-64328 from the Russian Foundation for Basic Research, and 1566.2003.05 from Ministry of Industry, Science and Technology.

REFERENCES

Anderson, G.M. 1996. *Thermodynamics of natural systems.* New York: John Wiley and Sons.

Belov, N.V. 1952. Geochemical accumulators. *Proc of. Inst. Kristallografii AN SSSR, issue 7*: 73–80. (in Russian).

Fripiat, I.I., Ielli, A., Polcelet, G., & Andre, I. 1965. Thermodynamic properties of absorbed water molecules and electrical conduction in montmorillonites and silicas. *J. Phys. Chem.* 69(7): 21–85.

Karpov, I.K. 1981. *Physicochemical computer modeling in geochemistry.* Novosibirsk: Nauka (in Russian).

Lebedev, V.I. 1954. Potential absorption of solar energy by crystalline matter of the Earth. *Izv. AN SSSR.Ser. Geol.*, 4: 50–74 (in Russian).

Letnikov, F.A. 1992. *Synergetics of geological systems.* Novosibirsk: Nauka (in Russian).

Molchanov, V.I. 1981. *Hydrogen generation in lithogenesis.* Novosibirsk: Nauka (in Russian).

Prigogine, I. & Stengers, I. 1984. *Order out of chaos: a new dialogue with nature.* London: Heineman.

Robie, R.A., & Hemingway, B.S. 1995. *Thermodynamics of minerals and related substances at 298.15K and 1 bar (10^5 pascals) pressure and at higher temperatures.*U.S. Geol. Surv. Bull. 2131. Washington: US Government Printing Office.

Sergeev, E.M., Golodkovskaya, G.A., Ziangirov R.S., Osipov V.I., Trofimov V.T. 1983. *Soil science*, Moscow: Izd-vo MGU (in Russian).

Shvartsev, S.L., Pinneker, E.V., Perel'man, A.I., Kononov V.I., Nazarov A.D., Rasskazov N.M., Udodov P.A., Schvetz V.M. 1982. *Fundamentals of hydrogeology. Hydrogeochemistry.* Novosibirsk: Nauka (in Russian).

Shvartsev, S.L. 1996. *General hydrogeology.* Moscow: Nedra (in Russian).

Shvartsev, S.L. 1998. *Hydrogeochemistry in zone of hypergenesis.* Moscow: Nedra (in Russian).
Shvartsev, S.L. 2001. The water-rock system synergy. *Earth Science Frontiers, N1*: 36–46.
Timofeev, P., Shcherbakov, A.V. & Il'in, V.A. 1989. *Energetics of sedimentation.* Moscow: Nauka (in Russian).
Zverev, V.P. 1983. *Energetics of hydrogeochemical processes of modem sedimentation.* Moscow: Nauka (in Russian).

Water–Rock Interaction, Wanty & Seal II (eds)

Chemical weathering and CO_2 fluxes in the Canadian Cordillera

J.D. Spence & K.H. Telmer
School of Earth and Ocean Sciences, University of Victoria, Victoria, British Columbia, Canada

ABSTRACT: Chemical weathering rates and associated CO_2 fluxes have been calculated for three major watersheds of the western Canadian Cordillera from major element concentrations and $\delta^{13}C$ of dissolved inorganic carbon (DIC) in river water. Weathering rates and sources determined from major elements are consistent with $\delta^{13}C_{DIC}$ values (−9.8 to −2.7‰ $_{vpdb}$) derived from the mixing of three sources of DIC: (i) silicate dissolution by carbonic acid (−17‰); (ii) carbonate dissolution by carbonic acid (−8.25‰); and (iii) carbonate dissolution by sulphuric acid derived by oxidative sulphide (pyrite) weathering (+0.5‰). DIC fluxes due to silicate weatheringand due to coupled pyrite-carbonate weathering are equivalent (60×10^3 mol C·km^{-2}·yr^{-1}). Coupled pyrite-carbonate weathering may therefore be a long-term source of atmospheric CO_2 offsetting silicate weathering CO_2 drawdown in the region.

1 BACKGROUND

To assess the chemical evolution of Earth's surface through geological time, including ocean, atmosphere and climate evolution, the controls on chemical weathering rates must be understood. Walker et al. (1981) proposed a negative feedback between silicate weathering rate and atmospheric CO_2 levels as a mechanism for regulating the greenhouse state of the atmosphere and maintaining climate stability.

There is still no consensus among researchers regarding the nature and the magnitude of the negative feedback between tectonic processes, atmospheric CO_2 and chemical weathering rates. For example, some claim that factors such as physical erosion rates driven by tectonic uplift, the geographic distribution of lithologies, frost shattering, and glaciation can have a greater influence on chemical weathering rates than global climate (cf. Ruddiman 1997). Others (e.g. Berner et al. 1983) claim seafloor spreading rate is the dominant control on atmospheric CO_2, driving chemical weathering rates through the resulting climate response. Notably, some of these processes occur on different time scales than the silicate weathering-climate feedback, and so linkages are difficult to make confidently.

In order to evaluate the linkages between chemical weathering and climate evolution, a detailed understanding of modern chemical weathering rates and fluxes to the oceans is required. Here we contribute to this effort by calculating chemical weathering rates and CO_2 fluxes for three large river systems (Fraser, Skeena and Nass) in the western Canadian Cordillera based on a new geochemical survey. We elaborate the importance of CO_2 emissions to the atmosphere through the weathering of carbonates by sulphuric acid derived by oxidative weathering of pyrite. If the rate of coupled pyrite-carbonate weathering exceeds the rate of pyrite deposition in marine sediments, the release of CO_2 to the atmosphere will offset CO_2 drawdown by silicate weathering. The geology of the area is similar to accretionary orogens found around the globe and so it is reasonable to expect this to be an important process globally.

2 WATERSHED CHARACTERISTICS

The portions of the Fraser, Skeena and Nass Watersheds considered by this study drain a combined area of 278,500 km^2, with cumulative annual discharge averaging 1.4×10^{11} m^3·yr^{-1} (Environment Canada 1994). The hydrograph of these basins is dominated by a pulse of water generated by snowmelt in late spring. Approximately 70% of the total discharge of these rivers occurs between May and August. The Fraser River, and its tributary, the Thompson River are the largest considered by this study, each traversing a transect of the geological belts that make up the Canadian Cordillera. The remaining rivers are representative of various geological terranes of the Canadian Cordillera.

2.1 *Bedrock geology*

The following geological descriptions are taken from: Gabrielse et al. (1991), Gabrielse & Yorath (1991) and Wheeler & McFeely (1991). The Canadian Cordillera is a geologically complex region making up the northern portion of the western margin of North America. It is typically described as a series of roughly north-south trending structural and lithological zones. From east to west these are: the Foreland, Omenica, Intermontane, and Coast Crystalline Complex.

The Foreland Belt is made up of ancestral passive margin carbonates and shales (Upper Proterozoic to Mesozoic) that have been thrust eastward onto the Continent. This is a region of very high relief, with many peaks exceeding 2500 m. The Omenica belt is made up of highly metamorphosed sedimentary and igneous rocks that are intruded by S-type granitic plutons, and is also a region of very high relief. The central regions of the Canadian Cordillera form the Intermontane Belt, meaning "between mountains". This belt has relatively low relief composed of a series of arc terranes of Late Triassic to Tertiary age that have been accreted onto the continental margin. The Intermontane Belt is composed of volcanic rocks of varying composition, and associated arc derived sediments, including clastics, shales, carbonates and cherts. This belt also contains significant Neogene back-arc flood basalts. The Coast Crystalline Complex is another high relief belt, dominated by I-type granitic intrusions and also contains significant volcanic and sedimentary sequences. In the studied watersheds, there are no known occurrences of evaporite minerals.

3 SAMPLE COLLECTION AND ANALYSIS

During the peak (June) and trough (October) of the 2002 hydrograph, 49 water samples (including field duplicates and blanks) were collected from 14 rivers draining the Fraser, Skeena and Nass watersheds. Dissolved oxygen, pH, temperature, and conductivity were measured in situ using a multi-meter probe. HCO_3^- was measured in the field using a digital titrator. Filtered (0.45 μm) water samples were transported to the University of Victoria in the dark at 4°C for analysis of SO_4^{2-}, Cl^-, NO_3^-, F^-, Ca^{2+}, Mg^{2+}, Na^+ and K^+ by ion chromatography (accuracy and precision better than ±5% for all except Mg which was ±10%); and $\delta^{13}C_{DIC}$ by continuous flow isotope ratio mass spectrometry. Accuracy and precision of the method, based on multiple analyses of field duplicates and sample splits was found to be better than ±1‰.

4 RESULTS AND DISCUSSION

The major element concentrations observed for the Cordilleran rivers follow the pattern $Ca^{2+} > Mg^{2+} > Na^+ > K^+$ for the cations, and $HCO_3^- >> SO_4^{2-} > Cl^- > NO_3^- > F^-$ for the anions. Charge balance is better than analytical uncertainty in all but 3 samples (probably due to organic anions), indicating the ionic constituents of these waters are well described by the major elements measured.

Measured $\delta^{13}C_{DIC}$ values range from −2.7 to −9.8‰$_{vpdb}$, with the October samples being ~2‰ heavier than the June samples on average. Greater seasonal variation was observed in the smaller tributaries. In the larger tributaries and in the Fraser, Skeena and Nass main stems, small seasonal $\delta^{13}C_{DIC}$ variation was observed (~1‰ heavier in October than June) reflecting the degree of buffering due to longer groundwater residence times in larger fluvial systems.

4.1 *Chemical weathering mass balance*

Chemical weathering rates were determined using generalized bedrock compositions using the following steps:

1. Concentrations are corrected for atmospheric input based on sea-salt ratios normalized to Cl^-.
2. Assign non-cyclic sulphate (SO_4^{2-*}) to pyrite oxidation (equation 1). Cyclic sulphate includes that released by evaporite dissolution but none is present in the watersheds (neither mapped nor detected by observed trends).

$$\frac{1}{2}FeS_2 + \frac{15}{8}O_2 + H_2O \rightarrow \frac{1}{4}Fe_2O_3 + 2H^+ + SO_4^{2-} \quad (1)$$

3. Assume the acidity produced by pyrite oxidation is dominantly consumed by the dissolution of Ca and Mg carbonates (equation 2), a process that has been proven important in glacial and alpine catchments (e.g. Drever & Hurcomb 1986; Anderson et al. 2000).

$$H_2SO_4 + 2CaCO_3 \rightarrow 2HCO_3^- + SO_4^{2-} + 2Ca^{2+} \quad (2)$$

The Ca^{2+} plus Mg^{2+} in the rivers resulting from sulphuric acid dissolution of carbonates ($[Ca^{2+} + Mg^{2+}]_{Cs}$) is calculated as:

$$[Ca^{2+} + Mg^{2+}]_{Cs} = 2 \times [SO_4^{2-*}] \quad (3)$$

The HCO_3^- resulting from the dissolution of carbonates by sulphuric acid ($[HCO_3^-]_{Cs}$) is calculated as:

$$[HCO_3^-]_{Cs} = 2 \times [SO_4^{2-*}] \quad (4)$$

4. Assume non-cyclic sodium (Na^{+*}) is dominantly derived from plagioclase dissolution, and non-cyclic potassium (K^{+*}) is dominantly from potassium feldspar dissolution. As was described in section 2.2 large portions of the Fraser, Skeena and Nass watersheds are composed of arc-derived rocks. These can be well represented by an intermediate plagioclase (An_{33}) with a Ca:Na ratio of 0.5. There are other possible sources of sodium and potassium, but this does not affect the stoichiometry of dissolution significantly, and the results are only moderately sensitive to the Ca:Na ratio chosen because Ca is much more concentrated in these rivers than Na. The Ca^{2+} resulting from silicate dissolution by carbonic acid ($[Ca^{2+}]_{Sc}$) is therefore calculated as:

$$[Ca^{2+}]_{Sc} = 0.5 \times [Na^{+*}] \tag{5}$$

Because charge balance must be maintained, anions and cations entering the water due to silicate dissolution by carbonic acid must be charge equivalent. Thus $[HCO_3^-]_{Sc}$ is calculated as:

$$[HCO_3^-]_{Sc} = [Na^{+*}] + [K^{+*}] + (2 \times [Ca^{2+}]_{Sc}) \tag{6}$$

5. Assume that any remaining Ca^{2+} and Mg^{2+} is derived from the dissolution of carbonates by carbonic acid ($[Ca^{2+} + Mg^{2+}]_{Cc}$). Equation 7 is the example of calcite dissolution by carbonic acid.

$$CaCO_3 + H_2CO_3 \rightarrow Ca^{2+} + 2HCO_3^- \tag{7}$$

The resulting HCO_3^- ($[HCO_3^-]_{Cc}$) is therefore:

$$[HCO_3^-]_{Cc} = 2 \times [Ca^{2+} + Mg^{2+}]_{Cc} \tag{8}$$

Figure 1 summarizes the mass balance results. Two main results are apparent: (i) that carbonate dissolution by carbonic acid is the dominant chemical weathering process for the three watersheds; and (ii) that chemical weathering of carbonates by sulphuric acid is occurring at roughly the same rate as the weathering of silicates by carbonic acid.

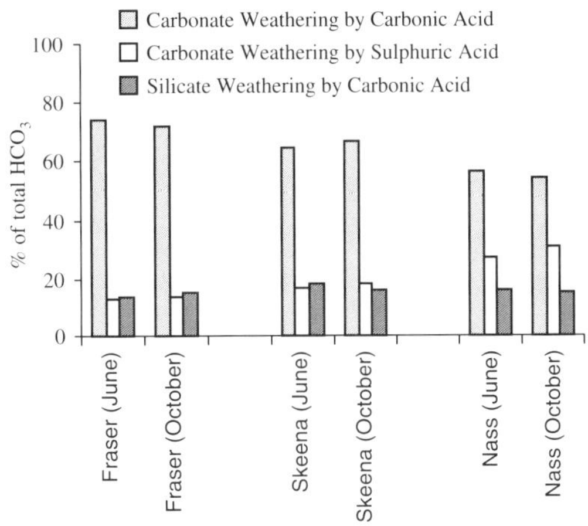

Figure 1. Chemical weathering sources of HCO_3^- ($\approx$DIC) fluxes calculated as percentages of measured HCO_3^- fluxes for the Fraser, Skeena and Nass Rivers.

4.2 *Carbon isotopes*

The isotopic signature of the HCO_3^- ($\approx\delta^{13}C_{DIC}$) offers an independent method of evaluating the mass balance results.

The two dominant sources of carbon in groundwater and in river water DIC are from organic matter decay and carbonate mineral dissolution. Carbonic acid forms in the soils by the dissolution of CO_2 derived from organic matter decay. Depending on the pH, the equilibrium dissolution of this CO_2 into groundwater gives the resulting DIC a $\delta^{13}C$ of $-17‰ \pm {\sim}5‰_{vpdb}$ (for a detailed description of this end-member definition see Telmer & Veizer 1999). Carbonate minerals inherit the $\delta^{13}C$ of the DIC in the water from which they precipitate. Marine carbonate rocks of Paleozoic to Tertiary age have an average $\delta^{13}C$ of ${\sim}0.5‰_{vpdb}$ (Veizer et al. 1999). This age range is representative of Cordilleran carbonates.

Using these $\delta^{13}C$ values and mineral dissolution stoichiometry, $\delta^{13}C_{DIC}$ end-members are defined. (i) The dissolution of one mole of calcium or magnesium carbonate by carbonic acid involves one mole of soil $CO_2(aq)$ ($\delta^{13}C = -17‰$), and one mole of carbonate ($\delta^{13}C = 0.5‰$), producing an end-member of $\delta^{13}C_{DIC} = -8.25‰$. (ii) The dissolution of silicate minerals by carbonic acid only involves carbon from soil $CO_{2(aq)}$producing an end-member of $\delta^{13}C_{DIC} = -17‰$. (iii) Carbonate dissolution by sulphuric acid involves only carbon in calcium or magnesium carbonate, resulting in a $\delta^{13}C_{DIC} = 0.5‰$.

Figure 2 shows $\delta^{13}C_{DIC}$ in relation to major element ratios. Many of the samples have $\delta^{13}C_{DIC}$ heavier than $-8‰$ confirming the importance of coupled pyrite-carbonate weathering. The agreement of isotopic based and major element based mass balance also supports the assumption that pyrite oxidation is the dominant source of dissolved SO_4^{2-} in the Cordilleran rivers.

4.3 *Chemical weathering CO_2 fluxes*

Based on the oceanic residence time of HCO_3^-, silicate weathering is proposed to provide a long-term sink for atmospheric CO_2 on timescales greater than

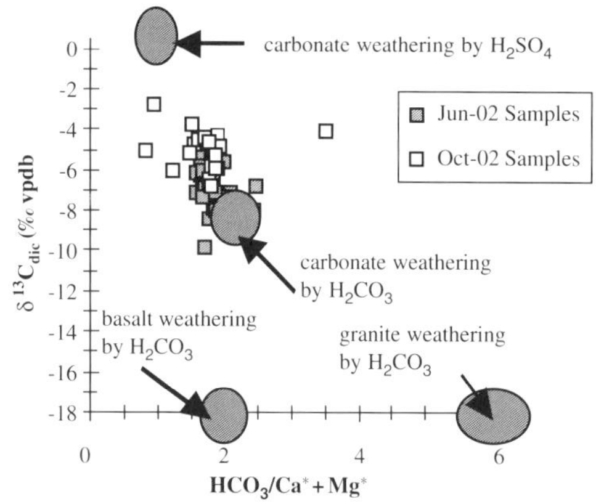

Figure 2. $\delta^{13}C_{DIC}$ versus $HCO_3/Ca^* + Mg^*$ for river water samples and for dominant weathering reaction end-members.

10^5 years by transferring it to the carbonate reservoir.

$$CaSiO_3 + CO_2 \rightarrow CaCO_3 + SiO_2 \tag{9}$$

We suggest that pyrite oxidation (equation 1)coupled with carbonate dissolution by sulphuric acid (equation 2), a process that is not currently included in geological carbon flux models, could be an important source of atmospheric CO_2 on equivalent timescales. The net coupled pyrite-carbonate reaction is:

$$\frac{1}{2}FeS_2 + CaCO_3 + \frac{15}{8}O_2 \rightarrow \frac{1}{4}Fe_2O_3 + Ca^{2+} + SO_4^{2-} + CO_2 \tag{10}$$

In order to maintain steady-state ocean and atmosphere composition, the pyrite and carbonate weathering that occurs on land must be reversed in the oceans by sedimentation processes (e.g. Garrels & Perry 1974). However, the residence time of HCO_3^- ($\sim 10^5$ years) is much shorter than the residence time of SO_4^{2-} ($\sim 10^7$ years) in seawater.

This implies that on timescales shorter than the oceanic residence time of SO_4^{2-}, if coupled pyrite-carbonate weathering on land exceeds the rate of pyrite-carbonate deposition in the oceans there will be a flux of CO_2 to the ocean-atmosphere system and vice versa.

To put this in perspective, based on current ocean and atmosphere composition, if coupled pyrite-carbonate weathering provided enough carbon to double the ocean-atmosphere inorganic carbon reservoir (3.3×10^{18} mol; Garrels & Perry 1974) marine concentrations of SO_4^{2-} and Ca^{2+} would increase from 28.9 to 31.2 mM and from 10.5 to 12.8 mM respectively; meanwhile, O_2 consumed by the oxidative pyrite weathering would decrease atmospheric O_2 concentration from 21 to 18%. This shows that a relatively small imbalance in the sulphur and oxygen cycles can lead to significant changes in atmospheric CO_2.

5 CONCLUSION

We show that carbon fluxes due to coupled pyrite-carbonate weathering are similar to those attributed to silicate weathering for the Canadian Cordillera. As sulphides are weathered more easily than silicates, the rate of physical erosion may be an important determinant in controlling pyrite oxidation rates. It is likely that coupled pyrite-carbonate weathering is greatest in regions of high physical erosion, such as tectonically and glacially active regions like the Canadian Cordillera. This suggests that coupled pyrite-carbonate weathering may provide a negative feedback to tectonically and glacially induced CO_2 drawdown and cooling. However, in order to evaluate the significance of this process: (i) global rates and controls on coupled pyrite-carbonate weathering will need to be evaluated; and (ii) rates of sulphide deposition through time must be considered.

REFERENCES

Anderson, S.P., Drever, J.I., Frost, C.D. & Holden, P. 2000. Chemical weathering in the foreland of a retreating glacier. *Geochim. Cosmochim. Acta* 64: 1173–1189.

Berner, R.A., Lasaga, A.C. & Garrels, R.M. 1983. The carbonate-silicate geochemical cycle and its effect on atmospheric carbon dioxide over the past 100 million years. *Am. J. Sci.* 283: 641–683.

Drever, J.I. & Hurcomb, D.R. 1986. Neutralization of atmospheric acidity by chemical weathering in an alpine drainage basin in the North Cascades Mountains. *Geology* 14: 221–224.

Environment Canada. 1994. *Surface Water and Sediment Data* [HYDAT CD-ROM]. Ottawa: Environment Canada.

Gabrielse, H. & Yorath, C.J. 1991. Tectonic Synthesis. In H. Gabrielse & C.J. Yorath (eds), *The Geology of the Cordilleran Orogen in Canada no. 4.* Ottawa: Geological Survey of Canada.

Gabrielse, H., Monger, J.W.H., Wheeler, J.O. & Yorath, C.J. 1991. Tectonic Framework: Part A. Morphogeological belts, tectonic assemblages and terranes. In H.Gabrielse & C.J. Yorath (eds), *The Geology of the Cordilleran Orogen in Canada no. 4.* Ottawa: Geological Survey of Canada.

Garrels, R.M. & Perry, E.A. 1974. Cycling of carbon, sulfur and oxygen through geologic time. In E.D. Goldberg (ed) *The Sea Vol 5.* New York: Wiley.

Ruddiman, W.F. (ed) 1997. *Tectonic uplift and climate change.* New York: Plenum.

Telmer, K. & Veizer, J. 1999. Carbon fluxes, pCO_2 and substrate weathering in a large northern river basin,

Canada: carbon isotope perspectives. *Chem. Geol.* 159: 61–86.
Veizer, J., Ala D., Azmy, K., Bruckschen, P., Buhl, D., Bruhn, F., Carden, G., Diener, A., Ebneth, S., Godderis, Y., Jasper, T., Korte, C., Pawallek, F., Podlala, G. & Strauss, H. 1999. 87Sr/86Sr, δ13C and δ18O evolution of Phanerozoic seawater. *Chem. Geol.* 161: 59–88.
Walker, J.C.G., Hays P.B. & Kasting, J.F. 1981. A negative feedback mechanism for the long-term stabilization of Earth's surface temperature. *J. Geophys. Res.* 86(c10): 9776–9782.
Wheeler, J.O. & McFeely, P. 1991. *Tectonic Assemblage Map of the Canadian Cordillera and Adjacent Parts of the United States of America (GSC Map 1712a)*. Ottawa: Geological Survey of Canada.

Water–Rock Interaction, Wanty & Seal II (eds)

Metal mobility, transport & fate during weathering of Devonian metalliferous black shale

Michele L.W. Tuttle & George N. Breit
U.S. Geological Survey, Denver Colorado, USA

ABSTRACT: Devonian metalliferous black shale crops out along the eastern Appalachian basin margin, Kentucky, USA and was investigated to evaluate processes producing high metal concentrations (As, Co, Cr, Mo, Ni, V, and Zn) in sediment from nearby streams. Unweathered and weathered shale, efflorescent salts, and soil were analyzed from New Albany black shale. Comparison of metal:sulfide ratios in pyrite and unweathered shale suggest multiple metal residences (pyrite, sphalerite, chalcopyrite, NiS, PbSe, clays and organic matter). Sulfide-mineral oxidation progressively removes sulfur, mobilizes metals, and creates low pH solutions. Evaporation of weathering solutions forms Fe- and Al-rich metal-sulfate salts containing high concentrations of trace metals. These soluble salts store metals and acidity until released by rainfall events. Mobilized metals then precipitate with, or sorb onto Fe oxides when runoff mixes with higher pH stream water. Metal:TiO_2 ratios show that mechanical transport of weathered shale particles and formation of metal-rich Fe oxides account for the stream sediment metal concentrations.

1 INTRODUCTION

Weathering of black shale recently has been a focus of global element cycling partly because black shale is enriched in environmentally sensitive trace metals. High concentrations of As, Co, Cr, Mo, Ni, V, and Zn were measured in 60 of 1200 NURE stream sediment samples from the Appalachian basin, Kentucky (Tuttle et al. 2001). Distribution of metal-rich sediment corresponds with outcrops of metalliferous Devonian black shale (New Albany, Ohio, and Chattanooga). Metal dispersion from these black shales during weathering was investigated from the chemical composition and metal residence in the samples from the Huron Member of the New Albany shale, including unweathered shale, variably weathered shale, efflorescent salts, soil, and nearby stream sediment. Results show that processes controlling redistribution of elements among secondary phases produce distinctly different element enrichments in soil and stream sediment.

2 METHODS

Shale, soil, and efflorescent salts from a New Albany shale outcrop (Pryor et al. 1981) and from a nearby core near the outcrop in Kentucky were analyzed. Analyses included bulk chemistry (ICP-AES and ICP-MS), sulfur speciation and isotopic composition, laser ablation-ICP/MS of pyrite, and characterization of mineral grains with SEM. Details of the sampling protocol, analytical procedures, and analytical results are in Tuttle et al. (2003). The chemical results for stream sediment samples are in Tuttle et al. (2001).

3 RESULTS AND DISCUSSION

Pyrite is the most abundant sulfide mineral in the Huron Member of the New Albany shale. It occurs as massive nodules and framboids (Fig. 1). The metal contents of these two sulfide phases as determined by laser ablation ICP/MS are different, with most metals being enriched in the framboids (Table 1). Metal:$S_{sulfide}$ ratios in the pyrite compared to those in bulk shale samples (Fig. 2) indicates that some metals reside in phases other than pyrite. SEM analyses identified sphalerite, chalcopyrite and NiS. Ripley et al. (1990) showed that most V resides in clays (likely the residence for most Cr and some Zn as well based on data presented in Fig. 2) and Mo is either in organic matter or an acid-volatile sulfide phase yet to be identified.

Sulfide minerals progressively oxidize across the weathering transect on the New Albany shale outcrop (Fig. 3). Oxidation results in acidic, metal-rich solutions. Efflorescent salts precipitated on the outcrop

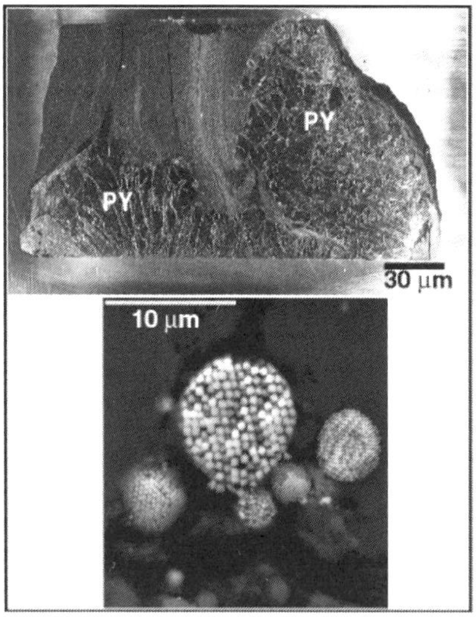

Figure 1. Photomicrographs of massive and framboidal pyrite in shale from the Huron Member.

Table 1. Average metal concentration in massive and framboidal pyrite from the Huron Member, New Albany shale.

Metal (ppm)	Massive*	Framboids*
As	420	480
Co	0.83	27
Cr	50	66
Cu	13	520
Mo	48	87
Ni	13	400
Pb	7.0	200
Se	2.5	23
V	2.8	15
Zn	3.3	60

* Averages based on $n = 75$ spot analyses for massive pyrite and $n = 64$ spot analyses for framboidal pyrite.

and in soil where solutions at depth had been wicked to the surface. (Although sulfate salts have not been positively identified in soil, high sulfate concentrations and pH's of <4 in laboratory 1:1 water extractions are consistent with their presence.) The efflorescent salts identified on the outcrop are similar to those found on mine-waste dumps, are predominately Fe- and Al-rich sulfates, and contain significant amounts of trace metals. Rainfall dissolves the metal-sulfate salts, and metals are released and transported to nearby streams. Stream surface water has a pH of about 8.0. Upon mixing with the stream water, Fe oxyhydroxides sorb or coprecipitate most metals. This process was quantified in the laboratory by dissolving efflorescent salts collected from the outcrop. The resultant solution had a pH of 3 and its relative composition was used as a proxy for the weathering solutions that are generated in the near surface during weathering. Titration of the solution to pH 8 (the pH of nearby streams) resulted in all metals being quantitatively removed from solution. The resulting Fe oxyhydroxides have about three times the concentration of metal (dry weight basis) as calculated in efflorescent salts (Table 2).

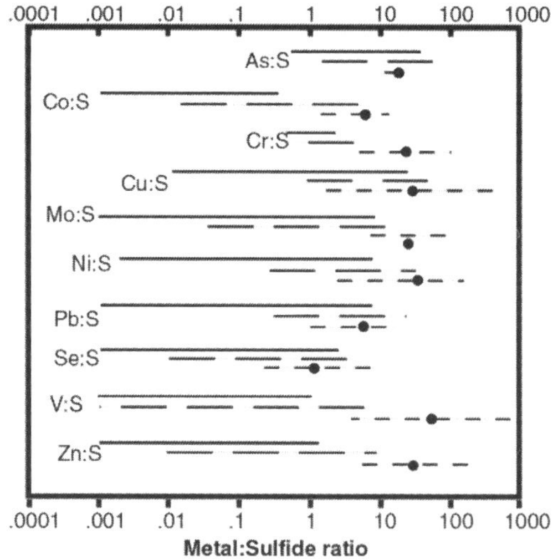

Figure 2. Range of metal:sulfide ($S_{sulfide}$) ratios (ppm/percent) in massive pyrite (solid line), range in framboids (long-dashed line), and range and geometric mean in unweathered shale (short-dashed lines with solid dot).

In addition to metals that enter the streams as a result of chemical weathering, metals are also transported during physical erosion of shale material (soil and disaggregated shale fragments). Metal:TiO_2 ratios (Table 3) were used to estimate the proportion of metals that enter stream sediment through chemical and mechanical transport.

Soil metal:TiO_2 ratios for the base metals Co, Ni, Pb, and Zn are either similar to or lower than weathered shale (Table 3). These metals originally reside in sulfide or selenide minerals and during weathering are mobilized into secondary phases. They are not strongly adsorbed at the low soil pH's (<4.0) (Dzombak & Morel 1990), and therefore, do not preferentially accumulate during soil-forming processes. Ratios in stream sediment are higher than in shale or soil, suggesting that solution transport of the metals to the stream in addition to transport of shale fragments and soil particles occurs, consistent with the mechanism of solution mixing outlined above.

Groundwater was also considered as a possible source of metals, but data from one nearby well

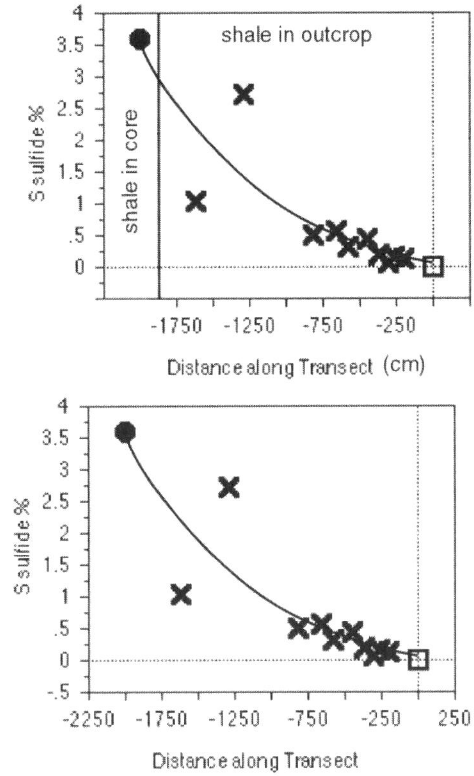

Figure 3. Plot showing the decrease in $S_{sulfide}$ across the outcrop weathering transect. Outcrop edge is to the right. Solid circle average content in unweathered core samples; **X**, content in outcrop shale; open square, content in soil ($n = 4$).

Table 2. Chemical composition of efflorescent salts ($n = 3$) that dissolved in water and of FeOOH precipitate (both ppm on dry weight basis), and percent precipitated from solution upon titration (% ppt'd).

Metal	Salt (ppm)	FeOOH (ppm)	% ppt'd
As	140	450	100
Co	160	490	95
Cr	28	78	89
Cu	150	510	97
Mo	35	100	93
Ni	360	1200	100
Pb	3.0	6.9	99
Se	1.9	6.2	100
V	41	130	100
Zn	310	950	96

completed at 30 feet below land surface in New Albany shale show the water to be anoxic with a pH of 7.8 (USGS NWIS database). Small amounts of ferrous iron and sulfate were measured in the groundwater, suggesting small amounts of pyrite oxidation, but the low iron concentrations suggest that trace-metal concentrations would be extremely low (concentrations were not measured). Therefore, the likely mechanism of release of these metals into the stream sediment is near-surface weathering, and the residence site is in iron oxyhydroxides that form as the solutions mix with higher pH stream water.

Table 3. Average metal:TiO_2 ratio in weathered shale ($n = 11$), soil ($n = 4$) and sediments from nearby streams ($n = 5$).

Metal:TiO_2 ratio*	Weathered shale	Soil	Stream sediment
As:TiO_2	65	61	41
Co:TiO_2	14	6.7	45
Cr:TiO_2	110	150	74
Cu:TiO_2	46	64	12
Fe_2O_3:TiO_2	8.9	9.4	12
Mo:TiO_2	87	120	<15
Ni:TiO_2	55	17	78
Pb:TiO_2	23	28	49
Se:TiO_2	2.1	4.6	nd
V:TiO_2	300	550	210
Zn:TiO_2	37	42	200

* ppm/percent except percent/percent for Fe_2O_3:TiO_2.

Ratios for metals that typically form oxyanions (As, Cr, Mo, and V) are high in soil and lower in the unweathered shale and stream sediment. Most of these elements reside in clays or organic matter as well as sulfide minerals in unweathered shale. During weathering, organic carbon decreases (Petsch et al. 2000; this study) and clays are weathered as evidenced by the Al in metal-sulfate salts). During these degradation processes, a portion of these metals are mobilized and then strongly adsorbed to mineral surfaces in the low pH soil (Dzombak & Morel 1990).

The trend for Cu:TiO_2 is very similar to those for the oxyanions. Cu forms very strong organic complexes in soil and it is likely these complexes, not adsorption to mineral surfaces, retain the metal in the soil.

4 SUMMARY

The metal anomalies in stream sediment along the western central Appalachian basin margin are attributed to weathering of the New Albany shale. Metals are differentially mobilized, primarily during oxidation of sulfide minerals but also during weathering of clays and organic matter. Evaporation of weathering solutions forms soluble metal-rich efflorescent metal-sulfate salts. These salts temporarily store metals and acidity in the shallowest subsurface until they dissolve during precipitation events. Base metals are preferentially carried in the acid weathering solutions and precipitate with iron oxyhydroxides when mixed with higher pH stream water. Elements that form oxyanions remain strongly bound to mineral surfaces in the soil and in weathering rinds and enter the streams when these materials are mechanically transported to the streambeds.

REFERENCES

Dzombak, D.A. & Morel, F.M.M. 1990. *SurfaceComplexation Modeling: Hydrous Ferric Oxide*. New York: Wiley & Sons.

Petsch, S.T., Berner, R.A. & Eglinton, T.I. 2000. A field study of the chemical weathering of ancient sedimentary organic matter: *Org. Geochem.* 31: 475–487.

Pryor, W.A., Fisk. H.N. & Kepferle, R.C. 1981. Road Log. In *Energy resources of Devonian-Mississippian shales of Eastern Kentucky, Annual Field Conference of the Geological Society of Kentucky, April 9–11, 1981*. Lexington: Kentucky Geological Survey.

Ripley, E.M., Shaffer, N.R. & Gilstrap, M.S. 1990. Distribution and geochemical characteristics of metal enrichment in the New Albany Shale (Devonian-Mississippian), Indiana. *Econ. Geol.* 85: 1790–1807.

Tuttle, M.L.W., Breit, G.N. & Goldhaber, M.B. 2003. Geochemical data from New Albany shale, Kentucky: A study in metal mobility during weathering of black shales: *U.S. Geological Survey Open-file Report* 03–207.

Tuttle, M.L.W., Goldhaber, M.B., Ruppert, L.F. & Hower, J.C. 2001. Arsenic in rocks and stream sediments of the Central Appalachian Basin, Kentucky: *U.S. Geological Survey Open-file Report* 02–28.

Water–Rock Interaction, Wanty & Seal II (eds)

Si cycling in a marine chronosequence demonstrated by pore-water Ge/Si ratios

Davison V. Vivit, Marjorie S. Schulz & Art F. White
US Geological Survey, Menlo Park, CA, USA

Suzanne P. Anderson
University of Colorado, Boulder, CO, USA

ABSTRACT: Si, Na, and Ge concentrations determined in pore-waters from a chronosequential flight of marine terraces in the vicinity of the Wilder Creek watershed (Santa Cruz County, CA, USA) show that silicate weathering controls solute fluxes from deep soils. At depths below the root zone (>2 meters), lithogenic Si, controlled by plagioclase dissolution, occurs. Ge/Si ratios indicate ongoing biogenic processes dominating in the root zone. Seasonal uptake of Si by native grasses brings about its incorporation into phytoliths which are amorphous opaline silica plant structures, and decay of grass litter during the wet season releases biogenic Si to the soil waters. Since cycling of both lithogenic and biogenic Si depends on dissolution, transport, deposition and seasonal climates; over time the pore-water Ge/Si ratio could vary at any terrace depth.

1 INTRODUCTION

Major and trace element constituent levels observed in soil pore waters from grasslands in the vicinity of Wilder Creek (Santa Cruz County, CA, USA) indicate prevailing silicate weathering processes. For instance, correlations between concentrations of dissolved sodium and dissolved silicon in deep soil waters beyond plant-dominated near-surface depths point to "lithogenic" processes, i.e. the weathering of specific primary minerals (Garrels & Mackenzie 1967). Furthermore, solid phase germanium-silicon fractionations observed for soils formed from Hawaiian basalts have yielded information about lithogenic secondary minerals derived from parent igneous rocks (Kurtz et al. 2002).

Since plant growth is a significant aspect of chemical weathering in watersheds at near-surface depths (Alexandre et al. 1997), application of the Ge-Si fractionation approach to the present investigation could indicate the degree to which "biogenic" silicate processes influence overall weathering. Larger Ge/Si ratios would seem likely to occur in the root-zone soil waters following Ge enrichment at the expense of Si uptake by vegetation.

2 EXPERIMENTAL METHODS

2.1 *Sampling*

Pore waters were collected over time (March 2001-present) from soil water suction sampler arrays installed at five field sites that are situated on five successive marine terraces comprising a chronosequence (age-dated from 65 to 226 ka). Terrace 1 (T1) is the youngest and Terrace 5 (T5) is the oldest. Soils have developed on terrace beach sands derived from regional granitic bedrock. Bradley (1957), Perg et al. (2001), and Schulz et al. (2002) describe the lithology and geochronology of the area (Fig. 1).

2.2 *Determination of Si*

Samples were filtered with 0.45-μm cellulose acetate syringe filters, diluted with distilled deionized H_2O, and acidified to 2.0% with Fisher Optima nitric acid, before determining Si and other cations by direct quantitative analysis using a Perkin-Elmer Sciex Elan 6000 ICP-MS spectrometer with associated Elan Version 2.2 software.

2.3 *Determination of Ge*

Measuring Ge near its detection limit by ICP-MS (~10 ng/L) required concentration of water samples by vacuum-evaporation prior to analysis. Figure 2 shows the setup assembled from commonplace laboratory equipment in a reasonably clean area without need of clean-room conditions.

Syringe-filtered sample aliquots of 50.00 ml were placed in polycarbonate flasks, which were connected to the house vacuum line by means of plastic luer manifolds, tubing, and rubber stoppers (see Figure 2). The samples were then warmed in a temperature range of 40–45°C while under low vacuum.

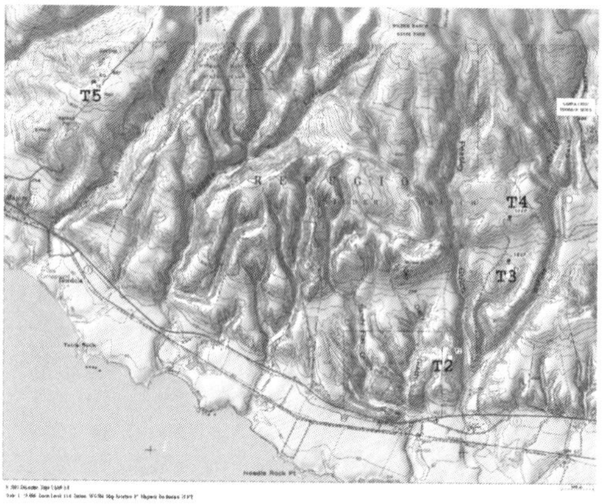

Figure 1. Topographic relief map of Santa Cruz marine terrace sites within and near Wilder Ranch State Park, Santa Cruz County, CA, USA. T1 is off the map (source: Topo USA 3.0, 2001, DeLorme).

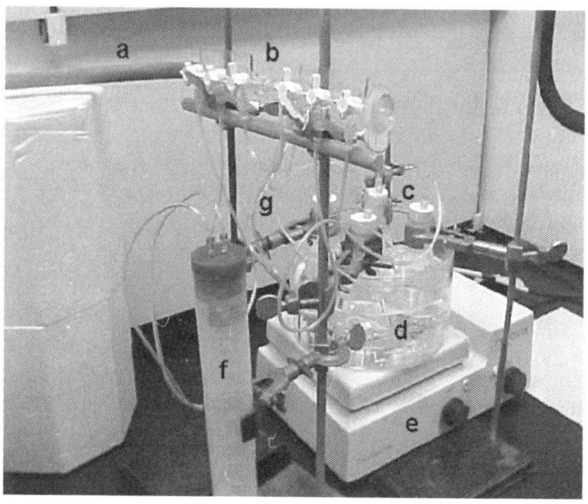

Figure 2. Setup for concentrating samples: (a) lab vacuum, (b) plastic manifold, (c) samples in polycarbonate flasks with stoppers, (d) water bath, (e) hotplate, (f) trap, (g) plastic luer tubing.

Concentrates were removed from the apparatus after 2 days when 65–70% of the water (130–140 ml with four samples) filled the trap, and individual concentration factors generally ranging from three- to four-fold were determined. The concentrates were then acidified to 2.0% nitric acid as previously described. Ge was determined at mass 74.

3 RESULTS AND DISCUSSION

3.1 *Processes controlling dissolved Si*

Figure 3 shows depth profiles for Si concentrations at Terraces 2 and 5. Down to two meters, higher Si concentrations occur at the shallowest depths due to evapotranspiration (E-T). Similar results were observed for Terraces 3 and 4 (not shown).

Minimal concentrations of dissolved Si occur between 60 and 180 cm before increasing suggesting the existence of biogenic controls on Si level.

Beyond 180 cm Si concentrations increase with depth to levels greater than those at shallower depths. From the near-surface to the saturated zone, Na levels (not shown) progressively increase. Comparable Si and Na trends exist at Terraces 2, 3, and 4 (not shown) implying that plagioclase dissolution controls Si levels in deeper soil waters (Garrels & Mackenzie 1967). Terrace 1, the youngest terrace, has Si and Na levels that exhibit no clear trends with depth due to its shallow unsaturated zone.

3.2 *Ge & Ge-Si fractionation*

The relative standard deviation (RSD) of the method used is 9.1% (n = 4) at the 5.7 ng/L level for a pore

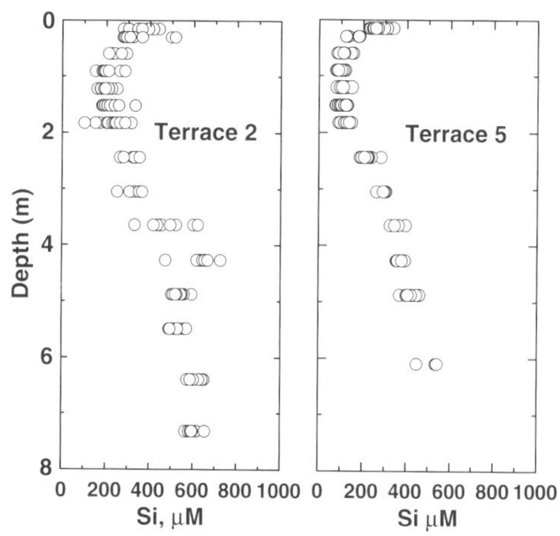

Figure 3. Variation of Si and Na concentrations in pore waters with depth at Terraces 5. Multiple points at a given depth show temporal variation.

water sampled at 122 cm depth at Terrace 4. The detection limit was 3 ng/L when the relative precision was established.

Table 1 shows the relative accuracy of the direct ICP-MS determination of Ge levels in the Terrace 3 pore water samples. Spike recoveries between 90% and 110% were satisfactory while those outside this range were acceptable taking considering that the detection limit for the tabulated results was 5 ng/L.

Figure 4 compares the Ge results obtained from vacuum-concentrated samples with those obtained from a germanium hydride generation-ICP-MS technique (Mortlock & Froelich 1996) which does not require sample concentration. The respective results between the two methods agreed favorably except for deeper samples at 426 cm (11.1 ng/L), 549 cm (47.5 ng/L and not shown in Fig. 4), and 640 cm (9.2 ng/L) which were 19%, 25%, and 25% higher, respectively, by the hydride method.

Altogether, measurable levels of Ge from Terraces 2, 3, 4, and 5 ranged between 3 and 48 ng/L. No Ge data is currently available for pore waters from Terrace 1. Figure 5 shows the Ge/Si mole ratio depth profiles for Terrace 5. Ge/Si ratios progressively decrease in the root zone down to ~2 meters, and the same trend is observed for Terraces 2, 3, and 4 (not shown). The higher Ge/Si ratios near the surface support the concept of root-zone uptake of Si by plants and its assimilation into phytoliths.

Phytoliths are amorphous silica structures inside or among plant cells (Fig. 6). Found mainly in above-ground foliage, the assimilated Si takes on a form of opal and plant litter decomposition liberates opaline Si into the watershed. Many plants physiologically incorporate silica; however, grasses are known to be prolific producers of phytoliths (Fearn 1999).

During plant growth, native grasses on the marine terraces deplete Si from soil water in the root zone while excluding Ge. Alexandre et al. (1997) observed Si depletion relative to Ge in an equatorial rainforest due to Si uptake by vegetation and its incorporation

Table 1. Germanium spike recoveries on select Terrace 3 pore waters sampled 12/12/01.

Depth cm	Added ng/L	Found ng/L	Recovered ng/L	Recovery %
122	–	27.3	–	–
122	10	36.4	9.4	94
122	25	47.5	20.9	84
243	–	14.8	–	–
243	10	25.6	11.0	110
243	25	40.3	25.9	104
732	–	27.2	–	–
732	10	38.0	11.1	111
732	25	55.2	28.7	115

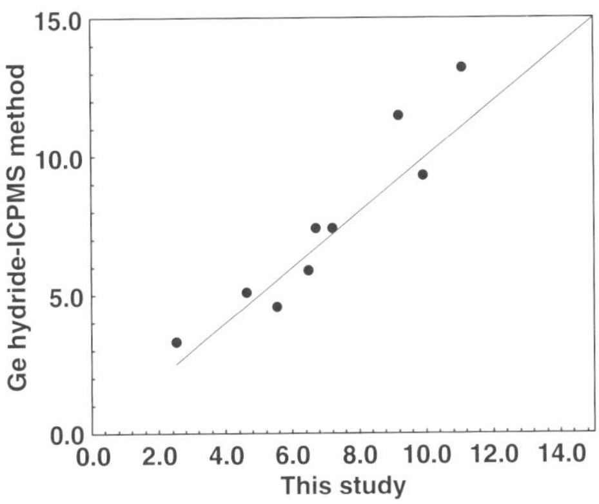

Figure 4. Germanium concentrations (ng/L) for Terrace 4 pore waters sampled 4/16/02 determined by two different methods.

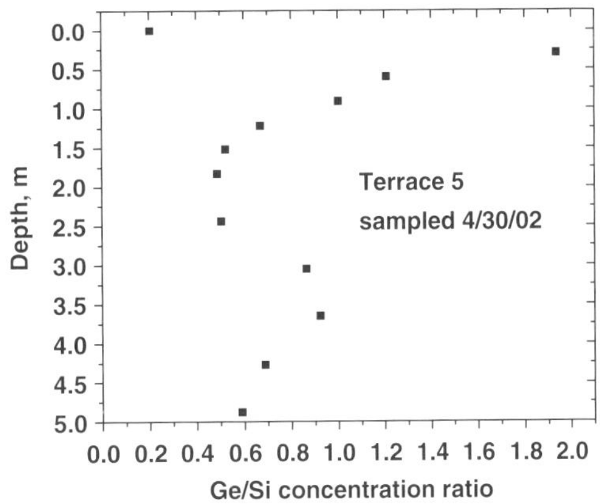

Figure 5. Variation of Ge/Si mole ratio with depth.

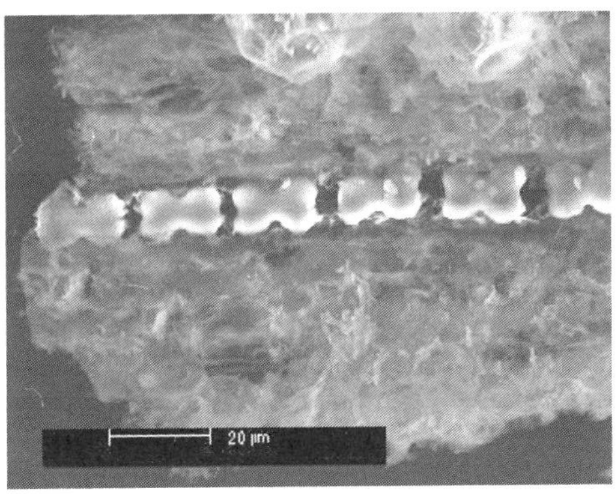

Figure 6. Electron micrograph of dumb-bell shaped phytoliths in a dry-ashed grass leaf (from Mulder 1999).

into phytoliths. At the marine terraces Si integrated into phytoliths becomes available for fluvial transport with the onset of seasonal rains coinciding with the end of the annual growth cycle for leaves. At greater depths the Ge/Si ratios follow no apparent trend and most likely result from lithogenic weathering of primary and secondary minerals. The same was also observed in soils derived from intensely weathered Hawaiian basalts (Kurtz et al. 2002).

These results have important implications for interpreting Si fluxes for estimating weathering rates in watersheds. Much of the shallow-depth, pore-water Si within the root zone originates from biogenic amorphous opaline silica, which weathers more rapidly than the lithogenic silicates (i.e. plagioclase, K-feldspar, and quartz). Derry et al. 2001, observed hydrologic fluxes in shallow soils exhibiting plant uptake and recycling of Si rather than mineral-water reactions controlling Si saturation. Hence, pore water sampled in the root zone at different times in the rainy season can yield varying Si concentrations as shown in Table 2.

Pore waters sampled in December, 2001 had high enough Ge levels to be analyzed directly by ICP-MS without need for sample pretreatment. Ge/Si values from December, being nearly two- to four-fold greater than those for April 2002, characterize pore waters originating from precipitation percolating into the soil early in the rainy season when phytoliths appear to be still intact. The April samples represent pore waters near the end of the wet period when Si levels are large relative to respective Ge levels. Extensive degradation of phytoliths late in the wet season might explain this relative increase with Si.

Table 2. Ge/Si molar ratios in Terrace 3 pore waters from two different dates.

Depth cm	Sample date	Ge/Si mole ratio × 10(−6)
91	12/12/2001	1.01
	4/16/2002	0.32
122	12/12/2001	1.28
	4/16/2002	0.36
243	12/12/2001	0.42
	4/16/2002	0.18
305	12/12/2001	0.45
	4/16/2002	0.25
427	12/12/2001	0.44
	4/16/2002	0.17
549	12/12/2001	1.66
	4/16/2002	0.49
610	12/12/2001	1.06
	4/16/2002	0.41
732	12/12/2001	0.45
	4/16/2002	0.44

However, Si uptake by concurrent rapid grass growth could retard the relative increase with Si.

The Ge/Si ratio variations seen here reflect the extent to which the formation and dissolution of biogenic Si from plant growth and decomposition might be controlling dissolved Si as seen in Figures 3 and 5. Furthermore, biogenic silica could remain in the soil column and recycle repeatedly through subsequent annual growth cycles of grass cover which would compound the difficulty in using Si for estimating weathering rates in watersheds.

ACKNOWLEDGEMENTS

Ge hydride-ICPMS analyses by S. King of Cornell University are gratefully acknowledged. This study is part of the USGS Water, Energy and Biogeochemical Budgets (WEBB) program.

REFERENCES

Alexandre, A., Meunier, J.-D., Colin, F. & Kudu, J.-M. 1997. Plant impact on the biogeochemical cycle of silicon and related weathering processes: *Geochim. Cosmochim. Acta* 61: 677–682.

Bradley, W.C. 1957. Origin of marine terrace deposits in the Santa Cruz area, CA: *Geol. Soc. Am. Bull.* 68: 421–444.

Derry, L.A., Kurtz, A.C., Ziegler, K., Chadwick, O.A. & Kelley, E.F. 2001. Plant phytolith source of dissolved silica in Hawaiian streams from Ge/Si ratios: 11th Annual V.M. Goldschmidt Conference (2001), Abstract 3805.pdf.

Fearn, M.C. 1999. Phytoliths. University of South Alabama, May 1999, URL: http://www.southalabama.edu/geography/ fearn/phyto.htm.

Garrels, R.M. & Mackenzie, T.T. 1967. Origin of the chemical composition of some springs and lakes. In W. Stumm (ed.), *Equilibrium concepts of natural water systems, Advances in Chemistry Series 67: 222–242,* American Chemical Society.

Kurtz, A.C., Derry, L.A. & Chadwick, O.A. 2002. Germanium-silicon fractionation in the weathering environment: *Geochim. Cosmochim. Acta* 66: 1525–1537.

Mortlock, R.A. & Froelich, P.N. 1996. Determination of germanium by isotope dilution-hydride generation inductively coupled plasma mass spectrometry: *Anal. Chim. Acta 332*: 277–284.

Mulder, C. 1999. Ecological significance of grass leaf phytoliths from southern Africa. Univ. of Utrecht, May 1999, URL: http://www.bio.uu.nl/~palaeo/Research/Namibia/namibia.htm.

Perg, L.A., Anderson, R.S. & Finkel, R.C. 2001. Use of a new ^{10}Be and ^{26}Al inventory method to date marine terraces, Santa Cruz, California, USA: *Geol.* 29: 879–882.

Schulz, M.S., White, A.F., Harden, J., Stonestrom, D., Anderson, S. & Vivit, D. 2002. Silicate weathering of marine terraces north of Santa Cruz, California: *Eos Trans. AGU, 83*, Fall Meet. Suppl., Abstract H61C-0808, 2002, American Geophysical Union.

Water–Rock Interaction, Wanty & Seal II (eds)
© 2004 Taylor & Francis Group, London, ISBN 90 5809 641 6

Hydrochemical evolution and contemporary weathering rates in a soil chronosequence, Merced, California, USA

A.F. White, M.S. Schulz, D. Vivit & A.E. Blum
US Geological Survey, Menlo Park, CA, USA

ABSTRACT: This study presents the first detailed description of the hydrochemical evolution and contemporary weathering rates in a soil chronosequence. During 3 million years of weathering, the soils become less permeable, progressively isolating pore water solutes from climatic perturbations and increasing the importance of mineral weathering. Mass balance models describe the elemental partitioning between minerals, pore water and exchange reservoirs and the elemental fluxes from precipitation, plants, discharge and weathering. Extrapolation of present day weathering fluxes reproduces observed long-term changes in soil compositions. Plagioclase reaction rates, based on solute weathering gradients, are similar to previously reported geologic rates and are several orders of magnitude slower than experimental rates due to near-equilibrium solute conditions.

1 INTRODUCTION

Chemical weathering at the Earth's surface can be addressed from two perspectives, either on a geologic time scale based on solid-state changes in soils and regoliths or else on a contemporary time scale based on present day solute compositions and fluxes (White 2002). Long-term weathering reflects the integrated effects of intrinsic changes in mineral properties and abundances as well extrinsic effects such as climate, biology and topography, commonly occurring over tens of thousands to millions of years of soil development. In contrast, contemporary weathering reflects current fluid residence times and the influences of seasonal to decadal changes in precipitation, temperature and biota.

Soil chronosequences represent ideal environments to characterize chemical weathering over both geologic and contemporary time scales. A significant body of research has used solid-state chemical, isotopic and physical characteristics of chronosequences to provide individual "snapshots in time" of geologic weathering processes and rates. In contrast, virtually no studies describe the corresponding hydrochemical characteristics of soil chronosequences and how these factors progressively evolve in soils of increasing age. The paper addresses specific solute controls on weathering and calculates contemporary mineral weathering fluxes and rates.

2 SITE CHARACTERIZATION

The Merced chronosequence, containing soils ranging in age from 3 to 3000 yrs., occurs on the eastern side of the Central Valley of California (White et al. 1996). The soil series, developed on terraces along the present-day Merced and Tuolumne Rivers, originally consisted mostly of granitic sands derived from glacial outwash from the Sierra Nevada Mountains to the east. The present day Mediterranean climate at Merced is relatively arid (average = 300 mm of annual precipitation), subjected to pronounced seasonal wetting and drying and has significant variability in total annual rainfall.

The most direct linkage between weathering and hydrology over 3 million years of soil development is centered on the progressive development of shallow argillic horizons (1–3 m deep). In these horizons, kaolinite progressively increases from 10 to 60% with concurrent decreases in hydraulic conductivities from 10^{-8} to 10^{-12} m s^{-1}. In younger, highly permeable soils, spatial and temporal solute variations are dominated by seasonal to decadal scale variations in rainfall and evapotranspiration (up to 99%). In the older soils, with the decreasing permeability, the climate signal is strongly damped at depth in the profiles (2–8 m) and solute distributions increasing reflect weathering inputs modified by the effects of heterogeneous flow.

3 MASS BALANCES

An example of contemporary mass balances is shown for the 250 ky old Riverbank soil in Figure 1. The soil is compartmentalized into the mineral, pore water and exchange reservoirs. The total elemental masses in the mineral reservoir are calculated from soil oxide percentages and bulk densities. The pore water reservoir is based on solute concentrations and gravimetric water contents and the exchange reservoir is calculated from acetate exchange measurements.

Solute fluxes into and out of the soil surface (arrows in Fig. 1) include precipitation $q_{j,precip}$, evapotranspiration (q_{ET}) and biological fluxes $q_{j,bio}$ which are calculated as the difference between the maximum and minimum masses in grassland vegetation measured during the year. The principal flux out of the base of the soil is pore water discharge $q_{j,dis}$ calculated as the difference between precipitation and ET across the soil surface. Elemental fluxes in discharge are calculated based on this net water flux and average solute compositions measured at the base of the soil.

The weathering flux, which in Figure 1 is considered a component of the discharge flux, is defined as:

$$q_{j,w} = q_{j,discharge} - q_{j,precip} \pm q_{j,bio} \pm q_{j,ex} \qquad (1)$$

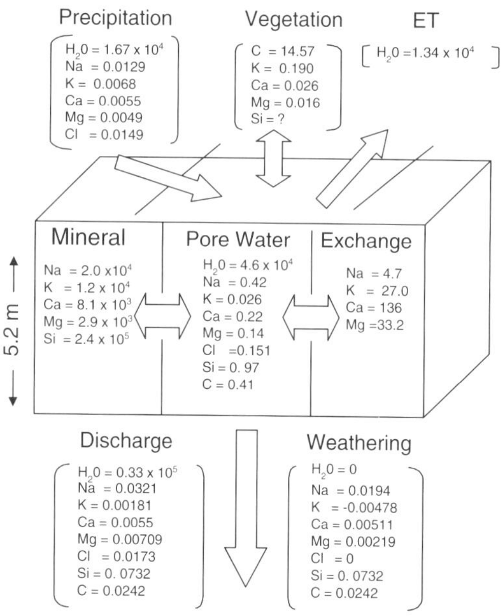

Figure 1. Schematic of contemporary mass balances in the 250 ky old Riverbank soil. Data in boxes are elemental and water masses in soil reservoirs (moles). Data in brackets are elemental fluxes (moles m^{-2}) between reservoirs as indicated by arrows.

As is frequently done in watershed studies (Paces 1986), Equation 1 is simplified such that the weathering flux $q_{j,w}$ for the Riverbank soil is calculated as the difference between precipitation and discharge, i.e. $q_{j,discharge} - q_{j,precip}$ with the net biologic and exchange fluxes set to zero (doubled headed arrows in Fig. 1). The upward biologic flux during the growing season is assumed to equal the downward flux resulting from decomposition during the dormant season. Similarly, exchange reactions on a contemporary time scale are assumed to be in equilibrium with no net contribution to or uptake from the pore water reservoir.

The degree which these assumptions accurately represent contemporary mass balances in the Riverbank soils is strongly dependent on the relative magnitudes of the various elemental reservoirs (Fig. 1). This is most obvious for K, the strong biological influence of which is evident in large seasonal variation in shallow pore waters. Small perturbations the biologic flux, i.e. $q_{j,bio} \neq 0$, strongly impact pore water solute concentrations and apparent weathering fluxes. Although K-feldspar and biotite weathering contribute to solute K, the calculated K weathering flux out of the soil is negative (Fig. 1), implying that during the course of the study, K was accumulating in the system due to antecedent decadal draught conditions and/or past land use.

In soils of increasing age, pore water solutes generally decrease in total concentration and increase in Na relative to K, Ca and Mg. These trends reflect changes with time in the amounts and relative proportions of residual minerals present as well of differences in kinetic reaction rates. In contrast, although Mg and Ca also decrease with soil age, their proportions remain constant with time at ratios of approximately 1:2, which are similar to that measured for sorbed species. These ratios produce an exchange selectivity coefficient defined such that:

$$K_{Mg/Ca} = \frac{\left[c_{Mg}/c_{Ca}\right]_{pw}}{\left[c_{Mg}/c_{Ca}\right]_{exchange}} \cong 1 \qquad (2)$$

which is consistent with a small smectite component in the soil with a high exchange capacity. The magnitude of the exchange reservoir is such that it strongly buffers perturbations in solute Mg and Ca on a contemporary time scale. However due to mass balance constraints, long term buffering by exchange reactions requires an addition input of Ca, probably introduced by open-system biological cycling and aerosol deposition.

The solid state mass change $\Delta m_{i,w}$, of an element j, produced by a constant weathering flux $q_{j,w}$ over contemporary time ($\Delta t_c = 1$ yr in Fig. 1), is defined such that:

$$\Delta m_{i,pw} = q_{j,w} \cdot \Delta t_c \tag{3}$$

Over such times, $\Delta m_{i,w}$ is very small compared to the total mass loss in the solid, i.e. $\Delta m_{i,w} << \Delta M_i$, However when total duration of soil weathering is considered, i.e. the geologic time scale Δt_g, the total elemental mass loss associated with the weathering flux becomes equal to the solid state loss and is defined as:

$$\Delta M_i = \int_0^{t_g} q_{j,w} dt \tag{4}$$

where $q_{j,w}$ may be a function of time. If the condition is true that contemporary weathering fluxes is comparable to past weathering, i.e., $q_{j,w}$ is constant, the following relationship is valid:

$$q_{j,w} = \frac{\Delta m_j}{\Delta t_c} = \frac{\Delta M_j}{\Delta t_g} \tag{5}$$

Weathering trajectories, defined by White et al. (1998), are produced by extrapolating the contemporary mass fluxes, expressed in Equation 3 to past geologic time scales (Eqn. 4). The resulting convex curves in Figure 2 show the progressive increases in $\Delta M_{i,w}$ for Na and Ca, released principally from plagioclase, as $\Delta t_c \rightarrow \Delta t_g$.

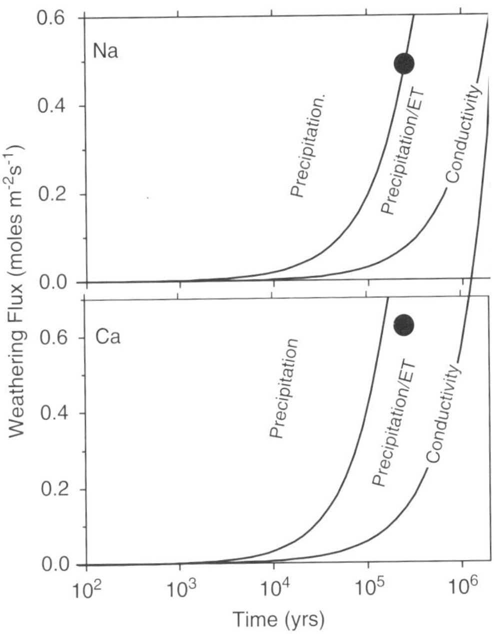

Figure 2. Mass losses with time in the Riverbank soil. Solid lines are weathering trajectories produced by extrapolating Equation 3 to geological time scales (Eqn. 4). Solid points are the measured mass loss in the soil.

The magnitudes of the solute weathering trajectories are constrained principally by the amount of water that physically moves through the soil, i.e., q_h. Fluid flow in the older soils is heterogeneous, thereby producing a range of trajectories. In Figure 2, the upper boundary of this continuum is constrained by precipitation, the maximum permissible water flux under current climatic conditions $q_h = q_p = 0.3$ m yr^{-1} (Eqn. 3). The lower boundary of the weathering trajectories corresponds to saturated conductivities in the Riverbank argillic horizon ($q_h = 0.01$ m yr^{-1}). The third trajectory, based on the precipitation-ET balance in Figure 1 ($q_h = 0.06$ m yr^{-1}) falls midway between the other two trajectories, a position, which may be argued, represents average integration at depth of fluid fluxes produced by rapid preferential flow around and slow flow through the argillic horizons. The precipitation/ET trajectory in Figure 2 essentially passes through the point representing total Na loss in the Riverbank soil after 250 kyrs of weathering, i.e., $\Delta M_{i,w} = 490$ kmoles versus 480 kmoles based on solid compositions. The corresponding solute losses of Ca are about 30% higher than the solid-state loss.

4 WEATHERING RATES

Conservative steady-state vertical infiltration, defined by the hydraulic flux q_h, is the simplest model for pore water movement through a soil. In such cases, increasing depth correlates with increasing fluid residence times. Adding a constant rate of chemical weathering to this scenario r_m (moles m^{-2} s^{-1}) results in the prediction that pore water solutes will increase consistently with depth in a soil (White 2002). The slope of this linear concentration/depth relationship is termed the weathering gradient b_p (m L mole^{-1}) and is directly proportional to the fluid flux and inversely proportional to the weathering rate.

Conversely, the weathering rate of a specific mineral r_m can be calculated from the solute flux divided by the weathering gradient such that (White 2002).

$$r_m = \frac{1}{10^6}\left(\frac{1}{\rho_w}\right)\left(\frac{1}{\phi \beta S}\right)\left(\frac{q_h}{b_p}\right) \tag{6}$$

This rate r_m is inversely proportional to the regolith bulk density ρ_w(g cm^{-3}) and a set of terms defining mineralogical properties i.e., the mineral fraction ϕ (moles/mole), the mineral stoichiometry β (mole/mole) and the specific mineral surface area S (m^2 g^{-1}).

The vertical distribution of solute Na, in the Turlock Lake soil, corrected for precipitation inputs, is

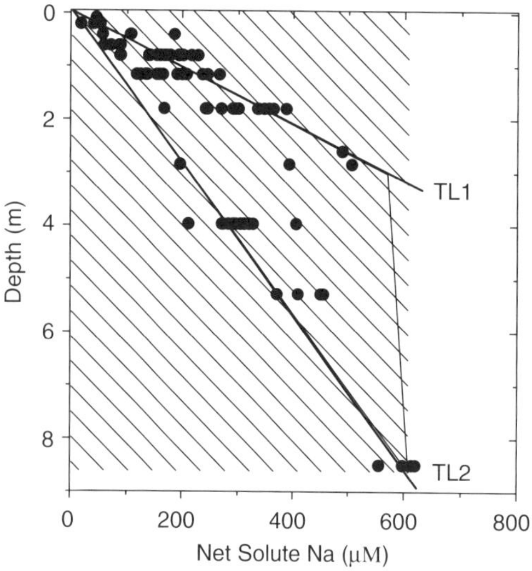

Figure 3. Net solute Na contributed from chemical weathering as a function of depth in the Turlock soil (600 kyr). Hashed area corresponds to the range in weathering gradients with specific lines defining those for the duripan (TL1) and for the deeper soils (TL2).

plotted in Figure 3. This distribution reflects the intensity of plagioclase weathering as a function of depth assuming that it is the sole source of Na. The shaded area in Figure 3 emphasizes the non-linear nature of the Na distributions, which is characterized by an array of weathering gradients. The upper limit is defined by the shallowest weathering gradient TL1, which pass through maximum Na concentrations in the argillic horizon. The lower limit is defined by steepest gradient TL2, which pass through solute concentrations in the deeper soil horizons. The presence of multi-gradients reflects the heterogeneous hydrology of the soils, specifically significant flow that bypasses the argillic horizon.

Contemporary plagioclase weathering rates r_m for the deep Riverbank and Turlock Lake soils are 7 to 3×10^{-16} moles m^{-2} s^{-1} respectively (Eqn. 6). These rates parallel the decrease with soil age of previously determined long-term rates based on changes in plagioclase content R_{plag} (White et al. 1998). The contemporary rates are consistently faster but the absolute differences for the deep soils are relatively small, i.e. $r_{plag}/R_{plag} =$ 4.3 and 3.4, suggesting the contemporary and long-term weathering rates are comparable in the Merced soils. These rates are also similar to plagioclase rates determined for other relatively old weathering environments (White & Brantley 2003). The present rates are many orders of magnitude slower than observed for experimental plagioclase dissolution and, based on solute thermodynamic stability calculations, reflect near-equilibrium weathering reactions.

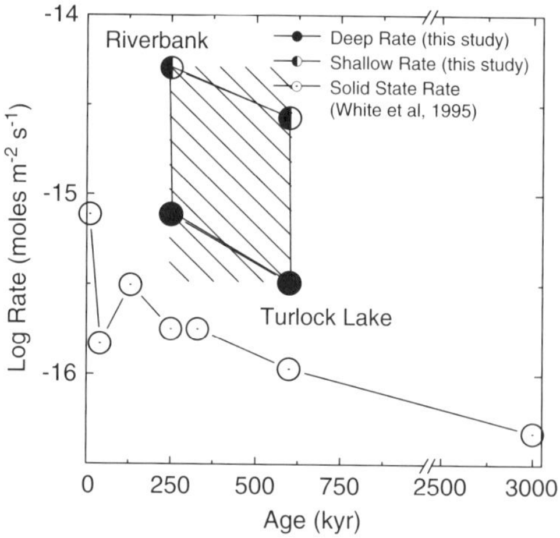

Figure 4. Contemporary weathering rates (Eqn. 4) compared to geologic rates based on changes in plagioclase contents in the Merced chronosequence (White et al. 1996).

REFERENCES

Paces, T. 1986. Rates of weathering and erosion derived from mass balance in small drainage basins. In S. Dethier & S. Colman (eds.), *Rates of Chemical Weathering of rocks and Minerals:* 531–550. New York: Academic Press.

White, A.F. 2002. Determining mineral weathering rates based on solid and solute weathering gradients: Application to biotite weathering in saprolites. *Chem. Geol.* 190: 69–89.

White, A.F., Blum, A.E., Schulz, M.S., Bullen, T.D., Harden, J.W. & Peterson, M.L. 1996. Chemical weathering of a soil chronosequence on granitic alluvium 1. Reaction rates based on changes in soil mineralogy. *Geochim. Cosmochim. Acta* 60: 2533–2550.

White, A.F., Blum, A.E., Schulz, M.S., Vivit, D.V., Larsen, M. & Murphy, S.F. 1998. Chemical weathering in a tropical watershed, Luquillo Mountains, Puerto Rico: I. Long-term versus short-term chemical fluxes. *Geochim. Cosmochim. Acta* 62: 209–226.

White, A.F. & Brantley, S.L. 2003. The effect of time on the weathering of silicate minerals: Why to weathering rates differ in the laboratory and field? *Chem. Geol.* 202: 23–41.

Water–Rock Interaction, Wanty & Seal II (eds)
© 2004 Taylor & Francis Group, London, ISBN 90 5809 641 6

Feldspar dissolution rates and clay precipitation in the Navajo aquifer at Black Mesa, Arizona, USA

Chen Zhu
Indiana University, Dept. of Geological Sciences, Bloomington, IN, USA

Alex Blum
US Geological Survey, Boulder, CO, USA

David Veblen
Department of Earth and Planetary Science, Johns Hopkins University, Baltimore, MD, USA

ABSTRACT: This study measured a plagioclase dissolution rate of $10^{-16.3}$ mol m^{-2} sec^{-1} and a K-feldspar dissolution rate of $10^{-17.7}$ mol m^{-2} sec^{-1} along up to 80 km long flow paths within the confined Navajo sandstone aquifer of the Black Mesa of Arizona, with residence times up to 35 Ky. These rates are among the slowest feldspar dissolution rates reported. However, evaluation of previously published water chemistry from other areas suggests these slow feldspar reaction rates are typical of long residence time, regional aquifers. The well constrained physical and chemical conditions of the Navajo aquifer, coupled with detailed FEG-SEM and HRTEM analysis of the compositions and textures of the primary and secondary phases, allowed the evaluation of most commonly proposed mechanisms for limiting feldspar dissolution rates, all of which are found to be inadequate to explain the observations. Rather, it is proposed that clay precipitation kinetics is a critical process in controlling overall feldspar reaction rates in regional aquifer systems.

1 INTRODUCTION

Numerous hypotheses have been proposed to explain the apparent retardation of feldspar dissolution in natural systems (Blum & Stillings 1995; Drever & Clow 1995). Nearly all previous *in situ* rates were derived from soil profiles and small watersheds. These unsaturated systems are fraught with issues of spatial and temporal variability, wetted versus unwetted surface areas, poor age control, channelized flow, and complex biological effects, all of which complicate understanding of the underlying geochemical processes. Nugent et al. (1998) also observed that the dissolution mechanism of fresh albite surfaces in soils appeared similar to that observed in laboratory experiments, not observed field dissolution rates. This has resulted in continuing doubts about the accuracy and origin of the slow feldspar dissolution rates measured at many field sites. Largely because of this uncertainty, some mathematical models used for performance assessment of high-level nuclear waste repositories (Spycher et al. 2003) and geological carbon sequestration, as a strategy to mitigate greenhouse gas effects on the global warming trend (Gunter et al. 1997) have opted to use laboratory rates.

To more clearly elucidate natural feldspar reaction mechanisms, we studied saturated aquifers where the many complexities of soils and watersheds are not present. The Navajo sandstone at Black Mesa, Arizona, has a) excellent age control, b) relatively simple hydrology and chemistry, c) more detailed and redundant measurements from different flow paths, time intervals, and solution chemistries d) alteration textures which are preserved, and e) groundwater residence time up to 35,000 years B.P.

2 WEATHERING IN THE NAVAJO SANDSTONE AT BLACK MESA

2.1 *Hydrogeology*

The Navajo sandstone is a well-sorted aeolian quartz sandstone with ~2% K-feldspar and plagioclase which forms a regional aquifer, and supplies most of the domestic and agricultural water supply in the area. Controversy over using this water resource for mining has generated abundant hydraulic, chemical, and isotopic data. The Navajo (or N) aquifer is recharged in the northwestern highlands and flows southeast (Fig. 1). In the recharge area the groundwater is $HCO_3^- - Ca^{2+} - Mg^{2+}$ type, with a pH of 7.5 to 8.0. After entering the confined portion of the aquifer, Ca^{2+} and Mg^{2+} concentrations decrease while Na^+ increases proportionally down gradient, and the pH rises up to ~10. ^{14}C dating of groundwater indicates the transit time along the 80 km flow path shown in Fig. 1 is ~35,000 years (Zhu 2000). The sedimentary basin has maintained the current flow regime for ~2 million years.

2.2 *In situ feldspar dissolution rates*

Dissolution rates (R) were calculated from 15 well pairs (Fig. 2), using ^{14}C travel times between wells.

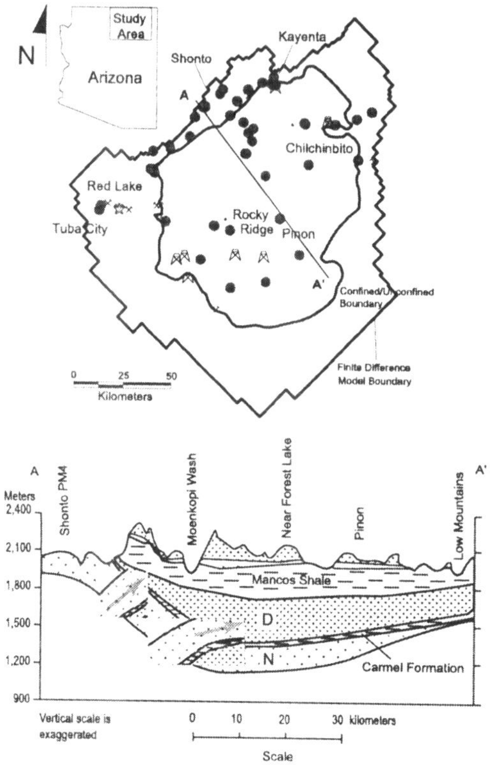

Figure 1. Top: Map of sample locations. Dots indicate groundwater samples; hammers, outcrop samples; well symbols, drill cuttings; stars, drill cores. Bottom: Cross-section of the Black Mesa basin. N and D stand for the Navajo and Dakota aquifers. Arrows indicate the direction of groundwater flow. The Carmel Formation hydraulically separates the N and D aquifers. (Modified from Cooley et al. 1969).

The mass of feldspar dissolved along each flow path was calculated from groundwater solute fluxes using an inverse mass balance approach (Garrels & Mackenzie, 1967). The specific surface area of the feldspars was assumed to be $0.7 \pm 0.2\ m^2/g$ (based on BET measurements). The *in situ* dissolution rates were $10^{-16.3 \pm 0.4}$ and $10^{-17.7 \pm 0.3}$ $(1\sigma)\ mol\ m^{-2}\ s^{-1}$ for plagioclase and K-feldspar, respectively. These rates are $\sim 10^5$ times slower than the laboratory rates at similar pH and temperature. To test the generality of these results, dissolution rates were also calculated for the Middendorf aquifer in the Atlantic Coastal Plain and the Lower San Pedro basin in southeastern Arizona. Similar apparent rates were obtained, although data in these aquifers are much less abundant (Fig. 2).

2.3 *Chemistry and morphology of the feldspar-clay interface*

The Navajo aquifer essentially provides an undisturbed natural flow-through reactor, preserving the morphology and phase relations of weathering reactions over geologic time scales.

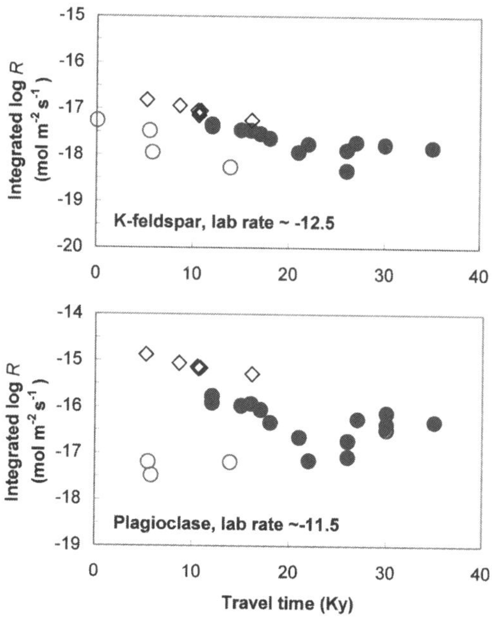

Figure 2. Derived *in situ* rates dissolution rates in aquifers for K-feldspar (top) and plagioclase (bottom). Solid circles are the Black Mesa basin; open circles the Middendorf aquifer, SC; and diamonds the Lower San Pedro basin, AZ. Rates from Black Mesa were normalized to measured BET surface areas. Rates for the other aquifers assumed Black Mesa surface areas as a first approximation.

Geochemists have argued for nearly a century about the existence and role of surface layer(s) on laboratory and naturally dissolved feldspars. High-resolution transmission electron microscopy (HRTEM) was performed on the Navajo sandstone, with an imaging information limit of 0.12 nm. The TEM and Field Emission Gun (FEG) Scanning Electron Microscopy (SEM) revealed two distinctly different surface coatings on K-feldspars. First, we observed a highly K^+ deficient, amorphous layer (no lattice fringes, Fig. 3), which continuously covers all feldspar surfaces. The amorphous layer remains thin, having a thickness <10 nm, even after reacting for hundreds of thousands to millions of years. The interface between the thin amorphous layer and completely unaltered feldspar is very sharp. This is similar to the observations of Hellmann et al. (2002) of experimentally altered albite, and suggests that diffusion processes within the feldspar are not an important factor in the dissolution kinetics. The amorphous layer, although ubiquitous, cannot be a diffusion barrier unless diffusion coefficients in the leach layer are unreasonably low.

Our HRTEM and FEG SEM observations also revealed the ubiquitous presence of clay coatings on feldspars within the Navajo sandstone. Feldspar surfaces are first covered with a small amount of patchy kaolinite in euhedral, oriented hexagonal plates. Outside of the kaolinite lies a continuous layer of oriented smectite 1–5 μm in thickness. These phase relationships are indicative of *in situ* alteration. In

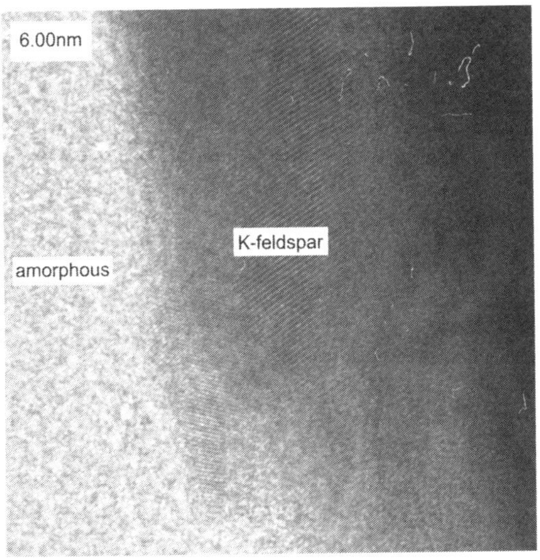

Figure 3. High-resolution TEM images showing K-feldspar with lattice fringes at atomic resolution and a K^+ deficient amorphous layer on the feldspar surface on the left. Chemical analysis indicates a sharp K^+gradient across the interface.

HRTEM observations, the thin amorphous layer is always present between crystalline feldspar and clay, and there is no structural inheritance between the feldspars and the amorphous surface layer or the surrounding clay minerals. An identical smectite layer occurs on quartz and Fe-Ti oxides, although kaolinite is not observed. We suspect that clay layers are widespread in undisturbed weathering environments, but that many studies remove the clay coatings before analysis. However, the clay rinds are also unlikely to be a significant diffusion barrier. Calculations indicate that the clay rind can only be a diffusion barrier using unrealistically low values for the effective porosity and tortuosity.

2.4 *The effect of solution chemistry on feldspar dissolution*

Probably the most currently favored explanation for slow feldspar dissolution in natural systems is an approach to equilibrium with respect to feldspars, whereas laboratory experiments are almost exclusively conducted far from equilibrium. A transition state based expression:

$$R_j = k_j(1 - e^{-A/RT}) \tag{1}$$

where k_j is the rate constant, is often used to approximate the effect of approach to equilibrium on dissolution rates. However, to reduce the dissolution rate by the observed factor of 10^5, the affinity (A) must be <0.025 J/mol, which is extremely close to equilibrium. This means that very little additional feldspar can dissolve before equilibrium with feldspar would be exceeded.

However, the observed geochemical evolution of groundwater at Black Mesa indicates that feldspar dissolution continues at a slow but nearly constant rate along the flow paths over thousands of years (Fig. 2). Thus, equilibrium must never be completely attained, since this would halt all further dissolution. Thus, affinity is most likely responsible for the slow feldspar dissolution rates, but can not completely explain the chemical evolution of the Black Mesa system.

Examination of other feldspar-containing regional aquifers, e.g., the Middendorf aquifer in the S. Carolina (Murphy & Schramke 1998), Lower San Pedro basin in SE Arizona (Robertson 1992), the Central Oklahoma aquifer (Parkhurst et al. 1995), and Santa Fe group in New Mexico (Zhu unpublished data), all show very slow feldspar dissolution, but still pronounced geochemical evolution over many kilometers. This requires continued, slow feldspar dissolution. Thus, it appears that slow feldspar dissolution rates in aquifers generally involve more than a simple approach to feldspar equilibrium.

2.5 *The role of clay precipitation in the overall reaction kinetics*

We conclude that the most important mechanism controlling the slow dissolution rates feldspars in aquifers is the slow precipitation kinetics of kaolinite and smectite. The precipitation of clays would remove solutes from solution, undersaturate the solution with respect to feldspars, and make additional feldspar dissolution possible. In our studies of the subsurface systems, *in situ* clay formation as a result of feldspar dissolution is ubiquitous. For reactions in series, the slowest reaction controls the overall rate. Thus, the overall rate of the feldspar dissolution along the flow path may be controlled by the precipitation rates of kaolinite and smectite if this is the slow step in the overall reaction.

3 REACTION PATH MODELING

This mechanism was demonstrated by reaction-path modelling. If feldspar initially dissolves at the laboratory rate and clay precipitates instantaneously or at the same rate as feldspar dissolution, feldspar dissolution rates decrease rapidly as A drops and the closed system approaches equilibrium. However, when clay precipitation rate constants are 10^2 to 10^4 slower than the feldspar dissolution rate constant, feldspar dissolution rate decreases rapidly due to the A effect, but reaches a steady state where groundwater is near equilibrium with feldspar. Therefore, slow clay precipitation effectively reduces K-feldspar dissolution rates by orders of magnitude, in a fashion consistent with laboratory rates, transition state theory and field observations, but with clay precipitation as the slow culprit. We believe that other factors may contribute to the slower feldspar dissolution rates in natural systems, such as surface chemistry and inhibitors, especially in soils. However, it is unlikely these factors can explain a five orders of magnitude difference.

The control of feldspar dissolution by the precipitation kinetics of the secondary minerals reconciles many of the apparent discrepancies between laboratory experimental rates and field measurements, and explains field observations that previously appeared inconsistent. It explains why smectite coatings occur on all sediment grains, not only on feldspars; a surface or diffusion controlled mechanism would result in higher concentrations of $SiO_{2(aq)}$ and $Al^{3+}_{(aq)}$ only in solutions near the feldspar surfaces. It also explains why kaolinite is preserved and coexists with smectite in natural systems over hundreds of thousands to millions years.

4 IMPLICATIONS OF A LINKED FELDSPAR DISSOLUTION-CLAY PRECIPITATION MECHANISM

A similar mechanism of clay precipitation controlling the overall feldspar weathering rates may also potentially operate in unsaturated soil systems. Isolated pores in soils may act as closed systems for weeks, months, or even years, and during these extended periods, the precipitation of clays may control the overall reaction rates within the pores.

A scheme in which clay precipitation kinetics controls the overall feldspar dissolution rate presents several dilemmas for the traditional application of kinetic theories to weathering systems. If the observed feldspar dissolution rate is not dependent on the surface area of the feldspar, then the typical approach of normalizing feldspar reaction rates to feldspar surface areas may be misleading. Feldspar surface chemistry, while critical to elucidating the atomic mechanisms of feldspar dissolution, may play a subsidiary role in quantitative models to simulate many large scale geological and environmental processes. It becomes more important to understand secondary mineral precipitation kinetics, which has been a difficult experimental and theoretical problem up to this time. Our hypothesis shifts the paradigm from the century-old debate about feldspar dissolution rates and mechanisms to the formation mechanisms of secondary phases, and opens up new possibilities for laboratory and field experiments to unravel the rates of overall feldspar weathering reactions.

ACKNOWLEDGEMENTS

This study was funded by NSF grant EAR 0003816.

REFERENCES

Blum, A.E. & Stillings, L.L. 1995. Feldspar dissolution kinetics. in A.F.White & S.L.Brantley (eds.) *Chemical Weathering Rates of Silicate Minerals:* 291–346 Washington, D.C.: Mineralogical Society of America.

Drever, J.I. & Clow, D.W. 1995. Weathering rates in catchments, in A.F.White & S.L.Brantley (eds.) in *Chemical Weathering Rates of Silicate Minerals*: 463–481. Washington, D.C.: Mineralogical Society of America.

Garrels, R.M. & Mackenzie, F.T. 1967. Origin of the chemical composition of some springs and systems. In *Equilibrium Concepts in Natural Water Systems*, Vol. 67, pp. 222–242.

Gunter, W.D., Wiwchar, B. & Perkins, E.H. 1997. Aquifer disposal of CO_2-rich greenhouse gases: extension of the time scale of experiment for CO_2-sequestering reactions by geochemical modeling. *Mineralogy and Petrology* 59: 121–140.

Hellmann, R., Penisson, J.M. & Thoomassin, J.H. 2002. Chemical weathering of minerals; do leached layers exist? *6th Int. Symp. Geochem. Earth's Surface* 6: 31–34.

Murphy, E.M. & Schramke, J.A. 1998. Estimate of microbial rates in groundwater by geochemical modelling constrained with stable isotopes. *Geochim. Cosmochim. Acta* 62: 3395–3406.

Nugent, M.A., Brantley, S.L., Pantano, C.G. & Maurice, P.A. 1998. The influence of natural mineral coatings on feldspar weathering. *Nature* 395: 588–591.

Parkhurst, D.L., Christenson, S. & Breit, G.N. 1995. Ground-water-quality assessment of the central Oklahoma aquifer, Oklahoma – Geochemical and geohydrologic investigations. *U.S. Geological Survey Water-Supply Paper 2357.*

Robertson, F.N. 1992. Radiocarbon dating of groundwater in a confined aquifer in southeast Arizona. *Radiocarbon* 34: 664–676.

Spycher, N., Apps, J. & Sonnenthal, E. 2003. Prediction of fluid flow and reactive transport around potential nuclear waste emplacement tunnels at Yucca Mountain, Nevada. *J. Contam. Hydrol.* 62–63, *653–673.*

Zhu, C. 2000. Estimate of recharge from radiocarbon dating of groundwater ages and numerical flow and transport modelling. *Water Resources Research* 36: 2607–2620.

Author index